BAYESIAN THEORY AND APPLICATIONS

Bayesian Theory and Applications

Edited by

PAUL DAMIEN

University of Texas, Austin

PETROS DELLAPORTAS

Athens University of Economics and Business

NICHOLAS G. POLSON

University of Chicago

DAVID A. STEPHENS

McGill University

Great Clarendon Street, Oxford, OX2 6DP,
United Kingdom

Oxford University Press is a department of the University of Oxford.
It furthers the University's objective of excellence in research, scholarship,
and education by publishing worldwide. Oxford is a registered trade mark of
Oxford University Press in the UK and in certain other countries

First published 2013
First published in paperback 2015

Impression: 1

Published in the United States of America by Oxford University Press
198 Madison Avenue, New York, NY 10016, United States of America

British Library Cataloguing in Publication Data
Data available

Library of Congress Cataloging in Publication Data
Data available

ISBN 978–0–19–969560–7 (Hbk.)
ISBN 978–0–19–873907–4 (Pbk.)

Printed and bound in Great Britain by
Clays Ltd, St Ives plc

Dedication

This volume is dedicated to Sir Adrian Smith, F.R.S.

Contents

Contributors

- C. K. Allen, Oak Ridge National Laboratory
- M. J. Bayarri, Universitat de Valencia
- J. O. Berger, Duke University
- Carlos M. Carvalho, University of Texas in Austin
- Sounak Chakraborty, University of Missouri-Columbia
- Christopher Challis, Duke University
- Siddhartha Chib, Washington University
- Hugh A. Chipman, Acadia University
- J. Andrés Christen, CIMAT, Mexico
- Merlise Clyde, Duke University
- A. Philip Dawid, University of Cambridge
- David Draper, University of California in Santa Cruz
- Colin Fox, University of Auckland
- Kassandra Fronczyk, M.D. Anderson Cancer Institute
- Dani Gamerman, UFRJ, Brazil
- Alan E. Gelfand, Duke University
- Edward I. George, University of Pennsylvania
- Malay Ghosh, University of Florida
- Souparno Ghosh, Duke University
- Michael Goldstein, Durham University
- Robert B. Gramacy, University of Chicago
- Jim E. Griffin, University of Kent
- E. Gutiérrez-Peña, IIMAS, UNAM
- Heikki Haario, Lappeenranta University of Technology
- Timothy E. Hanson, University of South Carolina
- D. Higdon, Los Alamos National Laboratory
- Gabriel Huerta, Indiana University
- Edwin S. Iversen, Duke University
- Eric Jacquier, MIT
- Alejandro Jara, Pontifica Universidad Catolica de Chile

- Jari Kaipio, University of Auckland
- George Karabatsos, University of Illinois
- Ville Kolehmainen, University of Eastern Finland
- Arthur Korteweg, Stanford University
- Athanasios Kottas, University of California in Santa Cruz
- Purushottam W. Laud, Medical College of Wisconsin
- Zesong Liu, University of Texas in Austin
- Hedibert F. Lopes, University of Chicago
- Udi Makov, University of Haifa
- Bani K. Mallick, Texas A & M University
- Robert E. McCulloch, University of Chicago
- Ramsés H. Mena, IIMAS, UNAM
- M. Mendoza, ITAM
- Daniel Merl, Lawrence Livermore National Laboratory
- Peter Müller, University of Texas in Austin
- C. Nakhleh, Sandia Labs
- Nicholas G. Polson, University of Chicago
- Ryan Prenger, Lawrence Livermore National Laboratory
- R. Ryne, Lawrence Berkeley National Laboratory
- Esther Salazar, Duke University
- Ana Paula Sales, Lawrence Livermore National Laboratory
- James G. Scott, University of Texas in Austin
- Siva Sivaganesan, University of Cincinnati
- Michael Stanley Smith, Melbourne Business School
- Glenn A. Stark, University of New Mexico
- David A. Stephens, McGill University
- Stephen G. Walker, University of Kent
- Mike West, Duke University
- Jesse Windle, University of Texas in Austin
- Sally Wood, Melbourne Business School

Introduction

At the outset, we would like to thank all the authors that have contributed to this volume. This book is dedicated to a statistician whose work in Bayesian statistics has forever changed the way in which statistical research and practice has been and will be carried out. Adrian Smith's accomplishments are documented at the end of this volume. Here, we simply note that three key ideas in this volume—hierarchical models, Markov chain Monte Carlo and sequential Monte Carlo—that have revolutionized Bayesian statistics are in large measure due to Adrian's contributions. These concepts are now ubiquitous wherever Bayesian models are used. In this volume, we have selected broad topic areas where these ideas come into play in a significant manner. Of course these topics are by no means exhaustive, but they serve to illustrate the impact that Adrian's research has had on Bayesian statistics in the last four decades or so.

When we conceived this volume, we wanted to position it somewhat differently from other tribute volumes. To accomplish this, based on our collective experiences, we felt that some of the basic ideas in modern Bayesian statistics with which Bayesian statisticians are familiar are foreign to some (if not many) colleagues and practitioners in other disciplines. Therefore, we felt that a volume that had a 'Bayesian textbook' flavour to it, and which also included application papers would prove useful in spreading modern Bayesian ideas. This is the modus operandi adopted in most of the chapters. We now discuss each part in turn.

Part I: Exchangeability Dawid and Goldstein explore the fundamental notion of Bayesian statistics, namely exchangeability.

Part II: Hierarchical Models The first key idea in modern Bayesian statistics is hierarchical models. Gelfand and Ghosh discuss the elementary ideas underlying such models. This is then followed up by Chakraborty, Mallick and Ghosh, and Kottas and Fronczyk's papers.

Part III: Markov Chain Monte Carlo The second key idea in modern Bayesian statistics is Markov chain Monte Carlo (MCMC). Chib reviews the key MCMC approaches to implementing full Bayesian analysis. Griffin and Stephens' contribution further describes and exemplifies advanced MCMC notions.

Part IV: Dynamic Models West describes the fundamentals of dynamic linear and nonlinear models. Papers by Gamerman and Salazar, and Huerta and Stark elaborate on these ideas via some novel applications.

Part V: Sequential Monte Carlo Carvalho and Lopes describe the use of SMC in a variety of Bayesian models, which is then followed up by an applications paper by Sales, Challis, Prenger and Merl.

Part VI: Nonparametrics Bayesian nonparametrics is embedded in the exchangeability ideas found in Part I. Walker discusses Bayesian nonparametric models and argues that to perform proper data analysis one must adopt nonparametric models at the outset. Two papers, one by Karabatsos and Walker, and the second by Ména complete this part.

Part VII: Spline Models and Copulas Part VI considers Bayesian nonparametrics using exchangeability as the basis. There are related approaches to nonparametrics but with some key differences. Two such classes of models are discussed in this part: Bayesian splines by Wood, and Bayesian copulas by Smith.

Part VIII: Model Elaboration and Prior Distributions Bayarri and Berger describe the fundamentals of Bayesian hypothesis testing, followed by three research papers by Draper; Liu, Windle and Scott; and Gutiérrez-Peña and Mendoza.

Part IX: Regressions and Model Averaging Chipman, George and McCulloch describe the correct way of doing regressions. This is further elaborated on in two papers by Clyde and Iversen, and Gramacy.

Part X: Finance and Actuarial Science Jacquier and Polson discuss the role of Bayes in financial applications. This is followed by a comprehensive review of Bayesian models in corporate finance by Korteweg. One area where Bayesian methods are only now beginning to gain popularity is actuarial science. Makov describes Bayesian models in this context.

Part XI: Medicine and Biostatistics It is safe to say that Bayesian methods have found most widespread use in biostatistics and bio-informatics. Mueller details the Bayesian models in these areas, followed by two key papers in biostatistics by Laud, Müller and Sivaganesan, and Hanson and Jara.

Part XII: Inverse Problems and Applications This is an exciting area of science where Bayesian methods are fast gaining in popularity. Fox, Haario and Christen provide a complete description of Bayesian ideas in this field, followed by two practical papers: one by Kaipio and Kolehmainen, and a second by Nakhleh, Higdon, Allen and Ryne.

Special thanks to Carlos Carvalho, Marcin Kacperczyk, Bani Mallick, Tom Shively, and Daniel Zantedeschi for helping review some of the papers.

Finally, we would like to thank Clare Charles, Elizabeth Hannon, Keith Mansfield, Viki Mortimer, Subramaniam Vengatakrishnan and their colleagues at Oxford University Press for their tireless efforts in ensuring that this book was completed in a timely and efficient manner.

Part I
Exchangeability

1 Observables and models: exchangeability and the inductive argument

MICHAEL GOLDSTEIN

1.1 Introduction

When quantifying uncertainty for large and complex systems, it is often considered helpful to regard such uncertainty as being of two kinds, epistemic and aleatory. Epistemic uncertainty is that which relates to our lack of knowledge, and could be reduced by receipt of further information. Aleatory uncertainty is that which relates to intrinsic chance variation in the system, and cannot be resolved except by direct observation. The distinction between aleatory and epistemic uncertainty is informal rather than precise, particularly within the view that all uncertainty stems from a lack of knowledge and understanding. Indeed, a basic activity in much of science is searching for explanatory structure within apparently random events, which corresponds to moving uncertainty from the aleatory to the epistemic form, where it can be better understood and, possibly, reduced.

The aleatory/epistemic distinction has a natural counterpart in much statistical analysis, where aleatory uncertainty is expressed through the likelihood function for the data given the population parameters, while epistemic uncertainty is expressed through the prior distribution over the parameters, within the Bayesian formulation, and is treated less formally within relative frequency based approaches. This division between uncertain model parameters and likelihoods conditional on the values of the parameters is helpful and constructive when modelling our uncertainty about a physical system. However, as with any other form of modelling, this does raise fundamental questions when we seek to apply the results of the model based analysis to actual real world inferences. All that we actually observe are individual measurements of real things. The parametric forms that we introduce to describe intrinsic chance variation are simply models whose meaning and justification remains to be established.

Within the subjectivist approach, there is a precise answer to the question of meaning for many statistical models. This meaning is rooted in the judgement of exchangeability. Exchangeability allows us to construct parametric statistical models purely on the basis of the uncertainty statements that we make about observable random quantities. Indeed, in many cases, the argument shows that we have no choice but to behave as though we consider that we are sampling from a parametric model (the aleatory uncertainty) given the true but unknown values of some population distribution (the epistemic uncertainty). Therefore, exchangeability is the logical bedrock to a large part of current statistical analysis. Beyond this, the distinction between aleatory and epistemic uncertainty pervades so much of current scientific analysis that the notion of exchangeability is a necessary conceptual tool to provide the underpinnings of meaning for uncertainty quantification in general and the inductive argument, namely the reasoning from particular cases to general principles, in particular.

Our aims in this chapter are two-fold. Firstly, we shall give an elementary and self-contained account of the notion of exchangeability and the derivation of de Finetti's representation theorem, which shows how we may construct operational statistical models based strictly on our judgements over observables. Secondly, we shall consider the relevance of this representation to real world inferences, and introduce a second collection of exchangeability judgements which are necessary in order that the inductive argument, when applied to inferences over models so constructed, also has an operational real world counterpart.

1.2 Finite population sampling

Finite population sampling gives a concrete illustration of the distinction between aleatory and epistemic uncertainty. Consider a simple version of this problem. We have a bucket, which contains a known large number, N, of counters, of which an unknown proportion q are red, and the remaining proportion $(1 - q)$ are blue. We intend to draw a counter at random from the bucket. (Here, and below, we use the term 'at random' as shorthand for the subjective judgement that each counter currently in the bucket is equally likely to be selected at each stage.) Let $Z = 1$ if this draw is red, and let $Z = 0$, otherwise. We are uncertain as to the value that Z will take. This uncertainty has two components. Firstly, we do not know the value of q. This is epistemic uncertainty. It can be quantified by consideration of what we know about the way that the population was formed, and will be further reduced if we take samples from the bucket. Different people will have different states of knowledge and so their epistemic uncertainty may differ. Secondly, even if we did know q, we still would not know the value of Z. This value would now be the realization of a Bernoulli random variable, parameter q, and this irreducible uncertainty is aleatory. The distinction between aleatory and epistemic uncertainty is most useful when there is a general consensus as to the representation of aleatory uncertainty, e.g. here, to the extent that there is general agreement that the draw from the bucket will be random, and no obvious way to impose more structure upon this variation.

The possible values of q are $q_i = i/n, i = 0, 1, \ldots, N$. In the subjective Bayes view, we may quantify our epistemic uncertainty for q by specifying our collection of probabilities $p_i = \mathrm{P}(q = q_i)$. Therefore, we can assess our probability that $Z = 1$, by the law of total probability, as

$$\mathrm{P}(Z = 1) = \sum_{i=0}^{N} p_i q_i \tag{1.1}$$

A useful way to rewrite (1.1) is

$$\mathrm{P}(Z = 1) = \int_0^1 q\, dF(q) \tag{1.2}$$

where F is the probability measure on [0,1] which assigns probability p_i to the point i/N.

Now suppose, instead, that we are going to take a random sample of size n, without replacement, from the bucket. Let X denote the number of red counters in the sample. Epistemic uncertainty is as before. Our aleatory uncertainty relates to the probability distribution for X if we know q, the proportion of red counters in the bucket. This distribution is hypergeometric, so that

$$P(X = k|q) = \frac{\binom{Nq}{k}\binom{N(1-q)}{n-k}}{\binom{N}{n}} \tag{1.3}$$

Therefore, the corresponding version of (1.2) is

$$P(X = k) = \int_0^1 \frac{\binom{Nq}{k}\binom{N(1-q)}{n-k}}{\binom{N}{n}} \, dF(q) \tag{1.4}$$

If n is small compared to N, then there is little difference between sampling with and without replacement, and so, approximately, we can rewrite (1.3) as

$$P(X = k|q) \approx \binom{n}{k} q^k (1-q)^{(n-k)} \tag{1.5}$$

so that (1.4) may be approximated as

$$P(X = k) \approx \int_0^1 \binom{n}{k} q^k (1-q)^{(n-k)} \, dF(q) \tag{1.6}$$

Representation (1.6) is familiar in the Bayesian context, and is often described by saying that X has a binomial likelihood, parameters n, q, where our prior measure for q is given by F. In the above examples, F was a discrete measure placing probabilities on each value i/N. As N increases, it is often helpful to approximate this discrete measure by a continuous pdf $f(q)$, so that

$$P(X = k) \approx \int_0^1 \binom{n}{k} q^k (1-q)^{(n-k)} f(q) \, dq \tag{1.7}$$

For example, most introductory treatments for Bayesian statistics deal with (1.7) by discussing the special case where $f(q)$ is a beta distribution, as this case has simplifying conjugacy properties. However, in our development, it is helpful to retain the possibility that F could have any form at all; for example F could be a mixture of discrete and continuous components if there were certain special choices for q. (Suppose, for example, that our bucket had been chosen by a coin flip between two buckets, for one of which we knew that q was 0.5, but we had no information about the value for q in the other bucket.)

As we increase the size of N as compared to n, the approximation (1.5), and so also (1.6), becomes increasingly precise. We can see this informally as (1.5) would be exact if we were sampling with replacement, and only removing a small number of counters from a large bucket will only change the proportion of red counters by a small amount. We can support this intuition by showing that the right-hand side of (1.3) tends uniformly to the right-hand side of (1.5) with N; for example, the most extreme change to the proportions in the bucket is to draw all counters of the same colour, and for any N, q and $n < Nq$, the probability of n successes when sampling without replacement is less than the probability when sampling with replacement, but greater than the probability for sampling with replacement if we first remove n red counters from the bucket, so that

$$\left(\frac{q-f}{1-f}\right)^n \leq P(X = n|q) \leq q^n$$

where f is the sampling fraction $f = n/N$, and so the approximation is very close for f near zero.

We have described this sampling problem from a Bayesian viewpoint. In the common situation where we have a large sample, n, from a much larger population, N, most inferential approaches will reach the same conclusion, namely that the proportion of red counters in the sample estimates the population proportion with high accuracy. When the sampling fraction f is not small, we must take more care in approximating the hypergeometric distribution, and if n is small, then our representation of epistemic uncertainty through F will be important. However, in all cases, the meaning of the analysis will be clear, in the sense that there is a true but unknown population parameter q, which is, in principle, observable, and a generally agreed aleatory description as to how the sample is drawn, given q.

Most statistical problems lack this logical bedrock. For example, if we spin a coin repeatedly, and would like to use our observed spins to revise our judgements about future spins, then we might represent our uncertainty by means of a model in which, given the value of q, the 'the true but unknown' value of the probability that the spun coin will land heads, the coin spins are independent Bernoulli variables with parameter q. This model is formally similar to the finite population problem that we have been discussing, but with the fundamental distinction that the quantity q over which we now express our uncertainty is only a model quantity, which is not observable even in principle and lacks even a real world definition. However, there is a bridge between such uncertainty models and the problem of finite population sampling and this comes through the concept of exchangeability, as we shall now describe.

1.3 Exchangeable samples

In the problem of sampling counters from the bucket that we described above, consider making an ordered series of draws, $X_1, X_2, \ldots, X_N$ from the bucket, without replacement, where $X_i = 1$ if the ith draw is red, $X_i = 0$ otherwise. For us, the sequence is not independent. Observing each draw alters both the aleatory uncertainty (each time we observe a red counter, then this reduces the proportion of red counters available for the next draw) and the epistemic uncertainty (each time we observe a red counter, this changes our state of knowledge about the true proportion of red counters in the bucket). However, the sequence does have certain probabilistic properties which are important for the general account that we shall develop.

Consider first a single draw X_i. For each draw i, given the initial proportion q of red counters, the probability of drawing a red counter is the same, namely q, as, on each draw, each individual counter has the same probability of being selected. Therefore, each X_i has the same probability distribution, namely Bernoulli, with parameter given by (1.2). Now consider any pair of draws, X_i, X_j. Given q, the ith draw has probability q of being red. If the ith draw is red, then the jth draw is a random draw from a bucket, size $N - 1$, with $qN - 1$ red counters. Therefore, the probability distribution of the number of red counters in two draws is given by (1.4), for the case where $n = 2$, and this is true for all pairs $i \neq j$. Continuing in this way, we have that the probability distribution of any collection of n elements $(X_{i_1}, \ldots, X_{i_n})$ from the series has the same probability distribution, however we select and permute the indices $i_1, \ldots, i_n$, as given by (1.4). This notion, that the probability distribution of any collection of n of the quantities depends only on the value of n, and not on the individual quantities selected, or the order in which they are arranged, is termed **exchangeability** and is fundamental to the subjectivist representation for epistemic and aleatory uncertainty.

While we have only discussed simple two-valued scalar quantities so far, the concept of exchangeability is quite general. We make the following definition.

Definition *A sequence $(Y_1, Y_2, \ldots)$ of random vectors $Y_i = (Y_{i1}, Y_{i2}, \ldots, Y_{im})$ taking values in some space Ω is said to be* **exchangeable** *if the joint probability distribution of each subcollection of n quantities $(Y_{i_1}, Y_{i_2}, \ldots, Y_{i_n})$ is the same.*

In our account of picking counters from the bucket, we deduced exchangeability of the sequence of selections from our views as to the physical description of the problem. The notion of exchangeability reverses the logic of this argument and allows us to deduce the structure of the problem directly from the judgement of exchangeability. This is usually termed **de Finetti's representation theorem for exchangeable sequences**. We will introduce this representation by discussing the example of spinning coins and then use a more general form of this argument to give the general form for the theorem. Our account builds on the treatment in [10].

1.4 The representation theorem for exchangeable binary sequences

Suppose that we spin a coin repeatedly. The familiar model, which treats coin spins as independent Bernoulli random variables each with a true but unknown probability q of landing heads, lacks an operational physical basis. However, we can retrieve a version of this model if we make the judgement that the coin spins are exchangeable.

Let $U_i = 1$ if the ith spin is heads, $U_i = 0$ otherwise. Suppose that we view the sequence $(U_1, U_2 \ldots)$ as exchangeable. To simplify our account, we will treat this sequence of coin spins as, in principle, infinite, which we can informally interpret as saying that we are able to consider as large a number of spins as we need when we construct our uncertainty judgements over the outcomes. (If we are restricted to a finite exchangeable sequence, then the results that we obtain will correspond to those deducible from finite population sampling, see for example, [3].)

Consider the following thought experiment. Imagine that we have an empty bucket, and a pile of counters, numbered, sequentially, 1 to N. Consider spinning the coin N times. We shall mark the outcome of the ith spin on the ith counter, so each counter is either marked 1 or 0. Each counter is added to the bucket.

As the spins are exchangeable, we must assign the same probability, q say, to the event that the first spin is heads, i.e. that $U_1 = 1$, as we do to the event that a randomly chosen counter from the bucket has value 1 (as the probability that the randomly chosen counter has value 1 is the average of the probabilities that each counter has value 1, which, by the exchangeability judgement, must all be equal to q). Again we may make a division into a notional epistemic uncertainty as to the value of the proportion of counters marked 1 in the bucket and an aleatory uncertainty for the value on the single counter that we pick, given this proportion. Therefore the probability, for the randomly selected counter in the thought experiment, can be constructed as in (1.1), by first considering the possible values $q_i = i/N, i = 0, 1, \ldots, N$ for the proportion of counters labelled 1, in the bucket, and assigning probabilities p_i to the outcomes for q as above. We have

$$P(U_1 = 1) = \int_0^1 q \, dF_N(q) \tag{1.8}$$

in the same way, where F_N assigns probability p_i to point i/N.

We can extend this argument to our judgement about the outcome of n tosses in the same way. If W_n is the number of heads in the first n spins, then our probability for observing $W_n = k$ is the same as the probability that we assign for this event in any sample of n spins, and so this probability must be equal to the probability of drawing k counters labelled 1 from the bucket in n random picks, which, by relation (1.4), is given, $\forall N \geq n$, as

$$P(W_n = k) = \int_0^1 \frac{\binom{Nq}{k}\binom{N(1-q)}{n-k}}{\binom{N}{n}} \, dF_N(q) \tag{1.9}$$

The simplest way to consider what happens as N increases is to invoke Helly's theorem (see, for example, [4], which also contains an insightful discussion of exchangeability) and a quite different derivation of the exchangeability representation theorem), which states that any infinite sequence of probability distributions G_N on a bounded interval contains a subsequence which converges in distribution to a limit, say G.

(Helly's theorem is a consequence of the result that any infinite sequence of numbers $a_1, a_2, \ldots$ on a bounded interval has a uniformly convergent subsequence. The result for number sequences can be shown as follows, where we suppose all numbers lie in [0,1]. Divide the sequence into ten subsequences, according to the first decimal place. At least one subsequence must be infinite. Keep one such subsequence and discard the rest. Let b_1 be the element a_{i_1} with the smallest subscript in this subsequence. Now divide this subsequence into ten subsequences according to the value of the second decimal place. At least one subsequence must be infinite. Keep one such subsequence and discard the rest. Let b_2 be the element a_{i_2} with the smallest subscript in this subsequence with $i_2 > i_1$. Continue in this way and the sequence $b_1, b_2, \ldots$ converges uniformly to a limit, as all values $b_j, b_{j+1}, \ldots$ agree in the first j decimal places, for each j. Helly's theorem follows by repeated application of this method. We first select an infinite subsequence of probability distributions which agree in the first decimal place for the probabilities that they assign to the intervals $[0, 0.5)$ and $[0.5, 1]$. From this subsequence, we select a subsequence which agrees to two decimal places for the probability assigned to intervals $[0, 0.25), [0.25, 0.5), [0.5, 0.75), [0.75, 1]$ and so forth. Choosing an element from each subsequence constructed in this way, we arrive at Helly's theorem.)

Applying Helly's theorem to the sequence F_N, there is a subsequence which converges in distribution to a limit F. Letting N tend to infinity, F_N tends to F and the hypergeometric integrand tends uniformly to the binomial, so that we have, for each k, n

$$P(W_n = k) = \int_0^1 \binom{n}{k} q^k (1-q)^{(n-k)}\, dF(q) \tag{1.10}$$

(1.10) is de Finetti's theorem for an infinite exchangeable sequence of binary outcomes, derived in [1]. The uniqueness of the distribution F satisfying (1.10) follows as a probability distribution on a bounded interval is uniquely determined by its moments: this is the Hausdorff moment problem (see [4] which contains a direct derivation of the exchangeability representation theorem based on this property). The theorem shows that the judgement of exchangeability, alone, is sufficient to ensure that our beliefs about the sequence are just as if we consider that there is a true but unknown quantity q given the value of which we view the sequence as a series of independent Bernoulli trials with probability q.

The convergence of the sequence F_N to F is uniform, as for any $N_1 < N_2 < N_3$ we may view F_{N_1}, F_{N_2} respectively as the distribution of q in buckets formed by draws of size N_1, N_2 respectively from a bucket formed by N_3 spins of the coin, so that F_{N_2} will be probabilistically closer than F_{N_1} to F_{N_3}, corresponding to the intuition that there are no features of a population that are better estimated by a small sample than by a large sample. In this way, we see that the exchangeability representation (1.10) is really a statement about our judgements over large finite collections of coin spins We invoke infinity simply to allow us to make a continuous approximation to the discrete process, to any order of accuracy that we require.

Notice, in particular, that we have constructed the measure $F(q)$ as the limit of the measures $F_N(q)$, namely the measures for the proportion of heads, $q_{[N]}$ in the first N tosses. This is another way of saying that the relative frequencies $q_{[N]}$ tend to a limit q in distribution. This is the subjectivist formulation of the notion of limiting relative frequency. The relative frequency approach to statistics uses the limiting relative frequency as the definition for the notion of probability but is unable to give a proper justification for this definition, or even a satisfactory explanation as to the way in which the limit should be understood. In contrast, the subjectivist approach constructs the limiting relative frequency as a subjective judgement which is implied by subjective exchangeability judgements over the sequence and deduces the limit as a necessary consequence of this judgement.

1.5 The general form for the exchangeability representation

The argument of the preceding section relates to coin spins, but it applies similarly to any infinite random exchangeable sequence of vectors, $(Y_1, Y_2, \ldots)$, over a space Ω. Just as before, we carry out the thought experiment of constructing a bucket with N counters, where the ith counter is marked with the value of Y_i. Let Q_N denote the empirical distribution of the counters in the bucket, so that Q_N assigns probability $1/N$ to the value on each counter. As the sequence Y is exchangeable, the first value, Y_1, has the same probability distribution as a draw according to the distribution Q_N. Therefore, we can split up our uncertainty as to the outcome of Y_1 into two parts. Firstly, we are uncertain as to the distribution Q_N, and secondly, given Q_N, we are uncertain as to the value of the observation Y_1. Denote our probability distribution for Q_N by F_N (so F_N assigns probabilities for all possible empirical distributions consisting of N selections from the space Ω). Then, analogously to (1.8), we have, for any $A_1 \in \Omega$,

$$P(Y_1 \in A_1) = \int Q_N(A_1)\, dF_N(Q_N) \tag{1.11}$$

where $Q_N(A_1)$ is the probability assigned to A_1 by the distribution Q_N (i.e. the proportion of the first N outcomes that are within A_1).

Now consider our probability distribution for the first n outcomes $(Y_1, Y_2, \ldots, Y_n)$. We can assess this distribution in two stages as above. First, we make a random choice for the empirical distribution Q_N according to F_N. Given the choice of Q_N, we now make n draws, without replacement, from the bucket consisting of N counters with this empirical distribution. We can evaluate this distribution exactly by a counting argument. If n is small compared to N, then each draw will only change the composition of the remaining counters in the bucket by a small amount, so that the draws will be almost independent. Therefore, we have that

$$P(Y_1 \in A_1, \ldots, Y_n \in A_n) \approx \int Q_N(A_1) \ldots Q_N(A_n) dF_N(Q_N) \tag{1.12}$$

As we let N increase, keeping n fixed, the exact form of the integrand in (1.12) tends uniformly to the product integrand. The distribution F_N tends to a limiting distribution F over the probability distributions Q over the space Ω. (The details of the limiting argument are technically more complicated than for the coin flips, due to the generality of the formulation, but the argument is the same, namely that the empirical distribution of a large sample, size N, from a much larger population, size M say, is close to the population distribution, by the standard arguments of finite population sampling, and therefore the sequence of distributions F_N must converge.)

Proceeding in this way, we have the generalization of de Finetti's result given by Hewitt and Savage, [11], which is as follows.

Theorem *Let $(Y_1, Y_2, \ldots)$ be an infinite exchangeable sequence of random quantities with values in Ω. Then there exists a probability measure F on the set of probability measures $Q(\Omega)$ on Ω, such that, for each n, and subsets $A_1, \ldots, A_n$ of Ω,*

$$P(Y_1 \in A_1, \ldots, Y_n \in A_n) = \int Q(A_1) \ldots Q(A_n) dF(Q) \tag{1.13}$$

F is the limiting distribution of the empirical measure, i.e. the probability assigned to any set A by F is given by the limit of the probability assigned to the proportion of the first N trials whose outcome is in A.

The exchangeability representation theorem is both surprising and prosaic. It is surprising, in the sense that the simple and natural symmetry judgement of exchangeability over observable quantities leads to such a strong result, namely that our beliefs must be as though we considered that we were making independent draws from a 'true but unknown' distribution Q for which we had assigned a prior measure F. This can be thought of as a version of the separation of our uncertainty into aleatory and epistemic components. Observation of a sample $Y_{[n]} = (Y_1, \ldots, Y_n)$ reduces our uncertainty about future elements of the sequence by applying the Bayes theorem to the prior measure F to update judgements about Q, as

$$P(Y_{i_1} \in A_1, \ldots, Y_{i_m} \in A_m | Y_{[n]}) = \int Q(A_{i_1}) \ldots Q(A_{i_m}) dF(Q|Y_{[n]}) \tag{1.14}$$

for all subsets $A_1, \ldots, A_m$ of Ω, and indices $i_1, \ldots, i_m$ all greater than n. Increasingly large samples tend to a 'relative frequency limit' eventually resolving all of our epistemic uncertainty, leaving the unresolvable aleatory uncertainty as to the outcomes of future draws from a known distribution, a posteriori. Compared to the conceptual confusion at the heart of traditional descriptions of statistical inference, this formulation is clear, unambiguous and logically compelling, building everything on natural belief statements about quantities which are, in principle, observable. The theory of exchangeability is rich and elegant and also of great practical and conceptual importance. This article has only focused on the most basic form for the representation. A characteristic example of the type of results that follow when we impose more structure on the exchangeability specifications is [12] which derives the additive model for log-odds in a two way table from natural exchangeability statements over rows and columns. (The discussion following that article contains some comments from me on the links between this result and the types of limiting finite population representations that we have described above.)

However, the representation theorem is also prosaic, as the population distribution is nothing more than the outcomes of all the possible future observations in the sequence, and the division into aleatory and epistemic components of uncertainty based on this structuring is just a partitioning of our judgements about such future observations. The bucket representation simply gives a concrete form to this identification with finite population sampling, and makes clear the role of exchangeability judgements in equating the observation of the members of the sequence with the random samples from the bucket.

1.6 Expectation as primitive

While the exchangeability representation is highly revealing, the real world implementation of the representation faces two difficulties, one in the construction of the representation and one in its inferential application.

The first difficulty is implicit in the derivation that we have described for the representation theorem. To construct the measure F in representation (1.13), we need to be able to quantify our beliefs for the outcome of the thought experiment comprising the composition of the large bucket with counters indexed by the vector outcomes of the first N members of the sequence. Specifying prior beliefs over the possible choices for this collection is both scientifically difficult, as we must consider questions at a level of detail beyond our ability to give scientifically meaningful answers, and technically difficult, because of the complexity of the objects over which we are aiming to develop a meaningful probabilistic representation. Therefore, one of the key advantages of the exchangeability representation, namely that it provides a method for us to restrict our belief statements to those related to observable quantities, in practice is usually unfeasible, and so the representation is rarely used in this way.

The second difficulty is as follows. The representation theorem appears to retrieve for us the familiar division into epistemic and aleatory uncertainties, but this division is itself based on an epistemic judgement, which is therefore subject to revision. We aim to use relation (1.14) to update beliefs about future outcomes given a current sample $Y_{[n]}$ by constructing the update $F(Q|Y_{[n]})$, and then deriving beliefs over future outcomes with respect to this distribution. However, the meaning that we ascribe to $F(Q)$ only holds for as long as we judge the sequence as exchangeable. We may change this judgement at any time. Bayesian statistics describes how to make inferences about quantities which have true but unknown values. There is no provision within the Bayesian approach (or any other approach to inference that I know of) for making operationally meaningful inferences about quantities which, at the time when we come to make the inference, may simply cease to exist.

What we need is both to simplify the specification requirements for the exchangeability representation, so that we may use it in practice, as well as in principle, and also to sharpen our formulation for inference to make sense of the issues raised when we make conditioning statements over evanescent quantities. We may address the first issue by changing the primitive for our theory from probability to expectation. To address the second issue requires us to augment our collection of exchangeability specifications, in ways that we shall describe below.

These are larger issues than we can do justice to in the space of this article. All that we will do here is to sketch the key steps that we must take to establish an operational meaning for our inferences over exchangeable quantities, building on ideas first outlined in [6] and [7].

Firstly, we shall discuss a simpler form of exchangeability, based on a different choice of primitive for the theory. Typically, the primitive of choice for the subjectivist theory is probability, but this is largely for historical reasons and to align the theory as closely as possible with its non-subjectivist counterparts. However, we do have a choice and we can, instead, choose expectation as the primitive for quantifying uncertainty. With this choice, we can make as many, or as few, expectation judgements as we wish, when treating a problem of uncertainty, including as many probability statements as we wish—these are simply expectation statements for the corresponding indicator variables. However, when probability is the primitive, we must make all possible probability statements before we can make any expectation statements. (For this reason, expectation was de Finetti's choice of primitive for the theory and the work which best summarized his views, [2], is actually a theory of expectation or, as he terms it, prevision.)

This is not an issue for non-subjectivist approaches—the probabilities all somehow exist separately from us and it is simply our task to learn about them. In the subjectivist theory, we are much more involved. Each uncertainty is a statement that we make, expressing our best judgements as to the likely outcomes. This is exactly the problem that we identified with the exchangeability representation. We need to specify so many probability judgements over observable quantities before we can construct the representation theorem that it is rarely used in this way. The theorem is drained of much of its power by the excessive demands of the probabilistic formalism. We shall now describe the second order version of the representation theorem, which does not suffer from this problem.

De Finetti makes expectation primitive under the operational definition in which $\mathrm{E}(X)$ is the value of x that you would choose if confronted with the penalty

$$L = k(X - x)^2$$

where k is a constant defining the units of loss and the penalty is paid in probability currency (i.e. tickets in a lottery with a single prize). The value of $E(X)$ is chosen directly, as a primitive, as is probability in the standard Bayesian account. De Finetti shows, under this definition, that $E(X)$ satisfies the usual properties of expectation, such as linearity. With this penalty scale, expectations are consistent with preferences, in the sense that preferring penalty A to B is equivalent to assigning $E(A) < E(B)$, as expectation for the penalty is equal to the probability of the reward.

Bayes linear analysis is a version of Bayesian analysis which follows when we take expectation as primitive; for a detailed account, see [9]. The particular features that are of concern for this article are the practical alternative for the exchangeability representation, which can actually be specified by judgements over observables in practice as well as in principle, and the linkage between this representation and an operationally meaningful form of inference for the evanescent model quantities expressed through the representation theorem. This formalism allows us to address the twin concerns that we have raised about current approaches to statistical induction (and we know of no alternative approach for so doing).

1.7 Second-order exchangeability representation theorem

We say that the sequence of random vectors $X_1, X_2, \ldots$, where $X_j = (X_{1j}, \ldots, X_{rj})$, is **Second-Order Exchangeable** (SOE), if each vector has the same mean and variance matrix and all pairwise covariance matrices are the same, i.e. if

$$E(X_i) = \mu, \ \mathrm{Var}(X_i) = \Sigma, \ \mathrm{Cov}(X_i, X_j) = \Gamma, \ \forall i \neq j \tag{1.15}$$

We suppose that all quantities in (1.15) are finite. We may separate our uncertainty about each X_i into aleatory and epistemic components, with corresponding second order specifications, according to the following representation theorem, derived in [5].

Theorem (Second-order exchangeability representation theorem) *If $X_1, X_2, \ldots$ is an infinite Second-Order Exchangeable sequence of random vectors, then, for each i,*

$$X_i = \mathcal{M}(X) \oplus \mathcal{R}_i(X) \tag{1.16}$$

where $\mathcal{R}_1(X), \mathcal{R}_2(X), \ldots$ is a mutually uncorrelated second-order exchangeable sequence, each with mean zero and uncorrelated with $\mathcal{M}(X)$.

(The notation $U \oplus W$ expresses the condition that all of the elements of the vector U are uncorrelated with all of the elements of the vector W.)

The proof of the representation theorem is similar to that for the full exchangeability representation. Our thought experiment is to construct a bucket containing N counters, marking the ith counter with the outcome for the ith case. Instead of considering the whole probability distribution of the counters in the bucket, we consider a single quantity, the average of the counters in the bucket, $\overline{X}_N = (\overline{X}_{1N}, \ldots, \overline{X}_{rN})$ where

$$\overline{X}_N = \frac{1}{N} \sum_{i=1}^{N} X_i$$

For the general exchangeability representation, we construct the population distribution from our beliefs relating to the limit of the sample distributions. For the second order theorem, we construct the population mean $\mathcal{M}(X)$ from our limiting beliefs about the sample means. We can do this because, from the specifications (1.15), the sequence $\overline{X}_i$ is a Cauchy sequence in mean square, as for $n < m$ and each i,

$$E((\overline{X}_{im} - \overline{X}_{in})^2) = (\frac{1}{n} - \frac{1}{m})(\Sigma_i - \Gamma_i) \tag{1.17}$$

where Σ_i, Γ_i are the ith diagonal terms of Σ, Γ. Therefore the sequence $\overline{X}_n$ tends to a limit, and this limit is the mean quantity $\mathcal{M}(X)$. The properties of the sequence $\mathcal{R}_i(X)$ follow by evaluating $\text{Cov}(\mathcal{R}_i(X), \mathcal{M}(X))$ as the limit of terms $\text{Cov}(X_i - \overline{X}_n, \overline{X}_n)$ and checking that this limit is zero, and similarly for $\text{Cov}(\mathcal{R}_i(X), \mathcal{R}_j(X))$.

We can formalize the construction of this limit, treating expectation as primitive, by constructing the inner product space $\mathcal{I}(X)$ whose vectors are linear combinations of all of the elements X_{ij}, with covariance as the inner product (we identify as equivalent all quantities which differ by a constant) and squared norm given by variance. $\mathcal{I}(X)$ is a pre-Hilbert space for which we may construct the minimal closure by adding limit points for all Cauchy sequences whose limits are not already elements of the space. The inner product over limit points is equal to the limit of the inner product for the associated Cauchy sequence. By (1.17), the sample means form Cauchy sequences, and therefore our specification is consistent with the existence of such limit points, which we identify with the elements of $\mathcal{M}(X)$.

The second-order exchangeability representation theorem is concerned with population mean quantities. It is our choice as to what elements we introduce into our base vectors and therefore what we may learn about from such specifications. For example, we may want to learn about population variation, in which case we must introduce appropriate squared terms into our base vectors, and make exchangeability statements over the corresponding fourth-order quantities. Details as to how to make the appropriate exchangeability specifications, the technicalities of the resulting inferences and the inter-relationship between the adjustment of means and variances are provided in [9].

1.8 Adjusted beliefs

The inner product space described above is the fundamental geometric construct underpinning the Bayes linear approach. The general form of this construction takes a collection U of random quantities, with covariance inner product, and constructs the closure of the inner product space $\mathcal{I}(U)$, denoted $[\mathcal{I}(U)]$. For any quantity $Y \in [\mathcal{I}(U)]$, the adjusted mean and variance of Y, given a data vector D, are defined to be, respectively, the orthogonal projection of Y into the subspace spanned by the elements of D and the orthogonal squared distance between Y and this subspace.

The explicit form for the *adjusted expectation* for a vector B given D, where $D = (D_0, D_1, \ldots, D_s)$, with $D_0 = 1$ is the linear combination $\overline{a}^T D$ where $\overline{a}$ is the value of a that you would choose if faced with the penalty

$$L = (B - a^T D)^2$$

It is given by

$$E_D(B) = E(B) + \text{Cov}(B, D)(\text{Var}(D))^{-1}(D - E(D))$$

(We may use an appropriate generalized inverse if $\text{Var}(D)$ is not invertible.)

The *adjusted variance matrix* for D given D, is

$$\begin{aligned}\mathrm{Var}_D(B) &= \mathrm{Var}(B - \mathrm{E}_D(D)) \\ &= \mathrm{Var}(B) - \mathrm{Cov}(B,D)(\mathrm{Var}(D))^{-1}\mathrm{Cov}(D,B)\end{aligned}$$

An important special choice for the belief adjustment occurs when D comprises the indicator functions for the elements of a partition, i.e. where each D_i takes value one or zero and precisely one element D_i will equal one. In this case adjusted expectation is equivalent to conditional expectation, e.g. if B is the indicator for an event, then

$$\mathrm{E}_D(B) = \sum_i \mathrm{P}(B|D_i)D_i$$

Therefore, the general inferential properties of belief adjustment that we shall describe below are inherited by full Bayes analysis, and this offers a formal interpretation of the real world inferential content of conditional probability arguments.

1.9 Temporal rationality

To understand how subjectivist theory can treat evanescent quantities such as population means, we must first discuss the inferential content of the standard Bayesian argument for observable quantities. This is a large and fundamental issue, which deserves far more space than we can give it here, where all that we will do is to sketch the outline of what is, in my view, the heart of the subjectivist argument.

Firstly, recall the precise meaning of a formal Bayesian inference. If A and B are both events, then $\mathrm{P}(B)$ is your betting rate on B (e.g. your fair price for a ticket that pays 1 if B occurs, and pays 0 otherwise) and $\mathrm{P}(B|A)$ is your current 'called off' betting rate on B (e.g. your fair price now for a ticket that pays 1 if B occurs, and pays 0 otherwise, if A occurs. If A doesn't occur your price is refunded).

This is not the same as the posterior probability that you will have for B if you find out that A occurs. There is no obvious relationship between the called off bet and the posterior judgement at all, and, in my view, no one has advanced an intellectually compelling argument as to why the two concepts should be conflated. The called off bet formulation can, however, be understood within the subjectivist theory as a model for the inference that you will make at the future time.

Models describe how system properties influence system behaviour. They involve two types of simplification, firstly, the description of system properties and secondly the rules by which system properties influence system behaviour. Good models capture enough features of the system that the insight and guidance they provide is sufficient to reduce our actual uncertainty as to system behaviour. This is valuable, provided that we do not commit the modeller's fallacy of considering that the analysis of the model is the same as the analysis of the system. A crucial condition for making good use of a model is to establish the relationship between the model and the actual system, as a basis for making real world inferences.

To derive such a relationship for the Bayesian model, we must make a link between our conditional judgements now and our actual future posterior judgements. This requires a meaningful notion of 'temporal rationality'. Our description is operational, based on preferences between random penalties, as assessed at different time points, considered as payoffs in probability currency.

Current preferences, even when constrained by current conditional preferences given possible future outcomes, cannot require you to hold certain future preferences; for example, you may obtain further, hitherto unsuspected, information or insights into the problem before you come to make your future judgements, and, always, the way in which you come to learn the information contained in any conditioning event will convey additional information that was not part of the formal conditioning.

These difficulties have no such force when considering whether future preferences should determine prior preferences. Suppose that you must choose between two random penalties, J and K. For your future preferences to influence your current preferences, you must know what your future preference will be. You have a **sure preference** for J over K at (future) time t, if you know now, as a matter of logic, that at time t you will not express a strict preference for penalty K over penalty J.

Our (extremely weak) temporal consistency principle is that future sure preferences are respected by preferences today. We call this

> **The temporal sure preference (TSP) principle** *Suppose that you have a sure preference for J over K at (future) time t. Then you should not have a strict preference for K over J now.*

At first sight, the temporal sure preference principle seems so weak that it can never be invoked, because we will never have a temporal sure preference. However, we actually have many such sure preferences and these are sufficient to determine the inferential content of the Bayesian model, provided we accept the temporal sure preference principle for the problem at hand. It is an interesting philosophical and practical question as to whether and when even this principle is too strong, but for our purposes here it is sufficient to note that this is the weakest principle which is sufficient to give a meaningful account of the content of a Bayesian inference. We will construct the argument for adjusted expectation, the argument for conditional expectation following as a special case, and then consider inference for exchangeable quantities under this formalism.

1.10 Prior inference

For a particular random vector B, suppose that you specify a current expectation $\mathrm{E}(B)$ and you intend to express a revised expectation $\mathrm{E}_t(B)$ at time t. As $\mathrm{E}_t(B)$ is unknown to you, you may express current beliefs about this quantity. Suppose that you will observe the vector D by time t. What information does the adjusted expectation, $\mathrm{E}_D(B)$, offer to you now about the posterior assessment $\mathrm{E}_t(B)$ that you will make having observed D?

We argue as follows. For any random quantity, Z, you can specify a current expectation for $(Z - \mathrm{E}_t(Z))^2$. Suppose that F is any further random quantity whose value you will surely know by time t. Suppose that you assess a current expectation for $(Z - F)^2$. From the definition of expectation, at future time t you will certainly prefer to receive penalty $(Z - \mathrm{E}_t(Z))^2$ to penalty $(Z - F)^2$. Therefore, by temporal sure preference, you should hold this preference now, and so you must now assign

$$\mathrm{E}((Z - \mathrm{E}_t(Z))^2) \leq \mathrm{E}((Z - F)^2) \tag{1.18}$$

Let D be a vector whose elements will surely be known by time t. Let $I(D, \mathrm{E}_t(Y))$ be the inner product space formed by adding the elements of $\mathrm{E}_t(B)$ to $I(D)$. From (1.18), $\mathrm{E}_t(B)$ is the orthogonal projection of B into $I(D, \mathrm{E}_t(B))$ and $\mathrm{E}_D(B)$ is the orthogonal projection of $\mathrm{E}_t(B)$ into $I(D)$.

Therefore, the temporal sure preference principle implies that your actual posterior expectation, $\mathrm{E}_t(B)$, at time t when you have observed D, satisfies the following prior assessments:

$$B = \mathrm{E}_t(B) \oplus S, \mathrm{E}_t(B) = \mathrm{E}_D(B) \oplus R \tag{1.19}$$

where S, R each have, a priori, zero expectation and are uncorrelated with each other and with D.

Therefore, evaluation of $\mathrm{E}_D(B)$ resolves some of your current uncertainty for $\mathrm{E}_t(B)$ which resolves some of your uncertainty for B. The actual amount of variance resolved is

$$\mathrm{Cov}(B, D)(\mathrm{Var}(D))^{-1}\mathrm{Cov}(D, B)$$

We say that $\mathrm{E}_D(B)$ is a **prior inference** for $\mathrm{E}_t(B)$, and therefore also for B. Relation (1.19) holds whatever the context in which the future judgements will be made. Adjusted expectation may be viewed as a model for such judgements which reduces, but does not eliminate, uncertainty about what those judgements should be. This argument is no different than that for the relationship between any real world quantity and a model for that quantity, except that, within a subjectivist analysis, we can rigorously derive the basis for this relationship, under very weak, plausible and testable assumptions.

Note that, if D represents a partition, then conditional and posterior judgements are related as

$$\mathrm{E}_t(B) = \mathrm{E}(B|D) \oplus R$$

where

$$\mathrm{E}(R|D_i) = 0, \forall i$$

with interpretation as above.

1.11 Prior inferences for exchangeable quantities

We now extend the notion of prior inference to the model quantities arising in the second order exchangeability representation, and thus provide an account of the inductive argument relating inferences about the population model and inferences about members of the population. To do this, we must first construct an operational meaning for posterior judgements over model quantities.

Suppose that the sequence of vectors $(X_1, X_2, \ldots)$ is infinite SOE. Suppose that you will observe a sample $X_{[n]} = (X_1, \ldots, X_n)$, by time t. You don't know whether you will still consider $(X_{n+1}, X_{n+2}, \ldots)$ to be SOE at time t. We would like to apply the posterior expectation operator $\mathrm{E}_t(.)$ directly to the exchangeability representation $X_i = \mathcal{M}(X) + \mathcal{R}_i(X)$, by the decomposition $\mathrm{E}_t(X_i) = \mathrm{E}_t(\mathcal{M}(X)) + \mathrm{E}_t(\mathcal{R}_i(X))$. In order to do this, we need to give a meaningful construction for the quantity $\mathrm{E}_t(\mathcal{M}(X))$. This cannot be done directly, as by time t there may be no vector $\mathcal{M}(X)$ to attach the posterior expectation to.

We construct an operational meaning for $\mathrm{E}_t(\mathcal{M}(X))$ by extending the thought experiment in which we construct a bucket marked with counters corresponding to the individual X_i values. For each $i > n$, we additionally record, on the counter marked with X_i, the value $\mathrm{E}_t(X_i)$, so that each counter is marked with a vector $U_i = (X_i, \mathrm{E}_t(X_i))$. Let us suppose that we currently view the sequence U_i as SOE, for $i > n$. This is a comparatively weak constraint. We do not now consider that, at time t, the sequence will necessarily still be exchangeable, but we cannot yet identify any future subsequences about which we already have reason to believe that our future judgements will be systematically different from our judgements over the rest of the sequence.

Therefore, we have the representation

$$U_i = \mathcal{M}(U) + \mathcal{R}_i(U)$$

The first half of the components of U_i consist of the elements of X_i. The remaining components consist of the elements of $\mathrm{E}_t(X_i)$, giving the representation

$$\mathrm{E}_t(X_i) = \mathcal{M}(\mathrm{E}_t(X)) + \mathcal{R}_i(\mathrm{E}_t(X))$$

For any $N > n$,

$$\frac{1}{N-n}\sum_{i=n+1}^{N} \mathrm{E}_t(X_i) = \mathrm{E}_t(\frac{1}{N-n}\sum_{i=n+1}^{N} X_i) \tag{1.20}$$

Taking the limit, in N, of the left hand side of (1.20) gives the quantity that we identify with $\mathcal{M}(\mathrm{E}_t(X))$. The corresponding limit in N of the right hand side of (1.20) is the limit of $\mathrm{E}_t(\overline{X}_n)$, which, as $\overline{X}_n$ tends to $\mathcal{M}(X)$, we equate with $\mathrm{E}_t(\mathcal{M}(X))$. Therefore, we can equate $\mathrm{E}_t(\mathcal{M}(X))$ with $\mathcal{M}(\mathrm{E}_t(X))$. By this construction, we can identify $\mathrm{E}_t(\mathcal{M}(X))$ as a quantity, derived through natural exchangeability judgements, which has the same logical status as the quantity $\mathcal{M}(X)$ itself.

We are now able to integrate model based assessments into our prior inference structure. We have the following theorem.

Theorem (Prior inferences for exchangeable models) *Suppose that, by time t, we will observe a sample $X_{[n]} = (X_1 \ldots, X_n)$ from an infinite SOE sequence of vectors. Suppose, also, that the sequence $U_i = (X_i, \mathrm{E}_t(X_i)), i = n+1, n+2, \ldots$ is a SOE sequence. We can construct the further vector, $\mathrm{E}_t(\mathcal{M}(X))$, which, given temporal sure preference, decomposes our judgements about any future outcome $X_j, j > n$ as*

$$X_j - \mathrm{E}(X) = \quad [\mathcal{M}(X) - \mathrm{E}_t(\mathcal{M}(X))] \tag{1.21}$$

$$\oplus[\mathrm{E}_t(\mathcal{M}(X)) - \mathrm{E}_{X_{[n]}}(\mathcal{M}(X))] \tag{1.22}$$

$$\oplus[\mathrm{E}_{X_{[n]}}(\mathcal{M}(X)) - \mathrm{E}(\mathcal{M}(X))] \tag{1.23}$$

$$\oplus[\mathcal{R}_j(X) - \mathrm{E}_t(\mathcal{R}_j(X))] \tag{1.24}$$

$$\oplus[\mathrm{E}_t(\mathcal{R}_j(X))] \tag{1.25}$$

(The orthogonal decomposition of (1.21), (1.22) and (1.23) follows by combining the construction for $(\mathcal{M}(X), \mathrm{E}_t(\mathcal{M}(X))$, as the limit of partial means of the quantities U_i, with the relationship (1.19) between each $X_i, \mathrm{E}_t(X_i)$ and $\mathrm{E}_{[n]}(X_i)$, derived from TSP. The orthogonal decomposition (1.24), (1.25) follows from TSP and the orthogonality between the two residual terms and the three mean terms follows as each covariance between an individual residual term and the mean terms must have the same value, by the SOE property of the sequence, and this covariance must therefore be zero, as the limiting average of the residual terms is equivalent to the zero random quantity.)

The above theorem shows that we may treat the vector $\mathcal{M}(X)$ as though it were, in principle, observable, allowing us to decompose our current uncertainty about each $X_j, j > n$, into five uncorrelated components of variation, as follows.

Firstly, our epistemic uncertainty is resolved into three components. We will be uncertain about the value of $\mathcal{M}(X)$, at time t, as expressed by the difference between the expectation, $\mathrm{E}_t(\mathcal{M}(X))$, that we will express for this quantity and the quantity itself, from (1.21). Secondly, part of our uncertainty (corresponding to (1.23)) about $E_t(\mathcal{M}(X))$ (and thus about $\mathcal{M}(X)$) will be resolved by the adjusted expectation for $\mathcal{M}(X)$ given $X_{[n]}$, but a part corresponding to (1.22), will be unresolved. Thirdly, this adjusted expectation given $X_{[n]}$ is uninformative for the uncertainty currently treated as aleatory, namely each $\mathcal{R}_j(X)$, about which our future expectation will reduce variation according to (1.25), leaving variation according to (1.24). Whether we will hold this variation to be aleatory at time t will be a subjective judgement that can only be made at that future time.

Each term in this decomposition raises basic practical, methodological, foundational and computational issues. As with the exchangeability representation itself, the prior inference theorem for the representation should be viewed as a starting point, establishing that such a formulation for inductive inference has a natural and operational meaning, based on the careful treatment of each of the five components of variation that we must account for. This is a part of the much wider issue as to the extent to which a Bayesian uncertainty analysis based on a complex scientific model may be informative for actual judgements about the real world; see the discussion in [8].

References

[1] de Finetti, B. (1931). Funzione caratteristica di un fenomeno aleatorio. Atti della R. Academia Nazionale dei Lincei, Serie 6. Memorie, Classe di Scienze Fisiche, Mathematice e Naturale, 4:251–299.

[2] de Finetti, B (1974, 1975). *Theory of Probability*, vol 1, 2, Wiley.

[3] Diaconis, P. (1977). Finite forms of de Finetti's theorem on exchangeability, *Synthese*, **36**, 271–281.

[4] Feller, W. (1966). *An Introduction to Probability Theory and its Applications*, vol II, Wiley, New York.

[5] Goldstein, M (1986). Exchangeable belief structures, *JASA*, **81**, 971–976.

[6] Goldstein, M. (1994). Revising exchangeable beliefs: subjectivist foundations for the inductive argument, in *Aspects of Uncertainty, A Tribute to D.V. Lindley*, 201–222.

[7] Goldstein, M. (1997). Prior inferences for posterior judgements in *Structures and Norms in Science*, M. C. D. Chiara *et al.*, eds, Kluwer, 55–71.

[8] Goldstein, M. (2011). External Bayesian analysis for computer simulators. In *Bayesian Statistics* 9. Bernardo, J. M. *et al.*, eds, Oxford University Press.

[9] Goldstein, M. and Woolf, D. (2007). *Bayes Linear Statistics: Theory and Methods*, Wiley.

[10] Heath, D. and Sudderth, W. (1976). de Finetti's theorem on exchangeable variables, *American Statistician*, 188–189.

[11] Hewitt, E. and Savage, L. J. (1955). Symmetric measures on Cartesian products. *Transactions of the American Mathematical Society*, **80**: 470–501.

[12] Lauritzen, S. L. (2003). Rasch models with exchangeable rows and columns, *Bayesian Statistics*, 7, Bernardo, J. M. *et al.*, eds, Oxford University Press.

2 Exchangeability and its ramifications

A. PHILIP DAWID

2.1 Introduction

Bruno de Finetti's concept of exchangeability [14], and its associated mathematics, form one of the cornerstones of subjectivist Bayesian inference, providing Bayesians with a good reason to take seriously the frequentist's model of independent observations from a common but unknown distribution. There is indeed a touch of magic about de Finetti's famous theorem: assuming nothing more than that our attitudes to a sequence of observations would be unchanged if they were arranged in a different order, it pulls the frequentist model out of a hat.

There are many ways of understanding de Finetti's theorem, and correspondingly many ways of generalizing it. One fruitful conception emphasizes the aspect of invariance under the action of a group: that of all finite rearrangements of the sequence of variables. This leads to generalizations in which we deal with other groups, acting on other structures, generating different kinds of statistical model. This chapter presents a brief survey of some of these generalizations and their applications. It is largely a summary of work that has previously been presented elsewhere [3, 5–7, 9, 10, 12, 13].

2.2 de Finetti's Theorem

At the purely mathematical level, de Finetti's theorem (henceforth dFT) supplies a characterization of those joint distributions P for an infinite sequence $X = (X_1, X_2, \ldots)$ of random variables, all defined on the same space $\mathcal{X}_0$, having the property of *exchangeability*: for any $n = 1, 2, \ldots$ and any permutation π of $(1, 2, \ldots, n)$, the joint distribution of the permuted sequence $(X_{\pi(1)}, X_{\pi(2)}, \ldots, X_{\pi(n)})$ is exactly the same as that of the unpermuted sequence $(X_1, X_2, \ldots, X_n)$. dFT shows that, for exchangeability to hold, it is necessary and sufficient that there exist a distribution (which, to avoid confusion, we shall sometimes term a *law*) ν over the space $\mathcal{Q}$ of distributions Q on $\mathcal{X}_0$, such that, for (suitably measurable) $A \subseteq \mathcal{X} := \mathcal{X}_0^\infty$,

$$P(X \in A) = \int_{\mathcal{Q}} Q^\infty(X \in A)\, d\nu(Q) \tag{2.1}$$

where Q^∞ is the distribution of $X = (X_1, X_2, \ldots)$ when the (X_i) are independent and identically distributed, each with distribution Q. The law ν can be uniquely recovered from P as the limit, as

$n \to \infty$, of the law of the empirical distribution, ν_n, of $(X_1, \ldots, X_n)$—see Goldstein's chapter in this volume [20].

For a binary sample space $\mathcal{X}_0 = \{0, 1\}$, dFT shows that any exchangeable distribution can be obtained from the model of Bernoulli trials with probability parameter q, by mixing with respect to a probability distribution ν for $q \in [0, 1]$. For more general $\mathcal{X}_0$ the mixing is over the set of distributions Q on $\mathcal{X}_0$, which will typically be a large nonparametric class. But—important mathematical niceties aside—the story is basically the same [21].

2.2.1 Intersubjective modelling

If your subjective joint distribution for $(X_1, X_2, \ldots)$ is exchangeable, you would not care if some demon rearranged the sequence in a different order: your opinions for the rearranged sequence would be exactly the same as they were for the original ordering. (In contrast, this would *not* hold if, for example, you thought there were some sort of time trend in the original sequence.)

We can regard this exchangeability property as the natural Bayesian explication of the intuitive concept of 'repeated trials of the same phenomenon under constant conditions', which forms the basis of the frequentist approach to probability and statistics. Indeed, this Bayesian approach, starting only with the very natural judgement of exchangeability as input, logically *derives*, via dFT, the model of independent and identically distributed trials as output. By providing a deeper justification for modelling assumptions that are typically—and thoughtlessly—simply taken as obvious and not in need of deeper analysis, it thus supplies a firmer basis for frequentist statistics than is available from that theory itself.

To expand on this point, consider a bevy of Bayesians who, while holding differing subjective probability distributions for X, all agree on exchangeability. Then they will all agree on the relevance of the independent and identically distributed model—the differences between them being relegated to their varying choices for the prior law for the 'parameter' Q of the model. We can thus justify the frequentist's statistical model of independent and identically distributed variables as an *intersubjective model* [5], conjured into existence by the simple and intuitive exchangeability judgement: that rearranging the variables should have no effect on judgements about them. Since this model comprises the common part of every exchangeable judgement, and since its parameter Q can be recovered from sufficiently extensive observation, it has at least as much claim to 'objectivity' as any other conception of that elusive term.

2.3 Group invariance

Exchangeability of a joint distribution P over $\mathcal{X} = \mathcal{X}_0^\infty$ can be restated as requiring that P be *invariant* under the group of all finite permutations acting on $\mathcal{X}$. dFT shows that the set of such invariant distributions forms a convex simplex, and that the extreme points of this simplex are the independent and identically distributed distributions—so that any member of the set has a unique representation as a convex combination of these extreme points.

Interesting extensions of this characterization arise when we apply it to other transformation groups, acting on sample spaces possibly other than those of infinite sequences. Thus let X be an uncertain quantity taking values in a sample space $\mathcal{X}$, let G be a group of transformations acting on $\mathcal{X}$, and let $\mathcal{P}$ denote the set of all distributions P for X that are invariant under G, so that $P(X \in A) = P(gX \in A)$ for all $g \in G$ and $A \subseteq \mathcal{X}$. Then $\mathcal{P}$ is a convex set. Let $\mathcal{T}$ be the set of extreme points of $\mathcal{P}$, i.e. those distributions in $\mathcal{T}$ that cannot be represented as a non-trivial mixture (convex combination) of distributions in $\mathcal{P}$. The extension of (2.1) to this more general context is then: for any $P \in \mathcal{P}$ there exists a unique law (distribution) ν over $\mathcal{T}$ such that, for $A \subseteq \mathcal{X}$,

$$P(X \in A) = \int_{\mathcal{T}} \theta(X \in A)\, d\nu(\theta). \tag{2.2}$$

Here $\theta \in \mathcal{T}$ is a distribution over $\mathcal{X}$. A friendlier notation renames this to P_θ, reinterpreting θ as a label for P_θ, with $\mathcal{T}$ the set of such labels, so yielding

$$P(X \in A) = \int_{\mathcal{T}} P_\theta(X \in A)\, d\nu(\theta). \tag{2.3}$$

The statistical interpretation of this mathematical property is as follows. Suppose You have made a personal judgement that Your opinions would not be affected if You were to be presented with gX ($g \in G$), rather than X. Then You must act as though You entertained the *statistical model* $\mathcal{P} = \{P_\theta : \theta \in \mathcal{T}\}$ for X, together with a prior distribution ν over its *parameter-space* $\mathcal{T}$. Again, if we consider the bevy of Bayesians all of whom agree in regarding invariance under G as appropriate, we can justify $\mathcal{P}$ as the intersubjective model engendered solely by this judgement of group invariance.

At this level of generality, the sample space $\mathcal{X}$ need not be a set of infinite sequences, and G need not be a permutation group. The link with frequentist understandings of modelling is then broken—but this is to very positive effect. By no longer insisting on any connection with the idea of 'repeated trials of the same phenomenon under constant conditions' this approach supplies a justification for statistical model building that is totally unavailable from the frequentist perspective.

2.3.1 Sufficiency

An added bonus of constructing statistical models through considerations of invariance is that we can use them to construct sufficient statistics: under suitable conditions a maximal invariant under the action of the group will be sufficient for the associated intersubjective model [17]. Thus for the case of exchangeability, where we observe just the first n variables $(X_1, \ldots, X_n)$, the maximal invariant under the permutation group is their *order statistic*—which is indeed sufficient for the model of all independent and identically distributed distributions for the (X_i). For a binary sample space, this reduces to the counts of 0s and 1s, which are sufficient for the Bernoulli model.

The above approach to model-building through invariance identifies the members of the intersubjective statistical model as the extreme points of the convex set of all invariant distributions. Another approach [18, 25, 26] starts by specifying what are the sufficient statistics, and suitably relating these, both algebraically and probabilistically, across different sample sizes. The set of all distributions consistent with these properties will again be a simplex, and its extreme points can be regarded as the associated statistical model—an *extreme point model*. When, as here, both approaches are possible they lead to the same model [5].

2.4 Other symmetry groups

Staying for the moment with the infinite product sample space $\mathcal{X} = \mathcal{X}_0^\infty$, we can entertain different groups acting on it.

2.4.1 Larger group

For a symmetry group G that contains the group of all finite permutations, the corresponding intersubjective model would still involve independent and identically distributed variables, but would impose additional structure on their common distribution.

One such larger symmetry group is that of all finite orthogonal transformations, where, for any $n = 1, 2, \ldots$, and any orthogonal transformation R of $\mathcal{R}^n$, the distributions of $X^n := (X_1, \ldots, X_n)^{\mathrm{T}}$ and of RX^n are judged to be the same. It was asserted by Freedman [19] and shown by Kingman [24] that a joint distribution has this property if and only if it can be expressed as a mixture, over some law for the parameter $\phi \geq 0$, of the joint distributions $X_i \overset{\text{i.i.d.}}{\sim} \mathrm{Norm}(0, \phi)$. That is, the model of independent and identically distributed normal variables with mean 0 arises from a judgement of rotational symmetry. With finitely many observations $(X_1, \ldots, X_n)$, the maximal invariant under this rotation group is $\sum_{i=1}^n X_i^2$, which is thus a sufficient statistic for this model.

Extending this result, Adrian Smith [31] showed that the independent and identically distributed normal model with both parameters unconstrained arises similarly from the assumption of invariance under the subgroup of finite orthogonal transformations that preserve the unit vector. The maximal invariant is now $(\sum_{i=1}^n X_i, \sum_{i=1}^n X_i^2)$, which is sufficient for this model.

2.5 Smaller group

Alternatively, we can consider 'restricted exchangeability', where we only require invariance under some smaller group of permutations. Often these will respect some additional structure in the sample space.

2.5.1 Partial exchangeability

Consider binary variables laid out in a semi-infinite two-way array: $(X_{ij} : i = 1, \ldots, k;$ $j = 1, 2, \ldots)$. An interpretation could be that we have a number of different coins, labelled by i, with possibly different biases, and can toss each of them over and over, with j labelling the toss. Then it could be appropriate to judge the problem invariant under any permutation of tosses j of the same coin (i.e. for fixed i), but not if we permute across the coins. This is the property of *partial exchangeability* [15]. The associated intersubjective statistical model has, as parameter, a vector $\mathbf{p} \in [0, 1]^k$, and then has all the (X_{ij}) independent, with $\mathrm{Prob}(X_{ij} = 1 | \mathbf{p}) = p_i$. That is, we simply assign different probabilities to the different coins.

As an intermediate position between full and partial exchangeability, we can permit further invariance of the problem under permutations of the label i of the coins. This might be appropriate if the coins were taken randomly from the output of a mint whose quality control is less than perfect, so that different coins might have different biases, but in an unsystematic way. We now allow a potentially infinite number of such coins (so $i = 1, 2, \ldots$), with invariance under all finite permutations of i.

The associated intersubjective statistical model can be expressed hierarchically [6]:

1. The parameter is a distribution Π on $[0, 1]$.
2. Given Π, variables $(P_i : i = 1, 2, \ldots)$ are independent and identically distributed according to Π.
3. Given (Π and) the (P_i), the X_{ij} are independent, with

$$\mathrm{Prob}(X_{ij} = 1) = P_i.$$

The model conditional on the P_i is thus exactly the same as for partial exchangeability; but now, corresponding to the new assumption of exchangeability of the coins, the (P_i), which were previously entirely arbitrary 'fixed effects', are themselves modelled as 'random effects', drawn independently from a common distribution.

2.5.2 Row–column exchangeability

In the above example the labelling j of the tosses has no objective meaning that carries across the coins: there is, for example, no special relationship between the third toss of coin 1 and the third toss of coin 2, and this indeed is why it can make sense to consider permuting the tosses of coin 1 while leaving those of coin 2 in place.

In other contexts there may be such correspondences of j-values across different i. Thus suppose a number of students (labelled by i) answer a number of questions (labelled by j). Then each of i and j has an intrinsic meaning across the levels of the other.

In such a case we might want to consider both students and questions as separately exchangeable. That is, our attitudes to the problem are considered unchanged when we replace the array (X_{ij}) by $X_{\rho(i)\sigma(j)}$, where ρ and σ are arbitrary finite permutations acting respectively on i and j. Note that such permutations preserve the integrity of individual students and questions.

Analysis of this problem is subtle. Aldous [1] showed that, for a doubly infinite array of binary variables, the associated intersubjective statistical model can be represented hierarchically as follows:

1. The parameter is a function $F : [0, 1]^2 \to [0, 1]$.
2. Given F, generate random (P_{ij}) in $[0, 1]$ as
$$P_{ij} = F(\alpha_i, \beta_j)$$
where the αs and βs are all independent and identically distributed with the uniform distribution over $[0, 1]$.
3. Given (F and) the (P_{ij}), generate the (X_{ij}) independently, with
$$\text{Prob}(X_{ij} = 1) = P_{ij}.$$

We can regard α_i as a measure of the quality of student i, and β_j as a measure of the difficulty of question j. The independent and identically distributed property reflects the assumed separate exchangeability of students and of questions. Note that, for any joint distribution of this form, the sets $(X_{ij} : i \in I, j \in J)$ and $(X_{ij} : i \in I', j \in J')$ are independent of each other whenever there is both no overlap of students ($I \cap I' = \emptyset$) and no overlap of questions ($J \cap J' = \emptyset$). Such an array is termed *dissociated* [27]. Moreover, any dissociated row–column exchangeable array can be represented in the above form.

However, this representation, while appealing in many ways, is not ideal, in that different choices for F can lead to identical joint distributions of the (X_{ij}), so that the parameter F is not identifiable (it is in fact identifiable modulo transformations of its arguments by separate functions each preserving the uniform distribution [22, 23]). Also, for a finite data array, the maximal invariant under the permutation group (which by, the general theory, is a sufficient statistic for the model) is hard to describe.

2.5.3 Spherical matrix models

We can combine the above generalizations of exchangeability—orthogonal tranformations rather than permutations, and two-way arrays rather than sequences. Thus suppose we have a doubly infinite array (X_{ij}) of real variables whose distribution can be regarded as *spherical*, i.e. the distribution of any finite submatrix is invariant under the actions of both left- and right-multiplication by an orthogonal matrix. Then the intersubjective model comprises those spherical distributions that are also dissociated. Moreover (assuming finite second moments), any such distribution can be expressed in the form

$$X_{ij} = \lambda_0 U_{ij} + \sum_{m=1}^{\infty} \lambda_m V_{im} W_{mk}$$

where the λs are non-negative real constants with $\lambda_1 \geq \lambda_2 \geq \ldots$ and $\sum_{m=1}^{\infty} \lambda_m^2 < \infty$, and all the Us, Vs and Ws are independent Norm$(0, 1)$ variables [1, 4]. It is rather easier to see that the group-induced sufficient statistic, based on a finite submatrix $(X_{ij} : 1 \leq i \leq I, 1 \leq j \leq J)$, is the unordered set of singular values of the matrix.

2.6 Second-order behaviour

For many purposes it is sufficient to confine attention to the joint first- and second-order moment structure of our variables. Thus if $(X_1, X_2, \ldots)$ is an exchangeable sequence with finite variance, then it is easy to see that there must exist constants μ, γ_0, γ_1 such that, for all $i \neq j$:

$$\mathbb{E}(X_i) = \mu \tag{2.4}$$

$$\text{var}(X_i) = \gamma_0 \tag{2.5}$$

$$\text{Cov}(X_i, X_j) = \gamma_1. \tag{2.6}$$

More generally, we call the sequence (X_i) *second-order exchangeable* if the properties (2.4)–(2.6) hold. This is equivalent to requiring that the mean and dispersion structure of the sequence be invariant under the group of finite permutations.

In this case, define $\phi_1 = \gamma_1$, $\phi_0 = \gamma_0 - \gamma_1$. It is readily checked that $\text{var}(\overline{X}_n) = \phi_0 + \phi_1/n$, whence, since n can be arbitrarily large, $\phi_0 \geq 0$. Also we find $\text{var}(X_n - \overline{X}_n) = (1 - 1/n)\phi_1$, so that $\phi_1 \geq 0$.

Now, for any real μ and $\phi_0, \phi_1 \geq 0$, consider uncorrelated variables $Z, Y_1, Y_2, \ldots$, with zero means, and

$$\text{var}(Z) = \phi_0$$

$$\text{var}(Y_i) = \phi_1$$

and define

$$X_i = \mu + Z + Y_i. \tag{2.7}$$

Then it is easy to see that the (X_i) satisfy (2.4)–(2.6). Moreover, we can recover Z as the mean square limit of $\overline{X}_n - \mu$ as $n \to \infty$, and then Y_i as $X_i - Z - \mu$. Hence an infinite sequence (X_i) is second-order exchangeable if and only if it can be represented as in (2.7), where $Z, Y_1, Y_2, \ldots$ are uncorrelated with zero means and $\text{var}(Y_i)$ is the same for all i. This can be regarded as a second-order (and indeed much simpler) variant of de Finetti's theorem: rather than the (X_i) being independent and identically distributed, after suitable conditioning, now they are uncorrelated with constant variance, after suitable partialling out of Z. See Goldstein [20, Section 7].

When we consider finite rather than infinite second-order exchangeable sequences, it need no longer be true that $\phi_0 \geq 0$, in which case there can be no real variable Z with $\text{var}(Z) = \phi_0$. Even so, for computing variances of linear functions of the (X_i) we can still proceed *as though* we had a representation of the form (2.7).

2.7 Extension to experimental layouts

Again, we can usefully extend the above ideas to other groups, acting on structures other than the sequence. We illustrate this here for the special case in which a number of workers, labelled by w, each operate a number of machines, labelled by m, for a number of different runs, labelled by r. However the theory extends straightforwardly to general distributive block structures [2, 11], which include most of the classical experimental layouts (in particular, all simple orthogonal block structures [28, 29]).

2.7.1 Fully random model

Let X_{mwr} be a measure of the quality of the rth run produced by worker w when operating machine m. In such a case it might sometimes be reasonable for attitudes about all the (X_{mwr}) to be invariant under the following permutations:

1. Permutations of r for a fixed combination of m and w—corresponding to exchangeability of the different runs made by a given worker on a given machine.
2. Permutations of w for fixed m—corresponding to a judgement of exchangeability of the workers with each other.
3. Permutations of m for fixed w—exchangeability of the machines one with another.

If we focus only on the dispersion structure, and assume this is invariant under the group generated by all the above permutations, then, analogous to (2.7), we obtain the following synthetic representation:

$$X_{mwr} = \mu + \alpha_m + \beta_w + \gamma_{mw} + \epsilon_{mwr} \tag{2.8}$$

where μ is a constant, and the other variables appearing on the right-hand side are mutually uncorrelated random variables, all terms involving the same Greek letter having the same variance. As long as the array is infinitely extendible in all directions, these variances will all be non-negative, and the representation (2.8) becomes a genuine equality, with the terms on the right-hand side identifiable as mean square limits of functions of the Xs.

The 'parameters' of the model (2.8) are μ and the respective variances $\phi_\alpha, \phi_\beta, \phi_\gamma, \phi_\epsilon$ of the random terms. We can thus consider this model, where these parameters can vary freely, as the intersubjective second-order model corresponding to the imposed symmetries. We see that our simple symmetry assumption delivers, for free, the usual 'random effects' model for this layout.

We can also use the symmetries (specifically, utilizing group representation theory) to deliver for free the appropriate decomposition of data (from a finite balanced layout) into main effects and interactions:

$$X_{mwr} = X_{...} + (X_{m..} - X_{...}) + (X_{.w.} - X_{...}) + (X_{mw.} - X_{m..} - X_{.w.} + X_{...}) + (X_{mwr} - X_{mw.}) \tag{2.9}$$

(where a dot indicates an average over the range of the replaced subscript in the data). Symmetry arguments can also be invoked to specify meaningful null hypotheses. For example, one possible interpretation of the assertion that there are 'no differences between the workers' is that the problem would remain invariant under the still larger group that (in addition to permutations of the machine label m) allowed permutations of *all* the runs on a given machine, irrespective of whether or not these were produced by the same worker. This delivers the model obtained from (2.8) by omitting the terms β_w and γ_{mw}. It thus corresponds to the 'null hypothesis' $\phi_\beta = \phi_\gamma = 0$, which in turn, in the context of the data-decomposition (2.9), is equivalent to the equality of the mean squares for workers, for machine–worker interaction, and for runs—thus suggesting appropriate statistical tests. See [9] for further details.

Another use of symmetry [3] is to make predictive inferences of various kinds—for example, for the production of a machine featuring in our experiment operated on by a new worker.

2.7.2 A mixed model

However, the symmetry approach is not just a different way of deriving and thinking about known results. It can also lead to new perspectives, models and analyses.

Thus suppose, in the above example, we have a finite number M of machines, and, recognizing that these are of various different kinds, are no longer willing to assume invariance under permutations of the index m—while still retaining (second-order) exchangeability between workers w, and between the runs r for any (m, w) combination. The intersubjective model generated by these reduced symmetry assumptions is now [8]

$$X_{mwr} = \mu_m + \alpha_{mw} + \epsilon_{mwr} \tag{2.10}$$

where:

1. $\mu_1, \ldots, \mu_M$ are arbitrary constants.
2. α_{mw} is $(\boldsymbol{\alpha}_w)_m$, where $\boldsymbol{\alpha}_w$ is a random $(M \times 1)$ vector, with mean $\mathbf{0}$ and arbitrary $(M \times M)$ dispersion matrix $\boldsymbol{\Phi}_\alpha$ (the same for all w).
3. ϵ_{mwr} is a real random variable, with variance ϕ_m depending on m alone.
4. The distinct vectors $\boldsymbol{\alpha}_w$ and scalars ϵ_{mwr} are all mutually uncorrelated.

In this model, the parameters are the 'fixed effect' machine means $\mu_1, \ldots, \mu_M$, the $(M \times M)$ dispersion matrix $\boldsymbol{\Phi}_\alpha$ of the 'random across-machine worker effects', $(\boldsymbol{\alpha}_w)$, and the within-machine variances ϕ_m $(m = 1, \ldots, M)$ of the 'random run effects', (ϵ_{mwr}). All these quantities can be consistently estimated from data on the M machines and indefinitely many workers and runs.

This model, fully justified by the symmetry assumptions from which it derives, differs from standard formulations of the 'mixed model'. Once again, the symmetries can be used to guide hypothesis generation, data-analysis and predictive inference. For example, the associated data-decomposition is

$$X_{mwr} = X_{m..} + (X_{mw.} - X_{m..}) + (X_{mwr} - X_{mw.})$$

where the component terms are independent, and have a special dispersion structure induced from that of (2.10). The symmetry analysis now leads to new tests of 'no worker effect', and even to a test of exchangeability between the machines (i.e. the model considered in Section 2.7.1).

2.8 Population genetics

An application of the symmetry approach to modelling arises in population genetics. We are interested in a collection of genetic markers (loci on the genome), labelled $m = 1, \ldots, M$. For simplicity, we suppose each of these has two possible variants (alleles), coded 0 and 1. We can sample data from a population of individuals, which itself is divided into many subpopulations (e.g. different ethnic groups), labelled $s = 1, \ldots, S$. We assume random mating within each subpopulation s. For each pair (s, m) there is a *gene pool*, comprising all the genes (with values 0 or 1) at marker m possessed by all the individuals in subpopulation s. We thus have an array (X_{smg}) of binary variables, where X_{smg} is the allele of the gth gene within the gene pool for marker m in subpopulation s.

Reasonable symmetry assumptions to impose on this array are:

1. Exchangeability of the genes within each gene pool (note that, because of random mating, there is no reason to expect the relationship between the two genes comprising the genotype of the same individual to differ in any way from the relationship between two genes belonging to different individuals).
2. Exchangeability of the subpopulations.

However we do not impose exchangeability of the markers.

Any joint distribution satisfying these symmetry assumptions can be represented in the following hierarchical way:

1. Generate, by some random process, a distribution Π over the space $[0, 1]^M$.
2. Given Π, generate vectors $(\mathbf{P}_s)$ in $[0, 1]^M$, independently from distribution Π.
3. Given (Π and) the $(\mathbf{P}_s)$, generate independent binary variables X_{smg} with $P(X_{smg} = 1) = P_{sm}$ (the m-entry of $\mathbf{P}_s$).

Thus $\mathbf{P}_s$ describes the frequency distributions of the different markers, within subpopulation s.

Also, although it does not follow directly from symmetry, it might be reasonable from scientific considerations (especially if the different markers are on different chromosomes) to assume that, for a random vector $\mathbf{P} \sim \Pi$, its components (P_m) are independent (though not necessarily identically distributed). For example, we might take, for Π, a distribution in which $P_m \sim \beta(a_{m0}, a_{m1})$, independently; then Π is determined by the (a_{mj}) $(m = 1, \ldots, M; j = 0, 1)$. However this extra specialization is inessential.

The intersubjective model corresponding to the assumed symmetry properties would take as its parameter the distribution Π (or, in the above specialization, the (a_{mj})), and so be described by levels 2 and 3 of the above hierarchy. The full hierarchical model of a single Bayesian would be obtained by adding in, as level 1, a subjective prior distribution (essentially unconstrained) for Π (or for the (a_{mj})).

Note that in this intersubjective model, because of the random nature of P_{ms} we will have exchangeability, but not independence, between the genes within a common gene pool,

However, the full hierarchical model can also be deconstructed in the following, equally valid, way. We take as our 'parameter' the quantities (P_{sm}), and as our 'model' the final stage 3 of the hierarchy. We flesh out the hierarchy with a joint prior distribution for the (P_{sm}), obtained by combining levels 1 and 2 of the hierarchy. (The first stage is again essentially arbitrary, but the second is now not, so imposing constraints on the form of the joint 'prior' distribution of the (P_{sm}).) In this description, the genes within a given gene pool are now independent.

We thus see that it is not meaningful to ask whether or not the genes in a gene-pool are *really* independent—the answer depends on a somewhat arbitrary choice we have to make as to where to draw the line between model and prior. This realization takes some of the heat out of the apparently discrepant modelling assumptions made in [16] and [30], and their application to criminal identification by means of DNA profiling [12, 13].

References

[1] Aldous, David J. (1981). Representations for partially exchangeable arrays of random variables. *Journal of Multivariate Analysis*, **11**, 581–598.

[2] Bailey, Rosemary A. (1981). Distributive block structures and their automorphisms. In *Combinatorial Mathematics VIII* (ed. K. L. McAveny), Volume 884 of *Lecture Notes in Mathematics*. Springer Verlag, Berlin.

[3] Dawid, A. Philip (1977). Invariant distributions and analysis of variance models. *Biometrika*, **64**, 291–297.

[4] Dawid, A. Philip (1978). Extendibility of spherical matrix distributions. *Journal of Multivariate Analysis*, **8**, 559–566.

[5] Dawid, A. Philip (1982). Intersubjective statistical models. In *Exchangeability in Probability and Statistics* (ed. G. Koch and F. Spizzichino), pp. 217–232. North-Holland Publishing Company, Amsterdam.

[6] Dawid, A. Philip (1985). Probability, symmetry and frequency. *British Journal for the Philosophy of Science*, **36**, 107–128.

[7] Dawid, A. Philip (1986). A Bayesian view of statistical modelling. In *Bayesian Inference and Decision Techniques* (ed. P. K. Goel and A. Zellner), Chapter 25, pp. 391–404. Elsevier Science Publishers B.V. (North-Holland), Amsterdam.

[8] Dawid, A. Philip (1986). Symmetry analysis of the mixed model. Research Report 53, Department of Statistical Science, University College London.

[9] Dawid, A. Philip (1988). Symmetry models and hypotheses for structured data layouts. *Journal of the Royal Statistical Society, Series B*, **50**, 1–34.

[10] Dawid, A. Philip (1993). Taking prediction seriously. *Bulletin of the International Statistical Institute*, **55**(3), 3–13.

[11] Dawid, A. Philip (1994). Distributive block structures: Mathematical properties. Research Report 133, Department of Statistical Science, University College London.

[12] Dawid, A. Philip (1997). Modelling issues in forensic inference. In *ASA Proceedings of the Section on Bayesian Statistical Science*, pp. 182–186. American Statistical Association (Alexandria, VA).

[13] Dawid, A. Philip and Pueschel, John (1999). Hierarchical models for DNA profiling using heterogeneous databases (with Discussion). In *Bayesian Statistics 6* (ed. J. M. Bernardo, J. O. Berger, A. P. Dawid, and A. F. M. Smith), Oxford, pp. 187–212. Oxford University Press.

[14] de Finetti, Bruno (1937). La prévision: Ses lois logiques, ses sources subjectives. *Annales de l'Institut Henri Poincaré*, **7**, 1–68. English translation "Foresight: Its logical laws, its subjective sources", in *Studies in Subjective Probability* (1964) (H. E. Kyburg and H. E. Smokler, eds.), pp. 93–158. Wiley, New York.

[15] de Finetti, Bruno (1938). Sur la condition d'équivalence partielle. In *Colloque Consacré à la Théorie des Probabilités*, Volume VI, pp. 5–18. Herman et Cie, Paris.

[16] Evett, Ian W., Foreman, Lindsey A., and Weir, Bruce S. (2000). Letter to the Editor (with responses by A. Stockmarr and B. Devlin). *Biometrics*, **56**, 1274–1275.

[17] Farrell, Roger H. (1962). Representation of invariant measures. *Illinois Journal of Mathematics*, **6**, 447–467.

[18] Freedman, David A. (1962). Invariants under mixing which generalize de Finetti's theorem. *The Annals of Mathematical Statistics*, **33**, 916–923.

[19] Freedman, David A. (1963). Invariants under mixing which generalize de Finetti's theorem: Continuous time parameter. *The Annals of Mathematical Statistics*, **34**, 1194–1216.

[20] Goldstein, Michael (2012). Observables and models: exchangeability and the inductive argument. In *Bayesian Theory and Applications* (P. Damien, P. Dellaportas, N. G. Polson and D. A. Stephens, eds.), pp. 3–18. Oxford University Press, Oxford.

[21] Hewitt, Edwin and Savage, Leonard J. (1955). Symmetric measures on Cartesian products. *Transactions of the American Mathematical Society*, **80**, 470–501.

[22] Hoover, Douglas N. (1979). Relations on probability spaces and arrays of random variables. Preprint, Institute for Advanced Study, Princeton, New Jersey.

[23] Hoover, Douglas N. (1982). Row-column exchangeability and a generalized model for exchangeability. In *Exchangeability in Probability and Statistics* (ed. G. Koch and F. Spizzichino), pp. 281–291. North-Holland, Amsterdam.

[24] Kingman, John F. C. (1972). On random sequences with spherical symmetry. *Biometrika*, **59**, 492–493.

[25] Lauritzen, Steffen L. (1988). *Extremal Families and Systems of Sufficient Statistics*. Number 49 in Lecture Notes in Statistics. Springer-Verlag, Heidelberg.

[26] Martin-Löf, Per (1974). Repetitive structures and the relation between canonical and microcanonical distributions in statistics and statistical mechanics. In *Proceedings of Conference on Foundational Questions in Statistical Inference* (ed. O. E. Barndorff-Nielsen, P. Blæsild, and G. Schou), Volume 1, Aarhus, pp. 271–294.

[27] McGinley, William G. and Sibson, Robin (1975). Dissociated random variables. *Mathematical Proceedings of the Cambridge Philosophical Society*, **77**, 185–188.

[28] Nelder, John A. (1965). The analysis of randomized experiments with orthogonal block structure: I. *Proceedings of the Royal Society of London, Series A*, **283**, 147–162.

[29] Nelder, John A. (1965). The analysis of randomized experiments with orthogonal block structure: II. *Proceedings of the Royal Society of London, Series A*, **283**, 163–178.

[30] Roeder, Kathryn, Escobar, Michael, Kadane, Joseph B., and Balazs, Ivan (1998). Measuring heterogeneity in forensic databases using hierarchical Bayes models. *Biometrika*, **85**, 269–287.

[31] Smith, Adrian F. M. (1981). On random sequences with centered spherical symmetry. *Journal of the Royal Statistical Society, Series B*, **43**, 208–209.

Part II

Hierarchical Models

3 Hierarchical modelling

ALAN E. GELFAND AND SOUPARNO GHOSH

3.1 Introduction

As we move into the second decade of the twenty-first century, we are witnessing a dramatic paradigm shift in the way that statisticians collaborate with researchers from other disciplines. Disappearing are the days when the statistician was called in at the end of a project to provide some routine data analysis and some summary displays. Now the statistician is an integral player in a research team, helping to formulate hypotheses, identify data needs, develop suitable stochastic models, and implement fitting of the resulting challenging models. Altogether, the statistician becomes sufficiently knowledgeable in the subject matter to 'walk the walk' and 'talk the talk', adding another scientific dimension to her/his skill set.

As part of this shift, there is increasing attention paid to bigger picture science, to looking at complex processes with an integrative perspective and to bringing a range of knowledge to this effort. Increasingly, we find researchers working with observational data, less with designed experiments, recognizing that the latter can help inform about the former, but the gathering of such experiments provides only one source of data for learning about the complex process. Other information sources, empirical, theoretical, physical, etc. will also be included in the synthesis.

The primary result of all of this is the development of a multilevel stochastic model which attempts to incorporate the foregoing knowledge, inserting it at various levels of the modelling, as appropriate. A key recognition in all of this is the importance of introducing uncertainty and how to do so. That is, as always, the stochastic models are only approximations to the complex process so error will always be introduced. Hence, in addition to the modelling challenge is the question of where and how to introduce error. In this regard, Michael Goldstein, in his presentation at the AFMS Conference, which motivated this volume, discussed a catalogue of such errors including: parameter uncertainty, functional uncertainty, model uncertainty, stochastic uncertainty measurement uncertainty and multiple model uncertainty.

Following the vision of Mark Berliner [8], we imagine a three-stage hierarchical specification:

First stage: [*data*|*process, parameters*]
Second stage: [*process*|*parameters*]
Third stage: [(*hyper*)*parameters*]

The simple form of this specification belies its breadth. The process component can include multiple levels. It can be dynamic, it can be spatial. The data can be conditioned on whatever aspects of the process are appropriate. The stochastic forms can be multivariate, perhaps infinite dimensional with parametric and/or nonparametric specifications. (In this volume, see the companion piece by

Chakraborty *et al.* for an example of the former and the companion piece by Kottas and Fronczyk for an example of the latter.) Moreover, the range of applications to which this generic specification has been applied runs the scientific gamut, e.g. biomedical and health sciences, economics and finance, environment and ecology, engineering and natural science, political and social science.

In view of the above, hierarchical modelling has taken over the landscape in contemporary stochastic modelling. Although analysis of such modelling can be attempted through non-Bayesian approaches, working within the Bayesian paradigm enables exact inference and proper uncertainty assessment within the given specification.

Finally, then, the objective of this chapter is to provide a representative review of the range of such modelling. In the development, we will also note the connections to Gibbs sampling, in particular, why Gibbs sampling and MCMC are ideally suited to fit such models. We also attempt to section the paper by type of model but acknowledge that, in fact, model types are not disjoint. The overarching *building block* is the notion of latent variables, e.g. random effects, missing data, labels. We will see that these variables introduce unobservable process features which will be of interest, as well as facilitating model-fitting. With regard to the latter, Gibbs sampling loops become natural, updating other parameters given the values of the latent variables and then updating the latent variables given the values of the other parameters. For instance, in modelling a complex process, we might have a dynamic model that introduces spatial random effects, temporal autoregressive effects, with errors in variables and which also accommodates missing data.

In any event, the format of the paper is as follows. In Section 3.2 we recall the basics of hierarchical forms, including random effects and missing data. Section 3.3 offers some scope for the introduction of other sorts of latent variables. Section 3.4 considers mixture models while Section 3.5 returns to random effects, primarily in the context of structured dependence. Section 3.6 looks at dynamic models while Section 3.7 considers relatively recent ideas in data fusion. We end with a brief summary in Section 3.8.

3.2 The basics

The earliest hierarchical models introduced Gaussian first and second stages [12]. Subsequently these models have been absorbed into the widely used class of hierarchical models popularized by Raudenbush and Bryk [54]. In particular, we can refer to these as standard hierarchical linear models with the specification:

$$\begin{aligned} \text{First stage: } & Y|X, \boldsymbol{\beta} \sim N(X\boldsymbol{\beta}, \Sigma_Y) \\ \text{Second stage: } & \boldsymbol{\beta}|Z, \boldsymbol{\alpha} \sim N(Z\boldsymbol{\alpha}, \Sigma_{\boldsymbol{\beta}}) \\ \text{Third stage: } & \boldsymbol{\alpha} \sim N(\boldsymbol{\alpha}_0, \Sigma_{\boldsymbol{\alpha}}) \end{aligned}$$

This simple version assumes all Σs known. If not, inverse Gamma or Wishart priors are introduced at the third stage. Model-fitting within the Bayesian framework using MCMC is routine. Indeed, due to the conjugacy, we have standard distributions for all full conditionals. Hence, we can implement a *vanilla* Gibbs sampler to perform the model-fitting [39].

This class of models encompasses what is generally characterized as the study of linear models [26]. If the first stage is changed to another member of the exponential family and connected to the second stage through a suitable link function, we obtain a hierarchical generalized linear model [32]. Now the conjugacy between the first and second stages is lost. Within MCMC, earlier, such models were fitted using adaptive rejection sampling [27] due to the log concavity of the resulting full conditional distributions. Nowadays, Metropolis–Hastings updating would more likely be used with some adaptive tuning of the acceptance rates.

In this vein, substantial effort was put into the investigation of so-called conditionally independent hierarchical models (CIHMs). Much of this work took place in the Statistics group at Carnegie Mellon University with model-fitting through Laplace approximation [35, 66]. This effort preceded the entrance of Gibbs sampling and MCMC as Bayesian computation tools. Interestingly, the approach is now enjoying a revival through the recent development of integrated nested Laplace approximation (INLA), led by Håvard Rue and his group at Trondheim in Norway [59].

The CIHM takes the basic form

$$\Pi_i[\mathbf{Y}_i|\boldsymbol{\theta}_i]\Pi_i[\boldsymbol{\theta}_i|\boldsymbol{\eta}][\boldsymbol{\eta}]$$

That is, exchangeable $\boldsymbol{\theta}_i$ are assumed. If $\boldsymbol{\eta}$ is fixed, then we can fit separate models for each i. In practice, this would not be the case and, with unknown $\boldsymbol{\eta}$, we observe the well-known phenomenon of shrinkage or borrowing strength across the is (see e.g. [19] and references therein). Evidently, the CIHM includes the hierarchical GLM, i.e. it allows a non-Gaussian first stage.

A more elaborate extension of the CIHM is the setting of dependent ARMA time series. Here, we model

$$Y_{it} = X_i^T \boldsymbol{\beta}_i + \sum_j \phi_{ij} Y_{i,t-j} + \sum_k \theta_{ik}\epsilon_{i,t-k} + \epsilon_{it}$$

At the second stage, we specify exchangeable $\boldsymbol{\beta}_i, \boldsymbol{\phi}_i, \boldsymbol{\theta}_i$. We adopt a usual vague Gaussian prior on $\boldsymbol{\beta}$, adding constrained priors on the ϕs and θs (to ensure stationarity). Finally, $\boldsymbol{\epsilon}_t \sim N(0, \Sigma)$.

3.2.1 Random effects and missing data

Arguably, the utilization of hierarchical models initially blossomed in the context of handling random effects and missing data, using the E-M algorithm [17] for likelihood analysis and Gibbs Sampling [24] for fully Bayesian analysis. In this subsection, we offer some elementary remarks, first on random effects, then on missing data.

With regard to random effects, both classical and frequentist modelling supply a stochastic specification for these effects, usually assumed to be a normal distribution with an associated variance component. These effects can be introduced at different levels of the modelling but, regardless, in much of the literature, they are assumed to be exchangeable, in fact i.i.d. [64]. More recently, we are seeing random effects with structured dependence in, e.g. dynamic, spatial and spatio-temporal models (see Section 3.5).

A typical linear version with i.i.d. effects takes the following form. At the first stage,

$$Y_{ij} = X_{ij}^T \boldsymbol{\beta} + \phi_i + \epsilon_{ij}.$$

At the second stage, $\boldsymbol{\beta}$ has a Gaussian prior while the ϕ_i are i.i.d. $\sim N(0, \sigma_\phi^2)$. The ϵ_{ij} are i.i.d. $\sim N(0, \sigma_\epsilon^2)$. The variance components become the third stage hyperparameters, i.e. we require prior specifications for $\sigma_\phi^2, \sigma_\epsilon^2$. As has been learned over recent years, care is required in these specifications. Improper priors can lead to improper posteriors [5–7]. The frequently employed inverse gamma priors, $IG(\epsilon, \epsilon)$ for small ϵ are *nearly* improper and result in *nearly* improper posteriors, as well as badly behaved MCMC in practice. A protective recommendation is an $IG(1, b)$ or $IG(2, b)$. Both are far from improper; the former has no integer moments, the latter has a mean but no variance. Evidently, we can revise the model to have a non-Gaussian first stage. Again, care is needed with prior specifications as well as in model-fitting.

In collecting information on, e.g. individuals, we often have vectors of data with one or more components of the components missing. It is unattractive to confine ourselves to analysing only the complete data cases. This may discard too much data and possibly introduce bias with regard to the ones retained. To use the individuals with missing data, we must *complete* them, the so-called imputation [38]. There is by now a very substantial literature on imputation but to do a fully model-based imputation in the Bayesian setting results in latent variables and Gibbs looping. In this sense, the Gibbs sampler extends the E-M algorithm to provide full posterior inference rather than an MLE with an asymptotic variance.

As a simple example, consider multivariate normal data, $\mathbf{Y}_i \sim N(\boldsymbol{\mu}_i, \Sigma)$ where the components of $\boldsymbol{\mu}_i$ may have regression forms in suitable covariates. Some components of some of the $\mathbf{Y}_i$s are missing. In order to perform the imputation, we do basic Gibbs sampling: we update the parameters given values for the missing data, then update the missing data given values for the parameters. Another standard example considers missing categorical counts within a multinomial model where the multinomial cell probabilities might be modelled using some sort of multivariate logit model [1]. For instance, some categories are aggregated/collapsed so counts for the disaggregated categories are missing. Again, we can envision a usual Gibbs loop: update the parameters given values for all the counts, update the missing counts given values for the parameters.

3.3 Latent variables

As noted above, latent variables are at the heart of most hierarchical modelling. Here, we provide examples which suggest they can be envisioned beyond random effects or missing data.[1] Latent variable models customarily result in a hierarchical specification of the form $[\mathbf{Y}|\mathbf{Z}][\mathbf{Z}|\boldsymbol{\theta}][\boldsymbol{\theta}]$. Here, the Ys are observed, the Zs are latent and the 'regression' modelling is shifted to the second stage.

An elementary version of a latent variable model arises with binary data models. In particular, the usual binary response model adopts a logit or probit link function. Illustrating with the probit, suppose $Y_i \sim \text{Bernoulli}(p(\mathbf{X}_i))$ (more generally, we can have $Y_i \sim Bin(n_i, p(\mathbf{X}_i))$). Specifically, let $\Phi^{-1}(p(\mathbf{X}_i)) = \mathbf{X}_i\boldsymbol{\beta}$ with a prior on $\boldsymbol{\beta}$. In fitting this model using MCMC computation, it is awkward to sample $\boldsymbol{\beta}$ using the likelihood in this form. So, let us introduce $Z_i \sim N(\mathbf{X}_i\boldsymbol{\beta}, 1)$. Immediately, $P(Y_i = 1) = \Phi(\mathbf{X}_i\boldsymbol{\beta}) = 1 - \Phi(-\mathbf{X}_i\boldsymbol{\beta}) = P(Z_i \geq 0)$. Once we bring in these Z_is, we achieve a routine Gibbs sampler: update the Zs given $\boldsymbol{\beta}, \mathbf{y}$ (this requires sampling from a truncated normal), update $\boldsymbol{\beta}$ given the Zs and $\mathbf{y}$ (this is the usual, typically conjugate normal updating). This approach was first articulated in the literature by Albert and Chib [2].

It is clear that this approach can be readily extended to general ordinal categorical data settings [13, 33]. In particular, for each i, the Bernoulli trial is replaced with a multinomial trial. There is still a latent Z_i, still following say a Gaussian linear regression. Now the multinomial outcomes are created by introducing cut points along the real line; the intervals determined by the cut points allocate probabilities to each of the multinomial outcomes. The cut points will be random as well, noting that, in order to identify the intercept in the regression, the smallest cut point can be taken to be 0, without loss of generality.

Another routine generalization of this approach accommodates censored or truncated data models. As a simple illustration, suppose we observe a variable with a point mass at 0 and the remainder of its mass spread over R^+ or perhaps $(0, 1]$. In the first case, such data arise when studying, for instance, daily precipitation at a location; in the second case when considering, for instance, the

[1] While, in a sense, all variables we can not observe are 'missing', in the previous subsection we take missing to mean some components of the data, while here they will be variables different from the data.

proportion of a particular land use classification over a region. Here, with observed Y_is, we can introduce Z_is such that $Y_i = g(Z_i)$ if $Z_i > 0$, $Y_i = 0$ if $Z_i \leq 0$, with $g(\cdot)$ a link function from R^1 to R^+ or perhaps to $(0, 1]$. Then, we could model the Z_i's using say a usual Gaussian linear regression.

Another setting for latent variables is change point problems [16]. Frequently, we observe a process over the course of time during which we would like to assess whether some sort of change in regime has occurred. Practically speaking, this requires the notion of a 'least' significant change. That is, we may be able to identify even a very small change with enough data but we will find challenging the case where the support for the change has 'no change' as a boundary point.

In the change point setting two sampling scenarios can be envisioned. In the first, we have a full set of data. Then we look, retrospectively, to try to find if change(s) occurred. In the second, we look at the data sequentially and we try to identify change(s) as the data collection proceeds. We illustrate with a simple version of the first scenario. Let $f_1(y|\theta_1)$ be the density for i.i.d. observations before the change point, $f_2(y|\theta_2)$ the density after the change point. With data Y_i, $i = 1, 2, \ldots, n$, let K be the change point indicator, i.e. $K \in \{1, 2, \ldots, n\}$ where $K = k$ means change at observation, $k + 1$; $k = n$ means 'no change'. Then, the model is

$$L(\theta_1, \theta_2, k; \mathbf{y}) = \Pi_{i=1}^{k} f_1(y_i|\theta_1)\Pi_{i=k+1}^{n} f_2(y_i|\theta_2)$$

Again, we have a hierarchical model, $[\mathbf{y}|k, \theta_1, \theta_2][K = k][\theta_1, \theta_2]$. (Note that we do not include any parameters in the prior for K; with only one change point, we could not hope to learn about such parameters.) With a prior on θ_1, θ_2, K, we have a full model specification. Again, a simple Gibbs sampler emerges for model fitting: update θs given $k, \mathbf{y}$ (so, we know exactly which observations are assigned to which density); update k given θs and $\mathbf{y}$ (this is just a discrete distribution, easily sampled). Obvious generalizations would allow the ys to be dependent [51], to have order restrictions on θs [70], to imagine multiple change points [29]. Also, in this version, time is discretized to the set of times when the measurements were collected. Extension to continuous time is available using point process models [53].

Errors in variables models [14, 23] offer another latent variables setting. Loosely stated, the objective is to learn about the relationship between say Y and X. Unfortunately, X is not observed. Rather, we observe say W instead of X. In some cases, W will be a version of X, subject to measurement error, i.e. W may be X_{obs} while X may be X_{true}. In other cases W may be a variable (variables) that plays the role of a surrogate for X. In any event, if we envision a joint distribution for W and X, we may condition in either direction. If we specify a model for $W|X$ we refer to this as a measurement error model [11, 44], imagining W to vary around the true or desired X; if we specify a model for $X|W$ we refer to this as a Berkson model [44, 57]. In fact, we can imagine further errors in variables component—perhaps we observe Z, a surrogate for Y. Altogether we have a hierarchical model with latent Xs, possibly Ys. In particular, for the measurement error case, we have

$$\Pi_i[Z_i|Y_i, \gamma][Y_i|X_i, \beta][W_i|X_i, \delta][X_i|\alpha]$$

while for the Berkson case we have

$$\Pi_i[Z_i|Y_i, \gamma][Y_i|X_i, \beta][X_i|W_i, \delta]$$

Typically, we will also have some *validation* data to inform about the components of the specification. We might have some X, Y pairs or perhaps some X, W pairs. It is noteworthy that the measurement error version requires a model (prior) for X while the Berkson model does not. In many applications the former will be more natural. However, in some contexts, model-fitting is only feasible with the latter [4]. In any event, what is most remarkable is that, within this hierarchical framework, using a full Bayesian specification, we can learn about the relationship between Y and X without ever observing X (and, possibly, without observing Y as well). This reveals the power of hierarchical modelling but, evidently, what we can learn depends upon the form of what we specify.

3.4 Mixture models

Mixture models have now become a staple of modern stochastic modelling [46, 67]. This has arisen on at least two accounts: (i) their flexibility to model unknown distributional shapes and (ii) their intuition in representing a population in terms of groups/clusters that may exist but are unidentified. Mixture models come in several flavours—parametric or nonparametric, incorporating finite, countable or uncountable mixing. In this regard, they are sometimes referred to as classification problems or discriminant analysis, reflecting a goal of assigning an individual to a population or assessing whether individuals belong to the same population.

The most rudimentary finite mixture version takes the form

$$\mathbf{Y} \sim \sum_{l=1}^{L} p_l f_l(\mathbf{Y}|\boldsymbol{\theta}_l).$$

Often the f_l are normal densities, whence we obtain a normal mixture. If we assume L is specified and we observe $\mathbf{Y}_i$, $i = 1, 2, \ldots, n$, then what is latent is a *label* for each $\mathbf{Y}_i$, i.e. an indicator of which component of the mixture Y_i was drawn from. These latent labels would be such that if $L_i = l$, then $\mathbf{Y}_i \sim f_l(\mathbf{Y}|\boldsymbol{\theta}_l)$. Upon introducing these labelling variables, the resulting hierarchical model becomes

$$\Pi_i[\mathbf{Y}_i|L_i, \boldsymbol{\theta}][\Pi_i[L_i|\{p_l\}][\boldsymbol{\theta}][\{p_l\}].$$

Here, $\boldsymbol{\theta}$ denotes the collection of $\boldsymbol{\theta}_l$. Once again, Gibbs sampling is routine to implement. We create a loop that updates $\boldsymbol{\theta}$, $\{p_l\}$ given the Ls and the data. With observations assigned to components, this becomes the equivalent of an analysis of variance problem to learn about the $\boldsymbol{\theta}_l$. To update the L_is given $\boldsymbol{\theta}$, $\{p_l\}$ and the data, requires sampling from an L-valued discrete distribution where, for a given i, the mass on the ls is determined by the relative likelihood for the observed $\mathbf{Y}_i$, as well as the prior on the p_ls (often a uniform). Hence, we obtain individual assignment probabilities as well as global assignment weights. Richer versions introduce covariates into the $\boldsymbol{\theta}_l$s and, possibly, into the p_ls [37, 60, 71]. Further challenge is added if L is unknown with a prior specification. Now, since the dimension of the model changes with L we may attempt reversible jump MCMC [30] to learn about L. An alternative might be to carry out model choice across a set of Ls.

It is evident that such mixture models are not identifiable, i.e. the subscripts can be permuted and the same mixture distribution results. This has led to substantial discussion in the literature [47, 55] with regard to introducing identifiability constraints or the possibility of fitting the MCMC, allowing multi-modality in the posterior in the absence of identifiability. The former path seems to be the most widely used, with order constraints on the means, imposed in some fashion, being the most common choice for achieving identifiability.

Next, we recall that many familiar distributional models can be developed through continuous mixing, i.e. closed forms are achieved by virtue of conjugacy between the mixand model and the mixing distribution. Well-known examples include scale mixing of normals to obtain t-distributions, as well as Poisson–Gamma (equivalently negative binomial) and beta-binomial models. In some modelling situations we may seek individual level mixing variables, whence we would introduce such variables as latent quantities. Again, a hierarchical model arises. An illustration is in the case of outlier detection through the use of suitable individual level gamma mixing of normals. Here outliers are 'detected' through the magnitudes of their associated mixing variables, i.e. the heavier the posterior tails, the more we are inclined to classify the observation as an outlier. See, e.g. [62, 68, 69].

We conclude with a brief discussion of nonparametric mixture models, illustrating with the Dirichlet process. That is, the foregoing finite models are all parametric in the sense that they are finite-dimensional specifications. The continuous mixture models are as well since they are characterized by the parameters of the mixing distribution as well as, perhaps, some parameters in the mixand distribution (that are not mixed). The so-called nonparametric setting envisions the mixing distribution to be unknown and drawn randomly from a family of mixing distributions. In particular, the family of all possible mixing distributions with regard to a parameter in the mixand would be a nonparametric specification, but it is not possible to assign a distribution over this entire family. Rather, in practice, we adopt a distribution over a subfamily of these distributions; the Dirichlet process (DP) is one example.

The literature on the use of DP priors has been growing, primarily because they are easy to specify, attractive to interpret and ideally suited for model-fitting within an MCMC framework. In particular, the stickbreaking representation of the DP [61] makes it most convenient for use. Let $\theta_1^*, \theta_2^*, \ldots$ be i.i.d. random elements independently and identically distributed according to the law G_0. G_0 can be a distribution over a general probability space, allowing the θ^*s to be random objects such as scalars, vectors, a stochastic process of random variables or even a distribution itself [58]. Let $q_1, q_2, \ldots$ be random variables independent of the θ^*s and i.i.d. among themselves with common distribution Beta$(1, \alpha)$. If we set $p_1 = q_1, p_2 = q_2\,(1 - q_1), \ldots, p_k = q_k \prod_{j=1}^{k-1}(1 - q_j), \ldots$, the random probability measure defined by

$$G(\cdot) = \sum_{k=1}^{\infty} p_k\, \delta_{\theta_k^*}(\cdot)$$

is distributed according to a DP. So, G, which is equivalent to $\{p_k\}$ and $\{\theta_k^*\}$, is a random distribution.

If we let G be the mixing distribution, with (mixand) kernel say $f(y; \theta)$, then the resulting mixture model becomes $f(y; G) = \int f(y; \theta)G(d\theta) = \sum_k p_k f(y; \theta_k^*)$, i.e. $f(y; G)$ is a countable mixture model. Again, with G random, so is $f(y : G)$. With data say $Y_i, i = 1, 2, .., n$, we immediately have the hierarchical form

$$\Pi_i[Y_i|\theta_i]\Pi_i[\theta_i|G][G|G_0, \alpha][G_0|\eta_G][\eta_G][\alpha].$$

Here G provides the latent structure, rather than assuming the $\theta_i \sim G_0$. Indeed, α is a precision parameter, reflecting how much G varies around G_0.

This mixture specification allows considerable technical development [49] as well as substantial extension [65], which we skip here. Rather, we note a few important features. First, we see that G is almost surely discrete, though the resulting $f(y; G)$ is not. Second, possibly paradoxical, though this is a countable mixture problem rather than finite, there are no identifiability issues. Third, this model allows ties, i.e. both θ_i and $\theta_{i'}$ can take the value θ_k^*. This suggests the introduction of labels, as above, to indicate which θ^* was drawn by each i. It also clarifies that there will be clustering for the Y_is. However, here the clustering is of a different type. There are no component models; the values of the θ^* don't really mean anything. They change with iteration and would not be saved, *rather*, we can only report the proportion of iterations when say Y_i and $Y_{i'}$ are in the same cluster, but nothing about a *common* distribution for them (clarifying why there are no identifiability problems). Finally, much can be said about fitting such models using MCMC (see, e.g. [42, 43, 52, 66]).

3.5 Random effects

Returning to the random effects setting, let us first consider individual level longitudinal data with interest in explanation through growth curves. A natural specification would model individual level curves centred around a population level curve. We would need the population level curve to see *average* behaviour of the process; we need individual level curves in order, for example, to prescribe *individual* level treatment. With parameters at each level and a third stage of hyperparameters, we again see a hierarchical form.

More precisely, if Y_{ij} is jth measurement for ith individual, let

$$Y_{ij} = g(\mathbf{X}_{ij}, \mathbf{Z}_i, \boldsymbol{\beta}_i) + \epsilon_{ij}$$

where $\epsilon_{ij} \sim N(0, \sigma_i^2)$. The form for g depends upon the application. It is often linear but need not be. We set $\boldsymbol{\beta}_i = \boldsymbol{\beta} + \eta_i$ where the η_i have mean 0 (or perhaps replace $\boldsymbol{\beta}$ with a regression in the $\mathbf{Z}_i$). Then the $\boldsymbol{\beta}_i$ (or the η_i) are the random effects. They provide the individual curves with $\boldsymbol{\beta}$ providing the global curve. Evidently, this specification falls under the heading of a CIHM as well. Learning with regard to any individual curve will borrow strength from the information about the other curves.

Customarily, random effects are modelled using normality. However, they need not be i.i.d. That is, if say the scalar ω_i is associated with individual i, we need not insist that the vector, $\boldsymbol{\omega}$, of ω_is, be distributed as say, $\boldsymbol{\omega} \sim N(\mathbf{0}, \sigma^2 I)$. We can replace $\sigma^2 I$ with $\Sigma(\boldsymbol{\theta})$ where $\Sigma(\boldsymbol{\theta})$ has structured dependence. That is, with say n individuals, we could not learn about an arbitrary positive definite $n \times n$ matrix Σ, but we could learn about Σ defined as a function of only a few parameters.

Structured dependence is at the heart of time series, spatial and spatio-temporal modelling and is frequently specified through a Gaussian process (GP). Here, we illustrate in the spatial setting, envisioning data in the form $Y(s_i), i = 1, 2, \ldots, n$, i.e. n observations at n different spatial locations s_i. We usually think of the s_i as being in R^2 but this is not necessary. We could imagine three-dimensional locations and, more generally, replacing geographic space with say covariate space. This takes us into the now very popular world of computer models [40, 41, 45, 50]. In any event, with a GP, we need only specify finite-dimensional (e.g. n) joint distributions with the joint dependence determined by a valid covariance function. Customarily, the covariance function assigns stronger association to variables that are closer to each other in geographic space. A common example is the exponential, $\text{cov}(Y(s), Y(s')) = \sigma^2 \exp(-\phi||s - s'||)$. With n locations, $\Sigma(\boldsymbol{\theta})_{ij} = \sigma^2 \exp(-\phi||s_i - s_j||)$.

Now, we can specify the standard univariate spatial model, incorporating spatial random effects [3]. Let

$$Y(s) = \boldsymbol{x}^T(s)\boldsymbol{\beta} + w(s) + \epsilon(s).$$

Here, the residual is partitioned into two pieces: one is spatial, $w(s)$, i.e. the $w(s)$ are spatial random effects and one is non-spatial, $\epsilon(s)$, i.e. the $\epsilon(s)$ are usual i.i.d. errors. $w(s)$ is from a Gaussian process, introducing say the stationary covariance function $\sigma^2 \rho(s - s'; \phi)$. $\epsilon(s)$ adds pure error with variance τ^2.

Interpretations that can be attached to $\epsilon(s)$ include: (i) a pure error term; the model is not perfectly spatial; τ^2, and σ^2 are variance components; (ii) measurement error or replication variability causing discontinuity in spatial surface $Y(s)$ (assuming $x(s)$ and $w(s)$ are continuous);

(iii) microscale structure; there may be dependence at a scale smaller than the smallest inter-location distance but in the absence of any data to learn about such dependence, independence is assumed.

Again, suppose we have data $Y(s_i), i = 1, \ldots, n$, and let $\boldsymbol{Y} = (Y(s_1), \ldots, Y(\boldsymbol{s}_n))^T$. The above model yields a marginal covariance matrix for $\boldsymbol{Y}$ of the form

$$\Sigma = \sigma^2 R(\phi) + \tau^2 I.$$

with $R_{ij} = \rho(s_i - s_j; \phi)$. The dependence incorporated into $R(\phi)$ enables us to learn about both variance components. Setting $\boldsymbol{\theta} = (\boldsymbol{\beta}, \sigma^2, \tau^2, \phi)^T$, we see that this is not a high-dimensional problem (perhaps half a dozen components in $\boldsymbol{\beta}$, three or four in Σ).

The likelihood is given by

$$\boldsymbol{Y}|\boldsymbol{\theta} \sim N(X\boldsymbol{\beta}, \sigma^2 R(\phi) + \tau^2 I)$$

Typically, independent priors are chosen for the parameters,

$$p(\boldsymbol{\theta}) = p(\boldsymbol{\beta})p(\sigma^2)p(\tau^2)p(\phi)$$

Common candidates are multivariate normal for $\boldsymbol{\beta}$, and inverse gamma for σ^2 and τ^2. Specification of $p(\phi)$ depends upon the choice of the correlation function, ρ; a uniform or discrete prior is usually selected. $p(\boldsymbol{\beta})$ can be 'flat' (improper). Care must be taken with regard to the parameters in Σ. First, results from [6] show that, for instance, if the prior on $\boldsymbol{\beta}, \sigma^2, \phi$ is of the form $\frac{\pi(\phi)}{(\sigma^2)^{a+1}}$ with $\pi(\cdot)$ proper, then, an improper posterior results if $a = 0$. This returns us to the point made in Section 3.2 regarding the problem with using $IG(\epsilon, \epsilon)$ priors for σ^2. Again, the posterior will be nearly improper. Again, it is safer to adopt an $IG(a, b)$ with $a \geq 1$. A further issue is that, without the pure error variance, τ^2, with say the exponential covariance function, we can not *identify* both σ^2 and ϕ [74]. We can only identify the product. (This is true for the more general Matérn class of covariance functions, as well.) So an informative prior on at least one of these parameters will be needed in order to achieve well-behaved MCMC model-fitting.

Of course, we may ask, "Where is the hierarchical modelling?" In fact, the foregoing is really a hierarchical setup by considering a first stage likelihood conditional on the spatial random effects $\boldsymbol{w} = (w(s_1), \ldots, w(s_n))$. That is, we have,

First stage: $\boldsymbol{Y}|\boldsymbol{\theta}, \boldsymbol{w} \sim N(X\boldsymbol{\beta} + \boldsymbol{w}, \tau^2 I)$.

The $Y(s_i)$ are conditionally independent given the $w(\boldsymbol{s}_i)$s.

Second stage: $\boldsymbol{w}|\sigma^2, \phi \sim N(O, \sigma^2 R(\phi))$

Here, **w** provides the process model.

Third stage: priors on $(\boldsymbol{\beta}, \tau^2, \sigma^2, \phi)$

With regard to model-fitting, we seek the marginal posterior $p(\boldsymbol{\theta}|\boldsymbol{y})$, which is the same under the marginal and hierarchical settings. That is, we can fit the model as $f(\boldsymbol{y}|\boldsymbol{\theta})p(\boldsymbol{\theta})$ or as $f(\boldsymbol{y}|\boldsymbol{\theta}, \boldsymbol{w})p(\boldsymbol{w}|\boldsymbol{\theta})p(\boldsymbol{\theta})$. Fitting the marginal model is usually computationally better behaved. We have a lower-dimensional MCMC (no $\boldsymbol{w}$s). Additionally, $\sigma^2 R(\phi) + \tau^2 I$ will be *diagonally dominant*, hence more stable than $\sigma^2 R(\phi)$ in terms of matrix inversion needed for sampling and likelihood evaluation.

Of course, interest will be in the spatial random effects (and, in fact, in the entire spatial surface of the ws) in order to see the pattern of spatial adjustment. We have not lost the $\boldsymbol{w}$s with the marginalized sampling. They are easily recovered, one-for-one with the posterior samples of $\boldsymbol{\theta}$, via familiar composition sampling:

$$p(\boldsymbol{w}|\boldsymbol{y}) = \int p(\boldsymbol{w}|\boldsymbol{\theta}, \boldsymbol{y})p(\boldsymbol{\theta}|\boldsymbol{y})d\boldsymbol{\theta}$$

In practice, we might have a non-Gaussian first stage. For instance, $Y(s)$ need not be a continuous variable. We can imagine a binary or two-colour map, i.e. at every s there is a light bulb and a realization of the map is a surface of 1s and 0s according to whether or not the bulb is illuminated at s. Specific examples include: presence/absence of a species at a location; abundance of a species at a location; was precipitation or deposition at a location measurable or not; the number of insurance claims by residents of a single family home at s; land use classification at a location (not ordinal).

To build models, we replace the Gaussian likelihood with an appropriate exponential family member, resulting in spatial generalized linear models [18]. The hierarchical model above recurs:

First stage: $Y(s_i)$ are conditionally independent given $\boldsymbol{\beta}$ and $w(s_i)$ with $f(y(s_i)|\boldsymbol{\beta}, w(s_i), \gamma)$ an appropriate non-Gaussian likelihood such that

$$g(E(Y(s_i))) = \eta(s_i) = \boldsymbol{x}^T(s_i)\boldsymbol{\beta} + w(s_i),$$

where η is a canonical link function (such as a logit) and γ is a dispersion parameter.

Second stage: Model $w(s)$ as a Gaussian process (GP), as above:

$$\boldsymbol{w} \sim N(\mathbf{0}, \sigma^2 H(\phi))$$

Third stage: Priors and hyperpriors according to the model specification.

Two points are worth noting here. First, we lose conjugacy between the first and second stage. We can not marginalize over the ws. We will have to generate them in the model-fitting; we will have to work with the hierarchical model itself. Second, it is not sensible to add a pure error term in the specification of g. We already have the equivalent of conditional independence due to the first stage exponential family mechanism that replaced the Gaussian choice. In fact, were we to include such ϵs in the model, poorly behaved MCMC will ensue.

In summary, if we introduce the spatial random effects in the transformed mean, then, with continuous covariates, this encourages the means of spatial variables at proximate locations to be close to each other. In spite of the conditional independence, marginal spatial dependence is induced between, say, $Y(s)$ and $Y(s')$, but the observed $Y(s)$ and $Y(s')$ need not be close to each other. In fact, there is no smoothness in the $Y(s)$ surface. In different terms, our second stage modelling is attractive for spatial explanation of the *process*, here in terms of the mean. It is not our intention to achieve smoothness in the observed surface.

3.6 Dynamic models

Dynamic models have now become a standard formulation for a wide variety of processes, including financial and environmental applications. Alternative names for them in the literature include Kalman filters, state space models and hidden Markov models [28, 48, 56]. They introduce a first stage (or observational model) and then a second stage (or transition model), with third stage hyperparameters. Again, the first stage provides the data model while the second stage provides a latent dynamic process model. See, e.g. West and Harrison [72] for a full development. In particular, there is a substantial modellling opportunity at the second stage, allowing evolution of process variables or process parameters, in either case, driven by covariate information.

The basic dynamic model takes the form,

$$\mathbf{Y}_t = g(\mathbf{X}_t, \boldsymbol{\theta}_1) + \boldsymbol{\epsilon}_t, \text{ the observation equation}$$

with

$$\mathbf{X}_t = h(\mathbf{X}_{t-1}; \boldsymbol{\theta}_2) + \boldsymbol{\eta}_t, \text{ the transition equation}$$

Evidently, time t is discrete and we are putting the dynamics in the mean. We illustrate with dynamic space–time models. Consider:

Stage 1: Measurement equation

$$Y(s,t) = \mu(s,t) + \epsilon(s,t)\,;\ \epsilon(s,t) \overset{ind}{\sim} N\left(0, \sigma_\epsilon^2\right).$$
$$\mu(s,t) = \mathbf{x}^T(s,t)\,\tilde{\boldsymbol{\beta}}(s,t).$$
$$\tilde{\boldsymbol{\beta}}(s,t) = \boldsymbol{\beta}_t + \boldsymbol{\beta}(s,t)$$

with

Stage 2: Transition equation

$$\boldsymbol{\beta}_t = \boldsymbol{\beta}_{t-1} + \boldsymbol{\eta}_t,\ \boldsymbol{\eta}_t \overset{ind}{\sim} N_p\left(0, \Sigma_\eta\right).$$
$$\boldsymbol{\beta}(s,t) = \boldsymbol{\beta}(s,t-1) + \mathbf{w}(s,t)$$

where $\mathbf{w}(s,t)$ are independent (over t) innovations of a spatial process (see, e.g., [25]). As noted above, this specification can be connected to a linear Kalman filter [36]. Thus, Bayesian model-fitting using the forward filter, backward sample (ffbs) algorithm [15, 21] becomes the customary approach.

Recently, Wikle and colleagues [31, 73] have adapted these dynamic forms to the fitting of models motivated by stochastic partial differential equations (SPDEs). The approach is to discretize time, raising the issue of sensitivity to temporal resolution as well as the fact that the limiting dependence structure from the discrete form need not be that associated with the SPDE. Nonetheless, the discretized specification stands as a model in its own right, worthy of fitting to enable inference.

In particular, there are many interesting ecological diffusions which can be applied to study the behaviour over time (and, perhaps space) of: (i) emerging diseases such as avian flu or H1N1 flu; (ii) exotic organisms, e.g. invasive plants and animals; (iii) the evolution of the distribution of size or age of a species; (iv) the dynamics explaining phenomena such as transformation of landscape, deforestation, land use classifications and urban growth.

Our objective for such processes is to forecast likely spread in space and time with associated uncertainty. We anticipate that the evolution will be nonlinear and nonhomogeneous in space and time, driven by explanatory covariates. We start with deterministic integro-differential equations or with partial differential equations, raising the question of how to add uncertainty. Loosely speaking, we provide theoretical models, with the data 'varying' around them, creating hierarchical specifications with second stage dynamics that lead us to the foregoing state space models. As remarked above, too much simplification would be required in order to obtain analytical solutions to the differential equations. So instead, we adopt discretization to fit models. Again, the issue is whether we care about the deterministic PDE or whether we should just work with the discrete time version, incorporating the behavioural features we want.

Hooten and Wikle [31] use data from the Breeding Bird Survey (BBS) to study the diffusion of the Eurasian collared dove. Gridding the eastern United States, let Z_{it} be the count in box i in year t, let n_{it} be the number of visits to cell i in year t and let λ_{it} be the *intensity* for box i in year t. Then, assume $Z_{it} \sim \text{Po}(n_{it}\lambda_{it})$ with $\log\lambda_{it} = w_{it} + \epsilon_{it}$. Here, the ϵ_{it} are i.i.d. (pure error or microscale variation). The focus is on the $\mathbf{w}_t$; they tell the diffusion story.

More precisely, the dynamics here are associated with continuous space and discrete time, i.e. $w_t(s)$. Without loss of generality, we take $\mathbf{t} = (1, 2, \ldots, T)$. We simplify $w_t(s)$ to be first-order Markov, i.e. for locations $s_1, s_2, \ldots, s_n$, let $\mathbf{w}_t = (w_t(s_1), w_t(s_2), \ldots, w_t(s_n))^T$. Then, $[\mathbf{w}_t|\mathbf{w}_0, \mathbf{w}_1, \ldots, \mathbf{w}_{t-1}] = [\mathbf{w}_t|\mathbf{w}_{t-1}]$. For example, we could set $\mathbf{w}_t = H\mathbf{w}_{t-1} + \boldsymbol{\eta}_t$ where $\eta_t(s)$ incorporates spatial structure. We have a vector AR(1) model and H is called the *propagator* matrix.

How might we specify H? The choice $H = I$ is not stationary. It leads to explosive behaviour with no interaction across space and time. Hence, it is not realistic for most dynamic processes of interest. The choice, $H = \text{Diag}(h)$ where $\text{Diag}(h)$ has diagonal elements $0 < h_i < 1$ avoids explosive behaviour but still offers no interactions. In fact, what we seek is integro-difference equation (IDE) dynamics over the space of locations:

$$w_t(s) = \int h(s, r; \phi) w_{t-1}(r) dr + \eta_t(s).$$

Here, h is a 'redistribution kernel' that determines the rate of diffusion and the advection. If required $w > 0$, then we could work with

$$\log w_t(s) = \log(\int h(s, r; \phi) w_{t-1}(r) dr) + \eta_t(s).$$

Alternatively, we could adopt

$$v_t(s) = \int h(s, r; \phi) v_{t-1}(r) dr$$

and

$$\log w_t(s) = \log v_t(s) + \eta_t(s).$$

Various forms can be considered for h, e.g. $h(s, r; \phi)$, $h(s, r; \phi(r))$, $h_t(s, r; \phi)$. Then, discretization of the spatial region will enable us to obtain H.

Lastly, we can reconnect to the PDE. In fact, recall the linear PDE, $\frac{dw(s,t)}{dt} = h(s)w(s, t)$. Finite differencing yields $w(s, t + \Delta t) - w(s, t) = h(s)w(s, t)\Delta t$, i.e. $w(s, t + 1) \approx \tilde{h}(s)w(s, t)$, suffering the same limitations as above. Hence, we need more general PDEs, in particular those that diffuse in space over time. Such PDEs can motivate IDEs, and therefore clarify H.

3.7 Data fusion

As a last example, we take up a modelling problem in data fusion. Data assimilation has some history in the meteorology community [34] but has only recently received serious attention in the statistics community. In particular, let us consider the Bayesian melding model of Fuentes and Raftery [22] which has gained considerable attention and has already been used in several applications [20, 63].

We present the model in the spatial setting, where we would be concerned with fusing a dataset consisting of measurements at monitoring stations, say exposure to ozone or particulate matter, with the output of a computer model for that exposure. The former is quite accurate but only sparsely available, often with missingness. The latter is uncalibrated but is available everywhere. The former is associated with point-referenced locations, the latter is supplied for grid cells, for example 12 km squares.

The melding or fusion model envisions a latent true exposure surface which is informed by both the station data and the computer model data. The hierarchical model arises with the two data sources providing the first-stage model. The latent true model provides a process specification at the second stage, with hyperparameters at the third stage, as is familiar by now. More precisely, let the $Y(s_i)$ be the observed station data at s_i, let $X(B_j)$ be the computer model output for grid cell B_j and let $Z(s)$ be the true exposure surface. We model the station data as a measurement error model, i.e.

$$Y(s_i) = Z(s_i) + \epsilon(s_i)$$

where the ϵ are a pure error specification. We model the computer output as a calibration specification, i.e. for grid cell B_j,

$$X(B_j) = \int_{B_j} (a(s) + b(s)Z(s) + \delta(s))ds$$

where $a(s)$ and $b(s)$ are Gaussian processes with the $\delta(s)$s being pure error. Here, the challenge for the melding approach emerges. The integral for $X(B_j)$ is stochastic because, for example, $a(s)$ is not a function but a realization of a stochastic process. So, the integral can not be computed explicitly; at best we can implement a Monte Carlo integration [3]. If we have to do many of these (and, in a practical situation we would have many grid cells), we would have an enormous number of Monte Carlo integrations to do at each iteration of an MCMC fitting algorithm.

Finally, we have the second-stage process model, say,

$$Z(s) = \mu(s) + \eta(s)$$

Here, the mean, $\mu(s)$, captures the large-scale structure, perhaps through covariates, perhaps through a trend surface, while $\eta(s)$ capture the small-scale structure or second-order dependence through a GP. Again, model-fitting is challenging; Fuentes and Raftery [22] observe that they were only able to successfully fit the model in the case that $b(s) = b$.

So, again, Bayesian melding has two important limitations. First, it is computationally intensive. Since computer model outputs usually cover large spatial domains, thereby introducing a very large number of grid cells, a very large number of stochastic integrals need to be computed. Secondly, as proposed, it does not incorporate a temporal dimension. Given the computational burden associated with Bayesian melding in its static version, a dynamic extension is, practically, infeasible. Fully model-based alternatives, so-called downscalers [9, 10], can address these limitations by using only a first-stage model for the relatively sparse station data.

3.8 Summary

We have argued that hierarchical models provide the stochastic framework within which to develop integrative process models. We have shown, with many examples, that these models typically share a common structure. There is a first-stage data model, there is a second-stage process model that is latent, i.e. it is endowed with a full model specification but it is unobserved, and a third stage which incorporates prior specifications for all of the remaining parameters in the model. We have noted that, in order to get the uncertainty right, these models should be fitted within the Bayesian framework. We have also noted that fitting of these models introduces familiar Gibbs looping. Hence, it is straightforward to envision how the MCMC model-fitting should be implemented. However, according to the size of the dataset and the complexity of the specifications, such model-fitting can be very challenging, perhaps infeasible. Indeed, this limitation will become more of a constraint as we continue to seek models which stretch the limits of our computing capabilities. Hence, we imagine a computing future built around simulation based model-fitting of these hierarchical forms but incorporating suitable approximation. Thus, the 'art' will encompass both specification (with comparison) and approximate fitting (to enable inference and comparison).

References

[1] Agresti, A. (2010). *Analysis of Ordinal Categorical Data* (2nd edn). Wiley, New York.

[2] Albert, J. H. and Chib, S. (1993). Bayesian analysis of binary and polychotomous response data. *Journal of the American Statistical Association*, **88**, 669–679.

[3] Banerjee, S., Carlin, B. P. and Gelfand, A. E. (2004). *Hierarchical Modelling and Analysis for Spatial Data*. Chapman & Hall/CRC, Boca Raton.

[4] Barber, J. J., Gelfand, A. E. and Silander, J. A. (2006). Modelling map positional error to infer true feature location. *Canadian Journal of Statistics*, **34**, 659–676.

[5] Berger, J. O. and Strawderman, W. E. (1996). Choice of hierarchical priors: admissibility in estimation of normal means. *Annals of Statistics*, **24**, 931–951.

[6] Berger, J. O., De Oliveira, V. and Sansó, B. (2001). Objective Bayesian analysis of spatially correlated data. *Journal of the American Statistical Association*, **96**, 1361–1374.

[7] Berger, J. O., Strawderman, W. E. and Tang, D. (2005). Posterior propriety and admissibility of hyperpriors in normal hierarchical models. *Annals of Statistics*, **33**, 606–646.

[8] Berliner, L. M. (1996). Hierarchical Bayesian time series models. In *Maximum Entropy and Bayesian Methods* (K. Hanson and R. Silver, eds.). Kluwer Academic Publishers, 15–22.

[9] Berrocal, V. J., Gelfand, A. E. and Holland, D. M. (2010). A spatio-temporal downscaler for output from numerical models. *Journal of Agricultural, Biological and Environmental Statistics*, **15**, 176–197.

[10] Berrocal, V. J., Gelfand, A. E. and Holland, D. M. (2010). A bivariate space-time downscaler under space and time misalignment, *Annals of Applied Statistics*, **4**, 1942–1975.

[11] Berry, S. M., Carroll, R. J. and Ruppert, D. (2002). Bayesian smoothing and regression splines for measurement error problems. *Journal of the American Statistical Association*, **97**, 160–169.

[12] Box, G. E. P. and Tiao, G. C. (1973). *Bayesian Inference in Statistical Analysis*. Addison-Wesley, Mass.

[13] Bradlow, E. T. and Zaslavsky, A. M. (1999). Hierarchical latent variable model for ordinal data from a customer satisfaction survey with "No Answer" responses. *Journal of the American Statistical Association*, **94**, 43–52.

[14] Carroll, R. J., Ruppert, D. and Stefanski, L. A. (1995). *Measurement Error in Nonlinear Models*. Chapman & Hall, Boca Raton, Fl.

[15] Carter, C. K. and Kohn, R. (1994). On Gibbs sampling for state space models. *Biometrika*, **81**, 541–553.

[16] Chib, S. (1998). Estimation and comparison of multiple change-point models. *Journal of Econometrics*, **86**, 221–241.

[17] Dempster, A. P., Laird, N. M. and Rubin, D. B. (1977). Maximum likelihood from incomplete data via the EM algorithm. *Journal of the Royal Statistical Society, Series B*, **39**, 1–38.

[18] Diggle, P. J., Moyeed, R. A. and Tawn, J. A. (1998). Model-based geostatistics (with discussion). *Applied Statistics*, **47**, 299–350.

[19] Fienberg, S. E. (2011). Bayesian models and methods in public policy and government settings. *Statistical Science*, **26**, 212–226.

[20] Foley, K. M. and Fuentes, M. (2008). A statistical framework to combine multivariate spatial data and physical models for hurricane wind prediction. *Journal of Agricultural, Biological and Environmental Statistics*, **13**, 37–59.

[21] Frühwirth-Schnatter, S. (1994). Data augmentation and dynamic linear models. *Journal of Time Series Analysis*, **15**, 183–202.

[22] Fuentes, M. and Raftery, A.E. (2005). Model evaluation and spatial interpolation by Bayesian combination of observations with outputs from numerical models. *Biometrics*, **61**, 36–45.

[23] Fuller, W. A. (1987). *Measurement Error Models*. Wiley, New York.

[24] Gelfand, A. E. and Smith, A. F. M. (1990). Sampling-based approaches to calculating marginal densities. *Journal of the American Statistical Association*, **85**, 398–409.

[25] Gelfand, A. E., Diggle, P. J., Fuentes, M. and Guttorp, P. (Eds.) (2010). *Handbook of Spatial Statistics*. CRC Press/Chapman & Hall, Boca Raton.

[26] Gelman, A., Carlin, J. B., Stern, H. S. and Rubin, D. B. (2004). *Bayesian Data Analysis* (2nd edn). Chapman & Hall/CRC, Boca Raton, FL.

[27] Gilks, W. R. and Wild, P. (1992). Adaptive rejection sampling for Gibbs sampling. *Applied Statistics*, **41**, 337–348.

[28] Girón, F. J. and Rojano, J. C. (1994). Bayesian Kalman filtering with elliptically contoured errors. *Biometrika*, **80**, 390–395.

[29] Girón, F. J., Moreno, E. and Casella, G. (2007). Objective Bayesian analysis of multiple changepoints for linear models (with discussion). In *Bayesian Statistics 8* (J. M. Bernardo, M. J. Bayarri, J. O. Berger, D. Heckerman, A. F. M. Smith and M. West, eds.). Oxford University Press, London, 227–252.

[30] Green, P. (1995). Reversible jump Markov chain Monte Carlo computation and Bayesian model determination. *Biometrika*, **82**, 711–732.

[31] Hooten, M. V. and Wikle, C. K. (2008). A hierarchical Bayesian non-linear spatio-temporal model for the spread of invasive species with application to the Eurasian Collared-Dove. *Environmental and Ecological Statistics*, **15**, 59–70.

[32] Ibrahim, J. G. and Laud, P. W. (1991). On Bayesian analysis of generalized linear models using Jeffreys's prior. *Journal of American Stististical Association*, **86**, 981–986.

[33] Johnson, V. E. and Albert, J. H. (1999). *Ordinal Data Modelling*. Springer, New York.

[34] Kalnay, E. (2003). *Atmospheric Modelling, Data Assimilation and Predictability*. Cambridge University Press.

[35] Kass, R. E. and Steffey, D. (1989). Approximate Bayesian inference in conditionally independent hierarchical models (parametric empirical Bayes models). *Journal of the American Statistical Association*, **84**, 717–726.

[36] Kent, J. T. and Mardia, K. V. (2002). Modelling strategies for spatial-temporal data. In *Spatial Cluster Modelling* (A. Lawson and D. Denison, eds.). Chapman & Hall, London, 214–226.

[37] Li, F., Villani, M. and Kohn, R. (2011). Modeling conditional densities using finite smooth mixtures. In *Mixtures: Estimation and Applications* (K. L. Mengersen, C. P. Robert and D. M. Titterington, eds.). John Wiley, Chichester, 123–144.

[38] Little, R. J. A. and Rubin, D. B. (1987). *Statistical Analysis with Missing Data*. John Wiley & Sons, New York.

[39] Lunn, D. J., Thomas, A., Best, N. and Spiegelhalter, D. (2000). WinBUGS – A Bayesian modelling framework: Concepts, structure, and extensibility. *Statistics and Computing*, **10**, 325–337.

[40] Lynch, P. (2006). *The Emergence of Numerical Weather Prediction: Richardson's Dream*. Cambridge University Press, Cambridge.

[41] Lynch, P. (2008). The origins of computer weather prediction and climate modelling. *Journal of Computational Physics*, **227**, 3431–3444.

[42] MacEachern, S. and Müller, P. (2000). Efficient MCMC schemes for robust model extensions using encompassing Dirichlet process mixture models. In *Robust Bayesian Analysis* (D. Rios Insua and F. Ruggeri, eds.). Springer-Verlag, New York, 295–315.

[43] MacEachern, S. N., Clyde, M. and Liu, J. S. (1999). Sequential importance sampling for Nonparametric Bayes Models: The Next Generation. *The Canadian Journal of Statistics*, **27**, 251–267.

[44] Mallick, B., Hoffman, F. O. and Carroll, R. J. (2002). Semiparametric regression modelling with mixtures of Berkson and classical error, with application to fallout from the Nevada test site. *Biometrics*, **58**, 13–20.

[45] McGuffie, K. and Henderson-Sellers, A. (2005). *A Climate Modelling Primer* (3rd edn). John Wiley, Chichester.

[46] McLachlan, G. J. and Peel, D. (2000). *Finite Mixture Models*. Wiley, New York.

[47] Mengersen, K. L. and Robert, C. P. (1996). Testing for mixtures: a Bayesian entropic approach (with discussion). In *Bayesian Statistics 5* (J. O. Berger, J. M. Bernardo, A. P. Dawid, D. V. Lindley and A. F. M. Smith, eds.). Oxford University Press, Oxford, 255–276.

[48] Minka, T. (2002). Bayesian inference in dynamic models: an overview. *Technical report*. Carnegie Mellon University.

[49] Neal, R. (2000). Markov chain sampling methods for Dirichlet process mixture models. *Journal of Computational and Graphical Statistics*, **9**, 249–265.

[50] Oey, L.-Y., Ezer, T. and Lee, H.-C. (2005). Loop Current, rings and related circulation in the Gulf of Mexico: A review of numerical models and future challenges. In *Circulation in the Gulf of Mexico: Observations and Models, Geophysical monograph series 161* (W. Sturges and A. Lugo-Fernandez, eds.). AGU, Washington D.C., 31–56.

[51] Perreault, L., Parent, E., Bernier, J., Bobée, B. and Slivitzky, M. (2000). Retrospective multivariate Bayesian change-point analysis: a simultaneous single change in the mean of several hydrological sequences. *Stochastic Environmental Research and Risk Assessment*, **14**, 243–261.

[52] Quintana, F. A. and Newton, M. A. (2000). Computational aspects of nonparametric Bayesian analysis with applications to the modelling of multiple binary sequences. *Journal of Computational and Graphical Statistics*, **9**, 711–737.

[53] Raftery, A. E. (1994). Change point and change curve modelling in stochastic processes and spatial statistics. *Journal of Applied Statistical Science*, **1**, 403–424.

[54] Raudenbush, S. W. and Bryk, A. S. (2002). *Hierarchical Linear Models* (2nd edn). Sage, Newbury Park, CA.

[55] Richardson, S. and Green, P. (1997). On Bayesian analysis of mixtures with an unknown number of components. *Journal of the Royal Statistical Society, Series B*, **59**, 731–792.

[56] Robert, C. P., Rydèn, T. and Titterington, D. M. (2000). Bayesian inference in hidden Markov models through the reversible jump Markov chain Monte Carlo method. *Journal of the Royal Statistical Society, Series B*, **62**, 57–75.

[57] Rodrigues, J. and Bolfarine, H. (2007). Bayesian inference for an extended simple regression measurement error model using skewed priors. *Bayesian Analysis*, **2**, 349–364.

[58] Rodriguez, A., Dunson, D. B. and Gelfand, A. E. (2008). The nested Dirichlet process. *Journal of the American Statistical Association*, **103**, 1131–1154.

[59] Rue, H., Martino, S. and Chopin, N. (2009). Approximate Bayesian inference for latent Gaussian models by using integrated nested Laplace approximation. *Journal of the Royal Statistical Society, Series B*, **71**, 1–35

[60] Scaccia, L. and Green, P. J. (2003). Bayesian growth curves using normal mixtures with nonparametric weights. *Journal of Computational and Graphical Statistics*, **12**, 308–331.

[61] Sethuraman, J. (1994). A constructive definition of Dirichlet priors. *Statistica Sinica*, **4**, 639–650.

[62] Sharples, L. D. (1990). Identification and accommodation of outliers in general hierarchical models. *Biometrika*, **77**, 445–453.

[63] Smith, B. J. and Cowles, M. K. (2007). Correlating point-referenced radon and areal uranium data arising from a common spatial process. *Applied Statistics*, **56**, 313–526.

[64] Spiegelhalter, D. J., Thomas, A., Best, N. and Gilks, W. R. (1995). BUGS: Bayesian inference using Gibbs sampling, Version 0.50. Technical report. Medical Research Council Biostatistics Unit, Institute of Public Health, Cambridge University.

[65] Teh, Y. W., Jordan, M. I., Beal, M. J. and Blei, D. M. (2006). Hierarchical Dirichlet processes. *Journal of the American Statistical Association*, **101**, 1566–1581.

[66] Tierney, L., Kass, R. E. and Kadane, J. B. (1989). Fully exponential Laplace approximations to expectations and variances of nonpositive functions. *Journal of the American Statistical Association*, **84**, 710–716.

[67] Titterington, D. M., Smith, A. F. M. and Makov, U. E. (1985). *Statistical Analysis of Finite Mixture Distributions*. John Wiley, Chichester.

[68] Verdinelli, I. and Wasserman, L. (1991). Bayesian analysis of outlier problems using the Gibbs sampler. *Statistics and Computing*, **1**, 105–117.

[69] Wakefield, J. C., Smith, A. F. M., Racine-Poon, A. and Gelfand, A. E. (1994). Bayesian analysis of linear and non-linear population models by using the Gibbs sampler. *Applied Statistics*, **43**, 201–221.

[70] Wang, J. and Zivot, E. (2000). A time series model of multiple structural changes in level, trend and variance. *Journal of Business and Economic Statistics*, **18**, 374–386.

[71] Wang, P. M., Cockburn, I. M. and Puterman, M. L. (1998). Analysis of patent data–A mixed-Poisson-regression-model approach. *Journal of Business and Economic Statistics*, **16**, 27–41.

[72] West, M. and Harrison, J. (1997). *Bayesian Forecasting and Dynamic Models* (2nd edn). Springer, New York.

[73] Wikle, C. K. (2003). Hierarchical Bayesian models for predicting the spread of ecological processes. *Ecology*, **84**, 1382–1394.

[74] Zhang, H. (2004). Inconsistent estimation and asymptotically equal interpolations in model-based geostatistics. *Journal of the American Statistical Association*, **99**, 250–61.

4 Bayesian hierarchical kernel machines for nonlinear regression and classification

SOUNAK CHAKRABORTY, BANI K. MALLICK AND MALAY GHOSH

4.1 Introduction

"One machine can do the work of fifty ordinary men. No machine can do the work of one extraordinary man."

Elbert Hubbard

Humans are capable of tackling extremely difficult problems without the benefit of an a priori solution. We learn from experience, and can often transfer knowledge acquired to novel instances or even whole new tasks. Are machines capable of similar problem-solving prowess? Machine learning is a direct descendant of an older discipline—statistical model-fitting. The goal in machine learning is to extract useful information from a corpus of data, by building good probabilistic models. The particular twist behind machine learning, however, is to automate this process as much as possible, often by using very flexible models characterized by large numbers of parameters, and to let the machine take care of the rest. Clearly, machine learning is driven by rapid technological progress in two areas: storage devices leading to large databases and datasets, and computing power enabling the use of more complex models.

Bayesian machine learning uses probability to express all forms of uncertainty. Learning is performed by simple application of the rules of probability. Results of Bayesian machine learning are expressed in terms of a probability distribution over all unknown quantities or parameters of the model. The probability distribution of a parameter quantifies the uncertainty of its value, and expresses our degree of belief in the various possibilities. In contrast, the conventional frequentist strategy takes the form of an estimator for unknown quantities or model parameters, with possibly some optimality criterion in mind.

Current technological trends inexorably lead to data flood. More data are generated from banking, telecom and other business transactions. More data are generated from scientific experiments in astronomy, chemistry, space exploration, biology, high-energy physics, etc. More data are created on the web, especially in text, image and other multimedia format. For example, Europe's Very Long

Baseline Interferometry produces 1 Gigabit/second of astronomical data in an observation window. Cell phone companies like AT&T handle so many calls per day that it is impossible to store all that data and analysis must be done 'on the fly'. DNA microarray allows measurement of gene expression levels for thousands of genes simultaneously. It is extremely complex to understand the relationship between these thousands of genes and a specific disease. To handle these massive dimensional complex datasets kernel based machine learning techniques are some of the best available options.

Support vector machines (SVM) are machine learning approaches, originally developed by Vapnik [27] and others who worked in this area. The SVMs are a system for efficiently training linear learning machines in kernel induced feature space. This has gained popularity due to its attractive, analytic and computational features, and promising empirical performance. The formulation embodies the Structural Risk Minimization (SRM) principle which has been shown to be superior to Empirical Risk Minimization (ERM), used by conventional neural networks. Wahba [28, 29] showed that SVM methodology can be cast as a regularization problem in terms of a reproducing kernel Hilbert space (RKHS) and hence SVMs and penalty methods, as used in the statistical theory of nonparametric regression, have a strong interrelationship. We have a training set $\{y_i, \boldsymbol{x}_i\}, i = 1, \ldots, n$, where y_i is the response variable and $\boldsymbol{x}_i$ is the vector of covariate of size p corresponding to y_i. Given the training data our goal is to find an appropriate function $f(\boldsymbol{x})$ to predict the responses y in the test set based on the covariates $\boldsymbol{x}$. Sollich [21] showed that SVM can be interpreted as a maximum a posterior solution to inference problems with Gaussian priors and an appropriate likelihood function based on a probabilistic interpretation. In the classification context, they are motivated by the geometric interpretation of maximizing the margin of discrimination, and are characterized by the use of a kernel function.

In recent years several authors have developed kernel machine models for regression and classification in the Bayesian framework. Law and Kwok [16] introduced a Bayesian formulation of SVM for regression, but they did not carry out a full Bayesian analysis and used instead certain approximations for the posterior. A similar remark applies to Sollich [23] who considered SVM in the classification context. As an alternative to SVM, Tipping [25, 26] and Bishop and Tipping [3] introduced relevance vector machines (RVMs). RVMs are suitable for both regression and classification, and are amenable to probabilistic interpretation. However, these authors did not perform a full Bayesian analysis. They obtained type II maximum likelihood (Good, [12]) estimates of prior parameters, which do not provide predictive distribution of future observations. Chakraborty *et al.* [6] developed full Bayesian SVM and RVM for single response and also multiple response correlated datasets. For classification problems, Mallick *et al.* [16] considered a Markov chain Monte Carlo (MCMC) based full Bayesian analysis for a binary classification scheme based on the RKHS theory, where the number of covariates was far greater than the sample size. Similarly, Chakraborty *et al.* [5] has developed a RKHS based multi-class kernel machine model for high-dimensional data and introduced a stochastic search variable selection scheme with it for variable selection.

The layout of this chapter is as follows: In Section 4.2, we describe the general regression and classification problem under the regularization framework and discuss RKHS and its related properties. In Section 4.3, we give a detailed description of Bayesian kernel machine regression models and the associated prior justification. In Section 4.4, we discuss the Bayesian kernel machine model for binary classification problems. Application to several real life datasets are also discussed in Section 4.3 and Section 4.4. Finally, some concluding remarks and future possibilities are discussed in Section 4.5.

4.2 RKHS and its related representation properties

In a regression or a classification problem, we have a training set $\{y_i, \boldsymbol{x}_i\}, i = 1, \ldots, n$, where y_i is the response variable (continuous for regression and categorical for classification problems) and $\boldsymbol{x}_i$ is the vector of covariate of size p corresponding to y_i. Given the training data our goal is to find an appropriate function $f(\boldsymbol{x})$ to predict the responses y in the test set based on the covariates $\boldsymbol{x}$. This can be viewed as a regularization problem of the form

$$\min_{f \in \mathcal{H}} \left[\sum_{i=1}^{n} L(y_i, f(\boldsymbol{x}_i)) + \lambda J(f) \right] \tag{4.1}$$

where $L(y, f(\boldsymbol{x}))$ is a loss function, $J(f)$ is a penalty functional, $\lambda > 0$ is the smoothing parameter and $\mathcal{H}$ is a space of functions on which $J(f)$ is defined. In this article, we consider $\mathcal{H}$ to be a reproducing kernel Hilbert space (RKHS) with kernel K, and we denote it by $\mathcal{H}_K$. A formal definition of RKHS in given in Aronszajn [1], Parzen [21], and Wahba [29].

When $h \in \mathcal{H}_K, f(\boldsymbol{x}) = \beta_0 + h(\boldsymbol{x})$ and $J(f) = \| h \|^2_{\mathcal{H}_K}$, (4.1) becomes

$$\min_{f \in \mathcal{H}_K} \left[\sum_{i=1}^{n} L(y_i, f(\boldsymbol{x}_i)) + \lambda \| h \|^2_{\mathcal{H}_K} \right] \tag{4.2}$$

The estimate of h is obtained as a solution of (4.2). It can be shown that the solution is finite-dimensional (Wahba, [28]) and leads to a representation of f (Kimeldorf and Wahba, [13]; Wahba [29]) as

$$f_\lambda(\boldsymbol{x}) = \beta_0 + \sum_{i=1}^{n} \beta_i K(\boldsymbol{x}, \boldsymbol{x}_i) \tag{4.3}$$

It is also a property of RKHS that

$$\| h \|^2_{\mathcal{H}_K} = \sum_{i,j=1}^{n} \beta_i \beta_j K(\boldsymbol{x}_i, \boldsymbol{x}_j) \tag{4.4}$$

Representation of f in the above form is of special interest to us, because in cases when the number of covariates p is much larger than the number of data points, we effectively reduce the dimension of covariates from p to n. To obtain the estimate of f_λ we substitute (4.3) and (4.4) in (4.2) and then minimize it with respect to $\beta = (\beta_0, \ldots, \beta_n)$ and the smoothing parameter λ. The other parameters in kernel K may be chosen by generalized approximate cross validation (GACV).

Similarly, for multivariate regression when $\boldsymbol{f}(\boldsymbol{x}) = (f_1(\boldsymbol{x}), \ldots, f_q(\boldsymbol{x}))$ is a q-tuple function we can have similar results based on RKHS as the univariate case. Here we consider $f(\boldsymbol{x}) \in \prod_{j=1}^{q}(\{1\} + \mathcal{H}_{K_j})$, the product space of q reproducing kernel Hilbert spaces $\mathcal{H}_{K_j}$ for $j = 1, \ldots, q$. In other words, each component can be expressed as $\beta_j + h_j(\boldsymbol{x})$ with $h_j \in \mathcal{H}_{K_j}$. Unless there is a compelling reason to believe that $\mathcal{H}_{K_j}$ should be different for $j = 1, \ldots, q$, we will assume that they are the same RKHS denoted by $\mathcal{H}_K$. All the results stated before hold in this framework as well. Detailed description is provided in Chakraborty *et al.* [6].

Regarding the loss $L(y, f(\boldsymbol{x}))$ as the negative of the log-likelihood, our problem is equivalent to maximization of the penalized log-likelihood

$$-\sum_{i=1}^{n} L(y_i, f(\boldsymbol{x}_i)) - \lambda \| h \|^2_{H_k} \tag{4.5}$$

This duality between 'loss' and 'likelihood', particularly viewing the loss as the negative of the log-likelihood, is referred to in the Bayesian literature as the 'logarithmic scoring rule' (Bernardo,

[2], p. 688). In Table 4.1, we summarize some popularly useful loss functions for kernel machine models in regression and classification. List of notations used in Table 4.1:

Table 4.1 Loss functions used in kernel machine models.

Loss function name	Formula	Use
Square error loss	$(y - f(\boldsymbol{x}))^2$	RVM regression
Hinge loss	$[1 - yf(\boldsymbol{x})]_+$	SVM binary classification
Multicategory hinge loss	$L(\boldsymbol{y}).[\boldsymbol{f}(\boldsymbol{x}) - \boldsymbol{y}]_+$	Multiclass SVM classification
Vapnik's ϵ-insensitive loss	$\lvert y - f(\boldsymbol{x})\rvert_\epsilon$	SVM regression
Multivariate ϵ-insensitive loss	$\parallel \boldsymbol{y} - \boldsymbol{f}(\boldsymbol{x}) \parallel_\epsilon$	Multiple response SVM regression
Multivariate quadratic loss	$(\boldsymbol{y} - \boldsymbol{f}(\boldsymbol{x}))^T(\boldsymbol{y} - \boldsymbol{f}(\boldsymbol{x}))$	Multiple response RVM regression

- $[a]_+ = a$ if $a > 0$ and is 0 otherwise.
- f and $\boldsymbol{f}$ (multivariate) is the unknown regression function that connects the response and the covariates.
- $L(\boldsymbol{y})$ is a J dimensional vector (for J class classification) with 0 in the jth component and 1 elsewhere when $\boldsymbol{y}$ correspond to class j. The . denotes the Euclidean inner product.
- $\lvert y - f(\boldsymbol{x})\rvert_\epsilon = 0$ if $\lvert y - f(\boldsymbol{x})\rvert \leq \epsilon$ and $\lvert y - f(\boldsymbol{x})\rvert - \epsilon$ otherwise.
- In a response vector with q components, $\parallel \boldsymbol{y} - \boldsymbol{f}(\boldsymbol{x}) \parallel_\epsilon = \sum_{j=1}^{q} \rho_j \lvert y_j - f_j\rvert_\epsilon$.

4.3 Bayesian kernel machine regression models

In this section we will discuss Bayesian kernel machine models for univariate regression problems and some real life applications. Let us consider that we have a dataset $\mathcal{D} = \{(y_i, \boldsymbol{x}_i) : i = 1, \ldots, n.\}$, where y_i is the response variable (continuous) and $\boldsymbol{x}_i$ is the $p \times 1$ covariate vector. Let us also assume that $p \gg n$ implying that the number of covariates far exceeds the sample size, hence there is an urgent need to reduce the dimension of the problem or the covariate space.

4.3.1 Hierarchical Bayes relevance vector machine

The Relevance Vector Machine (RVM) was first proposed by Tipping [25]. It is considered as a probabilistic model whose functional form is equivalent to the SVM with a least square loss function (also known as LS-SVM). Here we formulate the RVM in a complete hierarchical Bayesian setup based on the RKHS theory, as discussed in the previous section.

If we assume that f is generated from RKHS with the kernel function $K(.,.)$, using the representation theorem (4.3) we can express f as

$$f(\boldsymbol{x}_i) = \beta_0 + \sum_{j=1}^{n} \beta_j K(\boldsymbol{x}_i, \boldsymbol{x}_j | \theta) \tag{4.6}$$

where K is a positive definite function of the covariates $\boldsymbol{x}$ involving some unknown parameter θ. Hence,

$$y_i = f(\boldsymbol{x}_i) + \eta_i = \mathbf{K}_i^{\mathrm{T}} \beta + \eta_i \tag{4.7}$$

where $\eta_i \overset{iid}{\sim} \mathrm{N}(0,\sigma^2)$, $\beta = (\beta_0,\ldots,\beta_n)^{\mathrm{T}}$, and $\boldsymbol{K}_i = (1, K(\boldsymbol{x}_i,\boldsymbol{x}_1|\theta),\ldots,K(\boldsymbol{x}_i,\boldsymbol{x}_n|\theta))^{\mathrm{T}}$, $i = 1,\ldots,n$. We assign hierarchical priors to the unknown parameters β, θ and σ^2. A conjugate prior for this model is as follows.

$$\beta|\sigma^2 \sim \mathrm{N}_{n+1}(0,\sigma^2\boldsymbol{D}^{-1}) \tag{4.8}$$

where $\boldsymbol{D} = \mathrm{diag}(\lambda_0,\ldots,\lambda_n)$ is a $(n+1)\times(n+1)$ diagonal matrix and $\lambda = (\lambda_0,\ldots,\lambda_n)^{\mathrm{T}}$. λ_0 is fixed at a small value but other λs are kept unknown. We also assume that

$$\sigma^2 \sim \mathrm{IG}(\gamma_1,\gamma_2) \tag{4.9}$$

We further assume that

$$\theta \sim \mathrm{U}(a_L,a_U),\ \ \lambda_i \overset{iid}{\sim} \mathrm{Gamma}(c,d),\ \ i = 1,\ldots,n \tag{4.10}$$

where $\mathrm{U}(a_L,a_U)$ is the uniform distribution over (a_L,a_U).

The matrix with (i,j)th element $K_{ij} = \boldsymbol{K}(\boldsymbol{x}_i,\boldsymbol{x}_j|\boldsymbol{\theta})$ is known as the kernel matrix. The usual choices are (i) the Gaussian kernel $\boldsymbol{K}(\boldsymbol{x}_i,\boldsymbol{x}_j|\theta) = \exp\{-\frac{\|\boldsymbol{x}_i-\boldsymbol{x}_j\|^2}{2\theta}\}$ and (ii) the polynomial kernel, $\boldsymbol{K}(\boldsymbol{x}_i,\boldsymbol{x}_j|\theta) = (\boldsymbol{x}_i^{\mathrm{T}}\boldsymbol{x}_j + 1)^{\theta}$. Both kernels contain a single parameter θ. The joint posterior distribution is thus given by

$$\begin{aligned}\pi(\beta,\lambda,\sigma^2,\theta|\boldsymbol{y}) \propto\ & \frac{1}{(2\pi\sigma^2)^{n/2}}\exp\left\{-\sum_{i=1}^{n}\frac{(y_i-\boldsymbol{K}_i^{\mathrm{T}}\beta)^2}{2\sigma^2}\right\} \\ & \times \frac{1}{(2\pi)^{(n+1)/2}|\sigma^2\boldsymbol{D}^{-1}|^{1/2}}\exp(-\frac{1}{2\sigma^2}\beta^{\mathrm{T}}\boldsymbol{D}\beta) \\ & \times \exp(-\gamma_2/\sigma^2)(\sigma^2)^{-\gamma_1-1} \times \prod_{i=1}^{n}\exp(-d\lambda_i)\lambda_i^{c-1}\end{aligned} \tag{4.11}$$

This distribution is complex, and implementation of the Bayesian procedure requires MCMC sampling techniques, and in particular, the Gibbs sampling (Gelfand and Smith, [8]) and Metropolis–Hastings (MH) algorithm (Metropolis *et al.*, [19], Chib and Greenberg, [7]). The Gibbs sampler generates posterior samples using the conditional densities which we describe below.

Conditional distributions and posterior sampling of the parameters

The prior distributions in (4.8) and (4.9) are conjugate distributions for β and σ^2. The posterior density conditional on θ,λ is Normal-Inverse-Gamma

$$p(\beta,\sigma^2|\theta,\lambda) = \mathrm{N}_{n+1}(\beta|\tilde{\boldsymbol{m}},\sigma^2\tilde{\boldsymbol{V}})\mathrm{IG}(\sigma^2|\tilde{\gamma_1},\tilde{\gamma_2}) \tag{4.12}$$

where $\tilde{\boldsymbol{m}} = (\boldsymbol{K}_0^{\mathrm{T}}\boldsymbol{K}_0 + \boldsymbol{D})^{-1}(\boldsymbol{K}_0^{\mathrm{T}}\boldsymbol{y})$, $\tilde{\boldsymbol{V}} = (\boldsymbol{K}_0^{\mathrm{T}}\boldsymbol{K}_0 + \boldsymbol{D})^{-1}$, $\tilde{\gamma_1} = \gamma_1 + n/2$, and $\tilde{\gamma_2} = \gamma_2 + (\boldsymbol{y}^{\mathrm{T}}\boldsymbol{y} - \tilde{\boldsymbol{m}}^{\mathrm{T}}\tilde{\boldsymbol{V}}\tilde{\boldsymbol{m}})$. Here $\boldsymbol{K}_0^{\mathrm{T}} = (\boldsymbol{K}_1,\ldots,\boldsymbol{K}_n)$.

The conditional distribution for λ_i given β_i and σ^2 is given by

$$\lambda_i|\beta_i,\sigma^2 \overset{ind}{\sim} \mathrm{Gamma}\left(c+\frac{1}{2}, d+\frac{\beta_i^2}{2\sigma^2}\right),\ \ i = 1,\ldots,n. \tag{4.13}$$

We use the conditional distributions for constructing a Gibbs sampler through the following steps:

Step 1. Update β and σ^2 by sampling from their conditional distribution (4.12).

Step 2. Update λ by sampling from the conditional distribution (4.13) of λ_i in turn.

Step 3. Update of $\boldsymbol{K}$ is equivalent to the update of θ and we need a Metropolis–Hastings algorithm to sample from the marginal distribution of θ. We put a uniform prior (4.10) on θ so $p(\theta|\boldsymbol{y}) \propto p(\boldsymbol{y}|\theta)$, the marginal likelihood of θ. Let θ^* denote the proposed change. Accept this change with acceptance probability

$$\alpha = \min\left\{1, \frac{p(\boldsymbol{y}|\theta^*)}{p(\boldsymbol{y}|\theta)}\right\} \tag{4.14}$$

The ratio of the marginal likelihoods in (4.14) is given by

$$\frac{p(\boldsymbol{y}|\theta^*)}{p(\boldsymbol{y}|\theta)} = \frac{|\tilde{\boldsymbol{V}}^*|^{1/2}}{|\tilde{\boldsymbol{V}}|^{1/2}}\left(\frac{\tilde{\gamma}_2}{\tilde{\gamma}_2^*}\right)^{\tilde{\gamma}_1} \tag{4.15}$$

Using the marginal posterior of θ rather the full conditional posteriors improves the mixing by a large extent, also resulting in a much faster convergence of the chain. Stability of the chain is also increased without compromising the mixing. This way of using the marginal distribution of θ, the kernel parameter, for sampling, was also successfully used by Mallick *et al.* [18].

In the relevance vector machine discussed above, we effectively try to minimize the squared error loss function and use separate smoothing parameters λ_i for different β_i. The smoothing parameters determine the tradeoff between training accuracy and model complexity. We can also keep all λ_is the same by fixing $\lambda_i = \lambda$, for all $i = 1, \ldots, n$. This resulting approach will be referred to as the Bayesian relevance vector machine (BRVM). The multiple smoothing parameter is also used by Tipping [25]. By having multiple smoothing parameters over a single one we are effectively controlling each of the regression coefficients separately, thereby introducing sparseness in the model.

4.3.2 Hierarchical Bayes support vector machine

In this section, we show how SVM based on RKHS can be used in a complete Bayesian setup using Vapnik's ϵ-insensitive loss function. In the Bayesian SVM model proposed by Law and Kwok [16] they did not carry out a full hierarchical Bayesian analysis and used instead type II maximum likelihood to estimate the prior parameters.

The ϵ-insensitive loss function introduced by Vapnik [27] is as follows:

$$L(y,f(\boldsymbol{x})) = |y - f(\boldsymbol{x})|_\epsilon = \begin{cases} 0 & \text{if } |y - f(\boldsymbol{x})| \le \epsilon \\ |y - f(\boldsymbol{x})| - \epsilon & \text{otherwise} \end{cases} \tag{4.16}$$

This loss function (Figure 4.1) ignores errors of size less than ϵ, but penalizes in a linear fashion when the function deviates more than ϵ amount. This makes the fitting less sensitive to the outliers. It is interesting to note that, like other loss functions or error measures in robust regression (Huber, [13]), Vapnik's ϵ-insensitive loss function also has linear tails beyond ϵ. But in addition it flattens the contributions of those cases with small residuals. To construct a hierarchical model for regression using Vapnik's loss function we introduce n latent variables $z_1, \ldots, z_n$. Such that y_is are conditionally independent given z_i. Introduction of the latent variables makes the calculations

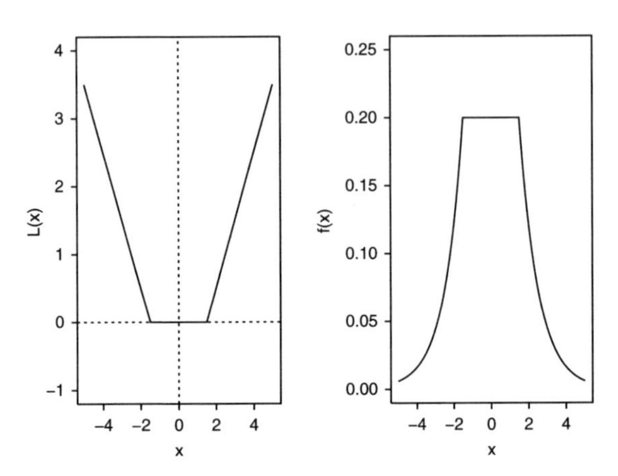

Figure 4.1 The figure on the left represents Vapnik's ϵ-insensitive ($\epsilon = 1$) loss function. The figure on the right represents the corresponding likelihood.

particularly simple. The likelihood of y_i conditional on z_i, corresponding to Vapnik's loss (4.16), as suggested by Law and Kwok [16], is given by

$$p(y_i|z_i) \propto \exp\{-\rho|y_i - z_i|_\epsilon\}, \;\; i = 1, \ldots, n. \tag{4.17}$$

It can be shown that this pdf figure can be written as a mixture of a truncated Laplace distribution and a uniform distribution as follows

$$\begin{aligned} p(y_i|z_i) &= \frac{\rho}{2(1+\epsilon\rho)} \exp\{-\rho|y_i - z_i|_\epsilon\} \\ &= p_1(\text{Truncated Laplace}(z_i, \rho)) + p_2(\text{Uniform}(z_i - \epsilon, z_i + \epsilon)) \end{aligned} \tag{4.18}$$

where $p_1 = \frac{1}{1+\epsilon\rho}$, $p_2 = \frac{\epsilon\rho}{1+\epsilon\rho}$. A Laplace$(\theta, \rho)$ distribution for a random variable U has pdf proportional to $\exp(-\rho|u - \theta|)$.

We connect z_i with $f(\boldsymbol{x}_i)$ by $z_i = f(\boldsymbol{x}_i) + \eta_i$, where η_i are the residual random effects that account for any unexplained source of variation not included in the model. We assume that f is generated from RKHS. By representation (4.6) the z_i are modelled as

$$z_i = \boldsymbol{K}_i^{\mathrm{T}} \beta + \eta_i \tag{4.19}$$

Where $\eta_i \overset{iid}{\sim} \mathrm{N}(0, \sigma^2)$ and β and $\boldsymbol{K}_i$ are defined as in Section 4.3.1. We assign hierarchical priors to the unknown parameters $\beta, \theta, \rho, \sigma^2, z_i, \lambda_i$ as follows.

$$z_i|\beta, \sigma^2, \theta, \lambda \overset{ind}{\sim} \mathrm{N}(\boldsymbol{K}_i^{\mathrm{T}} \beta, \sigma^2) \tag{4.20}$$

$$\beta|\sigma^2, \lambda \sim \mathrm{N}(0, \sigma^2 \boldsymbol{D}^{-1}); \boldsymbol{D} = \mathrm{diag}(\lambda_0, \lambda_1, \ldots, \lambda_n) \tag{4.21}$$

$$\sigma^2 \sim \mathrm{IG}(\gamma_1, \gamma_2) \tag{4.22}$$

$$\theta \sim \mathrm{U}(a_L, a_U), \;\; \lambda_i \overset{iid}{\sim} \mathrm{Gamma}(c, d), \;\; \rho \sim \mathrm{U}(r_L, r_U) \tag{4.23}$$

Asymptotically, if ϵ goes to infinity, we get the uniform likelihood, but if it goes to 0 we get the usual Laplace likelihood when ρ is fixed. So instead of keeping ϵ fixed, as in classical SVM we assigned prior to ϵ. However we observed that the performance of SVM decayed rapidly for priors spreading outside the range (0, 1). More detailed justifications are provided in Law and Kwok [16]. Hence we have considered

$$\epsilon \sim \text{Beta}(k_1, k_2) \tag{4.24}$$

This Bayesian SVM model bears some analogy to the classical SVM as the exponent of the Gaussian prior for β is equivalent to the quadratic penalty function, but with multiple smoothing parameters. The posterior is similar to the posterior for the RVM model, except now the Gaussian likelihood in RVM is changed to Vapnik's ϵ-insensitive loss based likelihood (4.17) as follows

$$\begin{aligned}
\pi(\beta, \lambda, \boldsymbol{z}, \sigma^2, \theta, \rho, \epsilon|\boldsymbol{y}) \propto\ & \frac{\rho^n}{2^n(1+\epsilon\rho)^n} \exp\left(-\rho \sum_{i=1}^{n} |y_i - z_i|_\epsilon\right) \\
& \times \frac{1}{(2\pi)^{n/2}(\sigma^2)^{n/2}} \exp\left(-\frac{1}{2\sigma^2}\sum_{i=1}^{n}(z_i - \boldsymbol{K}_i^{\mathrm{T}}\beta)^2\right) \\
& \times \frac{1}{(2\pi)^{(n+1)/2}|\sigma^2\boldsymbol{D}^{-1}|^{1/2}} \exp\left(-\frac{1}{2\sigma^2}\beta^{\mathrm{T}}\boldsymbol{D}\beta\right) \\
& \times \exp\left(-\frac{\gamma_2}{\sigma^2}\right)(\sigma^2)^{-\gamma_1-1} \times \prod_{i=1}^{n} \exp(-d\lambda_i)\lambda_i^{c-1} \times \epsilon^{k_1-1}(1-\epsilon)^{k_2-1}
\end{aligned} \tag{4.25}$$

As before, the implementation of Bayesian methods is done once again using MCMC. The conditional posterior distribution of λ_i is exactly the same as in the RVM case given by (4.13) in Section 4.3.1. The conditional distribution for β and σ^2 is also similar to that in the previous section, except that now $\boldsymbol{y}$ is replaced by the latent variable $\boldsymbol{z}$ in $\tilde{\boldsymbol{m}}$ and $\tilde{\gamma}_2$ in (4.12).

The distribution of z_i, conditional on $\boldsymbol{y}, \boldsymbol{K}_i, \beta, \theta, \rho, \lambda, \epsilon$, and $\boldsymbol{z}_{-i}$ ($\boldsymbol{z}_{-i}$ indicates the $\boldsymbol{z}$ vector with the ith element removed) does not have an explicit form. We thus resort to the Metropolis–Hastings (MH) algorithm with a proposal density $T(z_i^*|z_i)$ that generates moves from the current state z_i to a new state z_i^*. It is convenient to take the proposal distribution to be Gaussian with mean equal to $\boldsymbol{K}_i^{\mathrm{T}}\beta$ and variance σ^2 (Chib and Greenberg, [7]). The proposed updates then accepted with probabilities

$$\delta_i = \min\left\{1, \frac{\exp\left(-\rho|y_i - z_i^*|_\epsilon\right)}{\exp\left(-\rho|y_i - z_i|_\epsilon\right)}\right\} \tag{4.26}$$

The update of $\boldsymbol{K}$ or θ is done similarly as before using the MH algorithm. Instead of sampling from the full marginal of θ we sample from $p(\theta|\boldsymbol{y}, \boldsymbol{z}, \rho, \epsilon)$, the marginal posterior of θ integrating out λ, β, σ^2. If θ^* denotes the proposed change from current θ, we accept this θ^* with probability

$$\alpha = \min\left\{1, \frac{p(\theta^*|\boldsymbol{z}, \boldsymbol{y}, \rho, \epsilon)}{p(\theta|\boldsymbol{z}, \boldsymbol{y}, \rho, \epsilon)}\right\} \tag{4.27}$$

The ratios of the marginal posteriors are the same as in (4.15) except that $\boldsymbol{y}$ is now replaced by the latent variable $\boldsymbol{z}$ in the expression.

The conditional distribution of ρ depending only on the latent variable $\boldsymbol{z}$, ϵ and the data $\boldsymbol{y}$ is given by

$$p(\rho|\boldsymbol{y},\boldsymbol{z},\epsilon) \propto \frac{\rho^n}{(1+\epsilon\rho)^n} \exp\left(-\rho \sum_{i=1}^{n} |y_i - z_i|_\epsilon\right) \tag{4.28}$$

The conditional distribution of ϵ conditional on ρ, $\boldsymbol{y}$ and $\boldsymbol{z}$ is given by

$$p(\epsilon|\boldsymbol{y},\boldsymbol{z},\rho) \propto \frac{\rho^n}{(1+\epsilon\rho)^n} \exp\left[-\rho\epsilon \sum_{i=1}^{n} \mathrm{I}(|y_i - z_i| > \epsilon)\right] \times \epsilon^{k_1-1}(1-\epsilon)^{k_2-1} \tag{4.29}$$

We use the MH algorithm to generate ρ and ϵ respectively from (4.28) and (4.29) with the acceptance probabilities $\left\{1, \frac{p(\rho^*|\boldsymbol{y},\boldsymbol{z},\epsilon)}{p(\rho|\boldsymbol{y},\boldsymbol{z},\epsilon)}\right\}$ and $\left\{1, \frac{p(\epsilon^*|\boldsymbol{y},\boldsymbol{z},\rho)}{p(\epsilon|\boldsymbol{y},\boldsymbol{z},\rho)}\right\}$.

Using the above conditional distributions, we can construct a Gibbs sampler by following the same steps 1 to 3 as in RVM. The distributions of some of the conditionals are changed as we replace $\boldsymbol{y}$ by the latent variable $\boldsymbol{z}$. Three extra steps needed to generate the latent variables $\boldsymbol{z}$, ρ and ϵ are added as follows:

Step 4. We update each z_i in turn conditional on the rest using the Metropolis step discussed in (4.26).

Step 5. Update ρ using a Metropolis step involving (4.28).

Step 6. Update ϵ using a Metropolis step involving (4.29).

The resulting support vector machine discussed above is based on the likelihood corresponding to Vapnik's ϵ-insensitive loss function (4.16) with multiple smoothing parameters λ_i. This resulting approach will be referred as the Bayesian support vector machine (BSVM). As in BRVM, here also we can simplify our model by setting $\lambda_i = \lambda$ for all components of β.

4.3.3 Applications

Our Bayesian RVM (BRVM) and Bayesian SVM (BSVM) are applied on two simulated datasets and three real datasets. The three real datasets are (i) Blood sugar data (Spiegelman *et al.*, [24]); (ii) Gas data (Kalivas, [14]); (iii) Wheat data (Kalivas, [14]). All three real datasets are near infrared spectroscopy data. In all datasets the number of covariates exceeds the number of available data points in the training set. So they are all ideally large p small n scenarios. Each dataset is randomly split into a training set and a test set for a hundred times. In the training set we keep two-thirds of the available data and the rest of the data are kept in the test set. The 'out of sample' mean square errors of prediction (MSEP) are calculated on the test set for each split. In each dataset we centre and scale X and Y. In Table 4.2 and Table 4.3 we report the median out of sample MSEP and the corresponding standard deviations. The lowest attained MSEP is marked in bold. For our BRVM and BSVM models, we generate a MCMC sample of 10,000 with the first 2000 as burn in. To avoid the issue of multimodality and MCMC being stuck at one local mode we use five independent chains with different starting points. Final prediction is obtained after pooling samples from all five chains. The convergence of the MCMC chains is checked by monitoring the trace plots of the generated samples and calculating the Gelman–Rubin scale reduction factor (Gelman *et al.*, [10], p. 329).

To make our models less sensitive to the choice of hyperparameters of the priors, we have chosen near-diffuse but proper priors. Near-diffuse priors are proper priors but with large variance. Thus we can guarantee the propriety of the posterior and at the same time near diffuseness introduces some objectivity in our analysis objective. We have assigned a vague but proper prior to σ^2. For λ we select the values of the hyperparameters so that the mean is kept very small around 0.001, but the variance is large. This choice of hyperparameter produces a near diffuse proper prior for β also. In all the examples in this paper we have used the polynomial kernel. For kernel parameter θ we give a discrete uniform prior $U\{1, 2, \ldots, C\}$. In order to examine sensitivity in the choice of priors, we have considered several different combinations of near-diffuse but proper priors. The prediction error remains almost the same with all such choices. Here we report the results for two such choices (*i*) $\gamma_1 = 1, \gamma_2 = 10, c = 10^{-8}, d = 10^{-5}, C = 5$ and (*ii*) $\gamma_1 = 0.5, \gamma_2 = 1, c = 10^{-9}, d = 10^{-6}, C = 10$. In BSVM additional parameters ϵ and ρ are drawn from a large support uniform prior. For ρ two choices of its hyperparameters are considered (*i*) $r_L = 0, r_U = 100$, and (*ii*) $r_L = 0, r_U = 50$. Choice of prior for ϵ is made such that the fitting is less sensitive to outliers. The ϵ parameter controls the width of the ϵ-insensitive zone, used to fit the training data. The value of ϵ can affect the number of support vectors used to construct the regression function. The bigger ϵ, the fewer support vectors are selected. On the other hand, bigger ϵ-values results in more 'flat' estimates. We have done a simple cross-validation for finding out the range of ϵ where the SVM regression gives best results, and found out empirically that it works best in the range $(0, 0.5)$. Hence we have assigned a $\text{Beta}(k_1, k_2)$ distribution to ϵ . We have tried several combinations of (k_1, k_2) and the results are fairly similar. Here we report our results using (*i*) $k_1 = k_2 = 1$, i.e. we put a uniform $U(0, 1)$ prior on ϵ, and (*ii*) $k_1 = 1, k_2 = 5$. Law and Kwok [16] proposed a data dependent prior on ϵ, but sampling from the posterior under their prior becomes much more complicated. It may be noted that a better prediction accuracy can be attained by choosing a tight prior properly centred. However if this does not hold, the prediction will be highly inaccurate. The near-diffuse priors offer protection against this, and introduce some objectivity in our procedure.

We compare the performance of our procedures with some of the classical procedures which are specially equipped to handle high-dimensional regression cases. They are, partial least squares (PLS), principal component regression (PCR), as well as classical support vector machine (CSVM), and random forest (RF) (Breiman, [4]). The tuning parameters for all standard models are selected by five-fold cross-validation. We used the *mvr*() function in *R* to fit the PLS and PCR models. The CSVM model is fitted using the *svm*() function in *R*. For the CSVM models we used both the polynomial kernel (CSVM-P) and the radial basis function kernel (CSVM-R). The RF is fitted using the *randomForest*() function and 500 boosted trees.

Simulation study

Simulation 1: We simulate a vector of covariates $\boldsymbol{x}$ of length $p = 1000$ from a Gaussian process ($x_i \in (-5, 5)$) with covariance function $k(\boldsymbol{x}_r, \boldsymbol{x}_s) = \exp\left(\frac{-|\boldsymbol{x}_r - \boldsymbol{x}_s|^2}{2d}\right)$ and $d = 1.5$. We jitter around the realization from GP by adding normal noise and obtain 150 samples. Thus our feature matrix is $X_{150 \times 1000}$. We generate $\beta_{1000 \times 1}$ from an $Uniform[0, 1]^p$. The response Y is obtained by plugging in X and β in the equation $Y = X\beta + N(0, 1)$.

Simulation 2: Same as Simulation 1 except $d = 1$.

Both simulations mimic a typical high-dimensional regression problem where $p \gg n$. The results from simulation studies are tabulated in Table 4.2. In both simulation studies our BSVM model attains the lowest MSEP. In Simulation 1 the BSVM model has 7.83%, 7.16%, 2.32%, 4.23%, 3.86%, 8.47% and 10.91% lower MSEP than PLS, PCR, CSVM-P, CSVM-R and RF respectively. The BRVM model, though, has marginally higher MSEP than BSVM but it still improves considerably on the other standard methods. Similar results are also obtained for Simulation 2. Results under two different prior settings are very close to each other, suggesting our model is not very sensitive to prior choices.

Table 4.2 Simulation study. *The range of calculated GR diagnostics are reported in parenthesis in the first column. Results for CSVM-P are reported for degrees 1, 2, and 3 respectively. BRVM and BSVM results are reported under prior choice (i) and (ii).*

	Median MSEP	**SD**
Simulation 1		
PLS	0.9635	0.1629
PCR	0.9575	0.1843
CSVM-P	0.91425, 0.9313, 0.9280	0.1714, 0.1736, 0.1809
CSVM-R	0.9692	0.1898
RF	0.9910	0.1875
BRVM (0.98,1.05)	0.9041, 0.9039	0.1794, 0.1769
BSVM (0.99,1.11)	0.8935, **0.8897**	0.1483, 0.1521
Simulation 2		
PLS	0.9727	0.1662
PCR	0.9989	0.1663
CSVM-P	0.9685, 0.9593, 0.96974	0.1641, 0.1645, 0.1642
CSVM-R	0.9797	0.1620
RF	0.9853	0.1665
BRVM (0.96,1.06)	0.9661, 0.9670	0.1458, 0.1470
BSVM (0.95,1.10)	0.9460, **0.94398**	0.1524, 0.1546

Blood sugar data

Research in fluorescence based optics suggests that a less invasive measurement technique may be able to continuously monitor glucose levels in the body of diabetics. The data presented here are discussed in Spiegelman, Wikander, O'Neal and Coté [22]. Optical spectra are collected from an experiment measuring photon counts for 101 wavelengths in the range 509–625 nm and $n = 27$. Our response is the glucose concentration, and the covariates are the different photon counts corresponding to the $p = 101$ wavelengths. The target is to predict accurately the glucose concentration in the body of the diabetics in the test set on the basis of the available photon counts.

In Table 4.3 we report a comparative performance of various methods along with our BRVM and BSVM for prediction of blood glucose concentration on the basis of median MSEP. It is clear that our RKHS based Bayesian RVM and SVM perform much better than all other methods such as PLS, PCR which are often used by the chemometricians. Also Bayesian RVM and SVM leads to a better prediction than classical SVM. Introduction of multiple smoothing parameters also

Table 4.3 Real Data Analysis. *The range of calculated GR diagnostics are reported in parenthesis in the first column. Results for CSVM-P are reported for degrees 1, 2 and 3 respectively. BRVM and BSVM results are reported under prior choice (i) and (ii).*

	Median MSEP	SD
Blood sugar data		
PLS	0.5737	0.6602
PCR	0.5628	0.5419
CSVM-P	0.5848, 0.4465, 0.34122	2.6381, 6.1941, 4.1501
CSVM-R	0.3915	1.5244
RF	0.4820	0.2599
BRVM (0.94,1.02)	**0.3124**, 0.3148	0.7619, 0.7429
BSVM (0.97,1.08)	0.4507, 0.4500	1.7123, 1.7708
Gas data		
PLS	0.2300	0.0763
PCR	0.2294	0.0602
CSVM-P	0.0771, 0.0863, 0.0990	0.0229, 0.0328, 0.0651
CSVM-R	0.3756	0.2812
RF	0.35620	0.2645
BRVM (0.95,1.02)	0.04971, **0.0488**	0.0146, 0.0144
BSVM (0.98,1.16)	0.0509, 0.0506	0.0179, 0.0177
Wheat data		
PLS	0.5130	0.1177
PCR	0.3912	0.0809
CSVM-P	0.6407, 0.5992, 0.5303	0.1835, 0.1891, 0.1678
CSVM-R	0.8121	0.2444
RF	0.7091	0.2182
BRVM (0.98,1.03)	0.3148, 0.3177	0.1669, 0.1650
BSVM (0.93,1.13)	0.3096, **0.3090**	0.2202, 0.2230

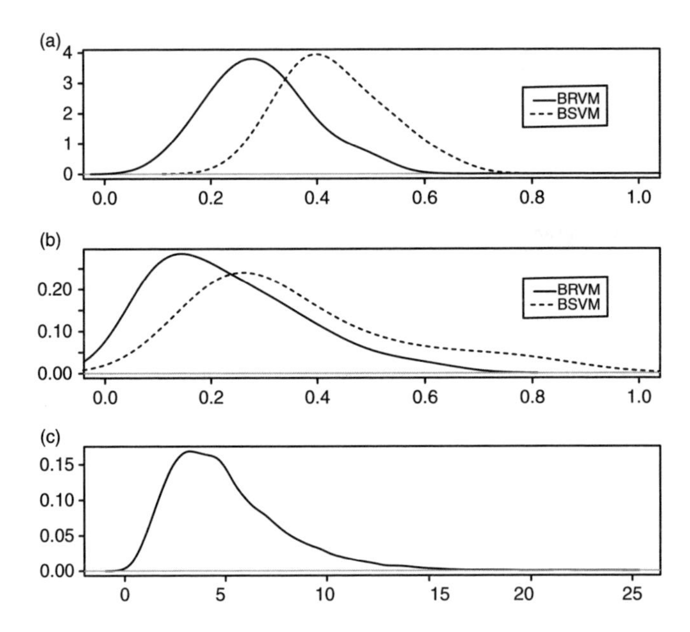

Figure 4.2 Blood sugar concentration dataset. (a) Distribution of MSEP under BRVM and BSVM. (b) Posterior distribution of the kernel parameter θ under BRVM and BSVM models. (c) Posterior distribution of ρ.

introduces sparsity in our model, as the β_is are then controlled individually. A number of these will then be shrunk to zero introducing the sparsity. Mallick *et al.* [18] have also provided a nice comparison of the benefits of a multiple smoothing parameter over a single smoothing parameter in a Bayesian SVM classification model. Introducing sparsity in this way also draws similarity with the Automatic Relevance Determination (ARD) models of Neal [20] and MacKay [17]. In terms of performance, BRVM with multiple smoothing parameters has made an improvement of 10% to 80% in prediction accuracy over all standard methods. Apart from the improvement in prediction accuracy, it enhances the flexibility of the model as well. In particular Bayesian BRVM with multiple smoothing parameters has the least average MSEP. Our model is not very sensitive to the choice of hyperparameters, as for both the choices we have nearly the same average MSEP. This is true as long as we stick to the class of near-diffuse but proper priors. In Figure 4.2 we plot the posterior distributions of MSEP, θ and ρ. The posterior plots of estimated kernel parameter θ indicate that nonlinear kernels of order 2 or 3 are required to attain the best prediction rate.

Gas data

This dataset is obtained from Kalivas [14]. In the dataset we have NIR spectra of 60 gasoline samples, measured in 2 nm intervals from 900 nm to 1700 nm as covariates. Our response is the octane number of each of the 60 gasoline samples. So here $p = 401$ and $n = 60$. From Table 4.3 we see that our BRVM and BSVM results in at least a five-fold decrease on MSEP over PLS, PCR, CSVM-R and RF. In this dataset BRVM is marginally better than BSVM. The CSVM-P is the main competitor in this data. Both BRVM and BSVM attain at least 50% lower MSEP. The prior sensitivity is also not an issue as long as we contain ourselves in the class of near-diffuse priors.

Wheat data

In this dataset we focus on a calibration NIR spectra dataset of wheat from Kalivas 1997 to predict the protein content. Here we have in total 100 (n) wheat samples. The NIR reflectance is measured from 1100–2500 nm in 2 nm intervals. So we have $p = 701$ covariates. The NIR data are used to predict protein content in the wheat samples. Measurements of protein were obtained by other standard reference methods. The goal is to accurately predict the protein content based on the NIR spectra data and to match it with the outcomes of the standard reference methods. In Table 4.3 we include the MSEP using our BRVM and BSVM along with other standard techniques. Both BRVM and BSVM reduces the MSEP over PLS, CSVM and RF by more than 50%. Our RKHS based kernel regression model improves upon the popularly used PCR by more than 25%. In this dataset BSVM comes out to be the overall winner.

In all simulation studies and real data analyses the calculated Gelman–Rubin scale reduction factor is close to the desired value of one. This ensures that our MCMC has converged satisfactorily and any inference drawn based on the MCMC will be appropriate.

4.4 Bayesian support vector machine model for binary data

For a binary classification problem, we have a training set $\{y_i, \boldsymbol{x}_i\}, i = 1, \ldots, n$, where y_i is the response variable indicating the class to which the ith observation belongs, and $\boldsymbol{x}_i$ is the vector of covariates of size p. The objective is to predict the posterior probability of belonging to one of the classes given a set of new covariates, based on the training data. Usually the response is coded as $y_i = 1$ for class 1 and $y_i = 0$ (or -1) for the other class. We utilize the training data $\boldsymbol{y} = (y_1, \cdots, y_n)^T$ and $\boldsymbol{X}^T = (\boldsymbol{x}_1, \cdots, \boldsymbol{x}_n)$ to fit a model $p(\boldsymbol{y}|\boldsymbol{x})$ and to use it to obtain $P(y_* = 1|\boldsymbol{y}, \boldsymbol{x}_*)$ for a future observation y_* with covariate $\boldsymbol{x}_*$. To simplify the structure, we introduce the latent variables $\boldsymbol{z} = (z_1, \ldots, z_n)$ and assume that $p(\boldsymbol{y}|\boldsymbol{z}) = \prod_{i=1}^{n} p(y_i|z_i)$, i.e. the y_i are conditionally independent given the z_i. In the next stage, the latent variables z_i are modelled as $z_i = f(\boldsymbol{x}_i) + \epsilon_i, i = 1, \ldots, n$, where $f \in \mathcal{H}_K$ the reproducing kernel Hilbert space spanned by the kernel $K()$. The ϵ_i, is the random residual effects, which are independent and identically distributed $N(0, \sigma^2)$. Therefore, we can construct a hierarchical model for classification as

$$p(y_i|z_i) \propto \exp\{-l(y_i, z_i)\}, i = 1, \ldots, n, \tag{4.30}$$

where the $y_1, y_2, \cdots, y_n$ are conditionally independent given $z_1, z_2, \cdots, z_n$ and l is any specific choice of the loss function, as explained in the previous section. We relate z_i to $f(\boldsymbol{x}_i)$ by $z_i = f(\boldsymbol{x}_i) + \epsilon_i$, where the ϵ_i are residual random effects. It is shown by Wahba [29], that a support vector machine can be fitted by finding $\boldsymbol{\beta}$ which minimizes the penalized loss function $\frac{1}{2} \parallel \boldsymbol{\beta} \parallel^2 + C\sum_{i=1}^{n}\{1 - y_i f(\boldsymbol{x}_i)\}_+$, where $[a]_+ = a$ if $a > 0$ and is 0 otherwise, $C \geq 0$ is a penalty term and f is the decision function.

In a Bayesian formulation, this optimization problem is equivalent to finding the posterior mode of $\boldsymbol{\beta}$, where the likelihood is given by $\exp[-\sum_{i=1}^{n}\{1 - y_i f(\boldsymbol{x}_i)\}_+]$, while $\boldsymbol{\beta}$ has the $N(\boldsymbol{0}, C\boldsymbol{I}_{n+1})$ prior. However, in our formulation with latent variables $\boldsymbol{z}$, we begin instead with the density

$$p(\mathbf{y}|\boldsymbol{z}) \propto \exp\left\{-\sum_{i=1}^{n}[1 - y_i z_i]_+\right\} \tag{4.31}$$

and assume independent $N(f(\boldsymbol{x}_i), \sigma^2)$ priors for the z_i.

If we use the density in (4.31), the normalizing constant may involve $\mathbf{z}$. Following Sollich [23], one may bypass this by assuming a distribution for $\boldsymbol{z}$ such that the normalizing constant cancels out. If the normalized likelihood is

$$p(\mathbf{y}|\boldsymbol{z}) = \exp\left\{-\sum_{i=1}^{n}[1 - y_i z_i]_+\right\}/c(\boldsymbol{z})$$

where $c(\cdot)$ is the normalizing constant, then choosing $p(\boldsymbol{z}) \propto Q(\boldsymbol{z})c(\boldsymbol{z})$, the joint distribution turns out to be

$$p(\boldsymbol{y}, \boldsymbol{z}) \propto \exp\left\{-\sum_{i=1}^{n}[1 - y_i z_i]_+\right\} Q(\boldsymbol{z}) \tag{4.32}$$

as the $c(\cdot)$ cancels from the expression. We will take $Q(\boldsymbol{z})$ as the product of independent normal pdfs with means $f(\boldsymbol{x}_i)$ and common variance σ^2. This method will be referred to as the Bayesian support vector machine (BSVM) classification.

Moreover, as explained in Section 4.2, we can express f as

$$f(\boldsymbol{x}_i) = \beta_0 + \sum_{j=1}^{n} \beta_j K(\boldsymbol{x}_i, \boldsymbol{x}_j|\boldsymbol{\theta}) \tag{4.33}$$

where K is a positive definite function of the covariates (inputs) $\boldsymbol{x}$ and we allow some unknown parameters $\boldsymbol{\theta}$ to enrich the class of kernels.

The random latent variable z_i is thus modelled as

$$z_i = \beta_0 + \sum_{j=1}^{n} \beta_j K(\boldsymbol{x}_i, \boldsymbol{x}_j|\boldsymbol{\theta}) + \epsilon_i = \mathbf{K}_i'\boldsymbol{\beta} + \epsilon_i, \tag{4.34}$$

where the ϵ_i are independent and identically distributed $N(0, \sigma^2)$ variables, and $\mathbf{K}_i' = (1, K(\mathbf{x}_i, \mathbf{x}_1|\boldsymbol{\theta}), \ldots, K(\mathbf{x}_i, \mathbf{x}_n|\boldsymbol{\theta})), i = 1, \ldots, n$.

To complete the hierarchical model, we need to assign priors to the unknown parameters $\boldsymbol{\beta}$, $\boldsymbol{\theta}$ and σ^2. We assign to $\boldsymbol{\beta}$ the Gaussian prior with mean $\mathbf{o}$ and variance $\sigma^2\mathbf{D}_*^{-1}$, where $\mathbf{D}_* \equiv \text{Diag}(\lambda_1, \lambda, \cdots, \lambda)$ is a $(n+1) \times (n+1)$ diagonal matrix, λ_1 being fixed at a small value, but λ is unknown. This amounts to a large variance for the intercept term. We will assign a proper uniform prior to $\boldsymbol{\theta}$, an inverse Gamma prior to σ^2 and a Gamma prior to λ. A Gamma(α, ξ) distribution for a random variable, say U, has probability density function proportional to $\exp(-\xi u)u^{\alpha-1}$, while the reciprocal of U will then be said to have a IG(α, ξ) distribution. Our model is thus given by

$$p(y_i|z_i) \propto \exp\left\{-\sum_{i=1}^{n}[1 - y_i z_i]_+\right\}$$

$$z_i|\boldsymbol{\beta}, \boldsymbol{\theta}, \sigma^2 \overset{\text{ind}}{\sim} \text{N}_1(z_i|\mathbf{K}_i'\boldsymbol{\beta}, \sigma^2) \tag{4.35}$$

$$\boldsymbol{\beta}, \sigma^2 \sim \text{N}_{n+1}(\boldsymbol{\beta}|\mathbf{o}, \sigma^2\mathbf{D}_*^{-1})\text{IG}(\sigma^2|\gamma_1, \gamma_2) \tag{4.36}$$

$$\boldsymbol{\theta} \sim \Pi_{q=1}^{p} U(a_{q1}, a_{q2}) \tag{4.37}$$

$$\lambda \sim \text{Gamma}(m, c) \tag{4.38}$$

where $U(a_{q1}, a_{q2})$ is the uniform probability density function over (a_{q1}, a_{q2}).

We can extend this model using multiple smoothing parameters so that the prior for $(\boldsymbol{\beta}, \sigma^2)$ is

$$\boldsymbol{\beta}, \sigma^2 \sim \mathrm{N}_{n+1}(\boldsymbol{\beta}|0, \sigma^2\mathbf{D}^{-1})\mathrm{IG}(\sigma^2|\gamma_1, \gamma_2), \tag{4.39}$$

where $\mathbf{D}$ is a diagonal matrix with diagonal elements $\lambda_1, \ldots, \lambda_{n+1}$. Once again, λ_1 is fixed at a small value, but all other λs are unknown. We assign independent Gamma(m, c) priors to them. Let $\boldsymbol{\lambda} = (\lambda_1, \ldots, \lambda_{n+1})'$.

Conditional distributions and posterior sampling of the parameters

The prior distributions given in (4.36) are conjugate for β and σ^2, whose posterior density conditional on $\mathbf{z}, \boldsymbol{\theta}, \boldsymbol{\lambda}$ is Normal-Inverse-Gamma,

$$p(\boldsymbol{\beta}, \sigma^2|\mathbf{z}, \boldsymbol{\theta}, \boldsymbol{\lambda}) = \mathrm{N}_{n+1}(\boldsymbol{\beta}|\tilde{\mathbf{m}}, \sigma^2\tilde{\mathbf{V}})\mathrm{IG}(\sigma^2|\tilde{\gamma_1}, \tilde{\gamma_2}), \tag{4.40}$$

where $\tilde{\mathbf{m}} = (\mathbf{K_o}'\mathbf{K_o} + \mathbf{D})^{-1}(\mathbf{K_o}'\mathbf{z}), \tilde{\mathbf{V}} = (\mathbf{K_o}'\mathbf{K_o} + \mathbf{D})^{-1}, \tilde{\gamma_1} = \gamma_1 + n/2,$ and $\tilde{\gamma_2} = \gamma_2 + \frac{1}{2}(\mathbf{z}'\mathbf{z} - \tilde{\mathbf{m}}'\tilde{\mathbf{V}}\tilde{\mathbf{m}})$. Here $\mathbf{K_o}' = (\mathbf{K}_1, \cdots, \mathbf{K}_n)$, where we recall that $\mathbf{K}_i = [K(\mathbf{x}_i, \mathbf{x}_1), \ldots, K(\mathbf{x}_i, \mathbf{x}_n)]'$.

The conditional distribution for the precision parameter λ_i given the coefficient β_i is Gamma and is given by

$$p(\lambda_i|\beta_i) = \mathrm{Gamma}\left(m + \frac{1}{2}, c + \frac{1}{2\sigma^2}{\beta_i}^2\right), i = 2, \ldots, n+1 \tag{4.41}$$

Finally, the full conditional density for z_i is

$$p(z_i|z_{-i}, \boldsymbol{\beta}, \sigma^2, \boldsymbol{\theta}, \boldsymbol{\lambda}) \propto \exp\left[-l(y_i, z_i) - \frac{1}{2\sigma^2}\{z_i - \sum_{j=1}^{n}\beta_j K(\mathbf{x}_i, \mathbf{x}_j)\}^2\right]$$

Similarly, the full conditionals are found when $\lambda_2 = \cdots = \lambda_{n+1} = \lambda$ from (4.38).

We make use of the above conditional distributions through a Gibbs sampler that iterates through the following steps: (i) update $\mathbf{z}$; (ii) update $\mathbf{K}, \boldsymbol{\beta}, \sigma^2$; (iii) update $\boldsymbol{\lambda}$.

For the update to $\mathbf{z}$, we propose to update each z_i in turn conditional on the rest. The conditional distribution of z_i does not have an explicit form; we thus resort to the Metropolis–Hastings procedure with a proposal density $T(z_i^*|z_i)$ that generates moves from the current state z_i to a new state z_i^*. The proposed updates are then accepted with probabilities

$$\alpha = \min\left\{1, \frac{p(y_i|z_i^*)p(z_i^*|\mathbf{z}_{-i}, \mathbf{K})T(z_i|z_i^*)}{p(y_i|z_i)p(z_i|\mathbf{z}_{-i}, \mathbf{K})T(z_i^*|z_i)}\right\} \tag{4.42}$$

We obtain $p(y_i|z_i)$ from (4.31) and

$$p(z_i|\mathbf{z}_{-i}, \mathbf{K}) \propto \exp\{-(z_i - \mathbf{K}_i\boldsymbol{\beta})^2/(2\sigma^2)\}$$

It is convenient to take the proposal distribution $T(z_i^*|z_i)$ to be a Gaussian with mean equal to the old value z_i and a prespecified standard deviation.

An update of $\mathbf{K}$ is equivalent to that of $\boldsymbol{\theta}$ and the marginal distribution of $\boldsymbol{\theta}$ conditional on $\mathbf{z}$ can be written as

$$p(\boldsymbol{\theta}|z) \propto p(z|\boldsymbol{\theta})p(\boldsymbol{\theta})$$

The new proposal $\boldsymbol{\theta}^*$ is then accepted with acceptance probability

$$\alpha = \min\left\{1, \frac{p(z|\boldsymbol{\theta}^*)}{p(z|\boldsymbol{\theta})}\right\} \tag{4.43}$$

The ratio of the marginal likelihoods is given by

$$\frac{p(z \mid \boldsymbol{\theta}^*)}{p(z \mid \boldsymbol{\theta})} = \frac{|\tilde{\mathbf{V}}^*|^{1/2}}{|\tilde{\mathbf{V}}|^{1/2}} \left(\frac{\tilde{\gamma}_2}{\tilde{\gamma}_2^*}\right)^{\tilde{\gamma}_1} \tag{4.44}$$

where $\tilde{\mathbf{V}}^*$ and $\tilde{\gamma}_2^*$ are similar to $\tilde{\mathbf{V}}$ and $\tilde{\gamma}_2$ with $\boldsymbol{\theta}^*$ replacing $\boldsymbol{\theta}$. Updating $\boldsymbol{\beta}$, σ^2 and $\boldsymbol{\lambda}$ is straightforward as they are generated from standard distributions.

4.4.1 Examples

We illustrate the methodology with several examples. For all examples, five models were fitted: (i) Bayesian support vector (BSVM) classification with a single penalty parameter; (ii) Bayesian support vector (BSVM) classification with multiple penalty parameters. We have used the SVM Matlab toolbox to obtain the classical SVM (SVM*) results. We obtained the RVM (Tipping, [25]) Matlab code from *http://research.microsoft.com/mlp/RVM/relevance.htm*. The neural network is fitted using the *nnet*() function in R.

Throughout the examples, we selected γ_1 and γ_2 to give a tight inverse gamma prior for σ^2 with mean 0.1. For $\boldsymbol{\lambda}$ we chose m and c so that the mean of the gamma distribution is small, say 10^{-3}, but with a large variance; a_{q1} and a_{q2}, the prior parameters of $\boldsymbol{\theta}$ are chosen using the $\boldsymbol{x}$ in such a way that computation of the kernel function does not over- or underflow. We performed the data analysis with both the Gaussian and polynomial kernels $\boldsymbol{K}$ as introduced in Section 4.3.1, and the results showed very little difference. The results reported here are based on Gaussian kernels. In all the examples we used a burn-in of 5000 samples, after which every 100th sample was retained in the next 50000 samples. The convergence and mixing of the chain were checked using two independent chains and the methods described in Gelman [9].

Benchmark comparisons

We analysed three well-known benchmark datasets and present the results in Table 4.4. The first two datasets are Pima Indians diabetes and Leptograpsus crabs (Ripley, [22]). The third is Wisconsin breast cancer data which contains ten basic features to classify two types of cancers, malignant and benign. We split the data randomly into training/testing partitions of sizes 300 and 269, and report average results over ten partitions. From Table 4.4, we see that all our multiple shrinkage models perform nearly as well as the best available alternatives.

Leukemia data

The leukemia dataset was described in Golub *et al.* [11]. Bone marrow or peripheral blood samples are taken from 72 patients with either myeloid leukemia (AML) or acute lymphoblastic leukemia (ALL). Following the experimental setup of the original paper, the data are split into training and test sets. The former consists of 38 samples, of which 27 are ALL and 11 are AML; the latter consists of 34 samples, 20 ALL and 14 AML. The dataset contains expression levels for 7129 human genes produced by Affymetrix high-density oligonucleotide microarrays.

We have provided our results in Table 4.5 with the modal or most frequent number of misclassification errors (the modal values) as well as the error bounds (maximum and minimum number of misclassifications).

Table 4.4 Modal classification error rates and 95% credible intervals for benchmark datasets. BSVM: Bayesian support vector machine; RVM: Relevance vector machine; SVM*: Classical support vector machine.

Method	Ripley's	Pima	Crabs
BSVM (single)	12.4(11.1,16.8)	21 (20,23.9)	4 (2,5)
BSVM (multiple)	8.8(8.4,11.6)	18.9 (18.3,20.6)	1 (0,4)
RVM	9.3	19.6	2
Neural Networks	N/A	22.5	3.0
SVM*	13.2	21.2	4

Table 4.5 Modal classification error rates and 95% credible intervals for Leukemia data. BSVM: Bayesian support vector machine; RVM: Relevance vector machine; CSVM: Bayesian SVM as proposed by Sollich [23]; SVM*: Frequentist support vector machine.

Model	Modal misclassification error	Error bound
BSVM (single)	4	(3,7)
BSVM (multiple)	1	(0,3)
CSVM (single)	5	(3,8)
CSVM (multiple)	2	(1,6)
SVM*	4	
RVM	2	

Table 4.5 shows that the results produced by the multiple shrinkage models are superior to the single precision models, as well as the classical SVM models. Although all the multiple shrinkage models performed well, the best performer among these appears to be the Bayesian support vector machine model.

The use of RKHS leads to a reduction in the dimension of the model, but the dimension can still be as high as the sample size. In the Bayesian hierarchical modelling framework, due to shrinkage priors, we obtain sparsity automatically (Tipping, [25]). Because of the presence of the unknown parameter θ in the expression of $\boldsymbol{K}$, this θ induces a posterior distribution for DF (rather than a fixed value). The posterior distributions of DF for all the three multiple shrinkage models were very similar.

4.5 Discussion

In this chapter, we have introduced Bayesian kernel based methods for regression and binary classification. Our kernel machine models can handle any large number of covariates. The regression and classification functions are not fixed, it is assumed to belong to the reproducing kernel Hilbert space or RKHS. RKHS encompasses a wide range of linear and nonlinear functions. In all the simulation studies and real data analysis, the number of covariates (p) we had was far greater than the available data points (n). Through our construction of the function through RKHS, the dimension of the problem is projected from higher-dimensional (p) covariate space to lower-dimensional (n) kernel space. Through posterior sampling our Bayesian models select the best choice of the kernel parameter θ. This enables our model to use multiple values of θ or a mixture of several kernels, linear and nonlinear. Moreover, multiple shrinkage models always appear to be superior to single parameter shrinkage models. With multiple shrinkage parameters, our Bayesian SVM model emerges as the winner in all the examples. Bayesian SVM also has the unique ability to quantify the prediction error, i.e. now we can have the entire posterior distribution of MSEP and can obtain a confidence interval. Rather than having just a point predictor now we can have the full posterior predictive probability distribution of a future observation. Bayesian kernel machine models are currently under constant development and are an active area of study. Their usefulness is also been explored in areas of multiple (correlated) response regression models, where the response is a vector with an unknown correlation structure, Chakraborty *et al.* [6]. They have also been successfully extended from binary classification to multiclass classification problems in high-dimensional microarray data (Chakraborty *et al.* [5]). A new direction may be applying the kernel trick for survival and longitudinal data with a large number of covariates. In the end we would like to view our methods as a new probability based way of modelling and prediction in high-dimensional problems. Much of their success and limitations will be revealed as more and more applications are made on various new datasets.

References

[1] Aronszajn, N. (1950). Theory of reproducing kernels. *Transactions of the American Mathematical Society*, **68**, 337–404.

[2] Bernardo, J. M. (1979). Expected information as expected utility, *Annals of Statistics*, **7**, 686–690.

[3] Bishop, C. and Tipping, M. (2000). Variational relevance vector machines. *Proceedings of the 16th Conference in Uncertainty and Artificial Intelligence* (eds C. Boutilier and M. Goldszmidt). Morgan Kauffman, San Francisco.

[4] Breiman, L. (2001). Random forests. *Machine Learning*, **45**, 5–32.

[5] Chakraborty, S., Ghosh, M., Mallick, B. K., Ghosh, D. and Dougherty E. (2007). Gene expression-based glioma classification using hierarchical Bayesian vector machines. *Sankhya*, **69**, 514–547.

[6] Chakraborty, S., Ghosh, M. and Mallick, B. (2011). Bayesian non-linear regression for large p small n problems. *Journal of Multivariate Analysis*.

[7] Chib, S. and Greenberg, E. (1995). Understanding the Metropolis–Hastings algorithm. *The American Statistician*, **49**, 327–335.

[8] Gelfand, A. and Smith, A. F. M. (1990). Sampling-based approaches to calculating marginal densities. *Journal of the American Statistical Association*, **85**, 398–409.

[9] Gelman, A. (1996). Inference and monitoring convergences, in *Markov Chain Monte Carlo in Practice* (eds Gilks, Richardson and Spiegelhalter), pp. 131–140, London: Chapman and Hall.

[10] Gelman, A., Carlin, J., Stern, H. and Rubin, D. (2003). *Bayesian Data Analysis*, Chapman and Hall.

[11] Golub, T. R., Slonim, D., Tamayo, P., Huard, C., Gaasenbeek, M., Mesirov, J., Coller, H., Loh, M., Downing, J., Caligiuri, M., Bloomfield, C. and Lander, E. (1999). Molecular classification

of cancer: Class discovery and class prediction by gene expression monitoring. *Science*, **286**, 531–537.
[12] Good, I. J. (1965). *The Estimation of Probabilities. An Essay on Modern Bayesian Methods.* MIT Press, MA.
[13] Huber, P. (1964). Robust estimation of a location parameter. *Annals of Mathematical Statistics*, **53**, 73–101.
[14] Kalivas, J. H. (1997). Two data sets of near infrared spectra. *Chemometrics and Intelligent Laboratory Systems*, **37**, 255–259.
[15] Kimeldorf, G. and Wahba, G. (1971). Some results on Tchebycheffian spline functions. *Journal of Mathematical Analysis and Applications*, **33**, 82–95.
[16] Law, M. H. and Kwok, J. T. (2001). Bayesian support vector regression. *Proceedings of the Eighth International Workshop on Artificial Intelligence and Statistics (AISTATS)*, 239–244. Key West, Florida, USA.
[17] MacKay, D. J. C. (1994). Bayesian non-linear modeling for the prediction competition. *ASHRAE Trans.*, **100**(2), 1053–1062.
[18] Mallick, B. K., Ghosh, D. and Ghosh, M. (2005). Bayesian classification of tumors using gene expression data. *Journal of the Royal Statistical Society, B*, **67**, 219–232.
[19] Metropolis, N., Rosenbluth, A. W., Rosenbluth, M. N., Teller, A. H. and Teller, E. (1953). Equations of state calculations by fast computing machines. *Journal of Chemical Physics*, **21**, 1087–1092.
[20] Neal, R. M. (1996). *Bayesian Learning for Neural Networks*, Springer-Verlag, New York.
[21] Parzen, E. (1970). Statistical inferences on time series by RKHS methods. *Proceedings of the 12th Biennial Seminar*, 1–37, Canadian Mathematical Congress, Montreal, Canada.
[22] Ripley, B. D. (1996). *Pattern Recognition and Neural Networks*, Cambridge University Press.
[23] Sollich, P. (2001). Bayesian methods for support vector machines: evidence and predictive class probabilities. *Machine Learning*, **46**, 21–52.
[24] Spiegelman, C., Wikander, J., O'Neal, P. and Coté, G. L. (2002). A simple method for linearizing nonlinear spectra for calibration. *Chemometrics and Intelligent Laboratory Systems*, **60**, 197–209.
[25] Tipping, M. (2000). The relevance vector machine. *Neural Information Processing Systems* Vol 12, S. Solla, T. Leen and K. Muller (eds), 652–658. MIT Press, Cambridge, MA.
[26] Tipping, M. (2001). Sparse Bayesian learning and the relevance vector machine. *J. Mach. Learn. Res.*, **1**, 211–244.
[27] Vapnik, V. N. (1995). *The Nature of Statistical Learning Theory*, 2nd edn. Springer: New York.
[28] Wahba, G. (1990). *Spline Models for Observational Data*. SIAM: Philadelphia.
[29] Wahba, G. (1999). Support vector machines, reproducing kernel Hilbert spaces and the randomized GACV, in *Advances in Kernel Methods*, B. Schölkopf, C. Burges and A. Smola (eds), 69–88. MIT Press, Cambridge, MA.

5 Flexible Bayesian modelling for clustered categorical responses in developmental toxicology

ATHANASIOS KOTTAS AND
KASSANDRA FRONCZYK

5.1 Introduction

Developmental toxicity studies investigate birth defects induced by toxic chemicals. In particular, under the standard Segment II developmental toxicity experiment, at each experimental dose level, $x_i, i = 1, \ldots, N$, a number, n_i, of pregnant laboratory animals (dams) are exposed to the toxin. Dam j at dose x_i has m_{ij} implants, of which the number of resorptions, that is, undeveloped embryos or very early foetal deaths, and prenatal deaths are recorded as R_{ij}, and the number of live pups at birth with a certain malformation are recorded as y_{ij}. Consequently, the number of viable foetuses for dam j at dose x_i is $m_{ij} - R_{ij}$. Additional continuous outcomes measured on each of the live pups may include body weight and length.

The main objective of developmental toxicity studies is to examine the relationship between the level of exposure to the toxin, which we generically refer to as the dose level, and the probability of the various responses of interest. We focus on the clustered categorical endpoints of embryolethality, that is, non-viable foetuses, and foetal malformation for live pups; thus, the data structure comprises $\{(m_{ij}, R_{ij}, y_{ij}) : i = 1, \ldots, N; j = 1, \ldots, n_i\}$. The corresponding dose–response curves are defined by the probability of the endpoints across dose levels. Also of interest is quantitative risk assessment, which evaluates the probability that adverse effects may occur as a result of the exposure to the substance.

Plotted in Figure 5.1 is a motivating dataset, available from the National Toxicology Program database, from an experiment that explored the effects of diethylhexalphthalate (DEHP), a commonly used plasticizing agent. The left and middle panels correspond to the endpoints of a non-viable foetus, that is, resorption or prenatal death, and malformation, that is, external, visceral or skeletal malformation of a live foetus. The right panel plots the proportions of combined negative outcomes, that is, adding the number of non-viable foetuses and malformations.

The number of dams is 30 for the control group, and 26, 26, 17 and 9 for doses 25, 50, 100 and 150 mg/kg $\times 1000$. The number of implants across all dams and dose levels ranges from 4 to 18, with 25th, 50th and 75th percentiles equal to 11, 13 and 14, respectively. Particularly noteworthy is the drop

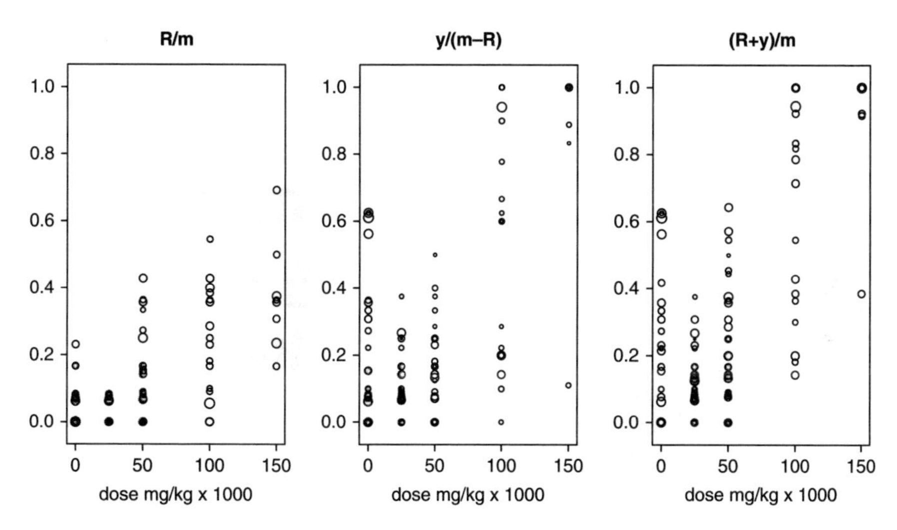

Figure 5.1 DEHP data. In each panel, a circle corresponds to a particular dam and the size of the circle is proportional to the number of implants. The coordinates of the circle are the dose level and the proportion of the specific endpoint: non-viable foetuses among implants (left panel); malformations among live pups (middle panel); combined negative outcomes among implants (right panel).

in the proportions of malformations, and combined negative outcomes, from dose 0 to 25 mg/kg $\times 1000$, which may correspond to a hormetic dose–response relationship. Hormesis refers to a dose–response phenomenon characterized by favourable biological responses to low exposures to toxins, and thus by opposite effects in small and large doses. For endpoints involving disease incidence, such as mutation, birth defects or cancer, hormesis results in a J-shaped dose–response curve. Although the possibility of different low dose effects is accepted, the suggestion of positive low dose effect is debated, hence, hormesis remains a controversial concept in the toxicological sciences [e.g. 1].

Notwithstanding the ultimate scientific conclusions, data such as those from the DEHP study motivate our modelling framework for the dose-dependent response distributions which enables rich inference for the implied, possibly non-monotonic, dose–response curves. Building flexible modelling for the response distribution is easy to justify for developmental toxicology data, which typically indicate vast departures from parametric models. This can be attributed to the inherent heterogeneity in the data due to the clustering of individuals within a group and the variability of the reaction of the individuals to the toxin. Note also that the typical toxicity experiment discussed above provides information on potentially different dose–response relationships for the distinct endpoints of embryolethality and foetal malformation, and it is thus biologically relevant to jointly analyse the clustered responses. This stands in contrast with the prevailing data structure found in the statistical literature, where the variables involved are the number of implants and the sum of all negative outcomes, as in the right panel of Figure 5.1.

We develop a Bayesian nonparametric mixture model for the joint distribution of the number of non-viable foetuses and malformations. We seek mixture modelling for response distributions that are related across doses with the level of dependence driven by the distance between the dose values. To this end, we consider a dependent Dirichlet process (DDP) prior [18] for the dose-dependent mixing distributions. Particular emphasis is placed on the choice of the mixture kernel and the DDP prior formulation to ensure an increasing trend in prior expectation for the implied dose–response curves, but without restricting prior realizations to be necessarily monotonic. This is key for the model's capacity to capture non-standard dose–response relationships. The nonparametric mixture model structure enables flexible inference for the response distributions at any observed dose level. Moreover, the dependence of the DDP prior across dose levels allows inference for the induced dose–response relationships through interpolation and extrapolation over any range of dose values of interest.

The modelling approach developed here extends work for the simpler data setting with combined negative outcomes [10]. To our knowledge, the literature does not include any Bayesian nonparametric approaches to modelling developmental toxicology data with a multicategory response classification. A Bayesian semiparametric model for the combined negative endpoints case, based on a product of Dirichlet process prior, was proposed by [5], and more recently extended in [19]. Examples of parametric Bayesian hierarchical models for toxicology data with discrete–continuous outcomes include [6] and [8]. Regarding the classical literature, a Dirichlet-trinomial model is presented in [2]; [24] develop an extended Dirichlet-multinomial model with Weibull dose–response functions; and [21] and [17] use quasi-likelihood and generalized estimating equations, respectively, to fit multinomial models which incorporate overdispersion.

The outline of the paper is as follows. Section 5.2 develops the DDP mixture model, including model properties and methods for prior specification and Markov chain Monte Carlo posterior inference. In Section 5.3, we present the application to the analysis of the DEHP data. Section 5.4 concludes with a summary.

5.2 Methods

5.2.1 The modelling approach

Under the Segment II toxicity study design, exposure occurs after implantation, and thus it is natural to treat the number of implants as a random quantity containing no information about the dose–response relationship. Hence, the modelling for the number of implants (m), the number of non-viable foetuses (R), and the number of malformations (y) is decomposed to $f(m, R, y) = f(m)f(R, y \mid m)$, where only the conditional distribution $f(R, y \mid m)$ will depend on dose level x. We assume a shifted Poisson distribution for $f(m)$ such that $m \geq 1$, although more general distributions can be readily utilized. Inference for the implant distribution is carried out separately from inference for $f(R, y \mid m)$, and is not discussed further.

To develop a flexible inference framework for risk assessment, we propose a nonparametric mixture model for the dose-dependent conditional distribution of the number of non-viable foetuses and malformations given the number of implants. Specifically, for a generic dose x,

$$f(R, y \mid m) \equiv f(R, y \mid m; G_x) = \int k(R, y \mid m; \varphi) \, \mathrm{d}G_x(\varphi)$$

where $k(R, y \mid m; \varphi)$ is a parametric kernel, with parameters φ, and G_x the dose-dependent mixing distribution. Placing a nonparametric prior on G_x results in a nonparametric mixture prior for $f(R, y \mid m; G_x)$. Nonparametric Bayesian mixture priors offer flexible modelling tools that, with the appropriate structure, can capture the complexity inherent in the data. These models can be viewed as extensions of finite mixture or continuous mixture models, where the random mixing distribution is not defined with a particular parametric family of distributions.

Regarding the mixture kernel, we take $k(R, y \mid m; \gamma, \theta) = \mathrm{Bin}(R; m, \pi(\gamma))\mathrm{Bin}(y; m - R, \pi(\theta))$, where $\pi(u) = \exp(u)/\{1 + \exp(u)\}$, $u \in \mathbb{R}$, will be used to denote the logistic function. Therefore, $\pi(\gamma)$ is the kernel probability of a non-viable foetus, and $\pi(\theta)$ the conditional probability of a malformation for a live pup. This formulation of the trinomial kernel distribution is natural as it highlights the nested nature of the count responses. Moreover, the logistic transformation for the Binomial probabilities is used to facilitate the formulation of the nonparametric prior model for the collection of mixing distributions, $G_{\mathcal{X}} = \{G_x : x \in \mathcal{X}\}$, where $\mathcal{X} \subseteq \mathbb{R}^+$.

As discussed in the Introduction, we seek modelling for the response distributions that allow nonparametric dependence structure across dose levels. We achieve such modelling by placing a DDP prior on the dose-dependent mixing distributions $\{G_x : x \in \mathcal{X}\}$. The DDP prior arises through extension of the Dirichlet process (DP) prior [9], the most widely used prior for mixing distributions in nonparametric or semiparametric mixture models. We use $\text{DP}(\alpha, G_0)$ to denote the DP prior defined in terms of a parametric centring distribution G_0, and precision parameter $\alpha > 0$.

To define the form of the DDP prior we use for $G_{\mathcal{X}}$, the almost sure discrete representation for the regular DP [22] is extended to

$$G_{\mathcal{X}}(\cdot) = \sum_{l=1}^{\infty} \omega_l \delta_{\eta_{l\mathcal{X}}}(\cdot) \tag{5.1}$$

where δ_a denotes a point mass at a, the $\eta_{l\mathcal{X}} = \{\eta_l(x) : x \in \mathcal{X}\}$ are independent realizations from a stochastic process $G_{0\mathcal{X}}$ over $\mathcal{X}$, and the weights are defined by a stickbreaking process: $\omega_1 = \zeta_1$, and $\omega_l = \zeta_l \prod_{r=1}^{l-1}(1 - \zeta_r)$ for $l \geq 2$, with ζ_l independent from a Beta(1,α) distribution, independently of the $\eta_{l\mathcal{X}}$. A key feature of the DDP prior is that for any finite collection of dose levels $(x_1, ..., x_k)$ it induces a multivariate DP prior for the corresponding collection of mixing distributions $(G_{x_1}, ..., G_{x_k})$. Therefore, the DDP prior model involves a countable mixture of realizations from stochastic process $G_{0\mathcal{X}}$ with weights matching those from the standard DP; this prior structure is referred to as single-p DDP prior. Single-p DDP mixture models have been applied to analysis of variance settings [4], spatial modelling [11, 15], dynamic density estimation [20], quantile regression [16] and survival regression [3].

Finally, the DDP prior mixture model is given by

$$f(R, y \mid m; G_{\mathcal{X}}) = \int \text{Bin}(R; m, \pi(\gamma))\text{Bin}(y; m - R, \pi(\theta))\, \text{d}G_{\mathcal{X}}(\gamma, \theta), \quad G_{\mathcal{X}} \sim \text{DDP}(\alpha, G_{0\mathcal{X}}) \tag{5.2}$$

Here, $\text{DDP}(\alpha, G_{0\mathcal{X}})$ denotes the DDP prior for $G_{\mathcal{X}} = \sum_{l=1}^{\infty} \omega_l \delta_{\eta_{l\mathcal{X}}}$, where $\eta_l(x) = (\gamma_l(x), \theta_l(x))$, for $x \in \mathcal{X}$, with precision parameter α and base stochastic process $G_{0\mathcal{X}}$. We define $G_{0\mathcal{X}}$ through two independent Gaussian processes, one driving each probability of response, each with a linear mean function, constant variance, and isotropic exponential correlation function. Hence, to introduce notation, we assume for all l, $\text{E}(\gamma_l(x) \mid \xi_0, \xi_1) = \xi_0 + \xi_1 x$, and $\text{E}(\theta_l(x) \mid \beta_0, \beta_1) = \beta_0 + \beta_1 x$; $\text{var}(\gamma_l(x) \mid \tau^2) = \tau^2$, and $\text{var}(\theta_l(x) \mid \sigma^2) = \sigma^2$; $\text{corr}(\gamma_l(x), \gamma_l(x') \mid \rho) = \exp(-\rho|x - x'|)$, and $\text{corr}(\theta_l(x), \theta_l(x') \mid \phi) = \exp(-\phi|x - x'|)$, with $\rho > 0$ and $\phi > 0$. As discussed in the next section, this specification for $G_{0\mathcal{X}}$ and, in particular, the linear mean functions, are key for flexible inference about the dose–response relationships implied by model (5.2). The full Bayesian model is implemented with priors on α and on the $G_{0\mathcal{X}}$ hyperparameters, $\psi = (\xi_0, \xi_1, \tau^2, \rho, \beta_0, \beta_1, \sigma^2, \phi)$.

5.2.2 Dose–response relationships

Here, we study the dose–response curves implied by DDP mixture model (5.2), including the probability of a non-viable foetus, the conditional probability of a malformation for a live pup and a risk function that combines the two endpoints.

To develop the dose–response curves, it is useful to note a connection of the mixture model with the clustered Binomial kernels with the model based on products of Bernoulli's kernel for the underlying binary responses. That is, for a generic dam at dose level x with m implants, let $R^* = \{R_k^* : k = 1, \ldots, m\}$ be the individual non-viable foetus indicators and denote by $y^* = \{y_s^* : s = 1, \ldots, m - \sum_{k=1}^{m} R_k^*\}$ the malformation indicators for the viable foetuses. Therefore, $R = \sum_{k=1}^{m} R_k^*$ and $y = \sum_{s=1}^{m-\sum_k R_k^*} y_s^*$. Then, a DDP mixture model for the clustered binary responses can be formulated as

$$f^* \left(R^*, y^* \mid m; G_{\mathcal{X}}\right) = \int \prod_{k=1}^{m} \text{Bern}(R_k^*; \pi(\gamma)) \prod_{s=1}^{m-\sum_k R_k^*} \text{Bern}(y_s^*; \pi(\theta)) \, \mathrm{d}G_{\mathcal{X}}(\gamma, \theta) \qquad (5.3)$$

where $G_{\mathcal{X}}$ is assigned the same DDP prior as the one for model (5.2). It is straightforward to show that mixture models (5.2) and (5.3) are equivalent with regard to the distribution for (R, y) conditional on m; in particular, the joint moment generating function for $(\sum_{k=1}^{m} R_k^*, \sum_{s=1}^{m-\sum_k R_k^*} y_s^*)$ under model (5.3) is equal to the joint moment generating function for (R, y) under model (5.2), in both cases, conditioning on m. Hence, we can define the dose–response curves under the DDP mixture model (5.2) working with probabilities of the two endpoints for a generic implant; this involves implicit conditioning on $m = 1$, which we suppress in the notation below.

The first risk assessment quantity of interest is the probability of embryolethality across effective dose levels, which is defined by

$$D(x) \equiv \text{pr}(R^* = 1; G_x) = \int \pi(\gamma) \, \mathrm{d}G_x(\gamma, \theta), \quad x \in \mathcal{X}$$

Risk assessment for the malformation endpoint is based on the conditional probability that a generic pup has a malformation given that it is a viable foetus, i.e.

$$M(x) \equiv \text{pr}(y^* = 1 \mid R^* = 0; G_x) = \frac{\text{pr}(R^* = 0, y^* = 1; G_x)}{\text{pr}(R^* = 0; G_x)}$$

$$= \frac{\int \{1 - \pi(\gamma)\}\pi(\theta)\mathrm{d}G_x(\gamma, \theta)}{\int \{1 - \pi(\gamma)\}\mathrm{d}G_x(\gamma, \theta)}, \quad x \in \mathcal{X}$$

Moreover, a full risk function at any given dose level can be defined through the combination of the probability of a non-viable foetus and the probability of a live, malformed pup; that is, the combined risk at dose level x is given by

$$r(x) \equiv \text{pr}(R^* = 1 \text{ or } y^* = 1; G_x) = \text{pr}(R^* = 0, y^* = 1; G_x) + \text{pr}(R^* = 1; G_x)$$

$$= \int \{1 - \pi(\gamma)\}\pi(\theta) \, \mathrm{d}G_x(\gamma, \theta) + \int \pi(\gamma) \, \mathrm{d}G_x(\gamma, \theta)$$

$$= 1 - \int \{1 - \pi(\gamma)\}\{1 - \pi(\theta)\} \, \mathrm{d}G_x(\gamma, \theta), \quad x \in \mathcal{X}$$

A key aspect of the modelling approach is that it does not force a non-decreasing shape restriction to the dose–response functions, which is the traditional assumption for more standard quantal bioassay experiments. As discussed in the Introduction and illustrated in Section 5.3 with the DEHP data, the model's capacity to capture non-standard, possibly non-monotonic, dose–response relationships is an asset of the proposed methodology. At the same time, given the relatively small number of observed dose levels in developmental toxicity studies, some structure is needed in the prior model in order to obtain meaningful interpolation and extrapolation posterior inference results for the dose–response curves. Under the specific formulation of the DDP prior for mixture model (5.2), such structure can be incorporated in the form of a non-decreasing trend in prior expectation for the dose–response curves.

Consider first the prior expectation for the dose-dependent probability of a non-viable foetus,

$$\mathrm{E}\{D(x)\} = \mathrm{E}\left\{\int \pi(\gamma)\mathrm{d}G_x(\gamma,\theta)\right\} = \int \pi(\gamma)\mathrm{d}G_{0x}(\gamma,\theta) = \int \pi(\gamma)\mathrm{dN}(\gamma;\xi_0+\xi_1 x,\tau^2)$$

that is, $\mathrm{E}\{D(x)\}$ is the expectation of the (increasing) logistic function with respect to the $\mathrm{N}(\xi_0+\xi_1 x,\tau^2)$ distribution, which is stochastically ordered in x when $\xi_1 > 0$. Hence, $D(x)$ is a non-decreasing function of x in prior expectation, under the $\xi_1 > 0$ prior restriction. Similarly,

$$\begin{aligned}\mathrm{E}\{r(x)\} &= 1 - \textstyle\int\{1-\pi(\gamma)\}\{1-\pi(\theta)\}\mathrm{d}G_{0x}(\gamma,\theta)\\ &= 1 - \left[\textstyle\int\{1-\pi(\gamma)\}\mathrm{dN}(\gamma;\xi_0+\xi_1 x,\tau^2)\right]\left[\textstyle\int\{1-\pi(\theta)\}\mathrm{dN}(\theta;\beta_0+\beta_1 x,\sigma^2)\right]\end{aligned}$$

Provided $\xi_1 > 0$ and $\beta_1 > 0$, distributions $\mathrm{N}(\xi_0+\xi_1 x,\tau^2)$ and $\mathrm{N}(\beta_0+\beta_1 x,\sigma^2)$ are stochastically ordered in x, which implies that both $\int\{1-\pi(\gamma)\}\mathrm{dN}(\gamma;\xi_0+\xi_1 x,\tau^2)$ and $\int\{1-\pi(\theta)\}\mathrm{dN}(\theta;\beta_0+\beta_1 x,\sigma^2)$ are decreasing functions of x, with values in the unit interval. Thus, $\mathrm{E}\{r(x)\}$ is non-decreasing in x when $\xi_1 > 0$ and $\beta_1 > 0$.

Therefore, with the $\xi_1 > 0$ and $\beta_1 > 0$ prior restrictions, we can build to both the probability of a non-viable foetus and to the combined risk function the non-decreasing trend in prior expectation. Although the same argument does not extend to the conditional probability of malformation, $M(x)$, the restriction $\xi_1 > 0$ and $\beta_1 > 0$ appears sufficient to provide the prior expectation non-decreasing trend for all three dose–response curves. In this respect, it is useful to note that, even though we develop inference about three dose–response relationships, there are only two endpoints and, consequently, the model is driven at any specific dose level by a bivariate random mixing distribution.

Note that the argument above relies on both the constant Gaussian process variances for the two components of $G_{0\mathcal{X}}$—which ensures the stochastic ordering of the induced normal distributions—and on the linear Gaussian process mean functions—which enables the non-decreasing trend through the restriction on the slope parameters. Indeed, the linear mean functions are crucial for practicable posterior inference. As suggested by Figure 5.2, if the model is applied using constant mean functions for the DDP prior centring Gaussian processes, that is, setting $\xi_1 = \beta_1 = 0$, we should not expect practically useful results outside the observed dose levels. For illustration, Figure 5.2 plots results from prior simulation for the embryolethality and malformation dose–response curves, and for the combined risk function, using fixed values for α $(= 1)$ and ψ. In particular, $(\xi_1 = 0.0085, \beta_1 = 0.12)$ and $(\xi_1 = 0.12, \beta_1 = 0.01)$ in the top and middle row, respectively. Although the relative magnitudes of ξ_1 and β_1 affect the rate of increase for the different curves, in all cases with $\xi_1 > 0$ and $\beta_1 > 0$, the non-decreasing trend in prior expectation is preserved.

Finally, smoothness properties of prior realizations for the dose–response curves relate directly to the respective properties of the centring process $G_{0\mathcal{X}}$. For details, we refer to the arguments in [18] and [11], extended and formalized by [12], but note briefly that the continuity of the realizations from the two Gaussian processes that define $G_{0\mathcal{X}}$ implies that as the distance between x and x' gets smaller, the difference between G_x and $G_{x'}$ gets smaller; moreover, it yields

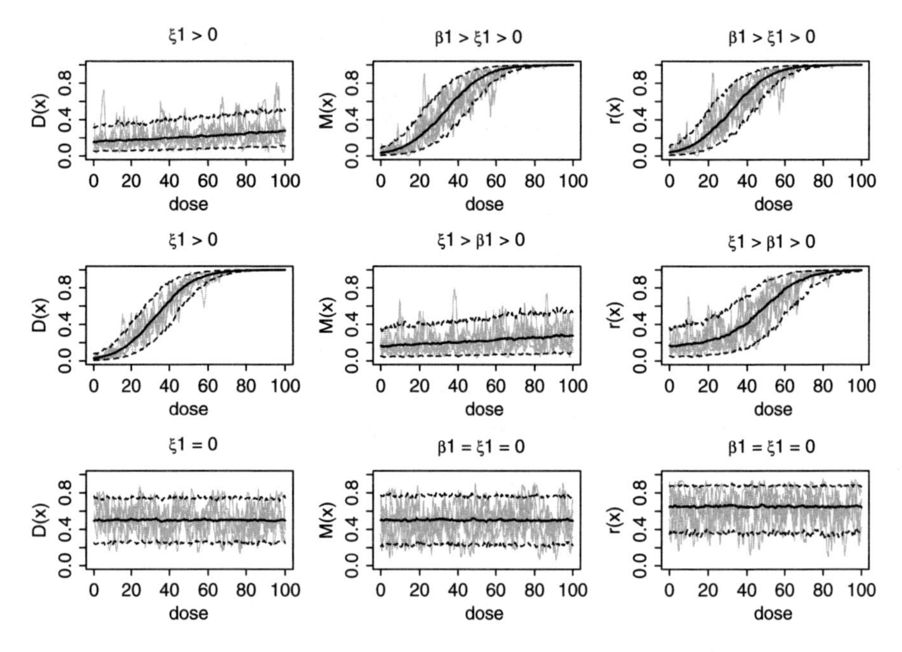

Figure 5.2 Prior mean and 90% interval estimates, along with five individual prior realizations, for the three dose–response curves. See Section 5.2.2 for details.

continuous prior realizations for the three dose–response functions defined above. The practical implication is that in prediction for the probability mass function $f(R, y \mid m; G_x)$ and for the corresponding dose–response curves, we learn more from dose levels x' nearby x than from more distant doses, a desirable property for distributions that are expected to evolve relatively smoothly with the dose level.

5.2.3 Implementation details

Markov chain Monte Carlo posterior simulation

Regarding the hierarchical model formulation for the data $= \{(m_{ij}, R_{ij}, y_{ij}) : i = 1, \ldots, N; j = 1, \ldots, n_i\}$, we observe that for the DEHP data (discussed in the Introduction) the dams are labelled and recorded in ascending numerical order across dose levels; that is, the smallest identification number corresponds to data from the first dam at the first dose level, the first dam at the second dose level has the next identification number, and so on. This is also the case for other datasets available from the database of the National Toxicology Program. Therefore, to write the model for the data, the animals are linked as a response vector across the dose levels with the conditional independence assumption built for the replicated response vectors. Hence, the data structure and corresponding hierarchical model are along the lines of the spatial DP [11] rather than, for instance, the ANOVA DDP [4].

Therefore, let $R_j = (R_{1j}, \ldots, R_{Nj})$, and $y_j = (y_{1j}, \ldots, y_{Nj})$ be the j-th response replicates with corresponding number of implants vector $m_j = (m_{1j}, \ldots, m_{Nj})$, for $j = 1, \ldots, n$, where $n = \max_i n_i$. Moreover, denote by $\gamma_j \equiv \gamma_j(x) = (\gamma_j(x_1), \ldots, \gamma_j(x_N))$, and $\theta_j \equiv \theta_j(x) = (\theta_j(x_1), \ldots, \theta_j(x_N))$ the latent mixing vector for R_j and y_j, respectively, where $x = (x_1, \ldots, x_N)$. We introduce missing value indicators, s_{ij}, such that $s_{ij} = 1$ if the j-th response replicates at dose level i are present and $s_{ij} = 0$ otherwise. Note that the s_{ij} are fixed for any particular dataset. Then, the first stage of the hierarchical model for the data can be written as

$$\{(R_{ij}, y_{ij})\} \mid \{m_{ij}\}, \{(\gamma_j, \theta_j)\} \sim \prod_{j=1}^{n} \prod_{i=1}^{N} \{\text{Bin}(R_{ij}; m_{ij}, \pi(\gamma_j(x_i)))\text{Bin}(y_{ij}; m_{ij} - R_{ij}, \pi(\theta_j(x_i)))\}^{s_{ij}}$$

where the (γ_j, θ_j), given G_x, are i.i.d. from G_x, which follows a DP(α, G_{0x}) prior implied by the DDP prior for $G_{\mathcal{X}}$. In particular, G_{0x} comprises two independent N-variate normal distributions, induced by the two Gaussian processes that define $G_{0\mathcal{X}}$; the first normal distribution has mean vector $(\xi_0 + \xi_1 x_1, \ldots, \xi_0 + \xi_1 x_N)'$ and covariance matrix with (i,j)-th element $T_{ij} = \tau^2 \exp(-\rho|x_i - x_j|)$; the mean of the second normal distribution is $(\beta_0 + \beta_1 x_1, \ldots, \beta_0 + \beta_1 x_N)'$ and its covariance matrix has (i,j)-th element $\Sigma_{ij} = \sigma^2 \exp(-\phi|x_i - x_j|)$.

Hence, the hierarchical model for the data is a DP mixture model induced by the DDP mixture prior. For Markov chain Monte Carlo posterior simulation, we use blocked Gibbs sampling [e.g. 13], which offers relatively ready implementation and also, in our context, can easily handle unbalanced response replicates. The approach is based on a finite truncation approximation of G_x such that $G_x \approx G_x^L = \sum_{l=1}^{L} p_l \delta_{(U_l(x), Z_l(x))}$, where the weights p_l arise from a truncated version of the stickbreaking construction: $p_1 = V_1, p_l = V_l \prod_{r=1}^{l-1}(1 - V_r), l = 2, \ldots, L-1$, and $p_L = 1 - \sum_{l=1}^{L-1} p_l$, with the V_l i.i.d., given α, from Beta$(1, \alpha)$. Moreover, $U_l(x) = (U_l(x_1), \ldots, U_l(x_N)) \equiv U_l$ and $Z_l(x) = (Z_l(x_1), \ldots, Z_l(x_N)) \equiv Z_l$, with the (U_l, Z_l) i.i.d., given ψ, from G_{0x}, for $l = 1, ..., L$. Hence, under the truncated version of mixing distribution G_x, $(\gamma_j, \theta_j) = (U_l, Z_l)$ with probability p_l, and $G_x^L \equiv (p, U, Z)$, where $p = (p_1, \ldots, p_L)$, $U = (U_1, \ldots, U_L)$ and $Z = (Z_1, \ldots, Z_L)$.

To represent the hierarchical model for the data under the DP truncation approximation, configuration variables $w = (w_1, \ldots, w_n)$ are introduced, such that $w_j = l$ if and only if $(\gamma_j, \theta_j) = (U_l, Z_l)$, for $l = 1, \ldots, L$ and $j = 1, \ldots, n$. Then, the model for the data can be expressed as

$$\{(R_j, y_j)\} \mid \{m_j\}, w, (U, Z) \sim \prod_{j=1}^{n} \prod_{i=1}^{N} \left\{\text{Bin}(R_{ij}; m_{ij}, \pi(U_{w_j}(x_i)))\text{Bin}(y_{ij}; m_{ij} - R_{ij}, \pi(Z_{w_j}(x_i)))\right\}^{s_{ij}}$$

$$w_j \mid p \sim \prod_{j=1}^{n} \sum_{l=1}^{L} p_l \delta_l(w_j)$$

$$p, (U, Z) \mid \alpha, \psi \sim f(p \mid \alpha) \times \prod_{l=1}^{L} G_{0x}(U_l, Z_l \mid \psi) \tag{5.4}$$

where $f(p \mid \alpha) = \alpha^{L-1} p_L^{\alpha-1}(1 - p_1)^{-1}\{1 - (p_1 + p_2)\}^{-1} \times \cdots \times (1 - \sum_{l=1}^{L-2} p_l)^{-1}$, a special case of the generalized Dirichlet distribution, is the prior for p, given α, induced by the truncated stickbreaking construction. The full Bayesian model is completed with independent hyperpriors for the DDP precision parameter α and the parameters ψ of the centring Gaussian processes. Specifically, we place a gamma(a_α, b_α) prior on α; normal priors N(m_ξ, s_ξ^2) and N(m_β, s_β^2) on ξ_0 and β_0; exponential priors Exp(b_ξ) and Exp(b_β) on ξ_1 and β_1 to promote the non-decreasing trend in prior expectation for the dose–response functions; inverse gamma priors inv-gamma(a_τ, b_τ) and inv-gamma(a_σ, b_σ) on the variance terms τ^2 and σ^2; and uniform priors Unif$(0, b_\rho)$ and Unif$(0, b_\phi)$ on the range parameters ρ and ϕ. Prior specification is discussed below.

Denote the n^* distinct values of vector w by $w_1^*, \ldots, w_{n^*}^*$, and let $M_k^* = |\{j : w_j = w_k^*\}|$, for $k = 1, \ldots, n^*$, and $M_l = |\{j : w_j = l\}|$, for $l = 1, \ldots, L$. Then, sampling from the posterior distribution $p(U, Z, w, p, \alpha, \psi \mid \text{data})$ corresponding to model (5.4) is based on simulation from the following posterior full conditional distributions.

The (U_l, Z_l) that correspond to $l \notin \{w_k^* : k = 1, \ldots, n^*\}$ are sampled from G_{0x} given its currently imputed parameters ψ. For $l = w_k^*, k = 1, \ldots, n^*$, the posterior full conditional for $U_{w_k^*}$ is proportional to $G_{0x}^{\gamma}(U_{w_k^*} \mid \psi) \prod_{\{j:w_j=w_k^*\}} \prod_{i=1}^{N} \{\text{Bin}(R_{ij}; m_{ij}, \pi(U_{w_k^*}(x_i)))\}^{s_{ij}}$, and the posterior full conditional for $Z_{w_k^*}$ to $G_{0x}^{\theta}(Z_{w_k^*} \mid \psi) \prod_{\{j:w_j=w_k^*\}} \prod_{i=1}^{N} \{\text{Bin}(y_{ij}; m_{ij} - R_{ij}, \pi(Z_{w_k^*}(x_i)))\}^{s_{ij}}$. Here, G_{0x}^{γ} and G_{0x}^{θ} denote the respective N-variate normal distributions arising from G_{0x}. Each of $U_{w_k^*}$ and $Z_{w_k^*}$ is updated using a random-walk Metropolis–Hastings step with an N-variate normal distribution as the proposal. The proposal covariance matrices were estimated dynamically, using initial runs based on normal proposals with scaled identity covariance matrices.

The posterior full conditional for each $w_j, j = 1, \ldots, n$, is given by a discrete distribution with values $l = 1, \ldots, L$ and corresponding probabilities

$$\tilde{p}_{lj} \propto p_l \prod_{i=1}^{N} \{\text{Bin}(R_{ij}; m_{ij}, \pi(U_l(x_i)))\text{Bin}(y_{ij}; m_{ij} - R_{ij}, \pi(Z_l(x_i)))\}^{s_{ij}}, \quad l = 1, \ldots, L$$

The updates for parameters α and p are the same with a generic DP mixture model [14]. Finally, the joint posterior full conditional for the hyperparameters ψ of the DDP prior centring Gaussian processes is proportional to

$$p(\xi_0)p(\xi_1)p(\tau^2)p(\rho)p(\beta_0)p(\beta_1)p(\sigma^2)p(\phi) \times \prod_{k=1}^{n^*} G_{0x}(U_{w_k^*}, Z_{w_k^*} \mid \psi)$$

where $p(\cdot)$ denotes the prior for each parameter. The form of G_{0x} and the parametric priors for the components for ψ result in normal posterior full conditionals for ξ_0 and β_0, and inverse gamma full conditionals for τ^2 and σ^2. We use Metropolis–Hastings updates for ξ_1 and β_1, and sample ρ and ϕ by discretizing their bounded support.

Inference for risk assessment

The samples from the posterior distribution of model (5.4) yield the mixing distribution G_x^L at all the observed dose levels through the posterior samples for (p, U, Z). To expand the inference over any range of doses of interest, we augment the N observed dose levels with M new doses, $\tilde{x} = (\tilde{x}_1, \ldots, \tilde{x}_M)$. Now, in the prior model, the $(U_l(x), U_l(\tilde{x}))$ and the $(Z_l(x), Z_l(\tilde{x}))$, for $l = 1, \ldots, L$, are independent realizations from two independent $(N + M)$-variate normal distributions, induced by the Gaussian processes that define $G_{0\mathcal{X}}$, with mean vectors and covariance matrices that are of the same form as above extending x to $(x, \tilde{x})$. But then, to sample from the conditional posterior distributions for each of the $U_l(\tilde{x})$ and $Z_l(\tilde{x})$, the additional sampling needed is from conditional M-variate normal distributions given the currently imputed $(U_l, Z_l), l = 1, \ldots, L$, and the parameters ψ.

Using the posterior samples from model (5.4), augmented with the posterior samples for $(U_l(\tilde{x}), Z_l(\tilde{x}))$, $l = 1, \ldots, L$, full inference for the response distributions and for risk assessment through the dose–response curves can be obtained by evaluating the relevant expressions developed in Sections 5.2.1 and 5.2.2. Under the DP truncation approximation used for posterior simulation, the integrals are replaced with sums. For instance, for any generic dose x_0 in $(x, \tilde{x})$, the posterior distribution for the probability of a non-viable foetus arises from $\sum_{l=1}^{L} p_l \pi(U_l(x_0))$, and for the combined risk function through $1 - \sum_{l=1}^{L} p_l\{1 - \pi(U_l(x_0))\}\{1 - \pi(Z_l(x_0))\}$. Moreover, for a specified number of implants

m_0, the conditional probability mass function for the number of malformations given R_0 non-viable foetuses, is evaluated through $\sum_{l=1}^{L} q_l(x_0)\text{Bin}(y; m_0 - R_0, \pi(Z_l(x_0)))$, where $q_l(x_0) = p_l\text{Bin}(R_0; m_0, \pi(U_l(x_0)))/\{\sum_t^L p_t\text{Bin}(R_0; m_0, \pi(U_t(x_0)))\}$. The posterior samples can be summarized with means and percentiles to provide posterior mean estimates and uncertainty bands for dose–response curves and probability mass functions for the response distributions; Section 5.3 reports such inferences for the DEHP data.

Prior specification

To specify the uniform priors for the range parameters ρ and ϕ, we consider the limiting case of the DDP model with $\alpha \to 0^+$, which yields the kernel of the mixture in (5.2) as the model's first stage with $G_{0\mathcal{X}}$ defining Gaussian process priors for the Binomial probabilities on the logistic scale. Then, under the exponential correlation function, $3/\rho$ is the range of dependence, that is, the distance between dose levels that yields correlation 0.05 for the Gaussian process realizations that define the probability of a non-viable foetus, and, analogously, for $3/\phi$. The range is usually assumed to be a fraction of the maximum interpoint distance over the index space. Let $D_{\max}$ be the maximum distance between observed doses. Since $3/b_\rho < 3/\rho$, we specify b_ρ such that $3/b_\rho = rD_{\max}$ for a small r; $r = 0.002$ was used for the DEHP data analysis in Section 5.3 leading to a Unif(0, 10) prior for ρ; the same uniform prior was used for ϕ. This approach to prior specification for ρ and ϕ is conservative, in particular, the posterior distributions for ρ and ϕ are concentrated on values substantially smaller than b_ρ and b_ϕ.

We set the prior means for ξ_0 and β_0 to 0, and the shape parameters of the inverse gamma priors for τ^2 and σ^2 to 2, implying infinite prior variance. The prior variances for ξ_0 and β_0, and the prior means for ξ_1, β_1, τ^2 and σ^2 are chosen by studying the induced prior distribution for the dose–response curves defined in Section 5.2.2. For the DEHP data, we placed a N(0, 10) prior on ξ_0 and β_0, an exponential prior with mean $b_\xi^{-1} = b_\beta^{-1} = 0.1$ on ξ_1 and β_1, and an inv-gamma(2, 10) on τ^2 and σ^2. Under this prior choice, the prior means for functions $D(x)$ and $M(x)$ have a relatively weak increasing trend starting around 0.5, with 90% uncertainty bands that cover almost the entire unit interval.

The DDP prior precision parameter, α, controls the number, n^*, of distinct mixture components [e.g. 7]. In particular, for moderate to large sample sizes, a useful approximation to the prior expectation $\text{E}(n^* \mid \alpha)$ is given by $\alpha \log\{(\alpha + n)\alpha^{-1}\}$. This expression can be averaged over the prior for α to obtain $\text{E}(n^*)$, thus selecting the gamma prior parameters to agree with a guess at the expected number of distinct mixture components. A gamma(2, 1) prior was used for the DEHP data example corresponding to $\text{E}(n^*) \approx 5$. Prior sensitivity analysis revealed robust posterior inference under more dispersed priors.

Finally, the level L for the DP truncation approximation can be chosen using standard distributional properties for the weights arising from the stickbreaking structure in (5.1). For instance, $\text{E}(\sum_{l=1}^{L} \omega_l \mid \alpha) = 1 - \{\alpha/(\alpha + 1)\}^L$, which can be averaged over the prior for α to estimate $\text{E}(\sum_{l=1}^{L} \omega_l)$. Given a tolerance level for the approximation, this expression is solved numerically to obtain the corresponding value L. For the analysis of the DEHP data, we used $L = 50$, which yields $\text{E}(\sum_{l=1}^{L} \omega_l) \approx 0.9999593$ under the gamma(2, 1) prior for α.

5.3 Data example

We illustrate the proposed DDP mixture modelling approach with the DEHP dataset discussed in the Introduction (Figure 5.1). It is known that plasticizers, such as the DEHP plasticizing agent, may leak in small quantities from plastic containers with various solvents such as food or milk.

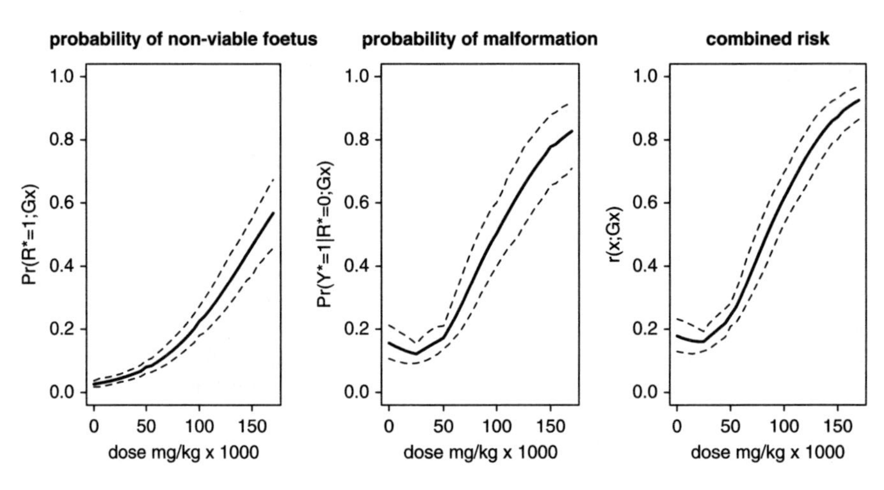

Figure 5.3 For the DEHP data, the posterior mean (solid lines) and 90% interval bands (dashed lines) for the risk assessment functions: probability of a non-viable foetus (left panel); conditional probability of malformation (middle panel); combined risk (right panel).

The possibility of toxic effects from these agents has been recognized and tested in developmental toxicity studies such as the one described in [23]. Recall that the two endpoints are non-viable foetus corresponding to resorption or actual prenatal death, and malformation involving external, visceral or skeletal malformation of a live foetus.

Figure 5.3 plots the posterior mean and 90% interval estimates for the three dose–response curves developed in Section 5.2.2 for risk assessment. The probability of a non-viable foetus across dose levels is a monotonically increasing function, with uncertainty bands around the posterior mean estimate that increase with increasing dose values, consistent with the decreasing number of dams for larger dose levels. The conditional probability of malformation, however, reveals a non-monotonic behaviour at the low dose levels, and this J-shaped pattern carries over to the combined risk which also exhibits a dip in the probability from the control through dose 25 mg/kg × 1000. The inference for the combined risk function agrees with the estimated dose–response curve for the combined negative outcomes version of the DEHP data, as obtained in [10] based on a DDP Binomial mixture model. The modelling approach presented in this paper is key to uncovering the malformation endpoint as the one that contributes to the non-monotonic, possibly hormetic, combined dose–response relationship.

Inference for response distributions is illustrated with posterior mean and 90% interval estimates for the probability mass function of the number of non-viable foetuses given $m = 12$ implants (Figure 5.4) and the number of malformations given $m = 12$ implants and $R = 3$ non-viable foetuses (Figure 5.5). Results are reported for the control group, the four effective dose levels, and a new dose at $x = 75$ mg/kg × 1000. As expected, there is more uncertainty in the estimation of the conditional response distributions for malformation. The interpolation at the new dose level appears to be influenced more by the distribution at dose 50, which can be attributed to the larger sample size relative to dose 100. The estimated distributions for the number of non-viable foetuses have relatively standard shapes, whereas there is some evidence of a bimodal shape at dose 100, and skewness in the estimated malformation distributions.

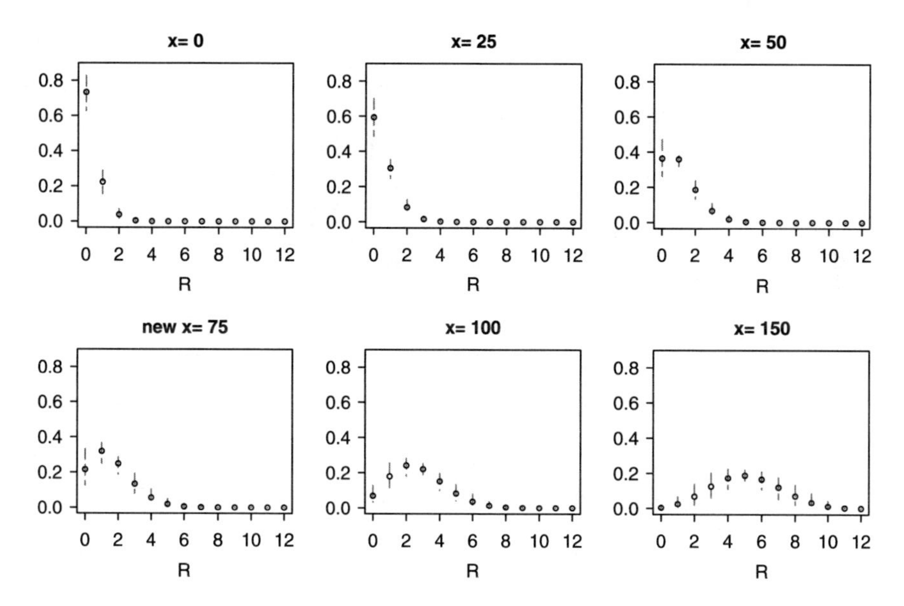

Figure 5.4 The posterior mean ("o") and 90% probability bands (dashed lines) of the probability mass functions for the number of non-viable foetuses given m = 12 implants.

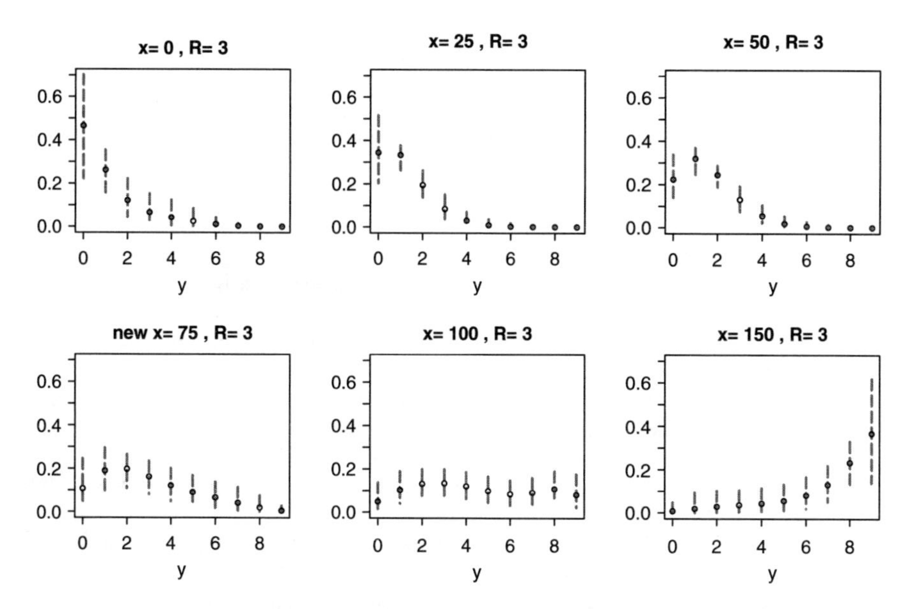

Figure 5.5 The posterior mean ("o") and 90% interval bands (dashed lines) of the probability mass functions for the number of malformations given m = 12 implants and R = 3 non-viable foetuses.

5.4 Summary

We have developed a Bayesian nonparametric modelling approach for risk assessment in developmental toxicity studies. The motivation for the proposed methodology is that it is critical to model flexibly the dose-dependent dam specific response distribution associated with the clustered categorical outcomes of a non-viable foetus and of malformation for a live pup. The model is built from a mixture with a product Binomial kernel, to capture the nested structure of the responses, and a dependent Dirichlet process prior for the dose-dependent mixing distributions. The resulting nonparametric DDP mixture model provides rich inference for the response distributions as well as for the dose–response curves. Data from a toxicity experiment involving a plasticizing agent were used to illustrate the scientifically relevant feature of the DDP mixture model with regard to estimation of different dose–response relationships for different endpoints, including non-monotonic dose–response curves.

Acknowledgements

This research is part of the PhD dissertation of Kassandra Fronczyk, completed at University of California, Santa Cruz, and was supported in part by the National Science Foundation under award DEB 0727543, and by a Special Research Grant awarded by the Committee on Research, University of California, Santa Cruz.

References

[1] Calabrese, E. J. (2005). Paradigm lost, paradigm found: The re-emergence of hormesis as a fundamental dose response model in the toxicological sciences. *Environmental Pollution*, **138**, 378–411.

[2] Chen, J. J., Kodell, R. L., Howe, R. B., and Gaylor, D. W. (1991). Analysis of trinomial responses from reproductive and developmental toxicity experiments. *Biometrics*, **47**, 1049–1058.

[3] DeIorio, M., Johnson, W. O., Müller, P., and Rosner, G. L. (2009). Bayesian nonparametric nonproportional hazards survival modeling. *Biometrics*, **65**, 762–771.

[4] DeIorio, M., Müller, P., Rosner, G. L., and MacEachern, S. N. (2004). An ANOVA model for dependent random measures. *Journal of the American Statistical Association*, **99**, 205–215.

[5] Dominici, F. and Parmigiani, G. (2001). Bayesian semiparametric analysis of developmental toxicology data. *Biometrics*, **57**, 150–157.

[6] Dunson, D., Chen, Z., and Harry, J. (2003). A Bayesian approach for joint modeling of cluster size and subunit-specific outcomes. *Biometrics*, **59**, 521–530.

[7] Escobar, M. D. and West, M. (1995). Bayesian density estimation and inference using mixtures. *Journal of the American Statistical Association*, **90**, 577–588.

[8] Faes, C., Geys, H., Aerts, M., and Molenberghs, G. (2006). A hierarchical modeling approach for risk assessment in developmental toxicity studies. *Computational Statistics & Data Analysis*, **51**, 1848–1861.

[9] Ferguson, T. S. (1973). A Bayesian analysis of some nonparametric problems. *The Annals of Statistics*, **1**, 209–230.

[10] Fronczyk, K. and Kottas, A. (2010). A Bayesian nonparametric modeling framework for developmental toxicity studies. Technical Report UCSC-SOE-10-11, University of California Santa Cruz, Department of Applied Mathematics and Statisics.

[11] Gelfand, A. E., Kottas, A., and MacEachern, S. (2005). Bayesian nonparametric spatial modeling with Dirichlet process mixing. *Journal of the American Statistical Association*, **100**, 1021–1035.

[12] Guindani, M. and Gelfand, A. E. (2006). Smoothness properties and gradient analysis under spatial Dirichlet process models. *Methodology and Computing in Applied Probability*, **8**, 159–189.

[13] Ishwaran, H. and James, L. (2001). Gibbs sampling methods for stick-breaking priors. *Journal of the American Statistical Association*, **96**(453), 161–173.

[14] Ishwaran, H. and Zarepour, M. (2000). Markov chain Monte Carlo in approximate Dirichlet and beta two-parameter process hierarchical models. *Biometrika*, **87**(2), 371–390.

[15] Kottas, A., Duan, J., and Gelfand, A. E. (2008). Modeling disease incidence data with spatial and spatio-temporal Dirichlet process mixtures. *Biometrical Journal*, **50**, 29–42.

[16] Kottas, A. and Krnjajić, M. (2009). Bayesian semiparametric modelling in quantile regression. *Scandinavian Journal of Statistics*, **36**, 297–319.

[17] Krewski, D. and Zhu, Y. (1994). Applications of multinomial dose–response models in developmental toxicity risk assessment. *Risk Analysis*, **14**, 613–627.

[18] MacEachern, S. N. (2000). Dependent Dirichlet processes. Technical report, Ohio State University, Department of Statistics.

[19] Nott, D. J. and Kuk, A. Y. C. (2009). Analysis of clustered binary data with unequal cluster sizes: A semiparametric Bayesian approach. *Journal of Agricultural, Biological, and Environmental Statistics*, **15**, 101–118.

[20] Rodriguez, A. and ter Horst, E. (2008). Bayesian dynamic density estimation. *Bayesian Analysis*, **3**, 339–366.

[21] Ryan, L. (1992). Quantitative risk assessment for developmental toxicity. *Biometrics*, **48**, 163–174.

[22] Sethuraman, J. (1994). A constructive definition of Dirichlet priors. *Statistica Sinica*, **4**, 639–650.

[23] Tyl, R. W., Jones-Price, C., Marr, M. C., and Kimmel, C. A. (1983). Teratologic evaluation of diethylhexyl phthalate (cas no. 111-81-7). Final Study Report for NCTR/NTP contract 222-80-2031(c), NITS PB85105674, National Technical Information Service, Springfield, Virginia.

[24] Zhu, Y., Krewski, D., and Ross, W. H. (1994). Dose-response models for correlated multinomial data from developmental toxicity studies. *Applied Statistics*, **43**, 583–598.

Part III

Markov Chain Monte Carlo

6 Markov chain Monte Carlo methods

SIDDHARTHA CHIB

6.1 Introduction

The growth of Bayesian thinking and practice over the past two decades has been in large part due to simulation-based computing methods, in particular, those based on Markov chain Monte Carlo (MCMC) methods. Starting with [44], MCMC methods had long been used in physics, for example, in computational statistical mechanics and quantum field theory to sample the coordinates of a point in phase space. The interest in MCMC methods in statistics was sparked by the paper of [23], where the Gibbs sampling method discussed in [24] was elaborated as a tool to generate marginal distributions of parameters of Bayesian models. Work on these methods since the Gelfand and Smith paper has been impressive. The theoretical underpinnings of MCMC methods have been clarified. Novel inference approaches that are inherently tied to MCMC-based computing have emerged. Applications of these methods in demanding applications across an increasingly diverse spectrum of scientific fields have become common.

MCMC sampling is a method for generating variates from a multivariate target probability distribution π^*, with support Θ, by simulating a Markov chain whose stationary distribution is π^*. Constructing such a chain is simpler, and in many instances the only viable strategy, compared to classical Monte Carlo methods that deliver independent and identically distributed (iid) draws from π^*. The fact that the variates from an MCMC simulation are correlated raises theoretical and practical issues that are different from classical Monte Carlo sampling. For example, there is the question about whether the effect of the starting value eventually wears off—this is the question related to the ergodicity of the chain; how long the chain should be allowed to run before one can suppose that the convergence to the invariant distribution has occurred—this is the question of the size of the burn-in; how the numerical accuracy of sample averages should be computed—this is the question of how the serial correlation in the chain should be incorporated.

Taking care of the preceding issues is important for the proper implementation of MCMC methods and can require considerable skill from the user when the model is high-dimensional and complex, especially in relation to the available sample information for conducting inferences. Nonetheless, provided the chain satisfies some weak theoretical conditions, the correlated sampled variates (beyond a suitably defined burn-in) are a surrogate for the target distribution. These can be used to estimate the probabilistic characteristics of π^*, just as in the case of iid draws. For example, the quantiles of π^* can be estimated by the quantiles of the sampled variates. The marginal density of any component can be estimated from the sample on that component, ignoring the sample on the other components. The expectation under π^* of any integrable function can be estimated as a sample-path average of that function.

The goal of this chapter is to provide a brief summary of MCMC methods, leaving further details for later chapters and the textbooks of [9], [40], and [50]. In Section 6.2 we describe the Metropolis–Hastings algorithm of [44] and its generalized version given by [30]. This is the original MCMC algorithm, and still the most important. The presentation borrows heavily from [54] and [12]. In Section 6.3 we consider the Gibbs sampling algorithm that arose from the work of [6], [24], [53] and [23], the last, in particular, responsible for elucidating its relevance for general Bayesian inference. Additional topics of importance, such as sampling with latent data, and calculation of the marginal likelihood, are discussed in Section 6.4. Section 6.5 has concluding remarks.

6.2 Metropolis–Hastings algorithm

Suppose that we are interested in sampling the target distribution π^* over a parameter vector $\boldsymbol{\theta} \in \Theta \subseteq \mathcal{R}^d$. Let $\mathbf{y}$ denote the sample data. The general idea behind the M-H approach is to sample a convenient Markov transition kernel $Q(\boldsymbol{\theta}, d\boldsymbol{\theta}'|\mathbf{y}) = \Pr(\boldsymbol{\theta} \in d\boldsymbol{\theta}'|\mathbf{y}, \boldsymbol{\theta})$ and to modify this transition kernel so that the modified transition kernel $P(\boldsymbol{\theta}, d\boldsymbol{\theta}'|\mathbf{y})$ has π^* as its unique stationary distribution. Assume that both the target distribution and the transition kernels are absolutely continuous with respect to a measure μ. For simplicity, one can suppose that μ is the Lebesgue measure. Letting π and q denote the densities of π^* and Q wrt to μ, respectively, we have that

$$\pi^*(d\boldsymbol{\theta}'|\mathbf{y}) = \pi(\boldsymbol{\theta}'|\mathbf{y})\mu(d\boldsymbol{\theta}')$$

and

$$Q(\boldsymbol{\theta}, d\boldsymbol{\theta}'|\mathbf{y}) = q(\boldsymbol{\theta}, \boldsymbol{\theta}'|\mathbf{y})\mu(d\boldsymbol{\theta}')$$

The distribution Q (equivalently, the density q) is used to generate candidate values or proposal values $\boldsymbol{\theta}'$. It is called the candidate generating density or proposal density.

In the M-H algorithm one constructs a Markov kernel $P(\boldsymbol{\theta}, d\boldsymbol{\theta}'|\mathbf{y})$ from $Q(\boldsymbol{\theta}, d\boldsymbol{\theta}'|\mathbf{y})$ with π^* as the stationary distribution. A key idea is that of reversibility. A Markov transition kernel $P(\boldsymbol{\theta}, d\boldsymbol{\theta}'|\mathbf{y})$ is reversible for π^* if

$$\pi^*(d\boldsymbol{\theta}|\mathbf{y})P(\boldsymbol{\theta}, d\boldsymbol{\theta}'|\mathbf{y}) = \pi^*(d\boldsymbol{\theta}'|\mathbf{y})P(\boldsymbol{\theta}', d\boldsymbol{\theta}|\mathbf{y}) \tag{6.1}$$

or, in terms of densities,

$$\pi(\boldsymbol{\theta}|\mathbf{y})p(\boldsymbol{\theta}, \boldsymbol{\theta}'|\mathbf{y}) = \pi(\boldsymbol{\theta}'|\mathbf{y})p(\boldsymbol{\theta}', \boldsymbol{\theta}|\mathbf{y})$$

for all $(\boldsymbol{\theta}, \boldsymbol{\theta}')$ in the support of π^*. Reversibility is an important restriction because a reversible Markov chain is automatically invariant. Invariance is the property that

$$\pi^*(d\boldsymbol{\theta}'|\mathbf{y}) = \int_{\Theta} P(\boldsymbol{\theta}, d\boldsymbol{\theta}'|\mathbf{y})\pi^*(d\boldsymbol{\theta}|\mathbf{y}) \tag{6.2}$$

which means intuitively that if $\boldsymbol{\theta} \sim \pi^*$, then the variate $\boldsymbol{\theta}'$ drawn from the transition kernel $P(\boldsymbol{\theta}, d\boldsymbol{\theta}'|\mathbf{y})$ is also from π^*. To see that reversibility implies invariance, one simply integrates both sides of (6.1) over $\boldsymbol{\theta}$. This leads to the invariance condition since $\int P(\boldsymbol{\theta}', d\boldsymbol{\theta}|\mathbf{y}) = 1$.

We can now follow [12] where an argumentation was introduced for deriving P from Q. Suppose one checks the reversibility condition for $Q(\boldsymbol{\theta}, d\boldsymbol{\theta}'|\mathbf{y})$. Since this transition kernel was chosen for its convenience, without necessarily any connection to the target distribution, it is unlikely to satisfy the reversibility condition. One possibility is that

$$\pi(\boldsymbol{\theta}|\mathbf{y})q(\boldsymbol{\theta}, \boldsymbol{\theta}'|\mathbf{y}) > \pi(\boldsymbol{\theta}'|\mathbf{y})q(\boldsymbol{\theta}', \boldsymbol{\theta}|\mathbf{y}) \tag{6.3}$$

which means informally that the process moves from $\boldsymbol{\theta}$ to $\boldsymbol{\theta}'$ too frequently and too rarely in the reverse direction. We can correct this situation by reducing the flow from $\boldsymbol{\theta}$ to $\boldsymbol{\theta}'$ by introducing probabilities $\alpha(\boldsymbol{\theta}, \boldsymbol{\theta}'|\mathbf{y})$ and $\alpha(\boldsymbol{\theta}', \boldsymbol{\theta}|\mathbf{y})$ of making the moves in either direction so that

$$\pi(\boldsymbol{\theta}|\mathbf{y})q(\boldsymbol{\theta}, \boldsymbol{\theta}'|\mathbf{y})\alpha(\boldsymbol{\theta}, \boldsymbol{\theta}'|\mathbf{y}) = \pi(\boldsymbol{\theta}'|\mathbf{y})q(\boldsymbol{\theta}', \boldsymbol{\theta}|\mathbf{y})\alpha(\boldsymbol{\theta}', \boldsymbol{\theta}|\mathbf{y}) \tag{6.4}$$

We now set $\alpha(\boldsymbol{\theta}', \boldsymbol{\theta}|\mathbf{y})$ to be as high as possible, namely equal to one. Solving for $\alpha(\boldsymbol{\theta}, \boldsymbol{\theta}'|\mathbf{y})$ we get that

$$\alpha(\boldsymbol{\theta}, \boldsymbol{\theta}'|\mathbf{y}) = \frac{\pi(\boldsymbol{\theta}'|\mathbf{y})}{\pi(\boldsymbol{\theta}|\mathbf{y})} \frac{q(\boldsymbol{\theta}', \boldsymbol{\theta}|\mathbf{y})}{q(\boldsymbol{\theta}, \boldsymbol{\theta}'|\mathbf{y})}$$

This quantity is less than one because we started from (6.3). The other possibility is that

$$\pi(\boldsymbol{\theta}|\mathbf{y})q(\boldsymbol{\theta}, \boldsymbol{\theta}'|\mathbf{y}) < \pi(\boldsymbol{\theta}'|\mathbf{y})q(\boldsymbol{\theta}', \boldsymbol{\theta}|\mathbf{y})$$

By the preceding argumentation it follows that $\alpha(\boldsymbol{\theta}, \boldsymbol{\theta}'|\mathbf{y})$ must now equal one. Therefore, on putting the two cases together, we get that

$$\alpha(\boldsymbol{\theta}, \boldsymbol{\theta}'|\mathbf{y}) = \min\left\{1, \frac{\pi(\boldsymbol{\theta}'|\mathbf{y})}{\pi(\boldsymbol{\theta}|\mathbf{y})} \frac{q(\boldsymbol{\theta}', \boldsymbol{\theta}|\mathbf{y})}{q(\boldsymbol{\theta}, \boldsymbol{\theta}'|\mathbf{y})}\right\} \tag{6.5}$$

The M-H algorithm is now in place. To find the next iterate of the Markov chain given the current value $\boldsymbol{\theta}$ to $\boldsymbol{\theta}'$, we first propose a value $\boldsymbol{\theta}'$ from $Q(\boldsymbol{\theta}, d\boldsymbol{\theta}'|y)$. With probability $\alpha(\boldsymbol{\theta}, \boldsymbol{\theta}'|\mathbf{y})$ we accept the proposed value. If rejected, the next iterate is the current value.

Algorithm. Metropolis–Hastings

Initialize the starting value $\boldsymbol{\theta}^{(0)}$, and specify the burn-in size n_0 and the required MCMC sample size G. Then, repeat the following two steps

- Sample $\boldsymbol{\theta}'$ from $Q(\boldsymbol{\theta}^{(g)}, d\boldsymbol{\theta}'|\mathbf{y})$
- Let the next iterate be

$$\boldsymbol{\theta}^{(g+1)} = \begin{cases} \boldsymbol{\theta}' & \text{with prob } \alpha(\boldsymbol{\theta}^{(g)}, \boldsymbol{\theta}'|\mathbf{y}) \\ \boldsymbol{\theta}^{(g)} & \text{with prob } 1 - \alpha(\boldsymbol{\theta}^{(g)}, \boldsymbol{\theta}'|\mathbf{y}) \end{cases}$$

Return the draws $\boldsymbol{\theta}^{(n_0+1)}, \ldots, \boldsymbol{\theta}^{(n_0+G)}$

It may be noted that the calculation of α does not require the norming constant of the target density. Another point is that if $q(\boldsymbol{\theta},\boldsymbol{\theta}'|\mathbf{y}) = q(\boldsymbol{\theta}',\boldsymbol{\theta}|\mathbf{y})$, which is the case when the proposal is sampled symmetrically around the current value, then

$$\alpha(\boldsymbol{\theta}',\boldsymbol{\theta}|\mathbf{y}) = \min\left\{1, \frac{\pi(\boldsymbol{\theta}'|\mathbf{y})}{\pi(\boldsymbol{\theta}|\mathbf{y})}\right\} \tag{6.6}$$

as in the original algorithm of Metropolis *et al.* [44].

6.2.1 Transition density of the M-H chain

An unusual aspect of the M-H algorithm is that it produces a chain in which values are repeated. As a result, the transition density of the M-H chain $P_{MH}(\boldsymbol{\theta}, d\boldsymbol{\theta}'|\mathbf{y})$ has two components—one for the move away from $\boldsymbol{\theta}$ given by

$$\alpha(\boldsymbol{\theta},\boldsymbol{\theta}'|\mathbf{y})Q(\boldsymbol{\theta}, d\boldsymbol{\theta}'|\mathbf{y})$$

and one for the probability of staying at $\boldsymbol{\theta}$ given by

$$r(\boldsymbol{\theta}|\mathbf{y}) = 1 - \int \alpha(\boldsymbol{\theta},\boldsymbol{\theta}'|\mathbf{y})Q(\boldsymbol{\theta}, d\boldsymbol{\theta}'|\mathbf{y})$$

In other words,

$$P_{MH}(\boldsymbol{\theta}, d\boldsymbol{\theta}'|\mathbf{y}) = \alpha(\boldsymbol{\theta},\boldsymbol{\theta}'|\mathbf{y})Q(\boldsymbol{\theta}, d\boldsymbol{\theta}'|\mathbf{y}) + \delta_{\boldsymbol{\theta}}(d\boldsymbol{\theta}')r(\boldsymbol{\theta}|\mathbf{y})$$

where $\delta_{\boldsymbol{\theta}}(d\boldsymbol{\theta}')$ is the Dirac measure, defined as equal to one whenever $\boldsymbol{\theta}'$ is equal to $\boldsymbol{\theta}$ and $\int \delta_{\boldsymbol{\theta}}(d\boldsymbol{\theta}') = 1$. It is easy to check that the integral of $P_{MH}(\boldsymbol{\theta}, d\boldsymbol{\theta}'|\mathbf{y})$ over all possible values of $\boldsymbol{\theta}'$ is one, as required.

6.2.2 MCMC convergence properties

We now digress by providing some theoretical properties related to MCMC simulations. This discussion has implications for the implementation of MCMC algorithms and for the analysis of the MCMC output. We begin by providing conditions under which the simulated sample path from an MCMC simulation leads to simulation-consistent estimates of posterior moments, posterior probabilities and other summaries of the target distribution. The definitions and results that follow are drawn from [54]. [50] provide further useful discussion.

Theorem 1 Suppose that the Markov chain $\{\boldsymbol{\theta}^{(g)}\}$ is π^*-irreducible and has invariant distribution $\pi^*(d\boldsymbol{\theta}|\mathbf{y})$. Then $\pi^*(d\boldsymbol{\theta}|\mathbf{y})$ is the unique invariant distribution. If the chain is π^*-irreducible, aperiodic and the invariant distribution is proper, then for π^*-every $\boldsymbol{\theta}^{(0)}$ and all measurable sets A

$$|\Pr(\boldsymbol{\theta}^{(g)} \in A|\mathbf{y}, \boldsymbol{\theta}^{(0)}) - \int_A \pi^*(d\boldsymbol{\theta}|\mathbf{y})| \to 0$$

as $g \to \infty$. If the chain is ergodic (π^*-irreducible, aperiodic and Harris recurrent), then for all functions $h(\boldsymbol{\theta})$ such that $\int_\Theta |h(\boldsymbol{\theta})|\pi^*(d\boldsymbol{\theta}|\mathbf{y}) < \infty$ and any initial distribution,

$$\hat{h}_G = G^{-1}\sum_{g=1}^{G} h(\boldsymbol{\theta}^{(g)}) \to \int_\Theta h(\boldsymbol{\theta})\pi^*(d\boldsymbol{\theta}|\mathbf{y})d\boldsymbol{\theta} \; as\; G \to \infty,\; a.s$$

These results hold under relatively weak conditions (for example, as discussed in [54], π^*-irreducibility of the chain is satisfied if the proposal density is everywhere positive in the support of the posterior density; it is Harris recurrent if it is π^*-irreducible, has π^* as it is unique invariant distribution and the transition kernel is absolutely continuous with respect to π^*.

One gets a central limit theorem for sample path averages if we further assume that the chain is uniformly ergodic.

Theorem 2 Suppose that the Markov chain $\{\boldsymbol{\theta}^{(g)}\}$ is uniformly ergodic and has invariant distribution $\pi^*(d\boldsymbol{\theta}|y)$. Then for functions $h(\boldsymbol{\theta})$ such that $\int_\Theta h(\boldsymbol{\theta})^2\pi^*(d\boldsymbol{\theta}|\mathbf{y}) < \infty$, and any initial distribution, the sample average $\hat{h}_G$ satisfies the ergodic limit theorem

$$\sqrt{G}\left(\hat{h}_G - \mathrm{E}_{\pi^*}h\right) \overset{d}{\to} \mathcal{N}(0, \sigma_h^2)$$

where

$$\mathrm{E}_{\pi^*}h = \int_\Theta h(\boldsymbol{\theta})\pi^*(d\boldsymbol{\theta}|\mathbf{y})$$

$$\sigma_h^2 = \lim_{G\to\infty} G\mathrm{Var}(\hat{h}_G) = \mathrm{Var}_{\pi^*}\left(h(\boldsymbol{\theta}^{(1)})\right) + 2\sum_{g=2}^{\infty}\mathrm{Cov}_{\pi^*}\left\{h(\boldsymbol{\theta}^{(1)}), h(\boldsymbol{\theta}^{(g)})\right\}$$

and the subscript π^* indicates that the expectations are calculated under the invariant distribution.

The next issue is how one should judge the success of the MCMC sampling strategy in estimating $\mathrm{E}_{\pi^*}h$. One way is to compare $\mathrm{Var}(\hat{h}_G) = \sigma_h^2/G$, where σ_h^2 is the variance that appears in Theorem 2, with the variance under (hypothetical) iid sampling. The square root of $\mathrm{Var}(\hat{h}_G)$ is called the numerical standard error. Under hypothetical iid sampling, $\mathrm{Var}(\hat{h}_G)$ is given by $G^{-1}\mathrm{Var}_{\pi^*}h(\boldsymbol{\theta}^{(1)})$. Therefore,

$$\tau_h^2 = \frac{\mathrm{Var}(\hat{h}_G)}{G^{-1}\mathrm{Var}_{\pi^*}h(\boldsymbol{\theta}^{(1)})} = \left\{1 + 2\sum_{g=1}^{\infty}\rho_{hg}\right\} \tag{6.7}$$

where ρ_{hg} is the sample autocorrelation at lag g. Because iid sampling produces an autocorrelation time that is theoretically equal to one, τ_h^2 will be close to one when the autocorrelations are declining quickly with lag g. Thus, the size of τ_h^2 reveals the inefficiency of the MCMC sampling procedure relative to iid sampling. In practice, the calculation of τ_h^2 is based on a windowing or truncation procedure that amounts to a down-weighting of autocorrelations at larger lags [27, 28]. The method of batch-means [49] is also used to estimate τ_h^2. First, we let $Z_g = h(\boldsymbol{\theta}^{(g)}), g = 1, 2, \ldots, G$. Next, we divide the data $\{Z_1, Z_2, \ldots, Z_G\}$ into k non-overlapping batches of length m with means

$$B_i = m^{-1}(Z_{(i-1)m+1} + \ldots + Z_{im}), i = 1, 2, \ldots, k$$

where the batch size m is chosen to ensure that the first-order serial correlation of the batch means is less than 0.05. The average of these batch means

$$\bar{B} = \frac{1}{k}\sum_{i=1}^{k} B_i$$

is of course $\hat{h}_G$ and the estimate of the sample variance of $\bar{B}$ by standard calculations is

$$\text{Var}\left(\bar{B}\right) = \frac{1}{k(k-1)}\sum_{i=1}^{k}(B_i - \bar{B})^2$$

In the batch means method, this variance estimate is taken to be the estimate of $\text{Var}(\hat{h}_G)$. [35] show that it is a consistent estimate of σ_h^2/G if k and m both increase with G.

6.2.3 Choice of proposal density

One significant practical problem in implementing the M-H algorithm is that there are many ways of specifying $Q(\boldsymbol{\theta}, d\boldsymbol{\theta}'|\mathbf{y})$. We present two that are popular in practice. In general, in each case we try to ensure that the chain makes large moves through the support of the invariant distribution, without staying at one place for many iterations.

Random walk proposals

In this version of the M-H algorithm, the proposal is drawn as

$$\boldsymbol{\theta}' = \boldsymbol{\theta} + \mathbf{z}$$

where $\mathbf{z}$ follows some symmetric distribution q such as the multivariate normal with mean of zero and covariance matrix $\mathbf{V}$. The covariance matrix has to be adjusted in pilot runs to reach some desired acceptance rate. Values in the range of 20% to 60% have been shown to be optimal under specific assumptions [50]. Because of the symmetry of the increment distribution, the M-H probability of move is a function of the target density, and is given by the expression in (6.6). A graphical view is shown in Figure 6.1. There has also been work on a more sophisticated version of this algorithm known as the Langevin M-H random-walk algorithm. In this case, the proposal value is generated with a drift term that is given by the first derivative of the log target density. In particular,

$$\boldsymbol{\theta}' = \boldsymbol{\theta} + \frac{c}{2}\frac{\partial \log \pi(\boldsymbol{\theta}|\mathbf{y})}{\partial \boldsymbol{\theta}} + \sqrt{c}\mathbf{z}$$

where $\mathbf{z}$ is distributed as $\text{N}(0, \mathbf{I})$. The tuning and efficiency of this algorithm is discussed in [51].

Although the random walk M-H algorithm is popular in applications, it is not always easy to tune, especially when the dimension of $\boldsymbol{\theta}$ is large. In such cases, it can be difficult to (simultaneously) generate large enough moves and get reasonable acceptance rates.

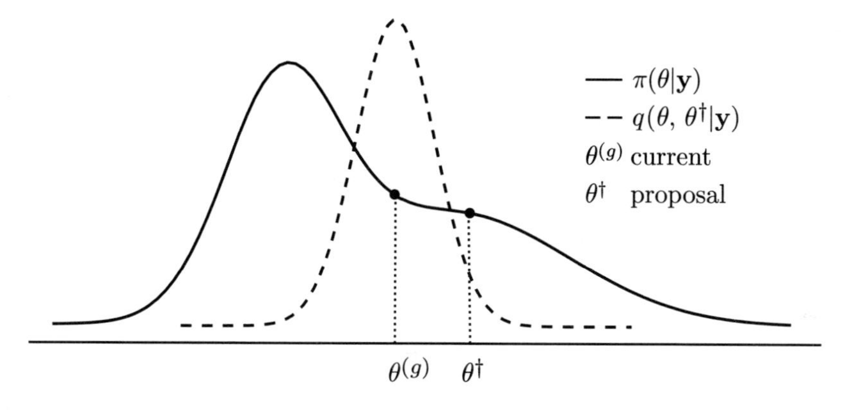

Figure 6.1 Random-walk M-H. The two points that determine the probability of move.

Independent proposals

Another possibility is to set $q(\boldsymbol{\theta}, \boldsymbol{\theta}'|\mathbf{y}) = q(\boldsymbol{\theta}'|\mathbf{y})$, an *independence M-H chain* in the terminology of [54]. In this case,

$$\alpha(\boldsymbol{\theta}, \boldsymbol{\theta}'|\mathbf{y}) = \min\left\{1, \frac{w(\boldsymbol{\theta}'|\mathbf{y})}{w(\boldsymbol{\theta}|\mathbf{y})}\right\}$$

where

$$w(\boldsymbol{\theta}|\mathbf{y}) = \frac{\pi(\boldsymbol{\theta}|\mathbf{y})}{q(\boldsymbol{\theta}|\mathbf{y})}$$

is the ratio of the target and proposal densities. [42] showed that the resulting MCMC chain is uniformly ergodic if $w(\boldsymbol{\theta}|\mathbf{y})$ is uniformly bounded.

It can easily be seen from the latter expression that if the target density can be expressed as $\pi(\boldsymbol{\theta}|\mathbf{y}) = w(\boldsymbol{\theta}|\mathbf{y})p(\boldsymbol{\theta}|\mathbf{y})$, where the first factor on the right is uniformly bounded and the second factor is a density that can be directly sampled, then one can just let $q(\boldsymbol{\theta}|\mathbf{y}) = p(\boldsymbol{\theta}|\mathbf{y})$ to produce a uniformly ergodic chain. Another way to implement independence chains is described by [11]. Under regularity conditions, the posterior density will be roughly quadratic around the posterior modal value. Then, a proposal density that is tailored to the target can be constructed by letting $q(\boldsymbol{\theta}|\mathbf{y}) = p(\boldsymbol{\theta}|\mathbf{m}, \mathbf{V})$, where p is some multivariate density and the parameters of the proposal density are

$$\mathbf{m} = \max_{\boldsymbol{\theta}} \log \pi(\boldsymbol{\theta}|\mathbf{y})$$

and

$$\mathbf{V} = c\left\{-\frac{\partial^2 \log \pi(\boldsymbol{\theta}|\mathbf{y})}{\partial\boldsymbol{\theta}\partial\boldsymbol{\theta}'}\right\}^{-1}_{\boldsymbol{\theta}=\hat{\boldsymbol{\theta}}} \tag{6.8}$$

where c is a tuning parameter that (along with $\mathbf{V}$) may be adjusted so that the tails of the proposal density are thicker than those of the target.

6.2.4 Multiple-block sampling

In applications it is not usually possible to sample $\boldsymbol{\theta}$ simultaneously in one block as in the basic M-H algorithm just presented. For such cases, [30] suggested revising the components of $\boldsymbol{\theta}$ one at a time. Each of these components can then be sampled by univariate M-H steps. This strategy can be generalized so that instead of revising one component at a time, the parameters are grouped more coarsely and revised in blocks. To explain this idea, suppose that $\boldsymbol{\theta}$ is grouped as $(\boldsymbol{\theta}_1, \boldsymbol{\theta}_2)$, with $\boldsymbol{\theta}_k \in \Omega_k \subseteq \Re^{d_k}$. The extension to more than two blocks is straightforward. Grouping in this way is also the hallmark of the Gibbs sampling algorithm. Often, this grouping is suggested by the model structure itself. For example, in a regression model, one block may consist of the regression coefficients and the other block of the error variance. The theoretical properties of this algorithm (in particular Harris-recurrence) is examined in [52].

Now let

$$Q_1(\boldsymbol{\theta}_1, d\boldsymbol{\theta}_1'|\mathbf{y}, \boldsymbol{\theta}_2) \ ; \ Q_2(\boldsymbol{\theta}_2, d\boldsymbol{\theta}_2'|\mathbf{y}, \boldsymbol{\theta}_1)$$

denote the proposal distributions, one for each block $\boldsymbol{\theta}_k$. There are many ways to specify these proposal distributions which can depend on the current value of the remaining block and on the data. For example, the random-walk and tailored approaches can be applied. As in the single block case, the proposed values are not necessarily accepted. The probabilities of acceptance are now defined as

$$\alpha_1(\boldsymbol{\theta}_1, \boldsymbol{\theta}_1'|\mathbf{y}, \boldsymbol{\theta}_2) = \min\left\{1, \frac{\pi_{1|2}(\boldsymbol{\theta}_1'|\mathbf{y}, \boldsymbol{\theta}_2) q_1(\boldsymbol{\theta}_1', \boldsymbol{\theta}_1|\mathbf{y}, \boldsymbol{\theta}_2)}{\pi_{1|2}(\boldsymbol{\theta}_1|\mathbf{y}, \boldsymbol{\theta}_2) q_1(\boldsymbol{\theta}_1, \boldsymbol{\theta}_1'|\mathbf{y}, \boldsymbol{\theta}_2)}\right\} \tag{6.9}$$

and

$$\alpha_2(\boldsymbol{\theta}_2, \boldsymbol{\theta}_2'|\mathbf{y}, \boldsymbol{\theta}_1) = \min\left\{1, \frac{\pi_{2|1}(\boldsymbol{\theta}_2'|\mathbf{y}, \boldsymbol{\theta}_1) q_2(\boldsymbol{\theta}_2', \boldsymbol{\theta}_2|\mathbf{y}, \boldsymbol{\theta}_1)}{\pi_{2|1}(\boldsymbol{\theta}_2|\mathbf{y}, \boldsymbol{\theta}_1) q_2(\boldsymbol{\theta}_2, \boldsymbol{\theta}_2'|\mathbf{y}, \boldsymbol{\theta}_1)}\right\} \tag{6.10}$$

where

$$\pi_{1|2}(\boldsymbol{\theta}_1|\mathbf{y}, \boldsymbol{\theta}_2) \text{ and } \pi_{2|1}(\boldsymbol{\theta}_2|\mathbf{y}, \boldsymbol{\theta}_1)$$

are called the *full conditional densities*. By Bayes' theorem, these are proportional to the joint posterior density. For instance,

$$\pi_{1|2}(\boldsymbol{\theta}_1|\mathbf{y}, \boldsymbol{\theta}_2) \propto \pi(\boldsymbol{\theta}_1, \boldsymbol{\theta}_2|\mathbf{y}).$$

Accordingly, the probabilities of move in (6.9) and (6.10) can be equivalently expressed in terms of the kernel of the joint posterior density $\pi(\boldsymbol{\theta}_1, \boldsymbol{\theta}_2|\mathbf{y})$ because the normalizing constant of the full conditional density (the norming constant in the latter expression) cancels in forming the ratio.

Given these ingredients, one sweep of the multiple-block algorithm is completed by updating each block, say sequentially in fixed order, where the proposed value for each block is accepted or rejected according to the M-H probabilities given above.

Algorithm. Multiple-block M-H

Step 1 Given $\boldsymbol{\theta}_2^{(g)}$ at the gth iteration, propose

$$\boldsymbol{\theta}_1' \sim q_1(\boldsymbol{\theta}_1^{(g)}, \boldsymbol{\theta}_1' | \mathbf{y}, \boldsymbol{\theta}_2^{(g)})$$

and move with probability

$$\alpha_1(\boldsymbol{\theta}_1^{(g)}, \boldsymbol{\theta}_1' | \mathbf{y}, \boldsymbol{\theta}_2^{(g)})$$

(otherwise stay at the current value) to produce the value $\boldsymbol{\theta}_1^{(g+1)}$

Step 2 Given this updated value of the first block, propose

$$\boldsymbol{\theta}_2' \sim q_2(\boldsymbol{\theta}_2^{(g)}, \boldsymbol{\theta}_2' | \mathbf{y}, \boldsymbol{\theta}_1^{(g+1)})$$

and move with probability

$$\alpha_2(\boldsymbol{\theta}_2^{(g)}, \boldsymbol{\theta}_2' | \mathbf{y}, \boldsymbol{\theta}_1^{(g+1)})$$

(otherwise stay at the current value) to produce the value $\boldsymbol{\theta}_2^{(g+1)}$

An illustrative schematic of these steps is given in Figure 6.2.

Given the sequential sampling of blocks, it is easy to see that the transition kernel of this Markov chain is given by the product of the two transition kernels

$$P(\boldsymbol{\theta}, d\boldsymbol{\theta}' | \mathbf{y}) = P_1(\boldsymbol{\theta}_1, d\boldsymbol{\theta}_1' | \mathbf{y}, \boldsymbol{\theta}_2) P_2(\boldsymbol{\theta}_2, d\boldsymbol{\theta}_2' | \mathbf{y}, \boldsymbol{\theta}_1') \tag{6.11}$$

This transition kernel is obviously not reversible because sampling in the reverse order never occurs. It is, however, invariant. To show this, we make use of the facts that each kernel satisfies invariance, conditioned on the value of the other block. In particular,

$$\pi_{1|2}^*(d\boldsymbol{\theta}_1' | \mathbf{y}, \boldsymbol{\theta}_2) = \int P_1(\boldsymbol{\theta}_1, d\boldsymbol{\theta}_1' | \mathbf{y}, \boldsymbol{\theta}_2) \pi_{1|2}^*(d\boldsymbol{\theta}_1 | \mathbf{y}, \boldsymbol{\theta}_2)$$

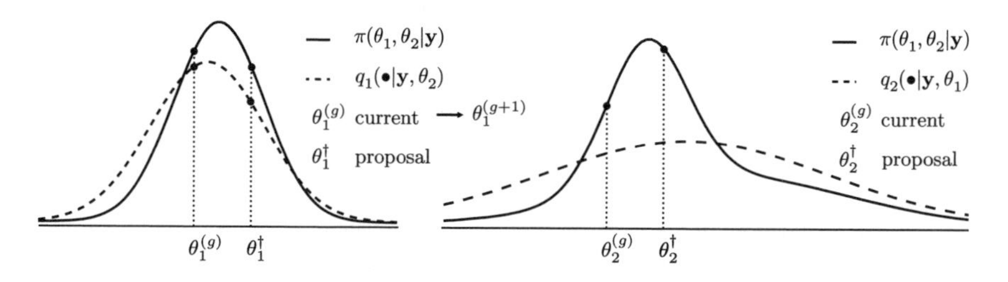

Figure 6.2 Multiple-block M-H: Left panel has the target and proposal densities of the first block and the four points that determine the probability of move; right panel has the same information for the second block.

and

$$\pi^*_{2|1}(d\boldsymbol{\theta}'_2|\mathbf{y},\boldsymbol{\theta}'_1) = \int P_2(\boldsymbol{\theta}_2, d\boldsymbol{\theta}'_2|\mathbf{y},\boldsymbol{\theta}'_1)\pi^*_{2|1}(d\boldsymbol{\theta}_2|\mathbf{y},\boldsymbol{\theta}'_1)$$

Following [12] one can now establish that the product kernel is invariant:

$$\begin{aligned}
&\int\int P_1(\boldsymbol{\theta}_1, d\boldsymbol{\theta}'_1|\mathbf{y},\boldsymbol{\theta}_2)P_2(\boldsymbol{\theta}_2, d\boldsymbol{\theta}'_2|\mathbf{y},\boldsymbol{\theta}'_1)\pi^*(d\boldsymbol{\theta}_1, d\boldsymbol{\theta}_2|\mathbf{y})\\
&= \int P_2(\boldsymbol{\theta}_2, d\boldsymbol{\theta}'_2|\mathbf{y},\boldsymbol{\theta}'_1)\left\{\int P_1(\boldsymbol{\theta}_1, d\boldsymbol{\theta}'_1|\mathbf{y},\boldsymbol{\theta}_2)\pi^*_{1|2}(d\boldsymbol{\theta}_1|\mathbf{y},\boldsymbol{\theta}_2)\right\}\pi^*_2(d\boldsymbol{\theta}_2|\mathbf{y})\\
&= \int P_2(\boldsymbol{\theta}_2, d\boldsymbol{\theta}'_2|\mathbf{y},\boldsymbol{\theta}'_1)\pi^*_{1|2}(d\boldsymbol{\theta}'_1|\mathbf{y},\boldsymbol{\theta}_2)\pi^*_2(d\boldsymbol{\theta}_2|\mathbf{y})\\
&= \int P_2(\boldsymbol{\theta}_2, d\boldsymbol{\theta}'_2|\mathbf{y},\boldsymbol{\theta}'_1)\frac{\pi^*_1(d\boldsymbol{\theta}'_1|\mathbf{y})\pi^*_{2|1}(d\boldsymbol{\theta}_2|\mathbf{y},\boldsymbol{\theta}'_1)\pi^*_1(d\psi'_1)}{\pi^*_2(d\boldsymbol{\theta}'_2|\mathbf{y})}\pi^*_2(d\boldsymbol{\theta}'_2|\mathbf{y})\\
&= \pi^*_1(d\boldsymbol{\theta}'_1|\mathbf{y})\int P_2(\boldsymbol{\theta}_2, d\boldsymbol{\theta}'_2|\mathbf{y},\boldsymbol{\theta}'_1)\pi^*_{2|1}(d\boldsymbol{\theta}_2|\mathbf{y},\boldsymbol{\theta}'_1)\\
&= \pi^*_1(d\boldsymbol{\theta}'_1|\mathbf{y})\,\pi^*_{2|1}(d\boldsymbol{\theta}'_2|\mathbf{y},\boldsymbol{\theta}'_1)\\
&= \pi^*(d\boldsymbol{\theta}'_1, d\boldsymbol{\theta}'_2|\mathbf{y}),
\end{aligned}$$

where the third line follows from the invariance of P_1, the fourth from the Bayes theorem, the sixth from the invariance of P_2, and the last from the law of total probability. The significance of this result is that we can take draws in succession from each of the kernels, instead of having to run each to convergence for every value of the conditioning variable.

6.3 Gibbs sampler

The Gibbs sampler is closely related to the multiple-block M-H algorithm. Just as in that algorithm, the parameters are sampled in blocks. The difference is that one uses the full conditional distribution of each block to update the blocks, rather than a general proposal distribution. As was orginally advertised by [23], the full conditional distributions in many Bayesian models are easily derived, and can be sampled directly. This ease of implementation led to numerous applications of the Gibbs sampler following the publication of that paper.

Formally, the transition kernel of the Markov chain is constructed from the set of full conditional distributions. For instance, suppose as above that $\boldsymbol{\theta}$ is grouped as $(\boldsymbol{\theta}_1, \boldsymbol{\theta}_2)$. Let

$$\pi^*_{1|2}(d\boldsymbol{\theta}_1|\boldsymbol{\theta}_2,\mathbf{y}) \text{ and } \pi^*_{2|1}(d\boldsymbol{\theta}_2|\boldsymbol{\theta}_1,\mathbf{y})$$

denote the full conditional distributions. Suppose that these distributions can be sampled directly. Sometimes, the strategic involvement of latent variables can lead to a tractable set of full conditional distributions, as discussed in the next section. Then, in the Gibbs algorithm, one samples the first distribution to produce a new value of $\boldsymbol{\theta}_1$. Given this new value, one then samples the second distribution to produce a new value of $\boldsymbol{\theta}_2$, and the process is repeated. The generalization to more blocks is immediate.

Algorithm. Gibbs sampling

In each iteration $g, g = 1, \ldots, n_0 + G$,

- Generate $\boldsymbol{\theta}_1^{(g+1)}$ from $\pi(\boldsymbol{\theta}_1|\mathbf{y}, \boldsymbol{\theta}_2^{(g)})\pi_{1|2}$
- Generate $\boldsymbol{\theta}_2^{(g+1)}$ from $\pi(\boldsymbol{\theta}_2|\mathbf{y}, \boldsymbol{\theta}_1^{(g+1)})\pi_{2|1}$

Return the values $\{\boldsymbol{\theta}^{(n_0+1)}, \boldsymbol{\theta}^{(n_0+2)}, ..., \boldsymbol{\theta}^{(n_0+G)}\}$.

Interestingly, one can show that the Gibbs algorithm is a particular instance of the multiple-block M-H algorithm. This requires showing that the M-H probability of accepting a block proposed from the full conditional distribution is one. In the notation of the multiple-block M-H algorithm, the proposal densities in the Gibbs sampler are

$$q_1(\boldsymbol{\theta}_1, \boldsymbol{\theta}_1'|\mathbf{y}, \boldsymbol{\theta}_2) = \pi_{1|2}(\boldsymbol{\theta}_1'|\mathbf{y}, \boldsymbol{\theta}_2),$$

and

$$q_2(\boldsymbol{\theta}_2, \boldsymbol{\theta}_2'|\mathbf{y}, \boldsymbol{\theta}_1) = \pi_{2|1}(\boldsymbol{\theta}_2'|\mathbf{y}, \boldsymbol{\theta}_1),$$

Then, for the first block, from (6.9) we get

$$\begin{aligned}
\alpha_1(\boldsymbol{\theta}_1, \boldsymbol{\theta}_1'|\mathbf{y}, \boldsymbol{\theta}_2) &= \min\left\{1, \frac{\pi_{1|2}(\boldsymbol{\theta}_1'|\mathbf{y},\boldsymbol{\theta}_2)q_1(\boldsymbol{\theta}_1', \boldsymbol{\theta}_1|\mathbf{y}, \boldsymbol{\theta}_2)}{\pi_{1|2}(\boldsymbol{\theta}_1|\mathbf{y},\boldsymbol{\theta}_2)q_1(\boldsymbol{\theta}_1, \boldsymbol{\theta}_1'|\mathbf{y}, \boldsymbol{\theta}_2)}\right\} \\
&= \min\left\{1, \frac{\pi_{1|2}(\boldsymbol{\theta}_1'|\mathbf{y},\boldsymbol{\theta}_2)\pi_{1|2}(\boldsymbol{\theta}_1|\mathbf{y},\boldsymbol{\theta}_2)}{\pi_{1|2}(\boldsymbol{\theta}_1|\mathbf{y},\boldsymbol{\theta}_2)\pi_{1|2}(\boldsymbol{\theta}_1'|\mathbf{y}, \boldsymbol{\theta}_2)}\right\} = 1
\end{aligned}$$

Thus, a variate proposed from the full conditional density is accepted without any rejection. With change of notation this is, of course, true for the second block as well. Thus, the updates from the multiple-block M-H algorithm, with proposals drawn from the full conditional distributions, correspond to those in the Gibbs sampling algorithm.

For the Gibbs sampler, the transition kernel is given by the product of the full conditional distributions:

$$P(\boldsymbol{\theta}, d\boldsymbol{\theta}'|\mathbf{y}) = \pi_{1|2}^*(d\boldsymbol{\theta}_1'|\boldsymbol{\theta}_2, \mathbf{y})\pi_{2|1}^*(d\boldsymbol{\theta}_2'|\boldsymbol{\theta}_1', \mathbf{y})$$

Invariance of this kernel can be established directly. One can also establish this fact indirectly by appealing to the invariance of the multiple-block M-H algorithm. [32] discuss the convergence properties of the Gibbs sampler that are independent of the selected version of the full conditional distributions. A key requirement is that the support of the target distribution be connected.

Metropolis-within-Gibbs

In some problems, not all the full conditional distributions are tractable. In that case, the blocks with the tractable full conditional distributions can be sampled directly and the blocks with the intractable full conditional distributions, by an M-H step. Such an algorithm is sometimes called the *Metropolis-within-Gibbs* algorithm.

6.4 Additional topics

6.4.1 Sampling with latent variables

In sampling $\pi^*(d\boldsymbol{\theta}|\mathbf{y})$ it is sometimes helpful to modify the target distribution by introducing latent variables or auxiliary variables into the sampling. To explain this idea, suppose that $\mathbf{z}$ are latent variables such that

$$\pi^*(d\boldsymbol{\theta}|\mathbf{y}) = \int_{\mathbf{z}} \pi^*(d\boldsymbol{\theta}, d\mathbf{z}|\mathbf{y}) \tag{6.12}$$

where $\pi^*(d\boldsymbol{\theta}, d\mathbf{z}|\mathbf{y})$ is the modified target distribution. Then, in many cases, the conditional distribution of $\boldsymbol{\theta}$ (or sub components of $\boldsymbol{\theta}$) given $\mathbf{z}$ are easy to derive. One can now sample the modified target distribution by (say) a multiple-block M-H algorithm to produce the sample

$$\left(\boldsymbol{\theta}^{(n_0+1)}, \mathbf{z}^{(n_0+1)}\right), \ldots, \left(\boldsymbol{\theta}^{(n_0+G)}, \mathbf{z}^{(n_0+G)}\right) \sim \pi^*(d\boldsymbol{\theta}, d\mathbf{z}|\mathbf{y})$$

By the usual Monte Carlo theory, the sampled draws on $\boldsymbol{\theta}$ are from the distribution $\pi^*(d\boldsymbol{\theta}, d\mathbf{z}|\mathbf{y})$ marginalized over $\mathbf{z}$. But since the condition (6.12) was assumed to hold, these draws are, therefore, from $\pi^*(d\boldsymbol{\theta}|\mathbf{y})$, the desired target of interest.

Sampling of $\boldsymbol{\theta}$ in this way is a powerful idea. [53] largely introduced this approach but in the context not of general latent data but in the setting of missing data problems and labelled the approach as data augmentation. In an early paper, [1], in the context of binary, ordinal and categorical outcomes, show vividly the usefulness of latent variables for Bayesian inference.

Example 6.1 (**Binary Categorical Data**) For binary (1, 0) outcomes y, where $\Pr(y = 1|\boldsymbol{\beta}) = \Phi(\mathbf{x}'\boldsymbol{\beta})$ and Φ is the distribution function of the standard normal, the posterior density under (say) a $\mathcal{N}(\boldsymbol{\beta}|\beta_0, \mathbf{B}_0)$ prior density and n independent observations $\mathbf{y} = (y_1, y_2, \ldots, y_n)$ is

$$\pi\left(\boldsymbol{\beta}|\mathbf{y}\right) \propto \mathcal{N}(\boldsymbol{\beta}|\beta_0, \mathbf{B}_0) \prod_{i=}^{n} \Phi(\mathbf{x}'\boldsymbol{\beta})^{y_i} \left(1 - \Phi(\mathbf{x}'\boldsymbol{\beta})\right)^{(1-y_i)}$$

which cannot be directly sampled. However, if one introduces latent data $z_i|\beta \sim \mathcal{N}(\mathbf{x}_i'\boldsymbol{\beta}, 1)$ such that $y_i = I[z_i > 0]$, $i \leq n$, then the posterior density $\pi\left(\boldsymbol{\beta}, \mathbf{z}|\mathbf{y}\right)$ as shown in [1] is

$$\begin{aligned}
\pi\left(\boldsymbol{\beta}, \mathbf{z}|\mathbf{y}\right) &\propto \pi\left(\boldsymbol{\beta}, \mathbf{z}, \mathbf{y}\right) \\
&\propto \mathcal{N}(\boldsymbol{\beta}|\beta_0, \mathbf{B}_0) \prod_{i=}^{n} \mathcal{N}(z_i|\mathbf{x}_i'\boldsymbol{\beta}, 1) \Pr\left(y_i|z_i, \boldsymbol{\beta}\right) \\
&\propto \mathcal{N}(\boldsymbol{\beta}|\beta_0, \mathbf{B}_0) \prod_{i=}^{n} \mathcal{N}(z_i|\mathbf{x}_i'\boldsymbol{\beta}, 1) \left\{I(z_i > 0)^{y_i} + I(z_i < 0)^{1-y_i}\right\}
\end{aligned}$$

where the term in curly braces is $\Pr\left(y_i|z_i, \boldsymbol{\beta}\right)$. It is easily checked that the integral of the latter density over $\{z_i\}$ leads to the correct target density. Thus, this is a valid modified target density. This target gives the right basis for developing an MCMC scheme because the full conditional distributions

$$\beta|\mathbf{y},\{z_i\}\,;\quad \{z_i\}|\mathbf{y},\beta$$

are both tractable. In particular, the distribution of β conditioned on the latent data becomes independent of the observed data and has the same form as in the Gaussian linear regression model with the response data given by $\{z_i\}$. It is multivariate normal with mean $\hat{\boldsymbol{\beta}} = \mathbf{B}(\mathbf{B}_0^{-1}\boldsymbol{\beta}_0 + \sum_{i=1}^{n}\mathbf{x}_i z_i)$ and variance matrix $\mathbf{B} = (\mathbf{B}_0^{-1} + \sum_{i=1}^{n}\mathbf{x}_i\mathbf{x}_i')^{-1}$. Next, the distribution of the latent data conditioned on the data and the parameters factor into a set of n independent truncated normal distributions, with each depending on the data through y_i

$$\{z_i\}|\mathbf{y},\boldsymbol{\beta} \propto \prod_{i=1}^{n}\mathcal{N}(z_i|\mathbf{x}_i'\boldsymbol{\beta},1)\left\{I(z_i > 0)^{y_i} + I(z_i < 0)^{1-y_i}\right\},$$

which are also easily sampled.

In some recent work, the idea of combining data augmentation with an expansion of the parameter space has been explored. [41], [42] and [31] provide theory and some applications. Latent variable augmentation is also central to the slice sampling method, for example see [18] and [45].

6.4.2 Choice of blocking

A crucial practical problem in both the multiple-block and Gibbs sampling algorithms is the composition of the blocks. As a general rule, sets of parameters that are highly correlated should be combined and sampled together. Otherwise, it becomes difficult to develop proposal densities that lead to large moves through the support of the target distribution. Beyond this advice, little in general can be said about how parameters should be grouped in practice. In some models it may not even be clear a priori which parameters are correlated. To deal with these difficulties, [14] propose a version of the multiple-block M-H algorithm in which the number of blocks and the components of the blocks are randomized in each MCMC iteration. The proposal distribution of each block is found by tailoring.

In designing a MCMC simulation one should be cognizant of opportunities for grouping parameters coarsely. For example, it is possible in some cases to reduce the number of blocks by the method of composition. For example, suppose that $\boldsymbol{\theta}_1, \boldsymbol{\theta}_2$ and $\boldsymbol{\theta}_3$ denote three blocks and that the distribution $\boldsymbol{\theta}_1|\mathbf{y},\boldsymbol{\theta}_3$ is tractable (i.e. can be sampled directly). Then, the blocks $(\boldsymbol{\theta}_1,\boldsymbol{\theta}_2)$ can be collapsed by first sampling $\boldsymbol{\theta}_1$ from $\boldsymbol{\theta}_1|\mathbf{y},\boldsymbol{\theta}_3$ followed by $\boldsymbol{\theta}_2$ from $\boldsymbol{\theta}_2|\mathbf{y},\boldsymbol{\theta}_1,\boldsymbol{\theta}_3$. This amounts to a two-block MCMC algorithm. In addition, if it is possible to sample $(\boldsymbol{\theta}_1,\boldsymbol{\theta}_2)$ marginalized over $\boldsymbol{\theta}_3$ then the number of blocks is reduced to one.

6.4.3 Estimation of density ordinates

If the full conditional densities are available, then the MCMC output can be used to estimate the posterior marginal density functions [23, 53]. By definition, the marginal density of a particular block $\boldsymbol{\theta}_k$ at the point $\boldsymbol{\theta}_k^*$ is

$$\pi(\boldsymbol{\theta}_k^*|\mathbf{y}) = \int \pi(\boldsymbol{\theta}_k^*|\mathbf{y},\boldsymbol{\theta}_{-k})\,\pi(\boldsymbol{\theta}_{-k}|\mathbf{y})d\boldsymbol{\theta}_{-k}$$

where $\boldsymbol{\theta}_{-k}$ is $\boldsymbol{\theta}$ excluding $\boldsymbol{\theta}_k$. Provided the normalizing constant of $\pi(\boldsymbol{\theta}_k^*|\mathbf{y},\boldsymbol{\theta}_{-k})$ is known, the marginal density can be estimated by the sample average

$$\hat{\pi}(\boldsymbol{\theta}_k^*|\mathbf{y}) = G^{-1}\sum_{g=1}^{G}\pi(\boldsymbol{\theta}_k^*|\mathbf{y}, \boldsymbol{\theta}_{-k}^{(g)})$$

[23] refer to this as the Rao–Blackwell method because of the connections with the Rao–Blackwell theorem in classical statistics. [10] extends this method for estimating the posterior density of $\boldsymbol{\theta}_k$ conditioned on one or more of the remaining blocks.

6.4.4 Comparison of models

In Bayesian statistics one is interested not just in summarizing the posterior distribution but also in comparing competing models, each defined by its own sampling distribution and prior [7]. MCMC methods have proved enormously useful for this purpose. In one approach, models and parameters are sampled jointly. The two leading MCMC methods in this category are the product space method [8] and the reversible jump method of [29]. Both methods have been widely applied. Recent applications and developments of these model space methods include [33], [19] and [34]. Specific versions of model space methods are particularly useful for the problem of variable selection [16, 25, 26, 35]. Another set of methods deals with the direct calculation of the model marginal likelihood. In this category, the method of [10] is both general and easy to implement. It provides an estimate of the marginal likelihood based on the output of the Gibbs sampling algorithm along with an estimate of the numerical standard error. [13] extend the framework for output from Metropolis–Hastings chains.

6.4.5 Output analysis

In implementing an MCMC method, it is important to assess the performance of the sampling algorithm to determine the rate of mixing and the size of the burn-in. A large literature is available on this topic, for example, [17], [36], [22], [50] and [20]. In some special cases, such as the hierarchical normal linear model, theoretical bounds on the burn-in time have been derived [37].

In practice, convergence (or more properly, lack of convergence) is assessed by empirical methods based on the sampled output, as for example those contained in the R CODA package which has the methods of Gelman and Rubin, Yu and Mykland, Raftery and Lewis and Geweke. One can also monitor the autocorrelation plots and the inefficiency factors. Slowly decaying correlations indicate problems with the mixing of the chain. It is also useful in connection with M-H Markov chains to monitor the acceptance rate of the proposal values with low rates implying 'stickiness' in the sampled values and thus a slower approach to the invariant distribution.

6.5 Concluding remarks

Work on MCMC methods continues at a rapid pace. Interesting recent developments include the particle filtering based sequential MCMC methods [2, 15, 21, 47, 48], primarily for the estimation of nonlinear state space models, and adaptive MCMC methods [1, 2, 5, 38, 45].

Over the last 20 years, MCMC methods have played a central role in the growth of Bayesian thinking. These methods have formed the basis for software programs such as WINBUGS and the many Bayesian packages in R. As a result, Bayesian applications across the sciences and social sciences have become common and the trends continue unabated.

References

[1] Albert, J. H. and Chib, S. (1993). Bayesian analysis of binary and polychotomous response data, *Journal of the American Statistical Association*, **88**, 669–679.

[2] Andrieu, C., Doucet, A. and Holenstein, R. (2010). Particle Markov chain Monte Carlo methods, *Journal of the Royal Statistical Society Series B – Statistical Methodology*, **72**, 269–342.

[3] Andrieu, C. and Moulines, E. (2006). On the ergodicity properties of some adaptive MCMC algorithms, *Annals of Applied Probability*, **16**, 1462–1505.

[4] Andrieu, C. and Thoms, J. (2008). A tutorial on adaptive MCMC, *Statistics and Computing*, **18**, 343–373.

[5] Atchade, Y. F. and Rosenthal, J. S. (2005). On Adaptive Markov chain Monte Carlo Algorithms, *Bernoulli*, **11**, 815–828.

[6] Besag, J. (1974). Spatial Interaction and Statistical-analysis of Lattice Systems, *Journal of the Royal Statistical Society Series B – Methodological*, **36**, 192–236.

[7] Carlin, B. and Louis, T. (2008). *Bayes and Empirical Bayes Methods for Data Analysis*, Boca Raton: Chapman & Hall, 3rd ed.

[8] Carlin, B. P. and Chib, S. (1995). Bayesian Model Choice via Markov Chain Monte Carlo Methods, *Journal of the Royal Statistical Society, Series B*, **57**, 473–484.

[9] Chen, M.-H., Shao, Q.-M. and Ibrahim, J. G. (2000). *Monte Carlo Methods in Bayesian Computation (Springer Series in Statistics)*, Springer.

[10] Chib, S. (1995). Marginal likelihood from the Gibbs output, *Journal of the American Statistical Association*, **90**, 1313–1321.

[11] Chib, S. and Greenberg, E. (1994). Bayes Inference in Regression Models with ARMA (p,q) Errors, *Journal of Econometrics*, **64**, 183–206.

[12] —— (1995). Understanding the Metropolis–Hastings algorithm, *The American Statistician*, **49**, 327–335.

[13] Chib, S. and Jeliazkov, I. (2001). Marginal likelihood from the Metropolis–Hastings output, *Journal of the American Statistical Association*, **96**, 270–281.

[14] Chib, S. and Ramamurthy, S. (2010). Tailored randomized block MCMC methods with application to DSGE models, *Journal of Econometrics*, **155**, 19–38.

[15] Chopin, N. (2004). Central limit theorem for sequential Monte Carlo methods and its application to Bayesian inference, *Annals of Statistics*, **32**, 2385–2411.

[16] Cottet, R., Kohn, R. J. and Nott, D. J. (2008). Variable Selection and Model Averaging in Semiparametric Overdispersed Generalized Linear Models, *Journal of the American Statistical Association*, **103**, 661–671.

[17] Cowles, M. K. and Rosenthal, J. S. (1998). A Simulation Approach to Convergence Rates for Markov chain Monte Carlo Algorithms, *Statistics and Computing*, **8**, 115–124.

[18] Damien, P., Wakefield, J. and Walker, S. (1999). Gibbs Sampling for Bayesian Non-conjugate and Hierarchical Models by Using Auxiliary Variables, *Journal of the Royal Statistical Society Series B – Statistical Methodology*, **61**, 331–344.

[19] Dellaportas, P., Friel, N. and Roberts, G. O. (2006). Bayesian Model Selection for Partially Observed Diffusion Models, *Biometrika*, **93**, 809–825.

[20] Fan, Y., Brooks, S. P. and Gelman, A. (2006). Output Assessment for Monte Carlo Simulations via the Score Statistic, *Journal of Computational and Graphical Statistics*, **15**, 178–206.

[21] Flury, T. and Shephard, N. (2011). Bayesian Inference Based Only On Simulated Likelihood: Particle Filter Analysis of Dynamic Economic Models, *Econometric Theory*, **27**, 933–956.

[22] Gamerman, D. and Lopes, H. F. (2006). *Markov Chain Monte Carlo: Stochastic Simulation for Bayesian Inference*, Boca Raton: Chapman and Hall/CRC, 2nd ed.

[23] Gelfand, A. E. and Smith, A. F. (1990). Sampling-Based Approaches to Calculating Marginal Densities, *Journal of the American Statistical Association*, **85**, 398–409.
[24] Geman, S. and Geman, D. (1984). Stochastic Relaxation, Gibbs Distribution and the Bayesian Restoration of Images, *IEEE Transactions, PAMI*, **6**, 721–741.
[25] George, E. I. and McCulloch, R. E. (1993). Variable selection via Gibbs sampling, *Journal of the American Statistical Association*, **88**, 881–889.
[26] —— (1997). Approaches for Bayesian variable selection, *Statistica Sinica*, **7**, 339–373.
[27] Geweke, J. (1992). Efficient simulation from the multivariate Normal and Student-t distributions subject to linear constraints, *Computing Science and Statistics: Proceedings of the Twenty-third Symposium*, 571–578.
[28] Geyer, C. J. (1992). Practical Markov chain Monte Carlo, *Statistical Science*, **4**, 473–483.
[29] Green, P. J. (1995). Reversible Jump Markov chain Monte Carlo Computation and Bayesian Model Determination, *Biometrika*, **82**, 711–732.
[30] Hastings, W. K. (1970). Monte-Carlo Sampling Methods Using Markov chains and their Applications, *Biometrika*, **57**, 97–109.
[31] Hobert, J. P. and Marchev, D. (2008). A theoretical comparison of the data augmentation, marginal augmentation and PX-DA algorithms, *Annals of Statistics*, **36**, 532–554.
[32] Hobert, J. P., Robert, C. P. and Goutis, C. (1997). Connectedness conditions for the convergence of the Gibbs sampler, *Statistics & Probability Letters*, **33**, 235–240.
[33] Holmes, C. C. and Mallick, B. K. (2003). Generalized Nonlinear Modeling with Multivariate Free-knot Regression Splines, *Journal of the American Statistical Association*, **98**, 352–368.
[34] Jasra, A., Stephens, D. A. and Holmes, C. C. (2007). Population-based Reversible Jump Markov chain Monte Carlo, *Biometrika*, **94**, 787–807.
[35] Jones, G. L., Haran, M., Caffo, B. S. and Neath, R. (2006). Fixed-width Output Analysis for Markov chain Monte Carlo, *Journal of the American Statistical Association*, **101**, 1537–1547.
[36] Jones, G. L. and Hobert, J. P. (2001). Honest exploration of intractable probability distributions via Markov chain Monte Carlo, *Statistical Science*, **16**, 312–334.
[37] —— (2004). Sufficient burn-in for Gibbs samplers for a hierarchical random effects model, *Annals of Statistics*, **32**, 784–817.
[38] Keith, J. M., Kroese, D. P. and Sofronov, G. Y. (2008). Adaptive independence samplers, *Statistics and Computing*, **18**, 409–420.
[39] Lamnisos, D., Griffin, J. E. and Steel, M. F. J. (2009). Transdimensional Sampling Algorithms for Bayesian Variable Selection in Classification Problems With Many More Variables Than Observations RID B-9845-2008, *Journal of Computational and Graphical Statistics*, **18**, 592–612.
[40] Liu, J. S. (2001). *Monte Carlo Strategies in Scientific Computing*, New York: Springer.
[41] Liu, J. S. and Wu, Y. N. (1999). Parameter expansion for data augmentation, *Journal of the American Statistical Association*, **94**, 1264–1274.
[42] Meng, X. L. and Van Dyk, D. A. (1999). Seeking efficient data augmentation schemes via conditional and marginal augmentation, *Biometrika*, **86**, 301–320.
[43] Mengersen, K. L. and Tweedie, R. L. (1996). Rates of convergence of the Hastings and Metropolis algorithms, *Annals of Statistics*, **24**, 101–121.
[44] Metropolis, N., Rosenbluth, A. W., Rosenbluth, M. N., Teller, A. H. and Teller, E. (1953). Equations of State Calculations by Fast Computing Machines, *Journal of Chemical Physics*, **21**, 1087–1092.
[45] Mira, A. and Tierney, L. (2002). Efficiency and Convergence Properties of Slice Samplers, *Scandinavian Journal of Statistics*, **29**, 1–12.
[46] Nott, D. J. and Kohn, R. (2005). Adaptive Sampling for Bayesian Variable Selection, *Biometrika*, **92**, 747–763.
[47] Pitt, M. K. and Shephard, N. (1999). Filtering via simulation: Auxiliary particle filters, *Journal of the American Statistical Association*, **94**, 590–599.

[48] Polson, N. G., Stroud, J. R. and Muller, P. (2008). Practical filtering with sequential parameter learning, *Journal of the Royal Statistical Society Series B – Statistical Methodology*, **70**, 413–428.
[49] Ripley, B. D. (1987). *Stochastic Simulation*, New York: Wiley.
[50] Robert, C. P. and Casella, G. (2004). *Monte Carlo Statistical Methods*, New York: Springer.
[51] Roberts, G. O. and Rosenthal, J. S. (2001). Optimal scaling for various Metropolis–Hastings algorithms, *Statistical Science*, **16**, 351–367.
[52] —— (2006). Harris Recurrence of Metropolis-within-Gibbs and Trans-dimensional Markov Chains, *Annals of Applied Probability*, **16**, 2123–2139.
[53] Tanner, M. A. and Wong, W. H. (1987). The Calculation of Posterior Distributions by Data Augmentation (with discussion), *Journal of the American Statistical Association*, **82**, 528–550.
[54] Tierney, L. (1994). Markov Chains for Exploring Posterior Distributions (with discussion), *The Annals of Statistics*, **21**, 1701–1762.

7 Advances in Markov chain Monte Carlo

JIM E. GRIFFIN AND DAVID A. STEPHENS

7.1 Introduction

7.1.1 Markov chain Monte Carlo: 1985–1995

The early work on Markov chain Monte Carlo (MCMC) throughout the 1980s, culminating with the publication of [12], led to the widespread use of Bayesian statistical inference in a broad range of application fields. The first half of the next decade saw many specific algorithmic developments, including the extension of the simple Gibbs sampling approaches of Gelfand and Smith to the use of the more general Metropolis–Hastings (MH) algorithm, and hybrid (mixture kernel) approaches in applications (see [55] for a summary of key theoretical concepts). There was also a huge growth in applications of Bayesian inference to challenging statistical problems that had previously proved intractable due to limitations of numerical approaches to integration.

By the mid 1990s, much was known about how MCMC should be implemented for optimal performance in standard statistical problems; see, for example, [20] for a summary of the state of knowledge at that time. Interest turned to addressing more complicated problems, such as model selection, and mixture problems, and a more in-depth theoretical study of classical algorithms such as the Metropolis algorithm. Model selection remains to this day an incompletely resolved issue in Bayesian inference, but in the mid 1990s specific focus fell on problems where a sequence of models was indexed by a discrete random variable, where models corresponding to different indices had different dimensions, mixture models with different numbers of components being a canonical example. Research into the Metropolis algorithm led to one of the most familiar results of MCMC folklore, The Goldilocks Principle, which relates to the optimal choice of the scale of Metropolis proposals [49]. This result, that suggests an optimal Metropolis acceptance rate of 0.234, continues to inform MCMC practitioners today, alerting them to the fact that they should focus on tuning their algorithm so that the acceptance rate is neither too high nor too low. It appeals to the logic that the MCMC kernels might be chosen in an adaptive fashion, in light of the observed performance of the algorithm. However, the result of [49] does not strictly relate to the construction of a truly adaptive algorithm.

In this chapter, we trace some of the key developments that further developed the underpinning theory, and potential applications, of MCMC since the mid 1990s. In particular, we review three main developments, namely reversible jump or transdimensional MCMC, population MCMC methods, and adaptive MCMC.

7.1.2 Notation

Following the notation introduced in the previous chapter, we initially consider the target distribution π^* over a parameter vector $\boldsymbol{\theta} \in \Theta \subseteq \mathcal{R}^d$ derived for observed data $\boldsymbol{y}$. Denote generic Markov transition kernel $Q(\boldsymbol{\theta}, d\boldsymbol{\theta}'|\boldsymbol{y}) = \Pr(\boldsymbol{\theta} \in d\boldsymbol{\theta}'|\boldsymbol{y}, \boldsymbol{\theta})$, and modified transition kernel $P(\boldsymbol{\theta}, d\boldsymbol{\theta}'|\boldsymbol{y})$ which has π^* as its unique stationary distribution. Let π and q denote the densities of π^* and Q wrt the Lebesgue measure. In the MH algorithm let $P(\boldsymbol{\theta}, d\boldsymbol{\theta}'|\boldsymbol{y})$ denote the reversible Markov transition kernel derived from $Q(\boldsymbol{\theta}, d\boldsymbol{\theta}'|\boldsymbol{y})$ with π^* as the stationary distribution. By the usual arguments, the acceptance probability for values generated from q is

$$\alpha(\boldsymbol{\theta}, \boldsymbol{\theta}'|\boldsymbol{y}) = \min\left\{1, \frac{\pi(\boldsymbol{\theta}'|\boldsymbol{y})}{\pi(\boldsymbol{\theta}|\boldsymbol{y})} \frac{q(\boldsymbol{\theta}', \boldsymbol{\theta}|\boldsymbol{y})}{q(\boldsymbol{\theta}, \boldsymbol{\theta}'|\boldsymbol{y})}\right\}.$$

Under mild conditions on $Q(.,.|\boldsymbol{y})$, variates generated from this Markov chain with transition kernel $P(.,.|\boldsymbol{y})$ form a dependent sample from π, and the ergodicity of the chain permits Monte Carlo estimation of functionals of π.

7.2 Reversible jump MCMC

After the foundational work on MCMC theory and applications, the next principal advance in MCMC theory was achieved by [26]. Around this time, interest in model selection via MCMC, and Markov chain methods operating on more complex parameter spaces, had become a major research field, and several authors had made significant contributions in this direction. In particular, the work of [27], [8] and [47] are notable. In [27] a particular continuous time Markov process, a Langevin jump diffusion, is constructed to sample the posterior distribution in a complex geometric and image analysis setting; a similar approach is used in [47], again in an image analysis setting. The use of a continuous time (birth–death) Markov process was, in itself, a novel contribution to Bayesian computation, although apart from a few specific areas of application (see for example [18, 54]), continuous time algorithms have remained under-utilized. In [8] a different approach is adopted; a more standard discrete time MCMC algorithm is constructed on an extended state space that encompasses all models considered simultaneously, with all parameters in all models being updated at each iteration, with model selection achieved during the MCMC run by sampling a discrete model index.

In the contributed Discussion of [27], Green laid out the basic construction that later formed the content of his paper [26]. He pointed out that whereas the Gibbs sampler construction utilized by [27] might only rarely be of use, the Metropolis–Hastings algorithm, they also suggested, could provide a general mechanism for jumping between subspaces of different dimension. The suggested general algorithm retained the reversibility and detailed balance properties of the usual MH algorithm, but required certain 'dimension-matching' terms to account for the differences in dimension of different subspaces.

We now study the reversible jump MCMC algorithm of [26]. Note that a formulation of MH algorithms on general state spaces was provided by [56], which includes Green's formulation as a special case.

7.2.1 Reversible jump MCMC: formulation

We consider a countable collection of Bayesian models, $\{M_k, k = 1, 2, \ldots\}$, where model M_k is parameterized via parameter $\boldsymbol{\theta}_k$ with parameter space $\Theta_k \subset \mathbb{R}^{d_k}$. We consider both the cases of a finite collection, where we present a series of possibly non-nested models and search for the most appropriate explanation of the data, and also cases of an infinite collection; the latter situation might arise if the index k corresponds to a non-negative integer-valued parameter. For data $\boldsymbol{y}$, the full posterior can be written

$$\pi(M_k, \theta_k|\boldsymbol{y}) = \frac{f(y|\theta_k, M_k)p(\theta_k|M_k)p(M_k)}{\sum_j \left\{\int f(y|\theta_j, M_j)p(\theta_j|M_j)d\theta_j\right\} p(M_j)}. \tag{7.1}$$

Our objective is to construct a Metropolis–Hastings Markov chain that is aperiodic and irreducible on the union parameter space $\Theta = \cup_k \Theta_k$. As usual, this requires the specification of a proposal transition density, $q(.,.|\boldsymbol{y})$, but, in contrast to the usual fixed-dimension case, we face the difficulty that the arguments of q are potentially of different dimensions, rendering the reversibility requirement difficult to meet. We now discuss the usual reversible jump solution.

At a specific iteration, suppose that the chain is in model M with parameter value $\boldsymbol{\theta}$ having dimension d, and the proposal is to move to model M' with parameter value $\boldsymbol{\theta}'$ having dimension d'. We envisage this move as first selecting a move between models, $M \longrightarrow M'$, and then the proposal of a $\boldsymbol{\theta}'$ possibly dependent on the current $\boldsymbol{\theta}$. Suppose that, when in model M, the move between the two models is selected with probability $r(M, M')$. To retain the possibility of a reversible proposal mechanism, with equally dimensioned arguments to $q(.,.|\boldsymbol{y})$, so consider the introduction of collections of latent variables $\boldsymbol{u}$ and $\boldsymbol{v}$ of dimension d_u and d_v respectively, so that $d + d_u = d' + d_v$. The proposal density q is then considered for the extended parameter vectors $(\boldsymbol{\theta}, \boldsymbol{u})$ and $(\boldsymbol{\theta}', \boldsymbol{v})$ such that $q((\boldsymbol{\theta}, \boldsymbol{u}), (\boldsymbol{\theta}', \boldsymbol{v})|\boldsymbol{y})$ is reversible; this is most easily constructed using a bijective differentiable mapping, that is

$$(\boldsymbol{\theta}', \boldsymbol{v}) = \boldsymbol{g}(\boldsymbol{\theta}, \boldsymbol{u}) \qquad \Longleftrightarrow \qquad (\boldsymbol{\theta}, \boldsymbol{u}) = \boldsymbol{h}(\boldsymbol{\theta}', \boldsymbol{v}).$$

Note that standard fixed-dimension moves also fall under this general proposal procedure; for example, for the ordinary Metropolis move, we might set

$$\boldsymbol{\theta}' = \boldsymbol{\theta} + \boldsymbol{u} \qquad \boldsymbol{u} \sim \mathrm{N}(0, \Sigma)$$

and for Metropolis–Hastings moves, we typically use a conditional generation $q(\boldsymbol{\theta}'|\boldsymbol{\theta}, \boldsymbol{y})$ taking the conditioning variable as a constant parameter in a suitably chosen density; the stochastic elements $\boldsymbol{u}$ in the MH move can be thought of as the *Uniform*(0, 1) variates used to perform basic random number generation from this conditional density.

To establish the acceptance probability for the reversible jump move, Green considers a 'hybrid' MH algorithm comprising a mixture of move types, with move indexed by m with transition proposal Q_m selected with probability r_m, some of which may be trans-dimensional, but each of which retains the detailed balance property. If ψ and ψ' represent the latent-augmented parameter vectors, and let the augmented posterior density be denoted

$$\tilde{\pi}_m(\psi|\boldsymbol{y}) = \tilde{\pi}_m(M, \boldsymbol{\theta}, \boldsymbol{u}|\boldsymbol{y}) = \pi(M, \boldsymbol{\theta}|\boldsymbol{y})p_u(\boldsymbol{u}).$$

Let $\alpha_m(\psi, \psi'|\boldsymbol{y})$ denote the acceptance probability for move type m, so that for arbitrary sets A and B

$$\int_A \tilde{\pi}^*_m(d\psi|\boldsymbol{y}) \int_B \alpha_m(\psi, \psi'|\boldsymbol{y})Q_m(\psi, d\psi'|\boldsymbol{y}) = \int_B \tilde{\pi}^*_m(d\psi'|\boldsymbol{y}) \int_A \alpha_m(\psi', \psi|\boldsymbol{y})Q_m(\psi', d\psi|\boldsymbol{y})$$

which implies as usual that

$$\alpha_m(\psi, \psi'|\mathbf{y})\tilde{\pi}_m(\psi|\mathbf{y})q_m(\psi, \psi'|\mathbf{y}) = \alpha_m(\psi', \psi|\mathbf{y})\tilde{\pi}_m(\psi'|\mathbf{y})q_m(\psi', \psi|\mathbf{y})$$

or, in Green's notation

$$\alpha_m(\psi, \psi'|\mathbf{y})f_m(\psi, \psi'|\mathbf{y}) = \alpha_m(\psi', \psi|\mathbf{y})f_m(\psi', \psi|\mathbf{y})$$

where f_m is defined with respect to a common, symmetric measure on the product space. Consider first the forward move; in this case $f_m(\psi, \psi'|\mathbf{y})$ is given by

$$f_m(\psi, \psi'|\mathbf{y}) = \tilde{\pi}_m(\psi|\mathbf{y})q_m(\psi, \psi'|\mathbf{y}) = \pi(M, \boldsymbol{\theta}|\mathbf{y})p_u(\boldsymbol{u})\overrightarrow{r}_m$$

say. For the reverse move, to preserve symmetry, we must set

$$f_m(\psi', \psi|\mathbf{y}) = \tilde{\pi}_m(\psi'|\mathbf{y})q_m(\psi', \psi|\mathbf{y}) = \pi(M', \boldsymbol{\theta}'|\mathbf{y})p_v(\boldsymbol{v})\left|\frac{\partial(\boldsymbol{\theta}', \boldsymbol{v})}{\partial(\boldsymbol{\theta}, \boldsymbol{u})}\right|\overleftarrow{r}_m$$

where the term

$$\left|\frac{\partial(\boldsymbol{\theta}', \boldsymbol{v})}{\partial(\boldsymbol{\theta}, \boldsymbol{u})}\right| = \left|\frac{\partial \boldsymbol{g}(\boldsymbol{t}_1, \boldsymbol{t}_2)}{\partial(\boldsymbol{\theta}, \boldsymbol{u})}\right|_{\boldsymbol{t}_1=\boldsymbol{\theta}, \boldsymbol{t}_2=\boldsymbol{u}}$$

is the Jacobian associated with the bijection $\boldsymbol{g} : (\boldsymbol{\theta}, \boldsymbol{u}) \mapsto (\boldsymbol{\theta}', \boldsymbol{v})$. This term arises as in the augmented posterior, we have

$$\tilde{\pi}(M', \boldsymbol{\theta}', \boldsymbol{v}|\mathbf{y}) = \tilde{\pi}(M, \boldsymbol{h}(\boldsymbol{\theta}', \boldsymbol{v})|\mathbf{y})|J(\boldsymbol{\theta}', \boldsymbol{v})| = \tilde{\pi}(M, \boldsymbol{\theta}, \boldsymbol{u})|\mathbf{y})|J(\boldsymbol{\theta}, \boldsymbol{u})|^{-1}$$

under the bijection; here

$$|J(\boldsymbol{\theta}', \boldsymbol{v})| = \left|\frac{\partial(\boldsymbol{\theta}, \boldsymbol{u})}{\partial(\boldsymbol{\theta}', \boldsymbol{v})}\right| = \left|\frac{\partial(\boldsymbol{\theta}', \boldsymbol{v})}{\partial(\boldsymbol{\theta}, \boldsymbol{u})}\right|^{-1} = |J(\boldsymbol{\theta}, \boldsymbol{u})|^{-1}$$

is the Jacobian. In the expressions above, the two terms $\overrightarrow{r}_m$ and $\overleftarrow{r}_m$ represent the probabilities of choosing to make move m (from M to M') and the probability of the reverse move (from M' to M). Thus, the acceptance probability for move type m is

$$\alpha_m((M, \boldsymbol{\theta}), (M', \boldsymbol{\theta}')|\mathbf{y}) = \min\left\{1, \frac{\pi(M', \boldsymbol{\theta}'|\mathbf{y})p_v(\boldsymbol{v})\overleftarrow{r}_m}{\pi(M, \boldsymbol{\theta}|\mathbf{y})p_u(\boldsymbol{u})\overrightarrow{r}_m}\left|\frac{\partial(\boldsymbol{\theta}', \boldsymbol{v})}{\partial(\boldsymbol{\theta}, \boldsymbol{u})}\right|\right\}.$$

Often, these general calculations simplify, as one of $\boldsymbol{u}$ or $\boldsymbol{v}$ is a null vector. If $\dim(\boldsymbol{\theta}) < \dim(\boldsymbol{\theta}')$, and the proposed move attempts to increase the dimension of the current model, then only the augmenting variables $\boldsymbol{u}$ are needed to match dimension, that is, the bijection can be constructed by setting $\boldsymbol{\theta}' = \boldsymbol{g}(\boldsymbol{\theta}, \boldsymbol{v})$. In this case

$$\alpha_m((M, \boldsymbol{\theta}), (M', \boldsymbol{\theta}')|\mathbf{y}) = \min\left\{1, \frac{\pi(M', \boldsymbol{\theta}'|\mathbf{y})\overleftarrow{r}_m}{\pi(M, \boldsymbol{\theta}|\mathbf{y})p_u(\boldsymbol{u})\overrightarrow{r}_m}\left|\frac{\partial(\boldsymbol{\theta}')}{\partial(\boldsymbol{\theta}, \boldsymbol{u})}\right|\right\}.$$

Conversely, if $\dim(\boldsymbol{\theta}) > \dim(\boldsymbol{\theta}')$, then only the augmenting variables $\boldsymbol{v}$ are needed, and

$$\alpha_m((M,\boldsymbol{\theta}),(M',\boldsymbol{\theta}')|\boldsymbol{y}) = \min\left\{1, \frac{\pi(M',\boldsymbol{\theta}'|\boldsymbol{y})p_v(\boldsymbol{v})\overleftarrow{r}_m}{\pi(M,\boldsymbol{\theta}|\boldsymbol{y})\overrightarrow{r}_m}\left|\frac{\partial(\boldsymbol{\theta}',\boldsymbol{v})}{\partial(\boldsymbol{\theta})}\right|\right\}.$$

As pointed out by [26, p. 717], these proposal mechanisms can be generalized by allowing the generation for $\boldsymbol{u}$ or $\boldsymbol{v}$ to depend on the values of $\boldsymbol{\theta}$ or $\boldsymbol{\theta}'$ respectively. Under this generalization, $p_u(\boldsymbol{u})$ and $p_v(\boldsymbol{v})$ are replaced by $p_u(\boldsymbol{u}|\boldsymbol{\theta})$ and $p_v(\boldsymbol{v}|\boldsymbol{\theta}')$ in the acceptance probabilities. This merely corresponds to an alternative construction of the augmented posterior $\tilde{\pi}$.

It is evident from the construction that each move type m comprises a pair of moves operating in each direction between models M and M'. Thus, if move m is a move from M to M', constructed by generation of augmenting variables $\boldsymbol{u}$ and transformation, there should exist in our collection of potential moves the reverse move, indexed m', say, which utilizes the augmenting variables $\boldsymbol{v}$, the two moves being selected with probabilities $\overrightarrow{r}_m$ and $\overleftarrow{r}_m$.

In some situations, the reversible jump acceptance probability simplifies even further. For example, in a move that increases dimension by d_u through the variables $\boldsymbol{u}$, it may be feasible to set the new parameters precisely equal to $\boldsymbol{u}$, that is, the proposed parameter vector $\boldsymbol{\theta}'$ is formed by concatenating $\boldsymbol{\theta}$ and $\boldsymbol{u}$. In this case, the Jacobian of the transformation is 1. Furthermore, it may be possible to generate $\boldsymbol{u}$ from a prior distribution, which facilitates cancellation in the Hastings ratio.

Example 7.1 (**Nested models**) Let $k = 1, 2, \ldots$ index models $M_1, M_2, \ldots$ under consideration, where k represents the dimension of $\boldsymbol{\theta}_k$. Let the elements of $\boldsymbol{\theta}_k$ be $(\theta_{k1}, \ldots, \theta_{kk})$. Suppose that the model specification is such that identical, independent priors are used for the components of the parameter vector

$$p(\boldsymbol{\theta}_k) = \prod_{l=1}^{k} p_0(\theta_{kl}).$$

For a proposed move from M_j to M_k with $j < k$, suppose that the elements of $\boldsymbol{u}$ are generated independently from p_0, so that

$$p_u(\boldsymbol{u}) = \prod_{l=1}^{k-j} p_0(u_l),$$

and suppose that $\boldsymbol{\theta}_k = (\boldsymbol{\theta}_l, \boldsymbol{u})$. Then

$$\alpha_j((M_j,\boldsymbol{\theta}_j),(M_k,\boldsymbol{\theta}_k)|\boldsymbol{y}) = \min\left\{1, \frac{\pi(M_k,\boldsymbol{\theta}_k|\boldsymbol{y})\overleftarrow{r}_j}{\pi(M_j,\boldsymbol{\theta}_j|\boldsymbol{y})p_u(\boldsymbol{u})\overrightarrow{r}_j}\left|\frac{\partial(\boldsymbol{\theta}_k)}{\partial(\boldsymbol{\theta},\boldsymbol{u})}\right|\right\}$$

$$= \min\left\{1, \frac{f(\boldsymbol{y}|M_k,\boldsymbol{\theta}_k)p(\boldsymbol{\theta}_k|M_k)p(M_k)\overleftarrow{r}_j}{f(\boldsymbol{y}|M_j,\boldsymbol{\theta}_j)p(\boldsymbol{\theta}_j|M_j)p(M_j)p_u(\boldsymbol{u})\overrightarrow{r}_j}\left|\frac{\partial(\boldsymbol{\theta}_k)}{\partial(\boldsymbol{\theta},\boldsymbol{u})}\right|\right\}$$

$$= \min\left\{1, \frac{f(\boldsymbol{y}|M_k,\boldsymbol{\theta}_k)p(M_k)\overleftarrow{r}_j}{f(\boldsymbol{y}|M_j,\boldsymbol{\theta}_j)p(M_j)\overrightarrow{r}_j}\right\}$$

as

$$p(\boldsymbol{\theta}_k|M_k) = \prod_{l=1}^{k} p_0(\theta_{kl}) = \left\{\prod_{l=1}^{j} p_0(\theta_{jl})\right\}\left\{\prod_{l=1}^{k-j} p_0(u_l)\right\} = p(\boldsymbol{\theta}_j|M_j)p_u(\boldsymbol{u}).$$

For the reverse move that attempts to decrease the model complexity by moving from model k to model j, the deterministic proposal that sets the last $k-j$ components of $\boldsymbol{\theta}_k$ to zero is used.

The approach described in this example can be useful in some applications, but can also lead to low acceptance rates.

7.2.2 Reversible jump MCMC: examples

Example 7.2 (**Moves between 1-d and 2-d models**) Consider two models M_1 and M_2 with parameters $\theta^{(1)} = \theta_1$ and $\theta^{(2)} = (\theta_{21}, \theta_{22})$, with all parameters taking values on $\mathbb{R}$. We consider four move types:

1. $m = 1$: move **within** Model M_1,
2. $m = 2$: move **within** Model M_2,
3. $m = 3$: move **from** Model M_1 **to** Model M_2,
4. $m = 4$: move **from** Model M_2 **to** Model M_1.

For the within-model moves, standard MH acceptance calculations proceed as usual. Moves 3 and 4 are a forward/reverse move pair. Clearly, if the current state of the chain is in M_1, only moves 1 or 3 can be selected, and similarly for M_2. We need only consider the relative magnitudes of the reversible jump moves selection probabilities; for example, if the probability of selecting move 3 is 0.3 and the probability of selecting move 4 is 0.1, then

$$\frac{\overleftarrow{r}_3}{\overrightarrow{r}_3} = \frac{0.1}{0.3} \qquad\qquad \frac{\overleftarrow{r}_4}{\overrightarrow{r}_4} = \frac{0.3}{0.1}$$

are the ratios that enter into the acceptance probability calculations for move types 3 and 4 respectively.

When the current state of the chain is in M_1, to propose a move to M_2 after selecting move 3, we must introduce a single random variate u; suppose that $u \sim Normal(0,1)$, and let

$$\theta_{21} = \theta_1 + u \qquad \theta_{22} = \theta_1 - u \qquad\qquad \Longleftrightarrow \qquad\qquad \theta_1 = \frac{\theta_{21}+\theta_{22}}{2}.$$

In this case the acceptance probability takes the form

$$\alpha_3((M_1,\theta_1),(M_2,(\theta_{21},\theta_{22}))|\boldsymbol{y}) = \min\left\{1, \frac{\pi(M_2,(\theta_{21},\theta_{22})|\boldsymbol{y})}{\pi(M_1,\theta_1|\boldsymbol{y})\phi(u)}\frac{\overleftarrow{r}_3}{\overrightarrow{r}_3}\left|\frac{\partial(\theta_{21},\theta_{22})}{\partial(\theta_1,u)}\right|\right\}$$

where $\phi(.)$ is the standard normal pdf, and

$$\left|\frac{\partial(\theta_{21},\theta_{22})}{\partial(\theta_1,u)}\right| = \begin{vmatrix} \dfrac{\partial\theta_{21}}{\partial\theta_1} & \dfrac{\partial\theta_{21}}{\partial u} \\ \dfrac{\partial\theta_{22}}{\partial\theta_1} & \dfrac{\partial\theta_{22}}{\partial u}\end{vmatrix} = \begin{vmatrix} 1 & 1 \\ 1 & -1\end{vmatrix} = 2.$$

For the reverse move 4, we have

$$\alpha_4((M_2,(\theta_{21},\theta_{22}),(M_1,\theta_1),)|\mathbf{y}) = \min\left\{1, \frac{\pi(M_1,\theta_1|\mathbf{y})\phi(u)}{\pi(M_2,(\theta_{21},\theta_{22})|\mathbf{y})}\frac{\overleftarrow{r}_4}{\overrightarrow{r}_4}\left|\frac{\partial(\theta_1,u)}{\partial(\theta_{21},\theta_{22})}\right|\right\}$$

where

$$\left|\frac{\partial(\theta_1,u)}{\partial(\theta_{21},\theta_{22})}\right| = \frac{1}{2}.$$

Example 7.3 (**Comparing nonlinear regression models**) Consider two competing pharmacokinetic (PK) models for scalar response data $y(t)$ measured at different time points $t_1,\ldots,t_n$:

1. **Model M_1:** one compartment, elimination only;

$$\mathbb{E}[Y(t)] = A_1\exp\{-\lambda_1 t\} \qquad t \geq 0$$

2. **Model M_2:** one compartment, absorption and elimination;

$$\mathbb{E}[Y(t)] = A_2\left(\exp\{-\lambda_{21}t\} - \exp\{-(\lambda_{21}+\lambda_{22})t\}\right) \qquad t \geq 0$$

where (A_1,λ_1) and $(A_2,\lambda_{21},\lambda_{22})$ are positive parameters. The data are displayed in Figure 7.1.

Under an assumption of additive, heteroscedastic Normal errors, we have two competing explanations for the observed data; both models can be fitted using ordinary least-squares, but model comparison is not straightforward as the models are nested but the nesting structure is complicated, as we require $\lambda_{22} \longrightarrow \infty$, that is, a boundary point in the parameter space, which leads to non-regular frequentist asymptotic theory. Bayesian model comparison can be carried out using Bayes factors, but this requires numerical integration.

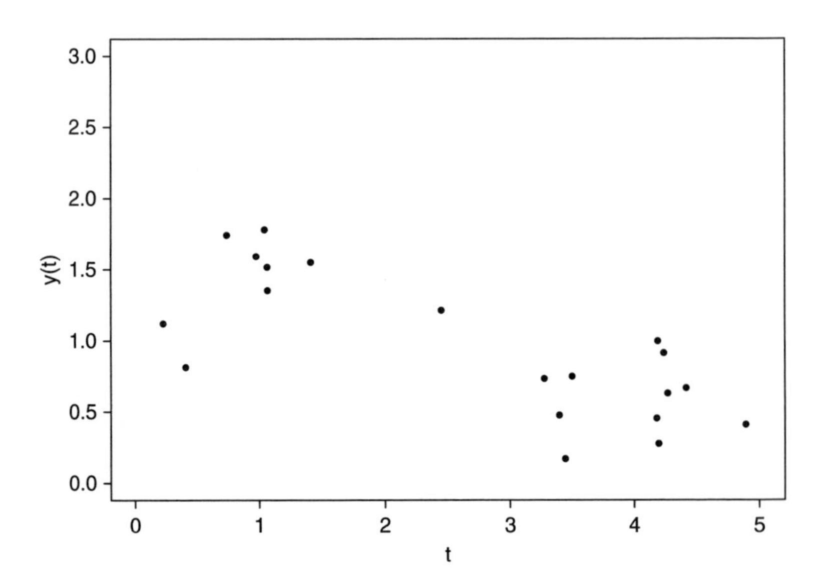

Figure 7.1 PK data from one compartment model.

For illustration, we consider a reversible jump MCMC solution. First, we use a log-scale parameterization and set

$$\boldsymbol{\theta}_1 = (\log A_1, \log \lambda_1) \qquad \boldsymbol{\theta}_2 = (\log A_2, \log \lambda_{21}, \log \lambda_{22}).$$

We place equal prior probabilities on M_1 and M_2, and then place independent $N(0, \tau)$ priors on the components of $\boldsymbol{\theta}_1$ and $\boldsymbol{\theta}_2$. The prior on residual error variance σ^2 is inverse Gamma with parameters 10 and 4. The ML estimates $\widehat{\boldsymbol{\theta}}_1$ and $\widehat{\boldsymbol{\theta}}_2$ can be computed easily, as can the Hessian matrices $\widehat{\mathbf{I}}_1$ and $\widehat{\mathbf{I}}_2$; these likelihood-based results yield reasonable approximations to the posterior distributions that can be used to produce independence MH algorithms. Specifically, at the ML estimates for σ under the two models, we may approximate the conditional posterior for $\boldsymbol{\theta}$ by the Normal density

$$p(\boldsymbol{\theta}_k|\widehat{\sigma}, \boldsymbol{y}) \simeq N(\widehat{\boldsymbol{\theta}}_k, n^{-1}\widehat{\sigma}_k^2 \widehat{\mathbf{I}}_k^{-1}). \tag{7.2}$$

On fitting using ML, the estimates of σ under the two models are found to be quite similar $(M_1 : \widehat{\sigma}_1 = 0.252, M_2 : \widehat{\sigma}_2 = 0.329)$.

A reversible jump MCMC algorithm can be constructed as follows: we again consider four move types:

1. $m = 1$: move **within** M_1; update $\boldsymbol{\theta}_1$ from $p(\boldsymbol{\theta}_1|M_1, \sigma, \boldsymbol{y})$.
2. $m = 2$: move **within** M_1; update $\boldsymbol{\theta}_2$ from $p(\boldsymbol{\theta}_2|M_2, \sigma, \boldsymbol{y})$.
3. $m = 3$: move **from** M_1 **to** M_2; propose a new $\boldsymbol{\theta}_2$, and carry out an accept/reject step.
4. $m = 4$: move **from** Model M_2 **to** Model M_1; propose a new $\boldsymbol{\theta}_1$, and carry out an accept/reject step.

with the remaining parameter σ^2 being updated in a Gibbs sampler algorithm at each iteration. Moves $m = 3, 4$ are a forward/reverse move pair. For move 3, several options are available; for example, we could adopt the strategy of Example 7.1, and generate a new variate u from the prior for the additional parameter, and then merely use the mapping

$$(\theta_{11}, \theta_{12}, u) \longmapsto (\theta_{21} = \theta_{11}, \theta_{22} = \theta_{12}, \theta_{23} = u)$$

with reverse move setting $\theta_{23} = 0$. This approach may be adequate, but more probably would not facilitate good mixing across the models. A perhaps better strategy is to consider a different augmentation, where we generate $\boldsymbol{u} = (u_1, u_2, u_3)$ from the model in (7.2) for $k = 2$, and use

$$(\theta_{11}, \theta_{12}, u_1, u_2, u_3) \longmapsto (\theta_{21} = u_1, \theta_{22} = u_2, \theta_{23} = u_3, v_1 = \theta_{11}, v_2 = \theta_{12})$$

with the paired reverse move being to generate $\boldsymbol{v} = (v_1, v_2)$ from the model in (7.2) for $k = 1$. This guarantees that the proposed value $\boldsymbol{\theta}_2$ lies in a region with reasonably high posterior support under model M_2, although it does not guarantee that the move will be accepted with high probability. In the Hastings ratio, the Jacobian of the transformation is 1, and under equal probabilities of forward/reverse moves, we have

$$\frac{\pi(M_2, \boldsymbol{\theta}_2|\boldsymbol{y})p_v(v_1, v_2)}{\pi(M_1, \boldsymbol{\theta}_1|\boldsymbol{y})p_u(u_1, u_2, u_3)} = \frac{f(\boldsymbol{y}|M_2, \boldsymbol{\theta}_2, \sigma)\left\{\prod_{j=1}^{3} \phi(\theta_{2j}/\tau)/\tau\right\}\phi_2(\theta_{11}, \theta_{12}; \widehat{\boldsymbol{\theta}}_1, \widehat{\mathbf{I}}_1)}{f(\boldsymbol{y}|M_1, \boldsymbol{\theta}_1, \sigma)\left\{\prod_{j=1}^{2} \phi(\theta_{1j}/\tau)/\tau\right\}\phi_3(\theta_{21}, \theta_{22}, \theta_{23}; \widehat{\boldsymbol{\theta}}_2, \widehat{\mathbf{I}}_2)}$$

where τ is the prior variance for the regression parameters. The logic of this construction is that numerically

$$f(\boldsymbol{y}|M_1,\boldsymbol{\theta}_1,\sigma) \approx \phi_2(\theta_{11},\theta_{12};\widehat{\boldsymbol{\theta}}_1,\widehat{\mathbf{I}}_1), \qquad f(\boldsymbol{y}|M_2,\boldsymbol{\theta}_2,\sigma) \approx \phi_3(\theta_{21},\theta_{22},\theta_{23};\widehat{\boldsymbol{\theta}}_2,\widehat{\mathbf{I}}_2).$$

A different approximation that incorporates the Normal asymptotic likelihood in equation (7.2) as well as the Normal prior distribution for the parameters can be constructed.

The algorithm was run for 100,000 iterations. In this run with $\tau = 4$, the chain spent about 66 % of the time in model M_1, indicating the posterior probabilities are

$$p(M_1|\boldsymbol{y}) \approx 0.66, \qquad p(M_2|\boldsymbol{y}) \approx 0.34.$$

The model posterior probabilities vary with the choice of τ; this is as expected, as the model probabilities are closely related to the marginal likelihood, or prior predictive distribution, which is the expected value of the likelihood for the observed data with respect to the prior distribution. It is evident from the discussion that the prior specification acts as a penalty for complexity. For illustration, if $\tau = 1$, the model probabilities change to $(0.80, 0.20)$; if $\tau = 10$, the model probabilities are $(0.26, 0.74)$.

Conditional on M_1 or M_2 being true, we can perform inference about the parameters of the two models, and also reconstruct estimates and posterior credible intervals for $\mathbb{E}[Y]$. Figure 7.2 displays the reconstructed posterior intervals for the two models.

It is evident that both models are plausible explanations of the observed data, but offer potentially different predictions of future responses, especially near $t = 0$. Note that the possible nesting structure, incorporated by the assumption that $\lambda_{22} \longrightarrow \infty$, is not supported by the priors studied. Finally, note that the BIC values for the two models, computed as

$$\text{BIC}_k = -2\log f(\boldsymbol{y}|M_k,\widehat{\boldsymbol{\theta}}_k,\widehat{\sigma}) + d_k \log n$$

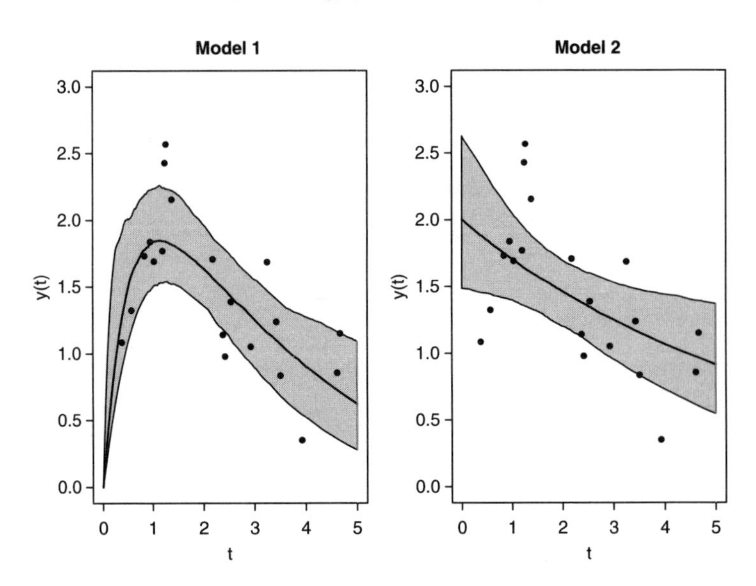

Figure 7.2 Example 7.3. Pointwise 95% credible intervals for $[Y]$ for Model 1 (left) and Model 2 (right).

where d_k is the total number of parameters fitted, are 25.553 and 33.227 respectively for models 1 and 2, indicating strong support for model 1.

Example 7.4 (**Finite mixture model**) The finite mixture model with K components has density

$$f(y|\boldsymbol{\omega},\boldsymbol{\theta},K) = \sum_{k=1}^{K} \omega_k f_k(y|\boldsymbol{\theta}_k)$$

where $\boldsymbol{\theta}_k$ are the parameters for component density k, and

$$0 < \omega_k < 1, \forall k \qquad \& \qquad \sum_{k=1}^{K} \omega_k = 1.$$

Most typically, although not exclusively, the component densities are from the same parametric location-scale family (say Gaussian) which differ in location and scale. For independent and identically distributed data $\boldsymbol{y} = (y_1, \ldots, y_n)$ from this model, the likelihood arising from this model does not permit analytical calculation of the posterior distribution, so MCMC is commonly used. If K is known, MCMC is typically implemented using a collection of auxiliary variables $z_1, \ldots, z_n$, and a completed data model

$$f(y, z|\boldsymbol{\omega},\boldsymbol{\theta},K) = \Pr[Z = z] f_z(y|\boldsymbol{\theta}_z) \qquad z \in \{1, 2, \ldots, K\}$$

and a Gibbs sampler strategy updating the $\boldsymbol{z}$ and $(\boldsymbol{\omega},\boldsymbol{\theta})$ from their full conditional posterior distributions. Conditional on $(\boldsymbol{\omega},\boldsymbol{\theta})$, the posterior distribution for each z_i is a discrete distribution on $\{1, 2, \ldots, K\}$, whereas conditional on the $\boldsymbol{z}$, the posterior for the mixture weights $\boldsymbol{\omega}$ is a density on the K-dimensional simplex, and the posterior for the components of $\boldsymbol{\theta}_k$ proceeds using only those y_i for which $z_i = k$, for $k = 1, \ldots, K$. Under a proper prior specification, the posterior is also proper, although inference is complicated by a permutation non-identifiability in the model specification (that is, the labels on the components may be permuted without changing the likelihood). Non-identifiability issues in inference can be resolved by, say, imposing constraints on the location parameters of the component densities, although this is not always completely satisfactory. See [30]. Perhaps the most satisfactory solution to MCMC mixing problems induced by lack of permutation identifiability is to introduce an independent random permutation of the component labels to supplement each MCMC iteration.

In addition to the difficulties of standard Bayesian analysis for this model, if K is also treated as an unknown parameter, then transdimensional MCMC is also needed; see [48] for an early influential paper. A discrete prior on K, typically on some finite set, completes the posterior specification, and yields a posterior distribution $\pi(K, \boldsymbol{\theta}^{(K)}, \boldsymbol{\omega}^{(K)}|\boldsymbol{y})$, where $\boldsymbol{\theta}^{(K)} = (\boldsymbol{\theta}_{K1}, \ldots, \boldsymbol{\theta}_{KK})$ and $\boldsymbol{\omega}^{(K)} = (\omega_{K1}, \ldots, \omega_{KK})$. In this model, dimension-changing moves correspond to changing the value of K, and in order to construct an efficient and tunable MCMC algorithm, moves which peturb K by ± 1 are typically considered. For example, moves of type $K \longrightarrow K + 1$ are termed 'birth' moves, and $K \longrightarrow K - 1$ are termed 'death' moves. The forward/reverse move pairs are then births from K and deaths from $K + 1$.

We consider the case of a Normal mixture model for simplicity. If y is one-dimensional, each component density has two parameters, so a birth/death pair requires the introduction of three latent variables $\boldsymbol{u}$, a new mean and variance, and also a parameter to introduce a new component weight. If y is D-dimensional, $D + D(D + 1)/2 + 1$ new scalar parameters are needed, that is, a new mean and covariance matrix, and a new component weight parameter.

- **New location/scale parameters.** Two strategies are used to propose a new set of location and scale parameters; either these new parameters are generated independently of the current model parameters from proposal distribution, or a subset of the current components is selected and used to generate a new additional component. In the latter case, a common strategy involves 'splitting' a currently existing component, that is, taking the mean and variance from component k, say, and by generating latent variables $\boldsymbol{u}$, and combining them with the parameters $\boldsymbol{\theta}_{Kk}$, using the approach described in the previous section. Perturbing a location parameter using a Normal increment is generally straightforward and may be sufficient to produce a reasonably functioning chain. Perturbing a scalar variance parameter is also reasonably easy, but perturbing a covariance matrix is more complex, although Wishart/Inverse Wishart generation can usually be implemented successfully in low dimensions. For example, if Σ_{Kk} is the covariance matrix for a selected component k, and $\boldsymbol{A}$ is a $D \times D$ positive definite symmetric matrix variate generated from a Wishart density, then $A\Sigma_{Kk}$ is also a positive definite symmetric matrix, so we may use it as the covariance matrix for a newly birthed component, and the Jacobian for this transform is straightforward to compute. The reverse 'merge' move can be defined using the rules outlined above.
- **New weight parameters.** Generating the component weight for a new component is straightforward, using a random split or rescaling of the currently existing weights.

In typical mixture applications, inference about the parameter K may be of primary interest in a given application, but perhaps more commonly the mixture model offers flexible modelling in the presence of heterogeneity. The Normal mixture case is the most commonly studied, but other mixtures of other parametric models have also been used.

It should be noted that MCMC computation for varying dimension mixture models is notoriously hard to implement successfully using single chain methods, even if the multiple modes caused by permutation invariance can be overcome. The likelihood surface is complicated and multi-modal, so it is easy for the chain to get trapped in local high probability regions of the parameter space. To overcome this, multi-chain (or population) MCMC methods are used; see Section 7.3.

The flexible modelling of multivariate density functions using finite mixture approximations has a long history in Bayesian inference, and has been extended to the use of nonparametric or infinite mixture models based on the Dirichlet Process. These Bayesian nonparametric approaches address dimension-changing issues from a different perspective, and typically through Polyá urn schemes; they essentially utilize a similar auxiliary variable scheme, but identify the 'clusters' of data having the same z value as observations from the same component mixture. *A priori*, the Polyá urn facilitates a countably infinite mixture model representation, although *a posteriori*, in light of a dataset of size n, a maximum of n components are supported. Working on this finite space of cluster allocations avoids the need for dimension-changing moves.

Example 7.5 (**Flexible curve-fitting via step functions**) Suppose the regression relationship between a scalar covariate x and scalar covariate y

$$y = g(x) + \epsilon$$

is the focus of interest, for independent homoscedastic Normal errors $\{\epsilon\}$. Typically, we seek the prediction $\widehat{y} = \widehat{g}(x)$. Suppose we seek to approximate g locally on a finite domain $\mathcal{X}$ by a series of step functions:

$$g_K(x; \boldsymbol{\beta}_K, \boldsymbol{\kappa}_K) = \sum_{k=1}^{K} \beta_{Kk} \mathbb{1}_{B_k}(x)$$

where $(B_1,\ldots,B_K)$ form a partition of $\mathcal{X}$ defined by a series of knots $\kappa_0,\kappa_1,\ldots,\kappa_{K-1},\kappa_K$ so that $B_k \equiv [\kappa_{k-1},\kappa_k)$. Without loss of generality, assume $\mathcal{X} \equiv [0,1)$, with $\kappa_0 = 0, \kappa_K = 1$. In this step-function approximation, $\boldsymbol{\beta}_K = (\beta_{K1},\ldots,\beta_{KK})$ are the parameters defining the piecewise constant levels of the function.

The likelihood for data (x_i, y_i), $i = 1,\ldots,n$ is a standard Normal likelihood

$$f(\mathbf{y}|M_K,\boldsymbol{\beta}_K,\sigma,\boldsymbol{\kappa}_K) = \prod_{i=1}^{n} \phi\left(\frac{y_i - g_K(x_i;\boldsymbol{\beta}_K,\boldsymbol{\kappa}_K)}{\sigma}\right).$$

Typically, within model M_K, a conjugate prior specification on $(\boldsymbol{\beta}_K,\sigma)$ is used so that the marginal likelihood $f(\mathbf{y}|M_K,\boldsymbol{\kappa}_K)$ can be computed analytically by integrating out $(\boldsymbol{\beta}_K,\sigma)$. The conjugate prior takes the form

$$p(\boldsymbol{\beta}_K,\sigma) = p(\boldsymbol{\beta}_K|\sigma)p(\sigma)$$

and $p(\boldsymbol{\beta}_K|\sigma)$ can be chosen to be some multivariate Normal density; a suitable choice might be to have a zero mean Gaussian process structure, where the covariance matrix depends on some hyperparameters and $\boldsymbol{\kappa}_K$.

A reversible jump MCMC algorithm can be constructed to explore the posterior $\pi(M_K,\boldsymbol{\kappa}_K|\mathbf{y})$ which, under the conjugate specification for $(\boldsymbol{\beta}_K,\sigma)$, depends on the marginal likelihood $f(\mathbf{y}|M_K,\boldsymbol{\kappa}_K)$, and the prior $p(M_K,\boldsymbol{\kappa}_K)$. One approach to the specification of this prior is to assume a Poisson process model for the change-in-level or jump locations of the step function. This amounts to setting $p(M_K)$ to be a Poisson distribution with rate parameter λ say, and then within model M_K, to regard $\boldsymbol{\kappa}_K$ as the order statistics derived from a Uniform random sample on (0,1), so that

$$p(\kappa_{K1},\ldots,\kappa_{K,K-1}|M_K) = (K-1)! \qquad 0 < \kappa_1 < \cdots < \kappa_{K,K-1} < 1.$$

Note that this prior depends only on the value of K and not the individual parameter values. A variant on this prior is given in [26], where a prior model based on the odd order statistics derived from a Uniform sample of size $2K-1$ is considered. This prior results in a larger expected separation between the κ values.

Dimension-changing moves in this model correspond to the addition or removal of a knot or knots. There are several mechanisms for doing this; a new knot can be proposed uniformly on (0,1), or uniformly in an interval selected in light of the current knot positions. In the former case, the reverse move corresponds to removing a knot uniformly at random from the current collection; this forward/reverse pair corresponds to the situation in Example 7.1, where the latent augmenting variable u is generated from the prior distribution, so the relevant terms cancel in the Hastings ratio. Model complexity in this model is controlled by the Poisson-model hyperparameter λ, and the covariance structure in the prior model for $\boldsymbol{\beta}_K$.

After the marginalized posterior $\pi(M_K,\boldsymbol{\kappa}_K|\mathbf{y})$ has been sampled at each iteration, it is trivial to sample from the conditional posterior on the remaining parameters, $\pi(\boldsymbol{\beta}_K,\sigma|M_K,\boldsymbol{\kappa}_K,\mathbf{y})$, which is available in closed form in the conjugate model. This then allows posterior samples of the function $g(x;\boldsymbol{\beta}_K,\boldsymbol{\kappa}_K)$ to be reconstructed. This methodology was exploited extensively for flexible modelling in the latter half of the 1990s to perform Bayesian versions of classical machine learning procedures such as classification and regression trees (CART), as it can be extended readily to the case of multiple predictors. In that case, the response function is modelled as a series of constants within hyper-rectangles defined by cut-points that partition predictor space. For a survey of the key techniques and literature, and examples, see [10].

Example 7.6 (**Flexible regression modelling and feature selection**) In the extended linear regression model, response y is modelled as a linear combination of parameters $\boldsymbol{\beta}$ plus error, that is

$$y = \boldsymbol{x}\boldsymbol{\beta} + \epsilon \tag{7.3}$$

where $\boldsymbol{x} = (x_1, \ldots, x_d)$ is a vector of predictor values. These predictors can correspond to the levels of a factor predictor, continuous covariates, nonlinear transforms of covariates, or functions combining any of these terms. With a fixed predictor set, there are $K = 2^d$ models that could be compared, and, although reversible jump MCMC could be used to explore the model space, it is not strictly necessary unless the marginal likelihood $f(\boldsymbol{y}|M_k)$ is not available analytically—this might be the case if ϵ is presumed to have a non-Normal distribution for example.

Using the basic regression model from equation (7.3), it is possible to broaden the modelling viewpoint to encompass flexible predictors. Flexible constructions include predictors based on splines, wavelets or other basis functions: these predictors allow the regression function to represent complicated regression relationships, and typically in such models it is the prediction, $\widehat{y} = \widehat{\boldsymbol{x}\boldsymbol{\beta}}$, that is of more interest than inference about the parameters.

To illustrate the use of reversible jump methodology in this field, consider the derived predictor, $\varphi(x)$, based on a truncated polynomial basis

$$\varphi(x) = (g(x) - \kappa)_+^\nu = \begin{cases} (g(x) - \kappa)^\nu & x > \kappa \\ 0 & x \leq \kappa \end{cases}$$

for some function $g(.)$ and parameters ν, κ. The simplest version is the linear form

$$\varphi(x) = (x - \kappa)_+ = \begin{cases} (x - \kappa) & x > \kappa \\ 0 & x \leq \kappa \end{cases}$$

which produces a piecewise linear continuous response function.

A series of such basis functions $\varphi_k(x)$, $k = 1, \ldots, K$, with corresponding $\kappa_1, \ldots, \kappa_K$ can be used to form the flexible regression model

$$y = \boldsymbol{\varphi}(x)\boldsymbol{\beta}_K + \epsilon = \sum_{k=1}^{K} \varphi_k(x)\beta_{Kk} + \epsilon$$

which may be augmented using standard polynomial terms, say

$$y = \boldsymbol{x}\boldsymbol{\alpha} + \boldsymbol{\varphi}(x)\boldsymbol{\beta}_K + \epsilon = \alpha_0 + \alpha_1 x + \alpha_2 x^2 + \sum_{k=1}^{K} \varphi_k(x)\beta_{Kk} + \epsilon.$$

In this latter case, setting $K = 0$, leaving only the polynomial terms, is legitimate. This model is a generalization of the model described in Example 7.5; the model, albeit nonlinear in x in general, is still linear in the parameters $(\boldsymbol{\alpha}, \boldsymbol{\beta}_K)$, so much of the previous conjugate analysis goes through as before.

Another strategy for flexible modelling is to use wavelets as the basis function set. These basis functions have compact support, are orthogonal, and typically do not rely on the specification of separate knot positions, as these are specified in an automatic fashion depending on the type of wavelet used. The standard classical procedure for implementing wavelet-based models is to select a wavelet family, which automatically specifies the discrete wavelet transform matrix $\boldsymbol{W}(\boldsymbol{x})$ such that $\boldsymbol{y} = \boldsymbol{W}(\boldsymbol{x})\boldsymbol{c}$ say, where $\boldsymbol{c}$ are the estimated wavelet coefficients. The Bayesian version of this model uses the $\boldsymbol{W}(\boldsymbol{x})$ matrix as the design matrix in a linear model; variable dimension MCMC then proceeds to simplify this model by omitting sets of columns of this matrix.

7.2.3 Reversible jump MCMC: summary

The reversible jump MCMC algorithm introduced by Green was an important development as it facilitated greater flexibility in Bayesian modelling. The algorithm itself relies on what are, in fact, standard Metropolis–Hastings moves that utilize latent or auxiliary variables and an extension to the state space on which the Markov chain operates. It is similar in spirit to the contemporaneous approach of [8], but does not retain the auxiliaries at all steps of the algorithm as in the Carlin and Chib method. Good mixing of the reversible jump chain can sometimes be hard to achieve in practice without expert tuning, but the approach offers an important contribution to Bayesian model selection.

7.3 Population MCMC

7.3.1 Motivation

In early statistical applications of MCMC, a common approach to implementation was to run multiple (and usually short) chains independently, and to combine the samples from the chains in order to produce posterior summaries (see for example the applications in [12]), with, on occasion, just a single draw from the posterior being obtained from each separate chain, and more commonly, the output from each chain being heavily thinned, so as to produce an independent Monte Carlo sample. In simple applications, short runs can be effective, as the chain quickly reaches its stationary distribution, but in more complicated multi-parameter settings there is a danger that the chain remains in its transient phase for many iterations. The *multi-chain* approach offers some protection against the chain getting trapped in localized high probability regions, but it can also be considered wasteful, as it prizes the (at equilibrium) independent draws from the target posterior above everything else. Once MCMC methods came into the statistical mainstream, most practitioners realized the utility of *single-chain* approaches, that could be used effectively to perform standard Monte Carlo calculations using dependent draws from the posterior, albeit at the cost of increased variance of the Monte Carlo estimators. The difficulty in implementing single-chain algorithms is that the chain may not explore the entire model space effectively; the usual Metropolis–Hastings algorithm is essentially local in nature, as proposed moves are usually highly dependent on the current value of the chain. Therefore, the practical application of MCMC methods using single chains can be challenging for complicated models.

An early multi-chain method, *adaptive direction sampling*, was devised by [21]. In this Markov chain algorithm, a population of chains with identical target distribution f is maintained, and values are updated by proposing new values in the direction connecting two randomly selected population members. Specifically, in d dimensions, if the population is $\{\boldsymbol{x}_1, \ldots, \boldsymbol{x}_N\}$, and $\boldsymbol{x}_j$ and $\boldsymbol{x}_k$ are selected at random, a change to $\boldsymbol{x}_j$ is proposed by setting

$$\boldsymbol{x}_j' = \boldsymbol{x}_j + u(\boldsymbol{x}_j - \boldsymbol{x}_k)$$

where u is a scalar random variate drawn from the density

$$f_U(u : \boldsymbol{x}_j, \boldsymbol{x}_k) \propto |u|^{d-1} f(\boldsymbol{x}_j + u(\boldsymbol{x}_j - \boldsymbol{x}_k)).$$

7.3.2 Simulated annealing

An early work that was influential on the statistical MCMC literature was the paper by Geman and Geman [15], which essentially popularized the Gibbs sampler, but which also utilized an optimization algorithm, *simulated annealing* [33], which operates as follows: to minimize the non-negative scalar function $g(.)$ of argument $\boldsymbol{x}$, consider the unnormalized probability density

$$p_T(\boldsymbol{x}) \propto \exp\left\{-\frac{1}{T} g(\boldsymbol{x})\right\}$$

for constant T. Simulated annealing proceeds by running a Markov chain sampling algorithm at a decreasing sequence of T values $T_1 > T_2 > \cdots$, with T_1 large. As T decreases, the function $p_T(\boldsymbol{x})$ becomes increasingly peaked around its mode. When T is large, $p_T(\boldsymbol{x})$ is relatively flat, so traversing the support is relatively straightforward, allowing the sampled points to locate the regions of high probability. In practice, T should be decreased slowly, so that the sampled points have sufficient chance to become concentrated around the mode.

Although simulated annealing was devised as an optimization algorithm, the core idea of running a chain or chains at different 'temperatures' was subsequently used to develop multiple-chain MCMC methods in which the chains, rather than operating in parallel, independently and under the same transition dynamics, instead have potentially different stationary distributions, but are allowed to interact and exchange information.

7.3.3 Parallel tempering

The key ideas in this section were developed by Geyer ([17, 19]), and form the basis of most current implementations of population MCMC methods. Consider a collection of related unnormalized (posterior) densities $\pi_1, \ldots, \pi_N$, defined for simplicity on a common support within $\mathbb{R}^d$. Consider also a related collection of MH proposal kernels $P_1, \ldots, P_N$ with proposal densities $q_1, q_2, \ldots, q_N$ that facilitate the implementation of MCMC for each density. Assume that the usual conditions of irreducibility and aperiodicity are met.

Consider the augmented posterior

$$\tilde{\pi}(\boldsymbol{\theta}_1, \ldots, \boldsymbol{\theta}_N) = \prod_{j=1}^{N} \pi_j(\boldsymbol{\theta}_j) \tag{7.4}$$

defined on the Cartesian product of the individual parameter spaces. An MCMC algorithm constructed by taking a fixed cycle (or random scan) across $P_1, \ldots, P_N$ is irreducible and aperiodic and has $\tilde{\pi}$ as its stationary distribution; of course, marginally, samples of $\boldsymbol{\theta}_j$ derived from the chain are distributed according to π_j.

In addition to the usual within-chain moves, consider the following 'exchange' move; in addition to the fixed scan, select two chains j and k at random from $\{1, \ldots, N\}$, and propose to exchange their values, that is, for chain j, propose to set $\boldsymbol{\theta}_j'$ equal to $\boldsymbol{\theta}_k$, the current value in chain k, and similarly propose $\boldsymbol{\theta}_k' = \boldsymbol{\theta}_j$. This move is then accepted with probability

$$\alpha(\boldsymbol{\theta}_j, \boldsymbol{\theta}_k|y) = \left\{1, \frac{\pi_j(\boldsymbol{\theta}_k)\pi_k(\boldsymbol{\theta}_j)}{\pi_j(\boldsymbol{\theta}_j)\pi_k(\boldsymbol{\theta}_k)}\right\}$$

This additional reversible move leaves the chain Markov on the product space, and does not affect the stationary distribution. However, the exchange move is likely to have a low acceptance rate unless the densities π_i and π_j are similar.

Combining the exchange move with temperature changing ideas inspired by simulated annealing, we have a mechanism for allowing parallel chains to exchange information in a useful fashion. Suppose that in a Bayesian inference problem, the target posterior $\pi(.)$ is indexed $j = 1$, and that

$$\pi_j(\boldsymbol{\theta}) \propto \{\pi(\boldsymbol{\theta})\}^{1/T_j}$$

say, for temperature constants $1 = T_1 < T_2 < \cdots$. Note that this construction does not necessarily leave π_j a proper density, so in the Bayesian setting the alternate construction

$$\pi_j(\boldsymbol{\theta}) \equiv \pi(\boldsymbol{\theta}_j|\mathbf{y}) \propto \{f(\mathbf{y}|\boldsymbol{\theta})\}^{1/T_j} p(\boldsymbol{\theta})$$

which ensures that a proper distribution might be considered. Using this approach, we ensure that if T_j and T_k are not separated by a large amount, π_j and π_k should be quite similar; it should be possible to propose and accept frequently if $k = j \pm 1$, say, and although moves between π_1 and π_N changes might only be accepted infrequently, the temperature 'ladder' allows for eventual passage of information across all chains. Applying the exchange move strategy produces an effective way of traversing the support of the target posterior, especially in the case of multi-modal densities, as densities that are severely multi-modal, with modes separated by low probability regions will be considerably more uniform (although still multi-modal) in their powered form.

At convergence, the stored MCMC samples represent a sample from the augmented joint posterior in equation (7.4), and consequently the samples of $\boldsymbol{\theta}_1$ values are variates generated from the corresponding marginal, that is, the target posterior.

Example 7.7 (**Tempering a mixture density**) Consider the two component mixture density

$$f(x) = \frac{3}{4}\phi(x+5) + \frac{1}{4}\phi(x-4)$$

with modes at -5 and 4. The effect of the tempering is evident in Figure 7.3; the two well-separated modes in the target ($T = 1$) density are linked by an intervening region of increasingly large probability content. Thus MH moves from one modal region to the other, which are difficult when $T = 1$, are much easier when $T = 10$.

Stochastic exchange moves

Simple deterministic exchange methods can be modified to exploit several different chains in a single move. For example, suppose a move within chain 1 (with stationary distribution $\pi_1 \equiv \pi$, the target posterior) is proposed as follows:

Algorithm 1 Stochastic exchange proposal

Current value of $\boldsymbol{\theta}_1$ is $\boldsymbol{u}_1$
for j in $2 : N$ **do**
  Propose $\boldsymbol{u}_j$ from $q_j(\boldsymbol{u}_{j-1}, .)$
end for
Propose $\boldsymbol{v}_N$ from $q_N(\boldsymbol{u}_N, .)$
for j in $(N-1) : 1$ **do**
  Propose $\boldsymbol{v}_j$ from $q_{j+1}(\boldsymbol{v}_{j+1}, .)$
end for
Take $\boldsymbol{v}_1$ as final proposed value for $\boldsymbol{\theta}'_1$.
Accept $\boldsymbol{\theta}'_1$ with probability

$$\alpha(\boldsymbol{\theta}_1, \boldsymbol{\theta}'_1) = \min\left\{1, \frac{\pi_1(\boldsymbol{\theta}'_1)}{\pi_1(\boldsymbol{\theta}_1)} q(\boldsymbol{u}, \boldsymbol{v})\right\}$$

where $q(\boldsymbol{u}, \boldsymbol{v})$ is the Hastings ratio for the proposed move.

It is straightforward to establish that

$$q(\boldsymbol{u}, \boldsymbol{v}) = \frac{\left\{\prod_{j=2}^{N} q_j(\boldsymbol{v}_{j-1}, \boldsymbol{v}_j)\right\} q_N(\boldsymbol{v}_N, \boldsymbol{u}_N) \left\{\prod_{j=1}^{N-1} q_{N-j+1}(\boldsymbol{u}_{N-j+1}, \boldsymbol{u}_{N-j})\right\}}{\left\{\prod_{j=2}^{N} q_j(\boldsymbol{u}_{j-1}, \boldsymbol{u}_j)\right\} q_N(\boldsymbol{u}_N, \boldsymbol{v}_N) \left\{\prod_{j=1}^{N-1} q_{N-j+1}(\boldsymbol{v}_{N-j+1}, \boldsymbol{v}_{N-j})\right\}}.$$

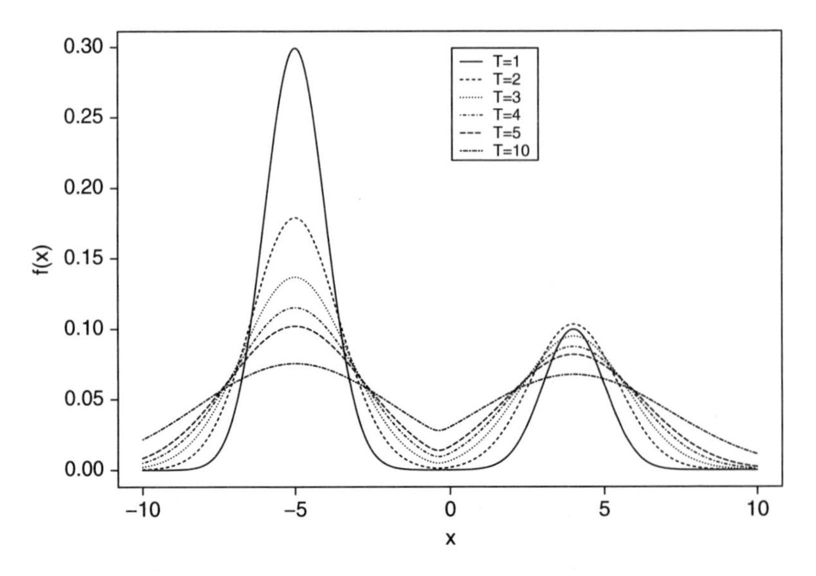

Figure 7.3 Example 7.7. Tempered mixture density with tempering parameters T = 1,2,3,4,5,10.

For example, if $N = 3$, we have

$$q(\boldsymbol{u}, \boldsymbol{v}) = \frac{q_2(\boldsymbol{v}_1, \boldsymbol{v}_2)q_3(\boldsymbol{v}_2, \boldsymbol{v}_3)q_3(\boldsymbol{v}_3, \boldsymbol{u}_3)q_3(\boldsymbol{u}_3, \boldsymbol{u}_2)q_2(\boldsymbol{u}_2, \boldsymbol{u}_1)}{q_2(\boldsymbol{u}_1, \boldsymbol{u}_2)q_3(\boldsymbol{u}_2, \boldsymbol{u}_3)q_3(\boldsymbol{u}_3, \boldsymbol{v}_3)q_2(\boldsymbol{v}_3, \boldsymbol{v}_2)q_3(\boldsymbol{v}_2, \boldsymbol{v}_1)}.$$

Recall that each $q_j(s, t)$ proposes a value t on the same space as s. If the proposal density is symmetric in its arguments, as in a standard Metropolis proposal, the expression simplifies to $q(\boldsymbol{u}, \boldsymbol{v}) = 1$.

The stochastic exchange approach is similar in spirit to the path and bridge sampling approaches to numerical integration in the computation or normalizing constants [13]. It has been used widely in Bayesian inference (see for example [30, 43] for examples in the context of mixture models).

Choosing the temperature values

It is clear that good choice of the number and values of the temperatures T_j is crucial to the adequate performance of the algorithm. In general, there are few results to give guidance as to the optimal choice for complex models where parallel tempering may be necessary. The most common strategy adopted is equal or geometric spacing, although several formulations have been presented suggesting alternative unequal spacing (see for example [4, 7, 19, 25],). Algorithms have also been developed that allow the temperatures to vary stochastically throughout the Markov chain runs, that is, the $\{T_j\}$ are treated as parameters in a global probability model. *Simulated tempering* [19, 41] is a parallel chain method that allows each T_j to vary according to stochastic updates, and allows choice of a pseudoprior for the temperatures. This can allow favourable choices of the temperatures to be discovered by the algorithm.

7.3.4 Evolutionary Monte Carlo

Evolutionary Monte Carlo (EMC) was introduced by [39]. It combined techniques from the literature on genetic algorithms with population Monte Carlo. A genetic algorithm is a general optimization approach that maintains a diverse collection of 'chromosomes' (parameter vectors) and searches for optimal configurations by quantifying and updating 'fitness' using pseudo-evolutionary modifications. The two basic genetic algorithm evolutionary steps acting on a set of chromosomes $\boldsymbol{x}_1, \ldots, \boldsymbol{x}_N$ include

- **Mutation.** Update $\boldsymbol{x}_j \longrightarrow \boldsymbol{x}_j'$
- **Crossover.** Recombine $\boldsymbol{x}_j$ and $\boldsymbol{x}_k$ by exchanging a portion of each chromosome. That is, if $\boldsymbol{x}_j = (\boldsymbol{x}_{j1}, \boldsymbol{x}_{j2})$ and $\boldsymbol{x}_k = (\boldsymbol{x}_{k1}, \boldsymbol{x}_{k2})$, set

$$\boldsymbol{x}_j' = (\boldsymbol{x}_{j1}, \boldsymbol{x}_{k2}) \qquad\qquad \boldsymbol{x}_k' = (\boldsymbol{x}_{k1}, \boldsymbol{x}_{j2}).$$

These two moves, in conjunction with the **Exchange** move from Section 7.3.3, form the basis of a stochastic version of the genetic algorithm, which, in line with the Metropolis–Hastings philosophy, proposes new candidate chromosomes, and then accepts or rejects the proposed move with the usual MH acceptance probability. In EMC, the chromosomes are parameter vectors in the joint probability model from equation 7.4 formed by combining tempered marginal densities $\pi_j, j = 1, \ldots, N$ at temperatures $T_j, j = 1, \ldots, N$.

It can readily be seen that stochastic versions of all three move types are reversible, and have standard accept/reject dynamics, and the acceptance probability can be computed in a straightforward fashion. For the crossover move, this can be carried out with a deterministic proposal (hence becoming a partial exchange move) or stochastically. To carry out a stochastic crossover move, an adaptive direction sampling or 'snooker' proposal [40] may be used; this proposes a new value for $\boldsymbol{x}_j$ in the direction connecting $\boldsymbol{x}_j$ and $\boldsymbol{x}_k$.

7.3.5 The Equi-energy sampler

The equi-energy sampler [34] is another population approach which utilizes other aspects of tempered posterior distributions. Suppose that the target posterior is

$$\pi(\boldsymbol{\theta}) \propto \exp\{-\psi(\boldsymbol{\theta})\}$$

and that the temperature ladder is given by $1 = T_1 < T_2 < \cdots < T_N < T_{N+1} = \infty$. Consider also a sequence of 'energy levels' $-\infty = E_1 < E_2 < \cdots < E_N$, and define the tempered distributions $\pi_1, \ldots, \pi_N$ by

$$\pi_j(\boldsymbol{\theta}) \propto \exp\left\{-\frac{1}{T_j}\max\{E_j, \psi(\boldsymbol{\theta})\}\right\} \qquad j = 1, \ldots, N.$$

Consider a partition of the parameter space $B_1, \ldots, B_N$ where B_j is the set $B_j \equiv \{\boldsymbol{\theta} : E_j \leq \psi(\boldsymbol{\theta}) < E_{j+1}\}$. The equi-energy sampler utilizes this partition to develop effective moves in the parameter space.

Algorithm 2 Equi-energy sampler

---

1. Chain N: Run an MCMC algorithm to sample π_N at the highest temperature; after a suitable period, use the samples to define the initial partition, $\widehat{B}_j^{(N)}, j = 1\ldots, N$.
2. After R iterations of Chain N, initiate a second chain, Chain $N-1$, to sample from π_{N-1} (whilst continuing to update Chain N) using two types of move:
   - **Move 1:** MH move within Chain $N-1$ (selected with probability $1-p$),
   - **Move 2:** Equi-Energy Jump into Chain $N-1$ (selected with probability p): suppose that the current value in Chain $N-1$, $\boldsymbol{\theta}_{N-1}$, lies within partition component $\widehat{B}_j^{(N)}$ say. Then propose a value $\boldsymbol{\theta}'_{N-1}$ uniformly from $\widehat{B}_j^{(N)}$, and accept the move with probability

   $$\alpha(\boldsymbol{\theta}_{N-1}, \boldsymbol{\theta}'_{N-1}) = \min\left\{1, \frac{\pi_{N-1}(\boldsymbol{\theta}'_{N-1})\pi_N(\boldsymbol{\theta}_{N-1})}{\pi_{N-1}(\boldsymbol{\theta}_{N-1})\pi_N(\boldsymbol{\theta}'_{N-1})}\right\}$$

   otherwise remain at the same point.
3. After a suitable period, use the samples from Chain $N-1$ to define a new partition, $\widehat{B}_j^{(N-1)}, j = 1\ldots, N$.
4. Repeat Steps 1 and 2, for new Chains $N-2, N-3, \ldots, 1$ etc., allowing equi-energy jumps between adjacent chains.

Under fairly general conditions, the equi-energy sampler can be shown to be ergodic at each level of the temperature ladder, such that Chain j has stationary distribution π_j [34]. The method is ingenious as it exploits both the temperature and location in the parameter space information.

7.3.6 Annealed importance sampling

In a related but distinct vein, Neal [44] developed a parallelized version of importance sampling that operated using a sequence of tempered distributions as in parallel tempering, and Markov transition kernels q_j with stationary distributions π_j. Instead of performing iterative sampling and exchanging information between chains, *annealed importance sampling* uses the tempered distributions to produce importance sampling weights that can be used to perform Monte Carlo integration.

Suppose that π_j is defined in terms of $\pi_1 \equiv \pi$, the (unnormalized) target posterior, and some (diffuse) density p_N that can be sampled in a straightforward manner, by

$$\pi_j(\boldsymbol{\theta}) = \{\pi_1(\boldsymbol{\theta})\}^{\eta_j} \left\{p_N(\boldsymbol{\theta})\right\}^{1-\eta_j}$$

for parameters $1 = \eta_1 > \eta_2 > \cdots > \eta_N = 0$. The annealed importance sampling algorithm proceeds as follows to estimate the quantity

$$\mathbb{E}[h(\boldsymbol{\theta})] = \int h(\boldsymbol{\theta})\pi_1(\boldsymbol{\theta})d\boldsymbol{\theta}.$$

Algorithm 3 Annealed importance sampling

1. **for** i in $1 : M$ **do**
2. Sample $\boldsymbol{\theta}_N$ from $\pi_N \equiv p_N$
3. **for** j in $(N-1) : 1$ **do**
4. Sample $\boldsymbol{\theta}_j$ from $q_{j+1}(\boldsymbol{\theta}_{j+1}, .)$
5. **end for**
6. Retain $\boldsymbol{\theta}_i = \boldsymbol{\theta}_1$, and form the importance sampling weight

$$w_i = \prod_{j=1}^{N-1} \frac{\pi_j(\boldsymbol{\theta}_j)}{\pi_{j+1}(\boldsymbol{\theta}_j)} \tag{7.5}$$

7. **end for**
8. Form the Monte Carlo estimate

$$\widehat{\mathbb{E}}[h(\boldsymbol{\theta})] = \frac{1}{M}\sum_{i=1}^{M} w_i h(\boldsymbol{\theta}_i)$$

Neal [44] demonstrates that this Monte Carlo estimator is unbiased for the estimated, as the importance sampling weights are correctly defined. Again, he used an augmented state space construction to demonstrate this. The first important note is that, for any π with associated Markov transition density q that has π as its invariant distribution, we have by definition of the invariance that

$$\pi(d\boldsymbol{\theta}) = \int q(\boldsymbol{u}, \boldsymbol{\theta})\pi(d\boldsymbol{u}).$$

Now, using the importance sampling identity with importance density p

$$\mathbb{E}_\pi[h(\boldsymbol{\theta})] = \int h(\boldsymbol{\theta})\pi(\boldsymbol{\theta})d\boldsymbol{\theta} = \int\int g(\boldsymbol{\theta})q(\boldsymbol{u},\boldsymbol{\theta})\pi(\boldsymbol{u})d\boldsymbol{u}d\boldsymbol{\theta}$$

$$= \int\int \frac{h(\boldsymbol{\theta})\pi(\boldsymbol{u})}{p(\boldsymbol{u})}p(\boldsymbol{u})q(\boldsymbol{u},\boldsymbol{\theta})d\boldsymbol{u}d\boldsymbol{\theta}$$

which suggests the estimator

$$\widehat{\mathbb{E}}_\pi[h(\boldsymbol{\theta})] = \frac{1}{M}\sum_{i=1}^{M}\frac{h(\boldsymbol{\theta}_i)\pi(\boldsymbol{u}_i)}{p(\boldsymbol{u}_i)}$$

where $\{u_i\}$ are independent draws from $p(.)$, and $\{\boldsymbol{\theta}_i\}$ are independent draws from $q(\boldsymbol{u}_i,.)$ for $i = 1,\ldots,n$. The annealed importance sampling algorithm adapts this result on the augmented space using the augmented posterior equation (7.4). We have that the augmented density can be written

$$\tilde{\pi}(\boldsymbol{\theta}_1,\boldsymbol{\theta}_2,\ldots,\boldsymbol{\theta}_N) = \pi_1(\boldsymbol{\theta}_1)\prod_{j=2}^{N}\tilde{q}_j(\boldsymbol{\theta}_{j-1},\boldsymbol{\theta}_j)$$

where

$$\tilde{q}_j(\boldsymbol{x},\boldsymbol{y}) = \frac{q_j(\boldsymbol{y},\boldsymbol{x})\pi_j(\boldsymbol{y})}{\pi_j(\boldsymbol{x})}$$

is the so-called reverse transition density. The proposal density is

$$p(\boldsymbol{\theta}_1,\boldsymbol{\theta}_2,\ldots,\boldsymbol{\theta}_N) = \pi_N(\boldsymbol{\theta}_N)\prod_{j=2}^{N}q_j(\boldsymbol{\theta}_j,\boldsymbol{\theta}_{j-1})$$

and thus the importance weight is given by

$$\frac{\tilde{p}(\boldsymbol{\theta}_1,\boldsymbol{\theta}_2,\ldots,\boldsymbol{\theta}_N)}{p(\boldsymbol{\theta}_1,\boldsymbol{\theta}_2,\ldots,\boldsymbol{\theta}_N)}$$

which reduces to (7.5).

7.3.7 Population MCMC: summary

Population MCMC methods facilitate effective exploration of the support of complicated and potentially multi-modal posterior distributions that single-chain methods often find difficult to explore. Usually population MCMC relies upon a series of temperature modulated versions of the target posterior, and on designed MH moves across the collection of Markov chains working with each version separately, allowing for exchange of sampled values between them. In this section, we have discussed the fixed-dimension case, but the methodology applies also to populations of variable-dimension chains [32]. Also, population approaches with exchange of information can be adopted in sequential Monte Carlo (see, for example, [31] for a comparison of population MCMC and sequential Monte Carlo). For a comprehensive survey of population MCMC methods, and other recent advances, see [38].

7.4 Adaptive MCMC

7.4.1 Introduction

The Metropolis–Hastings algorithm can be used to sample from any target distribution but at the price of having to choose a proposal distribution. This choice plays an important role in determining how well a sample drawn using the Metropolis–Hastings algorithm replicates the target distribution. Good choices of proposal distribution lead to algorithms which more closely replicate the target distribution and will usually depend on the shape of the distribution. In Bayesian statistics, the exact shape of the target distribution (the posterior distribution) will be determined by the choice of model (including the prior distribution) and data, and so will often change from model to model and from data to data. It would be desirable to have methods which automatically changes the proposal distribution so that the sample is a good representation of the target distribution. Such methods have been developed and are usually described as adaptive Markov chain Monte Carlo methods.

The idea of adaptation is appealing but leads to serious complications in the theory underlying non-adaptive MCMC methods. Allowing the proposal to depend on previous states stops the chains being Markovian and the usual theory for convergence of Metropolis–Hastings algorithms is no longer applicable. Therefore, new theory is needed to justify the use of such algorithms. An adaptive method was developed by [22] which used regeneration times to preserve the Markov properties. However, these methods have proved difficult to extend from their framework. An alternative approach was pursued by [28] who developed the Adaptive Metropolis (AM) algorithm which allows adaptation at every step of the algorithm and showed that it converges to the correct distribution. Allowing adaptation at every step of the algorithm (which is often known as infinite adaptation) has proved a fruitful idea. A fairly general theory has subsequently been developed which provides conditions under which adaptive MCMC algorithms can be shown to converge to the correct distribution. This has led to a general principle for the development of adaptive algorithms which can be applied to the full range of posterior distributions arising from Bayesian models.

Alternatively, a single chain can be used with the proposal at time j depending on the samples collected before time j, $\theta^{(1)}, \theta^{(2)}, \ldots, \theta^{(j-1)}$. Finite adaptive algorithms allow the proposal to be changed for the first K iteration, where it is specified before running the sampler, the same proposal is used after K. Clearly, samples are collected after K. Infinite adaptation algorithms continue to adapt the proposal for the whole length of the chain.

7.4.2 The Metropolis–Hastings algorithm and optimal proposals

Recall the basic Metropolis–Hastings algorithm: firstly, propose a new value $\boldsymbol{\theta}'$ from a transition density $q(\boldsymbol{\theta}^{(g)}, \boldsymbol{\theta}'|\mathbf{y})$ and calculate the acceptance probability

$$\alpha(\boldsymbol{\theta}^{(g)}, \boldsymbol{\theta}'|\mathbf{y}) = \min\left\{1, \frac{\pi(\boldsymbol{\theta}')q(\boldsymbol{\theta}', \boldsymbol{\theta}^{(g)}|\mathbf{y})}{\pi(\boldsymbol{\theta}^{(g)})q(\boldsymbol{\theta}^{(g)}, \boldsymbol{\theta}'|\mathbf{y})}\right\}.$$

Secondly, generate a uniform random variable u_g and set $\boldsymbol{\theta}^{(g+1)} = \boldsymbol{\theta}'$ if $u < \alpha(\boldsymbol{\theta}^{(g)}, \boldsymbol{\theta}'|\mathbf{y})$ or $\boldsymbol{\theta}^{(g+1)} = \boldsymbol{\theta}^{(g)}$ otherwise. This guarantees that $\boldsymbol{\theta}^{(1)}, \boldsymbol{\theta}^{(2)}, \ldots$ are a sample from π under weak conditions. This allows integrals of the form

$$I = \int h(\boldsymbol{\theta})\pi(\boldsymbol{\theta})\, d\boldsymbol{\theta}$$

to be approximated by

$$\widehat{I} = \frac{1}{G}\sum_{g=1}^{G} h(\boldsymbol{\theta}^{(g)})$$

since $\widehat{I}$ converges to I as $G \to \infty$. This guarantees the correctness of the approximation in the limit but, in practice, we will only run the chain for a finite number of iterations G. The accuracy of the approximation after this finite run will crucially depend on the choice of proposal distribution. This has led to interest in conditions for 'good' proposals.

The proposal in a Metropolis–Hastings algorithm is often chosen from a class which depends on some tuning parameters, $\boldsymbol{\zeta}$. There are two main classes of proposal. These are:

- The Metropolis–Hastings independence sampler, where $\boldsymbol{\theta}'$ is drawn from a particular distribution with parameters $\boldsymbol{\zeta}$. For example, a normal distribution with mean $\boldsymbol{\mu}$ and variance $\boldsymbol{\Sigma}$, where $\boldsymbol{\zeta} = (\boldsymbol{\mu}, \boldsymbol{\Sigma})$.
- The Metropolis–Hastings random walk sampler proposes $\boldsymbol{\theta}' = \boldsymbol{\theta} + \boldsymbol{\zeta}\boldsymbol{\epsilon}$, where ϵ is drawn from a zero mean distribution (such as a standard normal or t-distribution).

It is natural to ask whether certain choices of $\boldsymbol{\zeta}$ lead to good or bad mixing of the chain. In an independence sampler, [42] discussed the convergence of the Metropolis–Hastings independence sampler and showed that the optimal choice of proposal density is the target density $\pi(\boldsymbol{\theta})$. In practice, it will usually be impossible to sample from $\pi(\boldsymbol{\theta})$ (otherwise, we would just sample from $\pi(\boldsymbol{\theta})$ and there would be no need for MCMC), but good choices of proposal density approximate the target density as closely as possible. They showed that geometric ergodicity depends on the approximation having heavier tails than the target distribution. There are two main rules-of-thumb for Metropolis–Hastings random walk samplers. Firstly, [50] showed that choosing $\boldsymbol{\zeta}$ so that the average acceptance rate $\bar{\boldsymbol{\alpha}}$ (which is the average of the acceptance probabilities over the Metropolis–Hastings sampler) is equal to 0.234 leads to an optimal sampler in high-dimensional spaces. It has been shown that an average acceptance rate around this value can be optimal in many problems. Secondly, choosing $\boldsymbol{\zeta}\boldsymbol{\zeta}^T$ to be $(2.4)^2/d$ times the covariance of the target density, where d is the dimension of the target, has been shown to be optimal asymptotically in p [49, 50]. [14] showed that these results can also be useful in low-dimensional problems.

Example 7.8 A two-component mixture model

We will consider a one-dimensional two-component mixture distribution

$$\pi(\theta) = 0.3\,\mathrm{N}(\theta| - 1, 0.2^2) + 0.7\,\mathrm{N}(\theta|1, 0.2^2)$$

where $\mathrm{N}(\theta|\mu, \sigma^2)$ represents the probability density function of a normal distribution with mean μ and variance σ^2. Outputs from running the Metropolis–Hastings random walk sampler with different values of ζ are shown in Figure 7.4. The chain with $\zeta = 0.8$ tends to move often but short distances. As ζ increases, the chain tends to move less often but longer distances. The best approximation to the target distribution is provided by $\zeta = 1.4$ (middle row). This represent a compromise between the other values of ζ for which the chain either jumps too often ($\zeta = 0.8$) or too little ($\nu = 6$). The average acceptance rate for this chain is 0.234 and shows that a chain with this acceptance rate tends to mix well.

A useful measure of the mixing of a chain is the average squared jumped distance, which is the expected distance between consecutive values of the Markov chain, which is given by

$$\int \| \boldsymbol{\theta}' - \boldsymbol{\theta} \|^2 \, \alpha(\boldsymbol{\theta}, \boldsymbol{\theta}'|y) q(\boldsymbol{\theta}, \boldsymbol{\theta}'|y) \pi(\boldsymbol{\theta}) d\boldsymbol{\theta}' d\boldsymbol{\theta}$$

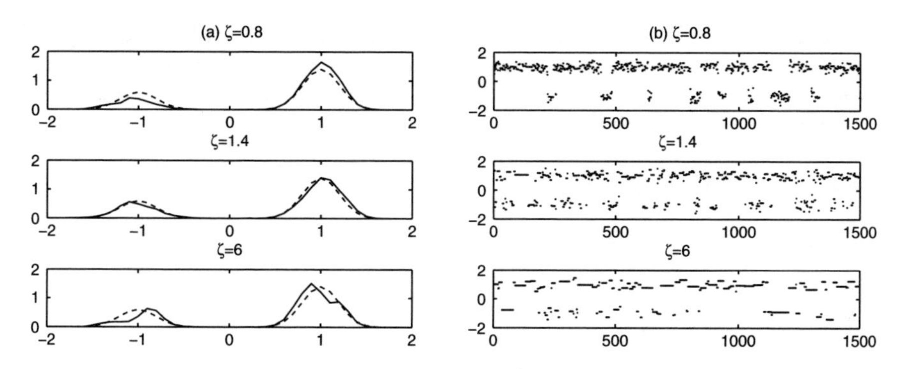

Figure 7.4 Example 7.8: (a) density estimate (solid line) and target density (dotted line) and (b) the trace plot for 1000 iterations of the Metropolis–Hastings random walk algorithm with different values of ν.

where $\| x \|$ represents the Euclidean norm. Larger values of the expected squared jumped distance suggest that the sampler is mixing better and so the mean squared error of estimating I by $\hat{I}$ will be smaller. It can be estimated from MCMC output by

$$\bar{J} = \frac{1}{G} \sum_{g=1}^{G} \| \boldsymbol{\theta}^{(g)} - \boldsymbol{\theta}^{(g-1)} \|^2 .$$

The relationship between ζ, the average acceptance rate $\bar{\alpha}$ and the average squared jumped distance $\bar{J}$ is illustrated in Figure 7.5. The average acceptance rate $\bar{\alpha}$ decreases as ζ increases (Panel (a)). Panel (b) shows that the average squared jumped distance $\bar{J}$ increases for small values of ζ but reaches an optimum when ζ is around 2 and then decreases. Small ζ is associated with small moves which are

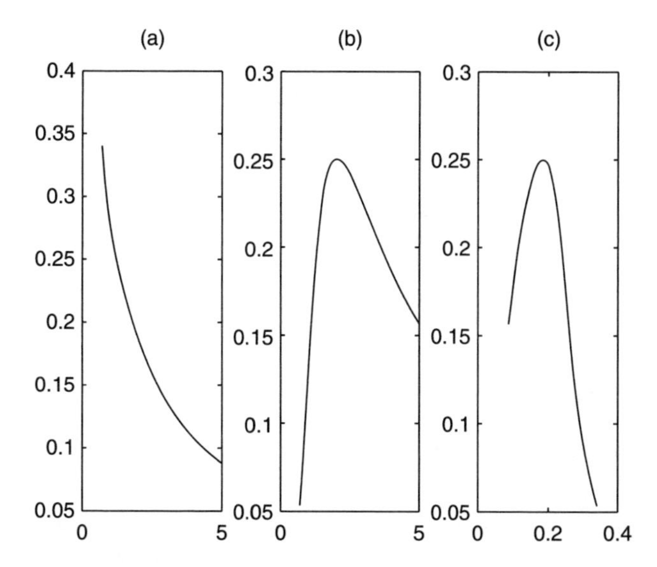

Figure 7.5 Example 7.8: (a) The average acceptance rate as a function of ν, (b) the average squared jumped distance as a function of ν, (c) the average squared jumped distance as a function of the average acceptance rate.

often accepted. Larger ζ leads to larger moves, but these often tend to be rejected if ζ becomes too large. Panel (c) shows that the average squared jumped distance is maximized by an acceptance rate around 0.19. The results fits roughly with the rule that $\bar{\alpha}$ around 0.234 leads to the largest average squared jumped distance $\bar{J}$ of about 0.25.

7.4.3 Adaptive samplers

In this section, we will look at different adaptive MCMC algorithms. We will make a slight change of notation to emphasize that both the proposal distribution and the acceptance probability depend on the tuning parameters $\boldsymbol{\zeta}$. The proposal density will be denoted by $q_{\boldsymbol{\zeta}}(\boldsymbol{\theta}, \boldsymbol{\theta}'|\mathbf{y})$ and the acceptance rate will be denoted by $\alpha_{\boldsymbol{\zeta}}(\boldsymbol{\theta}, \boldsymbol{\theta}'|\mathbf{y})$.

Adaptive Metropolis–Hastings random walk samplers

The first adaptive MCMC algorithm with infinite adaptivity was the Adaptive Metropolis (AM) algorithm introduced by [28]. They use the idea that the algorithm will be optimal if the variance–covariance matrix of the proposal distribution is chosen to be $s_d = 2.4^2/d$ times the variance–covariance matrix of the target. An estimate of the variance–covariance matrix at iteration g can be calculated using the sample covariance matrix of the previously generated samples $\boldsymbol{\theta}^{(1)}, \ldots, \boldsymbol{\theta}^{(g-1)}$. The sample covariance matrix may be a poor estimate when there are only a few samples. This is addressed by defining that the variance–covariance matrix is fixed at $\boldsymbol{\zeta}^{(0)}$ for the first g_0 iterations and taking a combination of the sample covariance matrix and the identity matrix for subsequent iterations.

Algorithm 4 Adaptive Metropolis–Hastings (AM) algorithm

1. Initialize the starting values $\boldsymbol{\theta}^{(0)}$ and $\boldsymbol{\zeta}^{(0)}$, the required MCMC sample size G, burn-in period n_0 and g_0.
2. **for** $g = 1, 2, \ldots$ **do**
3. Sample $\boldsymbol{\theta}' = \boldsymbol{\theta}^{(g)} + \boldsymbol{\epsilon}_g$ where $\boldsymbol{\epsilon}_g \sim \mathrm{N}(0, \boldsymbol{\zeta}^{(g)})$.
4. Let the next iterate be

$$\boldsymbol{\theta}^{(g+1)} = \begin{cases} \boldsymbol{\theta}' & \text{with prob } \alpha_{\boldsymbol{\zeta}^{(g)}}(\boldsymbol{\theta}^{(g)}, \boldsymbol{\theta}'|\mathbf{y}) \\ \boldsymbol{\theta}^{(g)} & \text{with prob } 1 - \alpha_{\boldsymbol{\zeta}^{(g)}}(\boldsymbol{\theta}^{(g)}, \boldsymbol{\theta}'|\mathbf{y}) \end{cases}$$

5. Compute

$$\boldsymbol{\zeta}^{(g+1)} = \begin{cases} \boldsymbol{\zeta}^{(0)} & \text{if } g \le g_0 \\ s_d \frac{1}{g-1}\left[\sum_{j=1}^{g} \boldsymbol{\theta}^{(j)} \boldsymbol{\theta}^{(j)T} - \frac{\left(\sum_{j=1}^{g} \boldsymbol{\theta}^{(j)}\right)\left(\sum_{j=1}^{g} \boldsymbol{\theta}^{(j)}\right)^T}{g}\right] + s_d \epsilon I_d & \text{if } g > g_0 \end{cases}.$$

   where ϵ is a small, positive constant and I_d is the d-dimensional identity matrix.
6. **end for**
7. Return the draws $\boldsymbol{\theta}^{(n_0+1)}, \ldots, \boldsymbol{\theta}^{(n_0+G)}$

It is shown in [28] that the samples come from the target distribution and that I can be approximated by $\hat{I}$. They also described how the covariance matrix can be calculated iteratively to avoid the method becoming computationally expensive. The algorithm was run on Example 7.8 with $\zeta^{(0)} = 1$ and $\epsilon = 0.01$ for 10 000 iterations. Output from the chain is shown in Figure 7.6. The variance converges fairly quickly to the posterior standard deviation.

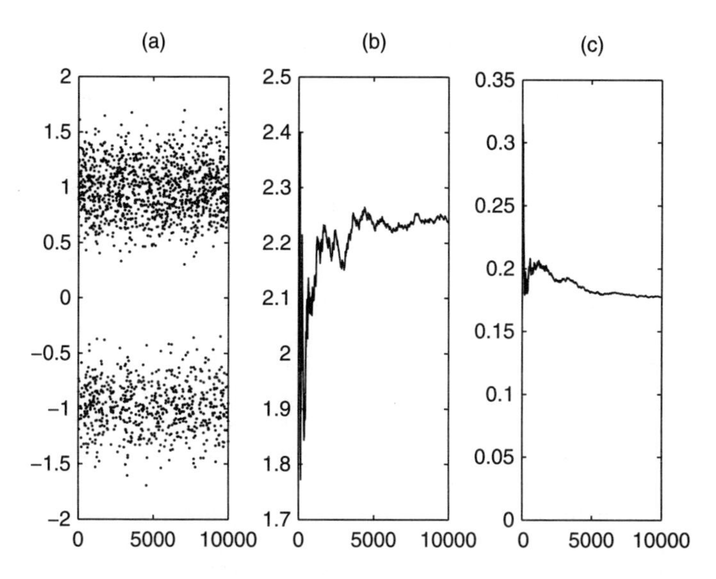

Figure 7.6 Example 7.8 – output for the AM algorithm: (a) trace plot of the samples, (b) trace plot of the standard deviation of the proposal and (c) trace plot of the average acceptance probability.

The average acceptance rate is 0.165 which is close to the optimal value of 0.19 for this example, showing that the rule used by AM works well here.

The idea that the chain should be adapted to achieve a particular average acceptable rate $\bar{\tau}$ is explored in [5]. A standard choice would be $\bar{\tau} = 0.234$, since this has been shown to be optimal in several situations. However, we have already seen in Example 7.8 that other values of $\bar{\tau}$ may lead to a better performing chain. It uses iterations of the Robbins–Monro algorithm from Stochastic Approximation which find the value ζ which solves $N(\zeta) = \mathrm{E}[M(x, \zeta)] = 0$, where x is a random variable whose distribution is unknown and M is a function. In our case $\zeta\, x$ are draws from the target distribution. The function that we wish to target $N(\zeta) = \bar{\alpha} - \bar{\tau}$ where $\bar{\alpha}$ depends on ζ. Finding a zero of this equation implies that we find the ζ for which $\bar{\alpha} = \bar{\tau}$.

Algorithm 5 Adaptive scale Metropolis–Hastings (ASM) algorithm

1. Initialize the starting values $\boldsymbol{\theta}^{(0)}$ and $\boldsymbol{\zeta}^{(0)}$, the required MCMC sample size G and the burn-in period n_0.
2. **for** $g = 1, 2, \ldots,$ **do**
3. Sample $\boldsymbol{\theta}' = \boldsymbol{\theta}^{(g)} + \boldsymbol{\epsilon}_g$ where $\boldsymbol{\epsilon}_g \sim \mathrm{N}(0, \boldsymbol{\zeta}^{(g)})$.
4. Let the next iterate be

$$\boldsymbol{\theta}^{(g+1)} = \begin{cases} \boldsymbol{\theta}' & \text{with prob } \alpha_{\zeta^{(g)}}(\boldsymbol{\theta}^{(g)}, \boldsymbol{\theta}'|y) \\ \boldsymbol{\theta}^{(g)} & \text{with prob } 1 - \alpha_{\zeta^{(g)}}(\boldsymbol{\theta}^{(g)}, \boldsymbol{\theta}'|y) \end{cases}$$

5. Compute

$$\zeta^{(g+1)} = \rho\left(\zeta^{(g)} + w^{(g)}\left(\alpha_{\zeta^{(g)}}(\boldsymbol{\theta}^{(g)}, \boldsymbol{\theta}'|\mathbf{y}) - \bar{\tau}\right)\right).$$

6. **end for**
7. Return the draws $\boldsymbol{\theta}^{(n_0+1)}, \ldots, \boldsymbol{\theta}^{(n_0+G)}$

The function ρ is used to stabilize the algorithm and defines a range $\Delta = \{\sigma : \delta_L \leq \sigma \leq \delta_U\}$ of possible values for ζ. The function ρ is designed to stop ζ moving outside Δ and has the form

$$\rho(\sigma) = \begin{cases} \delta_L \text{ if } \sigma < \delta_L \\ \sigma \text{ if } \sigma \in \Delta \\ \delta_U \text{ if } \sigma > \delta_U. \end{cases}$$

Ideas for relaxing this condition by allowing Δ to grow with the iterations are discussed by [1] and [53]. The sequence $w^{(g)}$ plays a crucial role in defining an algorithm for which ζ converges to the correct value and the chain converges to the correct distribution. [5] shows that this occur if $w^{(g)} = O(g^{-\lambda})$ for some constant $1/2 < \lambda_1 \leq 1$. The ASM algorithm was run on Example 7.8 with $w^{(g)} = g^{-0.7}$, $\zeta^{(0)} = 1$ and $\bar{\tau} = 0.234$ for 10 000 iterations. Output is shown in Figure 7.7. The algorithm converges well to the target average acceptance rate of 0.234 and the associated scale of 1.36. In this case, the choice of 0.234 is sub-optimal and the average squared jumped distance is 0.19, which is smaller than the average squared jumped distance of the AM algorithm, which was 0.24. The 0.234 value arises as a limit in the dimension of the target so it is surprising that it is not sub-optimal in a univariate example. However, this still represents a good value for this problem and supports the general applicability of this value.

The idea of the AM and ASM algorithms can be combined by adapting both the covariance matrix of the proposal to the covariance matrix of the target density (as in the AM) and the scale parameter s_d in the AM algorithm to achieve an average acceptance rate of 0.234 (as in the ASM). This is particularly useful if the target distribution is multivariate, in which case ASM is not directly applicable. This idea is discussed by [2] and [3] and will be referred to as the Adaptive Scaling within

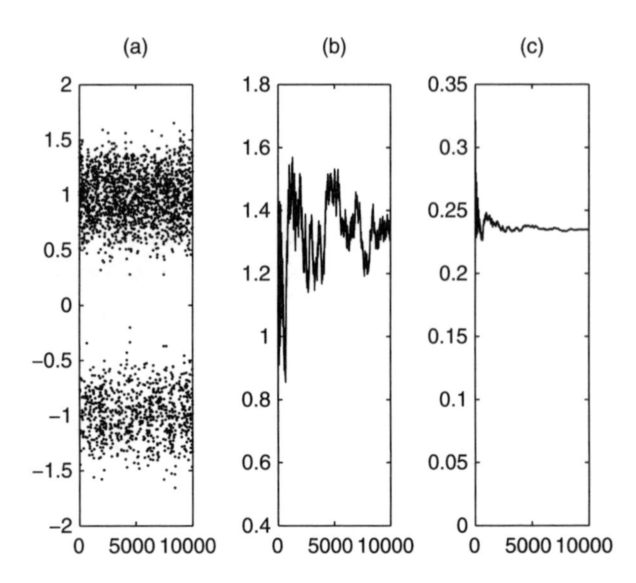

Figure 7.7 Example 7.8 – output for the ASM algorithm: (a) trace plot of the samples, (b) trace plot of the scale of the proposal, and (c) trace plot of the average acceptance probability.

the Adaptive Metropolis–Hastings (ASWAM) algorithm. The scale tuning parameter will now be written $s_d^{(g)}$.

Algorithm 6 Adaptive scaling within the Adaptive Metropolis–Hastings (ASWAM) algorithm

1. Initialize the starting values $\boldsymbol{\theta}^{(0)}$, $\boldsymbol{\zeta}^{(0)}$ and $s_d^{(0)}$, the required MCMC sample size G, burn-in period n_0 and g_0.
2. **for** $g = 1, 2, \ldots$ **do**
3. Sample $\boldsymbol{\theta}' = \boldsymbol{\theta}^{(g)} + \boldsymbol{\epsilon}_g$ where $\boldsymbol{\epsilon}_g \sim \mathrm{N}(0, \boldsymbol{\zeta}^{(g)})$.
4. Let the next iterate be

$$\boldsymbol{\theta}^{(g+1)} = \begin{cases} \boldsymbol{\theta}' & \text{with prob } \alpha_{\boldsymbol{\zeta}^{(g)}}(\boldsymbol{\theta}^{(g)}, \boldsymbol{\theta}'|\mathbf{y}) \\ \boldsymbol{\theta}^{(g)} & \text{with prob } 1 - \alpha_{\boldsymbol{\zeta}^{(g)}}(\boldsymbol{\theta}^{(g)}, \boldsymbol{\theta}'|\mathbf{y}) \end{cases}$$

5. Compute

$$s_d^{(g+1)} = \rho\left(s_d^{(g)} + w^{(g)}\left(\alpha_{\boldsymbol{\zeta}^{(g)}}(\boldsymbol{\theta}^{(g)}, \boldsymbol{\theta}'|\mathbf{y}) - \bar{\tau}\right)\right).$$

and

$$\boldsymbol{\zeta}^{(g+1)} = \begin{cases} \boldsymbol{\zeta}^{(0)} & \text{if } g \leq g_0 \\ s_d^{(g+1)} \frac{1}{g-1}\left[\sum_{j=1}^{g} \boldsymbol{\theta}^{(j)}\boldsymbol{\theta}^{(j)T} - \frac{\left(\sum_{j=1}^{g}\boldsymbol{\theta}^{(j)}\right)\left(\sum_{j=1}^{g}\boldsymbol{\theta}^{(j)}\right)^T}{g}\right] + s_d^{(g+1)}\epsilon I_d & \text{if } g > g_0 \end{cases}.$$

where ϵ is a small, positive constant and I_d is the d-dimensional identity matrix.

6. **end for**
7. Return the draws $\boldsymbol{\theta}^{(n_0+1)}, \ldots, \boldsymbol{\theta}^{(n_0+G)}$

The chain will converge under the same conditions on $w^{(g)}$ as ASM. The ASWAM algorithm was run on Example 7.8 with $\zeta_g = 0.1$, $\epsilon = 0.01$, $w^{(g)} = g^{-0.7}$ and $\tau = 0.234$ for 10 000 iterations. Figure 7.8 shows the output of the ASWAM algorithm in this case. The average acceptance rate converges to the targeted value of 0.234. The scale finishes around 1.5, which is similar to the value in the ASM algorithm (which is not surprising since both algorithms target the same average acceptance rate). The scale factor associated with this is scale is around 1.5 which is far from the 2.4 value used by the AM algorithm. This example illustrates the ability of the algorithm to generate the value of the covariance matrix and the scaling $s_d^{(g)}$ which both converge to the correct values.

The ASWAM algorithm is particularly suited to sampling from multivariate target distributions, since optimal scaling of the covariance matrix often produces a good proposal distribution. [58] suggests an alternative form of adaptation for multivariate target distribution which avoids the two-stage adaptation of ASWAM (adaptation of both the covariance matrix and the scaling parameter). The algorithm is termed the Robust Adaptive Metropolis–Hastings (RAM) algorithm and works directly on an adaptive scale matrix (so directly generalizing [5]).

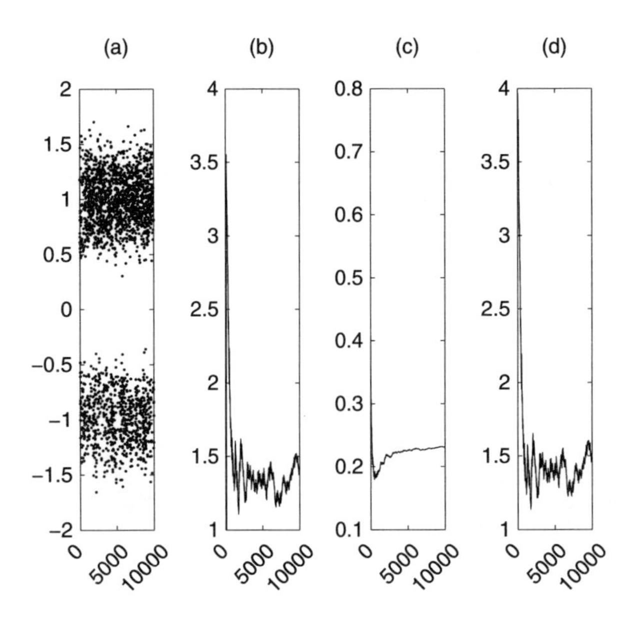

Figure 7.8 Example 7.8 – output for the ASWAM algorithm: (a) trace plot of the samples, (b) trace plot of the scale of the proposal, (d) trace plot of the average acceptance probability, and (d) trace plot of the scale parameter $s_d^{(g)}$.

Algorithm 7 Robust adaptive Metropolis–Hastings (RAM) algorithm

1. Initialize the starting value $\boldsymbol{\theta}^{(0)}$ and $\boldsymbol{\zeta}^{(0)}$, specify the burn-in size n_0 and the required MCMC sample size G.
2. **for** $g = 1, 2, \ldots,$ **do**
3. Sample $\boldsymbol{\theta}' = \boldsymbol{\theta}^{(g)} + \boldsymbol{\zeta}^{(g)}\epsilon_g$ where $\boldsymbol{\epsilon}_g \sim f$, some fixed distribution.
4. Let the next iterate be

$$\boldsymbol{\theta}^{(g+1)} = \begin{cases} \boldsymbol{\theta}' & \text{with prob } \alpha_{\zeta^{(g)}}(\boldsymbol{\theta}^{(g)}, \boldsymbol{\theta}'|y) \\ \boldsymbol{\theta}^{(g)} & \text{with prob } 1 - \alpha_{\zeta^{(g)}}(\boldsymbol{\theta}^{(g)}, \boldsymbol{\theta}'|y) \end{cases}$$

5. Compute

$$\boldsymbol{\zeta}^{(g+1)}\boldsymbol{\zeta}^{(g+1)^T} = \boldsymbol{\zeta}^{(g)}\left(I_d + w^{(g)}(\alpha_{\zeta^{(g)}}(\theta^{(g)}, \theta'|y) - \bar{\tau})\frac{1}{\| \epsilon_g \|^2}\epsilon_g\epsilon_g^T\right)\boldsymbol{\zeta}^{(g)^T}$$

where I_d is the d-dimensional identity matrix
6. **end for**
7. Return the draws $\boldsymbol{\theta}^{(n_0+1)}, \ldots, \boldsymbol{\theta}^{(n_0+G)}$.

---

[58] shows that the weights $w^{(g)}$ must obey the condition that

$$\sum_{g=1}^{\infty} g^{-1}w^{(g)} < \infty \qquad \sum_{g=1}^{\infty} w^{(g)} = \infty$$

for the algorithm to be ergodic. He also shows that the correct average acceptance rate will be targeted if the distribution is elliptically symmetric. This algorithm is very similar to the ASM algorithm for a one-dimension target and so we don't produce output from Example 7.8 with this algorithm.

Further examples

The example of a univariate mixture distribution is potentially challenging but simulation from multivariate target distributions is often needed in fitting Bayesian models. We will consider two examples: a bivariate mixture distribution and the posterior distribution of a logistic regression model.

Example 7.9 Bivariate mixture distribution
The univariate mixture distribution of Example 7.8 is extended to a bivariate mixture distribution which has a density of the form

$$\pi(\boldsymbol{\theta}) = 0.3\,\mathrm{N}_2\left(\boldsymbol{\theta}\,\middle|\,\begin{pmatrix}-1\\-1\end{pmatrix}, \begin{pmatrix}0.2^2 & 0\\ 0 & 0.2^2\end{pmatrix}\right) + 0.7\,\mathrm{N}_2\left(\boldsymbol{\theta}\,\middle|\,\begin{pmatrix}1\\1\end{pmatrix}, \begin{pmatrix}0.2^2 & 0\\ 0 & 0.2^2\end{pmatrix}\right)$$

where $\mathrm{N}_d(x|\boldsymbol{\mu}, \boldsymbol{\Sigma})$ is a d-dimensional multivariate normal distribution with mean $\boldsymbol{\mu}$ and variance–covariance matrix $\boldsymbol{\Sigma}$. The three algorithms suitable for multivariate target distributions (AM, ASWAM and RAM) were run for 10 000 iterations on this example. The initial setting used for ASWAM on Example 7.8 were used in this case. The RAM algorithm was run with $w^{(g)} = g^{-0.7}$. All algorithms performed well and converged to the correct distribution. Some results are reported in Table 7.1. All three algorithms lead to very similar average squared jumped distance. The average acceptance rate is 0.31, which is bigger than the target value of 0.234 used by both ASWAM and RAM algorithms. This illustrates the usefulness of all three algorithms for this example. This suggests that larger average squared jumped distances would be possible by targeting an acceptance rate between 0.24 and 0.31.

Example 7.10 Logistic regression model
The logistic regression model assumes that a binary response $y_i \in \{0, 1\}$ can be related to a p-dimensional vector of regressors $\mathbf{x}_i$ using an intercept α and a p-dimensional parameter vector of regression coefficients $\boldsymbol{\beta}$ by

$$y_i \sim \mathrm{Ber}(p_i), \qquad p_i = \frac{\exp\{\eta_i\}}{1 + \exp\{\eta_i\}}, \qquad \eta_i = \alpha + \mathbf{x}_i\boldsymbol{\beta}$$

where $\mathrm{Ber}(p)$ represents the Bernoulli distribution with probability p. We assume that each regressor has been centred and scaled to have sample variance 1. The model is completed by assuming a prior for the intercept α and regression coefficients $\boldsymbol{\beta}$. For illustration purposes, we will use $\alpha \sim \mathrm{N}(0, 1000^2)$ and $\boldsymbol{\beta} \sim \mathrm{N}(0, 10^2\, I_p)$. The target is the $(p + 1)$-dimensional posterior distribution and the problem is challenging for MCMC methods, since the full conditional distributions of α and $\boldsymbol{\beta}$ do not have a known form. The posterior distribution of α and β can also be highly

Table 7.1 Example 7.9: The average acceptance rate ($\bar{\alpha}$) and the average squared jumped distance ($\bar{J}$) for the AM, ASWAM and RAM algorithms using 10 000 iterations.

	AM	ASWAM	RAM
$\bar{\alpha}$	0.31	0.24	0.25
$\bar{J}$	0.036	0.036	0.035

Table 7.2 Example 7.10: The average acceptance rate ($\bar{\alpha}$) and the average squared jumped distance ($\bar{J}$) for the AM, ASWAM and RAM algorithms using 100 000 iterations.

	AM	ASWAM	RAM
$\bar{\alpha}$	0.24	0.23	0.21
$\bar{J}$	0.022	0.022	0.020

correlated, which suggests that a good algorithm will update α and β jointly. We will consider using the AM, ASWAM and RAM algorithms on this target distribution. The model was fitted to data on diabetes in Pima indians. This is a widely used dataset which is available in the MASS package of the statistical software program R. The response is the presence or absence of diabetes in a sample of 532 subjects. There are seven risk factors for diabetes which are used as the regressors in the logistic regression model. Further details are available from the MASS package. The results of running the three algorithms using the same setting as Example 7.9 are shown in Table 7.2. The average acceptance rates and average squared jumped distance are very similar across the three algorithms. This again shows the ability of these algorithms to automatically adapt to the target distribution

Adaptive Metropolis–Hastings independence samplers

In Metropolis–Hastings independence samplers, the proposal distribution is taken to be a parametric family of distributions such as a normal or a t-distribution. The Metropolis–Hastings algorithm will perform well if the proposal distribution is a good approximation to the target distribution. An adaptive version of the Metropolis–Hastings independence sampler was proposed by [11]. An iterative scheme is described for approximating the posterior distribution by estimating the parameters of the chosen family of distribution (such as the mean and variance of a normal distribution). The k-th iteration of the algorithm runs a Metropolis–Hastings independence sampler for G iterations with the approximation used as the proposal distribution. The approximations should improve as more approximations are calculated and a condition is described to decide when an approximation is 'good enough'. The MCMC sample to be used is generated conditional on a final approximation and so the algorithm has finite adaptation.

The quality of the approximation in the tails is important for the behaviour of the chain. [42] show that the sampler can only have geometric ergodicity if the tails of the proposal are at a slower rate than the tails of the target distribution. This suggests that the approximation should be flexible. A number of authors have considered a mixture of normal distributions as a flexible method for approximating the target distribution. [1] developing the idea of [11] into an infinitely adaptive algorithm where the Kullback–Leibler distance between the target distribution and the approximation is minimized using an EM algorithm. [24] also use a mixture of normal distributions for their target, but use the k-harmonic mean method for estimating the mixture of normals approximation as a faster alternative to the EM algorithm.

Adaptive Metropolis–Hastings random walk algorithms with multimodal targets

Adaptive MCMC algorithms provide a method for finding the optimal proposal distribution from a class of proposal distributions. However, they do not guarantee that the optimal proposal distribution will have good mixing properties. This is a particular problem if the target distribution is multi-modal. It is natural to ask the question, whether alternative methods can be developed which avoid this poor mixing.

One potential design for algorithms partitions the support of the target distribution into disjoint regions $\mathcal{S}_1, \ldots, \mathcal{S}_K$ for which $\cup_{i=1}^K \mathcal{S}_i = \theta$, and allow the proposal distributions to depend on the region containing $\boldsymbol{\theta}^{(g)}$. [9] introduced the Regional Adaptation (RAPT) algorithm and assumed that the target distribution in each region can be well-approximated by a normal distribution. It follows that the overall target distribution is approximated by a mixture of normal distributions. Their preferred algorithm is the Mixed RAPT algorithm which uses the following proposal

$$q(\boldsymbol{\theta}^{(g)}, \boldsymbol{\theta}'|y) = (1-\beta)\sum_{k=1}^{K} \lambda_k^{(i)} q_k(\boldsymbol{\theta}^{(g)}, \boldsymbol{\theta}'|y) + \beta q_{\text{whole}}(\boldsymbol{\theta}^{(g)}, \boldsymbol{\theta}'|y)$$

when $\boldsymbol{\theta}^{(g)}$ is in region $\mathcal{S}_i$ and $0 < \beta < 1$ is fixed. They allow adaptation of the $\lambda_k^{(i)}$s, the q_ks and q_{whole}. The q_ks are centred at $\boldsymbol{\theta}^{(g)}$ and have a covariance matrix which is $(2.4)^2/d$ times the covariance matrix of the $\boldsymbol{\theta}^{(1)}, \ldots, \boldsymbol{\theta}^{(g)}$ which fall in region $\mathcal{S}_k$. The $\lambda_j^{(t)}$ are adapted in the following way

$$\lambda_j^{(i)} = \begin{cases} \dfrac{d_j^{(i)}}{\sum_{k=1}^{K} d_k^{(i)}} & \text{if } \sum_{k=1}^{K} d_k^{(i)} > 0 \\ 1/2 & \text{otherwise} \end{cases}$$

where $d_k^{(i)}$ represents the average squared jumped distance sampled (over the current samples) from q_k when the chain was in region $\mathcal{S}_i$. The q_{whole} proposal is updated using the whole sample in the same way as the AM algorithm. This is included to provide some stability in the algorithm. [9] show the ergodicity of this algorithm. The Mixed RAPT algorithm assumes that the regions $\mathcal{S}_1, \ldots, \mathcal{S}_K$ are known before simulation (or through a pilot MCMC run) which limits the applicability of the method for general Bayesian inference. [6] extend this idea to the RAPT with Online Recursion (RAPTOR) algorithm. This allows the regions $\mathcal{S}_1, \ldots, \mathcal{S}_K$ to also adapt in the algorithm and uses a different method for estimation of the covariance matrix in each region.

7.4.4 Construction of adaptive algorithms

We have already seen a number of adaptive algorithms which have been shown to converge. Convergence results for the initially developed algorithms were proved on a case-by-case basis. However, it is natural to ask whether there are general conditions which show that adaptive algorithms work in general. There are two issues here:

(1) What criteria should the adaptive parameter $\boldsymbol{\zeta}$ be targeting? And how can we guarantee the sequence $\boldsymbol{\zeta}^{(1)}, \boldsymbol{\zeta}^{(2)}, \ldots$ will converge to the optimal value for a generated chain?
(2) Will the chain $\theta^{(1)}, \theta^{(2)}, \ldots$ be ergodic, i.e. will $\bar{I}$ converge to I as $G \to \infty$?

Answering these questions has been the central concern in the field. The lack of a Markovian structure means that standard MCMC theory does not apply to this class of methods and so (2) is hard to answer in general. Recently, there has been some success in defining general principles for checking (1) and (2). This section will not give a detailed description of these results but rather illustrate when these general principles can be applied to the algorithms in the previous section and how the general ideas can be applied to develop novel algorithms.

Question (1) can generally be answered using ideas from stochastic approximation. Most conditions discussed above can be expressed as expectations with respect to the target distribution (such as the average acceptance probability). Stochastic approximation concerns the developing of iterative methods for solving problems which consist of expectations with respect to a distribution which is only available through sampling and some tuning parameters. In particular, the Robbins–Monro algorithm has proved to be an important method which underlies both the AM and ASM algorithm [2].

An important condition for establishing (2) is the idea of *diminishing adaptation* [1, 51]. Intuitively, we want the difference in the adaptive parameter $\| \boldsymbol{\zeta}^{(j+1)} - \boldsymbol{\zeta}^{(j)} \|$ to become smaller as j tends to infinitely. Therefore, $\boldsymbol{\zeta}^{(j)}$ will 'settle down' to its optimal value. All the algorithms already described fit into this framework. The AM algorithm uses

$$\boldsymbol{\zeta}^{(g+1)} = \boldsymbol{\zeta}^{(g)} + \frac{1}{g-1}\left[\boldsymbol{\theta}^{(g)}\boldsymbol{\theta}^{(g)T} + \frac{\left(\sum_{j=1}^{g-1}\boldsymbol{\theta}^{(j)}\right)\left(\sum_{j=1}^{g-1}\boldsymbol{\theta}^{(j)}\right)^T}{g-1} - \frac{\left(\sum_{j=1}^{g}\boldsymbol{\theta}^{(j)}\right)\left(\sum_{j=1}^{g}\boldsymbol{\theta}^{(j)}\right)^T}{g} - \boldsymbol{\zeta}^{(g)}\right]$$

and the increments of $\boldsymbol{\zeta}^{(g+1)}$ are clearly getting smaller with g. The ASM algorithm uses

$$\zeta^{(g+1)} = \zeta^{(g)} + w^{(g)}(\alpha(\boldsymbol{\theta}^{(g)}, \boldsymbol{\theta}'|\mathbf{y}) - \bar{\tau})$$

since $0 < \alpha(\boldsymbol{\theta}^{(g)}, \boldsymbol{\theta}'|\mathbf{y}) < 1$ and $\bar{\tau}$ is fixed then the range of increments gets smaller with g if $w^{(g)}$ is a decreasing sequence (which leads to the condition that $w^{(g)}$ should be $O(g^{-\lambda})$ for $\lambda \in (1/2, 1]$). The RAM algorithm has a similar recursion to the ASM algorithm and has diminishing adaptation using similar arguments. The ASWAM algorithm combines the AM and ASM algorithms and so also has diminishing adaptation. Notice that many estimates calculated using the first g iterations of the chain will have diminishing adaptation (in a similar way to the AM algorithm). Therefore, any method which attempts to estimate the target distribution for the chain will have diminishing adaptation. The diminishing adaptation idea was made exact by [1] who assumed that $\| \boldsymbol{\zeta}^{(g+1)} - \boldsymbol{\zeta}^{g)} \|$ converges to zero. [51] introduced the weaker condition that

$$\lim_{g\to\infty} \sup_{\theta\in\Theta} \| Q_{\boldsymbol{\zeta}^{(g+1)}}(\boldsymbol{\theta}, \boldsymbol{\theta}'|\mathbf{y}) - Q_{\boldsymbol{\zeta}^{(g)}}(\boldsymbol{\theta}, \boldsymbol{\theta}'|\mathbf{y}) \|_{TV} = 0, \qquad \text{in probability}$$

where Θ is the state space of the target distribution. Notice that this condition does not assume that $\zeta^{(g)}$ will converge.

Once diminishing adaptation has been established, ergodicity can usually be proved, assuming that the chain has uniform behaviour for all possible MCMC kernels. For example, [51] show ergodicity under the assumption of 'simultaneous uniform ergodicity', which states that for all $\epsilon > 0$, there exists an $N(\epsilon)$ for which

$$\| Q_{\zeta}^{g}(\boldsymbol{\theta}, \boldsymbol{\theta}'|\mathbf{y}) - \pi(\boldsymbol{\theta}') \|_{TV} \leq \epsilon$$

for all $\theta \in \Theta$ and all possible values of ζ. Here, $Q_{\zeta}^{g}(\boldsymbol{\theta}, \boldsymbol{\theta}'|\mathbf{y})$ represents the distribution of the chain after g iterations starting at $\boldsymbol{\theta}$. This says that the chain should be converging at 'similar rates' for any starting point. This condition may be difficult to check in practice. They also show that the chain is ergodic if Θ and ζ are finite and that the chain is ergodic for all ζ. This is a looser condition on convergence and allows this result to be applied to most standard MCMC algorithms on finite spaces. They also show that if $q_\zeta(\boldsymbol{\theta}, \boldsymbol{\theta}'|\mathbf{y})$ is uniformly bounded on a space with finite measure and that it is continuous with respect to $\boldsymbol{\theta}$ and ζ, then the chain is ergodic. This covers many MCMC algorithms under the condition that the state space is finite (in practice, we can assume that the chain lives on a large compact space).

7.4.5 Metropolis-within-Gibbs

Much adaptive MCMC theory has been concerned with direct use of Metropolis–Hasting algorithms on a target distribution. This ignores a large area of application for Metropolis–Hastings algorithms in applied Bayesian work—as methods for sampling from full conditional distribution in a Gibbs sampler which have forms that cannot be easily sampled directly. This use of Metropolis–Hastings methods leads to Metropolis-within-Gibbs algorithms and it is interesting to ask whether these Metropolis–Hastings samplers can also be made adaptive and whether the limit theory extends to this more elaborate use. [29] provided initial work in this direction, introducing the Single Component Adaptive Metropolis (SCAM) algorithm. Suppose that our target distribution is d-dimensional, then a Metropolis-within-Gibbs scheme could be used to update each dimension consecutively using a Metropolis–Hastings random walk step. The SCAM algorithm is based on this scheme with the scale of the random walk in each dimension updated using the AM algorithm (Algorithm 1), and so defines a d-dimensional vector of variances $\boldsymbol{\zeta}^{(g)}$ which will be used in the update of each full conditional distribution at the g-th iteration.

Algorithm 8 Single component adaptive Metropolis (SCAM) algorithm.

1. Initialize the starting value $\boldsymbol{\theta}^{(0)}$ and $\boldsymbol{\zeta}^{(0)}$, the required MCMC sample size G, burn-in period n_0 and g_0.
2. **for** $g = 1, 2, \ldots,$ **do**
3.   **for** Component $j = 1, \ldots, d$ **do**
4.   Sample $\boldsymbol{\theta}'_j = \boldsymbol{\theta}_j^{(g)} + \boldsymbol{\epsilon}_{g,j}$ where $\boldsymbol{\epsilon}_{g,j} \sim \mathrm{N}(0, \boldsymbol{\zeta}_j^{(g)})$.
5.   Let the next iterate be

$$\boldsymbol{\theta}_j^{(g+1)} = \begin{cases} \boldsymbol{\theta}'_j & \text{with prob } \alpha^{(j)}_{\boldsymbol{\zeta}^{(g)}}(\boldsymbol{\theta}_j^{(g)}, \boldsymbol{\theta}'_j|\mathbf{y}) \\ \boldsymbol{\theta}_j^{(g)} & \text{with prob } 1 - \alpha^{(j)}_{\boldsymbol{\zeta}_j^{(g)}}(\boldsymbol{\theta}^{(g)}, \boldsymbol{\theta}'_j|\mathbf{y}) \end{cases}$$

6.   Compute

$$\boldsymbol{\zeta}_j^{(g+1)} = \begin{cases} \boldsymbol{\zeta}_j^{(0)} & \text{if } g \le g_0 \\ s_d \frac{1}{g-1}\left[\sum_{k=1}^{g} \boldsymbol{\theta}_j^{(k)} \boldsymbol{\theta}_j^{(k)T} - \frac{\left(\sum_{k=1}^{g} \boldsymbol{\theta}_j^{(k)}\right)\left(\sum_{k=1}^{g} \boldsymbol{\theta}_j^{(k)}\right)^T}{g}\right] + s_d \epsilon & \text{if } g > g_0 \end{cases}$$

7.   where ϵ is a small, positive constant.
8.   **end for**
9. **end for**
10. Return the draws $\boldsymbol{\theta}^{(n_0+1)}, \ldots, \boldsymbol{\theta}^{(n_0+G)}$

Here, $\alpha^{(j)}_{\boldsymbol{\zeta}^{(g)}}(\boldsymbol{\theta}_j^{(g)}, \boldsymbol{\theta}'_j|\mathbf{y})$ represents the acceptance probability of a Metropolis–Hastings algorithm applied to the j-th full conditional of the target distribution. The ergodicity of this algorithm is established by [29] under the same conditions as the AM algorithm.

Table 7.3 Example 7.9: The average acceptance rate ($\bar{\alpha}$) and the average squared jumped distance ($\bar{J}$) for the SCAM, ASM-SCAM and ASWAM-SCAM algorithms using 100 000 iterations.

	SCAM	ASM-SCAM	ASWAM-SCAM
$\bar{\alpha}_1$	0.41	0.23	0.24
$\bar{\alpha}_2$	0.35	0.23	0.23
$\bar{\alpha}_3$	0.42	0.23	0.24
$\bar{\alpha}_4$	0.40	0.23	0.24
$\bar{\alpha}_5$	0.36	0.23	0.24
$\bar{\alpha}_6$	0.35	0.23	0.24
$\bar{\alpha}_7$	0.43	0.23	0.24
$\bar{\alpha}_8$	0.34	0.23	0.24
$\bar{J}$	0.083	0.063	0.063

The basic approach of the SCAM algorithm is to adapt each Metropolis random walk step in a Gibbs sampler. This suggests that replacing the AM adaptation with a different adaptation scheme will not effect ergodicity. [52] suggest replacing the AM algorithm by an ASM algorithm in the SCAM algorithm and discuss its convergence. Similar arguments could also be made for the ASWAM and RAM algorithms. The SCAM algorithm can also be extended to samplers where some full conditional distributions are multivariate. This allows more sophisticated Gibbs samplers with blocking scheme to be made adaptive. Lastly, ergodicity will not be effected if some full conditional distributions are not updated adaptively using a Metropolis–Hastings step but can be directly sampled.

Example

We return to the logistic regression example (Example 7.10) and use the SCAM algorithm and versions using the ASM algorithm (ASM-SCAM) and the ASWAM algorithm (ASWAM-SCAM) in place of the AM algorithm for each dimension of the target distribution. The algorithms were implemented using the same algorithmic parameters as Example 7.8. Table 7.3 shows some results from running the algorithms for 100 000 iterations. All algorithms converge to the correct posterior distribution with the ASM-SCAM and ASWAM-SCAM converging to the correct average acceptance rate for each full conditional distribution. The average squared jumped distance is larger for SCAM than for the variation of SCAM that target the average acceptance probability. The average jumped distances are quite a bit larger than for the joint updating of the regression coefficients.

7.4.6 Other forms of adaptation

We have concentrated on an adaptation algorithm where the target distribution is defined on a subset of $\mathbb{R}^d$ and the scale (in a random walk proposal) or the proposal distribution (in the Metropolis–Hasings independence sampler) is adapted. Other forms of adaptation have been proposed in the literature. [37] considered adapting a random scan Gibbs sampler. In a random scan Gibbs sampler, one full conditional distribution is chosen to be updated at each iteration and is chosen at random. Standard implementations of random scan Gibbs sampling choose the full conditional distributions uniformly at random. [37] considered adapting these probabilities and showed that the chain can be ergodic under a wide range of updating schemes for the probabilities and Gibbs samplers.

In regression models, an important problem is variable selection where it assumed that only a subset of the regressors is needed to accurately predict the response. Each different subset of the regressors can be considered to be a model for the data. A standard Bayesian model for this problem assumes a prior on the space of all possible models and a prior on the regression coefficients given the model. The space of all models is a lattice, since each regressor is either included in the model or excluded. Therefore, an MCMC chain for this model will run on a lattice (if the regression coefficients can be marginalized from the posterior analytically) or on the union of subsets of $\mathbb{R}^{p+1}$ (where p is the number of regressors). For simplicity, we consider the case where the regression coefficients can be marginalized from the model. A standard Metropolis–Hastings algorithm [see e.g. 16] proposes to either add a variable to the model, remove a variable from the model, or combines those two moves (an addition and a deletion in the same move). This is markedly different to the Euclidean spaces considered previously. Several authors have proposed algorithms for this type of posterior distribution. [45] considered adapting the probabilities of including and excluding variables to the marginal posterior probability that a particular variable is included in the model. More recently, [35] extended the standard proposal in variable selection to allow more variables to be added, removed or swapped at each iteration and found that tuning this sampler to have an average acceptance rate of 0.234 leads to near optimal performance in the regression problems with many regressors. [36] considered automating the tuning of their algorithm by extending the ASM algorithm. They found that this leads to efficient methods for variable selection in a variety of Generalized Linear Models.

7.4.7 Summary

Adaptive Markov chain Monte Carlo methods have become an important tool in the Bayesian statistician's toolbox. They avoid the need to manually tune random walk Metropolis–Hastings algorithms or to carefully design Metropolis–Hastings independence samplers. This can be extremely challenging if the target distribution is not low-dimensional or multiple Metropolis–Hastings steps sitting inside a Gibbs sampler. Following the seminal work of [28], a general framework has been developed for verifying that an adaptive algorithm will converge correctly based around the idea of diminishing adaptation. This framework allows the development of algorithms with multiple forms of adaptation (such as the ASWAM and RAPT algorithms described in this chapter). This allows statisticians flexibility in designing sampling schemes and should allow these methods to be used creatively in the same way as other MCMC methods. The Grapham software [57] allows simulation from a hierarchical model using a Metropolis-within-Gibbs framework with a range of adaptive MCMC algorithms. Further work is still needed to apply these methods in posterior simulation of complicated Bayesian models.

There are still plenty of future directions for application of these methods. This chapter has concentrated on simulation in Euclidean spaces with algorithms which target an average acceptance rate of 0.234. An alternative approach would use directly the average squared jumped distance (or some other measure of the mixing of the chain) to find the optimal proposal. Initial work in the direction of using the average squared jumped distance is described by [46]. Work on non-Euclidean spaces has largely been restricted to variable selection problems where the chain works on a lattice. Other problems naturally lead themselves to adaptive algorithms. [23] consider learning about the location of structural breaks or outliers in dynamic models.

7.5 Chapter summary and discussion

In this chapter, we have outlined the key theoretical, methodological and algorithmic developments that sprang from the pioneering work on the application of MCMC to Bayesian statistical inference. The first and second topics, reversible jump and population methods, rely on application of the standard theory of MCMC, but apply the standard algorithms in an innovative fashion, by augmenting the usual Markov chain state space to include auxiliary variables that facilitate effective exploration of the posterior. The third topic, adaptive MCMC, required extension of the standard theory, but allowed for the construction of ergodic chains. All three approaches rely on more expertise on behalf of the algorithm developer than the standard algorithms, and are not completely automatic, however, typically population and adaptive algorithms are relatively easy to tune.

It is undoubtedly the case that the approaches discussed in this chapter equip the Bayesian statistician with MCMC tools that greatly enhance the chances of effective computational inference on a wide range of problems. The population MCMC methods from Section 7.3 dramatically increase the computational burden, but we feel that multi-chain methods are the most reliable way to ensure adequate posterior exploration. Adaptation can be achieved relatively cheaply, but as demonstrated in the examples from Section 7.4, it is often extremely advantageous. The conditions outlined in 7.4.4 are often straightforward to achieve and check, rendering adaptive MCMC a very promising approach that will become increasingly important in future applied work.

References

[1] Andrieu, C. and Moulines, E. (2006). On the ergodicity properties of some adaptive MCMC algorithms. *Ann. Appl. Prob.*, **16**, 1462–1505.

[2] Andrieu, C. and Thoms, J. (2008). A tutorial on adaptive MCMC. *Stat. Comp.*, **18**, 343–373.

[3] Atchadé, Y. and Fort, G. (2010). Limit theorems for some adaptive MCMC algorithms with subgeometric kernels. *Bernoulli*, **16**, 116–154.

[4] Atchadé, Y. F., Roberts, G. O. and Rosenthal, J. S. (2011). Towards optimal scaling of Metropolis-coupled Markov chain Monte Carlo. *Stat. Comp.*, **21**(4), 555–568.

[5] Atchadé, Y. F. and Rosenthal, J. S. (2005). On adaptive Markov chain Monte Carlo algorithms. *Bernoulli*, **11**, 815–828.

[6] Bai, Y., Craiu, R. V. and Di Narzo, A. F. (2011). Divide and conquer: A mixture-based approach to regional adaptation for MCMC. *J. Comp. Graph. Stat.*, **20**, 63–79.

[7] Behrens, G., Friel, N. and Hurn, M. (2012). Tuning tempered transitions. *Stat. Comp.*, **22**, 65–78.

[8] Carlin, B. P. and Chib, S. (1995). Bayesian model choice via Markov chain Monte Carlo. *J. R. Statist. Soc. B*, **57**, 473–484.

[9] Craiu, R. V., Rosenthal, J. S. and Yang, C. (2009). Learn from thy neighbor: Parallel-chain nad regional adaptive MCMC. *J. Amer. Statist. Assoc.*, **104**, 1454–1466.

[10] Denison, D. G. T., Holmes, C. C., Mallick, B. K. and Smith, A. F. M. (2002). *Bayesian Methods for Nonlinear Classification and Regression.* John Wiley.

[11] Gasemyr, J. (2003). An adaptive version of the Metropolis–Hastings algorithm with independent proposal distribution. *Scand. J. Stat.*, **30**, 159–173.

[12] Gelfand, A. E. and Smith, A. F. M. (1990). Sampling based approaches to calculating marginal densities. *J. Amer. Statist. Assoc.*, **85**, 398–409.

[13] Gelman, A. and Meng, X.-L. (1998). Simulating normalizing constants: From importance sampling to bridge sampling to path sampling. *Statistical Science*, **13**(2), 163–185.

[14] Gelman, A., Roberts, G. O. and Gilks, W. R. (1996). Efficient in Metropolis jumping rules. In *Bayesian Statistics 5* (ed. A. P. David, J. O. Berger, J. M. Bernardo and A. F. M. Smith), pp. 599–608. Oxford University Press, New York.

[15] Geman, S. and Geman, D. (1984). Stochastic relaxation, Gibbs distributions and the Bayesian restoration of images. *IEEE Trans. Patt. Anal. Mach. Intell.*, **6**, 721–741.

[16] George, E. I. and McCulloch, R. E. (1997). *Approaches for Bayesian variable selection*, Volume 7.

[17] Geyer, C. J. (1991). Monte Carlo maximum likelihood for dependent data. In *Comp. Sci. and Statis.: Proc. 23rd Symp. Interface*, (ed. E. Keramidas), pp. 156–163.

[18] Geyer, C. J. and Møller, J. (1994). Simulation procedures and likelihood inference for spatial point processes. *Scand. J. Stat.*, **21**(4), pp. 359–373.

[19] Geyer, C. J. and Thompson., E. A. (1995). Annealing Markov chain Monte Carlo with applications to ancestral inference. *J. Amer. Statist. Assoc.*, **90**, 909–920.

[20] Gilks, W. R., Richardson, S. and Spiegelhalter, D. J. (1995). *Markov Chain Monte Carlo in Practice: Interdisciplinary Statistics (Chapman & Hall/CRC Interdisciplinary Statistics)*. Chapman and Hall/CRC.

[21] Gilks, W. R., Roberts, G. O. and George, E. I. (1994). Adaptive direction sampling. *J. R. Statist. Soc. D (The Statistician)*, **43**(1), pp. 179–189.

[22] Gilks, W. R., Roberts, G. O. and Sahu, S. K. (1998). Adaptive Markov chain Monte Carlo through regeneration. *J. Amer. Statist. Assoc.*, **93**, 1045–1054.

[23] Giordani, P. and Kohn, R. (2008). Efficient Bayesian inference for multiple change-point and mixture innovation models. *J. Bus. Econ. Stat.*, **26**, 66–77.

[24] Giordani, P. and Kohn, R. (2010). Adaptive independent Metropolis–Hastings by fast estimation of mixtures of normals. *J. Comp. Graph. Stat.*, **19**, 243–259.

[25] Goswami, G. and Liu, J. S. (2007). On learning strategies for evolutionary Monte Carlo. *Stat. Comp.*, **17**(1), 23–38.

[26] Green, P. J. (1995). Reversible jump Markov chain Monte Carlo computation and Bayesian model determination. *Biometrika*, **82**(4), 711–732.

[27] Grenander, U. and Miller, M. (1994). Representations of knowledge in complex systems (with discussion). *J. R. Statist. Soc. B*, **56**(4), 549–603.

[28] Haario, H., Saksman, E. and Tamminen, J. (2001). An adaptive Metropolis algorithm. *Bernoulli*, **7**, 223–242.

[29] Haario, H., Saksman, E. and Tamminen, J. (2005). Componentwise adaptation for high dimensional MCMC. *Computat. Stat.*, **20**, 265–273.

[30] Jasra, A., Holmes, C. C. and Stephens, D. A. (2005). MCMC and the label switching problem in Bayesian mixture models. *Statistical Science*, **20**, 50–67.

[31] Jasra, A., Stephens, D. A. and Holmes, C. C. (2007). On population-based simulation for static inference. *Stat. Comp.*, **17**, 263–279. 10.1007/s11222-007-9028-9.

[32] Jasra, A., Stephens, D. A. and Holmes, C. C. (2007). Population-based reversible jump Markov chain Monte Carlo. *Biometrika*, **94**(4), 787–807.

[33] Kirkpatrick, S., Jr., Gelatt, C. D. and Vecchi, M. P. (1983). Optimization by simulated annealing. *Science*, **220**, 671–680.

[34] Kou, S. C., Zhou, Q. and Wong, W. H. (2006). Equi-energy sampler with applications in statistical inference and statistical mechanics. *Ann. Stat.*, **34**(4), pp. 1581–1619.

[35] Lamnisos, D., Griffin, J. E. and Steel, M. F. J. (2009). Transdimensional sampling algorithms for Bayesian variable selection in classification problems with many more variables than observations. *J. Comp. Graph. Stat.*, **18**, 592–612.

[36] Lamnisos, D., Griffin, J. E. and Steel, M. F. J. (2011). Adaptive Monte Carlo for Bayesian variable selection in regression models. *Technical Report*. University of Warwick.

[37] Latuszynski, K., Robert, G. O. and Rosenthal, J. S. (2012). Adaptive Gibbs samplers and related MCMC methods. *Ann. Appl. Prob.* (to appear).

[38] Liang, F., Liu, C. and Carroll, R. J. (2010). *Advanced Markov Chain Monte Carlo Methods: Learning from Past Samples.* Wiley Series in Computational Statistics. John Wiley & Sons.

[39] Liang, F. M. and Wong, W. H. (2001). Real-parameter evolutionary Monte Carlo with applications to Bayesian mixture models. *J. Amer. Statist. Assoc.*, **96**(454), 653–666.

[40] Liu, J. S., Liang, F. and Wong, W. H. (2000). The multiple-try method and local optimization in Metropolis sampling. *J. Amer. Statist. Assoc.*, **95**(449), 121–134.

[41] Marinari, E. and Parisi, G. (1992). Simulated Tempering: A New Monte Carlo Scheme. *Europhys. Lett.*, **19**(6), 451.

[42] Mengersen, K. L. and Tweedie, R. L. (1996). Rates of convergence of the Hastings and Metropolis algorithms. *Ann. Stat.*, **24**, 101–121.

[43] Neal, R. M. (1996). Sampling from multimodal distributions using tempered transitions. *Stat. Comp.*, **6**, 353–366. 10.1007/BF00143556.

[44] Neal, R. M. (2001). Annealed importance sampling. *Stat. Comp.*, **11**(2), 125–139.

[45] Nott, D. J. and Kohn, R. (2005). Adaptive sampling for Bayesian variable selection. *Biometrika*, **92**, 747–763.

[46] Pasarica, C. and Gelman, A. (2010). Adaptively scaling the Metropolis algorithm using expected squared jumped distance. *Stat. Sin.*, **20**, 343–364.

[47] Phillips, D. B. and Smith, A. F. M. (1995). Bayesian model comparison via jump diffusions. In *Markov Chain Monte Carlo in Practice*, (ed. W. R. Gilks, S. R. Richardson, and D. J. Spiegelhalter). Chapman and Hall, London.

[48] Richardson, S. and Green, P. J. (1997). On Bayesian Analysis of Mixtures with an Unknown Number of Components (with discussion). *J. R. Statist. Soc. B*, **59**(4), 731–792.

[49] Roberts, G. O., Gelman, A. and Gilks, W. R. (1997). Weak convergence and optimal scaling of random walk Metropolis algorithms. *Ann. Appl. Prob.*, **7**, 110–120.

[50] Roberts, G. O. and Rosenthal, J. S. (2001). Optimal scaling for various Metropolis–Hastings algorithms. *Statist. Sci.*, **16**, 351–367.

[51] Roberts, G. O. and Rosenthal, J. S. (2007). Coupling and ergodicity of adaptive MCMC. *J. Appl. Prob.*, **44**, 458–475.

[52] Roberts, G. O. and Rosenthal, J. S. (2009). Examples of adaptive MCMC. *J. Comp. Graph. Stat.*, **18**, 349–367.

[53] Saksman, E. and Vihola, M. (2010). On the ergodicity of the adaptive Metropolis algorithm on unbounded domains. *Ann. Appl. Prob.*, **20**, 2178–2203.

[54] Stephens, M. (2000). Bayesian analysis of mixture models with an unknown number of components - an alternative to reversible jump methods. *Ann. Stat.*, **28**(1), 40–74.

[55] Tierney, L. (1994). Markov chains for exploring posterior distributions. *Ann. Stat.*, **22**, 1701–1762.

[56] Tierney, L. (1998). A note on Metropolis–Hastings kernels for general state spaces. *Ann. Appl. Prob.*, **8**(1), pp. 1–9.

[57] Vihola, M. (2010). Grapham: Graphical models with adaptive random walk Metropolis algorithms. *Comput. Stat. Data Anal.*, **54**, 49–54.

[58] Vihola, M. (2012). Robust adaptive Metropolis algorithm with coerced acceptance rate. *Stat. Comp.* (to appear).

Part IV

Dynamic Models

8 Bayesian dynamic modelling

MIKE WEST

8.1 Introduction

Bayesian time series and forecasting is a very broad field and any attempt at other than a very selective and personal overview of core and recent areas would be foolhardy. This chapter therefore selectively notes some key models and ideas, leavened with extracts from a few time series analysis and forecasting examples. For definitive development of core theory and methodology of Bayesian state-space models, readers are referred to [46, 74] and might usefully read this chapter with one or both of the texts at hand for delving much further and deeper. The latter parts of the chapter link into and discuss a range of recent developments on specific modelling and applied topics in exciting and challenging areas of Bayesian time series analysis.

8.2 Core model context: Dynamic linear model

8.2.1 Introduction

Much of the theory and methodology of all dynamic modelling for time series analysis and forecasting builds on the theoretical core of linear, Gaussian model structures: the class of univariate normal dynamic linear models (DLMs or NDLMs). Here we extract some key elements, ideas and highlights of the detailed modelling approach, theory of model structure and specification, methodology and application.

Over a period of equally spaced discrete time, a univariate time series $y_{1:n}$ is a sample from a DLM with $p-$vector state θ_t when

$$y_t = x_t + \nu_t, \quad x_t = F_t'\theta_t, \quad \theta_t = G_t\theta_{t-1} + \omega_t, \qquad t = 1, 2, \ldots, \tag{8.1}$$

where: each F_t is a known regression $p-$vector; each G_t a $p \times p$ state transition matrix; ν_t is univariate normal with zero mean; ω_t is a zero-mean $p-$vector representing evolution noise, or innovations; the pre-initial state θ_0 has a normal prior; the sequences ν_t, ω_t are independent and mutually independent, and also independent of θ_0. DLMs are hidden Markov models; the state vector θ_t is a latent or hidden state, often containing values of underlying latent processes as well as time-varying parameters (Chapter 4 of [74]).

8.2.2 Core example DLMs

Key special cases are distinguished by the choice of elements F_t, G_t. This covers effectively all relevant dynamic linear models of fundamental theoretical and practical importance. Some key examples that underlie much of what is applied in forecasting and time series analysis are as follows.

Random walk in noise (Chapter 2 of [74]): $p = 1, F_t = 1, G_t = 1$ gives this first-order polynomial model in which the state $x_t \equiv \theta_{t1} \equiv \theta_t$ is the scalar local level of the time series, varying as a random walk itself.

Local trend/polynomial DLMs (Chapter 7 of [74]): $F_t = E_p = (1, 0, \cdots, 0)'$ and $G_t = J_p$, the $p \times p$ matrix with 1s on the diagonal and super-diagonal, and zeros elsewhere, define 'locally smooth trend' DLMs; elements of θ_t are the local level of the underlying mean of the series, local gradient and change in gradient etc., each undergoing stochastic changes in time as a random walk.

Dynamic regression (Chapter 9 of [74]): When $G_t = I_p$, the DLM is a time-varying regression parameter model in which regression parameters in θ_t evolve in time as a random walk.

Seasonal DLMs (Chapter 8 of [74]): $F_t = E_2$ and $G_t = rH(a)$ where $r \in (0, 1)$ and

$$H(a) = \begin{pmatrix} \cos(a) & \sin(a) \\ -\sin(a) & \cos(a) \end{pmatrix}$$

for any angle $a \in (0, 2\pi)$ defines a dynamic damped seasonal, or cyclical, DLM of period $2\pi/a$, with damping factor r per unit time.

Autoregressive and time-varying autoregressive DLMs (Chapter 5 of [46]): Here $F_t = E_p$ and G_t depends on a $p-$vector $\phi_t = (\phi_{t1}, \ldots, \phi_{tp})'$ as

$$G_t = \begin{pmatrix} \phi_{t1} & \phi_{t2} & \phi_{t3} & \cdots & \phi_{tp} \\ 1 & 0 & 0 & \cdots & 0 \\ 0 & 1 & 0 & \cdots & 0 \\ \vdots & & \ddots & \cdots & \vdots \\ 0 & 0 & \cdots & 1 & 0 \end{pmatrix},$$

with, typically, the evolution noise constrained as $\omega_t = (\omega_{t1}, 0, \ldots, 0)'$. Now $y_t = x_t + \nu_t$ where $x_t \equiv \theta_{t1}$ and $x_t = \sum_{j=1:p} \phi_{tj} x_{t-j} + \omega_{t1}$, a time-varying autoregressive process of order p, or TVAR(p). The data arise through additive noisy observations on this hidden or latent process.

If the ϕ_{tj} are constant over time, x_t is a standard AR(p) process; in this sense, the main class of traditional linear time series models is a special case of the class of DLMs.

8.2.3 Time series model composition

Fundamental to structuring applied models is the use of building blocks as components of an overall model—the principle of composition or *superposition* (Chapter 6 of [74]). DLMs do this naturally by collecting together components: given a set of individual DLMs, the larger model is composed by concatenating the individual component θ_t vectors into a longer state vector, correspondingly concatenating the individual F_t vectors, and building the associated state evolution matrix as the block diagonal of those of the component models. For example,

$$\begin{aligned} F' &= (1, f_t, E_2', E_2', E_2'), \\ G &= \text{block diag}\left\{1,\ 1,\ H(a_1),\ H(a_2),\ \begin{pmatrix}\phi_1 & \phi_2 \\ 1 & 0\end{pmatrix}\right\} \end{aligned} \tag{8.2}$$

defines the model for the signal as

$$x_t = \theta_{t1} + \theta_{t2} f_t + \rho_{t1} + \rho_{t2} + z_t$$

where:

- θ_{t1} is a local level/random walk intercept varying in time;
- θ_{t2} is a dynamic regression parameter in the regression on the univariate predictor/independent variable time series f_t;
- ρ_{tj} is a seasonal/periodic component of wavelength $2\pi/a_j$ for $j = 1, 2$, with time-varying amplitudes and phases—often an overall annual pattern in weekly or monthly data, for example, can be represented in terms of a set of harmonics of the fundamental frequency, such as would arise in the example here with $a_1 = \pi/6, a_2 = \pi/3$ yielding an annual cycle and a semi-annual (six month) cycle;
- z_t is an AR(2) process—a short-term correlated underlying latent process—that represents residual structure in the time series signal not already captured by the other components.

8.2.4 Sequential learning

Sequential model specification is inherent in time series, and Bayesian learning naturally proceeds with a sequential perspective (Chapter 4 of [46]). Under a specified normal prior for the latent initial state θ_0, the standard normal/linear sequential updates apply: at each time $t - 1$ a 'current' normal posterior evolves via the evolution equation to a 1-step ahead prior distribution for the next state θ_t; observing the data y_t then updates that to the time t posterior, and we progress further in time sequentially. Missing data in the time series is trivially dealt with: the prior-to-posterior update at any time point where the observation is missing involves no change. From the early days—in the 1950s—of so-called Kalman filtering in engineering and early applications of Bayesian forecasting in commercial settings (Chapter 1 of [74]), this framework of closed-form sequential updating analysis—or *forward filtering* of the time series—has been the centrepiece of the computational machinery. Though far more complex, elaborate, nonlinear and non-normal models are used routinely nowadays, based on advances in simulation-based computational methods, this normal/linear theory still plays central and critical roles in applied work and as components of more elaborate computational methods.

8.2.5 Forecasting

Forecasting follows from the sequential model specification via computation of predictive distributions. At any time t with the current normal posterior for the state θ_t based on data $y_{1:t}$, and any other information integrated into the analysis, we simply extrapolate by evolving the state through the state evolution equation into the future, with implied normal predictive distributions for sequences $\theta_{t+1:t+k}, y_{t+1:t+k}$ into the future any $k > 0$ steps ahead. Forecasting via simulation is also key to applied work: simulating the process into the future—to generate 'synthetic realities'—is often a useful adjunct to the theory, as visual inspection (and perhaps formal statistical summaries) of simulated futures can often aid in understanding aspects of model fit/misfit as well as formally elaborating on the predictive expectations defined by the model and fit to historical data; see Figures 8.2 and 8.3 for some aspects of this in the analysis of the climatological Southern Oscillation Index (SOI) time series, discussed later in Section 8.3.2. The concept is also illustrated in Figure 8.4 in a multivariate DLM analysis of a financial time series, discussed later in Section 8.4.1.

8.2.6 Retrospective time series analysis

Time series analysis—investigating posterior inferences and aspects of model assessment based on a model fitted to a fixed set of data—relies on the theory of *smoothing* or *retrospective filtering* that overlays forward-filtering, sequential analysis. Looking back over time from a current time t, this theory defines the revised posterior distributions for historical sequences of state vectors $\theta_{t-1:t-k}$ for $k > 0$ that complement the forward analysis (Chapter 4 of [74]).

8.2.7 Completing model specification: Variance components

The Bayesian analysis of the DLM for applied work is enabled by extensions of normal theory-based sequential analysis to incorporate learning on the observational variance parameters $V(\nu_t)$ and specification of the evolution variance matrices $V(\omega_t)$. For the former, analytic tractability is maintained in models where $V(\nu_t) = k_t v_t$, with known variance multipliers k_t, and $V(\omega_t) = v_t W_t$ with two variants: (i) constant, unknown $v_t = v$ (Section 4.3.2 of [46]) and (ii) time-varying observational variances in which v_t follows a stochastic volatility model based on variance discounting—a random walk-like model that underlies many applications where variances are expected to be locally stable but globally varying (Section 4.3.7 of [46]). Genesis and further developments are given in Chapter 10 of [74] and, with recent updates and new extensions, in Chapters 4,7 and 10 of [46].

The use of discount factors to structure evolution variance matrices has been and remains central to many applications (Chapter 6 of [74]). In models with non-trivial state vector dimension p, we must maintain control over the specification of W_t to avoid exploding the numbers of free parameters. In many cases, we are using W_t to reflect low levels of stochastic change in elements of the state. When the model is structured in terms of block components via superposition as described above, the W_t matrix is naturally structured in a corresponding block diagonal form; then the strategy of specifying these blocks in W_t using the discount factor approach is natural (Section 4.3.6 of [46]). This strategy describes the innovations for each component of the state vector as contributing a constant stochastic 'rate of loss of information' per time point, and these rates may be chosen as different for different components. In our example above, a dynamic regression parameter might be expected to vary less rapidly over time than, perhaps, the underlying local trend.

Central to many applications of Bayesian forecasting, especially in commercial and economic studies, is the role of 'open modelling'. That is, a model is often one of multiple ways of describing a problem, and as such should be open to modification over time, as well as integration with other formal descriptions of a forecasting problem (Chapter 1 of [74]). The role of statistical theory in guiding changes—interventions to adapt a model at any given time based on additional information—that maintain consistency with the model is then key. Formal sequential analysis in a DLM framework can often manage this via appropriate changes in the variance components. For example, treating a single observation as of poorer quality, or a likely outlier, can be done via an inflated variance multiplier k_t; feeding into the model new/external information that suggests increased chances of more abrupt change in one or more components of a state vector can be done via larger values of the corresponding elements of W_t, typically using a lower discount factor in the specification for just that time, or times, when larger changes are anticipated. Detailed development of a range of subjective monitoring and model adaptation methods of these forms, with examples, are given in chapters 10–12 of [74] and throughout [43]; see also Chapter 4 of [46] and earlier relevant papers [62, 72, 73].

8.2.8 Time series decomposition

Complementing the strategy of model construction by superposition of component DLMs is the theory and methodology of model decomposition that is far-reaching in its utility for retrospective time series analysis (Chapter 9 of [74]). Originally derived for the class of *time series DLMs* in which $F_t = F, G_t = G$ are constant for all time [66–70], the theory of decompositions applies also to time-varying models [44, 45, 47]. The context of DLM AR(p) and TVAR(p) models—alone or as components of a larger model—is key in terms of the interest in applications in engineering and the sciences, in particular (Chapter 5 of [46]).

Consider a DLM where one model component z_t follows a TVAR(p) model. The main idea comes from the central theoretical results that a DLM implies a decomposition of the form

$$z_t = \sum_{j=1:C} z^c_{tj} + \sum_{j=1:R} z^r_{tj}$$

where each z^*_{tj} is an underlying latent process: each z^r_{tj} is a TVAR(1) process and each z^c_{tj} is a quasi-cyclical time-varying process whose characteristics are effectively those of a TVAR(2) overlaid with low levels of additive noise, and which exhibits time-varying periodicities with stochastically varying amplitude, phase and period. In the special case of constant AR parameters, the periods of these quasi-cyclical z^c_{tj} processes are also constant.

This DLM decomposition theory underlies the use of these models—state-space models/ DLMs with AR and TVAR components—for problems in which we are interested in a potentially very complicated and dynamic autocorrelation structure, and aim to explore underlying contributions to the overall signal that may exhibit periodicities of a time-varying nature. Many examples appear in [46, 74] and references there, as well as the core papers referenced above. Figures 8.1, 8.2 and 8.3 exemplify some aspects of this in the analysis of the climatological Southern Oscillation Index (SOI) time series of Section 3.2.

8.3 Computation and model enrichment

8.3.1 Parameter learning and batch analysis via MCMC

Over the last couple of decades, methodology and applications of Bayesian time series analysis have massively expanded in non-Gaussian, nonlinear and more intricate conditionally linear models. The modelling concepts and features discussed above are all central to this increasingly rich field, while much has been driven by enabling computational methods.

Consider the example DLM of equation (8.2) and now suppose that $V(\nu_t) = k_t v$ with known weights k_t but uncertain v to be estimated, and the evolution variance matrix is

$$W_t \equiv W = \text{block diag}\left\{\tau_1,\ \tau_2,\ \tau_3 I_2,\ \tau_4 I_2,\ \begin{pmatrix} w & 0 \\ 0 & 0 \end{pmatrix}\right\}. \tag{8.3}$$

Also, write $\phi = (\phi_1, \phi_2)'$ for the AR parameters of the latent AR(2) model component. The DLM can be fitted using standard theory assuming the full set of model parameters $\mu = \{v, \phi, w, \tau_{1:4}\}$ to be known. Given these parameters, the forward filtering and smoothing based on normal/linear theory applies.

Markov chain Monte Carlo methods naturally open the path to a complete Bayesian analysis under any specified prior $p(\mu)$; see Chapter 15 of [74] and Section 4.5 of [46] for full details and copious references, as well as challenging applications in Chapter 7 of [46]. Given an observed data sequence $y_{1:n}$, MCMC iteratively re-simulates parameters and states from appropriate conditional distributions. This involves conditional simulations of elements of μ conditioning on current values of other parameters and a current set of states $\theta_{0:n}$ that often break down into tractable parallel simulators. The example above is a case in point under independent priors on ϕ and the variances v, τ_j, w, for example.

Central to application is the *forward filtering, backward sampling* (FFBS—[6, 18]) algorithm that arises naturally from the normal/linear theory of the DLM conditional on parameters μ. This builds on the sequential, forward filtering theory to run through the data, updating posterior distributions for states over time, and then steps back in time: at each point $t = n, t = n - 1, \ldots, t = 1, t = 0$ in turn, the retrospective distributional theory of this *conditionally linear, normal* model provides normal distributions for the states that are simulated. This builds up a sequence $\{\theta_n, \theta_{n-1}, \ldots, \theta_1, \theta_0\}$ that represents a draw—sampled via composition backwards in time—from the relevant conditional posterior $p(\theta_{0:n}|\mu, y_{1:n})$. The use of MCMC methods also naturally deals with missing data in a time series; missing values are, by definition, latent variables that can be simulated via appropriate conditional posteriors each step of the MCMC.

8.3.2 Example: SOI time series

Figures 8.1, 8.2 and 8.3 show aspects of an analysis of the climatological Southern Oscillation Index (SOI) time series. This is a series of 540 monthly observations computed as the 'difference of the departure from the long-term monthly mean sea level pressures' at Tahiti in the South Pacific and Darwin in Northern Australia. The index is one measure of the so-called 'El Nino-Southern Oscillation'—an event of critical importance and interest in climatological studies in recent decades which is generally understood to vary periodically with a very noisy 3–6 year period of quasi-cyclic pattern. As discussed in [24]— which also details the history of the data and prior analyses—one of several applied interests in this data is in improved understanding of these quasi-periodicities and also potential non-stationary trends, in the context of substantial levels of observational noise.

The DLM chosen here is $y_t = \theta_{t1} + z_t + \nu_t$ where θ_{t1} is a first-order polynomial local level/trend and z_t is an AR(12) process. The data is monthly data over the year, so the AR component provides opportunities to identify even quite subtle longer-term (multi-year) periodicities that may show quite high levels of stochastic variation over time in amplitude and phase. Extensions to TVAR components would also allow the associated periods to vary as discussed and referenced above. Here the model parameters include the 12-dimensional AR parameter ϕ that can be converted to autoregressive roots (Section 9.5 of [74]) to explore whether the AR component appears to be consistent with an underlying stationary process or not, as well as to make inferences on the periods/wavelengths of any identified quasi-periodic components. The analysis also defines posterior inferences for the time trajectories of all latent components z^r_{tj} and z^c_{tj} by applying the decomposition theory to each of the posterior simulation samples of the state vector sequence $\theta_{0:n}$.

Figure 8.1 shows approximate posteriors for the moduli of the 12 latent AR roots, all very likely positive and almost surely less than 1, indicating stationarity of z_t in this model description. The figure also shows the corresponding posterior for the wavelength of the latent process component z^c_{tj} having highest wavelength, indicating a dominant quasi-periodicity in the data with wavelength between 40–70 months—a noisy '4-year' phenomenon, consistent with expectations and prior studies. Figure 8.2 shows a few posterior samples of the time trajectory of the latent trend θ_{t1}, together with its approximate posterior mean, superimposed on the data. The inference is that of very limited change over time in the trend in the context of other model components. This figure also shows the data plotted together with a 'synthetic future' over the next three years: that is, a single draw from the posterior predictive distribution into the future. From the viewpoint of model fit, exploring such synthetic futures via repeat simulations studied by eye in comparison with the data can be most informative; they also feed into formal predictive evaluations for excursions away from (above/below) the mean, for example [24].

Additional aspects of the decomposition analysis are represented by Figure 8.3. The first frame shows the posterior mean of the fitted AR(12) component plotted over time (labelled as 'data' in the upper figure), together with the corresponding posterior mean trajectories of the three latent quasi-cyclical components having largest inferred periods, all plotted on the same vertical scale. Evidently, the dominant period component explains much of the structure in the AR(12) process, the second contributing much of the additional variation at a lower wavelength (a few months). The remaining components contribute to partitioning the noise in the series and have much lower amplitudes. The figure also shows several posterior draws for the z_t processes to give some indication of the levels of uncertainty about its form over the years.

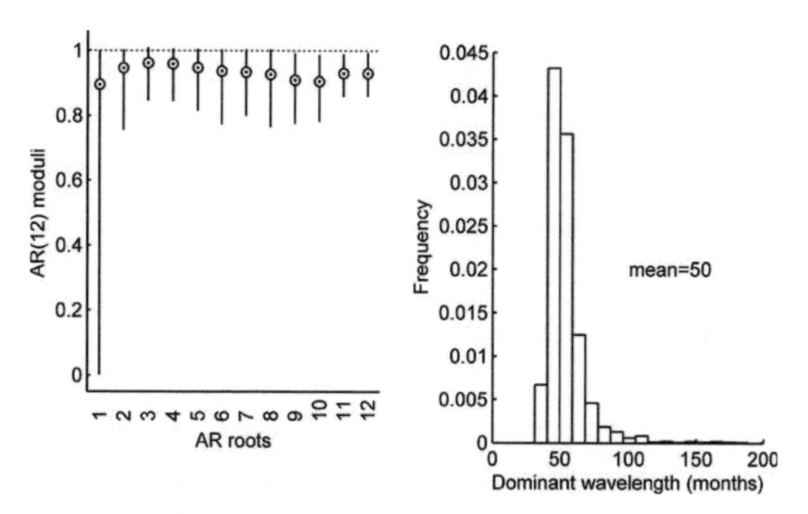

Figure 8.1 *Left frame:* Approximate posterior 95% credible intervals for the moduli of the 12 latent AR roots in the AR component of the model fitted to the SOI time series. *Right frame:* Approximate posterior for the wavelength of the latent process component z^c_{tj} with largest wavelength, indicating a dominant quasi-periodicity in the range 40–70 months.

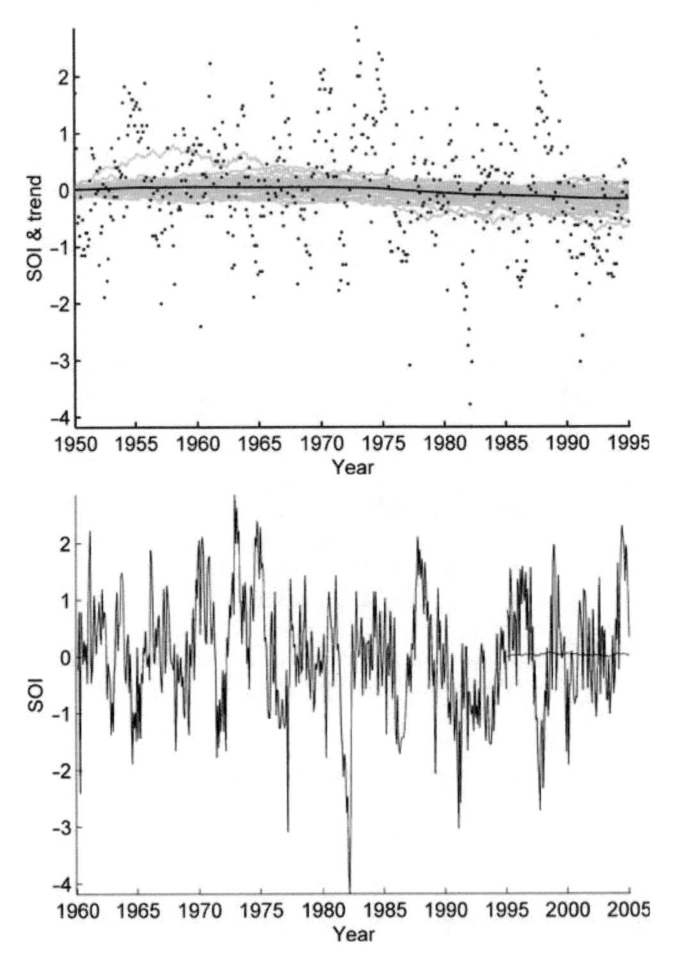

Figure 8.2 *Upper frame:* Scatter plot of the monthly SOI index time series superimposed on the trajectories of the posterior mean and a few posterior samples of the underlying trend. *Lower frame:* SOI time series followed by a single synthetic future—a sample from the posterior predictive distribution over the three or fours years following the end of the data in 1995; the corresponding sample of the predicted underlying trend is also shown.

8.3.3 Mixture model enrichment of DLMs

Mixture models have been widely used in dynamic modelling and remain a central theme in analyses of structural change, approaches to modelling non-Gaussian distributions via discrete mixture approximations, dealing with outlying observations, and others. Chapter 12 of [74] develops extensive theory and methodology of two classes of dynamic mixture models, building on seminal work by P.J. Harrison and others [23]. The first class relates to model uncertainty and learning model structure that has its roots in both commercial forecasting and engineering control systems applications of DLMs from the 1960s. Here a set of DLMs are analysed sequentially in parallel, being regarded as competing models, and sequentially updated 'model probabilities' track the data-based

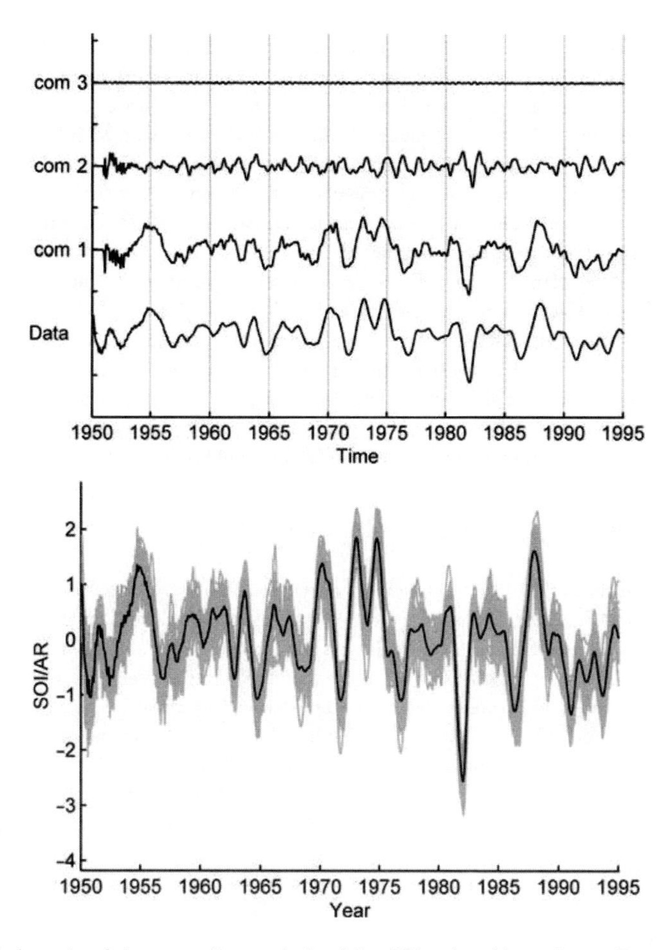

Figure 8.3 Aspects of decomposition analysis of the SOI series. *Upper frame:* Posterior means of (from the bottom up) the latent AR(12) component z_t (labelled as 'data'), followed by the three extracted component z^c_{tj} for $j = 1, 2, 3$, ordered in terms of decreasing estimated periods; all are plotted on the same vertical scale, and the AR(12) process is the direct sum of these three and subsidiary components. *Lower frame:* A few posterior samples (in grey) of the latent AR(12) process underlying the SOI series, with the approximate posterior mean superimposed.

evidence for each relative to the others, in what is nowadays a familiar model comparison and Bayesian model-averaging framework.

The second framework—adaptive *multi-process* models— entertains multiple possible models at each time point and aims to adaptively reweight sequentially over time; key examples are modelling outliers and change-points in subsets of the state vector as in applications in medical monitoring, for example [55, 56]. In the DLM of equation (8.2) with a 'standard' model having $V(\nu_t) = v$ and evolution variance matrix as in equation (8.3), a multi-process extension for outlier accommodation would consider a mixture prior induced by $V(\nu_t) = k_t v$ where, *at each time t*, k_t may take the value 1 or, say, 100, with some probability. Similarly, allowing for a larger stochastic change in the underlying latent AR(2) component z_t of the model would involve an extension so that the innovations variance w in equation (8.3) is replaced by $h_t w$, where now h_t may take the value 1 or 100, with some probability. These multi-process models clearly lead to a combinatorial explosion of the numbers of possible 'model states' as time progresses, and much attention has historically been placed on approximating the implied unwieldy sequential analysis. In the context of MCMC methods and batch analysis, this is resolved with simulation-based numerical approximations where the introduction of indicators of mixture component membership naturally and trivially opens the path to computation: models are reduced to *conditionally linear, normal DLMs* for conditional posterior simulations of states and parameters, and then the mixture component indicators are themselves re-simulated each step of the MCMC. Many more elaborate developments and applications appear in, and are referenced by, [19] and Chapter 7 of [46].

Another use of mixtures in DLMs is to define direct approximations to non-normal distributions, so enabling MCMC analysis based on conditionally normal models that they imply. One key example is the univariate stochastic volatility model pioneered by [26, 53] which is nowadays in routine use to define components of more elaborate dynamic models for multivariate stochastic volatility time series approaches [1, 2, 12, 36, 37, 41]; see also Chapter 7 of [46].

8.3.4 Sequential simulation methods of analysis

A further related use of mixtures is as numerical approximations to the sequentially updated posterior distributions for states in non-linear dynamic models when the conditionally linear strategy is not available. This use of mixtures of DLMs to define adaptive sequential approximations to the filtering analysis by 'mixing Kalman filters' [3, 11] has multiple forms, recently revisited with some recent extensions in [38]. Mixture models as direct posterior approximations, and as sequences of sequentially updated *importance sampling* distributions for nonlinear dynamic models were pioneered in [63–65], and some of the recent developments build on this.

The adaptive, sequential importance sampling methods of [65] represented an approach to sequential simulation-based analysis developed at the same time as the approach that became known as *particle filtering* [21]. Bayesian sequential analysis in state-space models using 'clouds of particles' in states and model parameters, evolving the particles through evolution equations that may be highly nonlinear and non-Gaussian, and appropriately updating weights associated with particles to define approximate posteriors, has defined a fundamental change in numerical methodology for time series. Particle filtering and related methods of sequential Monte Carlo (SMC) [7, 14], including problems of parameter learning combined with filtering on dynamic states [31], are reviewed in this book: see the chapter by H.F. Lopes and C.M. Carvalho, on *Online Bayesian learning....*

Recent methods have used variants and extensions of the so-called technique of *approximate Bayesian computation* [34, 54]. Combined with other SMC methods, this seems likely to emerge in coming years as a central approach to computational approximation for sequential analysis in increasingly complex dynamic models; some recent studies in dynamic modelling in systems biology [4, 35, 58] provide some initial examples using such approaches.

8.4 Multivariate time series

The basic DLM framework generalizes to multivariate time series in a number of ways, including multivariate non-Gaussian models for time series of counts, for example [5], as well as a range of model classes based on multi- and matrix-variate normal models (Chapter 10 of [46]). Financial and econometric applications have been key motivating areas, as touched on below, while multivariate

DLMs are applied in many other fields—as diverse as experimental neuroscience [1, 27, 28, 47], computer model emulation in engineering [30] and traffic flow forecasting [57]. Some specific model classes that are in mainstream application and underlie recent and current developments—especially to increasingly high-dimensional times series—are keyed out here.

8.4.1 Multivariate normal DLMs: Exchangeable time series

In modelling and forecasting a $q \times 1$ vector times series, a so-called *exchangeable time series DLM* has the form

$$\begin{aligned} y_t' &= F_t'\Theta_t + \nu_t', & \nu_t &\sim N(0, \Sigma_t) \\ \Theta_t &= G_t\Theta_{t-1} + \Omega_t, & \Omega_t &\sim N(0, W_t, \Sigma_t) \end{aligned} \tag{8.4}$$

where $N(\cdot,\cdot,\cdot)$ denotes a *matrix normal distribution* (Section 10.6 of [46]). Here the row vector y_t' follows a DLM with a matrix state Θ_t. The $q \times q$ time-varying variance matrix Σ_t determines patterns of co-changes in observation and the latent matrix state over time. These models are building blocks of larger (factor, hierarchical) models of increasing use in financial time series and econometrics; see, for example, [48, 49], Chapter 16 of [74] and Chapter 10 of [46].

Modelling multivariate stochastic volatility—the evolution over time of the variance matrix series Σ_t—is central to these multivariate extensions of DLMs. The first multivariate stochastic volatility models based on variance matrix discount learning [50, 51], later developed via matrix-beta evolution models [59, 60], and remain central to many implementations of Bayesian forecasting in finance. Here Σ_t evolves over one time interval via a nonlinear stochastic process model involving a matrix beta random innovation inducing priors and posteriors of conditional inverse Wishart forms. The conditionally conjugate structure of the exchangeable model form for $\{\Theta_t, \Sigma_t\}$, coupled with discount factor-based specification of the W_t evolution variance matrices, leads to a direct extension of the closed form sequential learning and retrospective sampling analysis of the univariate case (Chapter 10 of [46]). In multiple studies, these models have proven their value in adapting to short-term stochastic volatility fluctuations and leading to improved portfolio decisions as a result [48].

An example analysis of a time series of $q = 12$ daily closing prices (FX data) of international currencies relative to the US dollar, previously analysed using different models (Chapter 10 of [46]), generates some summaries including those in Figures 8.4 and 8.5. The model used here incorporates time-varying vector autoregressive (TV-VAR) models into the exchangeable time series structure. With y_t the logged values of the $12-$vector of currency prices at time t, we take F_t to be the $37-$dimensional vector having a leading 1 followed by the lagged values of all currencies over the last three days. The dynamic autoregression naturally anticipates the lag-1 prices to be the prime predictors of next time prices, while considering 3-day lags leads to the opportunity to integrate 'market momentum'. Figure 8.4 selects one currency, the Japanese Yen, and plots the data together with forecasts over the last several years. As the sequential updating analysis proceeds, forecasts on day t for day $t+1$ are made by direct simulation of the 1-step-ahead predictive distribution; each forecast vector y_{t+1} is then used in the model in order to use the same simulation strategy to sample the future at time $t+2$ from the current day t, and this is repeated to simulate day $t+3$. Thus we predict via the strategy of generating synthetic realities, and the figure shows a few sets of these 3-day-ahead forecasts made every day over three or four years, giving some indication of forecast uncertainty as well as accuracy.

Figure 8.5 displays some aspects of multivariate volatility over time as inferred by the analysis. Four images of the posterior mean of the *precision matrix* Σ_t^{-1} at four selected time points capture

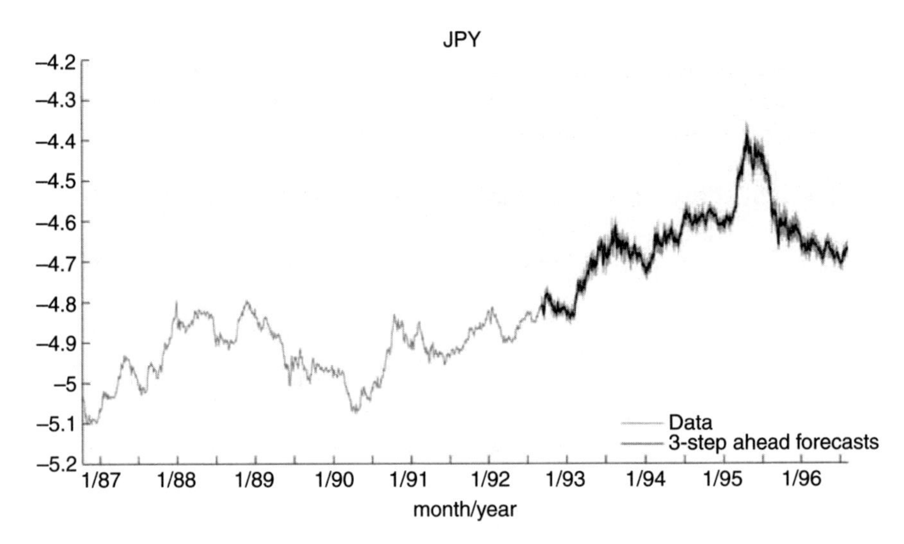

Figure 8.4 Daily prices of the Japanese Yen in US dollars over several years in the 1980s–1990s, followed by plots of forecasts from the multivariate TV-TVAR model with stochastic volatility fitted to a 12-dimensional FX time series of which the Yen is one element. The shaded forecast region is made up of 75 sets of 3-day-ahead forecasts based on the sequential analysis: on each day, the 'current' posterior for model states and volatility matrices is simulated to generate forecasts over the next three days.

some flavour of time variation, while the percentage of the total variation in the posterior mean of Σ_t explained by the first three dominant principal components at each t captures additional aspects.

8.4.2 Multivariate normal DLMs: Dynamic latent factor and TV-VAR models

Time-varying vector autoregressive (TV-VAR) models define a rich and flexible approach to modelling multivariate structure that allows the predictive relationships among individual scalar elements of the time series to evolve over time. The above section has already described the use of such a model within the exchangeable time series framework. Another way in which TV-VAR models are used is to represent the dynamic evolution of a vector of *latent factors* underlying structure in a higher-dimensional data series. One set of such *dynamic latent factor TV-VAR models* has the form

$$\begin{aligned} y_t &= F_t\theta_t + B_t x_t + \nu_t, & \nu_t &\sim N(0, \Psi), \\ x_t &= \textstyle\sum_{i=1:p} A_{ti} x_{t-i} + \omega_t, & \omega_t &\sim N(0, \Sigma_t). \end{aligned} \tag{8.5}$$

Here x_t is a latent $k-$vector state process following a TV-VAR(p) model with, typically, $k << q$ so that the common structure among the elements of y_t is heavily driven by a far lower-dimensional dynamic state. The set of p, $q \times q$ autoregressive coefficient matrices A_{ti} is often time-varying with elements modelled via, for example, sets of univariate $AR(1)$ processes or random walks. The factor

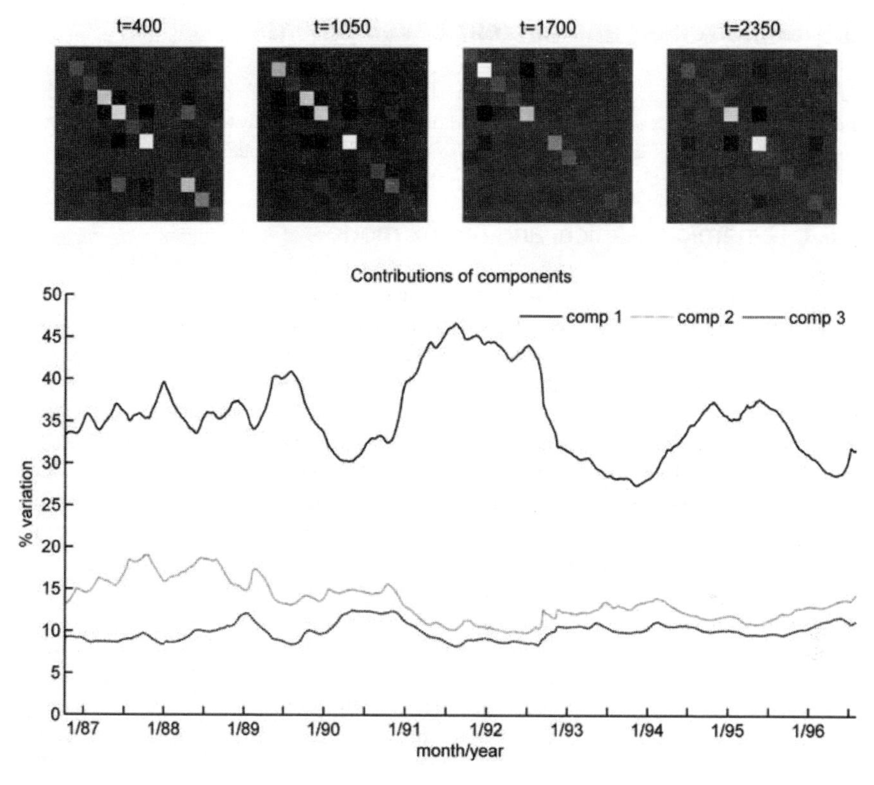

Figure 8.5 *Upper frames:* Images of posterior estimates of the 12 × 12 precision matrices Σ_t^{-1} in the analysis of the multivariate currency prices time series. The differences in patterns visually evident reflect the extent and nature of changes in the volatility structure across time, as represented by the four selected time points spaced apart by a few hundred days. *Lower frame:* Percentage variation explained by the first three dominant principal components of the posterior mean of Σ_t for each time $t = 1 : n$ over the FX time series, illustrating the nature of variation in the contribution of the main underlying 'common components' of volatility in the 12 currency price series over the ten year period.

loadings matrix B_t maps factors to data; in some models, including prior Bayesian factor analysis approaches [2], this will be taken as constant. The $k \times k$ dynamic covariance matrix Σ_t drives the innovations of the state evolution, and allowing for stochastic volatility here defines an enrichment of the TV-VAR structure. The additional component $F_t\theta_t$ superimposed may include dynamic regression and other terms, with relevant state evolution equations for θ_t. Such models are receiving increasing use in natural science and engineering applications [15, 16, 20, 45, 46, 75] as well as in econometrics and finance [36, 37]. Chapters 8 and 9 of [46] describe aspects of the theory and methodology of vector AR and TVAR models, connections with latent factor modelling in studies of multiple time series in the neurosciences, and discussion of multiple other creative applications of specialized variants of this rich class of models. The model contains traditional DLMs, VAR models, latent factor models as previously developed, as well as the more elaborate TV-VAR factor forms.

8.5 Some recent and current developments

Among a large number of recent and currently active research areas in Bayesian time series analysis and forecasting, a few specific modelling innovations that relate directly to the goals of addressing analysis of increasingly high-dimensional time series and nonlinear models are keyed out.

8.5.1 Dynamic graphical and matrix models

A focus on inducing parsimony in increasingly high-dimensional, time-varying variance matrices in dynamic models has led to the integration of Bayesian *graphical modelling* ideas into exchangeable time series DLMs [9, 10]. The standard theory of Gaussian graphical models using hyper-inverse Wishart distributions—the conjugate priors for variance matrices whose inverses Σ_t^{-1} have some off-diagonal elements at zero corresponding to an underlying conditional independence graph [13]—rather surprisingly extends directly to the time-varying case. The multivariate volatility model based on variance matrix discounting generalizes to define sequential analysis in which the posterior distributions for the $\{\Theta_t, \Sigma_t\}$ sequences are updated in closed multivariate normal, hyper-inverse Wishart forms. These theoretical innovations led to the development of dynamic graphical models, coupled with learning about graphical model structures based on existing model search methods [13, 25]. Applications in financial time series for predictive portfolio analysis show improvements in portfolio outcomes that illustrate the practical benefits of the parsimony induced via appropriate graphical model structuring in multivariate dynamic modelling [9, 10].

These developments have extended to contexts of *matrix time series* [61] for applications in econometrics and related areas. Building on Bayesian analyses of matrix-variate normal distributions, conditional independence graphical structuring of the characterizing variance matrix parameters of such distributions again opens the path to parsimonious structuring of models for increasingly high-dimensional problems. This is complemented by the development of a broad class of dynamic models for matrix-variate time series within which stochastic elements defining time series errors and structural changes over time are subject to graphical model structuring.

8.5.2 Dynamic matrix models for stochastic volatility

A number of recent innovations have aimed to define more highly structured, predictive stochastic process models for multivariate volatility matrices Σ_t, aiming to go beyond the neutral, random walk-like model that underlies the discounting approach. Among such approaches are multivariate extensions of the univariate construction method inspired by MCMC [42]; the such extension yields a class of stationary AR(1) stochastic process models for Σ_t that are reversible in time and in which the transition distributions give conditional means of the attractive form $E(\Sigma_t|\Sigma_{t-1}) = S + a(\Sigma_{t-1} - S)$ where $a \in (0, 1)$ is scalar and S an underlying mean variance matrix. This construction is, however, inherently limited in that there is no notion of multiple AR coefficients for flexible autocorrelation structures and the models do not allow time irreversibilty. Related approaches directly build transition distributions $p(\Sigma_t|\Sigma_{t-1})$ as inverse-Wisharts [39, 40] or define more empirical models representing Σ_t as an explicit function of sample covariance matrices of latent vector AR processes [22]. These are very interesting approaches, but are somewhat difficult to work with theoretically and model fitting is a challenge.

Recently, [32] used linear, normal AR(1) models for off-diagonal elements of the Cholesky of Σ_t and for the log-diagonal elements. This is a natural parallel of Bayesian factor models for multivariate volatility and defines an approach to building highly structured stochastic process models for time series of dynamic variance matrices with short-term predictive potential.

A related approach builds on theoretical properties of the family of inverse Wishart distributions to define new classes of stationary, *inverse Wishart autoregressive (IW-AR)* models for the series of $q \times q$ volatility matrices Σ_t [17]. One motivating goal is to maintain a defined inverse Wishart marginal distribution for the process for interpretation. Restricting discussion to the (practically most interesting) special case of a first-order model, the basic idea is to define an IW-AR(1) Markov process directly via transition densities $p(\Sigma_t|\Sigma_{t-1})$ that are the conditionals of a joint inverse Wishart on an augmented $2q \times 2q$ variance matrix and whose block diagonals are Σ_{t-1} and Σ_t. This yields

$$\Sigma_t = \Psi_t + \Upsilon_t \Sigma_{t-1} \Upsilon_t'$$

where the $q \times q$ random *innovations matrices* Υ_t and Ψ_t have joint matrix normal, inverse Wishart distributions independently over time. Conditional means have the form

$$E(\Sigma_t|\Sigma_{t-1}) = S + R(\Sigma_{t-1} - S)R' + C_t(\Sigma_{t-1})$$

where S is the mean variance matrix parameter of the stationary process, R is a $q \times q$ autoregressive parameter matrix and $C_t(\cdot)$ is a matrix naturally related to the skewness of the inverse Wishart model. This model has the potential to embody multiple aspects of conditional dependence through R as well as defining both reversible and irreversible special cases [17]. Some initial studies have explored use of special cases as models for volatility matrices of the innovations process driving a TV-TVAR model for multiple EEG time series from studies in experimental neuroscience. One small extract from an analysis of multi-channel EEG data [27] appears in Figure 8.6, showing aspects of the estimated time trajectories of volatility for one channel along with those of time--varying correlations from Σ_t across multiple channels. As with other models above, computational issues for model filtering, smoothing and posterior simulation analysis require customized MCMC and SMC methods, and represent some of the key current research challenges. The potential is clear, however, for these approaches to define improved representations of multivariate volatility processes of benefit when integrated into time series state space analysis.

8.5.3 Time-varying sparsity modelling

As time series dimension increases, the dimension of latent factor processes, time-varying parameter processes and volatility matrix processes in realistic dynamic models—such as special cases or variants of models of equation (8.5)—evidently increase very substantially. Much current interest then rests on modelling ideas that engender parsimonious structure and, in particular, on approaches to inducing data-informed *sparsity* via full shrinkage to zero of (many) parameters. Bayesian sparsity modelling ideas are well-developed in 'static' models, such as sparse latent factor and regression models [8, 71], but mapping over to time series raises new challenges of defining general approaches to *dynamic sparsity*. For example, with a dynamic latent factor component $B_t f_t$ of equation (8.5), a zero element $B_{t,(i,j)}$ in the factor loadings matrix B_t reflects lack of association of the *ith* series in y_t with the *jth* latent factor in f_t. The overall sparsity pattern of B_t—with potentially many zeros—reflects a model context in which each of the individual, univariate factor processes impacts on a subset of the output time series, but not all, and allows for complex patterns of cross-talk. The concept of dynamic sparsity is that these sparsity patterns will typically vary over time, so models are needed to allow time variation in the values of elements of B_t that can dynamically shrink completely to zero for some epochs, then reappear and evolve according to a specific stochastic model at others. A general approach has been introduced by [36, 37], referred to as *latent threshold modelling* (LTM).

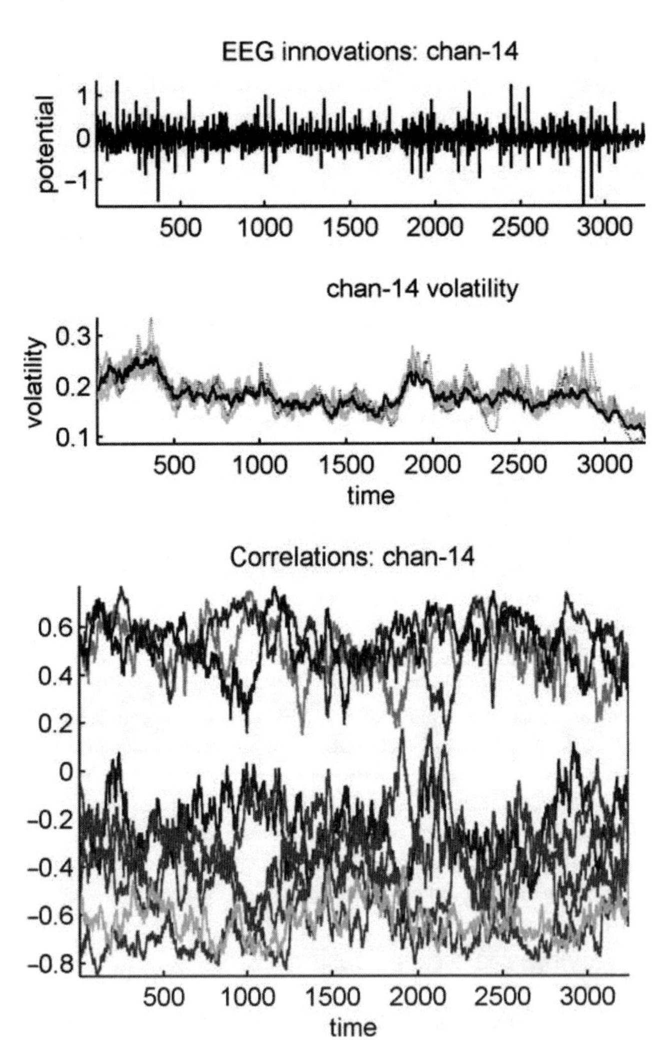

Figure 8.6 Aspects of results of approximate fitting of a $q \times q$ dimensional IW-AR(1) model to $q = 10$ EEG series from [27, 47]; here Σ_t is the volatility matrix of innovations driving a TV-VAR model for the potential fluctuations that the EEG signals represent. *Upper frame:* Estimated innovations time series for one EEG series/channel, labelled *chan-14*. *Centre frame:* Several posterior sample trajectories (grey) and approximate mean (black) for the standard deviation of channel 14. *Lower frame:* Corresponding estimates of time-varying correlations of *chan-14* with the other channels. Part of the applied interest is in patterns of change over time in these measures as the EEG channels are related spatially on the scalp of the test individual.

The basic idea of LTM for time series is to embed traditional time series model components into a larger model that thresholds the time trajectories, setting their realized values strictly to zero when they appear 'small'. For example, take one scalar coefficient process β_t, such as one element of the factor loadings matrix or a single dynamic regression parameter, and begin with a traditional evolution model of the AR(1) form $\beta_t = \mu + \rho(\beta_{t-1} - \mu) + \epsilon_t$. The LTM approach replaces the

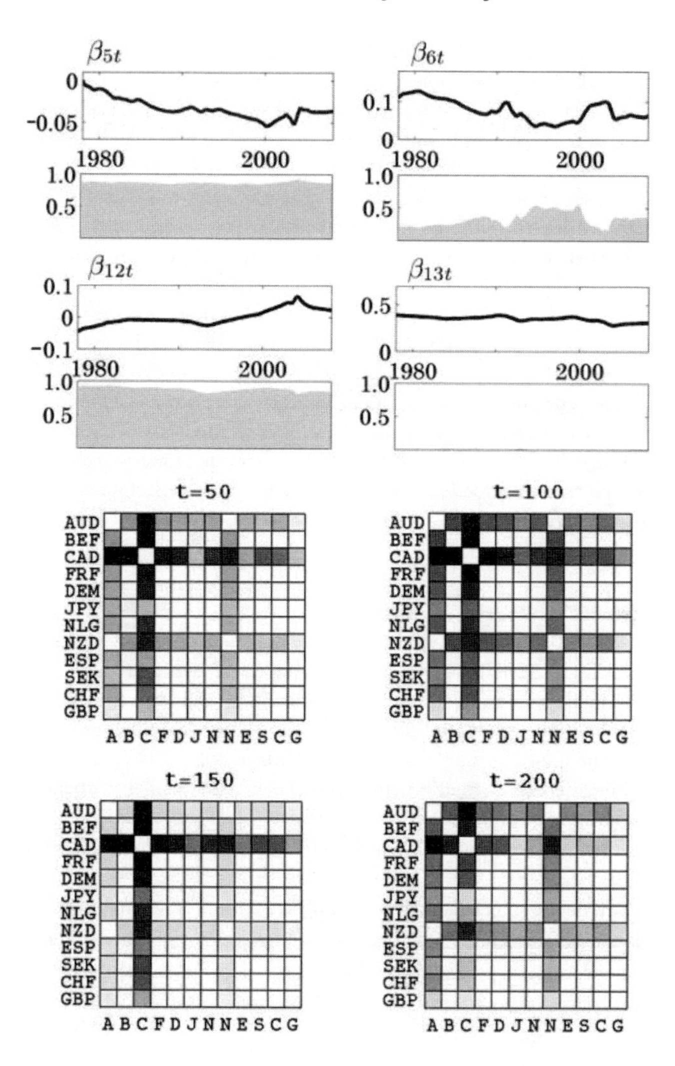

Figure 8.7 Examples of latent thresholding from an analysis of a three-dimensional economic time series using a TV-VAR(2) model (data from [36]). *Upper four frames:* Trajectories of approximate posterior means of 4 of the (18) *latent* time-varying autoregressive coefficients; the grey shading indicates estimated time-varying posterior probabilities of zero coefficients from the LTM construction. *Lower four images:* Images showing estimates of posterior probabilities (white = high, black = low) of non-zero entries in dynamic precision matrices Σ_t^{-1} modelled using an LTM extension of the Cholesky AR(1) model [32]. The data are time series on $q = 12$ daily international exchange rates (data from [46]) and the images show posterior sparsity probabilities for the 12×12 matrices at four selected time points, indicating both the ability of the LTM to identify zeros as well as how the sparsity pattern changes over time based on latent thresholding.

sequence $\beta_{1:n}$ with the thresholded version $b_{1:n}$ where $b_t = \beta_t I(|\beta_t| < \tau)$ for each t and based on some threshold τ. The concept is simple: the coefficient process is relevant, taking non-zero values, only when it beats the threshold, otherwise it is deemed insignificant and shrunk to zero. Extending MCMC analyses of multivariate DLMs to integrate the implied hierarchical model components, now embedding mutliple latent processes underlying the actual thresholded parameter processes, states and factors, requires substantial computational development, as detailed in [36]. The payoffs can be meaningful, as demonstrated in a series of financial time series and portfolio decision making examples in [37] where improved fit and parsimony feeds through to improvements in short-term forecasting and realized portfolio returns.

Figure 8.7 gives some flavour of the approach in extracts from two time series analysing: a TV-VAR(2) model of a $q = 3$-dimensional economic time series, and a multivariate stochastic volatility model analysis of a $q = 12$-dimensional financial time series. One key attraction of the LTM approach is its generality. Some of the model contexts addressed via LTM ideas in [36] include the following: *(i)* dynamic latent factor models; *(ii)* TV-VAR models, where the dynamic sparsity arises in collections of TV-VAR coefficient matrices A_{ti} of equation (8.5); *(iii)* dynamic regressions, where the approach can be regarded as a model for dynamic variable selection as well as a parsimony inducing strategy; and *(iv)* dynamic volatility modelling using extensions of the Cholesky volatility models of [32].

8.5.4 Nonlinear dynamical systems

The recent advances in Bayesian computational methods for dynamic models have come at a time when biotechnology and computing are also promoting significant advances in formal modelling in systems biology at molecular and cellular levels. In models of temporal development of components of gene networks and in studies of systems of cells evolving over time (and space, e.g. [33]), increasingly complex, multivariate nonlinear *mechanistic* models are being expanded and explored; these come from both the inherently stochastic biochemical modelling perspective and from the applied mathematical side using systems of coupled (ordinary or stochastic) differential equations [38, 76–78].

Statistical model development naturally involves discrete-time representations with components that realistically reflect stochastic noise and measurement error and inherently involve multiple underlying latent processes representing unobserved states that influence the network or cellular system. A specific class of models has a multivariate time series y_* modelled as

$$\begin{array}{rll|l} y_j & = x_{t_j} + \nu_j & & p(y_j|x_{t_j}, \Theta) \\ x_{t+h} & = x_t + G_h(x_t, \Theta)x_t + g_h(\Theta) + \omega_{t,h} & & p(x_{t+h}|x_t, \Theta) \end{array}$$

where the jth observation comes at real-time t_j and the spacings between consecutive observations are typically far greater than the fine time step h. Here x_t represents the underlying state vector of the systems (levels of gene or protein expression, numbers of cells, etc.), Θ all model parameters, ν_* measurement error and ω_* state evolution noise. The density forms to the right indicate more general model forms in which errors and noise may not be additive, when only partial observation is made on the state, and so forth.

The forefront research challenges in this area include development of efficient and effective computations for posterior inference and model comparison. Increasingly, SMC methods including SMC/importance sampling and ABC-SMC are being explored and evaluated, with models of increasing dimension and complexity [4, 35, 52, 58, 77]. In coming years, complex, multivariate dynamical systems studies in biology are sure to define a major growth area for Bayesian dynamic models and time series analysis.

Acknowledgement

The author is grateful to Ioanna Manolopoulou and Emily Fox for comments on this chapter. This work was partly supported by grant DMS-1106516 from the U.S. National Science Foundation (NSF), and grants P50-GM081883 and RC1-AI086032 of the U.S. National Institutes of Health. Any opinions, findings and conclusions or recommendations expressed in this work are those of the author and do not necessarily reflect the views of the NSF or the NIH.

References

[1] Aguilar, O., Prado, R., Huerta, G. and West, M. (1999). Bayesian inference on latent structure in time series (with discussion). In *Bayesian Statistics 6* (ed. J. M. Bernardo, J. O. Berger, A. P. Dawid and A. F. M. Smith), pp. 3–26. Oxford University Press, Oxford.

[2] Aguilar, O. and West, M. (2000). Bayesian dynamic factor models and portfolio allocation. *Journal of Business and Economic Statistics*, **18**, 338–357.

[3] Alspach, D. L. and Sorenson, H. W. (1972). Non-linear Bayesian estimation using Gaussian sum approximations. *IEEE Transactions on Automatic Control*, **AC-17**, 439–448.

[4] Bonassi, F. V., You, L. and West, M. (2011). Bayesian learning from marginal data in bionetwork models. *Statistical Applications in Genetics & Molecular Biology*, **10**, Art 49.

[5] Cargnoni, C., Müller, P. and West, M. (1997). Bayesian forecasting of multinational time series through conditionally Gaussian dynamic models. *Journal of the American Statistical Association*, **92**, 640–647.

[6] Carter, C. K. and Kohn, R. (1994). Gibbs sampling for state space models. *Biometrika*, **81**, 541–553.

[7] Carvalho, C. M., Johannes, M., Lopes, H. F. and Polson, N. G. (2010). Particle learning and smoothing. *Statistical Science*, **25**, 88–106.

[8] Carvalho, C. M., Lucas, J. E., Wang, Q., Chang, J., Nevins, J. R. and West, M. (2008). High-dimensional sparse factor modelling – Applications in gene expression genomics. *Journal of the American Statistical Association*, **103**, 1438–1456.

[9] Carvalho, C. M. and West, M. (2007). Dynamic matrix-variate graphical models. *Bayesian Analysis*, **2**, 69–98.

[10] Carvalho, C. M. and West, M. (2007). Dynamic matrix-variate graphical models – A synopsis. In *Bayesian Statistics 8* (ed. J. M. Bernardo, M. J. Bayarri, J. O. Berger, A. P. Dawid, D. Heckerman, A. F. M. Smith and M. West), pp. 585–590. Oxford University Press, Oxford.

[11] Chen, R. and Liu, J. S. (2000). Mixture Kalman filters. *Journal of the Royal Statistical Society, Series B*, **62**(3), 493–508.

[12] Chib, S., Omori, Y. and Asai, M. (2009). Multivariate stochastic volatility. In *Handbook of Financial Time Series* (ed. T. G. Andersen, R. A. Davis, J. P. Kreiss, and T. Mikosch), New York, pp. 365–400. Springer-Verlag.

[13] Dobra, A., Jones, B., Hans, C., Nevins, J. R. and West, M. (2004). Sparse graphical models for exploring gene expression data. *Journal of Multivariate Analysis*, **90**, 196–212.

[14] Doucet, A., de Freitas, N. and Gordon, N. J. (2001). *Sequential Monte Carlo Methods in Practice*. Springer-Verlag, New York.

[15] Doucet, A., Godsill, S. J. and West, M. (2000). Monte Carlo filtering and smoothing with application to time-varying spectral estimation. *Proceedings of the IEEE International Conference on Acoustics, Speech and Signal Processing*, **II**, 701–704.

[16] Fong, W., Godsill, S. J., Doucet, A. and West, M. (2002). Monte Carlo smoothing with application to speech enhancement. *IEEE Trans. Signal Processing*, **50**, 438–449.

[17] Fox, E. B. and West, M. (2011). Autoregressive models for variance matrices: Stationary inverse Wishart processes. Discussion Paper 11–15, Department of Statistical Science, Duke University. Submitted for publication.
[18] Frühwirth-Schnatter, S. (1994). Data augmentation and dynamic linear models. *Journal of Time Series Analysis*, **15**, 183–202.
[19] Frühwirth-Schnatter, S. (2006). *Finite Mixture and Markov Switching Models*. Springer-Verlag, New York.
[20] Godsill, S. J., Doucet, A. and West, M. (2004). Monte Carlo smoothing for non-linear time series. *Journal of the American Statistical Association*, **99**, 156–168.
[21] Gordon, N. J., Salmond, D. and Smith, A. F. M. (1993). Novel approach to nonlinear/non-Gaussian Bayesian state estimation. *Proc. IEE F*, **140**, 107–113.
[22] Gourieroux, C., Jasiak, J. and Sufana, R. (2009). The Wishart autoregressive process of multivariate stochastic volatility. *Journal of Econometrics*, **150**, 167–181.
[23] Harrison, P. J. and Stevens, C. (1976). Bayesian forecasting (with discussion). *Journal of the Royal Statististical Society, Series B*, **38**, 205–247.
[24] Huerta, G. and West, M. (1999). Bayesian inference on periodicities and component spectral structures in time series. *Journal of Time Series Analysis*, **20**, 401–416.
[25] Jones, B., Dobra, A., Carvalho, C. M., Hans, C., Carter, C. and West, M. (2005). Experiments in stochastic computation for high-dimensional graphical models. *Statistical Science*, **20**, 388–400.
[26] Kim, S., Shephard, N. and Chib, S. (1998). Stochastic volatility: Likelihood inference and comparison with ARCH models. *Review of Economic Studies*, **65**, 361–393.
[27] Krystal, A. D., Prado, R. and West, M. (1999). New methods of time series analysis for non-stationary EEG data: Eigenstructure decompositions of time varying autoregressions. *Clinical Neurophysiology*, **110**, 1–10.
[28] Krystal, A. D., Zoldi, S., Prado, R., Greenside, H. S. and West, M. (1999). The spatiotemporal dynamics of generalized tonic-clonic seizure EEG data: Relevance to the climinal practice of electroconvulsive therapy. In *Nonlinear Dynamics and Brain Functioning* (ed. N. Pradhan, P. Rapp and R. Sreenivasan). New York: Novascience.
[29] Lauritzen, S. L. (1996). *Graphical Models*. Clarendon Press, Oxford.
[30] Liu, F. and West, M. (2009). A dynamic modelling strategy for Bayesian computer model emulation. *Bayesian Analysis*, **4**, 393–412.
[31] Liu, J. and West, M. (2001). Combined parameter and state estimation in simulation-based filtering. In *Sequential Monte Carlo Methods in Practice* (ed. A. Doucet, J. D. Freitas and N. Gordon), pp. 197–217. New York: Springer-Verlag.
[32] Lopes, H. F., McCulloch, R. E. and Tsay, R. (2010). Cholesky stochastic volatility. Technical report, University of Chicago, Booth Business School.
[33] Manolopolou, I., Matheu, M. P., Cahalan, M. D., West, M. and Kepler, T. B. (2012). Bayesian spatio-dynamic modelling in cell motility studies: Learning nonlinear taxic fields guiding the immune response (with invited discussion). *Journal of the American Statistical Association*, **107**, doi: 10.1080/01621459.2012.655995.
[34] Marjoram, P., Molitor, J., Plagnol, V. and Tavaré, S. (2003). Markov chain Monte Carlo without likelihoods. *Proceedings of the National Academy of Sciences USA*, **100**, 15324–15328.
[35] Mukherjee, C. and West, M. (2009). Sequential Monte Carlo in model comparison: Example in cellular dynamics in systems biology. In *JSM Proceedings, Section on Bayesian Statistical Science. Alexandria, VA: American Statistical Association*, pp. 1274–1287.
[36] Nakajima, J. and West, M. (2012). Bayesian analysis of latent threshold dynamic models. *Journal of Business and Economic Statistics*, to appear.
[37] Nakajima, J. and West, M. (2012). Bayesian dynamic factor models: Latent threshold approach. *Journal of Financial Econometrics*, doi: 10.1093/jjfinec/nbs013.

[38] Niemi, J. B. and West, M. (2010). Adaptive mixture modelling Metropolis methods for Bayesian analysis of non-linear state-space models. *Journal of Computational and Graphical Statistics*, **19**, 260–280.

[39] Philipov, A. and Glickman, M. E. (2006). Factor multivariate stochastic volatility via Wishart processes. *Econometric Reviews*, **25**, 311–334.

[40] Philipov, A. and Glickman, M. E. (2006). Multivariate stochastic volatility via Wishart processes. *Journal of Business and Economic Statistics*, **24**, 313–328.

[41] Pitt, M. K. and Shephard, N. (1999). Time varying covariances: A factor stochastic volatility approach (with discussion). In *Bayesian Statistics VI* (ed. J. M. Bernardo, J. O. Berger, A. P. Dawid and A. F. M. Smith), pp. 547–570. Oxford University Press, Oxford.

[42] Pitt, M. K. and Walker, S. G. (2005). Constructing stationary time series models using auxilliary variables with applications. *Journal of the American Statistical Association*, **100**, 554–564.

[43] Pole, A., West, M. and Harrison, P. J. (1994). *Applied Bayesian Forecasting & Time Series Analysis*. Chapman-Hall.

[44] Prado, R., Huerta, G. and West, M. (2001). Bayesian time-varying autoregressions: Theory, methods and applications. *Resenhas*, **4**, 405–422.

[45] Prado, R. and West, M. (1997). Exploratory modelling of multiple non-stationary time series: Latent process structure and decompositions. In *Modelling Longitudinal and Spatially Correlated Data* (ed. T. Gregoire), pp. 349–362. Springer-Verlag.

[46] Prado, R. and West, M. (2010). *Time Series: Modeling, Computation & Inference*. Chapman & Hall/CRC Press.

[47] Prado, R., West, M. and Krystal, A. D. (2001). Multi-channel EEG analyses via dynamic regression models with time-varying lag/lead structure. *Journal of the Royal Statistical Society (Ser. C)*, **50**, 95–110.

[48] Quintana, J. M., Carvalho, C. M., Scott, J. and Costigliola, T. (2010). Futures markets, Bayesian forecasting and risk modeling. In *The Handbook of Applied Bayesian Analysis* (ed. A. O'Hagan and M. West), pp. 343–365. Oxford University Press, Oxford.

[49] Quintana, J. M., Lourdes, V., Aguilar, O. and Liu, J. (2003). Global gambling. In *Bayesian Statistics 7* (ed. J. M. Bernardo, M. J. Bayarri, J. O. Berger, A. P. Dawid, D. Heckerman, A. F. M. Smith and M. West), pp. 349–368. Oxford University Press, Oxford.

[50] Quintana, J. M. and West, M. (1987). An analysis of international exchange rates using multivariate DLMs. *The Statistician*, **36**, 275–281.

[51] Quintana, J. M. and West, M. (1988). Time series analysis of compositional data. In *Bayesian Statistics 3* (ed. J. M. Bernardo, M. H. DeGroot, D. V. Lindley and A. F. M. Smith), pp. 747–756. Oxford University Press, Oxford.

[52] Secrier, M., Toni, T. and Stumpf, M. P. H. (2009). The ABC of reverse engineering biological signalling systems. *Molecular Biosystems*, **5(12)**, 1925–1935.

[53] Shephard, N. (1994). Local scale models: State-space alternative to integrated GARCH models. *Journal of Econometrics*, **60**, 181–202.

[54] Sisson, S. A., Fan, Y. and Tanaka, M. M. (2007). Sequential Monte Carlo without likelihoods. *Proceedings of the National Academy of Sciences USA*, **104**, 1760–1765.

[55] Smith, A. F. M. and West, M. (1983). Monitoring renal transplants: An application of the multi-process Kalman filter. *Biometrics*, **39**, 867–878.

[56] Smith, A. F. M., West, M., Gordon, K., Knapp, M. S. and Trimble, I. (1983). Monitoring kidney transplant patients. *The Statistician*, **32**, 46–54.

[57] Tebaldi, C., West, M. and Karr, A. F. (2002). Statistical analyses of freeway traffic flows. *Journal of Forecasting*, **21**, 39–68.

[58] Toni, T. and Stumpf, M. P. H. (2010). Simulation-based model selection for dynamical systems in systems and population biology. *Bioinformatics*, **26**, 104–110.

[59] Uhlig, H. (1994). On singular Wishart and singular multivariate beta distributions. *Annals of Statistics*, **22**, 395–405.

[60] Uhlig, H. (1997). Bayesian vector autoregressions with stochastic volatility. *Econometrica*, **1**, 59–73.

[61] Wang, H. and West, M. (2009). Bayesian analysis of matrix normal graphical models. *Biometrika*, **96**, 821–834.

[62] West, M. (1986). Bayesian model monitoring. *Journal of the Royal Statistical Society (Ser. B)*, **48**, 70–78.

[63] West, M. (1992). Modelling with mixtures (with discussion). In *Bayesian Statistics 4* (ed. J. M. Bernardo, J. O. Berger, A. P. Dawid, and A. F. M. Smith), pp. 503–524. Oxford University Press.

[64] West, M. (1993). Approximating posterior distributions by mixtures. *Journal of the Royal Statistical Society (Ser. B)*, **54**, 553–568.

[65] West, M. (1993). Mixture models, Monte Carlo, Bayesian updating and dynamic models. *Computing Science and Statistics*, **24**, 325–333.

[66] West, M. (1995). Bayesian inference in cyclical component dynamic linear models. *Journal of the American Statistical Association*, **90**, 1301–1312.

[67] West, M. (1996). Bayesian time series: Models and computations for the analysis of time series in the physical sciences. In *Maximum Entropy and Bayesian Methods 15* (ed. K. Hanson and R. Silver), pp. 23–34. Kluwer.

[68] West, M. (1996). Some statistical issues in Palæoclimatology (with discussion). In *Bayesian Statistics 5* (ed. J. M. Bernardo, J. O. Berger, A. P. Dawid and A. F. M. Smith), pp. 461–486. Oxford University Press.

[69] West, M. (1997). Modelling and robustness issues in Bayesian time series analysis (with discussion). In *Bayesian Robustness* (ed. J. O. Berger, B. Betrò, E. Moreno, L. R. Pericchi, F. Ruggeri, G. Salinetti and L. Wasserman), IMS Monographs, pp. 231–252. Institute for Mathematical Statistics.

[70] West, M. (1997). Time series decomposition. *Biometrika*, **84**, 489–494.

[71] West, M. (2003). Bayesian factor regression models in the "large p, small n" paradigm. In *Bayesian Statistics 7* (ed. J. M. Bernardo, M. J. Bayarri, J. O. Berger, A. P. David, D. Heckerman, A. F. M. Smith and M. West), pp. 723–732. Oxford University Press.

[72] West, M. and Harrison, P. J. (1986). Monitoring and adaptation in Bayesian forecasting models. *Journal of the American Statistical Association*, **81**, 741–750.

[73] West, M. and Harrison, P. J. (1989). Subjective intervention in formal models. *Journal of Forecasting*, **8**, 33–53.

[74] West, M. and Harrison, P. J. (1997). *Bayesian Forecasting & Dynamic Models* (2nd edn). Springer Verlag.

[75] West, M., Prado, R. and Krystal, A. D. (1999). Evaluation and comparison of EEG traces: Latent structure in non-stationary time series. *Journal of the American Statistical Association*, **94**, 1083–1095.

[76] Wilkinson, D. J. (2006). *Stochastic Modelling for Systems Biology*. London: Chapman & Hall/CRC.

[77] Wilkinson, D. J. (2011). Parameter inference for stochastic kinetic models of bacterial gene regulation: a Bayesian approach to systems biology (with discussion). In *Bayesian Statistics 9* (ed. J. M. Bernardo, M. J. Bayarri, J. O. Berger, A. P. David, D. Heckerman, A. F. M. Smith and M. West), pp. 679–700. Oxford University Press.

[78] Yao, G., Tan, C., West, M., Nevins, J. R. and You, L. (2011). Origin of bistability underlying mammalian cell cycle entry. *Molecular Systems Biology*, **7**, 485.

9 Hierarchical modelling in time series: the factor analytic approach

DANI GAMERMAN AND ESTHER SALAZAR

9.1 Introduction

Adrian Smith's contribution to statistics spans across a very wide range of topics. One can single out his relentless effort towards making the Bayesian approach to inference applicable, in a series of computationally oriented papers with approximating methods. This effort reached its climax with his landmark paper [18], after which MCMC methods became famous and widespread.

Another line of contributions was more methodological in terms of proposing new routes for exploring more elaborate data structures. It led to another landmark paper [29]. This JRSSB discussion paper was devoted to explaining how information from different but related sources of information could be combined in a regression framework with a hierarchical structure. This paper was extended to the time-varying context in [16] with the use of dynamic models. One of us was fortunate to interact with Adrian in [17], where hierarchical and dynamic models were also used to combine information from different time series sources.

The idea of *borrowing information* from related sources is very powerful. It has proved to be very useful in past decades where complex data structures have begun to be tackled, as they required sophisticated modelling strategies. A vital element in such structured settings is the ability to extract from the data possible similarity patterns. This can be achieved in a number of ways, including hierarchical modelling, non-parametric components and factor analysis.

This chapter will address the issue of combining information from a possibly large time series with a factor analytic approach. Results obtained from this exercise are a (hopefully much) smaller number of latent time series that represent the main features of the complete dataset of time series originally available. Each combination of a time series and a factor gives rise to a weight or loading that informs in which ways the different original series were combined. These loadings are useful quantities as they allow the identification of common features and interpretation of the relationship or correlation structure between the different series.

These concepts will be discussed and combined in a number of forms in this chapter. Special attention will be devoted to the exploration of these ideas in the area of spatial statistics. It will be shown that this area is not only an area of application of these ideas but is one of the main beneficiaries of these developments. Spatial statistics is devoted to the analysis of a collection of processes that exhibit correlation due to their (geographic) location. The main goal there is to appropriately capture the spatial dependence in order to be able to extrapolate information from a few data sources to the whole region of interest. This inevitably leads to the need for parsimonious forms for representing the correlation structure. This chapter will show how the ideas behind dynamic factor models apply in this setting via illustrative examples with real data problems.

This chapter is organized as follows. Section 9.2 reviews the literature on factor analysis. Section 9.3 presents some basic factor model extensions for modelling high-dimensional multivariate time series. Section 9.4 describes applications of these ideas in the context of spatial analysis. Section 9.5 describes how regression ideas can be incorporated into the factor model setting. This is accomplished by enlarging the scope of the models to include explanation via covariate time series. Section 9.6 draws some concluding remarks and points at possible directions for further work.

9.2 A short review of factor analysis

Factor analysis is a useful statistical technique used widely for modelling multivariate data by a few unobserved set of variables called latent factors. More specifically, the observed variables are modelled as linear combinations of the latent factors plus an idiosyncratic error. In general, this approach is applied for the following purposes: (*i*) dimension reduction, (*ii*) identifying underlying structures, and (*iii*) modelling of sparse covariance structures. From a classical point of view, the term was first introduced in [49] and later discussed in [50], [1] and [22], among many others. In recent years, a fully Bayesian treatment of factor models became feasible due to the improvements in Bayesian computation, especially Markov chain Monte Carlo (MCMC) simulation methods. In this context, the Bayesian specifications proposed in [21], [42] and [3] can be mentioned.

The factor model is defined as follows

$$y_t = \boldsymbol{\beta} \boldsymbol{f}_t + \epsilon_t, \quad \epsilon_t \sim N(\mathbf{0}, \boldsymbol{\Sigma}) \tag{9.1}$$

where $t = 1, \ldots, T$, $\boldsymbol{y}_t$ is an n-dimensional observational vector, $\boldsymbol{\beta}$ is an $n \times k$ factor loadings matrix, $\boldsymbol{f}_t \sim N(\mathbf{0}, \boldsymbol{I}_k)$ are independent k-dimensional vectors called latent factors such that $k \ll n$ and $\boldsymbol{\Sigma} = \text{diag}(\sigma_1^2, \ldots, \sigma_n^2)$. This model implies that, given the factors, each $\boldsymbol{y}_t$ has independent components that are $\text{var}(y_{it}|\boldsymbol{f}_t) = \sigma_i^2$ and $\text{cov}(y_{it}, y_{jt}|\boldsymbol{f}_t) = 0\ (i \neq j)$. Moreover, dependence among components is induced by marginalizing over the distribution of the factors so $\boldsymbol{y}_t|\boldsymbol{\beta}, \boldsymbol{\Sigma} \sim N(\mathbf{0}, \boldsymbol{\Omega})$ where $\boldsymbol{\Omega} = \boldsymbol{\beta}\boldsymbol{\beta}^T + \boldsymbol{\Sigma}$. Note also that independence of the factors and the idiosyncratic terms $\boldsymbol{\epsilon}_t$ induces independence of the observations.

Two important issues have to be mentioned at this point. The first one is regarding identifiability problems related to the non-unique decomposition of $\boldsymbol{\Omega}$ and the inference about the number of factors. Many ways to handle this problem can found in the literature. Basically, the idea is to impose constraints on $\boldsymbol{\beta}$, as, for example the lower constraint defined in [21], [2] and [34]. However, in some applications, identifiability of the factor loadings is not required, especially for covariance matrix estimation, variable selection and prediction (see [5] for more details). This issue will be further discussed in the next section where structured priors for $\boldsymbol{\beta}$ can be used. On the other hand, uncertainty about the number of latent factors has been studied in different ways. The most common approach was fitting the factor model for different choices of k and then using a selection criterion like AIC or BIC for model selection. [34] proposed fully Bayesian inference on the number of factors through a reversible jump MCMC (RJMCMC) [23]. Their proposal was compared with a number of other alternatives based on bridge sampling [35]. Another recent approach relies on zeroing a subset of factor loadings using variable selection priors such as binary indicator δ_{ij} [19]. Based on this idea, interesting applications can be mentioned. See [8] and [15] for gene expression and financial modelling, respectively, are a few examples. In this chapter, we further discuss the RJMCMC scheme as a tool for model selection.

9.3 Dynamic factor models

9.3.1 Basic definitions

Dynamic factor models (DFMs) were developed in a number of ways and have become a useful tool for modelling high-dimensional multivariate time series. The core idea is to explain the common dynamic structure of the multivariate time series through a set of common (time series) factors. This is achieved by the introduction of flexible temporal correlation structures for the latent factors, previously assumed to be independent. This renders the DFM capable of assessing the complexity of time series data. Models along these lines were proposed in [4, 10, 13, 20, 36, 39, 45].

Earlier approaches have been primarily concerned with modelling multivariate stationary time series considering latent factors with a time-varying mean function. In this context, in [39] was proposed a methodology to identify the number of latent factors in a vector of stationary times series. Specifically, temporal correlation is introduced through a k-dimensional vector that follows an autoregressive moving average process (see [36, 4] and references therein for related ideas). For the nonstationary case, a methodology for building DFM for nonstationary time series in state space form was proposed in [40, 41] and, more recently, a new approach that allows nonstationary factors not necessarily driven by unit roots was introduced in [38].

In this section we focus on DFM for both stationary and nonstationary time series where the k-dimensional latent factor $\boldsymbol{f}_t$ (state vector) follows a general VARMA(p, q) representation

$$\boldsymbol{f}_t = \boldsymbol{\Gamma}_1\boldsymbol{f}_{t-1} + \ldots + \boldsymbol{\Gamma}_p\boldsymbol{f}_{t-p} + \boldsymbol{\omega}_t + \boldsymbol{\Xi}_1\boldsymbol{\omega}_{t-1} + \ldots + \boldsymbol{\Xi}_q\boldsymbol{\omega}_{t-q} \tag{9.2}$$

where $\boldsymbol{\omega}_t \sim N(\mathbf{0}, \boldsymbol{\Lambda})$, $\forall t$. The latent nature of the factors makes it difficult to precisely estimate this full model. It what follows, we will concentrate the presentation on a simplified VAR(1) version, obtained when $p = 1$ and $q = 0$. A number of features are more clearly understood in this setting and will be discussed below. This factor evolution is driven by the following transition equation

$$\boldsymbol{f}_t = \boldsymbol{\Gamma}\boldsymbol{f}_{t-1} + \boldsymbol{\omega}_t, \quad \boldsymbol{\omega}_t \sim N(\mathbf{0}, \boldsymbol{\Lambda}) \tag{9.3}$$

where $\boldsymbol{\Gamma}$ is a symmetric $k \times k$ autoregressive coefficient matrix characterizing the dynamic evolution of the common factors and $\boldsymbol{\Lambda}$ is a $k \times k$ covariance matrix with elements λ_{ij}, $i, j = 1, \ldots, k$. Note that $\boldsymbol{\Gamma}$ and $\boldsymbol{\Lambda}$ are not necessarily diagonal matrices so they can be defined to deal, for instance, with seasonal components and nonstationary common factors. Equations (9.1) and (9.2) or (9.3) define the dynamic factor model and, in a similar fashion to standard factor analysis, the latent factors $\boldsymbol{f}_t$ capture the time-varying correlation structure of the data.

Working within a Bayesian framework, some important issues related to model specification and posterior inference can be mentioned at this point.

Prior specification

The prior for the latent factor is given in equation (9.3) and completed by $\boldsymbol{f}_0 \sim N(m_0, C_0)$ with known hyperparameters m_0, C_0. As was mentioned before, many specifications for $\boldsymbol{\Gamma}$ can be considered. One possibility for the $\boldsymbol{\Lambda}$ matrix is a diagonal form with elements λ_i. In this case, a typical choice of prior for the λ_is is independent Gamma distributions. Similar independence assumptions can be made for the autoregressive matrix $\boldsymbol{\Gamma}$. One possibility is to consider $\boldsymbol{\Gamma} = \text{diag}(\gamma_1, \ldots, \gamma_k)$ such that, $\gamma_j \sim N(0, a)$ independent, for $j = 1, \ldots, k$, for some large value of a if one wants to represent vague prior information. If one is concerned with the possibility of unit roots and non-stationarity, the mixture prior $\gamma_j \sim \pi N_{(-1,1)}(0, a) + (1 - \pi)\delta_1(\gamma_j)$ may be assumed for the autoregressive coefficients, where $\pi \in (0, 1]$ and a are known hyperparameters, $N_{(l,u)}(\cdot, \cdot)$ denotes the normal distribution constrained to assumed values only in (l, u), $\delta_1(\gamma_j) = 1$ if $\gamma_j = 1$ and $\delta_j(\gamma_j) = 0$ if $\gamma_j \neq 1$ (see [26] for more details); for $\pi \neq 1$, the mixture prior allows the possibility that nonstationary factors be incorporated; if $\pi = 1$ we are in the stationary case.

Correlated factors can also be incorporated into the DFM. One example of that is the inclusion of h seasonal common factors to capture a possibly periodic behaviour of the time series. In that case, $\boldsymbol{\Gamma}$ could be specified as $\boldsymbol{\Gamma} = \text{diag}(\boldsymbol{\Gamma}_0, \boldsymbol{\Gamma}_1, \ldots, \boldsymbol{\Gamma}_h)$ where $\boldsymbol{\Gamma}_0 = \text{diag}(\gamma_{0,1}, \ldots, \gamma_{0,k})$,

$$\boldsymbol{\Gamma}_l = \begin{pmatrix} \cos(2\pi l/p) & \sin(2\pi l/p) \\ -\sin(2\pi l/p) & \cos(2\pi l/p) \end{pmatrix}, \quad l = 1, \ldots, h,$$

p is the seasonal period and $h = p/2$ is the *maximum* number of harmonics needed to capture the seasonal behaviour of the time series, (see [54], Chapter 8, for more details). As a consequence $\boldsymbol{\Lambda} = \text{diag}(\boldsymbol{\Lambda}_0, \boldsymbol{\Lambda}_1, \ldots, \boldsymbol{\Lambda}_h)$ and each $\boldsymbol{\Lambda}_l$ is no longer diagonal with *inverted Wishart* distribution as a prior.

Factor loadings specification

For the factor loadings one can take independent normal priors for each element of $\boldsymbol{\beta}$ such that $\beta_{ii} \sim N_{(0,\infty)}(0, b)$, $\beta_{ij} \sim N(0, b)$ only for $i > j$ (see [34] for more details), since identifiability constraints impose $\beta_{ij} = 0$ for $i < j$. However, in practice, one may also be interested in including conditional dependencies within the elements of $\boldsymbol{y}_t$. In order to do that, the underlying idea is to include a flexible correlation structure into the columns of $\boldsymbol{\beta}$, denoted by $\boldsymbol{\beta}_{(j)}$ $(j = 1, \ldots, k)$. In the context of spatial analysis, a number of papers have examined inducing dependencies through $\boldsymbol{\beta}_{(j)}$. For example, in [52] the columns of $\boldsymbol{\beta}$ are modelled as orthonormal basis functions and in [6] and [44] smoothed deterministic kernels are used to build $\boldsymbol{\beta}$. Alternatively, [32] introduced a spatial DFM where the columns of the factor loadings matrix follow independent Gaussian random fields. This idea will be discussed and illustrated in the next section for modelling space–time data. Additional developments on factor loadings specification include, for example, [8] and [5] for sparse factor analysis, and [30] for latent time-varying loadings, among others.

Posterior inference

Fully Bayesian treatment of the standard and dynamic factor models via MCMC methods is described in detail in [34] and [2], respectively. More specifically, the inference procedure is designed for two cases: known and unknown number of factors k. Considering that the number of factor k is known, the MCMC scheme described in [34] can easily be adapted where the common factors are jointly sampled via the well-known forward filtering backward sampling (FFBS) scheme [7, 14]. For the second case, model selection is performed by computing *posterior model probabilities* (PMPs) for different choices of k. In particular, the *reversible jump* MCMC algorithm, proposed/described in [32, 34] for DFM, can be used. The algorithm allows for a simple method of calculating the PMP from preliminary MCMC runs. As mentioned in the previous references, the Bayesian model search via RJMCMC penalizes over and under-parameterized factor models.

9.3.2 Hierarchical DFM

A common criticism in DFM is that the common latent factors are difficult to interpret. In large n settings and for multi-level datasets, the dimension reduction of the problem may involve loss of data structure. In this context, a hierarchical construction of the model to allow a progressive reduction in the dimensionality as the levels becomes higher may be desired. For example, the hierarchical construction for dynamic linear models (DLMs) proposed in [16] provides a general framework for analysis of multivariate time series. In accordance with this idea and following the same notation introduced in Subsection 9.3.1, the *3-level hierarchical dynamic factor model* (HDFM) can be written as

$$\boldsymbol{y}_t = \boldsymbol{\beta}_1 \boldsymbol{f}_{1t} + \boldsymbol{\epsilon}_{1t}, \quad \boldsymbol{\epsilon}_{1t} \sim N(\boldsymbol{0}, \boldsymbol{\Sigma}_1) \tag{9.4}$$

$$\boldsymbol{f}_{1t} = \boldsymbol{\beta}_2 \boldsymbol{f}_{2t} + \boldsymbol{\epsilon}_{2t}, \quad \boldsymbol{\epsilon}_{2t} \sim N(\boldsymbol{0}, \boldsymbol{\Sigma}_2) \tag{9.5}$$

$$\boldsymbol{f}_{2t} = \boldsymbol{\beta}_3 \boldsymbol{f}_{3t} + \boldsymbol{\epsilon}_{3t}, \quad \boldsymbol{\epsilon}_{3t} \sim N(\boldsymbol{0}, \boldsymbol{\Sigma}_3) \tag{9.6}$$

$$\boldsymbol{f}_{3t} = \boldsymbol{\Gamma} \boldsymbol{f}_{3,t-1} + \boldsymbol{\omega}_t, \quad \boldsymbol{\omega}_t \sim N(\boldsymbol{0}, \boldsymbol{\Lambda}) \tag{9.7}$$

where $\boldsymbol{f}_{it}, i = 1, 2, 3$, are k_i-dimensional vectors satisfying $k_1 > k_2 > k_3$, $\boldsymbol{\beta}_1$ is a $n \times k_1$ matrix, $\boldsymbol{\beta}_2$ and $\boldsymbol{\beta}_3$ are $k_1 \times k_2$ and $k_2 \times k_3$ matrices respectively, and $\boldsymbol{\Gamma}$ is a $k_3 \times k_3$ matrix. More specifically, eqn (9.4) represents the observation equation, eqns (9.5) and (9.6) the structural equations and eqn (9.7) the system equation. As mentioned in [16], the previous HDFM can be reduced to considering only two levels/stages of hierarchy by setting $\boldsymbol{\beta}_3 = \boldsymbol{I}_{k_3}$ and $\boldsymbol{\Sigma}_3 = \boldsymbol{0}$, a zero matrix. Again, further levels are easily induced but this would rarely be required.

9.3.3 Generalized DFM

The DFM can also be extended to allow for non-Gaussian observations. More specifically, the generalized DFM (GDFM) is a hierarchical model where the first level equation (observation equation) is given by

$$p(y_{ti}|\eta_{ti}, \psi) = \exp\{\psi(y_{ti}\eta_{ti} - b(\eta_{ti})) + c(y_{ti}, \psi)\} \tag{9.8}$$

where η_{ti} is the natural parameter and ψ is the dispersion parameter. The natural parameter η_{ti} is related to the temporal components through the link function ν such that $\eta_{ti} = \nu(\theta_{ti})$. Consequently, the model is completed by specifying the following two levels of hierarchy

$$\boldsymbol{\theta}_t = \boldsymbol{\mu}_t + \boldsymbol{\beta} \boldsymbol{f}_t \tag{9.9}$$

$$\boldsymbol{f}_t = \boldsymbol{\Gamma} \boldsymbol{f}_{t-1} + \boldsymbol{\omega}_t, \quad \boldsymbol{\omega}_t \sim N(\boldsymbol{0}, \boldsymbol{\Lambda}) \tag{9.10}$$

where $\boldsymbol{\theta}_t = (\theta_{t1}, \ldots, \theta_{tn})^T$, $\boldsymbol{\mu}_t$ is the mean level, and $\boldsymbol{\beta}$, $\boldsymbol{\Gamma}$ and $\boldsymbol{\Lambda}$ have the same specifications as the SDFM. See [53] for more details in the context of generalized DLMs.

Full Bayesian treatment for this new class is more challenging, specifically for MCMC sampling the common factor. In the previous cases, the full conditional distribution for joint sampling this component was normal and thus easily sampled from using, for example, the FFBS. This is no longer valid here and efficient proposal are very difficult to obtain, specially for large time series with large T. Componentwise sampling is also very inefficient. The solution here is a compromise with this component sampled in blocks. To this end, a block sampling scheme that combines techniques such as extended Kalman filter and block sampling was proposed in [31] with good performance in the applications.

9.4 Applications to spatial statistics

In this section, we discuss some applications of the above mentioned approaches for spatial and spatio-temporal processes.

The use of factor analysis to model multivariate spatial data has been treated in a number of ways. Here, we focus on the case in which factor analysis is used to identify clusters or groups of locations/regions (spatial dependence) whose temporal behaviour is driven by a set of common dynamic latent factors (temporal dependence). In previous works, either common dynamic factors or factor loadings matrices are restricted to be deterministic functions. Specifically, when the common factors are non-stochastic the space–time dynamic model proposed in [48] is obtained. On the other hand, when $\boldsymbol{\beta}$ is defined as a deterministic function the structure proposed in [52] and [6] is obtained.

In [32] was introduced a new class of models called the *spatial dynamic factor model* (SDFM), derived from the standard DFM. More specifically, the temporal dependence is modelled by the latent common factors and the spatial dependence is also modelled stochastically by the columns of the factor loadings $\boldsymbol{\beta}_{(j)}$. These are assumed to follow independent Gaussian processes. The role played by the stochastic structure is to allow further flexibility to the deterministic specification, that is restrictive by definition.

The SDFM is defined by eqns (9.1) and (9.3), where $\boldsymbol{y}_t = (y_{1t}, \ldots, y_{nt})^T$ such that y_{it} is an observation measured at time t and location $s_i \in \mathbb{R}^d$. Each column $\boldsymbol{\beta}_{(j)} = (\beta_j(s_1), \ldots, \beta_j(s_n))^T$ is defined as

$$\boldsymbol{\beta}_{(j)} \sim N(\boldsymbol{\theta}_j, \tau_j^2 \mathbf{R}_{\boldsymbol{\phi}_j}), \quad \text{for} \quad j = 1, .., k \tag{9.11}$$

where $\boldsymbol{\theta}_j$ is the n-dimensional mean vector, τ_j^2 is the common variance of the spatial process, $\mathbf{R}_{\boldsymbol{\phi}_j}$ is the matrix correlation function with (l, m)-element given by $\{\mathbf{R}_{\phi_j}\}_{l,m} = \rho_{\phi_j}(\| s_l - s_m \|)$, $l, m = 1, \ldots, n$ and $\rho_\phi(\cdot)$ represents a spatial correlation function like exponential or Matérn specified by the parameter ϕ.

As an illustration, Figure 9.1 shows a simulated SDFM with $k = 2$ common latent factors. Note that the surfaces for $\boldsymbol{y}_t$ are driven by the spatial behaviour of the $\boldsymbol{\beta}_{(j)}$s $(j = 1, 2)$ weighted by the values of the common factors. It is important to mention that the SDFM implies nonseparable forms of the covariance function when $k \geqslant 2$. In fact, if $\boldsymbol{\Gamma}$ and $\boldsymbol{\Lambda}$ are diagonal matrices, the covariance between two different sites at two different time indexes is given by

$$\text{cov}(y_{it}, y_{j,t+h}) = \sum_{l=1}^{k} \lambda_l \gamma_l^h (1 - \gamma_k^2)(\tau_j^2 \rho_{\phi_j} + \theta_{il}\theta_{jl})$$

That characteristic implies that the SDFM is able to model complex space–time interactions. In contrast, when $k = 1$ spatial and temporal covariance functions are identified separately.

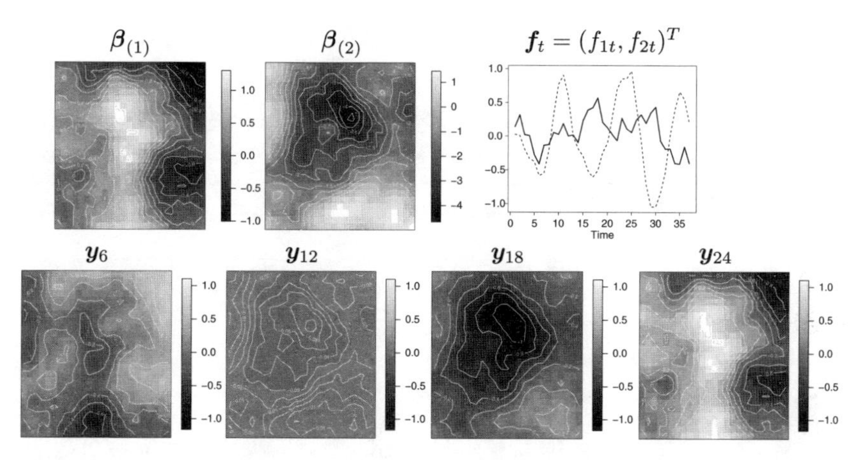

Figure 9.1 Simulated spatial dynamic 2-factor model. *First row*: Gaussian processes for the two columns of $\boldsymbol{\beta}$ and simulated dynamic factors (time series) $\boldsymbol{f}_t = (f_{1t}, f_{2t})^T$ for $t = 1, \ldots, 36$. The first factor (dashed line) has a seasonal behaviour with period $p = 12$. The second factor (solid line) follows an AR process with autoregressive parameter $\gamma_{22} = 0.9$. *Second row*: $\boldsymbol{y}_t$ processes following eqn (9.1) for $t = 6, 12, 18, 24$.

Example 9.1 The SDFM is used to examine the spatio-temporal variation in weekly concentration levels of nitrate (NO_3) across 22 monitoring stations located in eastern USA for $T = 312$ weeks (1st week of 1998 – 52nd week of 2003). The logarithm transformation was used to normalize the data and a seasonal common factor was considered to capture the yearly periodic behaviour repeated every 52 weeks (seasonal period). The SDFM with $k = 3$ regular factors and 1 seasonal factor was compared against other models and selected for fit to the data. Also, a Matérn correlation function is used to specify the spatial correlation structure of each $\boldsymbol{\beta}_{(j)}$ $(j = 1, \ldots, 4)$ with smoothness parameter equal to 1.

Figure 9.2 shows some posterior results of the fitted model. The four maps of the factor loadings (estimated via Bayesian interpolation) show distinct spatial patterns across the study area. Note that the temporal behaviour of the time series is directly related with the higher values of the interpolated surfaces (white areas). The loadings for the second factor are higher in the western region, specifically in some areas of Indiana, Ohio and Kentucky. This factor mimics, roughly, the spatial pattern of the NO_3 across time. Also, the results indicated that this factor is nonstationary. In addition, Figure 9.3, panel (a) shows a plot of observed versus fitted values (considering the original scale). The points roughly follow a straight line indicating good predictions. Panel (b) presents interpolation results for one of the out-of-sample monitoring stations (SPD station). The NO_3 interpolated values are very close to the real values, indicating the good interpolated performance of the model. Finally, panels (c) and (d) show forecast values for the 1st and 12th weeks of 2004. Note that the spatial pattern is almost preserved but with lowest values for the entire region for the 12th week. That behaviour is expected since high concentration levels of nitrate are expected in the first weeks of the year. ||

The SDFM can also be extended to allow for non-Gaussian observations. In [31] was introduced a new class of spatio-temporal models for multivariate exponential family data, called the *generalized spatial dynamic factor model* (GSDFM). In this formulation, the spatial and temporal components are modelled via a latent factor analysis of the canonical transformation of the mean function. The model is given by eqns (9.8)–(9.10) with the addition of the Gaussian prior (9.11) for the loadings. Note that this class of model also leads to a nonseparable spatio-temporal covariance structure. This characteristic is associated with the linear predictor $\boldsymbol{\theta}_t$ where, for $k > 1$, both spatial and temporal covariance structures can not be separately identified.

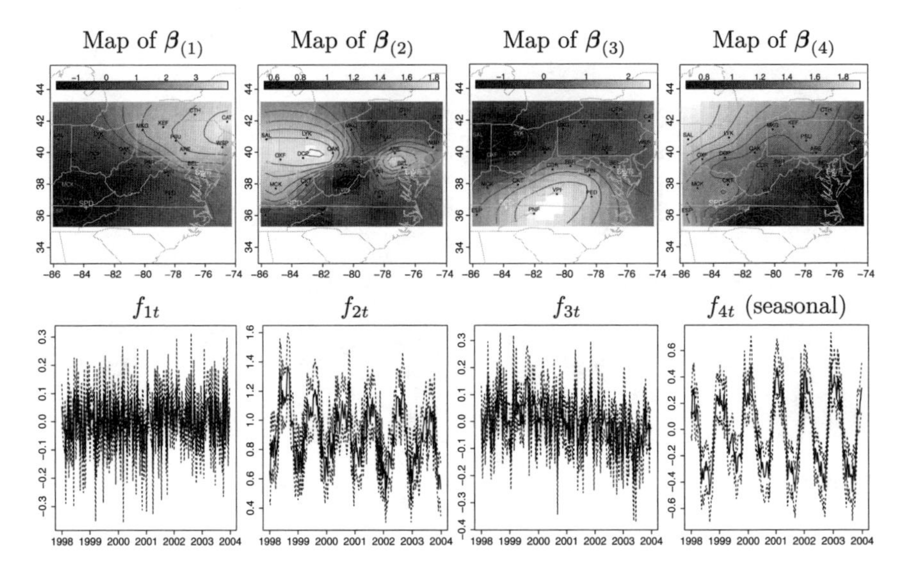

Figure 9.2 *First row*: Bayesian interpolation of the factor loadings. Values represent the range of the posterior means. *Second row*: Posterior means of the factors. Solid lines represent the posterior means and dashed lines the 95% credible intervals.

Example 9.2 We are interested in modelling daily rainfall occurrences (over 1 mm) in northern Oceania in 2001. The data contains $T = 365$ binary observations measured at 19 meteorological stations, 14 of them located in the Federated States of Micronesia and 5 in the Marshall Islands. Figure 9.4(a) shows the study area as well as the geographic location of the stations. We aim to identify microclimates over the study region and also fit rain probability maps for the whole area and across time. In addition, two stations were left out of the analysis for interpolation purposes. The model considered is given by eqns (9.8)–(9.10) such that $y_{ti} \sim \text{Bernoulli}(p_{ti})$, logistic link function $\theta_{ti} = \log(p_{ti}/(1 - p_{ti}))$, a Matérn correlation function for the specification of the factor loadings matrix and $\boldsymbol{\theta}_{(j)} = \mu_j \mathbf{1}_{17}$. The GSDFM was fitted considering $k = 1, 2, 3, 4$ common factors. Comparisons between models are based on the PMP. In this application, the model with three common factor shows the best results with PMP equal to 0.46.

Figure 9.5 shows the interpolated loading associated with the first, second and third factors as well as the estimated temporal behaviour of the common factors. The spatial loadings, interpolated via Bayesian kriging, indicate a smooth variation in different directions, especially for the first factor. These findings allow the recognition of microclimates, more specifically in the eastern part (around Marshall Islands), as shown in the map for $\boldsymbol{\beta}_{(3)}$. In addition, the results indicate the presence of one nonstationary factor (2nd factor) with posterior probability $\text{pr}(\gamma_2 = 1|\boldsymbol{y}^T) = 0.89$, where $\boldsymbol{y}^T = (\boldsymbol{y}_1, \ldots, \boldsymbol{y}_T)$. Other interesting results of the model are the rain probability time series for observed and interpolated stations, the latter interpolated via Bayesian kriging. Figure 9.4(b) shows those time series for two stations, FSM13 (included in the analysis) and MI5 (left out of the analysis). The posterior probabilities of rainfall occurrence seem to follow the general trend of the observed binary time series. Also, for station MI5, we tested the capability of the model in handling missing data, especifically for the days 121–170 (delimited by the vertical dashed lines). Note that the temporal behaviour of the probability is mainly driven by the third factor, which is expected, given the location of the station. ||

An alternative specification for the factor loadings matrix can be considered for the SDFM. For example, the discrete process convolution approach proposed in [25], or (more recently) the

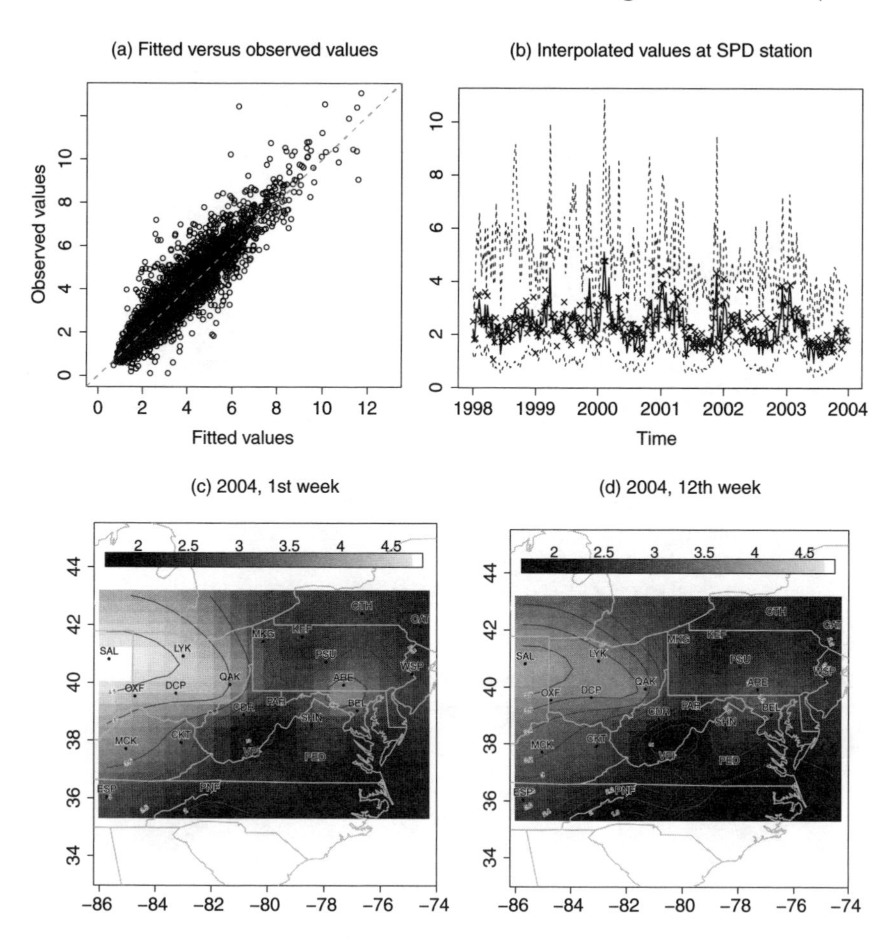

Figure 9.3 (a) Plot of fitted versus observed values of NO_3 for the whole period and for the 22 stations. (b) Interpolated values at SPD station left out from the sample used for fitting. Dashed lines represent the 95% credible intervals and the symbol × the observed NO_3 concentration level. (c)–(d) Forecast values for two different weeks in 2004.

spatial model considering compact support kernels as proposed in [28]. Here, we discuss a related approach that uses an approximate Gaussian process for $\boldsymbol{\beta}_{(j)}$s instead of deterministic kernels. This new model specification was recently proposed in [43] for comparing and blending regional climate model predictions. See the cited paper for more details related to the fully Bayesian treatment of the model.

More specifically, for each $y_t(s)$ measured at time t and location s we have that

$$y_t(s) = \mu_t(s) + \omega_t(s) + \epsilon_t(s), \quad \epsilon_t(s) \sim N(0, \sigma^2) \tag{9.12}$$

where $\mu_t(s)$ may represent a regression component and $\omega_t(s)$ is the space–time component that follows a Gaussian process. Many specifications can be considered for $\omega_t(s)$. Here we opted to use the *modified predictive process* (see [11] and the references therein), letting $\omega_t(s) = \tilde{\omega}_t(s) + \tilde{\epsilon}_t(s)$, where $\tilde{\epsilon}_t(s) \sim N(0, \tau^2 - \nu(s)^T \boldsymbol{H}^{-1} \nu(s))$ and $\tilde{\omega}_t(s)$ is represented on a set of k basis functions $B_l(s) = [\nu(s)^T \boldsymbol{H}^{-1}]_l$ where $\nu(s) = \tau^2(\rho_\phi(s, s_1^*), \ldots, \rho_\phi(s, s_k^*))$,

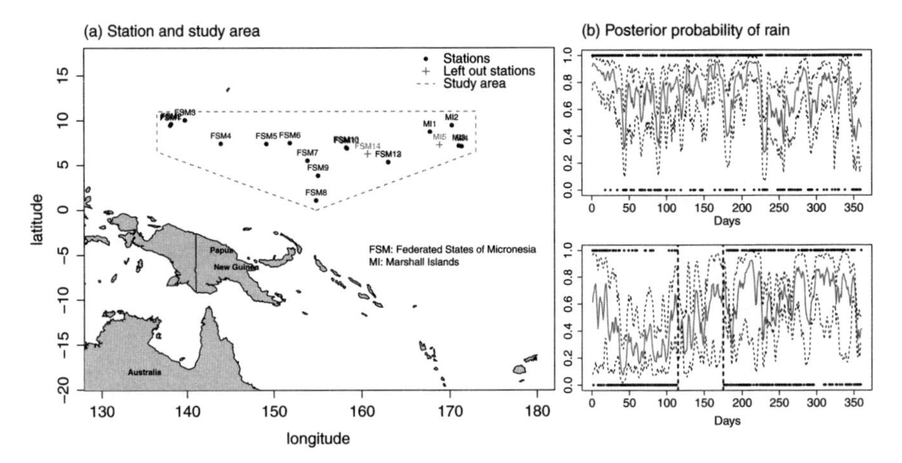

Figure 9.4 (a) Location of the monitoring stations. (b) Daily posterior probability of rainfall occurrence at FSM13 station (above) and MI5 station(below). The latter shows the results of the spatial interpolation since station MI5 was left out of the analysis for interpolation purposes. Dots are rain indicators, solid lines are rain mean probabilities and dashed lines are 95% credibility intervals.

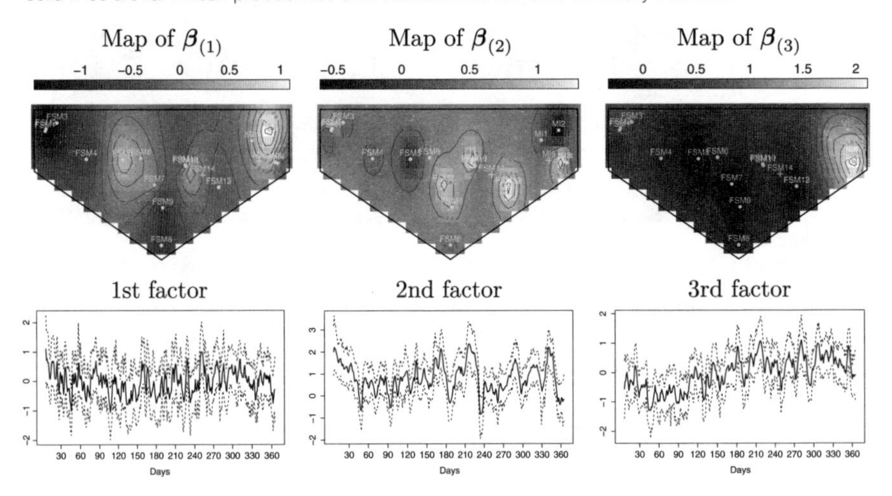

Figure 9.5 *First row*: Bayesian interpolation of the three columns of the factor loadings matrix. Values represent the range of the posterior means. *Second row*: Daily posterior means of the first, second and third dynamic factors. Solid lines represent the posterior means and dashed lines the 95% credible intervals.

$\{\boldsymbol{H}\}_{lm} = \tau^2 \rho_\phi(s_l^*, s_m^*)$ for $l, m = 1, \ldots, k$ and $\{s_l^*; l = 1, \ldots, k\}$ a set of *selected knots* and is given by $\tilde{\omega}_t(s) = \sum_{l=1}^{k} B_l(s)\gamma_{t,l} = \boldsymbol{B}(s)^T \boldsymbol{\gamma}_t$.

The temporal evolution of $\boldsymbol{\gamma}_t$ is specified as $\boldsymbol{\gamma}_t \sim N(\psi \boldsymbol{\gamma}_{t-1}, \boldsymbol{H})$ where $\psi \sim N_{(-1,1)}(\mu_\psi, \sigma_\psi)$. After a SVD decomposition of $\boldsymbol{H} = \boldsymbol{P}\boldsymbol{\Lambda}\boldsymbol{P}^T$ and letting $\boldsymbol{\gamma}_t = \boldsymbol{P}\boldsymbol{f}_t$ we can rewrite $\tilde{\omega}_t(s)$ as $\tilde{\omega}_t(s) = \boldsymbol{B}(s)^T \boldsymbol{P}\boldsymbol{f}_t = \boldsymbol{\beta}(s)^T \boldsymbol{f}_t$ and therefore $\boldsymbol{f}_t \sim N(\psi \boldsymbol{f}_{t-1}, \boldsymbol{\Lambda})$ (with independent elements given that $\boldsymbol{\Lambda}$ is a diagonal matrix).

If we conveniently rewrite eqn (9.12) in vector notation and by considering the previous specification we have

$$\boldsymbol{y}_t = \boldsymbol{\mu}_t + \boldsymbol{\beta}\boldsymbol{f}_t + \tilde{\boldsymbol{\epsilon}}_t + \boldsymbol{\epsilon}_t$$

$$\boldsymbol{f}_t = \psi \boldsymbol{f}_{t-1} + \boldsymbol{\omega}_t, \quad \boldsymbol{\omega}_t \sim N(\boldsymbol{0}, \boldsymbol{\Lambda})$$

The previous specification resembles the SDFM where $\boldsymbol{\beta}$ is an $n \times k$ matrix with k being the number of pre-selected knots (fixed), $\boldsymbol{\Lambda} = \mathrm{diag}(\lambda_1, \ldots, \lambda_k)$ with $\lambda_1 > \ldots > \lambda_k$, therefore $\boldsymbol{\beta}_{(1)}$ describes the main model of spatial variability, $\boldsymbol{\beta}_{(2)}$ the second, and so on.

Finally, in the following example we describe a *spatial hierarchical dynamic factor model* (SHDFM) for socio-economic multi-level measurements. The idea is to build a model-based vulnerability index that account for the different levels of hierarchy (for example, census tracts and capitals). The spatial version of this model was proposed in [33] to build Uruguayan vulnerability index at different geographical resolutions.

Example 9.3 Consider the p-dimensional vector of socio-economical variables $\boldsymbol{y}_{ijt} = (y_{ijt,1}, \ldots, y_{ijt,p})$ at capital i ($i = 1, \ldots, I$), census tract j ($j = 1, \ldots, n_i$) and time t. We aim to infer vulnerability indexes at two levels of resolution: capitals (coarse level) and census tracts (fine level). The proposed *two level SHDFM* can be written as

$$\boldsymbol{y}_{ijt} = \mu + \boldsymbol{\beta} f_{ijt}^{(1)} + \boldsymbol{\epsilon}_{ijt}^{(1)}, \quad \boldsymbol{\epsilon}_{ijt}^{(1)} \sim N(\boldsymbol{0}, \boldsymbol{\Sigma})$$

$$f_{ijt}^{(1)} = \theta_{it} + f_{ijt}^{(2)} + \epsilon_{ijt}^{(2)}, \quad \epsilon_{ijt}^{(2)} \sim N(0, \psi)$$

$$\boldsymbol{f}_{it}^{(2)} = \boldsymbol{f}_{i,t-1}^{(2)} + \boldsymbol{w}_{it}, \quad \boldsymbol{w}_{it} \sim N(\boldsymbol{0}, \tau_i^2 \boldsymbol{P}_i)$$

$$\boldsymbol{\theta}_t = \boldsymbol{\theta}_{t-1} + \boldsymbol{v}_t, \quad \boldsymbol{v}_t \sim N(\boldsymbol{0}, \delta^2 \boldsymbol{H})$$

where $\boldsymbol{\beta} = (1, \beta_2, \ldots, \beta_p)^T$ (that implies a 1-factor model for the first level), $\boldsymbol{\Sigma} = \mathrm{diag}(\sigma_1^2, \ldots, \sigma_p^2)$. Within capital i, the one dimensional factor $f_{ijt}^{(1)}$ is decomposed as the sum of two spatial components: θ_{it} (capital-level) and $f_{ijt}^{(2)}$ (census tract-level). Note that the $f_{ijt}^{(2)}$'s are conditionally independent and the joint vector $\boldsymbol{f}_{it}^{(2)} = (f_{i1t}^{(2)}, \ldots, f_{in_it}^{(2)})^T$ follows a Markovian evolution where the system innovation $\boldsymbol{w}_{it}$ follows a *proper Gaussian Markov random field*. More specifically, $\boldsymbol{P}_i = (\boldsymbol{I}_{n_i} + \phi \boldsymbol{M}_i)^{-1}$ where $\{\boldsymbol{M}_i\}_{lk} = m_k^{(i)}$ if $l = k$ and $\{\boldsymbol{M}_i\}_{lk} = -1/d_{lk}^{(i)}$ if sites l and k are neighbours (denoted by $l \sim k$) and zero otherwise, $m_k^{(i)} = \sum_{l \sim k} 1/d_{lk}^{(i)}$ and $d_{lk}^{(i)}$ is the Euclidean distance between centroids of regions l and k (see [51] for more details about this construction). An additional assumption is that the θ_{it}s are conditionally independent so the joint vector $\boldsymbol{\theta}_t = (\theta_{1t}, \ldots, \theta_{n_it})^T$ follows a Markovian evolution where the innovation $\boldsymbol{v}_t$ follows a zero mean Gaussian process with covariance structure $\boldsymbol{H}$ driven by the Euclidean distances between the centroids of the capitals. In this multi-level factor model, $f_{ijt}^{(1)}$ represents the vulnerability index at the census tract level of the capital i (fine level) and θ_{it} is the capital vulnerability index (coarse level). Note that the SHDFM takes full advantage of the multi-level data structure through the hierarchical specification of the common factor. ||

9.5 Regression with dynamic factor models

The ideas so far have been restricted to a single collection of time series. Even though any time series problem can be cast in a single collection of time series, the collections usually considered have a unified framework relating them. Typically they are measurements in a variety of settings of the same quantity. As such, they behave like a (random) sample of time series.

This section extends the scope of DFM beyond random samples by considering regression. The general idea of a regression is to explain a variable by a set of covariates. In time series context, this means explaining the behaviour of a (possibly multivariate) time series by a number of related explanatory time series. Although the approach is quite general, it is better explained without much loss of generality in the context of simple regression.

So from now on, we will restrict our attention to the situation where a collection of time series forming a multivariate time series $\boldsymbol{y}_t$ of a given variable is explained by another collection of time series forming a multivariate time series $\boldsymbol{x}_t$ of another given variable. This is a well-known setup in time series, sometimes referred to as transfer response models, covered in many standard time series books.

The idea of dimensionality reduction via factor models in the regression context is also not new and is also related to the basic factor model setup, as expected. Considering a set of multivariate observations $\boldsymbol{y}_t$ related to another collection of observations $\boldsymbol{x}_t$ at a latent level gives rise to the structural equation model (SEM)

$$\begin{aligned}
\boldsymbol{y}_t &= \boldsymbol{\beta}_y \boldsymbol{g}_t + \boldsymbol{\epsilon}_{yt}, \quad \boldsymbol{\epsilon}_{yt} \sim N(\mathbf{0}, \boldsymbol{\Sigma}_y) \\
\boldsymbol{x}_t &= \boldsymbol{\beta}_x \boldsymbol{f}_t + \boldsymbol{\epsilon}_{xt}, \quad \boldsymbol{\epsilon}_{xt} \sim N(\mathbf{0}, \boldsymbol{\Sigma}_x) \\
\boldsymbol{g}_t &= \boldsymbol{\Delta}_y \boldsymbol{g}_t + \boldsymbol{\Delta} \boldsymbol{f}_t + \boldsymbol{\varepsilon}_t, \quad \boldsymbol{\varepsilon}_{gt} \sim N(\mathbf{0}, \boldsymbol{\Sigma}_g)
\end{aligned}$$

The loading matrices $\boldsymbol{\beta}_x$ and $\boldsymbol{\beta}_y$ play exactly the same role as in factor models. The novelty here is the introduction of the relational matrices $\boldsymbol{\Delta}$ and $\boldsymbol{\Delta}_y$, establishing a regression relation between the set of variables $\boldsymbol{x}_t$ and $\boldsymbol{y}_t$ at a latent level. This basic SEM setup is described in detail in [46]. The Bayesian approach to SEM is described in [37]. It is worth pointing out that the standard factor model (9.1) is recovered after suitable concatenation of observables $(\boldsymbol{x}_t, \boldsymbol{y}_t)$ and factors $(\boldsymbol{f}_t, \boldsymbol{g}_t)$, respectively.

The extension towards time series problems is not difficult to obtain following the standard recipe of the previous sections of this chapter. Just like (9.2) establishes the dynamic of the factors for the time series settings, in [9] was proposed the dynamics of the two sets of factors as

$$\boldsymbol{g}(t) = \sum_{i=0}^{p} \boldsymbol{\Delta}_{yi} \boldsymbol{g}_{t-i} + \sum_{j=0}^{q} \boldsymbol{\Delta}_j \boldsymbol{f}_{t-j} + \boldsymbol{\varepsilon}_{gt}, \quad \boldsymbol{\varepsilon}_{gt} \sim N(\mathbf{0}, \boldsymbol{\Sigma}_g),$$

$$\boldsymbol{f}(t) = \sum_{j=1}^{s} \boldsymbol{\Delta}_{xj} \boldsymbol{f}_{t-j} + \boldsymbol{\varepsilon}_{ft}, \quad \boldsymbol{\varepsilon}_{ft} \sim N(\mathbf{0}, \boldsymbol{\Sigma}_f).$$

He refers to this model as dynamic SEM. [12] cast the dynamic SEM in state space form and applied it to the analysis of environmental problems. Once again, the DFM given in (9.1)–(9.2) can be recovered by appropriate concatenation of observables $(\boldsymbol{x}_t, \boldsymbol{y}_t)$ and factors $(\boldsymbol{f}_t, \boldsymbol{g}_t)$. Even more so than in DFM, it is very hard to estimate this model in its full expression for typical applications. The more natural simplifications can be obtained by restricting the order p, q, s of the auto- and cross-regressions to small values, say 1 or 2. Once again, the model can be written in state-space form by appropriately enlarging the state vector according to the order of lagged dependence of the latent factors.

These ideas were applied to the Spatial Statistics context in [27]. Once again, the columns of the loading matrices were assumed to follow independent Gaussian processes in order to impose stochastic similarity between neighbouring sites. The presence of two sets of variables introduces further possibilities beyond standard DFM. In particular, relationships between the loading matrices $\boldsymbol{\beta}_x$ and $\boldsymbol{\beta}_y$ may be introduced. For example, [27] use $\boldsymbol{\beta}_x$ as a (latent) design matrix for the mean of $\boldsymbol{\beta}_y$. Illustrative examples and further discussion about model specification and evaluation is provided in [27].

9.6 Concluding remarks

This chapter was concerned with a discussion on the use of factor models in the time series context via state space formulation. The key element of the approach is its ability to reduce the dimensionality in the multivariate time series context and at the same time to shed some light on the structure of the relationship between the different time series. Our presentation has focused exclusively on the discussion about model building. As a result, a number of other issues were not addressed. We will briefly comment upon them now.

By far the most important item not yet discussed is prediction. Time series are primarily concerned with forecasting into the future. The model-based approach of state space models followed here enables easy calculation of the predictive distributions $p(\boldsymbol{y}_{T+h}|\boldsymbol{y}^T)$, for all h. This is available approximately after obtaining the predictive distribution $p(\boldsymbol{f}_{T+h}|\boldsymbol{y}^T)$ for the latent factors and predictions can be approximated by samples. This exercise was made in Examples 9.2 and 9.3. Note also that the prediction exercise is very similar to the kriging exercise required for spatial extrapolation, and that was also illustrated in the examples above. Details are provided in [32].

Another important issue is generated by the large amount of possibilities rendered by these classes of models. There are a number of options provided by the choice of the number of regular and seasonal factors and the order of the factor dynamics. There are a few options available for model selection including AIC, BIC and DIC. These are mostly based on model fit after some penalization for complexity. One may also consider estimation of the number of factors in a RJMCMC algorithm. In this time series, we feel that model comparison should be more heavily based on predictions rather than fit. Even more so than in the other areas of statistics, given the relevance of prediction for the time series context. Standard practice in this area is based on cross-validation, where a portion of the data is left out of the fit. This portion typically consists of the last observed points to mimic the real exercise of forecasting into the future.

The description above illustrates some of the many possibilities for the use of factor models in the time series context. The presentation of spatial applications was entirely on data collected under continuous spatial variation. Similar ideas were applied to the context of discrete spatial variation or areal data in [47]. There are a number of extensions that can be envisaged by appropriately combining some of the model components described in this chapter. We are currently working on some of these and will be reporting them in the near future.

Acknowledgements

D. Gamerman was supported by CNPq-Brazil and *Fundação de Amparo à Pesquisa no Estado do Rio de Janeiro* (FAPERJ foundation). E. Salazar would like to thank the Department of Electrical and Computer Engineering at Duke University for their support.

Author's footnote

This book is very timely, right after Professor Adrian Smith's contributions to Science earned him a well-deserved knighthood. And we congratulate the editors for compiling this tribute.

References

[1] Anderson, T. W. (1963). The use of factor analysis in the statistical analysis of multiple time series. *Psychometrika*, **28**, 1–25.

[2] Aguilar, O. and West, M. (2000). Bayesian dynamic factor models and portfolio allocation. *Journal of Business and Economic Statistics*, **18**, 338–357.

[3] Arminger, G. and Muthén, B. O. (1998). A Bayesian approach to nonlinear latent variable models using the Gibbs sampler and the Metropolis–Hastings algorithm. *Psychometrika*, **63**, 271–300.

[4] Bai, J. and Ng, S. (2002). Determining the number of factors in approximate factor models. *Econometrica*, **70**, 191–222.

[5] Bhattacharya, A. and Dunson, D. B. (2011). Sparse Bayesian infinite factor models. *Biometrika*, **98**, 291–306.

[6] Calder, C. (2007). Dynamic factor process convolution models for multivariate space-time data with application to air quality assessment. *Environmental and Ecological Statistics*, **14**, 229–247.

[7] Carter, C. and Kohn, R. (1994). On Gibbs sampling for state space models. *Biometrika*, **81**, 541–553.

[8] Carvalho, C., Chang, J., Lucas, J., Nevins, J., Wang, Q. and West, M. (2008). High-dimensional sparse factor modelling: applications in gene expression genomics. *Journal of the American Statistical Association*, **103**, 1438–1456.

[9] Cziráky, D. (2004). Estimation of dynamic structural equation models with latent variables. *Metodološki zvezki*, **1**, 185–204.

[10] Engle, R. and Watson, M. (1981). A one-factor multivariate time series model of metropolitan wage rates. *Journal of the American Statistical Association*, **76**, 774–781.

[11] Finley, A. O., Sang, H., Banerjee, S. and Gelfand, A. E. (2009). Improving the performance of predictive process modelling for large datasets. *Computational Statistics and Data Analysis*, **53**, 2873–2884.

[12] Fontanella, L., Ippoliti, L. and Valentini, P. (2007). Environmental pollution analysis by dynamic structural equation models. *Environmetrics*, **18**, 265–283.

[13] Forni, M., Hallin, M., Lippi, M. and Reichlin, L. (2000). The generalized dynamic factor model: identification and estimation. *The Review of Economics and Statistics*, **82**, 540–554.

[14] Frühwirth-Schnatter, S. (1994). Data augmentation and dynamic linear models. *Journal of Time Series Analysis*, **15**, 183–202.

[15] Frühwirth-Schnatter, S. and Lopes, H. F. (2010). Parsimonious Bayesian factor analysis when the number of factors is unknown. *Technical Report*. The University of Chicago Booth School of Business.

[16] Gamerman, D. and Migon, H. S. (1993). Dynamic hierarchical models. *Journal of the Royal Statistical Society, Series B*, **55**, 629–642.

[17] Gamerman, D. and Smith, A. F. M. (1996). Bayesian analysis of longitudinal data studies, in *Bayesian Statistics* 5 (eds J. M. Bernardo *et al.*), Oxford University Press, Oxford, pp. 587–598.

[18] Gelfand, A. E. and Smith, A. F. M. (1990). Sampling-based approaches to calculating marginal densities. *Journal of the American Statistical Association*, **85**, 398–409.

[19] George, E. I. and McCulloch, R. (1993). Variable selection via Gibbs sampling. *Journal of the American Statistical Association*, **88**, 881–889.

[20] Geweke, J. (1977). The dynamic factor analysis of economic time series models. In *Latent Variables in Socio-Economic Models* (eds. D. J. Aigner and A. S. Goldberger). North Holland: Amsterdan, 365–383.

[21] Geweke, J. F. and Zhou, G. (1996). Measuring the pricing error of the arbitrage pricing theory. *The Review of Financial Studies*, **9**, 557–587.

[22] Gorsuch, R. L. (1983). *Factor Analysis*. 2nd edition, Hillsdale: Lawrence Erlbaum Associates.

[23] Green, P. (1995). Reversible jump Markov chain Monte Carlo computation and Bayesian model determination. *Biometrika*, **82**, 711–732.

[24] Higdon, D. (1998). A process-convolution approach to modelling temperatures in the north Atlantic Ocean. *Environmental and Ecological Statistics*, **5**, 173–190.

[25] Higdon, D. (2002). Space and space-time modelling using process convolutions. In *Quantitative Methods for Current Environmental Issues*, eds. C. Anderson, V. Barnett, P. C. Chatwin and A. H. El-haarawi, London: Springer Verlag, pp. 37–56.

[26] Huerta, G. and West, M. (1999). Priors and component structures in autoregressive time series models. *Journal of the Royal Statistical Society, Series B*, **61**, 881–899.

[27] Ippoliti, L., Valentini, P. and Gamerman, D. (2012). Space-time modelling of coupled spatio-temporal environmental variables. To appear in *Applied Statistics*.

[28] Lemos, R. T. and Sansó, B. (2009). A spatio-temporal model for mean, anomaly, and trend fields of north Atlantic sea surface temperature. *Journal of the American Statistical Association*, **104**, 5–25.

[29] Lindley, D. V. and Smith, A. F. M. (1972). Bayes estimates for the linear model (with discussion). *Journal of the Royal Statistical Society, Series B*, **34**, 1–41.

[30] Lopes, H. F. and Carvalho, C. M. (2007). Factor stochastic volatility with time varying loadings and Markov switching regimes. *Journal of Statistical Planning and Inference*, **137**, 3082–3091.

[31] Lopes, H. F., Gamerman, D. and Salazar, E. (2011) Generalized spatial dynamic factor analysis. *Computational Statistics and Data Analysis*, **55**, 1319–1330.

[32] Lopes, H. F., Salazar, E. and Gamerman, D. (2008). Spatial dynamic factor analysis. *Bayesian Analysis*, **3**(4), 759–792.

[33] Lopes, H. F., Schmidt, A. M., Salazar, E., Gomez, M. and Achkar, M. (2012) Measuring the vulnerability of the Uruguayan population to vector-borne diseases via spatially hierarchical factor model. To appear in *Annals of Applied Statistics*.

[34] Lopes, H. F. and West, M. (2004). Bayesian model assessment in factor analysis. *Statistica Sinica*, **14**, 41–67.

[35] Meng, X. L. and Wong, W. H. (1996). Simulating ratios of normalizing constants via a simple identity: a theoretical exploration. *Statistica Sinica*, **6**, 831–860.

[36] Molenaar, P. C. M. (1985). A dynamic factor model for the analysis of multivariate time series. *Psychometrika*, **50**, 181–202.

[37] Palomo, J., Dunson, D. B. and Bollen, K. (2007). Bayesian structural equation modelling. *Handbook of Latent Variable and Related Models*, Sik-Yum Lee (editor), Elsevier.

[38] Pan, J. and Yao, Q. (2008). Modelling multiple time series via common factors. *Biometrika*, **95**, 365–379.

[39] Peña, D. and Box, G. (1987). Identifying a simplifying structure in time series. *Journal of the American Statistical Association*, **82**, 836–843.

[40] Peña, D. and Poncela, P. (2004). Forecasting with nonstationary dynamic factor models. *Journal of Econometrics*, **119**, 291–321.

[41] Peña, D. and Poncela, P. (2006). Nonstationary dynamic factor analysis. *Journal of Statistical Planning and Inference*, **136**, 1237–1257.

[42] Polasek, W. (1997). Factor analysis and outliers: a Bayesian approach. Discussion paper, University of Basel.

[43] Salazar, E., Sansó, B., Finley, A., Hammerling, D., Steinsland, I., Wang, X. and Delamater, P. (2011). Comparing and blending regional climate model predictions for the American southwest. *Journal of Agricultural, Biological, and Environmental Statistics*, **16**, 586–605.

[44] Sansó, B., Schmidt, A. M. and Nobre, A. A. (2008). Bayesian spatio-temporal models based on discrete convolutions. *Canadian Journal of Statistics*, **36**, 239–258.
[45] Sargent, T. J. and Sims, C. A. (1977). Business cycle modelling without pretending to have too much a priori economic theory. In *New Methods in Business Research* (ed. C. A. Sims). Federal Reserve Bank of Minneapolis.
[46] Skrondal, A. and Rabe-Hesketh, S. (2004). *Generalized Latent Variable modelling*, Chapman & Hall, Boca Raton.
[47] Strickland, C., Simpson, D., Turner, I., Denham, R. and Mengersen, K. (2010). Fast Bayesian analysis of spatial dynamic factor models for multitemporal remotely sensed imagery. *Applied Statistics*, **60**, 109–124.
[48] Stroud, R., Müller, P. and Sansó, B. (2001). Dynamic models for spatiotemporal data. *Journal of the Royal Statistical Society, Series B*, **63**, 673–689.
[49] Thurstone, L. L. (1931). Multiple factor analysis. *Psychological Review*, **38**(5), 406–427.
[50] Thurstone, L. L. (1947). *Multiple Factor Analysis*. Chicago: University of Chicago Press.
[51] Vivar, J. C. and Ferreira, M. A. R. (2009). Spatiotemporal models for Gaussian areal data. *Journal of Computational and Graphical Statistics*, **18**(3), 658–674.
[52] Wikle, C. K. and Cressie, N. (1999). A dimension-reduced approach to space-time Kalman filtering. *Biometrika*, **86**, 815–829.
[53] West, M., Harrison, P. J. and Migon, H. S. (1985). Dynamic generalized linear models and Bayesian forecasting (with discussion). *Journal of the American Statistical Association*, **81**, 741–750.
[54] West, M. and Harrison, J. (1997). *Bayesian Forecasting and Dynamic Models*, Springer, New York.

10 Dynamic and spatial modelling of block maxima extremes

GABRIEL HUERTA AND
GLENN A. STARK

10.1 Introduction

Figure 10.1 shows the monthly maxima of precipitation at the Maíquetia-Simon Bolivar airport near Caracas, Venezuela within the period 1960–1999. The data only considers measurements at one site and has critical importance in extreme value analysis due to the catastrophic events that occurred near this location at the end of the year 1999. For example [5] carefully considered a Bayesian analysis of the annual maxima rainfall values at the same site, based on models that fully account for parameter uncertainties and non-stationarity in the data. Furthermore [16], analysed the monthly rainfall maxima of Figure 10.1 via dynamic regressions as in [29], [28] and in connection with a climatological index known as the *North Atlantic Oscillation* (NAO). Here we consider these rainfall observations as a starting point to study extreme events via the Generalized Extreme Value (GEV) distribution. Two key questions that arise from Figure 10.1 are: (1) How do we characterize these rainfall events? and (2) How do we assess for the non-stationary behaviour that is apparent in the data?

10.1.1 GEV distribution and likelihood

Let $y_{m,1}, y_{m,2}, \ldots, y_{m,n}$ be samples of extremes from m independent observations (i.e. block maxima), where m is the number of observations in each block and n is the number of blocks. For the precipitation maxima of Figure 10.1, a uniform block size across time was considered to produce the observations. In applications of block-maxima values it is typical to assume that $y_{m,i}, i = 1, 2, \ldots, n$, are independent and arise from a common GEV distribution as in [4] and [1]. If $y \sim GEV(\mu, \sigma, \xi)$, the cumulative distribution function for y is:

$$H(y) = \exp\left\{-\left[1 + \xi\left(\frac{y-\mu}{\sigma}\right)\right]_+^{-1/\xi}\right\} \tag{10.1}$$

where $-\infty < \mu < \infty$ is a location parameter, $\sigma > 0$ is a scale parameter and $-\infty < \xi < \infty$ is a shape parameter. The $+$ sign denotes the positive part of the argument so the support for $H(y)$

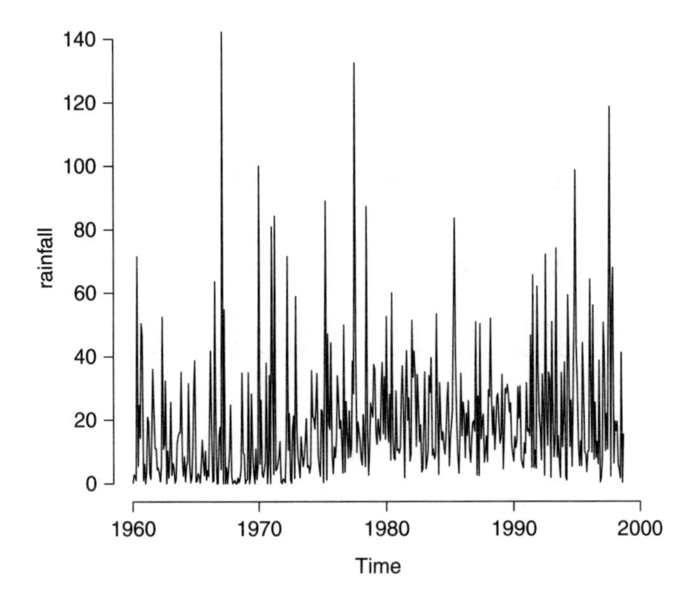

Figure 10.1 Monthly maxima of precipitation in Venezuela.

is given by the set $\{y : 1 + \xi(\frac{y-\mu}{\sigma}) > 0\}$. It is well known that different values of ξ imply different tail behaviours or *domains of attraction* for $H(y)$, namely Gumbel ($\xi \to 0$), Fréchet ($\xi > 0$) and Weibull ($\xi < 0$). $H(y)$ arises through a limiting argument for block maxima in the *Extremal Type* or *Fisher-Tippet* theorem in [10] and also presented in [4] and [1], so if the block size is large enough, one may assume that the observations $y_{m,i}$ follow a GEV distribution. Therefore, if we drop the dependence on the block size so that $y_i = y_{m,i}$, the GEV log-likelihood function based on n observations is

$$l(\mu, \sigma, \xi | \{y_t\}_{t=1}^n) = -n \log \sigma - (1 + 1/\xi) \sum_{i=1}^{n} \log\{1 + \xi(y_i - \mu)/\sigma\}$$
$$- \sum_{i=1}^{n} \{1 + \xi(y_i - \mu)/\sigma\}^{-1/\xi}. \tag{10.2}$$

10.1.2 Bayesian inference on the GEV

From a Bayesian point-of-view inferences on (μ, σ, ξ) can be directly obtained with Markov chain Monte Carlo (MCMC) methods based on a Gibbs sampling approach as in [13] with embedded Metropolis–Hastings (M-H) steps as in [15]. For instance, a prior distribution $p(\mu, \sigma, \xi)$ can be induced through a *trivariate* normal distribution on $(\mu, log(\sigma), \xi)$ which includes the case of an independence prior.

As described in [6], beta distributions for probability ratios or gamma distributions for quantile difference could alternatively be used to elicit $p(\mu, \sigma, \xi)$. In particular, the quantile-difference priors can be elicited via the $1 - p$ quantile of a GEV distribution which has a closed form in terms of the three parameters, $z_p = \mu - \frac{\sigma}{\xi}[1 - \{-\log(1 - p)\}^{-\xi}]$.

The posterior distribution for $(\mu, \log(\sigma), \xi)$ can be sampled with random walk proposals on each of the parameters as presented in [4] and [1],

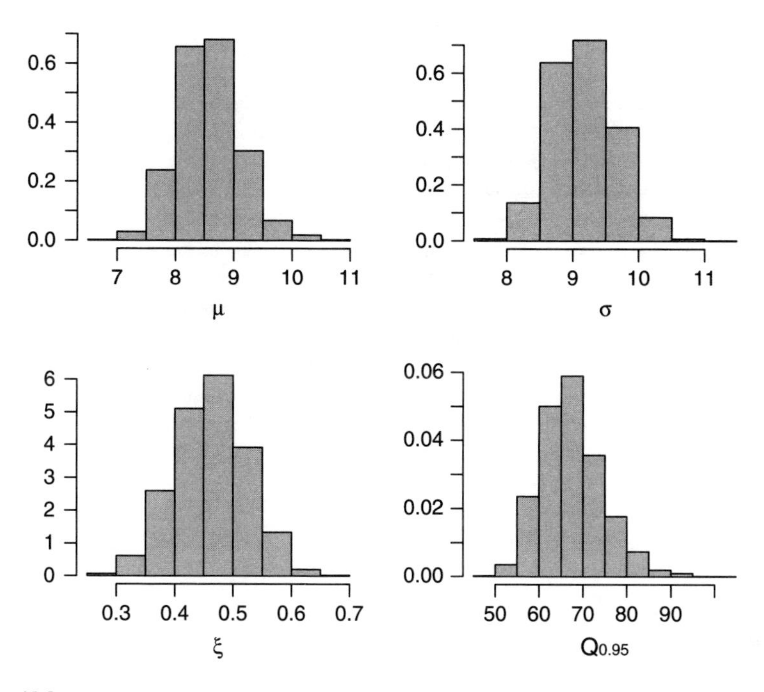

Figure 10.2 Marginal posterior distribution for (μ, σ, ξ) and 95% quantile of the GEV distribution.

$$log(\sigma^*) = log(\sigma^{(i)}) + \nu_\sigma \epsilon_1 \tag{10.3}$$

$$\mu^* = \mu^{(i)} + \nu_\mu \epsilon_2 \tag{10.4}$$

$$\xi^* = \xi^{(i)} + \nu_\xi \epsilon_3 \tag{10.5}$$

where $\epsilon_j \sim N(0,1); j = 1,2,3$ and ν_σ, ν_μ, ν_ξ denote the proposal tuning parameters. The proposed values are accepted or rejected as iterations of the MCMC are performed. To illustrate this method, Figure 10.2 shows histograms of posterior samples for μ, σ, ξ and $z_{0.05}$ corresponding to the data of Figure 10.1 and via this hybrid Gibbs sampling/Metropolis–Hastings method.

It is interesting to note that quantile estimation can also be achieved through the predictive distribution,

$$p(y^f|y) = \int_{-\infty}^{\infty} \int_0^{\infty} \int_{-\infty}^{\infty} f(y^f|\mu,\sigma,\xi)p(\mu,\sigma,\xi|y)d\mu d\sigma d\xi \tag{10.6}$$

where y^f denotes a future observation. Here $f(y^f|\mu,\sigma,\xi)$ represents the probability density function for this future observation based on the GEV distribution and conditional to model parameters. Through the *Method of Composition* it is possible to obtain samples of y^f from the predictive distribution. An empirical quantile based on these samples approximates the value y^* such that $P[Y^f \leq y^*|y] = 1 - p$, which fully takes into account all of the model parametric uncertainties. In fact, for the maximum monthly rainfall data of Figure 10.1, the 99% quantile based on the predictive distribution $(p = 0.01)$ is 162.87. On the other hand, if the GEV distribution is fitted with Maximum Likelihood Estimation (MLE), the 99%

quantile is estimated as 152.3. Furthermore, the posterior mean of the GEV distribution quantile, $z_{0.01}$, is 157.35.

Bayesian approaches for extreme values had been extensively studied from the beginning of Bayesian computation with MCMC and other numerical approximations. The paper by [25] is one of the first to illustrate the Bayesian modelling of extremes within the GEV framework. Also [7] offers some developments of Bayesian approaches to extreme value theory with applications to modelling areal rainfall extremes. A thorough review of the topic is offered by [6] where the authors studied the Bayesian approach from a variety of aspects, including how priors can be better elicited for extreme value data and how well a complete Bayesian analysis, that includes predictive-quantile estimation, performs compared to a pure likelihood based analysis as briefly illustrated here with the rainfall data at the Maíquetia/Simon Bolivar airport. The R-package *evdbayes* described in [26] and [27] implements the techniques in [6] including quantile-based prior specifications and peaks-over-threshold analysis. However, the modelling described in these references is mainly restricted to *constant* parameters or to parameters that may have deterministic trends in time.

10.2 Time-varying models for the GEV distribution

10.2.1 Introduction

To account for the type of non-stationarities that are present in the data of Figure 10.1, we can impose a time-dependent structure on any set of parameters of the GEV distribution. Here, we emphasize on a GEV distribution with a time-varying location parameter,

$$H(y_t) = \exp\left\{-\left[1+\xi\left(\frac{y_t-\mu_t}{\sigma}\right)\right]_+^{-1/\xi}\right\}, \quad t = 1,2,\ldots n \tag{10.7}$$

If a deterministic function is chosen to model time changes in μ_t, this can be expressed in a generalized form as

$$\mu_t = g(X^T\beta) \tag{10.8}$$

where g is a specific link function, β is a vector of parameters and X^T is a vector of covariates that involves time. In particular μ_t can follow a linear trend, $\mu_t = \beta_0 + \beta_1 t$, a higher degree polynomial such as $\mu_t = \beta_0 + \beta_1 t + \beta_2 t^2$ or a trend/seasonal model, $\mu_t = \beta_0 + \beta_1 t + \beta_2 S(t)$, where $S(t)$ represents the seasonal component. In this context, [4] discusses extensively these various models. On the other hand, μ_t can be treated as a stochastic process that depends on time through a hierarchical specification.

10.2.2 Dynamic linear model

We consider *Dynamic Linear Models* (DLMs) or state-space models as in [28] and [29] as our main choice for non-stationary modelling of the location parameter of the GEV distribution. We focus on DLMs to assess whether any short-term changes can occur in the extremal-type distribution. If $z_t, t = 1,2,\ldots$ represents a vector of observations of dimension r at time t, following the notation in [29], a DLM for z_t is specified as,

$$z_t = F_t'\theta_t + v_t, \quad v_t \sim N(0, V_t) \tag{10.9}$$

$$\theta_t = G_t \theta_{t-1} + w_t, \quad w_t \sim N(0, W_t) \tag{10.10}$$

$$\theta_0 | D_0 \sim N(m_0, C_0) \tag{10.11}$$

where F_t' is assumed to be a known $(r \times n)$ regression matrix and G_t is assumed as a known $(n \times n)$ state matrix. Equation 10.9 defines the observation equation of the DLM and the variance of the observation error is given by the $r \times r$ matrix V_t. Equation 10.10 is called the system or evolution equation and W_t is the $n \times n$ variance–covariance matrix of the evolution error. The errors v_t and w_t are generally assumed to be mutually independent. To obtain Bayesian inference on the state vector of the DLM, a prior distribution on the initial state is needed and is defined by equation 10.11, where m_0 defines the forecaster's initial belief about the level θ_0 and C_0 is the associated measure of uncertainty. At the lack of any true prior information, we adopt a non-informative prior where the mean level is fixed to an arbitrary constant and C_0 is given a large value. The quadruple (F_t, G_t, V_t, W_t) characterizes completely the DLM, so different quadruples define different subsets of the general class of DLMs. A special case of the quadruple is given by $(1, 1, V, W)$, which is referred to as a *first-order polynomial* DLM in [29].

10.2.3 DLMs and the GEV distribution

DLMs had been used by [12] and [16] for time-varying extreme value models. [12] outlines a semi parametric approach for smoothing extremes with applications to athletic records and temperature data. On the other hand, [16] consider DLMs for assessing dynamic trends and space–time structures for extreme values of ozone levels. Here we follow [16] and focus our presentation for the case of a time-varying location parameter.

Let $\{y_1, y_2, \ldots, y_n\}$ be independent realizations from a GEV distribution (μ_t, σ, ξ) conditional on model parameters. Assume that the temporal dependency on the location parameter is modelled through a DLM as just described in 10.2.2. For this case, the GEV likelihood function is:

$$L(\{\mu_t\}_{t=1}^n \xi, \sigma | \{y_t\}_{t=1}^n) = \prod_{t=1}^n \frac{1}{\sigma}\left[1 + \xi\left(\frac{y_t - \mu_t}{\sigma}\right)\right]_+^{-(1+\frac{1}{\xi})} \exp\left\{-\left[1 + \xi\left(\frac{y_t - \mu_t}{\sigma}\right)\right]_+^{-\frac{1}{\xi}}\right\}. \tag{10.12}$$

Assuming a *first-order polynomial* DLM $(1, 1, V, W)$ on μ_t, we have

$$\mu_t = \theta_t + v_t, \quad v_t \sim N(0, V) \tag{10.13}$$

$$\theta_t = \theta_{t-1} + w_t, \quad w_t \sim N(0, W) \tag{10.14}$$

In [7] the authors argue that prior eliciting in terms of quantiles or quantile differences is to be preferred over priors on GEV parameters for constant modelling situations. Given the complexities of a DLM-time-varying GEV model, we choose priors on the GEV parameters. The observational equation in 10.13 defines a prior for each μ_t conditional on θ_t and V, $\mu_t \sim N(\theta_t, V)$, $t = 1, \ldots, n$. We adopt independent normal priors for $\log \sigma$ and ξ,

$$\log \sigma \sim N(M_\sigma, V_\sigma), \quad \xi \sim N(M_\xi, V_\xi) \tag{10.15}$$

For V we adopt an inverse gamma prior, $V \sim IG(a, b)$, which provides a conditionally conjugate structure for this parameter. We adopt *discount factors* as in [29] to deal with the evolution variance W. The discount factor represents the change of information on state parameters from time $t-1$ to time t, which in [16] has proven its use for temporal and space-time modelling of extremes within the DLM framework.

10.2.4 MCMC for DLM-GEV distribution

The joint posterior distribution for all the model parameters, $(\{\mu_t\}_{t=1}^n, \xi, \sigma, \{\theta_t\}_{t=1}^n, V)$ given the data $\{y_t\}_{t=1}^n$ is

$$p(\{\mu_t\}_{t=1}^n, \xi, \sigma, \{\theta_t\}_{t=1}^n, V|\{y_t\}_{t=1}^n) \propto L(\{\mu_t\}_{t=1}^n\, \xi, \sigma|\, \{y_t\}_{t=1}^n)$$
$$\prod_{t=1}^n \left[\frac{1}{V^{1/2}} \exp\left\{-\frac{(\mu_t-\theta_t)^2}{2\,V}\right\} \exp\left\{-\frac{(\theta_t-\theta_{t-1})^2}{2\,W}\right\}\right]$$
$$\exp\left\{-\frac{(\xi - M_\xi)^2}{2\,V_\xi}\right\} \exp\left\{-\frac{(\log\sigma - M_\sigma)^2}{2\,V_\sigma}\right\} V^{-(a+1)} \exp(-b/V) \tag{10.16}$$

Posterior draws can be obtained through full conditional draws of each parameter based on the Metropolis or Metropolis–Hastings algorithm. For example, to draw the shape parameter ξ at iteration $i+1$,

1. We sample ξ^{i+1} from a normal distribution centred at ξ^i which defines a symmetric proposal distribution based on a *random walk.*
2. We compute $\alpha(\xi^i, \xi^{i+1}) = min\left\{1\ \frac{p(\xi^{i+1})}{p(\xi^i)}\right\}$, where $p(\xi^{i+1})$ denotes the full conditional posterior distribution from Equation 10.16 evaluated at the proposed value and $p(\xi^i)$ is the full conditional posterior evaluated at the previous sampled value.
3. Generate u from a $U(0,1)$ distribution. If $u < \alpha(\xi^i, \xi^{i+1})$, we accept the proposed value ξ^{i+1} as our current point in the chain. Otherwise, we reject ξ^{i+1} and keep the previous point ξ^i.

Sampling σ and μ_t follows similar steps. Specifically, we draw each μ_t individually and use the prior as the proposal distribution. This leads into a Metropolis ratio that exclusively depends on the full conditional distribution of μ_t. More details on this sampling strategy appear in [16]. To draw $\{\theta_t\}_{t=0}^n$ from its full conditional distribution, we use *Forward Filtering and Backward Simulation* (FFBS) as in [2] and [11] which, as mentioned in [28], is central to MCMC implementations of *conditionally linear* normal models. In our case and following Chapter 4 of [29], the state posterior distributions conditional on all other model parameters are computed sequentially in time using the following recursive equations. If δ denotes the DLM discount factor so that $W = \frac{(1-\delta)}{\delta} C_{t-1}$, and D_t represents all the information available up to time t, for $t = 1, 2, \ldots, n$

$$(\theta_t|D_{t-1}) = N(m_{t-1}, R_t), \quad R_t = C_{t-1} + W = C_{t-1}/\delta$$
$$(\mu_t|\theta_t, D_{t-1}) = N(m_{t-1}, Q_t), \quad Q_t = R_t + V$$
$$(\theta_t|D_t) = N(m_t, C_t), \quad m_t = m_{t-1} + A_t(\mu_t - m_{t-1}), C_t = A_t V, A_t = R_t/Q_t \tag{10.17}$$

To perform the backward simulation at time $t = n$, we draw θ_n from $(\theta_n|D_n)$ and draw the other θ_t parameters via *retrospective filtering*. So for $t = n-1, n-2, \ldots, 0$, θ_t is drawn conditionally on θ_{t+1} from a $N(h_t, H_t)$ where

$$h_t = m_t + \delta(\theta_{t+1} - m_t), \quad H_t = C_t(1 - \delta). \tag{10.18}$$

The posterior simulation for the DLM-GEV model proposed in (10.12–10.14) can be summarized as follows. At iteration $i + 1$, we

- Draw $\mu_t^{i+1} | y_t, \mu_t^i, \sigma^i, \xi^i, \theta_t^i, V^i$, for $t = 1, \ldots, n$ with individual Metropolis steps.
- Draw $\sigma^{i+1} | \{y_t\}_{t=1}^n, \{\mu_t^{i+1}\}_{t=1}^n, \sigma^i, \xi^i$ with a Metropolis step.
- Draw $\xi^{i+1} | \{y_t\}_{t=1}^n, \{\mu_t^{i+1}\}_{t=1}^n, \sigma^{i+1}, \xi^i$ with a Metropolis step.
- Draw $\{\theta_t^{i+1}\}_{t=1}^n | \{\mu_t^{i+1}\}_{t=1}^n, V^i$ via FFBS as described in (10.17–10.18).
- Draw $V^{i+1} | \{\mu_t^{i+1}\}_{t=1}^n, \{\theta_t^{i+1}\}_{t=1}^n$ from an inverse-gamma distribution.

Figure 10.3 shows the posterior mean of μ_t and θ_t with a 95% probability interval for θ_t based on our MCMC approach for the Maiquetía rainfall time series with a discount factor $\delta = 0.9$ and a burn-in period of 30000 iterations. The estimated parameters are consistent with a notion of more intense extremes in recent years. The changes in location parameter from the DLM-GEV are nonlinear and remarkably different from an estimated trend based only on a deterministic line where the intercept and slope area fitted via MLE. Furthermore, Figure 10.4 shows the posterior mean estimates of predictive GEV quantiles for four different probability levels: 95%, 75%, 50% and 5% based on our modelling approach and with the Maiquetía extreme rainfall measurements. These estimates appear more constant than the parameter estimates but clearly exhibit the non-stationary behaviour and skewness that is typically present in block maxima observations.

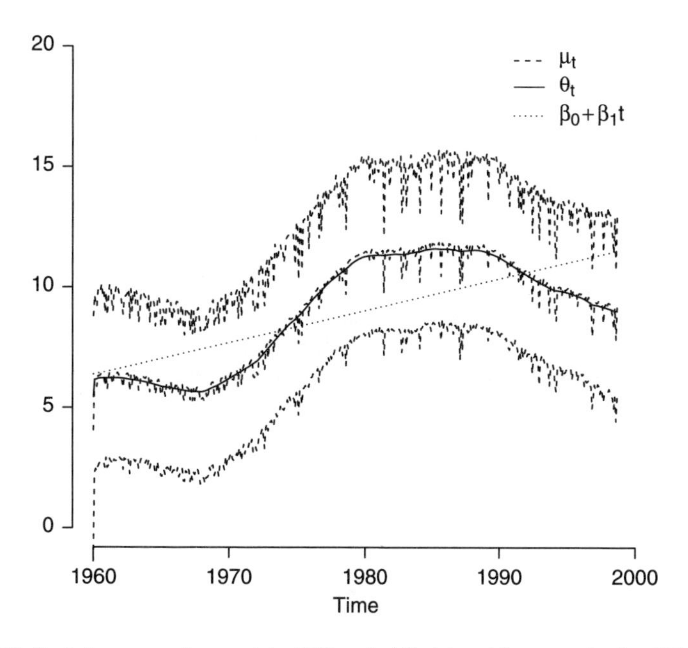

Figure 10.3 Posterior mean of μ_t and θ_t. 95% probability interval for μ_t under the GEV-DLM for Maiquetía rainfall data

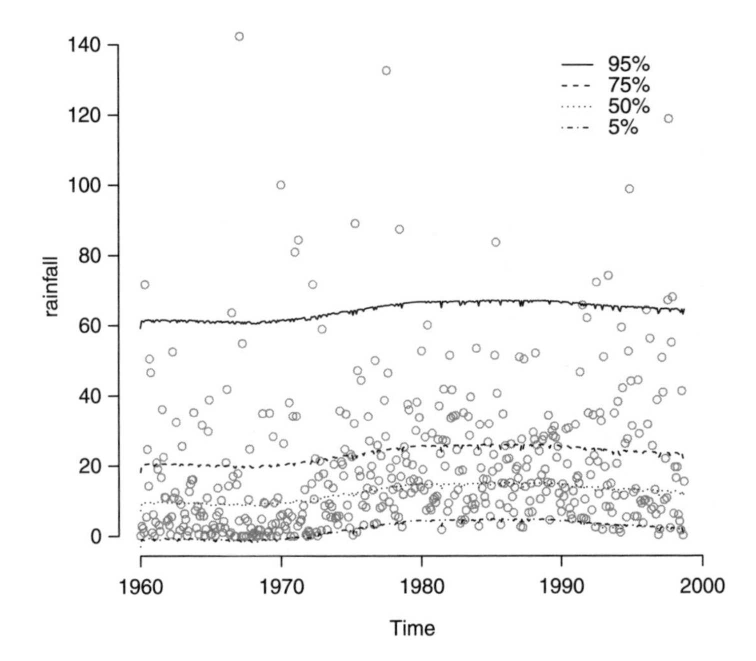

Figure 10.4 Posterior predictive quantiles based on GEV-DLM model for Maiquetía rainfall data

10.3 A spatial GEV distribution

10.3.1 Introduction

The analysis of extremes from a spatial perspective can arise as a natural extension to the time-varying GEV distribution-DLM models described in 10.2. The main scope of these spatial models considers the theory and applications of *Gauss Markov Random Fields* (GMRFs) as described in [21], which also discusses the connections of GRMFs to structural time series in the form of DLMs. Here we consider output of the Penn State/NCAR mesoscale (MM5) Regional Climate Model (RCM) which was driven by a NCAR/DOE parallel climate model. A similar output was previously analysed in [8]. The output contains extreme winter (December–January–February) precipitation of a 20-year control run that assumes current levels of greenhouse gases and begins in 1995. The spatial domain of the RCM includes 616 (28×22) grid points covering the western United States and southwestern Canada. Figure 10.5 shows maps of the output precipitation corresponding to two specific years. For year 2004, the extreme values are more intense in regions covering the Pacific coast, southwestern Canada and Arizona. Figure 10.6 shows the grid points over the spatial domain along with twenty-five locations that were held out and treated as missing in precipitation for all 20 years to assess the predictive ability of our models.

10.3.2 Objective and Gauss Markov random fields

The goal of our analysis is to treat the RCM output as data and to develop a hierarchical model around the GEV distribution that permits the characterization of the extremes through predictive quantiles, which may assist climate modellers to evaluate the output performance of the RCM. In this case, we are not comparing the results from our RCM analysis to real precipitation

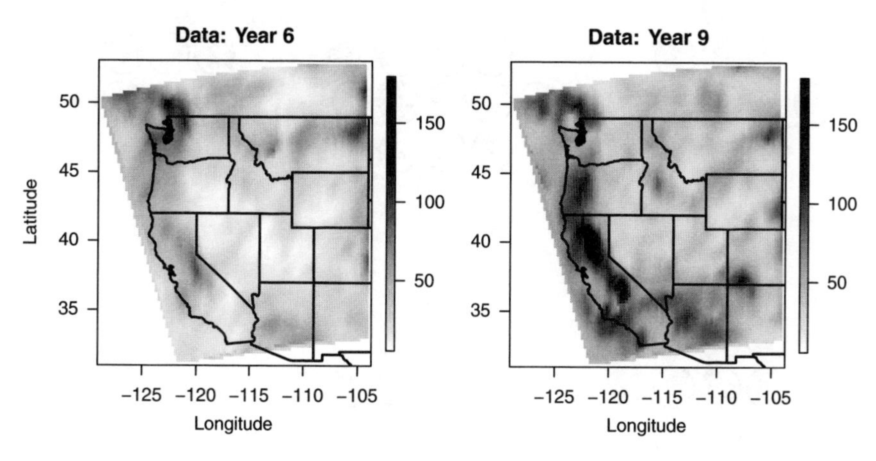

Figure 10.5 Two years of extreme precipitation for the MM5 Regional Climate Model.

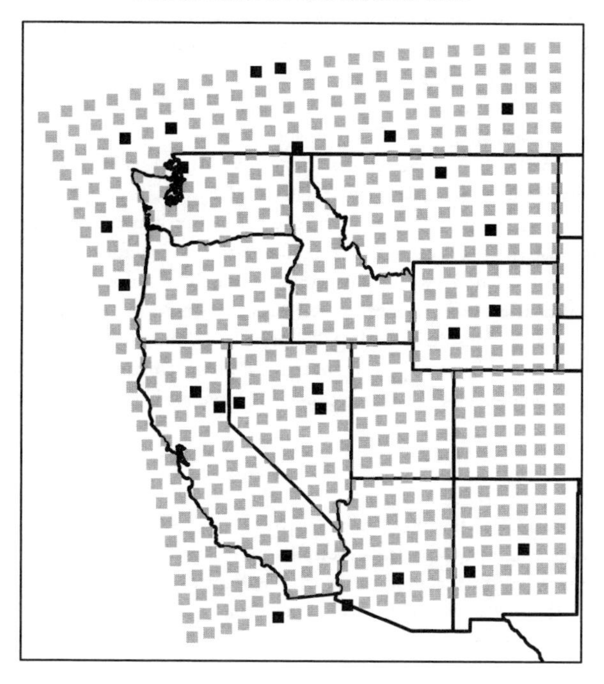

Figure 10.6 Grid points for the Regional Climate Model output and holdout locations.

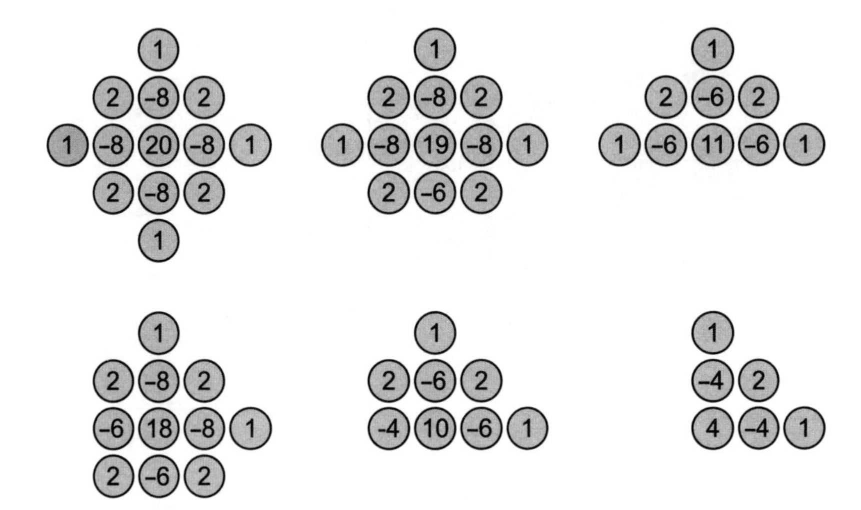

Figure 10.7 Stencils of the precision matrix for a IGMRF second-order neighbourhood structure.

measurements arising from station or satellite data. From a modelling standpoint, GMRFs provide a structure defined through neighbours and precision matrices that has a graphical representation and is attractive to represent spatial relationships on high-dimensional output from climate models. The *nodes* and *vertices* representing the graph of the GMRF correspond to points on a grid and neighbours. If $\mathbf{Q}$ represents the precision matrix of the GMRF, a point (node) i is connected to a point j or $i \sim j$, if and only if $\mathbf{Q}_{ij} \neq 0$. Different specifications of the matrix $\mathbf{Q}$ provide different spatial structures. In particular, our models consider a precision matrix from a *biharmonic difference operator* which corresponds to an *intrinsic* (improper) GMRF as in Chapter 3 of [21]. The schematic of Figure 10.7 gives this second-order neighbourhood structure where the point of reference is the node with the associated highest value and corrections are imposed for when this node is near a boundary. The hierarchical model we propose to analyse the RCM output depends on a likelihood based on the GEV distribution and a prior level on parameters that involve this second-order IGMRF. For example [8] considered an IGMRF prior based on a *first-order* neighbourhood. A potential advantage of the second-order prior over a first-order structure, is that it adds extra flexibility to represent dependencies for localized climate phenomena like regional storms.

10.3.3 Modelling strategy

More specifically, we assume that $Y_{st} \sim \text{GEV}(\mu^*_{st}, \sigma_s, \xi); s = 1, \ldots, n; t = 1, \ldots, 20$ are conditionally independent where Y_{st} represents precipitation from the RCM at the grid location s and at time t. In our case, $n = 616$, the number of grid point locations being considered in our analysis. Therefore the probability distribution for Y_{st} has the form,

$$H(y_{st}|\mu^*_{st}, \sigma_s, \xi) = exp\left\{-\left[1 + \xi\left(\frac{y_{st} - \mu^*_{st}}{\sigma_s}\right)\right]_+^{-1/\xi}\right\} \tag{10.19}$$

$$\mu^*_{st} = \mu_s + \phi_t \tag{10.20}$$

so μ^*_{st} has been additively decomposed in space and time. The scale parameter σ_s is allowed to vary in space while the shape parameter ξ is kept fixed across space and time with a prior distribution

$(\xi - 0.5) \sim \text{Beta}(9, 5)$ which guarantees that $-0.5 < \xi < 0.5$. This prior distribution was proposed in [17] and has been used as a penalization term in a GEV likelihood function in low sample size situations. This prior is not overly informative and is restricted to values that are sensible in studies of extreme precipitation.

Furthermore, we introduce vectors to represent a spatial component for the location and scale parameters respectively, $\boldsymbol{\mu} = (\mu_1, \ldots, \mu_n)$, $\boldsymbol{\sigma} = (\sigma_1, \ldots, \sigma_n)$ and $\boldsymbol{\eta} = \text{vec}(\boldsymbol{\mu}, log(\boldsymbol{\sigma}))$ is a vector that represents the concatenation of the elements in $\boldsymbol{\mu}$ and the logarithm of the elements in $\boldsymbol{\sigma}$. We model $\boldsymbol{\eta}$ as,

$$\boldsymbol{\eta} = (\boldsymbol{I}_2 \otimes \boldsymbol{X})\boldsymbol{B} + \boldsymbol{U} + \boldsymbol{\epsilon}, \tag{10.21}$$

$$\boldsymbol{\epsilon} \sim \text{N}(\boldsymbol{0}, \boldsymbol{T} \otimes \boldsymbol{I}_n), \tag{10.22}$$

$$\boldsymbol{U} \sim \text{GMRF}(\boldsymbol{0}, \boldsymbol{\theta} \otimes \boldsymbol{Q}), \tag{10.23}$$

$$\boldsymbol{T} \sim \text{Wishart}(n_T, \boldsymbol{V}_T), \tag{10.24}$$

$$\boldsymbol{\theta} \sim \text{Wishart}(n_\theta, \boldsymbol{V}_\theta), \tag{10.25}$$

$$\boldsymbol{B} \sim \text{N}(\boldsymbol{0}, \tau_{\boldsymbol{B}} \boldsymbol{I}_{10}). \tag{10.26}$$

For this specification both $\boldsymbol{T} \otimes \boldsymbol{I}_n$ and $\boldsymbol{\theta} \otimes \boldsymbol{Q}$ are precision matrices, where $\boldsymbol{Q}$ depends on the neighbourhood structure, $\otimes$ represents a *Kronecker* product and $\boldsymbol{I}_k$ is an identity matrix of dimension k. $\boldsymbol{X}$ is a matrix of covariates with $p = 5$ columns including an intercept, the longitudes, latitudes and altitudes of each grid point location and a vector indicating whether a grid point is located over the ocean or land. We expect that these covariates account for the variability as well as some of the spatial patterns in $\boldsymbol{\eta}$. $\boldsymbol{B}$ is a 10 $(2p)$-dimensional vector of regression coefficients where its first five elements are associated to $\boldsymbol{\mu}$ and the second five elements to $\boldsymbol{\sigma}$. $\boldsymbol{B}$ is assigned a 10-dimensional normal prior with precision $\tau_{\boldsymbol{B}}$. The spatial properties of $\boldsymbol{U}$ are defined through a GMRF prior which has a precision matrix $\boldsymbol{\theta} \otimes \boldsymbol{Q}$. Here $\boldsymbol{Q}$ is defined through the construction of a second-order intrinsic GMRF as described in Chapter 3 of [21] and illustrated in Figure 10.7. The second-order precision matrix $\boldsymbol{Q}$ is not a full rank matrix, and so the GMRF prior for $\boldsymbol{U}$ is not a proper probability distribution. Therefore, we constrain $\boldsymbol{U}$ so that $(\boldsymbol{I}_2 \otimes \boldsymbol{E})'\boldsymbol{U} = \boldsymbol{0}$ where $\boldsymbol{E}$ is a $n \times 3$ matrix whose columns are the eigenvectors of $\boldsymbol{Q}$ with zero eigenvalues. This constrain guarantees a proper prior on $\boldsymbol{U}$ and allows us to improve on the computational efficiencies of our MCMC simulations.

In addition, $\boldsymbol{\theta}$ is a 2×2 positive definite matrix that we model through a Wishart prior and provides a precision matrix for the blocks $\boldsymbol{\mu}$ and $\boldsymbol{\sigma}$. The term $\boldsymbol{\epsilon}$ represents global variability in $\boldsymbol{\eta}$ that is not captured by the model covariates. Every two elements of $\boldsymbol{\epsilon}$ are modelled independently at each grid location with a bivariate normal random variable with a 2×2 precision matrix $\boldsymbol{T}$, to which we assign a Wishart prior. The time term for μ^*_{st}, ϕ_t, follows a $\text{N}(\beta_1(t - \bar{t}), \nu)$, $t = 1, \ldots, 20$, so β_1 measures a potential annual shift in the GEV location parameter over the 20-year control run. We assign flat prior distributions on both β_1 and ν.

10.3.4 MCMC approach

The following steps describe the *ith* iteration of our MCMC method to sample model parameters and imputed values at held out locations in a Gibbs sampling scheme as in [13]. $\boldsymbol{y}$ represents the full set of observed and inputed data and $\boldsymbol{\phi} = (\phi_1, \ldots, \phi_{20})$.

- For a held out location s, we impute $Y_{s,t}$ from the GEV distribution, $Y_{st}^{(i)} \sim \text{GEV}(\mu_s^{(i-1)} + \phi_t^{(i-1)}, \sigma_s^{(i-1)}, \xi^{(i-1)})$ for $t = 1, \ldots, 20$.

- We draw $\boldsymbol{\eta}^{(i)}|\boldsymbol{y},\boldsymbol{B}^{(i-1)},\boldsymbol{U}^{(i-1)},\boldsymbol{\phi}^{(i-1)},\boldsymbol{T}^{(i-1)},\xi^{(i-1)},\boldsymbol{X}$ via Metropolis steps for the pair of values $\mu_s^{(i)}$ and $log(\sigma_s^{(i)})$ at each location $s = 1,\ldots,n$ where $n = 616$ is the number of grid point locations.
- We use a random walk Metropolis step to draw $\phi_t^{(i)}|\boldsymbol{y},\boldsymbol{\eta}^{(i)},\xi^{(i-1)},\beta^{(i-1)},\nu^{(i-1)}$ for $t = 1,\ldots,20$.
- We draw $\beta^{(i)}|\nu^{(i-1)},\boldsymbol{\phi}^{(i)}$ and $\nu^{(i)}|\beta^{(i)},\boldsymbol{\phi}^{(i)}$ from normal and inverse-gamma distributions respectively.
- We use a random walk Metropolis step to draw $\xi^{(i)}|\boldsymbol{y},\boldsymbol{\eta}^{(i)},\boldsymbol{\phi}^{(i)}$.
- We draw $\boldsymbol{B}^{(i-1)}|\boldsymbol{\eta}^{(i)},\boldsymbol{U}^{(i-1)},\boldsymbol{T}^{(i-1)},\boldsymbol{X}$ from a multivariate normal with mean vector $\mu_{\boldsymbol{B}} = \Sigma_{\boldsymbol{B}}(\boldsymbol{T}^{(i-1)} \otimes \boldsymbol{X}')(\boldsymbol{\eta}^{(i)} - \boldsymbol{U}^{(i-1)})$ and covariance matrix $\Sigma_{\boldsymbol{B}} = (\tau_{\boldsymbol{B}}\boldsymbol{I}_{\boldsymbol{10}} + \boldsymbol{T}^{(i-1)} \otimes \boldsymbol{X}'\boldsymbol{X})^{-1}$.
- We draw $\boldsymbol{U}^i|\boldsymbol{\eta}^{(i)},\boldsymbol{T}^{(i-1)},\boldsymbol{B}^{(i)},\boldsymbol{\theta}^{(i-1)},\boldsymbol{X}$ by first generating $\boldsymbol{U}^*$ from a $N_C(\boldsymbol{b_U},\boldsymbol{Q_U})$ where $\boldsymbol{b_U} = (\boldsymbol{T}^{(i-1)} \otimes \boldsymbol{I}_n)\boldsymbol{\eta}^{(i)} - (\boldsymbol{T}^{(i-1)} \otimes \boldsymbol{X})\boldsymbol{B}^{(i)}$ and $\boldsymbol{Q_U} = (\boldsymbol{T}^{(i-1)} \otimes \boldsymbol{I}_n) + (\boldsymbol{\theta}^{(i-1)} \otimes \boldsymbol{Q})$. N_C designates the canonical parameterization of a GMRF as in [21]. We then correct for the linear constraint $(\boldsymbol{I}_2 \otimes \boldsymbol{E})'\boldsymbol{U} = \boldsymbol{0}$, via the method of *conditioning by Kriging* as detailed in [21], page 37. This leads to the expression, $\boldsymbol{U}^{(i)} = (\boldsymbol{I}_2 \otimes (\boldsymbol{I}_n - \boldsymbol{E}\boldsymbol{E}'))\boldsymbol{U}^*$. The matrix $(\boldsymbol{I}_2 \otimes (\boldsymbol{I}_n - \boldsymbol{E}\boldsymbol{E}'))$ is the perpendicular projection operator that projects $\boldsymbol{U}^*$ into the column space of $\boldsymbol{Q}$.
- We draw $\boldsymbol{T}^{(i)}|\boldsymbol{\eta}^{(i)},\boldsymbol{B}^{(i)},\boldsymbol{U}^{(i)},\boldsymbol{X}$ from a Wishart$(n_{\boldsymbol{T}} + n, (\boldsymbol{V_T} + \boldsymbol{C}'\boldsymbol{C})^{-1})$ where $\boldsymbol{C}$ is a $n \times 2$ matrix with columns $\boldsymbol{\mu}^{(i)} - \boldsymbol{X}\boldsymbol{B}^{(i)}_{1,\ldots,p} - \boldsymbol{U}^{(i)}_{1,\ldots,n}$ and $log(\boldsymbol{\sigma}^{(i)}) - \boldsymbol{X}\boldsymbol{B}^{(i)}_{p+1,\ldots,2p} - \boldsymbol{U}^{(i)}_{n+1,\ldots,2n}$, and where $n = 616$ is the number of grid points and $p = 5$ is the number of model covariates, including an intercept. The notation $\boldsymbol{A}_{j,\ldots,k}$ with $j \leq k$ represents the vector formed with the consecutive entries of $\boldsymbol{A}$ starting from element j and ending in element k.
- Finally, we draw $\boldsymbol{\theta}^{(i)}|\boldsymbol{D},\boldsymbol{Q}$ from a Wishart$(n_{\boldsymbol{\theta}} + n, (\boldsymbol{V_\theta} + \boldsymbol{D}'\boldsymbol{Q}\boldsymbol{D})^{-1})$ where $n = 616$ and $\boldsymbol{D}$ is a $n \times 2$ matrix with columns formed by the vectors $\boldsymbol{U}^{(i)}_{1,\ldots,n}$ and $\boldsymbol{U}^{(i)}_{n+1,\ldots,2n}$ respectively.

10.3.5 Analysis of the RCM output

Figure 10.8 shows maps for the posterior mean estimates for the vector of location and scale parameters respectively based on the MCMC described in Section 10.3.4. The posterior estimates $\boldsymbol{\mu}$ are higher at most locations along the Pacific coast, from Canada to central California, as well as locations in central Arizona, suggesting that annual precipitation maxima are generally more extreme in these areas. The estimates of $\boldsymbol{\sigma}$ are higher along the Pacific coast of Washington state and British Columbia, along the Sacramento Valley of northern and central California, the coast of southern California, and in southern Arizona. This suggests that the distribution of annual precipitation maxima in these areas is more variable than in other areas.

Figure 10.9 shows the posterior distributions of ξ, the constant shape parameter, and of β_1, the slope of the location parameter trend. The posterior distribution of ξ has a mean of 0.063 and a standard deviation of 0.0083. This is very different compared to the prior distribution of ξ, which is a shifted beta distribution on $[-0.5, 0.5]$ with a mean of 0.1 and a standard deviation of 0.12. The posterior probability that $\xi > 0$ is close to one, which corresponds to a *Fréchet* case. However the range of values for ξ is lower than that traditionally obtained with GEV distribution fits to real measurements of precipitation. On the other hand, the posterior distribution of β_1 has a posterior mean of -0.016, but a standard deviation of 0.11, not showing any evidence that β_1 differs from zero. This indicates that over the 20 years of control runs for the RCM output, our statistical model is not able to detect any deterministic time changes in the GEV location parameters. Maps of posterior predictive quantiles of the distribution of annual precipitation maxima are shown in

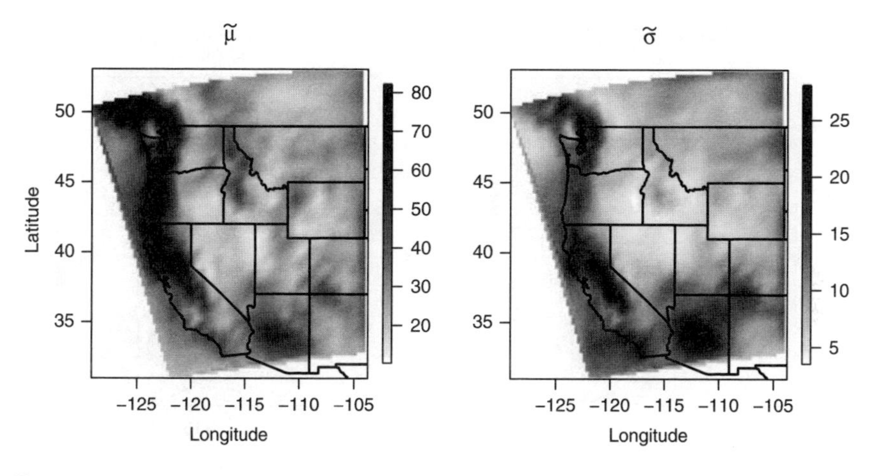

Figure 10.8 Posterior mean estimates for $\boldsymbol{\mu}$ and $\boldsymbol{\sigma}$ for the RCM output.

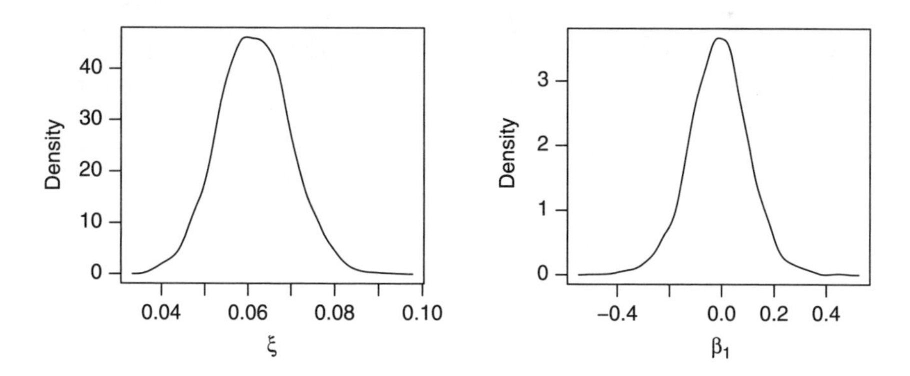

Figure 10.9 Posterior distributions for shape and slope parameters. RCM output.

Figure 10.10. These maps are based on posterior predictive samples and not on GEV quantiles or GEV *return levels*. All of these quantiles are relatively high along the Pacific coast from British Columbia through northern California, through northern and central California, and in central Arizona. The annual precipitation maxima in these areas are higher than in other areas of the spatial domain, but rarely, very extreme. In terms of posterior predictive evaluations, Figure 10.11 shows histograms of samples from the posterior predictive distribution corresponding to the held-out values of y_{st} at four different locations. The four locations were selected from the 25 originally held out locations and to be roughly representative of the northwest, northeast, southwest, and southeast regions of the study area. Vertical bars show the posterior predictive median and the 0.95 posterior predictive quantile based on the predictive samples. The small vertical lines along the x-axis show the actual 20 observed (output) values that were held out at each of the four locations. The observations are in accordance with our predictive distribution and were computed under the assumption of a zero-trend parameter. Similar figures were obtained for other held out

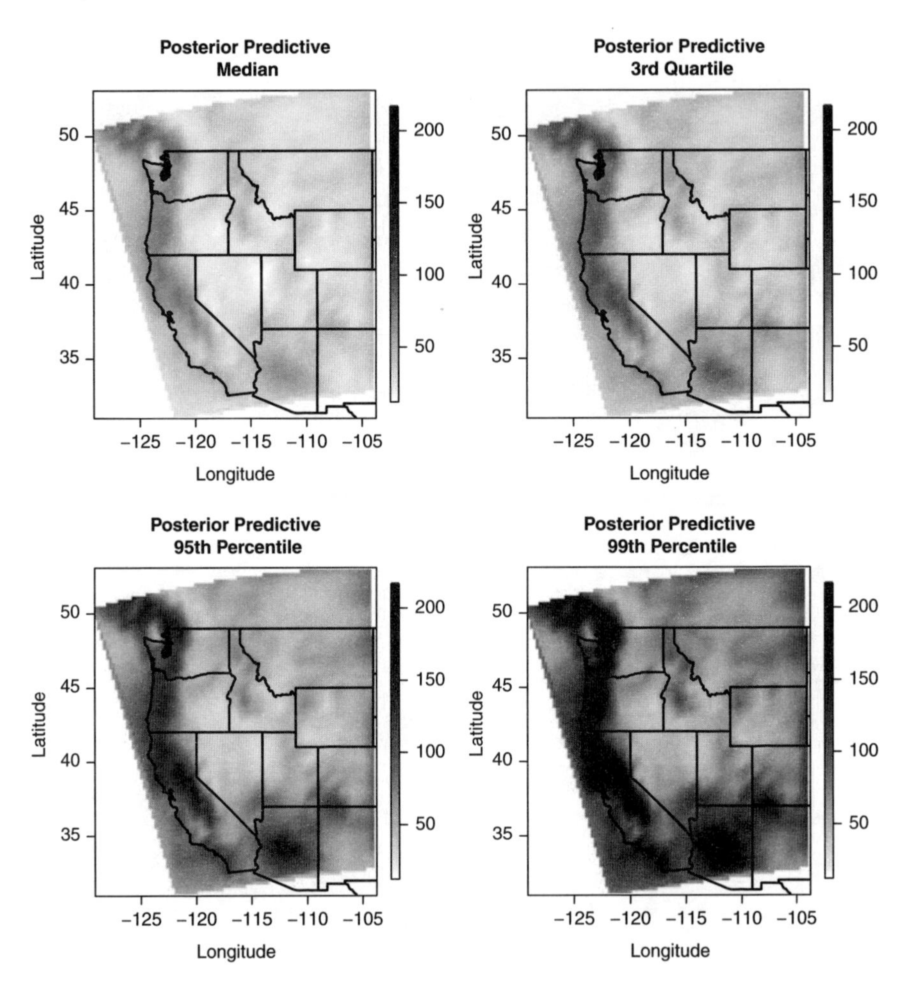

Figure 10.10 Posterior predictive percentiles at four levels: 50%,75%, 90% and 99% for the RCM output.

locations. Figure 10.12 shows the posterior distributions for each of the elements in $\boldsymbol{B}$. The first row of histograms corresponds to the parameters associated with $\boldsymbol{\mu}$ while the second row corresponds to $\boldsymbol{\sigma}$. The covariates that appear more relevant are those associated with longitude, latitude and relative position to ocean. The latitude coefficients show a negative change for both $\boldsymbol{\mu}$ and $\boldsymbol{\sigma}$, indicating that annual precipitation maxima become generally smaller and less variable with increasing latitude. The longitude coefficients also show a negative change with both $\boldsymbol{\mu}$ and $\boldsymbol{\sigma}$, indicating that annual precipitation maxima are generally smaller and less variable in the eastern portion of the study area than in the western portion. The elevation coefficients indicate a positive change on $\boldsymbol{\mu}$ and $\boldsymbol{\sigma}$, indicating that in general annual precipitation maxima are both more extreme and more variable at higher elevations, however, this change is anticipated to be rather small. The coefficients for the ocean indicator variable induce a negative change for $\boldsymbol{\mu}$ and $\boldsymbol{\sigma}$, so that annual precipitation maxima are both less extreme and less variable over the ocean than over land.

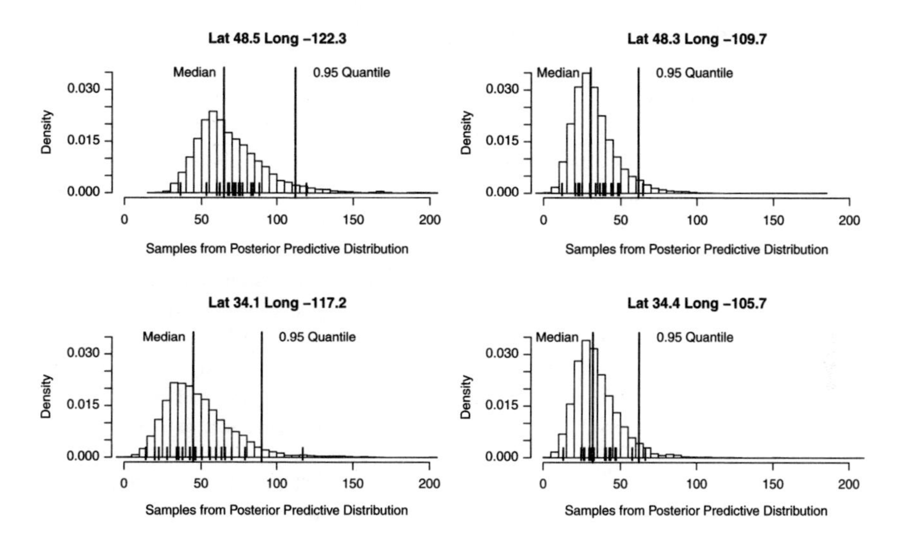

Figure 10.11 RCM output analysis. Predictive posterior distribution at four held out grid points.

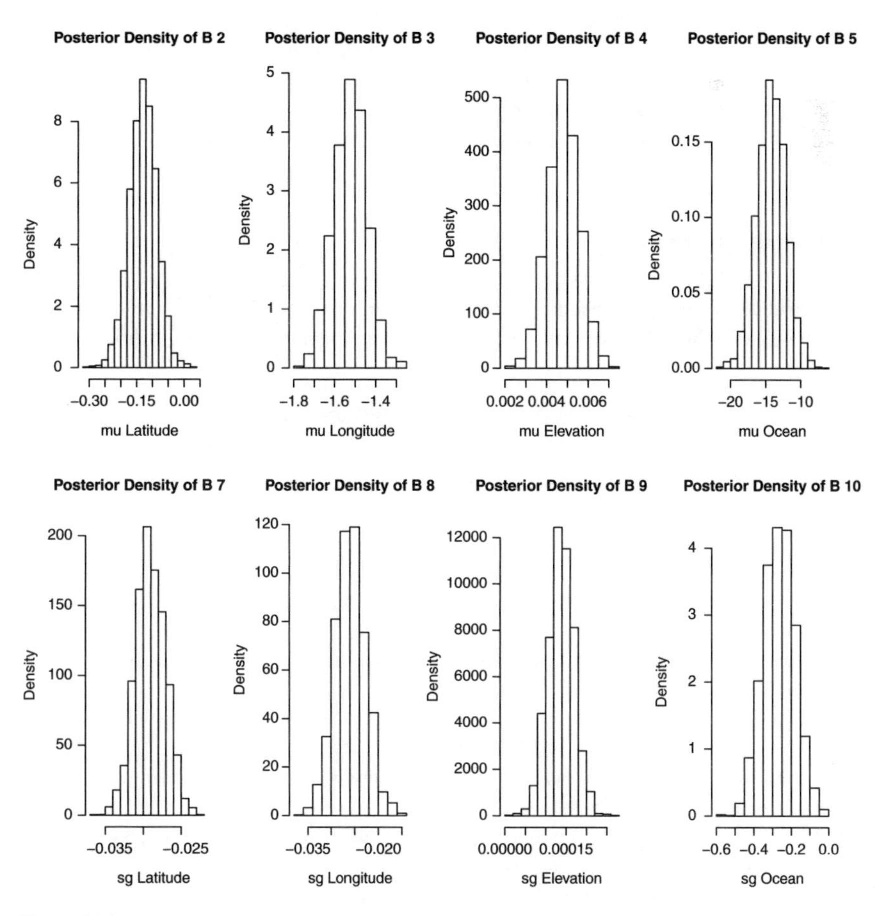

Figure 10.12 RCM output analysis. Posterior distributions for regression parameters.

10.4 Conclusions

This chapter presents a general class of models to study extreme values based on the GEV distribution and that rely on time domain and spatial latent components. These models had not been available until recently and their implementations are now available thanks to the developments of MCMC approaches. In particular, dynamic models as described in [29] and [28] provide a flexible approach to deal with time-varying extremes via MCMC algorithms based on FFBS, in contrast with the traditional deterministic parameter GEV regression models. Interesting developments in this area had arisen from *Particle Filter* methods starting from the work by [12] and more recently extended by [9], both emphasizing the analysis of the athletics dataset of [20]. In addition, [18] present a different framework where the time dependence in extremes is modelled through AR and MA with innovations arising from the Gumbel distribution and illustrated with extreme returns of daily stock data. From the spatial or spatial temporal perspective, some of the main model developments for the study of extremes arose from the Bayesian hierarchical perspective with MCMC methods initiated by [3] and also in for example, [16], [22] and [8]. As shown in this chapter, these modelling approaches had proven their value and flexibility in the representation of extremal phenomena from station data or from climate model output in high-dimensional situations. However, the main drawback of these approaches is that they are based on assumptions of *conditional independence* which may not be adequate to represent spatial dependencies of extremes. Developments based on *Copulas* as in [23] and [14] and the *Max stable* process as in [24] through *composite likelihood* methods combined with MCMC as in [19], provide some examples of the recent focus for modelling spatial extremes from a Bayesian point-of-view.

Acknowledgements

The rainfall data at Maíquetia was kindly provided by Bruno Sansó. The RCM output data was kindly provided by Steve Sain and Dan Cooley while the first author was visiting NCAR.

References

[1] Beirlant, J., Goegebeur, Y., Segers, J. and Teugels, J. (2004). *Statistics of Extremes*. Wiley, Chichester, England.

[2] Carter, C. K. and Kohn, R. (1994). On Gibbs sampling for state space models. *Biometrika*, **81**(3), 541–553.

[3] Casson, E. and Coles, S. (1999). Spatial regression models for extremes. *Extremes*, **1**(4), 449–468.

[4] Coles, S. (2001). *An Introduction to Statistical modelling of Extreme Values*. Springer-Verlag, New York, USA.

[5] Coles, S. and Pericchi, L. (2003). Anticipating catastrophes through extreme value modelling. *Journal of the Royal Statistical Society Series C, Applied Statistics*, **52**, 405–416.

[6] Coles, S. G. and Powell, E. A. (1996). Bayesian methods in extreme value modelling: A review and new developments. *International Statistical Review*, **64**(1), 119–136.

[7] Coles, S. G. and Tawn, J. A. (1996). Modelling extremes of the areal rainfall process. *Journal of Royal Statistics Society*, **B**(58), 329–347.

[8] Cooley, D. and Sain, S. (2010). Spatial hierarchical modelling of precipitation extremes from a regional climate model. *Journal of Agricultural Biological and Environmental Statistics*, **15**(3), 381–402.

[9] Fearnhead, P., Wyncoll, D. and Tawn, J. (2010). A sequential smoothing algorithm with linear computational cost. *Biometrika*, **97**(2), 447–464.

[10] Fisher, R. A. and Tippet, L. H. C. (1928). Limiting forms of the frequency distribution of the largest or smallest member of a sample. *Proceedings of the Cambridge Philosophical Society*, **24**, 180–190.

[11] Frühwirth-Schnatter, S. (1994). Data augmentation and dynamic linear models. *Journal of Time Series Analysis*, **15**, 183–202.

[12] Gaetan, C. and Grigoletto, M. (2004). Smoothing sample extremes with dynamic models. *Extremes*, **7**, 221–236.

[13] Gelfand, A. E. and Smith, A. F. M. (1990). Sampling-based approaches to calculating marginal densities. *Journal of the American Statistical Association*, **85**(410), 398–409.

[14] Ghosh, S. and Mallick, B. (2010). A hierarchical Bayesian spatio-temporal model for extreme precipitation events. *Environmetrics*, **22**, 192–204.

[15] Hastings, W. K. (1970). Monte Carlo sampling methods using Markov chains and their applications. *Biometrika*, **87**, 97–109.

[16] Huerta, G. and Sansó, B. (2007). Time-varying models for extreme values. *Environmental and Ecological Statistics*, **14**(3), 285–299.

[17] Martins, E. and Stedinger, J. (2000). Generalized maximum-likelihood generalized extreme-value quantile estimators for hydrologic data. *Water Resources Research*, **36**, 737–744.

[18] Nakajima, J., Kunihama, T., Omori, T. and Früwirth-Schnatter (2012). Generalized extreme value distribution with time-dependence using the ar and ma models in state space form. *Computational Statistics and Data Analysis*, **56**, 3241–3259.

[19] Ribatet, M., Cooley, D. and Davison, A. (2012). Bayesian inference for composite likelihood models and an application to spatial extremes. *Statistica Sinica*, **22**, 813–845.

[20] Robinson, M. E. and Tawn, J. A. (1995). Statistics for exceptional athletics records. *Journal of the Royal Statistical Society Series C, Applied Statistics*, **44**, 499–511.

[21] Rue, H. and Held, L. (2005). *Gaussian Markov Random Fields*. Chapman & Hall/CRC.

[22] Sang, H. and Gelfand, A. E. (2009). Hierarchical modelling for extreme values observed over space and time. *Environmental and Ecological Statistics*, **16**, 407–426.

[23] Sang, H. and Gelfand, A. E. (2010). Continuous spatial process models for spatial extreme values. *Journal of Agricultural Biological and Environmental Statistics*, **15**(1), 49–65.

[24] Smith, R. (1990). Max-stable processes and spatial extremes. Unpublished manuscript.

[25] Smith, R. L. and Naylor, J. C. (1988). A comparison of maximum likelihood and Bayesian estimators for the three parameter Weibull distribution. *Applied Statistics*, **36**, 358–369.

[26] Stephenson, A. G. (2002, September). *A User's Guide to the Evdbayes Package(Version 1.0)*. http://www.maths.lancs.ac.uk/~stephena/.

[27] Stephenson, A. G. and Gilleland, E. (2005). Software for the analysis of extreme events: the current state and future directions. *Extremes*, **8**(3), 87–109.

[28] West, M. (2012). Bayesian dynamic modelling. In *Bayesian Theory and Applications*, pp. 145–166. Oxford University Press, Oxford.

[29] West, M. and Harrison, J. (1997). *Bayesian Forecasting and Dynamic Models* (Second edn). Springer-Verlag, New York.

Part V

Sequential Monte Carlo

11 Online Bayesian learning in dynamic models: an illustrative introduction to particle methods

HEDIBERT F. LOPES AND
CARLOS M. CARVALHO

11.1 Introduction

In this chapter, we provide an introductory step-by-step review of Monte Carlo methods for filtering in general nonlinear and non-Gaussian dynamic models, also known as state-space models or hidden Markov models (see [60], [20], [5] and [25]). These MC methods are commonly referred to as *sequential Monte Carlo*, or simply *particle filters*. The standard Markovian dynamic model for observation y_t is

$$y_t \sim f(y_t|x_t, \theta), \tag{11.1}$$

$$x_t \sim g(x_t|x_{t-1}, \theta) \tag{11.2}$$

where, for $t = 1, \ldots, n$, x_t is the latent state of the dynamic system and θ is the set of fixed parameters defining the system. Equation (11.1) is referred to as the observation equation that relates the observed series y_t to the state vector x_t. Equation (11.2) is the state transition equation that governs the time evolution of the latent state. For didactical reasons, we assume throughout this chapter that y_t and x_t are both scalars. Multidimensional extensions are, in principle, straightforward and out of our scope.

The central problem in many state-space models, is the sequential derivation of the filtering distribution. By Bayes' theorem

$$p(x_t|y_{1:t}, \theta) = \frac{f(y_t|x_t, \theta)p(x_t|y_{1:t-1}, \theta)}{p(y_t|y_{1:t-1}, \theta)} \tag{11.3}$$

where $y_{1:t} = (y_1, \ldots, y_t)$ (the same for $x_{1:t}$). The problem translates, in part, to deriving the prior distribution of the latent state x_t given data up to time $t-1$:

$$p(x_t|y_{1:t-1},\theta) = \int g(x_t|x_{t-1},\theta)p(x_{t-1}|y_{1:t-1},\theta)dx_{t-1} \tag{11.4}$$

Even when θ is assumed to be known, sequential inference about x_t becomes analytically intractable, except when dealing with Gaussian dynamic linear models (DLM) (detailed in Section 11.2.1).

Most of the early contributions to the literature on the Bayesian estimation of state-space models boils down to the design of Markov chain Monte Carlo (MCMC) schemes that iteratively sample from states and parameters full conditional distributions:

$$p(x_{1:n}|y_{1:n},\theta) \quad \text{and} \quad p(\theta|x_{1:n},y_{1:n}). \tag{11.5}$$

The main references include, amongst others, [6], [7], [23] and [24]. See [45] for a thorough review of dynamic models.

On the one hand, MCMC methods gave researchers the means to free themselves from the (usually unrealistic) assumptions of normality and linearity for both observation equation (11.1) and state transition equation (11.2). On the other hand, however, they took from researchers the ability to sequentially learn about states and parameters.

Particle filters are Monte Carlo schemes designed to sequentially approximate the densities in equations (11.3) and (11.4) over time. The seminal *bootstrap filter* of Gordon, Salmond and Smith [29], for example, uses the sampling importance resampling algorithm to first propagate particles from time $t-1$, i.e. draws from $p(x_{t-1}|y_{1:t-1})$, via equation (11.4), and then to resample the discrete set of propagated particles with weights proportional to the likelihood (Bayes' theorem from equation (11.3)). Sections 11.3 and 11.4 provide additional details about the bootstrap filter as well as many other particles filters for state filtering or state and parameter learning.

The remainder of the chapter is organized as follows. Section 11.2 introduces the basic notation, results and references for the general class of Gaussian DLMs, the AR(1) plus noise model and for the standard stochastic volatility model with AR(1) dynamics. Particle filters for state learning with fixed parameters (also known as pure filtering) and particle filters for state and parameter learning are discussed in Sections 11.3 and 11.4, respectively. Section 11.5 deals with general issues, such as MC error, sequential model checking, particle smoothing and the interaction between particle filters and MCMC schemes.

11.2 Dynamic models

In what follows we provide basic notation and results, as well as key references, for the general class of Gaussian DLMs, the AR(1) plus noise model and for the standard stochastic volatility model with AR(1) dynamics.

11.2.1 Dynamic linear models

A Gaussian *dynamic linear model* (DLM) can be written as

$$y_t|x_t,\theta \sim N(\mu + F_t'x_t, \sigma_t^2) \tag{11.6}$$

$$x_t|x_{t-1},\theta \sim N(\alpha + G_t x_{t-1}, \tau_t^2) \tag{11.7}$$

where intercepts μ and α are added for notational reasons related to the stochastic volatility model of Section 11.2.3. Conditionally on $\theta = (F_{1:n}, G_{1:n}, \sigma^2_{1:n}, \tau^2_{1:n}, \mu, \alpha)$ and assuming the initial distribution $(x_0|y_0) \sim N(m_0, C_0)$, it is straightforward to show that

$$x_t|y_{1:t-1}, \theta \sim N(a_t, R_t) \quad (11.8)$$

$$y_t|y_{1:t-1}, \theta \sim N(f_t, Q_t) \quad (11.9)$$

$$x_t|y_{1:t}, \theta \sim N(m_t, C_t) \quad (11.10)$$

for $t = 1, \ldots, n$, where $N(a, b)$ denotes the normal distribution with mean a and variance b. The three densities in equations (11.8) to (11.10) are referred to as the *propagation density*, the *predictive density* and the *filtering density*, respectively. In fact, the propagation and filtering densities are the prior density of x_t given $y_{1:t-1}$ and the posterior density of x_t given $y_{1:t}$. The means and variances of the three densities are provided by the *Kalman recursions*:

$$a_t = \alpha + G_t m_{t-1} \text{ and } R_t = G_t C_{t-1} G_t' + \tau_t^2 \quad (11.11)$$

$$f_t = \mu + F_t' a_t \text{ and } Q_t = F_t' R_t F_t + \sigma_t^2 \quad (11.12)$$

$$m_t = a_t + A_t e_t \text{ and } C_t = R_t - A_t Q_t A_t' \quad (11.13)$$

where $e_t = y_t - f_t$ is the prediction error and $A_t = R_t F_t Q_t^{-1}$ is the *Kalman gain*. Two other useful densities are the conditional and marginal *smoothed densities*

$$x_t|x_{t+1}, y_t, \theta \sim N(h_t, H_t) \quad (11.14)$$

$$x_t|y_{1:n}, \theta \sim N(m_t^n, C_t^n) \quad (11.15)$$

where

$$h_t = m_t + B_t(x_{t+1} - a_{t+1}) \text{ and } H_t = C_t - B_t R_{t+1} B_t' \quad (11.16)$$

$$m_t^n = m_t + B_t(m_{t+1}^n - a_{t+1}) \text{ and } C_t^n = C_t + B_t^2(C_{t+1}^n - R_{t+1}) \quad (11.17)$$

and $B_t = C_t G_{t+1}' R_{t+1}^{-1}$ (see [60], Chapter 4, for additional details).

11.2.2 AR(1) plus noise model

The AR(1) plus noise model is a Gaussian DLM where the state follows a standard AR(1) process and y_t is observed with measurement error:

$$y_t|x_t, \theta \sim N(x_t, \sigma^2) \quad (11.18)$$

$$x_t|x_{t-1}, \theta \sim N(\alpha + \beta x_{t-1}, \tau^2) \quad (11.19)$$

Conditional on $\theta = (\sigma^2, \alpha, \beta, \tau^2)$, the whole state vector $x_{1:n}$ can be marginalized out analytically (see (11.9)):

$$p(y_{1:n}|\theta) = \prod_{t=1}^{n} p_N(y_t; f_t, Q_t) \quad (11.20)$$

where $p_N(x; \mu, \sigma^2)$ is the density of a normal random variable with mean μ and variance σ^2 evaluated at x. Notice that here f_t and Q_t are both nonlinear functions of θ. The density in equation (11.20) is commonly known as prior predictive density or integrated likelihood.

11.2.2.1 *MC sampling from the posterior*

Posterior draws from $p(x_{1:n}, \theta|y_{1:n})$ can be directly and jointly obtained:

Step (i): Draw $\{\theta^{(i)}\}_{i=1}^N$ from $p(\theta|y_{1:n}) \propto p(\theta)p(y_{1:n}|\theta)$. The likelihood $p(y_{1:n}|\theta)$ comes from (11.20). This can be performed by sampling importance resampling, acceptance–rejection algorithm or Metropolis–Hastings-type algorithms.

Step (ii): Draw $x_{1:n}^{(i)}$ from $p(x_{1:n}|\theta^{(i)}, y_{1:n})$, for $i = 1, \ldots, N$, by first computing forward moments via equations (11.11)–(11.13) and (11.16), and then sampling backwards x_t conditional on x_{t+1} and y_t via equations (11.14). This step is known as the *forward filtering, backward sampling* (FFBS) algorithm ([7]; [23]).

Alternatively, θ from step (i) could be sampled, via a Gibbs sampler step, for instance, from $p(\theta|y_{1:n}, x_{1:n})$. In this case, iterating between steps (i) and (ii) would lead to a MCMC scheme whose target, stationary distribution is the posterior distribution $p(x_{1:n}, \theta|y_{1:n})$.

11.2.2.2 *Prior specification and sufficient statistics*

Assume that the prior distribution of (α, β, τ^2) is decomposed into $\tau^2 \sim IG(\nu_0/2, \nu_0\tau_0^2/2)$ and $(\alpha, \beta)|\tau^2 \sim N(d_0, \tau^2 D_0)$, for known hyperparameters ν_0, τ_0^2, d_0 and D_0. It follows immediately, from basic Bayesian derivations for conditionally conjugate families, that $\tau^2|y_{1:t}, x_{1:t} \sim IG(\nu_t/2, \nu_t\tau_t^2/2)$ and $(\alpha, \beta)|\tau^2, y_{1:t}, x_{1:t} \sim N(d_t, \tau^2 D_t)$, where

$$
\begin{aligned}
D_t^{-1} &= D_{t-1}^{-1} + z_t z_t' \\
D_t^{-1} d_t &= D_{t-1}^{-1} d_{t-1} + z_t x_t \\
\nu_t &= \nu_{t-1} + 1 \\
\nu_t \tau_t^2 &= \nu_{t-1}\tau_{t-1}^2 + (x_t - z_t' d_t)x_t + (d_{t-1} - d_t)' D_{t-1}^{-1} d_{t-1}
\end{aligned}
\tag{11.21}
$$

and $z_t = (1, x_t)'$. The relevance of these conditional conjugacy results will become apparent when dealing with some of the particles filters with parameter learning in Section 11.4. See [52] for particle methods applied to AR models with structured priors.

11.2.3 SV-AR(1) model

Univariate stochastic volatility (SV) in asset price dynamics results from the movements of an equity index S_t and its stochastic volatility v_t via a continuous time diffusion by a Brownian motion: $d\log S_t = \mu dt + \sqrt{v_t} dB_t^P$ and $d\log v_t = \kappa(\gamma - \log v_t)dt + \tau dB_t^V$, where the parameters governing the volatility evolution are $(\mu, \kappa, \gamma, \tau)$ and (B_t^P, B_t^V) are (possibly correlated) Brownian motions ([54], [57], [31], [33]).

Data arises in discrete time so it is natural to take an Euler discretization of the above equations. This is then commonly referred to as the *stochastic volatility autoregressive*, SV-AR(1), model and is described by the following nonlinear dynamic model:

$$
y_t|x_t, \theta \sim N(0, \exp\{x_t/2\}) \tag{11.22}
$$

$$
x_t|x_{t-1}, \theta \sim N(\alpha + \beta x_{t-1}, \tau^2) \tag{11.23}
$$

where y_t are log-returns and x_t are log-variances. See [32] and [34] for the original Bayesian papers on MCMC estimation of the above SV-AR(1) model. In addition, [41] provides an extensive review of Bayesian inference in the SV-AR(1) model, as well as other univariate and multivariate SV models.

11.2.3.1 *Sampling parameters*

The SV model is completed with a conjugate prior distribution for $\theta = (\alpha, \beta, \tau^2)$, i.e. $p(\theta) = p(\alpha, \beta|\tau^2)p(\tau^2)$, where $(\alpha, \beta|\tau^2) \sim N(d_0, \tau^2 D_0)$ and $\tau^2 \sim IG(\nu_0/2, \nu_0\tau_0^2/2)$, for known hyperparameters d_0, D_0, ν_0 and τ_0^2. Apart from the nonlinear relationship between y_t and x_t in equation (11.22), notice the similarity between the above SV-AR(1) model and the AR(1) plus noise model of section 11.2.2. Therefore, sampling (α, β, τ^2) given $x_{1:t}$ can be done via equations (11.21).

11.2.3.2 *Sampling states*

Sampling from $x_{1:t}|y_{1:t}, \theta$ jointly is performed by a FFBS scheme introduced by [34] for the SV-AR(1) model. They approximate the distribution of $\log y_t^2$ by a carefully tuned mixture of normals with seven components. More precisely, the observation equation (11.22) is rewritten by $z_t = \log y_t^2 = x_t + \epsilon_t$, where $\epsilon_t = \log \varepsilon_t^2$ follows a $\log \chi_1^2$ distribution, a parameter-free left skewed distribution with mean -1.27 and variance 4.94. They argue that $\epsilon = \log \chi_1^2$ can be well approximated by $\sum_{i=1}^{7} \pi_i p_N(\epsilon_t; \mu_i, v_i^2)$, where

$$\begin{aligned}
\pi &= (0.0073, 0.10556, 0.00002, 0.04395, 0.34001, 0.24566, 0.2575) \\
\mu &= (-11.40039, -5.24321, -9.83726, 1.50746, -0.65098, 0.52478, -2.35859) \\
v^2 &= (5.79596, 2.61369, 5.17950, 0.16735, 0.64009, 0.34023, 1.26261)
\end{aligned}$$

Therefore, a standard data augmentation argument allows the mixture of normals to be transformed into individual normals, i.e. $(\epsilon_t|k_t) \sim N(\mu_{k_t}, v_{k_t}^2)$ and $Pr(k_t) = q_{k_t}$. Conditionally on $k_{1:t}$, the SV model can be rewritten as a standard Gaussian DLM:

$$(z_t|x_t, k_t, \theta) \sim N(\mu_{k_t} + x_t, v_{k_t}^2) \tag{11.24}$$

$$(x_t|x_{t-1}, \theta) \sim N(\beta_0 + \beta_1 x_{t-1}, \tau^2). \tag{11.25}$$

The FFBS algorithm is then used to sample from $p(x_{1:n}|y_{1:n}, k_{1:n}, \theta)$. Given $x_{1:n}$, k_t is sampled from $\{1, \ldots, 7\}$ with $Pr(\kappa_t = i|z_t) \propto \pi_i p_N(z_t; \mu_i + x_t, v_i^2)$, for $i = 1, \ldots, 7$ and $t = 1, \ldots, n$.

The above two steps, i.e. sampling parameters and sampling states, will both be very useful in the next two sections when deriving particle filters for both state and fixed parameters.

11.3 Particle filters

Particle filters use Monte Carlo methods, mainly the sampling importance resampling (SIR), to sequentially reweigh and resample draws from the propagation density. The nonlinear Kalman filter is summarized by the prior and posterior densities in equations (11.4) and (11.3):

$$\begin{aligned}
p(x_t|y_{1:t-1}) &= \int g(x_t|x_{t-1})p(x_{t-1}|y_{1:t-1})dx_{t-1} \\
p(x_t|y_{1:t}) &\propto f(y_t|x_t)p(x_t|y_{1:t-1})
\end{aligned}$$

where the vector of fixed parameters θ is assumed to be known and dropped from the notation, reappearing when necessary. The following joint densities will become useful in Sections 11.3.1 and 11.3.2:

$$p(x_t, x_{t-1}|y_{1:t-1}) = g(x_t|x_{t-1})p(x_{t-1}|y_{1:t-1}) \quad (11.26)$$

$$p(x_t, x_{t-1}|y_{1:t}) \propto f(y_t|x_t)g(x_t|x_{t-1})p(x_{t-1}|y_{1:t-1}). \quad (11.27)$$

Particle filters, loosely speaking, combine the sequential estimation nature of Kalman-like filters with the flexibility for modelling of MCMC samplers, while avoiding some of their shortcomings. On the one hand, like MCMC samplers and unlike Kalman-like filters, particle filters are designed to allow for more flexible observational and evolutional dynamics and distributions. On the other hand, like Kalman-like filters and unlike MCMC samplers, particle filters provide online filtering and smoothing distributions of states and parameters. Advanced readers are refereed to, for instance, [5], Chapters 7 to 9 for a more formal, theoretical discussions of sequential Monte Carlo methods.

The goal of most particle filters is to draw a set of i.i.d. particles $\{x_t^{(i)}\}_{i=1}^N$ that approximates $p(x_t|y_{1:t})$ by starting with a set of i.i.d. particles $\{x_{t-1}^{(i)}\}_{i=1}^N$ that approximates $p(x_{t-1}|y_{1:t-1})$. To simplify the notation, from now on we will simply refer to 'particles x_{t-1}' when describing a 'set of i.i.d. particles $\{x_{t-1}^{(i)}\}_{i=1}^N$'. The most popular filters are the bootstrap filter (BF), also known as sequential importance sampling with resampling (SISR) filter, proposed by [29], and the auxiliary particle filter (APF), also known as auxiliary SIR (ASIR) filter, proposed by [48]. We introduce both of them in the next section along with their optimal counterparts.

11.3.1 Bootstrap filter

The bootstrap filter (BF) is the seminal and perhaps the most implemented of the particle filters. It can be basically thought of as the repetition of the sampling importance resampling (SIR) over time. More precisely, let $p(x_{t-1}|y_{1:t-1})$ be the posterior density of the latent state x_{t-1} at time $t-1$. From equations (11.4) and (11.3) and Bayes' theorem, it is easy to verify that

$$p(x_t, x_{t-1}|y_{1:t}) \propto \underbrace{f(y_t|x_t)}_{\text{2. Resample}} \underbrace{g(x_t|x_{t-1})p(x_{t-1}|y_{1:t-1})}_{\text{1. Propagate}}. \quad (11.28)$$

In words, BF combines old particles x_{t-1}, generated from $p(x_{t-1}|y_{1:t-1})$, and new particles x_t, generated from $g(x_t|x_{t-1})$, so that the combined particles (x_t, x_{t-1}) are draws from $p(x_t, x_{t-1}|y_{1:t-1})$. This step is labelled '1. *Propagate*' in the above expression. BF then resamples the combined particles (x_t, x_{t-1}) with SIR weights proportional to the likelihood

$$\omega_t \propto \frac{f(y_t|x_t)g(x_t|x_{t-1})p(x_{t-1}|y_{1:t-1})}{g(x_t|x_{t-1})p(x_{t-1}|y_{1:t-1})} = f(y_t|x_t) \quad (11.29)$$

This step is labelled '2. *Resample*' in the above expression. These combined resampled particles approximate $p(x_{t-1}, x_t|y_{1:t})$ and, in particular, the marginal filtering density $p(x_t|y_{1:t})$.

11.3.1.1 *Particle impoverishment*

The overall SIR proposal density (the denominator of (11.29)) is $q(x_t, x_{t-1}|y_{1:t}) = p(x_t, x_{t-1}|y_{1:t-1}) = g(x_t|x_{t-1})p(x_{t-1}|y_{1:t-1})$. The particles x_t from (x_t, x_{t-1}) are, in fact, particles from the prior density $p(x_t|y_{1:t-1})$. It is well known that the SIR algorithm can perform badly when the prior is used as proposal density. The main reason is that in most cases either the prior is too flat relative to the likelihood or vice versa. Small overlap between the prior and the posterior leads to unbalanced weights, that is a small number of particles will have dominating weights and all other particles will have negligible weights. This decrease in particle representativeness, or *particle degeneracy*, is exacerbated when the SIR is carried over time.

11.3.1.2 *Adapted and fully adapted BF*

Instead of using the evolution density $g(x_t|x_{t-1})$ to propagate x_{t-1} to x_t, one could use an *unblinded proposal*, $q(x_t|x_{t-1}, y_t)$, i.e. a proposal that incorporates the information about the current observation y_t. These filters are commonly called *adapted filters*. In this case, $q(x_t, x_{t-1}|y_{1:t}) = q(x_t|x_{t-1}, y_t)p(x_{t-1}|y_{1:t-1})$ is the SIR proposal density, while the SIR weights are

$$\omega_t \propto \frac{f(y_t|x_t)g(x_t|x_{t-1})p(x_{t-1}|y_{1:t-1})}{q(x_t, x_{t-1}|y_{1:t})} = \frac{f(y_t|x_t)g(x_t|x_{t-1})}{q(x_t|x_{t-1}, y_t)} \tag{11.30}$$

Full adaptation occurs when one is able to sample from $p(x_t|x_{t-1}, y_t)$, in which case the SIR weights are proportional to the predictive density

$$\omega_t \propto p(y_t|x_{t-1}). \tag{11.31}$$

Even though full adaptation is rare, it can be used to guide the researcher in the selection of proposal densities $q(x_t|x_{t-1}, y_t)$. The closer $q(x_t|x_{t-1}, y_t)$ is to $p(x_t|x_{t-1}, y_t)$ the better. However, as [48] say, 'even fully adapted particle filters do not produce iid samples from $p(x_t|y_{1:t})$, due to their approximation of $p(x_t|y_{1:t-1})$ by a finite mixture distribution.' The AR(1) plus noise model of Section 11.2.2 and SV-AR(1) model of Section 11.2.3 can be implemented by fully adapted and adapted versions of the above BF.

11.3.2 Auxiliary particle filter

[52] noticed that writing Bayes' theorem from Equation (11.28) as

$$p(x_t, x_{t-1}|y_{1:t}) \propto \underbrace{p(x_t|x_{t-1}, y_{1:t})}_{2.Propagate} \underbrace{p(y_t|x_{t-1})p(x_{t-1}|y_{1:t-1})}_{1.Resample} \tag{11.32}$$

would lead to alternative ways of designing the SIR proposal density $q(x_t, x_{t-1}|y_{1:t})$. Since $p(y_t|x_{t-1})$ and $p(x_t|x_{t-1}, y_{1:t})$ are usually, respectively, unavailable for pointwise evaluation and sampling (see the discussion about fully adapted filters at the end of Section 11.3.1); they suggested a generic proposal

$$q(x_{t-1}, x_t|y_{1:t}) = g(x_t|x_{t-1})f(y_t|h(x_{t-1}))p(x_{t-1}|y_{1:t-1}), \tag{11.33}$$

where $h(.)$ is usually the expected value, median or mode of $g(x_t|x_{t-1})$. The SIR weights would then be written as

$$w_t \propto \frac{f(y_t|x_t)g(x_t|x_{t-1})p(x_{t-1}|y_{1:t-1})}{g(x_t|x_{t-1})f(y_t|h(x_{t-1}))p(x_{t-1}|y_{1:t-1})} = \frac{f(y_t|x_t)}{f(y_t|h(x_{t-1}))} \tag{11.34}$$

In words, APF would resample old particles x_{t-1} from $p(x_{t-1}|y_{1:t-1})$ with weights proportional to $f(y_t|h(x_{t-1}))$, which take into account the new observation y_t. These are usually called the *first-stage* weights. This step is labelled '1. *Resample*' in equation (11.32). Then, new particles x_t are sampled from $g(x_t|x_{t-1})$, such that the combined particles (x_{t-1}, x_t) are draws from $q(x_{t-1}, x_t|y_{1:t})$. These combined particles are then resampled with weights given by equation

(11.34). These are usually called the *second-stage* weights. This step is labelled '*1. Propagate*' in equation (11.32). The final, resampled combined particles approximate $p(x_{t-1}, x_t|y_{1:t})$ and, in particular, the marginal filtering density $p(x_t|y_{1:t})$. Comparing the above labels and their order of operation, we call the APF a *resample–sample* filter, while the BF is *sample–resample* filter.

11.3.2.1 *Fully adapted APF*

The above generic APF is a partially adapted filter by construction. However, the degree of adaptation depends on how close the first-stage weights $f(y_t|h(x_{t-1}))$ and the predictive $p(y_t|x_{t-1})$ are. For general adapted first-stage weights $q(x_{t-1}|y_t)$ and adapted resampling proposal $q(x_t|x_{t-1}, y_t)$, the SIR weights of equation (11.34) become

$$w_t \propto \frac{f(y_t|x_t)g(x_t|x_{t-1})}{q(x_t|x_{t-1}, y_t)q(x_{t-1}|y_t)}. \tag{11.35}$$

Similar to the fully adapted BF, the APF is fully adapted when $q(x_{t-1}|y_t) = p(y_t|x_{t-1})$ and $q(x_t|x_{t-1}, y_t) = p(x_t|x_{t-1}, y_t)$. In this case, the second-stage weights (equation (11.35)) are proportional to one (no resampling necessary).

11.3.2.2 *Local linearization*

[48] suggest, for more general settings, proposal density $q(x_t|x_{t-1}, y_t)$ that are based on local linearization of the observation equation via an extended Kalman filter-type approximation in order to better approximate $p(x_t|x_{t-1}, y_t)$. See [17] and [30], amongst others, for additional particle filters and discussion on proposals based on local linear approximations.

Another class of proposals, usually more efficient when available, is based on the *mixture Kalman filters* (MKF) of [11]. The MKF takes advantage of possible analytical integration of some components of the state vector by conditioning on some other components. Such filters are commonly referred to as *Rao-Blackwellized particle filter*. This is also acknowledged in [48] and many other references. See, for instance, [16], and [18], [8].

11.3.3 Marginal likelihood

The above filters can be used to approximate $p(y_{1:t})$, the marginal likelihood up to time t, as

$$\hat{p}(y_{1:t}) = \prod_{j=1}^{t} \hat{p}(y_j|y_{1:j-1}) = \frac{1}{N^t}\prod_{t=1}^{t}\sum_{i=1}^{N} f(y_j|x_j^{(i)}), \tag{11.36}$$

where x_t are particles from $p(x_t|y_{1:t-1})$. See [12] and [14] for further details and theoretical discussion.

11.3.4 Effective sample size

The quality of a particle filter can be measured by its ability to generate a 'diverse' particle set by drawing from proposals $q(x_t|x_{t-1}, y_t)$ and reweighting with densities $q(x_{t-1}|y_t)$. [35] suggest using the coefficient of variation $CV_t = (N\sum_{i=1}^{N}(\omega_t^{(i)} - 1/N)^2)^{1/2}$, where $\omega_t^{(i)} = w_t^{(i)}/\sum_{j=1}^{N} w_t^{(j)}$ are normalized weights, as a simple criterion to detect the weight degeneracy phenomenon. CV_t varies between 0 (equal weights) and $\sqrt{N-1}$ (N copies of a single particle). [37] and [36] propose tracking the *effective sample size* $N_{\text{eff}} = N/(1 + CV_t^2)$, which varies between 1 (N copies of a single particle) and N (equal weights). [5] tracks the Shannon entropy

$\text{Ent} = -\sum_{i=1}^{N} \omega_t^{(i)} \log_2 \omega_t^{(i)}$, which varies between 0 (N copies of a single particle) and $\log_2 N$ (equal weights).

11.3.5 Examples

11.3.5.1 *AR(1) plus noise model*

From Section 11.2.2 we can easily see that, given θ, the filtering densities $p(x_t|y_{1:t})$ are available in closed form and no particle filtering is necessary. However, we implement both BF and APF to this model and use the exact densities to assess their performances. It is easy to see that $p(y_t|x_{t-1})$ is normal with mean $h(x_{t-1}) = \alpha + \beta x_{t-1}$ and variance $\sigma^2 + \tau^2$, while $p(x_t|x_{t-1}, y_t)$ is normal with mean $Ay_t + (1 - A)h(x_{t-1})$ and variance $(1 - A)\tau^2$, where $A = \tau^2/(\tau^2 + \sigma^2)$. These results are used to implement fully adapted BF and APF, labelled here by OBF and OAPF (for optimal).

We simulate $S = 50$ datasets for each value of τ^2 in $\{0.05, 0.75, 1.0\}$ and all with $n = 100$ observations; a total of 150 datasets. The other parameters are $(\alpha, \beta, \sigma^2) = (0.05, 0.95, 1.0)$ and $x_0 = 1$. The prior for x_0 is $N(m_0, C_0)$ where $m_0 = 1$ and $C_0 = 10$. We run the four filters $R = 50$ times, each time based on $N = 500$ particles. A total of $150 \times 50 \times 4 = 30000$ combined runs. We then compute the logarithm of the mean square error of filter f and time t as $MSE_{ft} = \sum_{s=1}^{S} \sum_{r}^{R} (\hat{q}_{sftr}^{\alpha} - q_{st}^{\alpha})^2/RS$, where q_{st}^{α} and q_{sftr}^{α} are the true and approximated αth percentile of $p(x_t|y_{1:t})$, for dataset s, time period t, run r, percentile α in $\{5, 50, 95\}$ and filter f in $\{BF, APF, OBF, OAPF\}$.

Figure 11.1 summarizes our findings based on log relative MSEs of APF, OBF and OAPF relative to BF. It suggests that the optimal filters are better then their counterpart non-optimal filters. In addition, OAPF is uniformly superior to OBF (increasingly in τ^2), so favouring resampling–sampling filters over sampling–resampling filters. Finally, BF is usually better than APF for small τ^2/σ^2 (small signal-to-noise ratio). Similar results are found when MSEs are replaced by mean absolute errors (not shown here).

11.3.5.2 *SV-AR(1) model*

In this example we illustrate the performance of both the bootstrap filter and the auxiliary particle filter for the SV-AR(1) model of Section 11.2.3. The parameter vector $\theta = (\alpha, \beta, \tau^2)$ is assumed known (see Section 11.4.4 for the general case where θ is also learned sequentially). Let $\mu_t = \alpha + \beta x_{t-1}$. On the one hand, the BF propagates new particles x_t from $N(\mu_t, \tau^2)$, which are then resampled with weights proportional to $p_N(y_t; 0, e^{x_t})$. On the other hand, the APF resamples old particles x_{t-1} with weights proportional to $p_N(y_t; 0, e^{\mu_t})$. New particles x_t are then propagated from $N(\mu_t, \tau^2)$ and resampled with weights proportional to $p_N(y_t; 0, e^{x_t})/p_N(y_t; 0, e^{\mu_t})$.

Potentially better proposals can be obtained. One could, for instance, use the (rough) normal approximation $N(-1.27, 4.94)$ to $\log y_t^2$ presented in Section 11.2.3. This linearization leads to first-stage weights $q(x_{t-1}|y_t) = p_N(z_t; \mu_t, 4.94)$, where $z_t = \log y_t^2 + 1.27$, while the resampling proposal $q(x_t|x_{t-1}, y_t)$ is normal with mean $v(z_t/4.94 + \mu_t/\tau^2)$ and variance $v = 1/(1/4.94 + 1/\tau^2)$. Consequently, it can be shown that the second-stage weights are proportional to $p_N(y_t; 0, \exp\{x_t\})/p_N(z_t; x_t, 4.94)$. We call this APF filter simply APF1 in what follows.

A second example is based on [34]. They used, in a MCMC context, a first-order Taylor expansion of e^{-x_t} around μ_t to approximate the likelihood $p(y_t|x_t)$ by $\exp\{-0.5x_t(1 - y_t^2 e^{-\mu_t})\}$ (up to a proportionality constant). In this setting, the resampling proposal $q(x_t|x_{t-1}, y_t)$ is $N(\tilde{\mu}_t, \tau^2)$ with $\tilde{\mu}_t = \mu_t + 0.5\tau^2(y_t^2 e^{-\mu_t} - 1)$. First-stage weights are then $q(x_{t-1}|y_t) \propto \exp\{-0.5\tau^{-2}[(1 + \mu_t)\tau^2 y_t^2 e^{-\mu_t} + \mu_t^2 - \tilde{\mu}_t^2]\}$. We call this APF filter simply APF2 in what follows.

In a third, more involving example, inspired by [34], who use a seven-component mixture of normals to approximate $\log \chi_1^2$ (see equations (11.24) and (11.25) of Section 11.2.3), we obtain a fully adapted APF for the SV-AR(1) model. In this case, the first-stage weights are proportional to

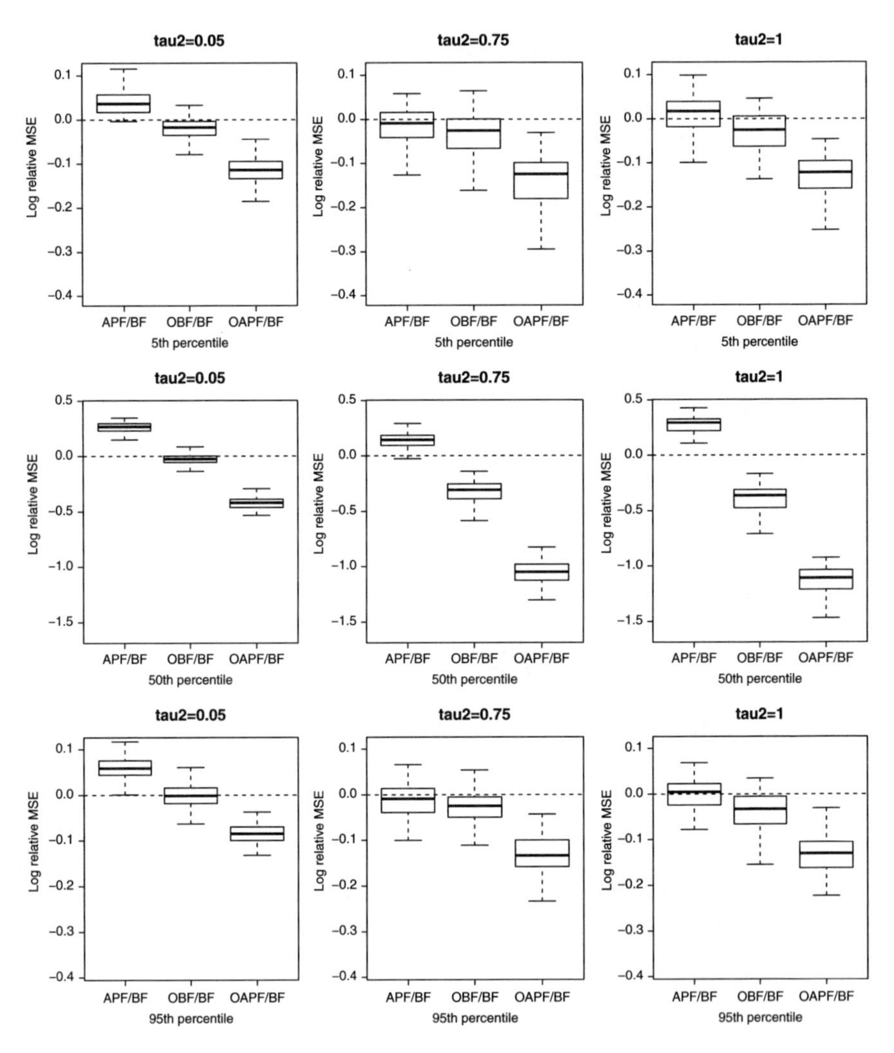

Figure 11.1 *AR(1) plus noise model (pure filter).* Relative mean square error performance (on the log-scale) of the four filters across $S = 50$ datasets of size $n = 100$ and $R = 50$ runs of each filter. Particle size for all filters is $N = 500$. Numbers below zero indicate a superior performance of the filter relative to the bootstrap filter (BF).

$\sum_{i=1}^{7} \pi_i p_N(\log y_t^2; \mu_i + \alpha + \beta m_{t-1}, v_i + \tau^2 + \beta^2 C_{t-1})$, where m_{t-1} and C_{t-1} are the Kalman moments from Section 11.2.1. By integrating out both states x_t and x_{t-1}, we expect the above weights to be flatter, more evenly balanced than the respective ones based on the BF, APF, APF1 and APF2. In addition, instead of sampling x_t, we first sample κ_t from $\{1, \ldots, 7\}$ with $Pr(\kappa_t = i) \propto \pi_i N(\log y_t^2; \mu_i + \alpha + \beta m_{t-1}, v_i + \tau^2 + \beta^2 C_{t-1})$, for $i = 1, \ldots, 7$, and then update m_t and C_t via equations (11.11) to (11.13) from Section 11.2.1. See the discussion in the last paragraph of Section 11.3.2. We call this APF filter simply FAAPF in what follows.

A total of $n = 200$ data points were simulated from $\alpha = -0.03052473$, $\beta = 0.9702$, $\tau^2 = 0.031684$ and $x_0 = -1.024320$. This is the specification used in one of the simulated exercises from [48] and is chosen to mimic the time series behaviour of financial returns. We assume that $x_0 \sim N(m_0, C_0)$ for $m_0 = -1.024320$ and $C_0 = 1$. We run the three filters for $R = 50$ times, each time and each one based on $N = 1000$ particles. We then compute their mean absolute error, $MAE = \sum_{t=1}^{n} |\hat{q}_{t,f}^{\alpha} - q_t^{\alpha}|/n$, where q_t^{α} and $q_{t,f}^{\alpha}$ are the true and approximated αth percentile of $p(x_t|y_{1:t})$, for $\alpha = (5, 50, 95)$ and f one of the filters.

Figure 11.2 summarizes our simulation exercise. The empirical findings suggest that the filters perform quite similarly, with the FAAPF, followed by the BF, being uniformly better than all other filters for all percentiles. This is probably partially due to the fact that the variability of the system equation ($\tau^2 = 0.02$) is much smaller than that of the observation equation. Recall, from Section 11.2.3, that the variance of the log χ_1^2 is around 4.94. In other words, $p_N(y_t; 0, \exp\{\alpha + \beta x_{t-1}\})$ does not seem to be a good SIR proposal for $p(y_t|x_{t-1})$. On one of their simulation exercises, [48] found similar results. They say that 'the auxiliary particle filter is more efficient than the plain

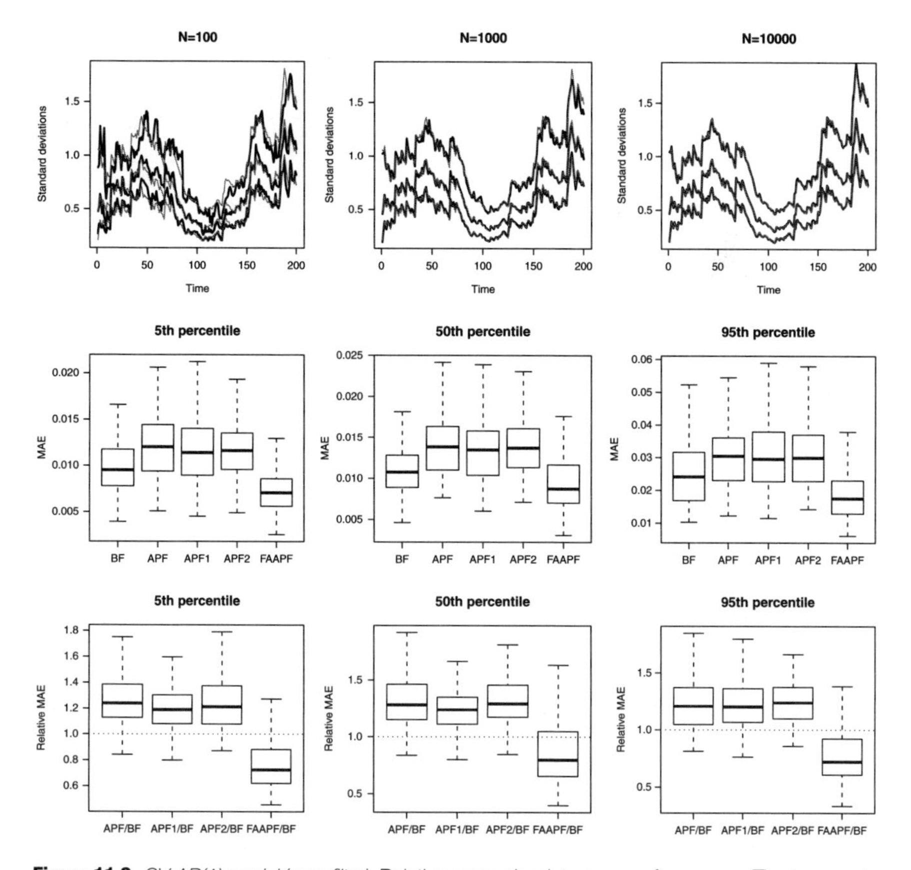

Figure 11.2 *SV-AR(1) model (pure filter).* Relative mean absolute error performances. The top panels show the trajectories of true (dark lines) and BF-based approximations (grey lines) for the αth percentiles $p(x_t|y_t)$, with α in $\{5, 50, 95\}$ and particle sizes N in $\{100, 1000, 10\,000\}$. True trajectories are basically BF with $N = 1\,000\,000$ (using APF or APF1 produced the same results). The middle panels show MAE based on $R = 100$ runs of each filter based on $N = 1000$ particles. APF1, APF2 and FAAPF are APF with first-stage weights $q(x_{t-1}|y_t)$ and resampling proposal $q(x_t|x_{t-1}, y_t)$ described in Section 11.3.5. Bottom panels are relative MAE of APF, APF1, APF2 and FAAPF relative to BF.

particle filter, but the difference is small, reflecting the fact that for the SV model, the conditional likelihood is not very sensitive to the state.'

11.4 Parameter learning

The particle filters introduced in Section 11.3, and illustrated in the examples of Section 11.3.5, assumed that θ, the vector of parameters governing both evolution and observation equations (see equations 11.1 and 11.2), is known. This was partially for didactical or pedagogical reasons and partially to emphasize the chronological order of appearance of the filters. Sequential estimation of fixed parameters θ is historically and notoriously difficult. Simply including θ in the particle set is a natural but unsuccessful solution, as the absence of a state evolution implies that we will be left with an ever-decreasing set of atoms in the particle approximation for $p(\theta|y_{1:t})$.

Important developments in the direction of sequentially updating $p(x_t, \theta|y_{1:t})$, instead of simply $p(x_t|y_{1:t}, \theta)$, have been made over the last decade and now sequential parameter learning is an important sub-area of research within the particle filter branch. [38], [55], [21], [50] and [8] are a good representation of the rapid developments in this area. We revisit several of these contributions here along with illustrations of their implementation in the AR(1) plus noise and SV-AR(1) models.

11.4.1 Liu and West's filter

[38] adapt the generic APF of Section 11.3.2 to sequentially resample and propagate particles associated with x_t and θ simultaneously. More specifically, equation (11.32) is rewritten as

$$p(x_t, x_{t-1}, \theta|y_{1:t}) \propto \underbrace{p(x_t, \theta|x_{t-1}, y_{1:t})}_{2.\ Propagate} \underbrace{p(y_t|x_{t-1}, \theta)p(x_{t-1}, \theta|y_{1:t-1})}_{1.\ Resample}. \tag{11.37}$$

Similarly to the APF's generic proposal (equation 11.33), Liu and West resample old particles (x_t, θ) with first-stage weights proportional to $p(y_t|h(x_{t-1}), m(\theta))$, with $h(\cdot)$ as before and $m(\theta) = a\theta + (1-a)\bar{\theta}$. Let $\tilde{\theta}$ and $\tilde{x}_{t-1}$ be the resampled particles. New particles θ are then propagated from the resampled particles via $N(m(\tilde{\theta}), h^2V)$, where $a^2 + h^2 = 1$, and new particles x_t are propagated from $g(x_t|\tilde{x}_{t-1}, \tilde{\theta})$. The second-stage weights are proportional to $p(y_t|x_t, \theta)/p(y_t|h(\tilde{x}_{t-1}), m(\tilde{\theta}))$. The quantities $\bar{\theta}$ and V are, respectively, the particle approximations to $E(\theta|y_{1:t})$ and $V(\theta|y_{1:t})$.

The key idea here is the choice of the proposal $q(x_t, \theta|x_{t-1}, y_{1:t})$ to approximate $p(x_t, \theta|x_{t-1}, y_{1:t})$. The proposal $q(x_t, \theta|x_{t-1}, y_{1:t})$ is decomposed into two parts: $q(x_t|\theta, x_{t-1}, y_{1:t}) = g(x_t|x_{t-1}, \theta)$ (blind propagation) and $q(\theta|x_{t-1}, y_{1:t})$, which is locally approximated by $N(m(\theta), h^2V)$. This *smooth kernel density* approximation ([58], [59]) literally adds an artificial evolution to θ, as suggested in [29], but it controls the inherent over-dispersion by locally shrinking the particles θ towards their mean $\bar{\theta}$. [38] use standard discount factor ideas from basic dynamic linear models to select the tuning constant a (or h). The constants a and h measure, respectively, the extent of the shrinkage and the degree of over dispersion of the mixture. The rule of thumb is to select a greater than or equal to, say, 0.99. The idea is to use the mixture approximation to generate fresh samples from the current posterior in a attempt to avoid particle degeneracy.

The main attraction of [38]'s filter is its generality as it can be implemented in any state-space model. It also takes advantage of APF's resample–propagate framework and can be considered a benchmark in the current literature. The steps of the LW algorithm are as follows:

Step 1 (Resample) $(\tilde{x}_{t-1}, \tilde{\theta})$ from (x_{t-1}, θ) with weights $w_t \propto p(y_t|h(x_{t-1}), m(\theta))$;
Step 2 (Propagate)

a) $\tilde{\theta}$ to $\hat{\theta}$ via $N(m(\tilde{\theta}), h^2V)$;
b) $\tilde{x}_{t-1}$ to $\hat{x}$ via $g(x_t|\tilde{x}_{t-1}, \hat{\theta})$;

Step 3 (Resample) (x_t, θ) from $(\hat{x}_t, \hat{\theta})$ with weights $w_{t+1} \propto p(y_t|\hat{x}_t, \hat{\theta})/p(y_t|h(\tilde{x}_{t-1}), m(\tilde{\theta}))$.

11.4.2 Storvik's filter

Storvik (2002) [55] (see also [21]) proposes a particle filter that sequentially updates states and parameters by focusing on the particular case where the posterior distribution of θ given $x_{1:t}$ and $y_{1:t}$ depends on a low-dimensional set of sufficient statistics, i.e. $p(\theta|y_{1:t}, x_{1:t}) = p(\theta|s_t)$, that can be recursively and deterministically updated via $s_t = \mathcal{S}(s_{t-1}, x_{t-1}, x_t, y_t)$.

We are using both models as illustrations in this chapter, i.e. the AR(1) plus noise and the SV-AR(1) models, allow sequential parameter learning via updating a set of sufficient statistics. Other, more general examples are the class of conditionally Gaussian DLMs and the class of discrete-state dynamic models, such as hidden Markov models (HMM), change-point models and generalized DLMs. The steps of the Storvik's algorithm are as follows:

Step 1 (Propagate) x_{t-1} to $\tilde{x}_t$ via $q(x_t|x_{t-1}, \theta, y_t)$;
Step 2 (Resample) (x_{t-1}, x_t, s_{t-1}) from $(x_{t-1}, \tilde{x}_t, s_{t-1})$ with weights $w_t \propto \frac{p(y_t|\tilde{x}_t, \theta)p(\tilde{x}_t|x_{t-1}, \theta)}{q(\tilde{x}_t|x_{t-1}, \theta, y_t)}$;
Step 3 (Propagate)

a) $s_t = \mathcal{S}(s_{t-1}, x_{t-1}, x_t, y_t)$;
b) θ from $p(\theta|s_t)$.

The resampling proposal density $q(x_t|x_{t-1}, \theta, y_t)$ plays the same role as it did in the BF and the APF.

11.4.3 Particle learning

[8, 9] present methods for sequential filtering, particle learning (PL) and smoothing for a rather general class of state space models. They extend Chen and Liu's (2000) [11] mixture Kalman filter (MKF) methods by allowing parameter learning and utilize a resample–propagate algorithm together with a particle set that includes state sufficient statistics. They also show via several simulation studies that PL outperforms both the LW and Storvik filters and is comparable to MCMC samplers, even when full adaptation is considered. The advantage is even more pronounced for large values of n.

Let s_t^x denote state sufficient statistics satisfying deterministic updating rule $s_t^x = \mathcal{K}(s_{t-1}^x, \theta, y_t)$, for $\mathcal{K}(\cdot)$ mimicking the Kalman filter recursions of Section 11.2.1. The steps of a generic PL algorithm are as follows:

Step 1 (Resample) $(\tilde{\theta}, \tilde{s}_{t-1}^x, \tilde{s}_{t-1})$ from $(\theta, s_{t-1}^x, s_{t-1})$ with weights $w_t \propto p(y_t|s_{t-1}^x, \theta)$;
Step 2 (Propagate)

a) (x_{t-1}, x_t) from $p(x_{t-1}, x_t|s_{t-1}^x, \theta, y_t)$;
b) $s_t = \mathcal{S}(\tilde{s}_{t-1}, x_{t-1}, x_t, y_t)$;

c) θ from $p(\theta|s_t)$;

d) $s_t^x = \mathcal{K}(\tilde{s}_{t-1}^x, \theta, y_t)$.

The reason for propagating x_{t-1} in step (2a) above, is that in the great majority of dynamic models used in practice, $\mathcal{S}$ is a function of x_{t-1}, and possibly several other lags x_t. The AR(1) plus noise model of Section 11.2.2 and the SV-AR(1) model of Section 11.2.3 fall into this category. In addition, it is worth mentioning that (x_{t-1}, x_t) is discarded after s_t is propagated.

11.4.4 Examples

We illustrate the various particle filters with parameter learning via the AR(1) plus noise model and the SV-AR(1) model as before. Then, the SV-AR(1) model is generalized to accommodate Student's t errors (Section 11.4.4.3), leverage effects (Section 11.4.4.4) and Markov switching (Section 11.4.4.5).

11.4.4.1 AR(1) plus noise model

We revisit the AR(1) plus noise model equations (11.18) and (11.19) from Section 11.2.2, but now assuming that $(\sigma^2, \tau^2) = (1, 0.05)$ and that the goal is to sequentially approximate $p(x_t, \alpha, \beta|y_{1:t})$. The priors of (α, β) and x_0 are, respectively, $N(a_0, \tau^2 A_0)$ and $N(m_0, C_0)$ (see Section 11.2.2.2), while parameter sufficient statistics s_t are defined by the set of equations (11.21). One dataset with $n = 100$ observations is simulated from $(\alpha, \beta, x_0) = (0.05, 0.95, 1.0)$. The prior hyperparameters are $(m_0, C_0) = (1.0, 10)$, $a_0 = (0, 1)$ and $A_0 = 2I_2$.

Figure 11.3 shows the true contours of $p(\alpha, \beta|y_{1:n}) \propto p(\alpha, \beta)p(y_{1:n}|\alpha, \beta)$ on a grid for the pair (α, β) along with approximate contours ($N = 1000$ particles) based on a OAPF approximation to $p(y_{1:n}|\alpha, \beta)$ (Section 11.3.2 and equation (11.20)). In practice, when (α, β) is replaced by larger parameter vectors, the use of grids could be replaced by a MCMC, SIR or rejection step. One can argue that approximating $p(y_{1:n}|\theta)$ by particle filters should be done with caution (see [47], and, more recently, [44], for further discussion).

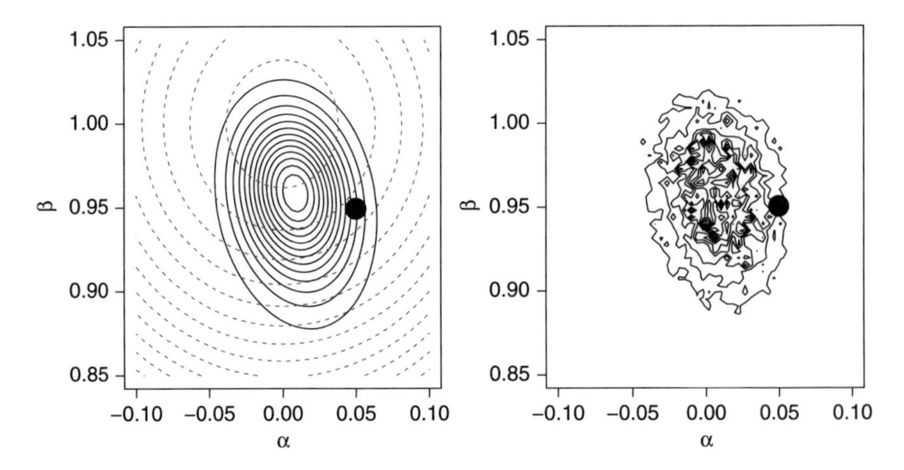

Figure 11.3 *AR(1) plus noise model (parameter learning).* Left panel: Contours of the prior distribution $p(\alpha, \beta)$ (dashed lines) and exact contours of the posterior distribution $p(\alpha, \beta|y_{1:n})$ (solid lines). Right panel: Contours of $\hat{p}(\alpha, \beta|y_{1:n}) \propto p(\alpha, \beta)\hat{p}(y_{1:n}|\alpha, \beta)$, where approximated integrated likelihood $\hat{p}(y_{1:n}|\alpha, \beta)$ is based on the OAPF of Section 11.3.2 and equation (11.20).

Figure 11.4 compares the performance of the LW filter (with $a = 0.995$) and PL to sequential (brute force) MCMC. The MCMC for this model is outlined in Section 11.2.2.1 and is run for 2000 iterations with the second half used for posterior summaries. The LW filter starts to show particle degeneracy around the 50th observation and moves away from the true percentiles.

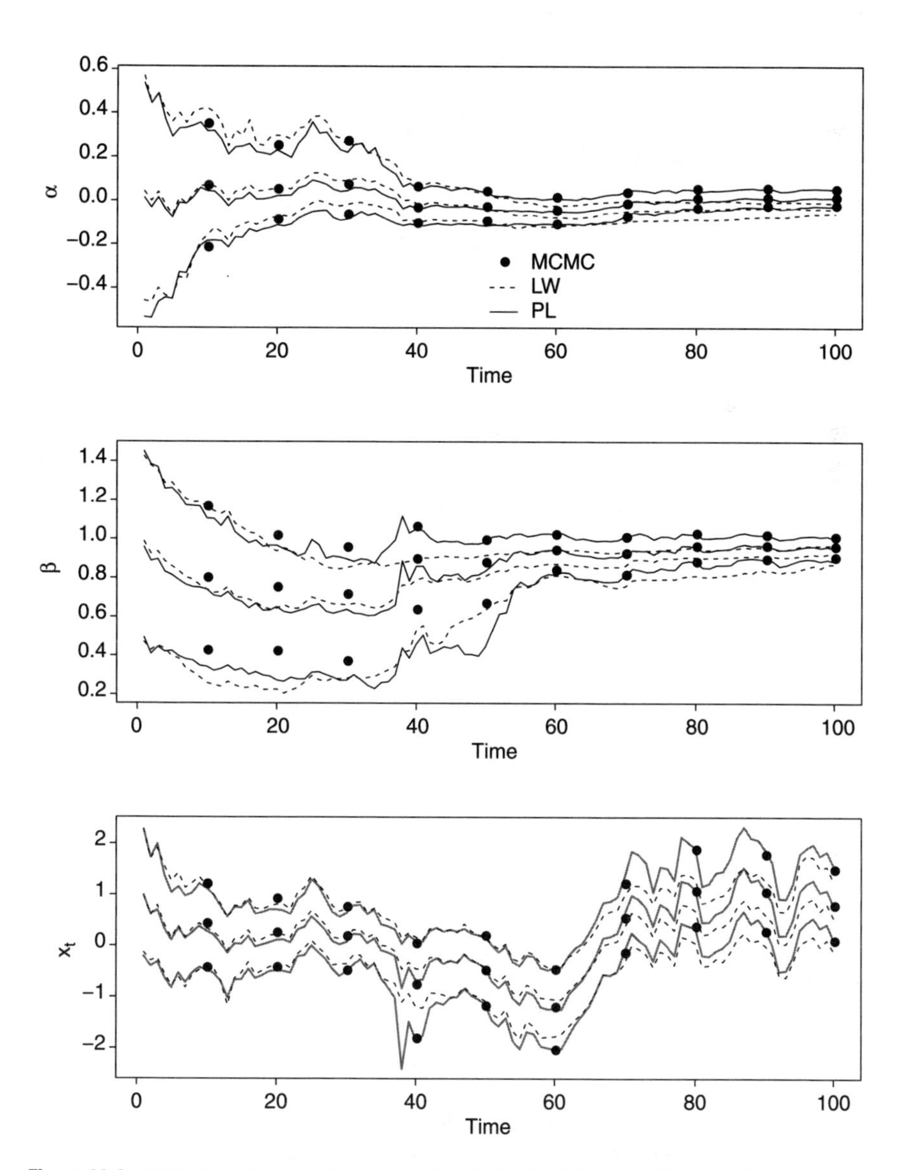

Figure 11.4 *AR(1) plus noise model (parameter learning).* 5th, 50th and 95th percentiles of $p(\alpha|y_{1:t})$ (top) and $p(\beta|y_{1:t})$ (middle) and $p(x_t|y_{1:t})$ (bottom) based on MCMC, LW filter and PL. MCMC is based on 1000 draws (after discarding the first 1000 draws). LW and PL are based on 1000 particles.

11.4.4.2 *SV-AR(1) model*

We revisit the SV-AR(1) plus noise model equations (11.22) and (11.23) from Section 11.2.3, but now assuming that $(\alpha, \beta|\tau^2)$, τ^2 and x_0 are, respectively, $N(a_0, \tau^2 A_0)$, $IG(\nu_0/2, \nu_0\tau_0^2/2)$ and $N(m_0, C_0)$. As in the illustration of Section 11.3.5.2, a total of $n = 200$ data points were simulated from $\alpha = -0.03$, $\beta = 0.97$, $\tau^2 = 0.03$ and $x_0 = -0.1$. We assume, as before, that $(m_0, C_0) = (-0.1, 1)$. The other hyperparameters are $a_0 = (-0.03, 0.97)$, $A_0 = 1.6I_2$ and $(\nu_0, \tau_0^2) = (10, 0.04)$.

The LW filter is based on 500 000 particles, while PL is based on 50 000. MCMC for the model (see Sections (11.2.3.1) and (11.2.3.2)) is implemented over time for comparison with both the LW filter and PL. MCMC, which starts at the true values, is based on 10 000 draws after the same number of draws is discarded as burn-in. Figure 11.5 summarizes the results. PL and MCMC produce fairly similar results, with LW slightly worse. Notice that LW is based on 10 times more particles than PL. We compared LW, PL and MCMC runs to a fine grid approximation of $p(\alpha, \beta, \tau^2|y_{1:n})$, with a 100-point grid for the log-volatilities x_t in $(-5, 2)$ and 50-point grids in the intervals $(-0.15, 0.1)$, $(0.85, 1.05)$ and $(0.01, 0.15)$, for α, β and τ^2, respectively. In this case, both LW and PL are based on 20000 particles and MCMC is based on 20000 draws after the same number of draws is discarded as burn-in.

Figure 11.6 summarizes the $R = 10$ replications of LW and PL, both based on $N = 10\ 000$ particles. LW has a larger Monte Carlo error when approximating the filtering distributions for all quantities, with particular emphasis on the volatility of the log-volatility τ^2 and, consequently, on the latent state x_t. Based on this simple exercise and running our code in R, it takes about 7 and 15 minutes to run the LW filter and PL, respectively. It takes about 8 minutes to run MCMC based on the whole time series of $n = 200$ observations. It takes about 13 hours to run MCMC based on $y_{1:t}$ for all $t \in \{1, \ldots, 200\}$, i.e. 50 times slower than PL and 100 times slower than the LW filter.

11.4.4.3 *SV-AR(1) model with t errors*

In order to illustrate particle filters' ability to approximate the predictive density $p(y_t|y_{1:t-1})$ via equation 11.36 from Section 11.3.3, we implement PL for the SV-AR(1) model and the SV-AR(1) model with Student's t error as in [42] on a simulated dataset with errors following t_ν for $\nu \in \{1, 2, 4, 30\}$. Figure 11.7 compares the Bayes factors (in the log scale) of the t_ν models against normality. For instance, when the data is t_1 or t_2, each additional outlier makes Bayes factors support t models more significantly. For additional discussion on sequential model comparison and model checking via particle methods see, for instance, [8, 9] and [39].

11.4.4.4 *SV-AR(1) model with leverage*

[46] introduce MCMC for posterior inference in the SV-AR(1) model with leverage. More precisely, log-volatility dynamics (equation (11.23)) is now $x_t|x_{t-1}, \theta \sim N(\alpha + \beta x_{t-1} + \tau\rho y_{t-1}\exp\{-x_{t-1}/2\}, \tau^2(1-\rho^2))$. Negative ρ captures the increase in (log-)volatility x_t that follows a drop in y_{t-1}. One of their examples, where $(\alpha, \beta, \tau^2, \rho) = (-0.026, 0.97, 0.0225, -0.3)$, is revisited here based on $n = 10\ 000$ observations (they use only $n = 1000$) in order to illustrate how a simple, generic LW filter performs relatively well, even when the sample size is fairly large. We use their prior specification, $(\beta + 1)/2 \sim Beta(20, 1.5)$, $\alpha|\beta \sim N(0, (1-\beta)^2)$, $\rho \sim U(-1, 1)$, and $\tau^2 \sim IG(5/2, 0.05/2)$, and run the LW filter based on $N = 500\ 000$ particles and tuning parameter $a = 0.995$. Figure 11.8 summarizes the results. This LW filter could easily be extended to fit the other SV models they considered, such as the SV-t model (see Section 11.4.4.3) and the superposition models.

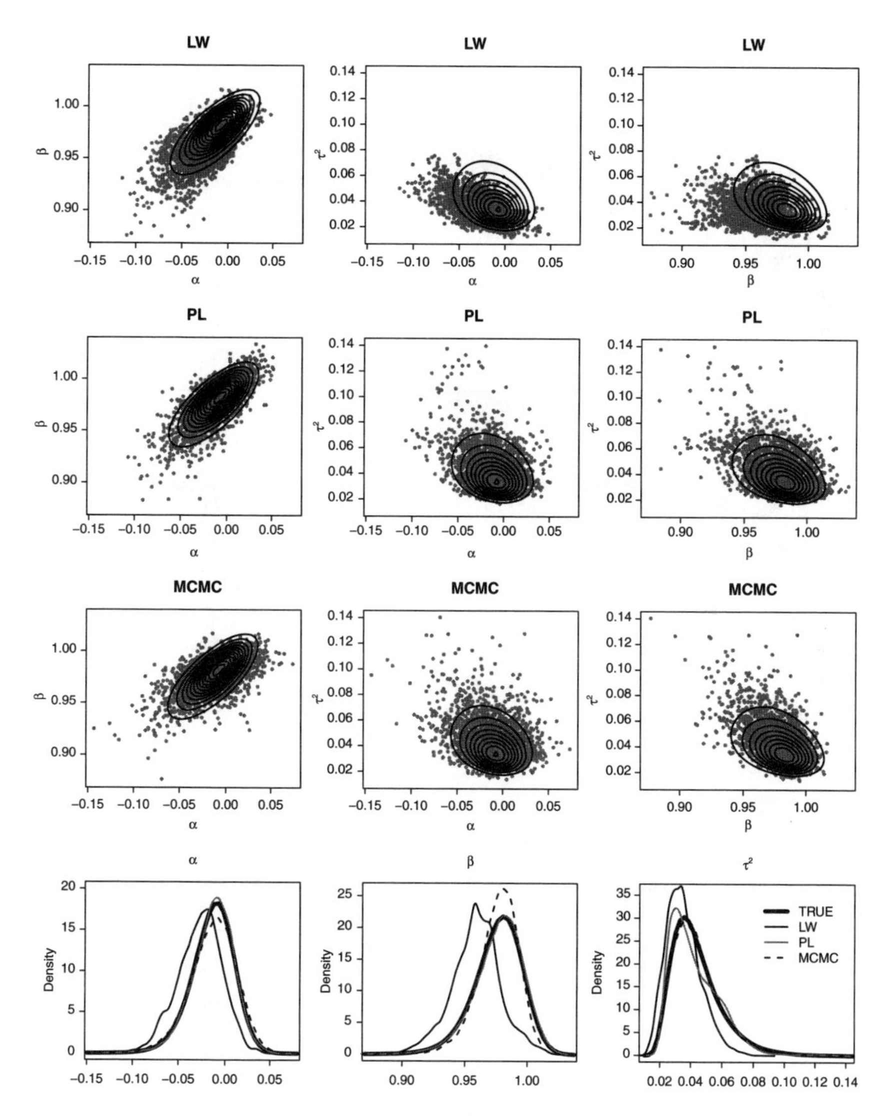

Figure 11.5 *SV-AR(1) model (parameter learning).* First three rows: True (contours) and approximated (dots) joint posterior distributions: $p(\alpha, \beta | y_{1:n})$ (1st row), $p(\alpha, \tau^2 | y_{1:n})$ (2nd row) and $p(\beta, \tau^2 | y_{1:n})$ (3rd row). Columns are based on LW filter, PL and MCMC. Fourth row: True and approximated marginal posterior distributions $p(\alpha | y_{1:n})$, $p(\beta | y_{1:n})$ and $p(\tau^2 | y_{1:n})$.

11.4.4.5 *SV-AR(1) model with regime switching*

[10] implement the LW filter for SV-AR(1) models with regime switching, where equation (11.23) becomes $x_t | x_{t-1}, s_t, \theta \sim N(\alpha + \beta x_{t-1} + \gamma s_t, \tau^2)$, for $\gamma > 0$ and latent regime switching variable $s_t \in \{0, 1\}$. We assume, for simplicity, that s_t obeys a two-regime homogeneous Markov model with $Pr(s_t = 0 | s_{t-1} = 0) = p$ and $Pr(s_t = 1 | s_{t-1} = 1) = q$. The vector of fixed parameters is $\theta = (\alpha, \beta, \tau^2, p, q)$ and the vector of latent states is (x_t, s_t). We revisit their analysis of the

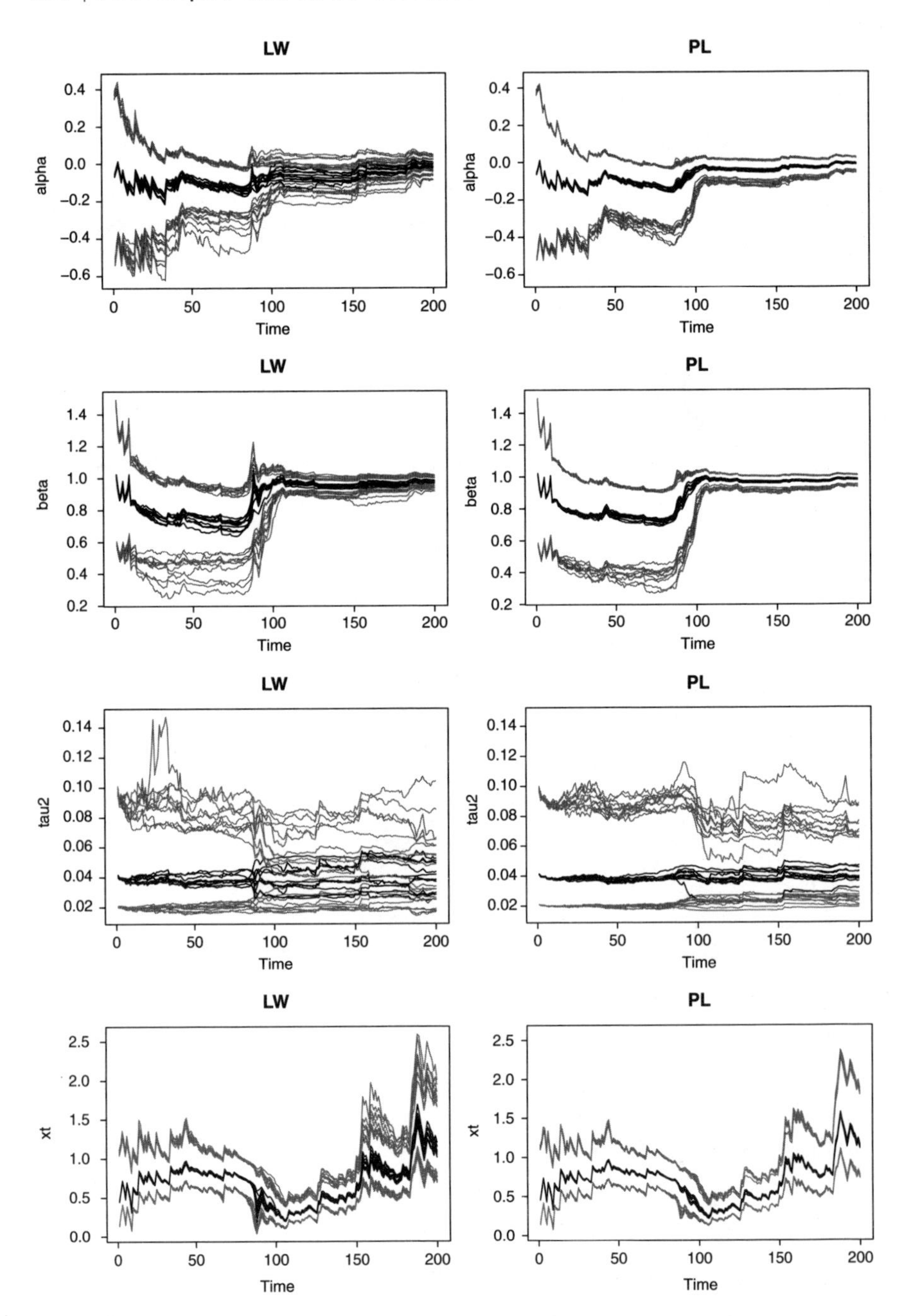

Figure 11.6 *SV-AR(1) model (parameter learning).* Approximate 5th, 50th and 95th percentiles of $p(\alpha|y_{1:t})$ (1st row), $p(\beta|y_{1:t})$ (2nd row), $p(\tau^2|y_{1:t})$ (3rd row) and $p(x_t|y_{1:t})$ (4th row) based on LW filter (left column) and PL (right column) for $R = 10$ replications of both filters and 10 000 particles.

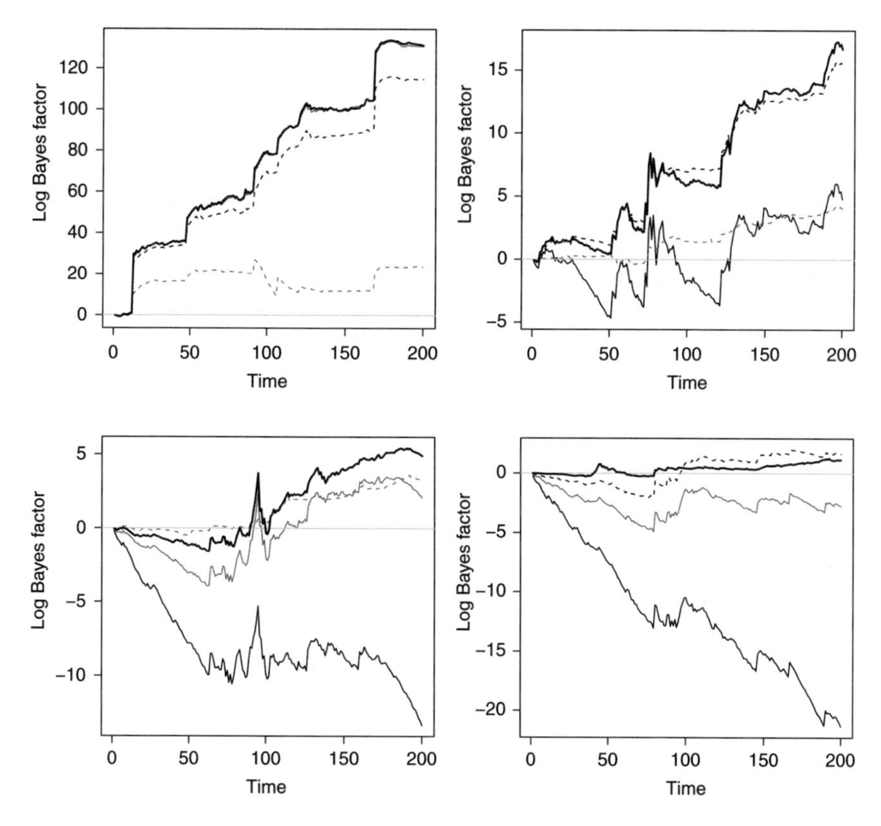

Figure 11.7 *SV-AR(1) model with t_ν error.* Bayes factors (in the log scale) of fitting t_ν models against normal models. The lines are t_1 (solid dark line), t_2 (solid grey line), t_4 (dashed dark line) and t_{30} (dashed grey line). Thicker solid lines correspond to the true data generating models.

IBOVESPA stock index (São Paulo Stock Exchange) but with a larger dataset spanning 01/02/1997 to 08/08/2011 ($n = 3612$ observations). The prior hyperparameters (Section 11.2.2.2) are $d_0 = (-0.25, 0.95, 0.05)$, $D_0 = 6I_3$, $\nu_0 = 10$ and $\tau_0^2 = 0.05$, with $p \sim Beta(50, 1)$, $q \sim Beta(1, 50)$, $x_0 \sim N(0, 1)$ and $s_0 \sim Ber(0.1)$. Figure 11.9 summarizes our findings. The model with regime switching captured the major 1997–1999 crisis listed in [11], as well as the more recent credit crunch crisis of 2008. It also captured the sharp drop on Monday, August 8th 2011, when the IBOVESPA (and most financial markets worldwide) suffered an 8% fall following worries about the weak US economy and the high levels of public debt in Europe. See [40] and [52] for further discussion and illustrations of particle methods in SV-AR(1) models with regime switching.

11.4.4.6 SV-AR(1) model with realized volatility.

In this final illustration, we revisit [56] who estimate SV models using daily returns and realized volatility simultaneously. Their most general model assumes that returns $y_{1t} \sim N(0, \exp\{x_t/2\})$ and that the log-volatility dynamics is $x_t|x_{t-1}, \theta \sim N(\alpha + \beta x_{t-1} +$

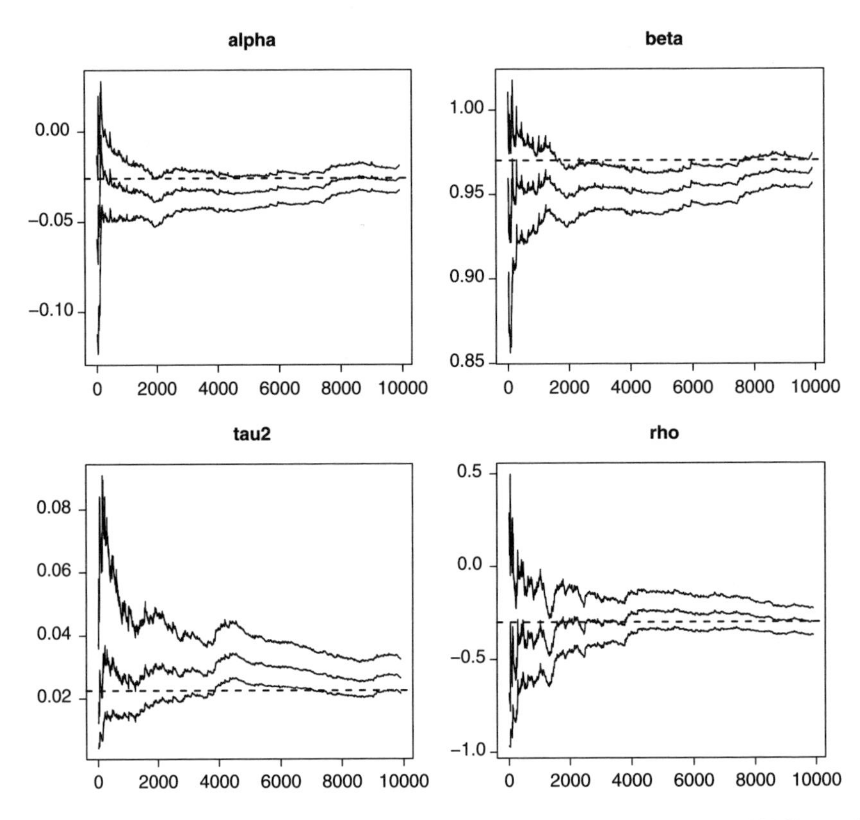

Figure 11.8 *SV-AR(1) model with leverage.* Sequential parameter learning based on LW filter and $N = 500\,000$ particles.

$\tau\rho y_{1,t-1}\exp\{-x_{t-1}/2\}, \tau^2(1-\rho^2))$ (as in Section 11.4.4.4). The model is completed with realized volatility $y_{2t} \sim N(\xi + x_t, \sigma^2)$, where ξ is the bias-correction term. We use high frequency data of the Tokyo price index (TOPIX) that what was kindly shared with us (the authors) for this illustration. In what follows y_{2t} is the logarithm of the scaled realized volatility based on one-minute intraday returns when the market is open during the 10-year period from April 1st, 1996 to March 31st, 2005 ($n = 2216$ trading days). Therefore, the vector of static parameters of the model is $\theta = (\alpha, \beta, \tau^2, \rho, \xi, \sigma^2)$. Implementation of the LW filter is fairly simple and we fit four models to the data: RV model, SV-AR(1) model, SV-AR(1) model with leverage and the current model. The RV model is basically an AR(1) plus noise model, in which case $\xi = 0$ for identification reasons. We label these four models RV, SV, ASV and ASV-RVC in what follows. The number of particles is $N = 100\,000$ and LW's tuning parameter is $a = 0.995$. Figure 11.10 shows posterior medians for time-varying standard deviations and their logarithms. The ASV model seems to be less sensitive to extremes when compared to the SV model. One can argue that the RV model is too adaptive when compared to the SV model. Similarly, the ASV-RVC is less sensitive to extremes when compared to the ASV model, while being less adaptive than the RV model. These results are corroborated by the marginal posterior densities for the models' parameters where the persistence parameter β and the leverage parameter ρ are smaller in the ASV-RVC model.

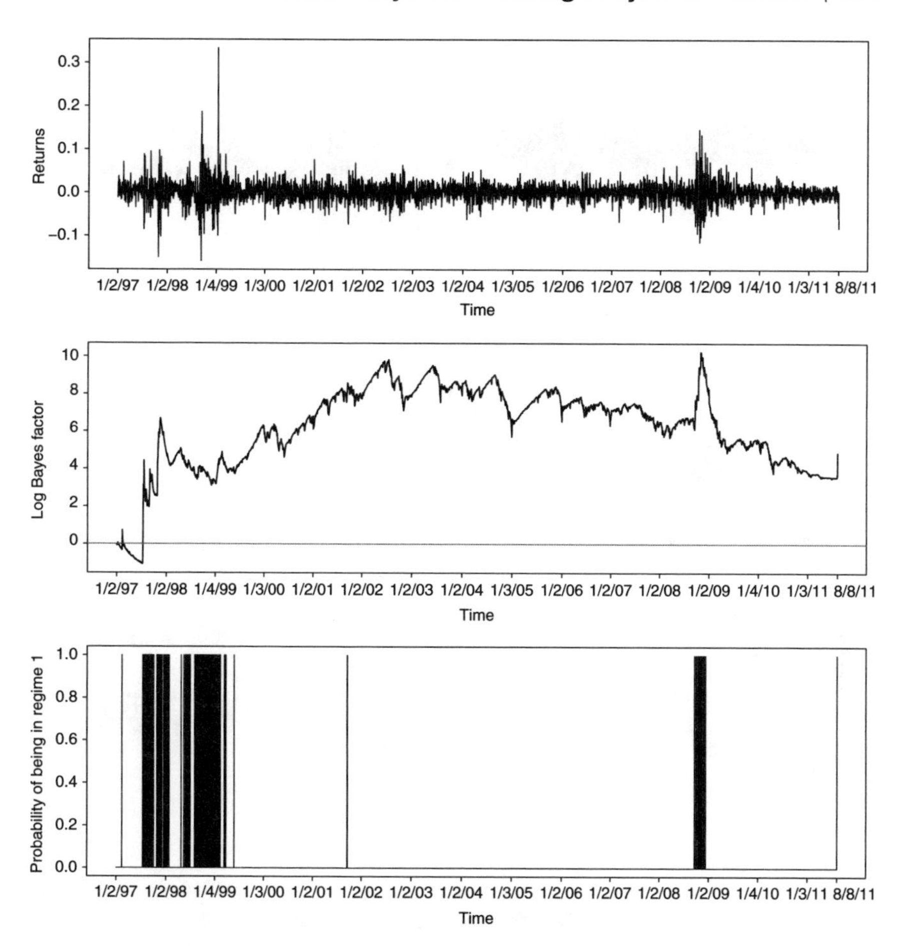

Figure 11.9 *SV-AR(1) model with regime switching.* IBOVESPA returns (top frame) from 01/02/1997 to 08/08/2011 ($n = 3612$ observations), Log Bayes factor (middle frame) and $Pr(s_t = 1|y_{1:t})$ (bottom frame). The LW filter is based on $N = 200\,000$ particles.

In addition, both parameters ξ and σ^2 are away from zero, suggesting that the biased-corrected realized volatility helps estimating daily log-volatilities x_t.

11.5 Discussion

This chapter reviews many of the important advances in the particle filter literature over the last two decades. Two relatively simple but fairly general models are used to guide the review: the AR(1) plus noise model and the SV-AR(1) model. We aim at a broad audience of researchers and practitioners and illustrate the benefits and the limitations of particle filters when estimating with dynamic models where sequentially learning of latent states and fixed parameters is the primary interest.

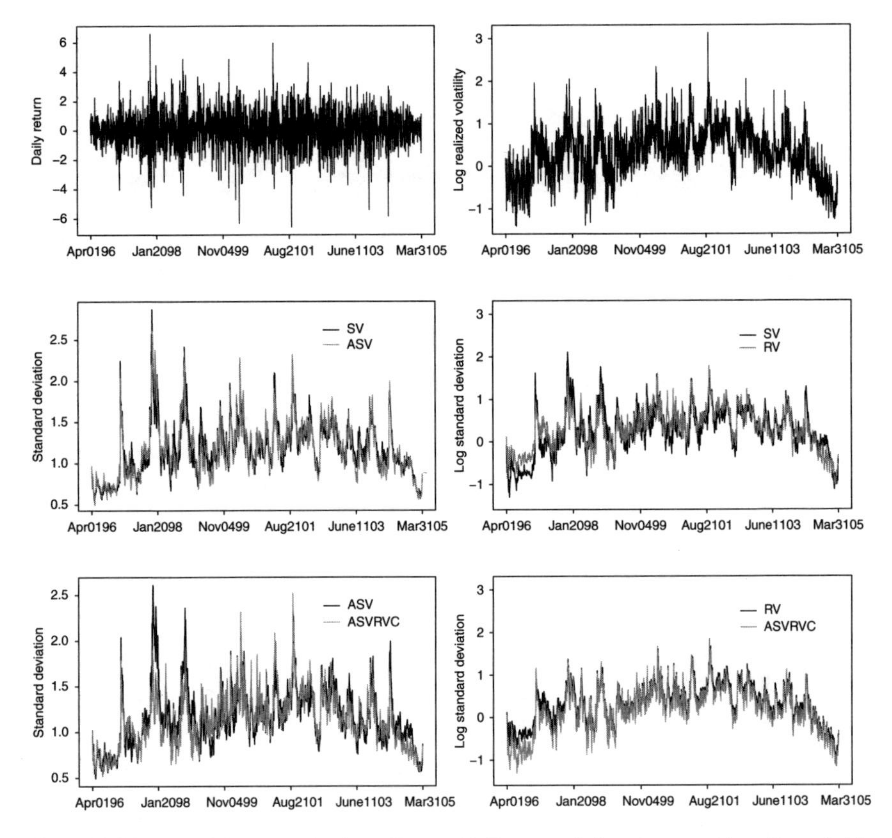

Figure 11.10 *SV-AR(1) plus realized volatility model.* Top row: Daily returns and logarithm of daily realized volatilities. Middle and bottom rows: Posterior medians of standard deviations and their logarithms based on four models: SV-AR(1) model (SV), SV with leverage (ASV), realized volatility (RV) and ASV-RV combined (ASV-SRVC).

The applications of Section 11.4.4 based on several (important) stochastic volatility models were intended to illustrate to the reader how relatively complex (despite univariate) models can be sequentially estimated via particle filters at relatively low computational cost. They are comparable in performance to the standard MCMC proposed in the references listed in each one of the examples. It is important to emphasize that this cost increases with the dimension of both latent state and static parameter vectors and that this is one of the leading sub areas of current theoretical and empirical research.

There are currently several review papers, chapter and books the reader should read after becoming fluent with the tools we introduce here. Amongst those are the earlier papers by [17], [2] and [13], books by [36], [17a] and [53] and the 2002 special issue of IEEE Transactions on Signal Processing on sequential Monte Carlo methods.

More recent reviews are [4], [19], [51] (chapter 6) and [43]. They carefully organize and highlight the fast development of the field over the last decade, such as parameter learning, more efficient particle smoothers, particle filters for highly dimensional dynamic systems and, perhaps the most recent, interconnections between MCMC and SMC methods.

Many important topics and issues were left out. Particle smoothers, for instance, are becoming a realistic alternative to MCMC in dynamic systems when the smoothed $p(x_{1:t}|y_{1:t})$, or simply $p(x_t|y_{1:t})$, is the distribution of interest. See [28], [22], [15] and [3], amongst others.

The interface between PF and MCMC methods is illustrated in our examples (see, for example, Section 11.3.3 and Figure 11.5). Hybrid schemes that combine particle methods and MCMC methods are abundant. [27] and [50], for instance, use MCMC steps to sample and replenish static parameters in dynamic systems. [1] introduce particle MCMC methods to efficiently construct proposal distributions in high dimension via SMC methods. See also [49].

Finally, particle filters have recently received a lot of attention in estimating non-dynamic models such as mixtures, Gaussian processes, tree models, etc. Important references are [39] and [9].

References

[1] Andrieu, C., Doucet, A. and Holenstein, R. (2010). Particle Markov chain Monte Carlo (with discussion). *Journal of the Royal Statistical Society, Series B*, 72, 269–342.

[2] Arulampalam, M. S., Maskell, S., Gordon, N. and Clapp, T. (2002). A Tutorial on Particle Filters for On-line Nonlinear/Non-Gaussian Bayesian Tracking. *IEEE Transactions on Signal Processing*, 50, 174–188.

[3] Briers, M., Doucet, A. and Maskell, S. (2010). *Annals of the Institute of Statistical Mathematics* Smoothing algorithms for state-space models, 6261–89.

[4] Cappé, O., Godsill, S. and Moulines, E. (2007). An overview of existing methods and recent advances in sequential Monte Carlo. *IEEE Proceedings in Signal Processing*, 95, 899–924.

[5] Cappé, O., Moulines, E. and Rydén, T. (2005). *Inference in Hidden Markov Models*. Springer, New York.

[6] Carlin, B. P., Polson, N. G. and Stoffer, D. S. (1992). A Monte Carlo approach to nonnormal and nonlinear state-space modelling. *Journal of the American Statistical Association*, 87, 493–500.

[7] Carter, C. K. and Kohn, R. (1994). On Gibbs sampling for state space models. *Biometrika*, 81, 541–53.

[8] Carvalho, C. M., Johannes, M., Lopes, H. F. and Polson, N. (2010). Particle learning and smoothing. *Statistical Science*, 25, 88–106.

[9] Carvalho, C. M., Lopes, H. F., Polson, N. and Taddy, M. (2010). Particle learning for general mixtures. *Bayesian Analysis*, 5.

[10] Carvalho, C. M. and Lopes, H. F. (2007). Simulation-based sequential analysis of Markov switching stochastic volatility models. *Computational Statistics & Data Analysis*, 51, 4526–4542.

[11] Chen, R. and Liu, J. S. (2000). Mixture Kalman filter. *Journal of the Royal Statistical Society, Series B*, 62, 493–508.

[12] Chopin, N. (2002). A sequential particle filter method for static models. *Biometrika*, 89, 539–52.

[13] Crisan, D. and Doucet, A. (2002). A survey of convergence results on particle filtering methods for practitioners. *IEEE Transations on Signal Processing*, 50, 736–746.

[14] Del Moral, P., Doucet, A. and Jasra, A. (2006). Sequential Monte Carlo samplers. *Journal of the Royal Statistical Society, Series B*, 68, 411–436.

[15] Douc, R., Garivier, E., Moulines, E. and Olsson, J. (2009). On the forward filtering backward smoothing particle approximations of the smoothing distribution in general state space models. *Annals of Applied Probability* (to appear).

[16] Douc, R., Moulines, E. and Olsson, J. (2009). Optimality of the auxiliary particle filter. *Probability and Mathematical Statistics*, 29, 1–28.
[17] Doucet, A., Godsill, S. J. and Andrieu, C. (2000). On sequential Monte Carlo sampling methods for Bayesian filtering. *Statistics and Computing*, 10, 197–208.
[17a] Doucet, A., de Freitas, N. and Gordon, N. (ed.) (2001). *Sequential Monte Carlo Methods in Practice*. New York: Springer-Verlag.
[18] Doucet, A. and Johansen, A. (2008). A Note on Auxiliary Particle Filters. *Statistics & Probability Letters*, 78, 1498–1504.
[19] Doucet, A. and Johansen, A. (2009). A Tutorial on Particle Filtering and Smoothing: Fifteen years Later. In D. Crisan and B. Rozovsky, editors, *Handbook of Nonlinear Filtering*. Oxford: Oxford University Press.
[20] Durbin, J. and Koopman, S. J. (2001). *Time Series Analysis by State Space Methods*. Oxford University Press.
[21] Fearnhead, P. (2002). Markov chain Monte Carlo, sufficient statistics and particle filter. *Journal of Computational and Graphical Statistics*, 11, 848–62.
[22] Fearnhead, P., Wyncoll, D. and Tawn, J. (2010). A sequential smoothing algorithm with linear computational cost. *Biometrika*, 97, 447–464.
[23] Frühwirth-Schnatter, S. (1994). Data augmentation and dynamic linear models. *Journal of Time Series Analysis*, 15, 183–202.
[24] Gamerman, D. (1998). Markov Chain Monte Carlo for dynamic generalized linear models. *Biometrika*, 85, 215–27.
[25] Gamerman, D. and Lopes, H. F. (2006). *Markov Chain Monte Carlo: Stochastic Simulation for Bayesian Inference*. Chapman & Hall/CRC.
[26] Ghysels, E., Harvey, A. C. and Renault, E. (1996). Stochastic Volatility. In C. R. Rao and G. S. Maddala (eds) *Handbook of Statistics: Statistical Methods in Finance*, 119–191. (Amsterdam: North-Holland.)
[27] Gilks, W. R. and Berzuini, C. (2001). Following a moving target-Monte Carlo inference for dynamic Bayesian models. *Journal of the Royal Statistical Society, Series B*, 63, 127–46.
[28] Godsill, S. J., Doucet, A. and West, M. (2004). Monte Carlo smoothing for non-linear time series. *Journal of the American Statistical Association*, 50, 438–449.
[29] Gordon, N., Salmond, D. and Smith, A. F. M. (1993). Novel approach to nonlinear/non-Gaussian Bayesian state estimation. *IEE Proceedings F. Radar Signal Process*, 140, 107–113.
[30] Guo, D., Wang, X. and Chen, R. (2005). New sequential Monte Carlo methods for nonlinear dynamic systems. *Statistics and Computing*, 15, 135–47.
[31] Hull, J., and White, A. (1987). The Pricing of Options on Assets with Stochastic Volatilities. *Journal of Finance*, 42, 281–300.
[32] Jacquier, E., Polson, N. G. and Rossi, P. E. (1994). Bayesian analysis of stochastic volatility models. *Journal of Business and Economic Statistics*, 20, 69–87.
[33] Johannes, M. and Polson, N. G. (2010). MCMC methods for continuous-time financial econometrics, in Ait-Sahalia, Y. and Hansen, L. P. eds, *Handbook of Financial Econometrics, Volume 2*. Princeton: University Press, 1–72.
[34] Kim, S., Shephard, N. and Chib, S. (1998). Stochastic Volatility: Likelihood Inference and Comparison with ARCH Models. *Review of Economic Studies*, 65, 361–393.
[35] Kong, A., Liu, J. S. and Wong, W. (1994). Sequential imputation and Bayesian missing data problems. *Journal of the American Statistical Association*, 89, 590–99.
[36] Liu, J. S. (2001). *Monte Carlo Strategies in Scientific Computing*. New York: Springer-Verlag.
[37] Liu, J. and Chen, R. (1995). Blind Deconvolution via Sequential Imputations. *Journal of the American Statistical Association*, 90, 567–76.

[38] Liu, J. and West, M. (2001). Combined parameters and state estimation in simulation-based filtering. In A. Doucet, N. de Freitas and N. Gordon, editors, *Sequential Monte Carlo Methods in Practice*. New York: Springer-Verlag.

[39] Lopes, H. F., Carvalho, C. M., Johannes, M. and Polson, N. G. (2011). Particle learning for sequential Bayesian computation (with discussion). In J. M. Bernardo, M. J. Bayarri, J. O. Berger, A. P. Dawid, D. Heckerman, A. F. M. Smith and M. West, editors, *Bayesian Statistics 9*. Oxford: Oxford University Press, 317–360.

[40] Lopes, H. F. and Polson, N. G. (2010). Extracting SP500 and NASDAQ volatility: The credit crisis of 2007-2008. In A. O'Hagan and M. West, editors, *The Oxford Handbook of Applied Bayesian Analysis*. Oxford: Oxford University Press, 319–42.

[41] Lopes, H. F. and Polson, N. G. (2010). Bayesian inference for stochastic volatility modelling. In Bocker, K. (Ed.) *Rethinking Risk Measurement and Reporting: Uncertainty, Bayesian Analysis and Expert Judgement*, 515–551.

[42] Lopes, H. F. and Polson, N. G. (2011). Particle Learning for Fat-tailed Distributions. Working Paper, The University of Chicago Booth School of Business.

[43] Lopes, H. F. and Tsay, R. (2011). Particle filters and Bayesian inference in financial econometrics. *Journal of Forecasting*, 30, 168–209.

[44] Malik, S. and Pitt, M. K. (2011). Particle filters for continuous likelihood evaluation and maximisation. *Journal of Econometrics* (in press).

[45] Migon, H. S., Gamerman, D., Lopes, H. F. and Ferreira, M. A. R. (2005). Dynamic models. In D. Dey, and C. R. Rao (Eds.), *Handbook of Statistics, Volume 25: Bayesian Thinking, Modelling and Computation*, 553–588.

[46] Omori, Y., Chib, S., Shephard, N. and Nakajima, J. (2009). Stochastic volatility with leverage: Fast and efficient likelihood inference. *Journal of Econometrics*, 140, 425–449.

[47] Pitt, M. K. (2002). Smooth particle filters for likelihood evaluation and maximisation. Technical Report Department of Economics, University of Warwick.

[48] Pitt, M. K. and Shephard, N. (1999). Filtering via simulation: Auxiliary particle filters. *Journal of the American Statistical Association*, 94, 590–99.

[49] Pitt, M. K., Silva, R. S., Giordani, P. and Kohn, R. (2012). On some properties of Markov chain Monte Carlo simulation methods based on the particle filter. *Journal of Econometrics*, 171 (2), 134–151.

[50] Polson, N. G., Stroud, J. R. and Müller, P. (2008). Practical filtering with sequential parameter learning. *Journal of the Royal Statistical Society, Series B*, 70, 413–28.

[51] Prado, R. and West, M. (2010). *Time Series: Modelling, Computation and Inference*. Chapman & Hall/CRC, The Taylor Francis Group.

[52] Rios, M. P. and Lopes, H. F. (2011). Sequential parameter estimation in stochastic volatility models. *Working Paper*. The University of Chicago Booth School of Business.

[53] Ristic, B., Arulampalam, S. and Gordon, N. (2004). *Beyond the Kalman Filter: Particle Filters for Tracking Applications*. Artech House Radar Library.

[54] Rosenberg, B. (1972). The Behaviour of Random Variables with Nonstationary Variance and the Distribution Of Security Prices. Working Paper No. 11, University of California, Berkeley, Institute of Business and Economic Research, Graduate School of Business Administration, Research Programme in Finance.

[55] Storvik, G. (2002). Particle filters for state-space models with the presence of unknown static parameters. *IEEE Transactions on Signal Processing*, 50, 281–89.

[56] Takahashi, M., Omori, Y. and Watanabe, T. (2009). Estimating stochastic volatility models using daily returns and realized volatility simultaneously. *Computational Statistics and Data Analysis*, 53, 2404–2426.

[57] Taylor, S. J. (1986). *Modelling Financial Time Series*. New York: John Wiley and Sons.
[58] West, M. (1993). Approximating posterior distributions by mixtures. *Journal of the Royal Statistical Society, Series B*, 54, 553–68.
[59] West, M. (1993). Mixture models, Monte Carlo, Bayesian updating and dynamic models. *Computing Science and Statistics*, 24, 325–33.
[60] West, M. and Harrison, J. (1997). *Bayesian Forecasting and Dynamic Models, 2nd Edition*. New York: Springer.

12 Semi-supervised classification of texts using particle learning for probabilistic automata

ANA PAULA SALES,
CHRISTOPHER CHALLIS,
RYAN PRENGER AND DANIEL MERL

12.1 Introduction

Some of the key statistical problems underlying many current national security applications stem from a common need for continuously deployable, self-adapting learning systems. Tracking and other signal processing tasks have long been the purview of dynamic Bayesian models [17] and sequential Monte Carlo techniques [8], however it is a relatively recent development that virtually all manner of inference tasks have been similarly relegated due to the staggering increase in data collection capabilities. In particular, many classic learning problems such as classification, regression and latent structure discovery, the inferential aspects of which were once dealt with in purely retrospective fashion by maximum likelihood or batch Bayes estimation, have become tracking problems of a sort as a result of the modern always-on data collection apparatus. In such settings, the target to be tracked is the joint posterior distribution of the parameters of the underlying probability model, for in a constant data collection setting, it can be difficult to justify model stationarity over long or indefinite periods of observation. This is especially true for adversarial classification problems, one example of which is spam detection, where intuitively the boundary between the positive class and negative class is constantly evolving as the adversary improves its attack vector. Therefore any robust solution to this type of detection problem must be similarly capable of adaptation, which is conceptually straightforward to accomplish when the associated inference is accomplished using sequential Monte Carlo methods.

The statistician will also recognize this as a natural setting for *semisupervised learning*, the statistical underpinnings of which were reviewed previously in Liang, Mukherjee, and West [11]. The salient characteristic of a semisupervised learning problem is the presence of a limited quantity of *labelled* data, and a usually much larger quantity of *unlabelled* data. In the context of a classification problem, the labels of course refer to the true class assignments associated with a set of observed predictor variables. Effective approaches for semisupervised learning is an important topic of research in the national security arena due to the fact that labelled data is laboriously generated by human analysts whose finite ground-truthing efforts must be efficiently utilized. Similarly, the feasibility of active learning [7] for intelligent tasking of human resources is another area of active research. Underlying both of these research areas, however, is the common need for an online learning framework conducive to making the sort of recommendations required of active learning and smoothly accommodating the sort of feedback obtained in a semisupervised setting.

In this chapter, we describe one such framework based upon the recent *particle learning* algorithm of Carvalho *et al.* [5, 6], which provides sequential parameter estimation for conjugate Bayesian models through a novel sequential Monte Carlo approach. The primary contributions of this work are the development of a new particle learning approach for efficient online estimation of simple text grammars as represented by probabilistic automata, and the development of a semisupervised classification system based on a flexible class of composite mixture models. Combined, these two components form the basis of a classification system for text-valued observations that does not require an exchangeability assumption on the tokens within the texts (e.g. the usual 'bag-of-words' representation).

The remainder of this chapter is organized as follows. In Section 12.2 we present a hierarchical model representation of a class of probabilistic automata and derive the particle learning algorithm for obtaining parameter estimates, highlighting the computational advantages of the probabilistic automata over more expressive grammars such as hidden Markov models. In Section 12.3 we describe a framework for semisupervised text classification problems based on a composite mixture model formulation. In Section 12.4 we describe an application of the classifier to a spam detection dataset in which we are able to accurately discriminate between spam and non-spam through directly modelling the generative grammars as mixtures of probabilistic automata. Finally, Section 12.5 concludes with a discussion of computational considerations and future work.

12.2 Particle learning for probabilistic deterministic finite automata

The seeming oxymoron probabilistic deterministic finite automata (PDFA) refers to a subclass of the greater family of probabilistic automata that includes many popular state space models such as hidden Markov models [9]. PDFA are characterized by probabilistically generated observables but *deterministic* state transitions conditional on *both* the current state *and* the current observation (Figure 12.1).

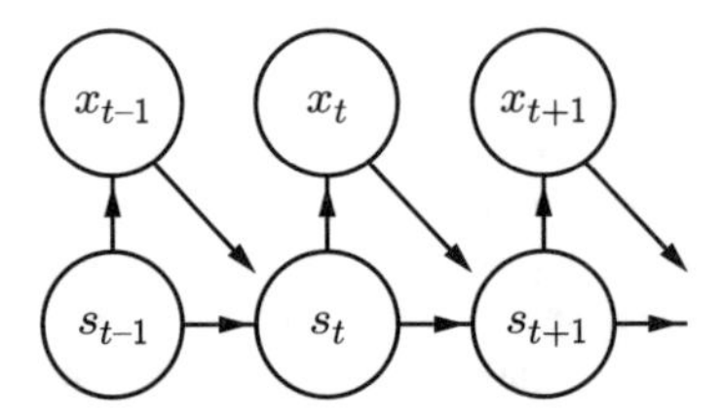

Figure 12.1 State transition diagram for a probabilistic deterministic finite automaton. The observables x_t follow emission distributions indexed by the underlying states s_t. The state transitions are deterministic given the previous state and previous observation.

PDFA have received recent attention as a possible computationally attractive substitute for HMMs for some natural language applications. Dupont *et al.* provide a thorough discussion of the connections between PDFA and HMMs, establishing PDFA to be a strict subclass of HMMs [9]. Pfau *et al.* recently introduced the probabilistic deterministic infinite automaton (PDIA), an infinite mixture of PDFA derived via a Dirichlet process prior on the emission distribution parameters [13]. Pfau *et al.* demonstrated the PDIA to have predictive performance exceeding that of an HMM on certain natural language problems.

Motivated by this favourable comparison to HMMs, in this work we consider only the fully parametric PDFA, and proceed to derive a particle learning approach to parameter estimation.

12.2.1 Model description

Let x_t denote a categorical observable at time t. The PDFA model can be written in hierarchical model form as follows:

$$x_t|s_t, \theta \sim \text{Multi}(x_t|1, \theta_{s_t}) \tag{12.1}$$

$$s_t|x_{t-1}, s_{t-1}, \pi \sim \delta_{\pi_{s_{t-1},x_{t-1}}}(s_t) \tag{12.2}$$

$$\pi_{s,x}|\alpha \sim \text{Multi}(\pi_{s,x}|1, \alpha) \tag{12.3}$$

$$\theta_s|\gamma \sim \text{Dirichlet}(\theta_s|\gamma) \tag{12.4}$$

where s_t denotes the associated hidden state at time t, θ_s parameterizes the emission distribution associated with state s, and $\delta_{\pi_{s,x}}(s_t)$ represents the degenerate distribution with mass at $\pi_{s,x}$. Thus $\pi_{s,x}$ specifies the transition out of state s after emitting x. The model is completed by specifying prior distributions for each $\pi_{s,x}$ and θ_s, parameterized by α and γ respectively.

The state transitions are deterministic given the previous observable and previous state. Thus, if we assume the initial state is some known starting state, then it is wholly artificial to represent the hidden state sequence, since every subsequent state is simply a function of the previous state and the previous observable. The only parameters of the model are those θ_s that characterize the emission distribution associated with each state, and the transition function, represented here as π. Finally, if θ_s is assigned a prior distribution that is conjugate to the form of the emission distribution, as we have done here with the Multinomial/Dirichlet combination, then the posterior distribution of each θ_s will have a closed form and it will be possible to effectively eliminate the parameter by integrating the likelihood with respect to this posterior. Thus we are left only with the task of conducting inference on the transition function π.

12.2.2 Parameter estimation using PL

In this section we derive a simple but highly effective particle learning algorithm for performing inference on the PDFA transition function π. A few preliminary remarks concerning the PDFA setting will be helpful. Consider now the process of estimating π: at time t only $\pi_{s,x}$ for which there was some $t' < t$ for which $x_{t'} = x$ and $s_{t'} = s$ need to be defined by each particle. All other $\pi_{s,x}$ are yet undefined, and are therefore implicitly represented by their common prior distribution (Eqn. 12.3). We will represent by π^t the set of transitions necessary to specify the state path up until time t. Note that this includes $\pi_{s_{t-1},x_{t-1}}$, but not necessarily π_{s_t,x_t}, as the transition from time t to time $t+1$ may yet be undefined.

As in any sequential Monte Carlo setting, our goal will be to produce samples from $p(\pi^t|x^t)$ through a combination of resampling and propagating a current set of posterior samples ('particles') from $p(\pi^{t-1}|x^{t-1})$ (where x^t denotes the vector of all observables $x_1, \ldots, x_t$ up to time t).

Assume that we begin with a uniformly weighted particle approximation $p(\pi^{t-1}|x^{t-1}) \approx \frac{1}{N}\sum_{i=1}^{N}\delta_{\pi_{(i)}^{t-1}}(\pi)$ where $\pi_{(i)}^{t-1}$ denotes a distinct sample from the target distribution. Upon observing x_t we have the following:

$$p(\pi^t|x^t) \propto p(x_t|\pi^{t-1}, x^{t-1})p(\pi|x^{t-1}) \tag{12.5}$$

This is the usual sequential importance sampling setting: given a particle approximation to $p(\pi^{t-1}|x^{t-1})$ we can produce a particle approximation to $p(\pi^t|x^t)$ by adjusting the weights of the current particles according to the posterior predictive density of x_t. The general form of the posterior predictive is as follows:

$$p(\pi^t|x^t) \propto p(\pi^t, x_t|x^{t-1}) \tag{12.6}$$

$$= p(\pi_{s_{t-1},x_{t-1}}|x^t, \pi^{t-1})p(x_t|x^{t-1}, \pi^{t-1})p(\pi^{t-1}|x^{t-1}) \tag{12.7}$$

Thus the weight for each particle is updated in this way:

$$\hat{w}_t = w_{t-1}\frac{p(\pi_{s_{t-1},x_{t-1}}|x^t, \pi^{t-1})p(x_t|x^{t-1}, \pi^{t-1})}{\phi(\pi_{s_{t-1},x_{t-1}}|x^t, \pi^{t-1})} \tag{12.8}$$

where $\hat{w}_t$ is the unnormalized weight at time t and $\phi(\pi_{s_{t-1},x_{t-1}}|x^t, \pi^{t-1})$ is a proposal distribution for $\pi_{s_{t-1},x_{t-1}}$. The natural choice for this proposal distribution is $p(\pi_{s_{t-1},x_{t-1}}|x^t, \pi^{t-1})$, which informs the proposed transition by the next data point. Thus the weight update simplifies to

$$\hat{w}_t = w_{t-1}p(x_t|x^{t-1}, \pi^{t-1}) \tag{12.9}$$

The particle filter then requires the tasks of computing the weight update $p(x_t|x^{t-1}, \pi^{t-1})$, and the proposal distribution $p(\pi_{x_{t-1},s_{t-1}}|x^t, \pi^{t-1})$.

We consider first the form of the weight update.

$$p(x_t|x^{t-1}, \pi^{t-1}) = \sum_{s_t} p(x_t, s_t|x^{t-1}, \pi^{t-1}) \tag{12.10}$$

$$= \sum_{s_t}\left\{p(x_t|s_t, x^{t-1}, \pi^{t-1})p(s_t|x^{t-1}, \pi^{t-1})\right\} \tag{12.11}$$

$$= \sum_{s_t}\left\{\left(\int p(x_t, \theta|s_t, x^{t-1}, \pi^{t-1})d\theta\right)p(s_t|x^{t-1}, \pi^{t-1})\right\} \tag{12.12}$$

$$= \sum_{s_t}\left\{\left(\int p(x_t|\theta, s_t)p(\theta|s_t, x^{t-1}, \pi^{t-1})d\theta\right)p(s_t|x^{t-1}, \pi^{t-1})\right\} \tag{12.13}$$

$$= \sum_{s_t}\left\{\left(\int \theta_{s_t,x_t}p(\theta|s_t, x^{t-1}, \pi^{t-1})d\theta\right)p(s_t|x^{t-1}, \pi^{t-1})\right\} \tag{12.14}$$

$$= \sum_{s_t}\frac{\gamma_{s_t,x_t}^{(t-1)}}{\gamma_{s_t}^{(t-1)}}p(s_t|x^{t-1}, \pi^{t-1}) \tag{12.15}$$

where $\gamma_{s,x}^{(t)} = \gamma_x + \sum_{k=1}^{t} \delta_s(s_k)\delta_x(x_k)$ and $\gamma_s^{(t)} = \sum_x \gamma_{s,x}^{(t)}$. Then $\gamma_{s,x}^{(t-1)}$ is the number of times the state-symbol pair (s, x) has been observed previously, and $\gamma_s^{(t-1)}$ is the total number of prior visits to state s_t. We can make this last step because the integral in (12.14) is the expectation of θ_{s_t,x_t} with respect to the posterior distribution of θ_{s_t} updated through time $t-1$. This posterior is a Dirichlet distribution due to multinomial Dirichlet conjugacy.

The proposal distribution for $\pi_{s_{t-1},x_{t-1}}$ requires a straightforward application of Bayes' rule.

$$p(\pi_{s_{t-1},x_{t-1}}|x^t, \pi^{t-1}) \propto p(x_t|x^{t-1}, \pi^t)p(\pi_{s_{t-1},x_{t-1}}|x^{t-1}, \pi^{t-1}) \tag{12.16}$$

$$= p(x_t|s_t, x^{t-1}, \pi^t)p(s_t|x^{t-1}, \pi^{t-1}) \tag{12.17}$$

$$= \frac{\gamma_{s_t,x_t}^{(t-1)}}{\gamma_{s_t}^{(t-1)}} p(s_t|x^{t-1}, \pi^{t-1}) \tag{12.18}$$

The first identity follows because s_t is determined by x^{t-1} and π^t, and the second from the preceding development of $p(x_t|x^{t-1}, \pi^{t-1})$. The distribution is then specified by computing (12.18) for each s_t and normalizing the values. The weight update in (12.15) can now be seen to be the normalizing constant of the transition proposal probabilities. This suggests the possibility of a simpler representation of $p(\pi^t|x^t)$. We have chosen to decompose the posterior in this way, however, in order to generate intelligent transition proposals.

12.2.2.1 *Algorithm*

In the previous section we developed general forms for the propagation distribution and weight update. These expressions simplify considerably when we consider the nature of the PDFA. Note first that when $\pi_{s_{t-1},x_{t-1}}$ is defined in π^{t-1}, $p(s_t|x^{t-1}, \pi^{t-1}) = 1$ when $s_t = \pi_{s_{t-1},x_{t-1}}$, and zero otherwise. Thus Eqn. 12.15 becomes simply $\gamma_{s_t,x_t}^{(t-1)}/\gamma_{s_t}^{(t-1)}$, and (12.18) leads us to propose $\pi_{s_{t-1},x_{t-1}}$ with probability 1.

In the case where $\pi_{s_{t-1},x_{t-1}}$ remains undefined at time $t-1$, the transition is still represented by the prior distribution and we have $p(s_t|x^{t-1}, \pi^{t-1}) = \alpha_{s_t}$. We can now completely specified the filtering algorithm.

Algorithm. Particle learning for PDFA

For each time period $t = 1, \ldots, T$:

For each particle $i = 1, \ldots, N$ (particle subscripts omitted):

(1) Update (and normalize) weights:

$$\hat{w}_t = \begin{cases} w_{t-1} \frac{\gamma_{s_t,x_t}^{(t-1)}}{\gamma_{s_t}^{(t-1)}} & \text{if } \pi_{x_{t-1},s_{t-1}}^{(i)} \text{ is already defined} \\ w_{t-1} \sum_{s_t} \frac{\gamma_{s_t,x_t}^{(t-1)}}{\gamma_{s_t}^{(t-1)}} \alpha_{s_t} & \text{if } \pi_{x_{t-1},s_{t-1}}^{(i)} \text{ is undefined} \end{cases} \tag{12.19}$$

(2) Resample if necessary.

(3) Sample new transitions. If not yet defined, sample a value s_t for $\pi_{s_{t-1},x_{t-1}}$ proportional to $\gamma_{s_t,x_t}^{(t-1)}/\gamma_{s_t}^{(t-1)}$.

(4) Increment γ: $\gamma_{s_t,x_t}^{(t)} = \gamma_{s_t,x_t}^{(t-1)} + 1$

The tasks of the algorithm are simply to track the number of observations of each state/symbol pair and calculate probabilities according to simple empirical ratios (adjusted by prior parameters). This simplicity allows for easy implementation, fast computation and widespread applicability.

12.2.2.2 *Multiple sequence inference*

There are many applications in which data consist of multiple sequences that can be treated as independently generated from the same model. The particle learning algorithm described above can be trivially modified to include a special character ω, signalling the end of a sequence by setting $\pi_{s,\omega} = s_0$, with s_0 a special initial state. This allows the model to be trained differently when natural divisions occur in the data, such as sentences in natural language.

12.3 Semisupervised classifiers using composite mixtures

In this section, we describe a simple computational framework for semisupervised learning tasks that leverages the sequential learning capabilities represented by particle learning and related methods. Figure 12.2 demonstrates the broad aim of this type of system. Data enters the system from the left; the underlying model can then be updated and/or the predicted dependent variables can be computed from the observed independent variables. The augmented observation is then considered according to various active learning criteria, and either the prediction is accepted (up arrow) or directed to a human analyst for ground-truthing and reincorporation into the data stream (down arrow).

The key task here lies in specifying a joint probability model on (y, x), where y is notionally a dependent variable (e.g. 'the label') and x is the independent/predictor variable, such that the parameters of $p(y, x)$ can be efficiently learned in sequential fashion, thereby facilitating the necessary feedback loop. Additionally, it will be useful if $p(y, x)$ is specified such that both $p(y|x)$ and $p(x)$ are analytically tractable, insofar as these derivations underlie our ability to perform prediction of the dependent variable. Borrowing from the literature of nonlinear regression via multivariate Gaussian mixtures [11, 12], we now specify a flexible family of composite mixture distributions for classification tasks with heterogeneous arrays of predictor variables.

12.3.1 Composite mixture models for classification

Let (y_t, x_t) denote the class label y_t associated with a corresponding p-dimensional array of predictor variables $x_t = [x_{t,1}, x_{t,2}, \ldots, x_{t,p}]$. Note that in general y_t can be multivariate, and not necessarily categorical. The usual strategy for specifying a joint probability model on a functional response (y_t) and a set of predictors (x_t) involves embedding y_t and x_t in a common space such that the full covariance structure can be modelled explicitly; this occurs by default in regression via Gaussian mixtures literature, since each mixture component is parameterized by a covariance matrix involving both response and predictors. In this application we are especially interested in

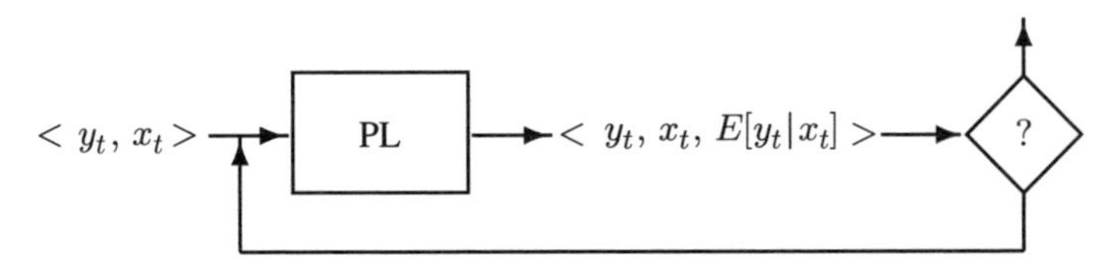

Figure 12.2 Flow diagram of semisupervised PL system capable of seamless switching between supervised updates, prediction and feedback.

cases where that embedding is non-trivial due to the disparity between the nature of the response and the various elements of the predictor array; for example the situation arising in text classification is that x_t may consist of one or more text strings, and y_t is a categorical value. In order to accomodate joint modelling of this type of response–predictor pair, rather than attempting to specify a common embedding via a GLM- or copula-based approach, we instead specify a *composite mixture model* (CMM) as follows:

$$p(y_t, x_t) = \sum_{k=1}^{K} \omega_k p_k(y_t, x_t) \tag{12.20}$$

$$= \sum_{k=1}^{K} \omega_k \left(\prod_{i=1}^{|y_t|} p_k(y_{t,i}) \prod_{j=1}^{|x_t|} p_k(x_{t,j}) \right) \tag{12.21}$$

In the CMM, each mixture component is a composite of independent distributions for each element of the response and predictor arrays. For text classification purposes we will be interested in CMM consisting of different compositions of multinomial and PDFA distributions, e.g. comparisons between $p_k(y_t, x_t) = Mn(y_t|\Omega_{k,y})PDFA(x_t|\Omega_{k,x})$ and $p_k(y_t, x_t) = Mn(y_t|\Omega_{k,y})Mn(x_t|\Omega_{k,x})$ would be used to investigate the effectiveness of the order-sensitive PDFA approach to modelling x_t over a standard multinomial bag-of-words model of x_t (here $\Omega_{k,\cdot}$ denotes the appropriate parameters for the given element within component k).

Although simplistic, the CMM formulation has several important properties. Firstly, it provides a mechanism for combining multiple disparate data types into a common probability model without resorting to complicated embeddings that would preclude sequential analysis. Secondly, the independence assumption within each mixture component results in the appealing property of analytically tractable conditional and marginal distributions, namely

$$p(y_t|x_t) = \sum_{k=1}^{K} \omega_k(x_t) p_k(y_t) \tag{12.22}$$

$$p(x_t) = \sum_{k=1}^{K} \omega_k p_k(x_t) \tag{12.23}$$

where $\omega_k(x_t) = \frac{\omega_k p_k(x_t)}{\sum_{\ell=1}^{K} \omega_\ell p_\ell(x_t)}$. Thirdly, although the elements of the response and predictor arrays are independent within each mixture component, it is still possible to recover complex correlations and dependencies between elements via the inclusion of a large number of mixture components. Although there may be more parsimonious ways of modelling the joint covariance structure, this is arguably the simplest and most generally applicable to a wide array of data types. Finally, the within-component independence allows computation associated with component updating to be parallelized across independent elements of the composite model.

12.3.2 Particle learning for composite mixtures

The basic particle learning algorithm for mixture models was developed in Carvalho *et al.* [6], and it is a straightforward extension of those ideas to derive a particle learning approach for CMM. Following the notation of [6] each particle *i* at time *t* is characterized by an essential state vector $\mathcal{Z}_t^{(i)}$ which contains the various sufficient statistics for posterior distributions of all model parameters. The key requirements of particle learning are that

1. Each $\mathcal{Z}_t^{(i)}$ allows computation of the posterior predictive distribution of the complete observation $p(y_{t+1}, x_{t+1} | \mathcal{Z}_t^{(i)})$,
2. Each $\mathcal{Z}_t^{(i)}$ allows sampling from the posterior distribution of sufficient statistics $p(\mathcal{Z}_{t+1}^{(i)} | \mathcal{Z}_t^{(i)}, y_t, x_t)$.

For mixture models, such $\{\mathcal{Z}_t^{(i)}\}$ arise through specifying a standard Dirichlet(α) prior on mixture weights and independent, conjugate composite priors for the model parameters of each mixture component. The state vectors are then characterized by the sufficient statistics associated with the parameters of mixture component *conditional* on a particular partition of the data amongst mixture components.

Furthermore, by modifying the PDFA model structure to accomodate multiple independent sequences (Section 12.2.2.2), we are able incorporate the particle propagation technique from the PL algorithm described above into a PLCMM setting, thus producing a sequential learning framework fully conducive to semisupervised text classification via CMM such as $p_k(y_t, x_t) = Mn(y_t|\theta_k)PDFA(x_t|\pi_k)$.

12.4 Application to spam detection

The CMM involving PDFA elements as described above provides a flexible framework with potential application to a variety of domains. We now demonstrate the utility of the approach in the context of classification of email messages into one of two categories: spam or non-spam. The task of spam filtering poses several interesting challenges that illustrate well some of the most appealing properties of our PLCMM approach.

Two primary aspects of spam data make spam filtering a non-trivial problem. Firstly, given the streaming nature of email, spam filtering datasets are large and constantly increasing in size. Secondly, spammers are continuously developing new ways to evade current spam filters, such that the distinguishing characteristics of spam messages change over time. Hence, in training a spam filter, one is faced with an infinite stream of ever-evolving data. Together, these two properties of email data effectively render infeasible the use of traditional batch Monte Carlo (MC) methods for spam filtering. Batch MC methods must be trained once; as such, in order to incorporate new data, the models must be retrained with an augmented or time shifted dataset. In contrast, particle learning (and SMC methods, in general) allows models to be repeatedly updated with individual observation; as such, it bypasses the issue of ever-increasing training dataset size faced by batch MC methods. Hence, particle filtering constitutes an elegant approach to accomodate and model this type of dynamical data.

Email text is generally composed in natural language following the rules of its grammar. As such, adjacent words are not independent of one another. For instance, given an adjective such as 'cold' in a text written in English, one would expect a noun such as 'weather' to follow. PDFA models account for this type of dependency among words by modelling both word frequency and order. This is in contrast with commonly used approaches to spam filtering (and text classification in general), such as multinomial factor models such as Latent Dirichlet Allocation [3]. In such models only the frequency with which words appear in the message is modeled. By modelling both word frequency and order, the PDFA model provides a natural representation for this type of data, without the recursive computations required of hidden Markov models.

Another relevant consideration regarding spam filtering involves the process by which labels are assigned to new training data. Given the large volume of incoming messages, the need for expensive and limited human resources becomes a bottleneck in continuous model updating. The PLCMM framework can reduce the need for human intervention by automatically triaging the data as follows.

For each new message that arrives, we compute the conditional expectation of the multinomial label given the observed text feature, and thus obtain predictive distribution of the label value. In cases in which the variance of this expectation is sufficiently low, the predicted label can be treated as the true label of the message, and the message along with its inputed label can be used to update the model. Only when the predicted label has sufficiently high variance is the message flagged for evaluation by a person. This allows all incoming messages to be used for both prediction and training, with minimal human intervention.

A final characterizing aspect of spam filtering is that misclassification costs (i.e. false positive cost and false negative cost) should not necessarily be treated as equal. It is generally preferable to misclassify a spam message as being a legitimate message (false positive) than to mislabel a legitimate message as being spam (false negative). In the PLCMM framework, these misclassification costs can easily be incorporated into the prediction step via the use of thresholds on the posterior probabilities that reflect this bias.

12.4.1 Description of spam data

In this application, we use the public email corpus PU1 [1].[2] This corpus consists of 1099 email messages, 481 of which are spam messages received over the course of 22 months, and 618 of which are legitimate email messages collected over the period of 36 months. All messages in the corpus have been preprocessed as follows. All fields of the header, except for the 'subject' field, have been removed, such that each message is composed of only two parts: 'subject' and 'body'. Additionally, all attachments and html tags have been removed. Finally, the messages in the corpus are encoded for privacy, with unique tokens being replaced with integer indices across all messages of the corpus.

The PU1 corpus is available in four different versions, obtained by enabling or disabling a lemmatizer and a stop-list. The lemmatizer converts words to their dictionary form, such that different forms of the same word are grouped together and can be treated as a single term. For example, the lemmatizer converts both 'selling' and 'sale' to 'sell', and both 'good' and 'better' to 'good'. The stop-list removes the 100 most frequent words of the British National Corpus from each message. Herein, these four versions of the data are referred to as follows: 'bare' (lemmatizer disabled, stop-list disabled), 'lemm' (lemmatizer enabled, stop-list disabled), 'stop' (lemmatizer disabled, stop-list enabled), and 'lemm+stop' (lemmatizer enabled, stop-list enabled).

12.4.2 Modelling strategy

In order to train a PLCMM spam filter we need to make a number of design choices regarding both the input data as well as several aspects of the model itself. Here we consider some of the most relevant aspects to demonstrate model flexibility and power.

Data considerations. The primary design question we address regarding the data is that of what parts of the email messages should be included in the model: subject, body or both. Additionally, we consider the benefits of processing the data with a lemmatizer and a stop-list, by comparing results obtained using the four datasets described above in Section 12.4.1: bare, lemm, stop and lemm+stop.

Model considerations. Many of the existing spam filters take into account only the distribution of words in the messages. By defining a CMM involving PDFA models for text-valued features we are able to incorporate both word distribution and word order. Hence, our foremost interest here is to assess whether including word order information into the model provides a

[2] Available at: http://labs-repos.iit.demokritos.gr/skel/i-config/.

performance improvement over using a model based only on aggregate word distributions. We accomplish this by comparing the predictive accuracies of CMMs in which the data fields (body and/or subject of message) are modelled alternately with a PDFA or a multinomial distribution (which is equivalent to a PDFA with a single state, as described in Section 12.3). In this analysis we do not perform extensive investigation of prior sensitivity, but rather focus solely on the impact of the number of hidden states in the PDFA and the number of components in the composite mixture model.

12.4.3 Experimental results

Here we present the results of performing predictions on PU1 data using PDFA models. All analyses were done via 10-fold cross-validation, where the dataset was partitioned into 10 parts, such that nine of them were used for training and the remaining one was used for testing. We explore two PDFA models, PDFA1 and PDFA2, which differ only in whether or not the body of the email is included in the model. In both models the email label, y, follows a multinomial distribution. In the PDFA1 model, the predictor variable x denotes only the email subject and is distributed according to a PDFA. Thus, in model PDFA1 the email data is represented by the composite mixture model,

$$p_{\mathrm{PDFA1}}(y,x) = \sum_{k=1}^{K} \omega_k Mn(y|\Omega_{k,y}) PDFA(x|\Omega_{k,x})$$

where once again the Ω terms indicate the appropriate parameter vectors for each element of the CMM. In the PDFA2 model, the predictor variable, (x_1, x_2), denotes both the email subject and the email body, with both following independent PDFAs,

$$p_{\mathrm{PDFA2}}(y,x_1,x_2) = \sum_{k=1}^{K} \omega_k Mn(y|\Omega_{k,y}) PDFA(x|\Omega_{k,x_1}) PDFA(x_2|\Omega_{k,x_2})$$

As described earlier, a simpler alternative to PDFAs is to model the subject and body of emails as multinomial distributions. In this case, only the frequency of each word is relevant, whereas with the PDFA both the frequency and order of words contribute to the model. In order to determine whether including the order of words leads to an improvement in model performance, we also consider two additional composite mixture models, Mn1 and Mn2,

$$p_{\mathrm{Mn1}}(y,x) = \sum_{k=1}^{K} \omega_k Mn(y|\Omega_{k,y}) Mn(x|\Omega_{k,x}),$$

$$p_{\mathrm{Mn2}}(y,x_1,x_2) = \sum_{k=1}^{K} \omega_k Mn(y|\Omega_{k,y}) Mn(x_1|\Omega_{k,x_1}) Mn(x_2|\Omega_{k,x_2}),$$

where, as in the PDFA models, x_1 denotes email subject and x_2 denotes email body. Table 12.1 presents a comparison of these four models. In all models, every multinomial random variable was given a Dirichlet prior with hyperparameter $\alpha = 0.01$. All simulations were performed using 30 mixture components, 100 particles and an effective sample size-based resampling strategy with a threshold of 75. In all PDFA elements, the maximum number of hidden states allowed was 20.

Table 12.1 Summary of models, showing the email features (email subject and email body) included in the model, as well as their distribution. Note that when both subject and body are included in the model, both follow the same distribution.

Model	Email subject	Email body	Subject/body distribution
Mn1	yes	no	Multinomial
Mn2	yes	yes	Multinomial
PDFA1	yes	no	PDFA
PDFA2	yes	yes	PDFA

Figures 12.3 and 12.4 summarize the results obtained with the four models, showing the ROC curves and their corresponding AUC measurements, respectively. There are several interesting observations to be made from these results. First and foremost, we address the primary question of interest, which is that of whether or not the PDFA model provides a better description of the data than a multinomial model. The short answer to this question is that it depends on the length and complexity of the variable being modelled. Specifically, using a PDFA to model email subject did not improve prediction performance in relation to a model using a multinomial distribution, as can be seen by comparing models Mn1 and PDFA1 in Figures 12.3 and 12.4. These results are, in fact, what one would expect given that email subjects are typically very short, with mean length of approximately five words. In contrast, modelling the body of emails with PDFAs leads to substantial improvements in prediction accuracy in comparison with modelling the body of emails with multinomial distributions. Indeed, the PDFA2 model produces significantly better ROC curves than the Mn2 model (see Figure 12.3). Once again, these results are expected since email bodies are sufficiently long for the order of the words to be informative of the email label, with the median email body length being over 160 words.

Perhaps the most surprising observation to be made from these results is the remarkably poor prediction performance of model Mn2, in which both email subject and body are modelled with multinomial distributions. This model performs worse than all other models we have tested, as can be seen in Figure 12.3. It is peculiar to observe that moving from model Mn1 to model Mn2, that is, simply adding a multinomial representation of the email body to a model in which the email subject follows a multinomial distribution, leads to substantially worse prediction accuracy (with AUC reducing from around 0.95 to around 0.55). This suggests that using a multinomial representation of the email body is not only uninformative for the email label prediction, but it is in fact detrimental to it, providing further validation to the usefulness of using PDFAs to model email text.

Our results also provides some insight into the impact of the different data pre-processing treatments on model performance. Lemmasterization of the text did not confer any obvious improvements to the performance of the classifiers. In contrast, the use of a stop-list, which removes the 100 most frequent words, led to improvements in classifier performance, both when words were modelled with a multinomial distribution, but primarily when words were modelled with a PDFA (see Figure 12.4). In fact, using a stop-list, regardless of whether or not a lemmasterizer was used, resulted in the best overall AUC measurements and accentuated the improvements of PDFAs over multinomial distributions.

Finally, we performed a parameter sensitivity analysis to determine the impact of the number of mixture components in the particles and of the number of states in the PDFA on the prediction performance of the PDFA1 and PDFA2 models. Figure 12.5 displays the mean AUC obtained from 10-fold cross-validation trials using the lemm+stop version of the PU1 dataset and the PDFA1

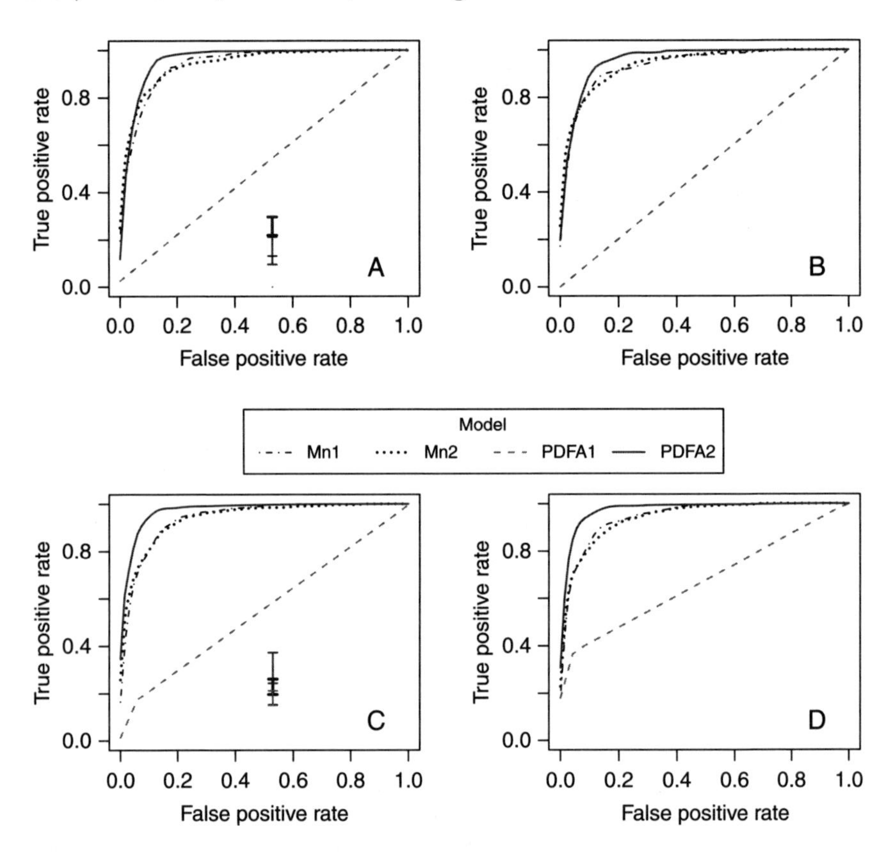

Figure 12.3 ROC curves for CMM-based classification on the four versions of the PU1 corpus: (A) bare, (B) lemm, (C) lemm+stop and (D) stop. Each line represents the vertical average of the ROC curves obtained by the 10 train–test partitions of the data used in the 10-fold cross-validation.

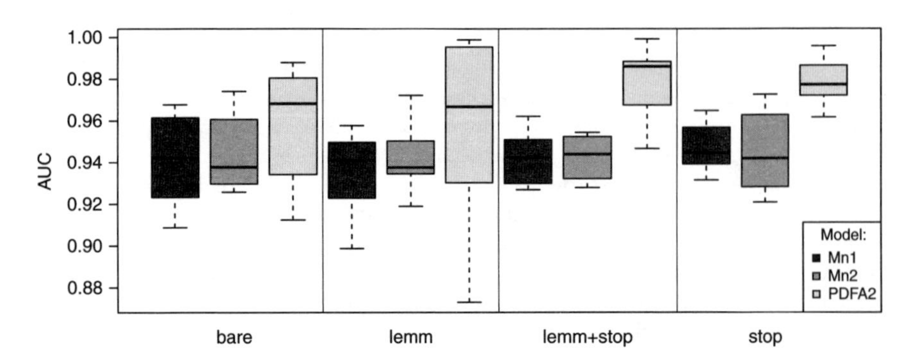

Figure 12.4 Distribution of AUC measurements corresponding to the ROC curves shown in Figure 12.3. The AUC measurements for the model Mn2 are now shown here, and their mean values were 0.5, 0.51, 0.56 and 0.69 for bare, lemm, lemm+stop and stop versions of the PU1 datasets, respectively.

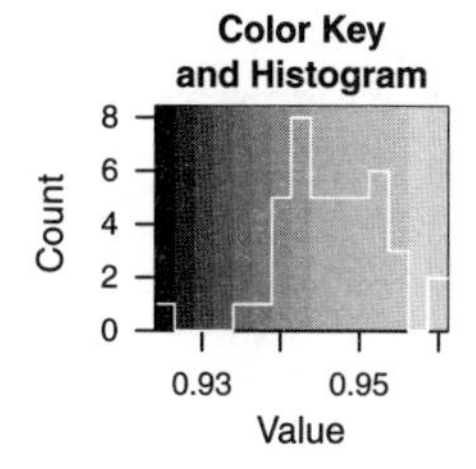

Number of components (K) \ Number of states (n)	2	5	10	20	25	30
3	0.011	0.017	0.016	0.019	0.017	0.028
5	0.019	0.015	0.016	0.013	0.02	0.016
7	0.018	0.02	0.012	0.018	0.01	0.022
10	0.015	0.012	0.013	0.017	0.021	0.017
20	0.014	0.015	0.018	0.015	0.013	0.015
30	0.012	0.02	0.012	0.018	0.016	0.016
40	0.017	0.02	0.014	0.021	0.017	0.018

Figure 12.5 Sensitivity analysis of the number of components in the mixtures and the number of states in the PDFA using the lemm+stop version of the PU1 dataset and including only the subject of the email messages. The grey levels denote the mean AUC from 10 fold cross validation trial, and the number within the cells denote one standard deviation from the mean. The grey level key to the left provides a mapping of the grey level to AUC value, and a histogram showing the frequency with which the values of AUC appear in the heatmap to the right.

model (results for the PDFA2 model are not shown, but are similar to the ones obtained with the PDFA1 model). In this particular example, the best AUC values were obtained using 10 mixture components per particle and five PDFA states. Similar AUC values were obtained with 30 mixture components and two PDFA states. It is somewhat surprising that the best results were obtained with such a small number of states. It would be interesting to investigate what words are emitted by each state. Unfortunately, this exercise is impossible for the PU1 data which, for privacy reasons, is made available in an encoded format, where we are not able to know what words actually correspond with codified values.

In general, all combinations of number of states and number of components led to high AUC values, ranging from 0.92 to 0.99. These are encouraging results, suggesting that this approach is reasonably robust to choices of number of components and number of states. Additionally, this observation could be particularly useful for situations where computational resources are limited; in this scenario, one could use small numbers of components and states, and still expect reasonable prediction accuracy.

12.5 Discussion

In recent years, interest in online learning systems has exploded as a result of the so-called 'big data' problem, which often precludes on computational grounds methods that require loading an entire dataset into a computer's memory. Online methods are by nature filters, operating on a reduced set of observations at a time and retaining state of much lower dimension than that of the entire data. An effective framework for online learning is also prerequisite for active learning and semisupervised learning tasks.

In this chapter, we have demonstrated a novel online learning system for classification of email texts based on particle learning for composite mixture models involving probabilistic automata. The composite mixture structure allows specification of a joint probability model for heterogeneous collections of independent variables without requiring complex embeddings via generalized linear models or copula techniques. In this sense, a primary advantage of the CMM is the speed with which new models can be instantiated simply by specifying the particular composite distribution characterizing each mixture component. Although we have only demonstrated several such composite distributions in this analysis, our software allows specification of arbitrary configurations of exponential family distributions, thus enabling generalized classification and regression through custom specification of heterogeneous collections of independent and dependent variables.

We have demonstrated the utility of the PL-based learning system in the context of spam detection, but the general approach is flexible and can be applied to a vast array of domains, including multiclass classification problems. One such application involves classification of newsgroup messages, in which the goal is to correctly associate a message with the newsgroup to which it was posted. In this type of application, a CMM classifier serves as a type of data validation by ensuring that message content is appropriate for the particular forum. Preliminary results indicate that the same CMM approaches involving PDFA elements demonstrated here in the context of spam will be similarly useful. In a reduced subset of the 20 newsgroup dataset [3] consisting of five newsgroups belonging to three subject matter categories (summarized in Table 12.2), we applied two CMMs: one in which

Table 12.2 Description of the newsgroup data subset.

Newsgroup name	Category	Train set size	Test set size
Medicine	Science	585	394
Cryptology	Science	594	395
Automobiles	Recreation	594	394
Religion_miscellanea	Religion	377	251
Atheism	Religion	480	319

[3] Available at: http://people.csail.mit.edu/jrennie/20Newsgroups/.

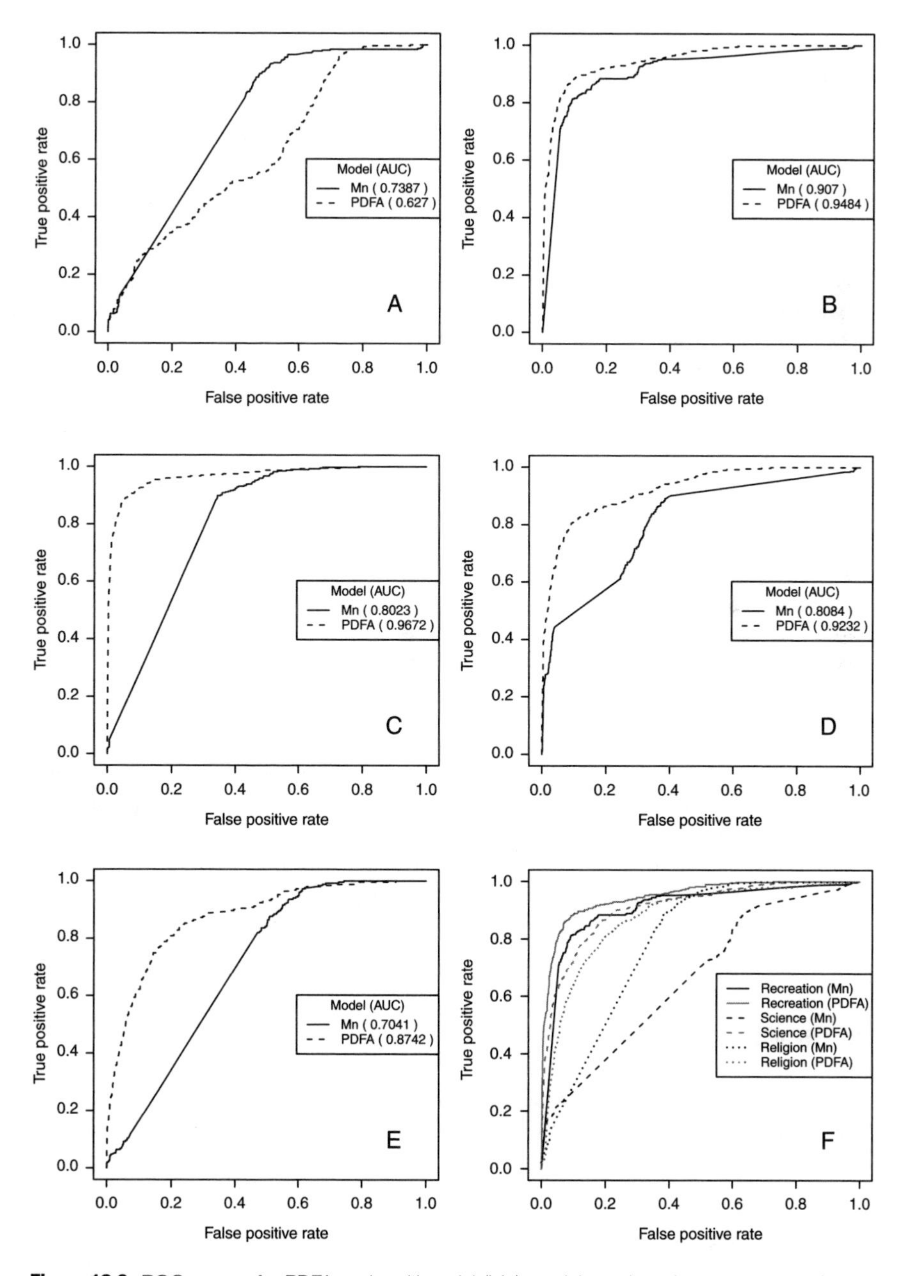

Figure 12.6 ROC curves for PDFA and multinomial (Mn) model used on the news group dataset in the prediction of new group name (A–E) and subject matter category (F). The new group ROC curves are (A) atheism, (B) automobiles, (C) cryptology, (D) medicine and (E) religion_miscellanea.

the message text follows a PDFA and another in which it follows a multinomial distribution. The prior parameters used in both models are the same as those described in Section 12.4.3. The results obtained with both models are shown in Figure 12.6, in which each ROC curve characterizes the ability of the underlying composite mixture model to correctly validate the actual newsgroup to which the message was posted. ROC curves obtained with the PDFA model are generally superior to those obtained with the multinomial, with the exception of group 'atheism'. In predicting subject matter category, the PDFA model was more accurate than the Mn model for all three groups (compare solid and dashed lines in Panel 12.6.F).

A key contribution of this work is the highly computationally efficient particle learning approach for probabilistic deterministic finite automata. The conditionally deterministic transition structure of the PDFA results in significantly reduced computational requirements relative to standard hidden Markov models, thus creating the opportunity for fast online learning, clustering and classification of high frequency observations of order-dependent categorical sequences such as text. Future work includes more direct investigation of the computational versus predictive tradeoff, in the context of classification and regression tasks, of the PDFA versus comparable filtering techniques for HMMs [2, 14]. Recent advances in online EM [4] and online variational methods [10, 16] may also have implications for composite mixture-based classifiers, especially for extremely high frequency data for which particle resampling produces a severe bottleneck.

It is currently an open topic for research how best to use the semisupervised learning capabilities afforded by a PL-based inference framework. In this analysis, our utilization of unlabelled data was restricted to performing prediction on the class label of the observation. Other possible semisupervised operations include performing marginal updates in the presence of missing data, or prioritizing labelling efforts through some measure capturing the uncertainty of the posterior conditional label prediction (e.g. active learning). In particular, in cyber security applications such as spam detection, there is a real need for active learning strategies suitable for adversarial classification problems in which the class distributions are not static but actually converging as the so-called "threat" class (e.g. spam) is deliberately made to look more and more similar to the benign class. Accommodating the adversarial situation clearly requires modification of the modelling strategy presented here to allow dynamic CMM, either by inclusion of discount factors iteratively applied to the essential state vectors, or by building the dynamism directly into the mixture structure in a manner very similar to that recently done for tree-based models in Taddy *et al.* [15].

Acknowledgement

This work was performed under the auspices of the U.S. Department of Energy by Lawrence Livermore National Laboratory under Contract DE-AC52-07NA27344; Document release number LLNL-JRNL-513511.

References

[1] Androutsopoulos, Ion, Koutsias, John, Cb, Konstantinos V. and Spyropoulos, Constantine D. (2000). An experimental comparison of naive Bayesian and keyword-based anti-spam filtering with personal e-mail messages. In *Proceedings of the 23rd annual international ACM SIGIR conference on Research and development in information retrieval*, pp. 160–167. ACM Press.

[2] Beal, Matthew, Ghahramani, Zoubin and Rasmussen, Carl (2002). The infinite hidden Markov model. In *Advances in Neural Information Processing Systems 14* (ed. T. Dietterich, S. Becker, and Z. Ghahramani), pp. 577–584. MIT Press, Cambridge, MA.

[3] Blei, D. M., Ng, A. Y. and Jordan, M. I. (2003). Latent Dirichlet allocation. *Journal of Machine Learning Research*, **3**, 993–1022.

[4] Cappé, Olivier (2011). Online EM algorithm for hidden Markov models. *Journal of Computational and Graphical Statistics*, **20**(3), 728–749.

[5] Carvalho, C., Johannes, M., Lopes, H. and Polson, N. (2010). Particle learning and smoothing. *Statistical Science*, **25**(1), 88–106.

[6] Carvalho, C., Lopes, H., Polson, N. and Taddy, M. (2010). Particle learning for general mixtures. *Bayesian Analysis*, **5**(4), 709–740.

[7] Cohn, David, Ghahramani, Zoubin and Jordan, Michael (1996). Active learning with statistical models. *Journal of Artificial Intelligence Research*, **4**, 129–145.

[8] Doucet, Arnaud, de Freitas, Nando and Gordon, Neil (ed.) (2001). *Sequential Monte Carlo Methods in Practice*. Springer.

[9] Dupont, P., Denis, F. and Esposito, Y. (2005). Links between probabilistic automata and hidden Markov models: probability distributions, learning models, and induction algorithms. *Pattern Recognition*, **38**, 1349–1371.

[10] Hoffman, Matthew, Blei, David and Bach, Francis (2010). Online learning for latent Dirichlet allocation. In *Advances in Neural Information Processing Systems* 23 (ed. J. Lafferty, C. K. I. Williams, J. Shawe-Taylor, R. Zemel, and A. Culotta), pp. 856–864. MIT Press, Cambridge, MA.

[11] Liang, Feng, Mukherjee, Sayan and West, Mike (2007). The use of unlabeled data in predictive modelling. *Statistical Science*, **22**(2), 189–205.

[12] Müller, Peter, Erkanli, Alaattin and West, Mike (1996). Bayesian curve fitting using multivariate normal mixtures. *Biometrika*, **83**(1), 67–79.

[13] Pfau, D., Bartlett, N. and Wood, F. (2010). Probabilistic deterministic infinite automata. In *Advances in Neural Information Processing Systems* 23 (ed. J. Lafferty, C. K. I. Williams, J. Shawe-Taylor, R. Zemel, and A. Culotta), pp. 1930–1938. MIT Press, Cambridge, MA.

[14] Rodriguez, Abel (2011). On-line learning for the infinite hidden Markov model. *Communications in Statistics–Simulation and Computation*, **40**(6), 879–893.

[15] Taddy, Matthew A., Gramacy, Robert B. and Polson, Nicholas G. (2011). Dynamic trees for learning and design. *Journal of the American Statistical Association*, **106**(493), 109–123.

[16] Wang, Chong, Paisley, John and Blei, David (2011). Online variational inference for the hierarchical Dirichlet process. In *Journal of Machine Learning Research Workshop and Conference Proceedings*, Volume 15, pp. 752–760.

[17] West, M. and Harrison, J. (1997). *Bayesian Forecasting and Dynamic Models*. Springer.

Part VI

Nonparametrics

13 Bayesian nonparametrics

STEPHEN G. WALKER

13.1 Introduction

Before talking about Bayesian nonparametrics specifically, it is worth discussing the Bayesian approach in general. There are a number of ways of introducing Bayes and one can find them in books such as [4]; [41]; [27]; and [37]. There are differences in the foundations, ranging from a rigorous development using notions of rational behaviour and axioms pertaining to such, to the mathematical formulation directly as the posterior being a product of the likelihood and prior.

There is a prevailing attitude that the foundations for Bayesian inference are resolved and practitioners should now 'just get on with it', without trying to restructure or reformulate ideas behind Bayesian thinking. This would be arguable were it not for the rapid advance in computation and computer power which allows the analyst to construct and estimate models of any size. Actions which took hold during an age when models were small,[4] out of necessity, and we can refer to [6] to see what could be achieved, may no longer be of any importance. And some actions are clearly at odds with Bayesian thinking. This is not to denigrate, but to point out that when models are small, one will necessarily behave differently compared to a situation where large models are available.

Let us expand on this point. I am trying to learn about something and have some current knowledge. My current knowledge is encapsulated in a small model. I learn through further observation that this small model is wrong or misguided. I must change it, whether the foundations for the inference I am undertaking allow me to do this or not. The current knowledge (knowledge prior to data) is being changed by the data. This cannot make any sense. Such knowledge should be updated, not changed then updated, by the data, and a framework should be established from the start that permits revisions of sufficiently large magnitude and variety. Large models, which are becoming increasingly better understood, can meet these needs.

With small models it must be expected to change them, to consider a range of possible models, or select a model post-data. It is necessary whether one accepts that this 'destroys' some aspect of Bayes or not. My opinion is that a methodology which relies on such strategies is not Bayes; it is something possibly close to Bayes, but lacks motivation. It is ad hoc. There is no single well-defined representation of current knowledge. And this current knowledge must be about something that

[4] For the purposes of this article, we can define a small model as one for which a number of assumptions about the data, e.g. symmetric, unimodal, have been made, and for which there is no external supporting evidence; a check of such assumptions would essentially be post-data.

exists. With a small model the only object that exists that I would be interested in learning about is the parameter value which takes my small model closest (in some sense) to the correct model.[5] I should be able to learn about this parameter because I am observing samples from this correct model. With this outlook there is no assertion that the small model is correct; one is simply trying to find the 'best parameter' one can from it. The question then is whether Bayes can be used in this scenario.

Entertaining a number of small models, trying to find one that is adequate, or even a perfect fit, is not satisfactory. There is, on the other hand, a proper Bayes approach of properly quantifying uncertainty with a large model. Indeed, if all the uncertainty has been adequately expressed, it must be an internal contradiction to then check that the data and the expressed uncertainty are compatible.

So small models could lead to undesirable issues; large models circumvent these issues and allow Bayes to be implemented in its purest form:

$$\text{uncertainty} \rightarrow \text{preliminary knowledge about the uncertainty} \rightarrow \\ \text{observation} \rightarrow \text{update knowledge}$$

Bayes provides the means to do this when knowledge is represented in probabilistic form.

13.1.1 Dependent models

I want to introduce Bayesian thinking by talking about the creation of dependence. I witness an observation and want to use this to predict another outcome, yet to be seen. In probability language, it is necessary to construct a dependence between the two outcomes; one seen, the other yet to be seen. This in no way implies that the outcome of one actually depends on the other; i.e. there is a physical dependence. Consider an example. An observer is watching a time trial in a cycling event. They have no idea what times to expect, and if it were possible they would construct a distribution describing the time to finish. The first rider finishes and provides a time. This provides information to the observer and they are able to revise the distribution of times. This new distribution must depend on the first outcome for it to be a revised distribution. And no one would pretend in reality that the times of the first and second rider physically depend on each other. The times are independent, and there is no reason not to assume that they are identically distributed.

To make this illustration more precise, it is quite feasible to think of a process by which person A is generating i.i.d outcomes from some machine and inviting person B to get at the distribution from which the observations arise. Persons A and B clearly have different knowledge and person B, needing to learn about where the samples are coming from, would be willing to set up a probability model in which there is a dependence structure between the observations. After every sample a new guess is made as to the distribution from which the observations arise. This must depend on the previous outcomes. There is no long-term principle involved here. This strategy can be invoked no matter how large the sample of outcomes is given. For any finite sample it cannot do any harm to assume that there is an underlying density generating the observations. Nor does it make any contribution to assume such a density does not exist. The key really being the notion that the order of the samples has no bearing on what the guess looks like after seeing all the observations.

The upshot is that person B creates a probability model which has a dependence structure connecting the outcomes so that learning can take place. The model is wrong, but it is very useful;

[5] This idea will be developed later. I am also aware that the notion of a correct model is something with which not all are comfortable. However, for now, I merely mention that without the notion of a correct model it is far from clear what a Bayesian is even trying to learn about.

[7] expressed a similar sentiment but in a different context. For me this is the essence of Bayes: creating dependent probabilistic models so that learning can take place. Without the dependence there is no learning. But what are these learning models and how are they comprised? While person B might have a sequence of guesses about where the outcomes are coming from, ultimately they do not believe the order of observation should have any impact on the model; so the order does not matter. When models are dependent and the order does not matter then the only selection is a Bayesian model ([11]; [22]).

In summary, therefore, if there is no physical dependence between observations and the order of observations is not relevant, then treating the data as i.i.d is appropriate. The Bayesian model creates variables out of the data and makes them dependent as well, for a good reason. To do this, there is no need to assume the data exchangeable. Neither is there any need to estimate properties of a model, such as consistency, using the notion of the data as coming from the model. (See for example [12], in the case of consistency.) This is unrealistic and would easily suggest the Bayesian model is better than it really is. However, to study the model assuming the observations are i.i.d is commonly termed the frequentist study of a Bayesian model. This is a misunderstanding.

13.1.2 Learning model

The Bayesian model is about learning. But learning about what? How does it start? One aspect of the learning is a sequence of guesses as to where the next outcome is coming from. So let $(X_1,\ldots,X_n)$ denote the sample. The sequence of guesses would be denoted by

$$m_n(x|x_1,\ldots,x_{n-1})$$

for $n = 1,2,\ldots$ The conditional notation is being used to emphasize that the guess depends on what has been seen to date. Given the symmetry, we can (as will be expanded on later) write the predictive guess as

$$m_n(x|x_1,\ldots,x_{n-1}) = \int f(x)\,\Pi(\mathrm{d}f|x_1,\ldots,x_{n-1})$$

where $\Pi(\mathrm{d}f|x_1,\ldots,x_n)$ is known as the posterior distribution and given by

$$\Pi(\mathrm{d}f|x_1,\ldots,x_n) = \frac{\prod_{i=1}^n f(x_i)\,\Pi(\mathrm{d}f)}{\int \prod_{i=1}^n f(x_i)\,\Pi(\mathrm{d}f)}$$

The starting point uses the prior distribution $\Pi(\mathrm{d}f)$; i.e.

$$m_1(x) = \int f(x)\,\Pi(\mathrm{d}f)$$

and so one sees that the only change as the observations come in is that Π gets updated. So we are learning about something called f. The only interpretation for this f is that it is the density which is generating the observations. Then it is quite clear that $\Pi(\mathrm{d}f)$ would at least represent what is known about the density generating the (X_i) based on evidence which does not include the (X_i).

So $\Pi(\mathrm{d}f|X_1,\ldots,X_n)$ must represent what is known about f with the evidence that yielded $\Pi(\mathrm{d}f)$ along with the evidence that is provided by $(X_i)_{i=1}^n$. And this is Bayes. It is a simple structure and the thinking is straightforward. It is an explicit and mathematical (using probability) evaluation of a real learning process.

Symmetry and the de Finetti representation [11] would imply Bayes. It is a misunderstanding that it is only possible to use the representation if one states the (X_i) are coming from the model itself. Symmetry holds for guesses as well; the guesses need to be symmetric and so there is a representation theorem for the guesses. If there is a representation theorem then there is Bayes. It would seem that Bayes is a more robust learning mechanism than previously given credit. This will be discussed further in Section 13.3. But at the outset, there are a couple of issues that must be dealt with sequentially:

(i) What is $\Pi(df)$ and how is it constructed?
(ii) What happens when $n \to \infty$?

13.2 Constructing the prior

Bayesian topics which have received prominence in the literature are due to the problems in how to deal with (i). An eagerness to make $\Pi(df)$ over simplistic leads to Bayesian model selection, Bayesian model checking, Bayes factors, and other ad hoc ideas. This issue is not about how to distribute mass within Π, which will be considered later, but rather how to get at the support of Π, denoted by Ω. The support requires a proper definition in terms of distances, but for now we can just leave it as the densities Π can generate. The choice of Ω can range from something very small, almost to a point mass at a specific density, to an Ω which includes all possible densities.

What is the biggest problem in this setup? It must be of the following type: I construct $\Pi(df)$ and this means that $\Pi(f \in \Omega) = 1$ for some set of densities Ω. So it has been stated that what is known without knowledge of the data is that the set of densities which could be generating the data lies in Ω. If that is what is known then that is it. How can one measure this certainty in practice? What is the test? The test is that the model, namely Ω is not checked off with the data. It is clearly an internal contradiction, and a demonstration of irrational behaviour, to specify Ω and then check it. Checking it indicates that there is more uncertainty around than is being acknowledged by the use of Ω. If this is the case, then Ω needs to be enlarged to appreciate the starting uncertainty properly.

But we cannot discount an error in judgement: Ω has been chosen and it becomes evident that after seeing a large enough sequence of the (X_i), that $f \notin \Omega$. What now? Is this the point where we need to throw away the simplicity and elegance of the Bayesian learning machine to accommodate a 'Bayesian' who has made an error in judgement? If, for example, Ω is too small then it could be quite possible that the discovery $f \notin \Omega$ is found.

The way to deal rigorously with the problem of an overly small Ω is not to have it in the first instance, and to make it as large as needed from the onset, even if this means making it as large as one can manage. Of course this refers to Ω rather than the number of parameters in the model. There are then subsequently no concerns about Ω being too small. This is the simple yet powerful message of Bayesian nonparametrics.

The typical Bayesian outlook is to start small and increase a model as and when it is deemed necessary, heading to something of an optimal fitting model. But as we have described earlier, this has all sorts of associated problems. Parsimony is an often cited objective; roughly speaking, if two models fit well, pick the one with the smallest number of parameters. To me, this is simply a recipe for underestimating uncertainty. It is an objective in direct conflict with Bayesian learning. I fail to see how such a perfect Bayesian model could have been obtained in a purely Bayesian way. To reiterate Ω is what one is looking for, which is a set of densities with particular properties or the set of all densities. The aim for parsimony is to put all the mass on Ω with the minimal of parameters. This is the critical difference of Bayes; parsimony with Ω, rather than the data. Because Bayes starts with fixed knowledge, which is Ω, the data rearranges the mass in Ω, the data are not to change Ω. Thus we need to describe how to construct $\Pi(df)$ so that Ω is as large as needed or as large as possible. This is not about adding more and more parameters into a model; rather it is about achieving a large Ω with the minimum of fuss.

13.2.1 A popular prior

What can we get for densities on the real line? The foundational model is the mixture of Dirichlet process model [31] and is constructed as

$$f(y) = \int k(y|\theta)\, \mathrm{d}P(\theta)$$

Here P is a mixing distribution and the original and remaining popular choice is the Dirichlet process [16].[6] This prior generates discrete random distribution functions and we will look at it in some detail. To fix ideas, assume $\theta = (\mu, \lambda)$ where μ denotes the mean of a normal distribution and $\sigma^2 = \lambda^{-1}$ is the corresponding variance. So k denotes the normal density function and P denotes a distribution function, and this is of the type

$$P = \sum_{j=1}^{\infty} w_j\, \delta_{\theta_j}$$

where the (w_j) are weights which sum to 1; the (θ_j) are a set of values from $(-\infty, +\infty) \times (0, \infty)$; and δ_θ denotes the measure with a point mass of 1 at θ. The prior distribution is assigned to (w_j, θ_j) and it is usual to allow the (θ_j) to have independent priors with the same distribution and the weights to have a stickbreaking prior construction, so that

$$w_1 = v_1 \quad \text{and for} \quad j > 1, w_j = v_j \prod_{l<j}(1 - v_l)$$

and the (v_j) are independent with beta distributions. For more on this type of construction, see [43] and [24]. Hence, $f(y)$ is an infinite mixture model of normal distributions. Any density function on the real line can be arbitrarily well approximated, with respect to the L_1 metric for example, by such a mixture of normal density functions.

Clearly then, with such a prior, it is possible to avoid the need to undertake anything like model selection. But there is still work to be done. The most notable is the problem of the choice of prior for (v_j, θ_j), which, if not careful, could have too influential a role in the learning process. But this problem of the influence of where the mass is placed in Ω is of course not unique to nonparametric models. To see how this issue can work in a nonparametric model we need go no further than the Dirichlet process. Recall the stickbreaking process whereby the (v_j) are independent beta$(1, c)$ variables, for some $c > 0$. If the (θ_j) are from the distribution function G, then for any set A we have

$$\mathrm{E}(P|X_1, \ldots, X_n) = \frac{c\,G + n\,P_n}{c + n}$$

[6] Typically, Bayesian nonparametric models resist in defining $\Pi(df)$ as a probability measure but rather describe how a random f can be taken from $\Pi(df)$.

where P_n is the empirical distribution function. By appropriate choices of c, this posterior expectation can clearly take any value from the empirical distribution to the prior expectation. Is this a cause for concern? But c and G have fairly straightforward roles from previous results: so

$$\mathrm{E}[P(A)] = G(A) \quad \text{and} \quad \mathrm{Var}[P(A)] = \frac{G(A)\, G(A^c)}{c+1}$$

for any set A, and where A^c is the complement set of A. But there are also a number of ideas to specify c based on the evaluation of $W_j = \sum_{l>j} w_l$ in terms of expectations and probabilities of outcomes. So we can easily make use of the fact that

$$\mathrm{E}(W_j) = \prod_{l=1}^{j} (1 - \mathrm{E}(v_l)) = \left(\frac{c}{1+c}\right)^j$$

There will then always be a way, however large the model, to be able to use knowledge about the problem to specify key parameters in the prior, usually based on prior expectations of quantities. Modern applications of the mixture of Dirichlet process model often assign a hyper-prior to c so as to mitigate the strength of any specific prior choices; see [15].

There are by now many ideas for estimating the model above using sampling based approaches which construct Markov chains with suitable stationary distributions. Such methods started with [14] and other ideas include [32]; [33]; [36]; [47]; [39]; and [25].

But for the Dirichlet process as well as other similar models and their use in mixture models, three issues need to be addressed. These are: identifiability and clustering; the use of the normal distribution as a kernel; and the infinite nature of the Dirichlet process.

13.2.2 Identifiability and clustering

For any density f there are, in the popular mixture of Dirichlet process model, many ways to construct the f. The basic idea is that we can achieve mass at a particular location, say θ, by having one of the θ_j at this location and the appropriate weight w_j. But we could also achieve this by largely ignoring the weights, making them simple, and we can achieve a certain weight at θ by placing an appropriate number of the (θ_j) close to θ. The Dirichlet process, and hence the mixture of Dirichlet process model, is not identifiable. When it comes to estimating the density f this may not be perceived as a problem, but it does mean that the number of mixtures will be overestimated when the model uses the latter plan to construct f; see the illustrations in [47]. The problem becomes more acute when mixture models are used for complex data structures such as regression and time series models.

It is far from easy to resolve this problem. If one adopts simple weights such as geometric weights, attractive since one removes a layer of parameters, namely the (w_j), then the support Ω is not diminished; see [38]. However, the estimate of the number of mixtures is not possible. An idea for modifying the model to provide both density estimation and to estimate components is now discussed which requires a change in kernel and model.

13.2.3 The choice of kernel

If there are mixtures in the model then it must be of interest to identify the mixtures. If the choice of kernel is the normal density function then one can hope to identify the normal components which make up the density. But this does not answer the question of number of clusters unless one has defined a cluster by a group of the data being modelled adequately by a normal density.

On the other hand, a unimodal density can be used to model a cluster adequately. One might believe that a bimodal density and higher number of modes model more than one cluster. Certainly this can be safely assumed in the absence of information beyond the data. Thus it becomes important, or at least of interest, to be able to model mixtures where the only assumption on each component is one of unimodality. However, in the context of mixture of Dirichlet process models, this is a difficult task. The components will already be nonparametric and to mix over these using a Dirichlet process would lead to unfathomable complications.

13.2.4 How to include infinity?

A solution to this problem is to use a variation on the nonparametric mixture model that involves modelling the number of components explicitly, as in [40]. So for every finite $k \geq 1$ a finite model is constructed as

$$f_k(x) = \sum_{j=1}^{k} w_{jk}\, p_j(y)$$

where the (p_j) are the components of the model. And to complete the model there would be a prior assigned to k, and each p_j would be assigned a prior which generated unimodal densities. One way to do this is via a mixture of uniform distributions. That is,

$$p(y) = \sum_{l=1}^{\infty} w_l\, \mathrm{Un}(y| - \theta_l, +\theta_l)$$

where the (θ_l) are positive. This would generate symmetric unimodal densities, and to incorporate skewness a number of ideas are possible. [42] use an idea from [17] that involves using a uniform component of the type

$$\mathrm{Un}\left(y| - e^{-\lambda}\theta_l, +\theta_l e^{\lambda}\right)$$

Hence,

$$f_k(y) = \sum_{j=1}^{k}\sum_{l=1}^{\infty} w_{jk}\, w_{lj}\, \mathrm{Un}(y| - e^{-\lambda_j}\theta_{lj}, +e^{\lambda_j}\theta_{lj}),$$

where

$$\sum_{j=1}^{k} w_{jk} = 1 \quad \text{and} \quad \sum_{l=1}^{\infty} w_{lj} = 1$$

Now k has the very real interpretation of being the number of clusters that can be modelled using unimodal densities.

13.3 What do we want as *n* grows?

We also consider another point which has to do with consistency. What happens to the posterior distributions as the sample size tends to infinity? This is the usual question. We can look at this from a slightly different perspective. My Bayesian model is learning about something, and so I identify an object to learn about. I collect observations from a source and, using the Bayesian learning machine, I think that, indeed, I am learning about this object. But how do I verify that this learning, via posterior distributions, is actually taking place? I only see this issue being resolved through an asymptotic study of the sequence of posterior distributions.

Before discussing the mathematics of consistency, it would be prudent to discuss what determines the asymptotic performance of the Bayesian model. We have already determined that the Bayesian model is wrong in the sense that no stochastic dependence of the type arising from exchangeability actually connects the observations. The stochastic dependence is used to construct a learning machine. If there is no physical dependence between observations, and they come from some identifiable similar source (such as the track cycling times described earlier), so the order of the sequence of observations does not ultimately matter, then the outcomes (once assigned to be random variables) must be treated as being i.i.d from some fixed but unknown distribution function, commonly written in the literature as F_0.

Briefly, if one is interested in $\psi(X)$ and $X \sim F_0$, then one studies probabilities such as $P(\psi(X) \in A) = F_0(\psi^{-1}(A))$. If there is a possibly infinite or arbitrarily large number of such X there is no sudden switch of the idea here to think about. Hence, the study of a Bayesian model based on the notion of the $(X_1, \ldots, X_n)$ being i.i.d from F_0^n, for any n, is not a frequentist setup; it is not even a Bayesian setup. It is a study of a Bayesian model with the appropriate assumption of how the data arrive.

So why would a Bayesian introduce a dependence structure to the outcomes when in reality there is none? There is a good reason to do this. It does not then mean a Bayesian must insist the data become dependent or assume they are dependent. The model creates a dependence for the reason we now elaborate on further.

A Bayesian is willing to provide a sequence of guesses as to where the next outcome is coming from. If this sequence is static, i.e. it is $m(x_n)$ for all n for some density function $m(x)$, then there seems little point in witnessing outcomes. So if $m_0(x_1)$ is the best guess for the density of x_1 then it would change to $m_1(x_2)$ once x_1 has been seen. The outcome must provide information which is helpful to the next best guess. So $m_1 \neq m_0$ and to highlight the point that it depends on x_1 we can write $m_1(x_2|x_1)$. Thus, since we have seen $(x_1, \ldots, x_{n-1})$, the best guess for x_n would be of the type $m_{n-1}(x_n|x_1, \ldots, x_{n-1})$.

But these guesses need some structure. The rule, namely the order does not matter, must result in

$$m(x_1, \ldots, x_n) = \prod_{i=1}^{n} m_{i-1}(x_i|x_1, \ldots, x_{i-1})$$

being symmetric, in the sense that for any permutation σ on the integers $(1, \ldots, n)$,

$$m(x_{\sigma(1)}, \ldots, x_{\sigma(n)}) = m(x_1, \ldots, x_n)$$

for any $n \geq 2$.

The consequences of the above are well known. To reiterate, the (x_i) are not dependent; the best guesses from the Bayesian about future outcomes force a model which attracts dependencies, given the desire to learn from experience. So $m(x_1, \ldots, x_n)$ can be viewed as the best guess for the first n outcomes taken one-by-one. It must be symmetric and therefore the best guess model must look like

$$m(x_1, \ldots, x_n) = \int \prod_{i=1}^{n} f(x_i)\, \Pi(\mathrm{d}f)$$

for some probability measure on a suitable and appropriate space of density functions.

This would be the model and the next outcome would be guessed as coming from

$$m_n(x_{n+1}|x_1, \ldots, x_n) = \frac{m(x_1, \ldots, x_{n+1})}{m(x_1, \ldots, x_n)}$$

and which is given by

$$m_n(x_{n+1}|x_1, \ldots, x_n) = \int f(x_{n+1})\, \Pi(\mathrm{d}f|x_1, \ldots, x_n)$$

where

$$\Pi(\mathrm{d}f|x_1, \ldots, x_n) = \frac{\prod_{i=1}^{n} f(x_i)\, \Pi(\mathrm{d}f)}{\int \prod_{i=1}^{n} f(x_i)\, \Pi(\mathrm{d}f)}$$

All that has changed due to the observations $(x_1, \ldots, x_n)$ is that $\Pi(\mathrm{d}f)$ has been updated into $\Pi(\mathrm{d}f|x_1, \ldots, x_n)$.

Hence, there is an important part to be played by $\Pi(\mathrm{d}f)$. So what exactly is it? This much is clear: it is expressing ideas about the location of a density function, in the sense that $\mathrm{P}(f \in A) = \Pi(A)$ and so on as the data arrive and $\Pi(\mathrm{d}f)$ is updated. It is also clear that f refers to the density or distribution function generating the data.

Now let us expand on $\Pi(\mathrm{d}f)$ a bit more. The sequence of guesses described in Section 13.1.2 and immediately above are precisely that, namely guesses as to the density generating the next piece of data; there is no certainty or the need for these guesses to actually ever coincide with the true density. For if $m_0(x)$ is to be the first best guess at the true density, which is what it is, then this tells us exactly what thought processes need to be adopted in order to construct $\Pi(\mathrm{d}f)$. The best f, call it f^*, once Ω has been decided, is the one which minimizes the Kullback–Leibler divergence [26] between f and f_0.[7] In this case, the best $\Pi(\mathrm{d}f)$ is the one with point mass one at f^*. From this we can see that one should be expressing beliefs about the location of f^* when one is constructing $\Pi(\mathrm{d}f)$. This must be the case to ensure that m_0 is the best guess. Thus (m_0, Π) form a pair; m_0 is to be a best guess for the density f_0 and, to attain this, it must be that Π is constructed with f^* as the target, since the best Π would be a point mass at f^*. Of course, f^* is a real density and therefore it is possible to write probability statements about its location. We are now discussing ideas where we are not differentiating between $f_0 \in \Omega$ or $f_0 \notin \Omega$. If $f_0 \in \Omega$ then $f^* = f_0$. The idea of thinking in terms of best guesses suggests that Bayes works either way, not through the Bayes theorem directly, but via de Finetti's representation theorem. Bayesians are asked to express uncertainty about densities, or parameters indexing density functions. But unless one has one of them as a target in mind, then probabilities such as $\Pi(f \in A)$ make no sense. For what is f when, as is the norm, $f_0 \notin \Omega$? The target is f^*. This is what needs to be learnt about. Hence, asymptotic studies need to establish that

[7] Other metrics or divergences are possible but the mathematics and the unique role the Kullback–Leibler divergence has with Bayesian theory make it the most suitable.

the sequence of posterior distributions move to f^*. In the literature this is typically done in two parts; the first to assume $f_0 \in \Omega$ and the second to assume $f_0 \notin \Omega$.

Formally, the statement is: if $f(x|\theta)$ is a model for $f_0(x)$ and $\pi(\theta)$ a prior for θ, then the update $\pi(\theta|X=x) \propto f(x|\theta)\,\pi(\theta)$, with $X \sim F_0$, is an appropriate update when interest is in learning about the value which minimizes $U(\theta) = -\int \log f(x|\theta)\, F_0(\mathrm{d}x)$. And this learning takes the process to θ^* which minimizes $U(\theta)$; see, for example, [3] and [8].

13.4 Non-i.i.d data

Our approach is to construct nonparametric models from parametric models in the same way as in the i.i.d case. So, we take a joint density, as it would arise in the non-i.i.d case, and construct a mixture model from it. For example, if we have a parametric regression model with y as the dependent variable and x as the independent variable, then we write a parametric joint density for (y, x) as $K_\theta(y, x)$.

The nonparametric version is then given by

$$f(y,x) = \sum_{j=1}^{\infty} w_j\, K_{\theta_j}(y,x)$$

from which the regression model is available as

$$f(y|x) = \frac{\sum_{j=1}^{\infty} w_j\, K_{\theta_j}(y,x)}{\sum_{j=1}^{\infty} w_j\, K_{\theta_j}(x)},$$

where $K_\theta(x)$ is the marginal for x from the joint $K_\theta(y, x)$. This, and a similar idea for time series, we believe leads to a well motivated class of nonparametric non-i.i.d models.

13.4.1 Time series

Since the work in [31], who introduced the mixture of Dirichlet process mixture model, and the advent of Bayesian posterior inference via simulation techniques (see [14], and [45]), Bayesian nonparametric methods have developed at a rapid pace and the Dirichlet mixture model is one of the most popular among these methods. The models have now moved away from the standard setup, namely i.i.d observations, to cover more complex data structures involving regression data. There are numerous works and papers over the last decade and it is therefore convenient to cite the book of [23] which contains references and discussions of many nonparametric models.

To set the notation, for the i.i.d case, assume $(y_1, \ldots, y_n)$ are the data. The mixture of Dirichlet process model takes the form

$$f(y) = \int k(y|\theta)\, \mathrm{d}P(\theta)$$

where $k(\cdot|\theta)$ is a density for all $\theta \in \Theta$ and P is a distribution function on Θ. If the prior for P is assigned as a Dirichlet process then, according to [43] we can construct

$$P = \sum_{j=1}^{\infty} w_j\,\delta_{\theta_j}$$

where the weights form a stickbreaking sequence and the (θ_j) are i.i.d from density function $g(\theta)$. And so, specifically for i.i.d (v_j) from a beta$(1, c)$ density, for some $c > 0$, $w_1 = v_1$ and, for $j > 1$, $w_j = v_j \prod_{l<j}(1 - v_l)$. Other stickbreaking constructions are allowed based on alternative beta distributions; see [24] for conditions.

Thus, the density model for the data arises as

$$f(y) = \sum_{j=1}^{\infty} w_j\,k(y|\theta_j)$$

which is an infinite mixture model; see [18] for a recent review of MCMC methods to implement these types of models.

The mixture of Dirichlet process model is now being regularly employed in regression problems. The idea now is to model dependent variable y on regression variable x, as

$$f(y|x) = \int k(y|\theta)\,\mathrm{d}P_x(\theta)$$

Here $P_x(\theta)$ is similar in construction to $P(\theta)$, but the weights and locations can both depend on x. That is

$$P_x = \sum_{j=1}^{\infty} w_j(x)\,\delta_{\theta_j(x)}$$

Exactly how to define the $(w_j(x), \theta_j(x))$ suitably is an interesting problem and a number of attempts have been tried; see [13] for a review. To some extent the complexity of how to construct these key functions over the x–space can be avoided by modelling the joint density as a mixture model:

$$f(y, x) = \int k(y, x|\theta)\,\mathrm{d}P(\theta)$$

But in this case the correct regression model is

$$f(y|x) = \frac{f(y, x)}{f(x)} = \frac{\int k(y, x|\theta)\,\mathrm{d}P(\theta)}{\int k(x|\theta)\,\mathrm{d}P(\theta)}$$

Such a model has not yet been entertained due to the difficulty of dealing with the denominator.

Perhaps one area that has not been fully exploited from a Bayesian nonparametric perspective and which involves the Dirichlet mixture model is time series data. For ease of exposition here we will only consider first-order time series data and models. The plan then is to construct a transition density $f(y|x)$ which would suitably capture the transition dynamics and then study the posterior distribution based on the likelihood function

$$\prod_{i=1}^{n} f(y_i|y_{i-1}).$$

A prior is assigned to the transition density and will be written as $\Pi(df)$. This will be based on the Dirichlet process mixture model.

The modelling of first-order time series data is also a vast area in the literature. Our aim is to use a Bayes nonparametric mixture model to construct $f(\cdot|\cdot)$; so, specifically, as with the regression setting,

$$f(y|x) = \int k(y|x,\theta)\,\mathrm{d}P_x(\theta),$$

where $k(y|x,\theta)$ is a density for every $\theta \in \Theta$ and P_x is a probability measure that depends on x. This type of model will be able to capture a wide class of transition functions, and the infinite-dimensional aspect to the model means that any surprise or change that arises in the future will be taken into account.

Let us start with a parametric first-order stationary time series model $k(y,x|\theta)$ which has identical marginals;

$$f(x|\theta) = \int k(y,x|\theta)\,\mathrm{d}y \quad \text{and} \quad f(y|\theta) = \int k(y,x|\theta)\,\mathrm{d}x.$$

Also, $f(y|\theta)$ is the stationary density:

$$f(y|\theta) = \int k(y|x,\theta) f(x|\theta)\,\mathrm{d}x.$$

This is now ready to be extended to the nonparametric setting by taking

$$k(y,x) = \int k(y,x|\theta)\,\mathrm{d}P(\theta) = \sum_{j=1}^{\infty} w_j\, k(y,x|\theta_j).$$

We can obtain the nonparametric model using the following specification for $P_x(\theta)$: the stationary density is

$$f(x) = \int k(x|\theta)\,\mathrm{d}P(\theta)$$

and the transition density is

$$f(y|x) = \frac{\int k(y|x,\theta)\,k(x|\theta)\,\mathrm{d}P(\theta)}{\int k(x|\theta)\,\mathrm{d}P(\theta)}$$

so

$$\mathrm{d}P_x(\theta) = \frac{k(x|\theta)\,\mathrm{d}P(\theta)}{\int k(x|\theta)\,\mathrm{d}P(\theta)}.$$

Equivalently,

$$w_j(x) = \frac{w_j\, k(x|\theta_j)}{\sum_{j=1}^{\infty} w_j\, k(x|\theta_j)}.$$

This model differs from recent approaches that have appeared in the literature; see for example [34]. These authors took the joint density $f(y, x)$ as

$$f(y, x) = \int p(x)\, p(y)\, \Pi(\mathrm{d}p)$$

where $\Pi(\mathrm{d}p)$ is a Bayesian nonparametric prior; specifically, [34] took it to be based on the Gaussian process prior of [30] and [28], [29]. This model results in a nonparametric transition density but only has a parametric stationary density, given by

$$f(x) = \int p(x)\, \Pi(\mathrm{d}p).$$

On the other hand the transition density is the predictive density function given a single observation from the Bayesian model, i.e.

$$f(y|x) = \int p(y)\, \Pi(\mathrm{d}p|x).$$

This can be nonparametric since the $\Pi(\mathrm{d}p|x)$ will be a probability measure that can accommodate two functions; one being the mean density $f(x)$ and another to do with the variance process, labelled $\tau(x)$, and which will be based on $\int p^2(x)\Pi(\mathrm{d}p)$. Then $f(y|x)$ will be a function of $(f(x), \tau(x))$. The current work is about obtaining nonparametric forms for both the stationary density and the transition density.

For the new model described above, we will need to estimate the parameters of k and P. To make this concrete we will present a particular model. Assume $k(x|\theta, \sigma^2)$ to be normal with mean θ and variance σ^2 and let

$$P(\theta) = \sum_{j=1}^{\infty} w_j\, \delta_{\theta_j}(\theta).$$

We will allow the means to change with component but keep the variance the same across components. Here the weights (w_j) sum to one and the (θ_j) are real numbers. We will then be interested in estimating $((w_j, \theta_j), \sigma)$. The prior for $\lambda = \sigma^{-2}$ will be denoted $\pi(\lambda)$.

Now it can be seen that the likelihood function based on a sample $(y_1, \ldots, y_n)$ is given by

$$\prod_{i=1}^{n} \frac{\sum_{j=1}^{\infty} w_j\, k(y_i|y_{i-1}, \theta_j, \lambda)\, k(y_{i-1}|\theta_j, \lambda)}{\sum_{j=1}^{\infty} w_j\, k(y_{i-1}|\theta_j, \lambda)}.$$

This looks an insurmountable likelihood to deal with. Our aim then is to show how to undertake Bayesian inference for this model using well designed latent variables which result in a viable latent model.

The numerator has a common and standard technique for simplification and this is to introduce the allocation variables (d_i) which lead to the latent model

$$\prod_{i=1}^{n} \frac{w_{d_i}\, k(y_i|y_{i-1}, \theta_{d_i}, \lambda)\, k(y_{i-1}|\theta_{d_i}, \lambda)}{\sum_{j=1}^{\infty} w_j\, k(y_{i-1}|\theta_j, \lambda)}.$$

Summing over the independent (d_i) returns the original likelihood. The issue now is to deal with the denominator.

First we can remove the λ from each k; so define

$$m(y|\theta,\lambda) = \exp\left\{-\tfrac{1}{2}\lambda(y-\theta)^2\right\}.$$

Now we write the latent likelihood model as

$$\lambda^{n/2}\prod_{i=1}^{n}\frac{w_{d_i}\, m(y_i|y_{i-1},\theta_{d_i},\lambda)\, m(y_{i-1}|\theta_{d_i},\lambda)}{\sum_{j=1}^{\infty} w_j\, m(y_{i-1}|\theta_j,\lambda)}.$$

We now focus on the denominator and the term

$$\frac{1}{\sum_{j=1}^{\infty} w_j\, m(y|\theta_j,\lambda)}.$$

This has been written with a generic y, and it is simpler to consider this first, and then put the product back together later.

Since the denominator is now between (0, 1) we can write it as

$$\sum_{k=0}^{\infty}\left[\sum_{j=1}^{\infty} w_j\,\left(1-m(y|\theta_j,\lambda)\right)\right]^k.$$

This suggests we should introduce the latent variable k yielding the latent model

$$\left[\sum_{j=1}^{\infty} w_j\,\left(1-m(y|\theta_j,\lambda)\right)\right]^k.$$

Finally, we can introduce the latent variables $(z_l : l = 1,\ldots,k)$ and the latent model

$$\prod_{l=1}^{k} w_{z_l}\left(1-m\left(y|\theta_{z_l},\lambda\right)\right).$$

Putting this with the latent model for the numerator, and recalling we have a k_i for each i, the final latent model is given by:

$$\lambda^{n/2}\prod_{i=1}^{n} w_{d_i} k\left(y_i|y_{i-1},\theta_{d_i},\lambda\right)\, k\left(y_{i-1}|\theta_{d_i},\lambda\right)\prod_{l=1}^{k_i} w_{z_{il}}\left(1-m\left(y_{i-1}|\theta_{z_{il}},\lambda\right)\right).$$

It is easy to see that summing over all the latent variables $((d_i),(k_i),(z_{il}))$ over their respective spaces returns the original likelihood.

We are now in a position where the latent model is similar to standard latent models for mixture of Dirichlet process models; see [25]. The form above suggests that there is a solution to the problem. Hence, we start to describe the sampling MCMC algorithm. As it stands, if we attempted to sample the d_i or the z_{il} we would face the problem that they are to be taken from the positive integers, and it would not be possible to evaluate all the relevant probabilities. We can therefore truncate this choice using ideas from [25] whereby we introduce latent variables δ_i and ζ_i which are combined with the latent model via

$$\mathbf{1}\left(\delta_i < e^{-\xi d_i}\right) e^{\xi d_i} \quad \text{and} \quad \mathbf{1}\left(\zeta_{il} < e^{-\xi z_{il}}\right) e^{\xi z_{il}}.$$

Here $\xi > 0$ and its value is not a modelling issue. A discussion on its choice and its role is given in [25]. Therefore,

$$\mathrm{P}(d_i = j|\cdots) \propto e^{\xi j}\, w_j\, k(y_i|y_{i-1}, \theta_j, \lambda)\, k(y_{i-1}|\theta_j, \lambda)\, \mathbf{1}(1 \le j \le N_i)$$

where $N_i = \lfloor -\xi^{-1} \log \delta_i \rfloor$. Also,

$$\mathrm{P}(z_{il} = j|\cdots) \propto e^{\xi j}\, w_j\, \left(1 - m(y_{i-1}|\theta_j, \lambda)\right)\, \mathbf{1}(1 \le j \le N_{il})$$

where $N_{il} = \lfloor -\xi^{-1} \log \zeta_{il} \rfloor$. The maximum value $N = \max\{N_i, N_{il}\}$ will then tell us exactly how many of the (θ_j, w_j) need to be sampled at each iteration of the MCMC algorithm. The weights are easy to sample and the conditional for each v_j is a straightforward extension of the usual mixture of Dirichlet process model, and is given by

$$v_j = \text{beta}\left(1 + \sum_{i=1}^{n} \mathbf{1}(d_i = j) + \sum_{i=1,l=1}^{n,k_i} \mathbf{1}(z_{il} = j),\; c + \sum_{i=1}^{n} \mathbf{1}(d_i > j) + \sum_{i=1,l=1}^{n,k_i} \mathbf{1}(z_{il} > j)\right)$$

These (v_j) can then be transformed to get the (w_j).

The (θ_j) are best sampled by introducing a latent variable u_{il} for each i and l. This enters the model via

$$\mathbf{1}(u_{il} < 1 - m(y_{i-1}|\theta_{z_{il}}, \lambda)).$$

These are standard slice random variables; see [9]. Hence,

$$p(\theta_j|\cdots) \propto \pi(\theta_j) \prod_{d_i=j} m(y_i|y_{i-1}, \theta_j, \lambda)\, m(y_{i-1}|\theta_j, \lambda) \prod_{z_{il}=j} \left(m(y_{i-1}|\theta_j, \lambda) < 1 - u_{il}\right).$$

The conditional for λ is given by

$$p(\lambda|\cdots) \propto \lambda^{n/2} \prod_{i=1}^{n} m(y_i|y_{i-1}, \theta_{d_i}, \lambda)\, m(y_{i-1}|\theta_{d_i}, \lambda) \prod_{l=1}^{k_i} \left(m(y_{i-1}|\theta_{z_{il}}, \lambda) < 1 - u_{il}\right).$$

Finally, we need to update each k_i. We do this independently and so we consider a generic k with relevant model part given by

$$\prod_{l=1}^{k} w_{z_l}\, \psi_{z_l}$$

where we have written $\psi_{z_l} = 1 - m(y|\theta_{z_l}, \lambda)$.

We deal with this apparent changing dimension part of the model using ideas in [20], which is based on the reversible jump MCMC methodology of [21]. If we write

$$p(k, z_1, \ldots, z_k) \propto \prod_{l=1}^{k} w_{z_l}\, \psi_{z_l}$$

then we extend the model to

$$p(k, z_1, \ldots, z_k, z_{k+1}, \ldots) \propto \left\{ \prod_{l=1}^{k} w_{z_l}\, \psi_{z_l} \right\} \prod_{l=k+1}^{\infty} w_{z_l}.$$

From k we can propose a move to $k+1$ with probability $\frac{1}{2}$, or to $k-1$ with probability $\frac{1}{2}$. The probability of accepting a move to $k+1$ is given by

$$\min\left\{1, \psi_{z_{k+1}}\right\}$$

where z_{k+1} has been sampled from the weights (w_j). On the other hand, the probability of accepting a move to $k-1$ is given by

$$\min\left\{1, \psi_{z_k}^{-1}\right\}.$$

13.4.2 Regression models

There has been a significant amount of recent research on Bayesian nonparametric regression models. This has primarily focused on developing models of the form

$$f(y|\mathbf{x}) = \sum_{j=1}^{\infty} w_j(\mathbf{x})\, K(y|\mathbf{x}, \theta_j(\mathbf{x}))$$

where $K(y|\mathbf{x}, \boldsymbol{\theta})$ is a chosen parametric density function. The $w_j(\mathbf{x})$ are mixture weights that sum to 1 at every value of the covariate vector $\mathbf{x} \in \mathcal{X}$, and with a prior distribution on weights $\{w_j(\mathbf{x})\}_{j=1,2,\ldots}$, and atoms $\{\boldsymbol{\theta}_j(\mathbf{x})\}_{j=1,2,\ldots}$, which are an infinite collection of processes indexed by $\mathcal{X}$. Our position is that it is a very difficult task to specify these components; i.e. the $w_j(\mathbf{x})$ and $K(y|\mathbf{x}, \boldsymbol{\theta})$. There are limitless possibilities and over-fitting and un-identifiability are serious issues. It is argued that some sort of guidance is needed in order to justify certain specifications.

An attractively simpler and intuitive approach to Bayesian nonparametric regression has been proposed by [35]. The idea is to specify a Dirichlet Process mixture model for the joint density $f(y, \mathbf{x})$ with mixture weights and atoms independent of $\mathbf{x}$. This would lead to a standard infinite mixture model, treating the $(y_i, \mathbf{x}_i)$ as i.i.d observations. This would not be a controversial choice for a model. In this case one would employ the likelihood function

$$\prod_i f(y_i, \mathbf{x}_i)$$

as used by [35].

However, the aim is regression rather than modelling the $(y, \mathbf{x})$ and hence the appropriate likelihood function is given by

$$\prod_i f(y_i, \mathbf{x}_i)/f(\mathbf{x}_i) = \prod_i f(y_i|\mathbf{x}_i).$$

To see this we simply note that there are two likelihood functions here and they are not the same. It is also clear that for regression purposes we are interested in the conditional density $f(y|\mathbf{x})$, and it is this that should form the basis of the likelihood function.

The reason why such a simple, motivated and useful regression model has not appeared in the literature is due to the fact that the posterior distribution has an intractable normalizing constant, i.e. $\prod_i f(\mathbf{x}_i)$. Inference is complicated by the need to evaluate the uncomputable integrals in this constant. It is possible to avoid this complication by proposing a modified prior distribution such that, when combined with the correct likelihood, it yields a posterior distribution that is identical to the posterior of the original model.

The aim here is to show that it is possible to use the correct likelihood for regression by showing how to deal with the problem of the normalizing constant. This uses ideas recently introduced in [48]. The details of the model and the MCMC used to estimate the model are provided in the companion article to this paper: [49].

13.5 Consistency

We will first discuss this issue assuming i.i.d observations. If we are operating with densities, then to confirm that Bayesian learning is about the true sampling density, we need to show that the posterior mass accumulates in suitable neighbourhoods about f_0. The appropriate metric to define neighbourhoods is the Hellinger distance, since the mathematics is amenable to this distance.

The Hellinger distance between densities f_1 and f_2 is defined as

$$d_H(f_1, f_2) = \left\{ \int \left(\sqrt{f_1} - \sqrt{f_2} \right)^2 \right\}^{1/2}$$

We will also use $d(f_1, f_2) = \frac{1}{2} d_H(f_1, f_2)^2$ which is bounded by 1, and specifically

$$d(f_1, f_2) = 1 - \int \sqrt{f_1 f_2}$$

The aim then is to find conditions on the prior Π which ensure that

$$\Pi_n(A_\epsilon) = \Pi(A_\epsilon | X_1, \ldots, X_n) = \frac{\int_{A_\epsilon} R_n(f)\, \Pi(df)}{\int R_n(f)\, \Pi(df)} \to 0 \text{ a.s.}$$

for all $\epsilon > 0$, where

$$R_n(f) = \prod_{i=1}^{n} f(X_i)/f_0(X_i)$$

and

$$A_\epsilon = \{f : d_H(f_1, f_2) > \epsilon\}$$

It was first shown in [44] that if the prior puts positive mass on all Kullback–Leibler neighbourhoods of f_0, then the posterior is consistent with respect to the weak topology. This result does not extend to neighbourhoods defined by the Hellinger metric since a counterexample is provided in [1]. The condition found by [51] deals with the denominator $I_n = \int R_n(f)\,\Pi(df)$. If $\Pi\{f : d_K(f_0,f) < \delta\} > 0$ for all $\delta > 0$, where $d_K(f,g) = \int f\,\log(f/g)$ is the Kullback–Leibler divergence, then $I_n > e^{-nc}$ a.s. for all large n for any $c > 0$.

To establish strong consistency, a condition is required for the numerator $L_n = \int_{A_\epsilon} R_n(f)\,\Pi(df)$ to ensure that $L_n < e^{-nd}$ a.s. for all large n for some $d > 0$. We can then establish posterior consistency. Let us first describe the popular approaches. The basic idea is to find a sieve with certain properties. If the sieve is denoted $\mathcal{F}_n$ then there are two conditions:

1. $\log N(\mathcal{F}_n, d_H, \delta) < nd$ for all large n, for some $d > 0$ for all $\delta > 0$. Here $N(A, d, \delta)$ denotes the number of balls of size δ with respect to metric d to cover set A.
2. $\Pi(\mathcal{F}_n^c) < e^{-nb}$ for all large n for some $b > 0$.

A proof of this is provided by [19]. [2] employs a slightly different version of condition 1. which uses a different measure of entropy. Hence, a specialist task of 'ball counting' is required. When counting balls is tricky, and imprecise bounds are used in the counting, then one can anticipate that stronger conditions on Π are suggested than are actually required.

An approach that avoids the need to count balls or work directly with entropies is as follows: There is a sieve and it does satisfy condition 1. Counting is obviated, because the sieve is constructed using the prior itself. So define the sieve

$$\mathcal{F}_n = \left\{A_j : \Pi(A_j) > e^{-nb}\right\}$$

for some $b > 0$, where the (A_j) are a partition of the set of densities such that two elements in the same A_j are no more than a Hellinger distance δ apart.

This sieve does the counting automatically; specifically, the Π does the counting, since

$$1 = \sum_j \Pi(A_j) > \sum_{A_j \in \mathcal{F}_n} \Pi(A_j) > |\mathcal{F}_n|\, e^{-nb}$$

and so $|\mathcal{F}_n| < e^{nb}$ as required. Thus there is only the need to establish condition 2. for this sieve. But

$$\Pi(\mathcal{F}_n^c) = \sum_{A_j \in \mathcal{F}_n^c} \Pi(A_j) = \sum_{A_j \in \mathcal{F}_n^c} \Pi(A_j)^\alpha\, \Pi(A_j)^{1-\alpha}$$

for any $0 < \alpha < 1$. Hence,

$$\Pi(\mathcal{F}_n^c) < e^{-n\alpha b} \sum_j \Pi(A_j)^{1-\alpha}$$

and so condition 2 is satisfied when the prior satisfies

$$\sum_j \Pi(A_j)^{1-\alpha} < +\infty$$

for some $\alpha \in (0, 1)$. An alternative proof choosing $\alpha = \frac{1}{2}$ is available. Hence this condition is only dealing with the prior mass on the complement of the sieve being suitably small.

One could, and should, think about the scenario when the prior does not put positive mass on all Kullback–Leibler neighbourhoods of f_0. One would have something like

$$\Pi \left\{ f : d_K(f_0, f) < \delta \right\} > 0$$

only for all $\delta > \delta_1$ for some $\delta_1 > 0$. With this, one can now only demonstrate that the denominator $I_n > e^{-nc}$ a.s. for all large n for $c > \delta_1$. The problem now is that the numerator, which works with the Hellinger distance, has to be bounded above by e^{-nd} for $d > \delta_1$. But d will be connected with a Hellinger distance and δ_1 will be connected with a Kullback–Leibler divergence. Thus, it makes sense to connect these up with a sequence of distances and these can be found in the family of α divergences:

$$d_\alpha(f_0, f) = \alpha^{-1} \left(1 - \int f_0^{1-\alpha} f^\alpha \right)$$

When $\alpha = \frac{1}{2}$, we recover the Hellinger distance. As $\alpha \to 0$ we will recover the Kullback–Leibler divergence. Working with these divergences was accomplished in [10].

13.6 Summary

Large models absorb all possible types of uncertainty that arise from data. This is pure Bayesian inference. There is no need to entertain a large number of small models in an attempt to determine which one performs best given the data. Statistical approaches that promote this latter idea are ad hoc, and are not Bayes.

But as discussed in this article, large models are not an unmixed blessing. Nonetheless, the point of Section 13.3 is that if interest is in the $f^* \in \Omega$ which minimizes

$$U(f) = -\int \log f(x)\, F_0(dx),$$

where F_0 denotes the true distribution function, then not only is f^* the appropriate density to target and learn about, but the Bayes machinery indeed works. Formally, there is a motivation to update

$$\Pi(df) \quad \text{to} \quad \Pi(df|X = x) \propto f(x)\, \Pi(df)$$

when $X \sim F_0$. Importantly, the mathematics demonstrated that learning is indeed about f^*.

This setup is also quite explicit in likelihood-based inference, since an estimate for f would follow from approximating $U(f)$, given a sample of size n, by

$$U_n(f) = -n^{-1} \sum_{i=1}^{n} \log f(X_i)$$

Minimizing this yields the Maximum Likelihood Estimator. Thus the MLE is also really about finding f^*. The problem with acknowledging this latter fact is that properties of estimators, such as unbiasedness, are unavailable.

We have discussed a certain type of large model for i.i.d data, namely the popular mixture model based on stickbreaking processes. We then showed how this structure can be extended to cover non-i.i.d data, such as time series and regression models. These latter extensions require the calculation of a troublesome and unavoidable normalizing constant in order to do full Bayesian inference. Using a novel combination of latent models and MCMC techniques, we showed that it is possible to satisfactorily provide complete Bayesian inference even in the non-i.i.d case.

References

[1] Barron, A. (1988). The exponential convergence of posterior probabilities with implications for Bayes estimators of density functions. Unpublished manuscript.

[2] Barron, A., Schervish, M. J. and Wasserman, L. (1999). The consistency of posterior distributions in nonparametric problems. *Annals of Statistics*, **27**, 536–561.

[3] Berk, R. H. (1966). Limiting behaviour of posterior distributions when the model is incorrect. *Annals of Mathematical Statistics*, **37**, 51–58. [Corrigendum **37**, 745–746.]

[4] Bernardo, J. M. and Smith, A. F. M. (1994). *Bayesian Theory*. Wiley.

[5] Besag, J. and Green, P. J. (1993). Spatial statistics and Bayesian computation. *Journal of the Royal Statistical Society, Series B*, **55**, 25–37.

[6] Box, G. E. P. and Tiao, G. C. (1973). *Bayesian Inference in Statistical Analysis*. Addison–Wesley.

[7] Box, G. E. P. (1980). Sampling and Bayes' inference in scientific modelling and robustness. *Journal of the Royal Statistical Society, Series A*, **143**, 383–430.

[8] Bunke, O. and Milhaud, X. (1998). Asymptotic behaviour of Bayes estimates under possibly incorrect models. *Annals of Statistics*, **26**, 617–644.

[9] Damien, P., Wakefield, J. C. and Walker, S. G. (1999). Gibbs sampling for Bayesian non-conjugate and hierarchical models using auxiliary variables. *Journal of the Royal Statistical Society, Series B*, **61**, 331–344.

[10] De Blasi, P. and Walker, S. G. (2012). Bayesian asymptotics with misspecified models. To appear in *Statistica Sinica*.

[11] de Finetti, B. (1937). La prévision: ses lois logiques, ses sources subjectives. *Ann. Inst. H. Poincaré*, **7**, 1–68.

[12] Doob, J. L. (1949). Application of the theory of martingales. In *Le Calcul des Probabilités et ses Applications*, Colloques Internationaux du Centre National de la Recherche Scientifique, **13**, 23–37. Paris: CNRS.

[13] Dunson, D. B. (2010). Nonparametric Bayes applications to biostatistics. In *Bayesian Nonparametrics*, Hjort *et al.* (Eds.), Cambridge University Press.

[14] Escobar, M. D. (1988). Estimating the means of several normal populations by nonparametric estimation of the distribution of the means. Unpublished PhD dissertation, Department of Statistics, Yale University.

[15] Escobar, M. D. and West, M. (1995). Bayesian density estimation and inference using mixtures. *Journal of the American Statistical Association*, **90**, 577–588.

[16] Ferguson, T. S. (1973). A Bayesian analysis of some nonparametric problems. *Annals of Statistics*, **1**, 209–230.

[17] Fernandez, C. and Steel, M. F. J. (1998). On Bayesian modelling of fat tails and skewness. *Journal of the American Statistical Association*, **93**, 359–371.

[18] Griffin, J. E. and Holmes, C. C. (2010). Computational issues arising in Bayesian nonparametric hierarchical models. In *Bayesian Nonparametrics*, Hjort *et al.* (Eds.), Cambridge University Press.

[19] Ghosal, S., Ghosh, J. K. and Ramamoorthi, R. V. (1999). Posterior consistency of Dirichlet mixtures in density estimation. *Annals of Statistics*, **27**, 143–158.

[20] Godsill, S. J. (2001). On the relationship between Markov chain Monte Carlo methods for model uncertainty. *Journal of Computational and Graphical Statistics*, **10**, 230–248.
[21] Green, P. J. (1995). Reversible jump Markov chain Monte Carlo computation and Bayesian model determination. *Biometrika*, **82**, 711–732.
[22] Hewitt, E. and Savage, L. J. (1955). Symmetric measures on Cartesian products. *Transactions of the American Mathematical Society*, **80**, 470–501.
[23] Hjort, N. L., Holmes, C. C., Müller, P. and Walker, S. G. (2010). *Bayesian Nonparametrics*. Cambridge University Press.
[24] Ishwaran, H. and James, L. F. (2001). Gibbs sampling methods for stick–breaking priors. *Journal of the American Statistical Association*, **96**, 161–173.
[25] Kalli, M., Griffin, J. E. and Walker, S. G. (2010). Slice sampling mixture models. *Statistics and Computing*, **21**, 93–105.
[26] Kullback, S. and Leibler, R. A. (1951). On information and sufficiency. *Annals of Mathematical Statistics*, **22**, 79–86.
[27] Lee, P. M. (2004). *Bayesian Statistics* (3rd Edition). Arnold.
[28] Lenk, P. J. (1988). The logistic normal distribution for Bayesian, nonparametric, predictive densities. *Journal of the American Statistical Association*, **83**, 509–516.
[29] Lenk, P. J. (1991). Towards a practicable Bayesian nonparametric density estimator. *Biometrika*, **78** 531–543.
[30] Leonard, T. (1978). Density estimation, stochastic processes and prior information (with discussion). *Journal of the Royal Statistical Society, Series B*, **40**, 113–146.
[31] Lo, A. Y. (1984). On a class of Bayesian nonparametric estimates I. Density estimates. *Annals of Statistics*, **12**, 351–357.
[32] MacEachern, S. N. (1994). Estimating normal means with a conjugate style Dirichlet process prior. *Communications in Statistics: Simulation and Computation*, **23**, 727–741.
[33] MacEachern, S. N. and Müller, P. (1998). Estimating mixture of Dirichlet process models. *Journal of Computational and Graphical Statistics*, **7**, 223–338.
[34] Mena, R. H. and Walker, S. G. (2005). Stationary models via a Bayesian nonparametric approach. *Journal of Time Series Analysis*, **26**, 789–805.
[35] Müller, P., Erkanli, A. and West, M. (1996). Bayesian curve fitting using multivariate normal mixtures. *Biometrika*, **83**, 67–79.
[36] Neal, R. M. (2000). Markov chain sampling methods for Dirichlet process mixture models. *Journal of Computational and Graphical Statistics*, **9**, 249–265.
[37] O'Hagan, A. and Forster, J. J. (2004). *Bayesian Inference* (2nd edition). Arnold, London.
[38] Ongaro, A. and Cattaneo, C. (2004). Discrete random probability measures: a general framework for nonparametric Bayesian inference. *Statistics and Probability Letters*, **67**, 33–45.
[39] Papaspiliopoulos, O. and Roberts, G. O. (2008). Retrospective Markov chain Monte Carlo methods for Dirichlet process hierarchical models. *Biometrika*, **95**, 169–186.
[40] Richardson, S. and Green, P. J. (1997). On Bayesian analysis of mixtures with an unknown number of components. *Journal of the Royal Statistical Society, Series B*, **59**, 731–758.
[41] Robert, C. P. (2001). *The Bayesian Choice*. Springer Texts in Statistics (2nd Edition).
[42] Rodriguez, C. E. and Walker, S. G. (2011). Bayesian nonparametric mixture modelling with unimodal kernels. Submitted.
[43] Sethuraman, J. (1994). A constructive definition of Dirichlet priors. *Statistica Sinica*, **4**, 639–650.
[44] Schwartz, L. (1965). On Bayes procedures. *Z. Wahrsch. Verw. Gebiete*, **4**, 10–26.
[45] Smith, A. F. M. and Roberts, G. O. (1993). Bayesian computations via the Gibbs sampler and related Markov chain Monte Carlo methods. *Journal of the Royal Statistical Society, Series B*, **55**, 3–23.

[46] Walker, S. G. and Hjort, N. L. (2001). On Bayesian consistency. *Journal of the Royal Statistical Society, Series B*, **63**, 811–821.
[47] Walker, S. G. (2007). Bayesian inference via a minimisation rule. *Sankhya*, **68**, 542–553.
[48] Walker, S. G. (2011). Posterior sampling when the normalizing constant is unknown. *Communications in Statistics: Simulation and Computation*, **40**, 784–792.
[49] Walker, S. G. and Karabatsos, G. (2012). Revisiting Bayesian curve fitting using multivariate normal mixtures. This book.

14 Geometric weight priors and their applications

RAMSÉS H. MENA

14.1 Introduction

The introduction by Walker, Chapter 13 in this volume, stresses the importance of having a dependent sample so that learning about the distribution generating the observations can take place. Further, if one assumes that the nature of the phenomenon under study generates independent and identically distributed (i.i.d.) observations, the dependence in the sample needed for Bayesian learning reduces to exchangeability. Whether one shares the notion of a correct model or prefers to think of the Bayesian approach as implied by assuming exchangeability among the observations, the role of such a dependence property is apparent.

From de Finetti's representation theorem, a set of random variables $(X_i)_{i\geq 1}$ taking values in a Polish space $\mathbb{X}$, endowed with the Borel σ-field $\mathscr{X}$, is exchangeable if and only if

$$\mathbb{P}(X_1 \in A_1, \dots, X_n \in A_n) = \int_{\mathcal{P}_{\mathbb{X}}} \prod_{i=1}^{n} \mathrm{P}(A_i)\, \mathrm{Q}(\mathrm{dP}), \qquad A_i \in \mathcal{X}, \tag{14.1}$$

for any $n \geq 1$ and where $\mathcal{P}_{\mathbb{X}}$ denotes the set of probability measures on $(\mathbb{X}, \mathcal{X})$. An interpretation follows directly from (14.1); the unknown, say P, that separates the joint law of $(X_i)_{i\geq 1}$ into conditional i.i.d. measures is random and uniquely driven by Q. Therefore its relation to Bayesian statistics, and the importance of specifying such a distribution, Q, for a random probability measure (r.p.m.) P, is evident.

There are various approaches to define r.p.m.s; namely, via extensions of finite dimensional distributions [20, 34]; via the normalization of stochastic processes [20, 47]; through predictive distributions [45]; or by virtue of stickbreaking constructions [30, 51], etc.

Up to date reviews are found in Walker *et al.* [54] and Hjort *et al.* [29]. Each of these constructions provides a different motive, i.e. analytic, numerical, or are useful for specific applications or in extensions to non-exchangeable contexts. The canonical example is Ferguson's [20] Dirichlet process, whose different constructions and representations have in part served as the gateway for the above more general approaches. For a partition $(B_1, \dots, B_k)$ of $\mathbb{X}$ and a finite non-atomic measure $\alpha > 0$ on $(\mathbb{X}, \mathcal{X})$, Ferguson [20] defined the Dirichlet process $(\mathcal{D}_\alpha)$ as the stochastic process having finite-dimensional distributions $(\mathrm{P}(B_1), \dots, \mathrm{P}(B_k)) \sim \mathrm{Dir}(\alpha(B_1), \dots, \alpha(B_k))$, where $\mathrm{Dir}(a_1, \dots, a_k)$ denotes the Dirichlet distribution over the $(k-1)$-dimensional simplex.

Although other r.p.m.s have made an impact, the Dirichlet process is the benchmark for applications. This is perhaps due to its mathematical tractability, which is in part due to its Pólya urn representation [7], which tells us that the $\mathcal{D}_{\theta \mathrm{P}_0}$ can be seen as the limit of predictive distributions

$$\mathbb{P}(X_{n+1} \in \cdot \mid X_1, \ldots, X_n) = \frac{\theta}{\theta + n} \mathrm{P}_0(\cdot) + \frac{n}{\theta + n} \mathrm{P}_n(\cdot), \qquad n \geq 1 \tag{14.2}$$

with $X_1 \sim \mathrm{P}_0$, $\mathrm{P}_0(\cdot) := \alpha(\cdot)/\theta$, $\theta := \alpha(\mathbb{X})$ and $\mathrm{P}_n(\cdot) := n^{-1} \sum_{i=1}^{n} \delta_{X_i}(\cdot)$. More general urn representations typically involve complex weights which in turn are cumbersome to incorporate in MCMC algorithms [see for instance 36]. Indeed, most constructive approaches of r.p.m.s seek to generalize the Dirichlet process resulting in richer models, but at the same time posing additional complications when applying or studying them.

A general approach to define a discrete r.p.m. on $(\mathbb{X}, \mathcal{X})$ is via

$$\mathrm{P}(B) = \sum_{i=1}^{\infty} \mathrm{w}_i \, \delta_{\mathrm{Z}_i}(B), \quad B \in \mathcal{X} \tag{14.3}$$

where the weights w_i and locations Z_i are random, i.e. $\sum_i \mathrm{w}_i = 1$ a.s., and independent of $\mathrm{Z}_i \stackrel{\text{iid}}{\sim} \mathrm{P}_0$, with P_0 a non-atomic distribution on $(\mathbb{X}, \mathcal{X})$. This class of r.p.m.s is termed proper species sampling models [see 45]. It follows that $\mathbb{E}[\mathrm{P}] = \mathrm{P}_0$, thus calling such a term the prior guess at the shape of P, a key component when applying and studying (14.3).

Sethuraman [51] proved that if the weights have the following stickbreaking form

$$\mathrm{w}_1 = \mathrm{v}_1 \quad \text{and} \quad \mathrm{w}_i = \mathrm{v}_i \prod_{j=1}^{i-1} (1 - \mathrm{v}_j), \quad i \geq 2 \tag{14.4}$$

with $\mathrm{v}_i \stackrel{\text{iid}}{\sim} \mathrm{Beta}(1, \theta)$, then $\mathrm{P} \sim \mathcal{D}_{\theta \mathrm{P}_0}$. If instead $\mathrm{v}_i \stackrel{\text{ind}}{\sim} \mathrm{Beta}(1 - \sigma, \theta + i\sigma)$ with $\theta > -\sigma$ and $\sigma \in [0, 1)$, then one has the two parameter Poisson–Dirichlet r.p.m. [30].

Here we undertake a different rationale; instead of searching for a further generalization, we present a simpler r.p.m. that results in a robust choice of nonparametric prior which, due to its simplicity, turns out to be appealing for practical implementations and also is easily extendable to the dependent processes settings. The idea centres around the following model:

Definition 14.1 *Let $\mathcal{P}_{\mathbb{X}}$ be the set of probability measures on $(\mathbb{X}, \mathcal{X})$. We term a r.p.m. $\mathrm{P} \in \mathcal{P}_{\mathbb{X}}$ a geometric weights prior, denoted by $\mathcal{GWP}(a, b)$, if*

$$\mathrm{P(B)} = \lambda \sum_{i=1}^{\infty} (1 - \lambda)^{i-1} \delta_{\mathrm{Z}_i}(B), \quad B \in \mathcal{X} \tag{14.5}$$

with $\mathrm{Z}_i \stackrel{iid}{\sim} \mathrm{P}_0$ independent of $\lambda \sim Be(a, b)$, $a, b > 0$.

Indeed, it is a simpler object than the Dirichlet process, as its weights depend on only one Beta random variable, instead of an infinite number. In other words, the stick is always broken with the same Beta random variable.

Before entering the discussion about these kinds of r.p.m.s let us first review some well-known facts about the Dirichlet process and some other nonparametric priors. This will then allow us to grasp the main idea and the appealing features of the model in Definition 14.1.

14.1.1 Disentangling the total mass parameter of the Dirichlet process

An appealing feature of the Dirichlet process, and other r.p.m.s constructed as in (14.3), is the almost sure discreteness [see 6]. Indeed, from (14.2), we see that $\mathbb{P}[X_i = X_j] > 0$ for any $i \neq j$. This implies that $X^{(n)} := (X_1, \dots, X_n)$ contains $K_n \leq n$ distinct observations $(X_1^*, \dots, X_{K_n}^*)$ with corresponding frequencies $\mathbf{N}_{K_n} = (N_1, \dots, N_{K_n})$ such that $\sum_{j=1}^{K_n} N_j = n$. Hence, selecting an exchangeable sample of size n driven by $\mathcal{D}_\theta\, P_0$ induces a partition into groups $\{\mathcal{G}_1, \dots, \mathcal{G}_{K_n}\}$, $\mathcal{G}_j := \{i : X_i = X_j^*\}, j = 1, \dots, K_n$, each with probability

$$\mathbb{P}[\{K_n = k\} \cap \{N_1 = n_1, \dots, N_{K_n} = n_k\}] = \frac{\theta^k}{(\theta)_n} \prod_{i=1}^{k} (n_i - 1)! \qquad (14.6)$$

where $n_j = \#\mathcal{G}_j$ and $(\theta)_n := \theta(\theta + 1) \cdots (\theta + n - 1)$. Notice that K_n and $\mathbf{N}_{K_n}$ depend only on the clustering structure among the X_is and not their actual values.

The support of (14.6) is in bijection with the set of partitions of a set with n elements, $[n] = \{1, \dots, n\}$, here denoted by $\mathcal{P}_{[n]}$, and whose cardinality is given by the nth Bell number, B_n. Because of their symmetry, these probabilities, denoted by $\Pi_k^{(n)}(n_1, \dots, n_k)$, are known in the literature as exchangeable partition probability functions (EPPFs), and are widely used to model random partitions in areas such as population genetics [17], combinatorics [28], economics [2] and excursion theory [45], among others [see also 18]. In Bayesian nonparametric inference they can be applied to depict the clustering structure among observations in hierarchical mixture models [see 38] or to give robust solutions to species sampling problems [see 35].

Marginalizing (14.6) one obtains the prior probability on the number of distinct values

$$\mathbb{P}[K_n = k] = \frac{\theta^k}{(\theta)_n} |s(n, k)|, \qquad k = 1, \dots, n \qquad (14.7)$$

where $s(n, k)$ is the Stirling number of the first kind. Figure 14.1 illustrates the above probabilities, and clearly θ is highly informative on the number of groups.

Remark 1 *The total mass parameter θ corresponding to a Dirichlet process is highly informative on the number of groups.*

A popular application of discrete r.p.m.s is as building blocks in hierarchical mixture models. Following Lo [38], a random density can be constructed as

$$\begin{aligned} Y_i \mid X_i &\overset{\text{ind}}{\sim} f(Y_i \mid X_i) \\ X_i \mid P &\overset{\text{iid}}{\sim} P \\ P &\sim Q \end{aligned} \qquad (14.8)$$

where $f(\cdot \mid X)$ is typically a Lebesgue density and thus used to model continuous data. This means that we can model a set of $\mathbb{Y}$-valued observables $(Y_i)_{i \geq 1}$, through the random density

$$f(y) = \int_{\mathbb{X}} f(y \mid x)\, P(dx)$$

When $P \sim \mathcal{D}_\theta P_0$, and using representation (14.2), Escobar [13, 14] and Escobar and West [15] devised a MCMC algorithm for this model. [See also 4, 33, for other numerical approaches in

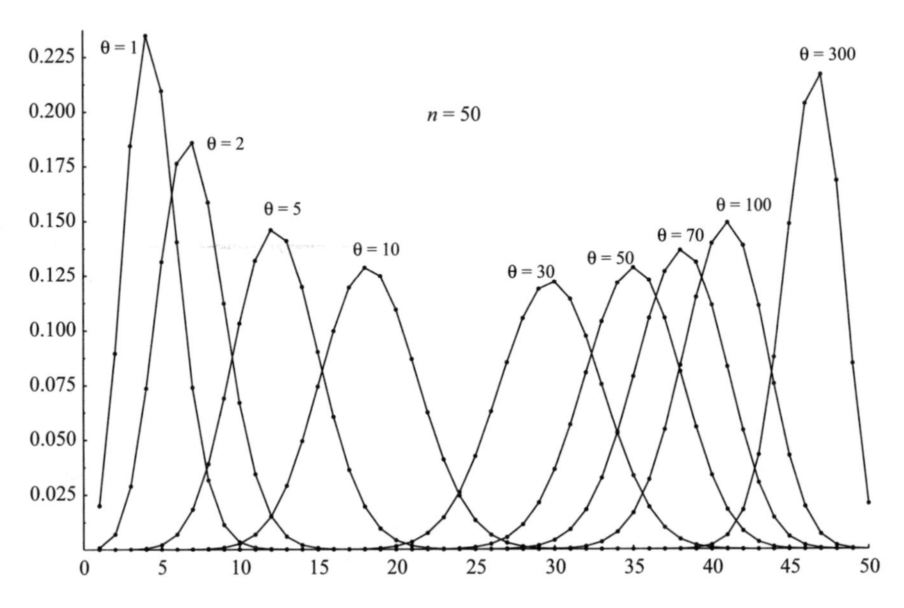

Figure 14.1 Prior probability on the number of different species corresponding to a sample of size 50 driven by $\mathcal{D}_{\theta P_0}$ as θ varies.

Bayesian nonparametrics.] Using representation (14.3) of a discrete r.p.m., the above model can be seen as the infinite mixture model

$$\mathrm{f}(y) = \int_{\mathbb{X}} f(y \mid x)\, \mathrm{P}(\mathrm{d}x) = \sum_{i=1}^{\infty} \mathrm{w}_i f(y \mid \mathrm{Z}_i) \tag{14.9}$$

with $\mathrm{Z}_i \overset{\text{iid}}{\sim} \mathrm{P}_0$. This clearly relates to the theory of mixture models used in model-based clustering [see 3] and density estimation problems. In fact, if we are interested in the clustering among the Y_is, it seems natural to think of it as being induced by a clustering at the latent level of the X_is. In particular, if we denote by $\mathrm{p}_n \in \mathcal{P}_{[n]}$ a given partition of $\mathrm{X}^{(n)}$, and thus of $\mathrm{Y}^{(n)}$, where each Y_i is i.i.d. from the random density (14.9), we can compute [see 37] the posterior probability

$$\mathbb{P}(\mathrm{p}_n \mid \mathrm{Y}^{(n)}) \propto \mathbb{P}(\mathrm{p}_n)\mathbb{P}(\mathrm{Y}^{(n)} \mid \mathrm{p}_n) \tag{14.10}$$

where

$$\mathbb{P}(\mathrm{p}_n) = \Pi_k^{(n)}(n_1, \ldots, n_k)$$

is the corresponding EPPF to P, e.g. expression (14.6) when $\mathrm{P} \sim \mathcal{D}_{\theta \mathrm{P}_0}$, and

$$\mathbb{P}(\mathrm{Y}^{(n)} \mid \mathrm{p}_n) = \prod_{j=1}^{k} \int_{\mathbb{X}} \prod_{i \in \mathcal{G}_j} f(y_i \mid x_j) \mathrm{P}_0(\mathrm{d}x_j), \tag{14.11}$$

sometimes termed the clustering likelihood of $\mathrm{Y}^{(n)}$ for a given partition $\{\mathcal{G}_1, \ldots, \mathcal{G}_k\}$. Although (14.10) simplifies for specific choices of f and P_0, direct evaluation through all its support, namely the subset $\mathcal{P}_{[n]}^k \subset \mathcal{P}_{[n]}$ of partitions of size k, is infeasible when n is large, since a Stirling number of the second type, $S_{n,k}$, of evaluations would be needed. Unlike the EPPF, the posterior distribution (14.10) is no longer exchangeable due to the effect of the kernel f in the observation's Y_is when evaluated in a particular group partition. Clearly, the problem amplifies when we are also interested in the number of groups as we would need to compute

$$\mathbb{P}[K_n = k \mid Y^{(n)}] \propto \sum_{p_n \in \mathcal{P}_{[n]}^k} \Pi_k^{(n)}(n_1, \ldots, n_k) \prod_{j=1}^{k} \int_{\mathbb{X}} \prod_{i \in \mathcal{G}_j} f(y_i \mid x_j) P_0(dx_j), \quad k = 1, \ldots, n. \tag{14.12}$$

Therefore, resorting to MCMC, or other kind of numerical methods, it is imperative for real applications. At the outset, building $\mathcal{P}_{[n]}$-valued Markov chains does not appear to be an easy task. But this is implicitly done with most MCMC methods based on Pólya urn schemes [cf. 15] by keeping track of the k and frequencies $(n_1, \ldots, n_k)$ at each iteration. See also Lau and Green [32] and the references therein for other approaches.

Note that when inferring about the clustering structure of $Y^{(n)}$ using quantities such as (14.10) and (14.12), the interpretation is different than that typically obtained in analyses based on finite mixture models [cf. 48]; under this latter approach the number of clusters is explicitly identified by the number of components in the mixture. Nonetheless, the former approach, based on the posterior probabilities on partitions, is also 'model based'.

Example 14.2 Consider the small dataset $y = (-1.522, -1.292, -0.856, -0.104, 2.388, 3.080, 3.313, 3.415, 3.922, 4.194)$, i.e. $n = 10$ with a $B_{10} = 115\,975$ possible groups and thus it is possible to evaluate (14.10) and (14.12). Assume the hierarchical model (14.8), with $P \sim \mathcal{D}_{\theta P_0}$,

$$f(\cdot \mid \mu, \xi) = N(\cdot \mid \mu, \xi^{-1}) \quad \text{and} \quad P_0(d\mu, d\xi) = N(\mu \mid 0, (\tau\,\xi)^{-1}) \mathrm{Ga}(\xi \mid \alpha, \beta) d\mu\, d\xi \tag{14.13}$$

If $S_j := \sum_{i \in \mathcal{G}_j} y_i^2 - n_j \bar{y}_j^2/(n_j + \tau)$ and $\bar{y}_j := n_j^{-1} \sum_{i \in \mathcal{G}_j} y_i$ then the likelihood (14.11) becomes

$$\prod_{j=1}^{k} \left\{ \frac{\tau}{n_j + \tau} \right\}^{\frac{1}{2}} \frac{\beta^{\alpha}\, \Gamma(\alpha + \frac{n_j}{2})}{(2\pi)^{\frac{n_j}{2}}\, \Gamma(\alpha) \left[\beta + \frac{S_j}{2}\right]^{\alpha + \frac{n_j}{2}}}.$$

Figure 14.2 shows that two groups is the intuitive choice. From Table 14.1 it is clear that θ also influences the posterior clustering probabilities. When $\theta = 0.5$ or $\theta = 1$, the posterior mode of (14.10) sits on $p = \{\{y_1, \ldots, y_4\}, \{y_5, \ldots, y_{10}\}\}$, with probabilities 0.522 and 0.332, respectively, whereas when $\theta = 5$ it sits on $p^* := \{\{y_1\}, \{y_2\}, \{y_3\}, \{y_4\}, \{y_5, \ldots, y_{10}\}\}$ with probability 0.0204. To some extent, these observations could have been predicted since $\mathbb{E}[K_n] = \sum_{i=1}^n \theta/(\theta + i - 1)$, which for $\theta = 0.5$, $\theta = 1$ and $\theta = 5$ yield 2.13, 2.93 and 5.84, respectively. Thus, slightly coinciding with the posterior results of Table 14.1. Clearly the effect of the prior could be diminished as the sample size n increases. ∎

Remark 2 *The total mass parameter θ corresponding to a Dirichlet process strongly influences the posterior inference on the number of groups in a nonparametric mixture model.*

Remarks 1 and 2 are not new, e.g. see Escobar and West [15]. Although such effects of an informative prior can be attenuated with other choices of r.p.m.s, such as the two-parameter Poisson–Dirichlet

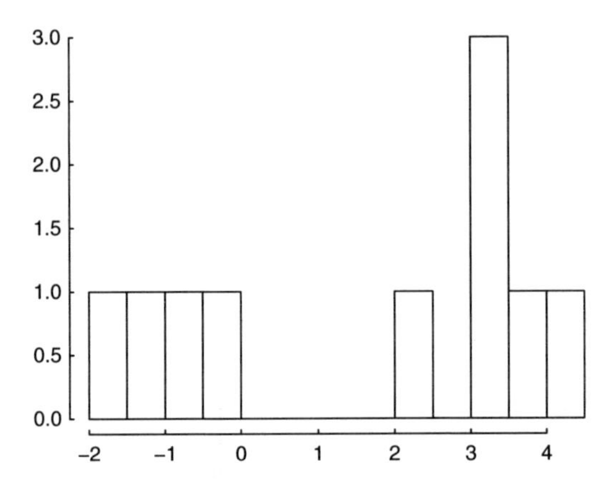

Figure 14.2 Small data set.

Table 14.1 Exact posterior probabilities (14.12) for K_n with $\tau = 0.1$ and $\alpha = \beta = 1$ and different choices of θ. Modal probabilities in bold font.

k	$\theta = 0.5$	$\theta = 1$	$\theta = 5$
1	0.019469	0.00619	0.000071
2	**0.591630**	0.37634	0.021504
3	0.312288	**0.39729**	0.113509
4	0.067986	0.17298	0.247113
5	0.008033	0.04088	**0.291972**
6	0.000568	0.00578	0.206592
7	0.000025	0.00051	0.090763
8	6.74 E−7	0.00003	0.024486
9	1.03 E−8	8.38 E−7	0.003740
10+	6.85 E−11	1.12 E−8	0.000249

process, or normalized generalized gamma processes [cf. 36], a similar effect still remains for their corresponding parameters.

The total mass parameter has different interpretations in the literature, e.g. as the scale parameter of the $\mathcal{D}_\theta \mathrm{P}_0$ that regulates the concentration mass around P_0; it can be thought of as the prior sample size [cf. 52]. It can also be used to match some moments of linear functionals of the Dirichlet process [cf. 55]. Furthermore, it arises in other areas such as population genetics, where it takes the role of the mutation rate in a Wright–Fisher-type population model [cf. 17].

Remark 3 *In practical implementations based on the Dirichlet process, the total mass parameter θ needs to be incorporated in the learning process.*

A technique to deal with stickbreaking priors is through truncation of representation (14.3),

$$\mathrm{P}_T(\cdot) = \sum_{i=1}^{T} \mathrm{w}_i \, \delta_{\mathrm{Z}_i}(\cdot), \qquad T < \infty \tag{14.14}$$

and thus relies on a L_1 error bound to determine the value of the truncation point T. Weights (14.4) are not necessarily ordered and therefore such a bound depends on their parameters. This, added to the required randomization of θ, implies that we would also need to incorporate T in the MCMC analysis, which is certainly neither commonly done nor an easy task.

Remark 4 *Unordered random weights in species sampling models might result in truncation methods that are highly dependent on their parameters.*

We have underlined some issues of the Dirichlet process that are in part a result of the randomness of its weights, and which is reflected on the parameter θ. The objective of this work is to present a somewhat easier r.p.m. that to some extent overcomes such issues. In Section 14.2 we study the geometric weights prior as one possible simpler model. Additionally, some estimation aspects and an intuitive derivation through a more general class of r.p.m.s are also presented. Being a simpler object it also makes it appealing to extensions to non-exchangeable contexts; in Section 14.3, some applications to dependent nonparametric processes are presented. In particular, its application to building models for Bayesian nonparametric regression and to construct probability measure-valued diffusion processes are explored. Examples illustrating the proposed models are also presented. Some concluding remarks are deferred to Section 14.4.

14.2 Geometric weights priors

A simple inspection of representation (14.3) of a discrete r.p.m. tell us that there are two sources of randomness, the weights and the locations. Furthermore, from Remark 3, we saw that we need to randomize the parameters of the weights, e.g. θ if $\mathrm{P} \sim \mathcal{D}_{\theta \mathrm{P}_0}$. As noted by Walker, Chapter 13 in this volume, the lack of order in the $\mathrm{P} \sim \mathcal{D}_{\theta \mathrm{P}_0}$ weights induces a non-identifiability problem. In other words, the fact that the stickbreaking weights are not deterministically ordered means that mass in a particular location $B \in \mathcal{X}$, i.e. $\mathrm{P}(B)$, can be attained by many different combinations of weights, w_is and locations Z_is. This issue is clearly inherited by mixtures based on such a prior. Hence, could one define a r.p.m. with simpler weights which makes better use of the availability of infinite locations to attain a particular mass, and eases the identifiability issue? The answer is yes, and is addressed below.

Instead of considering weights as in (14.4), consider

$$\omega_i := \mathbb{E}\left[\mathrm{w}_i\right] = \mathbb{E}\left[\mathrm{v}_i \prod_{j=1}^{i-1}(1-\mathrm{v}_j)\right] = \mu_i(\psi)\prod_{j=1}^{i-1}\left(1-\mu_j(\psi)\right)$$

with $\mu_i(\psi) := \mathbb{E}[\mathrm{v}_i]$ and where ψ here generically denotes the parameter (or parameters) corresponding to the distribution of the v_is. Next, assume that the v_is have been chosen such that $\omega_i > \omega_{i+1}$, that is $\mu_{i+1}(\psi) < \mu_i(\psi)(1-\mu_i(\psi))^{-1}$ for all i. Note also that when randomizing ψ the arguments in the right-hand side product of the above expression are no longer independent.

For the Dirichlet process we have $\mathrm{v}_i \stackrel{\text{iid}}{\sim} \mathrm{Be}(1,\theta)$; thus, setting $\lambda := \mu_i(\theta) = (1+\theta)^{-1}$, we have

$$\omega_i = \lambda(1-\lambda)^{i-1}\ : \tag{14.15}$$

geometric weights. Accordingly with the idea of randomizing θ, we could also assign a prior distribution to λ, say $\lambda \sim \mathrm{Be}(a,b)$. In other words, we recover the r.p.m. given in Definition 14.1.

Note that $\mathrm{P} \sim \mathcal{GWP}(a,b)$ is an almost sure discrete random probability measure and that, as in the Dirichlet process, and other species sampling models, the probability measure P_0 can be thought of as the prior guess at the shape of P, since $\mathbb{E}[\mathrm{P}] = \mathrm{P}_0$.

At first sight the random probability measure provided by (14.5) can be misinterpreted as a special case of the Dirichlet process, since the weights of the former can be obtained as in (14.4) by letting $(v_1, v_2, \dots)$ all be equal to the same realization of a Beta random variable, so that w_i, $i = 1, 2, \dots$ are mixed geometric with Beta(a, b) as the mixing distribution. However, this proves not to be the case. First, in the Dirichlet process case, the parameter a of the Beta distribution is constrained to be one. And, more significantly, because the Dirichlet process, $P \sim \mathcal{D}_{\theta P_0}$, is characterized by the distributional equation

$$P \stackrel{d}{=} v_1 \delta_{Z_1} + (1 - v_1)P \tag{14.16}$$

with $v_1 \sim \text{Be}(1, \theta)$ and $Z_1 \sim P_0$, stochastically independent of P [cf. 51]. The same procedure applied to $P^* \sim \mathcal{GWP}(a, b)$ yields

$$P^* \stackrel{d}{=} \lambda \, \delta_{Z_1} + (1 - \lambda)P^* \tag{14.17}$$

The crucial difference between these two cases is that in (14.16), P is independent of (v_1, Z_1), while in (14.17), P^* is independent of Z_1 but not of λ. Hence we are dealing with a different random probability measure.

A key issue that arises when considering whether to use a random probability measure of type (14.5) is whether it has full support. This could be a more relevant concern when one compares the Dirichlet process, or other more complex r.p.m.s, with the rather simplistic r.p.m. of Definition 14.1. The following proposition, whose proof easily follows from results in Ongaro and Cattaneo [43], justifies the use of the proposed model for inference purposes.

Proposition 14.1 *Let $\mathscr{P}$ be the probability distribution induced on $\mathcal{P}_{\mathbb{X}}$ by random probability measures $P \sim \mathcal{GWP}(a, b)$. Then the support of $\mathscr{P}$ in the topology of weak convergence on $\mathcal{P}_{\mathbb{X}}$ is given by all probability measures $G \in \mathcal{P}_{\mathbb{X}}$ such that the support of G is included in that of P_0.*

Therefore, the geometric weights prior constitutes a valid alternative to the Dirichlet process in Bayesian nonparametric applications.

14.2.1 An intuitive derivation

Walker [53] proposed an alternative method for posterior inference based on the nonparamteric mixture model (14.8) that overcomes the infinite summation in (14.9). By augmenting (14.9) through a latent uniform variable one can slice the corresponding infinite summation by considering the joint density

$$f(y, u) = \sum_{i=1}^{\infty} \mathbb{I}(u < w_i) f(y \mid Z_i) \tag{14.18}$$

with conditional density

$$f(y \mid u) = \frac{1}{|A_u|} \sum_{i \in A_u} f(y \mid Z_i), \tag{14.19}$$

where

$$A_u := \{j : u < w_j\} \tag{14.20}$$

and $|A_u| := \sum_{i=1}^{\infty} \mathbb{I}(u < w_i)$. Notice that given u the random set of component-indexes in the mixture, A_u, is finite, i.e. $|A_u| < \infty$. Using this idea Walker [53] proposed a Gibbs sampler for posterior analysis, based on the nonparametric mixture model (14.8), which avoids truncations such as (14.14). See also Kalli *et al.* [31] for more efficient implementations.

Here we are interested in the conditional distribution (14.19), as it can also be used to construct other kinds of random densities, namely

$$f(y \mid A) = \frac{1}{|A|} \sum_{i \in A} f(y \mid Z_i), \tag{14.21}$$

where A is a random finite subset of $\mathbb{N}_+$ and, as before, $Z_i \overset{\text{iid}}{\sim} P_0$. This resembles the approach undertaken in finite mixture models, but with some differences. First note that if we assume model (14.21) for each observation, Y_j, then there will be a random set A_j for each of them, whereas in the finite mixture model approach the number of components, N, suffices for all observations. In fact this is one of the reasons for having complex weight and location specifications in the finite mixture approach, i.e. to build a richer model which is able to allocate the required mass in a particular location. On the other hand, for Bayesian nonparametric mixtures such as those based on the Dirichlet process and other discrete r.p.m.s, there are an infinite number of locations, Z_is at our disposal to allocate a particular mass; therefore the complexity in the weights modulating a r.p.m. can be relaxed. In particular, note that the random set A corresponding to the Dirichlet process might have gaps, i.e. we can have realizations of the sort of $\{2, 5, 10, 345, 1004\}$; it is not a consecutive sequence of integers from 1 to N as it is typically for finite mixture models. In fact, having an infinite number of locations at our disposal, we do not see a clear need to have index sets with gaps.

Hence, an idea is to consider the random set $A := \{1, \ldots, N\}$ so (14.21) reduces to

$$f(y \mid N) = \frac{1}{N} \sum_{i=1}^{N} f(y \mid Z_i)$$

where N is random and modelled through a distribution supported on $\mathbb{N}_+$, namely $q_N(\cdot \mid \lambda)$, where for now λ denotes a generic parameter of the chosen distribution for N.

If we write out the model by marginalizing over N then we have

$$f(y) = \sum_{l=1}^{\infty} \frac{1}{l} \sum_{i=1}^{l} f(y \mid Z_i)\, q_N(l \mid \lambda)$$

which can also be written as

$$f(y) = \sum_{i=1}^{\infty} \omega_i f(y \mid Z_i) \tag{14.22}$$

where

$$\omega_i = \sum_{l=i}^{\infty} \frac{q_N(l \mid \lambda)}{l} \tag{14.23}$$

If we further randomize λ and assign a prior for it, e.g. $\lambda \sim \pi$, then (14.22) becomes a mixture based on a species sampling model (or simply a species sampling model if we take $f(\cdot \mid Z_i) = \delta_{Z_i}(\cdot)$).

Clearly the weights (14.23) sum up to one (a.s. if λ is random) and are in decreasing order as one sees that $\omega_{i+1} = \omega_i - q_N(i \mid \lambda)/i$.

Therefore (14.22) and (14.23) provide an alternative to the stickbreaking way of constructing species sampling models. By changing the choice of $q_N(\cdot \mid \lambda)$ and the prior for λ, we obtain different r.p.m.s.

Of particular interest here is when we assume $N \sim \text{Neg-Bin}(2, \lambda)$, that is

$$q_N(l \mid \lambda) = l\lambda^2(1-\lambda)^{l-1}\,\mathbb{I}(l \in \mathbb{N}_+)$$

which, by evaluating (14.23), allows us to recover the geometric weights, that is

$$\omega_i = \lambda(1-\lambda)^{i-1}$$

Hence, the r.p.m. of Definition 14.1 can be recovered as a r.p.m. of the sort

$$P(B) = \mathbb{E}_{q_N}\left[\frac{1}{N}\sum_{i=1}^{N}\delta_{Z_i}(B)\right] \tag{14.24}$$

where $Z_i \overset{\text{iid}}{\sim} P_0$, $q_N = \text{Neg-Bin}(2,\lambda)$ and $\lambda \sim \text{Be}(a,b)$.

14.2.2 Posterior inference

Based on the construction in the previous subsection, Fuentes *et al.* [22] proposed an algorithm for posterior inference under the following setting. Suppose we have a sample $Y^{(n)} := (Y_1, \ldots, Y_n)$ modelled by the nonparametric mixture model (14.8) where P follows a r.p.m. as in (14.24). Introducing a latent variable d_i which, given N_i, indicates where the component Y_i comes from: now, we can rewrite the nonparametric mixture model as

$$\begin{aligned} Y_i \mid Z^{(n)}, d_i, N_i &\overset{\text{ind}}{\sim} f(Y_i \mid Z_{d_i}) \\ d_i \mid N_i &\overset{\text{ind}}{\sim} U\{1,\ldots,N_i\} \\ N_i &\overset{\text{iid}}{\sim} q_N(\cdot \mid \lambda) \\ \lambda &\sim \pi, \end{aligned} \tag{14.25}$$

with $Z_i \overset{\text{iid}}{\sim} P_0$. Following [22] a Gibbs sampler algorithm, for the case $q_N = \text{Neg-Bin}\ (2,\lambda)$ and $\lambda \sim \text{Be}(a,b)$ (i.e. for a mixture model (14.8) when $P \sim \mathcal{GMP}(a,b)$) reduces to sampling from the following full conditional distributions:

$$f(Z_j \mid \cdots) \propto P_0(Z_j)\prod_{d_i=j} f(y_i \mid Z_j), \qquad \text{for } j = 1,\ldots,M \tag{14.26}$$

$$\mathbb{P}(d_i = l \mid \cdots) \propto f(Y_i \mid X_l)\,\mathbb{I}(l \in \{1,\ldots,N_i\})$$

$$\mathbb{P}(N_i = j \mid \cdots) = \lambda(1-\lambda)^{j-1}\mathbb{I}(j \geq d_i)$$

$$f(\lambda \mid \cdots) = \text{Be}\left(\lambda \mid a + 2n, b + \sum_{i=1}^{n} N_i - n\right) \tag{14.27}$$

for $i = 1, \ldots, n$ and where $M = \max\{N_1, \ldots, N_n\}$. Clearly, the Gibbs sampler simplifies when f and P_0 form a conjugate pair.

Example 14.3 In order to illustrate the performance of the above mixture of $\mathcal{GWP}(a, b)$, let us consider 240 data points coming from a mean-variance mixture of six normal distributions with weights $(0.17, 0.08, 0.125, 0.2, 0.125, 0.21)$ and mean-variance parameters given by $(-18, 2), (-5, 1), (0, 1), (6, 1), (14, 1)$ and $(23, 125)$. We assume the same kernel and prior guess at the shape specifications as in (14.13).

Figures 14.3 and 14.4 show the dynamics of the density estimator for the first 100 iterations based on mixtures of $\mathcal{GWP}(a, b)$ and on mixtures of $\mathcal{D}_{\theta P_0}$ respectively. From Figure 14.3 we note that the availability of an unlimited number of Z_js to represent a particular cluster location always results in an improvement in subsequent iterations, whereas for the Dirichlet process case, Figure 14.4, the algorithm requires several iterations to detect a good candidate for the Z_j representing a particular location. This feature is better appreciated in the mode around -18, which can be thought as being far from the overall mean of the data. It can also be observed at the tails of the density estimators in Figure 14.3, where for the initial iterations a larger mass than that shown for the Dirichlet process case is allocated. ■

It is probably worth mentioning that this drawback of nonparametric mixtures based on the Dirichlet process, and also on other discrete r.p.m.s, has received considerable attention in the Bayesian nonparametric literature, resulting in algorithms that aim to accelerate the identification of good candidates for the Z_js identifying particular cluster locations [see for instance 41]. However, in spite of these efforts, this issue is not fully resolved.

Figure 14.5 shows the estimates for both the Dirichlet process and the geometric weights prior cases after convergence. This figure also compares the true model that generated the observations;

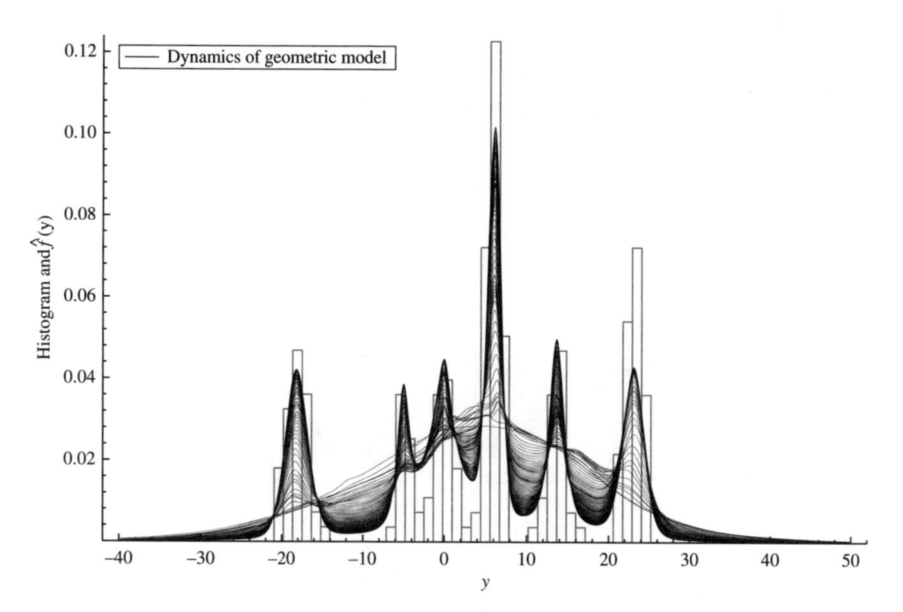

Figure 14.3 Dynamics of the density estimator, based on the geometric weights prior, through the first 100 iterations of the Gibbs sampler algorithm for the mean-scale mixtures dataset. The hyper-parameters are given by $(\tau, \alpha, \beta, a, b) = (100, 0.5, 0.5, 0.5, 0.5)$ and initial values $N_i = 10$ and $d_i \in \{1, \ldots, 10\}$ for all $i = 1, \ldots, 240$.

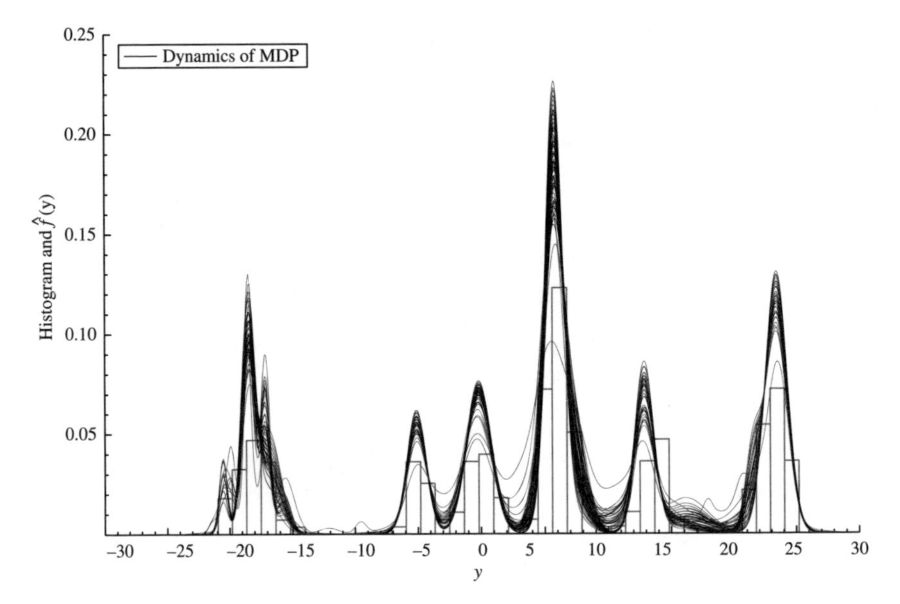

Figure 14.4 Dynamics of the density estimator for the mixture of Dirichlet process through the first 100 iterations of the Gibbs sampler algorithm for the mean-scale mixtures dataset.

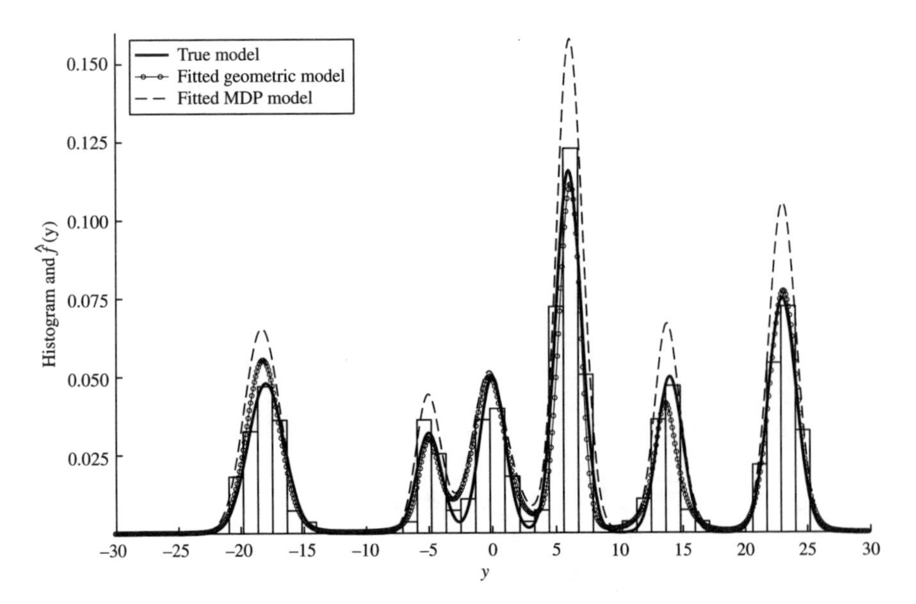

Figure 14.5 Density estimates for the 6 modes simulated dataset based on both $\mathcal{GWP}(a, b)$ and $\mathcal{D}_{\theta P_0}$. The estimates are based on 10 000 after a burn-in period of 2000 iterations. The hyperparameters are given by $(\mu, \tau, \alpha, \beta, a, b) = (0, 100, 0.5, 0.5, 0.5, 0.5)$ and $N_i = 10$ and $d_i \in \{1, \dots, 10\}$ for all $i = 1, \dots, 240$.

we can see both approaches are satisfactory, however the approach based on $\mathcal{GWP}(a, b)$ appears to be closer to the true model.

14.3 Dependent processes based on geometric weights priors

We noted earlier that the type of dependence driving Bayesian statistics is exchangeability. However, there are many phenomena where models with a more structured dependence are needed and would also benefit from a nonparametric approach, e.g. regression analysis, time series analysis, etc. This has motivated the notion of dependent nonparametric processes. The idea is to provide a family of random probability measures linked by a suitable dependence structure, e.g. by means of a set of covariates or a time parameter, and to use it for drawing inference on random phenomena in appropriate frameworks, usually with the aid of simulation techniques.

When this dependence structure can be indexed via a continuous time parameter, then we are effectively dealing with probability-measure-valued processes devised for nonparametric inference purposes. Apart from the theoretical developments offered by the probabilistic study of measure--valued processes, whose literature is certainly broad and well established [see for example 16], on the statistical side this is a relatively young area and to date the most productive ideas have involved the Dirichlet process.

Initiated in part by MacEachern [42] [see also 19] who introduced the notion of dependent Dirichlet processes, the current literature on the topic includes, among others, De Iorio *et al.* [10] who proposed a model with an ANOVA-type dependence structure, Gelfand *et al.* [23] who apply the dependent Dirichlet process to spatial modelling by using a Gaussian process for the atoms, Griffin and Steel [26] who let the dependence on the random masses be directed by a Poisson process, Caron *et al.* [8] who model nonparametrically the noise in a dynamic linear model, Dunson and Park [12] who construct an uncountable collection of dependent random probability measures based on a stickbreaking procedure with kernel-based weights. See also [11], [49], [44], [50] and [27] for other contributions.

The main idea is to construct a $\mathcal{P}_{\mathbb{X}}$-valued stochastic process $\{\mathrm{P}_z\}_{z\in\mathcal{Z}}$, which following representation (14.3) of a random probability measure, extends to

$$\mathrm{P}_z(B) = \sum_{i=1}^{\infty} \mathrm{w}_i(z)\,\delta_{\mathrm{Z}_i(z)}(B), \quad B \in \mathcal{X}, \tag{14.28}$$

where $\{\mathrm{w}_i(z)\}_{i=1}^{\infty}$ and $\{\mathrm{Z}_i(z)\}_{i=1}^{\infty}$ are infinite collections of stochastic processes, indexed by $z \in \mathcal{Z}$. Hence the dependence structure set in the weights and locations drives the dependence relations at the r.p.m. level. Clearly, while specifying a dependent nonparametric process, we could have both weights and locations dependent, or just one of them.

A natural idea when constructing dependent processes is to keep their marginal behaviour to be a known r.p.m., e.g. a Dirichlet process, a two-parameter Dirichlet process, a geometric weights prior, etc. This is easily done by setting strictly stationary processes with the desired marginal distribution. In what follows we will revise a couple of these examples based on geometric weights priors.

14.3.1 Covariate dependence for regression problems

A well-known problem in general regression analysis is to set a link function between observations and covariates. A Bayesian nonparametric approach to deal with this problem is to propose dependent random densities such as

$$f_z(y) = \int f(y \mid x) \, P_z(dx),$$

where $\{P_z\}_{z \in \mathcal{Z}}$ follows a covariate dependent nonparametric process.

Fuentes *et al.* [21] followed this idea and proposed a dependent nonparametric process with marginals $\mathcal{GWP}$ to model $\{P_z\}_{z \in \mathcal{Z}}$, that is

$$P_z(\cdot) = \sum_{i=1}^{\infty} \lambda(z)(1 - \lambda(z))^{i-1} \, \delta_{Z_i}(\cdot),$$

where $Z_i \overset{\text{iid}}{\sim} P_0$ and

$$\lambda(z) = \frac{e^{\xi(z)}}{1 + e^{\xi(z)}} \tag{14.29}$$

with $\xi := \{\xi(z)\}_{z \in \mathcal{Z}}$ a Gaussian process with continuous mean $\boldsymbol{\mu}$ and continuous covariance function $\boldsymbol{\sigma}$. Thus, we retain the iid locations and induce the dependence only through the simple structure of the geometric weights.

Some z-dependence properties can be studied directly from the geometric structure of the weights, for instance it is easy to see [see 21] that for $B \in \mathcal{X}$

$$\text{corr}(P_z(B), P_{z'}(B)) = \frac{\rho(z, z')}{\sqrt{\rho(z, z)}\sqrt{\rho(z', z')}} \tag{14.30}$$

where

$$\rho(z, z') = \mathbb{E}\left\{ \frac{\lambda(z)\lambda(z')}{1 - [1 - \lambda(z)][1 - \lambda(z')]} \right\}$$

When $\lambda(z) \sim \mathcal{LGP}(\boldsymbol{\mu}, \boldsymbol{\sigma})$, i.e. it follows the logistic Gaussian process (14.29), the above expression reduces to

$$\rho(z, z') = \mathbb{E}\left\{ \frac{e^{\xi(z)+\xi(z')}}{e^{\xi(z)} + e^{\xi(z')} + e^{\xi(z)+\xi(z')}} \right\}$$

Hence, returning to the dependent mixture model we can write

$$f_z(y) = \lambda(z) \sum_{i=1}^{\infty} (1 - \lambda(z))^{i-1} \, f(y \mid Z_i), \tag{14.31}$$

The logistic transformation (14.29) ensures that $0 < \lambda(z) < 1$ as required for the geometric weights prior.

Following the alternative derivation of geometric weights priors of Section 14.2.1 we can again make use of the latent variable d_i to build a Gibbs sampler algorithm, which is essentially based on the same full conditionals as in (14.40) but replacing (14.27) with the full conditional for $\xi^{(n)} = (\xi_1, \ldots, \xi_n)$, $\xi_i := \xi(z_i)$, which can be updated component-wise via

$$\mathbb{P}(\xi_i \mid \xi_{-i}) \propto \frac{1}{(1 + e^{\xi_i})^{N_i+1}} \, N\left(\xi_i; \mu_i - \frac{1}{c_{ii}} \sum_{j \neq i} (z_j - \mu_j)c_{ij} + \frac{1}{c_{ii}}, \frac{1}{c_{ii}} \right)$$

where $\mu_i = \boldsymbol{\mu}(z_i)$ and c_{ij} is the ij-term of the precision matrix $\boldsymbol{\Sigma}^{-1}$, $\boldsymbol{\Sigma} = \{\boldsymbol{\sigma}(z_i, z_j); i, j = 1, \ldots, n\}$. The above density is log-concave and thus can be easily sampled via the adaptive rejection sampling (ARS) algorithm of Gilks and Wild [25].

Example 14.5 Consider a simulated dataset with 61 observations coming from

$$Y_i = 0.2\, z_i^3 + \varepsilon_i$$

where $\varepsilon_i \overset{\text{iid}}{\sim} \mathrm{N}(0, 0.25)$ and $z = (-3, -2.9, \ldots, 2.9, 3)$. Hence assuming the dependent mixture model (14.31), with the same kernel and prior guess at the shape specifications as in (14.13), one can implement a MCMC algorithm, based on full conditionals (14.40) with the above component-wise update of the logistic Gaussian process, to infer about any random functional of the form

$$\boldsymbol{\eta}_z(h) = \int_{\mathbb{Y}} h(y)\, \mathrm{f}_z(y) \mathrm{d}y$$

This can be done through the Rao–Blackwellized MCMC estimator

$$\tilde{\boldsymbol{\eta}}_{z_i}(h) = \frac{1}{M} \sum_{l=1}^{M} \mathbb{E}_l[h(y) \mid z_i]$$

where M denotes the number of effective iterations in the MCMC. For example, one might be interested in the mean functional $\mathbb{E}_l[y \mid z_i] \approx \mu_{d_i}$, which in practice can be obtained as the updated mean value in the Gibbs sampler.

Figure 14.6 shows the MC estimator for the distribution of the mean functional $(h(y) = y)$ together with the observed data. For the corresponding Gaussian process we have set $\boldsymbol{\mu}(z) = -|z|$ and $\boldsymbol{\sigma}(z_i, z_j) = e^{-\phi||z_i - z_j||}$. ■

Just as in the marginal $\mathcal{GWP}$ case, simpler weights with an infinite number of locations, seems to be enough to allocate the required mass at a particular location. In particular, having the first and largest weight for each covariate point, z, combined with the choice of locations seems enough for regression purposes.

14.3.2 A nonparametric diffusion process based on geometric weights priors

Although there are currently many examples of dependent nonparametric priors, the current statistical literature devoted to the study of continuously dependent measures is sparse. Indeed, it is of interest to study this case from a statistical perspective as it would allow us to exploit the flexibility of a Bayesian nonparametric approach while enjoying desirable properties such as Markov, reversibility, regularity of sample paths for the constructed model, etc. These features are appealing in various modelling contexts and applications of stochastic processes such as finance or population genetics, where diffusion processes are typically used to model random phenomena evolving in time.

In this section we present an approach to constructing $\mathcal{P}_{\mathbb{X}}$-valued continuous time stochastic processes $(\mathrm{P}_t)_{t \geq 0}$ [see 40] by simply introducing the time dependence through the weights, while keeping the locations fixed (but random) over time, in the species sampling representation for the $\mathcal{GWP}(a, b)$. That is, we will consider

$$\mathrm{P}_t(B) = \lambda_t \sum_{i=1}^{\infty} (1 - \lambda_t)^{i-1}\, \delta_{Z_i}(B), \qquad B \in \mathcal{X}, \tag{14.32}$$

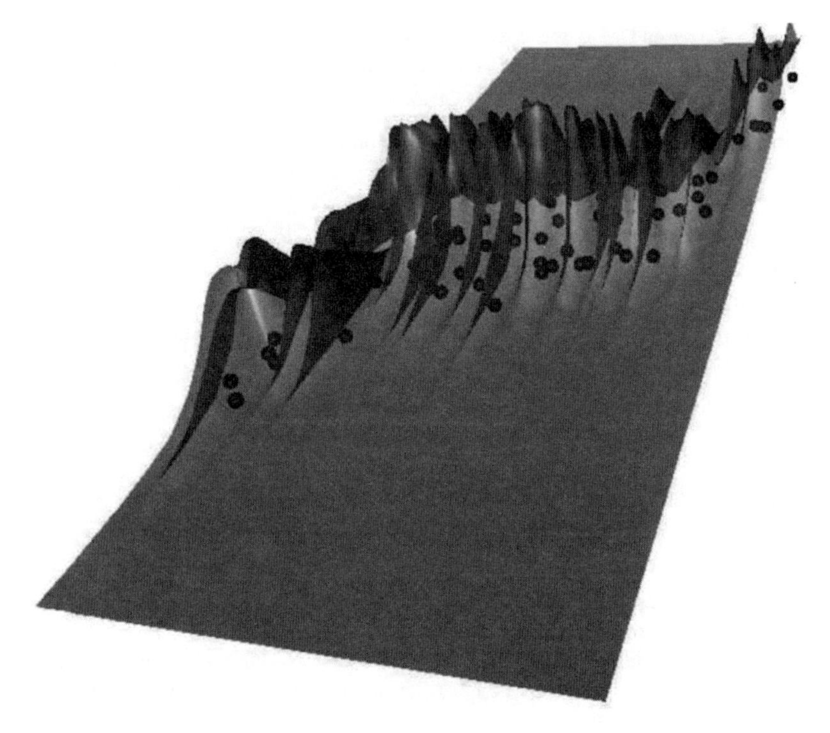

Figure 14.6 MC estimator for the density of $\eta_Z(y)$ for a simulated dataset. The spheres represent the observed data and the surface the Rao–Blackwellized MC estimator for $\eta_Z(y)$. The results are based on 10 000 iterations of the Gibbs sampler algorithm.

with $Z_i \overset{\text{iid}}{\sim} P_0$ and $(\lambda_t)_{t\geq 0}$ a suitable stochastic process. In order to keep the same marginal r.p.m. $\mathcal{GWP}(a,b)$, we assume $(\lambda_t)_{t\geq 0}$ follows a strictly stationary diffusion process with $\mathrm{Be}(a,b)$ invariant distributions. There are many choices for such a model, here we use a slight generalization of the well-known two-type Wright–Fisher diffusion process which can be described as the solution of the stochastic differential equation (SDE) on $[0,1]$ given by

$$d\lambda_t = \left[\frac{c}{a+b-1}(a-(a+b)\lambda_t)\right]dt + \sqrt{\frac{2c}{a+b-1}\lambda_t(1-\lambda_t)}\,dB_t \tag{14.33}$$

where $(B_t)_{t\geq 0}$ is a standard Brownian motion. The typical parameterization of the two-type Wright–Fisher with mutation SDE is found when $c = (a+b-1)/2$.

Definition 14.6 *[40] A geometric stickbreaking process with parameters $a, b, c > 0$ is a random process $(P_t)_{t\geq 0}$ taking values in $\mathcal{P}_{\mathbb{X}}$ defined at each $t \geq 0$ by (14.32) with $(\lambda_t)_{t\geq 0}$ a two-type Wright–Fisher diffusion and P_0 a nonatomic probability measure on $(\mathbb{X}, \mathcal{X})$. We denote it as $\mathcal{GSB}(a,b,c,P_0)$.*

It can be seen that the Wright–Fisher diffusion is time reversible, strictly stationary with invariant measure $\mathrm{Be}(a,b)$, so an immediate question is whether some of these properties are inherited by the $\mathcal{GSB}(a,b,c,P_0)$ process.

Before undertaking this problem, first let us note that constructing stationary one-dimensional diffusion processes with desired stationary distributions can be done via stochastic differential equations [see 5], however this is not entirely useful when, for estimation purposes, one needs to keep track of an analytical form or an exact representation of the corresponding transition density. Here we use an idea to construct continuous time Markov processes [see 39] that allows us to have a representation of the transition density of the Wright–Fisher diffusion process. The idea starts with a Gibbs sampler Markov process based on the joint density

$$f(y,x) = \text{Po}(y \mid \phi x)\, \text{Ga}(x \mid a, b)$$

from which we can construct a Markov process $(X_t)_{t\geq 0}$ with $\text{Ga}(a,b)$ marginals by the conditional updating $Y_t \mid X_0 \sim \text{Po}(\phi_t X_0)$ and $X_t \mid Y_t \sim \text{Ga}(a + Y_t, b + \phi_t)$. Indeed, such updating leads to the transition density

$$p(x_t \mid x_0) = \sum_{y=0}^{\infty} \text{Ga}(x_t \mid a+y, b+\phi_t)\text{Po}(y \mid \phi_t x_0) \tag{14.34}$$

$$= \frac{e^{-[\phi_t(x_t+x_0)+bx_t]}}{(\phi_t+b)^{-(a+1)/2}\,\phi_t^{(a-1)/2}} \left(\frac{x_t}{x_0}\right)^{\frac{a-1}{2}} \text{I}_{a-1}\left(2\sqrt{x_t\, x_0 \phi_t(\phi_t+b)}\right)$$

with $\phi_t := b(e^{ct}-1)^{-1}$. It can be verified that the above transition corresponds to the solution of a SDE given by

$$\text{d}X_t = c\left(\frac{a}{b} - X_t\right)\text{d}t + \sqrt{\frac{2c}{b} X_t}\, \text{d}B_t$$

known as the Cox–Ingersoll–Ross model for interest rates [see 9]. Therefore having a continuous-time Markov process with $\text{Ga}(a,b)$ invariant distribution suggests a simple transformation of two independent copies of these to obtain a diffusion with $\text{Be}(a,b)$ invariant densities. That is, $\lambda_t = X_{1t}/(X_{1t}+X_{2t})$, where $(X_{it})_{t\geq 0}$, $i = 1,2$ are independent Markov diffusion processes with $\text{Ga}(a,1)$ and $\text{Ga}(b,1)$ invariant densities respectively and transition probabilities (14.34). In fact, it easily follows [see 39] that the transition corresponding to this newly transformed process is given by

$$p(\lambda_t \mid \lambda_0) = \sum_{m=0}^{\infty} p_t(m) D(\lambda_t \mid m, \lambda_0) \tag{14.35}$$

where

$$p_t(m) = \frac{(a+b)_m\, e^{-mct}}{m!}(1-e^{-ct})^{a+b},$$

and

$$D(\lambda_t|m,\lambda_0) = \sum_{k=0}^{m} \text{Be}(\lambda_t|a+k, b+m-k)\, \text{Bin}(k|m,\lambda_0)$$

which again can be seen to correspond to the general class of Beta-binomial diffusion processes given by the solution of the SDE (14.33). Also, from the Gibbs sampler type construction, it easily follows that such a process is time reversible and has $\text{Be}(a,b)$ stationary distributions.

Having established our approach for constructing the diffusion process with $\mathrm{Be}(a,b)$ marginal distributions, we can go back to our question regarding the dependent nonparametric process of Definition 14.6. Let $\mathcal{P}_{\mathbb{X}}^{g} \subset \mathcal{P}_{\mathbb{X}}$ be the set of purely atomic probability measures with geometric weights as in Definition 14.1 and denote $\mathrm{C}_{\mathcal{P}_{\mathbb{X}}^{g}(\mathbb{X})}([0,\infty))$ the space of continuous functions from $[0,\infty)$ to $\mathcal{P}_{\mathbb{X}}^{g}$. Furthermore, for given locations $\mathrm{Z}=\{\mathrm{Z}_i\}_{i=1}^{\infty}$ define the continuous map $\Phi_{\mathrm{Z}}(\lambda) = \sum_{i=1}^{\infty}\lambda(1-\lambda)^{i-1}\delta_{\mathrm{Z}_i}$. Hence, because of the decreasing order of the geometric weights we can set $\Phi^{-1} = g_{\mathrm{Z}}$, where $g_{\mathrm{Z}}(\mathrm{P}) = \mathrm{P}(\{\mathrm{Z}_1\}) = \lambda$ and let $\tilde{\mathrm{B}}(a,b) = \mathrm{Be}(a,b)\circ\Phi_{\mathrm{X}}^{-1}$. Therefore we can state the following proposition whose proof can be found in Mena *et al.* [40].

Proposition 14.7 *Let $(\mathrm{P}_t)_{t\geq 0}$ be a $\mathcal{GSB}(a,b,c,\mathrm{P}_0)$ process on $\mathcal{P}_{\mathbb{X}}^{g}$. Then $(\mathrm{P}_t)_{t\geq 0}$ is reversible and stationary, with respect to $\tilde{\mathrm{B}}(a,b)$, Feller process with sample paths in $\mathrm{C}_{\mathcal{P}_{\mathbb{X}}^{g}(\mathbb{X})}([0,\infty))$.*

In other words the stability properties of the Wright–Fisher diffusion process are inherited by the $\mathcal{GSB}(a,b,c,\mathrm{P}_0)$ process at the probability measure-valued level. This is quite intuitive since the locations $(\mathrm{Z}_i)_{i\geq 1}$ are random but fixed across time. An immediate observation would be that letting the Z_is vary also leads to a more flexible model. While this is certainly true, we have two reasons for keeping the locations fixed. On the probabilistic side this would most likely tear apart the nice properties this process enjoys. But more importantly, on the inference side this is not even needed in view of Proposition 14.1 and as the example below will show.

All distributional properties of a diffusion process can be explained through its infinitesimal generator, in particular for the $\mathcal{GSB}(a,b,c,\mathrm{P}_0)$ process the generator can be found in Mena *et al.* [40], thus providing a valid alternative to other measure-valued processes found in the literature [cf. 16].

As in the previous section, interest might be on modelling a process taking values on the space of continuous densities. For this purpose the $\mathcal{GSB}(a,b,c,\mathrm{P}_0)$ process can also be incorporated into a dependent nonparametric mixture model given by

$$\mathrm{f}_t(y) = \int f(y\mid x)\,\mathrm{P}_t(\mathrm{d}x) = \lambda_t\sum_{l\geq 1}(1-\lambda_t)^{l-1} f(y\mid \mathrm{Z}_l) \tag{14.36}$$

with $\mathrm{Z}_l \overset{\text{iid}}{\sim} \mathrm{P}_0$. In fact, for a set of observations $\mathrm{Y}^{(n)}$ recorded at times $\{t_i\}_{i=1}^{n}$ and modelled through the above dependent density, Mena *et al.* [40] proposed a Gibbs sampler algorithm based on some slice sampler techniques that aid to overcome both the infinite summations, the one in (14.36) and the one in the representation of the Wright–Fisher transition density (14.35). The algorithm is a bit more demanding than those used in previous sections and so we briefly discuss it.

First, it is convenient to start by considering the part of the model related to the Wright–Fisher diffusion $(\lambda_t)_{t\geq 0}$. Hence to overcome the infinite summation in (14.36), we proceed as before and introduce the latent variable d_i that indicates the component $f(\cdot\mid \mathrm{Z}_i)$ from which the observation Y_i comes from, namely we have the augmented observations $(t_i,d_i)_{i=1}^{n}$ and the model can then be written as

$$d_i \mid \lambda_i \sim \mathrm{Geom}(\lambda_i)$$

with $\lambda_i := \lambda_{t_i}$ and corresponding transition density $p(\lambda_i \mid \lambda_{i-1})$ given as in (14.35), where t has to be replaced by $\tau_i = t_i - t_{i-1}$. Hence, to avoid the infinite summations needed for the transition (14.35), we introduce a further set of latent variables $(u_i,s_i,k_i)_{i=1}^{n}$ whereby the augmented transition density is given by

$$p(\lambda_i, u_i, s_i, k_i \mid \lambda_{i-1}) = \mathbb{I}(u_i < g(s_i))\frac{p_i(s_i)}{g(s_i)}\text{Be}(\lambda_i \mid a + k_i, b + s_i - k_i)\text{Bin}(k_i \mid s_i, \lambda_{i-1})$$

where g is a decreasing function with known inverse. Therefore the likelihood with the complete data is given by

$$l(a, b, c) = \text{Be}(\lambda_0 \mid a, b)\prod_{i=1}^{n} p(\lambda_i, u_i, s_i, k_i \mid \lambda_{i-1})\lambda_i(1-\lambda_i)^{s_i - 1}$$

Hence if, for instance, we assume independent standard exponential distributions as priors for a, b, c, we see that the full conditionals, e.g. $\pi(a \mid b, c, \ldots) \propto l(a, b, c)e^{-a}$ are log-concave and easily sampled through the ARS algorithm. For instance we have

$$\log \pi(c \mid a, b, \cdots) = \sum_{i=1}^{n}\{(a+b)\log(1-e^{-c\tau_i}) - s_i\, c\, \tau_i\} - c + O$$

where O is a constant which does not depend on c. The full conditionals for a and b follow similarly. The full conditional distribution for k_i is given by

$$\pi(k_i| \ldots) \propto \binom{s_i}{k_i}\frac{\mathbf{1}(k_i \in \{0, 1, \ldots, s_i\})}{\Gamma(a+k_i)\Gamma(b+s_i-k_i)}\left\{\frac{\lambda_i\lambda_{i-1}}{(1-\lambda_i)(1-\lambda_{i-1})}\right\}^{k_i}$$

which is clearly easy to sample since k_i can only take a finite number of values. The full conditional for u_i is simply a uniform distribution on $(0, g(s_i))$, where g is chosen for convenience, for example $g(s) = e^{-s}$ or $g(s) = s^{-2}$, so that g^{-1} is known. The benefit of this becomes apparent when we consider the full conditional for s_i. This is given by

$$\pi(s_i| \ldots) \propto \frac{p_i(s_i)}{g(s_i)}\binom{s_i}{k_i}\frac{\Gamma(a+b+s_i)}{\Gamma(b+s_i-k_i)}\{(1-\lambda_{i-1})(1-\lambda_i)\}^{s_i}\,\mathbf{1}(k_i \le s_i \le g^{-1}(u_i))$$

which by virtue of u_i is restricted to a finite set. The full conditional for λ_i, for $i \neq 0, n$, is given by

$$\pi(\lambda_i| \ldots) = \text{Beta}(1 + a + k_i + k_{i+1}, d_i - 1 + b + s_i + s_{i+1} - k_i - k_{i+1}) \tag{14.37}$$

whereas

$$\pi(\lambda_0| \ldots) = \text{Beta}(a + k_1, b + s_1 - k_1) \tag{14.38}$$

and

$$\pi(\lambda_n| \ldots) = \text{Beta}(1 + a + k_n, d_n - 1 + b + s_n - k_n). \tag{14.39}$$

This deals with the part of the model related to the Wright–Fisher process. For the remaining part of the model, which, for a given observation, is given by

$$y_i|t_i, \lambda_i, Z^{(n)} \sim \sum_{l=1}^{\infty}\lambda_i(1-\lambda_i)^{l-1}f(y_i|Z_l).$$

We proceed as before and introduce two latent variables (s_i, v_i) and a deterministic decreasing sequence of numbers (ψ_l) for which $\{l : \psi_l > v\}$ is a known set, such that

$$y_i, v_i, d_i|\lambda_i, Z^{(n)} \sim \psi_{d_i}^{-1} \mathbf{1}(v_i < \psi_{d_i}) \, \lambda_i(1-\lambda_i)^{d_i-1} f(y_i|Z_{d_i})$$

Note that we slice the infinite summation again. In order to complete the Gibbs sampler for the model we need to describe how to sample the s_i from their full conditionals and also the Z_ss. Now,

$$\pi(d_i|\ldots) \propto \psi_{d_i}^{-1} \lambda_i(1-\lambda_i)^{d_i-1} f(y_i|Z_{d_i}) \, \mathbf{1}(d_i \in \{l : \psi_l > v_i\})$$

and clearly the full conditional for v_i is the uniform distribution on $(0, \psi_{d_i})$. Since $\{l : \psi_l > v_i\}$ is a finite set this is easy to sample. Finally, we sample the Z_ls from

$$f(Z_j \mid \cdots) \propto \mathrm{P}_0(Z_j) \prod_{d_i=j} f(y_i \mid Z_j), \qquad \text{for } j = 1, \ldots, M \tag{14.40}$$

Note that, as before, we only need to consider a finite number of location updates $(Z_l)_{l=1}^M$ where $M = \max_i M_i$ and $\{1, \ldots, M_i\} = \{l : \psi_l > v_i\}$. We thus have all the full conditional distributions required to implement the Gibbs sampler needed for the estimation of model (14.36) given a discretely observed trajectory. The following algorithm summarizes the procedure.

Algorithm.

1. Select $g(\cdot)$ and $\psi(\cdot)$ functions, *e.g.* $g(x) = \psi(x) = e^{-x}$
2. Set initial values for:
    - Wright–Fisher diffusion parameters (a_0, b_0, c_0)
    - Parameters in the kernel K and possibly in P_0, *e.g.* θ^0
    - Latent variables needed to overcome infinite summations, $(u_i^0, s_i^0, k_i^0, d_i^0)_{i=1}^n$. For these an initial value for the augmented random probability measure is also needed, e.g. $M^0 = 20$
    - Use these values to initiate $\lambda^0 = (\lambda_i^0)_{i=0}^n$

    then for $j = 1, \ldots, I$
3. Update $v^j = (v_i^j)_{i=0}^n$, *i.e.* $v_i^j \sim \mathrm{U}[0, \psi_{s_i^{j-1}}]$, and compute

    $M^j = \max M_i^j$ with $\{1, \ldots, M_i^j\} = \{l : \psi(l) > v_i^j\}$
4. Update $\lambda^j = (\lambda_i^j)_{i=0}^n, \theta^j = (\theta_l^j)_{l=1}^M$ and $(u_i^j, s_i^j, k_i^j, d_i^j)_{i=1}^n$ using the corresponding full conditionals
5. Update (a_j, b_j, c_j), e.g. via ARS algorithm

The I iterations can then be used to build a Monte Carlo estimator for f_t or any desired functional of it.

Example 14.8 In order to illustrate how the modelling scheme described above is able to capture the dynamics of continuous time phenomena, we will consider data coming from 251 daily observations (corresponding to a financial year) from the adjusted close quotations of the S&P 500 index during the period 03.03.2008 to 27.02.2009 (the dataset can be found at http://finance.yahoo.com).

These types of data are typically modelled through parametric diffusion processes, however, one could argue to what extent such restrictive assumptions are justified. For example, in the case of interest rates one could choose among many existing models, such as the Cox–Ingersoll–Ross (CIR) diffusion, the Brennan–Schwartz diffusion or the Duffie–Kan diffusion (see Aït-Sahalia [1]). Adopting a nonparametric approach based on measure-valued processes provides enough flexibility to avoid such limiting assumptions.

As in our previous examples we use the kernel and prior guess at the shape specifications given by (14.13) and concentrate on the mean functional $\eta_t := \int y\, \mathrm{f}_t(y)\mathrm{d}y$, namely the evolution of the mean, which imitate that of a one-dimensional diffusion process. Figure 14.7 shows the MCMC estimates (heat contours) for the density process, $\hat{f}_t$, and the corresponding mean of the functional $\bar{\eta}_t$ (solid line) for the S&P 500 dataset (points). For both datasets the choice of hyperparameters was $\tau = 1000$ and $(\alpha, \beta) = (10, 1)$. The results are based on 100 000 iterations, after a 20 000 of burn-in (thinned each 10), enough to attain a satisfactory convergence of the sampler. Figure 14.8 shows the Markov chains corresponding to parameters (a, b) and the corresponding posterior densities; the results corresponding to c are proportional to those corresponding to b. A standard convergence analysis was performed, in particular the Gelman and Rubin [24] visual test and the Raftery and Lewis [46] diagnosis test were satisfactory. ■

It is apparent that the probability measure-valued approach here undertaken is able to capture the dependence induced by these datasets. Furthermore the model adapts well to drastic changes like those observed in the S&P 500 index and typically not captured by the (parametric) diffusion process. Note that the strict stationary and reversibility properties of the $\mathcal{GSB}(a, b, c, \mathrm{P}_0)$ process

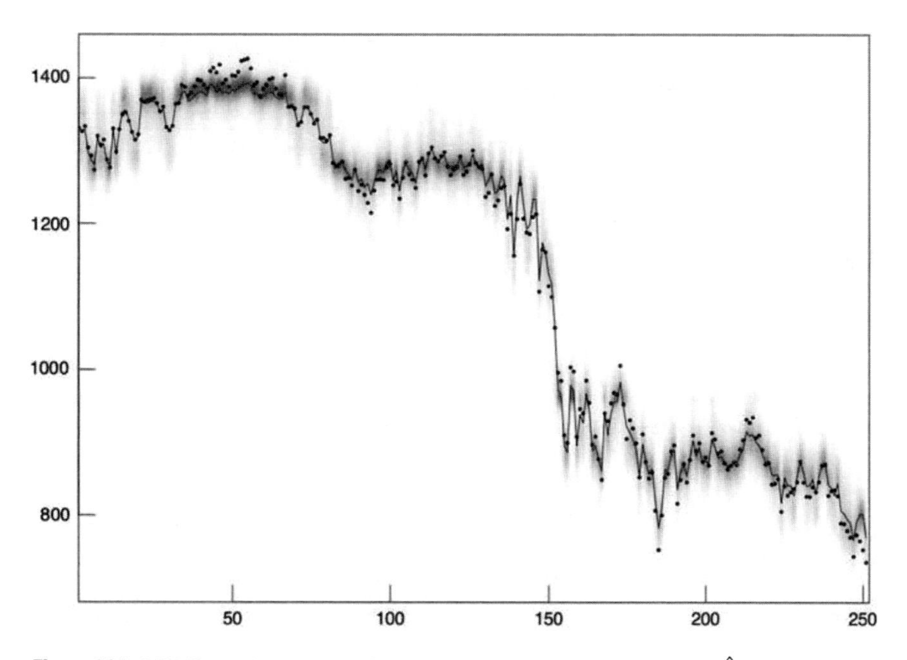

Figure 14.7 MCMC density estimator for the random density process (14.36), $\hat{f}_t$, (heat contour), mean of mean functional $\bar{\eta}_t$ (solid) for the S&P 500 dataset (dots). The estimates are based on 10 000 effective iterations, drawn from 100 000 iterations thinned each 10, of the Gibbs sampler algorithm after 20 000 iterations of burn-in.

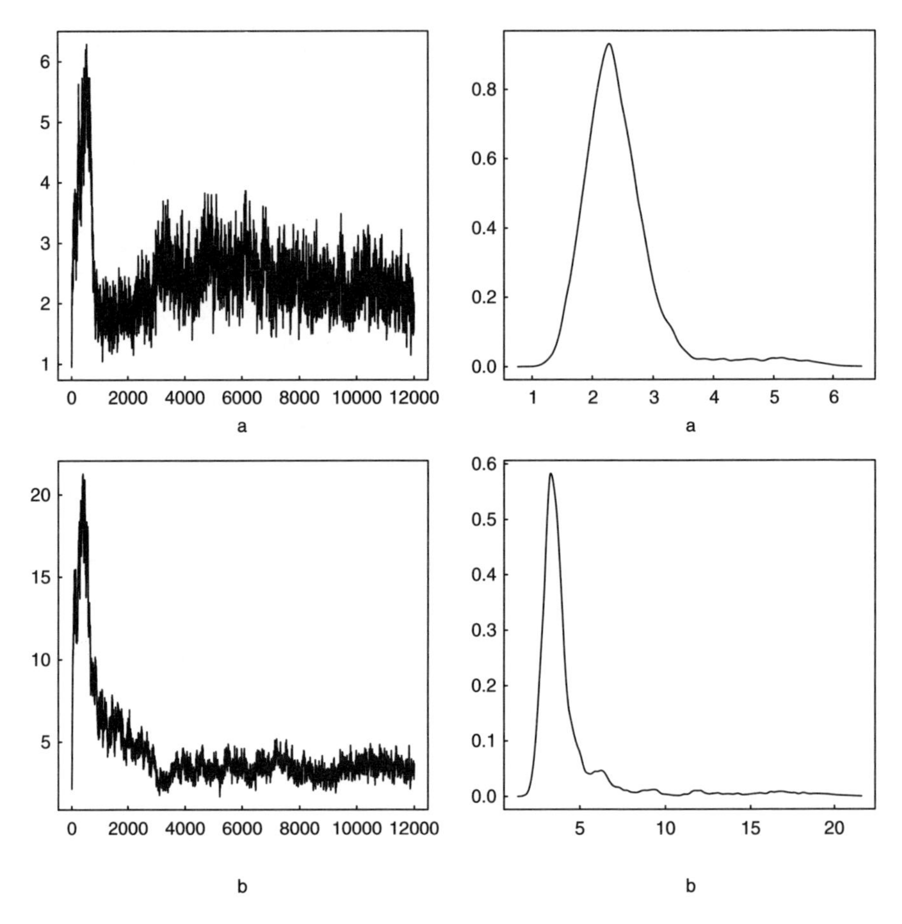

Figure 14.8 MCMC iterations and posterior densities for parameters (a, b). The estimates are based on 10 000 effective iterations, drawn from 100 000 iterations thinned each 10 iterations, of the Gibbs sampler algorithm after 20 000 iterations of burn-in.

are at the probability measure-valued level and not at the level of observations, therefore certain sudden changes of regime and unstable behaviours might be well captured by this model without compromising the measure-valued stationarity property.

14.4 Discussion

We have seen that r.p.m.s with simpler weights structure are able to provide good alternatives to more complex nonparametric priors. The key idea is that having simpler weights results in a more efficient use of the infinite collection of locations to assign the required mass to a particular set $B \in \mathcal{X}$. Having simpler weights also results in easier ways to estimate models and extend them to non-exchangeable contexts, as seen in Section 14.3.

Although for simplicity we have mainly concentrated on the geometric weights case, several generalizations are possible, for instance one could consider the expected weights corresponding to the two-parameter Poisson–Dirichlet process, i.e. with corresponding stickbreaking weights (14.4), with $v_i \overset{\text{ind}}{\sim} \text{Be}(1-\sigma, \theta + i\sigma)$, which implies

$$\omega_i = \frac{1-\sigma}{\theta + 1 + (i-1)\sigma} \prod_{j=1}^{i-1} \frac{\theta + j\sigma}{\theta + 1 + (j-1)\sigma}$$

and then $(\sigma, \theta) \sim \pi$. Furthermore, using the alternative derivation in Section 14.2.1 we could have a different choice of prior for N, e.g. $\text{N} \sim \text{Neg-Bin}(r, \lambda)$ which would lead to

$$w_l = \frac{1}{l} \binom{l+r-2}{r-1} \lambda^r (1-\lambda)^{l-1} \, {}_2F_1(1, l+r-1; l+1; \lambda) \tag{14.41}$$

where ${}_2F_1(a, b; c; \lambda)$ denotes the Gauss hypergeometric function. Both of the above possibilities are clearly more general than the geometric weights while keeping their decreasing feature, however some algorithmic modifications would be needed when randomizing their parameters.

We have also seen that from a numerical point of view having a simpler r.p.m. such as the $\mathcal{GWP}$ might lead to more efficient posterior inferences than those obtained via other nonparametric priors such as the Dirichlet process. Another appealing feature of the simplicity inherent to the $\mathcal{GWP}$ is that it is easily generalized to dependent nonparametric processes. This is in principle conceivable with other more complicated r.p.m.s such as the Dirichlet process; however, keeping a canonical construction of the resulting dependent process is not necessarily straightforward. In fact, to the best of our knowledge, Section 14.3.2 presents the first instance of a measure-valued Markovian process applied to model single trajectory phenomena (cf. Example 14.8).

Acknowledgements

The author wishes to thank the hospitality and support of Collegio Carlo Alberto during the preparation of this paper. The author was also partially supported by projects CONACyT 131179 and PAPIIT IN100411.

References

[1] Aït-Sahalia, Y. (1996). Nonparametric pricing of interest rate derivative securities. *Econometrica*, **64**, 527–560.

[2] Aoki, M. (2004). *Modeling Aggregate Behavior and Fluctuations in Economics: Stochastic Views of Interacting Agents*. Cambridge University Press.

[3] Banfield, J. D. and Raftery, A. (1993). Model-based Gaussian and non-Gaussian clustering. *Biometrics*, **49**, 803–821.

[4] Berry, D. A. and Christensen, R. (1979). Empirical Bayes estimation of a binomial parameter via mixtures of Dirichlet processes. *Annals of Statistics*, **7**, 558–568.

[5] Bibby, M., Skovgaard, M. and Sørensen, M. (2005). Diffusion-type models with given marginal distribution and autocorrelation function. *Bernoulli*, **11**, 191–220.

[6] Blackwell, D. (1973). Discreteness of Ferguson selections. *Annals of Statistics*, **1**, 356–358.

[7] Blackwell, D. and MacQueen, J. B. (1973). Ferguson distributions via Pólya urn schemes. *Annals of Statistics*, **1**, 353–355.

[8] Caron, F., Davy, M., Doucet, A., Duflos, E. and Vanheeghe, P. (2006). Bayesian inference for dynamic models with Dirichlet process mixtures. In *International Conference on Information Fusion*, pp. 1–8. Florence, Italy: INRIA-CCSd-CNRS.
[9] Cox, J. C., Ingersoll, J. E. and Ross, S. A. (1985). A theory of the term structure of interest rates. *Econometrica*, **53**, 385–407.
[10] De Iorio, M., Müller, P., Rosner, G. L. and MacEachern, S. N. (2004). An ANOVA model for dependent random measures. *Journal of the American Statistical Association*, **99**, 205–215.
[11] Dunson, D. B., Pillai, N. and Park, J. -H. (2007). Bayesian density regression. *Journal of the Royal Statistical Society. Series B*, **69**, 163–183.
[12] Dunson, D. B. and Park, J. -H. (2008). Kernel stick-breaking processes. *Biometrika*, **95**, 307–323.
[13] Escobar, M. D. (1988). Estimating the means of several normal populations by nonparametric estimation of the distribution of the means. Unpublished Ph.D. dissertation, Department of Statistics, Yale University.
[14] Escobar, M. D. (1994). Estimating normal means with a Dirichlet process prior. *Journal of the American Statistical Association*, **89**, 268–277.
[15] Escobar, M. D. and West, M. (1995). Bayesian density estimation and inference using mixtures. *Journal of the American Statistical Association*, **90**, 577–588.
[16] Ethier, S. N. and Kurtz, T. G. (1986). *Markov Processes: Characterization and Convergence.* Wiley Series in Probability and Mathematical Statistics. John Wiley & Sons Inc. New York.
[17] Ewens, W. (1972). The sampling theory of selectively neutral alleles. *Theoretical Population Biology*, **3**, 87–112.
[18] Ewens, W. J. and Tavarè, S. (1997). Multivariate Ewens distribution. In *Discrete Multivariate Distributions* (eds N. S. Johnson, S. Kotz and N. Balakrishnan), Wiley, New York.
[19] Feigin, P. D. and Tweedie, R. L. (1989). Linear functionals and Markov chains associated with Dirichlet processes. *Mathematical Proceedings of the Cambridge Philosophical Society*, **105**, 579–585.
[20] Ferguson, T. S. (1973). A Bayesian analysis of some nonparametric problems. *Annals of Statistics*, **1**, 209–230.
[21] Fuentes-García, R., Mena, R. H. and Walker, S. G. (2009). A nonparametric dependent process for Bayesian regression. *Statistics and Probability Letters*, **8**, 112–119.
[22] Fuentes-García, R., Mena, R. H. and Walker, S. G. (2010). A new Bayesian nonparametric mixture model. *Communications in Statistics – Simulation and Computation*, **39**, 669–682.
[23] Gelfand, A. E., Kottas, A. and MacEachern, S. N. (2005). Bayesian nonparametric spatial modeling with Dirichlet process mixing. *Journal of the American Statistical Association*, **100**, 1021–1035.
[24] Gelman, A. and Rubin, D. (1992). Inferences from iterative simulation using multiple sequences. *Statistical Inference*, **7**, 457–472.
[25] Gilks, W. R. and Wild, P. (1992). Adaptive rejection sampling for Gibbs sampling. *Applied Statistics*, **41**, 337–348.
[26] Griffin, J. E. and Steel, M. F. J. (2006). Order-based dependent Dirichlet processes. *Journal of the American Statistical Association*, **101**, 179–194.
[27] Griffin, J. E. and Steel, M. F. J. (2010). Stick-breaking autoregressive processes. *Journal of Econometrics*, **162**, 383–396.
[28] Hansen, J. C. (1994). Order statistics for decomposable combinatorial structures. *Random Structures and Algorithms*, **5**, 517–533.
[29] Hjort, N., Holmes, C., Müller, P. and Walker, S. G. (2010). *Bayesian Nonparametrics.* Cambridge University Press.
[30] Ishwaran, H. and James, L. F. (2001). Gibbs sampling methods for stick-breaking priors. *Journal of the American Statistical Association*, **96**, 161–173.

[31] Kalli, M., Griffin, J. E. and Walker, S. G. (2011). Slice sampling mixture models. *Statistics and Computing*, **21**, 93–105.

[32] Lau, W. L. and Green, P. L. (2007). Bayesian model-based clustering procedures. *Journal of Computational and Graphical Statistics*, **16**, 526–558.

[33] Lenk, P. (1988). The logistic normal distribution for Bayesian, nonparametric, predictive densities. *Journal of the American Statistical Association*, **83**, 509–516.

[34] Lijoi, A., Mena, R. H. and Prünster, I. (2005). Hierarchical mixture modelling with normalized inverse Gaussian priors. *Journal of the American Statistical Association*, **100**, 1278–1291.

[35] Lijoi, A., Mena, R. H. and Prünster, I. (2007). Bayesian nonparametric estimation of the probability of discovering new species. *Biometrika*, **94**, 769–786.

[36] Lijoi, A., Mena, R. H. and Prünster, I. (2007). Controlling the reinforcement in Bayesian non-parametric mixture models. *Journal of the Royal Statistical Society. Series B*, **69**, 715–740.

[37] Lijoi, A., Prünster, I. (2010). Models beyond the Dirichlet process. In *Bayesian Nonparametrics* (eds N. L. Hjort, C. C. Holmes, P. Müller, and S. G. Walker), Cambridge University Press.

[38] Lo, A. Y. (1984). On a class of Bayesian nonparametric estimates: I. Density estimates. *Annals of Statistics*, **12**, 351–357.

[39] Mena, R. H. and Walker, S. G. (2009). On a construction of Markov models in continuous time. *METRON – International Journal of Statistics*, **LXVII**, 303–323.

[40] Mena, R. H., Ruggiero, M. and Walker, S. G. (2011). Geometric stick-breaking processes for continuous-time Bayesian nonparametric modeling. *Journal of Statistical Planning and Inference*, **141**, 3217–3230.

[41] MacEachern, S. N. and Müller, P. (1998). Estimating mixtures of Dirichlet process models. *Journal of Computational and Graphical Statistics*, **7**, 223–238.

[42] MacEachern, S. N. (1999). Dependent nonparametric processes. In *ASA Proceedings of the Section on Bayesian Statistical Science*, pp. 50–5. Alexandria, VA: American Statistical Association.

[43] Ongaro, A. and Cattaneo, C. (2004). Discrete random probability measures: a general framework for nonparametric Bayesian inference.*Statistics and Probability Letters*, **67**, 33–45.

[44] Petrone, S., Guindani, M. and Gelfand, A. E. (2009). Hybrid Dirichlet mixture models for functional data. *Journal of the Royal Statistical Society. Series B*, **71**, 755–782.

[45] Pitman, J. (1996). Some developments of the Blackwell–MacQueen urn scheme. *Statistics, Probability and Game Theory. Papers in honor of David Blackwell*, **30**, 245–267.

[46] Raftery, A. and Lewis, S. (1992). One long run with diagnostics: Implementation strategies for Markov chain Monte Carlo. *Statistical Inference*, **7**, 493–497.

[47] Regazzini, E., Lijoi, A. and Prünster, I. (2003). Distributional results for means of random measures with independent increments. *Annals of Statistics*, **31**, 560–585.

[48] Richardson, S. and Green, P. J. (1997). On Bayesian analysis of mixtures with an unknown number of components (with discussion). *Journal of the Royal Statistical Society, Series B*, **59**, 731–792.

[49] Rodriguez, A. and Ter Horst, E. (2008). Bayesian dynamic density estimation. *Bayesian Analysis*, **3**, 339–366.

[50] Rodriguez, A. and Dunson, D. (2011). Nonparametric Bayesian models through probit stick-breaking processes. *Bayesian Analysis*, **6**, 145–178.

[51] Sethuraman, J. (1994). A constructive definition of Dirichlet priors. *Statistica Sinica*, **4**, 639–650.

[52] Sethuraman, J. and Tiwari, R. (1982). Convergence of Dirichlet measures and the interpretation of their parameter. *Proc. Third Purdue Symp. Statist. Decision Theory and Related Topics*, (eds S. S. Gupta and J. Berger), Academic Press NY.

[53] Walker, S. G. (2007). Sampling the Dirichlet mixture model with slices. *Communications in Statistics – Simulation and Computation*, **36**, 45–54.
[54] Walker, S. G., Damien, P., Laud, P. W. and Smith, A. F. M. (1999). Bayesian nonparametric inference for random distributions and related functions. *Journal of the Royal Statistical Society. Series B*, **61**, 485–527.
[55] Walker, S. G. and Mallick, B. (1997). A note of the scale parameter of the Dirichlet process. *The Canadian Journal of Statistics*, **25**, 473–479.

15 Revisiting Bayesian curve fitting using multivariate normal mixtures*

STEPHEN G. WALKER
AND GEORGE KARABATSOS

15.1 Introduction

There has been a significant amount of recent research on Bayesian nonparametric regression models. This has primarily focused on developing models of the form

$$f(y|\mathbf{x}) = \sum_{j=1}^{\infty} w_j(\mathbf{x})\, K(y|\mathbf{x}, \boldsymbol{\theta}_j(\mathbf{x}))$$

where $K(y|\mathbf{x}, \boldsymbol{\theta})$ is a chosen parametric density function, the $w_j(\mathbf{x})$ are mixture weights that sum to 1 at every value of the covariate vector $\mathbf{x} \in \mathcal{X}$, and with a prior distribution on weights $\{w_j(\mathbf{x})\}_{j=1,2,\ldots}$ and atoms $\{\boldsymbol{\theta}_j(\mathbf{x})\}_{j=1,2,\ldots}$, that are an infinite collection of processes indexed by $\mathcal{X}$. The literature on these models has exploded even in the last few years (see for example, [13]–[15], [2]–[8], [20], [11]).

Our thesis is that it is a very difficult task to specify these components; i.e. the $w_j(\mathbf{x})$ and $K(y|\mathbf{x}, \boldsymbol{\theta})$. It is almost limitless in possibilities and over-fitting and un-identifiability are serious issues. There needs to be some guide as to how to choose the components of the regression model.

An intuitive approach to Bayesian nonparametric regression was proposed by [17]. The idea is to specify a Dirichlet process mixture model for the joint density $f(y, \mathbf{x})$, with mixture weights and atoms independent of $\mathbf{x}$ ([12]). This would lead to a standard infinite mixture model treating the $(y_i, \mathbf{x}_i)$ as independent and identically distributed observations. This would not be a controversial choice of model.

In this case one would employ the likelihood function

$$\prod_i f(y_i, \mathbf{x}_i)$$

* This research is supported by National Science Foundation grant SES-1156372.

as used in [17]. However, when the aim is regression the appropriate likelihood function is given by

$$\prod_i f(y_i, \mathbf{x}_i)/f(\mathbf{x}_i) = \prod_i f(y_i|\mathbf{x}_i)$$

To see this we simply note that there are two likelihood functions here and they are not the same. It is also clear that for regression purposes we are interested in the conditional density $f(y|\mathbf{x})$ and it is this that should form the basis of the likelihood function.

It is then possible to note, and we will see this later, that the weights $w_j(\mathbf{x})$ take a particular form which has not appeared in the literature:

$$w_j(\mathbf{x}) = \frac{w_j\, K(\mathbf{x}|\boldsymbol{\theta}_j)}{\sum_j w_j\, K(\mathbf{x}|\boldsymbol{\theta}_j)}$$

The reason why such a simple, motivated and useful regression model has not appeared in the literature is due to the posterior distribution having an intractable normalizing constant. Inference is complicated by the uncomputable integrals in the normalizing constant. Previous authors ([18], Section 3.3) chose to avoid this complication by proposing a modified prior distribution such that, when combined with the correct likelihood, yields a posterior distribution that is identical to the posterior of the original model.

The aim in this paper is to show how it is possible to use the correct likelihood for regression by describing how to deal with the problem of the normalizing constant. This uses ideas recently introduced in [22] which shows how latent variables can be used to construct a latent model that can be studied using Markov chain Monte Carlo (MCMC) methods. The aim is not to compare the model with other models; we take it for granted, based on the work of [17], that the model is going to be useful.

The layout of the paper is as follows: Section 15.2 fully describes our regression model, and the methods for sampling the posterior distribution of the model. To obtain full posterior inference of the model, a reversible-jump sampling algorithm ([22]) is used to deal with the uncomputable normalizing constant. In Section 15.3 we illustrate our model through data analysis.

15.2 The regression model and modelling methods

15.2.1 The model

First we describe the model for $(y, \mathbf{x})$ which is a standard Bayesian nonparametric mixture model based on the mixture of Dirichlet process model. If we take the normal density as the kernel density; i.e.

$$\mathrm{n}(y|\boldsymbol{\mu}_j, \boldsymbol{\Sigma})$$

so

$$f(y, \mathbf{x}) = \sum_j w_j\, \mathrm{n}\,(y, \mathbf{x}|\boldsymbol{\mu}_j, \boldsymbol{\Sigma})$$

where the (w_j) are stick-breaking weights, that is, $w_j = \lambda_j \prod_{l=1}^{j-1}(1 - \lambda_l)$, $\lambda_j \in [0, 1]$, for $j \geq 1$ ([21]). As mentioned, such a modelling approach was taken by [17]. However, for regression modelling, the 'correct' likelihood function is

$$\prod_{i=1}^{n} f(y_i|\mathbf{x}_i)$$

Hence, given data $\mathcal{D}_n = \{(\mathbf{x}_i, y_i)\}_{i=1}^n$, we arrive at

$$f(y|\mathbf{x}) = \frac{f(y,\mathbf{x})}{p(\mathbf{x})} = \frac{\sum_j \mathrm{n}(y,\mathbf{x}|\boldsymbol{\mu}_j, \boldsymbol{\Sigma})\, w_j}{\sum_j \mathrm{n}(\mathbf{x}|\boldsymbol{\mu}_{\mathbf{x}j}, \Sigma_{\mathbf{x}})\, w_j} \tag{15.1}$$

To elaborate, the $f(y, \mathbf{x})$ has a numerator as though

$$\begin{pmatrix} y \\ \mathbf{x} \end{pmatrix} = \mathrm{n}\left(\begin{pmatrix} \mu_{yj} \\ \boldsymbol{\mu}_{\mathbf{x}j} \end{pmatrix}, \begin{pmatrix} \sigma_y^2 & \rho_{y\mathbf{x}} \\ \rho_{y\mathbf{x}} & \Sigma_{\mathbf{x}} \end{pmatrix}\right)$$

and $\boldsymbol{\mu}_j = (\mu_{yj}, \boldsymbol{\mu}_{\mathbf{x},j})$ and

$$\Sigma = \begin{pmatrix} \sigma_y^2 & \rho_{y\mathbf{x}} \\ \rho_{y\mathbf{x}} & \Sigma_{\mathbf{x}} \end{pmatrix}$$

But the likelihood does not assume that the **x**s are randomly generated. It is obviously a valid likelihood since if we integrate out the (y_i) we get 1. But it is strictly a regression model which has motivation when it is acknowledged that constructing $f(y|\mathbf{x})$ is problematic and guidelines are required. Of course, if the **x**s are randomly generated then the model is fully motivated.

The model is completed by the specification of prior densities $\lambda_j \sim_{ind} \mathrm{beta}(a_j, b_j)$ and $\boldsymbol{\mu}_j \sim_{ind} \pi_j(\boldsymbol{\mu}_j)$, $j = 1, 2, \ldots$, and $\Sigma \sim \pi(\Sigma)$. The choice of (a_j, b_j) will be discussed later but we point out now that the choice of $a_j = 1$ and $b_j = b > 0$ leads to the Dirichlet process. Also, a default choice of priors for the mean and covariance matrix $(\boldsymbol{\mu}_j, \Sigma)$ is given by a multivariate normal and inverted-Wishart prior densities, i.e. $\boldsymbol{\mu}_j \sim_{iid} \mathrm{n}_q(\boldsymbol{\mu}_{\boldsymbol{\mu}}, \Sigma_{\boldsymbol{\mu}})$ and $\Sigma \sim \mathrm{iw}_q(\nu_\Sigma, \mathbf{T})$, with $q = \dim(y, \mathbf{x})$, with $(1/\{\nu_\Sigma - q - 1\})\mathbf{T}$ as the mean of the inverted-Wishart density.

We need the denominator of the likelihood, as it contains parameters. It would appear that we would not be able to do inference due to the intractable nature of the denominator, but it turns out that inference is possible by defining a suitable latent model.

If we write

$$f(\mathbf{x}) = |\Sigma_{\mathbf{x}}|^{-1/2} \sum_{j=1}^{\infty} w_j\, m(\mathbf{x}; \boldsymbol{\mu}_{\mathbf{x}j}, \Sigma_{\mathbf{x}})$$

where

$$m(\mathbf{x}; \boldsymbol{\mu}_{\mathbf{x}j}, \Sigma_{\mathbf{x}}) = \exp\left\{-0.5(\mathbf{x} - \boldsymbol{\mu}_{\mathbf{x}j})^\top \Sigma_{\mathbf{x}}^{-1} (\mathbf{x} - \boldsymbol{\mu}_{\mathbf{x}j})\right\}$$

then we can easily note that

$$m(\mathbf{x}) = \sum_{j=1}^{\infty} w_j\, m(\mathbf{x}; \boldsymbol{\mu}_{\mathbf{x}j}, \Sigma_{\mathbf{x}}) < 1$$

And this will be the key to appropriately dealing with the denominator. The idea is that for any $0 < \zeta < 1$ it is that

$$\sum_{k=0}^{\infty}(1-\zeta)^k = \zeta^{-1}$$

Hence, we can represent the denominator at a generic $\mathbf{x}$ value as

$$|\Sigma_{\mathbf{x}}|^{1/2}\sum_{k=0}^{\infty}(1-m(\mathbf{x}))^k$$

and hence we can represent the $\prod_i f(\mathbf{x}_i)$ as

$$|\Sigma_{\mathbf{x}}|^{n/2}\prod_{i=1}^{n}\prod_{l=1}^{k_i} w_{d_{il}}(1-m(\mathbf{x}_i;\boldsymbol{\mu}_{\mathbf{x}_i\,d_{il}},\Sigma_{\mathbf{x}}))$$

For now if we sum over the d_{il}, i.e. for each i and $l \in \{1,\ldots,k_i\}$, we have $d_{il} \in \{1,2,\ldots\}$, then we recover

$$|\Sigma_{\mathbf{x}}|^{n/2}\prod_{i=1}^{n}(1-m(\mathbf{x}_i))^{k_i}$$

and if we now sum over each $k_i \in \{0,1,2,\ldots\}$ then we recover

$$|\Sigma_{\mathbf{x}}|^{n/2}\prod_{i=1}^{n} m^{-1}(\mathbf{x}_i)$$

which is precisely

$$\prod_{i=1}^{n} f^{-1}(\mathbf{x}_i)$$

Specifically, after introducing latent variables $(u_i, u_{il}, v_{il}, d_i, d_{il}, k_i)$, the likelihood becomes:

$$|\Sigma_{\mathbf{x}}|^{n/2}\prod_{i=1}^{n}\Big[\mathbf{1}\left(0 < u_i < \xi_{d_i}\right) w_{d_i}\xi_{d_i}^{-1}\mathrm{n}(y_i,\mathbf{x}_i;\boldsymbol{\mu}_{d_i},\Sigma) \tag{15.2}$$

$$\times\left\{\prod_{l=1}^{k_i}\mathbf{1}(0 < u_{il} < 1-\exp[-.5(\mathbf{x}_i-\boldsymbol{\mu}_{\mathbf{x}d_{il}})^{\top}\Sigma_{\mathbf{x}}^{-1}(\mathbf{x}_i-\boldsymbol{\mu}_{\mathbf{x}d_{il}})]),\right.$$

$$\left.\times\ \mathbf{1}(0 < v_{il} < \xi_{d_{il}})w_{d_{il}}\,\xi_{d_{il}}^{-1}\right\}\Big] \tag{15.3}$$

where $\mathbf{1}(\cdot)$ is the indicator function, and ξ_j is a fixed decreasing function, such as $\xi_j = \exp(-j)$. This slicing strategy is to force the (d_i), and also the (d_{il}) to be bounded and hence can be sampled in a MCMC algorithm. See [10] for a discussion on the choice of (ξ_j). It is easy to show that marginalizing over the latent variables yields the correct likelihood $f(y|\mathbf{x})$ of the regression model, as in (15.1). Importantly, the combined use of ξ with the latent variables facilitates MCMC sampling of the posterior distribution of the infinite-dimensional regression model.

15.2.2 MCMC sampling

For the regression model, an MCMC sampling algorithm is used to iteratively and repeatedly sample from its full conditional posterior distributions. Let $\mathbf{1}_N(j)$ be the function indicating whether $j \in \{1, \ldots, N\}$. Given the likelihood (15.2), the full conditional posterior distributions are given below, for $i = 1, \ldots, n$, $j = 1, \ldots, N$, and $l = 1, \ldots, k_i$.

$$\pi(u_i|\cdots) \propto \mathbf{1}(0 < u_i < \xi_{d_i});$$

$$\pi(u_{il}|\cdots) \propto \mathbf{1}(0 < u_{il} < 1 - \exp\{-.5(\mathbf{x}_i - \boldsymbol{\mu}_{\mathbf{x}d_{il}})^\top \Sigma_{\mathbf{x}}^{-1}(\mathbf{x}_i - \boldsymbol{\mu}_{\mathbf{x}d_{il}})\});$$

$$\pi(v_{il}|\cdots) \propto \mathbf{1}(0 < v_{il} < \xi_{d_{il}});$$

$$\pi(\lambda_j|\cdots) = \text{beta}\begin{pmatrix} a_j + \#(d_i = j) + \#(d_{il} = j), \\ b_j + \#(d_i > j) + \#(d_{il} > j) \end{pmatrix} \mathbf{1}_N(j)$$

$$\Pr\left(d_i = j|\cdots\right) \propto w_j \xi_j^{-1} \mathrm{n}(y_i, \mathbf{x}_i; \boldsymbol{\mu}_j, \Sigma)\mathbf{1}_{N_i}(j);$$

$$\Pr(d_{il} = j|\cdots) \propto w_j \xi_j^{-1}\{1 - \exp[-.5(\mathbf{x}_i - \boldsymbol{\mu}_{\mathbf{x}j})^\top \Sigma_{\mathbf{x}}^{-1}(\mathbf{x}_i - \boldsymbol{\mu}_{\mathbf{x}j})]\}\mathbf{1}_{N_{il}}(j);$$

$$\pi(\boldsymbol{\mu}_j|\ldots) \propto \pi_j(\boldsymbol{\mu}_j)\mathbf{1}_N(j) \prod_{d_i = j} \mathrm{n}(y_i, \mathbf{x}_i|\boldsymbol{\mu}_{d_i}, \Sigma)$$

$$\times \prod_{d_{il}=j} \mathbf{1}\left(u_{il} < 1 - \exp[-.5(\mathbf{x}_i - \boldsymbol{\mu}_{\mathbf{x}d_{il}})^\top \Sigma_{\mathbf{x}}^{-1}(\mathbf{x}_i - \boldsymbol{\mu}_{\mathbf{x}d_{il}})]\right);$$

$$\pi(\Sigma|\cdots) \propto \pi(\Sigma)|\Sigma_{\mathbf{x}}|^{n/2}\left[\prod_{i=1}^{n} \mathrm{n}(y_i, \mathbf{x}_i|\boldsymbol{\mu}_{d_i}, \Sigma)\right]$$

$$\times \prod_{i=1}^{n}\prod_{l=1}^{k_i} \mathbf{1}\left(u_{il} < 1 - \exp[-.5(\mathbf{x}_i - \boldsymbol{\mu}_{\mathbf{x}d_{il}})^\top \Sigma_{\mathbf{x}}^{-1}(\mathbf{x}_i - \boldsymbol{\mu}_{\mathbf{x}d_{il}})]\right)$$

Importantly, since ξ_j is a decreasing function, we can define finite values of $N = \max[\max_i N_i, \max_{i,l} N_{il}]$, $N_i = \max_j j\mathbf{1}(0 < u_i < \xi_j)$, $N_{il} = \max_j j\mathbf{1}(0 < v_{il} < \xi_j)$. Therefore, conditional on latent variables $(d_i, d_{il}, u_i, u_{il}, v_{il}, k_i)$, posterior sampling proceeds as if the infinite mixture model were a finite-dimensional model, as in [10]. All full conditionals can be easily sampled. The nonstandard full conditional densities $p(\boldsymbol{\mu}_j|\ldots)$ are each sampled using the random-walk Metropolis–Hastings algorithm, with normal proposal density $\mathrm{n}_q(\boldsymbol{\mu}_j, \mathrm{diag}(v_1, \ldots, v_q))$ having variances $(v_1, \ldots, v_q)$ automatically adapted to achieve the desired acceptance rate of 0.44 for each respective component of $\boldsymbol{\mu}$ over MCMC iterations, using a Robbins–Monro algorithm (see [1]). Also, when the model is assigned prior $\Sigma \sim \mathrm{iw}_q(\nu_\Sigma, \mathbf{T})$, the nonstandard full conditional posterior density $\pi(\Sigma|\cdots)$ can be sampled using an independent Metropolis–Hastings algorithm, with $\mathrm{iw}_q(\nu_\Sigma, c\mathbf{T})$ as the proposal density for some chosen constant $c > 0$.

The sampling of the full conditional distribution $\Pr(k_i|\cdots)$ requires reversible-jump MCMC methods ([22]), because the dimensionality of the model parameters changes with k_i. In particular, using the full set $(u_{il}, v_{il}, d_{il})_{l=1}^{\infty}$ we construct the following joint density

$$p(k_i, (u_{il}, v_{il}, d_{il})_{l=1}^{\infty} | \cdots) \propto \prod_{l=1}^{k_i} \mathbf{1}\left(0 < u_{id_{il}} < 1 - \exp\left[\begin{array}{c} -.5(\mathbf{x}_i - \boldsymbol{\mu}_{\mathbf{x}d_{il}})^\top \\ \times \Sigma_{\mathbf{x}}^{-1}(\mathbf{x}_i - \boldsymbol{\mu}_{\mathbf{x}d_{il}}) \end{array}\right]\right)$$
$$\times \prod_{l=1}^{k_i} \mathbf{1}_{N_{il}}(d_{il})\mathbf{1}(0 < v_{il} < \xi_{d_{il}})\, w_{d_{il}}\xi_{d_{il}}^{-1}$$
$$\times \prod_{l=k_i}^{\infty} p(u_{i,l+1}, v_{i,l+1}, d_{i,l+1} | u_{il}, v_{il}, d_{il}),$$

where the last term is a product of densities, each serving as a proposal density, which could be chosen as an independent (proposal) density:

$$p(u_{i,l+1}, v_{i,l+1}, d_{i,l+1} | u_{il}, v_{il}, d_{il}) = p(u_{i,l+1}, v_{i,l+1}, d_{i,l+1}) = \mathbf{1}_{N_{il}}(d_{i,l+1})\mathbf{1}(0 < v_{i,l+1} < \xi_{d_{i,l+1}})$$
$$\times \frac{\mathbf{1}(0 < u_{id_{i,l+1}} < 1 - \exp[-.5(\mathbf{x}_i - \boldsymbol{\mu}_{\mathbf{x}d_{i,l+1}})^\top \Sigma_{\mathbf{x}}^{-1}(\mathbf{x}_i - \boldsymbol{\mu}_{\mathbf{x}d_{i,l+1}})])}{1 - \exp[-.5(\mathbf{x}_i - \boldsymbol{\mu}_{\mathbf{x}d_{i,l+1}})^\top \Sigma_{\mathbf{x}}^{-1}(\mathbf{x}_i - \boldsymbol{\mu}_{\mathbf{x}d_{i,l+1}})]}.$$

Then given that the MCMC chain is at state k_i, a proposal is made to move to state $k_i + 1$ with probability $q(k_i + 1|k_i)$, and given a sample $(u_{i,k_i+1}, v_{i,k_i+1}, d_{i,k_i+1})$ from $p(u_{i,k_i+1}, v_{i,k_i+1}, d_{i,k_i+1})$, this proposal is accepted with probability

$$\min\left[1, \frac{w_{d_{i,k_i+1}}\xi_{d_{i,k_i+1}}^{-1}\{1 - \exp[-.5(\mathbf{x}_i - \boldsymbol{\mu}_{\mathbf{x}d_{i,k_i+1}})^\top \Sigma_{\mathbf{x}}^{-1}(\mathbf{x}_i - \boldsymbol{\mu}_{\mathbf{x}d_{i,k_i+1}})]\}}{q(k_i + 1|k_i)/q(k_i|k_i + 1)}\right]$$

Otherwise, with probability $q(k_i - 1|k_i)$, a proposal is made to move to state $k_i - 1$, and this proposal is accepted with probability

$$\min\left[1, \frac{q(k_i|k_i - 1)/q(k_i - 1|k_i)}{w_{d_{ik_i}}\xi_{d_{ik_i}}^{-1}\{1 - \exp[-.5(\mathbf{x}_i - \boldsymbol{\mu}_{\mathbf{x}d_{ik_i}})^\top \Sigma_{\mathbf{x}}^{-1}(\mathbf{x}_i - \boldsymbol{\mu}_{\mathbf{x}d_{ik_i}})]\}}\right].$$

We can define the proposal distribution by $q(1|0) = 1, q(0|1) = 0$, and $q(k'|k) = .5$ for $k > 0$ and for all $|k' - k| = 1$.

Finally, the full conditional posterior (predictive) density of Y_i ($i = 1, \ldots, n$) is:

$$\mathrm{n}(\mu_{yd_i} + \sum\nolimits_{s=1}^{p}(x_{si} - \mu_{x_s d_i})/\Delta_y, 1/\Delta_y),$$

with $\Delta = \Sigma^{-1}$, following known results involving a univariate conditional distribution of a multivariate normal density. Also, the predictive accuracy of the model can proceed via the evaluation of standardized residuals $(y_i - \mu_{yd_i})\Delta_y^{1/2}, i = 1, \ldots, n$.

To conduct MCMC sampling of the model, we wrote code in MATLAB (2011, The MathWorks, Natick, MA).

15.3 Illustrations

In this section we illustrate our model through the analysis of simulated and real data. In each illustration, we have rescaled y and each covariate x_s $(s = 1, \ldots, p)$ to have mean 0 and variance 1 prior to model fitting, assigned prior densities $\lambda_j \sim_{ind} \text{beta}(1/2, 1/2 + j/2)$, $\boldsymbol{\mu}_j \sim_{iid} \text{n}_q(\mathbf{0}, \mathbf{I})$ $(j = 1, 2, \ldots)$, and $\Sigma \sim \text{iw}_q(q + 2, \mathbf{I})$ to the parameters of the regression model, and chosen an inverted-Wishart $\text{iw}_q(\nu_\Sigma, .5\mathbf{T})$ proposal density for the independent Metropolis sampling of Σ in the MCMC algorithm. Hence, a Poisson–Dirichlet (Pitman–Yor) process prior was assigned to the mixture weights λ_j (e.g., [9]). Also, all results are reported on the original scale of y and $\mathbf{x}$. Finally, for each data illustration, we have estimated the regression model based on 20 000 samples of the MCMC algorithm, long after the predictions of the model seemed to stabilize over the MCMC iterations. We estimate the predictions of the model from every fifth MCMC sample of its predictive distribution.

15.3.1 Simulation

We first illustrate the model through the analysis of a simulated dataset with 61 observations, coming from $Y_i \sim \text{n}(0.2x_i^3, 0.25)$, with $x_1 = -3, x_2 = -2.9, \ldots, x_{n-1} = 2.9, x_n = 3$. Figure 15.1 presents the simulated data, and for the model, presents the estimate of the posterior predictive mean and interquartile range of Y conditional on each of the observed values of x. We see that the range captures the small number of simulated data points.

15.3.2 Crime Data

Here we present an analysis of a dataset described in [16], consisting of information on urban population percentage, the number of murder arrests, assault arrests, and rape arrests per

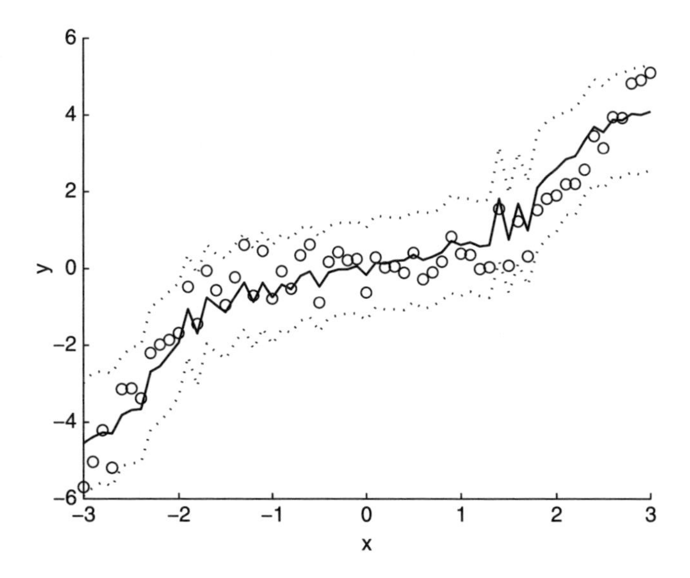

Figure 15.1 Simulated example. From the posterior predictive distribution of the model, the mean (solid line) and interquartile range (dashed lines) of Y, conditional on values of x.

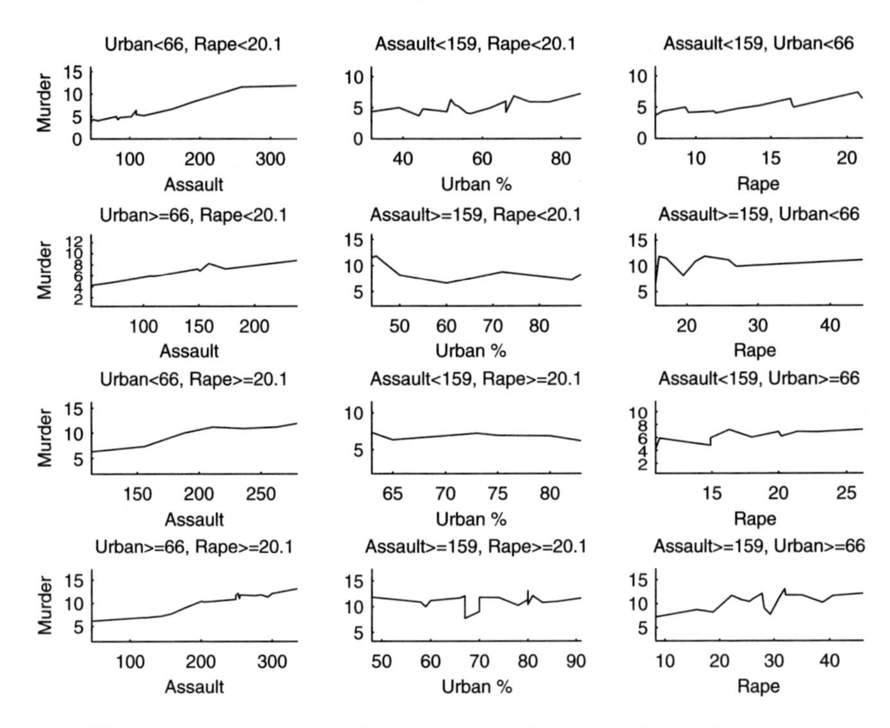

Figure 15.2 Crime data. Estimates of the posterior predictive mean of Y, conditional on values of the three covariates.

100 000 people, for each of the $n = 50$ U.S. states during the year 1973. This 'USArrests' dataset was obtained from the datasets package of the R software ([19]). We treat murder rate as the dependent variable, and the other three variables as covariates.

From the regression model, Figure 15.2 presents estimates of the posterior predictive mean of Y, conditional on values of the three covariates. The figure shows that there is generally a positive correlation between assault arrests and rape arrests with murder arrests, and that there are nonlinear and interactive relationships between each of the three covariates and murder arrests. Also, an inspection of the posterior predictive distribution of the standardized residuals revealed that there were no outliers in the model.

15.4 Discussion

In this paper, we have developed a Bayesian nonparametric regression model which relies on a standard Bayesian nonparametric form for the joint distribution of both the dependent and independent variables. The regression model then is available as a conditional density which can only be written as a ratio of two infinite-dimensional mixture models.

While this model has been acknowledged as desirable, its drawback has been the intractability of the likelihood, specifically the denominator of the likelihood. However, recent advances in both latent models for dealing with normalizing constants and simulation techniques linking slice sampling and infinite mixture models have now rendered this model tractable.

References

[1] Atchadé, Y. F. and Rosenthal, J. S. (2005). On adaptive Markov chain Monte Carlo algorithms. *Bernoulli*, **11**, 815–828.

[2] Chung, Y. and Dunson, D. B. (2009). Nonparametric Bayes conditional distribution modelling with variable selection. *Journal of the American Statistical Association*, **104**, 1646–1660.

[3] DeIorio, M., Müller, P., Rosner, G. L. and MacEachern, S. N. (2004). An ANOVA model for dependent random measures. *Journal of the American Statistical Association*, **99**, 205–215.

[4] Dunson, D. and Park, J.-H. (2008). Kernel stick breaking processes. *Biometrika*, **95**, 307–323.

[5] Fuentes-García, R., Mena, R. H. and Walker, S. G. (2010). A new Bayesian nonparametric mixture model. *Communications In Statistics*, **39**, 669–682.

[6] Gelfand, A. E., Kottas, A. and MacEachern, S. N. (2005). Bayesian nonparametric spatial modelling with Dirichlet processes mixing. *Journal of the American Statistical Association*, **100**, 1021–1035.

[7] Griffin, J. E. and Steel, M. F. J. (2006). Order-based dependent Dirichlet processes. *Journal of the American Statistical Association*, **101**, 179–194.

[8] Griffin, J. E. and Steel, M. F. J. (2010). Bayesian nonparametric modelling with the Dirichlet process regression smoother. *Statistica Sinica*, **20**, 1507–1527.

[9] Ishwaran, H. and James, L. F. (2001). Gibbs sampling methods for stick-breaking priors. *Journal of the American Statistical Association*, **96**, 161–173.

[10] Kalli, M., Griffin, J. and Walker, S. G. (2010). Slice sampling mixture models. *Statistics and Computing*, **21**, 93–105.

[11] Karabatsos, G. and Walker, S. G. (2011). Adaptive-modal Bayesian nonparametric regression. Technical report, University of Illinois, Chicago.

[12] Lo, A. Y. (1984). On a class of Bayesian nonparametric estimates. *Annals of Statistics*, **12**, 351–357.

[13] MacEachern, S. N. (1999). Dependent nonparametric processes. *Proceedings of the Bayesian Statistical Sciences Section of the American Statistical Association*, 50–55.

[14] MacEachern, S. N. (2000). Dependent Dirichlet Processes. Technical report, Department of Statistics, The Ohio State University.

[15] MacEachern, S. N. (2001). Decision theoretic aspects of dependent nonparametric processes. In *Bayesian Methods with Applications to Science, Policy and Official Statistics* (ed. E. George), Creta, pp. 551–560. International Society for Bayesian Analysis.

[16] McNeil, D. R. (1977). *Interactive Data Analysis*. John Wiley, New York.

[17] Müller, P., Erkanli, A. and West, M. (1996). Bayesian curve fitting using multivariate normal mixtures. *Biometrika*, **83**, 67–79.

[18] Müller, P., Quintana, F. A. and Rosner, G. (2004). A method for combining inference across related nonparametric Bayesian models. *Journal of the Royal Statistical Society, Series B*, **66**, 735–749.

[19] R Development Core Team (2011). *R: A Language and Environment for Statistical Computing*. R Foundation for Statistical Computing, Vienna, Austria.

[20] Rodriguez, A. and Dunson, D. B. (2011). Nonparametric Bayesian models through probit stick-breaking processes. *Bayesian Analysis*, **6**, 1–34.

[21] Sethuraman, J. (1994). A constructive definition of Dirichlet priors. *Statistica Sinica*, **4**, 639–650.

[22] Walker, S. G. (2011). Posterior sampling when the normalizing constant is unknown. *Communications in Statistics: Simulation and Computation*, **40**, 784–792.

Part VII

Spline Models and Copulas

16 Applications of Bayesian smoothing splines

SALLY WOOD

Over the last two decades there has been an explosion in the development and application of Bayesian smoothing techniques. The purpose of this chapter is to show how Bayesian smoothing splines together with the data augmentation techniques of [37] and [13] can be used to address several important practical problems. The chapter begins with a brief introduction to Bayesian smoothing splines and then shows how they can be used in a variety of applied settings, including medical diagnosis, climatology and astronomy.

16.1 Smoothing Splines. The basics

There are many ways to think about smoothing splines, [14] and [11] provide comprehensive and clear discussions on the topic, motivated from a roughness penalty approach. In this chapter we take a related but slightly different approach; we motivate the concept by asking what type of prior information do we need to impose on a function, so that, given some data, an estimate of this function will have desirable properties.

Suppose that an observation, y_i, is generated from a signal g which depends upon a covariate value, x_i, plus noise, e_i, so that

$$y_i = g(x_i) + e_i \tag{16.1}$$

In a Bayesian context a natural estimate of a function is its posterior mean, and we would like to select a prior for g so that this estimate has the following two properties:

1. The estimate of g is a smooth function, and
2. The prior is flexible enough to give good estimates for a large range of functions.

Choosing a parametric form for g, for example, taking g as a linear function satisfies the first requirement, but not the second.

One way to impose prior information on a function is to decompose the function into two components. The first component is a polynomial of degree $m - 1$ and the second is a Gaussian stochastic process, which represents departures from this polynomial, see [41], and [11] for details. One example is,

$$g(x_i) = \sum_{k=0}^{m-1} \alpha_k x_i^k + f(x_i)$$

$$f(x) = (\tau^2)^{1/2} \int_0^x \frac{(x-v)^{(m-1)}}{(m-1)!} dW(v) \tag{16.2}$$

where the notation $a^{(m)}$ means the mth derviative of a, W is a Wiener process with $W(0) = 0$, and $\text{var}(W(x)) = x$. This prior states that $f(x)$ has a normal distribution with a mean of zero and covariance matrix Ω_m, where $\tau^2\omega_{mij} = \text{cov}(f(x_i), f(x_j))$ is the $(i,j)th$ element of $\tau^2\Omega_m$. The values of ω_{mij} for $m = 1$ and $m = 2$ are,

$$\begin{aligned} \omega_{mij} &= x_i^2(x_j - x_i/3)/2 \text{ for } (x_i \le x_j) \text{ if } m = 2; \\ &= \min(x_i, x_j) \text{ if } m = 1. \end{aligned} \tag{16.3}$$

In (16.2) the parameter τ^2 controls the curvature of f and is called a smoothing parameter. The advantage of expressing the prior as a random function with a specific covariance structure is that many other forms of smoothing can be similarly expressed. For example kriging can be expressed as a smoothing spline, see [23] for details.

If the prior on $\boldsymbol{\alpha} = (\alpha_0, \ldots, \alpha_{m-1})$ is diffuse, for example $\boldsymbol{\alpha} \sim N(0, c_\alpha I_m)$, with $c_\alpha \to \infty$, then the prior in (16.3) has the following properties:

1. There is no prior information about g and its first $m-1$ derivatives at $x = 0$ because $\alpha_k = g^{(k)}(0)$ for $k = 0, \ldots, m-1$ and the prior on $\boldsymbol{\alpha}$ is diffuse.
2. The first $m-1$ derivatives of g are continuous because a Wiener process is continuous.
3. There is no prior information about the mth derivative of g, because $d^m g(x)/dx^m = dW(x)/dx$ and the first derivative of a Wiener process is infinite.

The prior in (16.2) is completed by specifying a prior for τ^2, which is usually taken to be an uninformative inverse gamma, $IG(a, b)$, with parameters a and b. If e_i in (16.1) is $N(0, \sigma^2)$ and $g(x)$ is estimated by its posterior mean, $E(g(x)|\boldsymbol{y})$ then, conditional on τ^2 and σ^2, this estimate has the following properties:

1. The estimate is a polynomial of degree $2m-1$ in each of the subintervals $(x_{i-1}, x_i), i = 2, ..., n$. This result can easily be shown by noting that $\boldsymbol{y} = (y_1, \ldots y_n)$ and $\boldsymbol{g} = (g(x_1), \ldots, g(x_n))$ are jointly normal and hence

$$E(g(x_i)|\boldsymbol{y}, \tau^2, \sigma^2) = E(\alpha_0 + \alpha_1 x_i|\boldsymbol{y}) + \omega_{mi.}\left(\Omega + I_n\frac{\sigma^2}{\tau^2}\right)^{-1}(\boldsymbol{y} - E(\boldsymbol{y}))$$

   where $\omega_{mi.}$ is a $1 \times n$ row vector with jth entry equal to ω_{mij} in (16.3), see [41] for a full discussion.
2. The estimate has continuous mth derivatives throughout its range.
3. The estimate is a polynomial of order $m-1$ for $x < x_1$ and $x > x_n$.
4. If $\tau^2 = 0$, the estimate is a polynomial of order $m-1$, and as $\tau^2 \to \infty$, $E(g(x_i)|\boldsymbol{y}) \to y_i$.

To compute the posterior mean and variance of $g(x_i)$ unconditional on σ^2 and τ^2, requires performing the multidimensional integrations

$$
\begin{aligned}
E[g(x_i)|\mathbf{y}] &= \int E\left[g(x_i)|\mathbf{y}, \sigma^2, \tau^2\right] p(\sigma^2, \tau^2|\mathbf{y}) d(\sigma^2, \tau^2) \\
\text{var}[g(x_i)|\mathbf{y}] &= E\left[g(x_i)^2|\mathbf{y}\right] - E\left[g(x_i)|\mathbf{y}\right]^2 \\
&= \int E\left[g(x_i)^2|\mathbf{y}, \tau^2, \sigma^2\right] d(\sigma^2, \tau^2) - E\left[g(x_i)|\mathbf{y}\right]^2
\end{aligned}
$$

respectively. Estimates of $E[g(x_i)|\mathbf{y}]$ and $\text{var}[g(x_i)|\mathbf{y}]$ are

$$
\begin{aligned}
\hat{g}(x_i) &= \frac{1}{M} \sum_{k=1}^{M} E\left[g(x_i)|\mathbf{y}, \sigma^{2[k]}, \tau^{2[k]}\right] \\
\hat{\sigma}_{g_i}^2 &= \frac{1}{M} \sum_{k=1}^{M} E\left[g(x_i)^2|\mathbf{y}, \sigma^{2[k]}, \tau^{2[k]}\right] - \hat{g}(x_i)^2
\end{aligned}
\tag{16.4}
$$

where the sequence $\sigma^{2[k]}, \tau^{2[k]}$ are drawn from $p(\sigma^2, \tau^2|\mathbf{y})$ using Markov chain Monte Carlo (MCMC). A typical sampling scheme to obtain these draws proceeds along the following lines;

Sampling scheme 16.1

1. Sample $\mathbf{g}$ conditional on $\mathbf{y}$, τ^2 and σ^2,
2. Sample τ^2 conditional on $\mathbf{g}$,
3. Sample σ^2 conditional on $\mathbf{y}$ and $\mathbf{g}$.

Steps (2) and (3) are straightforward and algorithms for the efficient sampling of $\mathbf{g}$ include [4], [12], [7] and [10].

16.1.1 Smoothing splines as linear combinations of basis functions

An equivalent representation of (16.2) is to write the unknown function $f(x)$ as a linear combination of basis functions

$$
\begin{aligned}
f(x) &= X_m \boldsymbol{\beta} \\
\Omega_m &= Q_m D_m Q_m' \text{ and } X_m = Q_m \\
\boldsymbol{\beta}_m &\sim N(0, \tau^2 D_m)
\end{aligned}
\tag{16.5}
$$

where $Q_m D_m Q_m'$ is an eigenvalue decomposition of the covariance matrix Ω_m, with Q_m the matrix of eigenvectors and D_m a diagonal matrix of eigenvalues. The columns of $X_m = Q_m$ are an orthogonal basis and similar to the Demmler Reinsch basis functions, see [8].[8] The subscript m indicates that different values of m will give different basis functions. The basis functions for $m = 1$ and $m = 2$ are shown in Figure 16.1 and, as can be seen, are very similar.

The prior on the coefficients of the basis functions, $\boldsymbol{\beta}_m \sim N(0, \tau^2 D_m)$, states that although the coefficients are different they have in common a distribution, which is normal with a zero mean and a variance $\tau^2 D_m$. The smoothness of the posterior mean of $g(x)$ is determined by the variance of

[8] The Demmler Reinsch basis functions are formed from the eigenvalue decomposition of the matrix, $\tau^2\Omega(\tau^2\Omega + \sigma^2 I)^{-1}$, which maps the data, $\mathbf{y}$, to the fitted values.

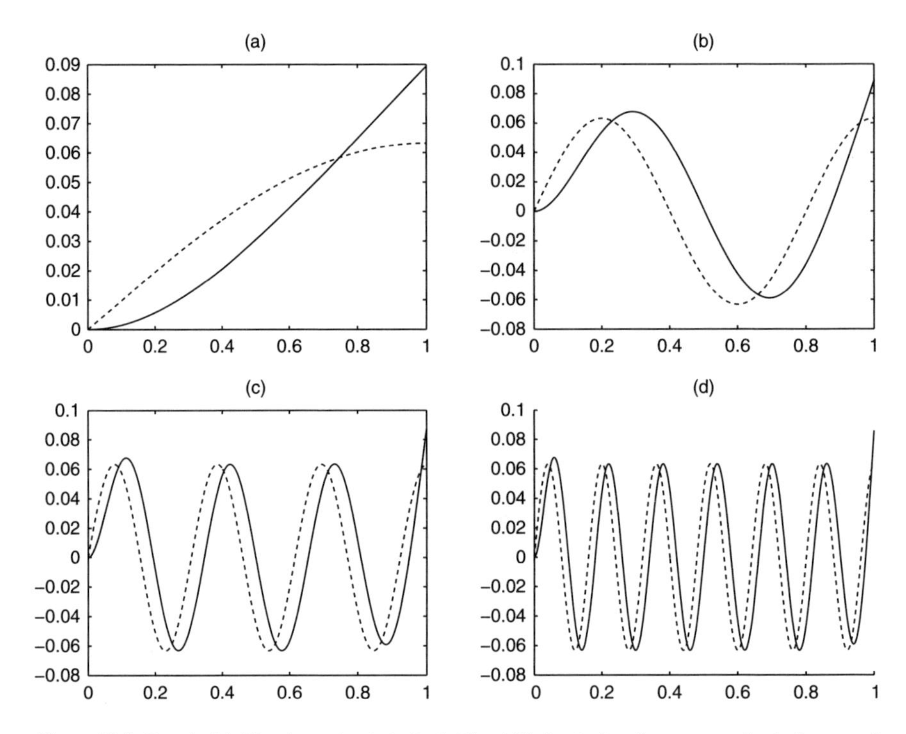

Figure 16.1 Panels (a)–(d), show the 1st, 2nd, 7th, 10th basis functions respectively for $m = 2$ (solid) and $m = 1$ (dashed), for equally spaced data on the interval [0,1], with $n = 500$.

these coefficients; a smaller variance shrinks the coefficients closer to zero, thus giving a smoother estimate. In the extreme if $\tau^2 = 0$, then all of the coefficients are identically zero and the posterior mean of $g(x)$ will be a polynomial of order $2m - 1$. Similarly as $\tau^2 \to \infty$ these coefficients have no restrictions and can lie anywhere in $\mathbb{R}^n$ and will therefore interpolate the data.

The value of m in the prior given by (16.2) also affects the variance of the coefficients via the diagonal matrix D_m, and hence the smoothness of the posterior mean of $g(x)$. In general the higher the value of m, the more prior information is imposed on the function and its derivatives and hence the smoother the value of $E(g(x)|\boldsymbol{y})$ will be. Figure 16.2 plots the diagonal elements of D_m, d_{jm} for $j = 1, \ldots, n$ against the index of the columns of X_m for $m = 1$ (dashed) and $m = 2$ (solid). For a given value of τ^2, Figure 16.2 shows that the variance of the coefficients β_{jm} decreases as the oscillation frequency of the basis functions increases. Figure 16.2 also shows that this rate of decrease is higher for $m = 2$ than $m = 1$ and therefore a priori we expect the estimate of a function based on a prior with $m = 1$ to be less smooth than one based on a prior with $m = 2$.

The advantage of using the representation in (16.5) is that the model in (16.1) is linear in the unknown parameters $\boldsymbol{\beta}_m$, and its posterior mean is the value of $\boldsymbol{\beta}_m$ which maximizes the penalized likelihood

$$p(\boldsymbol{\beta}^*_m|\boldsymbol{y}) \propto \exp -\frac{1}{2}\left\{\frac{(\boldsymbol{y} - X^*_m\boldsymbol{\beta}^*_m)'(\boldsymbol{y} - X^*_m\boldsymbol{\beta}^*_m)}{\sigma^2} + \frac{\boldsymbol{\beta}^{*'}_m D^{*-1}_m \boldsymbol{\beta}^*_m}{\tau^2}\right\}$$

where $X^*_m = [1_n, \boldsymbol{x}, \ldots, \boldsymbol{x}^{m-1}, X_m]$, and $\boldsymbol{x}^k = (x^k_1, \ldots, x^k_n)'$ for $k = 1, \ldots, m-1$; $\boldsymbol{\beta}^*_m = (\alpha_0, \ldots, \alpha_{m-1}, \boldsymbol{\beta}'_m)'$ and $D^*_m = \text{diag}(c_\alpha \mathbf{1}'_m, \boldsymbol{d}'_m)$ and $\boldsymbol{d}_m = (d_{1m}, \ldots, d_{nm})'$. As in Section 16.1 an

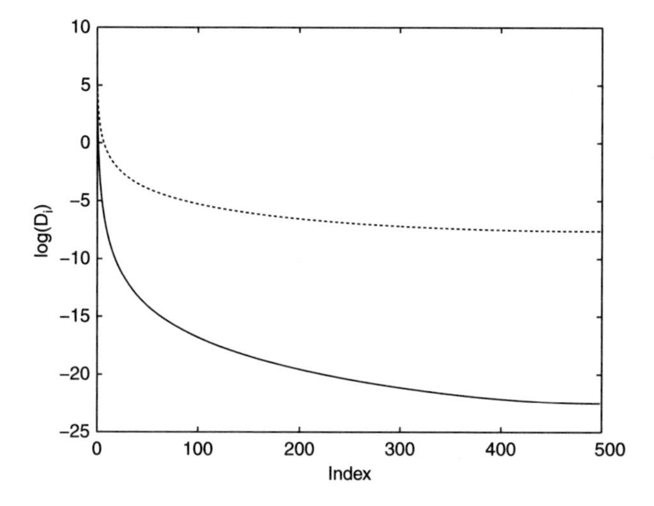

Figure 16.2 The log of the eigenvalues, $\log(D_{im})$ corresponding to the index of the eigenvectors, for $m = 2$ (solid) and $m = 1$ (dashed), for equally spaced data on the interval [0,1], with $n = 500$.

MCMC scheme to estimate the posterior mean $X_m^* \times E[\boldsymbol{\beta}_m^* | \mathbf{y}]$ usually proceeds along the following lines:

Sampling scheme 16.2

1. Sample $\boldsymbol{\beta}_m^*$ conditional on τ^2 and σ^2,
2. Sample τ^2 conditional on $\boldsymbol{\beta}_m^*$,
3. Sample σ^2 conditional on $\boldsymbol{\beta}_m^*$.

By representing smoothing splines as linear combinations of basis functions it becomes obvious that smoothing splines are similar to regression splines with a knot at each observation. If n is large then performing an eigenvalue decomposition and estimating $\boldsymbol{\beta}_m$ is computationally infeasible. However, truncating the number of basis functions to those eigenvectors corresponding to the largest p eigenvalues, with p of the order $n/10$ works well in practice.[9]

Example 16.1 *Modelling solar activity*
This example shows how smoothing splines can be used to model solar activity for the last 350 years. Sunspots are strong concentrations of magnetic flux on the sun's surface and appear as dark spots on the surface. They typically last for several days, although very large ones may live for several weeks. Sunspots have been measured by direct observation since 1610. The sunspot number is calculated by first counting the number of sunspot groups and then the number of individual sunspots. The sunspot number is then given by the sum of the number of individual sunspots and ten times the number of groups. The data are available at http://solarscience.msfc.nasa.gov/greenwch/spot_num.txt and appear in Figure 16.3. Our model for the number of sunspots is given by (16.1), where y_i is the square root of the sunspot number

[9] Although [17] point out that basis functions which have a small variance *a priori* do not necessarily have a posterior mean which is close to 0.

and x_i is the time at which the observation was recorded.[10] This figure shows that the number of sunspots has a cycle of approximately 11 years. The data were chosen because they show that the estimate of the posterior mean $E(\boldsymbol{g}|\boldsymbol{y})$ is insensitive to the number of truncated basis functions p, for $p > n/50$ even for functions with high oscillation frequency.

16.2 Extension to non-Gaussian data

The beauty of Bayesian statistics and MCMC methodology is having established estimation techniques for the plain vanilla version, techniques for estimating a richer class of models can often be achieved using data augmentation, as discussed by [37]. The purpose of data augmentation is to facilitate MCMC schemes which perform the multidimensional integration needed to estimate the required features of the marginal distributions, such as the posterior mean of a regression function. The type of data augmentation depends upon the problem at hand.

For example consider n binary observations $\boldsymbol{w} = (w_1, \ldots, w_n)$, where the goal is to estimate $\Pr(w_i = 1|x_i) = E(w_i = 1|x_i)$. One possible model for this expected value is

$$E(w_i|x_i) = H(g(x_i)) \tag{16.6}$$

where $H(.)$ is called a link function and g is some function of x. One choice for $H(.)$ in (16.6) is the standard normal cumulative distribution function (cdf), denoted by Φ. Estimating $\Phi\{g(x)\}$ by its posterior mean requires a sequence of draws from $p(\tau, \sigma^2|\boldsymbol{w})$. If the data are Gaussian (conditional on the covariates), then this is achieved by using Sampling schemes 16.1 or 16.2. If the data are non-Gaussian then this sequence of draws can be obtained by using data augmentation to modify these sampling schemes as discussed in [1].

The idea in [1] is to augment the data with a vector of latent variables $\boldsymbol{y} = (y_1, \ldots, y_n)$, where $y_i = g(x_i) + e_i$, with $e_i \sim N(0, 1)$. These latent variables are connected to the observations by requiring $y_i > 0$ if $w_i = 1$ and $y_i < 0$ otherwise. Then

$$\Pr(w_i = 1|g(x_i)) = \Pr(y_i > 0|g(x_i)) = \Phi\{g(x_i)\}$$

The posterior mean of $g(x_i)$ estimated by

$$\Phi\{\hat{g}(x_i)\} = \frac{1}{M}\sum_{j=1}^{M} E[\Phi\{g(x_i|\boldsymbol{w}, \boldsymbol{y}^{[j]}, \tau^{2[j]})\}]$$

where the sequence $\boldsymbol{y}^{[j]}, \tau^{2[j]}\, j = 1, \ldots, M$ is generated from the posterior distribution $p(\boldsymbol{y}, \tau^2|\boldsymbol{w})$. The sampling scheme implemented to obtain this sequence is:

Sampling scheme 16.3

1. Sample $\boldsymbol{y}$ conditional on $\boldsymbol{w}$ and $g(\boldsymbol{x})$,
2. Sample $g(\boldsymbol{x})$ conditional on τ^2 and $\boldsymbol{y}$,
3. Sample τ^2 conditional on $g(\boldsymbol{x})$.

[10] Note, data was transformed by taking the square root so that the error term in (16.1) is approximately *i.i.d* $\sim N(0, \sigma^2)$, despite that fact that it must always be positive.

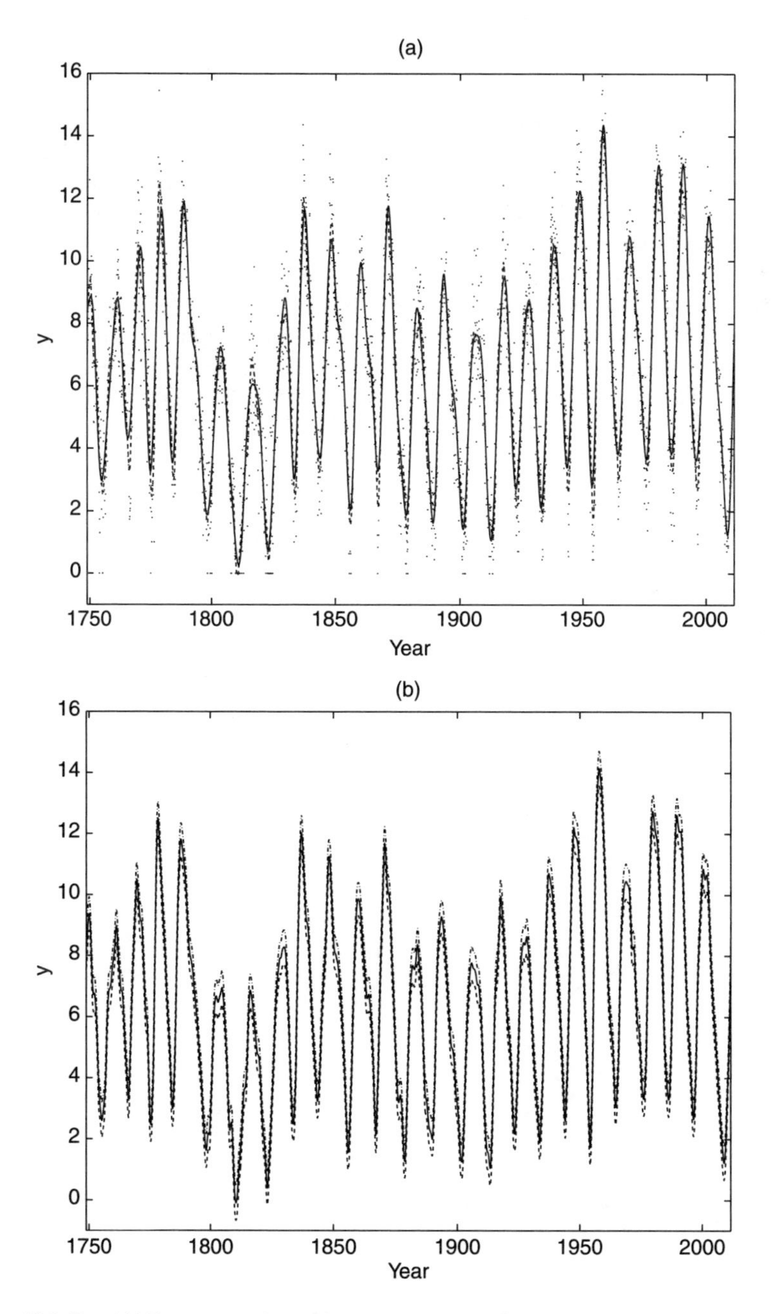

Figure 16.3 Panel (a) Square root of monthly sunspot numbers from 1741 to 2010. Three estimates of the posterior mean $E(\mathbf{g}|\mathbf{y})$, corresponding to the number of basis functions, p, equal to n, $n/10$ and $n/50$ are also plotted but are visually indistinguishable. Panel (b) Estimate of $E(\mathbf{g}|\mathbf{y})$ (solid) together with 95% posterior intervals (dashed) when $p = n/10$.

It should be noted that when the data are non-Gaussian the posterior mean is not necessarily the same as the posterior mode and therefore estimating $\Phi\{g(x_i)\}$ by maximizing the penalized likelihood as in [14] will give different results from estimating $\Phi\{g(x_i)\}$ by its posterior mean.

Example 16.2 *Modelling the probability of a heart attack*
The data augmentation technique is now applied to a dataset consisting of $n = 463$ observations on the dependent variable w, which equals 1 if the subject had a heart attack and 0 if he or she did not, and three risk factors; systolic blood pressure (BP), cholesterol ratio (CR), and age of patient (Age). These data have been described and analysed by [16] and [45].

The aim is to model the dependency between the probability of having a heart attack and these three risk factors in a flexible but robust manner. It is important that the method is made robust to outliers or miscoded observations because of the impact these observations can have on the inference regarding the effect of risk factors. To illustrate the data augmentation technique of [1] we first describe the sampling scheme omitting outlier detection and consider a single risk factor, CR, only. The model for the data in Figure 16.4 is

$$\Pr(w = 1|CR) = \Phi\{g(CR)\}$$

where

$$g(CR) = \alpha_0 + \alpha_1 CR + f(CR)$$

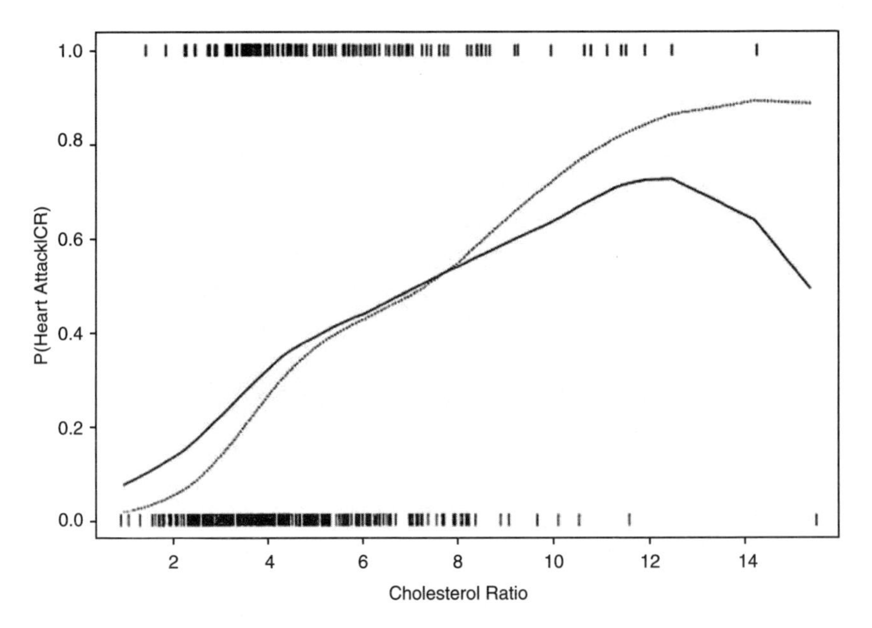

Figure 16.4 Plot of the incidence of heart attack versus cholestertol ratio (CR), with an estimate of the posterior mean of $\Phi\{(CR)\}$ when outlier detection is turned on (dashed) and when it is turned off (solid). There were many coincident points, so to show the distribution of the data the values of CR were perturbed by adding a small amount of Gaussian noise to their values. This figure was taken from [45].

The estimate of the posterior mean $E[\Phi\{g(CR)\}|\boldsymbol{w}]$ is

$$\Phi\{\hat{g}(CR)\} = \frac{1}{M}\sum_{k=1}^{M} E[\Phi\{g(CR)\}|\boldsymbol{w}, \boldsymbol{y}^{[k]}, \tau^{[k]})]$$

where the sequence $\boldsymbol{y}^{[k]}, \tau^{2[k]}$ is drawn from $p(\boldsymbol{y}^{[k]}, \tau^{[k]}|\boldsymbol{w})$ using Sampling scheme 16.3. Figure 16.4, solid line, shows that the probability of a heart attack increases as cholesterol ratio increases but drops sharply for high levels of cholesterol ratio. This sharp drop may be due to the effect the observation $(CR = 15.33, w = 0)$ has on the estimate of $\Phi\{\hat{g}(CR)\}$.

16.2.1 Robust binary nonparametric regression

To make the estimation procedure robust to outliers, data augmentation is used again. A vector of indicator variables is introduced where, $\boldsymbol{\gamma} = (\gamma_1, \ldots, \gamma_n)$, and $\gamma_i = 1$ if an observation is an outlier or miscoded and $\gamma_i = 0$ otherwise as in [39]. We assume a priori that the γ_i are independent with $\Pr(\gamma_i = 1) = \pi$. In this application we take $\pi = 0.05$ and find that this choice works well in practice. The sampling scheme now becomes

Sampling scheme 16.4

1. Sample $\boldsymbol{y}$ and $\boldsymbol{\gamma}$ jointly conditional on $\boldsymbol{w}$ and $\boldsymbol{g} = (g_1, \ldots, g_n)$ by drawing
   (a) $\boldsymbol{\gamma}$ from

$$p(\boldsymbol{\gamma}|\boldsymbol{w}, \boldsymbol{\alpha}, \boldsymbol{f}) = \prod_{i=1}^{n} p(\gamma_i|w_i, \boldsymbol{\alpha}, f(CR_i))$$

   where

$$\Pr(\gamma_i = 1|w_i, g(CR_i)) = \frac{\Pr(w_i|\gamma_i = 1, g(CR_i))\pi}{\Pr(w_i|\gamma_i = 1, g(CR_i))\pi + \Pr(w_i|\gamma_i = 0, g(CR_i))(1-\pi)}$$

   and then drawing
   (b) $\boldsymbol{y}$ as in Sampling Scheme 16.3.
2. Sample $\boldsymbol{g}$ and τ^2 as in Sampling Scheme 16.3.

Figure 16.4, dotted line, shows the estimate $\Phi\{\hat{g}(CR)\}$ when outlier detection is turned on. As can be seen, the sharp drop $\Phi\{\hat{g}(CR)\}$ disappears when the procedure is made robust to outliers. The posterior probability that the observation $(CR = 15.33, w = 0)$ is an outlier is approximately 0.50. The method also identifies outliers which occur at low levels of cholesterol ratio. For example the posterior probability that the observation $(CR = 1.74, w = 1)$ is an outlier is approximately 0.75. However, the observation $(CR = 1.74, w = 1)$ does not affect the estimate $\Phi\{\hat{g}(CR)\}$ to the same degree as does the observation $(CR = 15.33, w = 0)$, even though it has a higher posterior probability of being an outlier. This is because there are many observations for low levels of cholesterol ratio, while data is scarce in the vicinity of the observation $(CR = 15.33, w = 0)$. Hence the contribution of the observation $(CR = 1.74, w = 1)$ to the estimate is less than the contribution made by the observation $(CR = 15.33, w = 0)$.

16.2.2 Including other covariates

One method of including other risk factors is to assume that the effect of these risk factors on the probability of a heart attack is additive. Although this is not as flexible as modelling the three-dimensional surface nonparametrically, it is easy to implement and the discussion of modelling high-dimensional surfaces nonparametrically is left to Section 16.4. To model the probability of a heart attack as a function of the three risk factors, we assume that

$$\Pr(w = 1|BP, CR, Age) = \Phi(\{g(BP, CR, Age)\} \text{ where,}$$
$$g(BP, CR, Age) = \alpha_0 + \alpha_1 BP + \alpha_2 CR + \alpha_3 Age + f_1(BP) + f_2(CR) + f_3(Age)$$

and estimate $\Phi\{\hat{g}(BP, CR, Age)\} = \frac{1}{M}\sum_{k=1}^{M} E[\Phi\{g(CR, AGE, BP|\boldsymbol{w}, \boldsymbol{y}^{[k]}, \boldsymbol{\tau}^{2[k]})]$ using a similar sampling scheme to 16.3.

Sampling scheme 16.5

1. Sample $\boldsymbol{y}$ and $\boldsymbol{\gamma}$ jointly conditional on $\boldsymbol{w}$ and $\boldsymbol{\alpha} = (\alpha_0, \alpha_1, \alpha_2, \alpha_3)$ and $\boldsymbol{f}_1 = (f_1(BP_1), \ldots, f_1(BP_n))$, $\boldsymbol{f}_2 = (f_2(CR_1), \ldots, f_2(CR_n))$ and $\boldsymbol{f}_3 = (f_3(Age_1), \ldots, f_3(Age_n))$, by drawing
   (a) $\boldsymbol{\gamma}$ from $p(\boldsymbol{\gamma}|\boldsymbol{w}, \boldsymbol{\alpha}, \boldsymbol{f}_1, \boldsymbol{f}_2, \boldsymbol{f}_3)$ and then
   (b) $\boldsymbol{y}$ from $p(\boldsymbol{y}|\boldsymbol{\gamma}, \boldsymbol{w}, \boldsymbol{\alpha}, \boldsymbol{f}_1, \boldsymbol{f}_2, \boldsymbol{f}_3)$.
2. Sample $\boldsymbol{\alpha}, \boldsymbol{f}_1, \boldsymbol{f}_2$, and $\boldsymbol{f}_3$ conditional on $\boldsymbol{\tau}^2 = (\tau_1^2, \tau_2^2, \tau_3^2)$ and $\boldsymbol{y}$,
3. Sample $\boldsymbol{\tau}^2$ conditional on $\boldsymbol{f}_1, \boldsymbol{f}_2$, and $\boldsymbol{f}_3$.

Figure 16.5 shows that the probability of a heart attack increases monotonically with age and cholesterol ratio, when outlier detection is turned on. Interestingly, the posterior probability that the observation $(CR = 15.33, w = 0)$ is an outlier has decreased from 0.50, reported previously, to 0.30. This is because this patient had a systolic blood pressure of only 120 and was 49 years old. Thus after controlling for blood pressure and age, the likelihood that this observation is an outlier decreased. Conversely, the posterior probability that the observation $(CR = 1.74, w = 1)$ is an outlier increased from 0.75 to 0.86, because this patient was only 20 years old and had a systolic blood pressure of 106. The plot of blood pressure is interesting, because it shows that even with outlier detection turned on, the probability of having a heart attack increases as blood pressure decreases, for blood pressure less than 120. This is not a smoothing artefact but rather a real feature of the data. The graph shows a considerable number of patients who had heart attacks with a systolic blood pressure below 120.

16.3 Using smoothing splines to estimate spectral densities

16.3.1 Introduction

In this section we show how features of a time series such as the spectral density and unknown frequencies can be estimated simultaneously using smoothing splines together with data augmentation in a hierarchical Bayesian framework. Following [5], suppose that a time series

$$y_t = \beta_0 + v_t$$

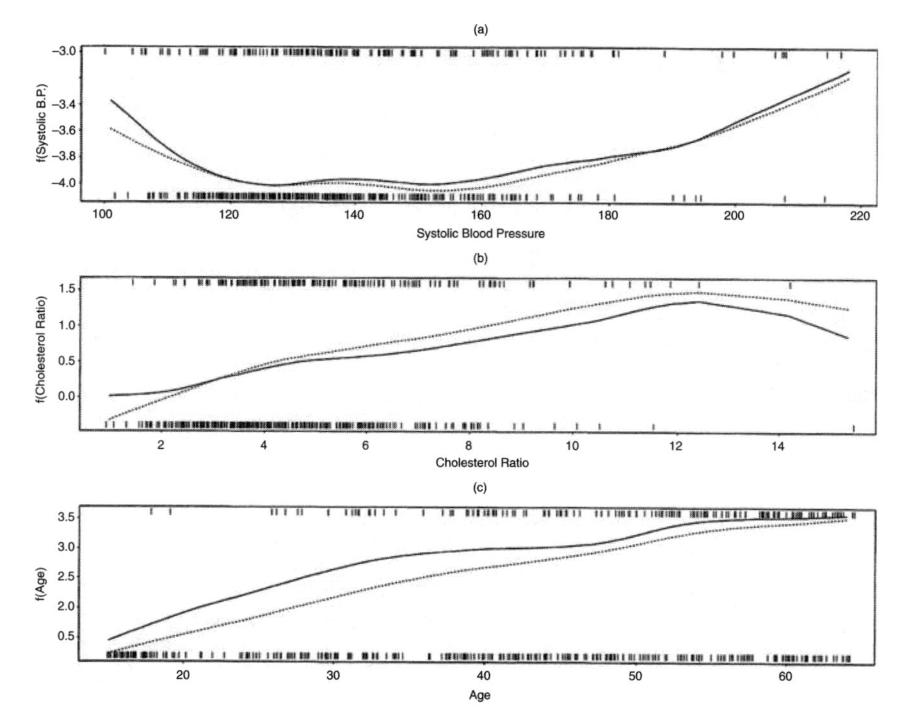

Figure 16.5 Heart attack data. Panels (a)–(c) are Bayesian estimates of $\alpha_0 + \alpha_1 BP + f_1(BP)$ (the effect of blood pressure), $\alpha_2 CR + f_1(CR)$ (the effect of the cholesterol ratio), and $\alpha_3 Age + f_3(Age)$ (the effect of age when outlier detection is turned on $(\cdots)$ and when it is not (—). For all three independent variables there were many coincident points. To show the distribution of the data, we perturbed the independent variables by adding a small amount of Gaussian noise to their values. This figure was taken from [45].

is modelled as the sum of a signal β_0 and noise v_t. The noise is a zero mean stationary Gaussian process with a spectral density given by $f(\nu)$, for $-1/2 < \nu \leq 1/2$. We assume that $f(\nu)$ is bounded and positive. Given a realization, $\boldsymbol{y} = (y_1, \ldots, y_n)$, the periodogram of the data at frequency ν is

$$I_{n,\beta_0}(\nu) = \frac{1}{n}\left|\sum_{t=1}^{n}(y_t - \beta_0)\exp(-i\nu t)\right|^2.$$

In what follows, the notation for the periodogram is simplified by omitting the dependence of I on n and β_0. Let $\nu_k = k/n$, for $k = 0, \ldots, n-1$, be the Fourier frequencies. [43] showed that, under appropriate conditions, for large n, the likelihood of $\boldsymbol{y}$ can be approximated as

$$p(\boldsymbol{y} \mid f, \beta) \propto \prod_{k=0}^{n-1} \exp\left\{-\frac{1}{2}\left[\log f(\nu_k) + I(\nu_k)/f(\nu_k)\right]\right\} \tag{16.7}$$

Let $x(\nu_k)$ be the log of the periodogram evaluated at the Fourier frequencies, then the representation in (16.7) suggests the log–linear model

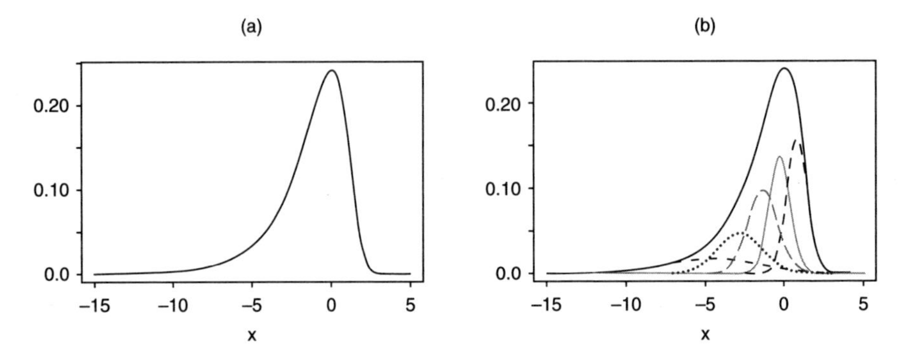

Figure 16.6 Panel (a) shows the density of a $\log(\chi_1^2)$ variable (solid line) and the normal mixture approximation (dotted line). Panel (b) shows the approximate density (solid line) and also the densities of the normal components (broken lines)

$$x(\nu_k) = \log f(\nu_k) + \epsilon_k \tag{16.8}$$

where $x(\nu_k) = \log I(\nu_k)$ for $k = 0, \ldots, [n/2]$, where $[n/2]$ is the largest integer less than or equal to $n/2$, the ϵ_ks are independent, $\epsilon_k \sim \log(\chi_2^2/2)$ for $k = 1, \ldots, [n/2] - 1$, and $\epsilon_k \sim \log(\chi_1^2)$ for $k = 0, [n/2]$. Note that in (16.7), there are only $[n/2] + 1$ distinct observations since the spectral density and the periodogram are both even functions of ν. For ease of notation, in what follows, we assume that n is even.

It is seen from (16.8) that the log spectral density can be estimated nonparametrically with the log periodogram as the dependent variable. [40] used a frequentist approach for estimating the log spectral density via cubic smoothing splines. This section shows how [5] used a Bayesian approach to model the log spectral density as cubic smoothing splines by approximating the error distribution in (16.8) by a mixture of five normal distributions and introduced latent component indicators.[11] Table 16.1 gives the weights, means, and variances of the five components in the mixture which approximate the density of $\log(\chi_1^2)$ variable (columns 1–3) and $\log(\chi_2^2)$ variable (columns 4–6). Figure 16.6, panel (a), shows plots of the density of a $\log(\chi_1^2)$ variable and its approximation, while panel (b) shows the approximation together with the densities of the components in the mixture.

Again data augmentation is used to facilitate the estimation of $g(\nu_k) = \log(f(\nu_k))$. Let γ_k determine the component of the mixture to which ϵ_k in (16.8) belongs, so that $\gamma_k = j$ if ϵ_k is generated from component j, and let $\boldsymbol{\gamma} = (\gamma_0, \ldots, \gamma_{n/2})'$. The prior probability of γ_k is given by columns 1 and 4 of Table 16.1. The γ_k are assumed to be independent a priori since the ϵ_k are independent.

Detecting signals in a time series by identifying spikes in the log periodogram has been studied extensively, see [27] for a discussion. In [5] detecting signals in a time series is achieved by allowing one of the components in the mixture to have a large variance and concluding that a spike occurs at frequency ν_k if the indicator variable γ_k has a high posterior probability corresponding to the mixture component with the large variance and the residual is positive. To this end a new mixture of five normal densities was used to approximate the $\log(\chi_2^2)$ density, with the variance of one of the distributions fixed to be very large and the other four components were selected so that the approximating mixture, given in Table 16.2, has the quality of the approximation similar to that in Table 16.1, except that it has a heavier right-hand tail.

[11] [30] provide a similar Bayesian technique to estimate the spectral matrix for multivariate time series.

Table 16.1 Five component approximations to $\log(\chi_1^2)$ and $\log(\chi_2^2/2)$

$\log(\chi_1^2)$			$\log(\chi_2^2/2)$		
probability	**mean**	**variance**	**probability**	**mean**	**variance**
0.13	−4.63	8.75	0.19	−2.20	1.93
0.16	−2.87	1.95	0.11	−0.80	1.01
0.23	−1.44	0.88	0.27	−0.55	0.69
0.22	−0.33	0.45	0.25	−0.035	0.60
0.25	0.76	0.41	0.18	0.48	0.29

Table 16.2 Five-component approximation to $\log(\chi_2^2/2)$ with one component having a variance of 25 and prior probability 0.02.

probability	mean	variance
0.13	−2.26	3.31
0.35	−0.91	0.92
0.02	−0.69	25
0.20	−0.32	0.63
0.30	0.34	0.38

Example 16.3 *Variable star data*

The technique is now demonstrated using observations on the variable star *S. Carinae*. The reported results are taken from [5]. *S. Carinae* is classified as a Mira type variable star and is located in the constellation *Carina*. *Carina* means keel and refers to the keel of the ship of the Argonauts in Greek mythology. A star is classified as variable if its apparent magnitude as seen from Earth changes over time. Mira type variable stars are characterized by very large pulsations with periods longer than 100 days. Data are available for *S Carinae* from the Royal New Zealand Society of Astronomy for 1189 ten-day periods.

Each observation is the average over the preceding 10 days and there are 40 missing observations. Missing data are distinguished from observed data by denoting them $\boldsymbol{y}_{\text{mis}}$ and $\boldsymbol{y}_{\text{obs}}$ respectively. For further details see [28]. The model for the log of the spectral density of the time series is given by (16.8) where the distribution of the error, ϵ_k, is modelled as a mixture of five normals with means, variances and prior probabilities given in Table 16.2. The posterior mean is,

$$E[g(\nu_k)|\boldsymbol{y})] = \int E[g(\nu_k)|\boldsymbol{y}_{\text{obs}}, \tau^2, \boldsymbol{y}_{\text{mis}}, \boldsymbol{f}]p(\boldsymbol{f}, \tau^2, \boldsymbol{y}_{\text{mis}}|\boldsymbol{y}_{\text{obs}})d\tau^2 d\boldsymbol{y}_{\text{mis}} d\boldsymbol{f}$$

and is estimated by

$$E[g(\nu_k)|\boldsymbol{y})] = \frac{1}{M}\sum_{k=1}^{M} E[g(\nu_k)|\boldsymbol{y}_{\text{obs}}, \tau^{2[k]}, \boldsymbol{y}_{\text{mis}}^{[k]}, \boldsymbol{f}^{[k]}]$$

Sampling scheme 16.6 is used to generate the sequence $\tau^{2[k]}, \boldsymbol{y}_{\text{mis}}^{[k]}, \boldsymbol{f}^{[k]}$ from $p(\tau^2, \boldsymbol{y}_{\text{mis}}, \boldsymbol{f}|\boldsymbol{y}_{\text{obs}})$.

Sampling scheme 16.6

1. Generate $\boldsymbol{y}_{\text{mis}}$ from $p(\boldsymbol{y}_{\text{mis}}|\boldsymbol{g}, \boldsymbol{y}_{\text{obs}}, \beta_0, \tau^2, \boldsymbol{\gamma})$ and compute $\boldsymbol{x} = (x(\nu_1), \ldots, x(\nu_n))$.
2. Generate τ^2 from $p(\tau^2|\boldsymbol{g})$
3. Generate $\boldsymbol{g}$ from $p(\boldsymbol{g}|\boldsymbol{x}, \beta_0, \tau^2, \boldsymbol{\gamma})$
4. Generate $\boldsymbol{\gamma}$ and β_0 from $p(\boldsymbol{\gamma}, \beta_0, |\boldsymbol{y}_{\text{mis}}, \boldsymbol{y}_{\text{obs}}, \boldsymbol{g})$

Figure 16.7 shows the results. Panel (a) shows the data; Panel (b) shows the log periodogram with the missing data replaced by the Markov chain Monte Carlo estimates; Panel (c) shows the Markov chain Monte Carlo estimate of the log spectral density of the error sequence; Panel (d) shows the posterior probability that γ_k belongs to the component in the mixture with the high variance. From panel (d) it can be seen that there are three frequencies at which the spectrum is likely to have a spike.

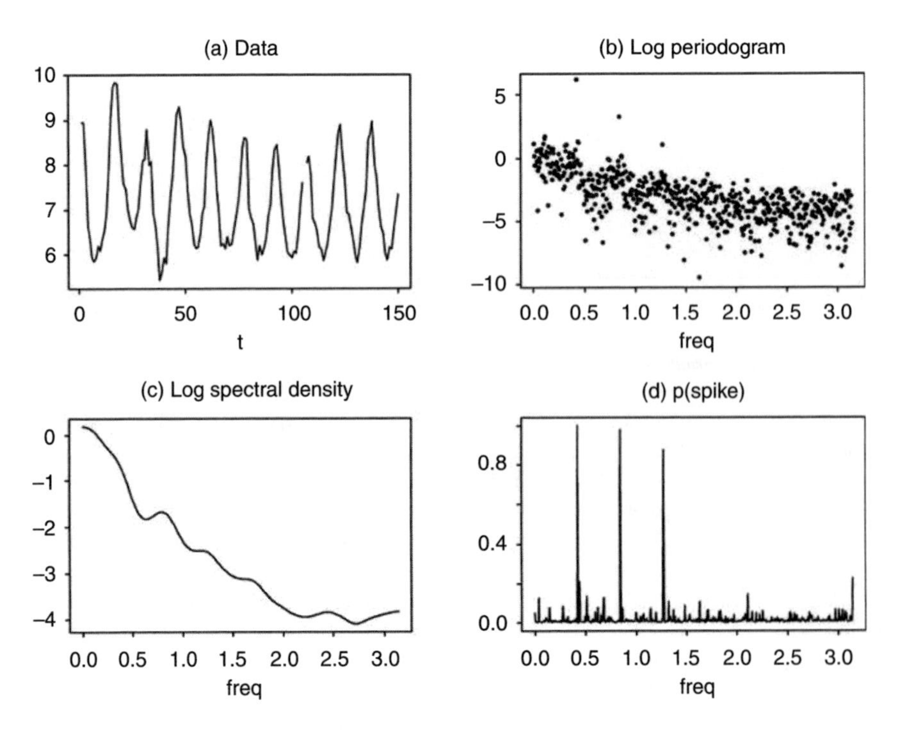

Figure 16.7 Panel (a) shows the first 150 observations on the varaible star *S Carinea*; Panel (b) shows the log periodogram; Panel (c) shows the Markov chain Monte Carlo estimate of the log spectral density of the error sequence (solid); Panel (d) shows the probabilities of a spike in the log periodogram. This figure is taken from [5].

The first spike occurs at frequency $\nu = 0.42$ and the two subsequent spikes occur at frequencies which are multiples of $\nu = 0.42$. The conclusion is that the time series contains a deterministic component, as discussed in Section 16.3.1. The frequency at which the spike first occurs is called the fundamental frequency. This value of $\nu = 0.42$ translates into 0.0669 cycles per 10 day period. In other words each cycle lasts 149.5 days.

16.4 Multidimensional smoothing splines

To extend the prior for g with a single covariate outlined in Section 16.1, we need to specify the prior covariance of the unknown function in d dimensional space. For example suppose that the signal g in (16.1) is a function of $d = 2$ covariates, x_1 and x_2, so that

$$y_i = g(x_{1i}, x_{2i}) + \epsilon_i \text{ with } \epsilon_i \sim N(0, \sigma^2) \tag{16.9}$$

$$g(x_{1i}, x_{2i}) = \alpha_0 + \sum_{k=1}^{m-1} \alpha_{k1} x_{1i}^k + \sum_{k=1}^{m-1} \alpha_{2k} x_{2i}^k + f(x_{1i}, x_{2i})$$

A prior for the covariance between $f(x_{1i}, x_{2i})$ and $f(x_{1j}, x_{2j})$, given by [42], is the thin-plate spline prior with $m = 2$ and

$$\text{cov}\{f(x_{1i}, x_{2i}), f(x_{1j}, x_{2j})\} = \tau^2 \Omega\{(x_i, x_{2i}), (x_j, x_{2j})\}.$$

where

$$\Omega\{(x_{1i}, x_{2i}), (x_{1j}, x_{2j})\} = A\{(x_{1i}, x_{2i}), (x_{1j}, x_{2j})\} - \sum_{k=1}^{3} p_k(x_{1j}, x_{2j}) A\{u_k, (x_{1i}, x_{2i})\}$$

$$- \sum_{k=1}^{3} p_k(x_{1i}, x_{2i}) A\{(x_{1j}, x_{2j}), u_k\} + \sum_{k=1}^{3} \sum_{l=1}^{3} p_k(x_{1i}, x_{2i}) p_l(x_{1j}, x_{2j}) A(u_k, u_l),$$

$$A\{(x_{1i}, x_{2i}), (x_{1j}, x_{2j})\} = r_{ij} \log(r_{ij}), \quad r_{ij} = \sqrt{\{(x_{1i} - x_{1j})^2 + (x_{2i} - x_{2j})^2\}},$$

$$\begin{aligned} p_1(x_1, x_2) &= 1 - 2x_1 - 2x_2, \\ p_2(x_1, x_2) &= 1 - 2x_1, \\ p_3(x_1, x_2) &= 1 - 2x_2 \\ u_1 &= (0, 0), \\ u_2 &= (\frac{1}{2}, 0), \\ u_3 &= (0, 2). \end{aligned}$$

As can be seen the kernel Ω is composed of radial basis functions and $M = \binom{d+m-1}{m}$ polynomials of order $\leq m - 1$. The polynomials ensure that the matrix is positive definite. Again we can write the prior on $g(x_1, x_2)$ as a linear combination of basis functions, by taking an eigenvalue decomposition of Ω so that $\Omega = QDQ'$ and letting $g(x_1, x_2) = X\beta$, where $X = Q$ and $\beta \sim N(0, \tau^2 D)$.

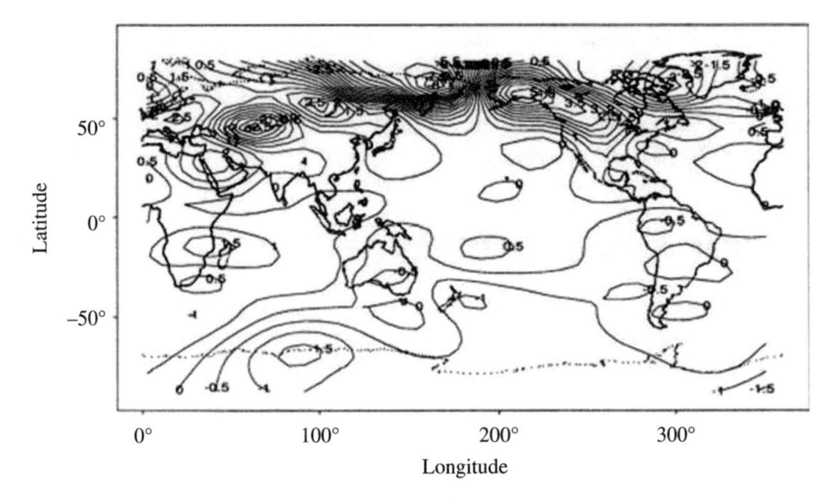

Figure 16.8 Global air temperature example. Isotherms of air temperature anomalies (°C) for December 1993 using a single smoothing spline. This figure is taken from [46].

Example 16.4 *Global air temperature anomalies*
This example demonstrates how smoothing splines can be used to model spatial data, such as global air temperature anomalies. The data were obtained from the air temperature anomaly archive developed by Jones (1994) and available at the web site *http://ingrid.ldeo.columbia.edu/SOURCES/.JONES/.landonly.cuf/*. The dataset contains monthly readings of air temperature anomalies at various points on the globe for December 1993. Air temperature anomalies are the deviations from a monthly mean temperature for a given latitude and longitude. The monthly mean temperature for a given latitude and longitude was defined to be the monthly mean temperature for that location for the period 1950–1979. Altogether 445 irregularly spaced observations across the entire globe were available for December 1993.

Let y_i be the December 1993 temperature anomaly recorded at latitude x_{1i} and longitude x_{2i}. The model for the data is given by (16.9) and an estimate of the posterior mean of $g(x_{1i}, x_{2i})$, is obtained using Sampling scheme 16.2. Figure 16.8 is taken from [46] and shows the isotherms for the model. One of the interesting features of this figure is that the isotherm fluctuations in the northern hemisphere are more pronounced than those in the southern hemisphere, indicating that the regression surface for northern hemisphere temperature anomalies requires less smoothing than the regression surface for southern hemisphere temperature anomalies. We return to this example in Section 16.5 where we discuss how to obtain an estimate of a regression surface where the degree of smoothness changes across the covariate space.

Example 16.5 *Union membership*
This example shows how the methodology discussed in Section 16.2 can be extended to higher dimensions by modelling the probability of union membership as a function of three continuous variables, years of education, wage, and age, and three dummy variables, south (1 = live in southern region of USA), female (1 = female), and married (1 = married). The data consist of 534 observations on US workers and can be found in [3] and at http://lib.stat.cmu.edu/datasets/CPS_85_Wages. [32] estimated the probability of union membership using a generalized additive model without interactions. In this example the three-dimensional surface of the continuous covariates is modelled

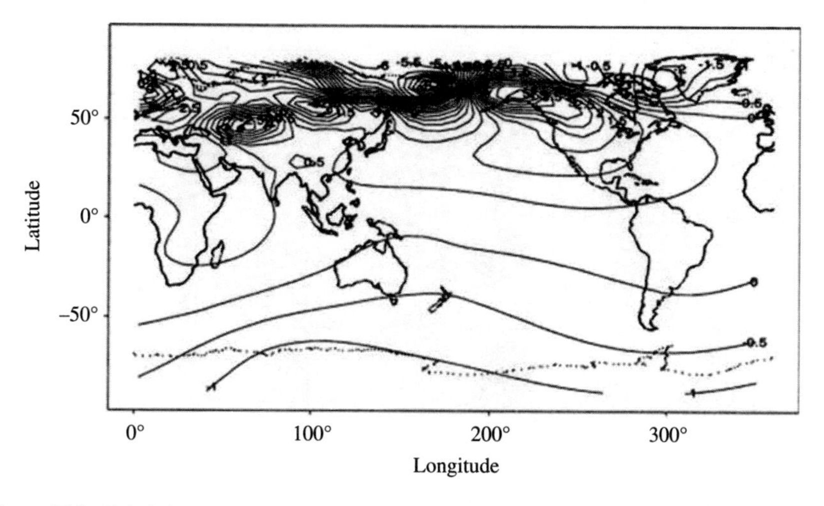

Figure 16.9 Global air temperature example. Isotherms of air temperature anomalies (°C) for December 1993 using a mixture of two smoothing splines. This figure is taken from [46].

using partial thin-plate spline basis functions, see [44] for details. The results suggest that this is more appropriate than an additive model. The model is

$$\Pr(w = 1|x) = \Phi\{g(x)\}$$

with

$$g(x) = \boldsymbol{\alpha} x + f(x^*)$$

where x=(years education, wage, age, south, female, married), $x^* =$ (years education,wage, age), wage is in US \$/hr and age is in years. The dependent variable, w, is 1 if the worker belongs to a union and 0 otherwise and $\boldsymbol{\alpha} = (\alpha_0, \alpha_1, \ldots, \alpha_6)$. The probabilities $\Pr(w_i = 1|x)$ are estimated by their posterior means and Sampling scheme 16.2 is used to perform the required multidimensional integration. Figure 16.10 panels (a) to (c) show the joint marginal effect of two continuous covariates at the mean of the third one with the dummy variables set to zero. These figures clearly show interactions among the continuous covariates. For example, Figure 16.10 panel (a) shows that for workers who did not finish high school ($<$ 12 years education) the probability of belonging to a union increases as wage increases. For workers who finished high school the probability of belonging to a union peaks at a wage of approximately \$15/hr, while for workers with tertiary eduction ($>$15 years) the probability of union membership is initially high and then decreases with increasing wage. Figure 16.10 panel (b) shows two modes. For workers with an average wage and who are older than 40 years, union membership peaks at 55 years, and between 8 and 10 years education (interestingly union membership peaks again at 55 years, and 18 years education, although this peak may be due to boundary effects). While for younger workers union membership increases as years of education increases. Figure 16.10 panel (c) shows that for workers whose age is less than 40, the probability of union membership initially increases with wage, before reaching a peak at a wage of about \$15/hr and then declines. For older workers this peak occurs at much lower wages, somewhere between \$5/hr and \$10/hr before declining sharply.

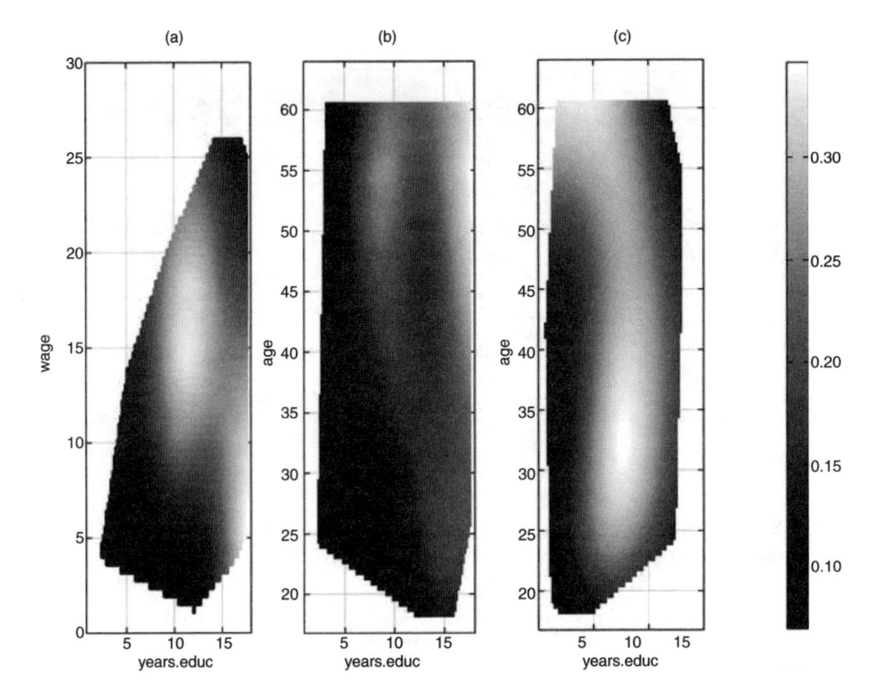

Figure 16.10 Panel (a) Plot of Pr(Union Member|wage, years education) at the mean age; Panel (b) Pr(Union Member|age, years education) at the mean wage; Panel (c) Pr(Union Member|wage, age) at the mean years education. In all plots the dummy variables are set to zero. This figure is taken from [44].

16.5 Locally adaptive smoothing splines

There are many Bayesian methods available which allow the degree of smoothness of the regression surface to change. This list includes, but it not limited to, [35], [18], and [9], who use regression splines and model averaging to obtain locally adaptive estimates; [21], [2], and [6] who use penalized splines; and [26] and [46] who use mixtures of smoothing splines. There is not enough space to do all these methods justice, but the reader is referred to [25] for an excellent discussion of these techniques. In this section the discussion is confined to smoothing splines and two applications of locally adaptive smoothing splines are described using the mixture-of-splines approach presented in [46] and [31].

Example 16.6 *Modelling global temperature anomalies adaptively*
To introduce the topic consider the example discussed in Section 16.4, where it was noted that the isotherms in the northern hemisphere were more wiggly than those in the southern hemisphere. In general it may be unrealistic to assume that temperature deviations from a historic average display the same variability across the globe. For example, areas close to the equator or on a coast may have less temperature variation than interior areas in the mid-latitudes. One method which allows the smoothness of a regression surface g to change across the globe is to model g as a mixture of J smoothing splines, where the weights in the mixture depend upon covariates. A model for temperature anomalies across the globe, given in [46], is

$$y(x_{1i}, x_{2i}) = g(x_{1i}, x_{2i}) + e_i$$

$$g(x_{1i}, x_{2i}) = \sum_{j=1}^{J} g_j(x_{1i}, x_{2i})\pi_j(x_{1i}, x_{2i})$$

$$\pi_j(x_{1i}, x_{2i}) = \frac{\exp(z_i\boldsymbol{\delta}_j)}{\sum_{k=1}^{J} \exp(z_i\boldsymbol{\delta}_k)} \quad \text{with} \quad \sum_{j=1}^{J} \pi_j(x_{1i}, x_{2i}) = 1 \tag{16.10}$$

where π_j is the weight given to the *jth* smoothing spline. This weight depends upon covariates $z_i = (1, x_{1i}, x_{2i})$ and is modelled by a multinomial logistic regression, with regression coefficients $\boldsymbol{\delta} = (\boldsymbol{\delta}_1, \ldots, \boldsymbol{\delta}_J)$, with $\boldsymbol{\delta}_j = (\delta_0, \delta_1, \delta_2)$ for $j = 1, \ldots J - 1$ and $\boldsymbol{\delta}_J = (0, 0, 0)$. In the neural network literature the formulation in (16.10) is known as a mixture-of-experts model (see [19] and [20]). Each of the g_j in (16.10) has the thin-plate spline prior given in (16.9) and is associated with a smoothing parameter, τ_j^2, which is defined over a local region of the covariate space, so that a smoothing spline with a given degree of smoothness is fitted to the data that fall into each region. The regions are allowed to overlap, such that individual data points may lie simultaneously in multiple regions. As such the regions are said to have 'soft' boundaries.

The posterior mean $E(g|\boldsymbol{y})$ is given by $\int E(g|\boldsymbol{y}, \Theta)p(\Theta|y)d\Theta$, where $\Theta = (\theta_1, \ldots, \theta_J, \boldsymbol{\delta}_1, \boldsymbol{\delta}_{J-1}, \sigma^2), \theta_j = (\alpha_j, \tau_j^2)$ and estimated by

$$\hat{g} = \frac{1}{M}\sum_{k=1}^{M}\sum_{j=1}^{J} E(\pi_j g_j|\boldsymbol{y}, \theta_j^{[k]}, \boldsymbol{\delta}_j^{[k]}, \sigma^{2[k]}) \tag{16.11}$$

where the sequence $\theta_1^{[k]}, \ldots, \theta_J^{[k]}, \boldsymbol{\delta}_1^{[k]}, \ldots, \boldsymbol{\delta}_{J-1}^{[k]}, \sigma^{2[k]}$ is generated from the joint posterior distribution $p(\theta_1, \ldots, \theta_J, \boldsymbol{\delta}_1, \ldots, \boldsymbol{\delta}_{J-1}, \sigma^2|\boldsymbol{y})$. Again data augmentation is used to generate this sequence by defining a vector of indicator variables $\boldsymbol{\gamma} = (\gamma_1, \ldots, \gamma_n)$ which identifies the component to which an observation belongs, s.t. $\Pr(\gamma_i = j|\boldsymbol{x}) = \pi_j(\boldsymbol{x})$. The sampling scheme is:

Sampling scheme 16.7

1. Sample $\boldsymbol{\gamma}$ conditional on $\boldsymbol{y}$, $\boldsymbol{g} = (g_1, \ldots, g_J)$, and σ^2.
2. Sample $\boldsymbol{\delta}_j$ for $j = 1, \ldots, J - 1$ conditional on $\boldsymbol{\gamma}$.
3. Conditional on a partition of the data, $\boldsymbol{g}$, τ_j^2 are sampled as in Sampling scheme 16.2.
4. Sample σ^2 conditional on $\boldsymbol{y}$, and $\boldsymbol{g}$.

Proper priors must be used for the parameters of the components in mixture models, because if no data are allocated to a component, then the parameters of that component are drawn from the prior distribution.

The choice of the number of components in the mixture, J is not trivial. Several model selection techniques exists such as BIC or AIC, while Reversible Jump MCMC (RJMCMC), can be used to perform model averaging, see [15] and [29]. For this example the BIC approach of [33] was used to select J, where the marginal likelihood $p(\boldsymbol{y}|J)$ is approximated by

$$p(\boldsymbol{y}|J) \approx p(\boldsymbol{y}|\hat{\theta}_1, \ldots, \hat{\theta}_J, \hat{\boldsymbol{\delta}}_1, \ldots, \hat{\boldsymbol{\delta}}_{J-1}, \hat{\sigma}^2, J)n^{-q_J/2}, \tag{16.12}$$

$\hat{\theta}_j$ and $\hat{\boldsymbol{\delta}}_j$ maximize the likelihood in (16.12) and q_j is the number of parameters. The quantity $p(\boldsymbol{y}|J)$ was maximized for $J = 2$ and Figure 16.9 presents an estimate of g if that data are modelled by (16.10) with $J = 2$. From Figure 16.9 one can see why a mixture of more than one thin-plate spline is required. The isotherms in the southern hemisphere and the lower latitudes of the northern hemisphere change very smoothly and show very little variation. In contrast, the isotherms in the higher latitudes of the northern hemisphere exhibit large fluctuations. In particular, the warmer than usual temperatures in Alaska, warmer by 10 celsius degrees, are related to the equally colder than usual temperatures in Siberia. Comparing Figure 16.9 with Figure 16.8 it can be seen that in order to achieve the degree of variation necessary in the northern hemisphere, extra variation is induced in the estimate of southern hemisphere isotherms.

16.6 Modelling non-stationary time series with a mixture of spectra

In many practical problems, time series are realizations of non-stationary random processes. In this section the approach presented in Section 16.3.1 is extended to modelling the log spectral density for non-stationary time series using a mixture of a finite but unknown number of individual log spectra. Non-stationarity is introduced to the model by allowing the mixing weights of the individual log spectra to change smoothly across partitions of time, thus allowing for slowly varying time series, as well as for piecewise stationary time series.

The model presented in this section is from [31]. In this section RJMCMC is used to perform model averaging, rather than model selection discussed in Section 16.6. This is achieved by allowing the number of underlying individual spectra to be unknown and vary from $j = 1, \ldots, J$. To do this it is assumed that the time series is a Dahlhaus-locally stationary process. [24] argue that Dahlhaus-locally stationary processes can be well approximated by piecewise stationary processes so that if the time series is partitioned into S small non-overlapping segments, $\boldsymbol{y} = (\boldsymbol{y}_1, \ldots, \boldsymbol{y}_S)$ then each segment $\boldsymbol{y}_s = (y_{s1}, \ldots, y_{n_s s})$ for $s = 1, \ldots, S$ is stationary with corresponding spectrum $f_s(\nu)$. The log spectrum in each segment $g_s(\nu) = \log(f_s(\nu))$ is modelled as a mixture of a maximum of R possible spectra so that

$$g_s(\nu) = \sum_{r=1}^{R} g_{rs}(\nu) \Pr(r) \tag{16.13}$$

where $\Pr(r)$ is the prior probability that the mixture contains r components, and $g_{rs}(\nu)$ is the log of the spectral density of a mixture of r components in segment s. For a given number of mixture components, r, and segment s, $g_{rs}(\nu)$ is modelled as

$$g_{rs}(\nu) = \sum_{j=1}^{r} \pi_{jsr} \log(f_{jr}(\nu) \tag{16.14}$$

where $f_{jr}(\nu)$ is the spectral density of the jth component in a mixture of r components. The prior for $\log(f_{jr}(\nu))$ is given by (16.3) with $m = 1$. The unknown weight assigned to the jth component in segment s, is π_{jrs} with $\sum_{j=1}^{r} \pi_{jsr} = 1$. Note that the spectral density $f_{jr}(\nu)$ is common to all segments, but the weight assigned to this component spectrum varies across segments, thus allowing the spectral density to change over time. As in Section 16.6 the π_{jsr} are modelled using a multinomial logistic regression given by (16.10).

In theory there are as many segments as observations, however a minimum number of observations in each segment is needed to estimate the spectral density and for the Whittle approximation to the likelihood to hold. [31] found that using a minimum of 64 observations in each segment gave reliable results assuming that the true local spectra have well separated peaks. It should be noted that the parameters of the mixing function π_{jsr} in (16.14) are of more importance than the number of segments because these parameters control the location and rate at which the time series moves from one stationary process to another.

The log of the spectral density in segment s, for $s = 1, \dots S$ is estimated by its posterior mean which is equal to

$$E[g_s(\nu)|\boldsymbol{y}] = \sum_{r=1}^{R} \int E\{g_s(\nu)|\boldsymbol{y}, \theta_r, r\} p(\boldsymbol{\theta}_r|\boldsymbol{y}, r) d\theta_r \Pr(r|\boldsymbol{y})$$

where $\theta_r = (\boldsymbol{f}_r(\nu), \boldsymbol{\tau}^2, \boldsymbol{\delta}_r^2), \boldsymbol{f}_r(\nu) = (f_{1r}(\nu), \dots, f_{rr}(\nu)), \boldsymbol{\tau}_r^2 = (\tau_{1j}^2, \dots, \tau_{rr}^2)$ and is estimated by

$$\hat{g}_s(\nu) = \frac{1}{M} \sum_{k=1}^{M} \sum_{r=1}^{R} E\{g_s(\nu)|\boldsymbol{x}_s(\nu), \theta_r^{[k]}, r^{[k]}\}$$

where $\boldsymbol{x}_s(\nu)$ is the periodogram of segment s and the sequence $\theta_r^{[k]}, r^{[k]}$ is drawn from the joint posterior $p(\boldsymbol{\theta}_r, r|\boldsymbol{y})$ using RJMCMC. Data augmentation is used to facilitate the RJMCMC scheme, by defining a vector of indicator variables identifying to which component a segment belongs, and a variable, r which indicates the number of the components in the mixture. Let $\gamma_{rs} = j$ if segment s is generated by component j in a mixture of r components. The sampling scheme used to obtain the necessary sequence of draws has two parts; a between-model move followed by a within-model move. The number of components r is first initialized, then the sampling scheme is

Sampling scheme 16.8

1. **Between model move**
   A new value of r is proposed, and conditional on this value, $\boldsymbol{f}_r$, $\boldsymbol{\tau}_r^2$ and $\boldsymbol{\delta}_r$ are proposed. These proposed values are then accepted or rejected using a Metropolis–Hastings step.
2. **Within model move**
   Given the new value of r, the parameters specific to a model of r components are then updated as follows.
   (a) Sample $\boldsymbol{\gamma}_r$ conditional on r, $\boldsymbol{f}_r$ and $\boldsymbol{\delta}_r$.
   (b) Sample $\boldsymbol{\delta}_r$ conditional on r and $\boldsymbol{\gamma}_r$.
   (c) Sample $\boldsymbol{f}_r$ conditional on r, $\boldsymbol{\gamma}_r$ and $\boldsymbol{\tau}_r^2$.
   (d) Sample $\boldsymbol{\tau}_r^2$ conditional on r and $\boldsymbol{f}_r$.

Example 16.7 *Southern Oscillation Index (SOI)*
One area of research in climate science over the last few decades is the El Niño/Southern Oscillation (ENSO) phenomenon. ENSO is an irregular low frequency oscillation between a warm El Niño state and a cold La Niña state. The Southern Oscillation Index (SOI) is an indicator of the ENSO phenomenon and is calculated to be the standardized anomaly of the mean sea level pressure difference between Tahiti and Darwin. In particular questions regarding the impact global warming may have had on the frequency of this phenomenon have been the source of much debate, see [38], [36] and [22].

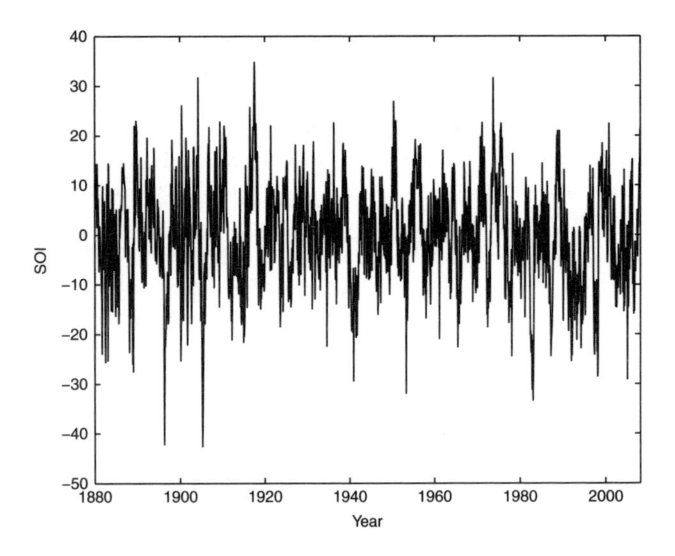

Figure 16.11 Monthly values of the Southern Oscillation Index (SOI) from January 1876 to April 2008.

The data, shown in Figure 16.11, are monthly values of the SOI from January 1876 to April 2008 and are available at http://www.bom.gov.au/climate/current/soihtm1.shtml.

The data were divided into 24 segments, each containing 64 observations (leaving out the first 29 observations), and the model given by equations (16.13) and (16.14) was fitted. Figure 16.12 shows the time-varying log spectrum for the SOI and indicates that the series is likely to be stationary given that the spectrum remains constant for the time period. To examine the effect that the prior for the number of components, $\Pr(r)$, might have on this finding, a number of choices for $\Pr(r)$ were used. The choice of these priors and the posterior probability $\Pr(r|\mathbf{y})$ appear in Table 16.3. Table 16.3 confirms that the series is most probably stationary, given that the modal number of components across all priors is $r = 1$. This finding is in contrast to [38] who concluded that the time series is non-stationary. There are a number of explanations for this disparity, see [31] for details.

Interestingly conditional on $r = 2$, which occurred with probability 0.25 when the uniform prior $\Pr(r) = 1/4$ was used, the change in the SOI spectrum occurred gradually between 1890 and 1910 and not in the 1970s as found by [38].

Example 16.8 *Distinguishing between earthquakes and nuclear explosions*
The second example of the model given by equations (16.13) and (16.14) relates to distinguishing between the seismic traces of an earthquake and a mining explosion. The data are from [34] and consist of 2048 measurements taken at a recording station in Scandinavia. The data have been analysed by several authors, and the methodology described here is taken from [31]. These particular time series both consist of two waves, the compression wave, also known as primary or P wave, which is the start of the series and the shear, or S wave, which arrives at the midpoint of the series. The analysis of such seismic data is one of critical importance for monitoring a comprehensive test-ban treaty. As argued by many authors, e.g. [34], distinguishing between the seismic traces of earthquakes and explosions is best accomplished in the frequency domain. The goal in this example is to estimate the time-varying spectrum of the process, in order to distinguish between the seismic traces of explosions and earthquakes. The top panels of Figures 16.14 and 16.15 show

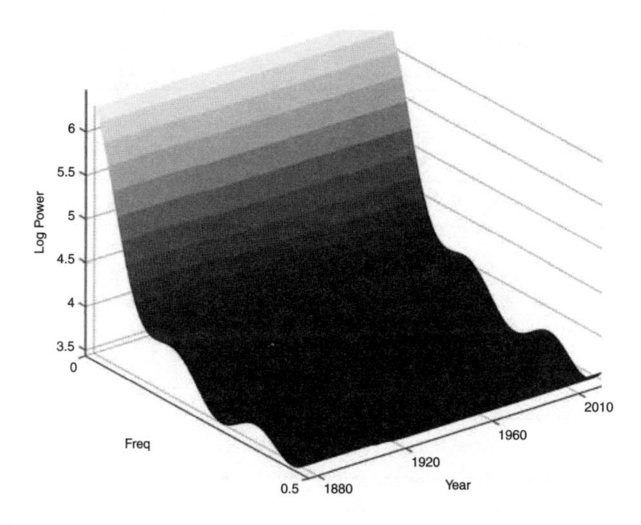

Figure 16.12 Time-varying log spectrum of the SOI index from 1876 to 2008.

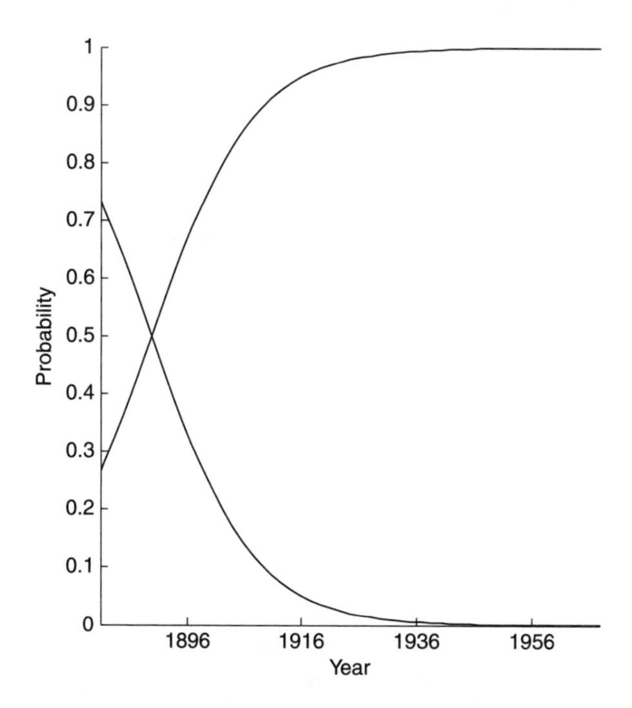

Figure 16.13 $\Pr(\gamma_S = j | r = 2, \mathbf{y})$, for $j = 1, 2$, for the SOI data.

Table 16.3 Posterior probabilities of the number of components in the mixture for the SOI data as a function of prior distribution.

Number of components	Prior			
	Uniform	Poisson		
	$P(j = k) = 0.05$	$\lambda = 1$	$\lambda = 2$	$\lambda = 5$
1	0.755	0.94	0.75	0.66
2	0.241	0.06	0.24	0.33
3	0.004	0.00	0.01	0.01
4	0.000	0.00	0.00	0.00

the recorded seismic trace of the explosion and earthquake respectively, while the bottom panels show the image plots of the time-varying log spectrum. As can be seen from Figures 16.14 and 16.15, both time series are clearly non-stationary. The difference between the two time series is the rate at which each moves from one locally stationary segment to the next. For the explosion data this occurs abruptly while the seismic trace from the earthquake moves more slowly. Other differences are that the S component for the earthquake shows power at the low frequencies only, and the power remains strong for a long time. In contrast, the explosion shows power at higher frequencies than the earthquake, and the power of the signals (P and S waves) does not last as long as in the case of the earthquake.

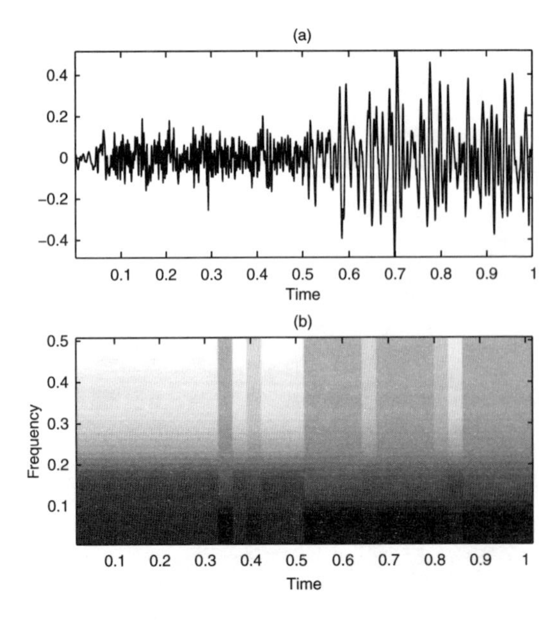

Figure 16.14 Earthquake data. Panel (a) seismic trace and panel (b) image plot for time-varying log spectrum.

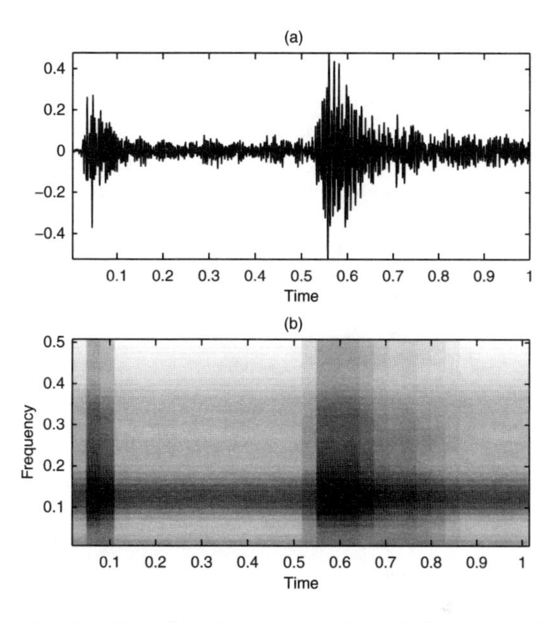

Figure 16.15 Explosion data. Panel (a) seismic trace and panel (b) image plot for time-varying log spectrum.

References

[1] Albert, J. H. and Chib, S. (1993). Bayesian analysis of binary and polychotomous response data, *Journal of the American Statistical Association*, **88**, pp. 669–679.

[2] Baladandayuthapani, V., Mallick, B. and Carroll, R. (2005). Spatially adaptive Bayesian penalized regression splines, *Journal of Computational and Graphical Statistics*, **14**, 378–394.

[3] Berndt, E. (1991). *The Practice of Econometrics: Classic and Contemporary*, Addison-Wesley Pub. Co.

[4] Carter, C. K. and Kohn, R. (1994). On Gibbs sampling for state space models, *Biometrika*, **81**, 541–553.

[5] —— (1996). Semiparametric Bayesian inference for time series with mixed spectra, *J. Royal Statist. Soc. Ser. B*, **59**, 255–268.

[6] Crainiceanu, C. M., Ruppert, D., Carroll, R., Joshi, A. and Goodner, B. (2007). Spatially adaptive Bayesian penalized splines with heteroscedastic Errors, *Journal of Computational and Graphical Statistics*, **16**, 265–288.

[7] De Jong, P. and Shephard, N. (1995). The simulation smoother for time series models, *Biometrika*, **82**, 339–350.

[8] Demmler, A. and Reinsch, C. (1975). Oscillation matrices with spline smoothing, *Numerische Mathematik*, **24**, 375–382, 10.1007/BF01437406.

[9] Denison, D. G. T., Mallick, B. K. and Smith, A. F. M. (1998). Automatic Bayesian curve fitting, *Journal of the Royal Statistical Society: Series B (Statistical Methodology)*, **60**, 333–350.

[10] Durbin, J. and Koopman, S. J. (2002). A simple and efficient simulation smoother for state space time series analysis, *Biometrika*, **89**, 603–616.

[11] Eubank, R. L. (1998). *Spline Smoothing and Nonparametric Regression*, New York: Marcel Dekker.

[12] Fruhwirth-Schnatter, S. (1994). Data augmentation and dynamic linear models, *Journal of Time Series Analysis*, **15**, 183–202.

[13] Gelfand, A. E. and Smith, A. F. M. (1990). Sampling-based approaches to calculating marginal densities, *Journal of the American Statistical Association*, **85**, pp. 398–409.

[14] Green, P. and Silverman, B. (1994). *Nonparametric Regression and Generalized Linear Models: A Roughness Penalty Approach*, Monographs on statistics and applied probability, Chapman & Hall.

[15] Green, P. J. (1995). Reversible jump Markov chain Monte Carlo computation and Bayesian model determination, *Biometrika*, **82**, 711–732.

[16] Hastie, T. and Tibshirani, R. (1987). Generalized additive models: some applications, *Journal of the American Statistical Association*, **82**, pp. 371–386.

[17] Holmes, C. and Mallick, B. (2003). Perfect simulation for Bayesian curve and surface fitting, Technical Report, Imperial College London.

[18] Holmes, C. C. and Mallick, B. K. (2001). Bayesian regression with multivariate linear splines, *Journal of the Royal Statistical Society: Series B (Statistical Methodology)*, 63.

[19] Jacobs, R., Jordan, M., Nowlan, S. and Hinton, G. (1991). Adaptive mixtures of local experts, *Neural Computation*, **3**, 79–87.

[20] Jordan, M. and Jacobs, R. (1994). Hierarchical mixtures of experts and the EM algorithm, *Neural Computation*, **6**, 181–214.

[21] Lang, S. and Brezger, A. (2004). Bayesian P-Splines, *Journal of Computational and Graphical Statistics*, **13**, 183–212.

[22] Nicholls, N. (2008). Recent trends in the seasonal and temporal behaviour of the El Niño–Southern Oscillation, *Geophysical Research Letters*, **35**, L19703.

[23] Nychka, D. (2000). *Spatial Process Estimates as Smoothers. Smoothing and Regression. Approaches, Computation and Application*, New York: John Wiley and Sons, pp. 393–424.

[24] Ombao, H. C., Raz, J. A., von Sachs, R. and Malow, B. A. (2001). Automatic statistical analysis of bivariate nonstationary time series, *Journal of the American Statistical Association*, **96**, 543–560.

[25] Panagiotelis, A. and Smith, M. (2008). Bayesian identification, selection and estimation of semiparametric functions in high-dimensional additive models, *Journal of Econometrics*, **143**, 291–316.

[26] Pintore, A., Speckman, P. and Holmes, C. (2006). Spatially adaptive smoothing splines, *Biometrika*, **93**, 113–125.

[27] Priestley, M. (1982). *Spectral Analysis and Time Series*, no. v. 1–2 in Probability and Mathematical Statistics, Academic Press.

[28] Quinn, B. G. and Thomson, P. J. (1991). Estimating the frequency of a periodic function, *Biometrika*, **78**, 65–74.

[29] Richardson, S. and Green, P. J. (1997). On Bayesian analysis of mixtures with an unknown number of components (with discussion), *Journal of the Royal Statistical Society: Series B (Statistical Methodology)*, **59**, 731–792.

[30] Rosen, O. and Stoffer D. S., (2007). Automatic estimation of multivariate spectra via smoothing splines, *Biometrika*, **94**, 335–345.

[31] Rosen, O., Stoffer, D. and Wood, S. A. (2009). Local spectral analysis via a Bayesian mixture of smoothing splines, *Journal of the American Statistical Association*, **104**, 249–262.

[32] Ruppert, D., Wand, M. and Carroll, R. (2003). *Semiparametric Regression*, Cambridge series on Statistical and Probabilistic Mathematics, Cambridge University Press.

[33] Schwarz, G. (1978). Estimating the dimension of a model, *The Annals of Statistics*, **6**, pp. 461–464.

[34] Shumway, R. and Stoffer, D. (2006). *Time Series Analysis and its Applications with R Examples, 2nd edn*, Springer.

[35] Smith, M. and Kohn, R. (1996). Nonparametric regression using Bayesian variable selection, *Journal of Econometrics*, **75**, 317–343.
[36] Solow, A. (2006). An ENSO shift revisited, *Geophysical Research Letters*, **33**, L22602.
[37] Tanner, M. A. and Wong, W. H. (1987). The calculation of posterior distributions by data augmentation, *Journal of the American Statistical Association*, **82**, pp. 528–540.
[38] Trenberth, K. E. and Hoar, T. J. (1996). The 1990–1995 El Niño-Southern oscillation event: longest on record, *Geophysical Research Letters*, **23**, 57–60.
[39] Verdinelli, I. and Wasserman, L. (1991). Bayesian analysis of outlier problems using the Gibbs sampler, *Statistics and Computing*, **1**, 105–117.
[40] Wahba, G. (1980). Automatic smoothing of the log periodogram, *Journal of the American Statistical Association*, **75**, pp. 122–132.
[41] —— (1983). Bayesian confidence intervals for the cross-validated smoothing spline, *Journal of the Royal Statistical Society B*, **45**, 133–150.
[42] —— (1990). *Spline models for observational data*, CBMS-NSF regional conference series in applied mathematics, Society for Industrial and Applied Mathematics.
[43] Whittle, P. (1957). Curve and Periodogram Smoothing, *Journal of the Royal Statistical Society. Series B (Methodological)*, **19**, pp. 38–63.
[44] Wood, S. A., Kohn, R., Cottet, R., Jiang, W. and Tanner, M. (2008). Locally adaptive nonparametric binary regression, *Journal of Computational and Graphical Statistics*, **17**.
[45] Wood, S. A. and Kohn, R. J. (1998). A Bayesian approach to robust binary nonparametric regression, *Journal of the American Statistical Association*, **93**, pp. 203–213.
[46] Wood, S. A., Wenxin, J. and Tanner, M. (2002). Bayesian mixture of splines for spatially adaptive nonparametric regression. *Biometrika*, **89**, 513–528.

17 Bayesian approaches to copula modelling

MICHAEL STANLEY SMITH

17.1 Introduction

Copula models are now used widely in the empirical analysis of multivariate data. For example, major areas of application include survival analysis, where much early work occurred [15, 51], actuarial science [24], finance [42, 12, 44], marketing [18], transport studies [7, 63], medical statistics [40, 49] and econometrics [61, 9, 53, 66]. Copula models are popular because they are flexible tools for the modelling of complex relationships between variables in a simple manner. They allow for the marginal distributions of data to be modelled separately in an initial step, and then dependence between variables is captured using a copula function.

However, the development of statistical inferential methodology for copula models has been limited. Most research has either been focused on the development and properties of copula functions (see [35] and [47] for excellent overviews), or their use in solving applied problems. Less attention has been given to the question of how to estimate the increasing variety of copula models in an effective manner. To date, the most popular estimation methods are full or two-stage maximum likelihood estimation [36] and method of moments style estimators in low dimensions [28]. There has been only limited work on developing Bayesian approaches to formulate and estimate copula models. This is surprising, given that Bayesian methods have proven successful in both formulating and estimating multivariate models elsewhere. The aim of this article is two-fold: (i) to introduce contemporary copula modelling to Bayesian statisticians, and (ii) to outline the advantages of Bayesian inference when applied to copula models. Therefore, there are two intended audiences: (i) Bayesians who are unfamiliar with the advances and features of copula models, and (ii) users of copula models who are unfamiliar with the advantages and features of modern Bayesian inferential methods.

Previous Bayesian work on copula modelling includes that of Huard, Évin and Favre [33], who suggest a method to select between different bivariate copulas, and Silva and Lopes [59] who use Markov chain Monte Carlo (MCMC) methods to estimate low-dimensional parametric copula functions. Gaussian copula regression models are estimated using MCMC methods in [54], [32] and [18]. Note that adopting a Gaussian copula does not mean the data are normally distributed. In [66] the work of [54] is extended to copulas derived by inversion from skew t distributions constructed by hidden conditioning. Methods to estimate so called 'vine' copulas with continuous margins using MCMC are proposed in [67] and [45, 46]. It is shown in [54] how Bayesian covariance selection approaches can be used in Gaussian copulas, while [67] and [46] also show how Bayesian selection ideas can be applied to determine whether, or not, the component 'pair-copulas' of a vine copula are equal to the bivariate independence copula. It is also shown in [67] that the D-vine copula provides a natural decomposition for serial dependence. Bayesian estimation of multivariate time series with copula-based time varying cross-sectional dependence is also considered in [4]. Last, [64] suggest efficient Bayesian data augmentation methodology for the estimation of copula models for multivariate discrete data, or a combination of discrete and continuous data. Their approach is for general copula functions, not just Gaussian copulas, or copulas constructed by inversion.

This article is divided into three main sections. The first provides an introduction to copula modelling. There are a number of excellent in-depth introductions to copulas and their properties; for example, see [35] and [47]. The purpose of this section is not to replicate any of these, but to introduce aspects that are important in Bayesian copula modelling. This includes an outline of what makes copula models so useful, how copulas models can be viewed as transformations, what are copulas constructed by inversion and vine copulas, and why the D-vine copula is an attractive model of serial dependence.

In the next two sections Bayesian approaches to formulating and estimating copula models are discussed separately for multivariate continuous and discrete data. This is because copula models, and associated methods, differ substantially in these two cases. In Section 17.3 the advantages of using Bayesian inference over maximum likelihood for the case of continuous data are discussed. For the Gaussian copula, a sampling scheme that can be used to evaluate the joint posterior distribution of the copula and any marginal model parameters is outlined in detail. Different priors for the correlation matrix of the Gaussian copula are considered, including priors based on a Cholseky factorization, the partial correlations as in [54], and the conditional correlations discussed in [36] and [19]. A new Bayesian selection approach using the latter is outlined, where the fitted copula model is a Bayesian model average over parsimonious representations of the dependence structure. Bayesian estimation and selection for D-vine copulas is also outlined. An interesting insight is that Bayesian selection of individual pair-copulas nests Bayesian selection of the conditional correlations for a Gaussian copula. Bayesian estimates of popular dependence metrics from the fitted copula are also discussed, where parameter uncertainty can be integrated out using the Monte Carlo iterates from the sampling scheme.

As noted by [21] and [27], popular method of moments style estimators based on ranks should not be used to estimate copula models for discrete data, making likelihood-based inference more important. However, the likelihood function differs substantially from that in the continuous case, and computational issues mean that maximum likelihood estimation is more difficult than in the continuous case. An effective solution is to employ Bayesian data augmentation, as outlined for a Gaussian copula in Section 17.4. The priors for the correlation matrix of the Gaussian copula, and also the Bayesian selection framework, are unaffected by whether the data is discrete or continuous. Last, it is discussed how measuring dependence in discrete data differs from that in the continuous case.

17.2 What are copula models?

17.2.1 The basic idea

Consider initially the bivariate case with two random variables, Y_1 and Y_2, with marginal distribution functions $F_1(y_1)$ and $F_2(y_2)$, respectively. A copula model is a way of constructing the joint distribution of (Y_1, Y_2). Sklar [60] shows that there always exists a bivariate function $C : [0, 1]^2 \to [0, 1]$, such that

$$F(y_1, y_2) = C(F_1(y_1), F_2(y_2))$$

The function C is itself a distribution function with uniform margins on [0, 1], and is labelled the 'copula function'. It binds together the univariate margins F_1 and F_2 to produce bivariate distribution F.

If both margins F_1 and F_2 are continuous distribution functions, then there is a unique copula function C for any given joint distribution function F. If either F_1 or F_2 are discrete-valued, then C is not unique. However, the objective of copula modelling is not to find the copula function(s) C that satisfy Sklar's representation, given knowledge of F_1, F_2 and F. Instead, the objective is to construct a joint distribution F from a copula function C and marginal models for F_1 and F_2. In this way, copula models can be used equally for discrete or continuous data, or a combination of both.

It is important to notice that the copula function C does not determine the marginal distributions of F, but accounts for dependence between Y_1 and Y_2. For example, in the case where Y_1 and Y_2 are independent, the copula function is $C(u_1, u_2) = u_1 u_2$, so that $F(y_1, y_2) = F_1(y_1)F_2(y_2)$. This copula function is called the 'independence copula'.

The copula model is easily generalized to m dimensions as follows. Let $Y = (Y_1, \ldots, Y_m) \in \mathcal{S}_Y$ be a random vector with elements that have marginal distribution functions $F_1, \ldots, F_m$, then the joint distribution function of Y is

$$F(y_1, \ldots, y_m) = C(F_1(y_1), \ldots, F_m(y_m)). \tag{17.1}$$

Again, the copula function $C : [0, 1]^m \to [0, 1]$ is itself a distribution function for random vector $U = (U_1, \ldots, U_m)'$ with uniform margins on [0, 1]. As before, if all elements of Y are continuous random variables, then there is a unique copula function C for any given F, but this is not the case if one or more elements are discrete-valued. Nevertheless, equation (17.1) can still be used to construct a well-defined joint distribution F, given $F_1, \ldots, F_m$ and C, just as in the bivariate case.

17.2.2 Why are copula models so useful?

A key feature of the copula representation of a joint distribution is that it allows for the margins to be modelled separately from the dependence structure. This promotes a 'bottom-up' modelling strategy, where models are first developed one-by-one for each univariate margin. Dependence is then introduced by an appropriate copula function C. Sklar's theorem reassures that this is not an ad-hoc approach, and that there should be at least one copula function C that correctly constructs the joint distribution F, as long as the marginal models $F_1, \ldots, F_m$ are accurate. Compare this to a more restrictive 'top-down' alternative, where the joint distribution function F is selected first, which then determines the form of the marginals. For example, if F is a multivariate t distribution with ν degrees of freedom, then each F_j is restricted to be univariate t with common degrees of freedom ν.

For much applied multivariate modelling, the flexibility that the bottom-up approach allows is compelling. The marginal models can be of the same form, or completely different, including any of the following:

(i) *Parametric distributions* A parametric distribution $F_j(y_j; \theta_j)$, with parameters θ_j. For example, F_j may be a t distribution with location μ_j, scale σ_j and degrees of freedom ν_j, so that $\theta_j = \{\mu_j, \sigma_j, \nu_j\}$. A copula model with t distributions for each margin is more flexible than a multivariate t distribution because the level of kurtosis can differ in each dimension [23]. For discrete data, F_j may be a negative binomial distribution with stopping parameter $r_j > 0$ and success parameter $p_j \in (0, 1)$, so that $\theta_j = \{r_j, p_j\}$. The negative binomial is a very popular model for count data that exhibit heterogeneity, and copula models provide flexible multivariate extensions [41, 50, 18].

(ii) *Nonparametric distributions* Approaches where each margin is modelled nonparametrically using the empirical distribution function (or a smoothed variant) have long been advocated in the copula literature; for example, see [26], [58] and [11]. Similarly, F_j can be modelled using Bayesian nonparametric methods; see [31] for recent accounts of these. Alternatively, rank likelihoods can be used for each marginal model as outlined by [32]. In all cases, copula models provide simple multivariate extensions of existing nonparametric methods.

(iii) *Regression models* Univariate regression models can be used for each margin, in which case the resulting copula model is called a 'copula regression model' [52, 54, 68]. The regression coefficients β_j can be pooled across margins $j = 1, \ldots, m$, so that $\beta_1 = \beta_2 = \ldots = \beta_m$, in which case the copula model is then an extension of the multivariate regression model. If the regression coefficients differ for each margin, then the copula model extends the 'seemingly unrelated regression' model popular in econometric analysis [74].

(iv) *Time series models* When observations are made on a multivariate vector over time, the marginal models can be parametric time series models, and contemporaneous dependence captured via the copula function [53, 10, 4]. Popular choices are GARCH or stochastic volatility models for the margins. As with copula regression models, marginal parameters can either be pooled or allowed to vary across the margin.

17.2.3 Copula functions and densities

The three conditions that C needs to meet to be an admissible copula function are listed in [47, p. 45], and are:

(i) For every $u = (u_1, \ldots, u_m) \in [0, 1]^m$, $C(u) = 0$ if at least one element $u_i = 0$.

(ii) If all elements of u are equal to one, except u_i, then $C(u) = u_i$.

(iii) For each $a = (a_1, \ldots, a_m), b = (b_1, \ldots, b_m) \in [0, 1]^m$, such that $a_i \leq b_i$ for all $i = 1, \ldots, m$,

$$\Delta_{a_m}^{b_m} \Delta_{a_{m-1}}^{b_{m-1}} \cdots \Delta_{a_1}^{b_1} C(v) \geq 0 .$$

Here, $\Delta_{a_k}^{b_k}$ is a differencing notation defined as

$$\Delta_{a_k}^{b_k} C(u_1, \ldots, u_{k-1}, v_k, u_{k+1}, \ldots, u_m) =$$
$$C(u_1, \ldots, u_{k-1}, b_k, u_{k+1}, \ldots, u_m) - C(u_1, \ldots, u_{k-1}, a_k, u_{k+1}, \ldots, u_m) ,$$

with v_k a variable of differencing, and $v = (v_1, \ldots, v_m)$. Notice that if $c(u) = \partial^m C(u)/\partial u_1 \ldots \partial u_m$ exists, then property (iii) is equivalent to

$$\int_{a_1}^{b_1} \cdots \int_{a_m}^{b_m} c(u) du \geq 0 .$$

Properties (i) and (iii) are satisfied if $C(u)$ is a distribution function on $[0, 1]^m$, while property (ii) is satisfied if C also has uniform margins. The density function $c(u)$ is commonly referred to as the 'copula density'.

Table 17.1 Copula functions, density functions and measures of dependence for the Frank, Clayton and Gumbel copulas. For the Frank copula, the function $D_1(\phi) = \frac{1}{\phi}\int_0^\phi t/(\exp(t)-1)dt$ is the Debye function; see [2; p. 998].

Frank ($\phi \in (-\infty, 0) \cup (0, \infty)$)
$C(u_1, u_2; \phi) = -\frac{1}{\phi}\log\left(1 + \frac{(\exp(-\phi u_1)-1)(\exp(-\phi u_2)-1)}{\exp(-\phi)-1}\right)$
$c(u_1, u_2; \phi) = \phi\left(\exp(\phi(1+u_1+u_2))(\exp(\phi)-1)\right) \times \left[\exp(\phi) - \exp(\phi(1+u_1)) - \exp(\phi(1+u_2)) + \exp(\phi(u_1+u_2))\right]^{-2}$
$\tau_{1,2}(\phi) = 1 + \frac{4}{\phi}(D_1(\phi)-1), \lambda^L_{1,2}(\phi) = \lambda^U_{1,2}(\phi) = 0$
Clayton ($\phi \in (-1, \infty)\backslash\{0\}$)
$C(u_1, u_2; \phi) = \max\left\{(u_1^{-\phi} + u_2^{-\phi} - 1)^{-1/\phi}, 0\right\}$
$c(u_1, u_2; \phi) = \max\left\{(1+\phi)(u_1 u_2)^{-1-\phi}\left(u_1^{-\phi} + u_2^{-\phi} - 1\right)^{-1/\phi-2}, 0\right\}$
$\tau_{1,2}(\phi) = \phi/(\phi+2), \lambda^L_{1,2}(\phi) = 2^{-1/\phi}$ and $\lambda^U_{1,2}(\phi) = 0$
Gumbel ($\phi \geq 1$)
$C(u_1, u_2; \phi) = \exp(-(\tilde{u}_1^\phi + \tilde{u}_2^\phi)^{1/\phi})$, where $\tilde{u}_j = -\log(u_j)$
$c(u_1, u_2; \phi) = C(u_1, u_2; \phi)\,(u_1 u_2)^{-1}(\tilde{u}_1^\phi + \tilde{u}_2^\phi)^{-2+2/\phi}(\tilde{u}_1\tilde{u}_2)^{\phi-1} \times \left[1 + (\phi-1)\left(\tilde{u}_1^\phi + \tilde{u}_2^\phi\right)^{-1/\phi}\right]$
$\tau_{1,2}(\phi) = 1 - \phi^{-1}, \lambda^L_{1,2}(\phi) = 0$ and $\lambda^U_{1,2}(\phi) = 2 - 2^{1/\phi}$

In the vast majority of cases parametric copula functions $C(u; \phi)$, with parameters ϕ, are used in applied analysis. There are a large number of choices for C, with [35] and [47] providing overviews of a wide range of copula functions and their properties. Particularly popular in the bivariate case are the family of Archimedean copulas; see [47; Chap. 4]. Three of the most popular Archimedean copulas are the Frank, Clayton and Gumbel. These are listed in Table 17.1, along with their densities and measures of dependence.

17.2.4 Constructing copulas by inversion (of Sklar's theorem)

Beyond the bivariate case, copulas that are constructed through inversion of Sklar's theorem are popular; see [47, Sect. 3.1]. To derive a copula function in this way, let $X = (X_1, \ldots, X_m) \in \mathcal{S}_X$ have distribution function $G(x; \phi)$, with parameters ϕ and strictly monotonic univariate marginal distribution functions $G_1(x_1; \phi), \ldots, G_m(x_m; \phi)$. By Sklar's theorem, there always exists a copula function C, such that

$$G(x; \phi) = C(G_1(x_1; \phi), \ldots, G_m(x_m; \phi)).$$

Denoting $u_j = G_j(x_j; \phi)$, then $x_j = G_j^{-1}(u_j; \phi)$, and substituting this into the equation above defines a copula function:

$$C(u_1, \ldots, u_m; \phi) = G(G_1^{-1}(u_1; \phi), \ldots, G_m^{-1}(u_m; \phi); \phi). \tag{17.2}$$

It is important to notice that the multivariate distribution G is only used to construct the copula function C, and is not the distribution function of the random vector Y, which remains F as given in equation (17.1). The parameters ϕ of the distribution of X are the parameters for copula function C.

Elliptical distributions are common choices for G [23], and the resulting copula functions are collectively called 'elliptical copulas'. The Gaussian copula [68] is the most popular of these, where G is the distribution function of a multivariate normal with zero mean, correlation matrix Γ and unit variances in each dimension. In this case, $\phi = \Gamma$, $G(x; \phi) = \Phi_m(x; \Gamma)$ and $G_j(x_j; \phi) = \Phi_1(x_j, 1)$, with $\Phi_k(\cdot; V)$ the distribution function of a k-dimensional $N(0, V)$ distribution. The Gaussian copula function is therefore

$$C(u_1, \ldots, u_m; \phi) = \Phi_m(\Phi_1^{-1}(u_1; 1), \ldots, \Phi_1^{-1}(u_m; 1); \Gamma). \tag{17.3}$$

The restrictions on the first and second moments of X are necessary to identify the copula parameters Γ in the likelihood.

When each marginal distribution F_j is univariate normal with mean μ_j and variance σ_j^2, then $u_j = \Phi_1(y_j - \mu_j; \sigma_j^2)$. If a Gaussian copula is also assumed, then the copula model for Y simplifies to a multivariate normal distribution with mean $\mu = (\mu_1, \ldots, \mu_m)$ and covariance matrix $D\Gamma D$, with $D = \text{diag}(\sigma_1, \ldots, \sigma_m)$.

Other choices for G include a multivariate t distribution, which results in the t copula [20], or a multivariate skew t distribution [66]. When selecting G, care has to be taken to consider any restrictions on ϕ that may be necessary to identify the parameters in the likelihood.

17.2.5 Copula models as transformations

Copula modelling can be interpreted as a transformation from the domain of the data, to another domain where the dependence is easier to model. The transformation is depicted in Figure 17.1. If the elements of Y are continuous-valued, the transformation $Y_j \mapsto U_j$ is one-to-one, as is the transformation $Y_j \mapsto X_j$ for inversion copulas.

The density of Y is given by

$$f(y) = \frac{\partial}{\partial y} C(F_1(y_1), \ldots, F_m(y_m)) = c(u) \prod_{j=1}^{m} f_j(y_j), \tag{17.4}$$

		$U_j = F_j(Y_j)$		$X_j = G_j^{-1}(U_j)$	
Variable	Y	$\longrightarrow$	U	$\longrightarrow$	X
Domain	S_Y	$\longrightarrow$	$[0, 1]^m$	$\longrightarrow$	S_X
Joint CDF	$F(y)$	$\longrightarrow$	$C(u)$	$\longrightarrow$	$G(x)$
Marginal CDFs	$F_j(y_j)$	$\longrightarrow$	Uniform	$\longrightarrow$	$G_j(x_j)$

Figure 17.1 Depiction of the transformation underlying a copula model. The right-hand column for variable X is for copulas constructed by inversion only. The transformations are given in the top row for Y_j continuous-valued.

with $u = (u_1, \ldots, u_m)$, $u_j = F_j(y_j)$, $f_j(y_j) = \frac{\partial}{\partial y_j} F_j(y_j)$ and $c(u) = \frac{\partial}{\partial u} C(u)$.

However, when the data are discrete-valued, the probability mass function is obtained by differencing the distribution function in equation (17.1), so that

$$\mathrm{pr}(Y = y) = \Delta_{a_m}^{b_m} \Delta_{a_{m-1}}^{b_{m-1}} \cdots \Delta_{a_1}^{b_1} C(v), \tag{17.5}$$

where $v = (v_1, \ldots, v_m)$ are indices of differencing. The upper bound $b_j = F_j(y_j)$ and lower bound $a_j = F_j(y_j^-)$ is the left-hand limit of F_j at y_j, with $F_j(y_j^-) = F_j(y_j - 1)$ when Y_j is ordinal-valued. In this case the transformations $Y_j \mapsto U_j$ and $Y_j \mapsto X_j$ are both one-to-many. This means that the elements $U_j|Y_j = y_j$ and $X_j|Y_j = y_j$ are only known up to bounds, with

$$F_j(y_j^-) \leq U_j < F_j(y_j) \text{ and,}$$

$$G_j^{-1}(F_j(y_j^-)) \leq X_j < G_j^{-1}(F_j(y_j)),$$

for $j = 1, \ldots, m$. Nevertheless, Y, U and X still have distribution functions F, C and G, respectively.

It is outlined later, in Section 17.4, how interpreting a copula model as a transformation allows for the construction of Bayesian data augmentation schemes to evaluate the posterior distribution when one or more margins are discrete.

17.2.6 Vine copulas

Much recent research in the copula literature has focused on building copulas in $m > 2$ dimensions. One popular family of copulas are called 'vines', which are constructed from sequences of bivariate copulas. Early examples of this approach are found in [34, 35], while [6] organize the different decompositions in a systematic way. The bivariate copulas are called 'pair-copulas' in [1], and vines are also known as pair-copula constructions (PCCs). Recent overviews are given by [30] and [17].

If the elements of Y are ordered in time, so that Y_t is observed before Y_{t+1}, then [67] point out that a vine labelled 'drawable' by [6] (or D-vine for short) proves a natural way of characterizing serial dependence; particularly Markovian serial dependence. This can be motivated by considering the following decomposition of the density of U,

$$c(u) = \prod_{t=2}^{m} f(u_t | u_{t-1}, \ldots, u_1),$$

where $f(u_1) = 1$ because the marginal distribution of u_1 is uniform on $[0, 1]$. The idea is to build a representation for each conditional distribution $f(u_t|u_{t-1}, \ldots, u_1)$ as follows. For $s < t$ there always exists a density $c_{t,s}$ on $[0, 1]^2$ such that

$$\begin{aligned} f(u_t, u_s | u_{t-1}, \ldots, u_{s+1}) &= f(u_t | u_{t-1}, \ldots, u_{s+1}) f(u_s | u_{t-1}, \ldots, u_{s+1}) \\ &\times c_{t,s}\left(F(u_t | u_{t-1}, \ldots, u_{s+1}), F(u_s | u_{t-1}, \ldots, u_{s+1}); u_{t-1}, \ldots, u_{s+1}\right). \end{aligned} \tag{17.6}$$

Here, $F(u_t|u_{t-1}, \ldots, u_{s+1})$ and $F(u_s|u_{t-1}, \ldots, u_{s+1})$ are conditional distribution functions of U_t and U_s, respectively. This is the theorem of Sklar applied conditional on $\{U_{t-1}, \ldots, U_{s+1}\}$. In a vine copula, $c_{t,s}$ is the density of a bivariate 'pair-copula' and it is simplified by dropping dependence on $(u_{t-1}, \ldots, u_{s+1})$; see [30] for a discussion of why this is often a good approximation. By setting $s = 1$, application of equation (17.6) gives

$$f(u_t|u_{t-1},\ldots,u_1) = c_{t,1}(F(u_t|u_{t-1},\ldots,u_2), F(u_1|u_{t-1},\ldots,u_2))f(u_t|u_{t-1},\ldots,u_2).$$

Denoting $u_{t|j} = F(u_t|u_{t-1},\ldots,u_j)$ and $u_{j|t} = F(u_j|u_t,\ldots,u_{j+1})$, for $j < t$,[12] repeated application of the above with $s = 2, 3, \ldots, t-1$ leads to the following:

$$f(u_t|u_{t-1},\ldots,u_1) = \prod_{s=1}^{t-1} c_{t,s}(u_{t|s+1}, u_{s|t-1}),$$

where the notation $u_{t|t} = u_t$, for $t = 1, \ldots, m$. Therefore, the D-vine copula is given by

$$c(u) = \prod_{t=2}^{m} \left\{ \prod_{s=1}^{t-1} c_{t,s}(u_{t|s+1}, u_{s|t-1}) \right\} \tag{17.7}$$

which is a product of $m(m-1)/2$ pair-copula densities, and $u = (u_{1|1}, \ldots, u_{m|m})$. If each pair-copula $c_{t,s}$ has copula parameter $\phi_{t,s}$, then the parameter vector of the D-vine is $\phi = \{\phi_{t,s}; t = 2, \ldots, m, s < t\}$. The hardest aspect of using the copula in equation (17.7) is the evaluation of the arguments of the component pair-copulas. An $O(m^2)$ recursive algorithm for the evaluation of these from u is given in [1], based on the identity in [34, p. 125]; see also Algorithm 1 in [67].[13]

Algorithm *Evaluation of the Arguments of a D-vine*

$k = 1, \ldots, m-1$ and $i = k+1, \ldots, m$:
Step 1: Compute $u_{i|i-k} = h_{i,i-k}(u_{i|i-k+1}|u_{i-k|i-1}; \phi_{i,i-k})$
Step 2: Compute $u_{i-k|i} = h_{i,i-k}(u_{i-k|i-1}|u_{i|i-k+1}; \phi_{i,i-k})$.

The functions $h_{t,s}(u_1|u_2; \phi_{t,s}) = \int_0^{u_1} c_{t,s}(v, u_2; \phi_{t,s})dv$ are the conditional distribution functions for the pair-copula with density $c_{t,s}$; see [1] and [67] for lists of these for some common bivariate copulas.

Because any combination of bivariate copula functions can be employed for the pair-copulas, the D-vine copula can be extremely flexible. Moreover, other vine copulas can be constructed using alternative sequences of pair-copulas; see [6] and [1]. However, the D-vine at Equation (17.7) is well-motivated when the elements of U are time-ordered.

17.2.7 Measures of dependence

Measures of dependence for copula models are discussed in [47; Chap. 5] and [35; Chap. 2]. In general, these are marginal pairwise dependencies between elements Y_i and Y_j. Kendall's tau and Spearman's rho are the two most popular measures of pairwise concordance, and empirical analysts are often familiar with sample versions based on ranked data. However, when Y_i and Y_j are continuous-valued, and Y follows the copula model at equation (17.1), the population equivalents can be expressed as

$$\tau_{i,j} = 4\left(\int_0^1 \int_0^1 C_{i,j}^B(u_i, u_j)dC_{i,j}^B(u_i, u_j)\right) - 1 = 4E(C_{i,j}^B(U_i, U_j)) - 1$$

[12] [67] denote $u_{t|j} = F(y_t|y_{t-1},\ldots,y_j)$ and $u_{j|t} = F(y_j|y_t,\ldots,y_{j+1})$ for $Y_1,\ldots,Y_m$ continuous random variables. However, this can be shown to be equivalent to the definition of $u_{t|j}$ and $u_{j|t}$ employed here.

[13] The algorithm here corrects a minor subscript typographical error in the algorithm in [67].

and

$$\rho^S_{i,j} = 12 \int_0^1 \int_0^1 u_i u_j dC^B_{i,j}(u_i, u_j) - 3 = 12E(U_i U_j) - 3\,. \tag{17.8}$$

In the above expressions, $C^B_{i,j}$ is the distribution function of (U_i, U_j) and is a bivariate margin of the m-dimensional copula function C. For some copulas $C^B_{i,j}$ can be computed in closed form, but for others this is not possible. Similarly, the expectations in the expressions for $\tau_{i,j}$ and $\rho^S_{i,j}$ can sometimes be computed in closed form, but for other choices of copulas they are computable only numerically, or by Monte Carlo simulation. Within a Bayesian MCMC framework the latter often proves straightforward; see Section 17.3.5.

In many situations high values of Y_i and Y_j exhibit different levels (or even directions) of dependence than low values of Y_i and Y_j; something that is called 'asymmetric (pairwise) dependence'. As noted by [47; Chap. 4], when Y_i and Y_j are continuous-valued, then the dependence properties of the bivariate margin in these two variables is characterized by the dependence properties between U_i and U_j. In this case, measures of asymmetric dependence are often based on the conditional probabilities

$$\lambda^{up}_{i,j}(\alpha) = \text{pr}(U_i > \alpha | U_j > \alpha)$$

$$\lambda^{low}_{i,j}(\alpha) = \text{pr}(U_i < \alpha | U_j < \alpha),$$

where $0 < \alpha < 1$. The limits of these are called the upper and lower tail dependencies [35, p. 33], and denoted as

$$\lambda^{up}_{i,j} = \lim_{\alpha \uparrow 1} \lambda^{up}_{i,j}(\alpha)\,, \text{ and } \lambda^{low}_{i,j} = \lim_{\alpha \downarrow 0} \lambda^{low}_{i,j}(\alpha)\,.$$

For bivariate copula models there is only a single pairwise combination, Y_1 and Y_2, and for many bivariate copula functions dependence measures are available in closed form. For example, Table 17.1 gives expressions for measures of dependence for the Frank, Gumbel and Clayton copulas; see [35], [47] and [33] for others. Pairwise dependence measures in multivariate m-dimensional elliptical copulas can also have closed form expressions. In particular, the Gaussian copula has zero tail dependence, with $\lambda^{up}_{i,j} = \lambda^{low}_{i,j} = 0$; whereas, the t copula has tail dependence that is non-zero, but is symmetric with $\lambda^{up}_{i,j} = \lambda^{low}_{i,j}$. When employing a copula model it is important to ensure that the copula has dependence properties that are consistent with those exhibited by the data.

17.3 Bayesian inference for continuous margins

When the data are continuous, the likelihood of n independent observations $y = \{y_1, \ldots, y_n\}$, each distributed as equation (17.1), is $f(y|\Theta, \phi) = \prod_{i=1}^n f(y_i|\Theta, \phi)$, where $y_i = (y_{i1}, \ldots, y_{im})'$ and

$$f(y_i|\Theta, \phi) = c(u_i; \phi) \prod_{j=1}^m f_j(y_{ij}; \theta_j)\,. \tag{17.9}$$

Here, $u_i = (u_{i1}, \ldots, u_{im})'$, $u_{ij} = F_j(y_{ij}; \theta_j)$, $\Theta = \{\theta_1, \ldots, \theta_m\}$ are any parameters of the marginal models, and $f_j(y_{ij}; \theta_j) = \frac{\partial}{\partial y_{ij}} F_j(y_{ij}; \theta_j)$ is the marginal density of y_{ij}. Initially, equation (17.9) appears separable in $\theta_1, \ldots, \theta_m$ and ϕ, but this is not the case because u_i depends on Θ. Most parametric copula functions have analytical expressions for the densities $c(u;\phi)$, so that maximum likelihood estimation is often straightforward. However, there are a number of circumstances where a Bayesian analysis can be preferable:

(i) For more complex marginal models $F_j(y_{ij};\theta_j)$ and/or copula functions $C(u;\phi)$, the likelihood can be hard to maximize directly. One solution is to use a two-stage estimator, where the marginal model parameters θ_j are estimated first, and then ϕ estimated conditional on these. In the copula literature, this is called 'inference for margins'; see [36] and references therein for a discussion. Another solution is to use to an iterative scoring algorithm to maximize the likelihood, as suggested by [69]. However, an attractive Bayesian alternative in this circumstance is to construct inference from the joint posterior $f(\Theta, \phi|y)$ evaluated in a Monte Carlo manner, with Θ and ϕ generated separately in a Gibbs style sampling scheme; see [54], [59] and [4] for discussions.

(ii) Bayesian hierarchical modelling has proven very successful for the modelling of multivariate data. This includes parsimonious modelling of covariance structures using Bayesian selection and model averaging; see [29], [65], [72] and [25] for examples. Bayesian selection can be extended to nonlinear dependence by considering priors with point mass components for ϕ. For example, [54] use a 'spike and slab' prior similar to [72] for the off-diagonal elements of the concentration matrix Γ^{-1} of a Gaussian copula. Similarly, [67] use Bayesian selection ideas to mix over independent and dependent pair-copulas in a vine copula. Hierarchical models can also be employed for the margins $F_j(y_j;\theta_j)$, and estimated jointly with the dependence structure captured by the copula function.

(iii) When estimating a copula model, the objective is often to construct inference on measures of dependence, quantiles and/or functionals of the random variable vector Y or parameters (Θ, ϕ). Evaluation of the posterior distribution of these quantities is often straightforward using MCMC methods.

17.3.1 The Gaussian copula model

To illustrate, Bayesian estimation of a Gaussian copula model for continuous margins is outlined as suggested by [54]. Following [68] and others, derivation of the copula density is straightforward by differentiation of equation (17.3), so that

$$c(u;\phi) = \frac{\partial}{\partial u} C(u;\phi) = |\Gamma|^{-1/2} \exp\left\{-\frac{1}{2} x'(\Gamma^{-1} - I)x\right\} \tag{17.10}$$

where $x = (\Phi_1^{-1}(u_1;1), \ldots, \Phi_1^{-1}(u_m;1))'$. Thus, the likelihood at equation (17.9) is a function of Θ and Γ, and can be written as

$$f(y|\Theta,\Gamma) = |\Gamma|^{-n/2}\left(\prod_{i=1}^{n} \exp\left\{-\frac{1}{2} x_i'(\Gamma^{-1} - I)x_i\right\} \prod_{j=1}^{m} f_j(y_{ij};\theta_j)\right), \tag{17.11}$$

where $x_i = (x_{i1}, \ldots, x_{im})'$, $x_{ij} = \Phi_1^{-1}(u_{ij};1)$ and $u_{ij} = F_j(y_{ij};\theta_j)$. Bayesian estimation can be undertaken using the following MCMC sampling scheme:

Sampling scheme. *Estimation of a Gaussian copula*

Step 1: Generate from $f(\theta_j|\{\Theta\backslash\theta_j\},\Gamma,y)$ for $j=1,\ldots,m$.
Step 2: Generate from $f(\Gamma|\Theta,y)$.

Here, $\{A\backslash B\}$ is notation for A with component B omitted. Steps 1 and 2 are repeated (in sequence) a large number of times, with each repeat usually called a 'sweep' in the Bayesian literature. The scheme requires an initial (feasible) state for the parameter values, which is denoted here as $(\Theta^{[0]},\phi^{[0]})$. The iterates from the scheme form a Markov chain, which can be shown to converge to draws from the joint posterior distribution $f(\Theta,\phi|y)$, which is the (unique) invariant distribution of the chain. After an initial number of sweeps, the chain is assumed to have converged and subsequent iterates form a Monte Carlo sample from which the parameters are estimated, and other Bayesian inference obtained as outlined in Section 17.3.5. For introductions to MCMC methods for computing Bayesian posterior inference see [70] and [55].

The posterior in Step 1 is given by

$$\begin{aligned} f(\theta_j|\{\Theta\backslash\theta_j\},\Gamma,y) &\propto f(y|\Theta,\Gamma)\pi(\theta_j) \\ &\propto |\Gamma|^{-n/2}\left(\prod_{i=1}^{n}\exp\left\{-\frac{1}{2}x_i'(\Gamma^{-1}-I)x_i\right\}f_j(y_{ij};\theta_j)\right)\pi(\theta_j), \end{aligned} \tag{17.12}$$

where $\pi(\theta_j)$ is the marginal prior for θ_j. In general, the density is unrecognizable because x_{ij} is a function of θ_j, so [54] suggest using a Metropolis–Hastings (MH) step with a multivariate t distribution as a proposal to generate θ_j in Step 1. The mean of the t distribution, $\hat{\theta}_j$, is the mode of equation (17.12), which is obtained via quasi-Newton–Raphson methods applied to the logarithm of the posterior density. The Hessian

$$H=\left.\frac{\partial^2\log(f(\theta_j|\{\Theta\backslash\theta_j\},\Gamma,y))}{\partial\theta_j\partial\theta_j'}\right|_{\theta_j=\hat{\theta}_j}$$

is calculated numerically using finite difference methods. The scale matrix of the MH proposal is $-H^{-1}$, and a low degrees of freedom, such as $\nu=5$ or $\nu=7$, is employed so that the proposal dominates the target density in the tails. If θ_j has too many elements for H to be evaluated in a numerically stable and computationally feasible fashion, θ_j can be partitioned and generated separately. Alternative MH steps are also possible, including those based on the widely employed random walk proposals.

The approach used to generate Γ in Step 2 varies depending on the prior and matrix parameterization adopted, of which there are several alternatives. For example, [54] consider a prior on the off-diagonal elements of Γ^{-1}, which is equivalent to assuming a prior for the partial correlations $\text{Corr}(X_t,X_s|X_{j\notin\{s,t\}})$ for $t=2,\ldots,m$; $s<t$. Alternatively, [32] suggests using a prior for Γ in a Gaussian copula that results from an inverse Wishart prior for a covariance matrix. However, because Γ is just a correlation matrix (for X), any prior for a correlation matrix can also be used; for example, see those suggested by [5], [43], [3], [19] and references therein.

17.3.1.1 *Prior based on a Choelsky factor*

One such prior for a correlation matrix is based on a Cholesky factorization, which is particularly suited to longitudinal data. This prior uses the decomposition

$$\Gamma=\text{diag}(\Sigma)^{-1/2}\Sigma\,\text{diag}(\Sigma)^{-1/2} \tag{17.13}$$

where Σ is a positive definite matrix, and $\text{diag}(\Sigma)$ is a diagonal matrix comprised of the leading diagonal of Σ. The matrix $\Sigma^{-1} = R'R$, with $R = \{r_{k,j}\}$ being an upper triangular Cholesky factor, and to ensure that the parameterization is unique, $r_{k,k} = 1$, for $k = 1, \ldots, m$. Generation of Γ in Step 2 is undertaken by generating the elements $\{r_{k,j}; j = 2, \ldots, m,\ k < j\}$ one at a time from the conditional posterior

$$f(r_{k,j}|\{R\backslash r_{k,j}\}, \Theta, y) \propto |\Gamma|^{-n/2} \left(\prod_{i=1}^{n} \exp\left\{ -\frac{1}{2} x_i'(\Gamma^{-1} - I)x_i \right\} \right) \pi(r_{k,j})$$

using random walk MH; see [70; p.177] for a discussion of this simulation tool. Once an iterate of R is obtained, the iterate of Γ can be computed using the relationship at equation (17.13). Using a different prior, [32] uses a similar approach to generate a correlation matrix for a Gaussian copula.

17.3.1.2 *Prior based on partial correlations*

It is suggested in [19] to parameterize a correlation matrix using the partial correlations

$$\lambda_{t,s} = \text{Corr}(X_t, X_s | X_{t-1}, \ldots, X_{s+1}), \text{ for } s < t. \tag{17.14}$$

This prior is based on the work of [37], who notes that these are unconstrained on $(-1, 1)$, and that $\Lambda = \{\lambda_{t,s}; t = 2, \ldots, m,\ s < t\}$ provides a unique parameterization of Γ. Note that $\lambda_{t,s}$ is sometimes called a 'semi-partial' correlation because it is not the correlation conditional on all other variables $\text{Corr}(X_t, X_s | X_{j \notin \{t,s\}})$, which is the 'full' partial correlation considered by [54]. One advantage is that the conditional distribution of $\lambda_{t,s}|\{\Lambda\backslash\lambda_{t,s}\}$ is only bounded to $(-1, 1)$, whereas the conditional distribution of the full partial correlations have more complex bounds. Beta or uniform priors for $\lambda_{t,s}$ are suggested by [19], which can be employed and Step 2 undertaken by generating the elements of Λ one at a time, again using MH with a random walk proposal. Once an iterate of Λ is obtained, Γ can be computed using the identity at equation (2) of [19].

There is an interesting link between the Gaussian copula parameterized by the partial correlations Λ, and the D-vine copula in equation (17.7). When the pair-copulas in the D-vine are bivariate Gaussian copulas, with densities

$$c_{t,s}(u_1, u_2; \phi_{t,s}) = \frac{1}{\sqrt{1 - \phi_{t,s}^2}} \exp\left\{ -\frac{\phi_{t,s}^2(x_1^2 + x_2^2) - 2\phi_{t,s} x_1 x_2}{2(1 - \phi_{t,s}^2)} \right\} \tag{17.15}$$

where $x_1 = \Phi_1^{-1}(u_1; 1)$ and $x_2 = \Phi_1^{-1}(u_2; 1)$, then the D-vine copula can be shown to be a Gaussian copula with copula density at equation (17.10); see [1] and [30]. In this case, the individual pair-copula parameters $\phi_{t,s}$ above are the partial correlations $\lambda_{t,s}$.

17.3.2 Bayesian selection in a Gaussian copula

Bayesian selection approaches can be employed to allow for parsimonious modelling of Γ in a Gaussian copula. It is well known that Bayesian selection can significantly improve estimates of a covariance matrix compared to maximum likelihood; see [73], [29], [65], [72], [25] and others for extensive evidence to this effect. As shown by [54], this is also the case when estimating the dependence structure of Y using a Gaussian copula model. They consider a selection prior with point mass probabilities on the off-diagonal elements of Γ^{-1}. In the Gaussian copula this is equivalent to identifying for which pairs (t, s) the full partial correlation $\text{Corr}(X_t, X_s|X_{j\notin\{s,t\}}) = 0$. This also corresponds to conditional independence between Y_t and Y_s, with the conditional density $f(y_t, y_s|y_{j\notin\{s,t\}}) = f(y_t|y_{j\notin\{s,t\}})f(y_s|y_{j\notin\{s,t\}})$.

17.3.2.1 *Priors for selection*

Bayesian selection can also be undertaken for the semi-partial correlations Λ defined in equation (17.14). In the Gaussian copula this is equivalent to determining for which pairs (t, s) there is conditional independence between elements of Y, with conditional density

$$f(y_t, y_s|y_{t-1}, \ldots, y_{s+1}) = f(y_t|y_{t-1}, \ldots, y_{s+1})f(y_s|y_{t-1}, \ldots, y_{s+1})$$

when $\lambda_{t,s} = 0$. To introduce a point mass probability for this value, binary indicator variables $\gamma = \{\gamma_{t,s}; t = 2, \ldots, m,\ s < t\}$ are introduced, such that

$$\lambda_{t,s} = 0 \text{ iff } \gamma_{t,s} = 0.$$

The non-zero partial correlations $\lambda_{t,s}|\gamma_{t,s} = 1$ are independently distributed with proper prior densities $\pi(\lambda_{t,s})$. It is noted by [37] that $\lambda_{t,s}|\{\Lambda\backslash\lambda_{t,s}\}$ are unconstrained on $[-1, 1]$, so that either independent uniform or Beta priors are simple choices for $\pi(\lambda_{t,s})$; see [19]. In comparison, each full partial correlation has bounds that are complex functions of the other full partial correlations and computationally demanding to evaluate. For this reason, Bayesian selection using the partial correlations Λ is computationally less burdensome than using the full partial correlations.

The prior on the indicators γ can be highly informative when the number of indicators $N = m(m-1)/2$ is large. For example, if $w_\gamma = \sum_{t,s}\gamma_{t,s}$ is the number of non-zero elements in Λ, then assuming flat marginal priors $\pi(\gamma_{t,s}) = 1/2$ puts high prior weight on values for $w_\gamma \approx N/2$. This problem has been noted widely in the variable selection literature; see [38], [75] and [8]. One solution is to employ the conditional prior

$$\pi(\gamma_{t,s} = 1|\{\gamma\backslash\gamma_{t,s}\}) \propto B(N - w_\gamma + 1, w_\gamma + 1)\,, \tag{17.16}$$

where $B(\cdot, \cdot)$ is the beta function. This prior has been used effectively in the Bayesian selection literature, with early uses in [62] and [65]. It corresponds to assuming the joint mass function

$$\pi(\gamma) = \frac{1}{N+1}\binom{N}{w_\gamma}^{-1}.$$

The implied prior for the total number of non-zero elements of Λ is uniform, with $\pi(w_\gamma) = 1/(1+N)$, while the marginal priors $\pi(\gamma_{t,s})$ are all equal; see [57] for a discussion. This prior is also equivalent to the uniform volume-based prior suggested by [72] and [16] on the model space.

17.3.2.2 *MCMC sampling scheme*

To evaluate the joint posterior distribution of the indicator variables and the partial correlations Λ, latent variables $\tilde{\lambda}_{t,s}$, for $t = 2, \ldots, m, s < t$, are introduced such that $\lambda_{t,s} = \tilde{\lambda}_{t,s}$ if $\gamma_{t,s} = 1$. Notice that $\lambda_{t,s}$ is known exactly given the pair $(\tilde{\lambda}_{t,s}, \gamma_{t,s})$, so it is sufficient to implement a sampling scheme to evaluate the joint posterior $f(\tilde{\Lambda}, \gamma, \Theta|y)$, where $\tilde{\Lambda} = \{\tilde{\lambda}_{t,s}; t = 2, \ldots, m,\ s < t\}$, as below.

Sampling scheme. *Bayesian selection for a Gaussian copula*

Step 1: Generate from $f(\theta_j|\{\Theta\backslash\theta_j\}, \Gamma, y)$ for $j = 1, \ldots, m$.
Step 2: Generate from $f(\tilde{\lambda}_{t,s}, \gamma_{t,s}|\Theta, \{\tilde{\Lambda}\backslash\tilde{\lambda}_{t,s}\}, \{\gamma\backslash\gamma_{t,s}\}, y)$ for $t = 2, \ldots, m$, $s < t$.
Step 3: Compute Λ from $(\tilde{\Lambda}, \gamma)$, and then Γ from Λ.

Step 1 is unchanged from that in Section 17.3.1, while Step 2 consists of MH steps to generate each pair $(\tilde{\lambda}_{t,s}, \gamma_{t,s})$, conditional on the others. The MH proposal density is

$$q(\tilde{\lambda}_{t,s}, \gamma_{t,s}) = q_1(\gamma_{t,s})q_2(\tilde{\lambda}_{t,s}).$$

To generate from the proposal q above, an indicator is generated from $q_1(\gamma_{t,s} = 0) = q_1(\gamma_{t,s} = 1) = 1/2$, and $\tilde{\lambda}_{t,s}$ from a symmetric random walk proposal q_2 constrained to $(-1, 1)$. For example, one such symmetric proposal for q_2 is to generate a new value of $\tilde{\lambda}_{t,s}$ from a normal distribution with mean equal to the old value, standard deviation 0.01, and constrained to $(-1, 1)$.

Temporarily dropping the subscripts (t, s) for convenience, a new iterate $(\tilde{\lambda}^{new}, \gamma^{new})$ generated from the proposal q is accepted over the old value $(\tilde{\lambda}^{old}, \gamma^{old})$ with probability

$$\min\left(1, \alpha\frac{\pi(\tilde{\lambda}^{new})}{\pi(\tilde{\lambda}^{old})}\kappa\right), \tag{17.17}$$

where κ is an adjustment due to the bounds $(-1, 1)$ on λ. If the symmetric density $q_2(\cdot)$ has distribution function $Q_2(\cdot)$, then this adjustment is

$$\kappa = \frac{Q_2(1 - \tilde{\lambda}^{old}) - Q_2(-1 - \tilde{\lambda}^{old})}{Q_2(1 - \tilde{\lambda}^{new}) - Q_2(-1 - \tilde{\lambda}^{new})}.$$

If a uniform prior is adopted for $\tilde{\lambda}_{t,s}$, as suggested in [19], then the ratio $\pi(\tilde{\lambda}^{new})/\pi(\tilde{\lambda}^{old}) = 1$ in equation (17.17). At each generation in Step 2, the likelihood in equation (17.11) is a function of $(\tilde{\lambda}, \gamma)$, so it can be written here as $L(\tilde{\lambda}, \gamma)$. Using this notation, the value α in equation (17.17) can be expressed separately for the four possible configurations of $(\gamma^{old}, \gamma^{new})$ as:

$$\alpha\left((\tilde{\lambda}^{old}, \gamma^{old} = 0) \rightarrow (\tilde{\lambda}^{new}, \gamma^{new} = 0)\right) = 1$$

$$\alpha\left((\tilde{\lambda}^{old}, \gamma^{old} = 0) \rightarrow (\tilde{\lambda}^{new}, \gamma^{new} = 1)\right) = \frac{L(\tilde{\lambda}^{new}, \gamma^{new} = 1)\delta_1}{L(0, \gamma^{old} = 0)\delta_0}$$

$$\alpha\left((\tilde{\lambda}^{old}, \gamma^{old} = 1) \rightarrow (\tilde{\lambda}^{new}, \gamma^{new} = 0)\right) = \frac{L(0, \gamma^{new} = 0)\delta_0}{L(\tilde{\lambda}^{old}, \gamma^{old} = 1)\delta_1}$$

$$\alpha\left((\tilde{\lambda}^{old}, \gamma^{old} = 1) \rightarrow (\tilde{\lambda}^{new}, \gamma^{new} = 1)\right) = \frac{L(\tilde{\lambda}^{new}, \gamma^{new} = 1)}{L(\tilde{\lambda}^{old}, \gamma^{old} = 1)}$$

where δ_0 and δ_1 are the conditional probabilities from equation (17.16) that $\gamma_{t,s} = 0$ and 1, respectively. Notice that when $(\gamma^{old} = 0) \rightarrow (\gamma^{new} = 0)$ the likelihood does not need computing to evaluate the acceptance ratio at Equation (17.17). This case will occur frequently whenever there is a high degree of sparsity in the dependence structure, so that each sweep of Step 2 will be much faster than if no selection were considered.

Reintroducing subscripts, Step 3 of the sampling scheme is straightforward, with each partial correlation

$$\lambda_{t,s} = \begin{cases} 0 & \text{if } \gamma_{t,s} = 0 \\ \tilde{\lambda}_{t,s} & \text{if } \gamma_{t,s} = 1 \end{cases}$$

and the correlation matrix Γ can be obtained directly from Λ using the relationship in [37] and [19].

17.3.3 Bayesian estimation and selection for a D-vine

Bayesian estimation for vine copulas is discussed in [45, 46] and [67]. The latter authors consider Bayesian selection and model averaging via the introduction of indicator variables in the tradition of Bayesian variable selection. It is this approach that is outlined here, although readers are referred to [67] for a full exposition.

The objective of Bayesian selection for a vine copula is to identify component pair-copulas that are equal to the bivariate independence copula. Recall that the bivariate independence copula has copula function $C(u_1, u_2) = u_1 u_2$, and corresponding copula density $c(u_1, u_2) = \partial C(u_1, u_2)/\partial u_1 \partial u_2 = 1$. This leads to a parsimonious representation because the independence copula is not a function of any parameters.

For the D-vine with copula density at equation (17.7), Bayesian selection introduces indicator variables $\gamma = \{\gamma_{t,s}; t = 2, \ldots, m,\ s < t\}$, where

$$c_{t,s}(u_1, u_2) = \begin{cases} 1 & \text{if } \gamma_{t,s} = 0 \\ c^{\star}_{t,s}(u_1, u_2; \phi_{t,s}) & \text{if } \gamma_{t,s} = 1. \end{cases} \tag{17.18}$$

In the above, $c^{\star}_{t,s}$ is a pre-specified bivariate copula density with parameter $\phi_{t,s}$.[14] The copula type can vary with (t, s), but for simplicity only the case where $c^{\star}_{t,s}(u_1, u_2; \phi_{t,s}) = c^{\star}(u_1, u_2; \phi_{t,s})$ is considered here. That is, each pair-copula $c_{t,s}$ is either an independence copula, or a bivariate copula of the same form for all pair-copulas, but with differing parameter values. From equation (17.6) it follows that when $c_{t,s}(u_1, u_2) = 1$, $f(u_t, u_s | u_{t-1}, \ldots, u_{s+1}) = f(u_t | u_{t-1}, \ldots, u_{s+1}) \times f(u_s | u_{t-1}, \ldots, u_{s+1})$, so that there is conditional independence between U_t and U_s.

The pre-specified bivariate copula can nest the independence copula, so that there exists a value ϕ^+, such that $c^{\star}(u_1, u_2; \phi^+) = 1$. In this case, the condition at equation (17.18) can be rewritten as $c_{t,s}(u_1, u_2) = c^{\star}(u_1, u_2; \phi_{t,s})$, with $\phi_{t,s} = \phi^+$ iff $\gamma_{t,s} = 0$. One example of such a copula is the Gumbel when $\phi^+ = 1$, which is easily seen by substituting the value into the copula density, as given in Table 17.1.

To estimate the joint posterior $f(\phi, \Theta | y)$, latent variables $\tilde{\phi}_{t,s}$, for $t = 2, \ldots, m$, $s < t$, are introduced such that $\phi_{t,s} = \tilde{\phi}_{t,s}$ if $\gamma_{t,s} = 1$. As with the partial correlations in the previous section, $\phi_{t,s}$ is known exactly given the pair $(\tilde{\phi}_{t,s}, \gamma_{t,s})$. Therefore, it is sufficient to implement a sampling scheme to evaluate the joint posterior $f(\tilde{\phi}, \gamma, \Theta | y)$, where $\tilde{\phi} = \{\tilde{\phi}_{t,s}; t = 2, \ldots, m,\ s < t\}$, as below.

[14] Note that this parameter is often a scalar, such as for an Archimedean or bivariate Gaussian copula. However, it can also be a vector, as in the case of a bivariate t copula, where both the degrees of freedom and correlation are parameters.

Sampling scheme. *Bayesian selection for a D-vine copula*

Step 1: Generate from $f(\theta_j|\{\Theta\backslash\theta_j\},\phi,y)$ for $j = 1,\ldots,m$.
Step 2: Generate from $f(\tilde{\phi}_{t,s},\gamma_{t,s}|\Theta,\{\tilde{\phi}\backslash\tilde{\phi}_{t,s}\},\{\gamma\backslash\gamma_{t,s}\},y)$ for $t = 2,\ldots,m$, $s < t$.
Step 3: Compute ϕ from $(\tilde{\phi},\gamma)$.

Generating the marginal parameters θ_j in Step 1 is undertaken using the same MH step outlined in Section 17.3.1, but where the conditional posterior is now

$$f(\theta_j|\{\Theta\backslash\theta_j\},\phi,y) \propto \left(\prod_{i=1}^{n} f(y_i|\Theta,\phi)\right)\pi(\theta_j)$$
$$\propto \left(\prod_{i=1}^{n} c(u_i;\phi)f_j(y_{ij};\theta_j)\right)\pi(\theta_j).$$

In the above, $c(u_i;\phi)$ is the D-vine copula density at equation (17.7), evaluated at observation $u_i = (F_1(y_{i1};\theta_1),\ldots,F_m(y_{im};\theta_m))$.[15] The algorithm in Section 17.2.6 is run separately for each observation u_i to evaluate the arguments of the component pair-copulas of $c(u_i;\phi)$. Interestingly, selection can speed up this algorithm substantially because $h_{t,s}(u_1|u_2;\phi_{t,s}) = u_1$ if $\gamma_{t,s} = 0$.

Generating the pair $(\tilde{\phi}_{t,s},\gamma_{t,s})$ follows the same MH step outlined in Section 17.3.2 for the partial correlations. The main difference is that whenever $\tilde{\phi}_{t,s}$ is vector-valued, each element is generated separately in the same manner. Also, for many bivariate copulas (particularly the Archimedean ones) proper non-uniform priors for $\tilde{\phi}_{t,s}$ are often preferred.

17.3.4 Equivalence of selection for Gaussian and D-vine copulas

It is worth highlighting here that the Bayesian selection approach for the D-vine nests that for the Gaussian copula, when the correlation matrix is parameterized by the semi-partial correlations Λ. If the pair-copula $c^\star$ is the bivariate Gaussian copula with density at equation (17.15), then $\phi_{t,s} = \lambda_{t,s}$ and $\phi = \Lambda$. In this case, the sampling schemes for Bayesian selection for D-vine and Gaussian copulas are identical.

17.3.5 Posterior inference

Estimation is based on the Monte Carlo iterates

$$\left\{(\phi^{[1]},\Theta^{[1]}),\ldots,(\phi^{[J]},\Theta^{[J]})\right\},$$

obtained from the sampling schemes after convergence to the joint posterior distribution, so that $(\phi^{[j]},\Theta^{[j]}) \sim f(\phi,\Theta|y)$. When Bayesian selection is undertaken, as in Sections 17.3.2 and 17.3.3, iterates $\{\gamma^{[1]},\ldots,\gamma^{[J]}\}$ are also obtained, with $\gamma^{[j]} \sim f(\gamma|y)$. Monte Carlo estimates of the posterior means can be used as point estimates. For example, the posterior means

$$E(\theta_k|y) \approx \frac{1}{J}\sum_{j=1}^{J}\theta_k^{[j]}, \text{ and } E(\phi|y) \approx \frac{1}{J}\sum_{j=1}^{J}\phi^{[j]},$$

[15] In the copula literature the n observations $\{u_1,\ldots,u_n\}$ are often called the 'copula data'.

are used as point estimates of the marginal model and copula parameters, respectively. Marginal $100(1-\alpha)\%$ posterior probability intervals can be constructed for any scalar parameter by simply ranking the iterates, and then counting off the $\alpha J/2$ lowest values, and the same number of the highest values.

When undertaking Bayesian selection for a Gaussian copula, the estimates

$$\text{pr}(\gamma_{t,s}=1|y) \approx \frac{1}{J}\sum_{j=1}^{J}\gamma_{t,s}^{[j]}\text{, and } E(\lambda_{t,s}|y) \approx \frac{1}{J}\sum_{j=1}^{J}\lambda_{t,s}^{[j]},$$

can be computed. The former gives the posterior probability that the pair Y_t, Y_s are dependent, conditional on $(Y_{s+1},\ldots,Y_{t-1})$, for $s<t$. The latter is the posterior mean of the semi-partial correlation. At each sweep of the sampling scheme, some elements of $\Lambda^{[j]}$ will be exactly equal to zero, as determined by $\gamma^{[j]}$. The estimate $E(\Gamma|y) \approx \frac{1}{J}\sum_{j=1}^{J}\Gamma^{[j]}$ is therefore often called a 'model average' because it is computed by averaging over these configurations of zero and non-zero semi-partial correlations in $\Lambda^{[j]}$.

Similar estimates can be computed when undertaking Bayesian selection for D-vine copulas. When the form of the component pair-copulas nests the independence copula, so that copula density $c^{\star}(u_1,u_2;\phi^{+})=1$, then it is possible to compute the posterior mean of the pair-copula parameters as $E(\phi_{t,s}|y) \approx \frac{1}{J}\sum_{j=1}^{J}\phi_{t,s}^{[j]}$, because $\phi_{t,s}^{[j]}=\phi^{+}$ when $\gamma_{t,s}^{[j]}=0$. However, when the pair-copulas do not nest the independence copula, $\phi_{t,s}$ is undefined when $\gamma_{t,s}=0$.

If the measures of pairwise dependence discussed in Section 17.2.7 have a closed form expression (or an accurate numerical approximation), then Monte Carlo estimates are straightforward to compute. For example, the estimate of Kendall's tau for continuous valued data is

$$E(\tau_{i,k}|y) = \int \tau_{i,k}(\phi)f(\phi|y)d\phi \approx \frac{1}{J}\sum_{j=1}^{J}\tau_{i,k}(\phi^{[j]}).$$

Posterior probability intervals are constructed using the iterates $\{\tau_{i,k}(\phi^{[1]}),\ldots,\tau_{i,k}(\phi^{[J]})\}$ in the same manner as for the model parameters. If the pairwise dependence measures are difficult to compute, then Kendall's tau and Spearman's rho can be obtained by evaluating the expectations at equation (17.8) via simulation as follows. At the end of each sweep of a sampling scheme, generate an iterate from the copula distribution $U^{[j]} \sim C(u;\phi^{[j]})$, and then compute

$$E(C_{i,k}^{B}(U_i,U_k)) \approx \frac{1}{J}\sum_{j=1}^{J}C_{i,k}^{B}(U_i^{[j]},U_k^{[j]})\text{, and } E(U_iU_k) \approx \frac{1}{J}\sum_{j=1}^{J}U_i^{[j]}U_k^{[j]}.$$

Simulating from most copula distributions is straightforward and fast; see [12; Chap. 6].

17.4 Bayesian inference for discrete margins

Estimation of copula models with one or more discrete marginal distributions differs substantially from those with continuous margins; see [27] for an extensive discussion on the differences. In this section, the case where all margins are discrete is considered, although extension to the case where some margins are discrete and others continuous is discussed in [64].

The likelihood of n independent observations $y = \{y_1, \ldots, y_n\}$, each distributed as equation (17.1) and with probability mass function at equation (17.5), is

$$L(\Theta, \phi) = \prod_{i=1}^{n} \Delta_{a_{im}}^{b_{im}} \Delta_{a_{im-1}}^{b_{im-1}} \cdots \Delta_{a_{i1}}^{b_{i1}} C(v; \phi). \tag{17.19}$$

Here, $v = (v_1, \ldots, v_m)$ are indices of differencing, each observation $y_i = (y_{i1}, \ldots, y_{im})$, the upper bound $b_{ij} = F_j(y_{ij}; \theta_j)$, and the lower bound $a_{ij} = F_j(y_{ij}^-; \theta_j)$ is the left-hand limit of F_j at y_{ij}. In general, computing the likelihood involves $O(n2^m)$ evaluations of C, which is prohibitive for high m. Moreover, even for low values of m, it can be difficult to maximize the likelihood for some copula and/or marginal model choices.

An alternative is to augment the likelihood with latent variables, and integrate them out in a Monte Carlo fashion. From a Bayesian perspective this involves evaluating the augmented posterior distribution by MCMC methods; an approach that is called Bayesian data augmentation [71]. It is shown in [64] how this can be undertaken by augmenting the posterior distribution with latent variables distributed as $U = (U_1, \ldots, U_m) \sim C(u; \phi)$. While their approach applies to all parametric copula functions, in the specific case of a copula constructed by inversion as at equation (17.2), latent variables distributed as $X \sim G(x; \phi)$, can also be used. This is proposed in [54] to estimate Gaussian copula models, and in [66] when G is the distribution function of the skew t of [56].

17.4.1 The Gaussian copula model

For the Gaussian copula, latent variables $x = \{x_1, \ldots, x_n\}$ are introduced, where $x_i = (x_{i1}, \ldots, x_{im}) \sim N(0, \Gamma)$. The augmented likelihood is $L(\Theta, \Gamma, x) = \prod_{i=1}^{n} f(y_i, x_i | \Theta, \Gamma)$, with mixed joint density

$$\begin{aligned} f(y_i, x_i | \Theta, \Gamma) &= \text{pr}(Y = y_i | x_i, \Theta) f_N(x_i; 0, \Gamma) \\ &\propto \left(\prod_{j=1}^{m} I(A_{ij} \leq x_{ij} < B_{ij}) \right) f_N(x_i; 0, \Gamma). \end{aligned}$$

Here, $f_N(x; \mu, V)$ is the density of a $N(\mu, V)$ distribution evaluated at x, $I(Z)$ is an indicator function equal to one if Z is true, and zero otherwise. The mass function

$$\text{pr}(Y_j = y_{ij} | x_{ij}, \theta_j) = \begin{cases} 1 & \text{if } A_{ij} \leq x_{ij} < B_{ij} \\ 0 & \text{otherwise} \end{cases},$$

where $A_{ij} = \Phi_1^{-1}(a_{ij}; 1)$ and $B_{ij} = \Phi_1^{-1}(b_{ij}; 1)$ as noted in Section 17.2.5, and $\Phi_1(\cdot; 1)$ is the distribution function of a standard normal.

The likelihood of the copula model in equation (17.19) is obtained by integrating over the latent variables, with $L(\Theta, \Gamma) = \int L(\Theta, \Gamma, x) \mathrm{d}x$. Let $x_{(j)} = \{x_{1j}, \ldots, x_{nj}\}$ be the latent variables corresponding to the jth margin, then the following sampling scheme can be used to evaluate the augmented posterior.

Sampling scheme. *Data augmentation for a Gaussian copula*

Step 1: For $j = 1, \ldots, m$:
1(a) Generate from $f(\theta_j|\{\Theta\backslash\theta_j\}, \{x\backslash x_{(j)}\}, \Gamma, y)$
1(b) Generate from $f(x_{(j)}|\Theta, \{x\backslash x_{(j)}\}, \Gamma, y)$
Step 2: Generate from $f(\Gamma|\Theta, x)$.

Steps 1(a) and 1(b) together produce an iterate from the density $f(\theta_j, x_{(j)}|\{\Theta\backslash\theta_j\}, \{x\backslash x_{(j)}\}, \Gamma, y)$. The conditional posterior at Step 1(b) can be derived as

$$f(x_{(j)}|\Theta, \{x\backslash x_{(j)}\}, \Gamma, y) \propto L(\Theta, \Gamma, x)$$
$$\propto \left(\prod_{i=1}^{n} I(A_{ij} \leq x_{ij} < B_{ij}) f_N(x_{ij}; \mu_{ij}, \sigma_{ij}^2)\right)$$

where μ_{ij} and σ_{ij}^2 are the mean and variance of the conditional distribution of $x_{ij}|\{x_i\backslash x_{ij}\}$ obtained from the joint distribution $x_i \sim N(0, \Gamma)$. Thus, $x_{(j)}$ can be generated element-by-element from independent constrained normal densities. In Step 1(a), θ_j is generated using the same MH approach as in the continuous case, but where the conditional density is now

$$f(\theta_j|\{\Theta\backslash\theta_j\}, \{x\backslash x_{(j)}\}, \Gamma, y) \propto \left(\prod_{i=1}^{n} \Phi_1\left(\frac{B_{ij} - \mu_{ij}}{\sigma_{ij}}; 1\right) - \Phi_1\left(\frac{A_{ij} - \mu_{ij}}{\sigma_{ij}}; 1\right)\right) \pi(\theta_j)$$

In Step 2 any of the existing methods for generating a correlation matrix Γ from its posterior distribution for Gaussian distributed data x can be used, as outlined in Section 17.3.1. Bayesian selection ideas can also be used as discussed in Section 17.3.2.

The efficiency of this sampling scheme is demonstrated empirically in [54], and [18] show it can be applied effectively to a problem with $m = 45$ dimensions. Alternative sampling schemes are proposed in [64] that can be used with the Gaussian copula, or with other copula models.

17.4.2 Measuring dependence

For continuous multivariate data, dependence between elements of Y is captured fully by the copula function C. In this case, the measures of dependence based on C discussed in Section 17.2.7 are adequate summaries. But when one or more margins are discrete-valued, in general, measures of concordance involve the marginal distributions; see [21], and [48]. Nevertheless, the dependence structure of the latent vector U (or the latent vector X for copulas constructed by inversion) is still informative concerning the level and type of dependence in the data. Moreover, estimation using nonparametric rank-based estimators becomes inaccurate [27] and likelihood-based inference, such as that outlined here, preferable.

17.4.3 Link with multivariate probit and latent variable models

Last, it is not widely appreciated that the multivariate probit model is a special case of the Gaussian copula model with univariate probit margins [68]. Data augmentation for a Gaussian copula therefore extends the approaches of [13], [22] and others for data augmentation for a multivariate probit model, to other Gaussian copula models. Similarly, the approach generalizes a number of Gaussian latent variable models for ordinal data, such as that of [14] and [39].

17.5 Discussion

The impact of copula modelling in multivariate analysis has been substantial in many fields. Yet, Bayesian inferential methods have been employed by only a few empirical analysts to date. Nevertheless, they show great potential for computing efficient likelihood-based inference in a number of contexts. One of these is in the modelling of multivariate discrete data, or data with a combination of discrete and continuous margins. Here, method of moments style estimators cannot be used effectively, and there can be computational difficulties in maximizing the likelihood, so that Bayesian data augmentation becomes attractive; see [64] for a full discussion. Another is in the use of hierarchical models, including varying parameter models [4] or hierarchical models for Bayesian selection and model averaging, as discussed here. Last, while this article has focused on the Gaussian and D-vine copulas, the Bayesian methods and ideas discussed here are applicable to a wide range of other copula models, and it seems likely that their usage will increase in the near future.

Acknowledgements

I would like to thank Robert Kohn, Claudia Czado, Anastasios Panagiotelis and particularly Mohamad Khaled, for their insightful comments on copula models and associated methods of inference. This research was funded by the Australian Research Council grants FT110100729 and DP1094289.

References

[1] Aas, K., Czado, C., Frigessi, A. and Bakken, H. (2009). Pair-copula constructions of multiple dependence. *Insurance: Mathematics and Economics*, **44**, 182–198.

[2] Abramowitz, M. and Stegun, I. A. (Eds.) (1965). *Handbook of Mathematical Functions*, New York: Dover Publications.

[3] Armstrong, H., Carter, C. K., Wong, K. F. K. and Kohn, R. (2009). Bayesian covariance matrix estimation using a mixture of decomposable graphical models. *Statistics and Computing*, **19**, 303–316.

[4] Ausin, M. C. and Lopes, H. F. (2010). Time-varying joint distribution through copulas. *Computational Statistics and Data Analysis*, **54**, 2383–2399.

[5] Barnard, J., McCulloch, R. and Meng, X. (2000). Modeling covariance matrices in terms of standard deviations and correlations, with application to shrinkage. *Statistica Sinica*, **10**, 1281–1311.

[6] Bedford, T. and Cooke, R. (2002). Vines–a new graphical model for dependent random variables. *Annals of Statistics*, **30**, 1031–1068.

[7] Bhat, C. R. and Eluru, N. (2009). A Copula-Based Approach to Accomodation Residential Self-Selection Effects in Travel Behavior Modeling. *Transportation Research Part B*, **43**, 749–765.

[8] Bottolo, L. and Richardson, S. (2010). Evolutionary Stochastic Search for Bayesian model exploration. *Bayesian Analysis*, **5**, 583–618.

[9] Cameron, A., Tong, L., Trivedi, P. and Zimmer, D. (2004). Modelling the differences in counted outcomes using bivariate copula models with application to mismeasured counts. *Econometrics Journal*, **7**, 566–584.

[10] Chen, X. and Fan, Y. (2006). Estimation and model selection of semiparametric copula-based multivariate dynamic models under copula misspecification. *Journal of Econometrics*, **135**, 125–154.

[11] Chen, X., Fan, Y. and Tsyrennikov, V. (2006). Efficient Estimation of Semiparametric Multivariate Copula Models. *Journal of the American Statistical Association*, **101**, 1228–1240.
[12] Cherubini, U., Luciano, E. and Vecchiato, W. (2004). *Copula Methods in Finance*, Wiley.
[13] Chib, S. and Greenberg, E. (1998). Analysis of multivariate probit models. *Biometrika*, **85**, 347–361.
[14] Chib, S. and Winkelmann, R. (2001). Markov chain Monte Carlo Analysis of Correlated Count Data. *Journal of Business and Economic Statistics*, **19**, 428–435.
[15] Clayton, D. (1978). A model for association in bivariate life tables and its application to epidemiological studies of family tendency in chronic disease incidence. *Biometrika*, **65**, 141–151.
[16] Cripps, E., Carter, C. and Kohn, R. (2005). Variable selection and covariance selection in multivariate regression models, in *Handbook of Statistics 25: Bayesian Thinking: Modeling and Computation*, D. Dey and C. Rao, (Eds.), Amsterdam: North-Holland, pp. 519–552.
[17] Czado, C. (2010). Pair-copula constructions of multivariate copulas. In *Workshop on Copula Theory and Its Applications*. Eds. F. Durante, W. Härdle, P. Jaworki and T. Rychlik, Dordrecht: Springer.
[18] Danaher, P. and Smith, M. (2011). Modeling multivariate distributions using copulas: applications in marketing, (with discussion). *Marketing Science*, **30**, 4–21.
[19] Daniels, M. and Pourahmadi, M. (2009). Modeling covariance matrices via partial autocorrelations. *Journal of Multivariate Analysis*, **100**, 2352–2363.
[20] Demarta, S. and McNeil, A. J. (2005). The t-copula and related copulas. *International Statistical Review*, **73**, 111–129.
[21] Denuit, M. and Lambert, P. (2005). Constraints on concordance measures in bivariate discrete data. *Journal of Multivariate Analysis*, **93**, 40–57.
[22] Edwards, Y. D. and Allenby, G. M. (2003). Multivariate analysis of multiple response data. *Journal of Marketing Research*, **40**, 321–334.
[23] Fang, H. B., Fang, K. T. and Kotz, S. (2002). The meta-elliptical distributions with given marginals. *Journal of Multivariate Analysis*, **82**, 1–16.
[24] Frees, E. W. and Valdez, E. A. (1998). Understanding relationships using copulas. *North American Actuarial Journal*, **2**, 1–25.
[25] Frühwirth-Schnatter, S. and Tüchler, R. (2008). Bayesian parsimonous covariance estimation for hierarchical linear mixed models. *Statistics and Computing*, **18**, 1–13.
[26] Genest, C., Ghoudi, K. and Rivest, L.-P. (1995). A semiparametric estimation procedure of dependence parameters in multivariate families of distributions. *Biometrika*, **82**, 543–552.
[27] Genest, C. and Nešlehová, J. (2007). A primer on copulas for count data. *The Astin Bulletin*, **37**, 475–515.
[28] Genest, C. and Rivest, L.-P. (1993). Statistical inference procedures for bivariate Archimedean copulas. *Journal of the American Statistical Association*, **88**, 1034–1043.
[29] Giudici, P. and Green, P. (1999). Decomposable graphical Gaussian model determination. *Biometrika*, **86**, 785–801.
[30] Haff, I., Aas, K. and Frigessi, A. (2010). On the simplified pair-copula construction- Simply useful or too simplistic? *Journal of Multivariate Analysis*, **101**, 1296–1310.
[31] Hjort, N. L., Holmes, C., Müller, P. and Walker, S. (2010). *Bayesian Nonparametrics*, CUP.
[32] Hoff, P. (2007). Extending the rank likelihood for semiparametric copula estimation. *The Annals of Applied Statistics*, **1**, 265–283.
[33] Huard, D., Évin, G. and Favre, A.-C. (2006). Bayesian copula selection. *Computational Statistics and Data Analysis*, **51**, 809–822.
[34] Joe, H. (1996). Families of m-variate distributions with given margins and m(m-1)/2 bivariate dependence parameters. In Rüschendorf, L., Schweizer, B., Taylor, M.D. (Eds.) *Distributions with Fixed Marginals and Related Topics.*
[35] Joe, H. (1997). *Multivariate Models and Dependence Concepts*, Chapman and Hall.

[36] Joe, H. (2005). Asymptotic efficiency of the two-stage estimation method for copula-based models. *Journal of Multivariate Analysis*, **94**, 401–419.

[37] Joe, H. (2006). Generating random correlation matrices based on partial correlations. *Journal of Multivariate Analysis*, **97**, 2177–2189.

[38] Kohn, R., Smith, M. and Chan, D. (2001). Nonparametric regression using linear combinations of basis functions. *Statistics and Computing*, **11**, 313–322.

[39] Kottas, A., Müller, P. and Quintana, F. (2005). Nonparametric Bayesian modeling for multivariate ordinal data. *Journal of Computational and Graphical Statistics*, **14**, 610–625.

[40] Lambert, P. and Vandenhende, F. (2002). A copula-based model for multivariate non-normal longitudinal data: analysis of a dose titration safety study on a new antidepressant. *Statistics in Medicine*, **21**, 3197–3217.

[41] Lee, A. (1999). Modelling rugby league data via bivariate negative binomial regression. *Australian and New Zealand Journal of Statistics*, **41**, 141–152.

[42] Li, D. X. (2000). On default correlation: a copula function approach. *The Journal of Fixed Income*, **9**, 43–54.

[43] Liechty, J. C., Liechty, M. W. and Müller, P. (2004). Bayesian correlation estimation. *Biometrika*, **91**, 1–14.

[44] McNeil, A. J., Frey, R. and Embrechts, R. (2005). *Quantitative Risk Management: Concepts, Techniques and Tools*, Princeton University Press, Princton: NJ.

[45] Min, A. and Czado, C. (2010). Bayesian inference for multivariate copulas using pair-copula constructions. *Journal of Financial Econometrics*, **8**, 511–546.

[46] Min, A. and Czado, C. (2011). Bayesian model selection for D-vine pair-copula constructions. *Canadian Journal of Statistics*, **39**, 239–258.

[47] Nelsen, R. B. (2006). *An Introduction to Copulas*, (2nd Ed.), New York: Springer.

[48] Nešlehová, J. (2007). On rank correlation measures for non-continuous random variables. *Journal of Multivariate Analysis*, **98**, 544–567.

[49] Nikoloulopoulos, A. and Karlis, D. (2008). Multivariate logit copula model with an application to dental data. *Statistics in Medicine*, **27**, 6393–6406.

[50] Nikoloulopoulos, A. and Karlis, D. (2010). Modeling multivariate count data using copulas. *Communications in Statistics- Simulation and Computation*, **39**, 172–187.

[51] Oakes, D. (1989). Bivariate survival models induced by frailties. *Journal of the American Statistical Association*, **84**, 487–493.

[52] Oakes, D. and Ritz, J. (2000). Regression in bivariate copula model. *Biometrika*, **87**, 345–352.

[53] Patton, A. (2006). Modelling asymmetric exchange rate dependence. *International Economic Review*, **47**, 527–556.

[54] Pitt, M., Chan, D. and Kohn, R. (2006). Efficient Bayesian inference for Gaussian copula regression models. *Biometrika*, **93**, 537–554.

[55] Robert, C. P. and Casella, G. (2004). *Monte Carlo Statistical Methods*, (2nd Ed.), New York: Springer

[56] Sahu, S. K., Dey, D. K. and Branco, M. D. (2003). A new class of multivariate skew distributions with applications to Bayesian regression models. *The Canadian Journal of Statistics*, **31**, 129–150.

[57] Scott, J. G. and Berger, J. O. (2010). Bayes and empirical-Bayes multiplicity adjustment in the variable-selection problem. *The Annals of Statistics*, **38**, 2587–2619.

[58] Shih, J. H. and Louis, T. A. (1995). Inferences on the association parameter in copula models for bivariate survival data. *Biometrics*, **51**, 1384–1399.

[59] Silva, R. and Lopes, H. (2008). Copula, marginal distributions and model selection: a Bayesian note. *Statistics and Computing*, **18**, 313–320.

[60] Sklar, A. (1959). Fonctions de répartition à n dimensions et leurs marges. *Publications de l'Institut de Statistique de L'Université de Paris*, **8**, 229–231.

[61] Smith, M. D. (2003). Modelling sample selection using Archimedean copulas. *Econometrics Journal*, **6**, 99–123.
[62] Smith, M. (2000). Modeling and short-term forecasting of New South Wales electricity system load. *Journal of Business and Economic Statistics*, **18**, 465–478.
[63] Smith, M. S. and Kauermann, G. (2011). Bicycle commuting in Melbourne during the 2000s energy crisis: A semiparametric analysis of intraday volumes. *Transportation Research Part B*, **45**, 1846–1862.
[64] Smith, M. S. and Khaled, M. A. (2012). Estimation of copula models with discrete margins via Bayesian data augmentation. *Journal of the American Statistical Association*, **107**, 290–303.
[65] Smith, M. and Kohn, R. (2002). Parsimonious covariance matrix estimation for longitudinal data, *Journal of the American Statistical Association*, **97**, 1141–1153.
[66] Smith, M. S., Gan, Q. and Kohn, R. J. (2012). Modelling dependence using skew t copulas: Bayesian inference and applications. *Journal of Applied Econometrics*, **27**, 500–522.
[67] Smith, M., Min, A., Almeida, C. and Czado, C. (2010). Modeling longitudinal data using a pair-copula decomposition of serial dependence. *Journal of the American Statistical Association*, **105**, 1467–1479.
[68] Song, P. (2000). Multivariate dispersion models generated from Gaussian copula. *Scandinavian Journal of Statistics*, **27**, 305–320.
[69] Song, P., Fan, Y. and Kalbfleisch, J. D. (2005). Maximization by Parts in Likelihood Inference. *Journal of the American Statistical Association*, **100**, 1145–1158.
[70] Tanner, M. A. (1996). *Tools for Statistical Inference: Methods for the Exploration of Posterior Distributions and Likelihood Functions*, 3rd Ed., New York: Springer.
[71] Tanner, M. A. and Wong, W. H. (1987). The calculation of posterior distributions by data augmentation. *Journal of the American Statistical Association*, **82**, 528–540.
[72] Wong, F., Carter, C. K. and Kohn, R. (2003). Efficient estimation of covariance selection models. *Biometrika*, **90**, 809–830.
[73] Yang, R. and Berger, J. O. (1994). Estimation of a covariance matrix using the reference prior. *Annals of Statistics*, **22**, 1195–1211.
[74] Zellner, A. (1962). An efficient method of estimating seemingly unrelated regressions and tests for aggregation bias. *Journal of the American Statistical Association*, **57**, 348–368.
[75] Zhang, Z., Dai, G. and Jordan, M. (2011). Bayesian generalized kernel mixed models. *Journal of Machine Learning Research*, **12**, 111–139.

Part VIII

Model Elaboration and Prior Distributions

18 Hypothesis testing and model uncertainty

M. J. BAYARRI AND J. O. BERGER

18.1 Introduction

This chapter gives a brief and mostly elementary review of Bayesian hypothesis testing and model uncertainty. In Section 18.2, the key concepts of Bayesian hypothesis testing are introduced. The Bayesian approach is compared with classical methods, such as use of p-values, in Section 18.3.

Section 18.4 introduces the basics of dealing with model uncertainty. The key difficulty is that of choosing prior distributions, especially when there are many models involved. General approaches to the development of suitable prior distributions for model uncertainty are discussed in Section 18.5. Other issues arising in model uncertainty are considered in Section 18.6, in particular, computation and search, approximation and asymptotics, multiplicity, and model criticism.

Much of the development in this area began with the work of Adrian Smith, especially the seminal papers (discussed later) [125], [126], and [57]. Overall, this is a huge methodological area, and many important techniques and approaches will be mentioned herein only briefly, if at all. Other useful reviews, with many more references, include [74], [14], [35], [90], [54], and [7].

18.2 Basics of Bayesian hypothesis testing

We introduce the basic elements of Bayesian hypothesis testing through a pedagogical example.

18.2.1 A pedagogical illustration

In hypothesis testing, one has to select among two possible models or explanations of data X; these are usually denoted by H_0 (the *null* hypothesis) and H_1 (the *alternative* hypothesis). We defer discussion of the important issue of correct formulation of hypotheses until Section 18.2.3, first presenting the key methodological ingredients of hypothesis testing.

One of the major scientific goals of the *Large Hadron Collider* (LHC) at CERN is to determine if the Higgs boson particle actually exists. A simplified version of the problem, that nevertheless contains many of its key features, presumes that the data X consists of the number of events observed in time T that are characteristic of Higgs boson production in LHC particle collisions. The statistical model assumes that the density of X is the Poisson density

$$\text{Poisson}(x \mid \theta + b) = \frac{(\theta + b)^x e^{-(\theta+b)}}{x!}$$

where θ is the mean rate of production of Higgs events in time T, and b is the (known) mean rate of production of events with the same characteristics from background sources in time T. One then wishes to test

$$H_0 : \theta = 0 \quad \text{versus} \quad H_1 : \theta > 0. \quad (H_0 \text{ corresponds to 'no Higgs.'})$$

As we present the Bayesian approach, it will be useful to simultaneously present the classical approach of using p-values for testing. For observed x, the p-value is given by

$$p = Pr(X \geq x \mid b, \theta = 0) = \sum_{m=x}^{\infty} \text{Poisson}(m \mid 0 + b)\,.$$

We will use two cases for illustration:

Case 1: $x = 7, b = 1.2$, which results in $p = 0.00025$;

Case 2: $x = 6, b = 2.2$, which results in $p = 0.025$.

18.2.2 Priors, Bayes factors and posteriors

18.2.2.1 *Prior distribution*

For Bayesian hypothesis testing, one needs to assess the prior probability of each hypothesis and, conditional on the truth of each of the hypotheses, the prior distribution of the unknown parameters in the statistical model corresponding to that hypothesis. In the pedagogical example, $Pr(H_i)$ will denote the prior probability that H_i is true, $i = 0, 1$; also, on $H_1 : \theta > 0, \pi(\theta)$ will denote the prior density for θ. (Note that H_0 does not involve any unknown parameters.) The choice of the prior distribution can be either subjective or objective:

> *Subjective Bayesian analysis* chooses the $Pr(H_i)$ and $\pi(\theta)$ based on personal beliefs and/or previous information about the problem (e.g. $\pi(\theta)$ could result from using the standard physics model predictions of the mass of the Higgs).
>
> *Objective Bayesian analysis* does not introduce any external or additional information into the problem (other than the data and the model). For instance, a natural objective choice of the prior probabilities of the hypotheses is $Pr(H_0) = Pr(H_1) = 1/2$. Making an objective choice for $\pi(\theta)$ is considerably more problematical, as will be discussed later. For illustration here we utilize the 'expected posterior prior' (discussed in Section 18.5.4.3), given by $\pi^E(\theta) = b(\theta + b)^{-2}$. Note that this prior is proper—essential as will be seen in Section 18.2.4—and has median b.

18.2.2.2 *Bayes factors*

An 'objective' alternative to choosing $Pr(H_0) = Pr(H_1) = 1/2$ is to report the *Bayes factor*. Informally, the Bayes factor of H_0 to H_1 is the *ratio of the average likelihood under* H_0 *to the average likelihood under* H_1, where the averages are with respect to the prior distributions of the parameters. (A more formal definition is deferred until Section 18.4.1.) In the pedagogical example,

$$B_{01} = \frac{\text{Poisson}(x \mid 0 + b)}{\int_0^\infty \text{Poisson}(x \mid \theta + b)\pi(\theta)\, d\theta} = \frac{b^x\, e^{-b}}{\int_0^\infty (\theta + b)^x\, e^{-(\theta+b)}\pi(\theta)\, d\theta}$$

For the prior $\pi^E(\theta) = b(\theta + b)^{-2}$, the Bayes factor is given by

$$B_{01} = \frac{b^x\, e^{-b}}{\int_0^\infty (\theta + b)^x\, e^{-(\theta+b)} b(\theta + b)^{-2}\, d\theta} = \frac{b^{(x-1)}\, e^{-b}}{\Gamma(x - 1, b)}$$

where Γ is the incomplete gamma function. In particular, in *Case 1*, $B_{01} = 0.0075$ (recall $p = 0.00025$), and, in *Case 2*, $B_{01} = 0.26$ (recall $p = 0.025$).

The common intuitive interpretation of the Bayes factor is simply as 'odds of H_0 to H_1'. Thus, in Case 2 above, $B_{01} = 0.26$ would be interpreted as saying that the data provide 1 to 4 odds in favour of the alternative hypothesis. It is interesting that the p-value here was $p = 0.025$, which many view as strong evidence against H_0, while 1 to 4 odds would hardly be viewed as strong evidence.

18.2.2.3 *Posterior distribution and Bayesian reporting*

The posterior probability of H_0, given the data, is given by

$$Pr(H_0 \mid x) = \frac{Pr(H_0)B_{01}}{Pr(H_1) + Pr(H_0)B_{01}}.$$

(Of course, $Pr(H_1 \mid x) = 1 - Pr(H_0 \mid x)$.) For the objective choice $Pr(H_0) = Pr(H_1) = 0.5$, this becomes

$$Pr(H_0 \mid x) = \frac{B_{01}}{1 + B_{01}}$$

For the pedagogical example, in *Case 1*, $Pr(H_0 \mid x) = 0.0075$ (recall $p = 0.00025$), and, in *Case 2*, $Pr(H_0 \mid x) = 0.21$ (recall $p = 0.025$).

The complete posterior distribution in this example can be summarized by

1. $Pr(H_0 \mid x)$, the posterior probability of the null hypothesis, and
2. $\pi(\theta \mid x, H_1)$, the posterior distribution of θ under H_1.

Both of these components are displayed in Figure 18.1 for the situation of Case 1; this type of figure is a very useful graphical presentation of the posterior. A useful numerical summary of the complete posterior is $Pr(H_0 \mid x)$ and C, a (say) 95% posterior credible set for θ under H_1. For the pedagogical example, these summaries are

Case 1: $Pr(H_0 \mid x) = 0.0075$; $C = (1.0, 10.5)$,

Case 2: $Pr(H_0 \mid x) = 0.21$; $C = (0.2, 8.2)$.

It is important to note that, for testing hypotheses as here, confidence intervals alone are *not* a satisfactory inferential summary. Thus, in Case 2 above, if one ignored the testing aspect of the problem and simply found a 95% credible set for θ, the result would be $C = (0.2, 8.2)$, which does not contain 0. In classical statistics, it is viewed as correct to say that, if a confidence interval does not contain the null parameter value, then one has 'significant' evidence against the null hypothesis. In Bayesian testing this is not so; here the credible interval if we had ignored the testing aspect

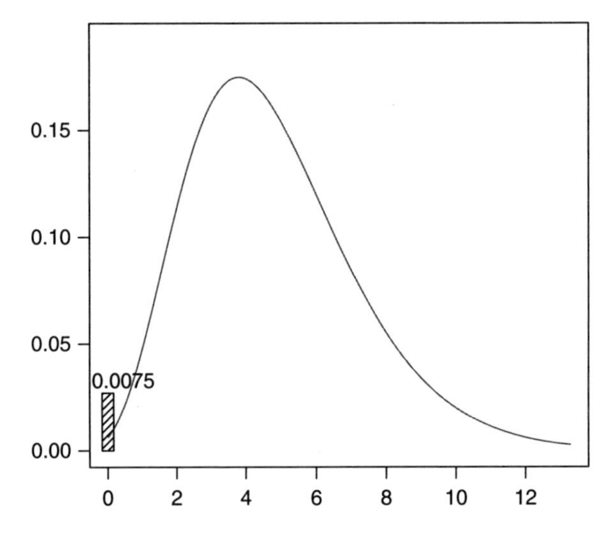

Figure 18.1 $Pr(H_0 \mid x)$ (the vertical bar) and the posterior density for θ given x and H_1 for Case 1 of the pedagogical example.

of the problem does not contain zero, yet the null hypothesis still has a reasonable probability of being true.

Finally, note that we can write

$$\underset{\text{(posterior odds)}}{\frac{Pr(H_0|x)}{Pr(H_1|x)}} = \underset{\text{(prior odds)}}{\frac{Pr(H_0)}{Pr(H_1)}} \times \underset{\text{(Bayes factor)}}{B_{01}}$$

Because prior probabilities of hypotheses often differ amongst individuals, it is common to primarily report Bayes factors, allowing any individual to convert to his/her own posterior probabilities. Note that we are not saying that Bayes factors are the correct final answer, but that they are an important component of the final answer and are very useful quantities for communication.

18.2.3 Formulating hypotheses when using Bayesian testing

18.2.3.1 *Precise versus non-precise hypotheses*

An important issue to consider is that of proper formulation of a hypothesis testing problem. The scientific literature is full of problems formulated as hypothesis testing problems which would better be addressed as estimation problems, and vice versa (although the latter is less frequent). A crucial issue is that the hypotheses must all be 'plausible,' i.e. they must have reasonable prior probability of being true. For instance, in our pedagogical illustration, the hypothesis H_0 : there is no Higgs boson particle, *is* a plausible hypothesis, as is the alternative hypothesis.

To better understand this crucial issue, consider the example of comparing the effects of two treatments A and B. Letting θ denote the difference between the mean treatment effects, one often sees tests of $H_0 : \theta = 0$ versus $H_1 : \theta \neq 0$. The key question is then whether H_0 is a plausible hypothesis. Consider, for instance, the following two possible scenarios involving treatment for cancer:

Scenario 1:
Treatment A = standard chemotherapy
Treatment B = standard chemotherapy + steroids

Scenario 2:
Treatment A = standard chemotherapy
Treatment B = a new radiation therapy

In Scenario 1, $H_0 : \theta = 0$ is plausible, in that the effect of steroids could well be essentially negligible. In Scenario 2, however, $H_0 : \theta = 0$ is *not* plausible; there is no plausible reason why the new radiation therapy would have essentially the same effect as the standard chemotherapy (assuming we can rule out both being completely ineffective or completely effective) and it would thus be much more appropriate to test $H_0 : \theta < 0$ versus $H_1 : \theta > 0$.

A final observation on this issue is that this is not an issue of one-sided versus two-sided testing. Indeed, in Scenario 1, it is quite possible to test $H_0 : \theta = 0$ versus $H_1 : \theta < 0$ versus $H_2 : \theta > 0$, which involves one-sided testing. The important point is rather whether $\theta = 0$ is a point that has non-negligible probability of being true; if so, $H_0 : \theta = 0$ must enter the formulation of the problem to properly capture the scientific situation; we call such hypotheses 'precise' hypotheses. Note that, if the scientific situation calls for a precise hypothesis having prior mass, it is an egregious mistake not to include such; it is not simply a modelling 'choice.'

18.2.3.2 *Approximating a believable null hypothesis by a precise null*

A precise null hypothesis, such as $H_0 : \theta = \theta_0$, is typically never true *exactly*; rather, it is used as a surrogate for an approximate precise hypothesis

$$H_0^\epsilon : |\theta - \theta_0| < \epsilon, \quad \epsilon \text{ small}$$

For instance, in the pedagogical example, while the Higgs boson may well not exist, the LHC experiment almost certainly has some very small bias ϵ, so that the real hypothesis being tested is $H_0^\epsilon : \theta < \epsilon$.

Luckily, it is typically the case that one obtains essentially the same answer whether using H_0 or H_0^ϵ. Indeed, in [10], it was shown that $Pr(H_0^\epsilon|\mathbf{x}) \approx Pr(H_0|\mathbf{x})$ if

$$\epsilon < \frac{1}{4}\sigma_{\hat{\theta}} \tag{18.1}$$

where $\sigma_{\hat{\theta}}$ is the standard error of the estimate of θ. Thus one simply needs to assess that ϵ satisfies this inequality to utilize the precise null. Note that, typically, $\sigma_{\hat{\theta}} \approx c/\sqrt{n}$, where n is the sample size so, for very large n, the above condition can be violated and using a precise null may not be appropriate, even if the real null (e.g. 'no Higgs') is believable.

18.2.4 Priors and paradoxes

When precise hypotheses are involved in the test, one needs to be quite careful in the choice of prior distributions. For instance, one cannot generally employ usual objective Bayes estimation priors. To see this, suppose $X \sim \mathrm{N}(x \mid \theta, 1)$, with density written $f(x \mid \theta)$. In estimation of the mean θ, it is perfectly reasonable to use the improper objective estimation prior $\pi(\theta) = c$, where c is any constant, in that the posterior density of θ is $\pi(\theta \mid x) = f(x \mid \theta)c / \int f(x \mid \theta)c d\theta = f(x \mid \theta)$ (since $\int f(x \mid \theta) d\theta = 1$), which does not depend on the choice of the constant c.

Consider now testing of $H_0 : \theta = 0$ versus $H_1 : \theta \neq 0$. The Bayes factor, if $\pi(\theta) = c$ were used, would be

$$B_{01}(c) = \frac{f(x \mid 0)}{\int_{-\infty}^{\infty} f(x \mid \theta)(c)d\theta} = \frac{f(x \mid 0)}{c}$$

and, hence, depends on the arbitrary choice of c. For this reason, one often hears it stated that improper priors cannot be used for hypothesis testing when there are precise hypotheses.

Even worse than use of improper priors here is use of *vague proper priors*, such as a *Uniform*$(-K, K)$ prior for θ, with K chosen very large; many Bayesians erroneously view the use of such vague proper priors to be better than the use of improper priors. But, for this prior in the normal example, the Bayes factor becomes

$$B_{01}(K) = \frac{f(x \mid 0)}{\int_{-K}^{K} f(x \mid \theta)(2K)^{-1}d\theta} \approx \frac{2Kf(x \mid 0)}{\int_{-\infty}^{\infty} f(x \mid \theta)d\theta} = 2Kf(x \mid 0)$$

which depends dramatically and inappropriately on the arbitrary choice of K. Indeed, the situation in practice is typically much worse with vague proper priors than with improper priors; the usual choice of c would be 1, which causes no dramatic problems, while K is usually chosen huge—e.g. $K = 1000$—leading to a nonsensical answer. This problem with vague proper priors is often referred to as Bartlett's Paradox (from [3]).

A closely related paradox is the Jeffreys–Lindley Paradox, from [79] and [91]. This is not really a paradox in the usual sense of the word, but simply describes a stark example of the contrasting behaviour of, say, Bayes factors and p-values. For instance, consider the normal example above but now with an i.i.d. sample of data $X_1, \ldots, X_n$. Reducing first to the sufficient statistic $\bar{X} = \sum_{i=1}^{n} X_i/n \sim \mathrm{N}(\bar{x} \mid \theta, n^{-1})$, and considering a fixed proper prior for θ, such as the $\mathrm{N}(\theta \mid 0, 1)$ prior, one obtains the Bayes factor

$$B_{01}(n) = \frac{\mathrm{N}(\bar{x} \mid 0, n^{-1})}{\int \mathrm{N}(\bar{x} \mid \theta, n^{-1})\mathrm{N}(\theta \mid 0, 1)d\theta} = \frac{\mathrm{N}(\bar{x} \mid 0, n^{-1})}{\mathrm{N}(\bar{x} \mid 0, n^{-1} + 1)} = \sqrt{n+1}\, e^{-\frac{n}{2(n+1)}z^2}$$

where $z = \sqrt{n}\bar{x}$ is the standardized test statistic. The 'paradox' is that z can be made arbitrarily large which makes the p-value arbitrarily small yet, for fixed z and large enough n, B_{01} will be very large. Then classical testing would state that there is very significant evidence against H_0, while Bayesian testing would state that H_0 has very strong odds of being true.

Again, this is not a paradox; if the Bayesian inputs are correct, the Bayesian answer is indisputably correct—it is just probability theory. Note, however, that the Bayesian formulation of an exact precise hypothesis test might well be wrong in practical contexts. If n is very large, (18.1) might well be violated, so that testing $H_0 : \theta = 0$ would be inappropriate and a Bayesian would need to do the more difficult analysis with H_0^{ϵ}.

We delay discussion of actual choice of priors until the model uncertainty sections, given that the same issues arise there. Note, however, that there is one completely general objective Bayesian procedure for testing two hypotheses with i.i.d. data; see Section 18.5.4.1.

18.2.5 Other testing scenarios

We have focused primarily on precise hypothesis testing in this section, although there are many other kinds of testing. For instance, in normal testing of $H_0 : \theta < 0$ versus $H_1 : \theta > 0$, with no precise hypothesis, one can legitimately use the ordinary estimation improper prior $\pi(\theta) = 1$ and, indeed, will find that the posterior probability of H_0 equals the p-value (see, e.g. [28]). There are many interesting generalizations of this idea (see [75]).

We do not focus on such problems, partly because of space constraints and partly because they can often just be approached as estimation problems, using standard objective estimation priors, obtaining the posterior distribution of the parameters, and computing the posterior probability of the regions corresponding to the non-precise hypotheses.

18.3 Understanding differences with classical testing

18.3.1 Discrepancy between p-values and Bayesian answers

In the pedagogical example for *Case 1*, the p-value was 0.00025 but the Bayes factor for the null was 0.0075, a quantity 30 times larger. Still, both would imply strong evidence against the null hypothesis while, in *Case 2*, the p-value was 0.025 and the Bayes factor was 0.26, a dramatic practical conflict; common understanding is that a p-value of 0.025 indicates strong evidence against the null hypothesis, but the Bayes factor suggests it is less than 1 to 4 odds against the null.

The discrepancy arises from two sources: the 'Ockham's razor' effect of the prior (see Section 18.4.6) and the fact that the p-value is a tail area of the sampling distribution, and not the likelihood of the data. As an aid to understanding this latter point, it is useful to first compute the lower bound on the Bayes factor over all possible prior distributions. Indeed,

$$B_{01} = \frac{\text{Poisson}(x \mid 0 + b)}{\int_0^\infty \text{Poisson}(x \mid \theta + b)\pi(\theta)\, d\theta} \geq \frac{\text{Poisson}(x \mid 0 + b)}{\text{Poisson}(x \mid \hat{\theta} + b)} = \min\{1, \left(\frac{b}{x}\right)^x e^{x-b}\}$$

the lower bound arising by choosing $\pi(\theta)$ to be a point mass at $\hat{\theta}$, an egregious favouring of the alternative hypothesis. In *Case 1*, this gives $B_{01} \geq 0.0014$ (recall $p = 0.00025$) and, in *Case 2*, $B_{01} \geq 0.11$ (recall $p = 0.025$). That the discrepancy holds for every prior, even that which is most favourable to the alternative hypothesis, highlights the serious error of replacing the likelihood of the actual data with the tail area of the data distribution.

The Ockham's razor effect of the prior is basically that the prior mass is spread over the a priori specified values of θ for the alternative, and many values of θ for the alternative are no more supportive of the data than the null value of θ. This is also the reason that Bayesian analysis seeks parsimony—a simpler model that explains the data is preferred to a complex one that does so.

Thus, in Case 1 of the pedagogical example, the difference, by a factor of 30, between the Bayes factor of 0.0075 and the p-value of 0.00025 arose as follows:

- A factor of $0.0014/0.00025 \approx 5.6$ is due to the difference between the tail area $\{X \geq 7\}$ and the actual observation $x = 7$.
- The remaining factor of roughly 5.4 in favour of the null results from the Ockham's razor penalty that Bayesian analysis automatically gives to a more complex model.

One way to help teach students that p-values are very different from error probabilities (both Bayesian and frequentist) is to utilize the *applet* (of German Molina) available at www.stat.duke.edu/~berger.

18.3.2 Conditional frequentist testing

To a strict frequentist, p-values are not relevant, since they cannot be given an interpretation as real error rates in repeated experimentation. Indeed, the difference between p-values and actual frequentist error probabilities can be seen from the *conditional frequentist viewpoint* ([88], [24]). This approach proceeds by

1. finding a statistic S that reflects the 'strength of evidence' in the data;
2. computing the frequentist measure of error conditional on the observed S.

Example 18.1 Observe X_1 and X_2 where

$$X_i = \begin{cases} \theta + 1 & \text{with probability } 1/2 \\ \theta - 1 & \text{with probability } 1/2 \end{cases}$$

It is easy to see that

$$C(X_1, X_2) = \begin{cases} (X_1 + X_2)/2 & \text{if } X_1 \neq X_2 \\ X_1 - 1 & \text{if } X_1 = X_2 \end{cases}$$

is a confidence set for θ, with coverage $Pr\{C(X_1, X_2) \text{ contains } \theta \mid \theta\} = 0.75$. This is a silly statement, however, since, when $x_1 \neq x_2$ it is a logical certainty that $C(X_1, X_2)$ is θ while, if $x_1 = x_2$, it is intuitively clear that $C(X_1, X_2)$ has only a 50% chance of being θ.

Here, the *strength of evidence* in the data is appropriately measured by $S = |X_1 - X_2|$ (which is either 0 or 2), so conditional frequentists compute instead the *conditional coverage*

$$Pr\{C(X_1, X_2) \text{ contains } \theta \mid \theta, S\} = \begin{cases} 0.5 & \text{if } S = 0 \\ 1.0 & \text{if } S = 2\,, \end{cases}$$

which appropriately agrees with intuition.

For the testing problem, [16] (continuous case) and [42] (discrete case) proposed the following conditional frequentist test:

1. Develop the 'strength of evidence' statistic S as follows:
   - let $p_i(x)$ be the p-value from testing H_i against the other hypothesis, recognizing that p-values are monotonically related to the strength of evidence against hypotheses (and any monotonic function of a conditioning statistic yields the same result);
   - define $S(x) = \max\{p_0(x), p_1(x)\}$; its use is based on deciding that data (in either the rejection or acceptance regions) with the same p-value has the same 'strength of evidence.'
2. Accept H_0 when $p_0(x) > p_1(x)$ and reject otherwise.
3. Compute Type I and Type II conditional error probabilities as

$$\alpha(s) = Pr(\text{rejecting } H_0 \mid S = s, H_0) \equiv Pr(p_0(X) \leq p_1(X) \mid S = s, H_0)$$

$$\beta(s) = Pr(\text{accepting } H_0 \mid S = s, H_1) \equiv Pr(p_0(X) > p_1(X) \mid S = s, H_1)\,.$$

The major surprise that results from this test is that the conditional frequentist Type I error probability exactly equals the Bayesian posterior probability of the null hypothesis when $Pr(H_0) = Pr(H_1) = 0.5$. Thus, in the pedagogical example, the conditional frequentist Type I error is $\alpha(s) = Pr(H_0 \mid x) = B_{01}/(1 + B_{01})$ (= 0.0075 in Case 1; = 0.21 in Case 2). So the discrepancy with the p-value is not a Bayesian versus frequentist issue, but rather an issue of conditioning on the actual data versus using a tail area.

18.3.3 When there is no alternative hypothesis

One commonly heard defence of p-values is that they can be used when there is only the null hypothesis, whereas Bayesian and frequentist testing requires an alternative hypothesis. Robust Bayesian theory suggests a way to proceed, however, even in the absence of an alternative hypothesis.

A *proper* p-value satisfies $H_0 : p(X) \sim \text{Uniform}(0, 1)$. [120] consider testing this versus $H_1 : p \sim f(p)$, where $Y = -\log(p)$ has a decreasing failure rate (a natural nonparametric alternative). Then,

Theorem 18.2 *If* $p < e^{-1}$, $B_{01}(p) \geq -e p \log(p)$. *An analogous lower bound on the conditional Type I frequentist error is*

$$\alpha(p) \geq (1 + [-e p \log(p)]^{-1})^{-1} \tag{18.2}$$

In the pedagogical example, $p = 0.00025$ results in $-ep \log p = 0.0057 \approx \alpha$.

Here are some common p-values and their associated 'calibrated conditional Type I frequentist error rates'.

p	0.2	0.1	0.05	0.01	0.005	0.001
$\alpha(p)$	0.465	0.385	0.289	0.111	0.067	0.0184

18.4 Basics of Bayesian model uncertainty

The Bayesian way of dealing with model uncertainty is not fundamentally different from the way of dealing with hypothesis testing. However, there can be so many models (if one is doing variable selection with p possible variables, there are 2^p models) that the nature of the problem changes, both in the methodology and the interpretation. In this section we highlight the basic issues.

18.4.1 Notation for model uncertainty

We assume that data, $\boldsymbol{X}$, can arise from one of q possible models $M_i, i = 1, \ldots, q$. Each model specifies a different density for $\boldsymbol{X}$ so that, under model M_i,

$$M_i\text{: } \boldsymbol{X} \text{ has density } f_i(\boldsymbol{x} \mid \boldsymbol{\theta}_i).$$

The prior distribution now consists of the model probabilities, $Pr(M_i)$, and the prior densities of each of the model parameters $\boldsymbol{\theta}_i$, to be denoted $\pi_i(\boldsymbol{\theta}_i)$. Because of the potentially large number of models involved, the specification of these quantities is difficult and will be the focus of much of the development.

Once the prior distributions have been assessed, the analysis proceeds by computing the marginal density of $\boldsymbol{X}$ under each model M_i, $i = 1, \ldots, q$, that is

$$m_i(\boldsymbol{x}) = \int f_i(\boldsymbol{x} \mid \boldsymbol{\theta}_i)\pi_i(\boldsymbol{\theta}_i)\, d\boldsymbol{\theta}_i$$

intuitively, this is the model likelihood, indicating how likely is $\boldsymbol{x}$ under M_i. The posterior density of $\boldsymbol{\theta}_i$ under model M_i is

$$\pi_i(\boldsymbol{\theta}_i \mid \boldsymbol{x}) = \frac{f_i(\boldsymbol{x}|\boldsymbol{\theta}_i)\pi_i(\boldsymbol{\theta}_i)}{m_i(\boldsymbol{x})}$$

and the posterior probability of each of the models is

$$Pr(M_i \mid \boldsymbol{x}) = \frac{Pr(M_i)\, m_i(\boldsymbol{x})}{\sum_{j=1}^{q} Pr(M_j)\, m_j(\boldsymbol{x})}$$

It is often more convenient to work with Bayes factors: the Bayes factor of model M_j to model M_i is

$$B_{ji} = \frac{m_j(\boldsymbol{x})}{m_i(\boldsymbol{x})}$$

since this is the ratio of the (integrated) likelihoods of model M_j to model M_i, it is typically interpreted as the odds of model M_j to model M_i provided by the data. It is often convenient to choose a base model M_1 (usually the null model if defined), and then report all the Bayes factors B_{j1}. From these, it is possible to reconstruct all the posterior model probabilities as

$$Pr(M_i \mid \boldsymbol{x}) = \frac{Pr(M_i)B_{i1}}{\sum_{j=1}^{q} Pr(M_j)B_{j1}}.$$

18.4.2 Assigning model prior probabilities

One must choose prior probabilities, $Pr(M_i)$, for each model. A common choice is equal probabilities, $Pr(M_i) = 1/q$, but this is often inappropriate, from several perspectives.

Consider the problem of variable selection, for instance. In this problem, there are variables (e.g. regression coefficients) $\theta_1, \ldots, \theta_m$ that could be included in the model, or could be excluded. Giving equal probability to all models is equivalent to saying that each variable is in or out of the model with probability $1/2$, which has the problem that there is no penalty for inclusion of variables. For instance, if one is testing 100 000 genes for gene expression related to some condition, allowing each gene to have probability 1/2 of being effective is typically ridiculous.

For this problem, the common solution is to allow each variable to be independently in the model with unknown probability p (called the *prior inclusion probability*), and then assign p a distribution. For instance, if p is given the uniform distribution, the prior probability of a model M_i that has k_i variables is

$$Pr(M_i) = \int_0^1 p^{k_i}(1-p)^{m-k_i} dp = Beta(1+k_i, 1+m-k_i) = \frac{k_i!(m-k_i)!}{(m+1)!}.$$

Curiously, this is the same probability assignment recommended by [79] from a different perspective. He suggested giving each potential model size an equal probability of $1/(m+1)$ (assuming the null model of no variables is also allowed). He then divided this probability up amongst all models of a given size, the result equalling the expression above.

For other discussions of the choice of prior probabilities of models, see [60, 61], [37], [124], [108], [118, 119], and [89].

18.4.3 Issues in assigning priors to model parameters

In hypothesis testing, the need for proper priors (as opposed to improper or vague proper priors) was discussed and applies equally well to choice of priors for model parameters in model uncertainty. Other complications can also arise; three of the most important are introduced in this section, namely, the meaning of parameters, predictive matching and information consistency. Section 18.5 considers particular strategies that have been proposed for selection of model parameter priors, especially those oriented to addressing the above issues.

18.4.3.1 *Meaning of parameters*

Parameters in different models typically have different meanings, even though they might be represented with the same symbol.

Example 18.3 Consider the variable selection in regression scenario with models

$$M_1 : Y = \theta_0 + \theta_1 X_1 + \sigma\,\epsilon$$
$$M_2 : Y = \theta_0 + \theta_1 X_1 + \theta_2 X_2 + \sigma\,\epsilon \qquad \epsilon \sim \mathrm{N}(0,1)$$

It is all too common to see assignment of the same prior, $\pi(\theta_i)$, to θ_i regardless of the model in which it occurs (implicitly also assuming independence of the parameters within a model). This is usually wrong, since the parameters can have radically different meanings in the two models. For instance, suppose the problem is predicting fuel consumption, Y, of a car from weight, X_1, and engine size, X_2. Clearly θ_1 has a very different meaning under M_1 than under M_2. (Regressing Y on X_1 alone produces larger θ_1 than regressing on both, due to the high correlation between X_1 and X_2.)

The right way to proceed is to formulate the model selection problem as:

$$M_1 : Y = \theta_0^{(1)} + \theta_1^{(1)} X_1 + \sigma^{(1)}\,\epsilon$$
$$M_2 : Y = \theta_0^{(2)} + \theta_1^{(2)} X_1 + \theta_2^{(2)} X_2 + \sigma^{(2)}\,\epsilon$$

and then assess *distinct priors* $\pi_1(\theta_0^{(1)}, \theta_1^{(1)}, \sigma^{(1)})$ and $\pi_2(\theta_0^{(2)}, \theta_1^{(2)}, \theta_2^{(2)}, \sigma^{(2)})$ under each of the models.

Of course, this only makes the problem more complicated, with the need for 2^p different priors in variable selection.

18.4.3.2 *Predictive matching*

Predictive matching is a major tool for deriving priors for model uncertainty, and is also very valuable as a method of evaluating prior choices. The idea is based on the observation that, if one makes a mistake of 2 in the prior scale for a one-parameter model compared to a null model, the Bayes factor is only wrong by a factor of 2. But, if one makes a mistake of a factor of 2 in the scale of a 50-dimensional model, the Bayes factor is off by a factor of 2^{50}. Hence it is imperative that priors across models be appropriately 'matched' to avoid major biases.

The most common type of matching is *predictive matching*, the idea being as follows. Suppose $\mathbf{X}^*$ are observables of interest, usually taken as some small set of possible observations. The pair $\{M_i, \pi_i\}$ would yield the marginal density

$$m_i(\mathbf{x}^*) = \int f_i(\mathbf{x}^* \mid \theta_i)\pi_i(\theta_i)d\theta_i\,.$$

We then say that $\{M_i, \pi_i\}$ is 'predictively matched' to $\{M_j, \pi_j\}$ if $m_i(\mathbf{x}^*) \approx m_j(\mathbf{x}^*)$. Numerous examples of this idea will be seen later.

18.4.3.3 *Information consistency*

Computational considerations often suggest use of conjugate prior distributions, but they can be problematical in situations of model uncertainty.

Example 18.4 We observe i.i.d. data, $X_1, \ldots, X_n$, from the $\mathrm{N}(x_i \mid \theta, \sigma^2)$ distribution, with both parameters unknown. The two models under consideration are $M_1 : \theta = 0$ and $M_2 : \theta \neq 0$. The 'natural' conjugate priors are

$$\pi_1(\sigma^2) = 1/\sigma^2$$
$$\pi_2(\theta \mid \sigma^2) = \mathrm{N}(\theta \mid 0, \sigma^2), \pi_2(\sigma^2) = 1/\sigma^2$$

Note that the improper $\pi_i(\sigma^2) = 1/\sigma^2$ can be used because they are both invariant scale parameters (see Section 18.5.1.1).

The Bayes factor can be computed to be

$$B_{21} = \sqrt{n+1}\left(\frac{1 + t^2/(n+1)}{1+t^2}\right)^{n/2}$$

where $t = \sqrt{n}\bar{x}/s$ is the usual t-statistic. Notice that B_{21} has the following strange behaviour: as $|t| \to \infty$, $B_{21} \to (n+1)^{-(n-1)/2} > 0$. For instance,

$$\text{if } n = 3, \quad B_{21} \to 1/4; \qquad \text{if } n = 5, \quad B_{21} \to 1/36.$$

While not conclusively damning, this behaviour is of enough concern that we recommend it be avoided for default procedures.

18.4.4 Model averaging

18.4.4.1 *Basics of model averaging*

Selecting a single model and using it for inference ignores model uncertainty, which can result in biased inferences and considerable overstatements of accuracy. The Bayesian approach to model uncertainty is *model averaging*, which simply views the uncertain model as another unknown, and treats uncertainties regarding the model probabilistically. Thus if, say, inference concerning ξ, a quantity which is common to all models (such as a future observation or the response mean) is desired, it would be based on

$$\pi(\xi \mid \boldsymbol{x}) = \sum_{i=1}^{q} Pr(M_i \mid \boldsymbol{x})\, \pi(\xi \mid \boldsymbol{x}, M_i),$$

where $\pi(\xi \mid \boldsymbol{x}, M_i)$ is the posterior distribution of ξ under model M_i alone. The reason for the name 'model averaging' is apparent.

Example 18.5 In [96], the problem of determining whether a vehicle type meets regulatory emission standards was considered. The emissions data $\boldsymbol{X} = (X_1, \ldots, X_n)$ was assumed to be i.i.d. from either the Weibull or lognormal distributions given, respectively, by

$$M_1 : f_W(x \mid \beta, \gamma) = \frac{\gamma}{\beta}\left(\frac{x}{\beta}\right)^{\gamma-1} \exp\left[-\left(\frac{x}{\beta}\right)^{\gamma}\right]$$

$$M_2 : f_L(x \mid \mu, \sigma^2) = \frac{1}{x\sqrt{2\pi\sigma^2}} \exp\left[\frac{-(\log x - \mu)^2}{2\sigma^2}\right]$$

The analysis from [43] proceeds by assigning each model prior probability 1/2. In Section 18.5.1.1, it will be shown that it is appropriate to use the model parameter priors $\pi_W(\beta, \gamma) = 1/(\beta\gamma)$ and $\pi_L(\mu, \sigma) = 1/\sigma$, respectively. (This is a case where improper priors can be utilized.) Then, the posterior probabilities (also equal to the conditional frequentist error probabilities) of the two models are

$$Pr(M_1 \mid \boldsymbol{x}) = 1 - Pr(M_2 \mid \boldsymbol{x}) = \frac{B(\boldsymbol{x})}{1 + B(\boldsymbol{x})}$$

where $z_i = \log x_i$, $\bar{z} = \frac{1}{n}\sum_{i=1}^{n} z_i$, $s_z^2 = \frac{1}{n}\sum_{i=1}^{n}(z_i - \bar{z})^2$, and

$$B(\boldsymbol{x}) = \frac{\int\int \prod_{i=1}^{n} f_W(x_i \mid \beta, \gamma)\,(\beta\gamma)^{-1} d\beta d\gamma}{\int\int \prod_{i=1}^{n} f_L(x_i \mid \mu, \sigma)\,(\sigma)^{-1} d\mu d\sigma}$$

$$= \frac{\Gamma(n) n^n \pi^{(n-1)/2}}{\Gamma(n - 1/2)} \int_0^{\infty} \left[\frac{y}{n}\sum_{i=1}^{n} e^{\left(\frac{z_i - \bar{z}}{s_z y}\right)}\right]^{-n} dy$$

For the data from [96], computation yields $Pr(M_1 \mid \boldsymbol{x}) = 0.712$. It thus follows that

$$Pr(\text{meeting regulatory standard} \mid \boldsymbol{x}) = 0.712\, Pr(\text{meeting standard} \mid \boldsymbol{x}, M_1) + 0.288\, Pr(\text{meeting standard} \mid \boldsymbol{x}, M_2)$$

This is superior to selecting one of the models and then computing the probability of meeting the standard under that model, as it accounts for the model uncertainty.

18.4.4.2 *Learning model structure*

One of the difficulties with model averaging is that one cannot easily understand model structure, since one cannot look at a single model to gain insight. One can, however, still look at model-averaged quantities that reflect model structure, although this is typically very context dependent.

An important example is when the parameters, θ_i, can be in or out of the model, reflecting the presence in the model of certain structure. In regression, for example, θ_i would be the coefficient of a covariate, and its presence or absence from the model reflects whether or not that covariate affects the response variable. In graphical models, θ_i could be associated with a link in the graph, its presence or absence determining whether that link is present in the graph. In all such situations key model-averaged structural quantities are the *posterior inclusion probabilities*

$$q_i = Pr(\theta_i \in \text{model} \mid \boldsymbol{x}) = \sum_{j:\theta_i \in M_j} Pr(M_j \mid \boldsymbol{x}), \tag{18.3}$$

i.e. the posterior probability that the *ith* feature of interest is in the model.

Example 18.6 This is a classical dataset that has been used by several authors (see [26] for references). The sample size of the data $\boldsymbol{Y}$ is $n = 13$ and there are four possible regressors: X_1, X_2, X_3, X_4. The full model for an observation is

$$Y = \theta_0 + \theta_1 X_1 + \theta_2 X_2 + \theta_3 X_3 + \theta_4 X_4 + \epsilon,\ \epsilon \sim \mathrm{N}(0, \sigma^2)$$

with σ^2 unknown. The models under consideration are all of the models that can be defined with subsets of regressors, with the intercept being present in all models. We use the following notation:

$$\text{Model } \{1,3,4\} \quad \text{denotes the model} \quad Y = \theta_0 + \theta_1 X_1 + \theta_3 X_3 + \theta_4 X_4 + \epsilon.$$

Table 18.1 reports the results of a model uncertainty analysis using the encompassing AIBF approach of [?]. The posterior probability of each model is reported, along with 'excess predictive risks', $\Delta R(M_i)$, that will be discussed in the next section.

The posterior inclusion probabilities here are

$$q_1 = \textstyle\sum_{j:\theta_1 \in M_j} Pr(M_j \mid \boldsymbol{Y}) = 0.95, \quad q_2 = \sum_{j:\theta_2 \in M_j} Pr(M_j \mid \boldsymbol{Y}) = 0.73,$$
$$q_3 = \textstyle\sum_{j:\theta_3 \in M_j} Pr(M_j \mid \boldsymbol{Y}) = 0.43, \quad q_4 = \sum_{j:\theta_4 \in M_j} Pr(M_j \mid \boldsymbol{Y}) = 0.55. \tag{18.4}$$

Thus there is a strong indication that X_1 is an important regressor, with lesser evidence for the importance of the other X_i as regressors.

Table 18.1 Posterior model probabilities and corresponding excess predictive risks for the Hald regression example.

Model	$Pr(M_i \mid Y)$	$\Delta R(M_i)$	Model	$Pr(M_i \mid Y)$	$\Delta R(M_i)$
null	0.000003	2652.44	{2,3}	0.000229	353.72
{1}	0.000012	1207.04	{2,4}	0.000018	821.15
{2}	0.000026	854.85	{3,4}	0.003785	118.59
{3}	0.000002	1864.41	{1,2,3}	0.170990	1.21
{4}	0.000058	838.20	{1,2,4}	0.190720	0.18
{1,2}	0.275484	8.19	{1,3,4}	0.159959	1.71
{1,3}	0.000006	1174.14	{2,3,4}	0.041323	20.42
{1,4}	0.107798	29.73	{1,2,3,4}	0.049587	0.47

Among the many references to model averaging are [62], [51], [108], and [33]. See also the Bayesian Model Averaging webpage http://www.research.att.com/volinsky/bma.html.

18.4.4.3 *Prediction*

When the goal is to predict a future Y, given $\boldsymbol{x}$, the optimal prediction is based on model averaging. If one of the M_i is indeed the true model (but unknown), the *posterior predictive distribution* of Y (its posterior distribution given $\boldsymbol{x}$) is

$$m(y \mid \boldsymbol{x}) = \sum_{i=1}^{q} Pr(M_i \mid \boldsymbol{x})\, m_i(y \mid \boldsymbol{x}, M_i),$$

where

$$m_i(y \mid \boldsymbol{x}, M_i) = \int f_i(y \mid \theta_i)\, \pi_i(\theta_i \mid \boldsymbol{x}) d\theta_i$$

is the posterior predictive distribution when M_i is true. For instance, the usual predictor of Y would be the posterior predictive mean, which would be $\hat{Y} = \sum_{i=1}^{q} Pr(M_i \mid \boldsymbol{x})\, \hat{Y}_i$, where $\hat{Y}_i$ is the posterior predictive mean arising from model M_i. This is, indeed, the optimal predictor of Y under squared error loss. For illustration of the excellent performance of model averaged prediction, see http://www.research.att.com/volinsky/bma.html.

It should be noted that the optimality of model averaged prediction implicitly assumes that the true model is among those being considered, which is often referred to as the *closed model* perspective. (See, e.g. [21].) Under the *open model perspective,* in which the true model is not assumed to be among those under consideration, it is not clear what is optimal. Nevertheless, model averaging would still seem intuitively to be valuable, at least to the extent of providing more accurate assessments of uncertainty in prediction. (See also [87].)

18.4.4.4 *Optimal single model prediction*

Computational or other considerations often require choice of a single model, which will then be used for prediction. Standard practice is to choose the maximum posterior probability model (leading, e.g. to the ubiquitous 'MAP' estimates). This practice is fine if only two models are being entertained (see [9]) and in problems such as variable selection in linear models with orthogonal design matrices (cf. [34, 35]), but is not generally optimal.

For instance, consider the scenario where models are defined by parameters θ_i that are in or out of the model, reflecting model features such as covariates. The optimal single predictive model is then often the *median probability model,* defined as the model consisting of those parameters whose posterior inclusion probability (18.3) is at least $1/2$.

Consider Example 18.6 for instance. From (18.4), it is clear that the median probability model is $\{1, 2, 4\}$, since the corresponding parameters have probability greater than $1/2$ of being in the model. Note from Table 18.1 that this is not the maximum posterior probability model, which is instead $\{1, 2\}$.

Conditions under which the median probability model is guaranteed to yield better predictive performance are given in [2]. Even when these conditions are not satisfied, however, the median probability model will often be optimal in practice. For instance, Example 18.6 is not a situation which satisfies the conditions in [2], yet the predictive risk of $\{1, 2, 4\}$ is seen in Table 18.1 to be smaller than the predictive risk of any other model. More precisely, the table reports $\Delta R(M_i) = E[(Y - \hat{Y}_i)^2] - E[(Y - \hat{Y})^2]$, the difference between the expected squared error loss of the single model (best) predictor, $\hat{Y}_i$, and the expected squared error loss of the optimal overall predictor, which is the model averaged predictor; expectation here is over the model averaged posterior predictive distribution of Y. Not only is $\{1,2,4\}$ clearly superior to $\{1,2\}$ (or any other model), but it does almost as well in this example as the optimal model-averaged predictor.

18.4.5 Consistency

A key feature of proper Bayesian model uncertainty analysis is that it is asymptotically *consistent*, which means that, if one of the entertained models is true, then the posterior probability of that model will go to one as the sample size grows to ∞ (assuming the model began with prior probability greater than zero, and the support of the parameter prior for that model is the entire parameter space). Even more remarkably, if the true model, M^*, is not among the models being considered, then the posterior probability will go to one for that model, among those being considered, that is closest to M^* in terms of Kullback–Leibler divergence. (See [19] and [50]; related references on consistency of Bayesian methods are [44], [17], [136] and [99].)

The paragraph above began with the phrase 'proper Bayesian model uncertainty analysis', because these consistency results are only guaranteed to hold when the analysis is fully Bayesian, with proper and non-zero priors. We will see that practical considerations often lead to a host of modifications or approximations to Bayesian model uncertainty analysis, and it is important to recognize that consistency is no longer guaranteed for these modifications. Indeed, verifying consistency then becomes one of the basic checks to see if the modification is satisfactory.

18.4.6 Ockham's razor

Ockham's razor is the long-believed scientific principle that, if two models are both compatible with the data, then the simplest one should be preferred. Simplest here does not refer to difficulty in understanding the model, but in the 'degrees of freedom' in the model.

For instance, [80] considered a historical problem in which two theories were advanced to explain an anomaly in the orbit of Mercury. One theory (model) was Einstein's general relativity, which was essentially a fully specified theory with no free parameters (and hence a 'simple' theory); the other theory (model) supposed the existence of a planet Vulcan circling too close to the sun to be seen, and this theory had free parameters (e.g. the mass of the planet). Both did a reasonable job of explaining the anomaly in Mercury's orbit, but Ockham's razor would say to favour the theory with fewer degrees of freedom, here general relativity.

One of the attractive features of Bayesian analysis is that it automatically follows Ockham's razor, because of the prior distributions placed on the parameters (the degrees of freedom) of models. Space precludes a demonstration here; see [80] for application of the Bayesian Ockham's razor to the historical problem.

18.5 Choosing priors for model uncertainty

In this section we review some frequently used methods for choosing priors for analysis of model uncertainty. More elaborate reviews of various of these methods can be found in [14], [64, 65], [90], and [7].

18.5.1 Choosing priors for 'common parameters'

It was seen, in Section 18.2.4, that, if a parameter occurs in one model but not another, then it must be assigned a proper (and non-vague) prior for the Bayes factor between the models to make sense. If the parameter occurs in both models, however, this is not necessarily the case; improper objective priors can then often be used. Here we discuss two situations in which this is reasonable.

18.5.1.1 *Invariance*

If there are parameters in each hypothesis that have what is called the same 'group invariance structure', then one can use the 'right-Haar priors' for those parameters, even when improper. Development of this is beyond the scope of the chapter, but the following illustration conveys the basic idea.

Example 18.7 Suppose i.i.d. data $X_1, \ldots, X_n$ arises from a density of the form $\sigma^{-1} g\left((x-\mu)/\sigma\right)$; such densities are called location-scale densities, with μ being the location parameter and σ the scale parameter. (This is an important class including, for instance, the normal distribution.) This class has a group invariance structure for which the right-Haar prior is $\pi^H(\mu, \sigma) = 1/\sigma$, which is clearly improper.

Suppose now that one has two different location scale models, determined by g_1 and g_2. The invariance theory then says that one can use the same right-Haar prior for both in computing the Bayes factor, leading to

$$B_{12} = \frac{\int\int \left[\prod_{i=1}^{n} \sigma^{-1} g_1\left((x_i-\mu)/\sigma\right)\right] \pi^H(\mu,\sigma) d\mu d\sigma}{\int\int \left[\prod_{i=1}^{n} \sigma^{-1} g_2\left((x_i-\mu)/\sigma\right)\right] \pi^H(\mu,\sigma) d\mu d\sigma}$$

This was the result used in Example 18.5 for the choices of priors there. ($\pi_L(\mu, \sigma) = 1/\sigma$ was clearly the right-Haar prior; $\pi_W(\theta, \gamma) = 1/(\theta\gamma)$ can be seen to be the right-Haar prior after transforming to $y = \log x$, $\mu = \log \beta$, and $\sigma = 1/\gamma$.)

One motivation for using the right-Haar prior here arises from the notion of predictive matching, as introduced in Section 18.4.3.2. The intuition, following similar arguments in [79], is that, if one only had two observations, X_1 and X_2, there should be no way to differentiate between g_1 and g_2, since the two observations can only learn about μ and σ; there are no 'degrees of freedom' left to learn about g. The argument proceeds that one should choose model parameter priors so that $B_{12} = 1$ if there are only two observations.

It is, indeed, shown in [18] that, for the right-Haar prior and any g,

$$m(x_1, x_2) = \int\int \left[\frac{1}{\sigma^2} g\left(\frac{x_1-\mu}{\sigma}\right) g\left(\frac{x_2-\mu}{\sigma}\right)\right] \frac{1}{\sigma} d\mu d\sigma = \frac{1}{2|x_1 - x_2|}$$

from which it is immediate that $B_{12} = 1$ for any two observations. [18] also generalize this to essentially any model with a group invariance structure, thus providing a wide class of problems for which we know that use of objective improper priors is fine.

18.5.1.2 *Parameters with the same scientific meaning*

If there are parameters in each model that have the same scientific meaning, reasonable default priors (e.g. the constant prior 1) can be used. For instance, in the pedagogical example of Section 18.2, if b were unknown it would have the same scientific meaning in both hypotheses, and could hence be assigned the same improper objective prior, e.g. a constant prior, or the Jeffreys prior $1/\sqrt{b}$ for a Poisson mean. Note that the difficulties with normalizing constants that were discussed in Section 18.2.4 do not apply when there is a common prior in both the numerator and denominator of a Bayes factor.

Often, similar parameters are present in various models, but they do not have the same meaning; see Section 18.4.3.1. A common strategy to obtain parameters with the same meaning—initiated by [79]—is to orthogonalize the parameters, i.e. to reparameterize so that the expected Fisher information matrix is diagonal. The idea is that the parameters are then 'independent in the likelihood' (at least asymptotically) so that a parameter then more plausibly has the same meaning across models.

There is still the issue of what objective prior to choose for the 'orthogonalized parameter,' but the notion is that usually one can then employ standard estimation objective priors. Discussions of this can be found in [85], [36], [41], [74] and [25].

18.5.2 Inducing model priors from a single prior

The number of models under consideration is often vast (e.g. 2^p in variable selection), so that separate determination of parameter priors for each model is typically not feasible. In this section we discuss the powerful idea of selecting only one prior, and inducing all others from this single prior.

18.5.2.1 *Basic idea*

Specify a prior $\pi_L(\theta_L)$ for the 'largest' model, and use this prior to induce priors on the other models. Possibilities for inducing priors on other models include

1. In variable selection, remove variables by conditioning them to be zero.
2. In variable selection, marginalize out the variables to be removed.
3. Projecting $\pi_L(\theta_L)$ onto the parameter space of other models through a minimization of Kullback–Leibler divergence. While this last is a potentially powerful tool (see [68] and [20] for illustration), it is often difficult to implement and will not be considered here.

Example 18.8 Suppose the largest model has prior $\pi_L(p_1,\ldots,p_m) \sim \textit{Dirichlet}(1,\ldots,1)$ (i.e. the uniform distribution on the simplex). If other models have parameters $(p_{i_1},\ldots,p_{i_l})$,

1. conditioning yields $\pi(p_{i_1},\ldots,p_{i_l},p^* \mid \text{other}\, p_j = 0) = \textit{Dirichlet}(1,\ldots,1)$, where $p^* = 1 - \sum_{j=1}^{l} p_{i_j}$;
2. marginalizing yields $\pi(p_{i_1},\ldots,p_{i_l},p^*) = \textit{Dirichlet}(1,\ldots,1,m-l)$.

This example clearly indicates that the conditional approach is the correct approach; the marginal approach would result in priors for $(p_{i_1},\ldots,p_{i_l})$ that are much too concentrated near zero if $m-l$ is large.

18.5.2.2 *Application to the linear model*

Analyses of model uncertainty for the linear model usually utilize so-called *g*-priors or Zellner–Siow priors ([135]). These and more modern alternatives can all be derived through the above conditional inducement from a full prior, as will be illustrated here.

The full linear model for data $\boldsymbol{Y} = (Y_1,\ldots,Y_n)'$ is given by

$$\boldsymbol{Y} = \boldsymbol{1}\,\theta_0 + \boldsymbol{X}\boldsymbol{\theta} + \boldsymbol{\varepsilon} \tag{18.5}$$

where $\boldsymbol{X}$ $(p \times n)$ is the matrix of covariates (full rank), $\boldsymbol{\theta} = (\theta_1,\ldots,\theta_p)$ is an unknown vector of regression coefficients; $\boldsymbol{1}$ is a vector of ones; θ_0 is the unknown mean level (intercept); and $\boldsymbol{\varepsilon}$ is $\mathrm{N}(\boldsymbol{0},\sigma^2\boldsymbol{I})$. For notational simplicity we assume that $\boldsymbol{1}^t\boldsymbol{X} = \boldsymbol{0}$, which results from 'centring' the columns of $\boldsymbol{X}$ by their means.

The models under consideration are all the submodels from (18.5) resulting from deleting components of $\boldsymbol{\theta}$ and corresponding columns in $\boldsymbol{X}$; we assume, however, that all models have the intercept. A convenient notation for such submodels is to let $\boldsymbol{\gamma} = (\gamma_1,\ldots,\gamma_p)$, where $\gamma_i = 1$ if variable X_i enters the model (that is, $\theta_i \neq 0$), and $\gamma_i = 0$ otherwise ($\theta_i = 0$). Then, the 2^p values for $\boldsymbol{\gamma}$ index the model space with elements $M_{\boldsymbol{\gamma}}$, with corresponding $\boldsymbol{\theta}_{\boldsymbol{\gamma}}$ and $\boldsymbol{X}_{\boldsymbol{\gamma}}$.

Note first that the parameters θ_0 and σ are common to all models. Furthermore, it can be shown that they are location-scale parameters and, hence, an adaptation of the argument in Section 18.5.1.1 justifies use of the common right-Haar prior $\pi_{\boldsymbol{\gamma}}(\theta_0, \sigma) \propto 1/\sigma$ for these parameters in all models. See [104] and [7] for further discussion.

The priors for $\boldsymbol{\theta}_{\boldsymbol{\gamma}}$ must be proper, since they are not common to all models. Considering first the full model in (18.5), an interesting class of priors is

$$\pi(\boldsymbol{\theta} \mid \sigma) = \int_0^\infty \mathrm{N}(\boldsymbol{\theta} \mid \mathbf{0}, g\,\sigma^2\,(\mathbf{X}^t\mathbf{X})^{-1})\pi(g)dg \tag{18.6}$$

where $\pi(g)$ is a density on $g > 0$. Centring the prior at $\mathbf{0}$ is a natural choice from an objective perspective. Choosing the covariance matrix to be proportional to $\sigma^2(\mathbf{X}^t\mathbf{X})^{-1}$, which is the covariance matrix of the least squares estimate of $\boldsymbol{\theta}$, has three purposes. The first is to ensure that the prior is invariant to change in the units of measurement of both $\boldsymbol{Y}$ and the covariates $\boldsymbol{X}$. The second is to satisfy a predictive matching criterion that is given in [7]; space limitations preclude discussion here. Finally, we will see that this choice of covariance matrix greatly simplifies the computation.

With a prior for the full model, we can now induce a prior on each submodel by conditioning. In particular, from multivariate normal theory it is immediate that, given g and σ^2, the distribution of $\boldsymbol{\theta}_{\boldsymbol{\gamma}}$ given that the other coordinates are set to zero is $\mathrm{N}(\boldsymbol{\theta}_{\boldsymbol{\gamma}} \mid \mathbf{0}, g\,\sigma^2\,(\mathbf{X}_{\boldsymbol{\gamma}}^t\mathbf{X}_{\boldsymbol{\gamma}})^{-1})$, from which it follows that the induced priors on the submodels are

$$\pi_{\boldsymbol{\gamma}}(\boldsymbol{\theta}_{\boldsymbol{\gamma}} \mid \sigma) = \int_0^\infty \mathrm{N}(\boldsymbol{\theta}_{\boldsymbol{\gamma}} \mid \mathbf{0}, g\,\sigma^2\,(\mathbf{X}_{\boldsymbol{\gamma}}^t\mathbf{X}_{\boldsymbol{\gamma}})^{-1})\pi(g)dg \tag{18.7}$$

Priors such as this are called *mixtures of g-priors*, and are extensively studied in [90] and [7]. The two most common choices of $\pi(g)$ are as follows.

g-priors (from [134]) use just a fixed g, typically $g = n$, the reason being that $n(\mathbf{X}_{\boldsymbol{\gamma}}^t\mathbf{X}_{\boldsymbol{\gamma}})^{-1}$ typically stabilizes as n grows. This choice allows closed form computation of Bayes factors but can be shown to violate information consistency (see Section 18.4.3.3).

Zellner–Siow priors (from [135]) utilize an inverse gamma density for g with scale parameter $n/2$ and shape parameter $1/2$; for this choice, (18.7) is actually the multivariate Cauchy density. The disadvantage of this prior is that it does not lead to closed form Bayes factors.

Recently, a g-mixture prior has been proposed that yields closed form answers and satisfies all the desiderata that have been suggested for priors for model uncertainty in the linear model. Indeed, [7] propose use of

$$\pi_{\boldsymbol{\gamma}}(g) = (0.5)\left[(1+n)/(1+k_{\boldsymbol{\gamma}})\right]^{1/2}(g+1)^{-3/2}1_{\{g>(1+n)/(1+k_{\boldsymbol{\gamma}})-1\}}$$

where $k_{\boldsymbol{\gamma}}$ is the dimension of $\boldsymbol{\theta}_{\boldsymbol{\gamma}}$. This results in the following closed form expression for the Bayes factor of model $M_{\boldsymbol{\gamma}}$ to the intercept only model M_0:

$$B_{\boldsymbol{\gamma}0} = \left[\frac{n+1}{k_{\boldsymbol{\gamma}}+1}\right]^{-\frac{k_{\boldsymbol{\gamma}}}{2}} \frac{Q_{\boldsymbol{\gamma}0}^{-\frac{n-1}{2}}}{k_{\boldsymbol{\gamma}}+1}\, {}_2F_1\left[\frac{k_{\boldsymbol{\gamma}}+1}{2}; \frac{n-1}{2}; \frac{k_{\boldsymbol{\gamma}}+3}{2}; \frac{(1-Q_{\boldsymbol{\gamma}0}^{-1})(k_{\boldsymbol{\gamma}}+1)}{(1+n)}\right]$$

where ${}_2F_1$ is the standard hypergeometric function (see [1]) and $Q_{\gamma 0} = SSE_\gamma / SSE_0$ is the ratio of the sum of squared errors of M_γ and M_0.

There is a huge literature on priors for linear and generalized linear models, with many other ideas and suggestions. A few of the references are [34], [59], [31], [81, 82], [39], [131], [93], [40], [70], [94], [95] and [114].

18.5.3 Predictive matching approaches

We have already seen an example of predictive matching in invariant situations utilizing the right-Haar prior. This section mentions three other examples of application of the idea. Other examples will be seen in Section 18.5.4.

18.5.3.1 *Subjective matching*

It is often asserted that, in subjective Bayesian statistics, one should elicit distributions of observables, not distributions of parameters (cf. [45], [83]). Thus suppose one begins the model uncertainty analysis by eliciting a distribution $m(\boldsymbol{x}^*)$ of observables $\boldsymbol{x}^*$; we will assume here that $\boldsymbol{x}^*$ is a potential sample of observations, and will usually be chosen to be a small sample size (so that the elicitation is easier).

If we now consider a model M_i, it is natural to choose a prior $\pi_i(\theta_i)$ so that this prior reflects the beliefs that have been encoded in $m(\boldsymbol{x}^*)$. In particular, if we were to interpret $m(\boldsymbol{x}^*)$ as the marginal density arising from our true model and prior, then $m(\boldsymbol{x}^*)$ should be close to $m_T(\boldsymbol{x}^*) = \int f_T(\boldsymbol{x}^* \mid \theta_T)\, \pi_T(\theta_T) d\theta_T$. It follows that, for any model M_i being considered, one should choose the prior $\pi_i(\theta_i)$ so that $m_i(\boldsymbol{x}^*) = \int f_i(\boldsymbol{x}^* \mid \theta_i)\, \pi_i(\theta_i) d\theta_i$ is as close as possible to $m(\boldsymbol{x}^*)$.

[121] established the following interesting result. Start with any prior $\pi_i^{(0)}(\theta_i)$ with full support (e.g. a constant prior), and iteratively define

$$\pi_i^{(l)}(\theta_i) = \int \pi_i^{(l-1)}(\theta_i \mid \boldsymbol{x}^*)\, m(\boldsymbol{x}^*) d(\boldsymbol{x}^*) \tag{18.8}$$

where $\pi_i^{(l-1)}(\theta_i \mid \boldsymbol{x}^*)$ is the posterior under model M_i if $\pi_i^{(l-1)}(\theta_i)$ were the prior. Then $\pi_i^{(l)}$ converges to that prior such that the resulting $m_i(\boldsymbol{x}^*)$ is closest to the elicited $m^*(\boldsymbol{x}^*)$ in Kullback–Leibler divergence. Furthermore, the convergence is extremely fast, so that often only two or three iterations are needed.

18.5.3.2 *Predictive moment matching*

This approach seeks to choose the $\pi_i(\theta_i)$ so that the $m_i(\mathbf{x}^*)$ are similar in terms of moments or other features for the various models. (See [76], [77].)

18.5.3.3 *'Calibrated' noninformative priors*

This approach utilizes standard improper objective estimation priors, but chooses specific multiplicative constants for the priors so as to obtain matching of the $m_i(\mathbf{x}_0^*)$ at some specified point (or points) $\mathbf{x}_0^*$. See [126] and [65] for examples and discussion.

18.5.4 Training sample methods of model selection

Efforts to utilize the data itself or imaginary data to help determine the prior have a long history, starting with [67], [79], and [125]. There have been very bad versions of this idea, such as repeated efforts to utilize the likelihood itself as the prior. The successful versions of this idea utilize only a small fraction of the data—or a small amount of imaginary data—to help determine the prior. Three variants of the idea are discussed in this section.

18.5.4.1 *Median intrinsic posterior probabilities*

This method utilizes subsets of the data to directly construct Bayes factors or posterior probabilities from objective improper priors. We only illustrate the idea here in the simplest setting, that of choosing between two models. Also, we discuss only one of a class of related methods, the *median intrinsic* method of [13].

Suppose that data $\boldsymbol{X} = \{X_1, X_2, \ldots, X_n\}$ are i.i.d. from either $M_1 : f_1(x \mid \theta_1)$ or $M_2 : f_2(x \mid \theta_2)$. The standard objective prior probabilities are $Pr(M_1) = Pr(M_2) = 1/2$, and we assume that objective estimation priors $\pi_j^{\mathrm{O}}(\theta_j)$ are available, leading to marginal likelihoods $m_j^{\mathrm{O}}(\boldsymbol{x}) = \int f_j(\boldsymbol{x} \mid \theta_j)\pi_j^{\mathrm{O}}(\theta_j)d\theta_j$. We know, of course, that we cannot use these marginal likelihoods directly for model comparison (unless the two models have the same group invariance structure).

A key notion is that of *minimal training sample size*, which is defined to be the smallest sample size for which $m_1^{\mathrm{O}}(\boldsymbol{x})$ and $m_2^{\mathrm{O}}(\boldsymbol{x})$ are finite. Let $\boldsymbol{x}(l)$ denote any subset of the data of minimal training sample size; such subsets are called *minimal training samples*. It is immediate that $\pi_j^{\mathrm{O}}(\theta \mid \boldsymbol{x}(l)) = f_j(\boldsymbol{x}(l) \mid \theta)\, \pi_j^{\mathrm{O}}(\theta_j)/m^{\mathrm{O}}(\boldsymbol{x}(l))$ are proper posterior distributions.

We can now utilize these posteriors as priors, together with the remaining data $\boldsymbol{x}(-l)$, to compute the posterior probabilities of M_1 and M_2; the result is

$$
\begin{aligned}
Pr(M_1 \,|\, \boldsymbol{x}(-l)) = 1 - Pr(M_2 \,|\, \boldsymbol{x}(-l)) &= \left(1 + \frac{\int f_2(\boldsymbol{x}(-l) \mid \theta_2)\pi_2^{\mathrm{O}}(\theta_2|\boldsymbol{x}(l))d\theta_2}{\int f_1(\boldsymbol{x}(-l) \mid \theta_1)\pi_1^{\mathrm{O}}(\theta_1|\boldsymbol{x}(l))d\theta_1}\right)^{-1} \\
&= \left(1 + \frac{m_2^{\mathrm{O}}(\boldsymbol{x})}{m_1^{\mathrm{O}}(\boldsymbol{x})} \cdot \frac{m_1^{\mathrm{O}}(\boldsymbol{x}(l))}{m_2^{\mathrm{O}}(\boldsymbol{x}(l))}\right)^{-1}
\end{aligned}
$$

Finally, the suggestion is to 'average' over the choice of the training samples $\boldsymbol{x}(l)$ by taking the median of the above values, leading to the *median intrinsic posterior probabilities*

$$
Pr^{med}(M_1 \mid \boldsymbol{x}) = 1 - Pr^{med}(M_2 \mid \boldsymbol{x}) = \left(1 + \frac{m_2^{\mathrm{O}}(\boldsymbol{x})}{m_1^{\mathrm{O}}(\boldsymbol{x})} \cdot \text{Median}\left\{\frac{m_1^{\mathrm{O}}(\boldsymbol{x}(l))}{m_2^{\mathrm{O}}(\boldsymbol{x}(l))}\right\}\right)^{-1}
$$

Note that this is a completely general prescription (assuming there are only two models under consideration), and requires only standard objective estimation priors (even a constant prior could be used). Furthermore, and most importantly, the procedure can be shown to correspond (at least for large n) to an actual Bayes procedure with sensible objective priors; the corresponding priors are called *intrinsic priors* and this property that the training sample methods correspond to actual priors motivates the name for the methodology. Among the many references related to this approach are [49], [55], [129], [50], [115], [12, 14, 15], [47], [78], [56], [97, 98] [87] [92], [116], [8], [110], [109], and [52].

18.5.4.2 *Fractional Bayes factors*

This idea was introduced in [101, 102]; instead of using a fraction of the data as a training sample, the notion is to use a fraction of the likelihood. The algorithm is as follows:

Algorithm

Step 1. Choose some 'fraction' $0 < b < 1$.

Step 2. For model M_j, use the prior $\pi_j^*(\boldsymbol{\theta}_j) \propto \left[f_j(\boldsymbol{x}|\boldsymbol{\theta}_j)\right]^b \cdot \pi_j^{\mathrm{O}}(\boldsymbol{\theta}_j)$, where $\pi_j^{\mathrm{O}}(\boldsymbol{\theta}_j)$ is an objective estimation prior (which could just be the constant prior).

Step 3. Compute Bayes factors using these priors and the 'remaining likelihoods' $[f_j(\boldsymbol{x}|\boldsymbol{\theta}_j)]^{(1-b)}$, resulting in

$$B_{ji} = \frac{\int \left[f_j(\boldsymbol{x}|\boldsymbol{\theta}_j)\right]^{(1-b)} \pi_j^*(\boldsymbol{\theta}_j) d\boldsymbol{\theta}_j}{\int \left[f_i(\boldsymbol{x}|\boldsymbol{\theta}_i)\right]^{(1-b)} \pi_i^*(\boldsymbol{\theta}_i) d\boldsymbol{\theta}_i} = \frac{\int f_j(\boldsymbol{x}|\boldsymbol{\theta}_j)\pi_j^{\mathrm{O}}(\boldsymbol{\theta}_j)d\boldsymbol{\theta}_j}{\int f_i(\boldsymbol{x}|\boldsymbol{\theta}_i)\pi_i^{\mathrm{O}}(\boldsymbol{\theta}_i)d\boldsymbol{\theta}_i} \cdot \frac{\int \left[f_i(\boldsymbol{x}|\boldsymbol{\theta}_i)\right]^b \pi_i^{\mathrm{O}}(\boldsymbol{\theta}_i)d\boldsymbol{\theta}_i}{\int \left[f_j(\boldsymbol{x}|\boldsymbol{\theta}_j)\right]^b \pi_j^{\mathrm{O}}(\boldsymbol{\theta}_j)d\boldsymbol{\theta}_j}.$$

The posterior probabilities are then computed from these Bayes factors.

This approach is particularly attractive when the Bayes factors for objective estimation priors are available in closed form. It is also broadly applicable (except for serious concerns in irregular problems, as observed in [13, 14]). The key issue is choosing the fraction b, and the typically recommended choice is $b = k_i/n$, where n is the sample size and k_i is the dimension of the parameter in model M_i; it has been noted, however, that specification of different fractions for different parts of the likelihood may be necessary ([48]).

18.5.4.3 *Expected posterior priors*

The name reflects the fact that the prior is computed as an appropriate expectation of posteriors (see [105, 106], [100]). The method proposes to train the initial objective estimation priors, $\pi_i^{\mathrm{O}}(\boldsymbol{\theta}_i)$, as follows:

1. Compute the 'initial marginals' $m_i^{\mathrm{O}}(\boldsymbol{x}) = \int f_i(\boldsymbol{x} \mid \boldsymbol{\theta}_i)\pi_i^{\mathrm{O}}(\boldsymbol{\theta}_i)d\boldsymbol{\theta}_i$; note that these will be improper if the initial priors are improper.
2. Next, consider a (random) training sample, $\boldsymbol{x}^*$, such that the posterior distributions $\pi_i^{\mathrm{O}}(\boldsymbol{\theta}_i \mid \boldsymbol{x}^*) = f_i(\boldsymbol{x}^* \mid \boldsymbol{\theta}_i)\pi_i^{\mathrm{O}}(\boldsymbol{\theta}_i)/m_i^{\mathrm{O}}(\boldsymbol{x}^*)$ exist, for $i = 1, \ldots, q$.
3. Specify a density $m^*(\boldsymbol{x}^*)$. The prior densities

$$\pi_i^*(\boldsymbol{\theta}_i) = \int \pi_i^{\mathrm{O}}(\boldsymbol{\theta}_i \mid \boldsymbol{x}^*)\, m^*(\boldsymbol{x}^*) d\boldsymbol{x}^*$$

are the *expected posterior priors* for the $\boldsymbol{\theta}_i$, with respect to $m^*(\cdot)$.

Note that the expected posterior priors, $\pi_i^*(\boldsymbol{\theta}_i)$, will not be proper unless m^* itself is proper, but are always properly 'calibrated' across models, which make them appropriate for objective Bayesian model selection.

An interesting choice for the mixing measure m^* is the *empirical distribution*. Specifically, given observations $x_1, \ldots, x_n$, let

$$m^*(\boldsymbol{x}^*) = \frac{1}{L} \sum_l I_{\{\boldsymbol{x}(l)\}}(\boldsymbol{x}^*),$$

where $\boldsymbol{x}(l) = (x_{l_1}, \ldots, x_{l_m})$ is a subsample of size $0 < m < n$ such that $\pi_i^{\text{O}}(\boldsymbol{\theta}_i \mid \boldsymbol{x}(l))$ exists for all models M_i, and L is the number of such subsamples of size m. For other reasonable choices of m^*, and discussion of computation with expected posterior priors, see [105, 106].

Example 18.9 For the pedagogical example in Section 18.2, suppose we choose the initial $\pi^{\text{O}}(\theta) = 1/(\theta + b)$. (Jeffreys prior, the square root of π^{O}, would probably be better, but leads to a much more difficult computation.) Following the ideas in [15], we represent the Poisson observation, X, over the time period T from the distribution in the example as a sum of i.i.d. observations from an exponential inter-arrival time process. Indeed, for $i = 1, \ldots,$ if we consider $Y_i \sim f(y_i \mid (\theta + b)/T) = (\theta + b)T^{-1} \exp\{-(\theta + b)\, y_i/T\}$, then $X \equiv \{\text{first } j \text{ such that } S_j = \sum_{i=1}^{j} Y_i > T\} - 1$. The minimal sample size for this exponential distribution can easily be seen to be 1. Computation then yields

$$\pi^E(\theta) = \int_0^\infty \pi^{\text{O}}(\theta \mid y_1) f(y_1 \mid b/T)\, dy_1 = \frac{b}{(\theta + b)^2}$$

which was the conventional proper prior used for Bayesian testing in the example.

18.5.5 Evaluating choices of priors for model uncertainty

In objective Bayesian estimation, there are a variety of criteria that objective priors should satisfy which often yield a unique (or at least, robust) objective prior. This is not the case with objective priors for model uncertainty. There are certainly important criteria; consistency, information consistency, and invariance have been discussed herein, and others can be found in the literature (see, e.g. [7]). But these criteria rarely yield a unique or robust answer, and even application of these criteria is far from universal (e.g. there are many proposed Bayesian model uncertainty procedures—such as DIC [127]—that are not even consistent).

One seemly obvious criterion is that objective model selection procedures should correspond, in some way, to actual Bayesian procedures arising from reasonable priors. Calling a procedure which fails to satisfy this criterion 'Bayesian' is not reasonable. Yet many so-called Bayesian procedures (DIC again is an example) fail this simple test. This basic point was first highlighted in [125], which showed how various model selection criteria were formal Bayesian procedures, but some corresponded to bizarre priors (e.g. prior variances $\to 0$ as $n \to \infty$). See [14] for additional discussion and examples.

Note, finally, that, if one is comparing real Bayesian model selection procedures, use of simulations for comparison is typically useless. By definition, a Bayes procedure will perform best for models and model parameters chosen from its prior distribution so, when one real Bayes procedure outperforms another, it means nothing more than that the priors used to construct the simulation were closer to the priors defining that Bayes procedure.

18.6 Other issues with Bayesian model uncertainty

18.6.1 Computation and search

There are two main computational difficulties encountered in dealing with Bayesian model uncertainty. The first is that computation of the marginal likelihoods $m_i(\boldsymbol{x}) = \int f_i(\boldsymbol{x} \mid \boldsymbol{\theta}_i)\pi_i(\boldsymbol{\theta}_i)\, d\boldsymbol{\theta}_i$ can be difficult, as they potentially involve high-dimensional numerical integration. There is a huge literature addressing such computations. One of the early key references was [57]. Others include [27], [69], [84], [130], [108], [33], [30], [46], [66], [72], [11], [29], [123], [122], [109], [113], [32].

The second computational difficulty is the potential size of the model space. If one has, say, 60 variables in a variable selection problem, the number of possible models is 2^{60}, which is far too large for enumeration; indeed, only a very small fraction of these models could ever be visited in any computational scheme. This is not an issue if the variables are orthogonal (see e.g. [36]), but otherwise can be highly problematic. For approaches to dealing with this problem, see [11], [73], [22], [132], and [38].

18.6.2 Approximations and asymptotics

An important approximation to a Bayes factor is the Laplace approximation, given by

$$B_{21}^{L} = \frac{\int f_2(\boldsymbol{x} \mid \boldsymbol{\theta}_2)\pi_2(\theta_2)d\theta_2}{\int f_1(\boldsymbol{x} \mid \boldsymbol{\theta}_1)\pi_1(\theta_1)d\theta_1} \approx \frac{f_2(\boldsymbol{x} \mid \widehat{\boldsymbol{\theta}}_2) \mid \widehat{\mathbf{I}}_2|^{-1/2}}{f_1(\boldsymbol{x} \mid \widehat{\boldsymbol{\theta}}_1) \mid \widehat{\mathbf{I}}_1|^{-1/2}} \cdot \frac{(2\pi)^{q_2/2}\pi_2(\widehat{\boldsymbol{\theta}}_2)}{(2\pi)^{q_1/2}\pi_1(\widehat{\boldsymbol{\theta}}_1)} \tag{18.9}$$

where $\widehat{\boldsymbol{\theta}}_1$ and $\widehat{\boldsymbol{\theta}}_2$ are the m.l.e.s for $\boldsymbol{\theta}_1$ and $\boldsymbol{\theta}_2$ (which have dimensions q_1 and q_2) and $\widehat{\mathbf{I}}_1$ and $\widehat{\mathbf{I}}_2$ are observed information matrices (slightly more accurate would be to base $\widehat{\boldsymbol{\theta}}_i$ and $\widehat{\mathbf{I}}_i$ on the full integrands in (18.9) rather than only the likelihoods). This is an asymptotically correct approximation as the sample size grows and, often, is surprisingly accurate for even small sample sizes. Indeed, this approximation has become the computational basis of a popular Bayesian analysis package INLA (see www.r-inla.org/ where software, references and useful material is provided).

Furthermore, this approximation is used to develop simple model selection tools such as BIC (the Bayes Information Criterion from [117]). Following [107], one route to BIC from (18.9) is to choose the $\pi_j(\boldsymbol{\theta}_j)$ to be $\mathrm{N}(\boldsymbol{\theta}_j \mid \widehat{\boldsymbol{\theta}}_j, n\widehat{\mathbf{I}}_j^{-1})$, where n is the sample size, in which case (18.9) becomes

$$B_{21} \approx \frac{f_2(\boldsymbol{x} \mid \widehat{\boldsymbol{\theta}}_2)}{f_1(\boldsymbol{x} \mid \widehat{\boldsymbol{\theta}}_1)} \cdot n^{\frac{1}{2}(q_2-q_1)}$$

which is the Bayes factor form of BIC and is very convenient to use. Unfortunately, this choice of prior is problematical (the prior being centred on the mle), and BIC can behave quite poorly when the sample size is not large relative to the dimensions of the model parameters. Further discussion of these issues, and generalizations, can be found in [86], [53], [85], [103], [14], [64], [17], [63], and [133].

18.6.3 Multiplicity

In classical statistics, dealing with multiple hypotheses requires a multiplicity correction to the error probabilities. In contrast, Bayesian analysis deals with multiplicity adjustment solely through the assignment of prior probabilities to models or hypotheses.

Example 18.10 Suppose one is testing mutually exclusive hypotheses $H_i,\ i = 1, \ldots, q$, so each hypothesis is a separate model. If the hypotheses are viewed as exchangeable, it is natural to choose $Pr(H_i) = 1/q$. For instance, suppose 1000 energy channels are searched for presence of a signal indicating the Higgs boson;

- if the signal is known to exist and occupy only one channel, but no channel is theoretically preferred, assign each channel prior probability 0.001;
- if the signal is not known to exist, prior probability 1/2 should be given to 'no signal,' and probability 0.0005 to each channel.

The key fact is that *this is the Bayesian solution regardless of the structure of the data.*

In contrast, frequentist control for multiplicity depends strongly on the structure of the data. To see this, further specialize the example so that, for each channel, one is testing $H_{0i} : \mu_i = 0$ versus $H_{1i} : \mu_i > 0$, $i = 1, \ldots, 1000$, based on data $X_i, i = 1, \ldots, 1000$, that are normally distributed with mean μ_i, variance 1, and correlation ρ.

If $\rho = 0$ and it is desired to obtain an overall error probability of $\alpha = 0.05$, one can just do the individual tests at level $0.05/1000 = 0.00005$; this 'Bonferonni correction' guarantees that the overall error probability of the multiple testing procedure is 0.05. If $\rho > 0$, however, the situation is more difficult. One natural way to proceed would be to choose the overall decision rule "declare μ_i to be a signal if X_i is the largest value and $X_i > K$," and then compute the corresponding frequentist type I error probability

$$\alpha = Pr(\max_i X_i > K \mid \mu_1 = \ldots = \mu_m = 0) = E^Z\left[1 - \Phi\left(\frac{K - \sqrt{\rho}Z}{\sqrt{1-\rho}}\right)^m\right]$$

where Φ is the standard normal cdf and Z is a standard normal random variable.

This gives (essentially) the Bonferroni correction when $\rho = 0$, but can be shown to converge to $1 - \Phi[K]$ as $\rho \to 1$, which is the type I error that would result from a *single test*. Thus the needed frequentist control for multiple testing ranges from the drastic Bonferroni correction to none, depending on the correlations among the data.

This highlights one of the ways in which dealing with model uncertainty or multiple testing from a Bayesian perspective introduces a major simplification; Bayesian accommodation of multiplicities does not depend on the error structure of the data, and the Bayesian adjustment is not highly conservative in the presence of dependent data.

At the same time, there is a common misconception among many Bayesians that Bayesian analysis need not concern itself with multiplicity adjustments; it is supposedly handled automatically by the Bayesian paradigm. This is indeed true if a thorough subjective Bayesian analysis is done, in the sense that choice of the subjective prior will spread the total prior probability of 1 among the possible models or hypotheses being considered. We have repeatedly argued, however, that a full subjective Bayes analysis is rarely possible in situations of model uncertainty (also, in multiple hypothesis testing), because of the large number of unknowns being considered. Furthermore, as mentioned in Section 18.4.2, naive and all-too-common choices, such as equal prior probabilities of models, will not typically control for multiplicity, so the issue is of major importance to Bayesians. Further general discussion of these issues can be found in [118, 119], which also refer to the previous Bayesian literature on multiplicity adjustment.

18.6.4 Model criticism

18.6.4.1 Introduction

Bayesian model uncertainty analysis is based on comparison of possible models. Often in the model development process, however, one is in the situation of entertaining a single specific model, and seeking to determine if the model is adequate or needs elaboration—i.e. should one even begin to develop and consider alternative models.

For the purpose of criticizing a postulated model, classical statisticians use p-values and a variety of diagnostic tools. Bayesians also use p-values and a variety of diagnostic tools. Because of space limitations, we limit discussion here to the use of p-values. (For review and references on alternative non-parametric tests see [128].)

Before beginning, an obvious question is why p-values are even being considered, given all the problems with p-values that were raised in Section 18.3.1. The answer is that we are not formally rejecting a model based on the p-value but, rather, deciding whether there is an indication that we should explore further. Indeed, a small p-value is an indication that something unusual has occurred (more precisely, a small value of (18.2) is such an indication), and when something unusual has occurred it is natural to look further.

Suppose that a statistical model $H_0 : \boldsymbol{X} \sim f(\boldsymbol{x} \mid \theta)$ is being entertained, data $\boldsymbol{x}_{obs}$ is observed, and it is desired to check the adequacy of the model. This is commonly done by choosing a statistic $T = t(\boldsymbol{X})$, where large values of T indicate incompatibility with the model. The classical p-value is then defined as

$$p = Pr\,(t(\boldsymbol{X}) \geq t(\boldsymbol{x}_{obs}) \mid \theta). \tag{18.10}$$

If θ is known, this probability computation is with respect to $f(\boldsymbol{x} \mid \theta)$; the question of interest here is what to do when θ is unknown, since computation of the p-value requires some way of 'eliminating' θ.

18.6.4.2 Plug-in p-value

There are many non-Bayesian approaches to this problem, some of which are reviewed in [5] and [111]. The most common method is to replace θ in (18.10) by its m.l.e., $\hat{\theta}$. The resulting p-value is called the *plug-in p-value* (p_{plug}), and is defined as

$$p_{\text{plug}} = Pr^{f(\cdot \mid \hat{\theta})}\,(t(\boldsymbol{X}) \geq t(\boldsymbol{x}_{obs})) \tag{18.11}$$

Although simple to use, there is a worrisome 'double use' of the data in p_{plug}—first to estimate θ and then to compute the tail area corresponding to $t(\boldsymbol{x}_{obs})$ in that distribution. Indeed, this difficulty was one of the motivations for creation of more sophisticated frequentist procedures such as the bootstrap.

18.6.4.3 Three Bayesian p-values

The principled Bayesian approach is to construct p-values, not from (18.10), but from the marginal (or predictive) density $m(\boldsymbol{x}) = \int f(\boldsymbol{x} \mid \theta)\pi(\theta)d\theta$, where $\pi(\theta)$ is a subjective proper prior. The resulting p-value was called the *predictive p-value* by [23].

Much of model checking, however, takes place in scenarios in which the model is quite tentative and, hence, for which serious subjective prior elicitation is not feasible. And, unfortunately, use of improper objective prior distributions is not directly possible with the predictive p-value, since the marginal distribution is itself then improper. (But see [4] for suggestions concerning the use of conditional marginal distributions based on improper priors.)

The 'Bayesian solution' that has become quite popular is to utilize objective initial priors, but then use the posterior distribution $\pi(\theta \mid \boldsymbol{x}_{obs})$, instead of the prior, to define the distribution used to compute the p-value. This leads to the *posterior predictive p-value*, originating with [71] and popularized in [112] and [58], given by

$$p_{\text{post}} = Pr^{m(\cdot \mid \boldsymbol{x}_{obs})}\,(T \geq t_{obs})\,, \quad m(t \mid \boldsymbol{x}_{obs}) = \int f(t \mid \theta)\pi(\theta \mid \boldsymbol{x}_{obs})\, d\theta \tag{18.12}$$

Note that, as with the plug-in p-value, there is a 'double use' of the data in p_{post}, first to convert the objective improper prior into a proper distribution for determining the predictive distribution, and then for computing the tail area corresponding to $t(\boldsymbol{x}_{obs})$ in that distribution.

To avoid a double use of the data, [5] proposed the following alternative way of eliminating θ, based on Bayesian methodology. Begin with an objective prior density $\pi(\theta)$. Next, define the *partial posterior density*

$$\pi(\theta \mid \boldsymbol{x}_{obs} \backslash t_{obs}) \propto f(\boldsymbol{x}_{obs} \mid t_{obs}, \theta)\pi(\theta) \propto \frac{f(\boldsymbol{x}_{obs} \mid \theta)\pi(\theta)}{f(t_{obs} \mid \theta)} \tag{18.13}$$

resulting in the *partial posterior predictive density* of T

$$m(t \mid \boldsymbol{x}_{obs} \backslash t_{obs}) = \int f(t \mid \theta)\pi(\theta \mid \boldsymbol{x}_{obs} \backslash t_{obs})\, d\theta \tag{18.14}$$

Since this density is free of θ, it can be used in (18.10) to compute the *partial posterior predictive p-value*

$$p_{\text{PPP}} = Pr^{m(\cdot \mid \boldsymbol{x}_{obs} \backslash t_{obs})}(T \geq t_{obs}). \tag{18.15}$$

Note that p_{PPP} uses only the information in $\boldsymbol{x}_{obs}$ that is *not* in $t_{obs} = t(\boldsymbol{x}_{obs})$ to 'train' the prior and eliminate θ. This avoids double use of the data because the contribution of t_{obs} to the posterior is removed before θ is eliminated by integration.

18.6.4.4 A hierarchical example

An interesting example comparing the above p-values follows, the example taken from [6].

Consider the hierarchical (or random effects) model

$$\begin{aligned} X_{ij} \mid \mu_i &\sim \text{N}(X_{ij} \mid \mu_i, \sigma_i^2) \text{ for } i = 1, \ldots, I, \quad j = 1, \ldots, n_i \\ \mu_i \mid \nu, \tau &\sim \text{N}(\mu_i \mid \nu, \tau^2) \quad \text{for } i = 1, \ldots, I, \end{aligned} \tag{18.16}$$

where all variables are independent and the variances, σ_i^2, at the first level are known. Of primary interest is whether the normality assumption for the means μ_i is compatible with the data. Consider the test statistic $T = \max\{\bar{X}_1, \ldots, \bar{X}_I\}$, where $\bar{X}_i$ denotes the group sample means. Since the μ_i are random effects, tests should be based on the marginal densities of the sufficient statistics $\bar{X}_i$, with the μ_i integrated out. The resulting null distribution is

$$\bar{X}_i \mid \nu, \tau \sim \text{N}(\bar{X}_i \mid \nu, \sigma_i^2 + \tau^2) \quad \text{for } i = 1, \ldots, I \tag{18.17}$$

Thus p_{plug} is computed with respect to this distribution, with the m.l.e.s, $\hat{\nu}, \hat{\tau}^2$ (numerically computed from (18.17)), inserted back into (18.17) and (18.16).

To compute the posterior predictive p-value and the partial posterior predictive p-value, we begin with a common objective prior for (ν, τ^2), namely $\pi(\nu, \tau^2) = 1/\tau$ (not $1/\tau^2$ as is sometimes done, which would result in an improper posterior). The computation of p_{post} and p_{PPP} require MCMC methods discussed in [6], though it should be noted that computation of p_{post} is much easier.

All three p-values were computed from a realization from the following simulated dataset, in which one of the groups comes from a distribution with a much larger mean than the other groups:

$$\begin{aligned} X_{ij} \mid \mu_i &\sim \text{N}(X_{ij} \mid \mu_i, 4) \text{ for } i = 1, \ldots, 5 \qquad j = 1, \ldots, 8, \\ \mu_i &\sim \text{N}(\mu_i \mid 1, 1) \quad \text{for } i = 1, \ldots, 4, \\ \mu_5 &\sim \text{N}(\mu_5 \mid 5, 1). \end{aligned} \tag{18.18}$$

The resulting sample means were 1.560, 0.641, 1.982, 0.014, 6.964. Crucial is that the sample mean of the 5th group is 6.65 standard deviations away from the mean of the other four groups. The results:

- $p_{\text{plug}} = 0.130$, failing to strongly indicate that the assumption of i.i.d. normality of the μ_i is wrong;
- $p_{\text{post}} = 0.409$, giving absolutely no indication that there is any problem with the assumption of i.i.d. normality of the μ_i;
- $p_{\text{ppp}} = 0.010$, properly indicating that more investigation of the model is in order.

There is much more that can be said here, and the articles referenced above contain many more details and arguments. But the take-home message is that double use of the data is harmful when done by frequentists, and seems to be even worse when done by Bayesians. More generally, there is a message that the extra power obtained from Bayesian analysis seems to accentuate logical errors, so that Bayesians must treat their methodology with care and respect, not giving in to 'intuitive flights of fancy.'

References

[1] Abramowitz, M. and Stegun, I. A. (1964). *Handbook of Mathematical Functions with Formulas, Graphs, and Mathematical Tables.* New York: Dover.

[2] Barbieri, M. M. and Berger, J. O. (2004). Optimal predictive model selection. *Annals of Statistics,* **32**, pp. 870–897.

[3] Bartlett, M. (1957). A Comment on D. V. Lindley's Statistical Paradox. *Biometrika* **44**, 533–534.

[4] Bayarri, M. J. and Berger, J. (1998). Quantifying surprise in the data and model verification [with discussion]. In *Bayesian Statistics 6* (Bernardo, J. M., Berger, J. O., Dawid , A. P. and Smith, A. F. M., eds.), 53–82, Oxford University Press, Oxford.

[5] Bayarri, M. J. and Berger, J. (2000). P-values for composite null models [with discussion]. *J. American Statist. Assoc,* **95**, 1127–1142.

[6] Bayarri, M. J. and Castellanos, M. E. (2007). Bayesian checking of the second level of hierarchical models (discussion paper). *Statistical Science* 22, 3, pp. 322–342; Rejoinder pp. 363–367.

[7] Bayarri, M. J., Berger, J. O., Forte, A. and García-Donato, G. (2012). Criteria for Bayesian Model Choice with Application to Variable Selection. *Annals of Statistics,* **40**, 1550–1570.

[8] Beattie, S. D., Fong, D. K. H. and Lin, D. K. J. (2002). A two-stage Bayesian model selection strategy for supersaturated designs. *Technometrics,* **44**, 55–63.

[9] Berger, J. O. (1997). Bayes factors. In *Encyclopedia of Statistical Sciences,* Update (S. Kotz, C. B. Read and D. L. Banks, eds.) 3 20–29. New York: Wiley.

[10] Berger, J. and Delampady, M. (1987). Testing Precise Hypotheses (with discussion). *Statistical Science,* **3**, 317–352.

[11] Berger, J. O. and Molina, G. (2003). Discussion of "Recognition of Faces versus Greebles: A Case Study in Model Selection" by Viele, K., Kass, R. E., Tarr, M. J., Behrmann, M., and Gauthier, I. In *Case Studies in Bayesian Statistics, Vol. VI* (Gatsonis, C., Carriquiry, A., Higdon, D., Kass, R.E., Pauler, D., and Verdinelli, I., eds.), 111–124. New-York: Springer-Verlag.

[12] Berger, J. and Pericchi, L. (1996). The intrinsic Bayes factor for model selection and prediction. *Journal of the American Statistical Association,* **91**, 109–122.

[13] Berger, J. O. and Pericchi, L. R. (1998). Accurate and stable Bayesian model selection: the median intrinsic Bayes factor. *Sankhya: The Indian Journal of Statistics. Series B,* **60**, 1–18.

[14] Berger, J. O. and Pericchi, R. L. (2001). Objective Bayesian Methods for Model Selection: Introduction and Comparison (with discussion). *Model Selection,* ed. P. Lahiri, Institute of Mathematical Statistics Lecture Notes – Monograph Series, volume 38, pp. 135–207.

[15] Berger, J. O. and Pericchi, L. R. (2004). Training samples in objective Bayesian model selection. *Annals of Statistics*, **32**, 841–869.
[16] Berger, J. O., Brown, L. D. and Wolpert, R. L. (1994). A unified conditional frequentist and Bayesian test for fixed and sequential simple hypothesis testing. *Annals of Statistics*, **22** (4), 1787–1807.
[17] Berger, J. O., Ghosh, J. K. and Mukhopadhyay, N. (2003). Approximations and consistency of Bayes factors as model dimension grows. *Journal of Statistical Planning and Inference*, **112**, pp. 241–258.
[18] Berger, J. O., Pericchi, L. R. and Varshavsky, J. A. (1998). Bayes factors and marginal distributions in invariant situations. *Sankhya: The Indian Journal of Statistics. Series A*, **60**, 307–321.
[19] Berk, R. (1966). Limiting Behavior of posterior distributions when the model is incorrect. *Annals of Mathematical Statistics*, **37**, 51–58.
[20] Bernardo, J. M. (1999). Nested hypothesis testing: the Bayesian reference criterion. In *Bayesian Statistics 6* (J. M. Bernardo, J. O. Berger, A. P. Dawid and A. F. M. Smith, eds.), pp. 101–130. London: Oxford University Press.
[21] Bernardo, J. M. and Smith, Adrian F. M. (1994). *Bayesian Theory*. John Wiley & Sons.
[22] Bottolo, L. and Richardson, S. (2010). Evolutionary stochastic search for Bayesian model exploration. *Bayesian Analysis*, **5**, 583–618.
[23] Box, G. E. P. (1980). Sampling and Bayes inference in scientific modelling and robustness. *Journal of the Royal Statistical Society, Series A*, **143**, 383–430.
[24] Brown, L. D. (1978). A contribution to Kiefer's theory of conditional confidence procedures. *Annals of Statistics*, **6**, 59–71.
[25] Buck, C. E. and Sahu, S. K. (2000). Bayesian models for relative archaeological chronology building. *Journal of the Royal Statistical Society: Series C (Applied Statistics)*, **49**, 423–440.
[26] Burnham, K. P. and Anderson, D. R. (1998). *Model Selection and Inference – A Practical Information-Theoretic Approach*. New York: Springer-Verlag.
[27] Carlin, B. P. and Chib, S. (1995). Bayesian model choice via Markov chain Monte Carlo. *Journal of Royal Statistical Society – Series B*, **57**, 473–484.
[28] Casella, G. and Berger, R. L. (1987). Reconciling Bayesian and frequentist evidence in the one-sided testing problem. *Journal of the American Statistical Association*, **82**, 106–111.
[29] Chen, M. H. (2005). Bayesian computations, from posterior densities to Bayes factors, marginal likelihoods and posterior model probabilities. In *Handbook of Statistics, Volume 25: Bayesian Thinking, Modeling and Computation* (Dipak K. Dey, C. R. Rao, eds.), 437–458. North Holland.
[30] Chib, S. and Jeliazkov, I. (2001). Marginal likelihood from the Metropolis–Hastings output. *Journal of the American Statistical Association*, **96**, 270–281.
[31] Chipman, H., George, E. I. and McCulloch, R. (2001). The practical implementation of Bayesian Model Selection. In *Model Selection* (P. Lahiri, ed.), 67–116. IMS Lecture Notes Volume 38.
[32] Chopin, N. and Robert, C. P. (2010). Properties of nested sampling. *Biometrika*, **97**, 741–755.
[33] Clyde, M. (1999). Bayesian Model Averaging and Model Search Strategies (with discussion). In *Bayesian Statistics 6*. (J. M. Bernardo, J. O. Berger, A. P. Dawid and A. F. M. Smith, eds.), pp. 157–185. Oxford University Press.
[34] Clyde, M. and George, E. I. (1999). Empirical Bayes Estimation in Wavelet Nonparametric Regression. In *Bayesian Inference in Wavelet-Based Models* (P. Muller and B. Vidakovic, eds.), 309–322. Springer-Verlag.
[35] Clyde, M. A. and George, E. (2004). Model uncertainty. *Stastical Science*, **19**, pp. 81–94.
[36] Clyde, M. and Parmigiani, G. (1996). Orthogonalizations and Prior Distributions for Orthogonalized Model Mixing. In *Modelling and Prediction: Honoring Seymour Geisser* (J. C. Lee, W. O. Johnson, and A. Zellner, eds.), 206–227, Springer-Verlag, New York.

[37] Clyde, M., DeSimone, H. and Parmigiani, G. (1996). Prediction via orthogonalized model mixing. *Journal of the American Statistical Association*, **91**, 1197–1208.

[38] Clyde, M., Ghosh, J. and Littman, M. (2011). Bayesian adaptive sampling for variable selection and model averaging. *Journal of Computational and Graphical Statistics*, **20**, 80–101.

[39] Cripps, E., Carter, C. and Kohn, R. (2005). Variable selection and covariance selection in multivariate regression models. In *Handbook of Statistics, Volume 25: Bayesian Thinking, Modeling and Computation* (Dipak K. Dey, C. R. Rao, eds.), 519–552. Elsevier.

[40] Cui, W. and George, E. I. (2008). Empirical Bayes vs. fully Bayes variable selection. *Journal of Statistical Planning and Inference*, **138**, 888–900.

[41] Czado, C. (1997). On selecting parametric link transformation families in generalized linear models. *Journal of Statistical Planning and Inference*, **61**, 125–139.

[42] Dass, S. C. (2001). Unified Bayesian and Conditional Frequentist Testing Procedures for Discrete Distributions, *Sankhya Ser. B*, **63**, 251–269.

[43] Dass, S. and Berger, J. (2003). Unified Bayesian and conditional frequentist testing of composite hypotheses. *Scandinavian Journal of Statistics*, **30**, 193–210.

[44] Dass, S. C. and Lee, J. (2004). A note on the consistency of Bayes factors for testing point null versus non-parametric alternatives. *Journal of Statistical Planning and Inference*, **119**, 143–152.

[45] de Finetti, B. (1975). *Theory of Probability* (English edition). New York: Wiley.

[46] Dellaportas, P., Forster, J. J. and Ntzoufras, I. (2002). On Bayesian Model and Variable Selection Using MCMC. *Statistics and Computing*, **12**, 27–36.

[47] De Santis, F. and Spezzaferri, F. (1997). Alternative Bayes factors for model selection. *Canadian Journal of Statistics* **25**, 503–515.

[48] De Santis, F. and Spezzaferri, F. (2001). Consistent fractional Bayes factor for linear models. *Journal of Statistical Planning and Inference*, **97**, 305–321.

[49] de Vos, A. F. (1993). A fair comparison between regression models of different dimension. Technical Report, The Free University, Amsterdam.

[50] Dmochowski, J. (1996). Intrinsic priors via Kullback–Liebler geometry. In *Bayesian Statistics 5* (J. M. Bernardo, J. O. Berger, A. P. Dawid and A. F. M. Smith, eds.), 543–550. London: Oxford University Press.

[51] Draper, D. (1995). Assessment and Propagation of Model Uncertainty. *Journal of the Royal Statistical Society, Ser. B*, **57**, 45–98.

[52] Draper, D. and Krnjajic, M. (2010). Calibration results for Bayesian model specification. *Bayesian Analysis*, **1**, 1–43.

[53] Dudley, R. M. and Haughton, D. (1997). Information criteria for multiple data sets and restricted parameters. *Statistica Sinica*, **7**, 265–284.

[54] Dutta, R., Bogdan, M. and Ghosh, J. K. (2012). Model Selection and Multiple Testing – A Bayesian and Empirical Bayes Overview and some New Results. *Journal of Indian Statistical Association*, Golden Jubilee Year (Volume 50).

[55] Gelfand, A. E. and Dey, D. (1994). Bayesian model choice: asymptotic and exact calculations. *Journal Royal Statistical Society B*, **56**, 501–514.

[56] Gelfand, A. E. and Ghosh, S. K. (1998). Model Choice: A Minimum Posterior Predictive Loss Approach. *Biometrika*, **85**, 1–11.

[57] Gelfand, Alan E. and Smith, Adrian F. M. (1990). Sampling-based approaches to calculating marginal densities. *Journal of the American Statistical Association*, **85**, 398–409.

[58] Gelman, A., Carlin, J. B., Stern, H. and Rubin, D. B. (1995). *Bayesian Data Analysis*. London: Chapman and Hall.

[59] George, E. I. and Foster, D. P. (2000). Calibration and Empirical Bayes Selection. *Biometrika*, **87**, 731–748.

[60] George, E. I. and McCulloch, R. E. (1993). Variable selection via Gibbs sampling. *Journal of the American Statistical Society*, **88**, 881–889.

[61] George, E. I. and McCulloch, R. (1997). Approaches for Bayesian Variable Selection. *Statistica Sinica*, **7**, 339–374.
[62] Geisser, S. (1993). *Predictive Inference: An Introduction*. New York: Chapman & Hall.
[63] Ghosh, J. K. and Ramamoorthi, R. V. (2003). *Bayesian Nonparametrics*. New York: Springer.
[64] Ghosh, J. K. and Samanta, T. (2001). Model selection – an overview. *Current Science*, **80**, 1135–1144.
[65] Ghosh, J. K. and Samanta, T. (2002). Nonsubjective Bayes testing – an overview. *Journal of Statistical Planning and Inference*, **103**, 205–223.
[66] Godsill, S. J. (2001). On the relationship between Markov chain Monte Carlo methods for model uncertainty. *Journal of Computational Graphics and Statistics*, **10**, 230–248.
[67] Good, I. J. (1950). *Probability and the Weighing of Evidence*. London: Charles Griffin.
[68] Goutis, C. and Robert, C. P. (1998). Model choice in generalized linear models: a Bayesian approach via Kullback–Leibler projections. *Biometrika*, **85**, 29–37.
[69] Green, P. (1995). Reversible jump Markov chain Monte Carlo computation and Bayesian model determination. *Biometrika*, **82**, 711–732.
[70] Gupta, M. and Ibrahim, J. (2009). An information matrix prior for Bayesian analysis in generalized linear models with high dimensional data. *Statistica Sinica*, **19**, 1641–1663.
[71] Guttman, I. (1967). The use of the concept of a future observation in goodness-of-fit problems. *Journal of the Royal Statistical Society Ser. B Stat. Methodol*, **29**, 83–100.
[72] Han, C. and Carlin, B. P. (2001). Markov Chain Monte Carlo Methods for Computing Bayes Factors: A Comparative Review. *Journal of the American Statistical Association*, **96**(455), 1122–1132.
[73] Hans, C., Dobra, A. and West, M. (2007). Shotgun stochastic search for "large p" regression. *Journal of the American Statistical Association*, **102**, 507–516.
[74] Hoeting, J. A., Madigan, D., Raftery, A. E. and Volinsky, C. T. (1999). Bayesian model averaging: a tutorial. *Statistical Science*, **14**(4), 382–417.
[75] Hoijtink, H., Klugkist, I. and Bollen, P. A. (2008). *Bayesian Evaluation of Informative Hypotheses*. New York: Springer.
[76] Ibrahim, J. and Laud, P. (1994). A predictive approach to the analysis of designed experiments. *Journal of the American Statistical Association*, **89**, 309–319.
[77] Ibrahim, J. G., Chen, M. H. and MacEachern, S. N. (1999). Bayesian variable selection for proportional hazards models. *Canadian Journal of Statistics*, **27**, 701–717.
[78] Iwaki, K. (1997). Posterior Expected Marginal Likelihood for Testing Hypotheses. *Journal of Economics, Asia University*, **21**, 105–134.
[79] Jeffreys, H. (1961). *Theory of Probability*. London: Oxford University Press.
[80] Jefferys, W. H. and Berger, J. O. (1992). Ockham's razor and Bayesian analysis. *American Scientist*, **80**, 64–72.
[81] Johnstone, I. M. and Silverman, B. W. (2004). Needles and hay in haystacks: empirical Bayes estimates of possibly sparse sequences. *Annals of Statistics*, **32**, 1594–1649.
[82] Johnstone, I. M. and Silverman, B. W. (2005). Empirical Bayes selection of wavelet thresholds. *Annals of Statistics*, **33**, 1700–1752.
[83] Kadane, J. B. (1980). Predictive and Structural Methods for Eliciting Prior Distributions. In *Studies in Bayesian Econometrics and Statistics in Honor of Harold Jeffreys* (A. Zellner, Eds.), 89–93. North Holland Publishing Company.
[84] Kass, R. E. and Raftery, A. (1995). Bayes factors. *Journal of the American Statistical Association*, **90**, 773–795.
[85] Kass, R. E. and Vaidyanathan, S. K. (1992). Approximate Bayes Factors and Orthogonal Parameters, with Application to Testing Equality of Two Binomial Proportions. *Journal of the Royal Statistical Society*, **54**, 129–144.

[86] Kass, R. E. and Wasserman, L. (1995). A Reference Bayesian Test for Nested Hypotheses and Its Relationship to the Schwarz Criterion. *Journal of the American Statistical Association*, **90**, 928–934.
[87] Key, J. T., Pericchi, L. R. and Smith, A. F. M. (1999). Bayesian Model Choice: What and Why? *Bayesian Statistics* 6 (J. M. Bernardo, J. O. Berger, A. P. Dawid, and A. F. M. Smith, eds.), pp. 343–370. Oxford University Press.
[88] Kiefer, J. (1977). Conditional confidence statements and confidence estimators (with discussion). *Journal of the American Statistical Association*, **72**, 789–827.
[89] Ley, E. and Steel, M. F. J. (2009). On the effect of prior assumptions in Bayesian model averaging with applications to growth regression. *Journal of Applied Econometrics*, **24**, 651–674. Wiley Online InterScience. DOI: 10.1002/jae.1057.
[90] Liang, F., Paulo, R., Molina, G., Clyde, M. A. and Berger, J. O. (2008). Mixtures of g Priors for Bayesian Variable Selection. *Journal of the American Statistical Association*, **103**, 410–423.
[91] Lindley, D. V. (1957). A statistical paradox. *Biometrika*, **44**, 187–192.
[92] Lingham, R. and Sivaganesan, S. (1999). Intrinsic Bayes factor approach to a test for the power law process. *Journal of Statistical Planning and Inference*, **77**, 195–220.
[93] Marin, J. M. and Robert, C. P. (2007). *Bayesian Core: A Practical Approach to Computational Bayesian Statistics*. New York: Springer.
[94] Maruyama, Y. and George, E. I. (2011). A fully Bayes factor with a generalized g-prior. *Annals of Statistics*, **5**, 2740–2765.
[95] Maruyama, Y. and Strawderman, W. E. (2011). Robust Bayesian variable selection with sub-harmonic priors. Technical report, Center for Spatial Information Science, University of Tokyo. arXiv:1009.1926v2 [stat.ME].
[96] McDonald, G. C., Vance, L. C. and Gibbons, D. I. (1995). Some tests for discriminating between lognormal and Weibull distributions – an application to emission data. In *Recent Advances in Life Testing and Reliability* (N. Balakrishnan, ed.), pp. 475–490 (Chapter 25), CRC Press. Inc. (Boca Raton).
[97] Moreno, E., Bertolino, F. and Racugno, W. (1998). An intrinsic limiting procedure for model selection and hypothesis testing. *Journal of the American Statistical Association*, **93**, 1451–1460.
[98] Moreno, E., Bertolino, F. and Racugno, W. (1999). Default Bayesian analysis of the Behrens–Fisher problem. *Journal of Statistical Planning and Inference*, **81**, 323–333.
[99] Moreno, E., Giron, F. J. and Casella, G. (2010). Consistency of objective Bayes factors as the model dimension grows. *Annals of Statistics*, **38**, 1937–1952.
[100] Neal, R. (2001). Transferring prior information between models using imaginary data. Technical Report 0108, Dept. Statistics, University of Toronto.
[101] O'Hagan, A. (1995). Fractional Bayes factors for model comparisons. *Journal of the Royal Statistical Society, Ser. B*, **57**, 99–138.
[102] O'Hagan, A. (1997). Properties of intrinsic and fractional Bayes factors. *Test*, **6**, 101–118.
[103] Pauler, D. (1998). The Schwarz Criterion and Related Methods for Normal Linear Models. *Biometrika*, **85**, 13–27.
[104] Paulo, R. *et al.* (2002). Notes on model selection for the normal multiple regression model. Model selection group of the 2002 stochastic computation SAMSI Program. SAMSI.
[105] Pérez, J. M. and Berger, J. (2001). Analysis of mixture models using expected posterior priors, with application to classification of gamma ray bursts. In *Bayesian Methods, with applications to science, policy and official statistics* (E. George and P. Nanopoulos, eds.), 401–410. Official Publications of the European Communities, Luxembourg.
[106] Pérez, J. M. and Berger, J. (2002). Expected posterior prior distributions for model selection. *Biometrika*, **89**, 491–512.
[107] Raftery, A. E. (1999). Bayes factors and BIC–Comment on "A critique of the Bayesian information criterion for model selection". *Sociological Methods and Research*, **27**, 411–427.

[108] Raftery, A. E., Madigan, D. and Hoeting, J. A. (1997). Bayesian model averaging for regression models. *Journal of the American Statistical Association*, **92**, 179–191.
[109] Raftery, A. E., Newton, M. A., Satagopan, J. and Krivitsky, P. (2007). Estimating the integrated likelihood via posterior simulation using the harmonic mean identity (with discussion), *Bayesian Statistics 8* (J. M. Bernardo, M. J. Bayarri, J. O. Berger, A. P. Dawid, D. Heckerman, A. F. M. Smith, and M. West, eds.), pp. 1–45. Oxford University Press.
[110] Robert, C. P. (2007). *The Bayesian Choice: From Decision-Theoretic Foundations to Computational Implementation.* Springer Texts in Statistics.
[111] Robins, J. M., van der Vaart, A. W. and Ventura, V. (2000). The asymptotic distribution of p-values in composite null models [with discussion]. *Journal of the American Statistical Association*, **95**, 1143–1172.
[112] Rubin, D. B. (1984). Bayesianly Justifiable and Relevant Frequency Calculations for the Applied Statistician. *The Annals of Statistics*, **12**, 1151–1172.
[113] Rue, H., Martino, S. and Chopin, N. (2009). Approximate Bayesian inference for latent Gaussian models by using integrated nested Laplace approximations. *Journal of the Royal Statistical Society B*, **71**, 319–392.
[114] Sabanés Bové, D. and Held, L. (2011). Hyper-g priors for generalized linear models. *Bayesian Analysis*, **6**, 1–24.
[115] Sansó, B., Pericchi, L. R. and Moreno, E. (1996). On the Robustness of the Intrinsic Bayes Factor for Nested Models. In *Bayesian Robustness* Volume 29 (J. Berger, F. Ruggeri, and L. Wasserman, eds.), pp. 157–176. Hayward: IMS Lecture Notes – Monograph series.
[116] Schluter, P. J., Deely, J. J. and Nicholson, A. J. (1999). The averaged Bayes factor: a new method for selecting between competing models. Technical Report, University of Canterbury.
[117] Schwarz, G. (1978). Estimating the dimension of a model. *Annals of Statistics*, **6**, 461–464.
[118] Scott, J. and Berger, J. (2005). An exploration of aspects of Bayesian multiple testing. *Journal of Statistical Planning and Inference*, **136**, 2144–2162.
[119] Scott, J. and Berger, J. (2010). Bayes and Empirical-Bayes multiplicity adjustment in the variable-selection problem. *Annals of Statistics*, **38**, 2587–2619.
[120] Sellke, T., Bayarri, M. J. and Berger, J. (2001). Calibration of p-values for testing precise null hypotheses. *The American Statistician*, **55**, 62–71.
[121] Shyamalkumar, D. N. (1996). Cyclic I_0 projections and its applications in Statistics. *Technical Report 96-24*, Purdue University, Department of Statistics.
[122] Sivia, D. S. and Skilling, J. (2006). *Data Analysis: A Bayesian Tutorial*, Second Edition. Oxford University Press.
[123] Skilling, J. (2006). Nested Sampling for General Bayesian Computation. *Bayesian Analysis 1*, pp. 833–860.
[124] Smith, M. and Kohn, R. (1996). Nonparametric regression using Bayesian variable selection. *Journal of Econometrics*, **75**, 317–344.
[125] Smith, A. F. M. and Spiegelhalter, D. J. (1980). Bayes Factors and Choice Criteria for Linear Models. *Journal of the Royal Statistical Society Ser. B*, **42**, 213–220.
[126] Spiegelhalter, D. J. and Smith, A. F. M. (1982). Bayes Factors for Linear and Log-Linear Models with Vague Prior Information. *Journal of the Royal Statistical Society, Ser. B*, **44**, 377–387.
[127] Spiegelhalter, D. J., Best, N. J., Carlin, B. P. and van der Linde, A. (2002). Bayesian measures of model complexity and fit (with discussion). *Journal of the Royal Statistical Society B*, **64**, 583–639.
[128] Tokdar, S. T., Chakrabarti, A. and Ghosh, J. K. (2010). Bayesian nonparametric goodness of fit. In *Frontier of Statistical Decision Making and Bayesian Analysis* (M. H. Chen, D. K. Dey, P. Muller, D. Sun and K. Ye, eds.), 185–193.

[129] Varshavsky, J. A. (1996). Intrinsic Bayes factors for model selection with autoregressive data. In *Bayesian Statistics 5* (J. M. Bernardo, J. O. Berger, A. P. Dawid, and A. F. M. Smith, eds.), pp. 757–763. London: Oxford University Press.

[130] Verdinelli, I. and Wasserman, L. (1996). Bayes Factors, Nuisance Parameters, and Imprecise Tests. In *Bayesian Statistics 5* (J. M. Bernardo, J. O. Berger, A. P. Dawid, and A. F. M. Smith, eds.), pp. 765–771. London: Oxford University Press.

[131] Wang, X. and George, E. I. (2007). Adaptive Bayesian criteria in variable selection for generalized linear models. *Statistica Sinica*, **17**, 667–690.

[132] Wilson, M. A., Iversen, E. S., Clyde, M. A., Schmidler, S. C. and Schildkraut, J. M. (2010). Bayesian model search and multilevel inference for SNP association studies. *Annals of Applied Statistics*, **4**, 1342–1364.

[133] Zak-Szatkowska, M. and Bogdan, M. (2011). Modified versions of Bayesian Information Criterion for sparse Generalized Linear Models. *Computational Statistics and Data Analysis*, **55**, 2908–2924.

[134] Zellner, A. (1986). On Assessing Prior Distributions and Bayesian Regression Analysis with g-prior Distributions. In A. Zellner, ed., *Bayesian Inference and Decision Techniques: Essays in Honor of Bruno de Finetti*, pp. 389–399. Edward Elgar Publishing Limited.

[135] Zellner, A. and Siow, A. (1980). Posterior odds for selected regression hypotheses. In *Bayesian Statistics* (J. M. Bernardo, M. H. DeGroot, D. V. Lindley and A. F. M. Smith, eds.), pp. 585–603. Valencia Univ. Press.

[136] Zhang, Z., Jordan, M. I. and Yeung, D. Y. (2009). Posterior consistency of the Silverman g-prior in Bayesian model choice. In *Proceedings of Advances in Neural Information Processing Systems (NIPS)* (Koller, D., Bengio, Y., Schuurmans, D., and Bottou, L., eds.), Volume 21, 1969–1976.

19 Proper and non-informative conjugate priors for exponential family models

E. GUTIÉRREZ-PEÑA AND M. MENDOZA

19.1 Introduction

Exponential families constitute an important class of probability models that occur, in one form or another, as part of more complex models widely used in applied statistics such as generalized linear models, hierarchical models and dynamic models. Therefore, it is of some importance to understand their properties. These families are related to the notion of sufficiency and can also be motivated as a set of solutions to certain maximum entropy problems ([1],[20],[17]). On the other hand, conjugate distributions play an important role in the Bayesian approach to parametric inference. One of the main features of such families is that they are closed under sampling, but a conjugate family often provides prior distributions which are tractable in various other respects. Indeed, conjugate families for exponential family models are themselves exponential families so, in particular, they can also be regarded as solutions to maximum entropy problems.

Maximum entropy (or, more generally, minimum relative entropy) is known to be closely related to the decision theoretical problem of minimizing a worst-case expected loss (see [17] and the references therein). Incidentally, Jeffreys' prior can be shown to be asymptotically least favourable under entropy risk [3]. In the context of exponential families, it is also noteworthy that, under certain conditions, Jeffreys' and other non-informative priors—including some forms of 'unbiased' priors—can be obtained as suitable limits of conjugate distributions. Moreover, there exists an interesting duality between unbiased estimators and optimal Bayes estimators that minimize expected risk ([27]). The aim of this paper is to discuss these various concepts and to highlight the relationship between them.

In the next section we briefly review some basic concepts concerning exponential families and information theory. Section 19.3 discusses Bayesian inference for exponential families based both on proper and certain non-informative, improper conjugate priors. In Section 19.4, we take a look at more general versions of these latter priors and discuss an interesting unbiasedness property of maximum likelihood estimators. Possible extensions of these and related results are also pointed out. Finally, Section 19.5 contains some concluding remarks.

19.2 Preliminaries

19.2.1 Exponential families

We first review some basic results concerning natural exponential families. See [1] for a comprehensive account of the properties of these models. Let η be a σ-finite positive measure on the Borel sets of $\mathbf{R}^d$, and consider the family $\mathcal{F}$ of probability measures whose density with respect to η is of the form

$$p(x|\theta) = b(x)\ \exp\{x^t\theta - M(\theta)\} \quad \theta \in \Xi$$

for some function $b(\cdot)$, where $M(\theta) = \log \int b(x)\exp\{x^t\theta\}\,\eta(dx)$ and $\Xi = \text{int}\,\Phi$, with $\Phi = \{\theta \in \mathbf{R}^d : M(\theta) < +\infty\}$. We assume that $b(x)\,\eta(dx)$ is not concentrated on an affine subspace of $\mathbf{R}^d$, and that Ξ is not empty. The family $\mathcal{F}$ is called a natural exponential family (with canonical parameter θ) and is said to be regular if Φ is an open subset of $\mathbf{R}^d$. In this paper we shall only be concerned with regular natural exponential families. The function $M(\theta)$, called the cumulant transform of $\mathcal{F}$, is convex and infinitely differentiable.

The mapping $\mu = \mu(\theta) = \partial M(\theta)/\partial\theta$ is one-to-one and differentiable, with inverse $\theta = \theta(\mu)$, and provides an alternative parameterization of $\mathcal{F}$, called the mean parameterization since $\mu = E(X|\theta)$. The set $\Omega = \mu(\Xi)$ is termed the mean parameter space. The function

$$V(\mu) = \frac{\partial^2 M\{\theta(\mu)\}}{\partial\theta^2} \qquad \mu \in \Omega$$

is called the variance function of $\mathcal{F}$. An important property of the variance function is that the pair $(V(\cdot), \Omega)$ characterizes $\mathcal{F}$ (see, for example, [26]).

The standard conjugate family for the natural exponential family $\mathcal{F}$ has densities (with respect to the Lebesgue measure on the Borel sets of $\mathbf{R}^d$) of the form

$$p(\theta|x_0, n_0) = H(x_0, n_0)\ \exp\{n_0 x_0^t\,\theta - n_0 M(\theta)\}$$

Some fundamental properties of this conjugate family are described by Diaconis and Ylvisaker ([5]). We shall refer to it as the DY-conjugate family.

19.2.2 Information theory and the minimum relative entropy principle

We now discuss some fundamental concepts of information theory (for details, see [4]). The entropy of a random variable X is defined by $H(X) = -\int_{\mathcal{X}} p(x)\log p(x)\eta(dx) = -E_p[\log p(X)]$. Intuitively, the entropy describes the uncertainty inherent in the distribution $p(x)$. Thus, for example, the entropy of a discrete distribution with finite support $p(x) = p_x$ $(p_x > 0, x = 1, 2, \ldots, k; \sum_x p_x = 1)$ is maximized when $p(x)$ is the uniform distribution, i.e. $p_x = 1/k$ for all $x = 1, 2, \ldots, k$. Unfortunately, in the continuous case the entropy (known as differential entropy in that case) does not share all the properties of the discrete version. The relative entropy (also known as Kullback–Leibler divergence [22]) addresses this issue by introducing an invariant measure $q(x)\eta(dx)$ with respect to which the entropy is computed:

$$D_{KL}(p(\cdot)\,||\,q(\cdot)) = \int p(x)\log\left\{\frac{p(x)}{q(x)}\right\}\eta(dx)$$

The maximum entropy principle ([20]) states that, subject to precisely stated prior information (usually in the form of moment constraints), the probability distribution that best represents the current state of knowledge about X is the one with the largest entropy. For continuous distributions, the relative entropy is minimized instead, leading to the minimum relative entropy principle[16] (also known as the Principle of Minimum Discrimination Information, [22]).

Another important quantity in information theory is the mutual information between the random variables X and Y:

$$\begin{aligned} I(X,Y) &= \int\int p(x,y)\log\left\{\frac{p(x,y)}{p(x)p(y)}\right\}\eta(dx)\eta(dy) \\ &= \int p(x)\left\{\int p(y|x)\log\left\{\frac{p(y|x)}{p(y)}\right\}\eta(dy)\right\}\eta(dx) \\ &= \int p(x)\left\{D_{KL}(p(y|x)\,||\,p(y))\right\}\eta(dx) \end{aligned}$$

$I(X,Y)$ describes the amount of information that the random variable Y carries about the random variable X (and vice versa). It is a measure of the dependence between the two random variables X and Y. From its definition, it is easy to see that $I(X,Y) = I(Y,X)$ and that $I(X,Y) = 0$ if and only if X and Y are independent. Another important property of $I(X,Y)$ is that it is a concave functional of $p(x)$ for fixed $p(y|x)$, and a convex functional of $p(y|x)$ for fixed $p(x)$ ([4]).

19.3 Bayesian parametric inference

From a Bayesian perspective, a problem of parametric inference is one where a phenomenon is to be described through the observation of a random variable X whose distribution function is completely specified up to the unknown value of a finite-dimensional parameter θ. Under such circumstances, the initial state of knowledge is described through a joint probability model $p(x,\theta)$, thus recognizing that uncertainty arises from two different sources: *variability* of X and *lack of information* regarding θ. This joint probability may be represented either as

$$p(x,\theta) = p(x|\theta)\,p(\theta)$$

or as

$$p(x,\theta) = p(x)\,p(\theta|x) \tag{19.1}$$

The first representation is the most common since then $p(x|\theta)$, the *sampling model*, describes the uncertainty about the observable X for a fixed value of θ, whereas the *prior*, $p(\theta)$, describes the state of knowledge regarding the value of the parameter. Thus, as a first step towards solving the inference problem, an appropriate model $p(x,\theta)$ must be chosen. This selection is usually accomplished sequentially: first the sampling model and then the prior distribution.

[16] We note that [17] define the relative entropy as $-D_{KL}(p(\cdot)\,||\,q(\cdot))$, so minimization of D_{KL} can be stated as the 'maximum relative entropy principle' in that case.

19.3.1 Sampling model

Following [2] (Section 4.5.4), let us assume that there exists a known measure $b(x)$ which might be regarded as a first rough approximation to the true sampling model, $q(x)$. Assume also that there are some conditions which are not satisfied by $b(x)$ but are useful to describe $q(x)$. If these conditions can be stated as moment constraints of the form

$$\int h_i(x)q(x)dx = \mu_i; \quad i = 1, \ldots, k$$

where the $h_i(\cdot)$ are known functions and the μ_i are given constants $(i = 1, \ldots, k)$, then the problem of selecting an appropriate sampling model (i.e. an approximation to the true model $q(x)$) may be posed as that of looking for a distribution, as close as possible to $b(x)$, but satisfying the above constraints. If, as a measure of closeness, we use the Kullback–Leibler divergence discussed in Section 19.2, then the optimal model belongs to the exponential family

$$p(x|\theta) = b(x) \exp\{h(x)^t\theta - M(\theta)\}$$

where θ is the k-dimensional canonical parameter and $h(x)$ is the corresponding vector of sufficient statistics. For the sake of simplicity, and without loss of generality, we can assume that the observation is the sufficient statistic so that the exponential family can be written in its natural form

$$p(x|\theta) = b(x) \exp\{x^t\theta - M(\theta)\} \tag{19.2}$$

Thus, we may conclude that under a variety of conditions an exponential family may be used as an optimal approximation to the sampling model.

Example. Suppose that a phenomenon will be observed, leading to a discrete random variable taking values in $\mathcal{X} = \{0, 1, 2, \ldots\}$. If, as a first rough approximation, we take the counting measure $b(x) = 1/x!$ and use the first moment to describe the true sampling model, then the appropriate constraint would be

$$\int x\, q(x)\, dx = \mu$$

Minimization of the relative entropy then leads to the solution

$$p(x|\theta) = \frac{1}{x!} \exp\{x\theta - \exp(\theta)\} \tag{19.3}$$

which corresponds to a Poisson distribution. This is usually parameterized in terms of the mean parameter $\mu = \exp(\theta)$.

19.3.2 Prior

19.3.2.1 *Conjugate priors*

Concerning the uncertainty about the parameter θ, it can be argued that, since θ can be thought of as the limit of a sequence of observable quantities, some information about the observables may be used to elicit a prior on θ. As before, we assume that this prior information can be roughly approximated by means of a base measure $B(\theta)$. If additional prior information can be expressed in terms of constraints regarding the expected value of the functions $g_1(\theta) = \theta$ and $g_2(\theta) = -M(\theta)$, then the optimal approximation to $B(\theta)$ satisfying the constraints is given by a member of the exponential family

$$p(\theta) \propto B(\theta) \exp\{n_0 x_0^t \theta - n_0 M(\theta)\} \tag{19.4}$$

where $\psi = (x_0, n_0)^t$ is the canonical *hyperparameter*. This family is conjugate for the sampling model (19.2). Conjugate families of priors have received a great deal of attention in the statistical literature since their introduction by [28]. One of the main features of such a family is that it is closed under sampling, in the sense that formal updating of a prior in the conjugate family via Bayes theorem yields a posterior distribution which also belongs to that family. In fact, given a sample $x_{(n)} = (x_1, \ldots, x_n)$ from $p(x|\theta)$, the posterior distribution is given by

$$p(\theta|x_{(n)}) \propto B(\theta) \exp\{(n\bar{x} + n_0 x_0)^t \theta - (n + n_0)M(\theta)\}$$

Besides this property, conjugate families often provide prior distributions which are tractable in at least two other respects: (i) for many exponential family likelihoods the normalizing constant of the conjugate density is readily found; (ii) it is often possible to express in convenient form the expectations (and, in some cases, higher-order moments) of some important functions of the parameters ([12]).

It is worth noting that [5] discuss conditions that ensure that the conjugate density (19.4) defines a proper distribution for θ (when the base measure $B(\theta)$ is taken as uniform). In Section 19.3.2.2 we shall return to the choice of $B(\theta)$, specifically in the context of *non-informative* priors.

Example. (*continued*). In the case of the Poisson model (19.3), we have $\theta = \log \mu$. Thus, the corresponding constraints for the prior are

$$\int \theta p(\theta) d\theta = m_1, \quad \int \exp(\theta) p(\theta) d\theta = m_2$$

Equivalently, in terms of μ,

$$\int \log \mu p(\mu) d\mu = m_1, \quad \int \mu p(\mu) d\mu = m_2$$

In this case (19.4) becomes

$$p(\theta) \propto B(\theta) \exp\{n_0 x_0 \theta - n_0 \exp(\theta)\}$$

which, under the corresponding transformation, leads to

$$p(\mu) \propto B_\mu(\mu) \mu^{s_0} \exp\{-n_0 \mu\}$$

with $s_0 = n_0 x_0$.

It is clear that if we use, as a first approximation, the improper base measure $B_\mu(\mu) = \mu^{-1}$, then we get the usual Gamma conjugate prior on μ. We note that, written in terms of θ, this measure is $B(\theta) \propto 1$.

19.3.2.2 *Some non-informative conjugate priors*

In the context of Bayesian inference, the need often arises to specify a prior distribution such that, even for moderate sample sizes, the information provided by the data dominates the prior because of the 'vague' nature of the prior knowledge about the parameter. Here we explore the case where the base measure $B(\theta)$ is chosen to be non-informative in some sense.

Uniform prior A (naive) prior often used as non-informative is the uniform prior $\pi_L(\theta) \propto 1$, also known as the Laplace prior. If we use $B(\theta) = \pi_L(\theta)$ as the base measure we get

$$p(\theta) \propto \exp\{n_0 x_0^t \theta - n_0 M(\theta)\}$$

the DY-conjugate family. This family has a number of important properties. In particular, if we reparameterize in terms of the mean parameter μ, it can be shown that

$$E(\mu | x_{(n)}) = \frac{n\bar{x} + n_0 x_0}{n + n_0}$$

so the Bayes estimator under squared error loss is a linear function of the sample mean. Moreover, $\hat{\mu} = E(\mu | x_{(n)})$ is also the Bayes estimator under the Kullback–Leibler loss ([11]). This result is particularly interesting since

$$E(\mu | x_{(n)}) = \bar{x}$$

when $x_0 = 0$ and $n_0 = 0$. The coincidence of the Bayes estimate with the usual frequentist unbiased estimator for the mean in this limiting case, shows that the idea of choosing the *equivalent sample size* (in the terminology of [28]) $n_0 = 0$ to define a non-informative conjugate prior, is most natural when the base measure is $\pi_L(\theta)$.

Jeffreys' prior A widely used non-informative prior is that obtained through the so-called Jeffreys' rule: $\pi_J(\theta) \propto i_\theta(\theta)^{1/2}$, where $i_\theta(\theta)$ denotes the Fisher information for θ. If we take $B(\theta) = \pi_J(\theta)$ we get

$$p(\theta) \propto i_\theta(\theta)^{1/2} \exp\{n_0 x_0^t \theta - n_0 M(\theta)\} \tag{19.5}$$

which may be regarded as a restricted reference prior in the sense of [2]. There is an interesting relationship between Jeffreys' prior and the DY-conjugate family. A parametrization $\lambda = \lambda(\theta)$ is said to be conjugate for θ (denoted $\lambda \smile \theta$), if the Jacobian $J_\lambda(\theta) = |\partial\lambda(\theta)/\partial\theta|$ is such that

$$J_\lambda(\theta) \propto \exp\{k_1^t \theta - k_2 M(\theta)\}$$

for some constants k_1, k_2 ([15]). In the case of the mean parameterization μ, we have $J_\mu(\theta) = i_\theta(\theta)$. Provided that $\mu \smile \theta$ (a sufficient condition for this to hold is that the sampling distribution be a natural exponential family having a quadratic variance function), it can be shown that (19.5) becomes

$$\begin{aligned} p(\theta) &\propto \exp\left\{\left(n_0 x_0 + \frac{k_1}{2}\right)^t \theta - \left(n_0 + \frac{k_2}{2}\right) M(\theta)\right\} \\ &= \exp\left\{\tilde{s}_0^t \theta - \tilde{n}_0 M(\theta)\right\} \end{aligned}$$

This is again a DY-conjugate prior. Thus, in this case Jeffreys' prior can be seen as a limiting case of the DY-conjugate prior as $\tilde{s}_0 \to k_1/2$ and $\tilde{n}_0 \to k_2/2$. Moreover, n_0 retains its interpretation as an equivalent sample size, only relative to the Jeffreys prior, so $n_0 = 0$ can still be regarded as non-informative in this sense ([16]).

Unbiased priors The idea of looking for a prior leading to a posterior mean for μ which equals the corresponding sufficient unbiased estimator, can be explored in more general settings. To start with, μ is just one possible transformation of θ. Suppose that we are interested in another parameterization, say λ, and that we want to use a non-informative prior for it. Specifically, if there exists a sufficient unbiased estimator $\hat{\lambda}(x_{(n)})$, we would like to choose a base measure $B(\theta)$ such that the prior is non-informative for λ in the sense that the corresponding posterior satisfies $E(\lambda|x_{(n)}) = \hat{\lambda}(x_{(n)})$.

The problem of finding these *unbiased priors* has been addressed by [18], [13], and [25]. For an ample class of transformations $\lambda = \lambda(\theta)$ such that $\lambda \smile \theta$, if the sampling distribution belongs to a natural exponential family with quadratic variance function then the required base measure $B(\theta)$ exists and is given by

$$B(\theta) = i_\theta(\theta)|J_\lambda(\theta)|^{-1} \tag{19.6}$$

In this case,

$$\begin{aligned} p(\theta) &\propto \exp\left\{[n_0 x_0 + (k_1/2) - r_1]^t \theta - [n_0 + (k_2/2) - r_2] M(\theta)\right\} \\ &= \exp\left\{\breve{s}_0^t \theta - \breve{n}_0 M(\theta)\right\} \end{aligned}$$

where r_1 and r_2 are constants that depend on the specific choice of $\lambda(\cdot)$. It is interesting to note that the base measure $B(\theta)$ in (19.6) can also be seen as a limiting case of the DY-conjugate family as $\breve{s}_0 \to [(k_1/2) - r_1]$ and $\breve{n}_0 \to [(k_2/2) - r_2]$. Written in terms of λ, the prior induced by (19.6) then takes the form $\pi_\lambda(\lambda) \propto i_\lambda(\lambda)$.

This prior was originally proposed by Hartigan ([18]). Thus, we shall denote the unbiased priors on θ and λ by

$$\pi_H(\theta) \propto i_\theta(\theta)|J_\lambda(\theta)|^{-1}$$

and

$$\pi_H(\lambda) \propto i_\lambda(\lambda)$$

respectively.

This latter representation is particularly evocative since, when we express Jeffreys' and Laplace priors in terms of λ, we get

$$\pi_J(\lambda) \propto i_\lambda(\lambda)^{1/2}$$

and

$$\pi_L(\lambda) \propto 1$$

thus suggesting that different powers (between 0 and 1) of the Fisher information may be regarded as non-informative in some sense.

As in the previous case, n_0 retains its interpretation as an equivalent sample size, this time relative to the unbiased prior $\pi_H(\theta)$, so $n_0 = 0$ can still be regarded as non-informative in the sense that it yields an unbiased prior for λ. Unbiasedness of priors will be further discussed in Section 19.4.

19.3.3 Hierarchical models

The sequential approach described at the beginning of this section to motivate the use of conjugate priors for exponential families, can be carried on in order to specify models with more than two levels. A structure of this kind is called a hierarchical model. A large amount of literature has been devoted to the analysis of hierarchical models since their introduction by [10] and further development by [24].

In this new setting we have k conditionally independent random samples such that $X_{(n_i)}$ is drawn from

$$p(x|\theta_i) = b(x)\exp\{x^t\theta_i - M(\theta_i)\}$$

for $i = 1, \ldots, k$. Thus, all sampling models have the same form but differ in the value of the corresponding parameter θ_i. If each sample is processed separately, then we have k instances of the same exponential conjugate structure discussed before. Alternatively, if a hierarchical model is adopted then $\theta_1, \theta_2, \ldots, \theta_k$ are assumed to be exchangeable (thus allowing for possible dependence among the parameters) and a conditional parametric model $p(\theta|\psi)$ can be used to describe the variability among these second level parameters. Following the ideas of the previous section, $p(\theta|\psi)$ can be approximated by a member of an exponential conjugate family. Thus, $\theta_1, \theta_2, \ldots, \theta_k$ may be assumed to be i.i.d. according to the model

$$p(\theta|\psi) \propto B(\theta)\exp\{s_0^t\theta - n_0 M(\theta)\}$$

In the hierarchical setting, a third level is defined by treating the canonical hyperparameter $\psi = (s_0, n_0)^t$ as unknown and eliciting a prior distribution for it. This structure allows the computation of a posterior for ψ which incorporates the information from all k samples. More importantly, the corresponding posterior for θ_i also involves the information from $X_{(n_j)}$ $(j \neq i)$ due to the dependence among the second level parameters. This mechanism is known as *borrowing strength*.

Here, the relevant issue is that one more probability model $p(\psi|\nu)$ must be elicited. We can recognize $\sum_{i=1}^k \theta_i$ and $\sum_{i=1}^k M(\theta_i)$ as the sufficient statistics for ψ, so the corresponding moment constraints may be defined in terms of s_0 and n_0. In this way, the minimum relative entropy principle leads to a conjugate exponential family

$$p(\psi|\nu) \propto G(\psi)\exp\{\psi^t\nu\}$$

with respect to a given approximating measure $G(\psi)$.

In a hierarchical model such as this there are three sources of uncertainty: variability within each sample $X_{(n_i)}$, variability among the second level parameters $\theta_1, \ldots, \theta_k$, and lack of information regarding the hyperparameter ψ. The joint model is then $p(x_{(n_1)}, \ldots, x_{(n_k)}, \theta_1, \ldots, \theta_k, \psi)$, which is represented as

$$p(x_{(n_1)}, \ldots, x_{(n_k)}|\theta_1, \ldots, \theta_k)\, p(\theta_1, \ldots, \theta_k|\psi)\, p(\psi)$$

using the fact that $\{x_{(n_1)}, \ldots, x_{(n_k)}\}$ and ψ are conditionally independent given $\{\theta_1, \ldots, \theta_k\}$. Once the data become available, interest focuses on the posterior distribution $p(\theta_1, \ldots, \theta_k, \psi|x_{(n_1)}, \ldots, x_{(n_k)})$ given by

$$p(\theta_1, \ldots, \theta_k|\psi, x_{(n_1)}, \ldots, x_{(n_k)})\, p(\psi|x_{(n_1)}, \ldots, x_{(n_k)})$$

or, alternatively

$$p(\psi | \theta_1, \ldots, \theta_k) \, p(\theta_1, \ldots, \theta_k | x_{(n_1)}, \ldots, x_{(n_k)})$$

From these expressions it is clear that, in both cases, the first factor is a conjugate posterior but the second is a mixture of conditional conjugate posteriors which will not be conjugate in general. This suggests that an alternative approach to conjugacy in hierarchical models might be to: first elicit the second level distribution $p(\theta | \psi)$ as an exponential conjugate prior for the sampling model as above; then, with this prior, calculate the predictive distribution $p(x_{(n_1)}, \ldots, x_{(n_k)} | \psi)$ as

$$\int p(x_{(n_1)}, \ldots, x_{(n_k)} | \theta_1, \ldots, \theta_k) \, p(\theta_1, \ldots, \theta_k | \psi) \, d\theta_1 \cdots d\theta_k;$$

finally, choose $p(\psi)$ to be conjugate for this predictive distribution. In the context discussed here, however, this approach is not useful since the predictive will not generally result in an exponential family.

Note that, even for those components of the posterior distribution which are conjugate (and hence closed under sampling), simplification of the calculations involved in the prior–posterior analysis is not guaranteed. This will be illustrated in the following example. In practice, most calculations for hierarchical models involve analytical and numerical approximations. For example, [21] are concerned with Laplace approximations to posterior moments, whereas [7] and [8] discuss the application of MCMC methods in this context. It is worth noting that the Gibbs sampler is particularly suitable in this case. Another potentially relevant contribution is [6], where it is shown that, for certain classes of exponential families, Laplace approximations to marginal posterior densities are exact up to a proportional constant. Moreover, in practice it is often the case that, even when the Laplace approximation to a marginal density cannot be shown analytically to be exact, it provides a remarkably accurate approximation. Hence, Laplace approximations to marginal densities may still prove useful in the analysis of some hierarchical models.

Example. (*continued*). We now assume that we have a set of k (conditionally independent) random samples such that $X_{(n_i)}$ is drawn from a Poisson sampling model

$$p(x|\mu_i) = \frac{1}{x!} \mu_i^x \exp\{-\mu_i\}$$

For the second level, let $\theta_1, \theta_2, \ldots, \theta_k$ be i.i.d. according to the Gamma distribution

$$p(\mu|\psi) = \frac{\beta^\alpha}{\Gamma(\alpha)} \mu^{\alpha-1} \exp\{-\beta\mu\}$$

where $\psi = (\alpha, \beta)^t$. If, for the third level, we approximate the corresponding prior starting from an initial base measure $G(\psi)$ with constraints on the expected values of α and β, we get the exponential conjugate family

$$p(\psi | \mu_0) \propto G(\psi) \left\{ \frac{\beta^\alpha}{\Gamma(\alpha)} \right\}^{k_0} \mu_0^{\alpha-1} \exp\{-\beta\mu_0\} \tag{19.7}$$

Analytic calculations with the posterior corresponding to (19.7) are no longer possible in this case. Even for the conjugate likelihood prior discussed by [7], we can proceed with the conditional distribution for β (given α), which turns out to be Gamma, but the marginal distribution for α

is analytically intractable. Thus, we still have a family of distributions for ψ which is closed under sampling (of the θs) but we do not obtain any advantage for analytic calculations.

We close this section by recalling that hierarchical models provide a simple way of describing the relationship among the parameters of the second level. It must be pointed out, however, that since we specify the prior through the components $p(\theta_1, \ldots, \theta_k|\psi)$ and $p(\psi)$, this dependence is only implicitly defined and described by

$$p(\theta_1, \ldots, \theta_k) = \int p(\theta_1, \ldots, \theta_k|\psi)p(\psi)d\psi$$

where the conditional density $p(\theta_1, \ldots, \theta_k|\psi)$ is given by $\prod_{i=1}^{k} p(\theta_i|\psi)$. Thus, dependence among $\theta_1, \ldots, \theta_k$ is generated by a mixture of these 'independent' priors using $p(\psi)$ as the weight function. Moreover, the posterior

$$p(\theta_1, \ldots, \theta_k|x_{(n_1)}, \ldots, x_{(n_k)}) \propto p(x_{(n_1)}, \ldots, x_{(n_k)}|\theta_1, \ldots, \theta_k)p(\theta_1, \ldots, \theta_k)$$

will always keep the same (prior) source for its dependence structure since $p(x_{(n_1)}, \ldots, x_{(n_k)}|\theta_1, \ldots, \theta_k)$ is given by the product $\prod_{i=1}^{k} p(x_{(n_i)}|\theta_i)$ which implies independence among the second level parameters.

19.4 Unbiasedness

19.4.1 A general definition of unbiasedness

In Section 19.3.2.2 we discussed the case where some Bayes estimators $\hat{\lambda} = \hat{\lambda}(x_{(n)})$ turned out to be unbiased for certain functions $\lambda = \lambda(\theta)$ of the canonical parameter, in the sense that $E(\hat{\lambda}|\lambda) = \lambda$. This definition of unbiasedness is related to the squared error loss $L^2_{\lambda}(\hat{\lambda}, \lambda) \equiv (\hat{\lambda} - \lambda)^2$, as is the Bayes estimator $\hat{\lambda} = E(\lambda|x_{(n)})$, i.e. the posterior expectation of λ.

[23] proposed a generalization of this notion of unbiasedness which takes into account the loss function being used. For a general loss function $L(\hat{\lambda}, \lambda)$, we shall say that the estimator $\hat{\lambda} = \hat{\lambda}(x_{(n)})$ is L-unbiased for the parameter λ if

$$E(L(\hat{\lambda}(X_{(n)}), \lambda)|\lambda) \leq E(L(\hat{\lambda}(X_{(n)}), \lambda')|\lambda)$$

for all $\lambda, \lambda' \in \Lambda$. Similarly, we shall say that the estimator $\hat{\lambda}$ is (L, p)-Bayes (for a loss function L and a prior p) if it minimizes the corresponding posterior expected loss

$$\int L(\hat{\lambda}, \lambda)\, p(\lambda|x_{(n)})\, d\lambda$$

Using this general definition of unbiasedness, [18] showed that the prior which minimizes the asymptotic bias of the Bayes estimator relative to the loss function $L(\hat{\lambda}, \lambda)$ is given by

$$\pi_H(\lambda) \propto E\left(\left\{\frac{\partial \log p(X_{(n)}|\lambda)}{\partial \lambda}\right\}^2 \middle| \lambda\right) \left[\frac{\partial^2 L(\lambda, \omega)}{\partial \omega^2}\right]^{-1/2}_{\omega=\lambda}$$

$$= i_{\lambda}(\lambda) \left[\frac{\partial^2 L(\lambda, \omega)}{\partial \omega^2}\right]^{-1/2}_{\omega=\lambda}$$

For the squared error loss, $L^2_\lambda(\hat{\lambda}, \lambda)$, this becomes

$$\pi_H(\lambda) \propto i_\lambda(\lambda) \tag{19.8}$$

[18] also showed that, for exponential family models and squared error loss with respect to the mean parameter (i.e. $L^2_\mu(\hat{\mu}, \mu) = (\hat{\mu} - \mu)^2$), the unbiasedness of $\pi_\mu(\mu) \propto i_\mu(\mu)$ is exact. This prior can be written as

$$\pi_\mu(\mu) \propto \frac{1}{V(\mu)}$$

while in terms of the canonical parameter it becomes

$$\pi_\theta(\theta) \propto 1$$

Hence, the (L^2_μ, π_μ)-Bayes estimator $\hat{\mu} = E(\mu|x_{(n)})$ is L^2_μ-unbiased and the corresponding unbiased prior is π_μ. As noted in Section 19.3.2.2, (19.8) also attains exact posterior unbiasedness (in the usual L^2_λ-sense) for certain other parameterizations $\lambda = \lambda(\theta)$.

19.4.2 Relative entropy loss

For an arbitrary parameterization λ, let

$$L^K(\hat{\lambda}, \lambda) = D_{KL}(p(x|\lambda) \,||\, p(x|\hat{\lambda}))$$

denote the relative entropy (Kullback–Leibler) loss, and let

$$L^{K*}(\hat{\lambda}, \lambda) = D_{KL}(p(x|\hat{\lambda}) \,||\, p(x|\lambda))$$

denote the corresponding dual loss. Note that, unlike the squared error loss L^2_λ, both of these loss functions are invariant under reparameterizations of the model and so we may drop the subscript λ from their notation.

With this notation, and as pointed out in Section 19.3.2.2, $\hat{\mu} = E(\mu|x_{(n)})$ is not only (L^2_μ, π_μ)-Bayes but also (L^K, π_μ)-Bayes. Therefore, *for the mean parameter of an exponential family, the (L^K, π_μ)-Bayes estimator is L^2_μ-unbiased.*

A partial dual result holds for the canonical parameter θ. In this case, the (L^{K*}, π_θ)-Bayes estimator is $\hat{\theta} = E(\theta|x_{(n)})$, which is also the (L^2_θ, π_θ)-Bayes estimator of θ. Thus, *if $\hat{\theta}$ happens to be unbiased in the usual sense,*[17] then we have that, *for the canonical parameter of an exponential family, the (L^{K*}, π_θ)-Bayes estimator is L^2_θ-unbiased.*

On the other hand, it is interesting that the (L^K, π_θ)-Bayes estimator of θ is its posterior mode which, since $\pi_\theta(\theta) \propto 1$, coincides with the maximum likelihood estimator (see [11]). In other words, $\pi_\theta(\theta) \propto 1$ is a 'maximum likelihood prior' as defined by [19].

Unlike (L^2, p)-Bayes estimators, both (L^K, p)- and (L^{K*}, p)-Bayes estimators are invariant under reparameterizations of the model. In a recent paper, [29] discuss a decomposition of the Kullback–Leibler risk which is analogous to the decomposition of the mean squared error into variance and (squared) bias.[18] More specifically, they introduce a general framework under which the

[17] Note that, under regularity conditions, $\hat{\mu}$ is always unbiased in the usual sense. See [5] and Section 19.3.2.2.

[18] Unlike us, [29] refer to L^{K*} as the Kullback–Leibler divergence and to L^K as the dual Kullback–Leibler divergence.

distribution that generates the data (and not just the mean or some other parameter) is estimated. Therefore, they define estimates to be probability distributions so that the analogues of bias and variance are parameter-free. Such estimates are compared using the Kullback–Leibler divergence as the loss function. They point out that the maximum likelihood estimator is parameter invariant and so is naturally described by a distribution.

In the case of exponential families, they show that the maximum likelihood estimator is L^{K*}-unbiased (regardless of the parameterization). We can rephrase this result to say that, *for exponential families, the* (L^K, π)*-Bayes estimator is* L^{K*}*-unbiased,* so it is unbiased in a dual sense. Here π is the prior induced by the uniform prior on the canonical parameter θ. It follows from the results mentioned at the begining of this subsection that, under regularity conditions, the L^{K*}-unbiased estimator for an arbitrary (one-to-one, continuous) transformation $\lambda = \lambda(\mu)$ of the mean parameter always exists and can be obtained by correspondingly transforming the usual L^2_μ-unbiased estimator $\hat{\mu} = E(\mu|x_{(n)})$, i.e. $\hat{\lambda} = \lambda(\hat{\mu})$, which in this case coincides with the maximum likelihood estimator. However, such $\hat{\lambda}$ is not necessarily unbiased in the usual L^2_λ-sense.

A partial dual result can be obtained for the canonical parameter. Using results from [29], it can be shown that an estimator is L^K-unbiased for θ if it is L^2_θ-unbiased. Recall that the (L^{K*}, π)-Bayes estimator is $\hat{\theta} = E(\theta|x_{(n)})$. Thus, *if* $\hat{\theta}$ *happens to be unbiased in the usual sense,* then we have that *the* (L^{K*}, π)*-Bayes estimator is* L^K*-unbiased.* Under these circumstances, the L^K-unbiased estimator for an arbitrary (one-to-one, continuous) transformation $\lambda = \lambda(\theta)$ of the canonical parameter can be obtained by correspondingly transforming the usual L^2_θ-unbiased estimator $\hat{\theta}$, i.e. $\hat{\lambda} = \lambda(\hat{\theta})$. As before, such $\hat{\lambda}$ is not necessarily unbiased in the usual L^2_λ-sense.

19.4.3 Extensions

[27] further extend Lehmann's definition of unbiasedness to take into account, besides a general loss function, the prior distribution of the parameter. They note an interesting duality between their definition of unbiased estimator and the usual definition of Bayes estimator given the same loss and prior. It would be interesting to explore this duality in the context of exponential families, both with respect to relative entropy loss and with respect to more general loss functions.

On the other hand, [17] found a close relationship between maximizing entropy and minimizing worst-case expected loss. They generalized this result to arbitrary decision problems and loss functions. This, in turn, allowed them to define generalized versions of entropy, divergence and exponential families. As an interesting particular case, mutual information $I(X, \Theta)$ can be regarded as a generalized entropy based on the loss function

$$L^{K*}(\hat{\theta}, \theta) = D_{KL}(p(x|\hat{\theta}) \,||\, p(x|\theta))$$

Recall the representation (19.1) and consider the distribution $p(x)$. It is shown in [14] that, for a certain class of distributions $p(x)$, minimization of $I(X, \Theta)$ with respect to $p(\theta|x)$, subject to the constraint

$$\int\int p(x)p(\theta|x)L(\hat{\theta}(x), \theta)\eta(dx)d\theta \leq l$$

leads to the conjugate posterior

$$p(\theta|x) \propto \exp\{(n\bar{x} + n_0 x_0)^t \theta - (n + n_0)M(\theta)\}$$

From this, we can derive the corresponding sampling model $p(x|\theta)$ and prior density $p(\theta)$.

Conversely, [3] have shown that fixing $p(x|\theta)$ and (asymptotically) maximizing $I(X, \Theta)$ with respect to $p(\theta)$, yields a Jeffreys prior.

19.5 Concluding remarks

Exponential families and maximization of entropy, subject to certain moment constraints, are closely related. Choosing appropriately such constraints on the canonical parameter leads to conjugacy. On the other hand, there exists a duality between maximization of entropy and minimization of worst-case expected loss. This relationship allows exponential families to be interpreted as robust Bayes acts against a class of distributions defined by mean-value constraints ([17]).

We have shown how several non-informative priors can be obtained as limiting cases of the DY-conjugate prior. A particular case is that where the posterior expected value of a parameter λ coincides with the corresponding unbiased estimator $\hat{\lambda}$. Another interesting duality holds between the classical notion of unbiasedness and minimization of expected squared error loss. We have seen that this result can be extended to account for more general loss functions. Further generalizations, using the theory developed by Grünwald and Dawid [17], may also be possible. Specifically, it would be interesting to explore the existence or conjugate priors for generalized exponential families derived from general discrepancies, as well as the corresponding definition of unbiasedness and its relationship to the resulting Bayes estimators. Finally, the existence of unbiased conjugate priors may also be of interest.

Acknowledgements

I would like to express my most sincere gratitude to Adrian, not only for his guidance, constructive criticism and openness during the course of my PhD studies, but also because he has been a constant source of inspiration throughout my career. (EGP)

This work was supported by Sistema Nacional de Investigadores, Mexico. The second author also wishes to acknowledge support from Asociación Mexicana de Cultura, A.C.

References

[1] Barndorff-Nielsen, O. (1978). *Information and Exponential Families in Statistical Theory*. Chichester: Wiley.

[2] Bernardo, J. M. and Smith, A. F. M. (2000). *Bayesian Theory*. Chichester: Wiley.

[3] Clarke, B. S. and Barron, A. R. (1994). Jeffreys' prior is asymptotically least favorable under entropy risk. *Jounal of Statistical Planning and Inference*, **41**, 37–60.

[4] Cover, T. M. and Thomas, J. A. (1991). *Elements of Information Theory*. New York: Wiley.

[5] Diaconis, P. and Ylvisaker, D. (1979). Conjugate priors for exponential exponential families. *Annals of Statistics*, **7**, 269–281.

[6] Efstathiou, M., Gutiérrez-Peña, E. and Smith, A. F. M. (1998). Laplace Approximations for natural exponential families with cuts. *Scandinavian Journal of Statistics*, **25**, 77–92.

[7] George, E., Makov, U. and Smith, A. F. M. (1993). Conjugate likelihood distributions. *Scandinavian Journal of Statistics*, **26**, 509–517.

[8] George, E., Makov, U. and Smith, A. F. M. (1994). Bayesian Hierarchical Analysis for exponential families via Markov Chain Monte Carlo. In *Aspects of Uncertainty: a Tribute to D. V. Lindley* (P. R. Freeman and A. F. M. Smith, eds.). Chichester: Wiley.

[9] Goel, P. K. and DeGroot, M. H. (1981). Information about hyperparameters in hierarchical models. *Journal of the American Statistical Association*, **76**, 140–147.
[10] Good, I. J. (1965). *The Estimation of Probabilities. An Essay on Modern Bayesian Methods.* Cambridge: MIT Press.
[11] Gutiérrez-Peña, E. (1992). Expected logarithmic divergence for exponential families. *Bayesian Statistics 4* (J. M. Bernardo, J. O. Berger, A. P. Dawid, A. F. M. Smith, eds.). Oxford: Oxford University Press, pp. 669–674.
[12] Gutiérrez-Peña, E. (1997). Moments for the canonical parameter of an exponential family under a conjugate distribution. *Biometrika*, **84**, 727–732.
[13] Gutiérrez-Peña, E. and Mendoza, M. (1999). A note on Bayes estimates for exponential families. *Revista de la Real Academia de Ciencias Exactas, Físicas y Naturales (España)*, **93**, 351–356.
[14] Gutiérrez-Peña, E. and Muliere, P. (2004). Conjugate priors represent strong pre-experimental assumptions. *Scandinavian Journal of Statistics*, **31**, 235–246.
[15] Gutiérrez-Peña, E. and Smith, A. F. M. (1995). Conjugate parameterizations for natural exponential families. *Journal of the American Statistical Association*, **90**, 1347–1356.
[16] Gutiérrez-Peña, E. and Smith, A. F. M. (1997). Exponential and Bayesian conjugate families: Review and extensions. *Test*, **6**, 1–90 (with discussion).
[17] Grünwald, P. D. and Dawid, A. P. (2004). Game theory, maximum entropy, minimum discrepancy and robust Bayesian decision theory. *Annals of Statistics*, **32**, 1367–1433.
[18] Hartigan, J. (1965). The asymptotically unbiased prior distribution. *Annals of Mathematical Statistics*, **36**, 1137–1152.
[19] Hartigan, J. (1998). The maximum likelihood prior. *Annals of Statistics*, **26**, 2083–2103.
[20] Jaynes, E. T. (1989). *Papers on Probability, Statistics and Statistical Physics* (2nd ed.). Dordrecht: Kluwer Academic.
[21] Kass, R. E. and Steffey, D. (1989). Approximate Bayesian inference in conditionally independent hierarchical models. (parametric empirical Bayes models). *Journal of the American Statistical Association*, **84**, 717–726.
[22] Kullback, S. (1959). *Information Theory and Statistics.* New York: Wiley.
[23] Lehmann, E. L. (1951). A general concept of unbiasedness. *The Annals of Mathematical Statistics*, **22**, 587–592.
[24] Lindley, D. V. and Smith, A. F. M. (1972). Bayesian estimates for the linear model. *Journal of the Royal Statistical Society B*, **34**, 1–41 (with discussion).
[25] Meng, X.-L. and Zaslavsky, A. M. (2002). Observation unbiased priors. *Annals of Statistics*, **30**, 1345–1375.
[26] Morris, C. N. (1982). Natural exponential families with quadratic variance functions. *Annals of Statistics*, **10**, 65–80.
[27] Noorbaloochi, S. and Meeden, G. (1983). Unbiasedness as the dual of being Bayes. *Journal of the American Statistical Association*, **78**, 619–623.
[28] Raiffa, H. and Schlaifer, R. (1961). *Applied Statistical Decision Theory.* Boston: Harvard University.
[29] Wu, Q. and Vos, P. (2012). Decomposition of Kullback–Leibler risk and unbiasedness for parameter free estimators. *Journal of Statistical Planning and Inference*, **142**, 1525–1536.

20 Bayesian model specification: heuristics and examples

DAVID DRAPER

20.1 Introduction

You (a person wishing to reason sensibly in the face of uncertainty: [12]) are about to begin work on a new scientific problem $\mathbb{P}$. You begin by identifying θ, the unknown aspect of $\mathbb{P}$ of principal interest; in the story I wish to tell here, θ could be just about anything (e.g., a map precisely locating the highest and lowest points on the surface of a newly-discovered Earth-like extra-solar planet), but (for concreteness) think of $\theta = (\theta_1, \ldots, \theta_k)$ as a vector in $\Re^k$ (all finite-dimensional unknowns can be expressed in this way). You take stock of Your resources and realize that it's possible to obtain a new dataset D to decrease Your uncertainty about θ; again, D could be just about anything (e.g., a surveillance-camera video record of a crime, offering a partial identification of the perpetrator), but (again, for concreteness) think of $D = (y_1, \ldots, y_n)$ as a vector in $\Re^n$ (all datasets can be expressed in this way). Your other source of information relevant to solving $\mathbb{P}$ is a set $\mathcal{B}$ of (true/false) propositions, all regarded by You as true, describing the scientific context of $\mathbb{P}$ and the nature of the data-gathering process. (An example of a proposition in $\mathcal{B}$ from (e.g.) the field of history is as follows: $\{(y_1, \ldots, y_n)$ is a random sample of size n from the population $\mathcal{P}$ of all words in essay 19 of the *Federalist Papers*$\}$ [20], with θ as the unknown author of the essay (among Alexander Hamilton, James Madison, and John Jay).) At design time (i.e., when You're still contemplating how to obtain D), You notice that the existence of D at analysis time (i.e., after D has arrived) partitions the overall information about θ into {information internal to D} and {information external to D}, and this means that (at analysis time) You'll face a fundamental question: how should the information about θ both internal and external to D be combined, to create an optimal summary of Your total information (and therefore an accurate audit of Your uncertainty) about θ?

Here's a simple but real example, to fix ideas. In 1962 and 1963 [11], two employees of the US *National Bureau of Standards* (now called the *National Institute of Standards and Technology*) made $n = 100$ weighings of a block of metal called *NB10*—given this name because it was supposed to weigh 10 grams—under conditions that were as close as humanly possible to the statistical ideal of independent, identically distributed (IID) sampling from the population $\mathcal{P}_{\text{NB10}} =$ {all possible weighings of *NB10* with the given apparatus}. Calling this problem $\mathbb{P}_{\text{NB10}}$, here θ is evidently the 'true' weight of *NB10*, by which I mean the average of all the potential data values in $\mathcal{P}_{\text{NB10}}$; D consists of the 100 weighings $y = (y_1, \ldots, y_n)$; and $\mathcal{B}$ contains the proposition {y is an IID sample from

$\mathcal{P}_{\text{NB10}}\}$ (along with background propositions known to be true from the context of $\mathbb{P}_{\text{NB10}}$, such as $\{\theta > 0\}$ and $\{\theta$ is close to 10 grams$\}$). In this problem the same fundamental question looms: how can an optimal summary of the total information about the weight of *NB10* be constructed?

The Bayesian approach to answering this *inferential* question, and to making *predictions* of future data D^* and *decisions* in the face of uncertainty, has been settled from a foundational perspective by de Finetti [2] and RT Cox [1]. Each of them proved a theorem, from different points of view about the meaning of probability: for de Finetti, probabilities arise from betting, and for Cox they're numerical expressions of information, in both cases about the truth status of propositions whose truth is unknown to You. The theorem says that if You specify two ingredients for inference and prediction—a probability distribution $p(D|\theta\,\mathcal{B})$ (usually referred to as Your *sampling distribution*) quantifying Your information about θ internal to D, and a probability distribution $p(\theta|\mathcal{B})$ (usually referred to as Your *prior distribution*) quantifying Your information about θ external to D—and two additional ingredients for decision-making—a set $\mathcal{A}$ of possible actions (usually referred to as Your *action space*) and a real-valued *utility function* $U(a, \theta^*)$ trading off the costs and benefits that will arise if You choose action a and θ takes the value θ^*—then (to obtain logically-internally-consistent inferences, predictions and decisions) You must combine the four ingredients according to the following three equations:

$$p(\theta|D\,\mathcal{B}) \propto p(\theta|\mathcal{B})\, p(D|\theta\,\mathcal{B}), \tag{20.1}$$

$$p(D^*|D\,\mathcal{B}) = \int_{\Theta} p(D^*|\theta\,D\,\mathcal{B})\, p(\theta|D\,\mathcal{B})\, d\theta, \tag{20.2}$$

$$a^* = \underset{a\in\mathcal{A}}{\operatorname{argmax}}\, E_{(\theta|D\,\mathcal{B})}\, U(a,\theta) = \underset{a\in\mathcal{A}}{\operatorname{argmax}} \int_{\Theta} U(a,\theta)\, p(\theta|D\,\mathcal{B})\, d\theta. \tag{20.3}$$

Here Θ is the set of possible values of θ; $p(\theta|D\,\mathcal{B})$ (usually referred to as Your *posterior distribution*) summarizes Your total information about θ and solves the inference problem; $p(D^*|D\,\mathcal{B})$, Your (posterior) *predictive distribution* for future data D^*, solves the prediction problem; and a^* solves the decision problem by *maximizing expected utility* (where the expectation is over Your posterior distribution $p(\theta|D\,\mathcal{B})$).

This is excellent, as far as it goes, but the original fundamental question has now been replaced by a new task that's almost as fundamental: how do You optimally specify the four ingredients {prior distribution, sampling distribution, action space, utility function} to be used in the three equations (20.1–20.3)? This task is *Bayesian model specification*, construed broadly. Sometimes this phrase is used more narrowly, to apply just to the sampling distribution, or just to {prior distribution, sampling distribution} if inference and/or prediction are the only goals. In the *NB10* problem, for instance, although there may be a subsequent decision with action space {replace *NB10* (because it doesn't actually weigh 10 grams), keep it}, I'll focus here on the inferential issue of the 'true' weight θ of *NB10*; in this problem let's call $M = \{p(\theta|\mathcal{B}), p(D|\theta\,\mathcal{B})\}$ Your *model* for (Your uncertainty about) θ.

To make the last paragraph meaningful I need to say what I mean by *optimal* Bayesian model specification, and this in turn depends on the following two-step argument:

- All Bayesian reasoning under uncertainty is based on $P(A|B) = \frac{P(A\,B)}{P(B)}$ for propositions A and B, and this is undefined if B is false; therefore
- **Rule 1:** You should try hard not to condition on propositions (a) that You know to be false and (b) that *may* be false.

This motivates the following terminology: in model specification, *optimal* = {to come as close as possible to the goal of [conditioning only on propositions rendered true by the context of the problem and the design of the data-gathering process, while at the same time ensuring that the set of conditioning propositions includes all relevant problem context]}.

Achieving this goal seems hard; for example, a popular method of Bayesian model specification involves looking at the data to specify $p(D|\theta\,\mathcal{B})$—for example, with the *NB10* data You could make a normal quantile plot of the 100 observations and assume $\{(y_i|\theta\,\sigma^2\,\mathcal{B}) \stackrel{\text{IID}}{\sim} N(\theta, \sigma^2)\}$ for Your sampling distribution if the plot indicated approximate normality—but if You do this You'll be conditioning on a proposition that *seems* true on the basis of Your data analysis (see Rule 1(b)) but was not compelled by the problem context or data-collecting design. This approach can be regarded as a kind of 'cheating' in the model-specification process: You peek at the data to help guide this process away from conditioning on obviously false propositions, but the something-for-nothing bell in Your head is probably ringing—the very fact that You peeked may be an action that should come with a price-tag.

In this chapter I examine three methods that may be helpful in moving toward the optimal-model-specification goal described above: an approach called *Calibration Cross-Validation* (CCV) that helps You to pay the right price for the data-peeking mentioned in the previous paragraph (Section 20.2), and Bayesian non-parametric methods for specifying sampling distributions (Section 20.3) and prior distributions (Section 20.4).

20.2 Calibration cross-validation

Two paragraphs ago I mentioned that a common method for specifying the model $M = \{p(\theta|\mathcal{B}), p(D|\theta\,\mathcal{B})\}$ involves (a) looking at the data to identify an apparently reasonable choice for the sampling distribution $p(D|\theta\,\mathcal{B})$—call this particular choice S^*—and then (b) acting as if S^* is something that can safely be conditioned on in drawing inferences about θ. This clearly doesn't satisfy the definition of optimal model specification introduced in Section 20.1, because S^* didn't arise from the problem context or data-gathering design, and it's also likely to be deficient from a *calibration* point of view (in this chapter, a *well-calibrated* inferential process is one that, informally, gets the right answer about as often as it claims to do so): the S^* approach uses the full dataset twice (once to find S^*, and again to draw inferential conclusions about θ based on S^*). The mis-calibration consequences of the S^* approach will generally be that Your nominal $100(1-\gamma)\%$ inferential intervals for (univariate components of) θ and predictive intervals for (univariate components of) future datasets D^* will include the actual values less than $100(1-\gamma)\%$ of the time.

A natural approach to improving on the calibration performance of the S^* method for sampling-distribution specification is two-component *cross-validation* (CV), undertaken in three steps: first (1) You partition D exchangeably (see Section 20.3) into (mutually exclusive and exhaustive) modelling and validation subsets—call them $\mathbb{M}$ and $\mathbb{V}$, respectively; then (2) You explore a variety of models with the data in $\mathbb{M}$, eventually settling on one or more that appear to fit the data well; and then finally (3) You see how well the model(s) from (2) validate on the data in $\mathbb{V}$, for example by constructing $100(1-\gamma)\%$ predictive intervals (based on the data in $\mathbb{M}$) for all of the data values in $\mathbb{V}$ and seeing what percentage of these intervals contain the actual observations. (The S^* approach could be considered a kind of one-component CV, in which modelling and validation take place on the same data.)

Two-component CV (2CV) is clearly a big improvement on the S^* method, but what happens if the model(s) in step (2) don't validate well in step (3)? This occurs more often than You would like it to, and is an embarrassment for 2CV. The natural thing to do is to go back to step (2), re-modelling and re-validating in step (3), iterating (2) and (3) until You finally *do* have one or more models that validate well in $\mathbb{V}$, but You now notice that You've painted Yourself into a corner: You don't have any pristine data values left to see how well the iterative modelling *process* calibrates on data not used in that process. This motivates *calibration cross-validation* (CCV; [4]): going out one more term in the Taylor series, so to speak, the idea is to

(a) partition the data into modelling ($\mathbb{M}$), validation ($\mathbb{V}$) and calibration ($\mathbb{C}$) subsets;

(b) use $\mathbb{M}$ to explore a variety of models until You've found one or more plausible candidates, which You can collect in an *ensemble* $\mathcal{M} = \{M_1, \ldots, M_m\}$;

(c) see how well the models in $\mathcal{M}$ predictively validate in $\mathbb{V}$;

(d) if none of them do, iterate (b) and (c) until You do get good validation, and

(e) fit the best model in $\mathcal{M}$ (or, better, use *Bayesian model averaging* (see, e.g., [19] and [3]) with the entire ensemble $\mathcal{M}$) on the data in ($\mathbb{M} \cup \mathbb{V}$), and report both (i) inferential conclusions based on this fit and (ii) the quality of predictive calibration of Your model/ensemble on the data in $\mathbb{C}$.

The goal with this method is both

(1) a good answer, to the main scientific question, that has paid a reasonable price for *model uncertainty* (the inferential answer is based only on ($\mathbb{M} \cup \mathbb{V}$), not the entire dataset, making Your uncertainty bands wider than those from an S^* analysis), and

(2) an indication of how well calibrated {the iterative fitting process yielding the answer in (1)} is in the calibration subset $\mathbb{C}$, which is a good proxy for future data.

You can use Bayesian decision theory [4] to decide how much data to put in each of $\mathbb{M}$, $\mathbb{V}$ and $\mathbb{C}$: the more important calibration is to You, the more data You want to put in $\mathbb{C}$, but only up to a point, because getting a good answer to the scientific question is also important. I've found that $(0.5, 0.25, 0.25)$ is often a reasonable allocation of data fractions into ($\mathbb{M}$, $\mathbb{V}$, $\mathbb{C}$), and that's what I'll use here. In the rest of this subsection I illustrate the use of CCV on the *NB10* dataset, which is summarized in the top part of Table 20.1 (values are expressed in micrograms below the nominal weight of 10 g).

I randomly partitioned the 100 *NB10* data values into the ($\mathbb{M}$, $\mathbb{V}$, $\mathbb{C}$) subsets of sizes $(50, 25, 25)$ given in the bottom part of Table 20.1 (for greatest stability of conclusions, this random partitioning should be repeated a number of times, with CCV performed in parallel on the repetitions and the results combined appropriately (see [4] for details); in the interests of brevity, here I only show results with the partition in Table 20.1). Step (b) of CCV now involves exploratory modelling with the data in $\mathbb{M}$.

Given the *NB10* problem context, it's natural to begin by fitting the parametric Gaussian model

$$M_1 : \left\{ \begin{array}{ccc} (\theta\, \sigma^2 | \mathcal{B}) & \sim & p(\theta\, \sigma^2 | \mathcal{B}) \\ (y_i | \theta\, \sigma^2\, \mathcal{B}) & \stackrel{\text{IID}}{\sim} & N(\theta, \sigma^2) \end{array} \right\} \tag{20.4}$$

for $i = 1, \ldots, n$. At the point at which these 100 weighings of *NB10* were performed in 1962–63, it's likely that workers at the National Bureau of Standards (NBS) already knew quite a bit about the weight of this block of metal, but here I'm going to illustrate the analysis from the viewpoint of someone (like me, and probably You) who has little information external to the present dataset D about the actual weight of *NB10* or the accuracy of the NBS weighing process. To make this state of information operational, I used the diffuse prior $p(\theta\, \sigma^2 | \mathcal{B}) = p(\theta | \mathcal{B})\, p(\sigma^2 | \mathcal{B})$, with $(\theta | \mathcal{B}) \sim N(0, 10^6)$ and $(\sigma^2 | \mathcal{B}) \sim \Gamma^{-1}(0.001, 0.001)$ (other diffuse prior specifications yielded nearly identical conclusions). With this prior, under the Gaussian model (20.4), (i) the marginal

Table 20.1 Top: A raw frequency distribution of n = 100 weighings of *NB10*; bottom: the random CCV partition illustrated here (with the data values in each component sorted).

Value	375	392	393	397	398	399	400	401
Frequency	1	1	1	1	2	7	4	12
Value	402	403	404	405	406	407	408	409
Frequency	8	6	9	5	12	8	5	5
Value	410	411	412	413	415	418	423	437
Frequency	4	1	3	1	1	1	1	1

M:	375	399	399	399	399	400	400	400	401	401	401	401	401
	401	402	402	402	402	402	402	403	403	403	403	403	404
	404	404	404	404	404	404	405	405	405	406	406	406	406
	406	407	407	407	408	408	408	409	410	411	437		
V:	393	397	398	399	400	401	401	402	403	404	405	406	406
	406	407	407	407	408	408	409	409	412	412	418	423	
C:	392	398	399	399	401	401	401	401	402	404	405	406	406
	406	406	407	407	409	409	410	410	410	412	413	415	

posterior for θ is approximately Gaussian with mean 403.8 and standard deviation (SD) 1.00, (ii) a 95% central posterior interval for θ runs from 401.8 to 405.8, (iii) the marginal posterior for σ has a moderately long right-hand tail (as You would expect for a scale parameter) with mean 7.06 and SD 0.730, (iv) the posterior predictive distribution for a future observation is approximately Gaussian with mean 403.8 and SD 7.17, and (v) the 95% central posterior predictive interval for the next data point runs from 389.7 to 417.9.

It's now interesting to see how well calibrated the Gaussian model is on the data set used to fit it. The left panel of Figure 20.1 presents a *calibration plot* based on the data in M, comparing nominal and actual coverage of $100(1-\gamma)\%$ predictive intervals for $\gamma = (0.01, 0.02, \ldots, 0.99)$. You can see that the Gaussian model produces predictive intervals that are sharply conservative; for example, at all nominal levels from 70% to 95%, the actual coverage is 96%. The right panel of Figure 20.1 gives a normal quantile plot of the data in M, which identifies the reason for the poor validation of the Gaussian model: the distribution is unimodal and close to symmetric but has substantially heavier tails than the Gaussian, and this has led in the Gaussian framework to a large estimate of σ. By way of a second, improved model this suggests a t sampling distribution, as in

$$M_2\colon \left\{ \begin{array}{lcl} (\theta\,\sigma^2\,\nu|\mathcal{B}) & \sim & p(\theta\,\sigma^2\,\nu|\mathcal{B}) \\ (y_i|\theta\,\sigma^2\,\nu\,\mathcal{B}) & \stackrel{\text{IID}}{\sim} & t_\nu(\theta,\sigma^2) \end{array} \right\}. \tag{20.5}$$

This model is easy to fit via MCMC with slice sampling; 100 000 monitoring iterations took 15 seconds at 3.3 GHz (this monitoring sample size produced Monte Carlo standard errors for all posterior summaries less than 0.01). Here I used the diffuse prior $p(\theta\,\sigma^2\,\nu|\mathcal{B}) = p(\theta|\mathcal{B})\,p(\sigma^2|\mathcal{B})\,p(\nu|\mathcal{B})$, with the same marginal priors as in the Gaussian model for θ and σ^2 and with $(\nu|\mathcal{B}) \sim \text{Uniform}(1.0, 10.0)$ (the right endpoint was chosen to be large enough to avoid truncation of the likelihood; other values that avoid truncation give similar results).

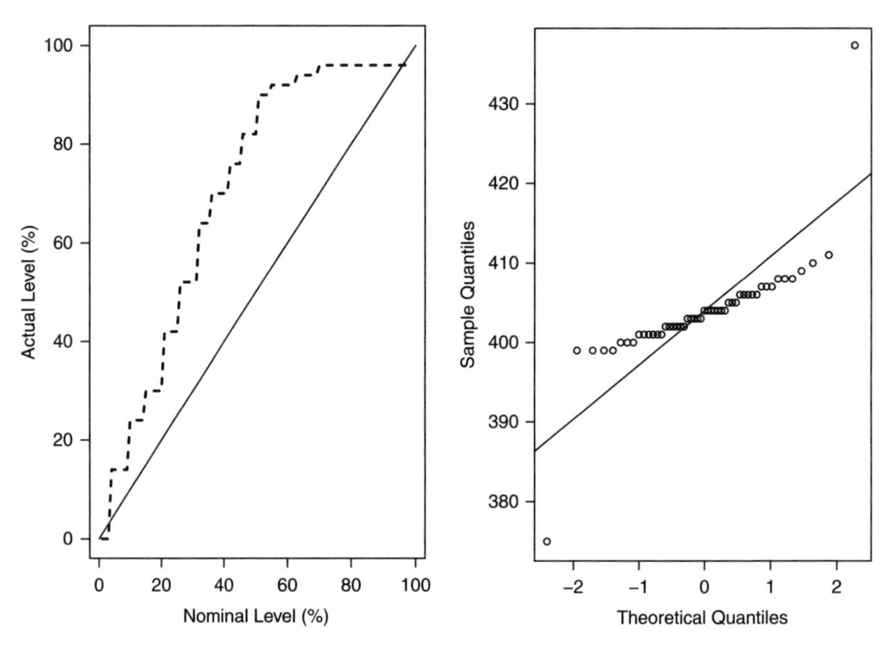

Figure 20.1 Left panel: calibration curve (dotted line) for the Gaussian model (20.4) fit to the *NB10* data in 𝕄 and validated in 𝕄 (the solid line represents the target behaviour under good calibration); right panel: normal quantile plot of the data in 𝕄.

The results from fitting the t model (20.5) to 𝕄 are as follows: (i) the marginal posterior for θ is again approximately Gaussian, but this time with mean 403.4 and SD 0.50; (ii) a 95% central posterior interval for θ runs from 402.5 to 404.4; (iii) the marginal posterior for σ again has moderate positive skew, this time with mean 2.73 and SD 0.46; (iv) the marginal posterior for ν has a substantial right-hand tail, with (mode, median, mean) = $(2.31, 2.44, 2.60)$ and SD 0.91; (v) the posterior predictive distribution for a future observation is approximately Gaussian with mean 403.4 and SD 2.80; and (v) the 95% central posterior predictive interval for the next data point runs from 397.9 to 409.0. Note how much smaller both the inferential uncertainty about θ and the predictive uncertainty about future observations are with the t model than with the Gaussian sampling distribution; this is a consequence of the Gaussian having minimal Fisher information for location among all symmetric unimodal sampling distributions on $\Re$. The much smaller values for σ are because observations from a $t_\nu(\theta, \sigma^2)$ sampling distribution have variance $\frac{\nu}{\nu-2}\sigma^2$ (i.e., scale and shape are confounded in this model).

Is the t model better than the Gaussian for the data in 𝕄? There are a number of ways to answer this question; the one I like best [5] involves *full-sample log scores*. The idea, with a univariate dataset $D = y = (y_1, \ldots, y_n)$ (such as the *NB10* weighings) and models M_j (here $j = 1, 2$) to be compared, involves computing

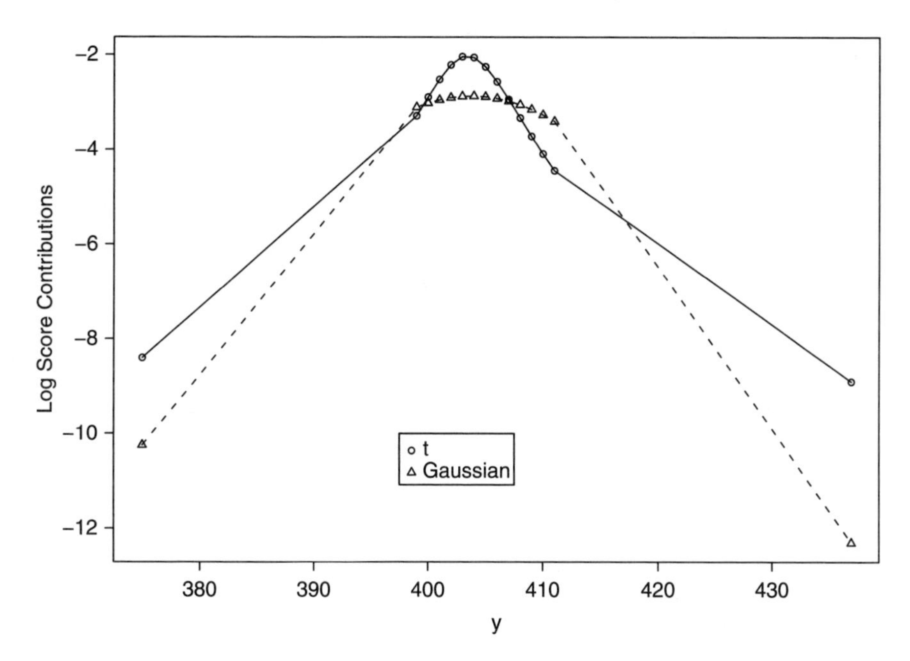

Figure 20.2 Contributions to the overall LS_{FS} values for each model from each observation; triangles (solid curve) and circles (dotted curve) track the Gaussian and *t* models, respectively.

$$LS_{FS}(M_j|D\,\mathcal{B}) = \frac{1}{n}\sum_{i=1}^{n}\log p(y_i|D\,M_j\,\mathcal{B}) \tag{20.6}$$

and favouring the model with the bigger log score LS_{FS}. Computation of LS_{FS} is straightforward; when parametric model M_j with parameter vector η_j is fit via MCMC, the predictive ordinate $p(y^*|D\,M_j\,\mathcal{B})$ in LS_{FS} can be approximated as follows. With m identically distributed (not necessarily independent) MCMC monitoring draws $(\eta_j)^*_k$ from $p(\eta_j|D\,M_j\,\mathcal{B})$,

$$\begin{aligned} p(y^*|D\,M_j\,\mathcal{B}) &= \int p(y^*|\eta_j\,M_j\,\mathcal{B})\,p(\eta_j|D\,M_j\,\mathcal{B})\,d\eta_j \\ &= E_{(\eta_j|D\,M_j\,\mathcal{B})}\,p(y^*|\eta_j\,M_j\mathcal{B}) \\ &\doteq \frac{1}{m}\sum_{k=1}^{m} p[y^*|(\eta_j)^*_k\,M_j\,\mathcal{B}]. \end{aligned} \tag{20.7}$$

Applying this method to models M_1 ((20.4), Gaussian) and M_2 ((20.5), *t*) with the data in $\mathbb{M}$ yields LS_{FS} values of -3.30 and -2.86, respectively; this represents a (sharp) preference for the *t* model. Figure 20.2 shows the individual contributions, from each data value in $\mathbb{M}$, to the overall LS_{FS} values from the Gaussian and *t* models. It's evident that the *t* model fits better both in the tails (where the most influential observations are from the Gaussian point of view) and in the centre (where most of the data values are); in fact, 80% of the data values in $\mathbb{M}$ are predicted better by the *t* model than by the Gaussian.

Next question: is the *t* model good enough to stop looking for a better model? The answer is complicated here by the small sample sizes in each of the $\mathbb{M}$ and $\mathbb{V}$ partition components. Figure 20.3

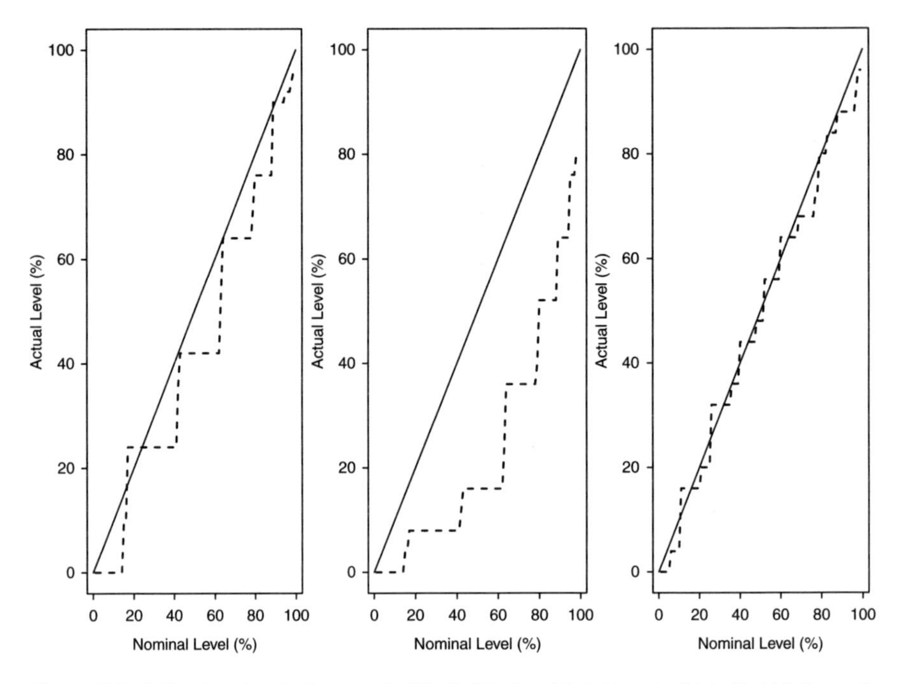

Figure 20.3 Calibration plots for the t model (20.5), fit to the data in $\mathbb{M}$ and validated in $\mathbb{M}$ (left panel); fit to $\mathbb{M}$ and validated in $\mathbb{V}$ (centre); and fit to $\mathbb{V}$ and validated in $\mathbb{V}$ (right).

gives calibration plots for the t model with three different data configurations: fit to the data in $\mathbb{M}$ and validated in $\mathbb{M}$ (left panel); fit to $\mathbb{M}$ and validated in $\mathbb{V}$ (centre; this corresponds to step (c) in the CCV algorithm); and fit to $\mathbb{V}$ and validated in $\mathbb{V}$ (right). *Internal* validation (evaluating the fit on the same dataset used to create the fit, as in {fit to $\mathbb{M}$, validate in $\mathbb{M}$}, which could be abbreviated $\mathbb{M} \rightarrow \mathbb{M}$, and similarly $\mathbb{V} \rightarrow \mathbb{V}$) ranges from barely adequate (the left panel) to excellent (the right display), but *external* validation (evaluating the fit on new data not used in the fitting process, as in $\mathbb{M} \rightarrow \mathbb{V}$, in the centre) is abysmal, with the opposite calibration problem (predictive intervals that are not wide enough) from that exhibited by the Gaussian $\mathbb{M} \rightarrow \mathbb{M}$ model in the left panel of Figure 20.1. With only 50 observations in $\mathbb{M}$ and 25 in $\mathbb{V}$, the parameter estimates from the t model are quite different when it's applied separately to $\mathbb{M}$ and $\mathbb{V}$ (see the first two rows in Table 20.2 below). Another way to put the difficulty, looking at the full *NB10* dataset in Table 20.1, is that there are 'only' 3 outliers in the entire dataset (namely, the observations 375, 423 and 437), and the presence or absence of any one of these outliers in $\mathbb{M}$ or $\mathbb{V}$ is, by its very nature, highly influential for the parameter estimates. As mentioned previously, [4] performs the obvious analysis to remedy this problem—repeat the CCV algorithm across many random ($\mathbb{M}, \mathbb{V}, \mathbb{C}$) partitions and average the results—which, in the interests of brevity, I do not reproduce here; the conclusion from this broader analysis is that the t model is a good basis for stopping the iterative step (d) in CCV and proceeding to the final step (e).

The third row in Table 20.2 gives posterior summaries from fitting the t model to the 75 observations in ($\mathbb{M} \cup \mathbb{V}$), and the left panel in Figure 20.4 gives the calibration plot for this fit when applied to $\mathbb{C}$. You can see that the model's validation is not perfect, again in part because of the small sample sizes: for instance, the nominal 95% predictive interval runs from 397.1 to 410.9 and includes only 88% of the data values in $\mathbb{C}$. Although it's not part of the CCV algorithm to do so, for comparison purposes the final row of Table 20.2 summarizes the fit of the t model to the entire

Table 20.2 Parameter and predictive summaries from fitting the t model separately to the $\mathbb{M}$ and $\mathbb{V}$ partition components, to the merged dataset ($\mathbb{M} \cup \mathbb{V}$), and to the entire dataset D; y^* is a future data value.

Data partition	Sample size	Posterior mean (SD)			
		θ	σ	ν	y^*
$\mathbb{M}$	50	403.4 (0.50)	2.73 (0.46)	2.60 (0.91)	403.4 (2.80)
$\mathbb{V}$	25	405.3 (1.23)	5.31 (1.12)	5.73 (2.39)	405.3 (5.54)
($\mathbb{M} \cup \mathbb{V}$)	75	404.0 (0.50)	3.42 (0.47)	2.88 (0.95)	404.0 (3.48)
D	100	404.3 (0.47)	3.85 (0.45)	3.56 (1.18)	404.3 (3.91)

dataset D, and the right panel in Figure 20.4 displays the calibration plot that results when the t model is fit to, and validated in, all of D; this shows what someone using the S^* approach would conclude, both about the parameters and about the quality of the model fit. The right panel of Figure 20.4 provides a somewhat rosier view of the quality of the t model than the left panel, and is therefore somewhat misleading about the calibration performance of the iterative modelling process leading to the results of the S^* method.

On the basis of the CCV approach to dealing with specification uncertainty about the sampling distribution in the *NB10* problem, I would draw the following conclusions:

(A) The block of metal called *NB10* weighed (in 1963) about 404.0 micrograms below the nominal weight of 10 grams, give or take about 0.50 micrograms, and a 95% interval for its weight runs from 403.0 to 404.9; and

(B) the iterative modelling process leading to the inferential conclusion in (A) is somewhat over-confident in its ability to predict future data values not used in the model-fitting, with nominal 95% predictive intervals for future observations including the actual data values about 88% of the time, give or take about $100\sqrt{\frac{(0.88)(0.12)}{25}}\,\% \doteq 6.5\%$.

20.3 Bayesian nonparametric sampling-distribution specification

In the *NB10* problem, at design time (before any data have been collected), and with no covariate information that would serve to distinguish one observation from another, upon reflection (following de Finetti [2]) You would notice that Your uncertainty about the *NB10* weighings $D = (y_1, \ldots, y_n)$ is *exchangeable*, in the usual sense that Your predictive distribution $p(y_1 \ldots y_n | \mathcal{B})$ is invariant under permutation of the order in which the data values are observed. Moreover, if the weighing process were to be continued indefinitely (still with no covariate information), yielding the entire population $\mathcal{P}_{\text{NB10}} = (y_1, y_2, \ldots)$, Your predictive distribution $p(y_1\, y_2 \ldots | \mathcal{B})$ would still be exchangeable (in the sense that exchangeability would hold for any finite subset of $\mathcal{P}_{\text{NB10}}$). In settings such as this, de Finetti [2] proved a celebrated theorem that says (slightly informally)

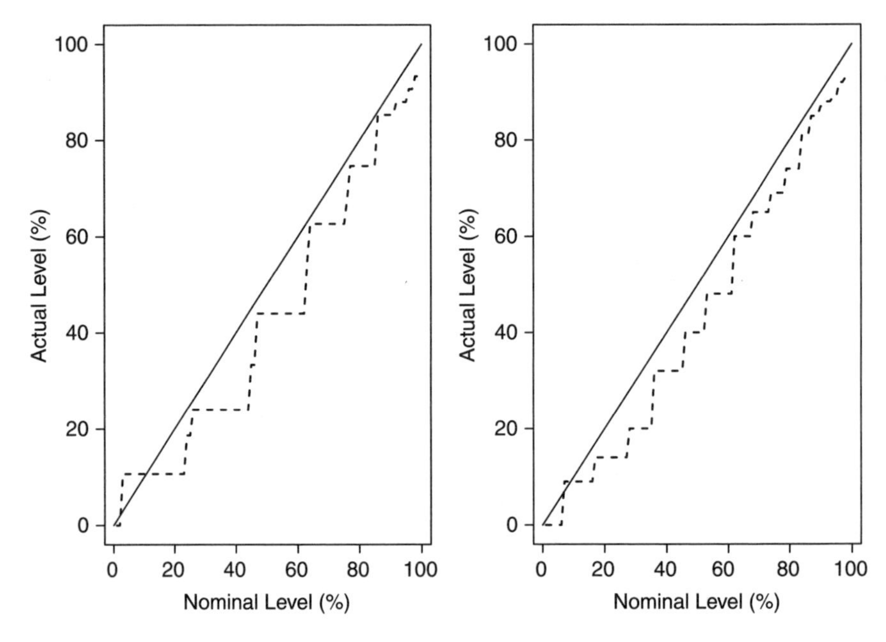

Figure 20.4 Calibration plots for the t model (20.5), fit to the data in ($\mathbb{M} \cup \mathbb{V}$) and validated in $\mathbb{C}$ (left panel); and fit to, and validated in, the entire data set (right panel).

that all logically-internally-consistent predictive distributions $p(y_1 \ldots y_n|\mathcal{B})$ are expressible hierarchically as

$$\left\{ \begin{array}{rcl} (G|\mathcal{B}) & \sim & p(G|\mathcal{B}) \\ (y_i|G\,\mathcal{B}) & \stackrel{\text{IID}}{\sim} & G \end{array} \right\}, \tag{20.8}$$

where G is the empirical CDF of $(y_1, y_2, \ldots)$; here $p(G|\mathcal{B})$ is a prior distribution on the set $\mathcal{G}$ of all CDFs on $\Re$. This theorem founded the sub-field of *Bayesian non-parametric (BNP) modelling*, which concerns inference on functions such as G (and also—not addressed in this chapter—functions such as regression surfaces). At the time he proved the theorem, de Finetti didn't know how to put a scientifically meaningful prior on $\mathcal{G}$, but progress toward this goal—started 40–50 years ago by Freedman [10] and Ferguson [9]—culminated, in the work of people such as Escobar and West [7] and Lavine [18], with MCMC-based approaches to extracting information from the posterior distribution $p(G|D\,\mathcal{B})$, and BNP modelling has become increasingly routine in the past 15 years. This approach offers the possibility of optimal model specification (in the definitional sense in Section 20.1) in the *NB10* problem, because the judgement of exchangeability leading to model (20.8) arises directly from problem context; the only remaining issue with this approach is how to specify $p(G|\mathcal{B})$ in a manner that is both (a) accurately driven by the nature of the data-gathering process and (b) well-calibrated.

Two approaches to specifying $p(G|\mathcal{B})$ have by now been developed to the point that they're both scientifically useful and computationally tractable: *Dirichlet-process (DP) mixture modelling* [7, 9] and *Pólya-tree (PT) mixture modelling* (e.g., [13]). I'll concentrate here on Pólya trees; see, e.g., [17] for practical examples of DP modelling with count data. For a univariate sample $D = y = (y_1, \ldots, y_n)$ such as the *NB10* dataset, a natural PT mixture model would take the following form:

$$\left\{\begin{array}{rcl} (y_i|G\mathcal{B}) & \overset{\text{IID}}{\sim} & G \qquad (i=1,\ldots,n) \\ (G|\alpha\theta\sigma^2\mathcal{B}) & \sim & PT\left[\Pi_{N(\theta,\sigma^2)},\mathcal{A}_\alpha\right] \\ (\alpha\theta\sigma^2|\mathcal{B}) & \sim & p(\alpha\theta\sigma^2|\mathcal{B}) \end{array}\right\}, \tag{20.9}$$

for an appropriately chosen prior distribution $p(\alpha\,\theta\,\sigma^2|\mathcal{B})$ on (α,θ,σ^2).

The meaning of the expression $PT\left[\Pi_{G_0(\eta)},\mathcal{A}_\alpha\right]$ is as follows. Rather generally in Bayesian work, prior distributions are specified through two main ingredients: a *prior estimate* of the thing receiving the prior distribution, and a *prior sample size* indicating how tightly concentrated the prior should be around the prior estimate. PT priors for a CDF G follow this pattern: $G_0(\eta)$ is the prior estimate or *centring distribution*, which will typically be a parametric family indexed (in this case) by the parameter vector η, and α acts like a prior sample size, in the sense that bigger (smaller) values of α lead to posterior distributions on G that are closer to (farther away from) the centring distribution. In model (20.9), $G_0(\eta)$ is the $N(\theta,\sigma^2)$ distribution; this is natural in the *NB10* problem for the same reason that the Gaussian sampling distribution appeared in model M_1 in Section 20.2.

This approach is referred to as PT mixture modelling because a point-mass prior on (α,θ,σ^2) of the form $(\alpha=\alpha_0,\theta=\theta_0,\sigma^2=\sigma_0^2)$ would correspond to fitting a single Pólya-tree prior for G, whereas a more realistic treatment of prior uncertainty about (α,θ,σ^2)—in which non-point-mass distributions are given to one or more elements of the (α,θ,σ^2) vector—amounts to mixing over individual Pólya trees. For univariate outcomes, PT priors are based on binary partitions of $\Re$ with 2^m partition sets at level m of the tree, and act like random histograms; to get PT priors to directly model continuous data, strictly speaking the number of histogram bars has to become countably infinite, but in practice finite Pólya trees (with 2^M bars, for finite M, at the bottom level) are all that's needed, because the real-world process of measuring conceptually continuous outcomes always discretizes them anyway.

These days it's relatively straightforward to fit model (20.9) via MCMC with a Metropolis-within-Gibbs approach: the full-conditional distribution $p(G|D\,\alpha\,\theta\,\sigma^2\,\mathcal{B})$ turns out to be another Pólya tree, and then You can Metropolis-sample the other full-conditionals (such as $p(\theta|D\,G\,\alpha\,\sigma^2\,\mathcal{B})$). The ensemble of R functions called `DPpackage` [14], available from `CRAN`, contains several functions that can fit model (20.9), including `PTlm` and `PTdensity`, and `WinBUGS` code for this model is available from Tim Hanson; this permits attention to shift away from the MCMC details and toward the modelling, where several surprises await (in relation to Your experience with parametric modelling).

It's possible to put a prior distribution on α, but—with an eye on calibration, as in Section 20.2—You can instead regard α as a kind of tuning constant that You can vary across a range of fixed values to achieve good out-of-sample calibration. In the *NB10* problem, I again use a diffuse prior on (θ,σ^2)—Gaussian with huge variance for θ, $\Gamma^{-1}(\epsilon,\epsilon)$ for σ^2 with small positive ϵ—to quantify the information base of someone who knows little, external to the *NB10* dataset, about the weight of *NB10* or the accuracy of the weighing process.

As a first example of the results from the BNP approach to dealing with specification uncertainty about sampling distributions, I used `PTlm` on the entire *NB10* dataset with $\alpha=1$ and $M=6$, employing a burn-in of 5000 iterations (from starting values for μ and σ that were not far from their likely posterior means) and a monitoring run of 10 000 saved values after thinning by a factor of 20. (In `PTlm`, by default θ is identified as the *median* of the population empirical CDF.) The resulting 205 000 iterations took 4.5 minutes at 3.3 GHz, and initially yielded poor acceptance rates for the Metropolis steps for θ and σ^2. Iterative tuning of the proposal distribution SDs eventually yielded near-optimal univariate acceptance rates of 44–49%, at which point I examined the Monte-Carlo accuracy achieved by this MCMC sampling strategy. The saved iterations for θ behaved like draws from an $AR_1(\rho_1)$ time series with a first-order autocorrelation $\hat{\rho}_1$ of +0.75, even after 20-fold thinning. From the usual expression

$$\widehat{MCSE}\left(\bar{\theta}^*\right) = \frac{\hat{\sigma}_\theta}{\sqrt{n^*}}\sqrt{\frac{1+\hat{\rho}_1}{1-\hat{\rho}_1}} \tag{20.10}$$

for the Monte Carlo standard error (MCSE) of the MCMC estimate $\bar{\theta}^* = \frac{1}{n^*}\sum_{j=1}^{n^*}\theta_j^*$ of the posterior mean of θ, where $\hat{\sigma}_\theta$ is the estimated posterior SD of θ and n^* is the number of saved monitoring iterations, it became clear that—with only 200 000 iterations going into the monitoring process—the MCSEs of the posterior mean and SD estimates were on the order of 0.08, which was too big for getting a good idea of the posterior SD of θ. To drive the MCSEs down to about 0.01, a monitoring run of 12 000 000 iterations (thinning by a factor of 200) was needed; this took about 3.9 hours at 3.3 GHz. The first surprise with BNP modelling is how much longer in clock time it can take to get results with decent Monte-Carlo accuracy, in relation to Your parametric-modelling experience; on reflection, this is perhaps not actually so surprising, for two reasons: (i) You're treating G as a nuisance parameter that has to be learned along with the main parameter(s) of interest (with $M = 6$ in Pólya trees, this is like learning an additional $2^6 = 64$ parameters (albeit rather highly correlated, so that the effective dimensionality of the learning process for G is probably on the order of a few dozen additional parameters)), and (ii) uncertainty about G is bound to create poorer mixing for θ and the other parameters You care about.

Figure 20.5 displays the marginal posterior distributions for θ (left panel) and σ (right panel) from this fitting of model (20.9), using default window-widths for the kernel density estimation. The second surprise with BNP modelling, when compared with parametric-modelling intuition, is how rough these posterior distributions are; on reflection this is once again perhaps not so

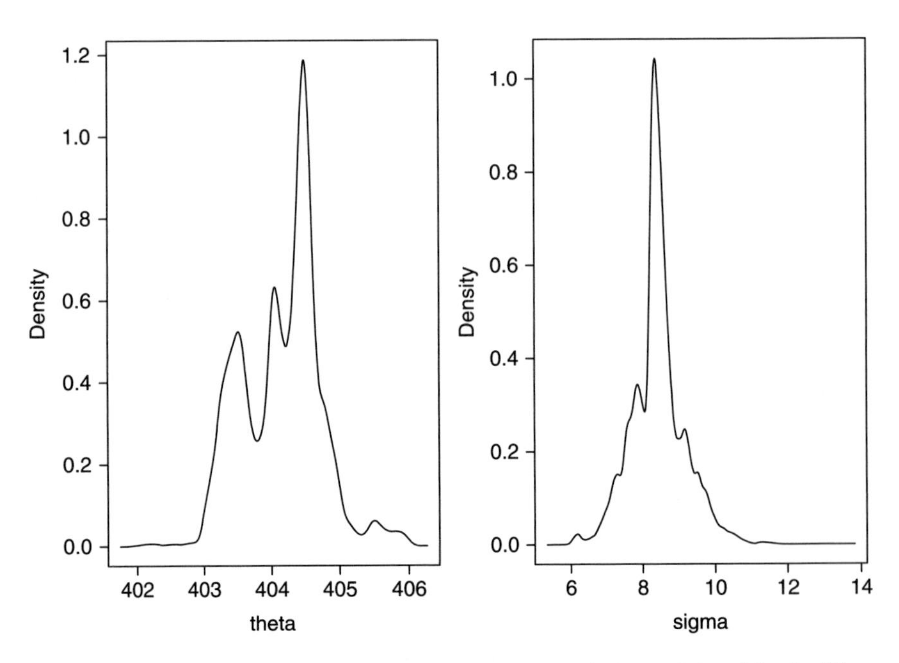

Figure 20.5 Marginal posterior distributions for θ (left panel) and σ (right panel), from fitting the Pólya-tree model (20.9).

startling (with the *NB10* data, the empirical CDF is itself quite rough, from the granularity of the observations). The marginal posterior for θ has a mean of 404.2 and an SD of 0.58, and the 95% central posterior interval runs from 403.1 to 405.4; these results are in reasonable agreement with those from the CCV approach in Section 20.2, with perhaps a bit more uncertainty about θ arising from what may be a better attempt to fully quantify uncertainty about G.

As a second example of the fitting of BNP models, to get a closer look at their calibration properties, I created an artificial dataset that had the same mean $\bar{y}$ and SD s as the *NB10* data but was (in a certain sense) as close to Gaussian as possible: observation i in this artificial dataset had the value $y'_i = \bar{y} + s\,\Phi^{-1}\left(\frac{i-\frac{1}{2}}{n}\right)$, where Φ is the standard normal CDF. I then fit model (20.9) to this artificial dataset, with 12 different priors on $(\alpha, \theta, \sigma^2)$; all these priors were of the form $p(\alpha, \theta, \sigma^2|\mathcal{B}) = p(\alpha|\mathcal{B})\,p(\theta|\mathcal{B})\,p(\sigma^2|\mathcal{B})$, with $(\theta|\mathcal{B}) \sim N(0, 10^6)$ and $(\sigma|\mathcal{B}) \sim U(0, 20)$ in each case (the upper limit on the uniform prior on σ was again chosen to avoid likelihood truncation). My goal in this work was (a) to represent (as in all of the results in this section) the information base, external to the *NB10* data, of someone who knows little about the weight of *NB10* or the accuracy of the NBS measuring process, and (b) to examine the resulting posterior inferences about (θ, σ) as a function of various priors on α, which is also (to put it mildly) not a quantity strongly pinned down by information external to the *NB10* dataset. I used Tim Hanson's `WinBUGS` code for these runs; in this code θ is identified as the *mean* of the population empirical CDF.

Consider a point-mass prior on α that sets $\alpha = \alpha^*$ (say). At each iteration of the Metropolis-within-Gibbs sampling used to fit model (20.9), at the point at which θ^* and σ^* values have been drawn from $p(\theta\,\sigma|G\,D\,\mathcal{B})$, imagine standardizing the artificial data values y'_i to create $y''_i = \frac{y'_i - \theta^*}{\sigma^*}$, and let G'' be the empirical CDF of the resulting y''_i values. To complete the current scan of the sampler, the final step is to draw a CDF G^* from the Pólya-tree distribution $PT\left[\Pi_{G^\dagger}, \mathcal{A}_{\alpha^*}\right]$, where $G^\dagger$ is a weighted average of the standard normal CDF Φ and G'' with weights given by α^* and n (respectively). Thus as α^* grows (with n fixed at 100), with the artificial dataset examined here (in which the empirical CDF is as close to Gaussian as possible), You would expect the Pólya-tree results to more and more closely resemble those from fitting the parametric Gaussian model (20.4), with the same diffuse prior on (θ, σ) as above; the question is how quickly (as α^* increases) this convergence will occur.

Table 20.3 presents the results of these calculations. By way of priors on α I used a popular choice in BNP modelling—a variety of $\Gamma(a, b)$ priors on α (almost all of which had $b = 1$)—and I compared these with point-mass priors having the same prior means as the Gamma distributions; the bottom row of the table gives the parametric Gaussian results for further comparison. (All of the Monte Carlo standard errors for the values in this table were 0.01 or smaller.) You can see that the expected convergence has indeed occurred, but the interesting thing (and this is a third surprise from BNP modelling) is how large α needs to be to get (calibrationally correct) results that are close to those from the parametric model. With small α, even with $n = 100$ observations, with diffuse priors on θ and σ, the uncertainty about those two parameters imposed upon the BNP modelling, above and beyond the uncertainty about G, makes the BNP inferences extremely conservative. (Of course, to really pin this down You would have to create a simulation environment in which many Gaussian datasets were generated at random, rather than simply using the one "super-Gaussian" artificial dataset I used here; I intend to report on results from this broader simulation experiment elsewhere.)

The reason for the inferential conservatism in Table 20.3 with small α is that, in the BNP formulation, (θ, σ) and G are correlated in the posterior (especially when α is small), and uncertainty about G is therefore propagated into uncertainty about the parameters. As an example of these correlations, I monitored the posterior for G on an equally spaced grid of 200 points in the range $(\bar{y} \pm 3.5s)$, obtaining a vector $(G^*_1, \ldots, G^*_{200})$ on each MCMC scan; with $\alpha = 1$, correlations

Table 20.3 Posterior summaries from fitting the Pólya-tree model (20.9) with an artificial Gaussian dataset having the same mean and SD as the *NB10* data, using a variety of prior distributions on α. In the first column, an integer k signifies $\alpha = k$, and $\Gamma_{a,b}$ is the $\Gamma(a, b)$ distribution. The last row gives results from fitting the parametric Gaussian model (20.4) to the same dataset, for comparison.

	Posterior summaries for					
	θ		σ		α	
α	Mean	SD	Mean	SD	Mean	SD
$\Gamma_{1,1}$	404.6	1.69	6.75	0.72	3.26	1.53
1	404.5	3.13	7.11	1.11	—	—
$\Gamma_{5,1}$	404.6	1.32	6.66	0.59	6.37	2.54
$\Gamma_{10,2}$	404.6	1.63	6.66	0.60	5.74	1.63
$\Gamma_{10,1}$	404.6	1.09	6.61	0.55	10.9	3.21
10	404.6	1.11	6.62	0.55	—	—
$\Gamma_{20,1}$	404.6	0.96	6.59	0.51	20.6	4.53
$\Gamma_{50,1}$	404.6	0.82	6.56	0.49	50.2	7.08
$\Gamma_{100,1}$	404.6	0.75	6.55	0.48	100.1	9.97
100	404.6	0.74	6.55	0.48	—	—
$\Gamma_{200,1}$	404.6	0.71	6.54	0.47	200.1	14.2
$\Gamma_{500,1}$	404.6	0.68	6.54	0.48	499.8	22.3
Parametric Gaussian	404.6	0.65	6.53	0.47	—	—

between θ and elements of this G^* vector ranged from -0.33 for G^*_1 to 0 for G^*_{100} to $+0.34$ for G^*_{200}, and correlations between σ and elements of the G^* vector ranged from $+0.67$ for G^*_1 to -0.06 for G^*_{100} to $+0.65$ for G^*_{200}.

The upshot of this inquiry is that if You know little, external to Your present dataset D, about the population empirical CDF G that gave rise to Your data (in a one-sample problem like that posed by the *NB10* dataset), and You express this uncertainty—in the Pólya-tree version of BNP modelling—with a prior on α that concentrates on small values (thereby ensuring that most of the information about G in the posterior comes from the empirical CDF based on Your sample), the resulting inferential answers for the parameters in Your model may not be well-calibrated, even if the centring distribution G_0 in Your Pólya-tree prior closely matches the actual data-generating mechanism. (Note that the conservatism in Table 20.3 is not present in the results summarized in Figure 20.5. I conjecture that this is because (a) θ was identified, in the modelling leading to Figure 20.5, as the median of G, whereas it was identified as the mean of G in the modelling that produced Table 20.3, and (b) the correlations noted in the previous paragraph are substantially smaller in the median modelling; this is a subject of continuing investigation.)

20.4 Bayesian nonparametric prior-distribution specification

Changing the focus now to specification of the prior distribution, it's common in Bayesian work to solve this specification problem with one member or another of a standard parametric family, sometimes chosen (e.g., for reasons of computational convenience) to be conjugate to Your sampling distribution. But this almost always goes beyond the optimal model-specification goal identified in Section 20.1; typically the sorts of propositions (relevant to Your prior distribution) that are rendered true by the context of the problem are (a) qualitative shape criteria such as monotonicity, convexity, or unimodality, and possibly also (b) one or more quantitative bounds on prior moments or percentiles. In such situations it would seem more satisfying to work with an infinite-dimensional non-parametric class $\mathcal{C}$ of prior densities satisfying the qualitative and quantitative criteria, for instance either (i) by sampling random members of this class and averaging over the implied uncertainty or (ii) by calculating upper and lower bounds over $\mathcal{C}$ for the posterior summaries of greatest interest (this is a form of sensitivity analysis).

Here's a case study in which to explore this idea. Suppose You're observing an IID Bernoulli(θ) process that has so far yielded n consecutive zeros, and the goal is to use the data to discriminate between two competing explanations for this outcome: $\theta = 0$ or $\theta > 0$. (In the real-world application on which this model is based, I once had occasion to buy n cups of tea over a several-week period from a machine that featured on its front a stick-on label announcing to customers that they might be lucky and get a free beverage, implying the existence of a device inside the machine that dispensed free drinks at random. After $n = 78$ consecutive fee-paying cups of tea, it was natural to speculate whether the makers of the machine had found it cheaper to attach the stick-on label, with no intent to offer free drinks at all, than to supply the machine with a randomization mechanism. Other applications of this problem arise, e.g., in medicine, when the first n patients screened in a particular sub-population all fail to have a disease that's rare in the overall population, and in process control, when the first n items manufactured have all been free of defects.)

Although most inferential settings involving observation of a Bernoulli process are more satisfyingly approached through interval estimation based on a model that treats $0 \le \theta \le 1$ continuously, with no individual value singled out for special treatment, this situation is a genuine sharp-null hypothesis-testing problem, and may be approached from the Bayesian point of view through the model

$$\left\{ \begin{array}{ccc} (\theta|\mathcal{B}) & \sim & p(\theta|\mathcal{B}) \\ (y_i|\theta\,\mathcal{B}) & \stackrel{\text{IID}}{\sim} & \text{Bernoulli}(\theta) \end{array} \right\} \tag{20.11}$$

($i = 1, \ldots, n$), with a prior of the form

$$p(\theta|\mathcal{B}) = \left\{ \begin{array}{ccc} 0 & \text{with probability} & \lambda \\ \pi(\theta|\mathcal{B}) & & (1-\lambda) \end{array} \right\} \tag{20.12}$$

for some $0 \le \lambda \le 1$. In the initial choice of $\pi(\theta|\mathcal{B})$ in the tea-machine case study, it seemed natural to quantify the following set of prior information about θ, conditional on θ being positive: (a) smaller values are more likely than bigger values, and (b) on substantive (economic) grounds, prior uncertainty about θ should be centred between two values (α_1, β_1), for instance $\left(\frac{1}{75}, \frac{1}{25}\right)$. (The upper bound in (b) arises because the makers of the tea machine would not wish to give away more free drinks than necessary, and the lower bound corresponds to the view that, if θ were too small, customers would not perceive a large enough reward from the possibility of a free cup of tea for the randomization strategy to be worthwhile. Unimodal priors with a positive mode are also worth considering in this problem; this possibility will be examined elsewhere.)

20.4.1 A conjugate parametric solution

The off-the-shelf choice for $\pi(\theta|\mathcal{B})$ is of course a member of the Beta (η_1, η_0) family chosen, in view of (a) and (b), to be monotonically decreasing (and possibly also convex) and to have a mean between α_1 and β_1. Examination of the qualitative behaviour of the Beta family reveals that the desired monotonicity and convexity correspond to the region within which $0 < \eta_1 \leq 1$ and $\eta_0 \geq 2$. The mean constraint (b) in the Beta family,

$$\alpha_1 \leq \frac{\eta_1}{\eta_1 + \eta_0} \leq \beta_1, \tag{20.13}$$

further restricts the appropriate subclass of parametric priors to those with

$$\eta_1 \left(\frac{1}{\beta_1} - 1 \right) \leq \eta_0 \leq \eta_1 \left(\frac{1}{\alpha_1} - 1 \right), \tag{20.14}$$

giving rise with $\alpha_1 = \frac{1}{75}$ and $\beta_1 = \frac{1}{25}$ to the roughly triangular admissible region in Figure 20.6 (the contours in this plot will be explained below).

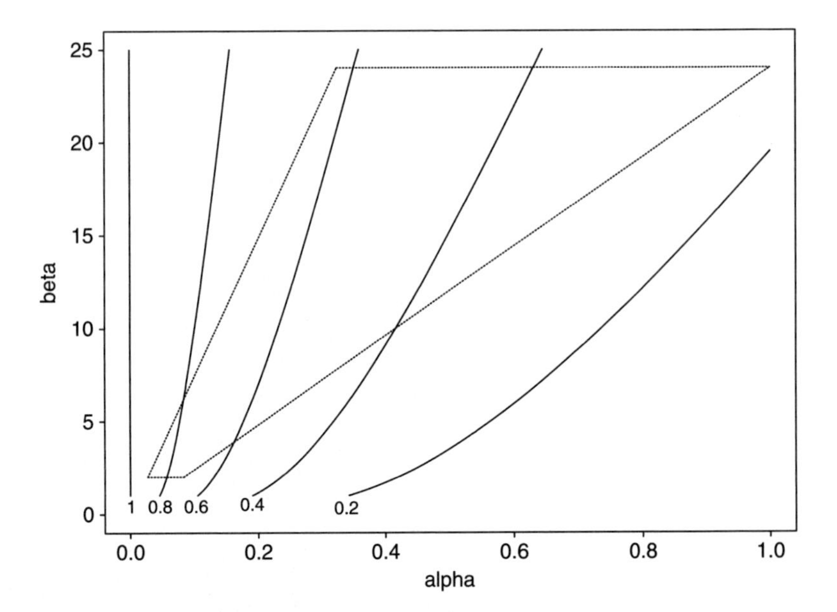

Figure 20.6 Admissible parametric priors given the substantive constraints (inside the dotted region), together with contours of $p(y|\theta > 0, \mathcal{B})$ with $n = 78$; see equation (20.16).

With data $y = (y_1, \ldots, y_n) = (0, \ldots, 0)$,

$$\begin{bmatrix} \text{posterior} \\ \text{odds} \end{bmatrix} = \begin{bmatrix} \text{prior} \\ \text{odds} \end{bmatrix} \cdot \begin{bmatrix} \text{Bayes} \\ \text{factor} \end{bmatrix}$$

$$\left[\frac{p(\theta = 0|y\,\mathcal{B})}{p(\theta > 0|y\,\mathcal{B})}\right] = \left[\frac{p(\theta = 0|\mathcal{B})}{p(\theta > 0|\mathcal{B})}\right] \cdot \left[\frac{p(y|\theta = 0, \mathcal{B})}{p(y|\theta > 0, \mathcal{B})}\right] \tag{20.15}$$

$$= \left(\frac{\lambda}{1-\lambda}\right) \cdot \left[\frac{1}{p(y|\theta > 0, \mathcal{B})}\right]$$

where

$$p(y|\theta > 0, \mathcal{B}) = \int_0^1 p(y|\theta, \theta > 0, \mathcal{B})\, p(\theta|\theta > 0, \mathcal{B})\, d\theta$$

$$= \int_0^1 (1-\theta)^n\, \pi(\theta|\mathcal{B})\, d\theta \equiv B^{-1}. \tag{20.16}$$

With the parametric choice $\pi(\theta|\mathcal{B}) = \text{Beta}(\eta_1, \eta_0)$, the Bayes factor in favour of $\theta = 0$ takes the form

$$B(\eta_1, \eta_0) = \frac{\Gamma(\eta_0)\,\Gamma(\eta_1 + \eta_0 + n)}{\Gamma(\eta_1 + \eta_0)\,\Gamma(\eta_0 + n)}. \tag{20.17}$$

$0 < B^{-1}(\eta_1, \eta_0) < 1$ is a probability and is easier to contour-plot than the Bayes factor; the contours in Figure 20.6 are values of $B^{-1}(\eta_1, \eta_0)$ with $n = 78$. From this it may be seen that in the admissible region B^{-1} takes its minimum value 0.235 at $(\eta_1, \eta_0) = (1, 24)$ and its maximum value of 0.899 at $(\eta_1, \eta_0) = (0.027, 2)$. Thus in the parametric Beta(η_1, η_0) class with the given prior specifications of monotonicity, convexity and bounds on the mean,

$$\frac{1}{0.899} = 1.11 \le \begin{pmatrix} \text{Bayes factor} \\ \text{in favor of} \\ \theta = 0 \end{pmatrix} \le 4.25 = \frac{1}{0.235}, \tag{20.18}$$

i.e., even with 78 consecutive zeros the strength of data evidence that $\theta = 0$ is surprisingly small. Using the informal guidelines of Jeffreys [15], as modified by Kass and Raftery [16], further calculation reveals that one would need more than 450 consecutive zeros for the evidence that $\theta = 0$ (as summarized by the upper bound on the Bayes factor) to pass from 'positive' to 'strong' with the prior specification examined here.

However, this conclusion is conditional on the Beta form of $\pi(\theta|\mathcal{B})$, which is not specified by the scientific context; how much bigger are the bounds when the calculation is made more appropriately over the nonparametric class $\mathcal{C}$ mentioned earlier? Answering this question involves finding the extreme values (here I mean supremum/infimum, which need not be attained) of the integral

$$I = I(\pi) = \int_0^1 (1-\theta)^n\, \pi(\theta|\mathcal{B})\, d\theta \tag{20.19}$$

when π ranges over $\mathcal{C}^*$, the set of functions $\pi(\theta|\mathcal{B})$: $[0, 1] \to \Re$ in the constraint set

$$\left\{ \begin{array}{c} \pi(\theta|\mathcal{B}) \geq 0, \quad \int_0^1 \pi(\theta|\mathcal{B})\, d\theta = 1, \\ (*)\ \pi \text{ is monotone nonincreasing} \\ 0 < \alpha_1 \leq \int_0^1 \theta\, \pi(\theta|\mathcal{B})\, d\theta \leq \beta_1 \leq \frac{1}{2} \end{array} \right\}, \tag{20.20}$$

or the set $\mathcal{C}^{**}$ of $\pi(\theta|\mathcal{B})$ in the same constraint set but with $(*)$ replaced by

$$(**)\ \pi \text{ is monotone nonincreasing and convex.} \tag{20.21}$$

20.4.2 A nonparametric solution

Draper and Toland [6] give solutions to the nonparametric specification problems detailed in the previous paragraph, using a method based on functional analysis that appears to be new to the literature; space constraints here permit only a sketch of these results, itemized as follows.

- Let $\mathcal{C}^*$ and $\mathcal{C}^{**}$ be as in (20.20) and (20.21) for $n > 1$. Implementation of the method detailed in [6] leads to the conclusion that

$$\sup_{\pi \in \mathcal{C}^*} \int_0^1 (1-\theta)^n\, \pi(\theta|\mathcal{B})\, d\theta = 1 - \frac{2\,\alpha_1\, n}{n+1}. \tag{20.22}$$

- It turns out that this supremum is not attained by any $\pi \in \mathcal{C}^*$, but instead occurs at the generalized function

$$\pi^*_{\text{sup}}(\theta|\mathcal{B}) = (1 - 2\,\alpha_1)\,\delta_0 + 2\,\alpha_1, \tag{20.23}$$

  where δ_0 is the Dirac delta measure at 0, i.e., the maximizing distribution has a point mass at 0 of size $(1 - 2\,\alpha_1)$ and is otherwise constant at height $2\,\alpha_1$ on $[0, 1]$.

- One of the main ideas in [6] is to (a) identify a relaxed version of the optimization problem and then (b) relate the solutions of the relaxed problem to those of the original problem. To this end, Toland rewrites the primary problem (20.19) and (20.20) as follows:

$$\sup / \inf \left\{ 1 + \int_0^1 \left[(1-\theta)^n - 1 \right] \pi(\theta|\mathcal{B})\, d\theta \right\} \tag{20.24}$$

  over all functions $\pi : [0, 1] \to \Re$ satisfying (20.20). This ensures that hypothesis (H1) in Section 2.1 of [6] holds, and the relaxed problem is then (20.24) over all functions $\pi : [0, 1] \to \Re$ in the relaxed constraint set

$$\left\{ \begin{array}{c} \pi(\theta|\mathcal{B}) \geq 0, \quad \int_0^1 \pi(\theta|\mathcal{B})\, d\theta \leq 1, \\ \pi \text{ is monotone nonincreasing} \\ 0 < \alpha_1 \leq \int_0^1 \theta\, \pi(\theta|\mathcal{B})\, d\theta \leq \beta_1 \leq \frac{1}{2} \end{array} \right\}. \tag{20.25}$$

Table 20.4 Bayes factor bounds as a function of how the prior is specified, with $n = 78, \alpha_1 = \frac{1}{75}$, and $\beta_1 = \frac{1}{25}$.

Specification	Bayes factor	
	Low	High
Parametric	1.11	4.25
Nonparametric C^*	1.03	6.33
Nonparametric C^{**}	1.03	5.29

The main point of the discussion in this case study is the observation that the supremum and infimum of the relaxed problem *are attained* at the constant function $\pi \equiv 2\alpha_1$, and coincide with the supremum and infimum of the primary problem (20.19, 20.20) (even though the latter supremum is *not attained*).

- The infimum of I over $\pi \in C^*$ turns out to be

$$\inf_{\pi \in C^*} \int_0^1 (1-\theta)^n \, \pi(\theta|\mathcal{B}) \, d\theta = \frac{1-(1-2\,\beta_1)^{n+1}}{2\,\beta_1\,(n+1)}; \tag{20.26}$$

the infimum is attained in C^* by

$$\pi^*_{\text{inf}}(\theta|\mathcal{B}) = \left\{ \begin{array}{cc} \frac{1}{2\,\beta_1} & \text{for } 0 \le \theta \le 2\,\beta_1 \\ 0 & 2\,\beta_1 < \theta \le 1 \end{array} \right\}, \tag{20.27}$$

i.e., a piecewise constant density (histogram).

- When convexity is added the supremum is unchanged, but the infimum of I over C^{**} is

$$2\left[\frac{1}{3\,\beta_1\,(n+1)} - \frac{1-(1-3\,\beta_1)^{n+2}}{(3\beta_1)^2\,(n+1)(n+2)}\right] \tag{20.28}$$

and occurs at the function

$$\pi^{**}_{\text{inf}}(\theta|\mathcal{B}) = \left\{ \begin{array}{cc} \frac{2}{(3\,\beta_1)^2}(3\,\beta_1-\theta) & \text{for } 0 \le \theta \le 3\,\beta_1 \\ 0 & 3\,\beta_1 < \theta \le 1 \end{array} \right\}, \tag{20.29}$$

i.e., a piecewise linear density (a frequency polygon).

With $n = 78, \alpha_1 = \frac{1}{75}$, and $\beta_1 = \frac{1}{25}$, the minimum and maximum values of I over C^* are 0.158 and 0.974, respectively, and over C^{**} the minimum rises to 0.189. Table 20.4 summarizes the numerical findings, and Figure 20.7 plots the optimizing densities. Without convexity the nonparametric limits are 8% lower and 49% higher than the parametric values, and even with convexity the corresponding figures are 8% and 24%; the casual adoption of a convenient parametric family satisfying the scientifically motivated monotonicity and convexity constraints has led to noticeably narrower sensitivity bounds than those that more appropriately arise from assuming only monotonicity and convexity.

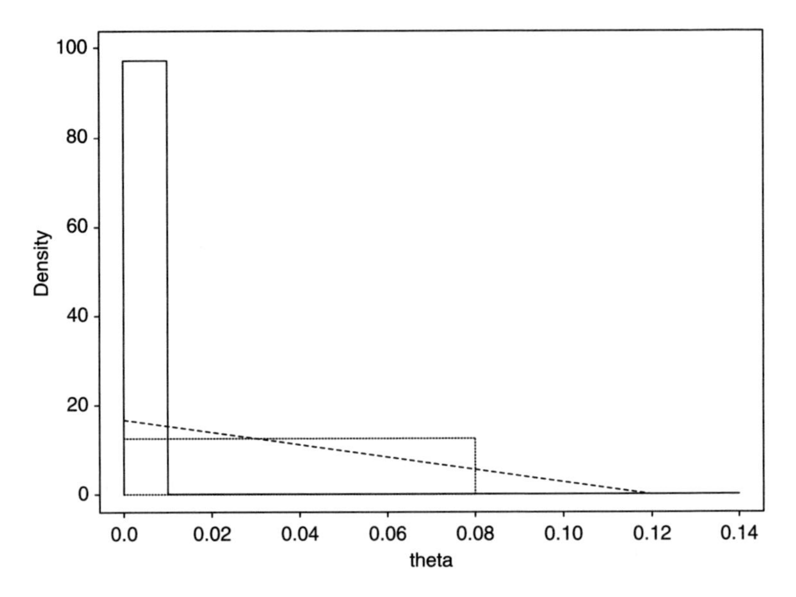

Figure 20.7 Optimal densities π^*_{inf} (long dotted line), π^{**}_{inf} (short dotted line), and approximation to π^*_{sup} in which the point mass at 0 is replaced by a histogram bar of width $\epsilon = 0.01$ (solid line).

20.4.3 A heuristic sketch of the method of proof

The goal is to optimize the integral $I = I(\pi)$ in (20.19) over functions π subject to monotonicity and/or convexity constraints, (20.20) and (20.21), respectively. The basic idea is to think of π as though it were a point in a convex subset of $\Re^k$ and to use intuitions from the geometry of such sets in $\Re^k$ to inform the proof.

(1) The set C^* of admissible functions π may be thought of as like a convex polygon in $\Re^k$, but with infinitely many vertices and edges. In this heuristic proof sketch, consider the situation with $k = 2$ and visualize a polygon with a finite number of vertices and edges. The boundary points are of two kinds: extreme points and all the other boundary points. An extreme point is not an interior point of any line segment in the set. (For example, the corners of a square are its only extreme points.)

(2) The function I in (20.19) is *linear* on π, analogous to a linear function in the plane. But it's important to realize that, because the set of functions satisfying (20.19) and (20.20) is infinite-dimensional, the extreme values may not be attained at points of the constraint set; and in fact the supremum is not attained but the infimum is. Because of this difficulty, Toland replaced the primary problem with a relaxed problem over a larger convex set and showed that in the relaxed problem the extreme values are attained and that they coincide with the extreme values of the original problem. Moreover he showed that the solution of the relaxed problem is finite-dimensional, although the original problem was not.

(3) The level sets of a linear function on $\Re^2$ are parallel straight lines. When the relaxed polygon lies on one side of a level line and touches it, the polygon and the line intersect in one of two ways: at single vertices (one for each of the minimizer and maximizer), or along an edge. In the latter case, even though the intersection set includes points that are not vertices, the set always includes vertices. Thus it suffices in optimizing a linear function to evaluate it at

the vertices of the polygon (this is the basic intuition behind linear programming). This is where the *Krein–Milman theorem* (e.g., [21]) comes into the infinite-dimensional argument (see Theorem 2 in Section 2 of [6]).

(4) Because of geometric constraints of monotonicity on π in the relaxed problem, the analogue of vertices in C^* turns out to be the class of *step functions* with at most three distinct values; thus the infinite set of extrema can be indexed with at most five parameters.

(5) A vertex π of the relaxed polygon has the property that no linear perturbation $(\pi \pm \phi\, \pi)$ away from it for small non-zero ϕ lies completely inside the admissible set. Toland was able to conjecture a particular form of ϕ, namely

$$\phi(\theta) = \int_0^\theta h(t)\, \pi'(t)\, dt \tag{20.30}$$

and show that if π is more complicated than a step function with two distinct values, a non-trivial $(\phi\, \pi \neq 0)$ h can always be found such that all of the constraints in C^* are satisfied. Therefore the vertices are histograms with two or three bars.

(6) Thinking of a two-bar histogram, in the limit as the left-hand bar becomes infinitely tall the maximizer π^*_{sup} over C^* results: a point mass at 0 plus a constant over the rest of $[0, 1]$. In the limit as the right-hand bar goes to 0 the minimizer over C^* is obtained.

(7) When convexity is added, the relaxed polygon vertices become the class of *convex piecewise linear functions* with exactly three distinct segments; here only six parameters are needed. The maximizer over C^{**} remains the same as in C^* because π^*_{sup} is already convex. The minimizer of the relaxed problem turns out to be a two–part piecewise linear function (frequency polygon) with the second segment 0.

(8) When unimodality is assumed instead of the other qualitative constraints examined, a modification of the method Toland employs for dealing with monotonicity is available, because unimodal densities on $[0, 1]$ are non-decreasing on $[0, d)$ and non-increasing on $(d, 1]$ for some $d \in [0, 1]$.

Results similar to the findings here under monotonicity have been obtained elsewhere in the Bayesian robustness literature by quite different means, through the use of Khintchine's Theorem (e.g., [8]) on generating unimodal densities as mixtures of uniform distributions. The approach sketched here, via functional analysis, both subsumes the condition of unimodality and yields new results under the alternative qualitative specifications of monotonicity and convexity.

20.4.4 Conclusions

As more moment constraints are added to the quantitative mean constraint examined here to increase realism, e.g.,

$$\sigma^2_{low} \leq \int_0^1 [\theta - E(\theta|\mathcal{B})]^2\, \pi(\theta|\mathcal{B})\, d\theta \leq \sigma^2_{high}, \tag{20.31}$$

the optimal nonparametric solutions become k–part piecewise linear functions (frequency polygons) with increasing k, approaching the smoothness built into parametric families like Beta (α, β). Thus continuous parametric assumptions are equivalent to infinite sets of moment constraints, i.e., when You choose continuous parametric priors You're probably assuming more than You think You are.

The set of practical problems in which

(a) the prior really matters and
(b) off-the-shelf parametric specifications are often used instead of qualitative descriptions involving shape (e.g., number of modes, monotonicity, convexity, smoothness) and substantive bounds on quantitative descriptions (e.g., moments or quantiles)

is larger than is generally acknowledged. The method of proof offered here shows promise to inform Bayesian sensitivity analysis in a wide variety of such problems, because all of the above characteristics may be enforced with linear constraints through derivatives and integrals of π.

Postscript On the 79th visit to the machine it yielded a free cup of tea.

Acknowledgements

I'm grateful to Tim Hanson and Alejandro Jara (a) for help with the programming that led to the results presented in Section 20.3 and (b) for comments that aided interpretation of the Bayesian nonparametric findings; to Milovan Krnjajić for helpful discussions about calibration cross-validation; to John Toland for doing all of the heavy lifting in the proofs underlying Section 20.4; and to Jim Berger, Brad Efron, Richard Olshen, Luc Tartar, and Stephen Walker for helpful comments and references. Membership on this list does not imply agreement with the conclusions drawn here, nor are any of these people responsible for any errors that may be present.

References

[1] Cox, R. T. (1946). Probability, frequency, and reasonable expectation. *American Journal of Physics*, **14**, 1–13.
[2] de Finetti, B. (1937). La prévision: ses lois logiques, ses sources subjectives. *Annales de l'Institut Henri Poincaré*, **7**, 1–68.
[3] Draper, D. (1995). Assessment and propagation of model uncertainty (with discussion). *Journal of the Royal Statistical Society (Series B)*, **57**, 45–97.
[4] Draper, D. (2012). Bayesian model specification: towards a theory of applied statistics. Submitted.
[5] Draper, D. and Krnjajić, M. (2012). Calibration results for Bayesian model specification. Submitted.
[6] Draper, D. and Toland, J. (2012). Bayesian non-parametric prior specification. Technical Report, Department of Applied Mathematics and Statistics, University of California, Santa Cruz.
[7] Escobar, M. and West, M. (1995). Bayesian density estimation and inference using mixtures. *Journal of the American Statistical Association*, **90**, 577–588.
[8] Feller, W. (1971). *An Introduction to Probability Theory and its Applications* (2nd edn), Volume 2. Wiley, New York.
[9] Ferguson, T. S. (1974). Prior distributions on spaces of probability measures. *Annals of Statistics*, **2**, 209–230.
[10] Freedman, D. (1963). On the asymptotic behavior of Bayes' estimates in the discrete case. *Annals of Mathematical Statistics*, **34**, 1194–1216.
[11] Freedman, D., Pisani, R. and Purves, R. (2007). *Statistics* (4th edn). Norton, New York.

[12] Good, I. J. (1950). *Probability and the Weighing of Evidence.* Charles Griffin, London.
[13] Hanson, T. (2006). Inference for mixtures of finite Pólya tree models. *Journal of the American Statistical Association,* **101**, 1548–1565.
[14] Jara, A., Hanson, T., Quintana, F., Müller, P. and Rosner, G. (2011). DPpackage: Bayesian semi- and nonparametric modeling in R. *Journal of Statistical Software,* **40**, 1–30.
[15] Jeffreys, H. (1967). *Theory of Probability* (3rd edn). Oxford University Press, Oxford.
[16] Kass, R. E. and Raftery, A. E. (1995). Bayes factors. *Journal of the American Statistical Association,* **90**, 773–795.
[17] Krnjajić, M., Kottas, A. and Draper, D. (2008). Parametric and non-parametric Bayesian model specification: a case study involving models for count data. *Computational Statistics and Data Analysis,* **52**, 2110–2128.
[18] Lavine, M. (1992). Some aspects of Pólya tree distributions for statistical modeling. *Annals of Statistics,* **20**, 1222–1235.
[19] Leamer, E. E. (1978). *Specification Searches: Ad Hoc Inference with Nonexperimental Data.* Wiley, New York.
[20] Mosteller, F. and Wallace, D. L. (1984). *Applied Bayesian and Classical Inference: The Case of The Federalist Papers.* Springer-Verlag, New York.
[21] Rudin, W. (1991). *Functional Analysis* (2nd edn). McGraw-Hill, New York.

21 Case studies in Bayesian screening for time-varying model structure: the partition problem

ZESONG LIU, JESSE WINDLE AND JAMES G. SCOTT

21.1 Introduction

Problems of model selection are often thought to be among the most intractable in modern Bayesian inference. They may involve difficult high-dimensional integrals, nonconcave solution surfaces or large discrete spaces that cannot be enumerated. All of these traits pose notorious computational and conceptual hurdles.

Yet model-choice problems—particularly those related to feature selection, large-scale simultaneous testing and inference of topological network structure—are also some of the most important. As modern datasets have become larger, they have also become *denser*—that is, richer with covariates, more deeply layered with underlying patterns, and indexed in ever more baroque ways (e.g. x_{ijkt}, rather than just x_{ij}). It is this complex structure, more than the mere tallying of terabytes, that defines the new normal in twenty-first-century statistical science. There is thus a critical need for Bayesian methodology that addresses the challenges posed by such datasets, which come with an especially compelling built-in case for sparsity.

The difficulties of model selection are further exacerbated when the structure of the model is not 'merely' unknown, but also changes as a function of auxiliary information. We call this the *partition problem*: the auxiliary information defines an unknown partition over sample space, with different unknown models obtaining within different elements of the partition. Here are three examples where model structure plausibly covaries with external predictors.

Subgroup analysis How can clinicians confront the multiplicity problem inherent in deciding whether a new cancer drug is effective for a specific subgroup of patients, even if it fails for the larger population? Here the model is simply a binary indicator for treatment effectiveness, while the partitions are defined by diagnostically relevant covariates—for example, age, sex or smoking status. No good approaches exist, Bayesian or otherwise, that are capable of systematically addressing this problem. The difficulty is that examining all possible partitions wastes power: many substantively meaningless or nonsensical partitions are considered, and must receive prior probability at the expense of the partitions we care about.

Partitioned variable selection A patient comes to hospital complaining of a migraine. The hospital wishes to use the patient's clinical history to diagnose whether the migraine may portend a subarchnoid haemorrhage, a catastrophic form of brain bleed. Such haemorrhages are thought to be etiologically distinct for children and adults. Thus the age of the patient influences which aspects of her clinical history (i.e. variables) should be included in the predictive model.

Network drift A delay-tolerant network (DTN) for communication devices has few instantaneous origin-to-destination paths. Instead, most messages are passed to their destination via a series of local steps (at close range) from device to device. In such a setting, it helps to know the underlying social network of users in the model—that is, who interacts with whom, and how often—in order to predict the likelihood of success for specific directed transmissions. Thus the time-varying topological structure of the *social* network has important implications for the efficient routing of traffic within the *device* network.

All three of these problems recall the literature on tree modelling, including [4], [6], [7] and [11]. Yet to our knowledge no one has studied tree models wherein one large discrete space (for example, predictors in or out of a linear model) is wrapped inside another large discrete space (trees).

This poses all the usual computational and modelling difficulties associated with tree structures, and large discrete spaces more generally. But it also poses a major, unique challenge. In all existing applications of Bayesian tree models we have encountered, the collapsed sampler (whereby node-level parameters are marginalized away and MCMC moves are made exclusively in tree space) is the computational tool of choice. But in the partition problem, the bottom level parameter in the terminal nodes of the tree is itself a model indicator, denoting an element of some potentially large discrete space. This makes it difficult to compute the marginal likelihood of a particular tree in closed form, since doing so would involve a sum with too many terms. The collapsed sampler therefore cannot always be used.

Trees are, of course, just one possible generative model for partitions of a sample space based on such auxiliary information. Others include species-sampling models, coalescent models or urn models (such as those that lie at the heart of Bayesian nonparametrics). But no matter what partitioning model is entertained, new models and algorithms are necessary to make inferences and quantify uncertainty for this very general class of problems.

This chapter presents two case studies of datasets within this class. For both data sets the auxiliary information is time. In the first case study, we identify time-varying graphical structure in the covariance of asset returns from major European equity indices from 2006–2010. This structure has important implications for quantifying the notion of financial contagion, a term often mentioned in the context of the European sovereign debt crisis of this period. In the second case study, we screen a large database of historical corporate performance in order to identify specific firms with impressively good (or bad) streaks of performance.

Our goal in these analyses is not to address all the substantive issues raised by each dataset. Rather, we intend: (1) to present our argument for the existence of non-trivial dynamic structure in each dataset, an argument that can be made using simple models; (2) to draw parallels between the case studies, both of which exemplify the partition problem quite well; and (3) to identify certain aspects of each model that must be generalized in future work if these case studies are to provide a useful template for other datasets.

21.2 Case study I: financial contagion and dynamic graphical structure

21.2.1 Overview

During times of financial crisis, such as the bursting of the US housing bubble in 2008 and the European sovereign-debt crisis in 2010, the co-movement of asset prices across global markets is hypothesized to diverge from its usual pattern. Many financial theorists predict, and many empiricists have documented, changes in market relationships after these large market shocks. It is important to track these large shocks and measure their impacts on the global economy so that we can better understand future market behaviours during times of crisis.

In general, the idea that market relationships change after large shocks is called contagion [9]. In the literature, there has been a lengthy debate over precise definition of this term. As a practical matter, we define contagion as significant change in the pattern of correlation in the residuals from an asset-pricing model during times of crisis, following in the tradition of previous authors [e.g. 1, 2, 10]. Focusing primarily on how the relationships between markets change, we want to determine whether large shocks have significant impact on the subsequent interactions between markets.

The standard way to study the co-movements and interdependent behaviour of markets is by looking at the covariance matrices of returns across different countries. In this case study, we explore ways of estimating this covariance structure to study the change of the market dynamic over time. Normally, when constructing covariance matrices, the algorithms applied are computationally identical to repeated applications of least squares regressions. Instead, we apply the ideas of Bayesian model selection, using the Bayesian information criterion, or BIC [16], to approximate the marginal likelihoods of different hypotheses, and a flat prior over model space.

In applying this method, we uncover many signs of contagion, which manifests itself as time-varying graphical structure in the covariance matrix of returns. For example, if we look at the relationship between Italy and Germany, the traditionally positive correlation between the countries changes sign during the sovereign debt crisis. This provides just one example of the evidence for contagion discovered in these investigations.

21.2.2 Contagion in factor asset-pricing models

The sovereign debt crisis started when Greece became in danger of defaulting on its debt. For years, Greece had been a rapidly growing economy with many foreign investors. This strong economy was able to withstand large government deficits that Greece had during that time. But after the worldwide 2008 financial crisis hit, two of the country's largest industries, tourism and shopping, were badly affected. This downturn caused panic in the Greek economy. Although Greece was not the only country that confronted debt problems, its debt-to-GDP ratio was judged excessively high by markets and ratings agencies, reaching 120% in 2010. Moreover, one of the major fears that arose during this period was that investors would lose faith in other similarly situated Euro-zone economies, which could cause something similar to a run on a bank.

One of the major events in this episode occurred on May 9, 2010. On that day the 27 member states of the EU agreed to create the European Financial Stability Facility, a legal instrument aimed at preserving financial stability in Europe by providing financial assistance to states in need. This legislation was upsetting to countries with large healthy economies such as Germany, whose electorate focused on the negative effects of the bailout. In light of these developments, not only do we want to show that the pattern of asset-price correlation changed, but we also want to make sense of these changes by linking them to the news headlines about bailouts.

Our raw data are daily market returns from equity indices corresponding to nine large European economies—Germany, the UK, Italy, Spain, France, Switzerland, Sweden, Belgium and the Netherlands—from December 2005 to October 2010. We do not include Greece because of its small size relative to the other economies of Europe, but we do include the Euro–Dollar exchange rate as a tenth column in the dataset.

By our definition of contagion, we need to examine the residuals of the returns within the context of an asset-pricing model. Specifically, we use a four-factor model where

$$E(y_{it} \mid EU, US) = \beta_i^{US} x_t^{US} + \beta_i^{EU} x_t^{EU} + \gamma_i^{US} \delta_t^{US} + \gamma_i^{EU} \delta_t^{EU}$$

where y_{it} is the daily excess return on index i; x_t^{US} is the daily excess return on a value-weighted portfolio of all US equities; x_t^{EU} is the daily excess return on the EU-wide index of Morgan Stanley Capital International; δ_t^{US} is the volatility shock to the US market; and δ_t^{EU} is the excess volatility shock to the European market. The excess shock is defined as the residual after regressing the EU volatility shock upon the US volatility shock. This is necessary to avoid marked collinearity, since the US volatility shock strongly predicts the EU volatility shock. These volatility factors are calculated using the particle-learning method from [14], not described here.

In this model, the loadings β_i^{US} and β_i^{EU} measure the usual betas relative to the US and Europe-wide equity markets. Thus we have controlled for regional and global market integration via an international CAPM-style model [2]. The loadings γ_i^{US} and γ_i^{EU}, meanwhile, measure country-level dependence upon global and regional volatility risk factors. As shown in [14], these loadings can be interpreted in the context of a joint model that postulates correlation between shocks to aggregate market volatility and shocks to contemporaneous country-level returns.

21.2.3 A graphical model for the residuals

The above model can be fitted using ordinary least squares, leading to an estimate of all model parameters along with a set of residuals ϵ_{it} for all indices. We now turn to the problem of imposing *graphical* restrictions on the covariance matrix of these residuals.

A Gaussian graphical model defines a set of pairwise conditional-independence relationships on a p-dimensional zero-mean, normally distributed random vector (here denoted x). The unknown covariance matrix Σ is restricted by its Markov properties; given $\Omega = \Sigma^{-1}$, elements x_i and x_j of the vector x are conditionally independent, given their neighbours, if and only if $\Omega_{ij} = 0$. If $G = (V, E)$ is an undirected graph whose nodes represent the individual components of the vector x, then $\Omega_{ij} = 0$ for all pairs $(i, j) \notin E$. The covariance matrix Σ is in $M^+(G)$, the set of all symmetric positive-definite matrices having elements in Σ^{-1} set to zero for all $(i, j) \notin E$.

We construct a graph by selecting a sparse regression model for each country's residuals in terms of all the other countries, as in [8] and [20]. We then cobble together the resulting set of conditional relationships into a graph to yield a valid joint distribution. Each sparse regression model is selected by enumerating all 2^9 possible models, and choosing the one that minimizes the BIC. This leads to a potentially sparse model for $E(\epsilon_{it} \mid \epsilon_{j,t}, j \neq i)$.

In this manner, an adjacency matrix can be constructed for the residuals from the four-factor model. The (i, j) element of the adjacency matrix is equal to 1 if the residuals for countries i and j both appear in each other's conditional regression models, and is equal to 0 otherwise. One may also assemble the pattern of coefficients from these sparse regressions to reconstruct the covariance matrix for all the residuals, denoted Σ.

We actually look at adjacency matrices and covariance matrices on a rolling basis, since we believe that there are changes in Σ over time. Each window involves a separate set of regressions for a period of 150 trading days, which is thirty weeks of trading, or about seven months. We shift the window in five-day increments, thereby spanning the whole five-year time period in our study. For each 150-day window, we refit the estimate for Σ_t using the entire graphical-model selection procedure. For the sake of comparison, we also include the estimates of Σ_t using OLS, ridge regression, and lasso regression.

21.2.4 Results

From Figure 21.1, which depicts the rolling estimates of the Italy–Germany and Spain–Germany regression coefficients, it is clear that there are nonrandom patterns that remain in the residuals. If the factor model fully explained the co-movements in the European market, we would expect the residuals to have no covariance and look like noise. We would also expect that the Bayesian variable-selection procedure would give regression coefficients of 0. To be sure, the data support the hypothesis that specific elements of the precision matrix $\Omega_t = \Sigma_t^{-1}$ are zero over certain time periods. An example of this is the ITA-DEU coefficient during much of 2008. Yet it is patently not the case that all such elements are zero for all time periods: the standard significance test of an empty graph is summarily rejected ($p < 10^{-6}$) at all time points. For details of this test, see [13] and Proposition 1 of [5].

There are several explanations for the observed correlation between the residuals. Firstly, it is highly probable that the factor models are imperfect. Regressing only on the market returns and market volatility, the four-factor model involves a substantial simplification of reality. For example, during the sovereign debt crisis, we could reasonably add an explanatory variable that takes into account the change in likelihood of a bailout. Secondly, even if we imagine that the four-factor model is the true model for this system of markets, we only have proxies for both the global and local returns and volatility. Specifically, using the US market return as a proxy for the world market return is a reasonable estimate, but it is far from perfect. Moreover, the factors that measure market volatility are at best a measure of the average market volatility over a short time span. This could potentially distort the residuals, since we cannot observe volatility spikes on a more granular scale. Most likely, the correlations in the residuals stems from some mixture of these two effects.

The same conclusion about significant residual correlation is also borne out by examining the time-varying topology of the graph itself. While it is difficult to visualize the time-varying nature of the network's structure in its entirety, we can look at quantities such as how the adjacency of a specific node in the graph—that is, how many neighbours it has—changes over time. Figure 21.2 shows the estimated adjaceny degrees for Sweden and the UK, two non-Euro countries. Again we see that the factor model is not perfect, as the residuals still exhibit correlation. On the other hand, we see that the degree of each vertex is not 9 at every time point. This means that shrinkage is often useful: it is clear that estimating $p(p-1)/2$ separate correlation parameters is a poor use of the data, and will lead to estimators with unnecessarily high variance. We can almost certainly reduce the required number of parameters while still obtaining a good estimate. This illustrates the utility of the graphical modelling approach.

Finally, the relationship between the Spain and Germany residuals is easily interpretable in terms of the underlying economic picture. In the summer of 2010, there is an apparent divergence from the historical norm, precisely coinciding with the Greek sovereign-debt crisis and associated bailout. The historically aberrant negative correlations between the residuals from Germany and the southern countries suggests that markets reacted very differently in these two countries to news of the period. A useful comparison is with the period of September and October 2008, when the global financial crisis associated with the bursting of the housing bubble was at its peak. These were global rather than EU-centric events, and no such divergence was observed involving the German-market residuals.

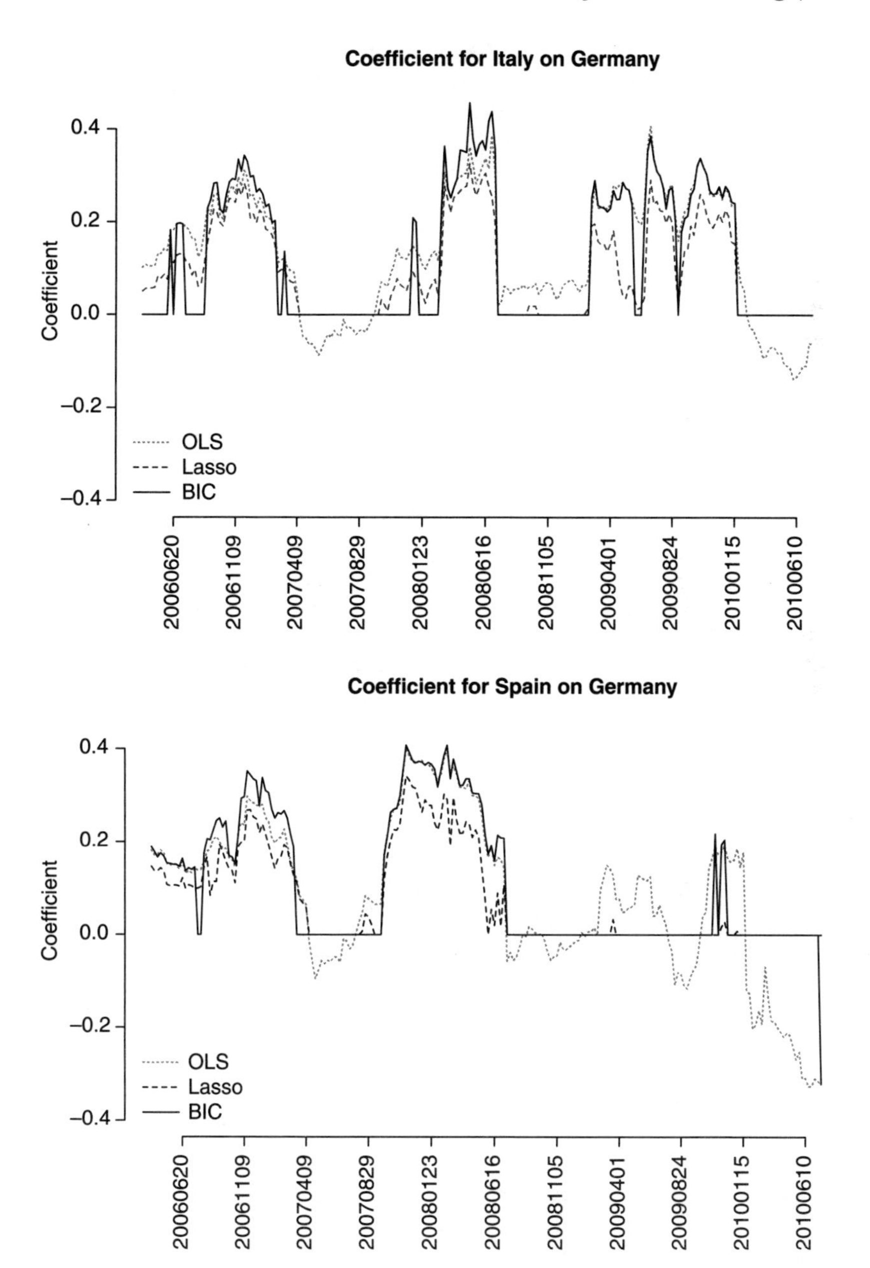

Figure 21.1 Estimated regression coefficients for Italy (top) and Spain (bottom) on Germany, where the coefficients are calculated on a rolling basis using Bayesian model selection via BIC.

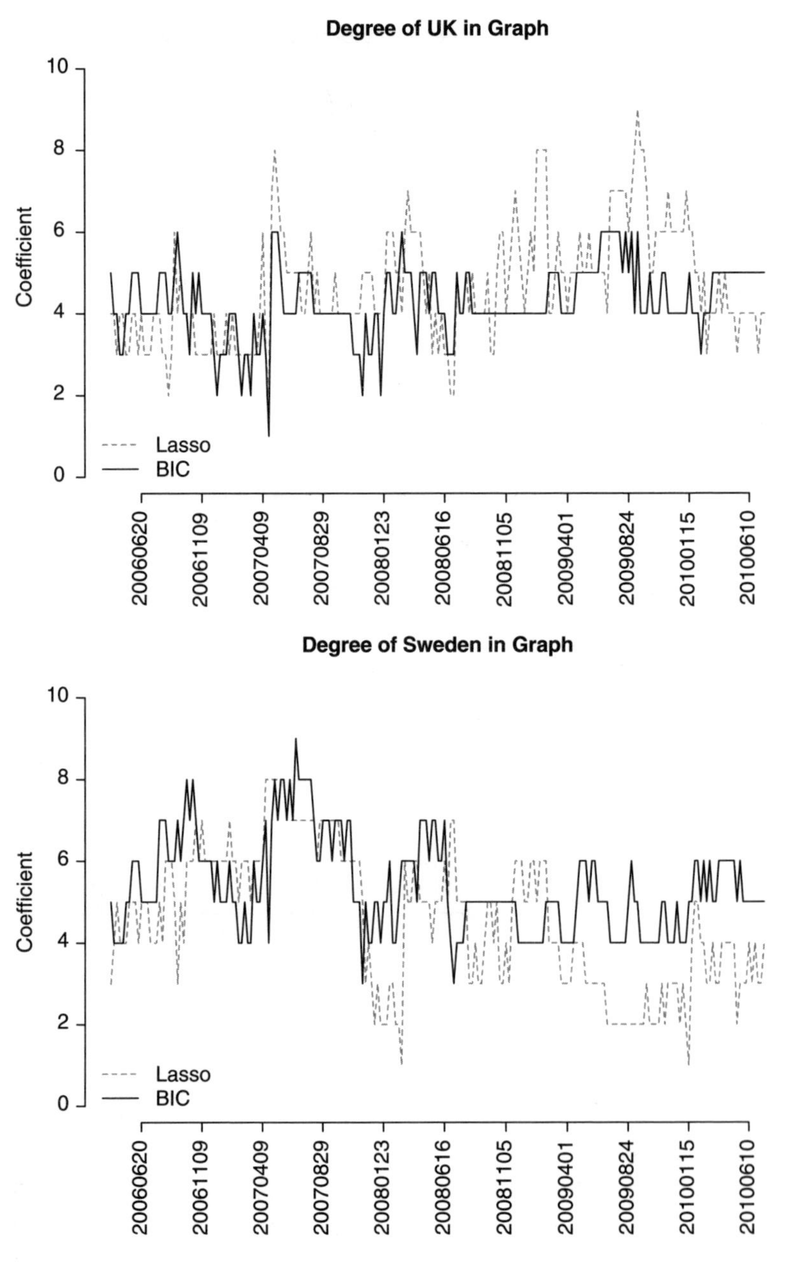

Figure 21.2 Estimated adjacency degree in the time-varying graph of residuals for the UK (top) and Sweden (bottom). There seems to be clear evidence of time-varying topological structure in the graph.

Our approach provides initial evidence for contagion effects, but has some important limitations. In particular, we have estimated time-varying graphical structure using a moving-window variable selection approach, which does not explicitly involve a dynamic model over graph space. Some authors have made initial efforts in studying dynamic graphs [e.g. 21, 22], but much further work remains to be done to operationalize the notion of contagion within this framework. The issue is that we expect contagion to be associated with a sharp change in the underlying graphical structure of the residual covariance matrix, as opposed to the locally drifting models considered by these other authors. Such sharp changes are likely obscured by our rolling-regression approach, in that only about 3% of the data changes in each new window.

21.3 Case study II: simultaneous change-point screening and corporate out-performance

21.3.1 Overview

In this case study, we compare publicly traded firms against their peer groups using a standard accounting metric known as ROA, or return on assets. The dataset comprises 645 456 company-year records from 53 038 companies in 93 different countries, spanning the years 1966–2008.

Just as in the previous example, we will attempt to uncover substantively meaningful changes over time in the underlying model for each time series. Let y_{it} denote the benchmarked ROA observation for company i at time t; let $\mathbf{y_i}$ denote the whole vector of observations for company i observed at times $\mathbf{t_i} = (\mathbf{t_1}, \ldots, \mathbf{t_{n_i}})$; and let $\mathbf{Y}$ denote the set of $\mathbf{y_i}$ for all i. We say that the observations have been 'benchmarked' to indicate that they have undergone a pre-processing step that removes the effects of a firm's size, industry and capital structure. For details, see [18].

The goal is to categorize each time series i as either signal or non-signal, both to be defined shortly. The model we consider takes the general form

$$\mathbf{y_i} = \mathbf{f_i} + \boldsymbol{\epsilon_i},\ \boldsymbol{\epsilon_i} \sim \text{Noise} \tag{21.1}$$

$$\mathbf{f_i} \sim \omega \cdot \text{Signal} + (1 - \omega) \cdot \text{Null} \tag{21.2}$$

where Signal is the distribution describing the signal and Null is the distribution describing the non-signal. In this case, 'noise' should not be conflated with actual measurement error. Instead, it represents short-term fluctuations in performance that are not indicative of any longer-term trends, and are thus not of particular relevance for understanding systematic out-performance.

We can rephrase this model as

$$\begin{aligned} \mathbf{y_i} &= \mathbf{f_i} + \boldsymbol{\epsilon_i},\ \boldsymbol{\epsilon_i} \sim \text{Noise} \\ \mathbf{f_i} &\sim \text{F}_{\gamma_i},\ \text{F}_1 = \text{Signal},\ \text{F}_0 = \text{Null} \\ \gamma_i &\sim \text{Bernoulli}(\omega) \end{aligned} \tag{21.3}$$

which provides us with the auxiliary variable γ_i that determines whether a time series i is either signal or noise. Thus the posterior distribution $p(\gamma_i = 1|\mathbf{Y})$ is a measure of how likely time series i is signal. One can sort the data points from most probable to least probable signal by simply ranking $p(\gamma_i = 1|\mathbf{Y})$. Importantly, these posterior probabilities will contain an automatic penalty for data dredging: as more unimpressive firms are thrown into the cohort, the posterior distribution for ω will favour increasingly smaller values, meaning that all observations have to work harder to overcome the prior bias in favour of the null [see, e.g. 3, 19].

Suppose, for example, that $\mathbf{f_i} = \mathbf{0}$ corresponds to the average performance of a company's peer group, and that we want deviations from zero to capture long-range fluctuations in $\mathbf{y_i}$ from this peer-group average—that is, changes in a company's fortunes or long-run trends that unfold over many years. Several previous authors have proposed methodology for the multiple-testing problem that arises in deciding whether $\mathbf{f}_i = 0$ for all firms simultaneously [12, 15, 17]. The origin of such a testing problem lies in the so-called 'corporate success study', very popular in the business world: begin with a population of firms, identify the successful ones, and then look for reproducible behaviours or business practices that explain their success.

The hypothesis that $\mathbf{f_i} = \mathbf{0}$, then, implies that a firm is no better, or worse, than its peer group over time. One way to test this hypothesis is by placing a Gaussian-process prior on those $\mathbf{f_i}$s that differ from zero. Suppose, for example, we have an i.i.d. error structure and some common prior inclusion probability:

$$\begin{aligned} \mathbf{y_i} &= \mathbf{f_i} + \epsilon_\mathbf{i},\ \epsilon_i \sim \mathrm{N}(\mathbf{0}, \sigma^2 \mathbf{I}) \\ \mathbf{f_i} &\sim \mathrm{F}_{\gamma_i},\ \mathrm{F}_1 = \mathrm{N}(0, \sigma^2 K(\mathbf{t_i})),\ \mathrm{F}_\mathbf{0} = \delta_\mathbf{0} \\ \gamma_i &\sim \mathrm{Bernoulli}(\omega) \end{aligned}$$

where $K(\mathbf{t})$ is the matrix produced by a covariance kernel $k(\cdot, \cdot)$, evaluated at the observed times $\mathbf{t} = (\mathbf{t_1}, \ldots, \mathbf{t_M})'$. The (i, j) entry of K is:

$$K_{i,j} = k(t_i, t_j).$$

Typically the covariance function will itself have hyperparameters that must be either fixed or estimated. A natural choice here is the squared-exponential kernel:

$$k(t_i, t_j) = \kappa_1 \exp\left\{ -\frac{(t_i - t_j)^2}{2\kappa} \right\} + \kappa_3 \delta_{t_i, t_j}$$

where δ_{t_i,t_j} is the Kronecker delta function. If the three κ hyperparameters are fixed, then this is just a special case of a conjugate multivariate Gaussian model, and the computation of the posterior model probabilities $p(\gamma_i = 1|\mathbf{Y})$ may proceed by Gibbs sampling. As shown in [17], this basic idea may be generalized to more complicated models involving autoregressive error structure and Dirichlet-process priors on non-zero trajectories.

21.3.2 Detecting regime changes

One shortcoming of this strategy is that $\mathbf{f}_i$ is assumed to be either globally zero or globally non-zero. The partition problem is different, and captures an important aspect of reality ignored by the model above: the signal of interest may not involve consistent performance, but rather a precipitous rise or fall in the fortunes of a company.

We therefore consider the possibility that each firm's history may, though not necessarily, be divided into two epochs. The separation of these epochs corresponds to some sort of significant schism in performance between time periods. For instance, a positive jump might arise by virtue of a drug patent or the tenure of an especially effective leader—someone like Steve Jobs of Apple, or Jack Welch of General Electric. There may also be periods of inferior performance when the jump is negative.

We therefore adapt model (21.3) in the following way. With each firm, we associate not a binary indicator of 'signal' or 'noise', but rather a multinomial indicator for the location in time of a major shift in performance. We then index all further parameters in the model by this multinomial indicator:

$$\begin{aligned} \mathbf{y_i} &= \mathbf{f_i} + \boldsymbol{\epsilon}_\mathbf{i},\ \epsilon_\mathbf{i} \sim \text{Noise} \\ \mathbf{f_i} &\sim \text{F}_{\gamma_i} \\ \gamma_i &\sim \text{Multinomial}(\boldsymbol{\omega}) \end{aligned} \tag{21.4}$$

with the convention that $\gamma_i = 0$ denotes the no-split case where $\mathbf{f}_i$ is globally zero, and $\gamma_i = n$ the no-split case where $\mathbf{f}_i$ is globally non-zero.

This differs from the traditional changepoint-detection problem in two respects: we are data-poor in the time direction, but data-rich in the cross-section. Indeed, this case study is the mirror image of the previous one, where the number of time series was moderate the number of observations per times series was large. This fact requires us to consider models that are simpler in the time domain, but also allows us to borrow cross-sectional information across firms for the purpose of estimating shared model parameters. This is particularly important for the multinomial parameter $\boldsymbol{\omega}$, which lives on the $(n + 1)$-dimensional simplex, and descibes the population-level distribution of changepoints.

There are many potential choices for F_k and Noise. For instance, we could choose

$$F_k = \text{N}(0, C_k) \text{ and Noise} = \text{N}(0, \Sigma)$$

where Σ describes the covariance structure of the noise and C_k describes a split in epochs at time k (again recalling the convention that C_0 is degenerate at zero and that C_n corresponds to no split at all). Of course, we need not limit ourselves to this interpretation, and may intead choose a collection of $\{C_k\}$ that embodies some other substantive meaning. For instance, we could consider the collection $C_{i,k}$ where $C_{1,k}$ corresponds to a small jump at time k and $C_{2,k}$ corresponds to a large jump at time k. We could also generalize to multiple regime changes per firm. For now, however, we consider the simpler case where there can be at most one shift, and where all shifts are exchangeable. Recall that we intend such an intentionally oversimplified model to be useful for high-dimensional screening, not nuanced modelling of an individual firm's history.

As before, both $\{C_k\}$ and Σ will include some sort of hyperparameters, which we will denote as $\boldsymbol{\theta}$ for now. Conditional on the hyperparameters, we may write the distribution of the data as

$$\mathbf{y_i}|\gamma_\mathbf{i} \sim \text{N}(\mathbf{0}, \mathbf{C}_{\gamma_\mathbf{i}} + \Sigma)$$

which are conditionally independent across i. Thus when calculating the posterior distribution using a Gibbs sampler, the conditional distribution $p(\boldsymbol{\gamma}|Y, \boldsymbol{\theta}, \boldsymbol{\omega})$ will conveniently decompose into a product over $p(\gamma_i|\mathbf{y_i}, \boldsymbol{\theta}, \boldsymbol{\omega})$. Moreover, when each time series has its own hyperparameter $\boldsymbol{\theta}_i$ and the only shared information across time series is the multinomial probability vector $\boldsymbol{\omega}$, the posterior calculation simplifies further. In particular, after marginalizing out each $\boldsymbol{\theta}_i$, the only quantities that must be sampled are $\{p(\gamma_i|\mathbf{y_i}, \boldsymbol{\omega})\}$ and $p(\boldsymbol{\omega}|\mathbf{Y}, \boldsymbol{\gamma})$.

Many of the details of such a model are encoded in particular choices for the covariance matrices describing signals and noise. As an illustration of this general approach, consider a simple model in which $\mathbf{f_i}$ is piecewise constant. This assumption is reasonable given the relatively short length of each time series; in any case, one may think of it as a locally constant approximation to the true model. Suppressing the index i for the moment, we write the model as

$$\begin{aligned}
y_{\mathbf{s}} &= f_{\mathbf{s}} + \epsilon_{\mathbf{s}}, \epsilon_{\mathbf{s}} \sim \mathrm{N}(0, \sigma_{\mathbf{s}}^2 I), \\
f_{\mathbf{s}} &\equiv \theta, \\
\theta &\sim \mathrm{N}(0, \sigma_{\mathbf{s}}^2 \tau^2), \\
\sigma_{\mathbf{s}}^2 &\sim \mathrm{IG}(a/2, b/2)
\end{aligned}$$

where $\mathbf{s}$ is some subsequence of the times $\{1, \ldots, n\}$. Marginalizing over θ and $\sigma_{\mathbf{s}}^2$ yields a multivariate-T marginal:

$$\begin{aligned}
y_{\mathbf{s}} &\sim \mathrm{T}_{a+|\mathbf{s}|}(0, R_{\mathbf{s}}) \\
R_{\mathbf{s}} &= \frac{a}{b}(I_{\mathbf{s}} + \tau^2 S_{\mathbf{s}}).
\end{aligned}$$

If we know $\theta = 0$, or in other words that $\theta \sim \delta_0$, then $R_{\mathbf{s}} = (a/b) I_{\mathbf{s}}$. Notice that this formulation automatically handles missing data, since one can simply exclude those times from $\mathbf{s}$.

This can be phrased in terms of model (21.4) by letting $S_k = \mathbf{1}_k \mathbf{1}_k'$ be the $k \times k$ matrix of ones, and defining

$$\mathrm{F}_k = \mathrm{N}(0, \tau C_k)$$

where

$$C_k | \sigma_{ij}^2 = \begin{bmatrix} \sigma_{i1}^2 S_k & 0 \\ 0 & \sigma_{i2}^2 S_{n-k} \end{bmatrix} \text{ for } k = 1, \cdots, n-1$$

$\mathrm{F}_n = N(0, \sigma_i^2 \tau S_n)$, and $\mathrm{F}_0 = \delta_0$ (that is $C_n = \sigma_i^2 S_n$ and $C_0 = 0$). Furthermore, we define the prior over the residuals so that it, too, depends on the index γ_i:

$$\text{Noise} = \mathrm{E}_{\gamma_i} = N(0, \Sigma_{\gamma_i})$$

where $\Sigma_k | \sigma_{ij}^2 = \sigma_{1i}^2 I$ for $k = 0$ and n, and where

$$\Sigma_k | \sigma_{ij}^2 = \begin{bmatrix} \sigma_{i1}^2 I & 0 \\ 0 & \sigma_{i2}^2 I \end{bmatrix} \text{ for } k = 1, \ldots, n-1$$

Finally, for $p(\boldsymbol{\omega})$ we assume a conjugate Dirichlet prior (details below).

This simple model has many advantages: it is analytically tractable; it handles missing data easily; it allows for the possibility of a sharp break between epochs; and it allows for preprocessing of all marginal likelihoods, which saves time in the Gibbs sampler. To see this, observe that the conditional posterior distribution for $\boldsymbol{\gamma}$ is

$$p(\boldsymbol{\gamma} | \mathbf{Y}, \boldsymbol{\omega}) \propto \prod_{\mathbf{i}} \mathbf{p}(\mathbf{y_i} | \gamma_{\mathbf{i}}, \boldsymbol{\omega}) \mathbf{p}(\gamma_{\mathbf{i}} | \boldsymbol{\omega})$$

so we can sample each $p(\gamma_i | \mathbf{y_i}, \boldsymbol{\omega})$ independently. In particular,

$$p(\gamma_i = k | \mathbf{Y}, \boldsymbol{\omega}) \propto \mathbf{p}(\mathbf{y_i} | \gamma_{\mathbf{i}} = \mathbf{k}) \mathbf{p}(\gamma_{\mathbf{i}} = \mathbf{k} | \boldsymbol{\omega})$$

Let ℓ_k be the observation times which are less than or equal to k and r_k be the observation times which are greater than k, both of which depend upon i. When $k > 0$,

$$p(\gamma_i = k|\mathbf{Y}, \boldsymbol{\omega}) \propto \mathrm{T}_{\mathbf{a}+|\ell_\mathbf{k}|}(\mathbf{y_i}(\ell_\mathbf{k}); \mathbf{o}, \mathbf{R}_{\ell_\mathbf{k}}) \cdot \mathrm{T}_{\mathbf{a}+|\mathbf{r_k}|}(\mathbf{y_i}(\mathbf{r_k}); \mathbf{o}, \mathbf{R}_{\mathbf{r_k}}) \cdot \mathbf{p}(\gamma_\mathbf{i} = \mathbf{k}|\boldsymbol{\omega})$$

with the convention that $r_k = \emptyset$ if $k = n$. When $k = 0$,

$$p(\gamma_i = k|\mathbf{Y}, \boldsymbol{\omega}) \propto \mathrm{T}_{\mathbf{a}+|\ell_\mathbf{k}|}(\mathbf{y_i}(\mathbf{t_i}); \mathbf{o}, \frac{\mathbf{a}}{\mathbf{b}}\mathbf{I}_{\mathbf{t_i}})\mathbf{p}(\gamma_\mathbf{i} = \mathbf{k}|\boldsymbol{\omega})$$

All of the T densities can be computed beforehand for each i and k, and the conditional posterior distribution of γ_i may be calculated directly over the simplex $\{0, \ldots, n\}$.

When implementing this model, we restrict our dataset to those firms which have at least 20 observations, leaving us with 6067 data points. Including firms with fewer data points would eliminate the possibility of survivorship bias, but would likely result in a significant decrease in power for the firms with longer histories, because of shared dependence of all firms on $p(\omega \mid \mathbf{Y})$.

Following the discussion above, the posterior distribution was simulated using a Gibbs sampler with 3000 steps. We set $\tau^2 = 10.0$ and chose the prior distributions as follows: $\mathrm{IG}(2.0/2, 2.0/2)$ for all noise variance parameters; and $\boldsymbol{\omega} \sim \mathrm{Dirichlet}(\alpha)$ with $\alpha_0 = 0.8$, $\alpha_n = 0.1$, and $\alpha_i = 0.1/(n-1)$ for $i = 1, \ldots, n-1$. This reflects the belief, borne out by prior studies, that most firms do not systematically over- or under-perform their peer groups over time. The choice of $\tau^2 = 10$ will result in increased power for detecting major shifts in performance, but also a significant Occam's-razor penalty against small shifts.

One must be careful in interpreting the posterior distribution $p(\boldsymbol{\gamma}|\mathbf{Y})$ that arises from this model. For instance, it may not be meaningful to categorize time series i in terms of the single time point j that maximizes $p(\gamma_i = j|\mathbf{Y})$, since it is entirely possible that $p(\gamma_i = j|\mathbf{Y})$ is of comparable magnitude for a range of different values of j. This would suggest either no split, or a split that cannot be localized very precisely. In such cases, it may be more appropriate to look at firms where the largest entry $p(\gamma_i \mid \mathbf{Y})$ is sufficiently large. This would be strong evidence that there is a split at time j for time series i.

Figure 21.3 provides intuition for the various scenarios we might encounter: the posterior of γ_i may strongly favour $\gamma_i = 0$ or $\gamma_i = n$, in which case no split occurs; it may show evidence of a split, but at an ambiguous time; it may show strong, but not decisive evidence of a split; or it may be flat, which does not tell us anything. The ideal case for interpretation, of course, would involve strong evidence of a split at a particular time, as seen in the last pane of Figure 21.3.

In examining these plots, it appears that the posterior mode

$$PM(\gamma_i) = \max_{0<j<n} p(\gamma_i = j|\mathbf{Y})$$

provides a good measure of a change in epochs as long as we choose a sufficiently high cutoff, such as requiring that the maximum satisfy $\max_{0<j<n}\{\gamma_i = j|\mathbf{Y}\} > 0.95$. To check that this is reasonable, we plot the histogram of $PM(\gamma_i)$ in Figure 21.4. Of the 3033 firms, only 58 have a value greater than or equal to 0.95. The largest twenty posterior modes were used to select the time series in Figure 21.5.

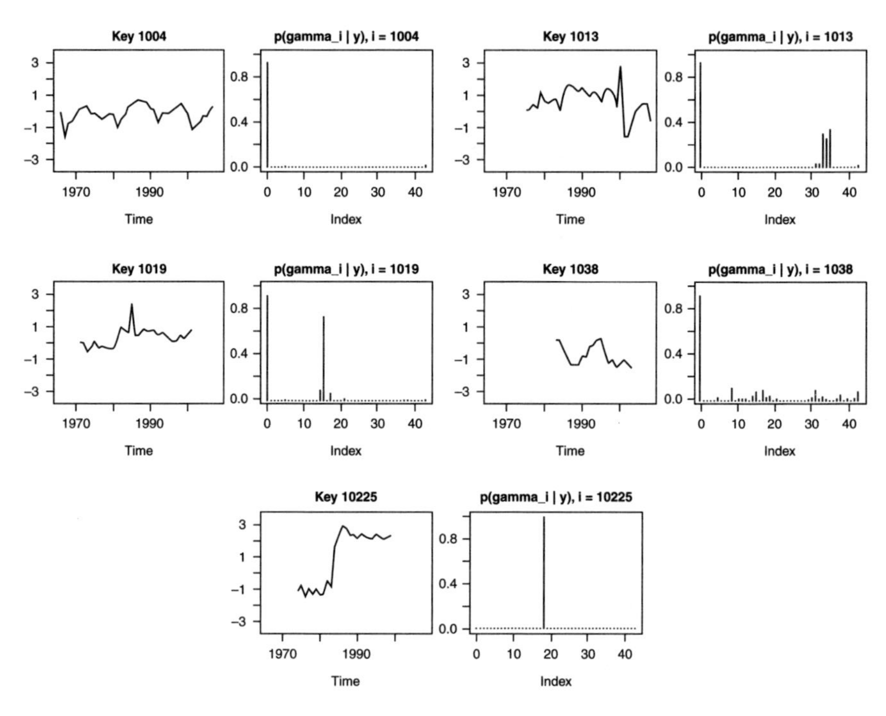

Figure 21.3 Five examples of a firm-level posterior distribution over possible changepoints, labelled by Compustat keys. The five panes embody a wide range of possible conclusions, with the data on the left and $p(\gamma_i \mid Y)$ on the right. From top to bottom, we see: strong evidence of a null case ($\gamma = 0$); evidence of a hard-to-localize changepoint; moderate evidence of a specific changepoint; near-total uncertainty over changepoints; and very strong evidence of a specific changepoint.

21.4 Discussion

In each of these two case studies, we have confronted a similar problem: time-varying model uncertainty for each of many different time series observed in parallel. In the first case, the model was a graph, encoding conditional independence relationships about residuals from country-level returns during the European sovereign debt crisis. In the second case, the model was an indicator of whether a firm's historical ROA trajectory was significantly different from its peer-group average. These were seen to be examples of a general class of problems wherein the model changes as a function of auxiliary information.

The models we have entertained are based upon fairly standard tools, and were chosen specifically to avoid the difficulties associated with the most general form of partitioning that were mentioned in the introduction. They are thus more appropriate for first-pass screening than for detailed analysis. Nonetheless, even these simple models were sufficient to support the general thrust of our argument: that each dataset exhibited non-trivial dynamic model structure. In each case, further work is clearly needed to build upon the limited, broad-brush conclusions that can be reached within the context of these simple models.

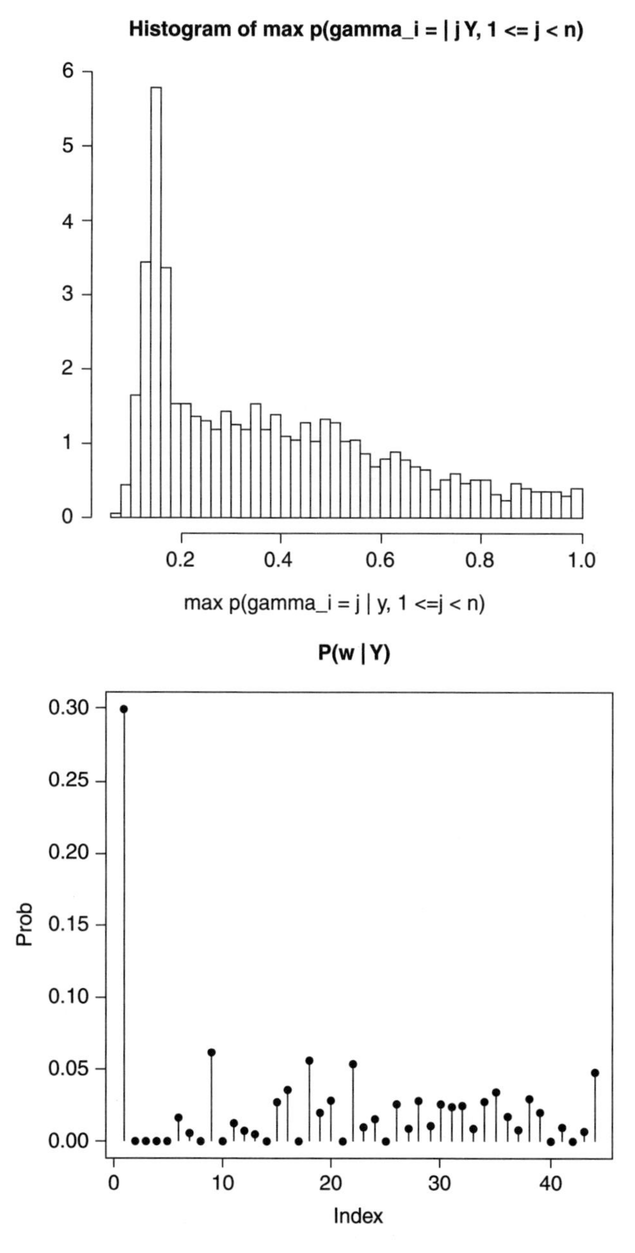

Figure 21.4 Top: the histogram of the largest entry in the posterior mode $PM(\gamma_i)$ across all firms. Bottom: the posterior mean of the multinomial probability vector ω. Notice that null cases ($\gamma_i = 0$) dominate the sample, but that some years clearly have more changepoints than others.

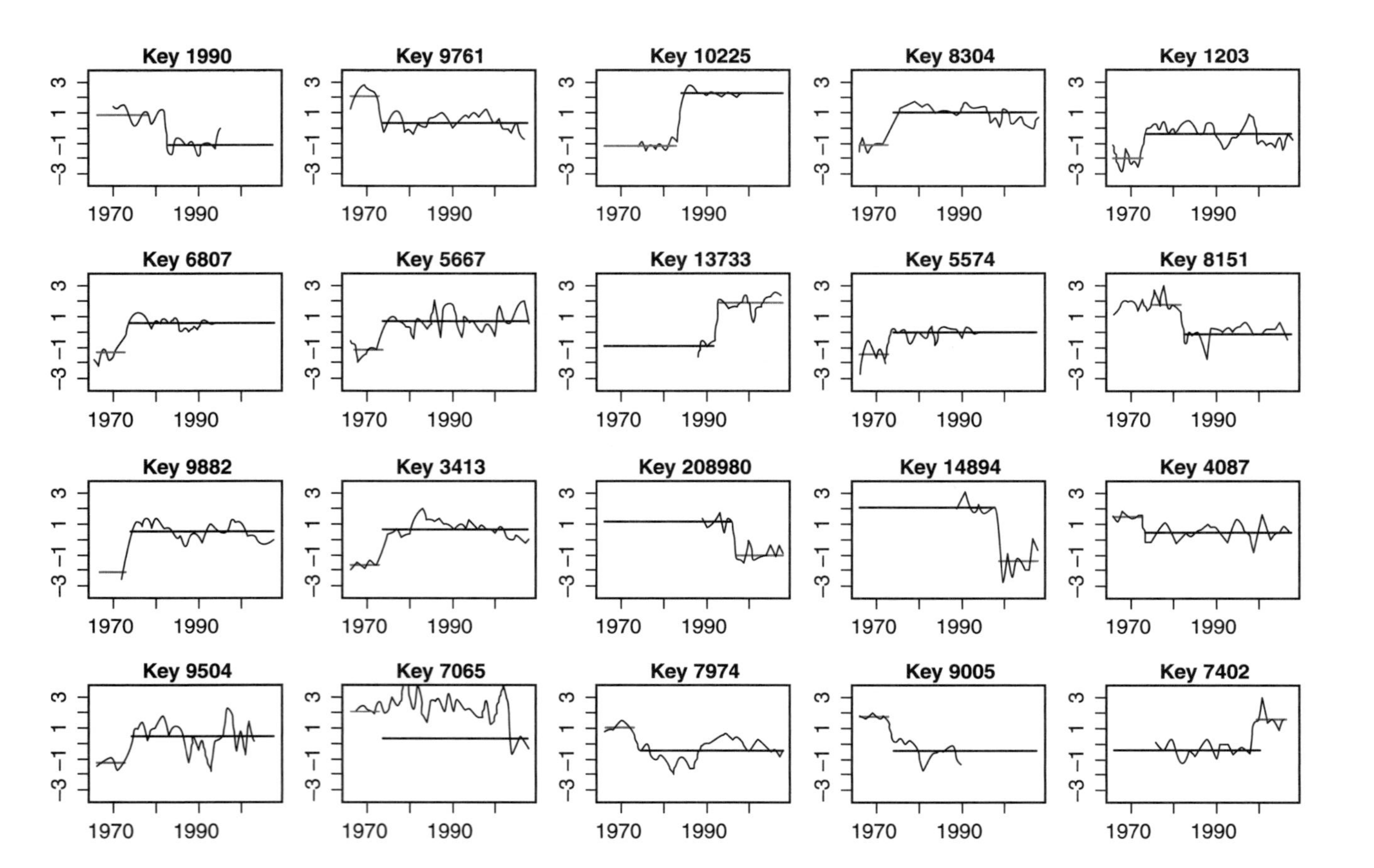

Figure 21.5 The firms with the 20 largest values for $PM(\gamma_i \mid Y)$. These are the 20 firms in the cohort where the evidence for a specific changepoint is the strongest.

References

[1] K. Bae, G. Karolyi and R. Stulz. (2003). A new approach to measuring financial contagion. *Review of Financial Studies*, **16**(3): 717–763.

[2] G. Bekaert, C. Harvey, and A. Ng. (2005). Market integration and contagion. *Journal of Business*, **78**(1): 39–69.

[3] M. Bogdan, A. Chakrabarti, F. Frommlet and J. K. Ghosh. (2011). Asymptotic Bayes-optimality under sparsity of some multiple testing procedures. *The Annals of Statistics*, **39**(3): 1551–1579.

[4] L. Breiman, J. H. Friedman, R. A. Olshen, and C. J. Stone. (1984). *Classification and Regression Trees*. Chapman and Hall/CRC.

[5] C. M. Carvalho and J. G. Scott. (2009). Objective Bayesian model selection in Gaussian graphical models. *Biometrika*, **96**(3): 497–512.

[6] H. A. Chipman, E. I. George, and R. E. McCulloch. (1998). Bayesian CART model search. *Journal of the American Statistical Association*, **93** (443): 935–948.

[7] D. G. Denison, B. K. Mallick, and A. F. Smith. (1998). A Bayesian CART algorithm. *Biometrika*, **85**(2): 363–377.

[8] A. Dobra, B. Jones, C. Hans, J. Nevins, and M. West. (2004). Sparse graphical models for exploring gene expression data. *Journal of Multivariate Analysis*, **90**: 196–212.

[9] R. Dornbusch, Y. C. Park, and S. Claessens. (2000). Contagion: Understanding how it spreads. *The World Bank Research Observer*, **15**(2): 177–197.

[10] K. J. Forbes and R. Rigobon. (2002). No contagion, only interdependence: Measuring stock market comovements. *The Journal of Finance*, **57**: 2223–2261.

[11] R. Gramacy and H. K. Lee. (2008). Bayesian treed Gaussian process models with an application to computer modeling. *Journal of the American Statistical Association*, **103** (483): 1119–1130.

[12] A. D. Henderson, M. E. Raynor, and M. Ahmed. (2009). How long must a firm be great to rule out luck? benchmarking sustained superior performance without being fooled by randomness. In *The Academy of Management Proceedings*.

[13] S. L. Lauritzen. (1996). *Graphical Models*. Clarendon Press, Oxford.

[14] N. G. Polson and J. G. Scott. (2011). An empirical test for eurozone contagion using an asset-pricing model with heavy-tailed stochastic volatility. Technical report, University of Texas at Austin, arXiv:1110.5789v2.

[15] N. G. Polson and J. G. Scott. (2012). Good, great, or lucky? Screening for firms with sustained superior performance using heavy-tailed priors. *The Annals of Applied Statistics*. (to appear).

[16] G. Schwarz. (1978). Estimating the dimension of a model. *Annals of Statistics*, **6**(2): 461–464.

[17] J. G. Scott. (2009). Nonparametric Bayesian multiple testing for longitudinal performance stratification. *The Annals of Applied Statistics*, **3**(4): 1655–1674.

[18] J. G. Scott. (2010). Benchmarking historical corporate performance. Technical report, University of Texas at Austin, http://arxiv.org/abs/0911.1768.

[19] J. G. Scott and J. O. Berger. (2006). An exploration of aspects of Bayesian multiple testing. *Journal of Statistical Planning and Inference*, **136**(7): 2144–2162.

[20] J. G. Scott and C. M. Carvalho. (2008). Feature-inclusion stochastic search for Gaussian graphical models. *Journal of Computational and Graphical Statistics*, **17** (790–808).

[21] M. Taddy, R. B. Gramacy, and N. G. Polson. (2011). Dynamic trees for learning and design. *Journal of the American Statistical Association*, **106** (493): 109–123.

[22] H. Wang, C. Reeson, and C. M. Carvalho. (2011). Dynamic financial index models: Modeling conditional dependencies via graphs. *Bayesian Analysis*.

Part IX

Regressions and Model Averaging

22 Bayesian regression structure discovery

HUGH A. CHIPMAN,
EDWARD I. GEORGE AND
ROBERT E. McCULLOCH

22.1 Introduction

The general problem of statistical regression is concerned with the discovery of a relationship between a variable of interest y and a set of potential predictors $x_1, \ldots, x_p$. It is usually realistic, especially when p is large, to consider that y may be related only to an unknown subset of the potential predictors, thus making variable selection an inherent part of the problem. In this paper, we describe two very different Bayesian approaches to this general problem. The feasible implementation of both of these approaches has been made possible by Markov chain Monte Carlo (MCMC) Bayesian posterior simulation, [7] and [14]. In particular, variations of the Gibbs sampler and the Metropolis–Hastings algorithms have allowed for the exploration of the otherwise intractable posteriors via simulation.

One approach, which dovetails with classical parametric approaches to variable selection, begins with an assumption that the relationship between y and $x_1, \ldots, x_p$ can be described by a full parametric model within which the actual subset model is nested. The most popular form used here is the normal linear model, in large part because of its appealing analytical tractability and because of its usefulness as an approximation to other forms, possibly after suitable transformations. A structured hierarchical mixture prior that captures all sources of remaining uncertainty is then used to obtain a posterior distribution which allocates more probability to the more promising subset models.

In contrast to the parametric approach, our second approach does not require an initial assumption about the nature of all the relationships between y and the subsets of $x_1, \ldots, x_p$. Nonparametric in nature, it begins with a very rich over parameterized functional form, a sum-of-trees model, that approximates a wide class of functions from R^p to R. However, with this more complex model it becomes more challenging to formulate useful priors and extract information about the relationship between y and x. A strong regularization prior over the multitude of parameters of the sum-of-trees model is used to obtain a posterior distribution over the possible relationships between y and $x_1, \ldots, x_p$. A variety of useful inferential summaries can be obtained by MCMC sampling from this posterior. In particular, by keeping track of how often each predictor is used in the sum-of-trees model, this approach allows for model-free variable selection, and further for model-free interaction detection, the discovery of when variables work together to influence the response.

22.2 Parametric Bayesian structure discovery

To illustrate the parametric Bayesian approach to structure discovery, we focus on the classical version of the problem which begins with the assumption of a normal linear model for every subset model, namely

$$Y = X_\gamma \beta_\gamma + \epsilon, \quad \epsilon \sim N_n(0, \sigma^2 I) \tag{22.1}$$

where Y is the $n \times 1$ vector of y observations, X_γ is the $n \times q_\gamma$ matrix whose columns correspond to the γth subset of $x_1, \ldots, x_p$, and β_γ is the $q_\gamma \times 1$ vector of unknown regression coefficients. For convenience, we have indexed each of the 2^p possible subset choices by

$$\gamma = (\gamma_1, \ldots, \gamma_p)' \tag{22.2}$$

where $\gamma_i = 1$ or 0 according to whether predictor x_i is included or excluded, respectively. The size (number of covariates) of the γth subset is thus $q_\gamma \equiv \gamma' 1$. The variable selection problem may then be regarded as how to use the data to choose γ. Particular Bayesian treatments of this formulation yield analytical reductions that allow for faster calculations as well as clearer insights into how the machinery works. Such Bayesian treatments also extend naturally to likelihoods that are a function of X_γ only through $X_\gamma \beta_\gamma$. There is by now a vast literature on Bayesian analyses for this formulation. See, for example, [9], [2] [4], and the references therein.

It should be noted that assumption (22.1) for every possible submodel γ is a strong assumption. Its strength is that it effectively turns the variable selection problem into a model selection problem which can be treated using variations of standard Bayesian parametric formulations. Its weakness is that a subset of predictors may be rejected because a normal linear submodel is inadequate rather than because Y is unrelated to the subset.

22.2.1 Prior formulations

The parametric problem formulation in (22.1), provides a likelihood $L(\beta_\gamma, \sigma, \gamma \,|Y)$. Thus, a Bayesian analysis proceeds with the choice of prior forms for

$$p(\beta_\gamma, \sigma, \gamma) = p(\beta_\gamma, \sigma \,|\gamma) p(\gamma) \tag{22.3}$$

For the specification of the model space prior $p(\gamma)$, many Bayesian variable selection implementations have used simple independence priors of the form

$$p(\gamma) = w^{q_\gamma} (1 - w)^{p - q_\gamma} \tag{22.4}$$

with a prespecified value for w, the expected proportion of x_i's in the submodel. Under this prior, each x_i enters the submodel independently with probability $p(\gamma_i = 1) = 1 - p(\gamma_i = 0) = w$. To avoid the fact that any such prior will be informative about the size of the model, a reasonable alternative is to margin out w in (22.4) with respect to a Beta prior to obtain

$$p(\gamma) = \frac{B(\alpha + q_\gamma, \beta + p - q_\gamma)}{B(\alpha, \beta)} \tag{22.5}$$

a special case of the more general form

$$p(\gamma) = \binom{p}{q_\gamma}^{-1} h(q_\gamma) \tag{22.6}$$

which is uniform over the set of submodels of a given size q_γ. See [8], [5] and [13].

For the specification of the parameter prior $p(\beta_\gamma, \sigma \mid \gamma) = p(\beta_\gamma \mid \sigma^2, \gamma)\, p(\sigma^2 \mid \gamma)$, an especially convenient choice is the conjugate normal-inverse-gamma prior

$$p(\beta_\gamma \mid \sigma^2, \gamma) = N_{q_\gamma}(0, \sigma^2 \Sigma_\gamma) \tag{22.7}$$

$$p(\sigma^2 \mid \gamma) = p(\sigma^2) = IG(\nu/2, \nu\lambda/2) \tag{22.8}$$

($p(\sigma^2)$ here is equivalent to $\nu\lambda/\sigma^2 \sim \chi^2_\nu$). A valuable feature of this prior is its analytical tractability; β_γ and σ^2 can be eliminated by routine integration to yield

$$p(Y \mid \gamma) \propto |X_\gamma' X_\gamma + \Sigma_\gamma^{-1}|^{-1/2} |\Sigma_\gamma|^{-1/2} (\nu\lambda + S_\gamma^2)^{-(n+\nu)/2} \tag{22.9}$$

where

$$S_\gamma^2 = Y'Y - Y'X_\gamma (X_\gamma' X_\gamma + \Sigma_\gamma^{-1})^{-1} X_\gamma' Y \tag{22.10}$$

The use of these closed form expressions can substantially speed up posterior evaluation and MCMC exploration, as we will see.

For choosing the prior covariance matrix Σ_γ that controls $p(\beta_\gamma \mid \sigma^2, \gamma)$, specification is substantially simplified by setting $\Sigma_\gamma = c\, V_\gamma$, where c is a scalar and V_γ is a preset form such as $V_\gamma = (X_\gamma' X_\gamma)^{-1}$ (as in the [16] g-prior) or $V_\gamma = I_{q_\gamma}$, the $q_\gamma \times q_\gamma$ identity matrix. Having fixed V_γ, the goal is then to choose c large enough so that $p(\beta_\gamma \mid \sigma^2, \gamma)$ is relatively flat over the region of plausible values of β_γ, thereby reducing prior influence. At the same time it is important to avoid excessively large values of c because the Bayes factors will eventually put increasing weight on the null model as $c \to \infty$, the Bartlett–Lindley paradox. For practical purposes, a rough guide is to choose c so that $p(\beta_\gamma \mid \sigma^2, \gamma)$ assigns substantial probability to the range of all plausible values for β_γ. A recent alternative of interest, are the hyper-g priors for β_γ which effectively integrate out c with respect to Beta prime distributions, Cui and George (2008) [5], [10] and [11].

In choosing values for the hyperparameters that control $p(\sigma^2)$, λ may be thought of as a prior estimate of σ^2, and ν may be thought of as the prior sample size associated with this estimate. Alternatively, one might use the data informally to choose λ and ν as follows. Let σ^2_{FULL} and σ^2_Y denote the traditional estimates of σ^2 based on the saturated and null models respectively. Treating σ^2_{FULL} and σ^2_Y as rough under- and over-estimates of σ^2, one might choose λ and ν so that $p(\sigma^2)$ assigns substantial probability to the interval $(\sigma^2_{FULL}, \sigma^2_Y)$. This should at least avoid gross misspecification. As a third option, the explicit choice of λ and ν can be avoided by using $p(\sigma^2) \propto 1/\sigma^2$, the limit of the inverse-gamma prior as $\nu \to 0$.

22.2.2 Posterior exploration and information extraction

The previous conjugate prior formulations allow for analytical margining out of β and σ^2 from $p(Y, \beta, \sigma^2 \mid \gamma)$ to yield a computable, closed form expression

$$g(\gamma) \propto p(Y \mid \gamma) p(\gamma) \propto p(\gamma \mid Y) \tag{22.11}$$

that can greatly facilitate posterior calculation and exploration. For example, when $\Sigma_\gamma = c\,(X'_\gamma X_\gamma)^{-1}$, we can obtain

$$g(\gamma) = (1+c)^{-q_\gamma/2}(\nu\lambda + Y'Y - (1+1/c)^{-1}W'W)^{-(n+\nu)/2}p(\gamma) \tag{22.12}$$

where $W = T'^{-1}X'_\gamma Y$ for upper triangular T such that $T'T = X'_\gamma X_\gamma$ (obtainable by the Cholesky decomposition). This representation allows for fast updating of T, and hence W and $g(\gamma)$, when γ is changed one component at a time, requiring $O(q_\gamma^2)$ operations per update, where γ is the changed value.

The availability of $g(\gamma) \propto p(\gamma \mid Y)$ allows for the flexible construction of MCMC algorithms that simulate a Markov chain

$$\gamma^{(1)}, \gamma^{(2)}, \gamma^{(3)}, \ldots \tag{22.13}$$

converging (in distribution) to $p(\gamma \mid Y)$. A variety of such MCMC algorithms can be conveniently obtained by applying the Gibbs sampler with $g(\gamma)$. For example, by generating each γ component from the full conditionals

$$p(\gamma_i \mid \gamma_{(i)}, Y) \tag{22.14}$$

($\gamma_{(i)} = \{\gamma_j : j \neq i\}$) where the γ_i may be drawn in any fixed or random order. The generation of such components can be obtained rapidly as a sequence of Bernoulli draws using simple functions of the ratio

$$\frac{p(\gamma_i = 1, \gamma_{(i)} \mid Y)}{p(\gamma_i = 0, \gamma_{(i)} \mid Y)} = \frac{g(\gamma_i = 1, \gamma_{(i)})}{g(\gamma_i = 0, \gamma_{(i)})} \tag{22.15}$$

The availability of such closed form $g(\gamma)$ also facilitates the use of MH algorithms. Because $g(\gamma)/g(\gamma') = p(\gamma \mid Y)/p(\gamma' \mid Y)$, these are of the form:

1. Simulate a candidate γ^* from a transition kernel $q(\gamma^* \mid \gamma^{(j)})$.
2. Set $\gamma^{(j+1)} = \gamma^*$ with probability

$$\alpha(\gamma^* \mid \gamma^{(j)}) = \min\left\{\frac{q(\gamma^{(j)} \mid \gamma^*)}{q(\gamma^* \mid \gamma^{(j)})}\frac{g(\gamma^*)}{g(\gamma^{(j)})}, 1\right\} \tag{22.16}$$

3. Otherwise, set $\gamma^{(j+1)} = \gamma^{(j)}$.

When available, fast updating schemes for $g(\gamma)$ can be exploited in all of these MCMC algorithms.

The simulated Markov chain sample $\gamma^{(1)}, \ldots, \gamma^{(K)}$ contains valuable information about the posterior $p(\gamma \mid Y)$. Empirical frequencies provide consistent estimates of individual model probabilities or characteristics such as $p(\beta_i \neq 0 \mid Y)$. When closed form $g(\gamma)$ are available, we can do better. For example, the exact relative probability of any two values γ^0 and γ^1 is obtained as $g(\gamma^0)/g(\gamma^1)$ in the sequence of simulated values. Such $g(\gamma)$ also facilitates estimation of the normalizing constant $p(\gamma \mid Y) = Cg(\gamma)$. Let A be a preselected subset of γ values and let $g(A) = \sum_{\gamma \in A} g(\gamma)$ so that $p(A \mid Y) = Cg(A)$. Then, a consistent estimate of C is

$$\hat{C} = \frac{1}{g(A)K} \sum_{k=1}^{K} I_A(\gamma^{(k)}) \tag{22.17}$$

where $I_A()$ is the indicator of the set A. This yields alternative estimates of the probability of individual γ values $\hat{p}(\gamma \mid Y) = \hat{C}g(\gamma)$, as well as an estimate of the total visited probability $\hat{p}(B \mid Y) = \hat{C}g(B)$, where B is the set of visited γ values.

22.3 Nonparametric Bayesian structure discovery

To illustrate the nonparametric Bayesian approach to structure discovery, we focus on an approach we call BART (Bayesian Additive Regression Trees), ([3]) which assumes only that y is related to $x = (x_1, \ldots, x_p)$ via a flexible sum-of-trees model of the form[19]

$$Y = \sum_{j=1}^{m} g(x; T_j, M_j) + \epsilon, \qquad \epsilon \sim N(0, \sigma^2) \tag{22.18}$$

where each T_j is a binary regression tree with a set M_j of associated terminal node constants μ_{ij}, and $g(x; T_j, M_j)$ is the function which assigns $\mu_{ij} \in M_j$ to x according to the sequence of decision rules in T_j. These decision rules are binary splits of the predictor space of the form $\{x \in A\}$ vs $\{x \notin A\}$ where A is a subset of the range of x. When $m = 1$, (22.18) reduces to the single tree model used by [1] for Bayesian CART.

Under (22.18), $E(Y \mid x)$ equals the sum of all the terminal node μ_{ij}s assigned to x by the $g(x; T_j, M_j)$'s. As these can be any values, it is easy to see that the sum-of-trees model (22.18) is a very flexible representation capable of representing a wide class of functions from R^n to R, especially when the number of trees m is large. Note also that the sum-of-trees representation is composed of many simple functions from R^p to R, namely the $g(x; T_j, M_j)$, rendering it much more manageable than a representation with more complicated basis elements such as multidimensional wavelets or multidimensional splines.

22.3.1 A regularization prior

We complete the BART model specification by imposing a prior over all the parameters of the sum-of-trees model, namely $(T_1, M_1), \ldots, (T_m, M_m)$ and σ. Note that these parameters entail all the bottom node parameters as well as the tree structures and decision rules, a very large number of parameters, especially when m is large. We do this using a prior that effectively regularizes the fit by keeping the individual tree effects from being unduly influential. Without such a regularizing influence, large tree components would overwhelm the rich structure of (22.18), thereby limiting its scope of approximation.

To begin with we simplify our prior specification task by restricting attention to prior formulations of the form

$$p((T_1, M_1), \ldots, (T_m, M_m), \sigma) = \left[\prod_j \left(\prod_i p(\mu_{ij} \mid T_j) \right) p(T_j) \right] p(\sigma) \tag{22.19}$$

[19] Note that here we use Y as a random scalar rather than an $n \times 1$ random vector as in Section 22.2.

where $\mu_{ij} \in M_j$. These independence restrictions simplify prior specification to the choice of prior forms for $p(T_j)$, $p(\mu_{ij} | T_j)$ and $p(\sigma)$, and to simplify matters further we consider for all of these, the same prior forms as those proposed by [1] for Bayesian CART. These forms are controlled by just a few interpretable hyperparameters which can be calibrated using the data to yield effective default specifications for regularization of the sum-of-trees model.

For $p(T_j)$, we use the [1] tree-generating process which is specified by three aspects: (i) the probability that a node at depth d ($= 0, 1, 2, \ldots$) is nonterminal, given by

$$\alpha(1+d)^{-\beta}, \qquad \alpha \in (0,1), \beta \in [0, \infty), \tag{22.20}$$

(ii) the distribution on the splitting variable assignments at each interior node, and (iii) the distribution on the splitting rule assignment in each interior node, conditional on the selected splitting variable. For (ii) and (iii) we use the simple defaults in [1] , namely a uniform prior on each set of possibilities

For $p(\mu_{ij} | T_j)$, we use the conjugate normal distribution $N(\mu_\mu, \sigma_\mu^2)$ which allows μ_{ij} to be margined out, greatly simplifying MCMC posterior calculations. Note that under this choice the prior distribution of $E(Y|x)$ is $N(m\,\mu_\mu, m\,\sigma_\mu^2)$, (because $E(Y|x)$ is the sum of m independent μ_{ij}s under the sum-of-trees model). Thus, it is highly probable that $E(Y|x)$ is between y_{min} and y_{max}, the observed minimum and maximum of y in the data, a fact which we can use to guide the specification of the hyperparameters μ_μ and σ_μ. The essence of our informal strategy is then to choose μ_μ and σ_μ so that $N(m\,\mu_\mu, m\,\sigma_\mu^2)$ assigns substantial probability to the interval (y_{min}, y_{max}). This can be conveniently done by choosing μ_μ and σ_μ so that $m\,\mu_\mu - k\sqrt{m}\,\sigma_\mu = y_{min}$ and $m\,\mu_\mu + k\sqrt{m}\,\sigma_\mu = y_{max}$ for some preselected value of k such 1,2 or 3. For example, $k = 2$ would yield a 95% prior probability that $E(Y|x)$ is in the interval (y_{min}, y_{max}). The goal of this specification strategy for μ_μ and σ_μ is to ensure that the implicit prior for $E(Y|x)$ is in the right 'ballpark' in the sense of assigning substantial probability to the entire region of plausible values of $E(Y|x)$ while avoiding overconcentration and overdispersion. As long as this goal is met, BART seems to be very robust to the variations of these specifications.

For $p(\sigma)$, we also use a conjugate prior, here the inverse chi-square distribution $\sigma^2 \sim \nu\lambda/\chi^2_\nu$, the same form we used for $p(\sigma)$ for the parametric variable selection problem previously. Here again, we use a data-informed prior approach, to guide the specification of the hyperparameters ν and λ, in this case to assign substantial probability to the entire region of plausible values of σ while avoiding overconcentration and overdispersion. Essentially, we calibrate the prior df ν and scale λ using a 'rough data-based overestimate' $\hat{\sigma}$ of σ. Two natural choices of where $\hat{\sigma}$ are (1) a 'naive' specification, the sample standard deviation of Y, or (2) a 'linear model' specification, the residual standard deviation from a least squares linear regression of Y on all the predictors. We then pick a value of ν between 3 and 10 to get an appropriate shape, and a value of λ so that the qth quantile of the prior on σ is located at $\hat{\sigma}$, that is $P(\sigma < \hat{\sigma}) = q$. We consider values of q such as 0.75, 0.90 or 0.99 to centre the distribution below $\hat{\sigma}$.

22.3.2 Posterior calculation and information extraction

Combining the regulation prior with the likelihood, $L((T_1, M_1), \ldots, (T_m, M_m), \sigma | y)$ induces a posterior distribution

$$p((T_1, M_1), \ldots, (T_m, M_m), \sigma | y) \tag{22.21}$$

over the full sum-of-trees model parameter space. Fortunately, the following backfitting MCMC algorithm can be used to simulate samples from this posterior.

We begin with a Gibbs sampler at the outer level. Let $T_{(j)}$ be the set of all trees in the sum *except* T_j, and similarly define $M_{(j)}$, so that $T_{(j)}$ will be a set of $m - 1$ trees, and $M_{(j)}$ the associated terminal node parameters. A Gibbs sampling strategy for sampling from (22.21) is obtained by m successive draws of (T_j, M_j) conditionally on $(T_{(j)}, M_{(j)}, \sigma)$:

$$(T_j, M_j)|T_{(j)}, M_{(j)}, \sigma, y \tag{22.22}$$

$j = 1, \ldots, m$, followed by a draw of σ from the full conditional:

$$\sigma|T_1, \ldots T_m, M_1, \ldots, M_m, y \tag{22.23}$$

The draw of σ in (22.23) is simply a draw from an inverse gamma distribution and so can be easily obtained by routine methods. More subtle is the implementation of the m draws of (T_j, M_j) in (22.22). This can be done by taking advantage of the following reductions. First, observe that the conditional distribution $p(T_j, M_j| \, T_{(j)}, M_{(j)}, \sigma, y)$ depends on $(T_{(j)}, M_{(j)}, y)$ only through

$$R_j \equiv y - \sum_{k \neq j} g(x; T_k, M_k), \tag{22.24}$$

the n—vector of partial residuals based on a fit that excludes the jth tree. Thus, the m draws of (T_j, M_j) given $(T_{(j)}, M_{(j)}, \sigma, y)$ in (22.22) are equivalent to m draws from

$$(T_j, M_j)|R_j, \sigma \tag{22.25}$$

$j = 1, \ldots, m$. Because we have used a conjugate prior for M_j, $p(T_j|R_j, \sigma)$ can be obtained in closed form up to a norming constant. This allows us to carry out each draw from (22.25) in two successive steps as

$$T_j|R_j, \sigma \tag{22.26}$$

$$M_j|T_j, R_j, \sigma. \tag{22.27}$$

The draw of T_j in (22.26), although somewhat elaborate, can be obtained using the Metropolis–Hastings (MH) algorithm of [1]. The draw of M_j in (22.27) is simply a set of independent draws of the terminal node μ_{ij}s from a normal distribution. The draw of M_j enables the calculation of the subsequent residual R_{j+1} which is critical for the next draw of T_j.

We initialize the chain with m simple single node trees, and then iterations are repeated until satisfactory convergence is obtained. Fortunately, this backfitting MCMC algorithm appears to mix very well as we have found that different restarts give remarkably similar results even in difficult problems. At each iteration, each tree may increase or decrease the number of terminal nodes by one, or change one or two decision rules. The sum-of-trees model, with its abundance of unidentified parameters, allows for 'fit' to be freely reallocated from one tree to another. Because each move makes only small incremental changes to the fit, we can imagine the algorithm as analogous to sculpting a complex figure by adding and subtracting small dabs of clay.

For inference based on our MCMC sample, we rely on the fact that our backfitting algorithm is ergodic. Thus, the induced sequence of sum-of-trees functions

$$f^*(\cdot) = \sum_{j=1}^{m} g(\cdot; T_j^*, M_j^*) \tag{22.28}$$

for the sequence of draws $(T_1^*, M_1^*), \ldots, (T_m^*, M_m^*)$, is converging to $p(f \mid y)$, the posterior distribution on the 'true' $f(\cdot)$. Thus, by running the algorithm long enough after a suitable burn-in period, the sequence of f^* draws, say $f_1^*, \ldots, f_K^*$, may be regarded as an approximate, dependent sample of size K from $p(f \mid y)$. Bayesian inferential quantities of interest can then be approximated with this sample as indicated below.

To estimate $f(x)$ or predict Y at a particular x, in-sample or out-of-sample, a natural choice is the average of the after burn-in sample $f_1^*, \ldots, f_K^*$,

$$\frac{1}{K}\sum_{k=1}^{K} f_k^*(x) \tag{22.29}$$

which approximates the posterior mean $E(f(x) \mid y)$. Posterior uncertainty about $f(x)$ may be gauged by the variation of $f_1^*(x), \ldots, f_K^*(x)$. For example, a natural and convenient $(1-\alpha)\%$ posterior interval for $f(x)$ is obtained as the interval between the upper and lower $\alpha/2$ quantiles of $f_1^*(x), \ldots, f_K^*(x)$.

Finally, BART provides a new approach to variable selection and interaction detection by identifying those variables or combination of variables that appear most often in the fitted sum-of-trees models. Interestingly, the variable selection strategy does not seem to work well when m is large because the redundancy offered by so many trees allows many irrelevant predictors to be mixed in with the relevant ones. However, as m is decreased and that redundancy is diminished, BART tends to heavily favour relevant predictors for its fit. In a sense, when m is small the predictors compete with each other to improve the fit. In contrast, interaction detection seems to work well with large m.

This model-free approach to variable selection is accomplished by observing what happens to the x component usage frequencies in a sequence of MCMC samples $f_1^*, \ldots, f_K^*$ as the number of trees m is set smaller and smaller. More precisely, for each simulated sum-of-trees model f_k^*, let z_{ik} be the proportion of all splitting rules that use the ith component of x. Then

$$v_i \equiv \frac{1}{K}\sum_{k=1}^{K} z_{ik} \tag{22.30}$$

is the average use per splitting rule for the ith component of x. As m is set smaller and smaller, the sum-of-trees models tend to more strongly favour inclusion of those x components which improve prediction of y and exclusion of those x components that are unrelated to y. In effect, smaller m seems to create a bottleneck that forces the x components to compete for entry into the sum-of-trees model. As we shall see in Section 22.4.2, the x components with the larger v_is will then be those that provide the most information for predicting y. A BART approach to model-free interaction detection proceeds in analogous fashion, for example, let z_{ijk} be the proportion of all trees in which both the ith and jth components of x appear.

22.4 Information extraction: details and examples

In Section (22.3) we outlined the BART model and discussed, in general terms, how it can be used to extract information about the relationship between y and x. In Section (22.4.1) we provide additional detail on the CART MCMC, highlighting the crucial aspects of our prior and algorithm that enable us to find structure. In Section (22.4.2) we give examples of information extraction. We show how we can extract information about what variables are important (using equation (22.30)) and which pairs of variables work together generating an 'interaction' effect.

22.4.1 Stochastic search in CART models

Perhaps the crucial model search step in Section (22.3) is the draw given in (22.26): $T_j|R_j, \sigma$. Our Gibbs sampler MCMC structure allows us to focus on one tree so that we are back to a CART problem. It is in this step, that we actually modify the structure of a tree. It is in this step, that a new variable may be introduced to our model.

This draw is done using a Metropolis-within-Gibbs proposal. The CART algorithm given in [1] uses several types of proposals (see also [15] for additional MCMC strategies).

The essential proposals[20] are a complementary BIRTH/DEATH pair of moves. In a BIRTH proposal, a bottom node of the current tree is chosen and we propose to give it a pair of children. A *nog* node of a tree is a tree which has children, but no grandchildren. Thus, both children of a nog node are bottom nodes. In a DEATH proposal, we choose a nog node from the current tree and we propose 'killing its children'. In order to make our general discussion more concrete and document some of the details, we give the acceptance probability for a BIRTH proposal.

The CART algorithm assumes a discrete set of possible split values for each component of x and integrates out the bottom node μ_{ij} so that our Metropolis search in tree space is over a large but discrete set of possible models. Let T_0 denote the *current* tree and T^* denote the *proposed* tree. Thus, T^* differs from T_0 only in that one of the bottom nodes of T_0 has given birth to a pair of children in T^*.

Since we are traversing a discrete space, we accept the proposal with Metropolis–Hastings probability

$$\alpha = \min\{1, \frac{P(T^*)\, P(T^* \to T_0)}{P(T_0)\, P(T_0 \to T^*)}\} \tag{22.31}$$

where $P(T_0)$ and $P(T^*)$ are the posterior probabilities of trees T_0 and T^* respectively, $P(P \to T_0)$ is the probability of proposing T_0 while at T^* (a DEATH), and $P(T_0 \to T^*)$ (a BIRTH) is the probability of proposing T^* while at T_0. $P(T_0)$ and $P(T^*)$ will depend on both the likelihood and our prior, while the transition probabilities depend on the mechanics of our proposal.

First we discuss the likelihood contribution. Let y_i denote the observed y in the *ith* bottom node given a tree T. Because the μ_{ij} are iid in our prior we have:

$$p(y \mid T) = \Pi p(y_i \mid T) \tag{22.32}$$

Thus the contribution of the likelihood to the ratio $P(P)/P(T_0)$ is just

$$\frac{p(y_l, y_r \mid T^*)}{p(y_{lr} \mid T_0)} = \frac{p(y_l \mid T^*)\, p(y_r \mid T^*)}{p(y_{lr} \mid T_0)} \tag{22.33}$$

where y_l denotes the observations in the new left child in T^*, y_r denotes the observation in the new right child in T^*, and y_{lr} denotes $\{y_l, y_r\}$. All other contributions to the likelihoods cancel out because of the product form of (22.32). Note that all three terms in the right-hand side of (22.33) are

[20] The other two proposals in CMG98 are CHANGE and SWAP.

just the predictive densities for a normal mean problem with known variance and normal prior on the mean.

As with the likelihood, much of the prior contributions to the posterior ratio cancel out since there is only one place where the trees differ and our stochastic tree growing prior draws tree components independently at different 'places' of the tree. Hence the prior contribution to the $P(T^*)/P(T_0)$ ratio is

$$\frac{(PG)\,(1 - PGl)\,(1 - PGr)\,P(rule)}{(1 - PG)} \tag{22.34}$$

where

- PG: prior probability of growing at chosen bottom node of T_0.
- PGl: prior probability of growing at new left child in T^*.
- PGr: prior probability of growing at new right child in T^*.
- $P(rule)$: prior probability of choosing the rule defining the new children in T^*.

Each of the PG quantities is obtained from (22.20). The prior $P(rule)$ places a uniform distribution on variables and then a uniform distribution on the discrete set of split values associated with the drawn variable.

Finally, the ratio $P(T^* \rightarrow T_0)/P(T_0 \rightarrow T^*)$, is given by

$$\frac{(PD)\,(Pnog)}{(PB)\,(Pbot)\,P(rule)} \tag{22.35}$$

where

- PD: probability of choosing the death proposal at tree T^*.
- $Pnog$: probability of choosing the nog node that gets you back T_0.
- PB: probability of choosing a birth proposal at T_0.
- $Pbot$: probability of choosing the T_0 bottom node such that a birth gets you to T^*.
- $P(rule)$: probability of drawing the new splitting rule to generate T^*'s children.

Our proposal draw of the new rule generating the two new bottom nodes is a draw from the prior. It is in this draw that variable selection (or, perhaps, variable proposal) occurs! Note that since our proposal for the *rule* is a draw from the prior, it cancels out in the ratio (22.31).

The formulas given above correspond very closely to the source code in the BayesTree package in R. However, there are still many details omitted. For example, a quantity PGl might be zero if we keep track of which variables in x have been 'used up' in that no further splits are possible.

22.4.2 Variable selection and interaction detection using BART

In this section we illustrate two forms of information extraction. The first is the variable selection approach given in (22.30). The second, interaction detection, uncovers which pairs of variables interact in analogous fashion by keeping track of the percentage of trees in the sum in which both variables occur. This exploits the fact that a sum-of-trees model captures an interaction between x_i and x_j by using them both for splitting rules in the same tree.

We illustrate the use of these methods in a simulated example and a real data example. Both examples are 'old chestnuts'. Since our goal is interpretation, rather than prediction, we hope the use of familiar examples eases the path of the reader.

22.4.2.1 *The Friedman simulation setup*

We simulate $n = 500$ observations from our basic model

$$y = f(x) + \sigma Z, \; Z \sim N(0, 1)$$

with x ten-dimensional and

$$f(x_1, x_2, \ldots, x_{10}) = 10 \sin(\pi x_1 x_2) + 20 (x_3 - 0.5)^2 + x_4 + x_5$$

The x_i are iid uniform on $(0, 1)$ and $\sigma = 1$.

[6] originally suggested this simulation setup to study the efficacy of nonlinear regression techniques. However, the setup is perfect for illustrating variable selection and the discovery of interaction. Only the first five of the ten x components matter. With ten xs there are 45 possible interaction pairs. Our simulated data has just one of these possibilities present: only x_1 and x_2 interact. In a real application it would be of tremendous interest to know that only these two variables interact, even without having further knowledge of the functional form.

Results for one simulated dataset are displayed in Figure (22.1). In panel (a) we have variable selection results. This panel corresponds closely to Figure (5) of [3]. For each variable, we plot the posterior mean of the percentage of rules (across all m tree) which use that variable. With $m = 20$, we very clearly identify the first five variables as being important.

Panel (b) gives the interaction detection results. With ten variables, there are $\binom{10}{2} = 45$ possible variable pairs. For each pair, we plot the posterior mean of the percentage of trees (out of m) which use both of the variables in splitting rules. We normalize the $m = 20$ and $m = 200$ results by dividing by each set of 45 posterior means by the maximum. Thus, the largest value displayed in each case is one. With both $m = 20$ and $m = 200$ we clearly identify the first pair (x_1 and x_2) as being of interest. With two variables involved, a pair is less likely to come in inconsequentially, so that the identification of interesting pairs is less sensitive to the choice of m than in the case of variable selection.

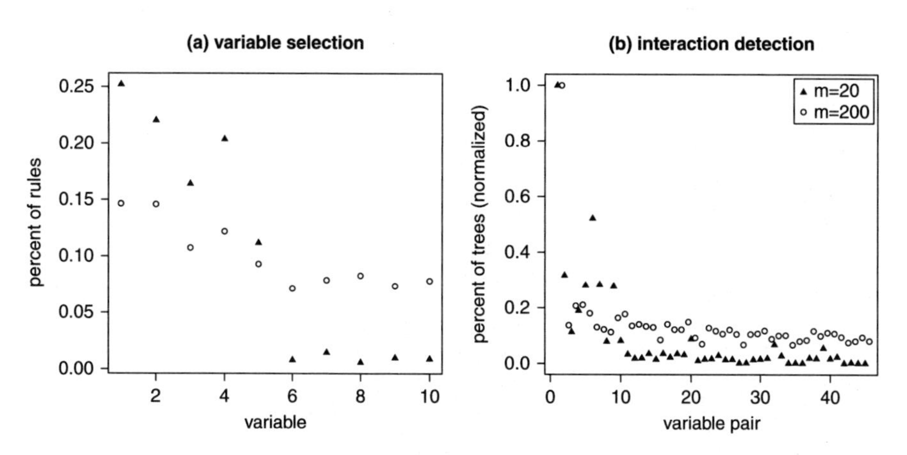

Figure 22.1 In panel (a) we correctly identify the first five variables as being important. In panel(b) we correctly identify the first interaction, which corresponds to variables x_1 and x_2.

22.4.2.2 *The Boston housing data*

For an example with real data we turn to our second 'old chestnut', the Boston housing data. The data were obtained from the R-package `mlbench` ([12]). There are 506 observations. Each observation corresponds to neighbourhood. The response is the median house price in the neighbourhood. There are 13 explanatory variables measuring characteristics of the neighbourhoods. We did a preliminary variable selection (using the approach illustrated in the previous section) and tossed out three of the *x*s. Fitted values (from BART) with and without the three *x*s are very similar.

Figure (22.2) displays the results of the interaction detection. The format is the same as in panel (b) of Figure (22.1). Several pairs of interest are identified. Our real data has more interesting structure than our simulated data! We will investigate the pair `dis` and `lstat` simply because these variables are more easily understood. `dis` is the 'weighted distances to five Boston employment centres'. `lstat` is the "percentage of lower status of the population".

In Figure (22.3) we attempt to see graphically the interaction between `dis` and `lstat` suggested by Figure (22.2). In panel (a) we plot `dis` vs. `lstat`. Four subsets of points are identified depending on whether `dis` and `lstat` are 'low' or 'high'. The points in the four subsets are plotted with a,b,c, or d while the rest of the points are plotted with a small circle. In the (b) panel we plot the fitted values from the BART run with $m = 200$. Before fitting we subtracted off the average response so the vertical axis is actually the amount the median value for a neighbourhood is above the average. The four boxplots correspond to the four data subsets indicated in panel (a).

So, for example, the first boxplot displays the fitted prices when both `ds` and `lstat` are low. The observations included here correspond to those highlighted in the bottom left corner of panel (a). The label '`dL_lL_a`' indicates that `ds` is Low and `lstat` is Low and the observations correspond to those plotted with a in panel (a). Similarly, the third boxplot is labelled '`dH_lL_c`', indicating that `ds` is High and `lstat` is Low and the points are plotted with the symbol c in panel (a).

The first pair of boxplots indicate the effect of increasing `lstat` when `ds` is low. The second pair of boxplots indicate the effect of increasing `lstat` when `ds` is high. Clearly, the boxplots indicate a strong interaction. For low `dl`, the effect of the change in `lstat` is much more pronounced. A nice neighbourhood close to the city centre is highly desirable whereas a bad neighbourhood close to the city centre may be very bad.

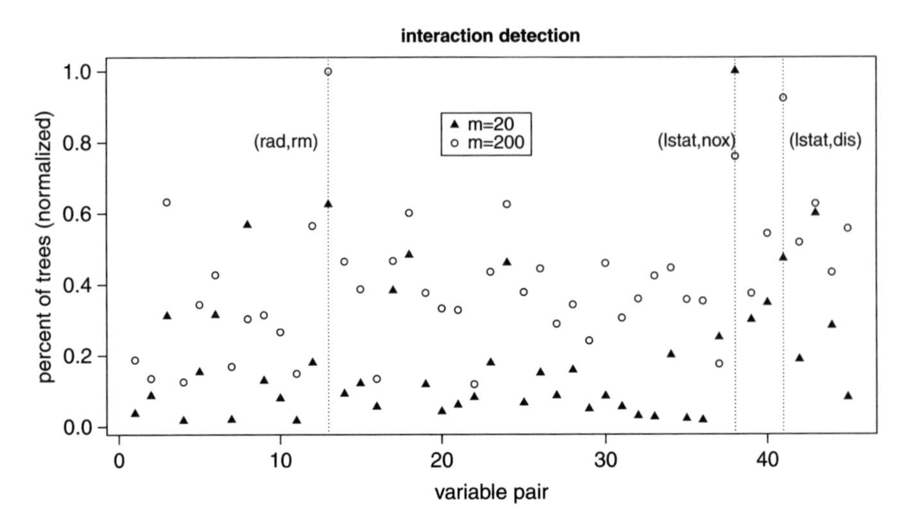

Figure 22.2 Interaction detection for the Boston housing data with ten explanatory variables.

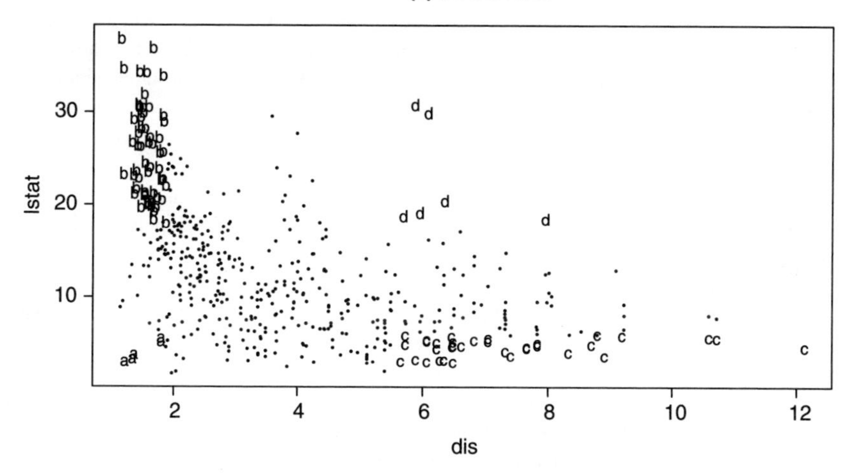

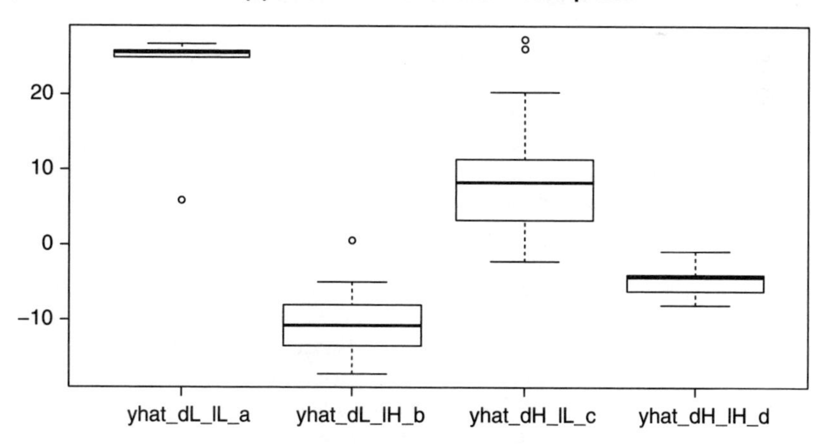

Figure 22.3 In panel (a) we identify four subsets of our data by whether each of `ds` and `lstat` are low or high. In panel (b) the boxplots display the fitted values (median house values) for the observations in the four subsets. The average of the dependent variable was subtracted off so that the vertical axis is the amount the median value of a neighbourhood is above average. The first pair of boxplots both have low values of `dis`. The first box has low values of `lstat` and the second box has high values of `lstat`. The second pair of boxplots again compare low and high `lstat` but now `ds` is high.

22.5 Discussion and beyond

The discovery of regression structure is an important and difficult problem for all approaches to data analysis. With our modern computational tools it has become even more important. However, in a sense, it has also become more difficult, as we struggle to grapple with complex, high-dimensional models.

In this article, we describe and contrast two different Bayesian approaches that illustrate the vast potential of Bayesian methods to extract information hidden in high-dimensional data. The first is based on the classical parametric form of the normal linear model, while the second is based on a rich overparameterized sum-of-trees model, nonparametric in nature. In our examples, we show that even though the overall BART sum-of-trees model is complex, the simple structure of the individual tree components enables us to uncover structure with inferential posterior summaries. In particular, we have shown how BART provides a novel approach to model-free variable selection, the search for interesting variables, and model-free interaction detection and the search for interesting pairs of variables. Going beyond what we have presented here, the companion pieces by Clyde and Iversen (Chapter 24 in this volume) and Gramacy (Chapter 23 in this volume) in this volume, shed new light on directions for the development of Bayesian methods for grappling with model uncertainty.

As laid out by Clyde and Iversen (see Chapter 24), the general Bayesian formulations for dealing with model uncertainly includes our parametric Bayesian formulation for variable selection as a special case. An important issue there is whether the class of models under consideration includes the actual data generating model. When it does, the so-called $\mathcal{M}$-closed setting, the implicit posterior model probabilities make sense. This will be the case in our parametric Bayesian variable selection framework when it is valid to assume that the complete data was generated by normal linear model (with possibly some zero coefficients). However, it may often be more realistic to allow that the unknown actual model is outside the class under consideration, the so-called $\mathcal{M}$-open setting. This is precisely the linear model assumption limitation which we alluded to earlier, that a subset of actually related predictors may be ignored simply because the relationship is not linear. Recognizing that the general limitation of using conventional Bayesian machinery for the $\mathcal{M}$-open setting, Clyde and Iversen consider alternatives to obtaining the weights corresponding to the conventional setting's posterior model probabilities. For this purpose they propose principled decision theoretic cross-validation approaches for selection of weights that optimize model averaged predictions.

A key message in Gramacy (see Chapter 23) is the central role played by the prior formulation in Bayesian variable selection. Indeed, the structured hierarchical mixture prior of Section 22.2.1 for the normal linear model, and the regularization prior of Section 22.3.1 for BART are essential for the effectiveness of these approaches. In both cases, it is necessary to use sensible hyperparameter values that balance 'null-versus-alternative' possibilities in a way that allows the variable selection information in the data to emerge. Gramacy speaks of this in discussing the prior allocations that must be balanced in a variety of structured hierarchical mixture priors for the linear model, priors with coefficient marginals that are *both* concentrated near 0 and heavy-tailed, similar in spirit to the lasso. By concentrating prior probability near zero, strong input from the data is needed to escape a neighbourhood about zero. However, once the estimate has escaped from zero, the heavy tails allow it to wander far. Going further, Gramacy shows how latent variable formulations allow extension of the approaches to non-normal errors and binary observations. Convenient and efficient Gibbs sampling algorithms for posterior computation are detailed throughout allowing Gramacy to argue persuasively that these Bayesian approaches are powerful tools in our modern data rich environment.

Acknowledgement

The authors are grateful to Adam Kapelner for valuable suggestions.

References

[1] Chipman, H. A., George, E. I. & McCulloch, R. E. (1998), Bayesian CART model search, *Journal of the American Statistical Association* **93**, 935–948.

[2] Chipman, H. A., George, E. I. & McCulloch, R. E. (2001), The practical implementation of Bayesian model selection, in P. Lahiri, ed., Model Selection, Vol. 38 of *IMS Lecture Notes – Monograph Series*, Institute of Mathematical Statistics, Beachwood, OH, pp. 65–116.

[3] Chipman, H. A., George, E. I. & McCulloch, R. E. (2010), BART: Bayesian additive regression trees, *Annals of Applied Statistics* **4**(1), 266–298.

[4] Clyde, M. & George, E. I. (2004), Model uncertainty, *Statistical Science* **19**, 81–94.

[5] Cui, W. & George, E. I. (2008), Empirical Bayes vs. fully Bayes variable selection, *Statist. Plann. Inference* **138**, 888–900.

[6] Friedman, J. H. (1991), Multivariate adaptive regression splines (Disc: P67-141), *The Annals of Statistics* **19**, 1–67.

[7] Gelfand, A. E. & Smith, A. (1990), Sampling-based approaches to calculating marginal densities, *Journal of the American Statistical Association* **85**, 398–409.

[8] George, E. I. & McCulloch, R. E. (1993), Variable selection via Gibbs sampling, *Journal of the American Statistical Association* **88**, 881–889.

[9] George, E. I. & McCulloch, R. E. (1997), Approaches for Bayesian variable selection, *Statistica Sinica* **7**, 339–374.

[10] Liang, F., Paulo, R., Molina, G., Clyde, M. A. & Berger, J. O. (2008), Mixtures of g-priors for Bayesian variable selection, *Journal of the American Statistical Association* **103**, 410–423.

[11] Maruyama, Y. & George, E. I. (2011), Fully Bayes factors with a generalized g-prior, *Annals of Statistics* **39**(5), 2740–2765.

[12] R Development Core Team (2011), *R: A Language and Environment for Statistical Computing*, R Foundation for Statistical Computing, Vienna, Austria. ISBN 3-900051-07-0.

[13] Scott, J. G. & Berger, J. O. (2010), Bayes and empirical Bayes multiplicity adjustment in the variable selection problem, *Ann. Statist* **38**, 2587–2619.

[14] Tierney, L. (1994), Markov chains for exploring posterior distributions, *Ann. Statist* **22**, 1701–1762.

[15] Wu, Y., Tjelmeland, H. & West, M. (2007), Bayesian CART: Prior specification and posterior simulation, *Journal of Computational and Graphical Statistics* **16**, 44–66.

[16] Zellner, A. (1986), On assessing prior distributions and Bayesian regression analysis with g-prior distributions, in *Bayesian Inference and Decision Techniques, Stud. Bayesian Econometrics Statist*, Vol. 6, North-Holland, Amsterdam, pp. 233–243.

23 Gibbs sampling for ordinary, robust and logistic regression with Laplace priors

ROBERT B. GRAMACY

23.1 Introduction

Whereas the advent of fast and cheap computation signalled a turning point in Bayesian statistical inference at the end of the last century, today a further technological advance is defining a new era. Data is now gathered on a massive scale. Examples include stock prices for thousands of assets at almost any resolution, text or images from web pages added to the internet in nearly continuous time, or genetics studies involving DNA sequencing and the like. Regression and classification, as ever the work horse of statistical inference, still play a major role. However, modern applications usually involve regressions or classifications where the number of predictors, p, in $x_i = (x_{i1}, \ldots, x_{ip})^\top$, is large. To help select relevant predictors, and also to obtain stable predictions, it is widely accepted that restrictions need to be placed on the parameters, e.g. the regression coefficients, in the model. In classical statistics this is what is meant by *regularization*. In Bayesian statistics it simply means deploying informative *priors*.

As one example, consider estimating the covariance of asset returns for balancing portfolios. Varying lengths of historical price information are available because some companies/indices have been publicly traded for longer than others. It has been illustrated that the appropriate estimators for this problem involve OLS regression as a subroutine [30]. However, scaling up to thousands of assets requires that the regressions be regularized [17]. Better still, a fully Bayesian approach with Laplace priors, discussed below, has been shown to produce superior regression estimators and more profitable portfolios [15]. These fully Bayesian estimators heavily leverage the technique of data augmentation and, thereby, Gibbs sampling [9]—historically, the first MCMC technique deployed for Bayesian inference.

Although such problems/applications are modern, the methods used to solve them were in fact so well established they had been forgotten (at least by some). They leveraged techniques from the 80s and 90s, making heavy use of Gibbs sampling algorithms that were abandoned when the sledge hammer of the Metropolis–Hastings (MH) sampler and other more generic algorithms came along. Logistic regression is another example. Many texts [e.g. 10] taught early twenty-first-century graduate students that MH and approximate methods were best suited to GLMs, implying that Gibbs sampling was not an option. But then [22] showed that a data augmentation/Gibbs sampling scheme, similar to the one for the Laplace and in the style popularized in the 90s, gave better MCMC performance than MH. There have since been many advances on these fronts. For example, a natural extension pairing the Gibbs samplers for logistic regression with those of Laplace priors for regularization [16] enables fully Bayesian inference for large classification problems like those needed for text classification [31].

This paper reviews the ideas behind the Gibbs samplers for both OLS and logistic regression under regularization, focusing on the Laplace prior. Section 23.2 considers OLS with extensions that allow for model selection and averaging, and heavy-tailed errors for robust estimation. Examples are provided using the implementation in the R package called `monomvn`, available on CRAN. Section 23.3 covers similar routines for logistic regression, with examples illustrated via the `reglogit` package. Finally, the paper concludes in Section 23.4 with references to further extensions to these methods.

23.2 Bayesian shrinkage regression

Consider the typical linear regression model:

$$\mathbf{y} = \beta_0 \mathbf{1}_n + \mathbf{X}\boldsymbol{\beta} + \boldsymbol{\epsilon}, \qquad \text{where} \qquad \epsilon \sim \mathcal{N}_n(\mathbf{0}, \sigma^2 \mathbf{I}_n) \tag{23.1}$$

To keep the discussion simple, assume a standardized $n \times p$ design matrix $\mathbf{X}$ where the columns are individually adjusted to have zero-mean and unit L_2-norm. This causes β_0 and $\boldsymbol{\beta}$ to be independent a posteriori and recognizes that regularized posterior summaries for $\boldsymbol{\beta}$ are not equivariant under a re-scaling of $\mathbf{X}$.

The *lasso* estimator [e.g. 21, Section 3.4.3] is the solution to an ordinary least squares criteria subject to an L_2 penalty on the regression coefficients:

$$\hat{\boldsymbol{\beta}} = argmin_{\boldsymbol{\beta}} \left\{ (\tilde{\mathbf{y}} - \mathbf{X}\boldsymbol{\beta})^\top (\tilde{\mathbf{y}} - \mathbf{X}\boldsymbol{\beta}) + \lambda \sum_{j=1}^{p} |\beta_j| \right\} \tag{23.2}$$

for some $\lambda \geq 0$. The intercept is excluded from penalization via $\tilde{\mathbf{y}} = \mathbf{y} - \bar{y}\mathbf{1}_n$. There is no closed form solution for $\hat{\boldsymbol{\beta}}$, but the entire path of solutions for all λ can be obtained iteratively via the LARS algorithm [6]. The estimator may be interpreted as the posterior mode under and i.i.d. Laplace (i.e. double-exponential) prior for each β_i: $\pi(\boldsymbol{\beta}^{(1)}|\sigma^2) = \prod_{j=1}^{p} \frac{\lambda}{2\sqrt{\sigma^2}} e^{-\lambda|\beta_j^{(1)}|/\sqrt{\sigma^2}}$. Estimators so obtained, either via the MAP or otherwise, are shrunk towards zero compared to their OLS alternatives by an amount that depends on the value of the penalty parameter, λ. A particular feature of the lasso, i.e. the MAP solution (23.2), is that $\hat{\boldsymbol{\beta}}$ may have many coefficients shrunk to exactly zero, which is convenient for variable selection. Often, λ is chosen via cross-validation (CV). Calculating the posterior mean estimator, or using the posterior distribution to choose the penalty parameter, λ, requires more work.

23.2.1 Hierarchical models for Bayesian shrinkage regression

For a fully Bayesian lasso, there is a latent variable formulation [3, 29] that represents the Laplace as a scale mixture of normals:

$$\begin{aligned} \mathbf{y}|\beta_0, \mathbf{X}, \boldsymbol{\beta}, \sigma^2 &\sim \mathcal{N}_n(\beta_0 \mathbf{1}_n + \mathbf{X}\boldsymbol{\beta}, \sigma^2 \mathbf{I}_n) & \mathbf{D}_\tau &= \mathrm{diag}(\tau_1^2, \ldots, \tau_p^2) \\ \boldsymbol{\beta}|\sigma^2, \tau_1^2, \ldots, \tau_p^2 &\sim \mathcal{N}_p(\mathbf{0}, \sigma^2 \mathbf{D}_\tau), \; \beta_0 \propto 1 & \sigma^2 &\sim \mathrm{IG}(a_\sigma/2, b_\sigma/2) \\ \tau_j^2|\lambda^2 &\stackrel{\text{iid}}{\sim} \mathrm{Exp}(\lambda^2/2) & \lambda^2 &\sim G(a_\lambda, b_\lambda) \end{aligned} \tag{23.3}$$

IG and G are the rate- and scale-parameterized inverse-gamma and gamma distributions, respectively. The default prior $\pi(\sigma^2) \propto \sigma^{-2}$ is obtained with $a_\sigma = b_\sigma = 0$.

Since the full conditionals for all of the parameters are of a standard form, Gibbs sampling (GS) is an obvious choice for MCMC.

$$\begin{aligned} \beta_0|\sigma^2, \mathbf{y} &\sim \mathcal{N}(\bar{y}, \sigma^2/n) \\ \boldsymbol{\beta}|\sigma^2, \{\tau_j^2\}_{j=1}^p, \mathbf{y} &\sim \mathcal{N}_p(\tilde{\boldsymbol{\beta}}, \sigma^2 \mathbf{A}^{-1}), & \mathbf{A} = \mathbf{X}^\top \mathbf{X} + \mathbf{D}_\tau^{-1}, \; \tilde{\boldsymbol{\beta}} = \mathbf{A}^{-1}\mathbf{X}^\top \tilde{\mathbf{y}} \\ \sigma^2|\boldsymbol{\beta}, \{\tau_j^2\}_{j=1}^p, \mathbf{y} &\sim \mathrm{IG}((a_\sigma + n - 1 + p)/2, (b_\sigma + \psi_\beta)/2), & \psi_\beta = ||\tilde{\mathbf{y}} - \mathbf{X}\boldsymbol{\beta}||^2 + \boldsymbol{\beta}^\top \mathbf{D}_\tau^{-1}\boldsymbol{\beta} \\ \tau_j^{-2}|\beta_j, \sigma^2, \lambda &\stackrel{\text{iid}}{\sim} \text{Inv-Gauss}(\sqrt{\lambda^2\sigma^2/\beta_j^2}, \lambda^2) \\ \lambda^2|\tau_1^2, \ldots, \tau_p^2 &\sim \mathrm{G}(a_\lambda + p\gamma, b_\lambda/\gamma + \textstyle\sum_{j=1}^p \tau_j^2/2) \end{aligned} \tag{23.4}$$

Using a marginal posterior conditional for σ^2 instead can help reduce autocorrelation in the Markov chain. Integrating over the posterior conditional for $\boldsymbol{\beta}$ gives

$$\sigma^2|\tau_1^2, \ldots, \tau_p^2, \mathbf{y} \sim \mathrm{IG}((a_\sigma + n - 1)/2, (b_\sigma + \psi_{\tilde{\beta}})/2),$$
$$\text{where } \psi_{\tilde{\beta}} = ||\tilde{\mathbf{y}} - \mathbf{X}\tilde{\boldsymbol{\beta}}||^2 + \tilde{\boldsymbol{\beta}}^\top \mathbf{D}_\tau^{-1}\tilde{\boldsymbol{\beta}} = \tilde{\mathbf{y}}^\top \tilde{\mathbf{y}} - \tilde{\boldsymbol{\beta}}^\top \mathbf{A}\tilde{\boldsymbol{\beta}}$$

An alternative hierarchical modelling framework for the Bayesian lasso is provided by [20]. While it does not require p latent τ_j^2 variables, the resulting GS procedure is not fully blocked, and rejection sampling is required for σ^2. 'Orthogonalizing' the sampler helps mitigate slow mixing of the un-blocked conditionals, and the resulting sampler is believed to be superior in such contexts. The current discussion focuses on the approach of [29] as it is more readily adaptable to some of the extensions, like heavy-tailed errors and model selection, described below.

Whereas the classical lasso has the property that the estimate $\hat{\boldsymbol{\beta}}$ may have components which are zero—in fact, it would never have more than $\min\{p, n-1\}$ non-zero components—samples of $\boldsymbol{\beta}$ from the posterior would never have zeros. So the Bayesian lasso is less useful for variable selection. We also note that when $p \geq n$—and without the ability to explicitly restrict $\boldsymbol{\beta}$ to having at most $\min\{p, n-1\}$ non-zero components—a proper prior must be used for σ^2 or the posterior will be improper. An empirical Bayes remedy that works well in this case is to take a small a_σ, say $a_\sigma = 3/2$, and then set b_σ so that the $(1-\alpha)$ part of the $\mathrm{IG}(a_\sigma, b_\sigma)$ distribution lies at the point $\tilde{\mathbf{y}}^\top \tilde{\mathbf{y}}$ (i.e. the MLE under the intercept model) via the incomplete gamma inverse function. Another remedy is Bayesian model averaging.

23.2.2 Bayesian model selection and averaging

Although the MAP lasso fit may indeed set some of the coordinates of $\hat{\boldsymbol{\beta}}^{(1)}$ to zero, this is more of a side effect of the solution space of the quadratic program (23.2) than the result of a deliberate prior modelling choice [20]. Bayesians rarely base inference on the MAP; it is more natural to select variables by inspecting the posterior model probabilities.

There are several standard ways of performing Bayesian variable selection in regression models that are amenable to GS. They essentially fall into two camps. Loosely, the first camp [e.g. 12, 14] employs a product-space wherein the prior for each β_j is augmented to include a point-mass at zero. Inference proceeds by GS on each of the conditionals $\beta_j|\boldsymbol{\beta}_{-j}, \mathbf{y}, \ldots, j = 1, \ldots, p$, which may flop between zero and non-zero values. [20] augmented this product space approach to variable selection under the Laplace prior by further conditioning on λ.

The second camp [e.g. 32] is transdimensional in that the $\boldsymbol{\beta}$-vector may vary in length while model space is traversed via Reversible Jump (RJ) MCMC [18]. An advantage of this approach is that, when $p \gg n$, it is implementationally more compact, only requiring memory for the (non-zero) $\boldsymbol{\beta}$-components. This can represent a huge saving in some contexts like large-scale portfolio balancing [15]. Another advantage is that it emits fully blocked sampling for the non-zero components of $\boldsymbol{\beta}$ for within-model moves.

Suppose that the transdimensional Markov chain is currently visiting a model with k non-zero regression coefficients $\boldsymbol{\beta}_k = (\beta_1, \ldots, \beta_k)$ using design matrix $\mathbf{X}_k$. The columns of $\mathbf{X}_k$ should come from a two-way partition (of k and $p - k$ elements) of the p columns of $\mathbf{X}$, but they need not coincide with the first k of the p columns. Now consider proposing to add a column to $\mathbf{X}_k$. Choose one of the $p - k$ columns of $\mathbf{X}$ not present in $\mathbf{X}_k$ for addition, thus creating $\mathbf{X}_{k+1}$. By considering the ratio of the marginal posterior distributions (integrating out $\boldsymbol{\beta}_k$ and $\boldsymbol{\beta}_{k+1}$) conditional on $\sigma^2, \tau_1^2, \ldots, \tau_k^2$ and a new proposed τ_{k+1}^2 (which can be taken from the prior), the move may be accepted with probability $\min\{1, A_{k\to k+1}\}$, where

$$A_{k\to k+1} = \frac{(\tau_{k+1}^{-2}|\mathbf{A}_{k+1}^{-1}|)^{\frac{1}{2}} \exp\left\{\frac{1}{2\sigma^2}\tilde{\boldsymbol{\beta}}_{k+1}^\top \mathbf{A}_{k+1}\tilde{\boldsymbol{\beta}}_{k+1}\right\}}{|\mathbf{A}_k^{-1}|^{\frac{1}{2}} \exp\left\{\frac{1}{2\sigma^2}\tilde{\boldsymbol{\beta}}_k^\top \mathbf{A}_k\tilde{\boldsymbol{\beta}}_k\right\} q(\tau_{k+1}^2)} \times \frac{\pi(k+1)q(k+1\to k)}{\pi(k)q(k\to k+1)} \tag{23.5}$$

and $\mathbf{A}_k = \mathbf{X}_k^\top\mathbf{X}_k + \mathbf{D}_{\tau_k}^{-1}$, $\tilde{\boldsymbol{\beta}}_k = \mathbf{A}_k^{-1}\mathbf{X}_k^\top\tilde{\mathbf{y}}$, with $\mathbf{D}_{\tau_k} = \text{diag}(\tau_1^2, \ldots, \tau_k^2)$. The reverse, of proposing to remove one of the columns of $\mathbf{X}_k$, may be accepted with probability $\min\{1, A_{k-1\to k}^{-1}\}$.

A uniform prior over all models with k non-zero components is typical, but there are other options [see, e.g. 12, 15, 20]. Movement throughout the 2^p-sized space is slow for large p, so a certain amount of thinning of the RJ-MCMC chain is appropriate. Collecting a sample from the posterior after p transdimensional moves approximates the model-level mixing (and computation burden) of the product-space approach. Throughout the RJ-MCMC the length of $\boldsymbol{\beta}$ varies, and the components shift to represent the partition of $\mathbf{X}$ stored in the columns of $\mathbf{X}_k$. The posterior probability that variable j, $j = 1, \ldots, p$, is relevant for predicting $\mathbf{y}$ can be approximated by the proportion of time that $\mathbf{X}_k$ contains variable j.

23.2.3 Student-*t* errors via scale-mixtures

The MVN assumption is not always appropriate. Instead, one may wish to consider the possibility that errors in $\mathbf{y}$ have a Student-t distribution with an unknown degrees of freedom ν: $\mathbf{y} = \beta_0\mathbf{1}_n + \mathbf{X}\boldsymbol{\beta} + \boldsymbol{\epsilon}$, $\{\epsilon_i\}_{i=1}^n \overset{\text{iid}}{\sim} \text{St}(0, \sigma^2; \nu)$. Following [4] and [13], it is convenient to represent the Student-t distribution as a scale mixture of normals with an $\text{IG}(\nu/2, \nu/2)$ mixing density.

We must redefine $\mathbf{X} = (\mathbf{1}_n, \mathbf{X})$ as a $n \times (p+1)$ matrix, $\boldsymbol{\beta} = (\beta_0, \boldsymbol{\beta}^\top)^\top = \{\beta_j\}_{j=0}^p$ so that the model becomes $\mathbf{y} = \mathbf{X}\boldsymbol{\beta} + \boldsymbol{\epsilon}$ since the posterior intercept β_0 is no longer independent of the other components of $\boldsymbol{\beta}$ in the presence of heavy-tailed errors. The setup is otherwise unchanged from Section 23.2.1. Upon assuming an exponential prior for the degrees of freedom parameter, ν, the modifications to the hierarchical model in Eq. (23.3) are:

$$
\begin{aligned}
\mathbf{y}|\mathbf{X},\boldsymbol{\beta},\sigma^2,\{\omega_i^2\}_{i=1}^n &\sim \mathcal{N}_n(\mathbf{X}\boldsymbol{\beta},\sigma^2\mathbf{D}_\omega) & \mathbf{D}_\omega &= \mathrm{diag}(\omega_1^2,\ldots,\omega_n^2) \\
\boldsymbol{\beta}|\sigma^2,\{\tau_j^2\}_{j=1}^p &\sim \mathcal{N}_{p+1}(\mathbf{0},\sigma^2\mathbf{D}_\tau) & \mathbf{D}_\tau &= \mathrm{diag}(\infty,\tau_1^2,\ldots,\tau_p^2) \\
\omega_i^2|\nu &\stackrel{\text{iid}}{\sim} \mathrm{IG}(\nu/2,\nu/2) & \nu|\theta &\sim \mathrm{Exp}(\theta)
\end{aligned}
\tag{23.6}
$$

Note that $\mathbf{D}_\tau$ is a $p+1$ diagonal matrix, and that the first component insures that β_0 is given a flat prior as before. After redefining $\mathbf{A} = \mathbf{X}^\top\mathbf{D}_\omega^{-1}\mathbf{X} + \mathbf{D}_\tau^{-1}$, $\tilde{\boldsymbol{\beta}} = \mathbf{A}^{-1}\mathbf{X}^\top\mathbf{D}_\omega^{-1}\mathbf{y}$ and $\psi_\beta = (\mathbf{y} - \mathbf{X}\boldsymbol{\beta})^\top\mathbf{D}_\omega^{-1}(\mathbf{y}-\mathbf{X}\boldsymbol{\beta}) + \boldsymbol{\beta}^\top\mathbf{D}_\tau^{-1}\boldsymbol{\beta}$, the modified full posterior conditionals follow:

$$
\begin{aligned}
\boldsymbol{\beta}|\sigma^2,\{\tau_j^2\}_{j=1}^p,\{\omega_i^2\}_{i=1}^n,\mathbf{y} &\sim \mathcal{N}_{p+1}(\tilde{\boldsymbol{\beta}},\sigma^2\mathbf{A}^{-1}) \\
\sigma^2|\boldsymbol{\beta},\{\tau_j^2\}_{j=1}^p,\{\omega_i^2\}_{i=1}^n,\mathbf{y} &\sim \mathrm{IG}\left(\frac{a_\sigma+n+p}{2},\frac{b_\sigma+\psi_\beta}{2}\right) \\
\omega_i^2|\boldsymbol{\beta},\sigma^2,\nu,\mathbf{y} &\stackrel{\text{iid}}{\sim} \mathrm{IG}\left(\frac{\nu+1}{2},\frac{\nu+\sigma^{-2}((\mathbf{y}-\mathbf{X}\boldsymbol{\beta})_i)^2}{2}\right) \\
p(\nu|\{\omega_i^2\}_{i=1}^n,\theta) &\propto \left(\frac{\nu}{2}\right)^{\frac{n\nu}{2}}\left(\Gamma\left(\frac{\nu}{2}\right)\right)^{-n}\exp(-\eta\nu)
\end{aligned}
\tag{23.7}
$$

where $\eta = \frac{1}{2}\sum_{i=1}^n(\log(\omega_i^2)+\omega_i^{-2})+\theta$

The conditional posterior of ν does not have a standard form, but there is an efficient rejection sampler [13]. A draw from $\nu \sim \mathrm{Exp}(\nu^*)$, where ν^* is chosen optimally as the root of $(n/2)[\log(\nu/2)+1-\Psi(\nu/2)]+\nu^{-1}-\eta$, may be retained with probability

$$
\min\left\{1,\left[\frac{\Gamma(\nu^*/2)}{\Gamma(\nu/2)}\right]^n\left[\frac{(\nu/2)^\nu}{(\nu^*/2)^{\nu^*}}\right]^{n/2}\exp[(\nu-\nu^*)((\nu^*)^{-1}-\eta)]\right\}
$$

As before, integrating out $\boldsymbol{\beta}$ gives $\sigma^2|\{\tau_j^2\}_{j=1}^p,\{\omega_i^2\}_{i=1}^n,\mathbf{y} \sim \mathrm{IG}((a_\sigma+n-1)/2,(b_\sigma+\psi_{\tilde{\beta}})/2)$ by redefining $\psi_{\tilde{\beta}} = (\mathbf{y}-\mathbf{X}\tilde{\boldsymbol{\beta}})^\top\mathbf{D}_\omega^{-1}(\mathbf{y}-\mathbf{X}\tilde{\boldsymbol{\beta}}) + \tilde{\boldsymbol{\beta}}^\top\mathbf{D}_\tau^{-1}\tilde{\boldsymbol{\beta}} = \mathbf{y}^\top\mathbf{D}_\omega^{-1}\mathbf{y} - \tilde{\boldsymbol{\beta}}^\top\mathbf{A}\tilde{\boldsymbol{\beta}}$, leading to a more efficient sampler. Finally, the Bayesian model selection and averaging method of Section 23.2.2, via equation (23.5), may be used with $\mathbf{X}_k = (\mathbf{1}_n,\mathbf{X}_k)$, $\boldsymbol{\beta}_k = (\beta_0,\boldsymbol{\beta}_k^\top)^\top$, $\tilde{\boldsymbol{\beta}}_k = \mathbf{A}_k^{-1}\mathbf{X}_k^\top\mathbf{D}_\omega^{-1}\mathbf{y}$ and $\mathbf{A}_k = \mathbf{X}_k^\top\mathbf{D}_\omega^{-1}\mathbf{X}_k + \mathbf{D}_{\tau_k}^{-1}$ and $\mathbf{D}_{\tau_k} = \mathrm{diag}(\infty,\tau_1^2,\ldots,\tau_k^2)$. The number of latent variables now grows with the sample size, so automatic $O(n)$ thinning from the Markov chain is sensible.

23.2.4 Empirical results on detecting fat tails

[20] offers a plethora of insights about the Bayesian lasso with comparison to the classical lasso. There is no need to re-produce these results here. Instead we offer a demonstration focusing on Student-*t* extensions. One can follow along with this example, on the diabetes data [6], via the help file for `blasso` in the `monomvn` package.

Consider testing the null hypothesis (model $\mathcal{M}_{\mathrm{N}}$) of normal errors versus the alternative (model $\mathcal{M}_{\mathrm{St}}$) that they follow a Student-*t* with $\nu \sim \mathrm{Exp}(\theta = 0.1)$ a priori. Jacquier *et al.* [23, Section 2.5.1] show how to exploit that the Student-*t* and normal models differ by just one parameter in the likelihood, ν, to calculate a Bayes factor (BF) for $\mathcal{M}_{\mathrm{N}}$ over $\mathcal{M}_{\mathrm{St}}$ as the expectation of the ratio of un-normalized posteriors using samples from the Student-*t* model. That is,

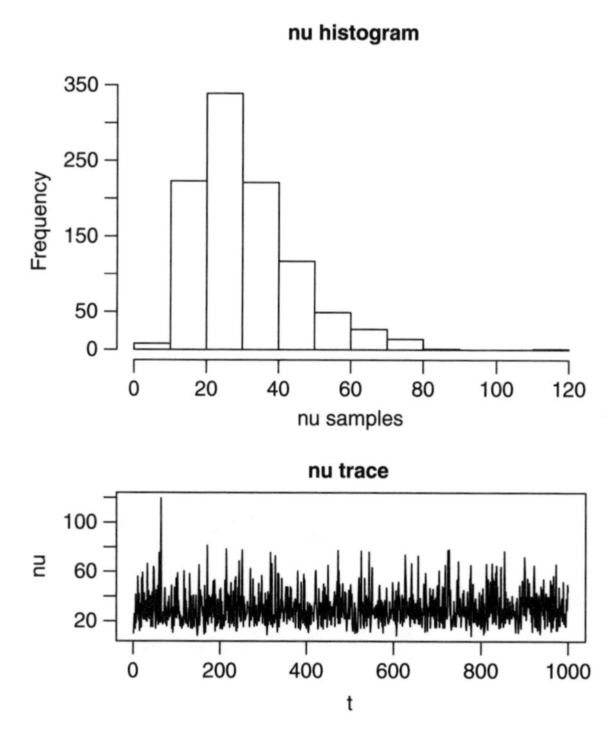

Figure 23.1 Histogram *(top)* and trace plot *(bottom)* for samples from marginal posterior for ν, the degrees of freedom parameter in the Student-*t* model for the `diabetes` data.

$$\mathbb{E}\left\{\frac{p(\mathbf{y}|\boldsymbol{\psi}, \mathcal{M}_{\mathrm{N}})}{p(\mathbf{y}|\boldsymbol{\psi}, \nu, \mathcal{M}_{\mathrm{St}})}\right\} \approx \frac{1}{T}\sum_{t=1}^{T} \frac{p(\mathbf{y}|\boldsymbol{\psi}^{(t)}, \mathcal{M}_{\mathrm{N}})}{p(\mathbf{y}|\boldsymbol{\psi}^{(t)}, \nu^{(t)}, \mathcal{M}_{\mathrm{St}})}, \quad \text{where} \quad (\boldsymbol{\psi}^{(t)}, \nu^{(t)}) \sim p(\boldsymbol{\psi}, \nu|\mathbf{y}, \mathcal{M}_{\mathrm{St}})$$

and where $\boldsymbol{\psi}$ collects the parameters shared by both models.

Doing this calculation on the `diabetes` data leads to a BF of essentially zero, showing 'decisive' evidence in favour of $\mathcal{M}_{\mathrm{N}}$ under reasonable priors [see 15]. Figure 23.1 shows that, after burn-in, the mixing was very good for ν (marginally) under $\mathcal{M}_{\mathrm{St}}$, starting at $\nu^{(0)} = 1/\theta$. It is easy to see now $\mathcal{M}_{\mathrm{N}}$ is favoured, since most of the posterior density is allocated to $\nu > 10$. Figure 23.2 shows that the mixing in model space (under $\mathcal{M}_{\mathrm{St}}$), through the model order k starting at $k^{(0)} = 0$, is also very good.

Figure 23.3 summarizes the posterior distribution of the coefficients of $\boldsymbol{\beta}$ and $\mu \equiv \beta_0$, offering a comparison to point estimates obtained under OLS and classical MLE/CV lasso, and the Bayesian MAP. Observe that for $\beta_1, \beta_2, \beta_6, \beta_8$, and β_{10} the classical lasso solution and MAP agree that the regression coefficient in question is zero. They disagree on β_5 and β_7 where one regards it as zero and the other does not. Finally, they agree that the rest (β_3, β_4, and β_9) are non-zero. Also observe that posterior distribution(s) are not symmetric, as expected under a Laplace prior and that, in the case of β_7, the sign of the OLS estimate is at odds with the posterior density. The probabilities that $\boldsymbol{\beta}|\mathbf{y}$ are non-zero are shown below the plot in Figure 23.3, demonstrating how the fully Bayesian approach, in contrast to the lasso–CV approach, offers full accounting in uncertainty in variable selection.

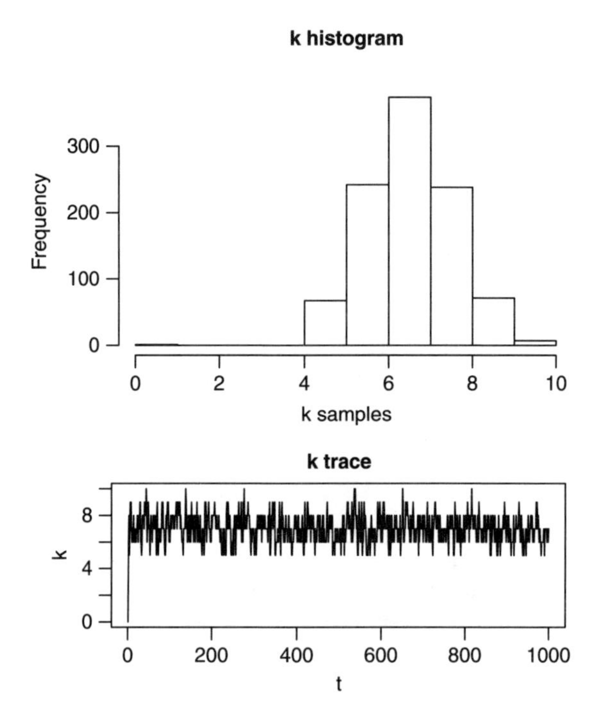

Figure 23.2 Histogram *(top)* and trace plot *(bottom)* for samples from the marginal posterior for the model order parameter $k \in \{0, 10\}$ for the **diabetes** data.

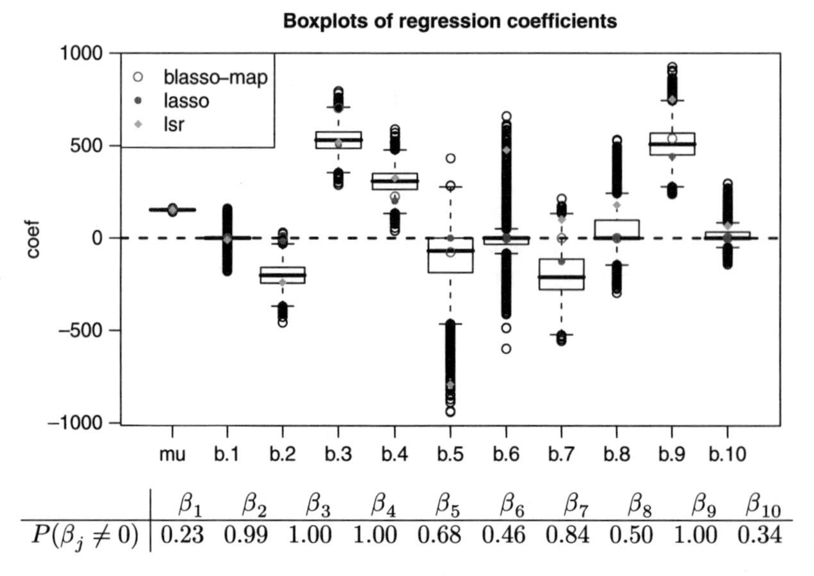

	β_1	β_2	β_3	β_4	β_5	β_6	β_7	β_8	β_9	β_{10}
$P(\beta_j \neq 0)$	0.23	0.99	1.00	1.00	0.68	0.46	0.84	0.50	1.00	0.34

Figure 23.3 Summarizing samples from the posterior distribution of $\boldsymbol{\beta}$ under the Bayesian lasso model averaging prior on the **diabetes** data. The *top* plot summarizes the marginal posterior of the components β_j, $j = 1, \ldots, 10$ and intercept $\mu \equiv \beta_0$ via boxplots; the *bottom* table shows the posterior probability that the components β_j are, marginally, non-zero.

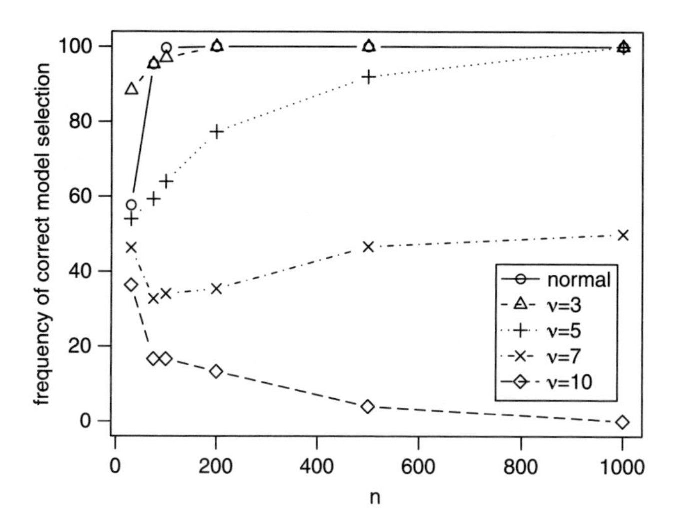

Figure 23.4 Frequency of correct model determinations as a function of the sample size, n, and the degrees of freedom parameter, ν, where 'normal' is interpreted as $\nu = \infty$.

To shed light on the 'selectability' of the Student-t model, consider synthetic data where $\boldsymbol{\beta} = (2, -3, 0, 0.75, 0, 0, -0.9)^\top$, $\mu \equiv \beta_0 = 1$, the rows of the $n \times 7$ design matrix $\mathbf{X}$ are uniformly distributed in $[0, 1]^7$, and $\epsilon_i \sim \mathrm{St}(0, \sigma^2 = 1; \nu)$, for $i = 1, \ldots, n$. Consider a Monte Carlo experiment where n and ν vary, with $n \in \{30, 75, 100, 200, 500, 1000\}$ and $\nu \in \{3, 5, 7, 10, \infty\}$, and the frequency of times that the BF indicated 'strong' preference for the correct model in repeated trials. In each trial, GS (23.7) was used to obtain 1200 samples from the posterior by thinning every $7n$ rounds, with the first 200 discarded as burn-in. For $n \leq 200$ we repeated the experiment with random data 300 times; when $n = 500$ we used 50 replications; and when $n = 1000$ we used 20. Figure 23.4 shows the relationships between n, ν and the frequency of correct model determinations (higher frequencies are better). In the case of normal errors, and Student-t errors with $\nu = 3$, the correct model can be determined with high accuracy when $n \geq 200$. When $\nu = 5$, a sample size of $n = 1000$ is needed; when $\nu = 7, 10$ we need $n \gg 1000$. Clearly for $10 \leq \nu < \infty$ the situation is hopeless unless n is very large. The conclusion is that the dataset must be huge, and simultaneously the degrees of freedom small, in order for the returns on investment in the extra latent variables in the Student-t model to be realized.

23.3 Logistic regression

Now consider a set of binary responses, y_i, encoded as ± 1, regressed on p-dimensional predictors $\mathbf{x}_i$ via the model $\mathbb{P}(y_i = \pm 1 | \mathbf{x}_i, \boldsymbol{\beta}) = (1 + \exp\{-y_i \mathbf{x}_i^\top \boldsymbol{\beta}\})^{-1}$, for $i = 1, \ldots, n$. For a review of Bayesian approaches to logistic regression see, e.g. [22] [HH hereafter] and [8]. As in the OLS setup, when p is large it is paramount to infer β under regularization or penalization. Mirroring (23.2), a common formulation [e.g., 11, 26, 28] in this context is

$$\hat{\boldsymbol{\beta}} = \mathrm{argmin}_{\boldsymbol{\beta}} \left\{ \sum_{i=1}^{n} \ln \left(1 + \exp\{-y_i \mathbf{x}_i^\top \boldsymbol{\beta}\} \right) + \nu^{-1} \sum_{j=1}^{p} |\beta_j| \right\} \tag{23.8}$$

The parameter ν dictates the amount of regularization, or the relative pull (ν^{-1}) or shrinkage of the β_js towards zero.[21] Borrowing from the OLS literature, [27] discuss how the LARS algorithm can be useful as a subroutine for the popular case of $\alpha = 1$. Again, it is typical to work with $\mathbf{x}_i$ pre-scaled, and not penalize the intercept.

We consider the following *power-posterior* distribution inspired by equation (23.8):

$$\pi_\kappa(\boldsymbol{\beta}|y,\nu,) = C_\kappa,(\nu)\exp\left\{-\kappa\left(\sum_{i=1}^n \ln\left(1+\exp\{-y_i\mathbf{x}_i^\top\boldsymbol{\beta}\}\right) + \nu^{-1}\sum_{j=1}^p|\beta_j|\right)\right\} \tag{23.9}$$

The placement of κ in the subscript in π_κ and $C_\kappa(\nu)$, a normalization factor, signals that it is user specified, not a parameter to be estimated. We call κ the *multiplicity parameter*, but it is known as the *thermodynamic parameter* the simulated annealing literature [see, e.g. 25]. It is a tool that facilitates several types of simulation based inference, as we shall describe. The discussion centres around the following likelihood–prior combination which, together with Bayes' rule, yields the expression in equation (23.9).

$$L_\kappa(y|\boldsymbol{\beta}) = \exp\left\{-\kappa\sum_{i=1}^n \ln\left(1+\exp\{-y_i\mathbf{x}_i^\top\boldsymbol{\beta}\}\right)\right\} = \prod_{i=1}^n\left(1+\exp\{-y_i\mathbf{x}_i^\top\boldsymbol{\beta}\}\right)^{-\kappa} \tag{23.10}$$

$$p_\kappa(\beta|\nu) \propto \exp\left(-\kappa\nu^{-1}\sum_{j=1}^p|\beta_j|\right) = \prod_{j=1}^p \exp\left\{-\kappa\left|\frac{\beta_j}{\nu}\right|\right\}.$$

Power-posterior analysis can be helpful for calculating modes and posterior means from complex optimization criteria, and marginal likelihoods for Bayesian estimators. See [16] for more details on these procedures. The focus here is on how κ can be used to obtain an efficient computational framework for binomial regression, where multiple binary responses are recorded for each predictor which is what motivates calling κ the multiplicity parameter. For the immediate discussion, however, let κ be fixed.

23.3.1 Representing the likelihood and prior

Extending a well-known result by [22], [16] recognized the likelihood (23.10) for $\boldsymbol{\beta}$ as a marginal quantity obtained after integrating over latent variables $(\mathbf{z},\boldsymbol{\lambda})$, where $\mathbf{z} = (z_1,\ldots,z_n)$ and $\boldsymbol{\lambda} = (\lambda_1,\ldots,\lambda_n)$. Briefly, one recognizes that each component $(1+\exp\{-y_i\mathbf{x}_i^\top\boldsymbol{\beta}\})^{-\kappa}$ of the likelihood can be written as the cumulative distribution function (cdf) evaluation (at zero) of a particular z-distribution [2]. For details see [16]. This implies a hierarchical model obtained by mixing

$$z_i|\boldsymbol{\beta},\lambda_i,y_i,\kappa \sim \mathcal{N}^+\left(y_i\mathbf{x}_i^\top\boldsymbol{\beta} + \frac{1}{2}(1-\kappa)\lambda_i,\lambda_i\right), \quad \text{over a particular } \lambda_i \sim q_{1,\kappa} \tag{23.11}$$

where $\mathcal{N}^+$ indicates the normal distribution truncated to the positive real line. The mixing density $q_{a,b}$ has a messy expression, but simple generative form:

[21] Note that $\nu = \lambda^{-1}$ from Section 23.2.1. We have chosen to stay closer to the notation in [16], who use λ for a different quantity.

$$\lambda \overset{D}{=} \sum_{k=0}^{\infty} 2\psi_k^{-1}\epsilon_k, \quad \text{where} \quad \epsilon_k \sim \text{Exp}(1), \quad \text{and} \quad \psi_k = (a+k)(b+k) \tag{23.12}$$

In more compact notation: $\mathbf{z}|\boldsymbol{\beta}, \boldsymbol{\lambda}, \mathbf{y}, \kappa \sim \mathcal{N}_n^+((\mathbf{y}.\mathbf{X})\boldsymbol{\beta} + \frac{1}{2}(1-\kappa)\boldsymbol{\lambda}, \boldsymbol{\Lambda})$, where $\mathbf{y} = (y_1, \ldots, y_n)^\top$, $\mathbf{y}.\mathbf{X} = \text{diag}(\mathbf{y})\mathbf{X}$, $\Lambda = \text{diag}(\lambda_1, \ldots, \lambda_n)$, and truncation is to the all-positive orthant. The $\kappa = 1$ the case is identical to the generative model described by HH, where

$$y_i = \text{sign}(z_i), \quad \text{where} \quad z_i \sim \mathcal{N}(\mathbf{x}_i^\top \boldsymbol{\beta}, \lambda_i) \quad \text{and} \quad \lambda_i = \sum_{k=1}^{\infty} \frac{2}{(1+k)^2}\epsilon_k, \quad \epsilon_k \overset{\text{iid}}{\sim} \text{Exp}(1). \tag{23.13}$$

When $\kappa > 1$, the asymmetry of the z-distribution makes it harder to extract y_i from $y_i\mathbf{x}_i^\top\boldsymbol{\beta} + \frac{1}{2}(1-\kappa)\lambda_i$, the mean of the truncated normal in equation (23.11). However, results in Section 23.3.3 indirectly suggest that one can interpret κy_i as a binomial response for integer κ.

The data augmentation scheme for the (power-) prior is very similar to the OLS setup. The essence is to consider $\beta_j = \frac{\nu}{\kappa}\sigma_j\sqrt{\tau_j}\epsilon_j$, where $\tau_j \sim p(\tau)$ and $\epsilon_j \overset{\text{iid}}{\sim} \mathcal{N}(0,1)$. Observe that small ν (i.e. heavy regularization) and large κ (i.e. heavy concentration of power-posterior density around the mode at the origin) *both* shrink β_j towards zero. Adapting a result from [33] to account for κ, we have that if $\tau_j \overset{\text{iid}}{\sim} \text{Exp}(2)$ then $p_\kappa(\boldsymbol{\beta}|\nu)$ is Laplace with a mean of zero and a scale of ν^2/κ^2. There are two reasonable choices for $\nu \sim p_\kappa(\nu)$ which lead to efficient inference by Gibbs sampling if we wish to fully account for its uncertainty, rather than fix it to a particular value, say via CV. One option is an inverse gamma (IG) prior for ν^2 with shape $r_\kappa = \kappa(r+1) - 1$ and scale $d_\kappa = \kappa d$, where $\kappa = 1$ yields a base $\text{IG}(\nu^2; r, d)$ prior. The second option is IG for ν, with identical powering-up identities. It has lighter tails in ν^{-1}, thus providing more aggressive shrinkage.

23.3.2 Simulation-based inference

We develop the posterior conditionals for sampling from power-posterior $p_\kappa(\boldsymbol{\beta}, \mathbf{z}, \boldsymbol{\tau}, \boldsymbol{\lambda}, \nu|\mathbf{y})$, for any κ, thereby describing a Gibbs sampling algorithm. When $\kappa = 1$ the marginal samples of $\boldsymbol{\beta}$ summarize the posterior distribution of the main parameters of interest. To obtain the MAP estimator or MLE requires establishing an inhomogeneous Markov chain, which is left to our references. Section 23.3.3 describes how a vectorized κ can be used for efficient inference with binomial responses.

By construction (23.11), the posterior full conditional for the latent $z_i|\lambda_i, \ldots$ is a truncated (non-negative) normal distribution. HH derive an expression for $\lambda_i|z_i, \ldots$ when $\kappa = 1$ and provide a rejection sampling algorithm by squeezing. Although this works for general κ, there is a simpler, yet radically different, Rao–Blackwellized approach which involves an alternate z-distribution result [see 16]. A proposal $\lambda_i' \sim q_{1,\kappa}(\lambda)$ from the mixing density with probability may be accepted with probability $\min\{1, A_i\}$ where

$$A_i = \frac{\Phi\{(-y_i\mathbf{x}_i^\top\boldsymbol{\beta} - \frac{1}{2}(1-\kappa)\lambda_i')/\sqrt{\lambda_i'}\}}{\Phi\{(-y_i\mathbf{x}_i^\top\boldsymbol{\beta} - \frac{1}{2}(1-\kappa)\lambda_i)/\sqrt{\lambda_i}\}} \tag{23.14}$$

The easiest way to simulate from $q_{1,\kappa}$ is approximately, by truncating the sum in equation (23.12) at $K = 100$ or so. There are many reasons to prefer a MH-within-Gibbs approach to the rejection/squeezing method of HH. But the most important reason, beyond implementational and computational simplicity, is that drawing λ_i unconditional on z_i yields lower autocorrelation in the overall joint MCMC sampling scheme.

The MVN mixing for $\mathbf{z}$ and priors for $\boldsymbol{\beta}$ combine to give that $\boldsymbol{\beta}|\mathbf{z}, \boldsymbol{\tau}, \boldsymbol{\lambda}, \nu, \kappa \sim \mathcal{N}_p(\tilde{\boldsymbol{\beta}}, \mathbf{V})$ where $\tilde{\boldsymbol{\beta}} = \mathbf{V}(\mathbf{y}.\mathbf{X})^\top \boldsymbol{\Lambda}^{-1}(\mathbf{z} - (1-\kappa)\boldsymbol{\lambda}/2)$, and $\mathbf{V}^{-1} = (\nu/\kappa^{1/\alpha})^{-2}\boldsymbol{\Sigma}^{-1}\mathbf{D}_\tau^{-1} + (\mathbf{y}.\mathbf{X})^\top\boldsymbol{\Lambda}^{-1}(\mathbf{y}.\mathbf{X})$. The full conditional distribution of τ_j is given by

$$p_\kappa(\tau_j|\beta_j, \nu) \propto \frac{1}{\sqrt{2\pi\tau_j}} \exp\left\{-\frac{1}{2}\left(\frac{\kappa^2\beta_j^2}{\nu^2\sigma_j^2\tau_j} + \tau_i\right)\right\} \equiv \mathrm{GIG}\left(\tau_j; \frac{1}{2}, 1, \frac{\kappa^2\beta_j^2}{\nu^2\sigma_j^2}\right)$$

which implies that $\tau_j^{-1} \sim \text{Inv-Gauss}\left(\frac{\nu}{\kappa}\left|\frac{\beta_j}{\sigma_j}\right|, 1\right)$. Finally, the IG priors for ν are both conditionally conjugate. We have,

$$\nu^2|\beta, \tau, \kappa \sim \mathrm{IG}\left(r_\kappa + \frac{\kappa p}{2}, d_\kappa + \frac{\kappa^2}{2}\sum_{j=1}^p \frac{\beta_j^2}{\sigma_j^2\tau_j}\right) \quad \text{or} \quad \nu|\beta, \kappa \sim \mathrm{IG}\left(r_\kappa + \kappa p, d_\kappa + \kappa\sum_{j=1}^p\left|\frac{\beta_j}{\sigma_j}\right|\right)$$

depending on whether the prior for ν^2 or ν is chosen, respectively. Observe that the latter leads to efficiency gains (in addition to better tail properties) since we do not need to condition on $\boldsymbol{\tau}$, and therefore this extends the analysis of [29].

23.3.3 Efficient handling of binomial data

Often, binary response data are collected repeatedly and independently for identical subjects, i.e. with the same covariates $\mathbf{x}$. For example, consider having observed $y_i|x_i \sim \mathrm{Bin}(n_i, \mu_i)$, where $\mu_i = e^{\eta_i}/(1+e^{\eta_i})$ and η_i is linear in $\mathbf{x}_i$. One way to work with this data is to *flatten* it, so that n_i components appear in the likelihood for each subject i: $\prod_{j=1}^{n_i}(1+\exp\{-y_{ij}\mathbf{x}_i^\top\boldsymbol{\beta}\})^\kappa$, where $y_{ij} \in \{-1, 1\}$ giving $|\sum_j^{n_i} y_{ij}| = n_i$. This allows inference to proceed as described in Section 23.3.2, but requires a lot of latents (n_i) for each individual (i).

A more efficient *multiplicity* representation is obtained by recognizing that the component of the likelihood for each observation, i, may be written instead with two terms as $(1+\exp\{-\mathbf{x}_i^\top\boldsymbol{\beta}\})^{\kappa y_i}(1+\exp\{\mathbf{x}_i^\top\boldsymbol{\beta}\})^{\kappa(n_i-y_i)}$. This suggests that the full likelihood, with m unique subjects, can be written by defining $\kappa_{i-} = \kappa(n_i - y_i)$ and $\kappa_{i+} = \kappa y_i$ as $\prod_{i=1}^m(1+\exp\{-\mathbf{x}_i^\top\boldsymbol{\beta}\})^{\kappa_{i+}}(1+\exp\{\mathbf{x}_i^\top\boldsymbol{\beta}\})^{\kappa_{i-}}$. In other words, this is a powered-up logistic likelihood where the first m terms use response 'data' $y_i' = +1$ with multiplicity parameter κ_{i+}, and the second m terms use $y_i' = -1$ with κ_{i-}. Forming vectors $\mathbf{y}'$ and $\boldsymbol{\kappa}'$, each of length $n = 2m$ in this way, allows the likelihood to be written as $\prod_{i=1}^n(1+\exp\{-y_i'\mathbf{x}_i^\top\boldsymbol{\beta}\})^{\kappa_i'}$. Usually $2m \ll \sum_{i=1}^m n_i$ yielding a much more efficient MCMC scheme, which is similar to that described in Section 23.3.2 with some vectorizing modifications. For example, the conditionals for z_i and λ_i would use κ_i' instead of κ. For $\boldsymbol{\beta}$, replace $\kappa 1_n$ with $\boldsymbol{\kappa}'$ in the expression for $\tilde{\boldsymbol{\beta}}$. The original, scalar, κ is used for the conditionals corresponding to the parameters of the prior. For example, the posterior conditional covariance $\mathbf{V}$ of $\boldsymbol{\beta}$ is unchanged.

23.3.4 Illustrations

Pima Indian data

The Pima Indian diabetes data is available from the UCI Machine Learning Repository [1]. It includes test outcomes for diabetes performed on $n = 768$ women of Pima heritage with eight real-valued predictors. Some of the predictors have many zeros, which may reasonably be interpreted as 'missing' values. However, to remain consistent with the treatment of this data by H&H, and other authors, we do not treat these values in any special way. In what follows we describe the estimators of $\boldsymbol{\beta} = (\beta_0 \equiv \mu, \beta_1, \ldots, \beta_8)$ obtained from our regularized logistic regression framework. Throughout, $T = 1000$ samples are taken from the posterior, discarding the first 100 as burn-in. This example is contained in the documentation for `reglogit` in the R package by the same name.

Figure 23.5 summarizes the marginal posterior for $\boldsymbol{\beta}$ with boxplots. Two settings of $\kappa \in \{1, 20\}$ (each panel) were used, and heavy regularization (fixing $\nu = 6$) was applied. The MLE, obtained from the `glm` command in R, and the MAP as estimated from the sample(s), are also shown. Shrinkage is apparent in the divergence between the MAP and MLE values in all panels. Observe how the quartiles and outliers converge on the MAP as κ is increased. The convergence is particularly rapid for the intercept term, and the two coefficients with considerable mass near zero (β_4 and β_5). In fact, the corresponding columns of X have the highest concentration of 'missing' values (30% and 49% respectively), so it is not surprising that the MAP estimator excludes them.

Synthetic binomial data

To illustrate the efficient handling of binomial data, consider the following simple binomial logistic regression problem. This example is also contained in the documentation for `reglogit`. The *true* linear predictor is $\eta_i = 1 + \mathbf{x}_i^\top \boldsymbol{\beta}$ where $\boldsymbol{\beta} = (2, -3, 2, -4, 0, 0, 0, 0, 0)^\top$, and the $p =$ 9-dimensional x_i are uniform in $[0, 1]^p$. The responses, $y_i \in \{0, \ldots, n_i\}$, are sampled with $y_i \sim$ Bin(μ_i, n_i) where $n_i = 20$ and $\mu_i = e^{\eta_i}/(1 + e^{\eta_i})$.

Two different implementations of regularized binomial logistic regression are compared based on the output of 100 repeated experiments with $\sum n_i = 2000$ (i.e. $m = 100$ distinct x_i predictors). The metrics for comparison are root mean squared error (RMSE) between the true and posterior mean $\boldsymbol{\beta}$s, and overall computing time of the respective MCMC samplers. In all cases we used $T =$ 1000 MCMC rounds with MH sampling of λ_i at thinning level(s) set by the effective $\boldsymbol{\kappa}'$ (i.e. via κ_i for each λ_i). The first 100 rounds were discarded as burn-in. The flattened version had a mean RMSE of 0.2117 (sd 0.0602), whereas the multiplicity version had 0.2120 (0.0606). So given the same number of MCMC iterations, there is apparently no benefit to one representation over the other, and moreover the extra/fewer latent variables do not seem to affect the MC error of the resulting estimators. But there is a big difference in CPU times. The flattened version took 570.4 s (37.8 s) whereas the multiplicity version was much (9x) faster at 64.6 s (0.82 s). Since this comes with no cost in accuracy (via RMSE), this implementation is much preferred over the flattened version.

A simulated $p \gg n$ experiment

Now consider a predictive comparison, including both fully Bayesian and full/joint MAP (including ν), benchmarked against other modern approaches to regularized logistic regression. Consider the following synthetic data experiment that uses the identical setup as above with the following exceptions. Here, $n_i = 5$ with 20 total samples giving $\sum n_i = 100$ total instances, and we consider three variations on the $\boldsymbol{\beta}$ describing the slope part of the linear predictor. In the first case let $p = 9$ and $\boldsymbol{\beta} = (2, -3, 0.74, -0.9, 0, 0, 0, 0)^\top$; in the second case $p = 100$ augmenting $\boldsymbol{\beta}$ from the first case with 91 more zeros; and in the third $p = 1000$ with 900 more zeros still. Random design matrices in the unit p-cube were used, yielding random training sets of (x, y) values, which was repeated 100 times for each of the three cases in a Monte Carlo fashion. Similarly created random test sets of size 1000, with $n_i' = 100$ so that $\sum n_i' = 10\,000$, were used to compare the methods by misclassification rate.

The fully Bayesian estimator (i.e. $\kappa = 1$) used priors/MCMC exactly as described in the preceding sections. MCMC rounds included (100, 1000), (500, 1500), (1000, 2000) burn-in and total samples in each of the cases $p = 9, 100, 1000$, respectively. The actual predictor used in the

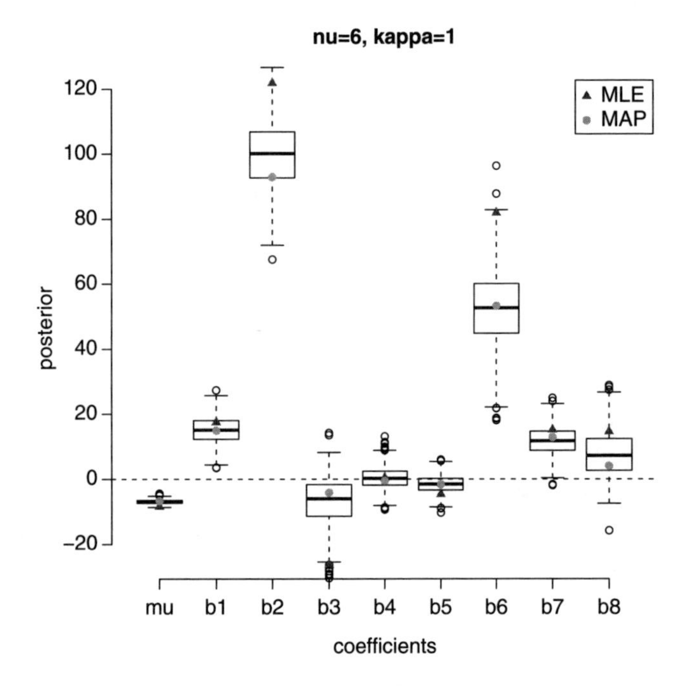

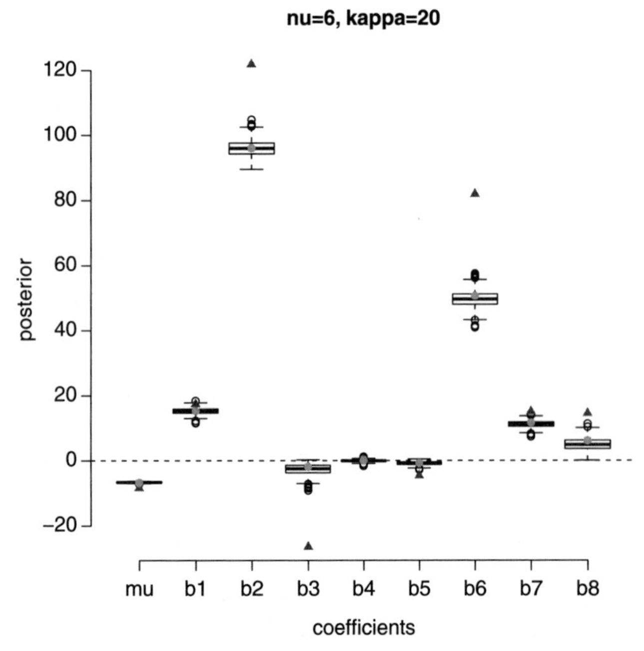

Figure 23.5 Illustrating shrinkage $\nu = 6$ in the power-posterior on the Pima Indian data for $\kappa \in \{1, 20\}$.

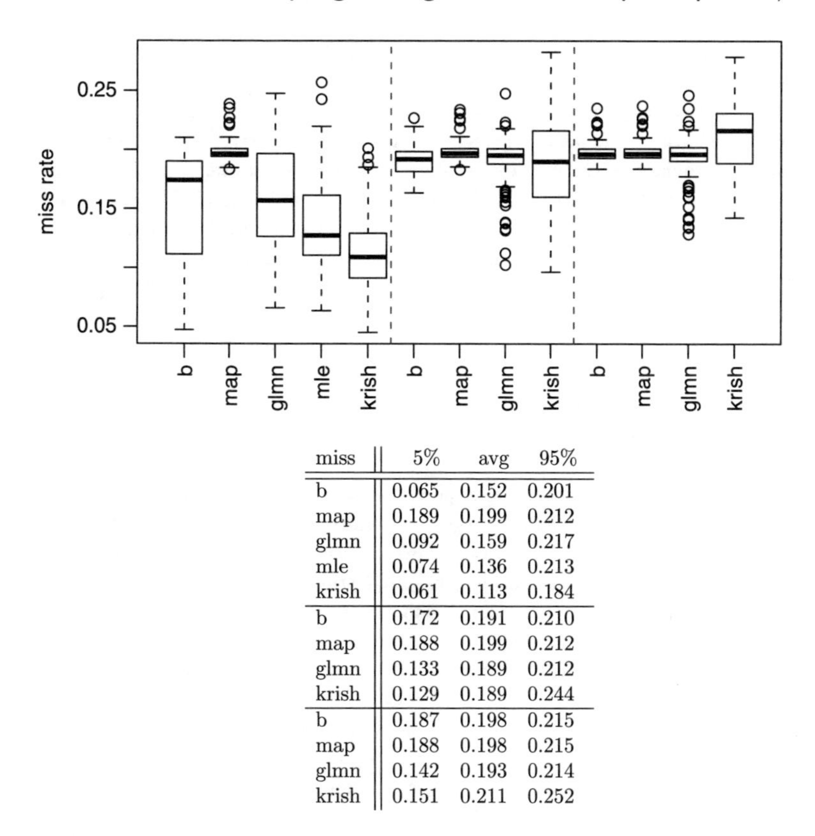

miss	5%	avg	95%
b	0.065	0.152	0.201
map	0.189	0.199	0.212
glmn	0.092	0.159	0.217
mle	0.074	0.136	0.213
krish	0.061	0.113	0.184
b	0.172	0.191	0.210
map	0.188	0.199	0.212
glmn	0.133	0.189	0.212
krish	0.129	0.189	0.244
b	0.187	0.198	0.215
map	0.188	0.198	0.215
glmn	0.142	0.193	0.214
krish	0.151	0.211	0.252

Figure 23.6 Misclassification rates in boxplot (*top*) and tabular (*bottom*) form. There are three sections, depending on the number of irrelevant predictors in the design matrix, wherein the same estimators are applied. The vertical dashed lines in the boxplots indicate the same demarkation as the horizontal lines in the tables.

comparisons was the mean of the posterior predictive after burn-in. The comparators include the standard MLE obtained via the `glm` command in R, a binomial fit from the `glmnet` package [7] for R, and the estimator of [26] ['krish' for short]. The MLE is unstable in all but the $p = 9$ case, so it is not considered for the $p = 100, 1000$ cases. CV was used to choose the penalty parameter in the $p = 9, 100$ cases for `glmnet`, via `cv.glmnet`. The same procedure gave fatal errors in the $p = 1000$ case so we plugged in the estimate obtained from the corresponding $p = 100$ run in for this final case. Reliably setting the penalty parameter for 'krish', via CV or otherwise, was too computationally intensive for the $p = 100, 1000$ cases so a setting was chosen by hand using the best out-of-sample simulations from the $p = 9$ case.

The results of the Monte Carlo experiment are summarized in Figure 23.6 by boxplots, and numerically. The best estimators have low miss rates with lower variability across the 100 repetitions. Lets consider the first, 'original', $p = 9$ part of the experiment, summarized in the *left*-hand region of the boxplots and the *top* region of the tables. The fully Bayesian and 'krish' methods come out on top. It is a close call, but 'krish' edges out the fully Bayesian estimation. This may not be surprising considering that its penalty parameter was, for all intents and purposes, set by oracle for this case in advance. The MLE is good on average, but has some extreme ELL and miss rate values. The glmnet and MAP estimators are in between, the former having better values of both metrics but with higher variability.

The distinctions in performance between the methods increases with p. See the *right*-hand regions of the boxplots and the *bottom* regions of the tables. The 'krish' method suffers from high variability due to the fixed choice of the penalty parameter. The glmnet variability is much lower, but there are many extreme outliers. The behaviour in both $p = 100$ and 1000 cases is qualitatively similar for this estimator even though the former used CV to set the penalty parameter and the latter used the same fixed value. The MAP and fully Bayesian estimators have similar average behaviour to other estimators, but with lower variability. Apparently, choosing the penalty parameter via the posterior is most reliable in high-dimensional settings. The fully Bayesian approach appears preferable to the MAP in all cases, but this distinction is harder to make out as p increases.

23.4 Discussion

The methods discussed here have many special cases, and extensions, that are outlined in our references. For example, Bayesian ridge regression, which involves an L_2-norm rather than L_1, is facilitated by straightforward simplifications. One extension would involve expanding the hierarchical model using a so-called normal-gamma (NG) prior for β [e.g. 19], a particular setting of which encodes the specific Laplace prior case that we handed in this paper. [15] give the details in the OLS framework, requiring an extra conditional in the Gibbs sampling. This is easily ported to logistic regression. Providing the option `case = "ng"` to `blasso` in the `monomvn` package invokes inference under the NG prior. A further extension which has been shown to improve on the Laplace and NG priors is the horseshoe prior [5]. Although we are unaware of any particular reference which outlines the Gibbs sampler for OLS (or logistic regression), the `monomvn` package provides an implementation through the (as yet undocumented) function `bhs`, which works similarly to `blasso`.

Handling polychotomous data, i.e. logistic classification with >2 classes, is straightforward. Following the setup in HH, one may introduce C collections of coefficients $\boldsymbol{\beta}^{(1)}, \ldots, \boldsymbol{\beta}^{(C)}$ for C classes with the convention that $\boldsymbol{\beta}^{(C)} = 0$ so that logistic regression is recovered in the $C = 2$ case. Then, the conditional likelihoods $L(\boldsymbol{\beta}^{(j)}|\mathbf{y}, \boldsymbol{\beta}^{(-j)})$ turn out to have exactly the form of a logistic regression likelihood for the class indicator for each $y_i = j$, independently for $i = 1, \ldots, n$. If there are $n_i > 1$ *trials* for predictors $\mathbf{x}_i$, then a vectorized multiplicity parameter can be used to get a fast implementation, as described in Section 23.3.3. Extending the methods to ordinal responses is even easier. Johnson and Albert [24, Chapter 4] describe a Bayesian probit model which is easily adapted for the logit case described here. Finally, adding variable selection via reversible jump into the regularized logistic regression arsenal would proceed similarly to the OLS setup described in Section 23.2.2.

References

[1] Asuncion, A. and Newman, D. (2007). *UCI Machine Learning Repository.*

[2] Barndorff-Nielsen, O., Kent, J. and Sorensen, M. (1982). Normal variance-mean mixtures and z-distributions. *International Statistical Review*, **50**, 145–159.

[3] Carlin, B. P. and Polson, N. G. (1991). inference for nonconjugate Bayesian models using the Gibbs sampler. *The Canadian Journal of Statistics*, **19**, 4, 399–405.

[4] Carlin, B. P., Polson, N. G. and Stoffer, D. S. (1992). A Monte Carlo approach to nonnormal and nonlinear state–space modeling. *Journal of the American Statistical Association*, **87**, 418, 493–500.

[5] Carvalho, C., Polson, N. and Scott, J. (2010). The horseshoe estimator for sparse signals. *Biometrika*, **97**, 2, 465–480.

[6] Efron, B., Hastie, T., Johnstone, I. and Tibshirani, R. (2004). Least angle regression (with discussion). *Annals of Statistics*, **32**, 2.

[7] Friedman, J. H., Hastie, T. and Tibshirani, R. (2010). Regularization paths for generalized linear models via coordinate descent. *Journal of Statistical Software*, **33**, 1, 1–22.

[8] Frühwirth-Schnatter, S. and Frühwirth, R. (2010). Data augmentation and MCMC for binary and multinomial logit models. In *Statistical Modelling and Regression Structures: Festschrift in Honour of Ludwig Fahrmeir*, eds. T. Kneib and G. Tutz, 111–132. Physica-Verlag.

[9] Gelfand, A. and Smith, A. (1990). Sampling-based approaches to calculating marginal densities. *Journal of the American Statistical Association*, **85**, 398–409.

[10] Gelman, A., Carlin, J., Stern, H. and Rubin, D. (2003). *Bayesian Data Analysis*. Chapman & Hall/CRC.

[11] Genkin, A., Lewis, D. and Madigan, D. (2007). Large-scale Bayesian logistic regression for text categorization. *Technometrics*, **49**, 3, 291–304.

[12] George, E. and McCulloch, R. (1993). Variable selection via Gibbs sampling. *Journal of the American Statistical Association*, **88**, 881–889.

[13] Geweke, J. (1993). Bayesian Treatment of the Independent Student–*t* Linear Model. *Journal of Applied Econometrics*, Vol. **8**, Supplement: Special Issue on Econometric Inference Using Simulation Techniques, S19–S40.

[14] —— (1996). Variable selection and model comparison in regression. In *Bayesian Statistics 5*, eds. J. Bernardo, J. Berger, A. Dawid, and A. Smith, 609–620. Oxford University Press.

[15] Gramacy, R. and Pantaleo, E. (2010). Shrinkage regression for multivariate inference with missing data, and an application to portfolio balancing. *Bayesian Analysis*, **5**, 2, 237–262.

[16] Gramacy, R. and Polson, N. (2010). Simulation-based regularized logisitic regression. Tech. Rep. arXiv:1005.3430, The University of Chicago, Booth School of Business.

[17] Gramacy, R. B., Lee, J. H. and Silva, R. (2008). On estimating covariances between many assets with histories of highly variable length. Tech. Rep. 0710.5837, arXiv. Url: http://arxiv.org/abs/0710.5837.

[18] Green, P. (1995). Reversible jump Markov chain Monte Carlo computation and Bayesian model determination. *Biometrika*, **82**, 711–732.

[19] Griffin, J. E. and Brown, P. J. (2010). Inference with Normal–Gamma prior distributions in regression problems. *Bayesian Analysis*, **5**, 1, 171–188.

[20] Hans, C. (2008). Bayesian lasso regression. Tech. Rep. 810, Department of Statistics, The Ohio State University, Columbus, OH 43210.

[21] Hastie, T., Tibshirani, R. and Friedman, J. (2001). *The Elements of Statistical Learning: Data Mining, Inference, and Prediction*. Springer-Verlag.

[22] Holmes, C. and Held, K. (2006). Bayesian auxilliary variable models for binary and multinomial regression. *Bayesian Analysis*, **1**, 1, 145–168.

[23] Jacquier, E., Polson, N. and Rossi, P. E. (2004). Bayesian analysis of stochastic volatility models with fat-tails and correlated errors. *J. of Econometrics*, **122**, 185–212.

[24] Johnson, V. and Albert, J. (1999). *Ordinal Data Modeling*. Springer.

[25] Kirkpatrick, S., Gelatt, C. and Vecci, M. (1983). Optimization by simulated annealing. *Science*, **220**, 671–680.

[26] Krishnapuram, B., Carin, L., Figueiredo, M. and Hartemink, A. (2005). Sparse multinomial logistic regression: fast algorithms and generalization bounds. *IEEE Pattern Analysis and Machine Intellegence*, **27**, 6, 957–969.

[27] Madigan, D. and Ridgeway, G. (2004). Discussion of 'least angle regression' by B. Efron, T. Hastie, I. Johnstone, and R. Tibshiran. *Annals of Statistics*, **32**, 2, 465–469.

[28] Park, M. and Hastie, T. (2008). Penalized logistic regression for detecting gene interactions. *Biostatistics*, **9**, 1, 30–50.
[29] Park, T. and Casella, G. (2008). The Bayesian Lasso. *Journal of the American Statistical Association*, **103**, 482, 681–686.
[30] Stambaugh, R. F. (1997). Analyzing investments whose histories differ in length. *Journal of Financial Economics*, **45**, 285–331.
[31] Taddy, M. A. (2011). Inverse regression for analysis of sentiment in text. Tech. Rep. arXiv:1012.2098, The University of Chicago, Booth School of Business.
[32] Troughton, P. T. and Godsill, S. J. (1997). A reversible jump sampler for autoregressive time series, employing full conditionals to achieve efficient model space moves. Tech. Rep. CUED/F-INFENG/TR.304, Cambridge University Engineering Department.
[33] West, M. (1987). On scale mixtures of normal distributions. *Biometrika*, **74**, 3, 646–648.

24 Bayesian model averaging in the M-open framework

MERLISE CLYDE
AND EDWIN S. IVERSEN

24.1 Introduction

Consideration of multiple models is ubiquitous in statistical practice. In Chapter 6, Bernardo & Smith [9] describe three distinct settings for the model comparison problem, denoted as $\mathcal{M}$*–closed* , $\mathcal{M}$*–complete* and $\mathcal{M}$*–open* which have far reaching consequences for how models should be compared, selected or combined.

The predominant perspective is the $\mathcal{M}$*–closed* view, where one entertains a collection of models $\mathcal{M} = \{\mathcal{M}_j, j = 1, \ldots J\}$, with the belief that one of the models in $\{\mathcal{M}_j\}$ is the 'true' generating model for the data, but that the true generating model is unknown. In this framework, a Bayesian would use probabilities $p(\mathcal{M}_j)$ to represent one's subjective (or objective) prior beliefs about the 'truth' of model $\mathcal{M}_j$. These beliefs combined with any prior beliefs about parameters within models are updated via Bayes theorem to obtain a joint posterior distribution for models and model specific parameters. The Bayesian paradigm provides a comprehensive framework for accounting for both parameter and model uncertainty, leading to the well-known Bayesian Model Averaging solution. For additional references, history, and examples we refer the reader to review articles by [24]. In conjunction with a decision theoretic approach, this joint posterior distribution can be used to construct optimal decision rules for selecting the 'best' model, make inferences about parameters or predicting future observations under selected utility functions. For additional references for the Bayesian approach to model choice we refer the reader to review articles by [12, 16, 17, 24] for history and examples.

In reality, the true process generating the data may be too complex to be used in practice or even to articulate as a probabilistic model, which leads to the $\mathcal{M}$*–complete* and $\mathcal{M}$*–open* perspectives discussed in [9]. In both of these formulations, the true generating model $\mathcal{M}_T$ is not included in the collection of models $\mathcal{M}$, rather the models in $\mathcal{M}$ are viewed as potential proxies available for comparison or model selection. With the belief that the true model is $\mathcal{M}_T$, assigning prior probabilities to models in $\mathcal{M}$ no longer makes sense, as the model is no longer part of the unknown specification of the data generating process. In the $\mathcal{M}$*–complete* specification, while one can specify $p(\mathbf{Y}|\boldsymbol{\theta}, \mathcal{M}_T)$, one may still wish to select a proxy model in $\mathcal{M}$ because of its attractive simplicity or ease of communication of results with others or computational tractability. In the more realistic

$\mathcal{M}$*–open* alternative, the list of models in $\mathcal{M}$ are also to be used in place of $\mathcal{M}_T$, however, there is no explicit specification of a belief model $p(\mathbf{Y}|\mathcal{M}_T)$.

Under the $\mathcal{M}$*–open* perspective, [9, 26] motivate the role of cross-validation to evaluate expected utility, leading to intrinsic Bayes factors for the model choice problem. Rather than model selection, our goal in this paper is optimal combination of multiple proxy models in the $\mathcal{M}$*–open* framework. In the next section, we review the standard $\mathcal{M}$*–closed* Bayesian Model Averaging approach and decision-theoretic methods for producing inferences and decisions. We then review model selection from the $\mathcal{M}$*–complete* and $\mathcal{M}$*–open* perspectives, before formulating a Bayesian solution to model averaging in the $\mathcal{M}$*–open* perspective. We construct optimal weights for MOMA: $\mathcal{M}$*–open* Model Averaging using a decision-theoretic framework, where models are treated as part of the 'action space' rather than unknown states of nature. We illustrate MOMA using 'incompatible' retrospective and prospective models for data from a case-control study and demonstrate that MOMA gives better predictive accuracy than using any of the proxy models. We conclude with open questions and future directions.

24.2 $\mathcal{M}$–*closed* model averaging

In the standard setup the joint distribution of the data and all unknowns may be described hierarchically, with $p(\mathbf{Y}|\boldsymbol{\theta}_j, \mathcal{M}_j)$ specifying the distribution of the data $\mathbf{Y} = (Y_1, \dots Y_n)^T$ given model specific parameters $\boldsymbol{\theta}_j$ in model $\mathcal{M}_j$, $p(\boldsymbol{\theta}_j|\mathcal{M}_j)$ reflecting prior uncertainty in the model specific parameters. A Bayesian would assign a prior probability, $p(\mathcal{M}_j)$, representing their belief (subjective or objective) that each model $\mathcal{M}_j$ is the true model.

In turn, posterior model uncertainty is represented by the posterior probabilities of models obtained via Bayes theorem

$$p(\mathcal{M}_j|\mathbf{Y}) = \frac{p(\mathbf{Y}|\mathcal{M}_j)p(\mathcal{M}_j)}{\sum_{j=1}^{J} p(\mathbf{Y}|\mathcal{M}_j)p(\mathcal{M}_j)} \tag{24.1}$$

where

$$p(\mathbf{Y}|\mathcal{M}_j) = \int p(\mathbf{Y}|\boldsymbol{\theta}_j, \mathcal{M}_j)p(\boldsymbol{\theta}_j|\mathcal{M}_j)d\boldsymbol{\theta}_j \tag{24.2}$$

is the marginal likelihood of $\mathcal{M}_j$. Given observed data $\mathbf{Y}$, the posterior probability of each model $p(\mathcal{M}_j|\mathbf{Y})$ represents a posterior measure that model $\mathcal{M}_j$ generated the data. The joint posterior distribution of $\boldsymbol{\theta}_j, \mathcal{M}_j$, $p(\boldsymbol{\theta}_j|\mathbf{Y}, \mathcal{M}_j)p(\mathcal{M}_j|\mathbf{Y})$, provides a complete post-data representation of parameter and model uncertainty that can be used for a variety of inferences and decisions. For example, the distribution of a future observation Y^*. Under the hierarchical model for the data, the Bayesian predictive distribution of Y^* is a mixture model

$$p(Y^*|\mathbf{Y}) = \sum_j p(Y^*|\mathcal{M}_j, \mathbf{Y})p(\mathcal{M}_j|\mathbf{Y}) \tag{24.3}$$

with components in the mixture the conditional predictive distributions

$$p(Y^*|\mathcal{M}_j, \mathbf{Y}) = \int p(Y^*|\boldsymbol{\theta}_j, \mathcal{M}_j)p(\boldsymbol{\theta}_j|\mathcal{M}_j, \mathbf{Y})d\boldsymbol{\theta}_j \tag{24.4}$$

and mixing weights given by the posterior probabilities of models from equation (24.1).

24.2.1 Optimal decisions

The joint posterior distribution of $\mathcal{M}_j$ and $\boldsymbol{\theta}_j$ provides a complete summary of one's beliefs after seeing the data. Combined with a decision-theoretic framework, this posterior permits making inferences or decisions that optimize one's utility. More formally, let $u(\omega, a)$ be a mapping from $A \times \Omega$ to $\mathbb{R}$ that reflects the utility of taking action a when the unknown state of nature is ω. Commonly used utility functions are negative quadratic loss for estimation or prediction or proper scoring rules if ω is a distribution. For a Bayesian, the optional action to take is the one that maximizes the posterior expected utility

$$a^* = \arg\sup_{a \in A} \int_{\Omega} u(\omega, a) p(\omega \mid \mathbf{Y}) d\omega \tag{24.5}$$

where $p(\omega \mid \mathbf{Y})$ is the posterior (predictive) distribution of ω given the data $\mathbf{Y}$.

Consider the decision problem of prediction under quadratic loss

$$u(Y^*, a) = -(Y^* - a)^2$$

where Y^* is the unknown 'state of nature', a is a possible action in action space $A = \mathbb{R}$ and u is the utility of taking action a when the future value is Y^*. Under quadratic loss for prediction, the optimal action for the point prediction of Y^* is $a^* = E(Y^*|\mathbf{Y})$, the (posterior predictive) mean of Y^* given Y, which under the $\mathcal{M}$-closed perspective, can be expressed as

$$E(Y^*|\mathbf{Y}) = \sum_{j=1}^{J} \mathrm{E}(Y^*|\mathcal{M}_j, \mathbf{Y}) p(\mathcal{M}_j|\mathbf{Y}) = \sum_{j=1}^{J} p(\mathcal{M}_j|\mathbf{Y}) \hat{Y}^*_{\mathcal{M}_j} \tag{24.6}$$

where $\hat{Y}^*_{\mathcal{M}_j}$ is the posterior mean under model $\mathcal{M}_j$. This is the well-known Bayesian Model Averaging solution, where the prediction is a weighted average of the model specific predictions $\hat{Y}^*_{\mathcal{M}_j}$ with weights that are given by the posterior model probabilities. Such model averaging or mixing procedures have been developed and advocated for by [27], [20], [17], [32] and [15], and are now widespread.

If the goal is to find the single model that leads to the best prediction under quadratic loss, then the set of actions consist of selecting a model and reporting the prediction under that model. The solution given by [9, Section 6.1] is the model that minimizes $(\hat{\mathbf{Y}}^*_{\mathcal{M}_j} - \mathrm{E}_{Y^*}(Y^* \mid \mathbf{Y}))^2$; the single model whose predictions are closest to the BMA solution. For the case of two models, this is the highest probability model, but in general the model closest to the BMA solution may not correspond to the highest probability model nor the median probability model of [1] except in special circumstances. While closed form expressions are generally unavailable, one can determine the best model by evaluating the distances between the models under consideration. When there is high correlation among the predictors this is preferable to the median probability model. [33] use a log scoring rule for selecting the model that is closest to model averaging solution in terms of predictive densities.

24.2.2 BMA is not a panacea

In problems where there are a large number of predictors (p), such as genome wide association studies, one might consider model averaging where the candidate models are each based on a single predictor, rather than approximating model averaging by stochastic search over the 2^p potential models. To illustrate that model averaging in such a setting may fail, consider the simplified setting with just two models in $\mathcal{M}$

$$\mathcal{M}_1 : \mathbf{Y} = \mathbf{X}_1\beta_1 + \mathbf{e} \tag{24.7}$$

$$\mathcal{M}_2 : \mathbf{Y} = \mathbf{X}_2\beta_2 + \mathbf{e} \tag{24.8}$$

leading to $\hat{\mathbf{Y}}^* = p(\mathcal{M}_1 \mid \mathbf{Y})\mathbf{X}_1\hat{\beta}_1 + p(\mathcal{M}_2 \mid \mathbf{Y})\mathbf{X}_2\hat{\beta}_2$. This seems appealing as the model averaging solution does contain all potential predictors, even though the full model was not included in the list of candidate models. If the 'true' model does in fact contain both predictors $\mathbf{Y} = \mathbf{X}_1\beta_{1T} + \mathbf{X}_2\beta_{2T} + \mathbf{e}$ then, under standard regularity conditions, the BMA model weights converge to 1 for the model that is 'closest' to the true model (in terms of Kullback–Leibler divergence); BMA only uses predictions from that model; and in the limit BMA is not consistent if $\mathcal{M}_T \notin \mathcal{M}$. The obvious solution is to add the full model to the list of models under consideration for this toy example. However, if the true model is some complex nonlinear function, one may need to use a richer set of basis vectors, such as in over-complete representations, for model averaging to lead to consistent results [38]. For ease of exposition, one may still wish to use simple proxy models when the true model is not included in $\mathcal{M}$, which leads to the $\mathcal{M}$*–closed* and $\mathcal{M}$*–open* perspectives.

24.3 Model comparison without the true model

When the true model is not in $\mathcal{M}$, we consider two cases

$\mathcal{M}$*–complete*: we know the true model, $\mathcal{M}_T$, and $p(Y^*|\mathbf{Y}) = p^C(Y^*|\mathcal{M}_T, \mathbf{Y})$ is available. We may, however, wish to use the models in $\mathcal{M}$ because of ease in communication of results, tractability of computations, reasonable proxies, etc.

$\mathcal{M}$*–open*: we know that the true model is NOT in $\mathcal{M}$, but we cannot specify $p(Y^*|\mathbf{Y}) = p^o(Y^*|\mathcal{M}_T, \mathbf{Y})$ because it is too difficult, we lack time to do so, or do not have the expertise, computational intractability, etc.

For the model comparison problem in the $\mathcal{M}$*–complete* setting, one simply finds the model in $\mathcal{M}$ which maximizes the expected utility, where now the expectation is with respect to the predictive distribution $p^c(Y^* \mid \mathcal{M}_T, \mathbf{Y})$. For the $\mathcal{M}$*–open* case, one again finds the optimal model and action $a^*(\mathbf{Y}, \mathcal{M}_j)$ under model $\mathcal{M}_j$ that maximizes expected utility,

$$\int u(y^*, a^*(\mathbf{Y}, \mathcal{M}_j))p^o(y^*|\mathcal{M}_T, \mathbf{Y})\, dy. \tag{24.9}$$

As the predictive distribution is not available in the $\mathcal{M}$*–open* setting [9, 26] argue that for exchangeable data and large n the expected utility can be approximated by

$$\frac{1}{n}\sum_{i=1}^{n} u(y_i, a^*(\mathbf{Y}_{(i)}, \mathcal{M}_j)) \tag{24.10}$$

based on partitioning $\mathbf{Y}^T = (y_i, \mathbf{Y}_{(i)}^T)$ into n partitions of the data where $\mathbf{Y}_{(i)}$ denotes the data without the ith observation and serves as a proxy for the observed data and y_i as a proxy for the future value Y^*. Randomly selecting from K of these partitions, they suggest a law of large numbers argument to justify that as $n, K \to \infty$

$$\left| \int u(Y^*, a(\mathbf{Y}, \mathcal{M}_j)) p^o(Y^* | \mathbf{Y}, \mathcal{M}_T)\, dY^* - \frac{1}{K} \sum_{k=1}^{K} u(y_k, a(Y_{(-k)}, \mathcal{M}_j)) \right| \to 0,$$

thereby justifying the use of cross-validation to approximate expected utility.

Walker & Gutiérrez-Peña [34] in the discussion of [26] provide an alternative justification for the above as an approximation based in a Bayesian nonparametric model. If we assume the data are exchangeable, coming from an unknown distribution F, then one may place a nonparametric prior on F, such as a Dirichlet process,

$$F \sim DP(\alpha_0, F_o)$$

with α_0 a scale or prior weight parameter and F_0 a parametric distribution that is the location parameter such that $E(F) = F_0$ [18]. Given a sample of size n, the posterior of F is again $DP(\alpha_n, F_n)$ where $\alpha_n = n + \alpha_0$, $F_n = (n\hat{F}_n + \alpha_0 F_0)/(n + \alpha_0)$ and $\hat{F}_n$ is the empirical distribution of the data. Using the nonparametric prior, the posterior predictive distribution for a new observation Y^* is F_n and the expected utility is expressed as

$$\int u(y^*, a^*(\mathbf{Y}, \mathcal{M}_j)) dF_n(y^*) = \frac{n}{n + \alpha_0} \int u(y^*, a^*(\mathbf{Y}, \mathcal{M}_j)) d\hat{F}_n(y^*) + \tag{24.11}$$

$$\frac{\alpha_0}{n + \alpha_0} u(y^*, a^*(\mathbf{Y}, \mathcal{M}_j)) dF_o(y^*) \tag{24.12}$$

[23] take F_0 to be centred at the $\mathcal{M}$*–closed* predictive distribution, so that as $\alpha_0 \to \infty$ one recovers the $\mathcal{M}$*–closed* solution, while as $\alpha_0 \to 0$

$$\int u(y^*, a^*(\mathbf{Y}, \mathcal{M}_j)) dF_n(y^*) \to \frac{1}{n} \sum_{i=1}^{n} u(y_i, a^*(\mathbf{Y}, \mathcal{M}_j)) \tag{24.13}$$

leads to their $\mathcal{M}$*–open* solution. The main difference between (24.13) and (24.10) is that the optional action under model $\mathcal{M}_j$ uses all data in constructing the distributions that go into $a^*(\mathbf{Y}, \mathcal{M}_j)$ in (24.13), while [26] use $a^*(\mathbf{Y}_{(i)}, \mathcal{M}_j)$ based on the training data in (24.10). Equation (24.13) provides internal rather than external validation as in (24.10).

In the case of prediction under quadratic loss under the limiting DP model, we would minimize over $\mathcal{M}$

$$\frac{1}{n} \sum_{i=1}^{n} (y_i - E(Y_i \mid \mathbf{M}_j, \mathbf{Y}))^2 \tag{24.14}$$

In the case of linear models $E[\mathbf{Y}] = \mathbf{X}_M \beta_{\mathcal{M}}$ with non-informative priors $p(\beta_{\mathcal{M}}) \propto 1$ assigned to parameters in $\mathbf{M}_j$, $E(Y_i \mid \mathbf{M}_j, \mathbf{Y}) = \mathbf{x}_i^T \hat{\beta}_{\mathcal{M}}$ where $\hat{\beta}_{\mathcal{M}}$ is the ordinary least squares estimate. The criterion would lead to picking the model with the smallest residual sum of squares regardless of model dimension (or highest R^2), which leads to poor predictive performance. In contrast, the CV-approach of Bernardo & Smith [9] chooses the model that minimizes

$$\frac{1}{n} \sum_{i=1}^{n} (y_i - E(Y_i \mid \mathbf{M}_j, \mathbf{Y}_{(i)}))^2 \tag{24.15}$$

over $\mathcal{M}$. This captures how well model $\mathcal{M}_j$ predicts, on average, a left-out observation given the remaining cases [21]. While we do not advocate non-informative priors in this setting, this highlights a potential problem of using the (limiting) DP prior.

Under the log scoring rule

$$\int \log(p(y \mid \mathcal{M}_j, \mathbf{Y}))p(y \mid \mathbf{Y})dy \tag{24.16}$$

[9, 26] propose the following approximation to the expected utility

$$\frac{1}{K}\sum_{k=1}^{k} \log(p(y_k \mid \mathbf{Y}_{(k)}, \mathcal{M}_j)) \tag{24.17}$$

which may be rearranged to form a criterion that implies that one would prefer model $\mathcal{M}_i$ to model M_0 if

$$\prod_{k=1}^{K} \left[\frac{p(y_k \mid \mathbf{Y}_{(k)}, \mathcal{M}_j)}{p(y_k \mid \mathbf{Y}_{(k)}, \mathcal{M}_0)}\right]^{1/K} > 1 \tag{24.18}$$

This corresponds to the Geometric Intrinsic Bayes factor criterion of [4–6] where $\mathbf{Y}_{(k)}$ represents a minimal training sample. This is related to the expression in [36] where $\mathbf{Y}$ replaces $\mathbf{Y}_{(k)}$.

Winkler [35] in the discussion of [26] raises the question of 'Why there is so much focus on model choice. If there is no 'true' model, why do we have to choose a single model?'. As George Box is often quoted 'Essentially all models are wrong, but some are useful.' Why should we restrict attention to just one of the proxy models if all are potentially useful? Rather selecting a model, we examine optimal weighted averages.

24.4 Combining models in the $\mathcal{M}$–*open* setting

In the $\mathcal{M}$–*closed* setting models are essentially an expansion of the 'parameter' space so that the unknown state of nature Ω is comprised of models and model specific parameters. The optimal weights for prediction $\sum w_j E(Y^* \mid \mathbf{Y}, \mathcal{M}_j)$ are the posterior probabilities of models, which are proportional to the prior model probabilities times the marginal distributions of the data $p(\mathbf{Y} \mid |M_j)$. In both the $\mathcal{M}$–*complete* and $\mathcal{M}$–*open* viewpoints, the assignment of prior probabilities $\{p(\mathcal{M}_i), i \in \mathcal{M}\}$ no longer makes sense as a measure of our degree of belief in model $\mathcal{M}_j$ if we really believe that $\mathcal{M}_T \notin \mathcal{M}$. Thus the standard BMA solution to combining models using weights that are posterior model probabilities is not applicable.

In the decision theoretic framework for the $\mathcal{M}$–*closed* or $\mathcal{M}$–*open* perspectives, models are not part of the state of unknowns Ω, but may be part of the decision or action space. Under this alternative viewpoint model weights w_j are solely part of the action space A, so that optimal weights to combine predictions or predictive distributions from the collection of proxy models in $\mathcal{M}$ becomes a decision problem.

24.4.1 Combining models as a decision problem

Let $\{w_j, j \in J\ w_j \in \Omega\}$ denote the weights $\mathbf{w}$ and consider decision rules of the form $a(\mathbf{Y}, \mathbf{w}) = \sum_j w_j E(y^* \mid \mathbf{Y}, \mathcal{M}_j)$ in the case of prediction or $a(\mathbf{Y}, \mathbf{w}) = \sum_j w_j p(y^* \mid \mathbf{Y}, \mathcal{M}_j)$ for predictive densities. For quadratic loss, the expected utility is

$$E_{Y^*}[u(y^*, a(\mathbf{Y}, w))|\mathbf{Y}] = -\int \| y^* - \sum_j w_j \hat{y}^*_{\mathcal{M}_j} \|^2 \, p(y^*|\mathbf{Y}, \mathcal{M}_t)\, dy^*$$

while for the log scoring rule,

$$E_{Y^*}[u(y^*, a(\mathbf{Y}, \mathbf{w}))|\mathbf{Y}] = \int \log(\sum_j w_j p(y^* \mid \mathcal{M}_j, \mathbf{Y}))p(y^* \mid \mathbf{Y}, \mathcal{M}_t)\, dy^*$$

In the $\mathcal{M}$*-complete* perspective, since we have $\mathcal{M}_T$, we can in principle solve the optimization problem.

In the $\mathcal{M}$*-open* formulation, as before, partition $\mathbf{Y}^T = (y_k^T, \mathbf{Y}_{(k)}^T)$ into future and training data vectors of size $n - m$ and m respectively. Randomly select K from these n choose m partitions to construct the approximate criterion

$$\hat{\mathbf{w}} = \arg\max_{\mathbf{w}} \frac{1}{K} \sum_{k=1}^{K} u(y_k, a(Y_{(-k)}, w)) \tag{24.19}$$

For the problem of prediction the problem may be stated as,

$$\text{Solve} \quad \hat{\mathbf{w}} = \arg\max_{\mathbf{w}} -\frac{1}{K} \sum_{k=1}^{K} \| y_k - \sum_{\mathcal{M}_j \in \mathcal{M}} w_j \hat{Y}_{(k),\mathcal{M}_j} \|^2 \tag{24.20}$$

subject to (24.21)

$$\sum_{j=1}^{J} w_j = 1 \tag{24.22}$$

$$w_j \geq 0 \quad \forall j \in \{1, \ldots, J\} \tag{24.23}$$

where the optimal solution may be found using a quadratic programming algorithm. This has an equivalent representation using Lagrangians:

$$-\frac{1}{K} \sum_{k=1}^{K} \| y_k - \sum_{j=1}^{J} w_j \hat{Y}_{(k),\mathcal{M}_j} \|^2 - \lambda_0 (\sum_{j=1}^{J} w_j - 1) + \sum_{j=1}^{J} \lambda_j w_j.$$

The two constraint functions ensure that the weights have the same support as posterior model probabilities in BMA (sum to one and non-negativity), however, the weights do not have an interpretation as posterior probabilities. While the expression above looks like a log likelihood from a Gaussian distribution, it is not the predictive log likelihood.

Similarly, under the log scoring rule,

$$\frac{1}{K} \sum_{k=1}^{K} \log \left(\sum_{j=1}^{J} w_j p(y_k \mid \mathbf{Y}_{(k),\mathcal{M}_j}, \mathcal{M}_j) \right) - \lambda_0 (\sum_{j=1}^{J} w_j - 1) + \sum_{j=1}^{J} \lambda_j w_j$$

the expected utility takes a form similar to a likelihood from a mixture model. This relationship permits an iterative solution for the weights as in estimation of mixture models, where starting with initial weights $\hat{w}_j^{(0)}$, we update the weights

$$\hat{w}_j^{(t)} \equiv \frac{1}{K} \sum_k \frac{\hat{w}_j^{(t-1)} p(y_k \mid \mathbf{Y}_{(k)}, \mathcal{M}_j)}{\sum_j \hat{w}_j^{(t-1)} p(y_k \mid \mathbf{Y}_{(k)}, \mathcal{M}_j)} \tag{24.24}$$

until convergence.

Remark on solution In the $\mathcal{M}$*–closed* setting the model weights may be obtained from Bayes factors for comparing any model to a base model, $P(\mathcal{M}_j \mid \mathbf{Y}) = w_j B(\mathcal{M}_j : \mathcal{M}_0) / \sum_j w_j B(\mathcal{M}_j : \mathcal{M}_0)$ where w_j is the prior probability of $\mathcal{M}_j$. In the $\mathcal{M}$*–open* framework, the geometric intrinsic Bayes factor (GIBF) leads to a model selection criterion under the log scoring rule[5], however, the optimal model weights in (24.24) for combining models under the log scoring rule are not equivalent to the renormalized GIBF nor the closely related arithmetic intrinsic Bayes Factors.

24.4.2 Restrictions on weights

As predictions may have support on $\mathbb{R}^K$, we may impose a range of restrictions on the weights by choice of λ_j. With $\lambda_0 = \lambda_1 = \ldots = \lambda_J = 0$ we obtain arbitrary weights ($w_j \in \mathbb{R}$). Setting $\lambda_1 = \ldots = \lambda_J = 0$ enforces the weights to sum to one, but allows positive and negative weights, while with all $\lambda_j >> 0$ the weights are constrained to be non-negative and sum to one. Because the weights are non-negative, the last penalty resembles a 'lasso' or L_1 penalty, which permits weights to be zero. For the log-scoring rule we must have both sets of constraints for the MOMA density estimate to be a valid density.

To illustrate the behaviour of the solutions under quadratic loss, we focus on the case where $m = n - 1$ and let $\hat{e} = [\hat{\epsilon}_{kj}] = [y_k - \hat{y}_{(-k)\mathcal{M}_j}]$ denote the $n \times J$ matrix of predicted residuals for predicting y_k under model $\mathcal{M}_j$ using data $\mathbf{Y}_{(k)}$.

Remark 1 With the sum to one constraint alone, $\hat{\mathbf{w}} \propto (\hat{\mathbf{e}}^T \hat{\mathbf{e}})^{-1}\mathbf{1}$. If residuals from models are uncorrelated, then weights are proportional to the inverse of the Predicted REsidual Sum of Squares for model $\mathcal{M}_j$, $\text{PRESS}_j = \sum_k \hat{\epsilon}_{kj}^2$. With non-informative priors in linear models, $\text{PRESS}_j = \sum e_{kj}^2/(1 - h_{kk})$ where e_{jk} is the ordinary residual for case k under model j and h_{kk} is the leverage of case k, putting a premium on models that are able to fit points with high leverage. In the more general case of correlated predicted residuals across models, the weights are adjusted for the other models.

Remark 2 With highly correlated residuals under similar proxies, weights with just the sum to one constraint may be negative and highly unstable. The non-negativity constraint induces a lasso-like L_1 penalty, which stabilizes weights and may drive the optimal weights to zero for redundant components.

Remark 3 The solution to (24.20) is in fact equivalent to the frequentist method of stacking [10, 11], although the Bayesian may prefer to include predictions under more robust prior distributions than using the non-informative prior distribution which leads to the least squares predictions.

24.5 Aeroplane failures examples

We compare the Dirichlet Process $\mathcal{M}$*–open* model averaging procedure of [37] and MOMA, which uses the predictive reuse approximation to the predictive distribution of future observations. The data are based on $n = 30$ intervals between air conditioning failure times for plane 7912 [30]. Walker *et al.* [37] consider two models: an exponential

$$p_E(y) = \theta \exp(\theta y) \tag{24.25}$$

$$\theta \sim G(a, b) \tag{24.26}$$

and a log-normal model

$$p_L(y) = (2\pi\sigma^2)^{-1/2} \frac{1}{y} \exp\left[-\frac{1}{2} \left(\frac{\log(y) - \mu}{\sigma} \right)^2 \right] \tag{24.27}$$

$$\mu \mid \sigma^2 \sim N(\mu_0, n_0\sigma^2) \tag{24.28}$$

$$(\sigma^2)^{-1} \sim G(\eta_0/2, \sigma_0^2\eta_0/2). \tag{24.29}$$

Using non-informative improper prior distributions, traditional model averaging leads to indeterminate Bayes factors due to potentially arbitrary constants in the improper priors. That is not a problem with the $\mathcal{M}$*-open* approach. Using non-informative priors $a = b = 0$ and $\mu_0 = n_0 = \sigma_0 = 0, \eta_0 = -1/2$, to obtain the predictive densities for a future observation under each model, the MOMA weight for the exponential model is 0.435 using equation (24.24). In contrast the optimal weight from [37] is 0.315 for the exponential model. The predictive distributions for both approaches are illustrated in Figure 24.1. While the distributions are very similar in the tails, the main difference is near the mode.

To compare the two methods, we generated 30 observations from a $G(2, 1)$ distribution and used Monte Carlo integration to evaluate the utility $\int \log \hat{p}(y \mid \mathbf{Y}) p_G(y) dy$ under the true Gamma model. Figure 24.2 illustrates one realization, where the utility under the MOMA estimate is -1.65 compared to -1.68 for the DP method. The estimate of the mixing weight for the exponential predictive distribution is 0.22 under MOMA, while it is 2.2×10^{-13}, virtually zero, under the DP

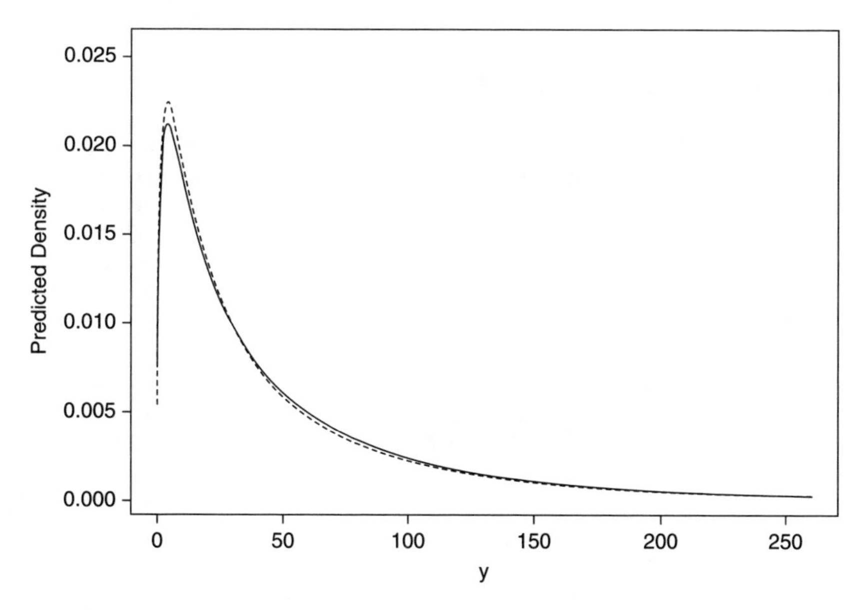

Figure 24.1 Predicted distributions under MOMA (solid line) and the DP process estimate of [37] (dashed line).

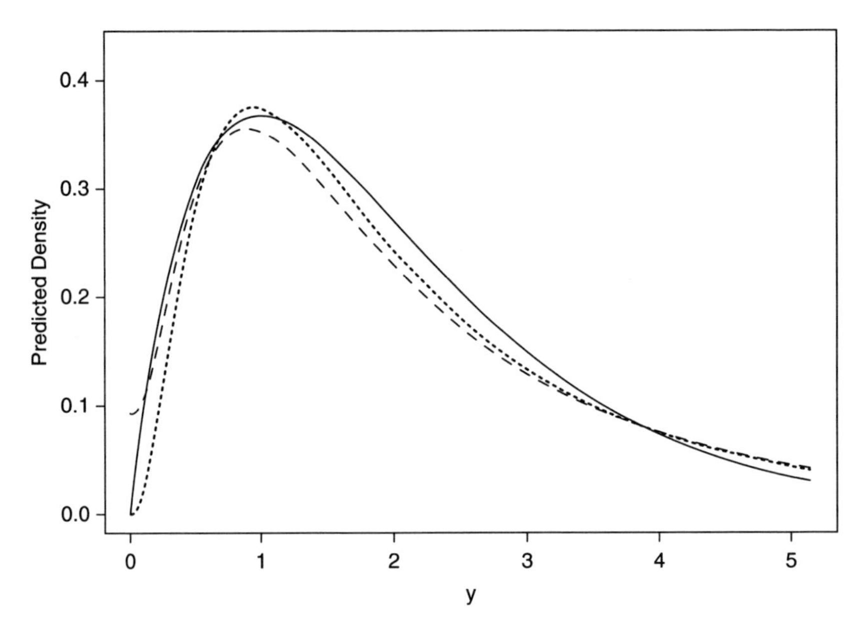

Figure 24.2 Predicted distributions under MOMA (long dash) and the DP process estimate of [37] (short dashed line), with the true Gamma(2,1) distribution (solid line).

mixture. The differences between the two approaches is most pronounced in small samples, while the methods yield very similar results for larger sample sizes.

24.6 Ovarian cancer example

Berchuck *et al.* [3] develop a model to predict binary survival status (short-term: < 3 years versus long-term: > 7 year) among patients diagnosed with advanced stage serous ovarian cancer using gene expression data from the primary tumor. The data consist of a retrospective sample with $n = 30$ short-term survivors, $n = 24$ long-term survivors and eleven early stage (I/II) cases. Expression was measured for 22, 283 targets using the Affymetrix U133a microarray. In addition to the tumor phenotype, six variables of clinical relevance (age, post-treatment CA125 levels, etc.) were also collected for each women.

Based on the retrospective sampling design, the likelihood would be proportional to the joint distribution of the 22, 283 expression and six clinical variables given survival status. The sample size and the dimensionality of the problem preclude undertaking serious joint modelling, and instead several proxy models are used to develop predictive models. We consider three classes of models:

Clinical trees (Five variants) Prospective Bayesian classification and regression tree models using only the six clinical variables;

Expression trees (Four variants) Prospective Bayesian classification and regression tree models using only expression data;

Expression LDA (Four variants) Retrospective discriminant models built using expression data given survival status.

Bayesian CART models using both clinical variables and expression variables lead to models that included only clinical variables, so this combination is excluded from the analysis as it would be redundant.

The clinical and expression tree models use a prospective likelihood and are based on Bayesian model averaged predictions from the Bayesian CART method of [29] with the different variants corresponding to different choices of the hyper-parameter settings. The LDA discriminant model is based on classification with a retrospective model described in [25]; predictions under this model are also model averaged. Because the data are retrospectively sampled, the prospective tree models provide simple proxies for prediction. The LDA method is a simple classification model that attempts to construct predictions assuming either conditional independence (labelled 'P1') or a sparse dependence structure (labelled 'P2') among the genes included in the analysis (either 100 or 200), and cannot reasonably be viewed as the true model. Traditional model averaging is not suitable for combining these predictions as there is no probability model that encompasses the two approaches employed in this example. Instead, we consider constructing optimal weights for MOMA estimates of the probability of being a long-term survivor under the quadratic loss criterion. This treats the 54 vectors of survival status, clinical data and expression data as being exchangeable in approximating the expected utility. A more realistic assumption would be that of partial exchangeability; given disease status the vectors of clinical and expression data are exchangeable. Population proportions of disease status could then be used for over or undersampling across the disease groups to construct the predictive distribution.

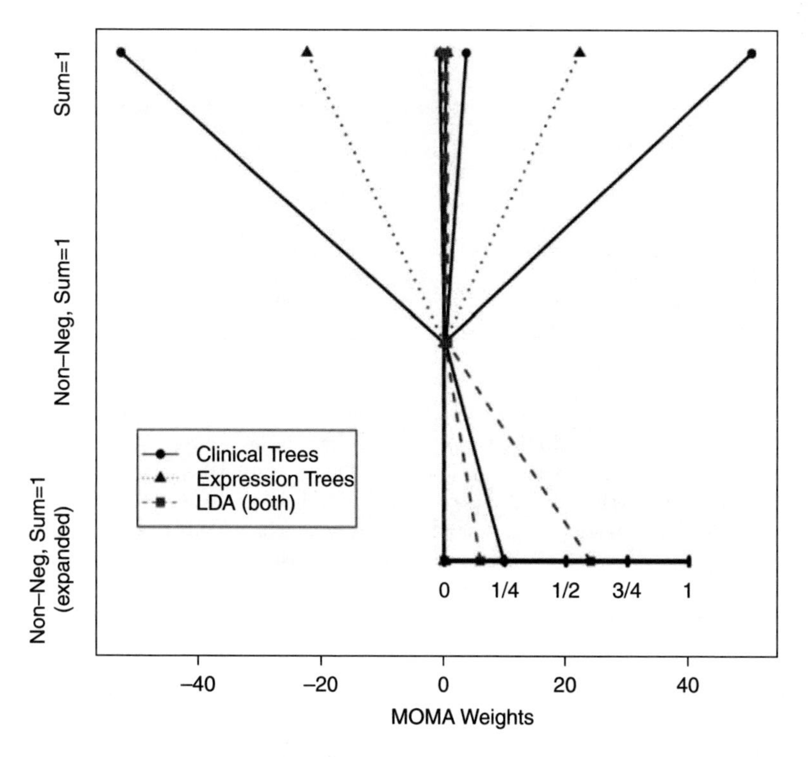

Figure 24.3 MOMA weights under the (top) sum to one constraint and (middle and bottom) the sum to one and non-negativity constraint. Solid with circles denotes the clinical trees, dotted line with triangles the expression trees and dashes with squares the LDA models.

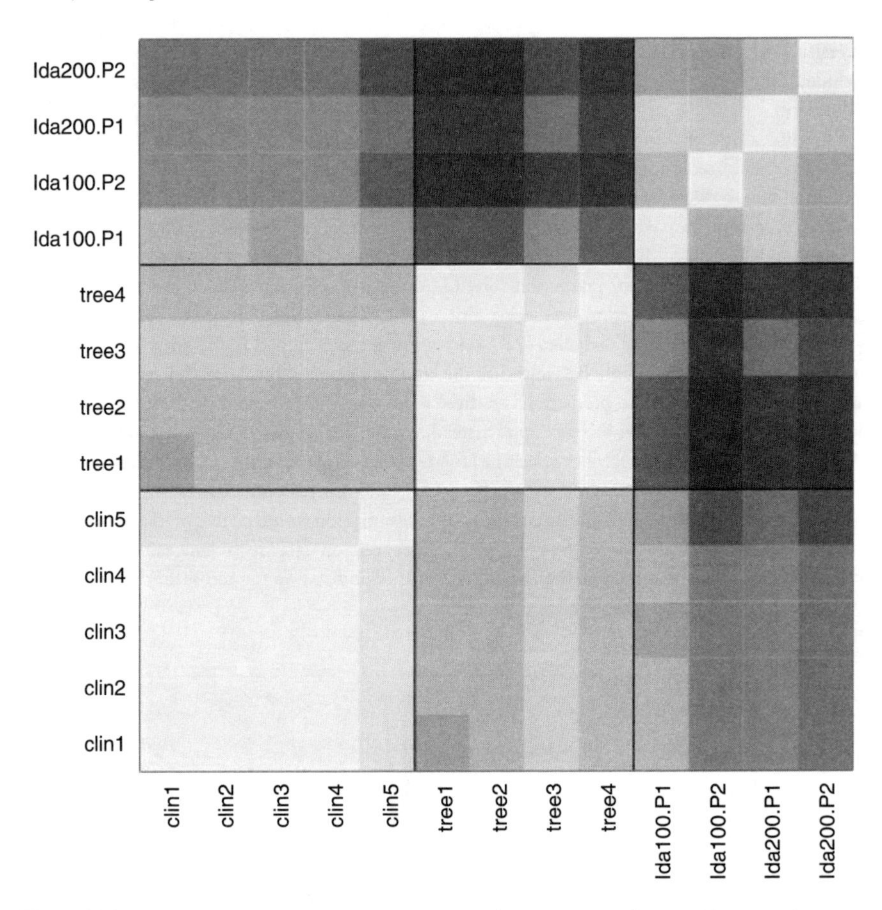

Figure 24.4 Correlation of predicted residuals between the clinical trees (bottom 5), expression trees (middle 4) and LDA models (top 4); the range is 0.9327–0.9999 for clinical trees, 0.9257–0.9999 for expression trees, and 0.7257–0.8424 for LDA models.

Figure 24.3 illustrates the effects of the constraints on the weights. Predictions from the expression trees are quite similar to each other leading to high correlation in the predicted residuals (Figure 24.4, middle block). This is also true in the case of the clinical trees and the LDA models, although less so, while correlations between residuals from models in different classes are weakly correlated. High positive correlations lead to large positive and negative weights within a class of models which in effect cancels the contribution from these models to the overall prediction. Under the additional non-negativity constraint, all of the weights of the expression trees are 0, with the resulting MOMA predictions based predominantly on a subset of one clinical tree and two LDA models.

24.6.1 Validation experiment

We evaluated the sensitivity of predictions to the form of constraint (sum-to-zero *versus* sum-to-zero and non-negativity) applied to the weight vector using five-fold cross-validation. We randomly split the data into a training set $\mathbf{Y}$ and a validation set $\mathbf{Y}^V$. Using the training data we obtained model weights $\hat{\mathbf{w}}$, using the Monte Carlo approximation to the expected utility in (24.20). We then

Table 24.1 MOMA weights and prediction accuracy for the ovarian cancer data for five randomly selected training and validation sets using the sum to one constraint.

	set1	set2	set3	set4	set5
clin1	53.08	−4.43	−0.01	−24.41	15.94
clin2	−79.92	−5.16	0.90	0.80	−4.63
clin3	−1.25	−0.24	−0.90	−0.01	5.35
clin4	27.36	10.14	−0.33	23.73	−17.24
clin5	1.13	0.27	0.27	0.36	0.55
tree1	−0.05	−0.55	−2.92	0.03	27.93
tree2	−0.12	−0.07	−3.21	−0.62	0.63
tree3	0.51	0.53	0.15	0.48	−3.35
tree4	−0.28	0.22	6.26	−0.04	−24.10
lda100.P1	−0.40	0.04	−0.01	0.02	−0.11
lda100.P2	0.44	−0.02	0.53	−0.06	−0.07
lda200.P1	0.30	0.17	−0.32	0.09	−0.03
lda200.P2	0.21	0.08	0.60	0.63	0.12
Accuracy	0.64	0.64	0.46	0.73	0.60

constructed the MOMA estimates of the probability of long-term survival $\hat{p}_j = \sum_i \hat{w}_i \hat{Y}^*_{\mathcal{M}_i}(\mathbf{Y})$ for the left-out validation samples and classified an individual as a long-term survivor if $\hat{p}_j \geq 1/2$. Finally, we computed the classification accuracy for the validation set, then repeated the procedure for the remaining partitions of the data. Tables 24.1 and 24.2 illustrate the variability of the weights across different training sets.

Overall the accuracy of the MOMA with non-negative weights and the sum to zero constraint is better than any of the individual models. The expression data, through the LDA models, appear to improve the predictions over the clinical trees alone. A followup study by [2] confirmed association of the top genes from the LDA models with long-term survival status.

24.7 Discussion

Expanding on the original work of [9, 26] we have presented a method for model averaging in the $\mathcal{M}$*−open* setting using sample re-use methods to approximate the predictive distribution of future observations [19, 20]. The solution of the mixing weights depends on the choice of utility functions as well as any constraints that are incorporated in the problem. The non-negativity constraint, which is natural, behaves like a lasso penalty and forces weights to be zero so that redundant predictions are not added. This is a striking difference to the traditional BMA solution where models with similar predictions often have similar marginal likelihoods, with the result that posterior mass is 'diluted'

Table 24.2 MOMA weights and prediction accuracy for the ovarian cancer data for five randomly selected training and validation sets using the non-negativity and sum to one constraints.

	set1	set2	set3	set4	set5
clin1	0.00	0.07	0.00	0.00	0.00
clin2	0.00	0.00	0.00	0.00	0.00
clin3	0.00	0.00	0.00	0.00	0.00
clin4	0.00	0.11	0.00	0.00	0.00
clin5	0.30	0.17	0.07	0.41	0.00
tree1	0.00	0.00	0.00	0.00	0.77
tree2	0.00	0.00	0.00	0.00	0.21
tree3	0.23	0.44	0.21	0.00	0.01
tree4	0.00	0.00	0.00	0.00	0.01
lda100.P1	0.00	0.00	0.00	0.00	0.00
lda100.P2	0.22	0.00	0.30	0.00	0.00
lda200.P1	0.00	0.00	0.00	0.00	0.00
lda200.P2	0.26	0.21	0.41	0.58	0.00
Accuracy	0.82	0.73	0.55	0.73	0.60

over similar models [14, 22]. MOMA using the log scoring rule is closely related to Ensemble BMA [31], which uses maximum likelihood to estimate the weights of a weighted average of bias corrected forecasts. The method of [37] uses a Dirichlet Process prior on the unknown distribution of the data to obtain the predictive distribution and leads to similar solutions to MOMA, although their method in the Gaussian case utilizes ordinary residuals, which may favour more complex models. The solutions in MOMA do not take into account the complexity of the models, however, the use of predicted residuals provides some penalization for poor out-of-sample prediction. If model complexity is an important criterion this can be incorporated as an additional constraint in the procedure.

An open question regards the selection of partitions of the data for MOMA. Rather than partition $\mathbf{Y}$ into $\{y_k, \mathbf{Y}_{(k)}\}$ where y_k is a single observation, [28] considered partitions $\{\mathbf{Y}_{-S}, \mathbf{Y}_S\}$ where $\mathbf{Y}_S$ is now a vector. In a simple but illustrative example, [28] showed that using the maximal training sample Y_S of dimension $n-1$ could lead to inconsistent selection of the true model, while use of the minimal training sample (dimension one) had more desirable asymptotic properties. As MOMA and the DP method may be viewed as using training samples of size $n-1$ and n, asymptotic properties of model averaging in this framework is an area that needs additional research.

Unlike traditional BMA, MOMA provides a formal mechanism for combining predictions from models where the likelihoods are not commensurate. We have focused on linear combinations of the predictions and predictive distributions as motivated by traditional BMA. However, there is a rich literature on the related problem of combining expert opinions in risk analysis (with beliefs represented as distributions) [13]. Rather than using the linear opinion pool (dating back to Laplace), alternative methods such as logarithmic pooling or Bayesian approaches may be useful in this context.

24.8 Acknowledgement

This work was supported by National Institute of Health grant 1-R01-HL-090559 and grant National Science Foundation DMS-1106891.

References

[1] Barbieri, Maria Maddalena and Berger, James O. (2004). Optimal predictive model selection. *Annals of Statistics*, **32**(3), 870–897.
[2] Berchuck, A., Iversen, E. S, Luo, J., Clarke, J. P., Levine, H. Horne D. A., Boyd, J., Alonso, M. A., Secord, A. A., Bernardini, M. Q., Barnett, J. C., Boren, T., Murphy, S. K., Dressman, H. K., Marks, J. R. and Lancaster, J. M. (2009). Microarray analysis of early stage serous ovarian cancers shows profiles predictive of favorable outcome. *Clinical Cancer Research*, **15**, 2448–2455.
[3] Berchuck, A., Jr., Iversen, E. S., Lancaster, J. M., Pittman, J., Luo, J., Lee, P., Murphy, S., Dressman, H. K., Febbo, P. G., West, M., Nevins, J. R. and Marks, J. R. (2005). Patterns of gene expression that characterize long-term survival in advanced stage serous ovarian cancers. *Clinical Cancer Research*, **11**(10), 3686–3696.
[4] Berger, James O. and Pericchi, Luis R. (1996). The intrinsic Bayes factor for linear models. See [7], pp. 25–44.
[5] Berger, James O. and Pericchi, Luis R. (1996). The intrinsic Bayes factor for model selection and prediction. *Journal of the American Statistical Association*, **91**, 109–122.
[6] Berger, James O. and Pericchi, Luis R. (2001). Objective Bayesian methods for model selection: Introduction and comparison. In *Model Selection* (ed. P. Lahiri), Volume 38 of *Lecture Notes in Statistics*, pp. 135–193. Institute of Mathematical Statistics, Hayward, CA.
[7] Bernardo, José Miguel, Berger, James O., Dawid, A. Phillip, and Smith, Adrian F. M. (ed.) (1996). *Bayesian Statistics 5*, Oxford, UK. Oxford Univ. Press.
[8] Bernardo, José Miguel, Berger, James O., Dawid, A. Phillip, and Smith, Adrian F. M. (ed.) (1999). *Bayesian Statistics 6*, Oxford, UK. Oxford Univ. Press.
[9] Bernardo, José M. and Smith, Adrian F. M. (1994). *Bayesian Theory*. Wiley, New York, NY.
[10] Breiman, Leo (1996). Heuristics of instability and stabilization in model selection. *Annals of Statistics*, **24**, 2350–2383.
[11] Breiman, Leo (1996). Stacked regressions. *Machine Learning*, **24**, 49–64.
[12] Chipman, Hugh A., George, Edward I., and McCulloch, Robert E. (2001). The practical implementation of Bayesian model selection. In *Model Selection* (ed. P. Lahiri), Volume 38 of *Lecture Notes in Statistics*, pp. 65–134. Institute of Mathematical Statistics, Hayward, CA.
[13] Clemen, Robert T. and Winkler, Robert L. (1999). Combining probability distributions from experts in risk analysis. *Risk Analysis*, **19**(2), 187–203.
[14] Clyde, Merlise (1999). Bayesian model averaging and model search strategies (with discussion). See [8], pp. 157–185.
[15] Clyde, Merlise, DeSimone, Heather and Parmigiani, Giovanni (1996). Prediction via orthogonalized model mixing. *Journal of the American Statistical Association*, **91**, 1197–1208.
[16] Clyde, Merlise and George, Edward I. (2004). Model uncertainty. *Statistical Science*, **19**(1), 81–94.

[17] Draper, David (1995). Assessment and propagation of model uncertainty (with discussion). *Journal of the Royal Statistical Society, Series B*, **57**, 45–70.
[18] Ferguson, Thomas S. (1974). Prior distributions on spaces of probability measures. *The Annals of Statistics*, **2**(4), 615–629.
[19] Geisser, Seymour (1975). A predictive sample reuse method with application. *jasa*, **70**, 320–328.
[20] Geisser, Seymour (1993). *Predictive Inference: An Introduction*. Chapman & Hall, New York, NY.
[21] Geisser, Seymour and Eddy, William F. (1979). A predictive approach to model selection (Corr: V75 p. 765). *Journal of the American Statistical Association*, **74**, 153–160.
[22] George, Edward I. (1999). Discussion of "Model averaging and model search strategies" by M. Clyde. See [8].
[23] Gutiérrez-Peña, E. and Walker, S.G. (2001). A Bayesian predictive approach to model selection. *Journal of Statistical Planning and Inference*, **93**(1–2), 259–276.
[24] Hoeting, Jennifer A., Madigan, David, Raftery, Adrian E., and Volinsky, Chris T. (1999). Bayesian model averaging: a tutorial (with discussion). *Statistical Science*, **14**(4), 382–401. Corrected version at http://www.stat.washington.edu/www/research/online/hoeting1999.pdf.
[25] Iversen, Edwin S. and Luo, Rosy J. (2003). Molecular and genetic modeling of disease risk. In *Proceedings of the American Statistical Association, Risk Section*, Alexandria, VA: American Statistical Association, pp. CD–ROM.
[26] Key, Jane T., Pericchi, Luis R. and Smith, Adrian F. M. (1999). Bayesian model choice: What and why? See [8], pp. 343–370.
[27] Leamer, Edward E. (1978). *Specification searches: Ad hoc Inference with Nonexperimental Data*. Wiley, New York, NY.
[28] Mukhopadhyay, Nitai, Ghosh, Jayanta K. and Berger, James O. (2005). Some bayesian predictive approaches to model selection. *Statistics & Probability Letters*, **73**, 369–379.
[29] Pittman, Jennifer, Huang, Erich, Nevins, Joseph, Wang, Quanli and West, Mike (2004). Bayesian analysis of binary prediction tree models for retrospectively sampled outcomes. *Biostatistics*, **5**(4), 587–601.
[30] Proschan, F. (1963). Theoretical explanation of observed decreasing failure rate. *Technometrics*, **5**, 375–383.
[31] Raftery, Adrian E., Gneiting, Tilmann, Balabdaoui, Fadoua and Polakowski, Michael (2005). Using Bayesian model averaging to calibrate forecast ensembles. *Monthly Weather Review*, **133**, 1155–1174.
[32] Raftery, Adrian E., Madigan, David and Volinsky, Chris T. (1996). Accounting for model uncertainty in survival analysis improves predictive performance. See [7], pp. 323–349.
[33] San Martini, A. and Spezzaferri, Fulvio (1984). A predictive model selection criterion. *Journal of the Royal Statistical Society, Series B*, **46**, 296–303.
[34] Winkler, R. L. (1999). "Bayesian Model Choice. What and Why?: Comment," in J. M. Bernardo, J. O. Berger, A. P. Dawid, and A. F. M. Smith, eds., *Bayesian Statistics 6*. Oxford: Oxford University Press, 367–368.
[35] Walker, Stephen G. and Gutiérrez-Pena, Eduardo (1999). Discussion of Bayesian model choice: What and why? See [8], pp. 367.
[36] Walker, Stephen G. and Gutiérrez-Pena, Eduardo (1999). Robustifying Bayesian procedures. See [8], pp. 685–710.
[37] Walker, Stephen G., Gutiérrez-Pena, Eduardo and Muliere, Pietro (2001). A decision theoretic approach to model averaging. *The Statistician*, **50**, 31–39.
[38] Wolpert, Robert L., Clyde, Merlise A. and Tu, Chong (2011). Stochastic expansions using continuous dictionaries: Lévy adaptive regression kernels. *Annals of Statistics*, **39**, 1916–1962.

Part X

Finance and Actuarial Science

25 Asset allocation in finance: a Bayesian perspective

ERIC JACQUIER
AND NICHOLAS G. POLSON

25.1 Introduction

Bayesian methods have long played a role in finance and asset allocation since the seminal work of de Finetti [12] and Markowitz (see [32]). In this paper, we show how the principle of maximum expected utility (MEU) [44], [45], [5] together with Stein's lemma for stochastic volatility distributions [19] solves for the optimal asset allocation. Stein's lemma provides the solution to the first-order condition that accompanies MEU. The optimal asset allocation problem couched in equilibrium then leads to models such as the Capital Asset Pricing Model (CAPM) or Merton's inter-temporal asset pricing model (ICAPM).

We consider an investor who wishes to invest in the risky asset in order to maximize the expected utility of their resulting wealth. Under logarithmic utility, this leads to the famous Kelly criterion which maximizes the expected long-run growth rate of the risky asset. We review the link between the Kelly rule [37] and the Merton optimal asset allocation [40]. We illustrate their implementation for a discrete binary setting and for the standard historical returns on the S&P500. Under a constant relative risk aversion utility (CRRA) we use Stein's lemma to derive fractional versions of the Kelly rules where the amount allocated is normalized by the investor's relative risk aversion.

The rest of the paper is as follows. Section 25.2 reviews the impact of Bayesian thinking in models of finance: asset pricing equilibrium models; how agents learn from prices; properties of returns data including predictability and stochastic volatility. Section 25.3 views asset allocation from a Bayesian decision theoretic perspective (see, for example, [5]). Section 25.3 studies maximization of the expected long-run growth rate and derives the classic Kelly and Merton allocation rules. Section 25.4 describes methods for estimating this long-run growth rate. Section 25.5 describes estimation methods for long-run asset allocation. Section 25.6 considers extensions to Bayesian dynamic learning [3] and time-varying investment opportunity sets [15]. When investors are faced with a return distribution that is an exchangeable Beta-Binomial process, the effect of dynamic learning makes investors willing to invest a small amount of capital to current returns that have a negative expectation, even though they are averse to risk. This is due to the fact that they might learn that the investment opportunity set improves in the future and this is taken account of in the Bayesian MEU solution. Finally, Section 25.7 concludes.

25.2 Bayesian methods in finance

Bayesian thinking underpins financial modelling in a number of empirical and theoretical ways. Bayesian MCMC and particle methods are prevalent in empirical finance, see [33], [34]. [35] discuss MCMC methods in financial econometrics, [9] describes portfolio choice problems and [43] provide an empirical analysis of the S&P500 stock index. [32] provide a recent survey of Bayesian methods in finance. For example, many of the theoretical developments in finance rely on Bayesian learning by agents. Here we discuss applications to learning from prices and the implications of return predictability and stochastic volatility of asset returns [50].

25.2.1 Learning from prices

Bayesian methods are designed for learning, and market equilibrium occurs after individuals with differences of opinion have an incentive to trade and finally agree on a price. The differences of opinion literature has a long history dating to [13], [17], [18], [25], [28] and [49, 50]. [42] and [2] discuss investor behaviour in financial markets.

A basic argument is as follows. Let y denote an observed signal. The insight is that observing prices $P(y)$ will change a trader's beliefs. The trader needs to be able to coherently update their probabilities via the Bayes theorem. [20, 21] summarizes the logic clearly as follows:

> *A smart trader t might even tell herself: let all the other traders naively use their own information. I will wait until the market clears, and after observing the current realization of $P(y)$, make my purchases of commodities to maximize $E\left[U(W)|y_t, P(y)\right]$. Since I am a price taker, I will expect to do better than by trading now and maximizing $E\left[U(W)|y_t\right]$*

Market efficiency arguments follow a similar route, see [22]. Let $\pi(y)$ be the vector of probabilities that the trader holds for the states. If the observed prices $P(\pi(y))$ are invertible in π, the trader can back out everybody else's beliefs. Then implementing the principle of maximum expected utility

$$E\left[U_t(W_t)|y_t, P(\pi(y))\right] \equiv E\left[U_t(W_t)|y\right]$$

The trader who only has information y_t can still act *as if* he had the full information set of all investors' signals y. Hence, in an efficient market prices are fully revealing of trader's beliefs.

25.2.2 Return predictability and stochastic volatility

Time varying stochastic volatility is a widely documented feature of financial series, Typically, volatility is time-varying, persistent, mean-reverting and in some instances has jumps [33, 34], [14]. [23, 24] shows that the eighteenth and twentieth century time series distributions of stock returns are statistically very similar, and both exhibit stochastic volatility. In particular, he analyses biweekly prices and returns for shares of the London stock market index and the Dutch East India company from 1723–1794 and shows empirical evidence of stochastic volatility.

Another, more controversial, feature of financial returns is predictability, typically modelled as a projection on a set of predetermined information variables. For example, the dividend yield, or net payout ratio, variable plays a central role, see [1] and [36]. [36] combine predictability and stochastic volatility. Rather than a constant optimal allocation as is the case with i.i.d. returns, the time-varying opportunity set arising from predictability induces a hedging-demand into the investor allocation. [1] shows how horizon effects can be dramatic when the investor engages in Bayesian learning about the mean return. In a continuous-time setting, [11] quantify the hedging demands from stochastic volatility. [38] provide a theoretical analysis of dynamic portfolio strategies when there is jump risk, an effect that is also present in a discrete time setting. Clearly, economically significant hedging demand can exist and sensitivity analysis to models and priors is an important issue. Finally, as the optimal allocation rule can lead to leverage, issues of short-sales and margin-based trading arise. [16] provide a framework for generalizing the CAPM when investors face margin-based constraints.

We now turn to the central problem of asset allocation.

25.3 Asset allocation

The asset allocation problem can be described as a Bayesian decision-theoretic problem [5] as follows. An investor has initial wealth W_0 that can be invested in a risk-free bond or a risky asset. The end of period wealth is $W = W_0 \left\{(1-\omega)r_f + \omega R\right\}$ where r_f is the return on the risk-free rate and R is the return on the risky asset. The Bayesian investor maximizes expected utility, namely $\max_\omega \mathbb{E}\left[U\left(W(\omega)\right)\right]$. This leads to a first-order condition of the form:

$$E\left[U'(W)(R-r_f)\right] = 0$$

Applying the definition of covariance yields

$$Cov\left[U'(W), R-r_f\right] + E\left[U'(W)\right]E\left[R-r_f\right] = 0$$
$$\omega E\left[U''(W)\right]var(R) + E\left[U'(W)\right]E\left[R-r_f\right] = 0$$

where we have used Stein's lemma for a differentiable function $g(X)$ where $\mathbb{E}|g'(X)| < \infty$ then $cov(g(X), X) = \mathbb{E}\left(g'(X)\right)var(X)$. This has a solution for the optimal weight

$$\omega^\star = \frac{1}{\gamma}\left(\frac{\mathbb{E}\left[R\right]-r_f}{Var\left[R\right]}\right) \tag{25.1}$$

where $\gamma = -W\mathbb{E}\left[U''(W)\right]/\mathbb{E}\left[U'(W)\right]$ is the agent's relative risk aversion. This is a form of the famous Merton optimal asset allocation result.

This result can be extended to allow for stochastic volatility in the risky asset return. Let $(X|V) \sim \mathcal{N}\left(0, V\right)$ where $V \sim p(V)$. The equivalent Stein result is

$$Cov\left[g\left(X\right), Y\right] = \mathbb{E}^Q\left[g'\left(X\right)\right]Cov\left[X, Y\right]$$

Here $\mathbb{E}^Q$ is taken with respect to the distribution $q(V) = Vp(V)/\mathbb{E}\left[V\right]$. This is referred to as the size-biasing of the original volatility distribution $p(V)$. If $Y = X$, we have $Cov\left[g\left(X\right), X\right] = \mathbb{E}^Q\left[g'\left(X\right)\right]var\left[X\right]$.The optimal allocation rule is then

$$\omega^\star_{SV} = \frac{1}{\Gamma^Q}\left(\frac{\mathbb{E}\left[R\right]-r_f}{Var\left[R\right]}\right) \tag{25.2}$$

where $\Gamma^Q = -\mathbb{E}^Q\left[U''\right]/\mathbb{E}\left[U'\right]$ is the volatility adjusted risk aversion. See [19] for an analysis of comparative statics.

In this one period setting, given that agents adopt the optimal allocation rule, one can then show that an equilibrium Capital Asset Pricing Model (CAPM) holds. Specifically, under stochastic volatility (SV), the expected return $\mathbb{E}(R_j)$ on the jth security is given by

$$\mathbb{E}(R_j) - r_f = \beta_j \left(\mathbb{E}(R_m) - r_f \right)$$

where r_f is the risk-free rate and $\beta_j = cov(R_j, R_m)/var(R_m)$ is the risk premium.

A commonly used utility function is constant relative risk aversion (CRRA). It has the advantage that the optimal rule is unaffected by wealth effects. The CRRA utility of wealth takes the form

$$U_\gamma(W) = \frac{W^{1-\gamma} - 1}{1 - \gamma}$$

The special case $U(W) = \log(W)$ for $\gamma = 1$ plays a central role in growth rate analysis. It leads to maximizing the expected long-run rate of growth, namely

$$\max_{\omega} \mathbb{E} \left(\log W_T | W_0 = x \right)$$

We now solve for this rule and derive the Kelly criterion and Merton's rule.

25.4 Maximizing expected long-run growth

25.4.1 Kelly rule and Merton's optimal allocation

The Kelly criterion corresponds to the following Bayesian decision problem under binary uncertainty. Consider a sequence of i.i.d. bets where

$$p(X_t = 1) = p \text{ and } p(X_t = -1) = q = 1 - p$$

The investor who maximises $\mathbb{E} \left(\log W_T | W_0 = x \right)$ uses a myopic-rule with weight

$$\omega^\star = p - q = 2p - 1$$

Indeed, we can see that maximizing the expected long-run growth rate leads to the solution

$$\begin{aligned} \max_{\omega} \mathbb{E} \left(\ln(1 + \omega W_T) \right) &= p \ln(1 + \omega) + (1 - p) \ln(1 - \omega) \\ &\leq p \ln p + q \ln q + \ln 2 \text{ and } \omega^\star = p - q \end{aligned}$$

If one believes the event is certain i.e. $p = 1$, then one bets all wealth and, a priori one is certain to double invested wealth. On the other hand, if one thinks the bet is fair, i.e. $p = \frac{1}{2}$, one bets nothing, $\omega^\star = 0$, due to risk-aversion.

We will use the following notation. Let p denote the probability of a gain and $O = (1 - p)/p$ the odds. We can generalize the rule to the case of asymmetric payouts (a, b) where

$$p(X_t = 1) = p \text{ and } p(X_t = -1) = q = 1 - p$$

Table 25.1 Kelly rule

Market	You	p	$\omega^\star$
4/1	3/1	1/4	1/16
12/1	9/1	1/10	1/40

Then the objective expected utility function is

$$p \ln(1 + b\omega) + (1 - p) \ln(1 - a\omega)$$

with optimal solution

$$\omega^\star = \frac{bp - aq}{ab} = \frac{p - q}{\sigma}$$

If $a = b = 1$ this reduces to the pure Kelly criterion.

A common case occurs when $a = 1$. We can now interpret b as the odds O that the market is willing to offer the invest if the event occurs and so we write $b = O$. The rule becomes

$$\omega^\star = \frac{p \cdot O - q}{O}$$

We now provide a counter-intuitive example.

Example. Betting Consider the following two betting situations described in Table 1. Assume two possible market opportunities: one where it offers you $4/1$ when you have personal odds of $3/1$ and a second one when it offers you $12/1$ while you think the odds are $9/1$. In expected return these two scenarios are identical both offering a 33% gain. In terms of maximizing long-run growth, however, they are not identical. From Table 25.1, the Kelly criteria advises an allocation that is twice as much capital to the lower odds proposition: $1/16$ weight versus $1/40$.

Specifically, we have the following optimal weight calculation $\omega^\star = (pO - q)/O$ with allocations

$$\frac{(1/4) \times 4 - (3/4)}{4} = \frac{1}{16} \text{ and } \frac{(1/10) \times 12 - (9/10)}{12} = \frac{1}{40}$$

respectively.

We now turn to the continuous-time setting and find Merton's rule.

Continuously compounded returns: In a continuous-time setting, let μ denote the expected return, σ the volatility and γ the risk aversion. Suppose that the risky asset follows a Black–Scholes geometric Brownian motion model

$$dS_t = S_t \left(\mu dt + \sigma dB_t\right)$$

for a constant volatility σ. Then the value of the asset at time T is

$$S_T = S_0 \exp\left\{\left(\mu - \frac{1}{2}\sigma^2\right) T + \sigma \sqrt{t} Z\right\}$$

where $Z \sim N(0, 1)$. This model implies that returns are log-normally distributed.

Now consider the evolution of wealth for an investor who keep a constantly rebalanced weight ω allocated to the risky asset and $(1-\omega)$ to the risk-free rate r_f. Their wealth W_t^{ω} now evolves according to $dW_t^{\omega}/W_t^{\omega} = \left\{(1-\omega)r_f + \omega\mu\right\} dt + \omega\sigma\, dB_t$. Let $\alpha = \mu + 0.5\sigma^2$ be the expected arithmetic return on the market. Then, using Itô's lemma, we can solve for wealth at time t with $Z \sim \mathcal{N}(0,1)$ as

$$W_t = W_0 \exp\left\{(r_f + \omega(\mu - r_f) - \frac{1}{2}\omega^2\sigma^2)t + \omega\sigma\sqrt{t}Z\right\}$$

Similarly, the expected utility of wealth is

$$\frac{1}{1-\gamma}\exp\left\{(1-\gamma)\left(r_f + \omega(\alpha - r_f) - \frac{1}{2}\omega^2\sigma^2 + \frac{1}{2}(1-\gamma)\omega^2\sigma^2\right)\right\}$$

Long-run growth rate: This leads to a natural definition of the growth rate as

$$G(\omega) \triangleq \omega(\alpha - r_f) - \frac{\gamma}{2}\omega^2\sigma^2$$

Maximizing the growth rate with respect to ω leads to the optimal allocation

$$\omega^\star = \frac{1}{\gamma}\frac{\alpha - r_f}{\sigma^2}$$

If $\gamma = 1$, this is known as the pure Kelly rule. Fractional Kelly rules are, as their name indicates, rules that allocate a fraction of the Kelly rule to the risky asset via the agents risk aversion γ.

This analysis can be seen as a utility interpretation of fractional Kelly rules and agrees with Stein's lemma approach in the previous section.

We can analytically compute the growth rate of wealth at the optimal allocation as:

$$G(\omega^\star) = \frac{1}{\gamma}\left(1 - \frac{1}{2\gamma}\right)\left(\frac{\alpha - r_f}{\sigma}\right)^2$$

and if $\gamma < \frac{1}{2}$, a case of an agent with extremely low risk aversion, then $G(\omega^\star) < 0$.

Estimation risk also affects the growth rate: if estimation error mistakenly leads to using twice the optimal rule, then we have lost all of the growth of our portfolio as $G(2\omega^\star) = 0$.

We now provide an equivalence with the Kelly criterion by scaling wealth in terms of volatility units. In the continuous case, let the investor be faced with a sequence of returns where the payout has expectation $E(R_T) = \mu$

$$p(R_T = \sigma) = p = \frac{1}{2} + \frac{1}{2}\frac{\alpha - r_f}{\sigma} \text{ and } p(R_T = -\sigma) = q = \frac{1}{2} - \frac{1}{2}\frac{\alpha - r_f}{\sigma}$$

Given (μ, σ), the optimal Kelly rule is

$$\omega^\star = \frac{p-q}{\sigma} = \frac{\alpha - r_f}{\sigma}\cdot\frac{1}{\sigma} = \frac{\mu}{\sigma^2}$$

This provides our equivalence with Merton's rule.

Example. S&P500: Consider a simple example of logarithmic utility (CRRA with $\gamma = 1$). This is a pure Kelly rule. We assume iid log-normal stock returns with an annualized expected excess return of 5.7% and a volatility of 16% which is consistent with long-run equity returns. In our continuous time formulation $\omega^\star = 0.057/0.16^2 = 2.22$ and the Kelly criterion which imply that the investor borrows 122% of wealth to invest a total of 220% in stocks. This is a the risk-profile of the Kelly criterion. One also sees that the allocation is highly sensitive to estimation error in $\hat{\mu}$. We consider dynamic learning in a later section and show how the long horizon and learning affect the allocation today.

The fractional Kelly rule leads to a more realistic allocation. Suppose that $\gamma = 3$. Then the Sharpe ratio is

$$\frac{\mu}{\sigma} = \frac{0.057}{0.16} = 0.357 \text{ and } \omega^\star = \frac{1}{3}\frac{0.057}{0.16^2} = 74.2\%$$

An investor with such a level of risk aversion then has a more reasonable 74.2% allocation.

This analysis ignores the equilibrium implications. If every investor acted this way, then this would drive up prices and drive down the equity premium of 5.7%.

25.5 Long run asset allocation

Discussions of long-term investment policy revolve around measures of the expected long-term return on a risky portfolio such as the global market index. In this section we review how optimal Bayesian asset allocation leads to an estimate of long run expected returns with very attractive properties. We first give a background on the estimates resulting from the classical literature, and their potential shortcoming.

25.5.1 Background on classical estimation

To concentrate on the main issue, that of the uncertainty in the mean, we assume that variance is known and that returns are not auto-correlated, see [31] and [29] for a robustness analysis. Recall that, as financial returns exhibit very little auto-correlation, there is no gain in observing the data more frequently. In this context, we therefore consider a sample of T annual i.i.d. log-normal returns $\log(1 + R) \sim N(\mu, \sigma^2)$. The reader will quickly see that they can adapt the discussion to any estimator of the mean return given its associated variance. The long-term, H-period return, or wealth return per dollar invested is log-normal. Long-term investors and policy-makers seek an estimate of its expected compound return

$$E(V_H) = e^{H(\mu + 0.5\sigma^2)} \tag{25.3}$$

The Maximum Likelihood estimate (MLE) of μ is the sample mean, it corresponds to the Bayesian posterior mean with no prior information. For long-term forecasts, practitioners used to choose a point estimate by compounding the sample geometric return $G = \frac{1}{T}\log\frac{P_T}{P_1}$. This amounts to estimating $E(V_H)$ by $e^{\hat{\mu}H}$. Academics, however, tended to substitute $\hat{\mu}$ in the theoretical expectation (25.3), often invoking a maximum likelihood justification, where the estimator of a function is approximated by the function of the estimator. This second suggestion is equivalent to the compounding H times of the arithmetic sample mean. The difference in these two estimates becomes very large in the long run. Using [47]'s geometric and arithmetic averages of 7% and 8.5%, the two approaches grow \$1 to \$160 versus \$454 over 75 years. [30] show that the ML approach makes little sense; as H is often of a magnitude comparable to T, the ML estimator suffers from an enormous upward bias due to the Jensen effect. An unbiased estimator is $U = \exp\{H(\hat{\mu} + 0.5\sigma^2(1 - \frac{H}{T}))\}$, a simple log-linear combination of the geometric and arithmetic estimators. Observe how the compounding factor is linearly decreasing in H. [31] argue that there is little justification for mere unbiasedness, and derive a minimum mean squared error classical estimator of $E(V_H)$:

$$M = e^{H(\widehat{\mu} + 0.5\sigma^2(1 - 3\frac{H}{T}))} \tag{25.4}$$

These estimators nest the ML (also known as arithmetic) estimator, only justified with extremely small $\frac{H}{T}$, and the geometric estimator, itself only justified when H happens to be equal to $\frac{T}{3}$. For both U and M, the compounding factor decreases linearly with the horizon H, with a downward penalty increasing with $\frac{H}{T}$. This is desirable because, (1) H affects the amount of compounding that magnifies (by the Jensen effect) the upward bias due to the estimation error in μ, and (2) T reduces the uncertainty in μ. For realistic values of μ, σ, T, H, the penalty is quite severe; longer-term investors must be far more modest in their predictions than shorter-term investors, given the same base estimator of μ.

However, since the compounding factor decreases linearly, it can lead to expected returns lower than the risk-free rate, even negative, for long enough horizons. This feature of the estimator makes no economic sense; any discount factor should tend to no lower than the risk-free rate as the horizon increases. We now turn to a simple Bayesian estimator based on optimal asset allocation, which naturally incorporates this desired feature.

25.5.2 Bayesian estimation of the long run expected return

Parameter uncertainty has an important impact on the optimal allocation. This has been recognized since [10]. Intuitively, the proper distribution for the investor to consider is the predictive density, which gets inflated by the uncertainty in the mean. Relative to the case with known parameters, the optimal allocation is then lower in the risky asset.

Consider first the classic [40] asset allocation, it is a framework with one risky asset with a i.i.d. normally distributed log-return $N(\mu, \sigma^2)$, a risk-free return r_f, and a power utility with risk aversion γ. If all parameters are known, [40] shows that, with continuous re-balancing, the horizon H is irrelevant and drops out of the computation. He derives the now well-known optimal allocation,

$$w^* = \frac{\alpha - r_f}{\gamma\sigma^2} \tag{25.5}$$

where $\alpha = \mu + 0.5\sigma^2$ is the expected arithmetic one-period return. To obtain this result, assuming constant rebalancing to an allocation w, it is easy to show first that the random H period return is log normally distributed as

$$\log(V_H|\alpha, \sigma) \sim N\left[(r_0(1 - w) + w\alpha - 0.5w^2\sigma^2)H, w^2\sigma^2 H\right] \tag{25.6}$$

The expected utility of this random wealth is then:

$$E[U(V_H)] = \frac{1}{1-\gamma}\exp\left[(1-\gamma)H(r_0 + w(\alpha - r_0) - 0.5w^2\sigma^2 + 0.5(1-\gamma)w^2\sigma^2)\right]. \tag{25.7}$$

The maximum likelihood estimator of μ is the standard average of past log-returns $\hat{\mu} \sim N(\mu, \sigma^2/T)$. With diffuse priors, the Bayesian posterior of μ is numerically equivalent, albeit with the well-known difference in interpretation. It is clear that simply substituting $\widehat{\alpha}$ in equation (25.7) or (25.5) is not the optimal solution. [29] derives the Bayesian optimal allocation with uncertainty in μ. The investor must consider the predictive utility, i.e. the utility of the predictive density of the H period return. Because the integrations over the parameter and over the distribution of returns can be exchanged, one can also view this as integrating μ out of the conditional expected utility in (25.7), using its posterior distribution. The expected (predictive) utility becomes:

$$E[U(V_H)] = \frac{1}{1-\gamma} \exp\left[(1-\gamma)H[r_0 + w(\hat{\alpha} - r_0) - 0.5w^2\sigma^2 + 0.5(1-\gamma)w^2\sigma^2(1+\frac{H}{T})]\right] \tag{25.8}$$

We observe that α has been replaced by its posterior mean $\hat{\alpha}$, and there is a new term in H/T at the end. Maximizing the expected utility in (25.8), [29] finds the optimal asset allocation under uncertainty:

$$w^* = \frac{\widehat{\alpha} - r_0}{\sigma^2\left[\gamma(1+\frac{H}{T}) - \frac{H}{T}\right]} \tag{25.9}$$

Figure 25.1 plots the optimal allocation versus the horizon for an investor with a relative risk aversion of 4 and standard values of the mean and variance of the market return. It shows that the effect of uncertainty on the optimal allocation is important for realistic values of μ, r_f, σ, γ.

This optimal allocation in (25.9) implies an estimate of the expected long-term return under uncertainty. For this investor with risk aversion γ, denote α^* this *risk-adjusted* future expected return, where the risk is estimation risk. For this risk-adjusted return, the optimal allocation would be the Merton allocation in (25.5). Equating the two allocations, we find that this risk-adjusted estimate is

$$\alpha^\star - r_f = \frac{\widehat{\alpha} - r_f}{1 + \frac{H}{T}(1-\frac{1}{\gamma})} \tag{25.10}$$

It can be rewritten in terms of μ if desired. Figure 25.2 shows this estimate versus the horizon. Even with a moderate risk aversion of 4, it reinforces the stark (low expected return) prediction of the minimum mean squared error estimate in (25.4).

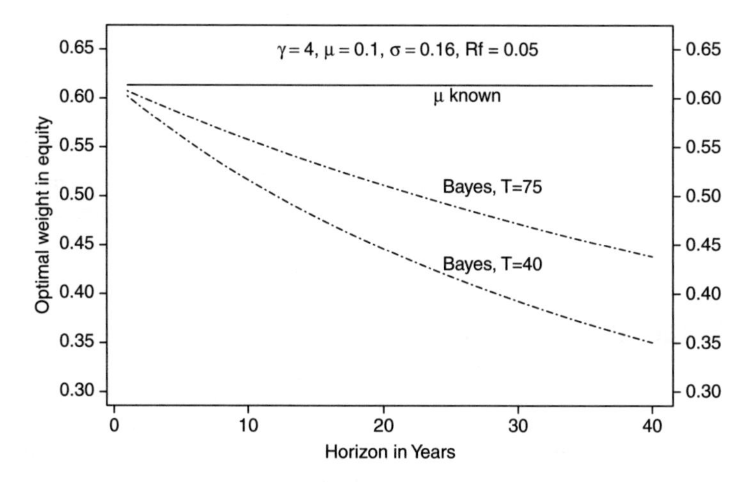

Figure 25.1 Bayesian long-term asset allocation under uncertainty.

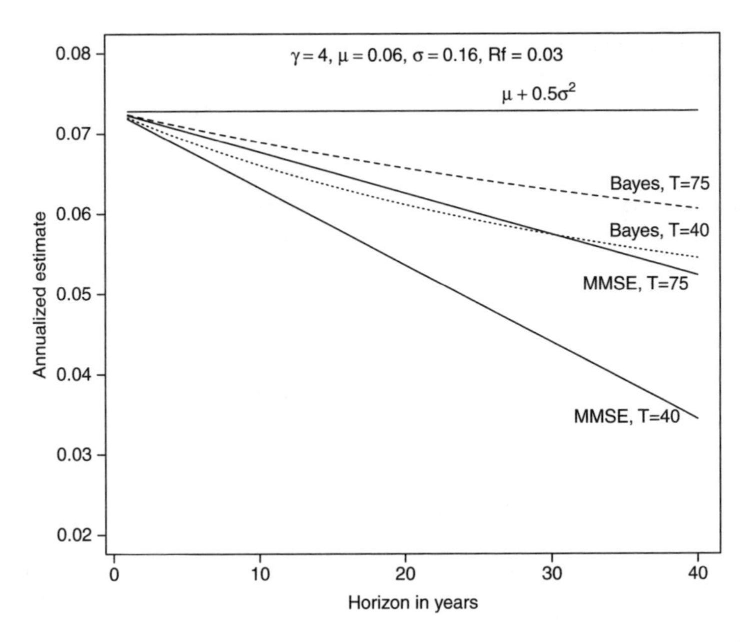

Figure 25.2 Diffuse prior on μ. Bayesian and classical minimum mean-squared estimates of annual compound factor.

This estimate is optimal for a given investor with risk aversion γ. Now, not only investors with different horizons but also investors with different risk aversion, formulate different optimal point forecasts. Note that the estimate in (25.10) can never be below the risk-free rate. This is to be expected since it is consistent with the Bayesian optimal allocation which has a lower bound of zero for the very long run.

Another related approach is the classic Black and Litterman [6, 7] framework for combining investor views with market equilibrium. In a multivariate returns setting the optimal allocation rule is $\omega^\star = \frac{1}{\gamma}\Sigma^{-1}\mu$. The question is how to specify (μ, Σ) pairs? For example, given $\hat{\Sigma}$, BL derive Bayesian inference for μ given market equilibrium model and a priori views on the returns of pre-specified portfolios which take the form

$$(\hat{\mu}|\mu) \sim \mathcal{N}\left(\mu, \tau\hat{\Sigma}\right) \text{ and } (Q|\mu) \sim \mathcal{N}\left(P\mu, \hat{\Omega}\right)$$

Combining views, the implied posterior is $(\mu|\hat{\mu}, Q) \sim \mathcal{N}(Bb, B)$ with mean and variance specified by $B = (\tau\hat{\Sigma})^{-1} + P'\hat{\Omega}^{-1}P$ and $b = (\tau\hat{\Sigma})^{-1}\hat{\mu} + P'\Omega^{-1}Q$. These posterior moments are then used in the optimal allocation rule.

25.6 Dynamic Bayes learning

When the opportunity set varies in time, the agent who maximizes long run growth will take this into account in her decision rule today. The Bayesian dynamic portfolio problem dates to [3, 4]. Here we will follow the analysis of [15] and consider the extensions to CRRA utility and to exchangeable return distributions where the agent will engage in dynamic learning.

Remember that in an i.i.d. discrete setting, the classic Kelly rule is for $p > \frac{1}{2}$,

$$\omega^\star = p - q = 2p - 1$$

and zero otherwise. The investor never bets on outcomes which have unfavourable (given their views) odds, or even fair odds due to risk aversion.

Now if returns are still independent but have time-varying known probabilities p_t we can still solve for the optimal allocation: for $p_t > \frac{1}{2}$ invest

$$\omega_t^\star = \frac{p_{t-1}^{\gamma^{-1}} - q_{t-1}^{\gamma^{-1}}}{p_{t-1}^{\gamma^{-1}} + q_{t-1}^{\gamma^{-1}}}$$

If p_t is unknown, a myopic Bayes rule is to track the sufficient statistics under a Beta prior $Be(\alpha, \beta)$ distribution with T observations $y = (y_1, \ldots, y_T)$ as

$$\mathbb{E}(p|y) = \frac{\alpha \sum_{t=1}^{T} y_t}{T + \alpha + \beta}$$

and uses a myopic plug-in rule by replacing p with $E(p|y^T)$.

To be fully Bayesian, we consider the exchangeable case where the investor can learn from experience. The solution of maximizing discounted expected utility has a very different solution. We need to solve for the value function using Bellman's equation and take account of the fact that *in the future* they will be using Bayes rule and learning from prices.

Bayes learning

Consider a conjugate Bernoulli-Beta model where $(y|p) \sim Ber(p)$ and prior distribution $(p|a, b) \sim Be(a, b)$ for given hyperparameters.

Value function

If we consider the case of power utility, the agent solves

$$\max_{\omega} \mathbb{E}\left[U(W_T)|W_0 = x\right] = x^{1-\gamma} V_T(a, b)$$

where V_T is the value function at period n.

Let $f_T(x|\alpha, \beta)$ denote the optimal value of $\mathbb{E}\left(W_T^{1-\gamma}|W_0 = x\right)$ when the prior distribution is $p \sim Be(\alpha, \beta)$. The posterior is then $Be\,(\alpha + 1, \beta)$ given a success and $Be\,(\alpha, \beta + 1)$ given a failure. At the initial condition, we have $f_0(x|\alpha, \beta) = x^{1-\gamma}$. We can then solve for the value function recursively using the identity

$$\begin{aligned} f_1(x|\alpha, \beta) &= \max_{0 \le b \le x} \left\{ \frac{\alpha}{\alpha + \beta}(x + b)^{1-\gamma} + \frac{\beta}{\alpha + \beta}(x - b)^{1-\gamma} \right\} \\ &= \max_{0 \le c \le 1} \left\{ \frac{\alpha}{\alpha + \beta}(1 + c)^{1-\gamma} + \frac{\beta}{\alpha + \beta}(1 - c)^{1-\gamma} \right\} \\ &= x^{1-\gamma} V_1(\alpha, \beta) \end{aligned}$$

Following recursively, we find

$$V_k(a,b) = \max_{0 \le c \le 1} \left\{ \frac{a}{a+b} V_{k-1}(a+1,b)(1+c)^{1-\gamma} + \frac{b}{a+b} V_{k-1}(a,b+1)(1-c)^{1-\gamma} \right\}$$

The optimal allocation is then

$$\omega_k(a,b) = \left(\frac{1-w}{1+w} \right)^+ \text{ where } w = \left(\frac{bV_{k-1}(a,b+1)}{aV_{k-1}(a+1,b)} \right)^{\gamma^{-1}}, \gamma > 1$$

In the myopic Kelly case where $\gamma = 1$, we have the rule $\omega_k(\alpha, \beta) = 0$ for $\alpha < \beta$ as one might expect. You will not invest if your prior mean says the bet is unfair.

This is not true in general though due to the learning effect. For example, with a uniform prior

$$w_{10}(1,1) = 0.875 \quad \text{and} \quad w_9(1,2) = 0$$

$$w_{10}(1,1) = 0.973 \quad \text{and} \quad w_9(1,2) = 0.117$$

and the investors allocate a large portion to the risky asset. Even in the case where they see a failure and they think that the odds are two to one against them, they are still willing to invest $w_9(1,2) = 0.117$ that is 11.7% of their wealth.

We therefore have the important effect of Bayesian learning on asset allocation, that the agent will invest in the first period even though the expected return is negative and they are risk averse!

25.7 Discussion

Bayesian thinking is central to finance and asset allocation. Stein's lemma provides a useful tool for analysing first-order maximum expected utility conditions. The classic Kelly and Merton allocation rules correspond to Bayes rules where the investor maximizes the expected long-run growth rate of accumulation. These rules are sensitive to estimation risk and Bayesian estimation methods provide the necessary inputs for expected returns and volatility.

Under the maximum growth condition when returns are independent these rules are myopic. Investors will not be willing to allocate any capital to unfavourable bets. However, when the invest opportunity set is extended to allow for Bayesian learning with exchangeable return distributions, we see that risk-averse investors are willing to hold a small amount of the risky asset in the hope that they will learn in the future that conditions are improved.

There are many avenues for future research. One currently active area is to use learning methods to explain bubbles and speculative behaviour (see [26], [46] and [48]). Early models in this literature 'explained' bubbles by incorporating agents who are persistently 'naive' and do not update beliefs coherently. The recent literature has focused more on the dynamics of Bayesian learning and belief structures. For example, [41] allows agents to do some beta-binomial updating of their beliefs, and [27] use prices to learn whether one is smart or dumb money in a market.

An interesting vignette comes from the South Sea bubble of 1720. The English parliament passed the Bubble Act in the following year 1721 trying to ban future bubbles. At the height of the bubble, the price-earnings multiple of the stock was between 150 and 190 based on current earnings. In the first six month of 1720 the South sea stock rose 500 per cent. The major 'news' being only the large demand for their shares. A number of investors lost large sums of money. Sir Isaac Newton (1642–1729) who was Master of the Mint from 1699 to 1729 and who put England on the Gold standard in 1717, reportedly lost the equivalent of $20,000 in the Bubble. He was quoted as saying *I can predict the motion of Heavenly bodies but not the behaviour of the stock-market.*

25.8 Acknowledgement

Polson is Professor of Econometrics and Statistics at the Chicago Booth School of Business. email: ngp@chicagobooth.edu. Jacquier is visiting professor of finance at MIT Sloan School of Management, on leave from HEC Montreal, email: jacquier@mit.edu. Jacquier acknowledges support from the HEC Montreal professorship in derivative securities and Sloan research fund.

References

[1] Barberis, N. (2000). Investing in the long run when returns are predictable. *Journal of Finance*, **55**, 225–264.

[2] Barberis, N. and R. Thaler (2003). A Survey of behavioral finance. *Handbook of the Economics of Finance* (eds Constantinides *et al.*), 1053–1128.

[3] Bellman, R. and R. Kalaba (1956). On the role of dynamic programming in statistical communication theory. *IRE Transactions on Information Theory*, IT-3, 197–203.

[4] Bellman, R. and R. Kalaba (1958). On communication processes involving learning and random duration. *IRE National Convention Record*, **4**, 16–20.

[5] Bernardo, J. and A. F. M. Smith (2000). *Bayesian Theory*. Wiley.

[6] Black, F. (1976). Studies of Stock Market Volatility Changes. *Proceedings of Journal of American Statistical Association*, 177–181.

[7] Black, F. and R. Litterman (1991). Asset allocation: combining investor views with market equilibrium. *Journal of Fixed Income*, **1**, 7–18.

[8] Black, F. and R. Litterman (1992). Global portfolio optimization. *Financial Analysts Journal*, **48**(5), 28–43.

[9] Borel, E. (1924). Apropos of a treatise on probability. *Revue Philosophique*. Reprinted in *Studies in Subjective Probability*, Kyburg and Smokler (1980) (eds).

[10] Brandt, M. (2009). Portfolio choice problems. *Handbook of Financial Econometrics* (eds Ait-Sahalia and L. P. Hansen), 269–336.

[11] Brown, S. (1976). Optimal Portfolio Choice Under Uncertainty: A Bayesian Approach. *Ph.D, University of Chicago*.

[12] Campbell, J. Y. and L. M. Viceira (2002). *Strategic Asset Allocation*. Oxford University Press.

[13] de Finetti, B. (1941). Il problema dei Pieni. Reprinted: *Journal of Investment Management*, **4**(3), 19–43.

[14] DeGroot, M. H. (1973). Reaching a consensus. *Journal of the American Statistical Association*, **69**, 118–121.

[15] Eraker, B., M. Johannes and N. G. Polson (2003). The impact of jumps in volatility and returns. *Journal of Finance*, **58**(3), 1269–1300.

[16] Ferguson, T. S. and C. Z. Gilstein (1985). A General Investment Model. *Technical Report, UCLA*.

[17] Garleanu, N. and L. H. Pedersen (2011). Margin-based asset pricing and deviations from the law of one price. *Review of Financial Studies*, **24**(6), 1980–2022.

[18] Geanakopolos, J. and H. Polemarchakis (1982). We can't disagree forever. *Journal of Economic Theory*, 192–200.

[19] Geanakopolos, J. and J. Sebenius (1983). Don't bet on it: a note on contingent agreements with asymmetric information. *Journal of the American Statistical Association*, **78**, 224–226.
[20] Gron, A., B. Jorgensen and N. G. Polson (2011). Optimal portfolio choice and stochastic volatility. *Applied Stochastic Models.*
[21] Grossman, S. (1976). On the efficiency of competitive stock markets where traders have diverse information. *Journal of Finance*, **31**, 573–585.
[22] Grossman, S. (1978). Further results on the informational efficiency of competitive stock markets. *Journal of Economic Theory*, **18**, 81–101.
[23] Grossman, S. J. and J. E. Stiglitz (1980). On the impossibility of informationally efficient markets. *American Economic Review*, **70**, 393–408.
[24] Harrison, P. (1998). Similarities in the distribution of stock market price changes between the eighteenth and twentieth centuries. *Journal of Business*, **71**(1), 55–79.
[25] Harrison, P. (2001). Rational equity valuation at the time of the South Sea Bubble. *History of Political Economy*, **33**(2), 269–281.
[26] Harris, M. and A. Raviv (1993). Differences of opinion make a horse race. *Review of Financial Studies*, **6**(3), 473–506.
[27] Harrison, J. M. and D. Kreps (1978). Speculative behavior in a stock market with heterogeneous expectations. *Quarterly Journal of Economics*, **92**(2), 323–336.
[28] Heaton, J. B. and N. G. Polson (2011). Smart Money, Dumb Money and Learning from Prices. *Working Paper, University of Chicago.*
[29] Hong, H. and J. C. Stein (2003). Differences of opinion, short-sales constraints, and market crashes. *Review of Financial Studies*, **16**, 487–525.
[30] Jacquier, E. (2008). Long-term forecasts of mean returns: Statistical versus economic rationales. *Working paper, HEC Montreal.*
[31] Jacquier, E., Kane, A. and A. Marcus (2003). Geometric or arithmetic mean: a reconsideration. *Financial Analysts Journal*, **59**(6), 46–53.
[32] Jacquier, E., Kane, A. and A. Marcus (2005). Optimal estimation of the risk premium for the long-term and asset allocation. *Journal of Financial Econometrics*, **3**, 37–56.
[33] Jacquier, E. and N. G. Polson (2011). Bayesian Methods in Finance. *Handbook of Bayesian Econometrics*, H. van Dyk *et al.* (eds).
[34] Jacquier, E., N. G. Polson and P. Rossi (1994). Bayesian analysis of stochastic volatility models. *Journal of Business and Economic Statistics*, **12**(4), 371–89.
[35] Jacquier, E., N. G. Polson and P. Rossi (2005). Bayesian analysis of stochastic volatility models with fat-tails and correlated errors. *Journal of Econometrics*, **122**(1), 185–212.
[36] Johannes, M. and N. G. Polson (2009). MCMC Methods for continuous-time financial econometrics. *Handbook of Financial Econometrics* (eds Ait-Sahalia and L. P. Hansen), 1–72.
[37] Johannes, M., A. Korteweg and N. G. Polson (2011). Sequential Learning, Predictive Regressions and Optimal Portfolio Allocation. *Working paper.*
[38] Kelly, J. R. (1956). A new interpretation of the information rate. *Bell System Technical Journal*, **35**, 917–926.
[39] Liu, J. and J. Pan (2003). Dynamic derivatives strategies. *Journal of Financial Economics*, **69**, 401–430.
[40] Markowitz, H. (2006). de Finetti scoops Markowitz. *Journal of Investment Management*, **4**(3), 5–18.
[41] Merton, R. C. (1969). Lifetime portfolio selection under uncertainty: the continuous time case. *Review of Economics and Statistics*, **50**, 247–257.
[42] Morris, S. (1996). Speculative investor behaviour and learning. *Quarterly Journal of Economics*, **111**(4), 1111–1133.
[43] Odean, T. (1998). Volume, volatility, price, and profit when all traders are above average. *Journal of Finance*, **53**, 1887–1934.

[44] Polson, N. G. and B. V. Tew (2000). Bayesian portfolio selection: an empirical analysis of the S&P500 indices 1970–1996. *Journal of Business and Economic Statistics*, 164–173.
[45] Ramsey, F. P. (1926). Truth and Probability. Reprinted in *Studies in Subjective Probability*, Kyburg and Smokler (1980) (eds).
[46] Savage, L. J. (1954). *Foundations of Statistics*. John Wiley and Sons.
[47] Scheinkman, J. A. and W. Xiong (2003). Overconfidence and speculative bubbles. *Journal of Political Economy*, **111**(6), 1183–1220.
[48] Siegel, J. J. (1994). *Stocks for the Long Run*. McGraw-Hill.
[49] Stein, J. C. (2009). Sophisticated investors and market efficiency. *Journal of Finance*, **64**(4), 1517–1548.
[50] Varian, H. R. (1985). Divergence of opinion in complete markets: A Note. *Journal of Finance*, **40**(1), 309–317.
[51] Varian, H. R. (1989). Differences of Opinion in Financial Markets. In *Financial Risk: Theory, Evidence and Implications*. Courtnet C. Stone (ed).

26 Markov chain Monte Carlo methods in corporate finance

ARTHUR KORTEWEG

This chapter introduces Markov chain Monte Carlo (MCMC) methods for empirical corporate finance. These methods are very useful for researchers interested in capital structure, investment policy, financial intermediation, corporate governance, structural models of the firm and other areas of corporate finance. In particular, MCMC can be used to estimate models that are difficult to tackle with standard tools such as OLS, Instrumental Variables regressions and Maximum Likelihood. Starting from simple examples, this chapter exploits the modularity of MCMC to build sophisticated discrete choice, self-selection, panel data and structural models that can be applied to a variety of topics. Emphasis is placed on cases for which estimation by MCMC has distinct benefits compared to the standard methods in the field. I conclude with a list of suggested applications. Matlab code for the examples in this chapter is available on the author's personal homepage.

26.1 Introduction

In the last two decades the field of empirical corporate finance has made great strides in employing sophisticated statistical tools to achieve identification, such as instrumental variables, propensity scoring and regression discontinuity methods. The application of Bayesian econometrics and in particular Markov chain Monte Carlo (MCMC) methods, however, has been lagging other fields of finance such as fixed income and asset pricing, as well as other areas of scientific inquiry such as marketing, biology, and statistics. This chapter explores some of the many potential applications of this powerful methodology to important research questions in corporate finance.

With the current trend in the corporate finance literature towards more complex empirical models, MCMC methods provide a viable and attractive means of estimating and evaluating models for which classical methods such as least squares regressions, GMM, Maximum Likelihood and their simulated counterparts are too cumbersome or computationally demanding to apply. In particular, MCMC is very useful for estimating nonlinear models with high-dimensional integrals in the likelihood (such as models with many latent variables), or a hierarchical structure. This includes, but is not limited to, discrete-choice, matching and other self-selection models, duration, panel data and structural models, encompassing a large collection of topics in corporate finance such as capital structure and security issuance, financial intermediation, corporate governance, bankruptcy, and structural models of the firm. The MCMC approach thus opens the door to estimating more realistic and insightful models to address questions that have thus far been out of reach of empirical corporate finance.

To illustrate the method, I consider the effect of firm attrition on the coefficient estimates in typical capital structure panel data regressions. Firms disappear from the sample for non-random reasons such as through bankruptcy, mergers or acquisitions, and controlling for this non-random selection problem alters the estimated coefficients dramatically. For example, the coefficient on profitability changes by about 25%, and the coefficient on asset tangibility drops roughly in half. Whereas estimating this selection correction model is difficult with classical methods, the MCMC estimation is not particularly complex, requiring no more than standard probability distributions and standard regressions. I provide the Matlab code for this model on my personal website.[22]

The goals of this chapter are two-fold. First, I want to introduce MCMC methods and provide a hands-on guide to writing algorithms. The second goal is to illustrate some of the many applications of MCMC in corporate finance. However, these goals come with a good deal of tension. Most sections in this chapter start with developing MCMC estimators for simple problems that have standard frequentist solutions that come pre-packaged in most popular software packages. The reason I discuss these examples is not because I think that researchers should spend their time coding up and running their own MCMC versions of these standard estimators, but rather to illustrate certain core principles and ideas. These simple examples then function as a stepping stone to more complex problems for which MCMC has a distinct advantage over the standard approaches such as least squares regressions, GMM, and Maximum Likelihood (or where such approaches are simply not feasible). For example, Section 26.4 starts with a standard probit model, focusing on the core concept of data augmentation. The modularity of MCMC allows us to extend this model to build a dynamic selection model in Section 26.4.3 that is nearly impossible to estimate by Maximum Likelihood or other classical methods.

To aid readers interested in applying MCMC methods, I have provided Matlab code for all the numbered algorithms in this chapter on my personal webpage.[23] Apart from educational purposes, these examples can also be used as building blocks to estimate more complex models, thanks to the inherent modularity of MCMC. It is, for example, quite straightforward to add a missing data feature to a model by adding one or two steps to the algorithm, without the need to rewrite the entire estimation code.

At the core of MCMC lies the Hammersley–Clifford theorem, by which one can break up a complex estimation problem into bite-size pieces that usually require no more than standard regression tools and sampling from simple distributions. Moreover, MCMC methods do not rely on asymptotic results but instead provide exact small-sample inference of parameters (and nonlinear functions of parameters), and do not require optimization algorithms that often make Maximum Likelihood and GMM cumbersome to use.

This chapter is organized by modelling approach. Section 26.2 introduces MCMC estimation through a simple regression example. Section 26.3 introduces the concept of data augmentation through a missing data problem. Section 26.4 discusses limited dependent variable and sample selection models, currently the most widely used application of MCMC in corporate finance. Section 26.5 addresses panel data models, introduces the powerful tool of hierarchical modelling, and presents the application to capital structure regressions with attrition. Section 26.6 describes the estimation of structural models by MCMC, and in particular the concepts of Metropolis–Hastings sampling and Forward Filtering and Backward Sampling. Section 26.7 suggests a number of further

[22] https://faculty-gsb.stanford.edu/korteweg/pages/DataCode.html.

[23] Other popular packages for running MCMC estimations are R, WinBugs and Octave (all can be downloaded free of charge). Many code examples for these packages can be found online.

applications in corporate finance for which MCMC is preferable to classical methods. Section 26.8 concludes.

26.2 Regression by Markov chain Monte Carlo

A simple introduction to MCMC is found by considering the standard linear model

$$y = X\beta + \varepsilon,$$

where y is an $N \times 1$ vector of dependent variables, X is an $N \times k$ matrix of predictor variables, and the error term is distributed iid Normal, $\varepsilon \sim \mathcal{N}\left(0, \sigma^2 I_N\right)$. With conjugate Normal-Inverse Gamma priors, $\sigma^2 \sim \mathcal{IG}(a, b)$ and $\beta|\sigma^2 \sim \mathcal{N}\left(\mu, \sigma^2 \cdot A^{-1}\right)$, this problem has well-known analytical expressions for the posterior distribution of the parameters of interest, β and σ^2. Alternatively, one can learn about the joint posterior of β and σ^2, by drawing a sample from it through Monte Carlo simulation. Algorithm 1 explains the simulation steps in detail (for ease of notation I do not write the conditioning on the observed data, X and y, in the algorithm). First, draw a realization of σ^2 from its *posterior* Inverse Gamma distribution. Next, draw a realization of β from its posterior distribution, using the draw of σ^2 from the first step. Together, these two draws form one draw of the *joint* posterior distribution $p(\beta, \sigma^2|X, y)$. Repeating these two steps many times then results in a sequence of draws from the joint posterior distribution. This sequence forms a Markov chain (in fact the chain is independent here), hence the name Markov chain Monte Carlo [31, 33]. This particular MCMC algorithm, in which we can draw from exact distributions, is also known as *Gibbs sampling*.

Algorithm 1 Regression

---

1. Draw $\sigma^2 \sim \mathcal{IG}(a + N, b + S)$,
   where $S = \left(y - Xm\right)' \left(y - Xm\right) + (m - \mu)'A(m - \mu)$, and $m = (X'X + A)^{-1}(X'y + A\mu)$
2. Draw $\beta|\sigma^2 \sim \mathcal{N}\left(m, \sigma^2 \cdot (X'X + A)^{-1}\right)$
3. Go back to step 1, repeat.

To illustrate the algorithm, I simulate a simple linear model with $N = 100$ observations and one independent variable, X, drawn from a standard Normal distribution, and no intercept. I set the true $\beta = 1$ and $\sigma = 0.25$. For the priors I set $a = 2.1$ and $b = 1$, corresponding to a prior mean of σ^2 of 0.91 and a variance of 8.26, and I set $\mu = 0$ and $A = I_K \cdot 1/10\,000$, such that the prior standard deviation on β is one hundred times σ^2. Unless otherwise noted, I use the same priors throughout this chapter. Using Matlab it takes about 3.5 seconds on a standard desktop PC to run 25 000 iterations of Algorithm 1. Figure 26.1 shows the histogram of the draws of β and σ. These histograms are the marginal posterior distributions of each parameter, numerically integrating out the other. In other words, the histogram of the 25 000 draws of β on the left-hand plot represents $p\left(\beta|X, y\right)$, and with enough draws converges to the analytical t-distribution. In the right-hand plot, the draws of σ^2 are first transformed to draws of σ by taking square roots before plotting the histogram. This highlights the ease of computing the posterior distribution of nonlinear functions of parameters. This example may appear trivial here, but this transformation principle will prove very useful later.

The vertical lines in Figure 26.1 indicate the true parameter values. Note that the point estimates are close to the true parameters, despite the small sample and the *prior* means being centred far from the true parameter values. Moreover, the simulated posterior means coincide with the analytical

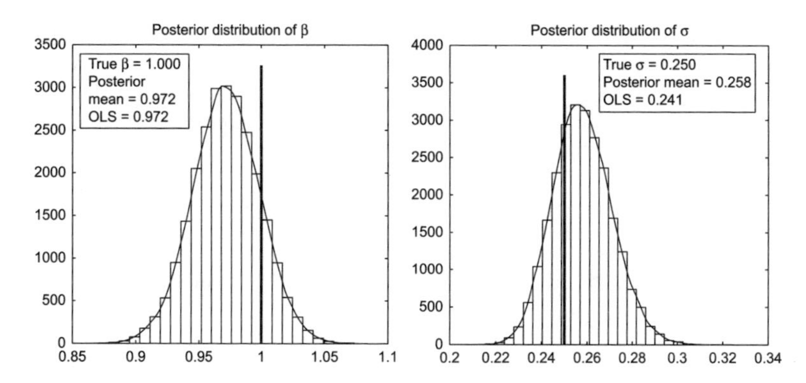

Figure 26.1 Posterior distribution of regression parameters. Histograms of 25 000 parameter draws of the standard regression model estimated by MCMC on a simulated dataset of 100 observations. The vertical lines indicate the true parameter values that were used to generate the data.

Bayesian solutions. The (1, 99%) credible intervals are $[0.910, 1.034]$ for β, and $[0.231, 0.290]$ for σ. The posterior standard deviation of β is 0.026, compared to a standard error of 0.025 from OLS regression. The difference is due to the MCMC estimates being small-sample estimates that do not rely on asymptotic approximations, unlike standard errors. After all, the very definition of the posterior distribution implies that the estimates are conditioned on the observed data only, not an imaginary infinitely large dataset. This allows for exact inference, which may be quite different from the asymptotic inference of classical methods, especially in smaller datasets.

26.3 Missing data

To make the inference problem more interesting, suppose that some of the observations in y are missing at random (I postpone the problem of non-randomly missing data until the next section). The problem of missing data is widespread in corporate finance, even for key variables such as investment and leverage. For example, for the 392,469 firm-year observations in Compustat between 1950 and 2010, capital expenditure is missing for 13.9% of the observations. Debt issuance has been collected since 1971 but 14.1% of the 348,228 firm-year observations are missing, whereas market leverage (book debt divided by book debt plus market value of equity) is missing 22.4% of the time. For R&D expenditures the missing rate is over 50%. Even a canonical scaling variable such as total assets is missing around 5% of the time. As I will illustrate below, MCMC provides a convenient way of dealing with the missing data problem.

With missing data one loses the Bayesian analytical solution to the posterior distribution. However, the sampling algorithm from the previous section needs only one, relatively minor, modification, based on the important concept of data augmentation [94], in order to deal with this issue. Think of the missing observations, denoted by y^*, as parameters to be estimated along with the regression parameters. In other words, we augment the parameter vector with the latent y^*, and sample from the joint posterior distribution $p\left(\beta, \sigma^2, y^* | X, y\right)$.

The key to sampling from this augmented posterior distribution is the Hammersley–Clifford theorem. For our purposes, this theorem implies that the complete set of conditional distributions $p\left(\beta, \sigma^2 | y^*, X, y\right)$ and $p\left(y^* | \beta, \sigma^2, X, y\right)$ completely characterizes the joint distribution. Algorithm 2 shows that, unlike the joint distribution, the complete conditionals are very straightforward to sample from: $p\left(y^* | \beta, \sigma^2, X, y\right)$ is a Normal distribution (and each missing observation can be sampled independently since the error terms are iid), and $p\left(\beta,\sigma^2|y^*,X,y\right)$ is simply Algorithm 1, a Bayesian regression treating the missing y^* as observed data. This gives a first taste of the modularity of the MCMC approach: we go from a standard regression model to a regression with missing data by adding an extra step to the algorithm. Note again that I suppress the conditioning on the observed data in the algorithm.

Algorithm 2 Missing data

1. Draw the missing y_i^* for all i with missing y_i, treating β and σ^2 as known:

$$y_i^*|\beta,\sigma^2 \sim \mathcal{N}\left(X'\beta,\sigma^2\right)$$

2. Draw β,σ^2 from a Bayesian regression with Normal-IG priors, treating the y^*as observed data. The posterior distributions are:

$$\sigma^2|y^* \sim \mathcal{IG}(a+N,b+S)$$
$$\beta|\sigma^2,y^* \sim \mathcal{N}\left(m,\sigma^2\cdot(X'X+A)^{-1}\right)$$

3. Go back to step 1, repeat.

Denoting by $\left\{\sigma^2\right\}^{(g)}$ the draw of σ^2 in cycle g of the MCMC algorithm, the MCMC algorithm thus starts from initial values $\{\beta\}^{(0)}$ and $\left\{\sigma^2\right\}^{(0)}$, and cycles between drawing y^*, σ^2 and β, conditioning on the latest draw of the other parameters:

$$\{\beta\}^{(0)},\left\{\sigma^2\right\}^{(0)} \rightarrow \left\{y^*\right\}^{(1)} \rightarrow \left\{\sigma^2\right\}^{(1)} \rightarrow \{\beta\}^{(1)} \rightarrow \left\{y^*\right\}^{(2)} \rightarrow \ldots$$

The resulting sequence of draws is a Markov chain with the attractive property that, under mild conditions, it converges to a stationary distribution that is exactly the augmented joint posterior distribution. This is the essence of MCMC. Figure 26.2 shows the first 50 draws of Algorithm 2, using the same regression model as the previous section (with true $\beta = 1$ and $\sigma^2 = 0.25$), but randomly dropping half of the observations of y. The convergence to the stationary distribution is most noticeable for σ^2, and quite rapid for this particular model. The period of convergence is called the 'burn-in' period and should be dropped before calculating parameter estimates and other properties of the posterior distribution.

For many problems the likelihood function is not globally concave and has multiple local maxima. For such problems the chain needs to run for a larger number of cycles in order to fully explore the posterior distribution. As a general rule, the MCMC algorithm is more hands-off than Maximum Likelihood, which requires a great deal of manual work by the researcher to make sure that a global optimum is reached, for example through the use of different starting values, applying a variety of maximization algorithms, or simulated annealing routines. On rare occasions, it is possible even for the MCMC chain to get 'stuck' in a local maximum, so it is still good practice to try different starting values. This is also helpful in determining when the chain has converged (Gelman and Rubin [32] develop a convergence test based on the within and between variance of multiple MCMC chains), and does not waste much computing time because the post-convergence draws of the different chains can be combined in order to obtain a better approximation to the posterior distribution.

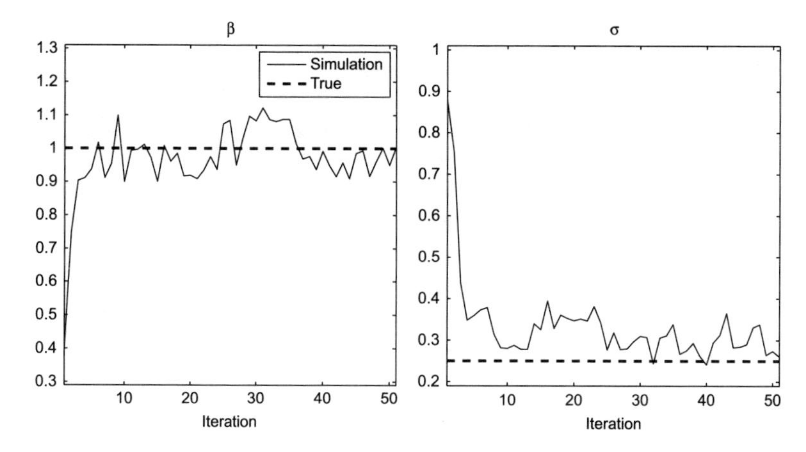

Figure 26.2 Convergence of the MCMC chain. Plot of the first 50 iterations of the Markov chain from the estimation of the missing data regression model in Algorithm 2, estimated on a simulated dataset of 100 observations, of which 50 are dropped at random. The dashed horizontal lines indicate the true parameter values that were used to generate the data.

Table 26.1 shows parameter estimates from two OLS regression approaches to the missing data problem, as well as MCMC estimates. The first OLS results drop the observations for which y is unobserved altogether. The second set of OLS estimates fill in the unobserved y^* using the point estimates of β from the dropped observations. In other words, $y_i^* = x_i'\hat{\beta}$. Unlike other common fill-in schemes such as using the sample average, this results in unbiased estimates of β.

The key issue here is one of statistical efficiency: dropping the unobserved data altogether ignores the information in X that is contained in the dropped observations, while filling in the missing data with point estimates understates standard errors by ignoring the prediction variance in the filled-in data. The latter problem is evident from the fact that the standard errors, as well as the estimates of the residual standard deviations of the filled-in OLS regressions, are considerably lower compared to the OLS regressions that drop the data with missing observations. The MCMC algorithm solves both these problems by using all observations on X while accounting for prediction variance by filling in different values for y^* every time we take a draw of β and σ^2. For comparison, the first MCMC column shows the MCMC version of the regression dropping the observations with missing ys. The posterior standard deviations are larger than the OLS standard errors because they are small-sample rather than asymptotic estimates. The last column shows the results from Algorithm 2, accounting for the information in the dropped X while also accounting for the uncertainty of the missing y^*.

It is evident from Table 26.1 that integrating out the missing y^* is not particularly helpful in this example. However, Table 26.2 shows that the benefits are more substantial when allowing for missing observations on the *explanatory* variables in a multiple regression. Simulating and integrating out the missing variables leads to parameter estimates that are generally closer to the true parameters than the other methods, even for a relatively low correlation between the explanatory variables of 0.15. Moreover, since the full information about the dependent and non-missing explanatory variables is exploited, the estimates have lower posterior standard deviations, i.e. they are more precise, compared to the MCMC estimates that drop the observations with some missing data altogether. The algorithm for tackling this problem is very similar to Algorithm 2, essentially simulating the missing explanatory variables from a regression of the variable with missing data on the other variables. An example of a corporate finance application of such a problem is

Table 26.1 Missing dependent variable regressions. Estimates of a missing data regression model based on a simulated dataset of 100 observations, randomly dropping 50% of the dependent variable data (but not the regressor). The true coefficients are shown in the first column (there is no intercept). The OLS columns estimate the model by dropping the missing observations altogether (the column labelled 'Drop'), and filling in the missing data using fitted values from the 'Drop' regression (the column labelled 'Filled'). The MCMC estimates in the 'Drop' column uses the standard Bayesian regression of Algorithm 1, dropping the missing observations. The final column uses Algorithm 2, which simulates and integrates out the missing observations. The MCMC estimates use 1 000 burn-in cycles followed by 25 000 cycles to sample the posterior distribution. Standard errors for the OLS estimates and posterior standard deviations for the MCMC estimates are in brackets.

	True	OLS		MCMC	
		Drop	Filled	Drop	Alg 2
β	1	1.012	1.012	1.013	1.011
		(0.047)	(0.025)	(0.058)	(0.057)
σ	0.25	0.239	0.171	0.298	0.297
		–	–	(0.029)	(0.032)

in Frank and Goyal [29], who impute missing factors in leverage regressions using an MCMC algorithm.

Other, more complex, cases also promise better results, for example if y follows a time-series process but has missing data gaps in the series. An illustration of this kind is found in Korteweg [56], who uses panel data for corporate bond and equity values to estimate the net benefits to leverage, where a non-trivial fraction of the corporate bond values are unobserved. Another avenue for improving performance is to sample the empirical distribution of the residuals, instead of imposing the Normal distribution, to obtain results that are more robust to distributional assumptions. For further information on using Bayesian methods for missing data, see Rubin [88] and Graham and Hirano [36].

26.4 Limited dependent variable and selection models

In the previous section I assumed that data is missing at random. If data is instead missing for a reason, it becomes necessary to specify the model by which observations are selected in order to obtain estimates of the parameters of interest. More generally, selection models are useful for addressing many questions in corporate finance such as the effect of underwriter choice on bond issue yields, the diversification discount due to conglomeration choice, and the impact of bank versus pubic debt financing on cost of capital (see Li and Prabhala [68] for a comprehensive overview and references).

I start this section with a simple probit example. The probit model serves as a basis to developing an estimation algorithm for the Heckman selection model. Since both probit and Heckman have canned Maximum Likelihood-based modules in popular statistics packages such as Stata, this may not sound very exciting. However, in Section 26.4.3 and beyond, I introduce extensions to

Table 26.2 Missing explanatory variable regressions. Estimates of a missing data regression model based on a simulated dataset of 100 observations with two explanatory variables, $y = x_1\beta_1 + x_2\beta_2 + \varepsilon$, and randomly dropping 50% of the observations on x_2. The true coefficients are shown in the first column. The explanatory variables have a correlation coefficient of 0.15. The first column estimates the model by dropping the missing observations altogether, and the second column uses the Griliches [37] GLS method to fill in the missing data, using a regression of the variable with missing data on the other explanatory variable. The MCMC estimates in the 'Drop' column use the standard Bayesian regression of Algorithm 1, dropping the missing observations, whereas the final column uses a version of Algorithm 2 to simulate and integrate out the missing observations. The MCMC estimates use 1 000 burn-in cycles followed by 25 000 cycles to sample the posterior distribution. Standard errors for the OLS estimates and posterior standard deviations for the MCMC estimates are in brackets.

	True	OLS	GLS	MCMC	
		Drop	Filled	Drop	Alg 2
β_1	0.5	0.493	0.416	0.493	0.476
		(0.038)	(0.002)	(0.042)	(0.039)
β_2	−0.5	−0.425	−0.426	−0.424	−0.474
		(0.044)	(0.003)	(0.049)	(0.039)
σ	0.25	0.251	0.262	0.280	0.282
		–	–	(0.020)	(0.025)

dynamic selection, matching models and switching regressions, that are very difficult to estimate with Maximum Likelihood, but are quite feasible with MCMC, both from an ease of implementation as well as from a computational perspective.

26.4.1 Probit model

In the standard probit model, y_i has two possible outcomes, zero and one. The probability of observing $y_i = 1$ is:

$$pr(y_i = 1) = \Phi(x_i\beta)$$

where $\Phi(\cdot)$ is the standard Normal cdf. Observations are assumed to be iid. Probit models have been used in corporate finance to estimate, for example, the probability of issuing debt or equity [49], takeovers [4], bankruptcy [84] and the firing of CEOs [53].

The estimation goal is to find the posterior distribution of β given y and X. It will prove convenient to rewrite the model in the following way:

$$y_i = \mathbb{I}_{\{w_i \geq 0\}}$$
$$w_i = x_i\beta + \eta_i$$

with $\eta_i \sim \mathcal{N}(0,1)$, iid. The auxiliary selection variable, w, is unobserved. If $w \geq 0$ then y equals one, otherwise y equals zero. Augmenting the posterior distribution with w, MCMC Algorithm 3 (from Albert and Chib [3]) shows how to sample from the joint posterior distribution of β and w, conditional on the observed data. The algorithm cycles between drawing from the complete conditionals $w|\beta$ and $\beta|w$, which by the Hammersley–Clifford theorem fully characterize the joint posterior distribution.

Algorithm 3 Probit model

1. Draw $w_i|\beta$ for all i:

   (a) for $y_i = 1$:

   $$w_i|\beta \sim \mathcal{LTN}\,(x_i\beta, 1)$$

   (b) for $y_i = 0$:

   $$w_i|\beta \sim \mathcal{UTN}\,(x_i\beta, 1)$$

2. Draw $\beta|w$ from a Bayesian regression of w on X, with Normal priors on β and known variance equal to one:

   $$\beta|w \sim \mathcal{N}\left((X'X+A)^{-1}(X'w+A\mu),(X'X+A)^{-1}\right)$$

3. Go back to step 1, repeat.

In step 1, when y equals one, w must be greater than zero, and the distribution of w is therefore truncated from below at zero. I denote this lower-truncated Normal distribution by $\mathcal{LTN}$. Similarly, $\mathcal{UTN}$ is the upper truncated Normal distribution, again truncated at zero. Step 2 draws from the posterior distribution of coefficients in a Bayesian regression, as in Algorithm 1 but fixing the variance of the error term to unity.

An advantage of MCMC for the probit model is in the calculation of nonlinear functions of parameters. Researchers are often interested in the partial effects, $\partial pr(y=1|x)/\partial x = \phi(x'\beta)\beta$, which are highly nonlinear functions of β. With the standard Maximum Likelihood approach the asymptotic distribution of the partial effects has to be approximated using the Delta method. With MCMC we obtain a sample from the exact posterior distribution without relying on asymptotics or approximations by simply calculating the partial effect for each draw of β (discarding the burn-in draws). We can then compute means, variances, intervals etc.

Many extensions of Algorithm 3 have been developed, for example to the multivariate probit model with several outcome variables [14], the multinomial probit that allows more than two discrete outcomes [73], and the multinomial-t and multinomial logit models [15]. In the next section I discuss another extension of the probit model, the classic Heckman selection model.

26.4.2 Heckman selection model

In the Heckman (also known as Tobit type 2) selection model, y_i is no longer a binary variable but can take on continuous outcomes:

$$y_i = x_i\beta + \varepsilon_i$$

$$w_i = z_i\gamma + \eta_i$$

The outcome variable y_i is observed only when $w_i \geq 0$. The error terms are distributed iid jointly Normal with zero means, $var(\varepsilon_i) = \sigma^2$, $var(\eta_i) = 1$, and correlation ρ. The top equation is referred to as the outcome equation, and the bottom equation is the selection equation.

The Heckman selection model can be used to control for endogenously missing data and self-selection by firms. For example, the choice to issue equity may depend on the type of firm. If there is an unobserved component to firm type, then selection cannot be controlled for by covariates in the outcome equation alone. In panel data applications it is common practice to use fixed effects to control for selection, but these do not control for the fact that the reasons for selection can (and often do) change over time. Selection models do allow for that possibility.

To estimate the Heckman model by MCMC, we decompose ε_i into a component that loads on η_i and an orthogonal component ξ_i:

$$
\begin{aligned}
y_i &= x_i\beta + \delta \cdot \eta_i + \sigma_\xi \cdot \xi_i \\
w_i &= z_i\gamma + \eta_i
\end{aligned}
$$

where $\delta = \sigma \cdot \rho$ is the covariance between ε and η, and $\sigma_\xi = \sigma \cdot \sqrt{1-\rho^2}$ is the conditional standard deviation of $\varepsilon|\eta$. From this representation it follows immediately that the selection equation cannot be ignored if ρ (and hence δ) is not equal to zero. Consider the expected value of y_i if it is observed:

$$
E\left[y_i | w_i \geq 0, data\right] = x_i\beta + \delta E\left[\eta_i | \eta_i \geq -z_i\gamma\right]
$$

Ignoring the selection equation (dropping $\delta \cdot \eta$ in the observation equation) thus introduces an omitted variables bias if $\rho \neq 0$ [43]. The omitted variable, $E[\eta_i|\eta_i \geq -z_i\gamma] = \phi(-z_i\gamma)/\Phi(-z_i\gamma)$, is called the inverse Mills ratio.

MCMC Algorithm 4 (based on Li [65]) shows how to draw from the posterior distribution of the parameters augmented with the latent selection variable, w, and the missing observations, y^*. The algorithm essentially combines the randomly missing data routine with the probit estimation.

From the sampled parameters it is straightforward to use nonlinear transformations to recover the posterior distribution of ρ and σ, as well as treatment effects, analogous to the calculation of partial effects in the probit model.

The typical approach to estimating Heckman models is the two-step estimator [43]: In the first stage we estimate the selection equation using a probit model, and in the second stage we plug the fitted inverse Mills ratio from the first stage into the outcome equation to correct for the omitted variable bias. This estimator generates inconsistent estimates for the covariance matrix in the second stage, and correct standard errors have to be computed from an asymptotic approximation or through a bootstrap. Full information estimators (MCMC and the Maximum Likelihood estimator) exhibit better statistical properties but are often criticized for their sensitivity to the Normality assumption. Robust and nonparametric estimators have been proposed to deal with this issue (e.g. [45, 71, 72]), but tend to be limited in the types of models they can estimate. For example, Manski's [72] model is limited to two regressors. In contrast, the MCMC algorithm is more flexible. Van Hasselt [97] extends the algorithm to accomodate a mixture of Normals in the error terms without losing the generality of the model. Mixtures of Normals are able to generate many shapes of distributions such as skewness, kurtosis and multimodality, and the algorithm lets the data tell you what the shape of the error distribution is. Van Hasselt also shows how to estimate the number of mixture components, something that is very difficult to do with frequentist methods.[24]

[24] A common concern with the standard selection model is that it is identified from distributional assumptions only, unless one employs an instrument that exogenously changes the probability of being selected but is

Algorithm 4 Heckman selection model

1. Draw $w_i, y_i^* | \beta, \gamma, \delta, \sigma_\xi^2$

   (a) for y_i observed:

$$w_i | \beta, \gamma, \delta, \sigma_\xi^2 \sim \mathcal{LTN}\left(z_i\gamma + \rho \cdot \left[\frac{y_i - x_i\beta}{\sqrt{\delta^2 + \sigma_\xi^2}} \right], 1 - \rho^2 \right)$$

$$\text{where} \rho = \delta / \sqrt{\delta^2 + \sigma_\xi^2}$$

   (b) for y_i not observed:

$$w_i | \beta, \gamma, \delta, \sigma_\xi^2 \sim \mathcal{UTN}\left(z_i\gamma, 1\right)$$

$$y_i^* | w_i, \beta, \gamma, \delta, \sigma_\xi^2 \sim \mathcal{N}\left(x_i\beta + \delta\left[w_i - z_i\gamma\right], \sigma_\xi^2 \right)$$

2. Draw $\beta, \gamma | w, y^*, \delta, \sigma_\xi^2$ from a Bayesian Seemingly Unrelated Regression ([99]) of $[y; w]$ on $[X; Z]$, with Normal priors on β and γ and known covariance matrix Ω:

$$\beta, \gamma | w \sim \mathcal{N}\left(\left(C'\Omega^{-1}C + A\right)^{-1} \left(C'\Omega^{-1}\left[y; w\right] + A\mu\right), \left(C'\Omega^{-1}C + A\right)^{-1} \right)$$

   where

$$\Omega = \begin{bmatrix} \sigma_\xi^2 + \delta^2 & \delta \\ \delta & 1 \end{bmatrix} \otimes I_N \quad \text{and} \quad C = \begin{bmatrix} X & 0 \\ 0 & Z \end{bmatrix}$$

3. Draw $\delta, \sigma_\xi^2 | \beta, \gamma, w, y^*$ from a Bayesian regression of $y - X\beta$ on $w - Z\gamma$, with Normal-IG priors (see Algorithm 1).
4. Go back to step 1, repeat.

The Heckman selection algorithm outlined above can be adapted to estimate many related models. For example, Chib [13] develops an MCMC algorithm to estimate the Tobit censoring model where y takes on only non-negative values. Tobit censoring is relevant in corporate finance applications because many variables are naturally non-negative, such as gross equity issuance, cash balances and investment, and ignoring any censoring (for example from irreversible investment bounding capital expenditures at zero) may mask certain causal relationships of interest. Double--censoring can also be accommodated, to account for the fact that leverage (measured gross of cash) is bounded between zero and one, which is important for estimating speed-of-adjustment models [10, 51]. Similarly, the outcome variable may be qualitative (exactly zero or one) to model binary outcomes such as default versus no default, or merger versus no merger.

orthogonal to ε, the shock in the observation equation [45]. Relaxing the Normality assumption may loosen the distributional assumptions somewhat, but should not be seen as a substitute for an instrument.

The standard selection model has only two possible outcomes from selection: a data point is either observed or not observed. In many corporate finance applications there are multiple possible outcomes. For example, a company can choose not to raise debt financing, choose to issue public bonds or to obtain a bank loan. Obrizan [81] shows how to estimate the selection model with a multinomial probit in the selection equation by MCMC. Classical estimation of this model is possible as well, but requires numerical integration of the likelihood using quadrature or simulation. I will return to this issue below. In the remainder of this section, I introduce three other extensions to the Heckman model that have many potential applications in corporate finance but are very difficult to estimate with traditional methods.

26.4.3 Dynamic selection

The standard Heckman model assumes that the error terms in the selection equation are independent across units of observation. Under this assumption the unobserved data points are only informative about γ, the parameters of the selection equation. They carry no further information about the parameters of the outcome equation, β. In many corporate finance applications this is not the case. For example, a firm may be more inclined to issue equity because a peer firm is doing the same, or because some common unobserved factor induces both firms to issue. For the purpose of illustration, consider the case of two firms. The outcome of the first company is unobserved and the outcome of the second is observed. The expected value of the outcome of the second firm is

$$E\left[y_2|w_2 \geq 0, w_1 < 0, data\right] = x_2\beta + \delta E\left[\eta_2|\eta_2 \geq -z_2\gamma, \eta_1 < -z_1\gamma\right]$$

Unlike the standard Heckman model, if the ηs are correlated then firm 1 carries information about the conditional mean of y_2 despite firm 1's outcome being unobserved. In other words, the fact that firm 1 is unobserved matters for the conditional mean of firm 2. With two firms the expectation is a two-dimensional integral. With thousands of firms the integral becomes very high-dimensional, and with the current state of computing power it is too time-consuming to evaluate in a Maximum Likelihood estimation, or even to compute the inverse Mills ratio in the two-step procedure.[25]

A similar estimation issue arises when the outcome variable follows an autoregressive distributed lag (ADL) model. For example, consider the following ADL process for an individual firm:

$$y_t = \lambda y_{t-1} + x_t\beta + \varepsilon_t$$

Assume that the error terms are temporally independent (but ε and η are still contemporanesouly correlated so that $\delta \neq 0$). In a two-period setting, if the outcome at time 1 is unobserved but observed at time 2, the conditional mean of the outcome at time 2 is

$$E\left[y_2|w_2 \geq 0, w_1 < 0, data\right] = x_2\beta + \lambda E\left[y_1|w_2 \geq 0, w_1 < 0, data\right] + \delta E\left[\eta_2|\eta_2 \geq -z_2\gamma\right]$$

With $\lambda = 0$ this works out to the standard Heckman correction. With non-zero λ we get a similar integration issue as in the cross-sectional case, because the value of y_1 depends on the realized value of y_2, due to the ADL process. The resulting model is thus a dynamic generalization of the

[25] The usual way to numerically integrate out the latent variables in Maximum Likelihood is through quadrature methods, which are effective only when the dimension of the integral is small, preferably less than five [96]. The alternative, Simulated Maximum Likelihood (SML), is less computationally efficient than MCMC. I will discuss this in more detail in Section 26.5.

standard selection model. Autocorrelation in the error term results in similar estimation problems, even without the lagged dependent variable.

Korteweg and Sørensen [59] tackle the estimation problem of the dynamic selection model using an MCMC algorithm, and apply it to estimate the risk and return to venture capital funded entrepreneural firms. The outcome variable is the natural logarithm of a start-up's market value, which follows a random walk (the ADL process above with $\lambda = 1$). However, the valuation is only observed when the company obtains a new round of funding. The probability of funding depends strongly on the firm's valuation, and this gives rise to the selection problem. Korteweg and Sørensen develop an MCMC algorithm to estimate the model in a computationally efficient way.[26] In a follow-up paper, Korteweg and Sørensen [60] apply the model to estimating loan-to-value ratios for single-family homes, as well as sales and foreclosure behaviour and home price indices. They extend the model to include multiple selection equations, in order to capture the probability of a regular sale versus a foreclosure sale.

The dynamic selection model is very versatile, and can be applied to virtually any linear asset pricing model with endogenous trading. The model is also applicable to a variety of corporate finance problems, since many of the standard variables follow ADL processes, and, in addition, autocorrelation in the error term is a common occurrence. Moreover, the estimation problem is not unique to the selection model, but generalizes to the Tobit model and other censoring problems. For example, investment follows an ADL but is censored at zero (if irreversible), leading to similar integration issues as the dynamic selection model. Other examples include cash balances, default or M&A intensity, and leverage, where the latter is subject to double censoring as mentioned above.

26.4.4 Switching regressions

In a switching regression, y is always observed, but the parameters of the outcome equation depend on whether w is above or below zero:

$$y_i = \begin{cases} y_{i0} = x_{i0}\beta_0 + \varepsilon_{i0} & \text{if} \quad w_i > 0 \\ y_{i1} = x_{i1}\beta_1 + \varepsilon_{i1} & \text{if} \quad w_i \leq 0 \end{cases}$$

For example, Scruggs [89] considers the market reaction to calls of convertible bonds, which can be 'naked' (without protection) or with the assistance of an underwriter that guarantees conversion at the end of the call period. We observe the announcement reaction, y, under either call method but the choice of method is endogenous through the correlations between η, ε_0 and ε_1. Unlike standard average announcement effect studies, the β_0 and β_1 parameters in the switching regressions reflect announcement effects *conditional* on the endogenously chosen call method, as advocated by Acharya [1], Eckbo, Maksimovic, and Williams [25], Maddala [70], and Prabhala [83]. In other words, the parameters capture the counterfactual of what would have happened if a firm had chosen the unobserved alternative. As such, we can use the model to analyse treatment effects. As in the Heckman model, without an instrument the model is identified from parametric assumptions alone, and imposing an exclusion restriction is helpful to achieve nonparametric identification. Scruggs assumes that the error terms have a fat-tailed multivariate t-distribution and estimates the model using an MCMC algorithm.

Switching models have many potential applications in corporate finance. For example firm's cash and investment policies may be different in growth versus recession regimes, or hiring and firing intensities of CEOs may vary depending on the state of the firm.

[26] A detailed description of the algorithm as well as Matlab and C++ code to implement it can be found on my personal webpage.

It is possible to estimate a switching regression using classical methods [43, 61], but this is cumbersome at best. The MCMC approach is flexible and easy to extend to more complex models. For example, Li and McNally [67] use MCMC to estimate a switching regression with multiple outcome equations within each regime. They apply their model to the choice of share repurchase method, modelling the percentage of shares bought, the announcement effects and the tender premium (if the repurchase is by tender offer). This model can in turn be extended to outcome equations that are not all continuous, but instead may be composed of a mix of continuous, truncated and discrete outcomes. Another logical extension is to have more than two regimes since in many cases corporate managers face a decision between multiple options (similarly to the multinomial selection model). Such models become increasingly intractable with classical methods, but are quite manageable using MCMC.

26.4.5 Matching models

A prevalent form of matching in corporate finance is the endogenous two-sided matching between two entities. For example, firms match with banks for their financing needs, and CEOs match with the firms they run. Sørensen [91] develops a model in which venture capitalists (VCs) match with entrepreneurs. He asks whether the better performance of experienced VCs is driven by sorting (more experienced VCs pick better firms) or influence (more experienced VCs add more value). Since some of the dimensions along which sorting occurs are unobserved, the resulting endogeneity problem makes identification more tricky. The economics of the problem makes finding an instrument very difficult, so Sørensen develops a structural model that exploits the fact that investors' decision to invest depends on the other agents in the market, whereas the outcome of the investment does not. This provides the exogenous variation needed for identification.

The resulting model is prohibitively time-consuming to estimate by Maximum Likelihood because investment decisions interact. If one investor invests in a start-up, then other investors cannot. This implies that the error terms in the model are not independent and have to be integrated jointly in order to compute the likelihood function. Given that there are thousands of investments, such an extremely high-dimensional integral is computationally infeasible at present. Sørensen develops a feasible MCMC procedure to estimate the model, which is computationally much quicker than Maximum Likelihood.

Later studies [9, 82] use a similar MCMC methodology to study the matching of targets and acquirers in M&A, and the matching between banks and firms.

The next section on panel data dives deeper into the benefits of MCMC methods for formulating feasible estimators that perform high-dimensional integration in a computationally efficient way.

26.5 Panel data

In corporate finance one often observes the actions of a set of agents (companies, CEOs etc.) over time. Such panel datasets are a rich source of identification, but also come with certain empirical challenges. The standard issues in classical estimation of panel data models are the assumptions regarding asymptotics (whether we assume that N or T approaches infinity) and the related incidental parameters problem [78],[27] the initial values problem [44], and the Hurwicz asymptotic bias for

[27] The incidental parameters problem in the panel data context states that individual fixed effects are not estimated consistently for fixed T, which results in inconsistent estimates of the parameters of interest. In some cases a transformation (such as first-differencing to cancel out the fixed effects) can resolve the problem, but these are rarely found outside of the linear and logit models.

ADL type models (also known as Nickell bias or, when applied to predictive regressions, Stambaugh bias). The Bayesian approach avoids many of these pitfalls. For example, asymptotic assumptions are unnecessary in the Bayesian paradigm since one conditions on the observed data only, and the initial values problem is easier to handle since we can treat it like a missing data problem. Moreover, MCMC methods allow for the estimation of a wider variety of panel data models, as I will discuss below.

26.5.1 Random effects probit

Consider the panel data extension of the probit model with random effects (RE):

$$\begin{aligned} y_{it} &= \mathbb{I}_{\{w_{it} \geq 0\}} \\ w_{it} &= x_{it}\beta + \alpha_i + \eta_{it} \end{aligned}$$

For example, $pr(y_{it} = 1)$ could represent the probability that firm i goes bankrupt at time t. The unit-specific intercept, α_i, is assumed to be randomly generated from a Normal distribution with mean zero and variance τ^2, and is uncorrelated with $\eta_{it} \sim \mathcal{N}\left(0, 1 - \tau^2\right)$. This 'random effect' controls for time-invariant, unobserved heterogeneity across units.[28] The parameters, β, are therefore identified from the time-series variation within firms.[29]

It is useful to think of the structure of the panel probit model as a hierarchy, where each level builds upon the previous levels:

$$\begin{aligned} \tau^2 &\sim \mathcal{IG}\left(a, b\right) \\ \alpha_i|\tau^2 &\sim \mathcal{N}\left(0, I_N \cdot \left(1 - \tau^2\right)/\tau^2\right) \\ w_{it}|\alpha_i, \tau^2 &\sim \mathcal{N}\left(x_{it}\beta + \alpha_i, 1 - \tau^2\right) \end{aligned}$$

This hierarchy can be extended to as many levels as desired (e.g. industry-company-executive-year data). Hierarchical models [69] are useful in many corporate finance settings, and MCMC methods are very well suited for estimating these models, due to the complete conditional structure of the algorithm. By breaking up the problem into simple regression steps based on its hierarchical structure, one can estimate models for which even the act of writing down the likelihood function becomes an arduous task. This allows one to compute correct standard errors and perform hypothesis testing without resorting to standard shortcuts such as two-stage estimators.

Algorithm 5 shows how to extend the probit Algorithm 3 to estimate the panel probit model. Steps 1 and 2 follow straight from Algorithm 3. Step 3 jointly draws a set of αs by regressing $w_{it} - x_{it}\beta$ on a set of dummies, one for each firm (note that the prior means are zero). Step 4 estimates the variance of the αs, again in regression form. Note that the Algorithm follows the hierarchy of the model.

28 Alternatively, one can think of the random effects as a form of error clustering. Note that in Stata the 'cluster' command gives larger standard errors than the RE estimates, because Stata only considers the residual idiosyncratic error after removing the group error component.

29 The random effects estimator is different from the fixed effects estimator, which is typically estimated using dummy variables. The random effects estimator dominates the fixed effects estimator in mean-squared error [26, 92], whereas the benefit of the fixed effects estimator is that it allows the unit-specific means to be correlated with the other explanatory variables. Mundlak [76] develops a correlated random effects model by specifying $\alpha_i = \bar{x}_i\gamma + u_i$, where $\bar{x}_i$ is the time-series average of x_{it}, and u_i is an othogonal error term. Chamberlain [8] extends the approach to a more flexible specification of α as a function of x.

Algorithm 5 Panel random effects probit

1. Draw $w_{it}|\beta, \alpha, \tau^2$ for all i and t:

   (a) for $y_{it} = 1$:

$$w_{it}|\beta, \alpha \sim \mathcal{LTN}(x_{it}\beta + \alpha_i, 1)$$

   (b) for $y_{it} = 0$:

$$w_{it}|\beta, \alpha \sim \mathcal{UTN}(x_{it}\beta + \alpha_i, 1)$$

2. Draw $\beta|w, \alpha, \tau^2$ from a Bayesian regression of $w - \alpha$ on x_{it}, with Normal priors on β and known variance $1 - \tau^2$:

$$\beta|w \sim \mathcal{N}\left((X'X + A)^{-1}(X'(w - \alpha) + A\mu), \left(1 - \tau^2\right) \cdot (X'X + A)^{-1}\right)$$

   where $w - \alpha$is the stacked vector $\{w_{it} - \alpha_i\}$ across i and t, corresponding to the matrix X.
3. Draw $\alpha|\beta, w, \tau^2$ from a Bayesian regression of $w - X\beta$ on a $NTxN$ matrix of firm dummies D, using a $\mathcal{N}\left(0, I_N \cdot \left(1 - \tau^2\right)/\tau^2\right)$ prior:

$$\alpha|\beta, w, \tau^2 \sim \mathcal{N}\left(\left(D'D + I_N \cdot \left(1 - \tau^2\right)/\tau^2\right)^{-1} \cdot D'(w - X\beta), \left(1 - \tau^2\right) \cdot \left(D'D + I_N \cdot \left(1 - \tau^2\right)/\tau^2\right)^{-1}\right)$$

4. Draw $\tau^2|\alpha, \beta, w$, using an $\mathcal{IG}(a, b)$ prior:

$$\tau^2|\alpha, \beta, w \sim \mathcal{IG}\left(a + N, b + \sum_{i=1}^{N} \alpha_i^2\right)$$

5. Go back to step 1, repeat.

Besides the relative ease of programming,[30] there is also a computational advantage to MCMC in models with high-dimensional integrals over many latent variables. To appreciate why nonlinear panel data models such as the panel RE probit are difficult to estimate by Maximum Likelihood, consider the likelihood function:

$$L = \prod_{i=1}^{N} \int_{-\infty}^{\infty} \left[\prod_{t=1}^{T} \Phi\left((2y_{it} - 1) \cdot \frac{x_{it}\beta + \alpha_i}{\sqrt{1 - \tau^2}}\right)\right] \phi\left(\frac{\alpha_i}{\tau}\right) d\alpha_i$$

where, as before, $\phi(\cdot)$ is the pdf of the standard Normal distribution, and $\Phi(\cdot)$ is the cdf. The term in square brackets is the standard probit likelihood, conditional on α_i. Because of the nonlinearity of $\Phi(\cdot)$, the expectation over α cannot be solved analytically, so numerical methods are required to evaluate the integral. In order to calculate the likelihood for one set of parameters, we need to eval-

[30] The core (steps 1 through 5) of the panel probit routine of Algorithm 5 requires only about 20 lines of code in Matlab.

uate N unidimensional integrals (in addition to the integrals required to evaluate the standard Normal cdf in the inner term of the likelihood). This can be done quite efficiently by Gauss–Hermite quadrature (this is how Stata estimates this model). However, even with small changes to the model the integral becomes of high dimension, at which point quadrature quickly loses its effectiveness (even for as few as five dimensions [96]). For example, allowing for auto-correlation in the η_{it}s, the inner term, $\prod_{t=1}^{T} \Phi\left((2y_{it} - 1) \cdot \frac{x_{it}\beta + \alpha_i}{\sqrt{1-\tau^2}}\right)$, becomes a T-dimensional integral with no analytical expression.[31] Allowing instead for cross-correlation (but no auto-correlation) in the error terms, rewrite the likelihood as

$$\prod_{t=1}^{T} \int f\left(y_{1t} \ldots y_{Nt}\right) g\left(\alpha_1 \ldots \alpha_N\right) d(\alpha_1 \ldots \alpha_N)$$

where $f(\cdot)$ is the joint likelihood of $y_{1t} \ldots y_{Nt}$ conditional on the αs, and $g(\cdot)$ is the joint probability density of the REs. Even conditional on the αs, the N-dimensional joint likelihood $f(\cdot)$ has no analytical solution, and on top of that one needs to integrate over the distribution of the REs. The classical alternative to quadrature, Simulated Maximum Likelihood, thus requires a large simulation exercise to evaluate the likelihood, covering the entire joint distribution of all latent variables. This simulation has to be repeated for every guess of the parameters vector. MCMC on the other hand switches between drawing new parameters and drawing the latent states. The integration over the distribution of the latent states only needs to be done once, after the simulation is finished. This speeds up estimation considerably. For example, Jeliazkov and Lee [53] extend MCMC Algorithm 5 and estimate a random effects panel probit model in which the η_{it} are serially correlated, and apply it to women's labour force participation. The estimation problem extends to many related models for which the likelihood is non-linear in the parameters, and Algorithm 5 can be adapted to estimate these models as well. For example, Bruno [6] develops an algorithm for a panel RE Tobit model. Such models have wide applicability in corporate finance for the same reasons as mentioned above: many standard variables, such as leverage, investment or the decision to fire a CEO, are of a binary or truncated nature, and fixed or random effects help control for time-invariant unobserved heterogeneity in a panel data setting. Another useful extension is to deal with unbalanced panel data by combining Algorithm 5 with the randomly missing data Algorithm 2.

26.5.2 Panel data with selection/attrition

In this section I combine the Heckman model from Section 26.4.2 with the panel RE model from the previous section. In other words, I allow for non-random missing data in a random effects panel model. This model is useful for controlling for non-random attrition, for example firms disappearing through bankruptcy or merger/acquistion. In spite of the wide range of potential applications, no canned estimators are currently available in popular software packages.[32]

The model is

$$\begin{aligned} y_{it} &= \alpha_i + x_{it}\beta + \delta \cdot \eta_{it} + \sigma_\xi \cdot \xi_{it} \\ w_{it} &= \theta_i + z_{it}\gamma + \eta_{it} \end{aligned}$$

[31] Note that the problem of integration with auto-correlated errors is closely related to the estimation of the dynamic selection model of Section 26.4.3.

[32] Running a panel Heckman in Stata and using the 'cluster' option to allow for random effects does not lead to the same result, as the error clustering is not true Maximum Likelihood and does not allow for random effects in the selection equation.

where δ and σ_ξ are as defined in Section 26.4.2. The random effects are iid, $\alpha_i \sim \mathcal{N}(\mu, \tau^2)$, and $\theta_i \sim \mathcal{N}(\kappa, \omega^2)$. The error terms are iid, $\eta_{it} \sim \mathcal{N}(0, 1-\omega^2)$, and $\xi_{it} \sim \mathcal{N}\left(0, \sigma_\xi^2\right)$, and are uncorrelated with each other and with the random effects. As in the standard Heckman model in Section 26.4.2, selection enters through $\delta \neq 0$. Hausman and Wise [41] were the first to consider this model. Other Maximum Likelihood approaches have been developed by Ridder [85], Nijman and Verbeek [79] and Vella and Verbeek [98]. These models impose strong assumptions and are computationally burdensome as they require the evaluation of multiple integrals. To circumvent the computational burden, two-step estimators are typically employed, but these understate the standard errors of the parameters in the observation equation by not accounting for the estimation error in the inverse Mills ratios. MCMC is useful in these models because it is a computationally more efficient estimation method, and it leads to correct inference, taking into account all sources of estimation error.

Algorithm 6 shows the MCMC procedure, which is essentially the standard Heckman model of Algorithm 4, augmented with the RE components as in Algorithm 5. Dropping Steps 3 through 6 and setting μ, κ, τ and ω to zero collapses the algorithm back to the standard Heckman model. Similarly, forcing $\rho = 0$ collapses the algorithm down to a standard RE panel regression without selection. Although Algorithm 6 is slightly longer than the algorithms shown thus far, it is not much more complicated as it is still essentially a sequence of regressions.

To illustrate the importance of selection effects in corporate finance, I run a RE panel regression of quasi-market leverage (defined as the book value of debt divided by the book value of debt plus the market value of equity) on lagged profitability (operating income divided by the book value of assets), tangibility (net property plant and equipment divided by book assets), market-to-book ratio (market value divided by book value of equity) and the natural logarithm of book assets. Regressions of this nature are quite common in the literature (e.g. [64]), although researchers typically use fixed effects rather than random effects. I use data for a random sample of 1 000 firms from Compustat between 1950 and 2010, for a total of 11 431 firm-years. The first column in Table 26.3 shows the standard GLS estimates of the regression coefficients. To gauge the effect of the priors and the additional distributional assumptions of the MCMC algorithm in this example, the second column of the table reports the MCMC estimates of the same regression model, ignoring the selection issue (i.e. forcing $\rho = 0$). The coefficient estimates are very close, suggesting that MCMC indeed replicates the standard GLS regression when ignoring selection. The random effects are important as they account for 57% of the variance (result not reported in the table).

An important concern is that leverage is missing for 1 572, or 14%, of firm-years. If observations are not missing at random, for example if firms drop out of the sample due to a merger or a bankruptcy, this can result in misleading estimates. Unreported results reveal that the missing firm-years are characterized by higher profitability, higher tangibility and lower market-to-book ratios, suggesting that they are indeed different in observable respects. This is in principle not a problem, unless there is also selection on *unobservable* variables, which would manifest itself as a non-zero correlation between the error terms in the selection and observation equations, ρ. The last column of Table 26.3 shows the estimates from MCMC Algorithm 6. The posterior mean of the correlation ρ is 0.367, the first percentile of the posterior distribution is 0.237 and the 1/10th percentile is 0.070. This finding indicates that ρ is not zero and there is indeed non-random selectivity on unobserved variables. Moreover, the selection correction has a large impact on parameter estimates. The coefficient on profitability drops from −0.186 in the GLS to −0.237 after correcting for selection, a change of about 25%. Moreover, the coefficient on tangibility roughly drops in half, from 0.207 to 0.101. The estimates of market-to-book and firm size are only marginally affected by the selection issue. Note also the well-known result that the estimate of residual variance (and hence standard errors) is biased downwards when ignoring selection [42], hence the larger posterior standard deviations of the coefficients in the selection correction model. For brevity I do not report the coefficients

Algorithm 6 Panel random effects with selection

1. Draw $w_{it}, y^*_{it}|\alpha, \theta, \beta, \gamma, \delta, \sigma_\xi, \mu, \tau^2, \kappa, \omega^2$

   (a) for y_{it} observed:

$$w_{it} \sim \mathcal{LTN}\left(\theta_i + z_{it}\gamma + \rho\sqrt{1-\omega^2} \cdot \left[\frac{y_{it} - \alpha_i - x_{it}\beta}{\sqrt{\delta^2 + \sigma_\xi^2}}\right], (1-\rho^2) \cdot (1-\omega^2)\right)$$

   where $\rho = \delta / \sqrt{\delta^2 + \sigma_\xi^2}$.

   (b) for y_{it} not observed:

$$w_{it} \sim \mathcal{UTN}\left(\theta_i + z_{it}\gamma, 1 - \omega^2\right)$$

$$y^*_{it}|w_{it} \sim \mathcal{N}\left(\alpha_i + x_{it}\beta + \delta\left[w_{it} - \theta_i - z_{it}\gamma\right], \sigma_\xi^2\right)$$

2. Draw $\beta, \gamma|w, y^*, \alpha, \theta, \delta, \sigma_\xi, \mu, \tau^2, \kappa, \omega^2$ from a Bayesian Seemingly Unrelated Regression of $[y - \alpha; w - \theta]$ on $[X; Z]$, with Normal priors on β and γ and known covariance matrix Ω as in Algorithm 4 Step 2, where

$$\Omega = \begin{bmatrix} \sigma_\xi^2 + \delta^2 & \delta\sqrt{1-\omega^2} \\ \delta\sqrt{1-\omega^2} & 1 - \omega^2 \end{bmatrix} \otimes I_N$$

3. Draw $\alpha|w, y^*, \theta, \beta, \gamma, \delta, \sigma_\xi, \mu, \tau^2, \kappa, \omega^2$ from a Bayesian regression of $y - X\beta$ on a $NTxN$ matrix of firm dummies D, using a $\mathcal{N}\left(\mu, I_N \cdot \left(\delta^2 + \sigma_\xi^2\right)/\tau^2\right)$ prior.
4. Draw $\theta|w, y^*, \alpha, \beta, \gamma, \delta, \sigma_\xi, \mu, \tau^2, \kappa, \omega^2$ from a Bayesian regression of $w - Z\gamma$ on a $NTxN$ matrix of firm dummies D, using a $\mathcal{N}\left(\kappa, I_N \cdot \left(1 - \omega^2\right)/\omega^2\right)$ prior.
5. Draw $\mu, \tau^2|w, y^*, \alpha, \theta, \beta, \gamma, \delta, \sigma_\xi, \kappa, \omega^2$ from a Bayesian regression.
6. Draw $\kappa, \omega^2|w, y^*, \alpha, \theta, \beta, \gamma, \delta, \sigma_\xi, \mu, \tau^2$ from a Bayesian regression.
7. Draw $\delta, \sigma_\xi^2|w, y^*, \alpha, \theta, \beta, \gamma, \mu, \tau^2, \kappa, \omega^2$ from a Bayesian regression of $y - \alpha - X\beta$ on $w - \theta - Z\gamma$, with Normal-IG priors (see Algorithm 1).
8. Go back to step 1, repeat.

of the selection equation, although these could be interesting in their own right as they convey information about the reasons for selection.

It is important to note that the results in Table 26.3 should be taken as suggestive only. There are other observable variables that can be included in the observation and selection equations that may alleviate the omitted variable problem. Moreover, I include the same variables in the selection equation as in the observation equation, so the model is identified from distributional assumptions only. A more thorough analysis requires an instrument that changes the probability of observing leverage but does not drive leverage itself. In other words, one needs a variable in the selection equation that does not appear in the observation equation. This requires delving deeper into the economic reasons for selection. For example, the fact that more profitable firms are more likely to disappear from the data suggests that the main source of attrition is mergers and acquisitions rather than bankruptcy. One would then look for an instrument that drives M&A but not leverage. Other

Table 26.3 Random effects panel regressions. Random effects panel regression estimates of quasi--market leverage (defined as the book value of debt divided by the book value of debt plus the market value of equity) on profitability (operating income divided by the book value of assets), tangibility (net property plant and equipment divided by book assets), market-to-book ratio (market value divided by book value of equity) and the natural logarithm of book assets, based on a random sample of 1 000 firms spanning 11 431 firm-years from Compustat between 1950 and 2010. All explanatory variables are lagged by one period. The first column uses the standard GLS method to estimating a random effects panel model. The 'No Selection' column estimates the same random effects model, but using MCMC (Algorithm 6 with ρ forced to zero). The column labelled 'Selection' uses MCMC Algorithm 6 to correct for sample selection. The MCMC estimates use 1 000 burn-in cycles followed by 10 000 cycles to sample the posterior distribution. Standard errors for the GLS estimates and posterior standard deviations for the MCMC estimates are in brackets.

Dependent variable: Quasi-market leverage			
	GLS	**MCMC**	
		No selection correction	**Selection correction**
Operating income / assets	−0.183	−0.186	−0.237
	(0.015)	(0.015)	(0.017)
Tangibility	0.209	0.207	0.101
	(0.016)	(0.016)	(0.021)
Market-to-book	−0.029	−0.029	−0.028
	(0.002)	(0.002)	(0.002)
Log(assets)	0.020	0.020	0.019
	(0.002)	(0.002)	(0.002)
ρ	–	–	0.367
	–	–	(0.049)

possible extensions are to use two selection equations to separately model bankruptcy and M&A as different reasons for attrition, essentially producing a joint model of capital structure, M&A and bankruptcy, or to use the Korteweg and Sørensen [59] approach and model leverage as an AR(1) process. Note also that the above example, given the use of random effects, only considers selection in the time series. There could be important cross-sectional sample selection issues over and above the time-series selection problem.

It is fairly straightforward to generalize Algorithm 6 to make the dependent variable a binary (probit) or truncated (Tobit) variable. For example, Hamilton [38] specifies a RE panel Tobit selection model with t-distributed error terms and uses MCMC to estimate the effect of HMO choice on healthcare costs. A different extension to the model is in Cowles, Carlin, and Connett [19] who estimate a panel selection model allowing for two observation equations. Selection is multinomial and the observation equations are observed at different cutoffs of the selection variable. They apply their model to a longitudinal clinical trial measuring the effect of smoking cessation paired with an inhaler treatment on lung function. The selection problem is that patients endogenously do not show up for follow-up visits, or show up but fail to return canisters with the inhaler.

26.6 Structural models

Structural models have many potential applications in empirical corporate finance, and have been used in areas such as capital structure (e.g. [46, 47, 93]) and corporate governance (e.g. [95]). These models are typically estimated by simulated method of moments (SMM) or, in some cases, Simulated Maximum Likelihood (SML, see e.g. [5, 80]). In this section I will illustrate the benefits of estimating these models by MCMC, especially models with latent state variables.

Consider Merton's [75] model of the firm in state-space representation:

$$
\begin{aligned}
v_t &= v_{t-1} + \mu + \sigma \varepsilon_t \\
E_t &= P\left(v_t, \sigma; \theta\right) + \eta_t
\end{aligned}
$$

The state variable of the model is the natural logarithm of the market value of the firm's assets, v_t, which follows a random walk with drift μ. The shocks to firm value, ε_t, have a standard Normal distribution. The observed market value of equity, E_t, is a function of v_t, volatility, σ, and other parameters, θ. In Merton's model the pricing function, $P(\cdot)$, is the Black–Scholes European call option function, and the parameters vector θ consists of the maturity of debt (time-to-expiration), the face value of debt (strike price), and the risk-free interest rate, all of which I will assume are observed. The pricing error, η_t, allows for unmodelled features such as market microstructure noise [50, 58] that may force observed prices away from the model.[33]

I assume for now that, in addition to θ, the time series of firm value, $v^T = \{v_1 \ldots v_T\}$ as well as the equity values are all observed. The inference problem is then to obtain the posterior distribution $p\left(\sigma^2, \mu | E^T, v^T\right)$. As usual, an MCMC algorithm cycles between drawing from the complete conditionals, $p\left(\mu | \sigma^2, E^T, v^T\right)$ and $p\left(\sigma^2 | \mu, E^T, v^T\right)$. The first distribution is a simple Bayesian regression and poses no problems. The second distribution is more tricky because σ is present in both equations of the state-space, and although the conditional posterior can be evaluated without much trouble, it is not a known distribution that is easily sampled. The solution to this problem is an accept–reject type algorithm known as Metropolis–Hastings.[34] Denote by ς the current draw of $\left\{\sigma^2\right\}^{(g)}$ in cycle g of the MCMC algorithm. To sample the next draw, $\left\{\sigma^2\right\}^{(g+1)}$, one proceeds as follows:

1. Draw s from a proposal density $f\left(s; \varsigma\right)$.
2. Compute $\alpha = \min\left(1, \frac{p\left(s|\mu, E^T, v^T\right) \cdot f\left(\varsigma; s\right)}{p\left(\varsigma|\mu, E^T, v^T\right) \cdot f\left(s; \varsigma\right)}\right)$.
3. Set $\left\{\sigma^2\right\}^{(g+1)} = \begin{cases} s & \text{w/ probability } \alpha \\ \varsigma & \text{w/ probability } 1 - \alpha \end{cases}$

Note that we use the latest draw for μ when computing α. The proposal density is key for the performance of the algorithm. Ideally, it is a density that is close to the target density, $p\left(\sigma^2 | \mu, E^T, v^T\right)$, yet easy to sample from. The proposals can be iid (commonly known as 'independence Metropolis') or proposals may depend on the current draw of σ^2. A popular example of the latter type is the 'random

[33] An alternative motivation for the pricing errors is simply to avoid a stochastic singularity problem in the sense that the model makes predictions about more observable variables than there are structural shocks and hence could be rejected with the observation of as few as two time periods.

[34] The Metropolis–Hastings algorithm is derived from the detailed balance condition of the stationary distribution of the Markov chain, which ensures that the chain will be reversible. I will not go into the theoretical details here, but instead refer the reader to Johannes and Polson [54], Robert and Casella [86], or Rossi *et al.* [87] for a comprehensive treatment.

walk Metropolis' in which s equals the current draw, ς, plus a random increment. Note that if we are able to sample from the target density, i.e. $f(\cdot) = p(\cdot)$, then we are back to a Gibbs sampler step: $\alpha = 1$ and we always accept the proposal. In this sense we can think of the Metropolis–Hastings algorithm as a generalization of the Gibbs sampler. As in the Merton model example, an MCMC algorithm can be a mixture of Gibbs steps and Metropolis–Hastings steps.

It is important to 'tune' the proposal density to get reasonable performance from the sampler. Clearly an acceptance rate near zero is bad because the sampler will be slow to converge to, and explore, the posterior distribution due to the many rejected proposals. An acceptance rate near 100% may seem great at first sight, but one may worry that the chain is moving very slowly because the incremental steps in a random walk Metropolis are too small, or because the tails of the proposal are too thin relative to the target distribution in an independence Metropolis sampler. Either way, we may be undersampling an important part of the posterior distribution, leading to poor inference. There is no generic theoretical advice that can be given as to how best to pick a proposal distribution. Typical advice in practice is to aim for a 50–80% acceptance rate, but it should be noted that the acceptance rate alone does not determine whether the sampler does a good job. For more details on the Metropolis–Hastings algorithm and tuning, see Johannes and Polson [54], Robert and Casella [86], or Rossi *et al.* [87].

In many structural models inference is complicated further by the fact that the underlying state variable is not observed (e.g. [46, 47, 63]). MCMC deals with this by augmenting the posterior distribution with the state variable, $p\left(\sigma, \mu, v^T | E^T\right)$, which can subsequently be integrated out to obtain the marginal distribution of the model parameters. Conveniently, the complete conditionals for μ and σ do not need any modification (they are already conditioned on a draw of v^T), so the only extra step that is required is to sample from the conditional distribution of the state variable $p\left(v^T | \mu, \sigma, E^T\right)$. In the Merton model a linear Kalman filter provides us with the distribution of $v_1|E_1$ through $v_T|E_T$, dropping the conditioning on the parameters for ease of exposition. Since we need to draw from $v_1|E^T$ through $v_T|E^T$, one more step is required: smoothing. This essentially involves running the Kalman filter again, but going backwards starting at time T, and sampling the v_ts as we go along. This procedure is called 'Forward Filtering, Backwards Sampling' (FFBS) [7, 30].

Korteweg and Polson [58] describe the MCMC algorithm for a structural model of the firm in detail, with FFBS and Metropolis–Hastings sampling for σ^2. They estimate Leland's [62] model for a panel of firms and compute corporate bond credit spreads, taking into account the effect of parameter and latent state uncertainty. Using the robust pricing framework of Hansen and Sargent [40], the bond price under uncertainty, P_t, is the expectation of the model price over the distribution of latent state and parameters given the data up to time t:

$$
\begin{aligned}
P_t &= \mathbb{E}_{\sigma, v_t | E^t}\left(B(v_t, \sigma; \theta)\right) \\
&\approx \frac{1}{G}\sum_{i=1}^{G} B\left(v_t^{(i)}, \sigma^{(i)}; \theta\right)
\end{aligned}
$$

where $B(\cdot)$ is the model's bond pricing function. Bayesian methods are very well suited for problems of learning and uncertainty, since the posterior distribution captures the degree of parameter and state uncertainty based on the observed data. The second line shows the approximation to the bond price based on G cycles of the MCMC algorithm (after dropping the initial burn-in cycles). The expectation in the top line becomes a simple average over the draws of the algorithm. The concavity of the bond pricing formula in v and σ results in larger credit spreads for bonds compared to estimates from standard methods (e.g. [27]).

Apart from the usefulness for learning and uncertainty problems, MCMC methods have the same advantage over SMM and SML in speed of computation as described in Section 26.5.1: instead of simulating the latent state variables for each set of parameters as is required in SMM/SML, the MCMC algorithm bounces between parameter draws and draws of the state vector, leading to faster convergence. Moreover, the MCMC and other likelihood-based estimates incorporate all available information, unlike SMM which only considers a set of moments chosen by the researcher and therefore results in less powerful tests. However, this efficiency gain does come at the expense of making stronger distributional assumptions on the observation error, η (see Korteweg and Lemmon [57] for a detailed discussion of structural model testing). Finally, as shown earlier, MCMC gives correct small-sample inference and is amenable to computing nonlinear functions of parameters and states. This is a particularly useful feature for structural models where the observables are often nonlinear in both the parameters and states.

The MCMC algorithm for structural models can be extended in various important directions, including but not limited to panel data, autocorrelation in the error structure, handling missing data, and adding stochastic volatility and jumps to the state process, as well as more flexible observation error distributions, for example through mixtures of Normals. The inherent modularity of MCMC makes this task more convenient to handle than with classical methods.

There are other applications of state space models outside of strict structural modelling. One example is Korteweg [56], who uses MCMC to estimate the net benefits to leverage from a state space model in which the state vector is composed of individual firms' unlevered asset values and industry asset betas. The observations are the market values of firms' debt and equity and their riskiness (i.e. betas) with respect to the market portfolio. Under the identifying assumption that all firms within the industry share the same (unlevered) asset beta, Korteweg identifies the present value of the net benefits to leverage as a function of leverage and other firm characteristics such as profitability, market-to-book ratio, and asset tangibility. The MCMC algorithm for this model is similar to the algorithms described above, employing the Kalman filter and FFBS to integrate out the latent states.

26.7 Extensions and other applications

In this section I will highlight some further potential applications of MCMC methods in corporate finance, focusing on cases which are difficult to estimate using classical methods.

Hierarchical models are a particularly useful type of model that has to date seen little application in corporate finance. Hierarchical models were introduced in Section 26.5.1 in the context of estimating random effects in a panel probit model, but the basic concept of the hierarchical model has much broader applicability. For example, consider a model of the decision to issue equity. The typical approach would be to run a probit or hazard model that specifies the probability of issuing equity as a function of a number of covariates, most importantly the financing deficit, one of the key variables in the pecking order theory of capital structure. However, a common concern is that the financing deficit and the decision to issue equity are jointly determined in the sense that they are both driven by (unobserved) investment opportunities. Lacking a good instrument or natural experiment, one example of a hierarchical modelling solution to this problem is to model the deficit conditional on an underlying (latent) investment opportunity variable, and to specify the probability of issuing equity conditional on both the deficit and the latent variable, potentially allowing for correlation between the error terms.[35]

It is also useful to add hierarchical model features to structural models. For example, it would be interesting to analyse the cross-sectional distribution of firms' bankruptcy costs, or the distribution of CEO talent in structural models of the firm. However, there is usually not enough data to reliably pin down the estimate on a firm by firm basis. One way to overcome this issue is to use a hierarchical

[35] Of course an instrument that shocks the deficit (but not the equity issuance probability) will help the identification of the model. The intuition is similar to the argument for having an instrument in the Heckman selection model.

setup and specify a firm's bankruptcy cost or a CEO's talent as being generated from a particular distribution, using the observed data to estimate the mean and variance of this distribution, as was done in the random effects model.

Bayesian methods are also very useful for duration (hazard) models. For example, one could model the probability of attrition in Section 26.5.2 through a hazard rate instead of a probit model. This would do a better job of capturing the essence of liquidation or an acquisition in the sense that, unlike in the probit model, a firm could not come back to life the next period. Li [66] estimates the duration of Chapter 11 bankruptcy spells from a Bayesian perspective, using an approximation to the posterior density, accounting for parameter and model uncertainty as well as providing correct sample properties based on the small dataset. Horny, Mendes and Van den Berg [48] estimate a hazard model of job duration with worker and firm-specific random effects. They use random effects because fixed effects are not feasible due to right-censoring of the unemployment spells, and because the random effects allow them to decompose the variation of job durations into the relative contributions of workers' and firms' characteristics. In addition, they allow for multiple job spells per worker and for correlation between the worker and firm random effects. They estimate the model by MCMC because the joint dependence of the random effects makes classical likelihood-based estimation very difficult for both computational speed and the difficulty of finding the appropriate asymptotic distribution in order to calculate standard errors. Chen, Guo and Lin [11] develop a model of switching hazards which they use to estimate the probability of IPO withdrawal in relation to its subsequent survival hazard, which in turn depends on the IPO decision. They use MCMC because of the computational advantage in both estimation and performing model selection and cross-validation, as well as the small-sample inference properties. Fahrmeir and Knorr Held [28] develop a nonparametric MCMC method for a duration model with multiple outcomes, such as unemployment ending in a full-time versus a part-time job. Such a model could, for example, be applied to firms' survival spells ending in either bankruptcy or an acquisition, or CEO's tenure spells ending in forced or voluntary retirement.

Another area where MCMC can be applied is count data. These models are related to duration models, but instead of a time spell the dependent variable is a non-negative integer that counts the number of events that have occurred within a given time period. Examples include the number of takeover bids received by a target firm [52], the number of failed banks [22] or the number of defaults in a portfolio of assets [55]. Chib, Greenberg and Winkelmann [16] develop an MCMC algorithm for panel count data models with random effects. They argue that Maximum Likelihood is not a viable estimator for this model due to the presence of random effects paired with the nonlinearity of count data. Chib and Winkelmann [17] generalize this model to allow for multivariate correlated count data, represented by correlated latent effects. Munkin and Trivedi [77] embed a count model within a self-selection model with two correlated outcome equations, one of which is a count and the other a continuous variable. The authors strongly motivate their MCMC approach by the computational difficulties encountered when attempting to estimate the model by SML. Deb, Munkin and Trivedi [23] extend this selection model by allowing the entire outcome function to be different among the treated and untreated groups, essentially turning the model into a switching regression with count outcomes. They apply their MCMC estimation to separate the incentive and selection effects in private insurance on the number of doctor visits, using a multiyear sample of the US adult non-Medicare population. This could be applied in corporate finance, for example, to identify the incentive and selection effects in public versus private firms undertaking acquisitions, using the count of M&A as the dependent variable.

A class of models that is very closely related to the switching regressions of Section 26.4.4, and which is also related to the state space models of Section 26.6 is the set of hidden Markov models (HMM). Unlike switching regressions, there are typically no covariates that drive the selection into a specific state, but they generally allow for multiple states (in some cases a continuum of states) of the system that switch according to a Markov process. In economics, HMM are usually applied to regime switching settings, where the economy may switch between expansion and recession. Albert and Chib [2] argue that the two-step Maximum Likelihood approach to estimating HMM of Hamilton [39] does not give correct standard errors as the uncertainty about the parameters in the first step is not incorporated in the second step. In addition, the ML approach does not provide a complete description of the likelihood, such as bimodality or asymmetry. Albert and Chib develop an MCMC algorithm that deals with both issues. McCullogh and Tsay [74] use MCMC to estimate an HMM that is very close to the switching regressions of Section 26.4.4 and apply it to GNP data. Ghysels, McCullogh and Tsay [35] estimate a nonlinear regime switching model and apply their MCMC estimator to two examples, one using housing starts data while the other employs industrial production data. They argue that the MCMC approach is particularly suitable to their model because classical estimation of periodic Markov chain models often results in parameter estimates at the boundary.

Recent work has started exploring the uses of MCMC for Instrumental Variables, with some success (see Sims [90] for a discussion and further references). In particular, when the error distributions are non-Normal, the Bayesian estimates may be more efficient than classical methods [18]. On a different note, two recent papers by Davies and Taillard [20, 21] propose an MCMC approach to dealing with omitted variables that does not use an instrument at all, but instead achieves identification by assuming that any common variation in the residuals is due solely to the omitted variables.

Finally, thanks to the modularity of MCMC, models can be mixed and extended in various ways that are quite straightforward. For example, one can deal with missing data by adding the Algorithm 2 steps to any of the other algorithms. Error clustering can be handled by adding random effects to the models. Last but not least, one can deal with heteroskedasticity and non-Normal error terms by using mixtures of Normals (see Diebolt and Robert [24] and Geweke [34] for MCMC mixture models, Chen and Liu [12] for Kalman filters with mixtures of Normals, and Korteweg and Sørensen [59] for an application). These mixture models can be added to MCMC algorithms with relative ease, are very flexible in generating both skewness and kurtosis in the error distributions, and make the MCMC estimates more robust to outliers.

26.8 Conclusion

With the current trend towards more complex empirical models in the corporate finance literature, Markov chain Monte Carlo methods provide a viable and attractive means of estimating and evaluating models where classical methods such as least squares regression, GMM and Maximum Likelihood and their simulated counterparts are difficult or too computationally demanding to apply. In particular, this includes nonlinear models with many latent variables that require high-dimensional integration to evaluate the likelihood, or models that have a hierarchical structure. Examples of such models include panel limited dependent variable models, matching and other self-selection models, and structural models of the firm. The potential application of these types of models in corporate finance is vast, including such diverse areas as capital structure, financial intermediation, bankruptcy, and corporate governance.

The core feature of the method is the Hammersley–Clifford theorem, which breaks up the problem into its complete conditional distributions. These are usually relatively easy to sample from, requiring no more than standard regression tools. Another benefit of the MCMC approach is that it is modular, so that, for example, one can add a missing data module to a panel probit algorithm with relative ease. Moreover, the method allows for exact small-sample inference of parameters and nonlinear functions of parameters (the latter being helpful, for example, when calculating marginal effects in a probit or logit model), and does not require optimization algorithms, simulated annealing or other methods that can make Maximum Likelihood and GMM cumbersome to use.

Every introductory text necessarily has to focus on certain ideas at the expense of others, and I have chosen to focus on applications rather than to discuss some of the more technical details of the MCMC method, the role of priors, and convergence diagnostics. Most introductory textbooks thoroughly discuss these and other topics, and I refer the interested reader to Rossi *et al.* [87] and Johannes and Polson [54] for a particularly lucid treatise and further reading.

This chapter has only touched the tip of the iceberg of possibilities that MCMC has to offer for empirical corporate finance research, and I hope to have convinced the reader that the potential benefits of the method are plentiful. Given that learning MCMC does come with some fixed cost, I hope that this chapter and the accompanying code samples help to lower the cost of adoption, and inspire and motivate corporate finance researchers to dive deeper into MCMC methods. With time, hopefully this methodology will become part of the standard toolkit in finance, as it already is in many other areas of scientific inquiry.

Acknowledgements

I thank the editors, and Dirk Jenter, Kai Li, Michael Roberts, Morten Sørensen and Toni Whited for helpful comments and suggestions. All errors are my own.

References

[1] Acharya, S. (1988). A generalized econometric model and tests of a signalling hypothesis with two discrete signals. *Journal of Finance*, **43**, 412–429.

[2] Albert, J. and Chib, S. (1993). Bayes inference via Gibbs sampling of autoregressive time series subject to Markov mean and variance shifts. *Journal of Business and Economic Statistics*, **11**, 1–15.

[3] Albert, J. and Chib, S. (1993). Bayesian analysis of binary and polychotomous response data. *Journal of the American Statistical Association*, **88**, 669–679.

[4] Billet, M. T. and Xue, H. (2007). The takeover deterrent effect of open market share repurchases. *Journal of Finance*, **62**, 1827–1850.

[5] Bruche, M. (2007). Estimating structural models of corporate bond prices. Working paper, CEMFI Madrid.

[6] Bruno, G. (2004). Limited dependent panel data models: A comparative analysis of classical and Bayesian inference among econometric packages. Working paper, Bank of Italy.

[7] Carter, C. K. and Kohn, R. J. (1994). On Gibbs sampling for state space models. *Biometrika*, **81**, 541–553.

[8] Chamberlain, G. (1984). Panel data. In *Handbook of Econometrics* (ed. Z. Griliches and M. Intriligator). Elsevier, Amsterdam: North Holland.

[9] Chen, J. (2010). Two-sided matching and spread determinants in the loan market. Working paper, UC Irvine.

[10] Chen, L. and Zhao, X. (2007). Mechanical mean reversion of leverage ratios. *Economics Letters*, **95**, 223–229.

[11] Chen, R., Guo, R.-J. and Lin, M. (2010). Self-selectivity in firm's decision to withdraw IPO: Bayesian inference for hazard models of bankruptcy with feedback. Working paper, Rutgers University.

[12] Chen, R. and Liu, J. S. (2000). Mixture Kalman filters. *Journal of the Royal Statistical Society Series B*, **62**, 493–508.

[13] Chib, S. (1992). Bayes inference in the Tobit censored regression model. *Journal of Econometrics*, **51**, 79–99.

[14] Chib, S. and Greenberg, E. (1998). Analysis of multivariate probit models. *Biometrika*, **85**, 347–361.

[15] Chib, S., Greenberg, E. and Chen, Y. (1998). MCMC methods for fitting and comparing multinomial response models. Working paper, Washington University in St. Louis.
[16] Chib, S., Greenberg, E. and Winkelmann, R. (1998). Posterior simulation and Bayes factors in panel count data models. *Journal of Econometrics*, **86**, 33–54.
[17] Chib, S. and Winkelmann, R. (2001). Markov chain Monte Carlo analysis of correlated count data. *Journal of Business and Economic Statistics*, **19**, 428–435.
[18] Conley, T. G., Hansen, C. B., McCulloch, R. E. and Rossi, P. E. (2008). A semi-parametric Bayesian approach to the instrumental variable problem. *Journal of Econometrics*, **144**, 276–305.
[19] Cowles, M. K., Carlin, B. P. and Connett, J. E. (1996). Bayesian Tobit modelling of longitudinal ordinal clinical trial compliance data with nonignorable missingness. *Journal of the American Statistical Association*, **91**, 86–98.
[20] Davies, P. and Taillard, J. (2010). Omitted variables, endogeneity, and the link between managerial ownership and firm performance. Working paper, University of Iowa and Boston College.
[21] Davies, P. and Taillard, J. (2011). Estimating the return to education: An instrument-free approach. Working paper, University of Iowa and Boston College.
[22] Davutyan, N. (1989). Bank failures as Poisson variates. *Economics Letters*, **29**, 333–338.
[23] Deb, P., Munkin, M. K. and Trivedi, P. K. (2006). Private insurance, selection, and health care use: A Bayesian analysis of a Roy-type model. *Journal of Business and Economic Statistics*, **24**, 403–415.
[24] Diebolt, J. and Robert, C. P. (1994). Estimation of finite mixture distributions through Bayesian sampling. *Journal of the Royal Statistical Society Series B*, **56**, 363–375.
[25] Eckbo, B. E., Maksimovic, V. and Williams, J. (1990). Consistent estimation of cross-sectional models in event studies. *Review of Financial Studies*, **3**, 343–365.
[26] Efron, B. and Morris, C. (1975). Data analysis using Stein's estimator and its generalizations. *Journal of the American Statistical Association*, **70**, 311–319.
[27] Eom, Y. H., Helwege, J. and Huang, J. (2004). Structural models of corporate bond pricing: An empirical analysis. *Review of Financial Studies*, **17**, 499–544.
[28] Fahrmeir, L. and Knorr Held, L. (1997). Dynamic discrete-time duration models. Working paper, University of Munich.
[29] Frank, M. Z. and Goyal, V. K. (2009). Capital structure decisions: Which factors are reliably important? *Financial Management*, **38**, 1–37.
[30] Fruhwirth-Schnatter, S. (1994). Data augmentation and dynamic linear models. *Journal of Time Series Analysis*, **15**, 183–202.
[31] Gelfand, A. and Smith, A. F. M. (1990). Sampling based approaches to calculating marginal densities. *Journal of the American Statistical Association*, **85**, 398–409.
[32] Gelman, A. and Rubin, D. B. (1992). Inference from iterative simulation using multiple sequences. *Statistical Science*, **7**, 457–511.
[33] Geman, S. and Geman, D. (1984). Stochastic relaxation, Gibbs distributions, and the Bayesian restoration of images. *IEEE Transactions on Pattern Analysis and Machine Intelligence*, **6**, 721–741.
[34] Geweke, J. (2006). Interpretation and inference in mixture models: Simple MCMC works. Working paper, University of Iowa.
[35] Ghysels, E., McCulloch, R. E. and Tsay, R. S. (1998). Bayesian inference for periodic regime-switching models. *Journal of Applied Econometrics*, **13**, 129–143.
[36] Graham, B. S. and Hirano, K. (2011). Robustness to parametric assumptions in missing data models. *American Economic Review Papers & Proceedings*, **101**, 538–543.
[37] Griliches, Z. (1986). Economic data issues. In *Handbook of Econometrics, Volume 3* (ed. Z. Griliches and M. Intriligator), pp. 1466–1514. Elsevier, Amsterdam: North Holland.

[38] Hamilton, B. (1999). HMO selection and medicare costs: Bayesian MCMC estimation of a robust panel data Tobit model with survival. *Health Economics*, **8**, 403–414.
[39] Hamilton, J. D. (1989). A new approach to the economic analysis of nonstationary time series and the business cycle. *Econometrica*, **57**, 357–384.
[40] Hansen, L. and Sargent, T. (2010). Fragile beliefs and the price of uncertainty. *Quantitative Economics*, **1**, 129–162.
[41] Hausman, J. A. and Wise, D. A. (1979). Attrition bias in experimental and panel data: The Gary income maintenance experiment. *Econometrica*, **47**, 455–473.
[42] Heckman, J. J. (1976). The common structure of statistical models of truncation, sample selection and limited dependent variables and a simple estimator for such models. *Annals of Economic and Social Measurement*, **5**, 475–492.
[43] Heckman, J. J. (1979). Sample selection bias as a specification error. *Econometrica*, **47**, 153–161.
[44] Heckman, J. J. (1981). The incidental parameters problem and the problem of initial conditions in estimating a discrete time–discrete data stochastic process. In *Structural Analysis of Discrete Data with Econometric Applications* (ed. C. Manski and D. McFadden). MIT Press, Cambridge.
[45] Heckman, J. J. (1990). Varieties of selection bias. *American Economic Review*, **80**, 313–318.
[46] Hennessy, C. A. and Whited, T. M. (2005). Debt dynamics. *Journal of Finance*, **60**, 1129–1165.
[47] Hennessy, C. A. and Whited, T. M. (2007). How costly is external financing? Evidence from a structural estimation. *Journal of Finance*, **62**, 1705–1745.
[48] Horny, G., Mendes, R. and Van den Berg, G. J. (2009). Job durations with worker and firm specific effects: MCMC estimation with longitudinal employer–employee data. Working paper, IZA Institute for the Study of Labor, Bonn.
[49] Hovakimian, A., Hovakimian, G. and Tehranian, H. (2004). Determinants of target capital structure: The case of dual debt and equity issues. *Journal of Financial Economics*, **71**, 517–540.
[50] Huang, S. and Yu, J. (2010). Bayesian analysis of structural credit risk models with microstructure noises. *Journal of Economic Dynamics and Control*, **34**, 2259–2272.
[51] Iliev, P. and Welch, I. (2010). Reconciling estimates of the speed of adjustment of leverage ratios. Working paper, UCLA.
[52] Jaggia, S. and Thosar, S. (1995). Contested tender offers: An estimate of the hazard function. *Journal of Business and Economic Statistics*, **13**, 113–119.
[53] Jeliazkov, I. and Lee, E. H. (2010). MCMC perspectives on simulated likelihood estimation. In *Advances in Econometrics* (ed. W. Green and R. Hill), Volume 26, pp. 3–39. Emerald Group Publishing Limited.
[54] Johannes, M. and Polson, N. (2012). *Computational Methods for Bayesian Inference: MCMC methods and Particle Filtering*. Unpublished manuscript.
[55] Koopman, S. J., Lucas, A. and Schwab, B. (2010). Macro, industry and frailty effects in defaults: The 2008 credit crisis in perspective. Working paper, Tinbergen Institute.
[56] Korteweg, A. (2010). The net benefits to leverage. *Journal of Finance*, **65**, 2137–2170.
[57] Korteweg, A. and Lemmon, M. (2012). Structural models of capital structure: A framework for model evaluation and testing. Working paper, Stanford University.
[58] Korteweg, A. and Polson, N. G. (2010). Corporate credit spreads under uncertainty. Working paper, University of Chicago.
[59] Korteweg, A. and Sørensen, M. (2010). Risk and return characteristics of venture capital-backed entrepreneurial companies. *Review of Financial Studies*, **23**, 3738–3772.
[60] Korteweg, A. and Sørensen, M. (2012). Estimating loan-to-value and foreclosure behavior. Working paper, Stanford University.
[61] Lee, L.-F. (1979). Identification and estimation in binary choice models with limited (censored) dependent variables. *Econometrica*, **47**, 977–996.
[62] Leland, H. (1994). Bond prices, yield spreads, and optimal capital structure with default risk. Working paper, UC Berkeley.

[63] Leland, H. (1994). Risky debt, bond covenants and optimal capital structure. *Journal of Finance*, **49**, 1213–1252.
[64] Lemmon, M. L., Roberts, M. R. and Zender, J. F. (2008). Back to the beginning: Persistence and the cross-section of corporate capital structure. *Journal of Finance*, **63**, 1575–1608.
[65] Li, K. (1998). Bayesian inference in a simultaneous equation model with limited dependent variables. *Journal of Econometrics*, **85**, 387–400.
[66] Li, K. (1999). Bayesian analysis of duration models: An application to Chapter 11 bankruptcy. *Economics Letters*, **63**, 305–312.
[67] Li, K. and McNally, W. (2004). Open market versus tender offer share repurchases: A conditional event study. Working paper, University of British Columbia.
[68] Li, K. and Prabhala, N. R. (2007). Self-selection models in corporate finance. In *Handbook of Corporate Finance: Empirical Corporate Finance* (ed. B. Eckbo), Volume I, Chapter 2, pp. 37–86. Elsevier, Amsterdam: North Holland.
[69] Lindley, D. V. and Smith, A. F. M. (1972). Bayes estimates for the linear model. *Journal of the Royal Statistical Society Series B*, **34**, 1–41.
[70] Maddala, G. S. (1996). Applications of limited dependent variable models in finance. In *Handbook of Statistics* (ed. G. Maddala and G. Rao), Volume 14, pp. 553–566. Elsevier, Amsterdam: North Holland.
[71] Manski, C. (1989). Anatomy of the selection problem. *Journal of Human Resources*, **24**, 343–360.
[72] Manski, C. (1990). Nonparametric bounds on treatment effects. *American Economic Review*, **80**, 319–323.
[73] McCulloch, R. E., Polson, N. G. and Rossi, P. E. (2000). A Bayesian analysis of the multinomial probit model with fully identified parameters. *Journal of Econometrics*, **99**, 173–193.
[74] McCulloch, R. E. and Tsay, R. S. (1994). Statistical analysis of economic time series via Markov switching models. *Journal of Time Series Analysis*, **15**, 523–539.
[75] Merton, R. C. (1974). On the pricing of corporate debt: The risk structure of interest rates. *Journal of Finance*, **29**, 449–470.
[76] Mundlak, Y. (1978). On the pooling of time series and cross section data. *Econometrica*, **46**, 69–85.
[77] Munkin, M. K. and Trivedi, P. K. (2003). Bayesian analysis of a self-selection model with multiple outcomes using simulation-based estimation: An application to the demand for healthcare. *Journal of Econometrics*, **114**, 197–220.
[78] Neyman, J. and Scott, E. L. (1948). Consistent estimates based on partially consistent observations. *Econometrica*, **16**, 1–32.
[79] Nijman, T. and Verbeek, M. (1992). Nonresponse in panel data: The impact on estimates of the life cycle consumption function. *Journal of Applied Econometrics*, **7**, 243–257.
[80] Nikolov, B., Morellec, E. and Schuerhoff, N. (2012). Corporate governance and capital structure dynamics. *Journal of Finance*, **67**, 803–848.
[81] Obrizan, M. (2011). A Bayesian model of sample selection with a discrete outcome variable: Detecting depression in older adults. Working paper, Kyiv School of Economics.
[82] Park, M. (2008). An empirical two-sided matching model of acquisitions: Understanding merger incentives and outcomes in the mutual fund industry. Working paper, UC Berkeley.
[83] Prabhala, N. R. (1997). Conditional methods in event studies and an equilibrium justification for standard event-study procedures. *Review of Financial Studies*, **10**, 1–38.
[84] Pulvino, T. (1999). Effects of bankruptcy court protection on asset sales. *Journal of Financial Economics*, **52**, 151–186.
[85] Ridder, G. (1990). Attrition in multi-wave panel data. In *Panel Data and Labor Market Studies* (ed. G. R. J. Hartog and J. Theeuwes). Elsevier, Amsterdam: North Holland.

[86] Robert, C. P. and Casella, G. (2004). *Monte Carlo Statistical Methods* (2nd edn). Springer, New York.

[87] Rossi, P. E., Allenby, G. M. and McCulloch, R. (2005). *Bayesian Statistics and Marketing* (1st edn). John Wiley and Sons, Hoboken, NJ.

[88] Rubin, D. (1983). Some applications of Bayesian statistics to educational data. *The Statistician*, **32**, 55–68.

[89] Scruggs, J. T. (2007). Estimating the cross-sectional market response to an endogenous event: Naked vs. underwritten calls of convertible bonds. *Journal of Empirical Finance*, **14**, 220–247.

[90] Sims, C. A. (2007). Thinking about instrumental variables. Working paper, Princeton University.

[91] Sørensen, M. (2007). How smart is smart money? A two-sided matching model of venture capital. *Journal of Finance*, **62**, 2725–2762.

[92] Stein, C. (1955). Inadmissibility of the usual estimator for the mean of a multivariate normal distribution. In *Proceedings of the third Berkeley symposium*.

[93] Strebulaev, I. A. (2007). Do tests of capital structure theory mean what they say? *Journal of Finance*, **62**, 1747–1787.

[94] Tanner, M. A. and Wong, W. H. (1987). The calculation of posterior distributions by data augmentation. *Journal of the American Statistical Association*, **82**, 528–549.

[95] Taylor, L. A. (2010). Why are CEOs rarely fired? Evidence from structural estimation. *Journal of Finance*, **65**, 2051–2087.

[96] Train, K. (2003). *Discrete Choice Methods with Simulation* (1st edn). Cambridge University Press, New York.

[97] van Hasselt, M. (2009). Bayesian inference in a sample selection model. Working paper, University of Western Ontario.

[98] Vella, F. and Verbeek, M. (1994). Two-step estimation of simultaneous equation panel data models with censored endogenous variables. Working paper, Tilburg University.

[99] Zellner, A. (1971). *An introduction to Bayesian inference in econometrics*. Wiley, New York.

27 Actuarial credibility theory and Bayesian statistics—the story of a special evolution

UDI MAKOV

27.1 Introduction

Actuarial practitioners and researchers have clearly been more open to Bayesian thinking than their counterparts in the statistical world. This is much in evidence in the context of *credibility theory* which has adopted Bayesian influences almost from the start and is to date solidly founded on empirical Bayes philosophy. The paper is not devoted to the Bayesian impact on actuarial science, but rather to the evolution of *credibility theory*, the early Bayesian impact and the current divergence between *credibility theory* and Bayesian statistics.

According to [38] "The word *credibility* was originally introduced into actuarial science as a measure of the credence that the actuary believes should be attached to a particular body of experience for ratemaking purposes". Since 1918 to this very day, credibility is used to assess premiums by considering the way to merge two sources of information: the individual experience of the insured (the data) and the existing information on the entire portfolio or class the individual belongs to (prior information). While the problem is clearly tailored for a Bayesian framework, solutions are shown to be constrained by limiting practices deeply rooted in the actuarial profession.

Section 27.2 traces the early days of *credibility theory* and the influence of Bayesian thoughts. Section 27.3 describes the empirical Bayes solution to the problem, a solution which has dominated the actuarial profession until today. Section 27.4 is devoted to approximate Bayesian credibility solutions for special loss distributions. These approximations are inspired by the motivation to enrich *credibility theory* with Bayesian inputs while maintaining the constraints imposed by the profession.

For surveys of *credibility theory* see [22], [54], [18] and [43]. For surveys of Bayesian statistics in actuarial science, with some emphasis on credibility models, see [49], [40] and [39].

27.2 *Credibility theory*—early Bayesian thoughts

Bayesian thoughts first entered into actuarial science (life insurance in particular) with Richard Price (1723–1791), who was influenced by the writing of Thomas Bayes (1702–1761) and who published Bayes' essay ('An essay towards solving a problem in the doctrine of chances') posthumously in the *Philosophical Transactions of the Royal Society of London* (1763). Incidentally, the first life insurance policy in North America was issued in 1761, the year Bayes died. *Credibility theory* was conceived by American actuaries before World War I as the body of knowledge aimed at generating practical solutions to insurance ratemaking problems. [41] was concerned with the amount of individual risk exposure needed for the estimated mean claim to be regarded as reliable. In particular, if given θ, the risk parameter, annual claims amounts $X_1, \ldots, X_n$ are i.i.d with mean $\mu(\theta)$ (otherwise known as *risk* or *fair premium*) and variance $\sigma^2(\theta)$, how large should n be for $\bar{x}$ to be a *fully credible* estimator of $\mu(\theta)$? Mowbray relied on $p[|\ \bar{x} - \mu(\theta)\ | \leq k\mu(\theta)] \geq 1 - \varepsilon$ (k is somewhat arbitrary) and suggested that $n \geq \frac{z_{1-\varepsilon}^2 s^2}{k^2 \bar{x}}$, where $\bar{x}$, and s^2 are, respectively, the sample mean and variance of the n claims. The philosophy at the time, termed *limited fluctuation credibility,* was that if n is sufficiently large, $\bar{x}$ is a *fully credible* estimator of $\mu(\theta)$.

[52] suggested to replace *full credibility* with *partial credibility*. In particular, he proposed that the risk premium be estimated by *credibility premium* $\hat{\mu}$ via a *credibility formula,* a weighted average of the individual experience $\bar{x}$ and $m = E[\mu(\theta)]$, the industry-wide premium rate charged for a particular class, or the overall mean in the insurance portfolio:

$$\hat{\mu}(\theta) = z\bar{x} + (1 - z)m \tag{27.1}$$

where the weight z is the *credibility factor,* a factor which was regarded as equal to one in *limited fluctuation credibility*. This was the first step in what was called the *greatest accuracy credibility theory.*

[52] justified (27.1) by assuming a binomial distribution for the number of claims with a normal prior distribution for the parameter p, the probability of making a claim. Although his derivation was erroneous, it was a pioneering attempt to adopt Bayesian thinking for the solution of a practical actuarial problem. This has to be taken in historical perspective: At a time when Bayesian statistics was under the adverse dominance of Fisherian and frequentist approaches, actuaries suggested Bayesian-driven credibility models. [27] suggested a Poisson distribution with Gamma prior, and [3] investigated the Beta-Binomial and the Normal-Normal models. For Bailey, a Bayesian at heart and, unfortunately, unknown in Bayesian statistics circles, it was almost inconceivable to calculate premiums without the use of priors [4]:

> The statistical methods, developed by the mathematicians and available in the standard textbooks on statistical procedures, deal with the evaluation of the indications of a group of observations, but under the tacit or implicit assumption that no knowledge existed prior to the making of those particular observations. The credibility procedures, used in the revisions of casualty rates, have been developed by casualty actuaries to give consistent weightings to additional knowledge in its combination with already existing knowledge.

Similarly, [19] reflected:

> What was amazing was that, at the same time, several philosophers of science and statisticians were saying almost exactly the same thing. This group included Jeffreys, Barnard, Ramsey, de Finetti (also a contributor to actuarial science), Savage, and

Lindley. What made Bailey remarkable was that he came to his criticism of sampling theory and support of the Bayesian approach because of intensive study of a very practical business problem.

In an earlier paper, [2] introduced a generalized theory of credibility which contained the seeds of empirical Bayes thinking to be developed a decade later by [46].

27.3 The empirical Bayes phase

The early empirical Bayes ideas of Bailey were finally sprouting in the late 1960s against the background of dramatic advancement in Bayesian statistics, empirical Bayes and decision theory. In his classical model [7, 8] aimed at minimizing the *mean square error*

$$E[\mu(\theta) - \hat{\mu}(\theta)]^2 \tag{27.2}$$

within the linear form

$$\hat{\mu}(\theta) = a + b\bar{x}. \tag{27.3}$$

He derived the credibility formula (27.1) by using

$$z = \frac{\lambda n}{\lambda n + \eta}, \tag{27.4}$$

where

$$\lambda = V[m(\theta)] \tag{27.5}$$

and

$$\eta = E[\sigma^2(\theta], \tag{27.6}$$

with the unknown parameters λ and η estimated using the data.

In what has become a seminal paper, [9] modify the risk variance to take the form

$$V[X_j \mid \theta] = \frac{\sigma^2(\theta)}{p_j}, \tag{27.7}$$

to allow for different amounts of risk exposure (or volume) p_j. This led to a modified credibility formula with

$$z = \frac{\sum p_j \lambda}{\sum p_j \lambda + \eta} \tag{27.8}$$

and

$$\bar{x} = \frac{\sum p_j x_j}{\sum p_j}. \tag{27.9}$$

Here, too, in the spirit of empirical Bayes, the unknown parameters, akin to Bayesian hyperparameters, were estimated using the data, taking into account the exposures p_j. A decade later, [42] summarized this state of affairs as follows:

> A credibility estimator is Bayes in the restricted class of linear estimators and may be viewed as a linear approximation to the unrestricted Bayes estimator. When the structural parameters occurring in a credibility formula are replaced by consistent estimators based on data from the collective of similar risks, we obtain an empirical estimator, which is a credibility counterpart of empirical Bayes estimators.

[9]'s results swept the actuarial world and were universally adopted as the ultimate methodology for premium calculations. Since it was first published, practising actuaries are using these results unquestionably, although new research inroads into *credibility theory* have demonstrated that there is life beyond empirical Bayes techniques. This conservatism is, in part, the responsibility of professional bodies, like the Institute of Actuaries in the UK and the Society of Actuaries & the Casualty Actuarial Society in the USA, whose syllabuses have failed to bridge the gap between *credibility theory* and Bayesian statistics. This is in spite of the fact that the link between the two has long been established. [3] was the first to identify *exact credibility*, the credibility formula which coincides with a Bayesian solution. This was further investigated by [20, 21], [17], [28], [48], [10] in the case of the exponential family, [53] in a nonparametric setup and [30, 31] for the exponential dispersion family. In these contexts, the credibility formula is equivalent to $E[X_{n+1} \mid x_1, \ldots x_n]$, i.e. setting the premium at a value equal to the mean predicted risk, which has such an intuitive appeal. Unfortunately, most actuaries are unable to calculate this mean when the conditional expectation does not have a linear form.

27.4 Constrained Bayesian solutions

Over the years *credibility theory* has attracted a lot of attention in various directions, Bayesian and otherwise. Since it is not the aim of this paper to provide an extensive review of the topic, only a sample of recent contributions are now mentioned: [12], [47], [11], [16], [37], [1], [51], [14, 15], [44], [50] and [45].

The main aim of this section is to provide a brief description of the published approximate-Bayesian papers attempting to adapt the credibility formula (27.1) to other families of distributions, for which the predictive mean of the risk is no longer linear. The intention of these papers was to equip practising actuaries with useful approximate Bayes estimators which could only be accepted if they agreed with the structure of the classical credibility formula. Two approaches were introduced: the derivation of approximate credibility formula using second-order Bayes estimators [29] and the use of stochastic approximation techniques, which is briefly outlined.

We first make the observation [30, 31] that for the exponential dispersion family (EDF)

$$dP_{\theta,\lambda} = f(x|\theta,\lambda)\,dx = e^{\lambda(x\theta - k(\theta))} q_\lambda(x)dx, \quad \theta \in \Theta \subset R^1, \; \lambda \in \Lambda \in R^+, \tag{27.10}$$

with a conjugate prior for θ

$$\Pi(\theta) \propto e^{n_0(x_0\theta - k(\theta))}, \tag{27.11}$$

the following credibility formula exists:

$$E[X_{n+1} \mid x_1, \ldots, x_n, \lambda] = E[\mu(\theta) \mid x_1, \ldots, x_n, \lambda] = \frac{n\lambda}{n\lambda + n_0}\bar{x} + \frac{n_0}{n\lambda + n_0}m. \tag{27.12}$$

(27.12) can be written as a stochastic approximation recursion,

$$[\widehat{\mu}_n = \widehat{\mu}_{n-1} - a_n (\widehat{\mu}_{n-1} - x_n), \quad (27.13)$$

where $a_n = \frac{\lambda}{n_0 + n\lambda}$ is the gain function fully defined by the prior distribution's hyperparameters. Using stochastic approximation theory, (27.13) can be shown to converge to the true mean risk w.p.1. (For additional Bayesian analysis of credibility models for the EDF, see [34, 35].)

[32] exploited this observation to generate credibility formulas for other families of loss distribution. The first family is the location dispersion family (LDF) (see [23, 24, 25]),

$$dP_{\mu,\lambda} = f(x|\theta,\lambda)\, dx = a(\lambda) \exp(\lambda u(x-\theta)))dx, \;\; x \in R, \quad (27.14)$$

$$\theta \in R, \; \lambda \in R_+, \quad (27.15)$$

which includes, amongst other distributions, the log-gamma distribution, the Brandorff–Nielsen hyperbolic distribution and the log generalized inverse Gaussian distribution. For this family the stochastic approximation takes the sequential credibility form

$$\widehat{\mu}_n = \widehat{\mu}_{n-1} - a_n \lambda u'(x_n - \widehat{\mu}_{n-1} + \mu_0(\lambda)), \quad (27.16)$$

where $\mu_0(\lambda)$ is the expectation of (27.14) when $\theta = 0$. Regularity conditions are provided for the convergence with probability one of (27.16).

The gain function in (27.16) cannot automatically rely on the hyperparameters since the Bayes estimator is nonlinear. Instead, a gain function, the *first-order optimal gain function*, which minimizes the Bayes risk, $E_\lambda[\widehat{\mu}_n - \mu]^2$, is used

$$a_n^* = \frac{R_o}{n\kappa R_0 + 1} \quad (27.17)$$

where $R_0 = E_\lambda(\mu - m)^2$ is the variance of μ with respect to $\pi(\mu)\, d\mu$ and where $k = B_\lambda / I_\lambda$. Here $I_\lambda = \lambda^2 a(\lambda) \int_{-\infty}^{\infty} u'(x)^2 \exp(\lambda u(x))\, dx = -\lambda a(\lambda) \int_{-\infty}^{\infty} u''(x) \exp(\lambda u(x))\, dx$, is Fisher Information about parameter μ, and $B_\lambda = \lambda^2 a(\lambda) \int_{-\infty}^{\infty} u''(x)^2 \exp(\lambda u(x))\, dx$. The resulting credibility formula is called the *generalized sequential credibility formula.* For example, for losses following the log-gamma distribution, $\kappa = I_\lambda / B_\lambda = \lambda + 1$ and $a_n^* = \frac{R_o}{n(\lambda+1)R_0+1}$.

Special attention is given in [33] to the *symmetric location dispersion family* which includes two important members: the *exponential power family* $f(x|\theta,\lambda) = \frac{\delta\lambda^{1/\delta}}{2\Gamma(1/\delta)} \exp(-\lambda|x-\theta|^\delta)$, which attracted considerable attention in the context of robust Bayesian statistics [6], and the *generalized student-t family*, discussed in [5], $f(x|\theta,p,\sigma) = c_p\left[1 + \frac{(x-\theta)^2}{k\sigma^2}\right]^{-p}$, $-\infty < x < \infty$, where $c_p = \frac{1}{\sigma\sqrt{k}\beta(0.5,\lambda-0.5)}$ and for $p \geq 2, k = 2p-3$, and for $1 \leq p < 2, k = 1$. $\beta(.,.)$ is the beta function.

[36] extended the sequential approach to a problem of credibility evaluation of a *scale-dispersion family* (SDF) ([26], Section 1.4.2),

$$dP_{\theta,\lambda} = a(\lambda)\, x^{-1} \exp(\lambda u(\theta x)))dx, \;\; x \in R_+. \quad (27.18)$$

Here $\theta \in R_+$ is the risk parameter of $P_{\theta,\lambda}$, the distribution of the claim size X, and $\lambda \in R_+$ is a dispersion parameter. SDF contains important distributions. For example, the lognormal family, for which $u(x) = -\frac{1}{2}\ln^2 x$, frequently used successfully to represent loss data. A considerably rich three-parameter class of SDF was introduced by [13], $dP_{\theta,\lambda,\gamma} = \frac{|\gamma|\lambda^{\lambda}}{\Gamma(\lambda)} \frac{1}{x} \exp(\lambda\left(\gamma \ln(x\theta) - (x\theta)^{\gamma}\right))dx$.

[36] modified, respectively, the Bayesian risk sequences and the first-order stepwise optimal gain:

$$R_n^* = \frac{R_o}{nkR_0/(c_\lambda V_\lambda(X)) + 1} \tag{27.19}$$

$$a_n^* = \frac{\mu_0(\lambda)}{nk + c_\lambda V_\lambda(X)/R_0}, \tag{27.20}$$

$$n = 1, 2, \ldots \tag{27.21}$$

where

$$c_\lambda = \frac{\mu_0(\lambda)^2}{V_{\mu_0(\lambda),\lambda}(X)}, \tag{27.22}$$

and provided regularity conditions for the convergence w.p.1 of the sequential credibility formula

$$\widehat{\mu}_n = \widehat{\mu}_{n-1} - a_n^* \lambda x_n u' \left(x_n \frac{\mu_0(\lambda)}{\widehat{\mu}_{n-1}} \right). \tag{27.23}$$

27.5 Epilogue

The actuarial profession is disadvantaged by its failure to adopt modern Bayesian methodologies. This unfortunately affects us all as risk management tools do not exploit the rich Bayesian advancements continuously reported in the statistical literature. MCMC for example, while reported in actuarial journals to a limited extent, has failed to enter the professional guidelines provided to practicing actuaries. Given the conservatism of this profession, changes should be introduced at a slow pace, centred on gradual modifications of exiting actuarial techniques with an increasing Bayesian content. In this way, as demonstrated above, the common linear empirical Bayes can be modified to cater for a wide range of loss distributions within the constraint of linearity. While the focus of this paper was on *credibility theory*, the observations made here would apply to other major problems in actuarial science.

References

[1] Atanasiu, V. (2007). Theoretical aspects of credibility theory. *Romai J.*, **3**, 1–14.

[2] Bailey, A. (1945). A generalized theory of credibility. *Proc. of the Casualty Act. Soc.*, **32**, 13–20.

[3] Bailey, A. (1950). Credibility procedures: LaPlace's generalization of Bayes' rule and the combination of collateral knowledge with observed data. *Proc. of the Casualty Act. Soc.*, **37**, 7–23.

[4] Bailey, A. (1950). Discussion of a paper of Carson. *Journal of American Teachers of Insurance* (now *The Journal of Risk and Insurance*), **17**, 17–24.

[5] Bian, G. & Tiku, M. (1997). Bayesian inference based on robust priors and MML estimators: Part 1, symmetric location-scale distributions. *Statistics*, **29**, 317–345.

[6] Box, G. & Tiao, G. (1973). *Bayesian Inference in Statistical Analysis*. Addison-Wesley, Reading, MA.

[7] Bühlmann, H. (1967). Experience and credibility. *Astin bulletin*, **4**, 199–207.

[8] Bühlmann, H. (1969). Experience rating and credibility. *Astin bulletin*, **5**, 157–165.
[9] Bühlmann, H. & Straub, E. (1970). Glaubwürdigkeit für Schadensätze. *Mitt. SVVM*, **70**, 111–133.
[10] Diaconis, P. & Ylvisaker, D. (1979). Conjugate priors for exponential families. *Annals of Statistics*, **7**, 269–281.
[11] Ebegil, M. (2006). A study to examine Bühlmann–Straub credibility model in generalized linear models. *Commun. Fac. Sci. Univ. Ank.* Series A1, **55**, 9–16.
[12] Frees, E. (2003). Multivariate credibility for aggregate loss models. *North American Actuarial Journal*, **7**, 13–37.
[13] Fergusson, T. (1962). Location and scale parameters in exponential families of distributions. *Ann. Math. Stat.*, **33**, 986–1001.
[14] Gomez-Deniz, E. (2008). A Generalization of the credibility theory obtained by using the weighted balanced loss function. *Insurance: Mathematics & Economics*, **42**, 850–854.
[15] Gomez-Deniz, E. (2008). Deriving credibility premiums under different Bayesian methodology. *Advances in Mathematical and Statistical Modelling. Statistics for Industry and Technology*, **6**, 219–229.
[16] Goulet, V., Forgues, A. & Lu, J. (2006). Credibility for severity revisited. *North American Actuarial Journal*, **10**, 49–62.
[17] Herzog, T. (1989). Credibility: the Bayesian model versus Bühlmann's model. *Transactions of Society of Actuaries*, **41**, 43–88.
[18] Herzog, T. (1994). *Introduction to Credibility Theory*. ACTEX Publications, Winsted.
[19] Hickman, J. & Heacox, L. (1999). Credibility theory: The cornerstone of actuarial science. *North American Actuarial Journal*, **3**, 1–8.
[20] Jewell, W. (1974). Credible means are exact Bayesian for exponential families. *Astin bulletin*, **8**, 77–90.
[21] Jewell, W. (1975). The use of collateral data in credibility theory: a hierarchical model. *Giornale dell' Istituto Italianio degh Attuari*, **38**, 1–16.
[22] Jewell, W. (1976). A survey of credibility theory. Operation Research Center, University of California, Berkeley.
[23] Jorgensen, B. (1983). Maximal likelihood estimation and large-sample inference for generalized linear and nonlinear regression models. *Biometrika*, **70**, 19–28.
[24] Jorgensen, B. (1987). Exponential dispersion models (with discussion). *J. Roy. Statist. Soc.* Ser.B, **49**, 127–162.
[25] Jorgensen, B. (1992). Exponential dispersion models and extensions: A review. *Internat. Statist. Rev.*, **60**, 5–20.
[26] Jorgensen, B. (1997). *The Theory of Dispersion Models*. Chapman and Hall, London.
[27] Keffer, R. (1929). An experience rating formula. *Transactions of the Actuarial Society of America*, **30**, 130–139.
[28] Klugman, S. (1992). *Bayesian Statistics in Actuarial Science with Emphasis on Credibility*. Kluwer Academic Publishers.
[29] Landsman, Z. (2002). Credibility theory: a new view from the theory of second order optimal statistics. *Insurance: Mathematics & Economics*, **30**, 351–362.
[30] Landsman, Z. & Makov, U. (1998). Exponential dispersion models and credibility. *Scandinavian Actuarial Journal*, **1**, 89–96.
[31] Landsman, Z. & Makov, U. (1999). Credibility evaluations for exponential dispersion families. *Insurance: Mathematics & Economics*, **24**, 33–39.
[32] Landsman, Z. & Makov, U. (1999). On stochastic approximation and credibility. *Scandinavian Actuarial Journal*, **1**, 15–31.
[33] Landsman, Z. & Makov, U. (1999). Sequential credibility evaluation for symmetric location claim distributions. *Insurance: Mathematics & Economics*, **24**, 291–300.

[34] Landsman, Z. & Makov, U. (2001). Bayesian prediction in the exponential dispersion family with an application to actuarial credibility. *Monographs of Official Statistics*, Eurostat, 283–289.
[35] Landsman, Z. & Makov, U. (2001). On credibility evaluation and the tail area of exponential dispersion family. *Insurance: Mathematics & Economics*, **27**, 277–283.
[36] Landsman, Z. & Makov, U. (2003). Sequential quasi credibility for scale dispersion models. *Scandinavian Actuarial Journal*, **2**, 119–135.
[37] Lau, J., Siu, T. K. & Yang, H. (2006). On Bayesian mixture credibility. *Astin bulletin*, **36**, 573–588.
[38] Longley-Cook, L. (1962). An introduction to credibility theory. *Proc. of the Casualty Act. Soc.*, **49**, 194–226.
[39] Makov, U. (2001). Principal applications of Bayesian methods in actuarial science: A perspective. (with discussion). *North American Actuarial Journal*, **5**, 53–73.
[40] Makov, U., Smith, A. & Liu, Y. (1996). Bayesian methods in insurance: a review. *The Statistician*, **45**, 503–515.
[41] Mowbray, A. (1914). How extensive a payroll exposure is necessary to give a dependable pure premium. *Proc. of the Casualty Act. Soc.*, **1**, 24–30.
[42] Norberg, R. (1980). Empirical Bayes credibility. *Scandinavian Actuarial Journal*, **4**, 177–194.
[43] Norberg, R. (2006). Credibility theory. *Encyclopedia of Actuarial Science*, John Wiley & Sons.
[44] Ohlsson, E. (2008). Combining generalized linear models and credibility models in practice. *Scandinavian Actuarial Journal*, **4**, 301–314.
[45] Payandeh Najafabadi, Amir T. (2010). A new approach to the credibility formula. *Insurance: Mathematics & Economics*, **46**, 334–338.
[46] Robbins, H. (1955). An empirical Bayes approach to statistics. *Proceeding of the Third Berkeley Symposium on Mathematical Statistics and Probability*, **1**, University of California Press, Berkeley, 157–163.
[47] Ronka-Chmielowiec, W. & Poprawska, E. (2005). Selected methods of credibility theory and its application to calculating insurance premium in heterogeneous insurance portfolios. in *Innovations in Classification, Data Science, and Information Systems*, **VI**, 490–497.
[48] Schmidt, K. (1990). Convergence of Bayes and credibility premiums. *Astin bulletin*, **20**, 167–172.
[49] Schmidt, K. (1998). Bayesian models in actuarial mathematics. *Math. Meth. Oper. Res.*, **48**, 117–146.
[50] Siu, T. K. & Yang, H. (2009). Nonparametric Bayesian credibility. *Australian Actuarial Journal*, **15**, 209–230.
[51] Van Der Merwe, A. & Bekker, K. (2007). A computational Bayesian approach to the balanced Bühlmann credibility model. *South African Statistical Journal*, **41**, 65–103.
[52] Whitney, A. (1918). The theory of experience rating. *Proc. of the Casualty Act. Soc.*, **4**, 274–292.
[53] Zehnwirth, B. (1977). The mean credibility formula is a Bayes rule. *Scandinavian Actuarial Journal*, **4**, 212–216.
[54] Zehnwirth, B. (1983). Credibility theory: a concise partial survey. *Austral. J. Statist.*, **25**, 402–411.

Part XI

Medicine and Biostatistics

28 Bayesian models in biostatistics and medicine

PETER MÜLLER

28.1 Introduction

Biomedical studies provide many outstanding opportunities for Bayesian thinking. The principled and coherent nature of Bayesian approaches often leads to more efficient, more ethical and more intuitive solutions. In many problems the increasingly complex nature of experiments and the ever increasing demands for higher efficiency and ethical standards leads to challenging research questions.

In this chapter we introduce some typical examples. Perhaps the biggest Bayesian success stories in biostatistics are hierarchical models. We will start the review with a dicussion of hierarchical models. Arguably the most tightly regulated and well controlled applications of statistical inference in biomedical research are the design and analysis of clinical trials, that is, experiments with human subjects. While far from being an accepted standard, Bayesian methods can contribute significantly to improving trial designs and to constructing designs for complex experimental layouts. We will discuss some areas of related current developments. Another good example of how the Bayesian paradigm can provide coherent and principled answers to complex inference problems are problems related to the control of multiplicities and massive multiple comparisons. We will conclude this overview with a brief review of related research.

28.2 Hierarchical models

28.2.1 Borrowing strength in hierarchical models

A recurring theme in biomedical inference is the need to borrow strength across related subpopulations. Typical examples are inference in related clinical trials, data from high throughput genomic experiments using multiple platforms to measure the same underlying biologic signal, inference on dose–concentration curves for multiple patients etc. The generic hierarchical model includes multiple levels of experimental units. Say

$$y_{ki} \mid \boldsymbol{\theta}_k, \phi \sim p(y_{ki} \mid w_{ki}, \boldsymbol{\theta}_k, \phi)$$
$$\boldsymbol{\theta}_k \mid \phi \sim p(\boldsymbol{\theta}_k \mid \boldsymbol{x}_k, \phi),$$
$$\phi \sim p(\phi) \tag{28.1}$$

Without loss of generality we will refer to experimental units k as 'studies' and to experimental units i as 'patients', keeping in mind an application where $\boldsymbol{y}_k = (y_{ki}, i = 1, \ldots, n_k)$ are the responses recorded on n_k patients in the k-th study of a set of related biomedical studies. In that case $\boldsymbol{w}_k = (w_{ki}, i = 1, \ldots, n_k)$ could be patient-specific covariates, $\boldsymbol{\theta}_k$ are study specific parameters, typically including a study-specific treatment effect, and $\boldsymbol{x}_k$ might be study-specific covariates. We will use these terms to refer to elements of the hierarchical model, simply for the sake of easier presentation, but keeping in mind that the model structure is perfectly general.

In a pharmacokinetic (PK) study k could index patients and $\boldsymbol{y}_k = (y_{ki}, i = 1, \ldots, n_k)$ could be drug concentrations for patient k observed at n_k time points, $w_{ki}, i = 1, \ldots, n_k$. In that case $\boldsymbol{\theta}_k$ are the PK parameters that characterize how patient k metabolizes the drug. In a pharacodynamic (PD) study y_{ki} could be repeat measurements on blood pressure, blood counts, etc.

In another example, k could index different platforms for high throughput genomic experiments, for example $k = 1$ for RPPA data that records protein activation and $k = 2$ for microarray data that measures gene expression. In that case $i, i = 1, \ldots, n$, could index different genes and proteins and $\boldsymbol{\theta}_k = (\theta_{ki}, i = 1, \ldots, n)$ could code differential gene expression and protein activation.

In [40] we use a hierarchical model for small area estimation. We borrow strength across states to estimate the rate of mammography usage θ_k in each state in the USA, $k = 1, \ldots, K$. The data were collected at a national level, leaving very small or zero sample sizes in some states. By borrowing strength across states we can still report inference for all states. Figure 28.1 shows the posterior means $E(\theta_k \mid data)$.

In [54] a hierarchical model is used to define a clinical trial design for sarcoma patients. Sarcoma is a very heterogeneous disease. The subpopulations $k = 1, \ldots, K$ index $K = 12$ different sarcoma types, and $i = 1, \ldots, n_k$ indexes enrolled patients who are diagnosed with sarcoma subtype k. The hierarchical model borrows strength across sarcoma types. Some sarcoma types are very rare. Even in a large cancer hospital it would be difficult to accrue sufficiently many patients for a trial for one subtype alone. Only by borrowing strength across subtypes does it become feasible to investigate such rare subtypes. The other extreme of pooling all patients would be equally inappropriate, as it would ignore the heterogeneity and varying prognosis across different subtypes. The hierarchical model allows for a compromise of borrowing strength at a level between pooling the data and running separate analyses. One limitation, however, remains. The hierarchical model (28.1) assumes that all subtypes are a priori exchangeable. That is not quite appropriate for the sarcoma subtypes. There are likely to be some known differences. [38] develop a variation of hierarchical models that allows for exchangeability of patients across subsets of subpopulations. In the case of the sarcoma study this implies that patients within some sarcoma subtypes are pooled. The selection of these subsets itself is random, with an appropriate prior.

In all five examples the second level of the hierarchical model formalizes the borrowing of strength across the submodels. Most applications include conditional independence at all levels, with $\boldsymbol{\theta}_k$ independent across k conditional on ϕ and y_{ki} independent across i conditional on $\boldsymbol{\theta}_k, \phi$. All five examples happen to use hierarchical models with two levels. Extensions to more than two levels are conceptually straightforward.

The power of the Bayesian approach to inference in hierarchical models is the propagation of uncertainties and information across submodels. For example, when $k = 1, \ldots, K$ indexes related clinical trials then inference for the k-th trial borrows strength from patients enrolled in the other $K - 1$ trials. Let $\boldsymbol{y}_{-k} = (\boldsymbol{y}_\ell, \ell \neq k)$ denote all data excluding the k-th study. We can rewrite the implied model for the k-th study as

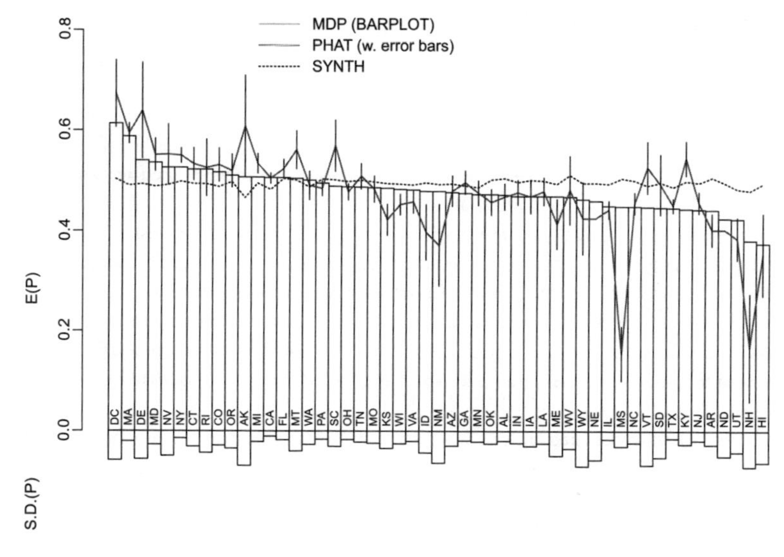

Figure 28.1 Posterior estimates $E(\theta_k \mid data)$ for each US state in a hierarchical model for mammography usage. The barplot shows the posterior means. The barplot below the x-axis shows the posterior standard deviations $SD(\theta_k \mid data)$. The solid line shows the data (empirical frequency of mammography usage). The dotted line shows the estimate $\widehat{\theta_k}$ in a regression on state-specific demographic summaries.

$$p(\mathbf{y}_k \mid \boldsymbol{\theta}_k, \phi) \text{ and } p(\boldsymbol{\theta}_k, \phi \mid \mathbf{y}_{-k})$$

This highlights the nature of borrowing information across studies. The original prior is replaced by the posterior conditional on data from the other studies. We could describe this aspect of hierarchical models as a principled construction of informative priors based on related studies. The important feature in the process is that this borrowing of strength is carried out in a coherent fashion as dictated by probability calculus, rather than ad-hoc plug-in approaches. The implication is a coherent propagation of uncertainties and information.

Besides the pragmatic aspect of borrowing strength, hierarchical models can also be introduced from first principles. Essentially, if observations within each subpopulation are judged as arising from an infinitely exchangeable sequence of random quantities, and the subpopulations themselves are judged to be exchangeable a priori, then model (28.1) is implied [5, Chapter 4].

28.2.2 Posterior computation

One of the early pathbreaking discussions that introduced Bayesian thinking for hierarchical models appears in [39]. The paper appeared long before the routine use of posterior Markov chain Monte Carlo simulation, when computational implementation of Bayesian inference in complex models was still a formidable challenge. One of the important contributions of Lindley and Smith's paper was to highlight the simple analytic nature of posterior inference when all levels of the hierarchical model are normal linear models.

The restriction to models that allow analytic results severely limited the practical use of Bayesian inference in biomedical applications. This radically changed with the introduction of Markov chain Monte Carlo posterior simulation in [21]. In fact, hierarchical models were one of the illustrative examples in [21] and the companion paper [19] with more illustrative applications.

28.2.3 Related studies and multiple platforms

One of the strengths of the Bayesian approach is the coherent and principled combination of evidence from various sources. A critical need for this feature arises in the combination of evidence across related studies. While many studies are still planned and published as if they were stand-alone experiments, as if they were carried out in total isolation from other related research, this is an incresingly unreasonable simplification of reality.

One simple approach to borrow information across related studies that investigate the same condition is post-processing of results from these studies. This is known as meta-analysis [11, Chapter 3]. A typical example appears in [24], who analyse evidence from eight different trials that all investigated the use of intravenous magnesium sulphate for acute myocardial infarction patients. The discussion in [24] shows how a Bayesian hierarchical model with a suitably sceptic prior could have anticipated the results of a later large-scale study that failed to show any benefit of the magnesium treatment.

Multiple related studies need not always refer to clinical trials carried out by different investigators. An increasingly important case is the use of multiple experimental platforms to measure the same underlying biologic signal. This occurs frequently in high throughput genomic studies. In a recent review paper [29] argue for the need of hierarchical modelling to obtain improved inference by pooling different data sources.

28.2.4 Population models

In [63] the authors discuss population models as an important special case of hierarchical models and Bayesian inference in biostatistics. Typically each submodel corresponds to one patient, with repeated measurements y_{ki}, and a sampling model that is indexed by patient-specific parameters θ_k and perhaps additional fixed effects ϕ. The first-level prior in the general hierarchial model (28.1) now takes the interpretation of the distribution of patient-specific parameters θ_k across the entire patient population.

In population PK/PD models the hierarchical prior $p(\boldsymbol{\theta}_k \mid \boldsymbol{x}_k, \phi)$ represents the distribution of PK (or PD) parameters across the population. One of the typical characteristics of patient populations is heterogeneity. There is usually no good reason beyond technical convenience to justify standard priors like a multivariate normal. While the population distribution $p(\boldsymbol{\theta}_k \mid \boldsymbol{x}_k, \phi)$ is usually not of interest in itself, good modelling is important, mainly for prediction and inference for future patients. Let $i = n + 1$ index a future patient and let $\boldsymbol{Y} = (\boldsymbol{y}_1, \ldots, \boldsymbol{y}_n)$ denote the observed data. Inference for a future patient is driven by

$$p(\boldsymbol{\theta}_{n+1} \mid \boldsymbol{x}_{n+1}, \boldsymbol{Y}) = \int p(\boldsymbol{\theta}_{n+1} \mid \boldsymbol{x}_{n+1}, \phi)\, \mathrm{d}p(\phi \mid \boldsymbol{Y})$$

assuming that the patient-specific PK parameters $\boldsymbol{\theta}_k$ are conditionally independent given ϕ. The expression for the posterior for $\boldsymbol{\theta}_{n+1}$ highlights the critical dependence of prediction on the parametric form of $p(\boldsymbol{\theta}_k \mid \boldsymbol{x}_k, \phi)$. For example, assuming a normal distribution might severely underestimate the probability of patients with unusual PK parameters. Several authors have investigated the use of more general population models in Bayesian population PK/PD models. Let $N(x; m, S)$ denote a multivariate normal distribution for the random variable x with moments m and S. For example, a mixture of normals

$$p(\boldsymbol{\theta}_k \mid \phi) = \sum_{\ell=1}^{L} w_\ell N(\boldsymbol{\theta}_k; \boldsymbol{\mu}_\ell, \Sigma_\ell) \tag{28.2}$$

could be used to generalize a normal population model without substantially complicating posterior simulation. Here $\phi = (L, w_\ell, \boldsymbol{\mu}_\ell, \Sigma_\ell, \ell = 1, \ldots, L)$ indexes the population model. This and related models have been used, for example, in [31, 43] and others. The mixture model needs to be completed with a prior for the parameters ϕ. This is easiest done by writing the mixture as $\int N(\boldsymbol{\theta}_k; \boldsymbol{\mu}, \Sigma) dG(\boldsymbol{\mu}, \Sigma)$ for a discrete probability measure $G = \sum_\ell w_\ell \delta_{(\boldsymbol{\mu}_\ell, \Sigma_\ell)}$. Here δ_x indicates a point mass at x. The prior specification then becomes the problem of constructing a probability model for the random probability measure G. We will discuss such models in more detail below, in Section 28.7.

28.3 Bayes in clinical trials: phase I studies

Few other scientific studies are as tightly regulated and controlled as clinical trials. However, most regulatory constraints apply for phase III studies that compare the new experimental therapy against standard of care and clinically relevant standards. For early phase studies the only constraint is that they be carried out in scientifically and ethically responsible ways. This is usually controlled by internal review boards (IRB) that have to approve the study.

Phase I studies aim to establish safety. In oncology a typical problem is to find the maximum tolerable dose (MTD) of a new chemotherapy agent (or combination of agents). Let $\boldsymbol{z} = (z_j, j = 1, \ldots, J)$ denote a grid of available doses. Many designs assume that the formal aim of the study is to find a dose with toxicity closest to a pre-determined maximum tolerable target level $\pi^\star$. Typically the outcome is a binary indicator, $y \in \{0, 1\}$ for a dose-limiting toxicity. Let $\pi_j = p(y = 1 \mid z_j)$ denote the probability of toxicity at dose level j. One of the still most widely used designs is the so called 3+3 design. It is simply a rule for escalating the dose in subsequent patient cohorts until we observe a certain maximum number of toxicities in a cohort. The design is an example of a rule-based design. In contrast to model-based designs that are based on inference with respect to underlying statistical models, rule-based designs simply follow a reasonable but otherwise ad hoc algorithm. There is no probabilistic guarantee that the reported MTD is in fact a good approximation of the unknown truth. Rule-based designs are popular due to the ease of implementation, but they are also known to be inefficient [36]. Here, efficiency is judged by frequentist summaries under repeated use of a design. Summaries include the average sample size and the average true probability of toxicity at the reported MTD.

[47] proposed one of the first model-based Bayesian designs to address some of the limitiations of traditionally used rule-based designs. The method is known as continual reassessment method (CRM). The underlying model is quite straightforward. Let d_j, $j = 1, \ldots, J$ denote a grid of available dose levels and recall that π_j is the probability of toxicity at dose j. For a given skeleton $\boldsymbol{d} = (d_1, \ldots, d_J)$ of doses the model is indexed with only one more parameter as $\pi_j = d_j^a$. Several variations with alternative one-parameter models are in use. The algorithm uses a target dose, say $\pi^\star = 0.30$ and proceeds by sequentially updating posterior inference on a and assigning the respective next patient cohort to the dose with $\widehat{\pi}_j = d_j^{\widehat{a}}$ closest to the target toxicity $\pi^\star$. Here $\widehat{a} = E(a \mid data)$ is the posterior mean conditional on all currently available data and $\widehat{\pi}_j$ is a plug in estimate of π_j. In a minor variation one could imagine to replace $\widehat{\pi}_j$ by the posterior mean $\widehat{\pi}_j = E(d^a \mid data)$. The coding of the doses d_j is part of the model, but is fixed up-front. The values d_j are not raw dose values, but chosen to achieve desired prior means $\pi_j = E(d_j^a)$. Here the

expectation is with respect to the prior on a. The initially proposed CRM gave rise to serious safety concerns. The main issue is that the algorithm could jump to inappropriately high doses, skipping intermediate and yet untried doses.

Several later modifications have improved the original CRM, including the modified CRM of [22] to address the safety concerns, and the TITE-CRM of [9] for time to event outcomes and many more. The TITE-CRM still uses essentially binary outcomes, but allows for weighting to accomodate early responses and enter patients in a staggered fashion. Let U_i denote the event time for patient i, for example, time to toxicity. Let y_i denote a binary outcome for patient i, defined as U_i beyond a certain horizon T, i.e, $y_i = I(U_i > T)$. Let y_{in} denote the toxicity status of patient i just before the n-th patient enters the trial. When U_i is already observed then $y_{in} = y_i$. Also when U_i is censored beyond T, i.e. U_i is known to be beyond the horizon T, then $y_{in} = y_i = 0$. Only when U_i is censored before T, then $y_{in} = 0$, while y_i would still be considered censored. The TITE-CRM replaces the binary response y_i by y_{in} and uses an additional weight $w_i = \min(U_i/T, 1)$ to replace π_j in the likelihood by $g_j = \pi_j w_i$. This allows use to be made of early responses implied by censored $U_i < T$ and can significantly reduce the trial duration. The approach of [55] goes a step further using a parametric model for the time to event endpoint. [3] generalize the TITE-CRM with a probit model for discretized event times to allow for lack of monotonicity. Some authors have suggested alternative model-based approaches. The EWOC (escalation with overdose control) method of [1] uses a cleverly parameterized logistic regression of outcome on dose. A common theme of these model-based methods is the use of very parsimonious models. This is important in the context of the small sample sizes in phase I trials.

Another common feature is the use of a target toxicity level $\pi^\star$. Several recent authors argue that the assumption of a single value $\pi^\star$ is unrealistic, and replace the notion of a target dose by target toxicity intervals. This view is taken, for example, in [30]. Keeping the model ultimately simple they use independent Beta/Binomial models for each dose. Only post-processing with isotonic regression ensures monotonicity. A related approach is proposed in [45] who go a step further and introduce ordinal toxicity intervals, including intervals for underdosing, target toxicity, excessive toxicity, and unacceptable toxicity. The underlying probability model is a logistic regression centred at a reference dose. Sequential posterior updating after every patient-cohort includes updated posterior probabilities for the target toxicity intervals at each dose. The respective next dose is assigned by trading off the posterior probabilities of these intervals.

The designs and models mentioned so far are all exclusively phase I designs. Inference is only concerned with toxicity outcomes, entirely ignoring possible efficacy outcomes. The EffTox model introduced in [56] explicitly considers both, toxicity and efficacy. Thall and Cook develop a design that trades off target levels in both endpoints. The probability model is based on two marginal logistic regressions for a binary toxicity and a binary efficacy outcome, and one additional parameter that induces dependence of the two outcomes. The design includes sequential posterior updating and a desirability function that is used like a utility function to select acceptable doses and eventually an optimal dose.

28.4 Phase II studies

Phase II studies aim to establish some evidence for a treatment effect, still short of a formal comparison with a control or standard of care in the following final phase III study. Larger sample sizes and more structure, compared with phase I, allow for more impact of model-based Bayesian design. Opportunities for innovative Bayesian designs arise in sequential stopping, adaptive design and the use of problem-specific utility functions.

28.4.1 Sequential stopping

Some sequential stopping designs use posterior predictive probabilities to decide upon continuation versus termination of a trial. Let PP denote the posterior predictive probability of a positive result at the end of the trial. For example, assume that the efficacy outcome is a binary indicator for tumour response, and that a probability of tumour response $\pi > \pi_0$ is considered a clinically meaningful response. Let y denote the currently available data, i.e. outcomes for already enrolled patients, and let y^o denote the still unobserved responses for future patients if the trial were to run until some maximum sample size. Also assume that evidence for efficacy is formalized as the event $\{p(\pi > \pi_0 \mid data) > 1 - \epsilon\}$. The posterior predictive probability PP $= p\{p(\pi > \pi_0 \mid y, y^o) > 1 - \epsilon \mid y\}$ is the probability of a successful trial if the study continues until the end. Continuous monitoring of such predictive probabilities facilitates the implementation of flexible stopping rules based on the chance of future success. This is implemented in [28] who define a sequential stopping rule based on the posterior predictive probability of (future) conclusive evidence in favour of one of two competing treatments. Similarly, also [37] argue for the use of posterior predictive probabilties to implement sequential stopping in phase II trials. See also the discussion in [7].

Similar in spirit, many recently proposed clinical trial designs use continuously updated posterior probabilities of clinically meaningful events to define stopping rules. The use of posterior predictive probabilities can be seen as a special case of this general principle, using a particular type of posterior inference related to future posterior probabilities. In general, one could consider any event of interest, usually related to some comparison of success probabilities or other parameters under one versus the other therapy. [11, Chapter 6] refers to such approaches as 'proper Bayes' design. A class of such designs that use continuously updated posterior probabilities for sequential stopping for futility and efficacy was introduced in [58] and [57]. For example, Let $p_n = p(\theta_E > \theta_S + \delta \mid data)$ denote the posterior probability that the response rate θ_E under the experimental therapy is larger than the response rate θ_S under standard of care by more than δ. A design could stop for futility when $p_n < L$ and stop for efficacy when $p_n > U$, where U and L are fixed thresholds. [58] include a model for multiple outcomes, including indicators for toxicity and tumour response. In that case the posterior probabilities of appropriate combinations of outcomes under competing treatments can be used to define stopping. Figure 28.2 summarizes a possible history of a trial using such stopping rules. The two curves $p_n(EFF)$ and $p_n(TOX)$ plot the continuously updated posterior probabilities $p_n(\cdot) = p(\cdot \mid data)$ of a toxicity event (TOX) and an efficacy event (EFF). The two horizontal lines are thresholds. When $p_n(TOX) > U$ the trial stops for toxicity. When $p_n(EFF) < L$ the trial stops for futility, i.e. lack of efficacy. This particular trial had no sequential stopping for efficacy. The figure shows a simulated trial history under a very favourable assumed truth. Neither curve crosses the corresponding stopping boundary.

28.4.2 Adaptive allocation

Sequential stopping is one (important) example of outcome adaptive designs. Those are clinical trial designs that use interim analysis during the course of the trial to change some design element as a function of already observed outcomes. The implementation and careful planning of such experiments is far more natural and straightforward under a Bayesian approach than under the classical paradigm [6].

Besides sequential stopping another commonly used type of adaptive design is adaptive treatment allocation for multi-arm trials, i.e. trials that recruit patients for more than one therapy. The intent of most adaptive allocation designs is to favour allocation to the better therapy, but to keep some minimum level of randomization. Adaptive allocation is thus different from deterministic rules like play-the-winner. The designs are usually rule-based, i.e. there is no notion of optimality.

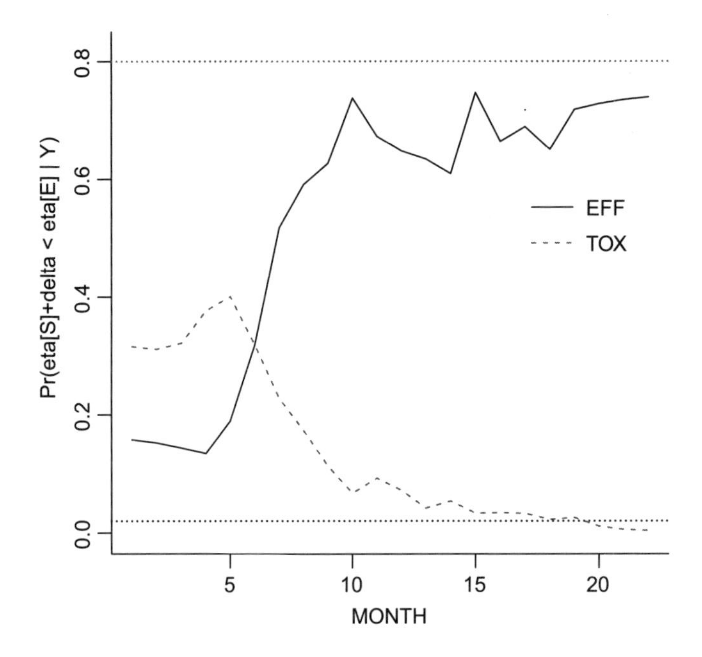

Figure 28.2 Sequentially updated posterior estimates for toxicity and efficacy and stopping by decision boundaries.

The adaptive allocation is carried out following some reasonable, but ad-hoc rule. The idea goes back to at least [59]. A recent review appears in [53]. Consider a study that allocates patients to two treatments, say T_1 and T_2, and let θ_1 and θ_2 denote treatment-specific parameters that can be interpreted as efficacy of treatment T_1 and T_2, respectively. For example, θ_j might be the probability of tumour response under treatment T_j. Let $r_j(\boldsymbol{y})$ denote the probability of allocating the next patient (or patient cohort) to treatment T_j. Importantly, $r_j(\cdot)$ is allowed to depend on the observed outcomes $\boldsymbol{y}$. A popular rule is

$$r_j(\boldsymbol{y}) \propto \left\{ p(\theta_j = \max_i \theta_i \mid \boldsymbol{y}) \right\}^c$$

The two probabilities, the randomization probability r_j and the posterior probability of T_j being the best treatment, are defined on entirely different spaces, and are entirely unrelated in the probability model. Only the adaptive allocation rule links them, following the vague notion that it is preferable to assign more patients to the better treatment. The power c is a tuning parameter. It is usually chosen as $c = 1$ or $c = 1/2$, [53] recommend $c = n/2N$ for current sample size n and maximum sample size N. The recommendation is based on empirical evidence only.

Finally, for a fair discussion of adaptive trial design we should note that Bayesian adaptive designs are far from routine. In fact, in a recent guidance document the US Food and Drug Administration (FDA) does not encourage the use of Bayesian methods for adaptive designs.

28.4.3 Delayed response

One of the big white elephants in drug development is the high failure rate of phase III confirmatory studies, even after successful phase II trials. In cancer more than 50% of phase III studies fail [32].

One of the possible causes for this is that phase II studies typically use a binary indicator for tumour response, for example an indicator for tumour shrinkage, as a proxy for survival, which is a typical phase III endpoint. Ideally one would want to see the same endpoint being used in phase II and III. But the delayed nature of a survival response would make phase II trials unacceptably slow. See [26] for a discussion. They use model-based Bayesian inference to develop a phase II design that mitigates the problem. The solution is simple and principled. Let $S_i \in \{1, 2, 3, 4\}$ denote an indicator for tumour response for the i-th patient. They code tumour response as a categorical outcome with four levels. Let T_i denote survival time. They use a joint sampling model $p(S_i, T_i \mid \boldsymbol{\lambda}, \boldsymbol{p})$, where $\boldsymbol{p}$ are the probabilities of S_i under different treatments and $\boldsymbol{\lambda}$ indexes an exponential regression of T_i on S_i and treatment assignment. Under this model it now becomes meaningful to report posterior probabilities for events related to survival. As a welcome side effect, inference about tumour response is improved by borrowing strength from the survival outcome.

28.5 Two Bayesian success stories

One of the big opportunities and challenges for Bayesian approaches in clinical trial design is the possibility to match treatments with patients in a coherent fashion. Two recent successful studies illustrate these features.

28.5.1 ISPY-2

ISPY-2 [2] uses a Bayesian adaptive phase II clinical trial design. The trial considers neoadjuvant treatments for women with locally advanced breast cancer. The study simultaneously considers five different experimental therapies. All five treatments are given in combination with standard chemotherapy, before surgery (neoadjuvant). ISPY-2 defines an adaptive trial design, i.e. design elements are changed in response to the observed data. Adaptation in ISPY-2 includes changing probabilities of assigning patients to the treatment arms and the possibility of dropping arms early for futility or efficacy. In the latter case the protocol recommends a following small phase III study. The treatment is 'graduated.'

In addition to these more standard adaptive design elements, the most innovative and important feature of the trial is explicit consideration of population heterogeneity by defining subpopulations based on biomarkers and a process that allows different treatments to be recommended for each subpopulation. The important detail is that both, the identification of subpopulations and the treatment recommendation happen in the same study. With this feature ISPY-2 might be able to break the so-called biomarker barrier. Many previous studies have attempted to identify subpopulations, but only very few stood the test of time and proved useful in later clinical trials. In ISPY-2, the process of biomarker discovery starts with a list of biomarkers that define up to 256 different subpopulations, although only about 14 remain as practically interesting, due to prevelance and biologic constraints. For each patient we record presence or absence of these biomarkers, including presence of hormone receptors (estrogen and progesterone), human epidermal growth factor receptor 2 (HER2) and MammaPrint risk score. The recorded biomarkers determine the relevant subgroup, which in turn determines the allocation probabilities.

The trial is a collaboration of the US National Cancer Institute, the US Food and Drug Administration, pharmaceutical companies and academic investibators. Please see http://www.ispy2.org for more details.

28.5.2 BATTLE

The BATTLE (Biomarker-integrated approaches of targeted therapy of lung cancer elimination) trial is in many ways similar to ISPY-2. The study is described in [66]. BATTLE is a phase II trial for patients with advanced non-small cell lung cancer (NSCLC). The design considers five subpopulations defined by biomarker profiles, including EGFR mutation/amplification, K-ras and B-raf mutation, VEGF and VEGFR expression and more. The primary outcome is progression free survival beyond eight weeks, reported as a binary response. We refer to the binary outcome as disease control. Similarly to ISPY-2, the design allows allocation of treatments with different probabilities in each subpopulation, and to report subpopulation-specific treatment recommendations upon conclusion of the trial. Treatment allocation is adaptive, with probabilities proportional to the probabilities of disease control. Let γ_{jk} denote the current posterior probability of disease control for a patient in biomarker group k under treatment j. The next patient in biomarker group k is assigned to treatment j with probability proportional to γ_{jk}. Posterior probabilities are with respect to a hierarchical probit model. The probit model is written in terms of latent probit scores z_{jki} for patient i under treatment j in biomarker group k. The model assumes a hierarchical normal/normal model for z_{jki}. The model includes mean effects μ_{jk} of treatment j in biomarker group k, and mean effects ϕ_j for treatment j.

The model is also used to define early stopping of a treatment arm j for the k-th disease group. An arm is dropped for futility when the posterior predictive probability of the posterior probability for disease control being beyond θ_1 is less than δ_L. Here the latter posterior probability refers to inference conditional on future data that could be observed if the treatment were not dropped. And θ_1 would naturally be chosen to be the probability of disease control under standard of care. Similarly, treatment j is recommended for biomarker group k if the posterior predictive probability of (future) posterior probability of disease control being greater than θ_0 is greater than δ_U. Here $\theta_0 > \theta_1$ would be some clinically meaningful improvement over θ_1.

28.6 Decision problems

Some biomedical inference problems are best characterized as decision problems. A Bayesian decision problem is described by a probability model $p(\theta, y)$ on parameters θ and data y, a set of possible actions $d \in \mathcal{A}$, and a utility function $u(d, \theta, y)$. The probability model is usually factored as a prior $p(\theta)$ and a sampling model $p(y \mid \theta)$. It is helpful to further partition the data into (y^o, y) for data y^o that is already observed at the time of decision making, and future data y. The utility function describes the decision maker's relative preferences for actions d under assumed values of the parameters θ and hypothetical data y. In this framework the optimal decision is described as

$$d^\star = \arg\max_{d \in \mathcal{A}} U(d, y^o) \text{ with } U(d, y^o) = \int u(d, \theta, y)\, dp(\theta, y \mid y^o) \tag{28.3}$$

In words, the optimal decision maximizes the utility, in expectation over all random quantities that are unknown at the time of decision making, and conditional on all known data. The expectation $U(d, \cdot)$ is the expected utility.

In [48] we find an optimal schedule for stem cell collection (apheresis) for high-dose chemo-radiotherapy patients under two alternative treatments, $x \in \{1, 2\}$. We develop a hierarchical model to represent stem cell counts for each patient over time. Stem cell counts are measured by CD34 antigen levels per unit volume. We do not need details of the model for the upcoming discussion. The model includes a sampling model $p(y_{ij} \mid t_{ij}, \boldsymbol{\theta}_i)$ for the CD34 counts y_{ij} that is recorded for

Table 28.1 Optimal (and 2nd and 3rd best) design for two future patients. The first column reports the estimated value $\widehat{U}(\boldsymbol{d}_x^\star, x)$.

U	d_1	d_2	d_3	d_4	d_5	d_6
			treatment $x = 1$			
2.03	1	0	0	0	0	0
3.00	1	1	0	0	0	0
4.00	1	1	1	0	0	0
			treatment $x = 2$			
5.14	0	1	1	0	0	0
5.18	0	1	1	1	0	0
5.47	1	0	1	1	0	0

patient i at (known) time t_{ij}, a random effects distribution $p(\boldsymbol{\theta}_i \mid x_i, \phi)$ for patient-specific parameters of a patient under treatment $x_i \in \{1, 2\}$, and a hyperprior $p(\phi)$. The decision is a vector of indicators $\boldsymbol{d} = (d_1, \ldots, d_N)$, with $d_j = 1$ when an apheresis for a future patient is scheduled on day j of an N day period. The action space $\mathcal{A}$ is the set of all binary N-tuples. The utility function $u(\cdot)$ is a combination of sampling cost for $n_d = \sum_j d_j$ stem cell collections and a reward for collecting a target volume $y^\star$ of stem cells. Let L denote the volume collected at each collection and let $\boldsymbol{y} = (\boldsymbol{y}_1, \ldots, \boldsymbol{y}_n, \boldsymbol{y}_{n+1})$ denote the observed CD34 counts, including the future patient $n + 1$. Then

$$u(\boldsymbol{d}, \boldsymbol{\theta}, \boldsymbol{y}, x) = n_d + \lambda I(\sum_j y_{n+1,j} L < y^\star),$$

for a future patient $i = n + 1$ assigned to treatment $x_i = x$. The optimal design $\boldsymbol{d}_x^\star$ is found by maximizing

$$U(\boldsymbol{d}, x) = \int u(\boldsymbol{d}, \boldsymbol{\theta}, \boldsymbol{y})\, dp(\boldsymbol{\theta}, \boldsymbol{y}_{n+1} \mid x_{n+1} = x, \boldsymbol{y}_1, \ldots, \boldsymbol{y}_n)$$

Table 28.1 summarizes the optimal design for treatments $x = 1$ and $x = 2$.

Another interesting example of the benefit of decision theoretic thinking occurs in phase II clinical trials with sequential stopping decisions. After each cohort of patients we make a decision $d \in \{0, 1, 2\}$, where $d = 0$ indicates stopping enrolment and recommending against further development (failure), $d = 2$ indicates stopping enrolment recommending for a following phase III confirmatory trial (victory), and $d = 1$ indicates continued enrolment. A useful utility function in this context formalizes the considerations that feature in this decision. Enroling patients is expensive, say c units per patient. If the following confirmatory trial shows a statistically (frequentist) significant effect then the drug can be marketed and the developer collects a substantial reward C. A significant effect in a follow-up trial is usually easy to consider. The beauty of frequentist tests is that they are usually quite simple. Let $\boldsymbol{z}$ denote the (future) responses in the follow-up trial. All that is needed for the utility function is the posterior predictive probability of the (future) test statistic $S(\boldsymbol{z})$ falling in the rejection region R, using a sample size n_3 based on the usual desiderata of type-I error and power under a current estimate of the treatment effect.

For the statement of the utility function we do not need any details of the probability model, short of an assurance that there is a joint probability model for data and parameters. Let $\boldsymbol{y} = (y_1, \ldots, y_n)$ denote the currently observed data. Let $n_3(\boldsymbol{y})$ denote the sample size for a future phase III trial.

The sample size can depend on current estimates of the treatment effect and other parameters. Let z denote the (future) data in the phase III study, let $S(z)$ denote the test statistic that will be evaluated upon conclusion of the phase III study, and let R denote the rejection region for that test. Also, without loss of generality we assume that patients are recruited in cohorts of size 1, i.e. we consider stopping after each new patient. Formally, the utility function becomes

$$u(d,\boldsymbol{\theta},\boldsymbol{y}) = \begin{cases} c + E\{U(d^\star,\boldsymbol{y},y_{n+1}) \mid \boldsymbol{y}\} & \text{if } d = 1 \\ 0 & \text{if } d = 0 \\ cn_3(\boldsymbol{y}) + C\,p\{S(z) \in R \mid \boldsymbol{y}\} & \text{if } d = 2 \end{cases}$$

Here $U(d^\star, \ldots)$ is the expected utility under the optimal action in the next period. This appears in the utility function because of the sequential nature of the decision problem. Discussing strategies for the solution of sequential decision problems is beyond the scope of this discussion. Here we only want to highlight how relatively straightforward it is to incorporate a stylized description of the following phase III study in the utility function. Variations of this utility function are used in [41, 49, 62].

There is a caveat about the use of decision theoretic arguments in biomedical research problems. Many investigators are reluctant to use formal decision theoretic approaches in biomedical problems. The main reason is perhaps the nature of the optimal decision $d^\star$ as an implicitly defined solution. There is no guarantee that the solution $d^\star$ looks intuitively sensible. Technical details of the probability model and the typically stylized utility function determine the solution in a very implicit way, as the solution to the optimization problem (28.3). However, many details are usually chosen for technical convenience rather than based on substantive prior information. Of course, when counter-intuitive solutions are found, one could always go back and include additional constraints in $\mathcal{A}$. But this is a post-hoc fix, and not all counter-intuitive features are readily spotted.

28.7 Nonparametric Bayes

28.7.1 BNP priors

Nonparametric Bayesian (BNP) models are priors for random probability models or functions. A common technical definition of BNP models is probability models for infinite-dimensional random quantities. However, traditionally most applications of BNP involve random distributions only. So for this discussion we will restrict attention to BNP priors for random probability measures. Let G denote an (unknown) probability model of interest. For example, G could be the distribution of event times in a survival analysis, or the random effects distribution in a mixed effects model, or the probability model for residuals in a regression problem. Generically, let $p(y \mid G, \eta)$ denote the sampling model for the observable data given G and possibly other parameters η. To be specific we assume a density estimation problem for a random sample $\boldsymbol{y} = (y_1, \ldots, y_n)$ with

$$p(\boldsymbol{y} \mid G) = \prod_{i=1}^{n} G(y_i) \tag{28.4}$$

Bayesian inference requires that the model be completed with a prior for the unknown quantities G and η. When the prior $p(G)$ restricts G to a family of probability models $\{G_\theta, \theta \in R^p\}$ that

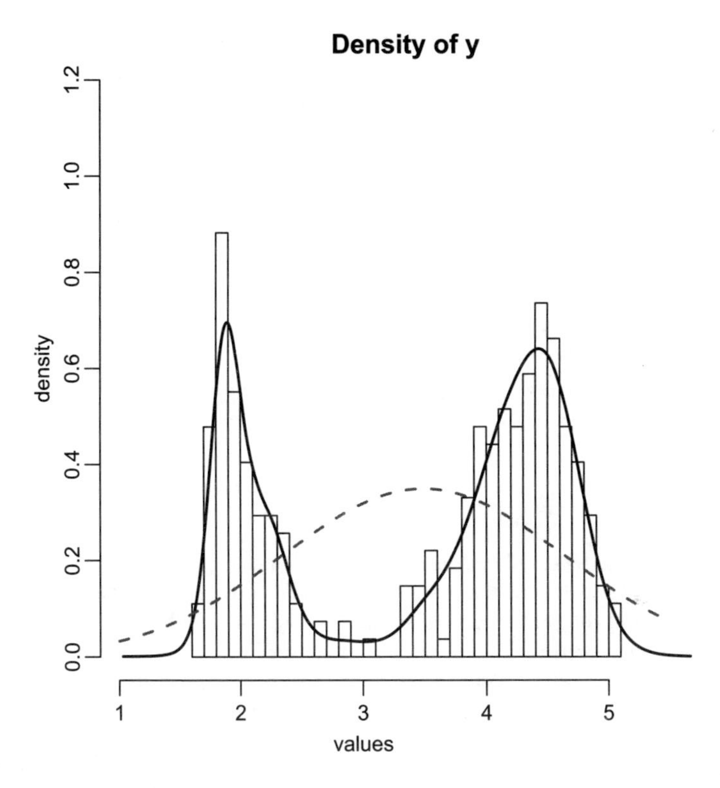

Figure 28.3 Inference on the unknown distribution G under a BNP model (solid line) and a parametric model (dashed line).

is indexed by a finite-dimensional parameter vector θ, then we are back to standard parametric inference. For example, assuming a normal sampling model $G(\cdot) = N(\cdot; \mu, \sigma)$ reduces the problem to inference for the unknown normal moments $\theta = (\mu, \sigma)$. If, however, the investigator is not willing or not able to restrict attention to a parametric family then we require a prior for the infinite dimensional quantity G,

$$G \sim p(G)$$

Good recent reviews of BNP appear in [65] and [25]. Figure 28.3 illustrates the flexibility of BNP in a simple density estimation problem. For a good recent discussion of nonparametric Bayes in biostatistics and bioinformatics we refer to [13].

28.7.2 Survival analysis

One of the most common applications of BNP in biostatistics is to survival analysis, i.e. density estimation (28.4) when the data are event times, and typically include extensive censoring. [52] and [18] developed BNP estimates of survival functions in the presence of censoring, still relying on analytic results. They used the Dirichlet process (DP) prior [16, 17]. The DP prior and variations is to date by far the most commonly used BNP model. With the introduction of

Gibbs sampling the modelling options greatly improved. [34] discuss Gibbs sampling for right, left and interval censored data. They assume that data is observed at discrete times only, reducing the DP prior to a Dirichlet distribution of the quantiles over finitely many time intervals. The mentioned approaches all use variations of the basic model

$$T_i \sim F \text{ and } F \sim \mathrm{DP}(F^\star, \alpha),$$

i.e. a DP prior on the unknown event time distribution. The DP requires two parameters. The base measure $F^\star$ and the total mass parmaeter α. The base measure $F^\star$ fixes the prior mean, $E(F) = F^\star$, Among other implications, α determines the variation, $F(A) \sim \mathrm{Be}[\alpha F^\star(A), \alpha\{1 - F^\star(A)\}]$. An important property of the DP prior is that it generates a.s. discrete random probability measures. This is awkward and often avoided by using an additional convolution with a continuous kernel $k(x; \theta)$, for example, a normal kernel $N(x; \theta, \sigma)$ with known scale.

$$F(T_i) = \int k(T_i; \theta)\, dG(\theta) \text{ and } G \sim \mathrm{DP}(G^\star, \alpha). \tag{28.5}$$

[10] propose an accelerated failure time model using a Dirichlet process prior. Similarly, [33] and [20] develop accelerated failure time models based on DP mixtures. These models are semiparametric. The model component that implements the regression on baseline covariates remains parametric. A fully nonparametric survival regression based on DP priors is proposed in [12]. In principle any nonparametric density regression model, i.e. a nonparametric prior for families of random distributions $\{G_x\}$ indexed by covariates x, could be used. Many such models have been proposed in the recent literature, including, for example, [14] and [15].

Besides the DP model and variations, many other BNP models have been proposed for survival data. Many approaches are based on the Polya tree (PT) prior. See [35] for a discussion of these priors and references. A fully Bayesian semiparametric survival model using the PT was proposed by [64]. More recently, [23, 60] proposed mixtures of Polya Tree priors for use in semiparametric proportional hazards, accelerated failure time and proportional odds models. They assume a mixture of PT priors for a baseline survival distribution. A fully nonparametric extension of the PT model for survival data to include a nonparametric regression on covariates, is proposed in [61]. We refer to [27] and the following chapter in this volume for a thorough review of nonparametric Bayesian methods in survival analysis.

28.8 Multiplicities and error control

28.8.1 Posterior inference accounts for multiplicities

Many important scientific problems naturally lead to massive multiple comparisons. Typical examples are experiments that record gene expression under different biologic conditions, simultaneously for massively many genes. The problem is to compare for each gene relative gene expression across biologic conditions and report those genes that show significant differences across conditions. The number of comparisons can be thousands or tens of thousands.

Formally, let $\delta_i \in \{0, 1\}$, $i = 1, \ldots, n$ denote the unknown truth for n comparisons. A stylized example of differential gene expression experiments could involve a sampling model $p(y_i \mid \theta_i) = N(\theta_i, \sigma)$, $i = 1, \ldots, n$, for a gene-specific difference score y_i. The sampling model is indexed with a gene-specific parameter θ_i that can be interpreted as the level of differential expression. The variance σ is assumed to be common across all genes. The i-th comparison is $\theta_i = 0$ versus $\theta_i \neq 0$, i.e.

no differential expression versus non-zero differential expression. In this context $\delta_i = I(\theta_i \neq 0)$ is an indicator of (true) differential expression. The model is completed with a hierarchical prior on θ_i. For the moment, we need not worry about the details of this prior model. For a meaningful discussion we only need to assume that the prior includes a positive prior probability $p(\delta_i = 0) > 0$ of non-differential expression. We focus on inference of δ_i. Under a Bayesian approach one might report, for example, $\overline{p}_i = p(\delta_i = 1 \mid y)$, the posterior probability of differential expression for the i-th gene.

A popular folk theorem asserts that Bayesians need not worry about multiplicities. Posterior probabilities are already (automatically) adjusted for multiplicities. This is illustrated, for example in [50]. Under some assumptions on the hierarchical prior for δ_i, the statement is correct. The posterior probabilities $\overline{p}_i$ adjust for multiplicities, in the following sense. Focus on one particular comparison, say the i-th comparison with a particular difference score y_i and consider two scenarios with the same value y_i, but different values for other $y_j, j \neq i$. When all other observed difference scores $y_j, j \neq i$, are closer to zero, then $\overline{p}_i$ will be shrunken more towards zero than what would be the case for larger y_j. In other words, posterior probabilities adjust for multiplicities by increased shrinkage in $\overline{p}_i$ when the data suggest a high level of noise in the data. Of course, this statement depends on the details of the model, but it holds for any reasonable hierarchical model. See [50] for details.

However, reporting posterior probabilities $\overline{p}_i$ 'is only half the solution' [8]. The remaining part of the solution is the decision to select which genes should be reported as differentially expressed. The magic automatic adjustment for multiplicities only happens for the probabilities, not for the decision. Let $d_i(y) \in \{0, 1\}, i = 1, \ldots, n$, denote this inference summary. In a classical framework, this could simply be a test of $H_{0i} : \theta_i = 0$. Under a Bayesian perspective one could consider, for example, $d_i = I(\overline{p}_i > c)$. This seems an innocent and reasonable decision rule. It will turn out to actually be the Bayes rule under certain assumptions. In any case, it is a plausible rule to consider. In both cases, fequentist and Bayesian, we are left with the decision of where to draw the line, i.e. how to choose the threshold c? Clearly, traditional type I error control for each comparison is meaningless. In most applications one would end up reporting way too many false positives. The other extreme of controlling experiment-wide error rates, on the other hand, is way too conservative in most applications.

28.8.2 False discovery rate (FDR)

A commonly used compromise between the two extremes of comparison-wise and experiment-wide error control is the control of false discovery rate, the relative fraction of false positives, relative to the number of positives. Let $D = \sum d_i$ denote the number of reported genes. The false discovery rate is defined as

$$\text{FDR} = \frac{1}{D} \sum_{i=1}^{n} (1 - \delta_i) d_i.$$

Often the denominator is replaced by $(D + \epsilon)$ to avoid zero division. At this moment the FDR is neither frequentist nor Bayesian. It is a function of both, the data, indirectly through $d_i(y)$, and the parameters δ_i. Marginalizing with respect to the data leads to a frequentist expectation of the FDR. Most references to FDR in the literature refer to this frequentist expectation. We refer to it as $\widehat{\text{FDR}}$, to distinguish it from the posterior expected $\overline{\text{FDR}} = E(\text{FDR} \mid y)$, conditional on the data.

A clever procedure proposed and popularized by [4] allows straightforward control of $\widehat{\text{FDR}}$. In contrast, the evaluation of $\overline{\text{FDR}}$ requires no clever tricks. Conditional on the data the only unknown remains δ_i and we find $\overline{\text{FDR}} = 1/D \sum_{i=1}^{n} (1 - \overline{p}_i) d_i$. One could use a desired bound on $\overline{\text{FDR}}$ to fix the threshold c in the decision rule $d_i = I(\overline{p}_i > c)$. This is proposed, for example, in [46]. Under the Bayesian paradigm, FDR control turns out not only to be easy, it is even the correct thing to do. The decision rule $d_i = I(\overline{p}_i > c)$ can be justified as a Bayes rule under several loss functions combining false discovery counts (or proportions) and false negative counts (or proportions). See [42] for a discussion.

An important special case of multiplicity adjustments arises in subgroup analysis. Many clinical studies fail to show the hoped for effect for the target patient population. It is then tempting to consider subgroups of patients. Often it is possible to make clinically or biologically reasonable arguments of why the treatment should be more specifically suitable for certain subgroups. With the increased use of biomarkers it is becoming more common to consider interesting subpopulations defined by various relevant biomarkers. For example, a targeted therapy that aims to address a pathological disruption of some molecular network can only be effective when the disease is in fact caused by a disruption of this network. Naturally, the investigation of possible subgroup effects needs to proceed in a controlled fashion to avoid concerns related to multiplicities. The problem is naturally approached as a decision problem, with a probability model for all unknowns, including the unknown subgroup effects, and a utility function that spells out the decision criterion. Recent discussions of a Bayesian decision theoretic approach appear in [51] and [44].

28.9 Conclusion

We have discussed some prominent applications of Bayesian theory in biostatistics. The judgement of what is prominent, of course, remains subjective. Surely many important applications of Bayesian methods have been missed. A good recent discussion of some applications of more sophisticated Bayesian models in biostatistics appears in [13]. An extensive review of Bayesian methods specifically in clinical trials and healthcare evaluations appears in [11].

References

[1] Babb, James, Rogatko, André and Zacks, Shelemyahu (1998). Cancer phase 1 clinical trials: efficient dose escalation with overdose control. *Statistics in Medicine*, **17**(10), 1103–1120.

[2] Barker, A. D., Sigman, C. C., Kelloff, G. J., Hylton, N. M., Berry, D. A., and Esserman, L. J. (2009). I-spy 2: An adaptive breast cancer trial design in the setting of neoadjuvant chemotherapy. *Clinical Pharmacology*, **86**, 97–100.

[3] Bekele, B. N., Ji, Y., Shen, Y. and Thall, P. F. (2008). Monitoring late-onset toxicities in phase I trials using predicted risks. *Biostatistics*, **9**, 442–47.

[4] Benjamini, Yoav and Hochberg, Yosef (1995). Controlling the false discovery rate: A practical and powerful approach to multiple testing. *Journal of the Royal Statistical Society, Series B: Methodological*, **57**, 289–300.

[5] Bernardo, J. M. and Smith, Adrian F. M. (1994). *Bayesian Theory*. John Wiley & Sons.

[6] Berry, Donald A. (1987). Statistical inference, designing clinical trials, and pharmaceutical company decisions. *The Statistician: Journal of the Institute of Statisticians*, **36**, 181–189.

[7] Berry, Donald A. (2004). Bayesian statistics and the efficiency and ethics of clinical trials. *Statistical Science*, **19**(1), 175–187.

[8] Berry, Scott M. and Berry, Donald A. (2004). Accounting for multiplicities in assessing drug safety: A three-level hierarchical mixture model. *Biometrics*, **60**(2), 418–426.

[9] Cheung, Ying Kuen and Chappell, Rick (2000). Sequential designs for phase I clinical trials with late-onset toxicities. *Biometrics*, **56**(4), 1177–1182.

[10] Christensen, Ronald and Johnson, Wesley (1988). Modelling accelerated failure time with a Dirichlet process. *Biometrika*, **75**, 693–704.
[11] Spiegelhalter, D. J., Abrams, Keith R. and Myles, Jonathan P. (2003). *Bayesian Approaches to Clinical Trials and Health-care Evaluation.* Wiley.
[12] De Iorio, M., Johnson, W., Müller, P., Rosner, G. Trippa, L., Müller, P. and Johnson, W. (2009). A ddp model for survival regression. *Biometrics*, **65**, 762–771.
[13] Dunson, David (2010). Nonparametric Bayes applications to biostatistics. In *Bayesian Nonparametrics* (ed. N. L. Hjort, C. Holmes, P. Müller, and S. G. Walker), pp. 235–291. Cambridge University Press.
[14] Dunson, D. B. and Park, J. H. (2008). Kernel stick-breaking processes. *Biometrika*, **95**(2), 307–323.
[15] Dunson, David B., Pillai, Natesh and Park, Ju-Hyun (2007). Bayesian density regression. *Journal of the Royal Statistical Society: Series B (Statistical Methodology)*, **69**(2), 163–183.
[16] Ferguson, Thomas S. (1973, April). A Bayesian analysis of some nonparametric problems. *Annals of Statistics*, **1**(2), 209–230.
[17] Ferguson, Thomas S. (1974, August). Prior distributions on spaces of probability measures. *Annals of Statistics*, **2**(4), 615–629.
[18] Ferguson, Thomas S. and Phadia, Eswar G. (1979, January). Bayesian nonparametric estimation based on censored data. *Annals of Statistics*, **7**(1), 163–186.
[19] Gelfand, Alan E., Hills, Susan E., Racine-Poon, Amy and Smith, Adrian F. M. (1990). Illustration of Bayesian inference in normal data models using Gibbs sampling. *Journal of the American Statistical Association*, **85**, 972–985.
[20] Gelfand, Alan E. and Kottas, Athanasios (2003). Bayesian Semiparametric Regression for Median Residual Life. *Scandinavian Journal of Statistics*, **30**(4), 651–665.
[21] Gelfand, Alan E. and Smith, Adrian F. M. (1990). Sampling-based approaches to calculating marginal densities. *Journal of the American Statistical Association*, **85**, 398–409.
[22] Goodman, S., Zahurak, M. and Piantadosi, S. (1995). Some practical improvements in the continual reassessment method for phase I studies. *Statistics in Medicine*, **14**, 1149–1161.
[23] Hanson, Timothy and Johnson, Wesley O. (2002). Modelling regression error with a mixture of polya trees. *Journal of the American Statistical Association*, **97**, 1020–1033.
[24] Higgins, Julian P. T. and Spiegelhalter, David J. (2002). Being sceptical about meta-analyses: a Bayesian perspective on magnesium trials in myocardial infarction. *International Journal of Epidemiology*, **31**(1), 96–104.
[25] Hjort, Nils Lid (ed.), Holmes, Chris (ed.) Müller, Peter (ed.), and Walker, Stephen G. (ed.) (2010). *Bayesian Nonparametrics.* Cambridge University Press.
[26] Huang, X., Ning, J., Li, Y., Estey, E., Issa, J. P. and Berry, D. A. (2009). Using short-term response information to facilitate adaptive randomization for survival clinical trials. *Statistics in Medicine*, **28**(12), 1680–1689.
[27] Ibrahim, Joseph G., Chen, Ming-Hui and Sinha, Debajyoti (2001). *Bayesian Survival Analysis.* Springer-Verlag, New York, NY, USA.
[28] Inoue, Lurdes Y. T., Thall, Peter F. and Berry, Donald A. (2002). Seamlessly expanding a randomized phase II trial to phase III. *Biometrics*, **58**(4), 823–831.
[29] Ji, H. and Liu, X. S. (2010). Analyzing 'omics data using hierarchical models. *Nature Biotechnology*, **28**(4), 337–340.
[30] Ji, Y., Li, Y., and Bekele, N. (2007). Dose-finding in oncology clinical trials based on toxicity probability intervals. *Clinical Trials*, **4**, 235–244.
[31] Kleinman, K. P. and Ibrahim, J. G. (1998). A semi-parametric Bayesian approach to the random effects model. *Biometrics*, **54**, 921–938.
[32] Kola, Ismail and Landis, John (2004). Can the pharmaceutical industry reduce attrition rates? *Nat Rev Drug Discov*, **3**(8), 711–716.

[33] Kuo, L. and Mallick, B. (1997). Bayesian semiparametric inference for the accelerated failure--time model. *Canadian Journal of Statistics*, **25**, 457–472.

[34] Kuo, L. and Smith, A. F. M. (1992). Bayesian computations in survival models via the Gibbs sampler. In *Survival Analysis: State of the Art* (ed. J. P. Klein and P. K. Goel), pp. 11–24. Boston: Kluwer Academic.

[35] Lavine, Michael (1992). Some aspects of Polya tree distributions for statistical modelling. *Annals of Statistics*, **20**, 1222–1235.

[36] Le Tourneau, Christophe, Lee, J. Jack and Siu, Lillian L. (2009). Dose escalation methods in phase I cancer clinical trials. *Journal of the National Cancer Institute*, **101**(10), 708–720.

[37] Lee, J. J. and Liu, D. D. (2008). A predictive probability design for phase II cancer clinical trials. *Clinical Trials*, **5**(2), 93–106.

[38] Leon-Novelo, L., Bekele, B. N., Müller, P., Quintana, F. and Wathen, K. (2011). Borrowing strength with non-exchangeable priors over subpopulations. *Biometrics*, to appear.

[39] Lindley, D. V. and Smith, A. F. M. (1972). Bayes estimates for the linear model. *Journal of the Royal Statistical Society. Series B (Methodological)*, **34**(1), pp. 1–41.

[40] Malec, D. and Müller, P. (2008). A Bayesian semi-parametric model for small area estimation. In *Festschrift in Honor of J.K. Ghosh* (ed. S. Ghoshal and B. Clarke), pp. 223–236. IMS.

[41] Müller, Peter, Berry, Don A., Grieve, Andy P., Smith, Michael, and Krams, Michael (2007). Simulation-based sequential Bayesian design. *Journal of Statistical Planning and Inference*, **137**(10), 3140–3150.

[42] Müller, P., Parmigiani, G., Robert, C. and Rousseau, J. (2004). Optimal sample size for multiple testing: the case of gene expression microarrays. *Journal of the American Statistical Association*, **99**(468), 990–1001.

[43] Müller, P. and Rosner, G. (1997). A Bayesian population model with hierarchical mixture priors applied to blood count data. *Journal of the American Statistical Association*, **92**, 1279–1292.

[44] Müller, P., Sivaganesan, S. and Laud, P. W. (2010). A Bayes rule for subgroup reporting. In *Frontiers of Statistical Decision Making and Bayesian Analysis* (ed. M.-H. Chen, D. Dey, P. Mueller, D. Sun, and K. Ye), pp. 277– 284. Springer-Verlag.

[45] Neuenschwander, B., Branson, M. and Gsponer, T. (2008). Critical aspects of the Bayesian approach to phase I cancer trials. *Statistics in Medicine*, **27**, 2420–2439.

[46] Newton, Michael A. (2004). Detecting differential gene expression with a semiparametric hierarchical mixture method. *Biostatistics (Oxford)*, **5**(2), 155–176.

[47] O'Quigley, John, Pepe, Margaret and Fisher, Lloyd (1990). Continual reassessment method: A practical design for Phase 1 clinical trials in cancer. *Biometrics*, **46**, 33–48.

[48] Palmer, J. and Müller, P. (1998). Bayesian optimal design in population models of hematologic data. *statmed*, **17**, 1613–1622.

[49] Rossell, D., Müller, P. and Rosner, G. (2007). Screening designs for drug development. *Biostatistics*, **8**, 595–608.

[50] Scott, James G. and Berger, James O. (2006). An exploration of aspects of Bayesian multiple testing. *Journal of Statistical Planning and Inference*, **136**(7), 2144–2162.

[51] Sivaganesan, S., Laud, P. and Müller, P. (2011). A Bayesian subgroup analysis with a zero-enriched polya urn scheme. *Statistics in Medicine*, to appear.

[52] Susarla, V. and Ryzin, J. Van (1976). Nonparametric Bayesian estimation of survival curves from incomplete observations. *Journal of the American Statistical Association*, **71**, 897–902.

[53] Thall, P. F. and Wathen, J. K. (2007). Practical Bayesian adaptive randomisation in clinical trials. *European Journal of Cancer*, **43(5)**, 859–866.

[54] Thall, P. F., Wathen, J. K., Bekele, B. N., Champlin, R. E., Baker, L. H. and Benjamin, R. S. (2003). Hierarchical Bayesian approaches to phase II trials in diseases with multiple subtypes. *Statistics in Medicine*, **22**, 763–780.

[55] Thall, P. F., Wooten, L. H. and Tannir, N. M. (2005). Monitoring event times in early phase clinical trials: some practical issues. *Clinical Trials*, **2**, 467–478.

[56] Thall, Peter F. and Cook, John D. (2004). Dose-finding based on efficacy-toxicity trade-offs. *Biometrics*, **60**(3), 684–693.

[57] Thall, P. F. and Simon, R. (1994). Practical Bayesian guidelines for phase IIB clinical trials. *Biometrics*, **50**, 337–349.

[58] Thall, Peter F., Simon, Richard M. and Estey, Elihu H. (1995). Bayesian sequential monitoring designs for single-arm clinical trials with multiple outcomes. *Statistics in Medicine*, **14**, 357–379.

[59] Thompson, William R. (1933). On the likelihood that one unknown probability exceeds another in view of the evidence of two samples. *Biometrika*, **25**(3/4), pp. 285–294.

[60] Hanson, Timothy, Johnson Wesley O. and Laud, Purashotam (2009). Semiparametric inference for survival models with step process covariates. *Canadian Journal of Statistics*, **37**, 60–79.

[61] Trippa, L., Müller, P. and Johnson, W. (2011). The multivariate beta process and an extension of the polya tree model. *Biometrika*, **98**(1), 17–34.

[62] Trippa, Lorenzo, Rosner, Gary L. and Müller, Peter (2012). Bayesian enrichment strategies for randomized discontinuation trials. *Biometrics*, **68**(1), 203–11.

[63] Wakefield, J. C., Smith, A. F. M., Racine-Poon, A. and Gelfand, A. E. (1994). Bayesian analysis of linear and non-linear population models by using the Gibbs sampler. *Applied Statistics*, **43**, 201–221.

[64] Walker, Stephen and Mallick, Bani K. (1999). A Bayesian semiparametric accelerated failure time model. *Biometrics*, **55**, 477–483.

[65] Walker, Stephen G., Damien, Paul, Laud, Purushottam W. and Smith, Adrian F. M. (1999). Bayesian nonparametric inference for random distributions and related functions (Disc: P510–527). *Journal of the Royal Statistical Society, Series B: Statistical Methodology*, **61**, 485–509.

[66] Zhou, Xian, Liu, Suyu, Kim, Edward S., Herbst, Roy S. and Lee, J. Jack (2008). Bayesian adaptive design for targeted therapy development in lung cancer–a step toward personalized medicine. *Clinical Trials*, **5**, 181–193.

29 Subgroup analysis

PURUSHOTTAM W. LAUD, SIVA SIVAGANESAN AND PETER MÜLLER

29.1 Introduction

Subgroup analysis is the comparison of treatment efficacies in a clinical trial among subgroups defined by baseline patient characteristics such as gender and age. Such analyses are commonplace and often not carried out with statistically sound techniques. Lagakos [15], in an article addressing medical researchers in the New England Journal of Medicine, discusses in a clear and direct manner the challenges of reporting subgroup analyses. He also suggests some simple and minimal remedies to avoid distorted reporting. In a Special Report in the same journal, Lagakos and coauthors [31] present empirical evidence of then-current reporting practices. Subgroup analyses were reported in 59 of the 97 primary reports of clinical trials in the same prestigious journal in the year July 2005–June 2006. The vast majority of these subgroup analysis reports were substandard. The Special Report then provides guidelines for good reporting practice. Other authors have also addressed such issues; see, for example, [3, 4, 33].

Subgroup analysis is carried out for different purposes. One of these is the identification of particular subgroups where the treatment is highly efficacious (or is ineffective or even harmful). Another is to look for consistency of effect across various subgroups [12], and it is also recommended for risk-stratified subgroups [24]. All of these are important considerations leading to direct clinical implications. Improvement in subgroup analysis techniques can, therefore, have a substantial effect on better extraction of information from clinical trials and lead to improved clinical trial designs.

29.1.1 A dementia trial

We introduce an example of a clinical trial to focus attention on subgroup analysis. Kovach *et al.* [14] reported on a double-blinded randomized experiment to study the effectiveness of a new treatment called Serial Trial Intervention (STI) compared to standard treatment (control). The treatments are intended to affect positively the comfort and behaviour of patients with late-stage dementia and are initiated after certain behavioural symptoms are observed. The study was conducted in 14 nursing homes on 112 subjects with late-stage dementia. STI is an innovative clinical protocol for assessment and management of unmet needs in people with late-stage dementia. The outcome variable of interest was the difference, pre and post treatment, in Discomfort-DAT (Discomfort-Dementia of the Alzheimer's Type scale), a measure of discomfort felt by the subjects. The treatment group contained 55 subjects and the control group 57. The investigators were interested in the subgroups defined by two covariates, Functional Assessment Staging of Dementia (FAST) score (covariate X_1) and presence/absence of vocalization (MVOCAL, covariate X_2) in

Table 29.1 Dementia trial sample sizes

	$X_1 = 0$	$X_1 = 1$	Total	$X_2 = 0$	$X_2 = 1$
Control	31	26	57	19	38
Treated	35	20	55	19	36
Total	66	46	112	38	74

the behavioural symptoms that initiated treatment, standard or STI. Two subgroups were of interest based on X_1, defined by $X_1 = 1$ when FAST Score ≥ 7, and, $= 0$ if FAST score < 7. The presence $(X_2 = 1)$ and absence $(X_2 = 0)$ of vocalization in behavioural symptoms initiating treatment also define two additional subgroups of interest. Subgroup sample sizes are shown in Table 29.1.

29.1.2 Different approaches to subgroup analysis

The main concern in conducting tests of hypotheses for subgroup effects is that such multiple testing has a detrimental effect on Type I error rate. Most clinical trials are designed to address the overall effect in the entire population with careful attention to this error rate and to the power to detect a prespecified effect size of the treatment over control. Thus all recommended methods require an a priori specification of subgroups to be tested. Pocock *et al.* (2002) [23] and Wang *et al.* (2007) [31], among others, argue for the use of interaction tests rather than the reporting of treatment effects separately for each level of the subgroup. A test of interaction between treatment groups (STI versus Control) and subgroups (presence or absence of vocalizaton) examines the evidence for inconsistency of treatment effects across subgroups. Standard multiplicity adjustments such as the Bonferroni method are then used with these interaction tests to account for testing of multiple pre-specified subgroups. In the Dementia Trial example, there are two interaction tests and the overall effect test. Thus the adjustment would require conducting each test at a level obtained by dividing the originally intended Type I error rate by 3. Without multiplicity control, subgroup analyses—even when identified using an interaction test—are considered exploratory rather than confirmatory. Proschan and Waclawiw (2000) [25] advocate testing for a subgroup effect only if the overall treatment effect has been established. This controls the Type I error based on the principle of fixed sequence testing. Alternatively, the overall Type I error can simply be split across the overall and subgroup comparisons in a possibly unequal allocation. Other authors have advocated fallback type procedures, in which the subgroup can still be tested even if the overall treatment effect is not significant ([2, 30, 32]). These utilize the correlation between the overall and the subgroup specific test statistics as well as the anticipated directional consistency of the treatment effect across subgroups to improve efficiency. Alpha-splitting and fallback type procedures require a smaller significance level for the test of an overall effect in order to allocate some Type I error to the subgroup test and still maintain the overall Type I error rate at the prespecified level α.

There are also some recommendations from the Bayesian viewpoint in the literature. These fall mainly into two categories. The first approach is estimation based. It assumes that the treatment–subgroup interactions are exchangeable [9, 11, 13, 17, 28], and has essentially focused on post hoc analysis. This approach reports shrinkage estimates of the treatment–subgroup interactions, and estimates of the variance components representing the inter–subgroup variability. This inference is the basis for making judgements about the existence and the extent of subgroup effects. Various forms of exchangeability for treatment-subgroup interactions have been studied.

The second approach is the one we take here and in [20, 29]. It focuses on evaluating subgroup effects, i.e. treatment–subgroup interactions, only for pre-specified subgroups. Subgroups are defined by the different values of pre-specified categorical covariates. The main goal is to determine whether there is a subgroup effect, and if so, which of the subgroups have different effect sizes. Various configurations of subgroup effects are represented by different models, and Bayesian model selection is used to choose among models using predetermined threshold values for posterior probabilities. Multiplicity adjustment is incorporated with both the Bayesian and frequentist approaches, by a suitable choice of priors and by controlling Type I and other error rates of interest via a suitable choice of threshold values in the model selection procedure.

This Bayesian model selection approach that we take here has several advantages. Firstly, with appropriate choice of priors, an automatic multiple comparison adjustment of the reported probabilities is possible [26, 27]. Secondly, coherence and decision theoretic justification [20] makes the approach attractive from the viewpoint of optimality. Thirdly, it is possible and desirable to evaluate the operating characteristics of the procedure over repeated experiments. Indeed results of such evaluations can be used in the design phase to quantify the price of including subgroup analysis in the protocol of the clinical trial.

In Section 29.2 we describe the models under consideration, followed by definitions of various error types of interest in Section 29.3. Section 29.4 details the priors to be used, on the model space as well as on the parameters of each model. Posterior computation and a model selection procedure are described in Section 29.5. We revisit the Dementia Trial in Section 29.6, outline a decision theoretic approach in Section 29.7 and conclude with a discussion of possible further developments in Section 29.8.

29.2 Models for subgroup analysis

To set the stage, consider a two-arm clinical trial, where a treatment is compared with a control. To be specific, suppose that a continuous response variable is observed on independent random samples of subjects under treatment and control. Let $Y_{0j}, j = 1, \ldots, J_0$ and $Y_{1j}, j = 1, \ldots, J_1$ denote the responses under control and treatment, respectively, and take Y_{tj}, $t = 0, 1$ to be $N(\mu_t, \sigma^2)$ where μ_t may depend on some covariates. To begin, suppose that the values of a covariate X are available on each subject, and that the values are classified into S categories. The main question of interest is whether the treatment is effective overall, i.e. $\mu_1 = \mu_0$ or not. In subgroup analysis, interest lies in the effectiveness of the treatment among the subgroups of patients as defined by each covariate. In particular, we want to determine whether the treatment is effective within any of the S subgroups and, if it is found effective in two or more subgroups, whether the sizes of effects differ between these subgroups.

To cast this problem as a model selection problem, first consider two competing models

$$M_{00} : \delta = \mu_1 - \mu_0 = 0 \quad M_{01} : \delta = \mu_1 - \mu_0 \neq 0$$

and define the model space at this stage as $\mathcal{M}_0 = \{M_{00}, M_{01}\}$. Now let μ_{0s}, μ_{1s} denote the mean outcome under control and treatment, respectively, in subgroup s, $s = 1, \ldots, S$ so that $\delta_s = \mu_{1s} - \mu_{0s}$ represents treatment efficacy in subgroup s. The goal is to identify subgroups which have no treatment effect and, among those having treatment effects, to characterize how the treatment effects differ across the corresponding levels of the covariate. To this end, we will consider several models representing all such configurations of subgroup effects, and use $\mathcal{M}_{\mathcal{X}}$ to represent the set of all such models.

29.2.1 Indexing and counting models

We index the models in $\mathcal{M}_{\mathcal{X}}$ using a vector, $\boldsymbol{\gamma} = (\gamma_1, \ldots, \gamma_S)$, of length S, where γ_i is a non-negative integer assigned to the i-th subgroup. A zero in any position indicates no treatment effect in that subgroup. The non-zero elements are integers ranging from 1 to K, where K is the number of *distinct* non-zero treatment effects among all subgroups. The integers are assigned by order of appearance, such that subgroups with common values of treatment effect receive the same integer. For example, taking $S = 3$, the model with non-zero and distinct treatment effects in the three subgroups is denoted by $\boldsymbol{\gamma} = (1, 2, 3)$; the model with equal but non-zero effect in the first and third subgroups and a distinct non-zero effect in subgroup 2 is denoted as $\boldsymbol{\gamma} = (1, 2, 1)$. Taking $S = 5$ subgroups, $\boldsymbol{\gamma} = (1, 0, 2, 1, 3)$ represents the case $\delta_2 = 0, \delta_1 = \delta_4 \neq 0, \delta_3 \neq 0, \delta_5 \neq 0, \delta_1 \neq \delta_3, \delta_1 \neq \delta_5, \delta_3 \neq \delta_5$. Thus $\gamma_s, s = 1, .., S$, can be regarded as the cluster membership indicator for the S subgroup effects corresponding to the covariate X. We also note that $\boldsymbol{\gamma}$ with all elements equalling zero represents the overall null, and with all elements equalling 1 corresponds to the hypothesis of overall effect in the population.

With $S = 2$ subgroups, there are five models in $\mathcal{M}_{\mathcal{X}}$ enumerated as:

$$(0, 0),\ (0, 1),\ (1, 0),\ (1, 2),\ (1, 1)$$

and $S = 3$ yields the 15 models:

$$(0, 0, 0),\ (0, 0, 1),\ (0, 1, 0),\ (1, 0, 0),\ (0, 1, 1),\ (1, 0, 1),\ (1, 1, 0),\ (0, 1, 2),\ (1, 0, 2),\ (1, 2, 0),$$
$$(1, 2, 2),\ (1, 2, 1),\ (1, 1, 2),\ (1, 2, 3),\ (1, 1, 1).$$

The number of models grows rapidly with S. To count all possible models, note that a model puts the S subgroups into clusters and assigns each cluster a distinct treatment effect δ_k including, possibly, zero. The number of distinct ways S distinguishable objects can be put in R clusters (or indistinguishable cells) equals $St(S, R)$, the Stirling number of the second kind, as in Chapter 24.1.4 of [1]. In each case, all of the R clusters may correspond to R distinct non-zero treatment effects; or one of the R clusters to a zero treatment effect, and all others to $R - 1$ distinct non-zero effects; leading to $R + 1$ possibilities. In other words, each of the $St(S, R)$ partitions corresponds to $(R + 1)$ models obtained by labelling one of the R clusters with $\delta = 0$, plus the model that does not label any of the R clusters with $\delta = 0$. Thus, the number of models, H, is given by

$$H = \sum_{R=1}^{S} (R + 1) St(S, R) = \sum_{R=1}^{S} \frac{(R + 1)}{R!} \sum_{i=0}^{R} (-1)^i \binom{R}{i} (R - i)^S$$

Actual counting can be accomplished more readily by the recursion

$$St(1, 1) = 1,\ \ St(S, R) = 0 \text{ if } R < 1,\ \text{ and } St(S, R) = R\, St(S - 1, R) + St(S - 1, R - 1)$$

yielding:

S	2	3	4	5	6
H	5	15	52	203	877

When subgroups are made with multiple covariates $X_1, \ldots, X_I$, we denote models in $\mathcal{M}_{\mathcal{X}_i}$ by $M_{ih}, h = 0, \ldots, H_i$ reserving $h = 0$ for the null model $\gamma = (0, \ldots, 0)$ and $h = H_i$ for the overall effect model $\gamma = (1, \ldots, 1)$. More details are given in a later subsection as needed.

29.3 Error rates of interest

As the procedure to be described below selects a model from many competing models, several error rates of interest can be defined. Of prime importance among these is the Type I Error (TIE) defined as the probability, in repeated experiments, of rejecting the overall null (M_{00}) by selecting any other model when the true model is M_{00}. Other error rates pertain to probabilities under the overall effect model and each subgroup model. While there are many possibilities, we focus on the following definitions, where P_f represents probability under repeated experiments.

TIE : Type I Error	$P_f(M_{00}$ not selected$\vert M_{00})$
FNR : False Negative Rate	$P_f(M_{00}$ selected$\vert M_{01})$
TPR : True Positive Rate	$P_f(M_{01}$ selected$\vert M_{01})$
FSR : False Subgroup Rate	$P_f($some $M_{ih}, i \neq 0, h \neq 0, h \neq H_i$ selected$\vert M_{01})$
TSR : True Subgroup Rate	$P_f(M_{ih}$ selected$\vert M_{ih}, i \neq 0, h \neq 0, h \neq H_i)$
FPR : False (overall) Positive Rate	$P_f(M_{01}$ selected$\vert M_{ih}, i \neq 0, h \neq 0, h \neq H_i)$

It is important to note that all but the TIE require additional specifications for proper definitions. An effect size is required for FNR, TPR and FSR. Moreover, TSR and FPR further depend on i, h. These definitions lead to Table 29.2, summarizing the various possibilities.

These error rates and the operating characteristics of the procedure can be evaluated via simulation over repeated datasets. For the Dementia Trial example we show the results of such simulations in Section 29.6.

29.4 Priors

To carry out Bayesian inference we need a prior on the model space $\mathcal{M}_{\mathcal{X}}$ and priors on the parameters in each model in $\mathcal{M}_{\mathcal{X}}$. We specify the latter first, considering and restating the simple model for observables mentioned at the beginning of Section 29.2. With independent normally distributed samples of J_0 subjects under control and J_1 under treatment, with measurements denoted by $Y_{0j}, j = 1, \ldots, J_0$ and $Y_{1j}, j = 1, \ldots, J_1$, respectively. More precisely,

$$Y_{0j} \sim N(\mu_{0s}, \sigma^2), \;\; s = x_{0j}; \quad Y_{1j} \sim N(\mu_{1s}, \sigma^2), \;\; s = x_{1j}$$

where $x_{0j} = s$, if the covariate X equals s for the j-th subject in the control group and similarly for x_{1j}.

Table 29.2 Error rates

Rates		Overall null model M_{00}	Truth subgroup effect model M_{ih}	Overall alternative model M_{01}
	M_{00}	1-TIE		FNR
Decision	M_{ih}		TSR_{ih}	FSR
	M_{01}		FPR_{ih}	TPR

29.4.1 Priors for model parameters

The unknown parameters under a model M are the vector of S control means, $\boldsymbol{\mu}_0 = (\mu_{01}, \ldots, \mu_{0S})$, the K non-zero distinct treatment effects vector $\boldsymbol{\delta}^\star = (\delta_1^\star, \ldots, \delta_K^\star)$, and σ^2. We assign mixture g-priors for the $\delta^\star$'s, and noninformative priors for the other parameters which are common to all models. Mixture g-priors have been found to be reasonable non-informative priors in linear model settings, and also computationally easy to work with [16]. The choice of prior is critical for a fair model comparison and selection. For a good discussion of these issues, see [5, 10] and the references cited there. The priors we use here are commonly used as good non-informative priors.

Using generic notation $P(\cdot)$ to denote a probability distribution, conditionally on g, $\boldsymbol{\mu}_0$ and σ^2, we take

$$\boldsymbol{\delta}^\star \sim N_K(\mathbf{0}, g\sigma^2 I_K)$$

and

$$P(g, \boldsymbol{\mu}_0, \sigma^2) = P(g) \cdot P(\boldsymbol{\mu}_0, \sigma^2),$$

where

$$P(g) = \frac{1}{(1+g)^2} \text{ and } P(\boldsymbol{\mu}_0, \sigma^2) = \frac{1}{\sigma^2};\ g > 0, \boldsymbol{\mu}_0 \in \mathcal{R}^S, \sigma^2 > 0.$$

29.4.2 Prior on the model space

The prior distribution on the space $\mathcal{M}_\mathcal{X}$ can be specified by what we call a zero-enriched Polya urn. Before spelling out the details below, we first give an outline emphasizing construction of the prior. Consider the following useful version of the Polya urn noted by Blackwell & MacQueen (see [6, 22]) in the context of a Dirichlet process. The urn consists of balls of various colours. At each step, a ball is added to the urn with the following scheme. At the beginning of step $n+1$, the urn contains n balls of $k \le n$ colours. With probability $n/(\alpha+n)$ a ball is randomly drawn from this urn, and returned to it along with another ball of the same colour. With probability $\alpha/(\alpha+n)$ a ball of a new colour is added, the colour determined by a random draw from a bagful of colours. A prior on $\mathcal{M}_\mathcal{X}$ is specified by the following procedure for generating the model index $\boldsymbol{\gamma} = (\gamma_1, \ldots, \gamma_S)$. Given $\gamma_1, \ldots, \gamma_s$, the next element γ_{s+1} equals 0 with probability p. Then, given $\gamma_1, \ldots, \gamma_s$ and $\gamma_{s+1} \neq 0$, γ_{s+1} is generated from the above described Polya urn where the non-zero γs serve as colours. As an example, the probability that γ equals $(1, 2, 1)$ is given by

$$\{(1-p)\} \times \{(1-p) \times \frac{\alpha}{\alpha+1}\} \times \{(1-p) \times \frac{2}{\alpha+2} \times \frac{1}{2}\}.$$

We reparameterize α to $q = 1/(\alpha+1)$ and take p, q to be independent beta random variables with parameters (α_1, β_1) and (α_2, β_2), respectively.

For a complete specification, let $K_s = \max\{\gamma_{s'} \neq 0; s' \le s\}$ denote the number of distinct non-zero treatment effects among the first s covariate levels, with $K_S = K$ for the full set of γ_ss. Also, when $K_s \ge 1$ and $1 \le k \le K_s$, let $N_{sk} = \#\{s' : \gamma_{s'} = k, s' \le s\}$ and $L_s = \#\{s' : \gamma_{s'} \neq 0, s' \le s\}$ denote, respectively, the number of treatment effects that match the k-th distinct non-zero effect and the total number of non-zero treatment effects, among the first s levels of the covariate. Note that $\sum_{k=1}^{s} N_{sk} = L_s$. We use two parameters (p, q) to define the probabilities:

$$\begin{aligned} P(\gamma_{s+1} = 0 \mid \gamma_1, \ldots, \gamma_s) &= p \\ P(\gamma_{s+1} = k \mid \gamma_1, \ldots, \gamma_s) &= (1-p)\,\tfrac{N_{sk}q}{1-q+L_s q} \quad \text{for } k = 1, \ldots, K_s \geq 1 \\ P(\gamma_{s+1} = K_s + 1 \mid \gamma_1, \ldots, \gamma_s) &= (1-p)\,\tfrac{1-q}{1-q+L_s q} \quad \text{for } K_s \geq 0 \end{aligned} \tag{29.1}$$

Conditional on non-zero treatment effects, the last two lines define a Polya urn with total mass parameter $\alpha = (1-q)/q$. With probability proportional to $q \cdot L_s$ the treatment effect for the $(s+1)$-st level of the covariate is tied with an earlier level. With probability proportional to $1-q$ the treatment effect is distinct.

We assign independent Beta priors for the parameters (p, q),

$$P(p, q) = Beta(p; \alpha_1, \delta_1) Beta(q; \alpha_2, \delta_2) \tag{29.2}$$

where $Beta(x; a, b)$ indicates the probability density function of a Beta distribution with parameters a and b.

The equation (29.1) implies the prior probability of a model $M \in \mathcal{M}_X$, given p and q. To give a closed form expression to this probability, let, for $0 \leq k \leq K \leq S$

$$N_k = \#\{s : \gamma_s = k, 1 \leq s \leq S\} \tag{29.3}$$

so that $\sum_{k=0}^{K} N_k = S$ and $N_k = N_{Sk}$. For example, N_0 is the number of subgroups with zero treatment effect, and N_1 is the number of subgroups with treatment effects that are equal to the first non-zero subgroup effect, and so on. The process of specifying the probability of a model M indexed by $\boldsymbol{\gamma}$ can be thought of as filling in each of the S positions in $\boldsymbol{\gamma}$ by 0 with probability p, or by a positive integer with probability $1-p$. The positive integers, which identify the configuration of the non-zero subgroup effects, are chosen successively according to the probability specification in (29.1).

Then, the prior probability of a model M, given p and q is

$$P(M \mid p, q) = c(p, q)\, P(N_0, \ldots, N_K \mid p, q) \tag{29.4}$$

where

$$P(N_0, \ldots, N_K \mid p, q) = p^{N_0} (1-p)^{S-N_0} \frac{\alpha^K \prod_{k=1}^{K} [N_k - 1]!}{\prod_{s=1}^{S-N_0} \{\alpha + (s-1)\}} \tag{29.5}$$

$[x] = \max\{x, 0\}$, N_ks are as in (29.3), $c(p, q)$ is the normalizing constant, and a product over an empty set is equal to 1.

The probability model 29.1 is a zero-enriched Polya urn scheme. Each integer in the index vector $\boldsymbol{\gamma}$ of a model M is allowed to be zero with probability p; and each non-zero integer is either equal to a previously selected value or to the subsequent (hitherto unselected) value.

The normalizing constant $c(p, q)$ is determined by counting the number of different models corresponding to a given K and $(N_0, \ldots, N_K)$,

$$c(p, q)^{-1} = \sum_{K=0}^{S} \sum \binom{S}{N_0} P(N_0, \ldots, N_K \mid p, q)$$

Table 29.3 Conditional prior on model space, $S = 3$.

Model(s)	$P(M\|p, q)$	$P(M), \alpha_1 = \beta_1 = \alpha_2 = \beta_2 = 1/2$
$(0, 0, 0)$	p^3	$10/32$
$(0, 0, 1), (0, 1, 0), (1, 0, 0)$	$p^2(1-p)$	$6/32$
$(0, 1, 1), (1, 0, 1), (1, 1, 0)$	$p(1-p)^2q$	$3/32$
$(0, 1, 2), (1, 0, 2), (1, 2, 0)$	$p(1-p)^2(1-q)$	$3/32$
$(1, 2, 2), (1, 1, 2), (1, 2, 1)$	$(1-p)^3q(1-q)/(1+q)$	$0.255(10/32)$
$(1, 2, 3)$	$(1-p)^3(1-q)^2/(1+q)$	$0.320(10/32)$
$(1, 1, 1)$	$2(1-p)^3q^2/(1+q)$	$0.425(10/32)$

where the inside summation spans over all integers $N_0 \geq 0, N_1 > 0, \ldots, N_K > 0$ satisfying the condition $N_0 + N_1 + \ldots + N_K = S$ and $P(N_0, \ldots, N_K \mid p, q)$ is as in 29.5.

As an example, for $S = 3$, the assigned probabilities for the 15 models are displayed in Table 29.3 conditioned on p and q as well as marginally for one choice of parameters for the two beta distributions. These unconditional probabilities can be controlled by modifying the choice of $\alpha_1, \beta_1, \alpha_2, \beta_2$.

29.5 Posterior probabilities and a subgroup reporting procedure

Calculation of posterior model probabilities is straightforward. With $\vec{y}$ denoting all observed data, we have $P(\vec{y}|M, p, q) = P(\vec{y}|M)$ as a consequence of conditional independence. The posterior probability of a model M, therefore, is

$$P(M \mid \vec{y}) = \frac{P(\vec{y}|M)P(M)}{\sum_{M' \in \mathcal{M}} P(\vec{y}|M')P(M')}$$

where

$$P(M) = \int P(M|p, q)P(p, q)dpdq \tag{29.6}$$

Further,

$$P(\vec{y}|M) = \int P(\vec{y}|M, g)P(g)dg$$

where

$$P(\vec{y}|M, g) = \int P(\vec{y} \mid \mu_0, \delta^*, \sigma^2)P(\delta^*|g, \mu_0, \sigma^2)P(\mu_0, \sigma^2)d\mu_0 d\delta^* d\sigma^2,$$

where $\boldsymbol{\delta}^*$ is the vector of $(S - N_0)$ non-zero treatment effects. For the normal models and priors specified in Section 29.4, $P(\vec{y}|M, g)$ has a closed form expression, which allows the calculation of $P(\vec{y}|M)$ using a one-dimensional integral over g.

29.5.1 Multiple covariates

Suppose there are I covariates of interest $X_1,\ldots,X_I$, with X_i classified into S_i categories. Now define a model space $\mathcal{M}_{X_i} = \mathcal{M}_i$ for each covariate X_i, leading to I distinct probability models. The primary hypotheses of interest are the overall null and alternative models M_{00} and M_{01}, which constitute the space $\mathcal{M}_0$ in subsection 4.2.1. This overall-effect model space $\mathcal{M}_0$, along with $\mathcal{M}_i, i = 1, \ldots, I$ define the collection of models that are of interest in the subgroup analysis. List the models in $\mathcal{M}_i$ in sequence, labelling the hth model by M_{ih}, $0 \leq h \leq H_i$. Often, the index i is understood from the context in which case we omit it and write simply H. The labelling of the models in $\mathcal{M}_i$ is done so that the first model $(h = 0)$ corresponds to the absence of subgroup effects in all s_i subgroups and is represented by $\boldsymbol{\gamma} = (0, \ldots, 0)$. The last model $(h = H)$ corresponds to the presence of subgroup effects of equal size in all subgroups and is represented by $\boldsymbol{\gamma} = (1, \ldots, 1)$. These two models correspond to the overall null and alternative models M_{00} and M_{01}, respectively. Thus, each of the remaining models in $\mathcal{M}_i$ represents the presence of a 'bona-fide' subgroup effect.

29.5.2 Procedure for reporting subgroup effects

For each model space $\mathcal{M}_i$, first calculate posterior model probabilities denoted $\bar{P}_i(M_{ih}), i = 1, \ldots, H_i$. The reporting procedure is described in full step-wise detail in [29]. Here we describe the essence of it. In each model space, the first and last models correspond to M_{00} and M_{01}. Using thresholds c_0 and c_1 to compare with these respectively, we proceed as follows.

Start: Choose M_{01} and stop if M_{01} is better than M_{00}, i.e., $\frac{\bar{P}(M_{01})}{\bar{P}(M_{00})} > c_0$, and in each model space, no subgroup model is better than the last model M_{iH_i}, i.e. $\frac{\bar{P}(M_{ih})}{\bar{P}(M_{iH_i})} \leq c_1$ for all h.

Continue: In each model space, choose the highest probability model if it is better than both end-models with the respective thresholds.

Finish: Choose M_{00} if no subgroup effect model is picked in any model space.

The overall effect model is chosen, only if there is strong evidence in its favour in comparison to the overall null model, and if there is no strong evidence against it in comparison to any single subgroup effect model. This serves to control two types of errors. When a model representing a subgroup effect (in $\mathcal{M}_i, i > 0$) is true, simply comparing M_{00} and M_{01} alone could result in a very high posterior probability for M_{01} (if this is the one 'closer' to the true model) and lead to choosing the wrong model M_{01}. Comparing the overall effect model with the subgroup effect models (in addition to the null model) protects against this error. On the other hand, it is also important to control the error of choosing a subgroup effect model when in fact the overall effect model is true; this can be achieved by requiring sufficiently large evidence against the overall effect model, i.e. by choosing a large value for c_1.

The values of c_0 and c_1 can be chosen so that the overall Type-I error is equal to a pre-determined value, and the probability of choosing a subgroup effect model when the overall effect model (with a specified effect size) is true is also small. This can be done using simulation. The value of c_0 relates to the amount of evidence required for rejecting the null model in favour of either the overall or a subgroup effect model. The value of c_1 represents the threshold for protecting the overall alternative against wrongly deciding in favour of a subgroup effect model. Adjusting for multiple testing, and controlling the error of choosing a subgroup effect when there is an overall effect are important goals; the framework here is specifically geared to achieving these goals.

29.6 Error rate evaluation by simulation

Let us return to the Dementia Trial. We used the sample sizes in Table 29.1 and independent $Beta(0.5, 0.5)$ priors for p and q. For each of the many settings needed to construct Figures 29.1 and 29.2, one thousand datasets were generated.

The two lower curves in Figure 29.1(a) show the estimated True Positive Rate (TPR) and the False Subgroup Rate (FSR), both calculated by simulating data from a specific overall effect (true) model. The standardized effect size for this model was set at 0.8. The uppermost curve marked TPR0 results from a simple procedure without subgroup analysis, i.e. this shows the traditional power at various levels of significance. The Type I Error (TIE) was varied by setting $c_1 = c_0$, and varying the common value from 19 down to 1, resulting in increasing values of TIE shown in the figure. At TIE of 0.05, TPR is 0.85, FSR is 0.10 and TPR0 is 0.98. These quantify the tradeoffs involved in planning a subgroup analysis as opposed to a simple overall effect analysis without considering subgroups.

The curves in Figure 29.1(b) were obtained with data from a particular subgroup model, namely, $M_{11} = (1, 0)$ representing an effect in the subgroup with FAST score less than 7. The standardized effect size was set to 0.8 again, i.e. this was the effect size in the subgroup with a non-zero treatment effect while the other subgroup had effect size zero. The TSR curve represents the rate of correctly choosing the subgroup effect model while the FPR curve represents the rate of incorrectly choosing the overall effect model M_{01}. Note that $1 - (\text{TSR} + \text{FPR})$ is the rate of incorrectly choosing the overall null or choosing a different subgroup model.

Figure 29.2 addresses issues similar to those in traditional power curves, plotting rates against various standardized effect sizes. We calculated TIE for a range of common threshold values for $c_0 = c_1$. For $c_0 = c_1 = 1.9$ we found TIE $= 0.05$. We thus fixed TIE at 0.05 by choosing $c_0 = c_1 = 1.9$. In Figure 29.2(a) the true model is the overall effect model M_{01}. As in Figure 29.1(a), TPR and FSR are plotted along with the traditional power curve TPR0 for comparison. The curves in Figure 29.2(b) are computed under the same true subgroup model as in Figure 29.1(b), i.e. $M_{11} = (1, 0)$.

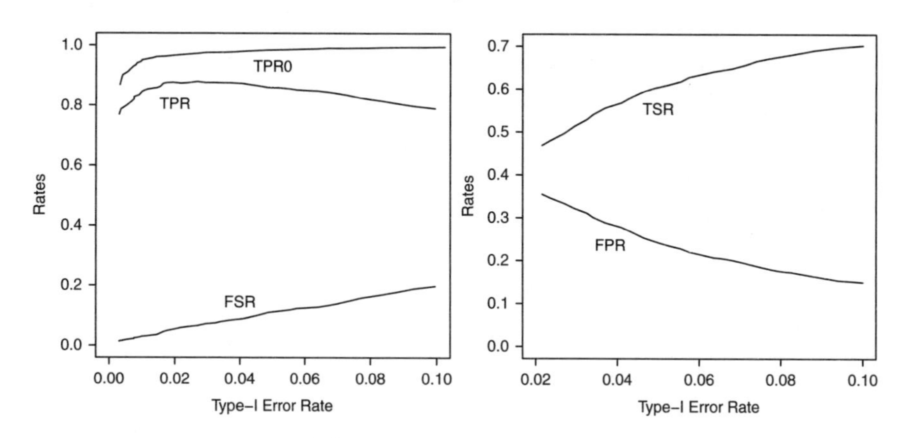

Figure 29.1 Panel (a) on left, panel (b) on right.

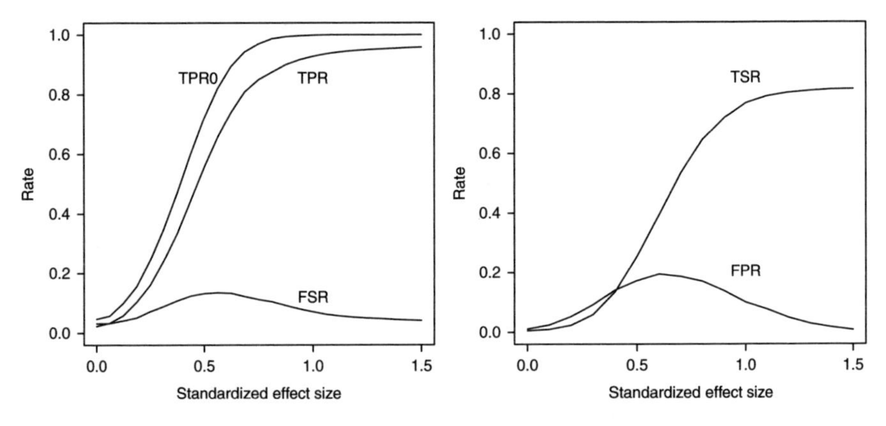

Figure 29.2 Panel (a) on left, panel (b) on right.

With these operating characteristics of the procedure, we used the experimental data to calculate posterior model probabilities. As noted above, the choice $c_1 = c_2 = 1.9$ can achieve a Type I error rate of 0.05.

To address the FSR, one would need to study it as a function of the overall effect size under the assumption of no subgroup effects. However, it is possible to define and control the value of average FSR, averaged over a reasonable effect size distribution. We used the normal distribution for the overall effect size with mean 0 and standard deviation equal to that of the data with simulation carried out under zero subgroup-effects. By choosing a range of values of c_0 and c_1, a range of values for TIE and average FSR can be obtained and suitably small values for these rates may then be chosen.

Setting both c_0 and c_1 at 1.9, the average FSR was also found to be 0.05. we also calculated the values of TIE on a rectangular grid of values for (c_0, c_1) and found that the choice $(c_0, c_1) = (1.9, 5.7)$ also corresponds to a TIE of 0.05. The average FSR for this choice was 0.01. Thus the tuning parameters in the procedure were chosen to control the probability of incorrectly rejecting the overall null and the probability of incorrectly picking a subgroup model when the overall effect model is true.

Implementation of the stepwise procedure resulted in the posterior probabilities as given in Table 29.4, and selection of the models, as given in Table 29.5. The result shows that, when TIE is 0.05 and average FSR is 0.05, the procedure selects the subgroup effect for MVOCAL; the treatment effect is absent when MVOCAL=0 and present when MVOCAL=1. When TIE is 0.05 and average FSR is 0.01, the procedure selects the overall effect.

29.7 Subgroup reporting as a decision problem

It is natural to cast subgroup reporting as a Bayesian decision problem. To describe how, let Γ_i denote all models except M_{i0} and M_{iH_i} (corresponding to the overall null and alternative hypotheses) in the model space $\mathcal{M}_{\mathcal{X}_i}$ for subgroups made by the covariate X_i. Then take $\mathcal{M} = \mathcal{M}_0 \cup \Gamma_1 \cup \cdots \cup \Gamma_I$. Let δ denote the desired decision. Possible decisions are to report an overall treatment effect, $\delta = M_{01}$; to report no evidence for any treatment effects, $\delta = M_{00}$; or to report subgroup effects. When reporting subgroup effects we allow reporting of multiple subgroups, one for each covariate i. Thus reporting subgroups involves the identification of a set of covariates,

Table 29.4 Posterior probabilities of overall and subgroup effects models for Dementia Trial data. Model spaces $\mathcal{M}_1$ and $\mathcal{M}_2$ correspond to the covariates FAST and MVOCAL, respectively.

Model Space			$\mathcal{M}_0$		
Models	M_{00}				M_{01}
Posterior Probability	0.011				0.989
Model Space (FAST)			$\mathcal{M}_1$		
Models	M_{10}	M_{11}	M_{12}	M_{13}	M_{14}
Posterior Probability	0.0105	0.0032	0.2089	0.3235	0.4539
Model Space (MVOCAL)			$\mathcal{M}_2$		
Models	M_{20}	M_{21}	M_{22}	M_{23}	M_{24}
Posterior Probability	0.0023	0.0003	0.5714	0.2843	0.1417

Table 29.5 Overall or Subgroup effect model as selected Dementia Trial data.

c_0	c_1	TIE		Model(s) Selected	Average FSR
1.9	1.9	0.05	M_{22}	No effect if MVOCAL = 0, non-zero effect if MVOCAL = 1	0.05
1.9	5.7	0.05	M_{01}	Overall Alternative	0.01

$A_I \equiv \{i_1, \ldots, i_m\} \subset \{1, \ldots, I\}$ together with a subgroup model index $\gamma_i \in \Gamma$ for each covariate. Let $\Gamma_I = \{\gamma_i, i \in A_I\}$ and let $A = (A_I, \Gamma_I)$ denote the pair of covariate indices and list of subgroups for each chosen covariate. Reporting subgroups is thus characterized as $\delta = A$. In summary

$$\delta \in \{M_0, M_1, A\} \equiv \mathcal{D} \text{ with } A = (A_I, \Gamma_I)$$

Note that the action space $\mathcal{D}$ differs from the model space $\mathcal{M}$ because we allow reporting of multiple subgroups simultaneously, but the probability model $p(M)$ allows only for one true subgroup model at a time.

A utility function $u(\delta, M, y)$ represents the investigator's relative preferences for the alternative actions under an assumed true model M and data y. Let $n_A = |A_I|$ denote the number of reported subgroups when $\delta = A$. We assume a natural generalization of a traditional o/c utility function for testing problems:

$$u(\delta, M, y) = \begin{cases} u_0\, I(M = M_0) & \text{if } \delta = M_0 \\ u_1\, I(M = M_1) & \text{if } \delta = M_1 \\ u_2\, I(M \in A_I) - (n_A - 1) & \text{if } \delta = A \end{cases} \tag{29.7}$$

In short, we realize a reward when the correct model is reported, and we pay a price for reporting more than one subgroup. As in many decision problems, the specific choice of u includes some arbitrary and simplifying choices. In particular, we assume that the data enters the utility function only indirectly through the decision rule $\delta = \delta(y)$. This is typical for inference problems.

Let $p(M)$ denote a probability model over the model space and let

$$U(\delta, y) \propto \sum u(\delta, M, y)\, p(M \mid y)$$

denote the posterior expected utility. The optimal decision δ^* is the action with maximum expected utility. It is easy to show that the optimal decision under 29.7 can be characterized as follows. Assume $\delta^*(y) = A$. If $i \in A_I$, i.e. we report subgroups for covariate i, then the reported subgroups for i are the subgroups with highest posterior probability,

$$\gamma_i^* = \arg\max_{\gamma \in \Gamma} \{p(M_{i\gamma} \mid y)\}$$

And

$$A_I^* = \{i : p(M_{i\gamma_i^*} \mid y) \geq 1/u_2\}$$

i.e. we report subgroups for all covariates that include a model $M_{i\gamma_i^*}$ with posterior probability greater than $1/u_2$. Let $\Gamma_I^* = \{\gamma_i^*, i \in A_I^*\}$. In summary, if $\delta^* = A$, then it must be $A^* = (A_I^*, \Gamma_I^*)$. Thus, to determine the Bayes rule δ^* we only need to compare expected utilities for $\delta = M_0, M_1$ and A^*:

$$U(\delta, y) = \begin{cases} u_0\, p(M_0 \mid y) & \text{if } \delta = M_0 \\ u_1\, p(M_1 \mid y) & \text{if } \delta = M_1 \\ u_2 \sum_{i \in A_I^*} p(M_{i\gamma *_i} \mid y) - (n_{A^*} - 1) & \text{if } \delta = A^* \end{cases}$$

We can now see the Bayes rule. Write $\overline{p}(M)$ as short for $p(M \mid y)$, let $M_i^* = M_{i\gamma_i^*}$ denote the maximum posterior subgroup model with covariate x_i, let $M^* = \arg\max\{\overline{p}(M_i^*)\}$ denote the highest posterior probability subgroup model and let $i^* = \arg\max_i \overline{p}(M_i^*)$ denote the index of the covariate that defines M^*. Then

$$\delta^* = \begin{cases} M_1 & \text{if } \frac{\overline{p}(M_1)}{\overline{p}(M_0)} > \frac{u_0}{u_1} \text{ and } \frac{\overline{p}(M_1)}{\overline{p}(M^*)} > \frac{u_2}{u_1} + \sum_{A_I^* \setminus i^*} \frac{u_2 \overline{p}(M_i^*) - 1}{u_1\, \overline{p}(M^*)} \\ A^* & \text{if } \frac{\overline{p}(M_1)}{\overline{p}(M^*)} < \frac{u_2}{u_1} + \sum_{A_I^* \setminus i^*} \frac{u_2 \overline{p}(M_i^*) - 1}{u_1 \overline{p}(M^*)} \text{ and } \frac{\overline{p}(M_0)}{\overline{p}(M^*)} < \frac{u_2}{u_0} + \sum_{A_I^* \setminus i^*} \frac{u_2 \overline{p}(M_i^*) - 1}{u_0 \overline{p}(M^*)} \\ M_0 & \text{otherwise} \end{cases}$$

Note that $u_2\, \overline{p}(M_i^*) > 1$ for all $i \in A_I^*$, i.e. the terms in the sums are all strictly positive (although some can be very small).

It is possible to relate this Bayes rule to the procedure described in Section 29.5. To do this, we depart from a strictly decision theoretic implementation, and take the form of the Bayes rule δ^* as a motivation for a slightly simplified rule. We drop the sum over $A^* \setminus i^*$ in the conditions for reporting A^*, i.e. we are slightly more conservative in reporting subgroups. Then

$$\delta^* = \begin{cases} M_1 & \text{if } \frac{\overline{p}(M_1)}{\overline{p}(M_0)} > \frac{u_0}{u_1} \text{ and } \frac{\overline{p}(M_1)}{\overline{p}(M^*)} > \frac{u_2}{u_1} \\ A^* & \text{if } \frac{\overline{p}(M_1)}{\overline{p}(M^*)} < \frac{u_2}{u_1} \text{ and } \frac{\overline{p}(M_0)}{\overline{p}(M^*)} < \frac{u_2}{u_0} \\ M_0 & \text{otherwise} \end{cases}$$

and finally, we replace A^* by $\{M_{i\gamma_i^*} : \overline{p}(M_{i\gamma_i^*}) > \frac{1}{u_2} \max\{\overline{p}(M_1)u_1, \overline{p}(M_0)u_0\}\}$. This enables us to describe the final rule in terms of thresholds on odds $\overline{p}(M_1)/\overline{p}(M_0)$, $\overline{p}(M_i^*)/\overline{p}(M_0)$ and $\overline{p}(M_i^*)/\overline{p}(M_1)$ only. Noting that $\overline{p}(M^*) < x \Leftrightarrow \overline{p}(M_i^*) < x \forall i$ we get

$$\delta^* = \begin{cases} M_1 & \text{if } \frac{\overline{p}(M_1)}{\overline{p}(M_0)} > t_0 \text{ and } \frac{\overline{p}(M_i^*)}{\overline{p}(M_1)} < t_1 \text{ for all } i \\ A^* & \text{if for some } i\text{: } \frac{\overline{p}(M_i^*)}{\overline{p}(M_0)} > t_0 t_1 \text{ and } \frac{\overline{p}(M_i^*)}{\overline{p}(M_1)} > t_1 \\ & \text{and } A^* = \{i : \text{ above holds}\} \\ M_0 & \text{otherwise} \end{cases} \tag{29.8}$$

This is very nearly the procedure described in Section 29.5. The main difference is that $t_0 t_1$ is replaced by t_0.

For an implementation of the proposed rule we need to specify a probability model $p(M)$ over the space of all models $\mathcal{M}$. Let $\mathcal{M}_i = \{M_{i\gamma_i}, \gamma_i \in \Gamma\} \cup \{M_{00}, M_{01}\}$ denote the subspace defined by groupings based on covariate x_i, including the overall null and alternative. We define probability models $p_i(M)$ for $M \in \mathcal{M}_i$, $i = 1, \ldots, I$. We construct p_i such that $\pi_0 \equiv p_i(M_{00})$ and $\pi_1 \equiv p_i(M_{01})$ are common across i. The we piece the sub-models together by

$$p(M) = \begin{cases} \pi_0 & \text{for } M = M_{00} \\ \pi_1 & \text{for } M = M_{01} \\ p_i(M)\,\frac{1}{I} & \text{for } M = M_{i\gamma},\ \gamma \in \Gamma \end{cases}$$

We simplify the construction by using the same zero-enriched Polya urn model with common parameter values (p, q). If $S_i = 2$ for all covariates, the above definition works as stated. Otherwise we need to rescale $p_i(M)$ in the last line of the above display as

$$p_i(M)\frac{1 - \pi_0 - \pi_1}{1 - \pi_{i0} - \pi_{iH_i}}$$

where π_{i0} is the probability assigned by the zero-enriched Polya urn to model M_{i0} and similarly π_{iH_i} to the last model (equal and non-zero effect in all subgroups) M_{iH_I}.

More details and results for the Dementia Trial are reported in [20].

29.8 Discussion

The methods presented here cast subgroup analysis as a model selection procedure that can also be seen as a decision problem. Bayesian thinking allows one readily to formulate the probability model and to implement inference and decision making. We have used a simple model here for the observables (namely, conditionally independent normals) in order to focus more on various aspects of model selection. These aspects include a careful specification of the model space, priors on this space and on the parameters in each model, and repeated-data frequency evaluation or operating characteristics. Beyond these aspects, the Bayesian formulation makes it fairly straightforward to proceed to post-data calculations and inference for other models for the observables such as binary, count or time-to-event outcomes. The material presented here can therefore be seen as a framework in which to carry out subgroup analysis rather than a specific instance of it.

We have used low-information priors for illustration. We have not addressed how to incorporate substantive contextual information that might be available, external to the data from the clinical trial. One example is post-approval studies of a drug's safety and effectiveness. Typically, information gathered during the clinical trials that led to the approval of the drug is available when designing such studies. The framework presented here should allow the use of this information. However, methods need to be developed for this purpose. Also, investigation is needed into the flexibility of the priors to represent the available external information.

In this article, we have limited our goals to determining the presence or absence of a subgroup effect (as well as overall effect) and which subgroups show equal or differing effects. It is also of interest to estimate the effect sizes with interval estimates for making clinical judgements. It would be desirable to extend the method presented here to provide estimate(s) based on the selected model(s), or to provide smoothed estimates in the spirit of [28] and [8].

The framework presented here treats each subgroup-making covariate separately, not accomodating any possible interactions such as due to age–gender combinations. It is possible to form a single covariate using these combinations and proceed as before. However, a better approach could be formulated to include and estimate interactions and to find optimal combinations. The work in [21] and [7] is relevant in this context.

The current implementation of the calculations employs a full enumeration of models. Sampling based evaluation of the posterior probabilities of models using reversible jump MCMC would be much more efficient. Such methods have been typically implemented in the variable selection context. The model space here is somewhat distinct and some features of it could offer opportunities for added efficiencies by minimizing transdimensional MCMC. A sparse prior and the expectation of an effect for only a few subgroups could be exploited for efficiency as well. It is also possible to adapt the posterior MCMC simulation to discover the optimal decision theoretic solution as in [18] and [19].

Finally, it would be interesting to explore the possibility of employing Bayesian nonparametric models that allow flexible clustering, for post-hoc subgroup analysis. Carrying out the clustering without covariate information and then aligning any discovered clusters with the available covariates may mitigate concerns related to multiplicities. It would be important to evaluate repeated-data operating characteristics of such methods.

Acknowledgements

We thank Christine Kovach, PhD, RN of the University of Wisconsin-Milwaukee and Brent Logan, PhD of the Medical College of Wisconsin for providing advice and the data from their study. This research was initiated during a research program on multiplicities and repeatability at SAMSI (Statistical and Mathematical Sciences Institute, NC).

References

[1] Abramowitz, M. and Stegun, I. A. (1972). *Stirling Numbers of the Second Kind* (9th printing edn)., Chapter 24.1.4, pp. 824–825. Handbook of Mathematical Functions with Formulas, Graphs and Mathematical Tables. Dover, New York.

[2] Alosh, M. and Huque, M. F. (2009). A flexible strategy for testing subgroups and overall population. *Statistics in Medicine*, **28**(1), 3–23.

[3] Assmann, S. F., Pocock, S. J., Enos, L. E. and Kasten, L. E. (2000). Subgroup analysis and other (mis)uses of baseline data in clinical trials. *Lancet*, **355**(9209), 1064–1069.

[4] Bailar, J. C. and Mosteller, F. (1992). *Medical Uses of Statistics* (2nd edn). NEJM Books, Waltham, MA.

[5] Berger, J. O. (2006). The case for objective Bayesian analysis. *Bayesian Analysis*, **1**(3), 385–402.

[6] Blackwell, D. and MacQueen, J. B. (1973). Ferguson distributions via polya urn schemes. *Annals of Statistics*, **1**, 353–355.

[7] Brinkley, J., Tsiatis, A. and Anstrom, K. J. (2010, Jun). A generalized estimator of the attributable benefit of an optimal treatment regime. *Biometrics*, **66**(2), 512–522.

[8] Chipman, Hugh A., George, Edward I. and McCulloch, Robert E. (2010). Bart: Bayesian additive regression trees. *The Annals of Applied Statistics*, **4**(1), 266–298.

[9] Dixon, D. O. and Simon, R. (1991, Sep). Bayesian subset analysis. *Biometrics*, **47**(3), 871–881.

[10] Goldstein, M. (2006). Subjective Bayesian analysis: principles and practice. *Bayesian Analysis*, **1**(3), 403.

[11] Hodges, J. S., Cui, Y., Sargent, D. J. and Carlin, B. P. (2007). Smoothing balanced single-error-term analysis of variance. *Technometrics*, **49**(1), 1225.

[12] Jackson, R. D., LaCroix, A. Z., Gass, M., Wallace, R. B., Robbins, J., Lewis, C. E., Bassford, T., Beresford, S. A., Black, H. R., Blanchette, P., Bonds, D. E., Brunner, R. L., Brzyski, R. G., Caan, B., Cauley, J. A., Chlebowski, R. T., Cummings, S. R., Granek, I., Hays, J., Heiss, G., Hendrix, S. L., Howard, B. V., Hsia, J., Hubbell, F. A., Johnson, K. C., Judd, H., Kotchen, J. M., Kuller, L. H., Langer, R. D., Lasser, N. L., Limacher, M. C., Ludlam, S., Manson, J. E., Margolis, K. L., McGowan, J., Ockene, J. K., O'Sullivan, M. J., Phillips, L., Prentice, R. L., Sarto, G. E., Stefanick, M. L., Horn, L. Van, Wactawski-Wende, J., Whitlock, E., Anderson, G. L., Assaf, A. R., Barad, D., and Investigators, Women's Health Initiative (2006, Feb 16). Calcium plus vitamin d supplementation and the risk of fractures. *The New England Journal of Medicine*, **354**(7), 669–683.

[13] Jones, H. E., Ohlssen, D. I., Neuenschwander, B., Racine, A. and Branson, M. (2011). Bayesian models for subgroup analysis in clinical trials. *Clinical Trials*, **8**, 129–143.

[14] Kovach, C. R., Logan, B. R., Noonan, P. E., Schlidt, A. M., Smerz, J., Simpson, M. and Wells, T. (2006, Jun–Jul). Effects of the serial trial intervention on discomfort and behaviour of nursing home residents with dementia. *American Journal of Alzheimer's Disease and Other Dementias*, **21**(3), 147–155.

[15] Lagakos, S. W. (2006). The challenge of subgroup analyses–reporting without distorting. *The New England Journal of Medicine*, **354**(16), 1667–1669.

[16] Liang, F., Paulo, R., Moina, G., Clyde, M. A. and Berger, J. O. (2008). Mixtures of g-priors for Bayesian variable selection. *Journal of the American Statistical Association*, **103**(481), 410–423.

[17] Louis, T. A. (1984). Estimating a population of parameter values using Bayes and empirical Bayes methods. *Journal of the American Statistical Association*, **79**, 393–398.

[18] Müller, P. (1999). Simulation based optimal design, pp. 459–474. *Bayesian Statistics* 6. Oxford University Press.

[19] Müller, P., Sansó, B. and DeIorio, M. (2004). Optimal Bayesian design by inhomogeneous Markov Chain Monte Carlo. *Journal of the American Statistical Association*, **99**(467), 788–798.

[20] Müller, Peter, Sivaganesan, Siva, and Laud, Purushottam W. (2010). A Bayes rule for subgroup reporting. In *Frontiers of Statistical Decision Making and Bayesian Analysis: In Honor of James O. Berger* (ed. M.-H. Chen, D. K. Dey, P. Müller, D. Sun, and K. Ye), pp. 277–284. Springer-Verlag Inc.

[21] Nobile, A. and Green, P. J. (2000). Bayesian analysis of factorial experiments by mixture modeling. *Biometrika*, **66**(2), 512–522.

[22] Pitman, J. (2006). *Some developments of the Blackwell-MacQueen urn scheme*, Volume 30 of *Statistics, Probability and Game Theory*, pp. 245–267. IMS Lecture Notes.

[23] Pocock, S. J., Assmann, S. E., Enos, L. E. and Kasten, L. E. (2002). Subgroup analysis, covariate adjustment and baseline comparisons in clinical trial reporting: current practice and problems. *Statistics in Medicine*, **21**(19), 2917–2930.

[24] Pocock, S. J. and Lubsen, J. (2008). More on subgroup analyses in clinical trials. *The New England Journal of Medicine*, **358**(19), 2076; author reply 2076–7.

[25] Proschan, M. A. and Waclawiw, M. A. (2000). Practical guidelines for multiplicity adjustment in Clinical Trials. *Controlled Clinical Trials*, **21**(6), 527–539.
[26] Scott, J. G. and Berger, J. O. (2006). An exploration of aspects of Bayesian multiple testing. *Journal of Statistical Planning and Inference*, **136**, 2144–2162.
[27] Scott, J. G. and Berger, J. O. (2010). Bayes and empirical Bayes multiplicity adjustment in the variable selection problem. *The Annals of Statistics*, **38**(5), 2587–2619.
[28] Simon, R. (2002). Bayesian subset analysis: application to studying treatment-by-gender interactions. *Statistics in Medicine*, **21**(19), 2909–2916.
[29] Sivaganesan, S., Laud, P. W. and Müller, P. (2011). A Bayesian subgroup analysis with a zero-enriched polya urn scheme. *Statistics in Medicine*, **30**(4), 312–323.
[30] Song, Y. and Chi, G. Y. (2007). A method for testing a prespecified subgroup in clinical trials. *Statistics in Medicine*, **26**(19), 3535–3549.
[31] Wang, R., Lagakos, S. W., Ware, J. H., Hunter, D. J. and Drazen, J. M. (2007). Statistics in medicine – reporting of subgroup analyses in clinical trials. *The New England Journal of Medicine*, **357**(21), 2189–2194.
[32] Wiens, B. L. and Dmitrienko, A. (2005). The fallback procedure for evaluating a single family of hypotheses. *Journal of Biopharmaceutical Statistics*, **15**(6), 929–942.
[33] Yusuf, S., Wittes, J., Probstfield, J. and Tyroler, H. A. (1991). Analysis and interpretation of treatment effects in subgroups of patients in randomized clinical trials. *JAMA : The Journal of the American Medical Association*, **266**(1), 93–98.

30 Surviving fully Bayesian nonparametric regression models

TIMOTHY E. HANSON AND
ALEJANDRO JARA

30.1 Introduction

Semiparametric survival models split the model, and hence inference, into two parts: parametric and nonparametric. The parametric portion of the model provides a succinct summary relating patient survival to a relatively small number of regression coefficients: risk factors, acceleration factors, odds ratios, etcetera. The nonparametric part of the model—the baseline hazard, cumulative hazard, or survival function—is modelled as flexibly as possible, so inference does not depend on a particular parametric form such as log-normal or Weibull.

This chapter compares two Bayesian nonparametric models that generalize the accelerated failure time model, based on recent work concerning probability models for predictor-dependent probability distributions. Both models allow for crossing hazards for different covariates $\mathbf{x}_1$ and $\mathbf{x}_2$, as well as crossing cumulative hazards, and hence crossing survival curves, and thus convey substantial flexibility over standard accelerated failure time models. Furthermore, the entire density is modelled at every covariate level $\mathbf{x} \in \mathcal{X}$, so full density and hazard estimates are available, accompanied by reliable interval estimates, unlike many median (and other quantile) regression models. Perhaps most importantly, both models are implemented as user-friendly functions calling compiled FORTRAN in DPpackage for R [45].

The chapter is organized as follows. Commonly used semiparametric survival models are reviewed in Section 30.2. A discussion about the Bayesian nonparametric priors used in the generalizations of the accelerated failure time model is given in Section 30.3. The two generalizations of the accelerated failure time model are introduced in Section 30.4. The two models are compared by means of real-life data analyses in Section 30.5. The chapter concludes with a short discussion in Section 30.6.

30.2 Semiparametric survival models

A common starting point in the specification of a regression model for time-to-event data is the definition of a baseline survival function, S_0, that is modified (either directly or indirectly) by subject-specific covariates $\mathbf{x}$. Let T_0 be a random survival time from the baseline group (with all covariates equal to zero). The baseline survival function is defined by $S_0(t) = P(T_0 > t)$. Continuous survival is assumed throughout. Thus, the baseline density and hazard functions are defined by $f_0(t) = -\frac{d}{dt}S_0(t)$ and $h_0(t) = f_0(t)/S_0(t)$, respectively. The survival, density and hazard functions for a member of the population with covariates $\mathbf{x}$ will be denoted by $S_{\mathbf{x}}(t)$, $f_{\mathbf{x}}(t)$, and $h_{\mathbf{x}}(t)$, respectively.

30.2.1 Proportional hazards

A proportional hazards (PH) model [20], for continuous data, is obtained by expressing the covariate-dependent survival function $S_{\mathbf{x}}(t)$ as

$$S_{\mathbf{x}}(t) = S_0(t)^{\exp(\mathbf{x}'\boldsymbol{\beta})} \tag{30.1}$$

In terms of hazards, this model reduces to

$$h_{\mathbf{x}}(t) = \exp(\mathbf{x}'\boldsymbol{\beta})h_0(t)$$

Note then that for two individuals with covariates $\mathbf{x}_1$ and $\mathbf{x}_2$, the ratio of hazard curves is constant and proportional to $\frac{h_{\mathbf{x}_1}(t)}{h_{\mathbf{x}_2}(t)} = \exp\{(\mathbf{x}_1 - \mathbf{x}_2)'\boldsymbol{\beta}\}$, hence the name 'proportional hazards.' Cox [20] is the second most cited statistical paper of all time [74], and the proportional hazards model is easily the most popular semiparametric survival model in statistics, to the point where medical researchers tend to compare different populations' survival in terms of instantaneous risk (hazard) rather than mean or median survival as in common regression models. Part of the populariy of the model has to do with the incredible momentum the model has gained from how easy it is to fit the model through partial likelihood [21] and its implementation in SAS in the procedure `proc phreg`. The use of partial likelihood and subsequent counting process formulation [4] of the model has allowed ready extension to stratified analysis, proportional intensity models, frailty models, and so on [81].

The first Bayesian semiparametric approach to PH models posits a gamma process as a prior on the baseline cumulative hazard $H_0(t) = \int_0^t h_0(s)ds$ [48]; partial likelihood emerges as a limiting case (of the marginal likelihood as the precision approaches zero). The use of the gamma process prior in PH models, as well as the beta process prior [41], piecewise exponential priors, and correlated increments priors are covered in [43] (pp. 47–94) and [78]. Other approaches include what are essentially Bernstein polynomials [11, 29] and penalized B-splines [40, 51]. The last two models are available in a free program called BayesX [7].

30.2.2 Accelerated failure time

An accelerated failure time (AFT) model is obtained by expressing the covariate-dependent survival function $S_{\mathbf{x}}(t)$ as

$$S_{\mathbf{x}}(t) = S_0\{\exp(-\mathbf{x}'\boldsymbol{\beta})t\} \tag{30.2}$$

This is equivalent to a linear model for the log transformation of the corresponding time-to-event response variable, T,

$$\log T = \mathbf{x}'\boldsymbol{\beta} + \epsilon \tag{30.3}$$

where $\exp(\epsilon) \sim S_0$. The mean, median, and any quantile of survival for an individual with covariates $\mathbf{x}_1$ is changed by a factor of $\exp\{(\mathbf{x}_1 - \mathbf{x}_2)'\boldsymbol{\beta}\}$ relative to those with covariates $\mathbf{x}_2$.

An early frequentist least squares teatment of the AFT model with right-censored data is due to Buckley and James [10]; the Buckley–James estimator is implemented in Frank Harrell's `Design` library for R [3]. More refined estimators followed in the 1990s [86, 89] focusing on median-regression.

From a Bayesian nonparametric perspective, the first approach, based on a Dirichlet process prior, obtained approximate marginal inferences to the AFT model [18]; a full Bayesian treatment using the Dirichlet process is not practically possible [47]. Approaches based on Dirichlet process mixture models have been considered by Kuo and Mallick [56], Kottas and Gelfand [55] and Hanson [34]. Dirichlet process mixtures 'fix' the discrete nature of the Dirichlet process, as do other discrete mixtures of continuous kernels. We refer the reader to Komarek and Lesaffre [54], for an alternative approach based on mixtures of normal distributions. Tailfree priors that have continuous densities can directly model the distribution of ϵ in expression (30.3) [33, 37, 83, 93].

Although PH is by far the most commonly-used semiparametric survival model, several studies have shown vastly superior fit and interpretation from AFT models [33, 38, 42, 49, 67, 71]. Many further argue for alternatives to hazard ratios in reporting the results of survival analyses [9, 39, 49, 50, 67, 68, 75, 80, 85, 94]. Cox pointed out himself [72] "*... the physical or substantive basis for ... proportional hazards models ... is one of its weaknesses ... accelerated failure time models are in many ways more appealing because of their quite direct physical interpretation ...* ". Additionally, Keene [50] points out difficulties with PH analysis for use in meta-analytic analysis designed to combine information across several studies and further comments "*Accelerated failure time models are a valuable and are often a more realistic alternative to proportional hazards models.*"

Since the AFT model is a log-linear model, one can obtain a point estimate of survival for covariates $\mathbf{x}$ as simply $\exp(\mathbf{x}'\hat{\boldsymbol{\beta}})$, where $\hat{\boldsymbol{\beta}}$ is an estimate of $\boldsymbol{\beta}$. Prediction is impossible within the PH model framework without an estimate of the baseline hazard function, so reporting only coefficients—which is common—disallows others to predict survival.

30.2.3 Proportional odds

The proportional odds (PO) model has recently gained attention as an alternative to the PH and AFT models. PO defines the survival function $S_{\mathbf{x}}(t)$ for an individual with covariate vector $\mathbf{x}$ through the relation

$$\frac{S_{\mathbf{x}}(t)}{1 - S_{\mathbf{x}}(t)} = \exp\{-\mathbf{x}'\boldsymbol{\beta}\} \left(\frac{S_0(t)}{1 - S_0(t)} \right) \tag{30.4}$$

The odds of dying before any time t are $\exp\{(\mathbf{x}_1 - \mathbf{x}_2)'\boldsymbol{\beta}\}$ times greater for those with covariates $\mathbf{x}_1$ versus $\mathbf{x}_2$.

The first semiparametric approaches to proportional odds models involving covariates are due to Cheng *et al.* [17], Murphy *et al.* [65], and Yang and Prentice [87]. A semiparametric frequentist implementation of the proportional odds model is available in Martinussen and Scheike's `timereg` package [62] for R. Bayesian nonparametric approaches for the PO model have been based on Bernstein polynomials [5], B-splines [84], and Polya trees [33, 36, 38, 93].

The PH, AFT, and PO models all make overarching assumptions about the data generating mechanism for the sake of obtaining succinct data summaries. An important aspect associated with the Bayesian nonparametric formulation of these models is that, by assuming the *same, flexible model* for the baseline survival function, they are placed on a common ground [33, 36, 38, 92, 93]. Furthermore, parametric models are special cases of the nonparametric models. Differences in fit and/or predictive performance can therefore be attributed to the *survival* models only, rather than to additional possible differences in quite different nonparametric models or estimation methods.

Of the Bayesian approaches based on Polya trees considered by Hanson [33], Hanson and Yang [38], Zhao *et al.* [93] and Hanson *et al.* [36], the PO model was chosen over PH and AFT according to the log-pseudo marginal likelihood (LPML) criterion [28]. In three of these works, the parametric log-logistic model, a special case of PO that also has the AFT property, was chosen. This may be due to the fact that the PO assumption implies that hazard ratios $\lim_{t\to\infty} \frac{h_{\mathbf{x}_1}(t)}{h_{\mathbf{x}_2}(t)} = 1$, that is, eventually everyone has the same risk of dying tomorrow. These authors also found that, everything else being equal, the actual semiparametric model chosen (PO, PH or AFT) affects prediction far more than whether the baseline is modelled nonparametrically. It is worth noting that none of these papers favoured the semiparametric PH model in actual applications.

30.2.4 Other models

PH, AFT and PO are only three of many semiparametric survival models used in practice. There are a few more hazard-based models including the additive hazards (AH) model [1, 2], given by

$$h_{\mathbf{x}}(t) = h_0(t) + \mathbf{x}'\boldsymbol{\beta}$$

which is implemented in Martinussen and Scheike's `timereg` package for R. An empirical Bayes approach to this model based on the gamma process was implemented by [79]. Fully Bayesian approaches require elaborate model specification to incorporate the rather awkward constraint $h_0(t) + \mathbf{x}'\boldsymbol{\beta} \geq 0$ for $t > 0$ [23, 88].

Recently, there has been some interest in the accelerated hazards model [16, 91], given by

$$h_{\mathbf{x}}(t) = h_0\{\exp(-\mathbf{x}'\boldsymbol{\beta})t\}$$

This model allows hazard and survival curves to cross. A highly interpretable model that relates covariates to the residual life function, m_0, which is defined by $m_0(t) = E(T_0 - t|T_0 > t)$, is the proportional mean residual life model [15]. Under this model, the residual life function for a subject with covariates $\mathbf{x}$ is given by

$$m_{\mathbf{x}}(t) = \exp(\mathbf{x}'\boldsymbol{\beta})m_0(t).$$

It is important to stress that there have been no Bayesian approaches to these two models to date. Certainly there are other semiparametric models we are omitting here, but these round out several available methods.

Finally, several interesting 'super models' have been proposed in the literature, including transformation models that include PH and PO as special cases [61, 76], transformation and extended regression models that include PH and AH as special cases [62, 88] and hazard regression models that include both PH and AFT as special cases [14]. While highly flexible, all these models suffer in that, once fit, the resulting regression parameters lose any simple interpretability.

30.2.5 Extensions

There are several generalizations that have been made to the semiparametric models presented here. A standard approach for dealing with correlated data has been the introduction of frailty terms to the linear predictor (e.g., $\mathbf{x}_{ij}'\boldsymbol{\beta} + \gamma_i$ for the jth subject in cluster i). Frailty models have been widely discussed in the literature and correspond to particular cases of hierarchical models.

Hazard-based models (proportional, additive and accelerated) naturally accommodate time-dependent covariates; the linear predictor is simply augmented to be $\mathbf{x}(t)'\boldsymbol{\beta}$. Similarly, hazard-based models can also include time-dependent regression effects via $\mathbf{x}'\boldsymbol{\beta}(t)$ or even $\mathbf{x}(t)'\boldsymbol{\beta}(t)$. A traditional 'quick fix' for nonproportional hazards is to introduce an interaction between a covariate x and time, e.g. $h_{\mathbf{x}}(t) = \exp(x\beta_1 + xt\beta_2)h_0(t)$, yielding a particular focused deviation from PH. After implementing time-dependent regression effects, model inference is essentially reduced to examining plots, much like additive models.

These extensions allow one to continue using the familiar proportional hazards model is situations where proportional hazards does not hold. Such is the mindset of many people involved in analysing survival data that other, potentially more parsimonious models, are never even considered. Therneau and Grambsch [81] discuss the Cox model including many generalizations. When proportional hazards fails they recommend, (a) stratification *within the Cox model,* (b) partitioning the time axis so that proportional hazards may hold over shorter time periods *within the Cox model,* (c) time-varying effects $\boldsymbol{\beta}(t)$ *within the Cox model,* and (d) as a last resort the consideration other models, e.g. AFT or AH.

Other model modifications include cure rate models, joint longitudinal/survival models, recurrent events models, multistate models, competing risks models, and multivariate models that incorporate dependence more flexibly than frailty models.

30.3 Two Bayesian nonparametric priors used in survival analysis

Ultimately, we generalize AFT models based on extension of Dirichlet process mixture models and Polya tree models. Therefore, we briefly review both of these priors in this section. Many other priors for baseline hazard, cumulative hazard, or survival functions have been successfully employed over the last 20 years. These include the gamma process [48], the beta process [41], Bernstein polynomials [12, 29, 69, 70], piecewise exponential models [43], penalized B-splines [40, 51, 84] and extensions of these approaches. The literature is too vast to attempt even a moderate review, and we instead refer the interested reader to Sinha and Dey [78], Ibrahim *et al.* [43], Müller and Quintana [64], and Hanson *et al.* [35] for general overviews.

30.3.1 The Dirichlet process mixture model

Convolving a Dirichlet process (DP) [26] with a parametric kernel, such as the normal, gives a DP mixture (DPM) model [25, 59]. A simple DPM of Gaussian densities for continuous data $\epsilon_1, \ldots, \epsilon_n$ is given by

$$\epsilon_i | G \overset{iid}{\sim} \int N(\mu, \sigma^2) dG(\mu, \sigma^2), \tag{30.5}$$

where the mixing distribution, G, is a random probability measure defined on $\mathbb{R} \times \mathbb{R}^+$, following a DP. A random probability measure G follows a DP with parameters (α, G_0), where $\alpha \in \mathbb{R}^+$ and G_0 is an appropriate probability measure defined on $\mathbb{R} \times \mathbb{R}^+$, written as

$$G \mid \alpha, G_0 \sim DP(\alpha G_0) \tag{30.6}$$

if for any measurable non-trivial partition $\{B_l : 1 \leq l \leq k\}$ of $\mathbb{R} \times \mathbb{R}^+$ the vector $\{G(B_l) : 1 \leq l \leq k\}$ has a Dirichlet distribution with parameters $(\alpha G_0(B_1), \ldots, \alpha G_0(B_k))$. It follows that

$$G(B_l) \mid \alpha, G_0 \sim \text{Beta}(\alpha G_0(B_l), \alpha G_0(B_l^c)),$$

and therefore, $E[G(B_l) \mid \alpha, G_0] = G_0(B_l)$ and

$$Var[G(B_l) \mid \alpha, G_0] = G_0(B_l)G_0(B_l^c)/(\alpha + 1).$$

These results show the role of G_0 and α, namely, that G is centred around G_0 and that α is a precision parameter. If $G \mid \alpha, G_0 \sim DP(\alpha G_0)$, then the trajectories of the process can be represented by the following stickbreaking representation, due to Sethuraman [77]:

$$G(\cdot) = \sum_{i=1}^{\infty} w_i \delta_{(\mu_i, \sigma_i^2)}(\cdot)$$

where $\delta_\theta(\cdot)$ is the Dirac measure at θ, $w_i = V_i \prod_{j<i}(1 - V_j)$, with $V_i \mid \alpha \overset{iid}{\sim} Beta(1, \alpha)$, and $(\mu_i, \sigma_i^2) \mid G_0 \overset{iid}{\sim} G_0$.

The stickbreaking representation of the DP allows formulating (30.5) as a countably infinite mixture of normals given by

$$\epsilon_i | G \overset{iid}{\sim} \sum_{j=1}^{\infty} \left[V_j \prod_{k=1}^{j-1} (1 - V_k) \right] N(\mu_j, \sigma_j^2) \tag{30.7}$$

Note that $E(w_j) > E(w_{j+1})$ for all j, so the weights are stochastically ordered. The prior distribution on ϵ_i is centred at the normal distribution; Griffin [30] discusses prior specifications that control the 'non-normalness' of this distribution.

30.3.2 The Polya tree

A Polya tree (PT) successively partitions the reals into finer and finer partitions; each refinement of a partition doubles the number of partition sets by cutting each of the previous level's sets into two pieces; there are two sets at level 1, four sets at level 2, eight sets at 3, and so on. We focus on a PT centred at the standard normal density, that is, where $N(0, 1)$ is the *centring distribution* for the Polya tree. At level j, the Polya tree partitions the real line into 2^j intervals $B_{j,k} = (\Phi^{-1}((k - 1)2^{-j}), \Phi^{-1}(k2^{-j}))$ of probability 2^{-j} under Φ, $k = 1, \ldots, 2^j$. Note that $B_{j,k} = B_{j+1,2k-1} \cap B_{j+1,2k}$. Given an observation ϵ is in set k at level j, i.e. $B_{j,k}$, it could then be in either of the two offspring sets $B_{j+1,2k-1}$ or $B_{j+1,2k}$ at level $j + 1$. The conditional probabilities associated with these sets will be denoted by $Y_{j+1,2k-1}$ and $Y_{j+1,2k}$. Clearly they must sum to one, and so a common prior for either of these probabilities is a beta distribution [27, 33, 37, 57, 58, 82, 83, 93], given by

$$Y_{j,2k-1}|c \overset{ind.}{\sim} \text{beta}(cj^2, cj^2),\ j = 1, \ldots, J;\ k = 1, \ldots, 2^{j-1}$$

where $c \in \mathbb{R}^+$, which ensures that every realization of the process has a density, allowing the modelling of continuous data without the need of convolutions with continuous kernels. The resulting model for data $\epsilon_1, \ldots, \epsilon_n$ is given by

$$\epsilon_i|G \overset{iid}{\sim} G \tag{30.8}$$

where

$$G \sim PT_J(c, N(0, 1)) \tag{30.9}$$

The user-specified weight c controls how closely the posterior follows $N(0,1)$ in terms of L_1 distance [32], with larger values forcing G closer to $N(0,1)$; often a prior is placed on c, e.g. $c \sim \Gamma(a,b)$. The PT is stopped at level J (typically $J = 5,6,7$); within the sets $\{B_{J,k} : k = 1,\ldots,2^J\}$ at the level J G follows $N(0,1)$ [33]. The resulting density is given by

$$p(\epsilon|\{Y_{j,k}\}) = \phi(\epsilon)\prod_{j=1}^{J} 2Y_{j,\lceil 2^j\phi(\epsilon)\rceil} \tag{30.10}$$

where $\lceil\cdot\rceil$ is the ceiling function, and so a likelihood can be formed. For the simple model, the PT is conjugate. Let $\boldsymbol{\epsilon} = (\epsilon_1,\ldots,\epsilon_n)$. Then

$$Y_{j,2k-1}|\boldsymbol{\epsilon} \overset{ind.}{\sim} \text{beta}\left(cj^2 + \sum_{i=1}^{n} I\{\lceil 2^j\phi(\epsilon)\rceil = 2k-1\}, cj^2 + \sum_{i=1}^{n} I\{\lceil 2^j\phi(\epsilon)\rceil = 2k\}\right)$$

and $Y_{j,2k} = 1 - Y_{j,2k-1}$.

Location μ and spread σ parameters are melded with expression (30.8) and the PT prior (30.9) to make a median-μ location-scale family for data $y_1,\ldots,y_n$, given by

$$y_i = \mu + \sigma\epsilon_i,$$

where the $\epsilon_i \mid G \overset{iid}{\sim} G$ and G follows a PT prior as in expression (30.9), with the restriction $Y_{1,1} = Y_{1,2} = 0.5$. Allowing μ and σ to be random induces a mixture of PT (MPT) model for $y_1,\ldots,y_n$, smoothing out predictive inference [37, 57]. Note that Jeffreys' prior under the normal model is a reasonable choice here [8], and leads to a proper posterior [33].

30.3.3 Which approach is better?

In recent years, there has been a dramatic increase in research and applications of Bayesian nonparametric models, such as DPM and MPT, motivated largely by the availability of simple and efficient methods for posterior computation. However, much of the published research has been concentrated on the proposal of new models with insufficient emphasis given to the real practical advantage of the new proposals. The overriding problem is the choice of what method to use in a given practical context. In general, the full support of the models and the extremely weak conditions under which the different models have been shown to have consistent posteriors might well trap the unwary into a false sense of security, by suggesting that good estimates of the probability models can be obtained in a wide range of settings. More research seems to be needed in that direction.

Early papers on PTs [57, 82, 83] show 'spiky' and 'irregular' density estimates. However, these papers picked a very small precision c and used a fixed centring distribution (without a random scale σ) that was often much more spread out than what the data warranted. The MPT prior automatically centres the prior at a reasonable centring distribution and smooths over partition boundaries. We argue that both MPT and DPM are competitor models, with appealing properties regarding support and posterior consistency, and that performance of each should be evaluated in real-life applications with finite sample sizes. We will illustrate this point by means of the analyses of simulated data.

We compared MPT and DPM estimates using 'perfect samples' (data are percentiles of equal probability from the distribution, approximating expected order statistics) from four densities, motivated by Figures 2.1 and 2.3 in [24]. Both models were fitted under more or less standard prior specifications for samples of size $n = 500$. Specifically, the MPT model was fitted in `DPpackage`

[45] using the PTdensity function with an $N(\mu, \sigma^2)$ baseline measure, and the following prior settings: $J = 6$, $c \sim \Gamma(10, 1)$, and $p(\mu, \sigma) \propto \sigma^{-1}$ (Jeffreys' prior under the normal model). The DPM model was fit using the DPdensity function included in DPpackage [45]. This function fits the DPM model considered by Escobar and West [25], which is given by

$$y_i|\mu_i, \tau_i \sim N(\mu_i, \tau_i^{-1}), \;\; (\mu_i, \tau_i)|G \sim G, \;\; G|\alpha, G_0 \sim DP(\alpha G_0),$$

where the centring distribution, G_0, corresponds to the conjugate normal / gamma distribution, i.e. $G_0 \equiv N(\mu|m, (k\tau)^{-1}) \times \Gamma(\tau_1, \tau_2)$. The model was fitted by assuming $\tau_1 = 2$ and $\tau_2 = 1$ and, $m \sim N(0, 10^5)$, $k \sim \Gamma(0.5, 50)$ and $\alpha \sim \Gamma(1, 1)$.

Figure 30.1 shows the true models and the density estimates under the DPM and MPT models, along with the histogram of the data. Although it is difficult to see differences between the estimates under the DPM and MPT models across true models, the density estimates under the MPT are a

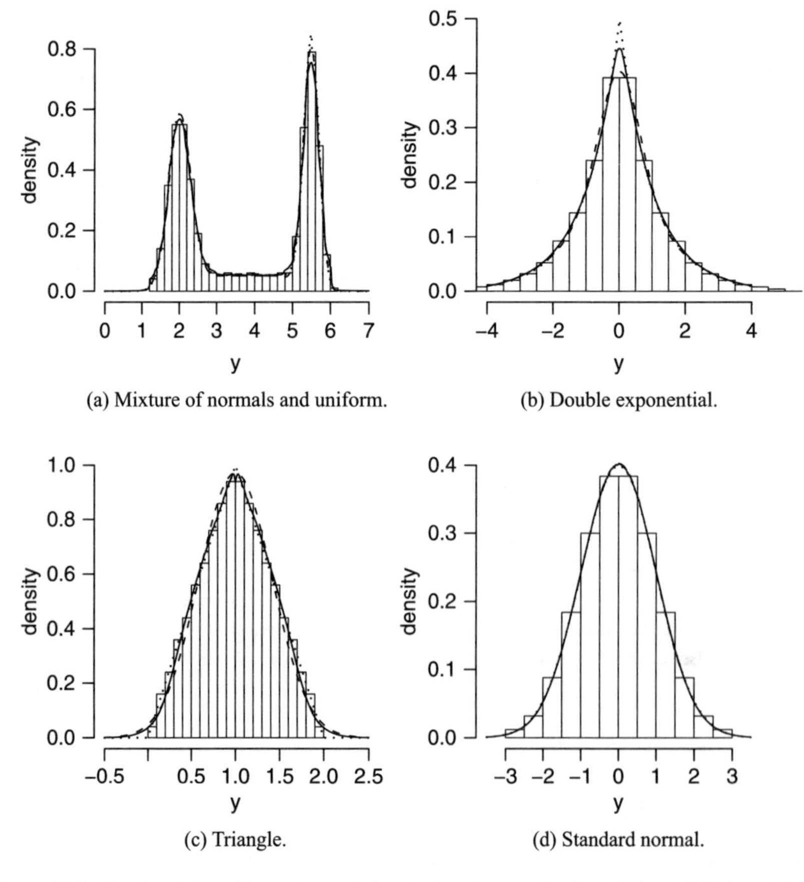

Figure 30.1 Simulated data. The estimated density functions under the MPT and DPM models are displayed as solid and dashed lines, respectively. The data generating model is represented as dotted lines. Panels (a), (b), (c) and (d) display the density estimates, true model and the histogram of the simulated data under a mixture of normals and uniform, double exponential, triangle and standard normal distribution, respectively.

bit rougher than under DPM, but there are no obvious partitioning effects or unruly spikes where there should not be. More importantly, either method can perform better than the other depending on the data generating mechanism; they both can do a good job.

When the true model is a mixture of two normals and a uniform distribution, the L_1 distance between the MPT estimates and the true model was 0.056, while for the DPM $L_1 = 0.034$. When the true model was a double exponential distribution, the MPT model outperformed the DPM. In this case $L_1 = 0.025$ and $L_1 = 0.060$ for the MPT and DPM model, respectively. A similar behaviour was observed when the model under consideration had a triangular shape. In this case, $L_1 = 0.051$ and $L_1 = 0.096$ for the MPT and DPM model, respectively. Finally, when the data generating mechanism was a standard normal distribution, both models performed equally well; $L_1 = 0.009$ for both MPT and DPM. From the plots in Figure 30.1 and the L_1 distances, the MPT appears to be a serious competitor to the DPM model, doing (two times) better or as well in three out of four cases.

30.4 Two generalizations of the AFT model

Section 30.2 overviewed several useful semiparametric models for analysing survival data. We mentioned that the PH model can be augmented with time-varying effects, stratification, etc., to handle non-proportional hazards, but that these fixes destroy any easy interpretability from the model. In this section we discuss two generalizations of the AFT model to handle data that do not follow AFT assumptions. Similarly to augmenting hazard models with time-varying effects, the AFT generalizations allow for crossing survival and hazard curves, but still allow straightforward interpretability. Furthermore, both models boil down to simple DPM and PT models for baseline $\mathbf{x} = \mathbf{0}$ groups, thus enabling the placement of all competing models on common ground, and facilitating comparisons among quite different assumptions on how covariates affect survival. Both augmentations are examples of 'density regression', allowing the entire density $f_{\mathbf{x}}(\cdot)$ to change smoothly with predictors.

The two approaches are the 'linear dependent Dirichlet process mixture' (LDDP) and the 'linear dependent tailfree process' (LDTFP). The former can be interpreted as a mixture of parametric AFT models, and the latter is an AFT model with very general heteroscedastic error terms. In this section we introduce the models and their properties, in the next we compare them using datasets that have been examined in the literature.

Note that unlike traditional linear models that focus on the trend (mean or median), we are interested in modelling the entire density as well; otherwise we could directly use quantile regression models, such as the ones implemented in Koenker's `quantreg` package in R [52]. In what follows, consider standard interval-censored failure time data $\{(l_i, u_i, \tilde{\mathbf{x}}_i)\}_{i=1}^n$, where the responses are known up to an interval, $T_i \in (l_i, u_i]$, and $\tilde{\mathbf{x}}_i$ are covariates for subject i, without the intercept term. The AFT model for the failure time response is given by the log-linear model

$$y_i = \log(T_i) = \tilde{\mathbf{x}}_i'\boldsymbol{\beta} + \epsilon_i \tag{30.11}$$

30.4.1 Linear dependent Dirichlet process mixture

A natural semiparametric specification of the AFT model would consider a nonparametric model for the error distribution of the errors in (30.11). By considering a DPM of normal distributions for the errors, the distribution for the log failure time is the distribution of ϵ_i, given by (30.7), shifted by the covariates $\tilde{\mathbf{x}}_i'\boldsymbol{\beta}$. Specifically,

$$y_i|\boldsymbol{\beta}, G \overset{ind.}{\sim} \sum_{j=1}^{\infty} w_j N(\mu_j + \tilde{\mathbf{x}}_i'\boldsymbol{\beta}, \sigma_j^2)$$

The interpretation of the components of $\boldsymbol{\beta}$ are as usual and the model can be fitted using standard algorithms for Dirichlet process mixture models [66].

The LDDP [22, 45, 46] can be interpreted as a generalization of the previous model, which arises by additionally mixing over the regression coefficients, yielding a mixture of log-normal AFT models. Set $\mathbf{x}_i = (1, \tilde{\mathbf{x}}_i')'$. The LDDP model is given by

$$y_i|G \overset{ind.}{\sim} \sum_{j=1}^{\infty} w_j N(\mathbf{x}_i'\boldsymbol{\beta}_j, \sigma_j^2) \tag{30.12}$$

where, as before, $w_i = V_i \prod_{j<i}(1 - V_j)$, with $V_i \mid \alpha \overset{iid}{\sim} Beta(1, \alpha)$, and $\boldsymbol{\beta}_j \overset{iid}{\sim} N(\mathbf{m}_0, \mathbf{V}_0)$ and $\sigma_j^2 \overset{iid}{\sim} \Gamma^{-1}(a, b)$.

The model trades easy interpretability offered by a single $\boldsymbol{\beta}$ for greatly increased flexibility. In particular, the LDDP model does not stochastically order survival curves from different predictors $\mathbf{x}_{i_1}$ and $\mathbf{x}_{i_2}$, and both the survival and hazard curves can cross. However, if the data warrant only a few weights from $\{w_1, w_2, \dots\}$ with non-negligible mass, the model can be re-fitted using a simple, finite mixture of log-normal distributions. The number of components J in the finite mixture can be estimated from the posterior number of components from a fit of (30.12) yielding

$$y_i|\mathbf{w}, \boldsymbol{\beta}, \boldsymbol{\tau} \overset{ind.}{\sim} \sum_{j=1}^{J} w_j N(\mathbf{x}_i'\boldsymbol{\beta}_j, \sigma_j^2) \tag{30.13}$$

where $\mathbf{w} = (w_1, \dots, w_J)$, $\boldsymbol{\beta} = (\boldsymbol{\beta}_1, \dots, \boldsymbol{\beta}_J)$ and $\boldsymbol{\tau} = (\sigma_1^2, \dots, \sigma_J^2)$. This model defines J homogeneous subpopulations with simple *unimodal* survival densities $LN(\beta_{j1}, \tau_j)$ and accompanying acceleration factors given through $(\beta_{j2}, \dots, \beta_{j,p})$. These can be viewed as homogeneous subpopulations corresponding to an omitted variable with J levels. The model is also a mixture of experts model with a gating mechanism that is independent of the covariates. Generalization of this model, where weights also depend on covariates can be found in, for instance, Müller *et al.* [63] and Chung and Dunson [19].

30.4.2 Mixture of linear dependent tailfree processes

A PT defines the conditional probabilities $Y_{j+1,2k-1}$ and $Y_{j+1,2k}$ as beta. However, we can instead define a logistic regression for each of these probabilities, allowing the *entire* shape of the density to change with predictors; this is the approach considered by Jara and Hanson [44]. Given covariates $\mathbf{x}$,

$$(Y_{j+1,2k-1}, Y_{j+1,2k})$$

are modelled through logistic regressions

$$\log\{Y_{j+1,2k-1}(\mathbf{x})/Y_{j+1,2k}(\mathbf{x})\} = \mathbf{x}'\boldsymbol{\tau}_{j,k}$$

There are $2^J - 1$ covariate vectors $\boldsymbol{\tau} = \{\boldsymbol{\tau}_{j,k}\}$. For instance, for $J = 3$, $\{\boldsymbol{\tau}_{0,1}, \boldsymbol{\tau}_{1,1}, \boldsymbol{\tau}_{1,2}, \boldsymbol{\tau}_{2,1}, \boldsymbol{\tau}_{2,2}, \boldsymbol{\tau}_{2,3}, \boldsymbol{\tau}_{2,4}\}$. Let $\mathbf{X} = [\mathbf{x}_1 \cdots \mathbf{x}_n]'$ be the $n \times p$ design matrix. Following Jara and Hanson [44], each is assigned an independent normal prior, $\tau_{j,k} \sim N_p\left(\mathbf{0}, \frac{2}{c(j+1)^2}\Psi\right)$. Several options could be considered for Ψ. Jara and Hanson [44] discussed in detail the case where $\Psi = n(\mathbf{X}'\mathbf{X})^{-1}$, generating a g-prior [90] for the tailfree regression coefficients.

Augmenting (30.10), the random density is given by

$$g_{\mathbf{x}}(\epsilon) = \phi(\epsilon) 2^J \prod_{i=1}^{J} Y_{j,\lceil 2^j \Phi(\epsilon) \rceil}(\mathbf{x})$$

The parameter $c \in \mathbb{R}^+$ controls how non-normal $g_{\mathbf{x}}(e)$ is, and can be interpreted as a measure of the random L_1 distance $||g_{\mathbf{x}} - \phi||$ [32]. Since the $\{Y_{j,k}\}$ are modelled as logistic-normal instead of beta, the resulting random density is called a tailfree process. The final linear dependent tailfree process AFT model is given by

$$y_i = \mathbf{x}_i'\boldsymbol{\beta} + \sigma\epsilon_i, \ \ \epsilon_i | \boldsymbol{\tau} \overset{ind.}{\sim} g_{\mathbf{x}_i}.$$

Unlike the LDDP, the LDTFP separates survival into one distinct trend $\mathbf{x}'\boldsymbol{\beta}$ and an evolving log-baseline survival density $g_{\mathbf{x}}$. By setting $g_{\mathbf{x}}$ to have median-zero, e^{β_j} gives a factor by how median survival changes when x_j is increased *just as in standard AFT models*. This heightened interpretability in terms of median-regression in the presence of heteroscedastic error allows a fit of the LDTFP model to easily relate covariates $\mathbf{x}$ to median survival.

The LDTFP models the probability of falling above or below quantiles of the $N(\mathbf{x}'\boldsymbol{\beta}, \sigma^2)$ distribution, but in terms of conditional probabilities. This model can be viewed as a particular kind of quantile regression model. Koenker and Hallock [53] suggest that '... instead of estimating linear conditional quantile models, we could instead estimate a family of binary response models for the probability that the response variable exceeded some prespecified cutoff values.' However, Koenker and Hallock [53] prefer the linear (in covariates) quantile specification because '... it nests within it the independent and identically distributed error location shift model of classical linear regression.' By augmenting a median-zero tailfree process with a general trend $\mathbf{x}'\boldsymbol{\beta}$ we accomplish the same objective, nesting the ubiquitous normal-errors linear model within a highly flexible median regression model, but with heteroscedastic error that changes shape with covariate levels $\mathbf{x} \in \mathcal{X}$.

30.5 Illustrations

The two generalizations of the AFT model are illustrated using real-life datasets. The generalized AFT models were fitted using the `LDDPsurvival` and `LDTFPsurvival` functions, which are available in version 1.1–4 of `DPpackage` [45]. The models were compared in terms of the LPML [28].

30.5.1 Breast cancer data

We consider a dataset involving time to cosmetic deterioration of the breast for women with stage 1 breast cancer who have undergone a lumpectomy [6]. The data come from a retrospective study designed to compare the cosmetic effects of radiotherapy versus radiotherapy plus chemotherapy on women with early breast cancer. Both treatments are alternatives to a mastectomy that preserve (and thus enhance the appearance of) the breast. It is postulated that chemotherapy in addition to radiotherapy (treatment A) reduces the cosmetic effect of the procedure by inducing breast retraction more quickly than radiotherapy alone (treatment B).

There are 46 radiation only and 48 radiation plus chemotherapy patients. Patients were typically observed every 4 to 6 months, at which point a clinician graded the level of breast retraction as none, moderate, or severe. The event is moderate or severe breast retraction, and thus the event times are interval censored, with interval endpoints occurring at clinic visits. These data were analysed using a traditional (homoscedastic) AFT model considering baseline distributions modelled as a mixture of DP by Hanson and Johnson [31].

The LDDP and LDTFP models were fitted, including as predictor the treatment indicator. For the LDTFP model we set $J = 4$ and $\Psi = n(\mathbf{X}'\mathbf{X})^{-1}$, where n is the sample size. The median function parameters β_0 and β_1 were given a Zellner's g-prior [90], $g(\mathbf{X}'\mathbf{X})^{-1}$, with $g = 2n$, $\sigma^{-2} \sim \Gamma(5.01, 2.01)$, and $c \sim \Gamma(10, 1)$. For the LDDP model, we assume $\mathbf{m}_0 \sim N_2(\mathbf{0}_2, 100 \times \mathbf{I}_2)$, $\mathbf{V}_0^{-1} \sim \text{Wishart}(4, \mathbf{I}_2)$, $a = 3.01$, $b \sim \Gamma(3.01, 1.01)$ and $\alpha \sim \Gamma(10, 1)$. For all models, a burn-in of 20 000 iterates was followed by a run of 100 000 thinned down to 10 000 iterates.

The two models based on dependent process priors outperformed a classical semiparametric analysis based in the AFT assumption. Rounded to the nearest integer, the LPML for the LDDP and LDTFP model was -147 and -149, respectively, better than -159 obtained using the mixture of DP model [31], fixing the total mass parameter $\alpha = 5$ and using a $N(\gamma, \theta^2)$ centring distribution. Figure 30.2 shows the estimated survival curves for the two treatment groups under the different models, evaluated in a grid of 200 equally-spaced points. The survival curves are similar to the ones reported by Hanson and Johnson [31], with the exception that the estimated survival curves are initially indistinguishable before 15 months under the LDDP and LDTFP model; the AFT model forces a more pronounced stochastic ordering of the survival curves. Although the LDDP model shows marginally better predictive performance than the LDTFP model for these data, the survival point estimates obtained under the two models are qualitatively similar. The better predictive performance of the LDDP models is explained by its lower posterior variability.

Under the LDTFP model, the estimated treatment effect was $\hat{\beta}_1 = 0.30$ and nonsignificant with a 95% highest posterior density (HPD) interval of $(-0.07, 0.68)$. The median time to retraction from treatment B is estimated to be $e^{0.30} \approx 1.38$ times longer than treatment A with 95% HPD interval $(0.90, 1.91)$. Priors favouring smaller values of c yielded qualitatively similar inferences, although estimated point estimates of the survival curves cross at about 15 months.

The results of the AFT analyses with homoscedastic error show a significant regression effect, indicating lower times to retraction under treatment A as expected, somewhat contradicting the LDTFP analysis where no significant difference in median survival was found. However, a glance at Figure 30.2 shows marginal evidence of different median lifetimes given the large variability of the survival curves across the groups. Under the homoscedastic AFT model the regression parameter affects all quantiles simultaneously and indicates a net scale shift in probability; under the LDTFP model the conditional probabilities change beyond the median function. The significant effect under the homoscedastic model can be viewed as an averaging of the overall warping of the density across treatment levels, embodied in the parameters $\{\boldsymbol{\tau}_{j,k}\}$.

30.5.2 Cancer clinical trial data

We consider data arising from a cancer clinical trial, described in Rosner [73], and analyzed by De Iorio *et al.* [22] using a LDDP mixture of normals model. The data records the event-free survival time in months for 761 women, i.e. the response of interest corresponds to the time until death, relapse or treatment-related cancer. Researchers are interested in determining whether high doses of the treatment are more effective for treating the cancer compared to lower doses. High doses of the treatment are known to be associated with a high risk of treatment-related mortality. The

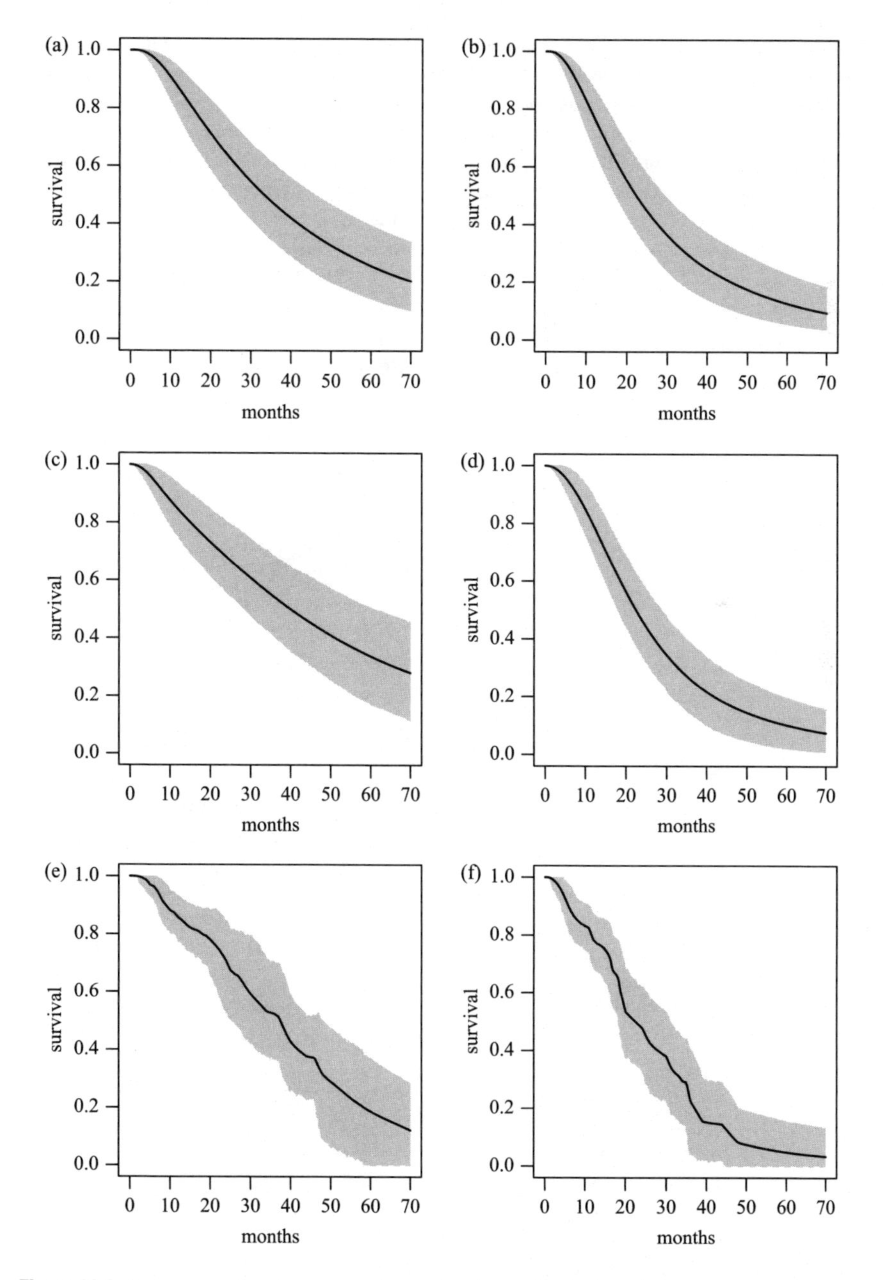

Figure 30.2 Breast retraction data. Panels (a) and (b) show estimated survival curves for treatments A and B, respectively, under the LDTFP model. Panels (c) and (d) display survival curves for treatments A and B, respectively, under the LDDP model. Panels (e) and (f) display survival curves for treatments A and B, respectively, under the mixture of Dirichlet process model for comparison purposes. In all cases, the pointwise 95% credible bands are also displayed as a grey area.

clinicians hope that this initial risk is offset by a substantial reduction in mortality and disease recurrence or relapse, consequently justifying more aggressive therapy. Thus the primary reason for carrying out the clinical trial was to compare low versus high dose. Following [22], we consider two categorical covariates, one continuous covariate, and one interaction term: treatment dose (low or high), estrogen receptor (ER) status (negative or positive), the size of the tumour (standardized to zero mean and unit variance), and the treatment dose and ER status interaction.

The LDDP and LDTFP models were fitted to the data. For the LDTFP, we set $J = 5$ and $\Psi = 10^3\mathbf{I}_5$, and the median function parameters were assigned independent normal priors $\boldsymbol{\beta} \sim N_5(\mathbf{0}_5, 10^3\mathbf{I}_5)$, $\sigma^{-2} \sim \Gamma(1.5, 6.0)$, and $c \sim \Gamma(7.0, 0.1)$. For the LDDP model, we assume $\mathbf{m}_0 \sim N_5(\mathbf{0}_5, 100 \times \mathbf{I}_5)$, $\mathbf{V}_0^{-1} \sim \text{Wishart}(7, \mathbf{I}_5)$, $a = 1.01$, $b \sim \Gamma(1.51, 3.01)$ and $\alpha \sim \Gamma(5, 1)$. For both models, a burn-in of 20 000 iterates was followed by a run of 100 000 thinned down to 10 000 iterates. Qualitatively similar inferences were obtained under the two models.

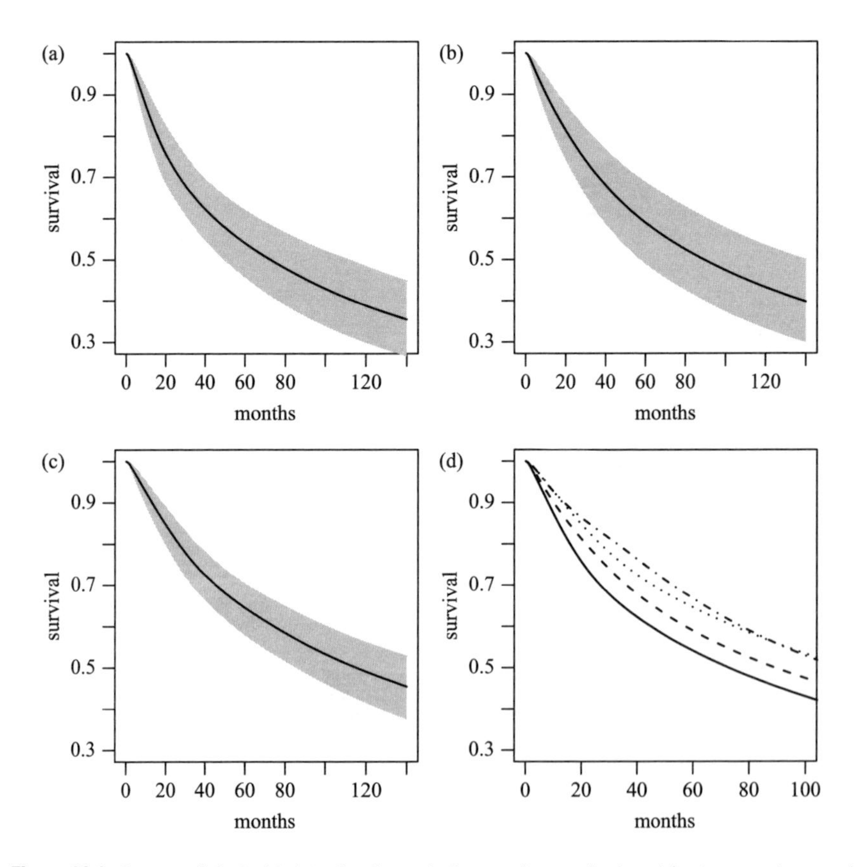

Figure 30.3 Cancer clinical trial data. In all panels the results are displayed for tumour size equal to 2.0 cm (first quartile) under the LDTFP model. Panels (a), (b) and (c) show the posterior mean (and pointwise 95% HPD band in grey) for the survival curves for low treatment dose—negative ER status, high treatment dose—negative ER status and low treatment dose—positive ER status, respectively. Panel (d) shows the posterior mean for the four combinations of treatment dose and ER status. The low treatment dose—negative ER status, high treatment dose—negative ER status, low treatment dose—positive ER status and high treatment dose—positive ER status is represented as a continuous, dashed, dotted and dotdash lines, respectively.

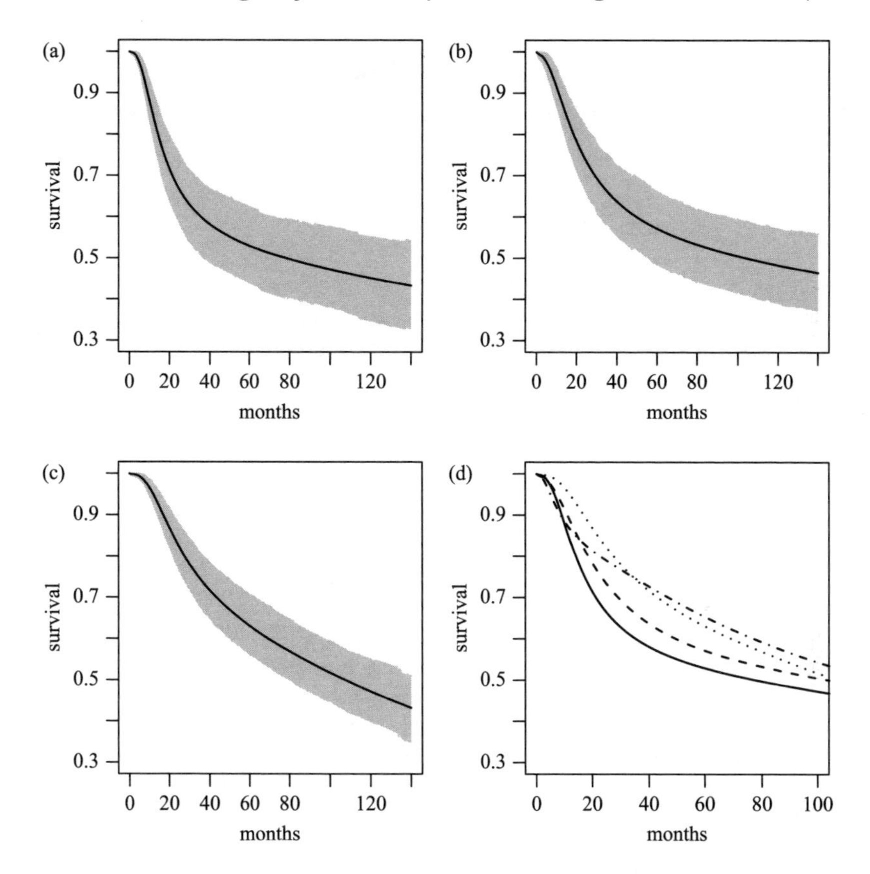

Figure 30.4 Cancer clinical trial data. In all panels the results are displayed for tumour size equal to 2.0 cm (first quartile) under the LDDP model. Panels (a), (b) and (c) show the posterior mean (and pointwise 95% HPD band in grey) for the survival curves for low treatment dose—negative ER status, high treatment dose—negative ER status and low treatment dose—positive ER status, respectively. Panel (d) shows the posterior mean for the four combinations of treatment dose and ER status. The low treatment dose—negative ER status, high treatment dose—negative ER status, low treatment dose—positive ER status and high treatment dose—positive ER status is represented as a continuous, dashed, dotted and dotdash lines, respectively.

Figures 30.3 and 30.4 show the estimated survival curves, and corresponding posterior uncertainty, for ER positive patients with tumour size 2.0 cm (the first quartile) under the LDTFP and LDDP model, respectively. The two models based on dependent process priors outperformed a classical semiparametric analysis based in the AFT assumption. Rounded to the nearest integer, the LPML for the LDDP and LDTFP model was -2048 and -2052, respectively, better than -2063 obtained using a parametric AFT lognormal regression model.

An advantage of the LDTFP model over the LDDP model is that direct inferences can be made on the median survival time. In order to evaluate the posterior evidence against the hypothesis of null effect of the covariates on the median survival function, the pseudo contour probability (PsCP) was evaluated for each hypothesis. The PsCP was computed based on the highest posterior density (HPD) intervals, which were estimated using the method proposed by Chen and Shao [13]. The PsCP is defined as one minus the smallest credible level for which the null hypothesis is contained in the corresponding HDP. The results suggest a non-important effect of the treatment dose (PsCP $= 0.55$) and its interaction with ER status (PsCP $= 0.5$), and an important effect of the ER status (PsCP < 0.01) and a negative effect of the tumour size (PsCP < 0.01) on the median survival time.

30.5.3 Lung cancer data

We consider data presented in Maksymiuk *et al.* [60] on the treatment of limited-stage small-cell lung cancer in $n = 121$ patients. The data have been analysed in the literature by using median-regression models [33, 55, 83, 86, 89]. In the study, it was of interest to determine which sequencing of the drugs cisplaten and etoposide increased the lifetime from time of diagnosis, measured in days, of those with limited-stage small-cell lung cancer. Treatment A applied cisplaten followed by etoposide, whereas treatment B applied etoposide followed by cisplaten. The patients' ages in years at entry into the study were also included as a concomitant variable. The LDTFP model was fitted to the data assuming $J = 5$ and $\Psi = 10^3\mathbf{I}_2$. The median function parameters were assigned independent normal priors $\boldsymbol{\beta} \sim N_2(\mathbf{0}_2, 10^3\mathbf{I}_2), \sigma^{-2} \sim \Gamma(3, 1.5)$, and $c \sim \Gamma(1.0, 1.0)$. For the LDDP model, we assume $\mathbf{m}_0 \sim N_2(\mathbf{0}_2, 100\mathbf{I}_2), \mathbf{V}_0^{-1} \sim \text{Wishart}(5, \mathbf{I}_2), a = 3.01, b \sim \Gamma(3.01, 3.01)$ and $\alpha \sim \Gamma(5, 1)$. For both models, a burn-in of 20 000 iterates was followed by a run of 100 000 thinned down to 10 000 samples.

The LPML measures for the LDDP and LDTFP models were -732 and -733, respectively. These results suggest that both dependent models slightly outperform from a predictive point of view alternative parametric and semiparametric survival models. In fact, the LPML for the Weibull, log-logistic, and PO, PH, and AFT models, using a MPT prior for the baseline survival function, were -747, -735, -734, -737, and -734, respectively [33]. The LDDP and LDTFP models in some sense predict the data 'best', but there is little real predictive difference among the LDDP, LDTFP, PO and AFT models. The Weibull model is clearly inferior, whereas the LDDP and LDTFP models have a pseudo Bayes factor of about 50 relative to the PH model. The similar predictive behaviour of the dependent model is confirmed by the density plots in Figures 30.5 and 30.6.

Table 30.1 presents posterior regression parameter inferences for the MPT AFT, PO and PH models and for the LDTFP model. Holding age fixed, patients typically survive $e^{0.345} \approx 1.4$ times longer under treatment A versus treatment B under the AFT assumption. The PO model indicates that the odds of surviving past any time t is $e^{0.93} \approx 2.5$ greater for treatment A versus treatment B. Similarly to observations under the MPT AFT model, the results of the LDTFP suggest that the median survival time for patients under treatment A is $e^{0.407} \approx 1.5$ times the median survival time for patients under treatment B.

30.6 Concluding remarks

We have discussed, compared and illustrated flexible nonparametric models that can be used to introduce categorical and continuous covariates in the context of time-to-event data. The models correspond to generalizations of AFT models based on dependent extensions of the DP and PT priors. Advantages of the induced survival regression models include ease of interpretability and computational tractability. An important property of the proposed models is that the complete distribution of survival times is allowed to change with values of the predictors (including properties such as skewness, multimodality, quantiles, etc.) instead of just one or two characteristics, as implied for many commonly used survival models.

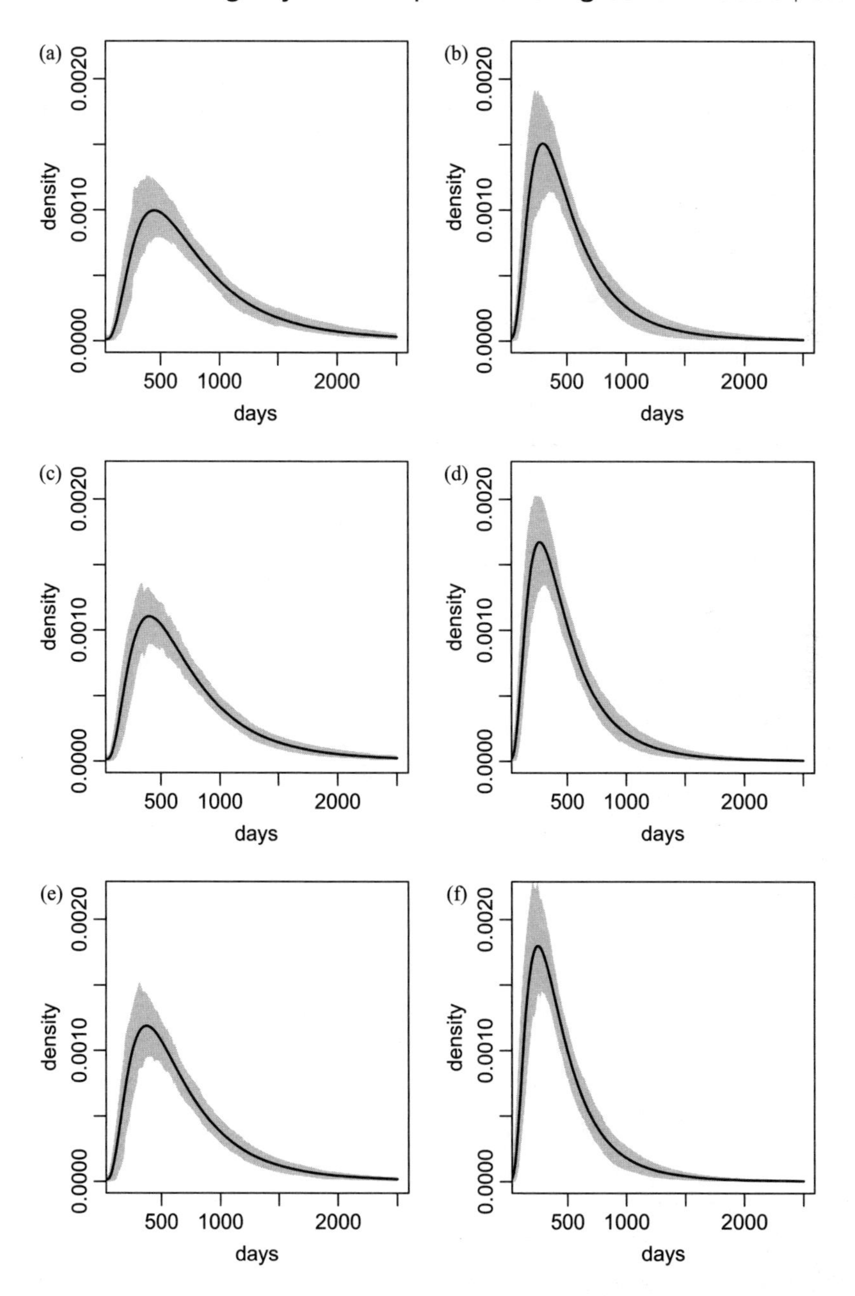

Figure 30.5 Lung cancer data. Panels (a), (c) and (e) show the posterior mean (and pointwise 95% HPD band in grey) for the densities at age 56, 61.1 and 68 for treatment A under the LDDP model. Panels (b), (d) and (f) show the posterior mean (and pointwise 95% HPD band in grey) for the densities at age 56, 61.1 and 68 for treatment B under the LDDP model.

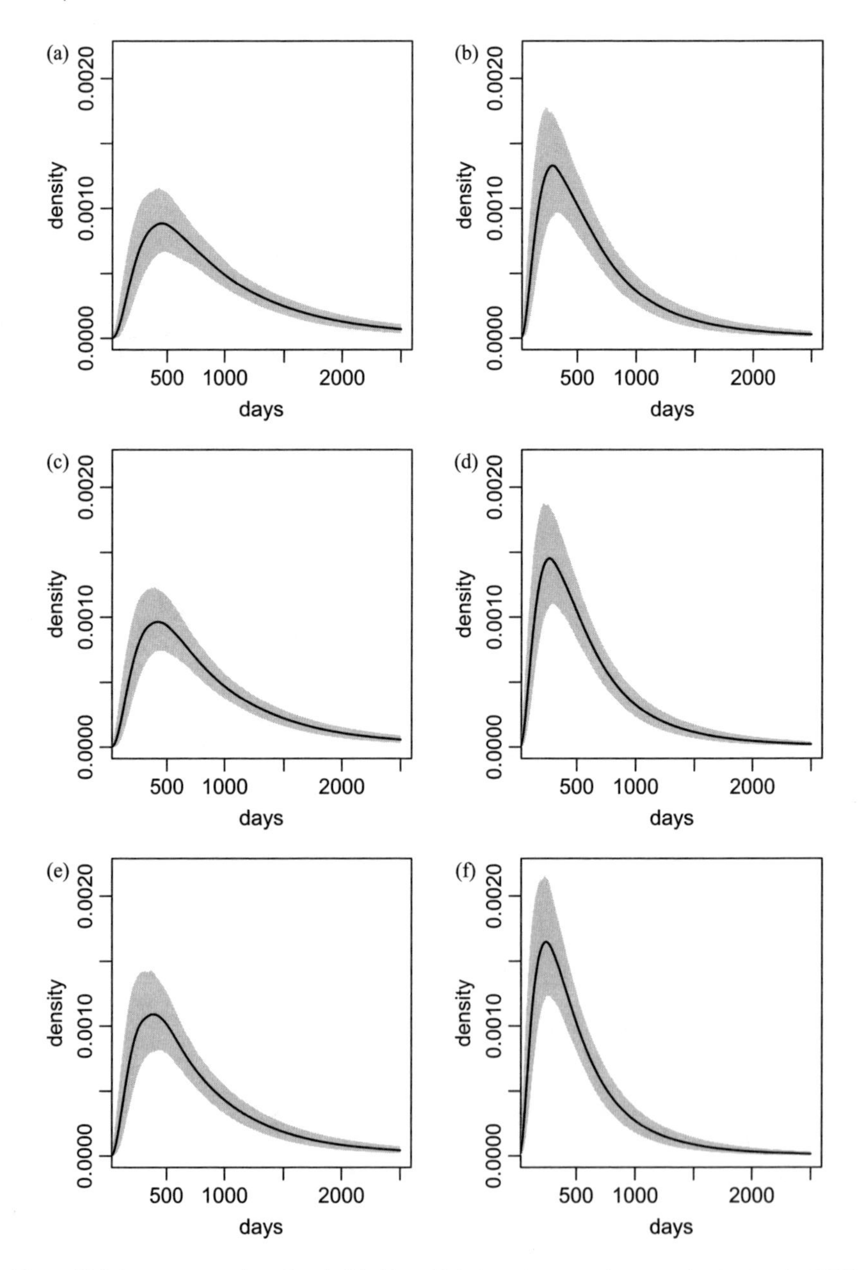

Figure 30.6 Lung cancer data. Panels (a), (c) and (e) show the posterior mean (and pointwise 95% HPD band in grey) for the densities at age 56, 61.1 and 68 for treatment A under the LDTFP model. Panels (b), (d) and (f) show the posterior mean (and pointwise 95% HPD band in grey) for the densities at age 56, 61.1 and 68 for treatment B under the LDTFP model.

Table 30.1 Lung cancer data Posterior mean (95% credible interval) for the regression coefficients.

Coefficient	MPT AFT	MPT PO	MPT PH	LDTFP
β_1 (Age)	0.007	0.034	0.028	−0.019
	(−0.004, 0.036)	(−0.001, 0.071)	(0.003, 0.054)	(−0.037, −0.001)
β_2 (Treatment)	0.345	0.930	0.533	0.407
	(0.157, 0.533)	(0.292, 1.568)	(0.130, 0.926)	(0.130, 0.691)

Acknowledgements

The work of the first author was supported in part by NSF grant CMMI-0855329. The second author was supported by Fondecyt 1100144 grant.

References

[1] Aalen, O O (1980). A model for nonparametric regression analysis of counting processes. In *Lecture Notes in Statistics, Vol. 2*, pp. 1–25. Springer-Verlag.

[2] Aalen, O O (1989). A linear regression model for the analysis of life times. *Statistics in Medicine*, **8**, 907–925.

[3] Alzola, C and Harrell, F (2006). *An introduction to S and the Hmisc and Design libraries.* Online manuscript available at http://biostat.mc.vanderbilt.edu/wiki/pub/Main/RS/sintro.pdf.

[4] Andersen, P K and Gill, R D (1982). Cox's regression model for counting processes: A large sample study. *The Annals of Statistics*, **10**, 1100–1120.

[5] Banerjee, S and Dey, D K (2005). Semi-parametric proportional odds models for spatially correlated survival data. *Lifetime Data Analysis*, **11**, 175–191.

[6] Beadle, G, Harris, J, Silver, B, Botnick, L, and Hellman, S (1984). Cosmetic results following primary radiation therapy for early breast cancer. *Cancer*, **54**, 2911–2918.

[7] Belitz, C, Brezger, A, Kneib, T, and Lang, S (2009). *BayesX – Software for Bayesian inference in structured additive regression models. Version 2.00.* Available from http://www.stat.uni-muenchen.de/~bayesx.

[8] Berger, J O and Guglielmi, A (2001). Bayesian testing of a parametric model versus nonparametric alternatives. *Journal of the American Statistical Association*, **96**, 174–184.

[9] Bradburn, M J, Clark, T G, Love, S B, and Altman, D G (2003). Survival analysis part II: Multivariate data analysis – an introduction to concepts and methods. *British Journal of Cancer*, **89**, 431–436.

[10] Buckley, J and James, I (1979). Linear regression with censored data. *Biometrics*, **8**, 907–925.

[11] Carlin, B P and Hodges, J S (1999). Hierarchical proportional hazards regression models for highly stratified data. *Biometrics*, **55**, 1162–1170.

[12] Chang, I S, Hsiung, C A, Wu, Y J, and Yang, C C (2005). Bayesian survival analysis using Bernstein polynomials. *Scandinavian Journal of Statistics*, **32**, 447–466.

[13] Chen, M H and Shao, Q M (1999). Monte Carlo estimation of Bayesian credible and HPD intervals. *Journal of Computational Graphical Statistics*, **8**, 69–92.

[14] Chen, Y Q and Jewell, N P (2001). On a general class of semiparametric hazards regression models. *Biometrika*, **88**, 687–702.

[15] Chen, Y Q, Jewell, N P, Lei, X, and Cheng, S C (2005). Semiparametric estimation of proportional mean residual life model in presence of censoring. *Biometrics*, **61**, 170–178.

[16] Chen, Y Q and Wang, M C (2000). Analysis of accelerated hazards models. *Journal of the American Statistical Association*, **95**, 608–618.

[17] Cheng, S C, Wei, L J, and Ying, Z (1995). Analysis of transformation models with censored data. *Biometrika*, **82**, 835–845.

[18] Christensen, R and Johnson, W O (1988). Modeling accelerated failure time with a Dirichlet process. *Biometrika*, **75**, 693–704.

[19] Chung, Y and Dunson, D B (2009). Nonparametric Bayes conditional distribution modeling with variable selection. *Journal of the American Statistical Association*, **104**, 1646–1660.

[20] Cox, D R (1972). Regression models and life-tables (with discussion). *Journal of the Royal Statistical Society, Series B: Methodological*, **34**, 187–220.

[21] Cox, D R (1975). Partial likelihood. *Biometrika*, **62**, 269–276.

[22] De Iorio, M, Johnson, W O, Müller, P, and Rosner, G L (2009). Bayesian nonparametric nonproportional hazards survival modeling. *Biometrics*, **65**, 762–771.

[23] Dunson, D B and Herring, A H (2005). Bayesian model selection and averaging in additive and proportional hazards. *Lifetime Data Analysis*, **11**, 213–232.

[24] Efromovich, S (1999). *Nonparametric Curve Estimation: Methods, Theory and Applications.* Springer-Verlag.

[25] Escobar, M D and West, M (1995). Bayesian density estimation and inference using mixtures. *Journal of the American Statistical Association*, **90**, 577–588.

[26] Ferguson, T S (1973). A Bayesian analysis of some nonparametric problems. *The Annals of Statistics*, **1**, 209–230.

[27] Ferguson, T S (1974). Prior distributions on spaces of probability measures. *The Annals of Statistics*, **2**, 615–629.

[28] Geisser, S and Eddy, W F (1979). A predictive approach to model selection. *Journal of the American Statistical Association*, **74**, 153–160.

[29] Gelfand, A E and Mallick, B K (1995). Bayesian analysis of proportional hazards models built from monotone functions. *Biometrics*, **51**, 843–852.

[30] Griffin, J (2010). Default priors for density estimation with mixture models. *Bayesian Analysis*, **5**, 45–64.

[31] Hanson, T and Johnson, W O (2004). A Bayesian semiparametric AFT model for interval-censored data. *Journal of Computational and Graphical Statistics*, **13**, 341–361.

[32] Hanson, T, Kottas, A, and Branscum, A (2008). Modelling stochastic order in the analysis of receiver operating characteristic data: Bayesian nonparametric approaches. *Journal of the Royal Statistical Society, Series C*, **57**, 207–225.

[33] Hanson, T E (2006). Inference for mixtures of finite Polya tree models. *Journal of the American Statistical Association*, **101**, 1548–1565.

[34] Hanson, T E (2006). Modeling censored lifetime data using a mixture of gammas baseline. *Bayesian Analysis*, **1**, 575–594.

[35] Hanson, T E, Branscum, A, and Johnson, W O (2005). Bayesian nonparametric modeling and data analysis: An introduction. In *Bayesian Thinking: Modeling and Computation (Handbook of Statistics, volume 25)* (ed. D. Dey and C. Rao), pp. 245–278. Elsevier: Amsterdam.

[36] Hanson, T E, Branscum, A, and Johnson, W O (2011). Predictive comparison of joint longitudinal–survival modeling: a case study illustrating competing approaches. *Lifetime Data Analysis*, **17**, 3–28.

[37] Hanson, T E and Johnson, W O (2002). Modeling regression error with a mixture of Polya trees. *Journal of the American Statistical Association*, **97**, 1020–1033.

[38] Hanson, T E and Yang, M (2007). Bayesian semiparametric proportional odds models. *Biometrics*, **63**, 88–95.

[39] Heller, G and Simonoff, J S (1992). Prediction in censored survival data: A comparison of the proportional hazards and linear regression models. *Biometrics*, **48**, 101–115.

[40] Hennerfeind, A, Brezger, A, and Fahrmeir, L (2006). Geoadditive survival models. *Journal of the American Statistical Association*, **101**, 1065–1075.

[41] Hjort, N L (1990). Nonparametric Bayes estimators based on beta processes in models for life history data. *The Annals of Statistics*, **18**, 1259–1294.

[42] Hutton, J L and Monaghan, P F (2002). Choice of parametric accelerated life and proportional hazards models for survival data: Asymptotic results. *Lifetime Data Analysis*, **8**, 375–393.

[43] Ibrahim, J G, Chen, M H, and Sinha, D (2001). *Bayesian Survival Analysis*. Springer-Verlag.

[44] Jara, A and Hanson, T E (2011). A class of mixtures of dependent tailfree processes. *Biometrika*, **98**, 553–566.

[45] Jara, A, Hanson, T E, Quintana, F A, Müller, P, and Rosner, G L (2011). DPpackage: Bayesian semi- and nonparametric modeling in R. *Journal of Statistical Software*, **40**, 1–30.

[46] Jara, A, Lesaffre, E, De Iorio, M, and Quitana, F (2010). Bayesian semiparametric inference for multivariate doubly-interval-censored data. *The Annals of Applied Statistics*, **4**, 2126–2149.

[47] Johnson, W O and Christensen, R (1989). Nonparametric Bayesian analysis of the accelerated failure time model. *Statistics and Probability Letters*, **8**, 179–184.

[48] Kalbfleisch, J D (1978). Nonparametric Bayesian analysis of survival time data. *Journal of the Royal Statistical Society, Series B: Methodological*, **40**, 214–221.

[49] Kay, R and Kinnersley, N (2002). On the use of the accelerated failure time model as an alternative to the proportional hazards model in the treatment of time to event data: A case study in influenza. *Drug Information Journal*, **36**, 571–579.

[50] Keene, O N (2002). Alternatives to the hazard ratio in summarizing efficacy in time-to-event studies: an example from influenza trials. *Statistics in Medicine*, **21**, 3687–3700.

[51] Kneib, T and Fahrmeir, L (2007). A mixed model approach for geoadditive hazard regression. *Scandinavian Journal of Statistics*, **34**, 207–228.

[52] Koenker, R (2008). Censored quantile regression redux. *Journal of Statistical Software*, **27**(6), 1–25.

[53] Koenker, R and Hallock, K F (2001). Quantile regression. *Journal of Economic Perspectives*, **15**, 143–156.

[54] Komárek, A and Lesaffre, E (2008). Bayesian accelerated failure time model with multivariate doubly-interval-censored data and flexible distributional assumptions. *Journal of the American Statistical Association*, **103**, 523–533.

[55] Kottas, A and Gelfand, A E (2001). Bayesian semiparametric median regression modeling. *Journal of the American Statistical Association*, **95**, 1458–1468.

[56] Kuo, L and Mallick, B (1997). Bayesian semiparametric inference for the accelerated failure-time model. *Canadian Journal of Statistics*, **25**, 457–472.

[57] Lavine, M (1992). Some aspects of Polya tree distributions for statistical modelling. *The Annals of Statistics*, **20**, 1222–1235.

[58] Lavine, M (1994). More aspects of Polya tree distributions for statistical modelling. *The Annals of Statistics*, **22**, 1161–1176.

[59] Lo, A Y (1984). On a class of Bayesian nonparametric estimates: I. Density estimates. *Annals of Statistics*, **12**, 351–357.

[60] Maksymiuk, A W, Jett, J R, Earle, J D, Su, J Q, Diegert, F A, Mailliard, J A, Kardinal, C G, Krook, J E, Veeder, M H, Wiesenfeld, M, Tschetter, L K, and Levitt, R (1994). Sequencing and schedule effects of cisplatin plus etoposide in small cell lung cancer results of a north central cancer treatment group randomized clinical trial. *Journal of Clinical Oncology*, **12**, 70–76.

[61] Mallick, B K and Walker, S G (2003). A Bayesian semiparametric transformation model incorporating frailties. *Journal of Statistical Planning and Inference*, **112**, 159–174.

[62] Martinussen, T and Scheike, T H (2006). *Dynamic Regression Models for Survival Data.* Springer-Verlag.

[63] Müller, P, Erkanli, A, and West, M (1996). Bayesian curve fitting using multivariate normal mixtures. *Biometrika,* **83**, 67–79.

[64] Müller, P and Quintana, F A (2004). Nonparametric Bayesian data analysis. *Statistical Science,* **19**, 95–110.

[65] Murphy, S A, Rossini, A J, and van der Vaart, A W (1997). Maximum likelihood estimation in the proportional odds model. *Journal of the American Statistical Association,* **92**, 968–976.

[66] Neal, R M (2000). Markov chain sampling methods for Dirichlet process mixture models. *Journal of Computational and Graphical Statistics,* **9**, 249–265.

[67] Orbe, J, Ferreira, E, and Núñez Antón, V (2002). Comparing proportional hazards and accelerated failure time models for survival analysis. *Statistics in Medicine,* **21**, 3493–3510.

[68] Orbe, J and Núñez Antón, V (2006). Alternative approaches to study lifetime data under different scenarios: from the PH to the modified semiparametric AFT model. *Computational Statistics and Data Analysis,* **50**, 1565–1582.

[69] Petrone, S (1999). Bayesian density estimation using Bernstein polynomials. *The Canadian Journal of Statistics,* **27**, 105–126.

[70] Petrone, S (1999). Random Bernstein polynomials. *Scandinavian Journal of Statistics,* **26**, 373–393.

[71] Portnoy, S (2003). Censored regression quantiles. *Journal of the American Statistical Association,* **98**, 1001–1012.

[72] Reid, N (1994). A conversation with Sir David Cox. *Statistical Science,* **9**, 439–455.

[73] Rosner, G L (2005). Bayesian monitoring of clinical trials with failure–time endpoints. *Biometrics,* **61**, 239–245.

[74] Ryan, T and Woodall, W (2005). The most-cited statistical papers. *Journal of Applied Statistics,* **32**, 461–474.

[75] Sayehmiri, K, Eshraghian, M R, Mohammad, K, Alimoghaddam, K, Foroushani, A R, Zeraati, H, Golestan, B, and Ghavamzadeh, A (2008). Prognostic factors of survival time after heatopoietic stem cell transplant in acute lymphoblastic leukemia patients: Cox proportional hazard versus accelerated failure time models. *Journal of Experimental & Clinical Cancer Research,* **27**, 74.

[76] Scharfstein, D O, Tsiatis, A A, and Gilbert, P B (1998). Efficient estimation in the generalized odds-rate class of regression models for right-censored time-to-event data. *Lifetime Data Analysis,* **4**, 355–391.

[77] Sethuraman, J (1994). A constructive definition of Dirichlet priors. *Statistica Sinica,* **4**, 639–650.

[78] Sinha, D and Dey, D K (1997). Semiparametric Bayesian analysis of survival data. *Journal of the American Statistical Association,* **92**, 1195–1212.

[79] Sinha, D, McHenry, M B, Lipsitz, S R, and Ghosh, M (2009). Empirical Bayes estimation for additive hazards regression models. *Biometrika,* **96**, 545–558.

[80] Swindell, W R (2009). Accelerated failure time models provide a useful statistical framework for aging research. *Experimental Gerontology,* **44**, 190–200.

[81] Therneau, T M and Grambsch, P M (2000). *Modeling Survival Data: Extending the Cox Model.* Springer-Verlag Inc.

[82] Walker, S G and Mallick, B K (1997). Hierarchical generalized linear models and frailty models with Bayesian nonparametric mixing. *Journal of the Royal Statistical Society, Series B: Methodological,* **59**, 845–860.

[83] Walker, S G and Mallick, B K (1999). A Bayesian semiparametric accelerated failure time model. *Biometrics,* **55**, 477–483.

[84] Wang, L and Dunson, D B (2011). Semiparametric Bayes' proportional odds models for current status data with underreporting. *Biometrics*, **67**, 1111–1118.
[85] Wei, L J (1992). The accelerated failure time model: a useful alternative to the Cox regression model in survival analysis. *Statistics in Medicine*, **11**, 1871–1879.
[86] Yang, S (1999). Censored median regression using weighted empirical survival and hazard functions. *Journal of the American Statistical Association*, **94**, 137–145.
[87] Yang, S and Prentice, R L (1999). Semiparametric inference in the proportional odds regression model. *Journal of the American Statistical Association*, **94**, 125–136.
[88] Yin, G and Ibrahim, J G (2005). A class of Bayesian shared gamma frailty models with multivariate failure time data. *Biometrics*, **61**, 208–216.
[89] Ying, Z, Jung, S H, and Wei, L J (1995). Survival analysis with median regression models. *Journal of the American Statistical Association*, **90**, 178–184.
[90] Zellner, A (1983). Applications of Bayesian analysis in econometrics. *The Statistician*, **32**, 23–34.
[91] Zhang, J, Peng, Y, and Zhao, O (2011). A new semiparametric estimation method for accelerated hazard model. *Biometrics*, **67**, 1352–1360.
[92] Zhang, M and Davidian, M (2008). "Smooth" semiparametric regression analysis for arbitrarily censored time-to-event data. *Biometrics*, **64**, 567–576.
[93] Zhao, L, Hanson, T E, and Carlin, B P (2009). Mixtures of Polya trees for flexible spatial frailty survival modelling. *Biometrika*, **96**, 263–276.
[94] Zhou, M and Li, G (2008). Empirical likelihood analysis of the Buckley-James estimator. *Journal of Multivariate Analysis*, **99**, 649–664.

Part XII

Inverse Problems and Applications

31 Inverse problems

COLIN FOX, HEIKKI HAARIO
AND J. ANDRÉS CHRISTEN

31.1 Introduction

The aim of collecting data from a physical system is to gain meaningful information about the system or phenomenon of interest. However, in many situations the quantities that we wish to determine are different from the ones which we are able to measure, or have measured. Starting with the data that we have measured, the problem of trying to reconstruct the quantities that we really want is called an *inverse problem*. Loosely speaking, we say an inverse problem is where we measure an *effect* and want to determine the *cause*.

Most science and statistics is data-driven in this way, though not always called an 'inverse problem'. Here we want to discuss the features that are characteristic for the problems most typically treated under the umbrella of inverse problems. The quintessential setting is where the measurement process is a complex physical relationship, and inversion presents analytic difficulties.

In a mathematical setting, we represent the measurement process by a family of models parameterized by x, where all necessary physical parameters are contained in x, including nuisance parameters. In the language of inverse problems, simulation of the model for given x defines the *forward map* $A : x \mapsto d$ giving data d in the absence of errors. Determining and simulating the map $A : x \mapsto d$ is the *forward problem*, whereas inferring x from d is the *inverse problem*.

A mathematical model of the forward map A is usually based on some physical theory. For many physical models the mathematical analysis of the forward map is well developed; indeed, many areas of mathematics have been developed precisely to understand the structure of these mappings. Computer evaluation of $A(x)$ is typically the subject of computational science, and again, much of numerical computation has been developed to simulate these problems. For example, solving large-scale partial differential equations arising as models of physical systems drives a great deal of computational science and engineering. Thus, distinctive features of inverse problems are that the forward map is based on physics, mathematical analysis of the forward map is well developed, and evaluation of the forward map uses advanced numerical computation.

Bayesian methods are well suited to incorporating these mathematical and computational models, and for accounting for errors or uncertainties in each of these steps. In this chapter we present methodology and algorithms that are currently used for the Bayesian analysis of inverse problems.

A diverse range of researchers and practitioners work on inverse problems. There are probably as many notions of what it means to *solve* an inverse problem as there are communities of people working on inverse problems. Our notion of an inverse problem and the methods we use to solve them has been influenced by the problems in front of us, and the shared experience of trying to achieve solutions with quantified accuracy in industrial and scientific contexts. That has led us to reformulate the inverse problem in the Bayesian (probabilistic) framework, and to employ sample-based inference to evaluate summary posterior statistics. In doing so we are outliers in the wider inverse-problems community in which deterministic 'regularization' methods (discussed in Section 31.2) are overwhelmingly the most popular. Bayesian methods have the reputation of providing the 'gold standard' amongst solutions, but also of being computationally impractical. Perhaps for those reasons, and also because regularization has a Bayesian *interpretation*, it is common to see analyses of inverse problems under the title of 'Bayesian' that amount to nothing more than regularization. While regularized solutions can be very useful, actually regularization is not a Bayesian method and our view is that scientific accuracy is served by making a linguistic distinction.

A recent development is the focus on *uncertainty quantification* (UQ) within computational models, particularly in the computational science and engineering community. We see this development as very heartening, as we are already seeing a renewed vigour in research into methods for tackling the sizable computational tasks involved in Bayesian analysis of inverse problems.

Inverse problems are often high dimensional in the sense of many unknowns and many data. When using low-level representations it is common to work with 10^3 or 10^4 unknown parameters, which we call *high* dimensional. For example, in impedance tomography about 10^3 elements are needed in an unstructured finite element mesh to ensure that the computed forward map accurately simulates the physics. A global climate model contains upwards of 10^7 unknowns, which we call *very* high dimensional. Mid-level representations, such as representations of surfaces, can effectively *reduce* the number of unknowns. In inverse problems, this reduction often leads to a *more difficult* sampling problem, that we attribute to the geometry of state space becoming more complex. Of order 10 unknowns is *low* dimensional for inverse problems, and usually arises when using parametric representations. Such problems can be very difficult when the system response is chaotic, as occurs in weather and chemical systems.

The remainder of this chapter is organized as follows. This introductory section continues with a list of representative examples of inverse problems followed by a discussion of the the key mathematical property of ill-posedness. We further discuss deterministic and regularization methods in Section 31.2. Some history of Bayesian analysis, as viewed from physics, is presented in Section 31.3. We present the framework for current methodology in Section 31.4, in the context of case studies. We also present some of the recent advances in MCMC algorithms in Section 31.4. We conclude with a glimpse of future directions in Section 31.5.

31.1.1 Examples of inverse problems

- **Compton scattering** The inelastic scattering of photons in matter can be used to probe the wave function of electrons in matter. The forward problem is to predict the angle and energy of scattered photons given the electron structure; the inverse problem is to determine electron structure from measurements of the scattering.
- **Computer axial tomography** X-rays are partially transmitted through the body, with various internal structures having different opacity to X-rays. CAT scans display a picture of that variation *in vivo*. Non-invasive measurements are made of the *total* absorption along lines through the body. Given measurement of such line integrals, how do we reconstruct the absorption as a function of position in the body?
- **Model fitting** A common task in science and engineering is to 'fit' parameters θ of a model

$$d = f(x, \theta) + \epsilon$$

for a given set of measured points $\{x_i, d_i\}_{i=1}^n$. The unknown vector θ may be low dimensional, and the fit routinely done by suitable optimization routines. But even here, with a nonlinear model and possibly non-ideal data, only the rather recent advent of efficient Bayesian sampling algorithms has enabled us to properly analyse the reliability of parameter values and model predictions.

- **Radio-astronomical imaging** When using a multi-element interferometer as a radio telescope, the measured data is not the distribution of radio sources in the sky (called the 'sky brightness' function) but is approximately the Fourier transform of the sky brightness. It is not possible to measure the entire Fourier transform, but only to sample this transform on a collection of irregular curves in Fourier space. From such data, how is it possible to reconstruct the desired distribution of sky brightness?
- **Measuring bulk flow** Many industrial processes transport mixed phase fluids in closed pipes. Control of the process is often improved by real-time measurement of total flow of one or more of the phases. Soft-field imaging, that uses diffusive or highly scattering fields, provides a suitable non-invasive measurement that is sensitive to bulk properties. The image recovery problem is ill-posed, while the determination of bulk flow corresponds to image analysis or segmentation.
- **Geophysics** Inverse problems have always played an important role in geophysics as the interior of the Earth is not directly observable yet the surface manifestation of waves that propagate through its interior is measurable. Like many classes of inverse problems, 'inverse eigenvalue problems' were first investigated in geophysics when, in 1959, the normal modes of vibration of the Earth were first recorded and the modal frequencies and shapes were used to learn about the structure of the Earth in the large.

From this short and incomplete list, it is apparent that inverse problems occur in a myriad of settings.

31.1.2 Ill-posed and ill-conditioned

The problem of solving

$$A(x) = d \tag{31.1}$$

for x given d is called *well-posed* (in the sense of Hadamard) [42] if:

1. a solution *exists* for any data d,
2. the solution is *unique*, and
3. the inverse mapping $d \mapsto x$ is *continuous*.

Conditions 1 and 2 are equivalent to saying that the operator A is onto and one-to-one. Condition 3 is a necessary but not sufficient condition for stability of the solution.

A problem that is not well-posed is said to be *ill-posed*. So an ill-posed problem is one where an inverse does not exist because the data is outside the range of A, or the inverse is not unique because more than one value of x is mapped to the same data d, or because an arbitrarily small change in the data can cause an arbitrarily large change in the solution. Most correctly stated inverse problems turn out to be ill-posed, including all of the examples listed above.

For a well-posed problem, relative error propagation from the data to the solution is controlled by the *condition number* of A, denoted cond (A). If Δd is a variation of d and Δx the corresponding variation of x, then

$$\frac{||\Delta x||}{||x||} \leq \text{cond}\,(A)\,\frac{||\Delta d||}{||d||} \tag{31.2}$$

where (for linear forward problems) $\text{cond}\,(A) = ||A||\,\left|\left|A^{-1}\right|\right|$. When the 2-norm is used, cond (A) is just the ratio of largest to smallest singular values of A. It is possible to find a variation in data Δd for which eqn (31.2) is arbitrarily close to equality, so we usually think of eqn (31.2) with equality since the worst case behaviour will dominate the inverse.

Smaller values of cond(A) give more stable problems. If cond(A) is not too large, the problem in eqn (31.1) is said to be *well-conditioned*, otherwise the problem is said to be *ill-conditioned*. The separation between well-conditioned and ill-conditioned problems is not very sharp and depends on the computational environment. Strictly speaking, a problem that is ill-posed because it fails condition 3 must be infinite dimensional—otherwise the ratio $||\Delta x||/||\Delta d||$ is bounded. However, for ill-conditioned problems the ratio can become very large and we refer to such problems as (discrete) ill-posed problems [22].

The classical example of an ill-posed problem is a Fredholm integral equation of the first kind

$$\int_a^b k\,(t,s)\,x\,(s)\;\mathrm{d}s = d\,(t)\,, \qquad a \leq t \leq b \tag{31.3}$$

with a square integrable, or Hilbert–Schmidt, kernel k. If the solution x is perturbed by $\Delta x\,(s) = \epsilon\,\sin(2\pi ps)$, ϵ a constant, and $\Delta d\,(t)$ is the corresponding perturbation of $d\,(t)$, it follows from the Riemann–Lebesgue lemma that $\Delta d \to 0$ as $p \to \infty$. Hence, the ratio $||\Delta x||/||\Delta d||$ can become arbitrarily large by choosing the frequency p large enough, showing that eqn (31.3) is an ill-posed problem because it fails condition 3. In particular, this calculation shows that inverses of Hilbert–Schmidt integral equations are extremely sensitive to high-frequency perturbations.

Hilbert–Schmidt operators are examples of *compact* operators [45] that commonly arise in inverse problems. Since the inverse of a compact operator cannot be continuous (in standard topologies), all such inverse problems are ill-posed. Many forward problems, especially those that probe an object by the propagation of energy, are also *smoothing* operators. That is, the energy fields throughout the domain have a higher order of differentiability than the imposed excitation. It follows that the singular values of the forward map are summable to some power [3], again ensuring that the inverse is unbounded and the inverse problem is ill-posed. These considerations also explain why best-fit and maximum likelihood estimates are unreliable.

The properties of compact and smoothing both imply that the forward map is arbitrarily well approximated by a *finite-dimensional* operator, even though the spaces for parameters and data could be arbitrarily high dimensional. This means that, in the presence of uncertainty, the physical measurement process conveys only a finite amount of information about the unknowns, even when many more data are measured. Commonly the effective (local) rank can be of the order of 10 to 100. Then the physically possible data lies on a manifold of much lower dimension than data space. This explains the extreme sensitivity that inverse problems display to measurement error or model error, since measurement error will easily put data out of the range of the forward map, while modelling error will mean that the range of the model does not coincide with the physical process.

31.2 Deterministic approaches

The deterministic inverse problem is to invert the function A to obtain unknowns x as a function of data d. Mathematical studies in inverse problems typically focus on the idealized inverse problem in which *all* data is measured, and are concerned with invertability of the forward map and to what degree the inverse problem is ill-posed.

In the absence of an inverse, a solution that achieves a *best fit* to data can be computed as

$$\hat{x}_0 = \arg\min_x C(x), \qquad \text{where} \qquad C(x) = ||d - Ax||^2$$

is the *data misfit* functional, in this case the square of the norm of the residual. When A is invertible the minimum misfit is $C(\hat{x}_0) = 0$ for $\hat{x}_0 = A^{-1}d$.

However, choosing $\hat{x}$ that minimizes $C(x)$ almost always gives a poor solution. In the presence of noise, finding the (possibly non-unique) minimum of C leads to amplification of the noise because of the ill-posedness. Instead, deterministic studies often regard the data as defining a *feasible set* of solutions for which $C(x) \leq C_m$ where C_m depends on the 'level' of the noise.

The primary difficulty in deterministic solutions to ill-posed inverse problems is due to small singular values of the linearized forward map. Actually, the situation is a little worse in practice since the forward map A never models the measurement process precisely. If we consider measurement error e and model error ΔA and the simple observation model $d = (A + \Delta A)x + e$ then direct inversion may be written symbolically as

$$\hat{x} = \frac{d}{A} = x + \frac{\Delta Ax + e}{A}$$

Using the bases of singular vectors makes this formula precise, and shows that the direct inverse will be dominated by model error and measurement noise in the directions of singular vectors of A corresponding to small singular values.

31.2.1 Regularization methods

The most common resolution in the deterministic setting is to formulate and apply a *regular* operator that approximates the singular inverse operator A^{-1}. That is most commonly performed using the *method of regularization* introduced by Tikhonov [42], via the variational statement

$$\hat{x}_\lambda = \arg\min_x \left\{C(x) + \lambda^2 R(x)\right\} \tag{31.4}$$

Here $R(\cdot)$ is a *regularizing functional* that represents our aversion to a particular solution, with larger values being larger aversion, and λ is the *regularizing parameter*.

There are many ways of arriving at this variational form. One way is to think of minimizing the regularizing functional $R(x)$ over the set of solutions satisfying $C(x) = C_m$, for some C_m. Introducing the Lagrange multiplier $1/\lambda^2$ gives the form in eqn (31.4).

The most common regularizing functional is *Tikhonov regularization*

$$R(x) = ||x||_2^2$$

Sometimes, there is a preference for solutions which are close to some *default solution* x_∞ which can be accommodated by choosing

$$R(x) = ||x - x_\infty||^2 \tag{31.5}$$

More generally, it may not be the norm of $x - x_\infty$ which needs to be small, but some linear operator acting on this difference. Introducing the operator L for this purpose, we can set

$$R(x) = ||L(x - x_\infty)||^2 = (x - x_\infty)^{\mathrm{T}} L^{\mathrm{T}} L (x - x_\infty) \tag{31.6}$$

In discrete problems the matrix L is of size $p \times n$ where $p \leq n$. Typically, L is a banded matrix approximation to the $(n-p)$th derivative. For example, when data and unknowns are one-dimensional functions discretized with interval h, approximations to the first and second derivatives are given by the matrices

$$L_1 = \frac{1}{h}\begin{pmatrix} -1 & 1 & & & \\ & -1 & 1 & & \\ & & \ddots & \ddots & \\ & & & -1 & 1 \end{pmatrix} \quad \text{and} \quad L_2 = \frac{1}{h^2}\begin{pmatrix} 1 & -2 & 1 & & & \\ & 1 & -2 & 1 & & \\ & & \ddots & \ddots & \ddots & \\ & & & 1 & -2 & 1 \end{pmatrix}$$

Use of the second derivative, also called *Laplacian regularization*, penalizes curvature in the solution and is commonly used when making contour maps.

In other cases, it may be appropriate to minimize some combination of the derivatives such as

$$R(x) = \alpha_0 \left\|x - x_\infty\right\|^2 + \sum_{k=1}^{q} \alpha_k \left\|L_k (x - x_\infty)\right\|^2$$

where L_k is a matrix which approximates the kth derivative, and α_k are non-negative constants. Such a quantity is the square of a *Sobolev norm* that may also be written in the form of eqn (31.6).

Equation (31.4) provides a family of solutions parameterized by the regularization parameter λ. If λ is very large, the data misfit term $C(x)$ is negligible compared to $R(x)$ with $\lim_{\lambda\to\infty} \hat{x}_\lambda = x_\infty$. We effectively ignore the data (and any noise on the data) and minimize the solution seminorm by choosing the default solution. On the other hand, if λ is small, the weighting placed on the solution seminorm is small and the data misfit at the solution becomes more important. If λ is reduced to zero, the solution reduces to the least squares case.

When A is linear and the regularizing functional has the quadratic form in eqn (31.6), a solution to eqn (31.4) may readily be found by solving

$$\left(A^{\mathrm{T}}A + \lambda^2 L^{\mathrm{T}}L\right)\hat{x}_\lambda = \lambda^2 L^{\mathrm{T}} L x_\infty + A^{\mathrm{T}} d \tag{31.7}$$

Computing the regularized solution is thus reduced to solving a (large) system of simultaneous equations with a symmetric positive definite coefficient matrix, for which there are many efficient algorithms. In stationary time-series problems sequential solutions may sometimes be implemented by repeated action of a linear operator, or *filter*. Examples are the Wiener filter and the Kalman filter.

The regularization functionals we have discussed are norms or *seminorms* on the space of solutions, as is typically the data misfit functional. There are many other regularizing functionals in common use, many designed to overcome the observation that regularization can over smooth solutions, especially at transitions in images. For example, *total variation* regularization is often used to encourage 'blocky' images [22]. Other norms are also used such as the 0-norm that penalizes the number of non-zero components and hence prefers sparse solutions.

31.2.1.1 *Truncated singular value decomposition*

A linear operator A with rank r has the singular value decomposition (SVD)

$$A = \sum_{l=1}^{r} \sigma_l u_l^{\mathrm{t}} v_l \tag{31.8}$$

for some bases of left and right singular vectors $\{u_l\}$ and $\{v_l\}$, respectively, and singular values σ_l. The truncated SVD method is based on the observation that the components of the solution for singular vectors associated with the larger singular values of A are well determined by the data, whereas the components corresponding to smaller singular values are not. When the singular values up to $k \leq n$ are deemed to be significant the *truncated SVD* solution is

$$x'_k = \sum_{l=1}^{k} \left(\frac{u_l^{\mathrm{T}} d}{\sigma_l} \right) v_l \tag{31.9}$$

The integer k takes the role of regularizing parameter.

31.2.1.2 *Filter factors*

For the case of linear forward maps, the filter factor representation displays the solutions to the regularization problem for all values of λ in a convenient form. Here we analyse Tikhonov regularization since the SVD in eqn (31.8) suffices. Equivalent results for more general L are available using the generalized SVD.

Writing $d_l = u_l^{\mathrm{T}} d$, $\hat{x}_l = v_l^{\mathrm{T}} \hat{x}$, and $x_{\infty l} = v_l^{\mathrm{T}} x_\infty$, i.e. resolving each vector into the bases of singular vectors, the regularized solution can be written

$$\hat{x}_l = \begin{cases} \dfrac{\sigma_l^2}{\lambda^2 + \sigma_l^2} \left(\dfrac{d_l}{\sigma_l} \right) + \dfrac{\lambda^2}{\lambda^2 + \sigma_l^2} x_{\infty l} & \text{for } l = 1, 2, \ldots, r, \\ x_{\infty l} & \text{for } l = r + 1, \ldots, n. \end{cases} \tag{31.10}$$

The terms d_l/σ_l and $x_{\infty l}$ give the solution coefficient in the extreme cases of no regularization ($\lambda = 0$) and no data ($\lambda = \infty$), respectively. The coefficients of these terms are the *filter factors*.

Notice how the filter factors sum to one, and the first filter factor smoothly decreases to zero as the singular values gets smaller, or as λ increases. The value of λ sets the boundary between 'small' and 'large' singular values. In contrast, the filter factors for the truncated SVD method are equal to unity for those singular values which are deemed to be non-negligible ($l \leq k$) and to zero for those singular values which are negligible ($l > k$). That sharp cutoff typically leads to ringing[36] in solutions. Thus, Tikhonov regularization may be viewed as a type of *windowing* as employed in signal processing.

31.2.1.3 *Choosing the regularization parameter*

We have seen that λ sets the balance between minimizing the residual norm $||d - Ax||$ and minimizing the solution seminorm $||L(x - x_\infty)||$. There is no single rule for selecting λ that works in all cases. Perhaps the most convenient graphical tool is the *L-curve* [22], that is a parametric plot of log of the solution seminorm versus log of the data misfit. One of the simplest methods is the *Morozov discrepancy principle* that sets λ so that the data misfit equals the measurement error 'level'. Another method is *generalized cross validation* (GCV) for selecting the parameter in ridge regression [15], which is equivalent to regularized inversion.

[36] More formally known as Gibbs' phenomenon.

31.3 A subjective history of subjective probability

For the many physicists and astronomers who were applying Bayesian analysis to inverse problems in the 1980s, the history of Bayesian methods is synonymous with the development of probabilistic methods in the physical sciences. This viewpoint is supported by many key components in Bayesian methodology being developed in response to problems arising in physics, including the Metropolis algorithm. This section presents a history of Bayesian methods as imbibed by one of us (CF) while studying inverse problems amongst *Bayesian physicists*.[37,38]

The name of Bayes was attached to Bayes' theorem by Poincaré around 1886, in his own work on probability. Bayes never wrote Bayes' theorem in the modern form. He did, however, give a method for finding *inverse probability* while solving an unfinished problem stated by Bernoulli. That method was reasoned by lengthy arguments and appeared in a paper published in 1763, after Bayes' death in 1760.

The first clear statement and use of Bayes' theorem was given by Laplace in almost his first published work in 1774. Laplace rediscovered Bayes' principle in greater clarity and generality, and then for the next 40 years applied it to scientific and civic problems. Laplace published in 1812 his two-volume treatise *Théorie Analytique des Probabilités* in which the analytical techniques for Bayesian calculations were developed. The second volume contains Laplace's definition of probability, Bayes' theorem, remarks on moral and mathematical hope (or expectation), a discussion of the method of least squares, Buffon's needle problem, and inverse probability. Later editions also contain supplements which consider applications in physics and astronomy. Laplace was mainly concerned with overdetermined problems (many observations and few unknowns) and solely used the *principle of insufficient reason*[39] to determine prior probabilities.

Laplace's calculus of probability was soon applied to explaining physical phenomena. The physicist James Clerk Maxwell said in 1850 [32],

> the true logic for this world is the calculus of Probabilities, which takes account of the magnitude of the probability which is, or ought to be, in a reasonable man's mind.

Even though Maxwell was only 19 years old at the time, he was already a formidable scientist and these principles remained in Maxwell's later work. In his kinetic theory of gases Maxwell determined the distribution over molecular velocities, effectively determining a prior probability distribution by 'pure thought' [27]. Experimental verification promoted the Maxwell distribution to the status of physical *law*, founding the subject of statistical physics.

However, among those looking to develop a theory of uncertain events the concept of probability as representing *a state of knowledge* was rejected, from about 1850, and replaced by the notion that probability must refer to *frequency in a random experiment*. Largely that rejection took place when it was realized that the notion of equiprobable, encapsulated in Laplace's principle of insufficient reason, gave results that depended on the parameterization chosen, and since Laplace had based his notion of *probable* on the more fundamental notion of *equiprobable* the whole theory was rejected. Bertrand constructed his paradox in 1889, as a transformation of Buffon's needle problem, to demonstrate the difficulties.

By the beginning of the twentieth century, application of Bayes' theorem was severely criticized with a growing tendency to avoid its application [9]. The new statistics[40] was connected with

[37] This term was apparently coined by Brian Ripley as a pejorative.
[38] A more extensive early history can be found in the first section of [28].
[39] Renamed the *principle of indifference* by Keynes [32].
[40] Interestingly, Cramér referred to Bayesian methods as 'classical' in 1945.

the theory of *fiducial probabilities* due to R. A. Fisher and the theory of *confidence intervals* due to J. Neyman. These methods became so dominant that for half a century from 1930 a student of statistics could easily not know that any other conception had existed. In that period, von Mises said that Bertrand's paradox did not even belong to the field of probability, apparently unaware of the Boltzmann (including Maxwell) distributions in physics that resolve problems of the same type.

In the 1930s, Harold Jeffreys found himself unconvinced by Fisher's arguments and rediscovered Laplace's rationale while working on 'extracting signals from noise' in geophysics. In 1939 he published his *Theory of Probability* in which he extended Bayesian inference, explaining the theory much more clearly than did Laplace. In the 1948 edition Jeffreys gave a much more general *invariance theory* for determining ignorance priors, which remains of importance today in the form of *reference priors*.

For many physicists the question of whether one can or cannot use Bayes' theorem to quantify uncertainty was answered by the physicist Richard T. Cox in 1946 and 1961 [7]. Instead of asking whether or not Laplace gave us the right 'calculus of inductive reasoning', he raised the question of what such a calculus must look like. Supposing that degrees of plausibility are to be represented by real numbers, he found the functional conditions that such a calculus be consistent and showed that the general solution uniquely determines the product and sum rules for probability to within a change of variables. An immediate consequence is Bayes' theorem. This does not answer the question of how to assign probabilities, but it does determine how they must be manipulated once assigned.

The reappearance of Bayesian methods in the physical sciences from about 1970 can in many cases be traced to the physicist Edwin T. Jaynes who, from the 1960s to 1980s, championed Bayesian methods as an inductive extension of deductive logic. While looking to unify statistical physics and Shannon's new theory of communication, he observed that methods that were experimentally verified in statistical physics appeared to be derided in statistics, and set about formalizing the basis of those methods. This led Jaynes to formulate the *maximum entropy* principle for prior distributions [28], as an extension of Jeffreys' uninformative prior. Jaynes also adapted the group invariance methods, that are standard in physics for deriving the mathematical form of physical laws, to the method of *transformation groups* for determining prior probabilities. Notably, this resolved Bertrand's 'paradox', showing that it is actually well posed [27]. Jaynes had rephrased Laplace's *indifference between events* to an *indifference between problems*. An anonymous poet celebrated this contribution in the lines:

> So, are you faced with problems you can barely understand?
> Do you have to make decisions, though the facts are not in hand?
> Perhaps you'd like to win a game you don't know how to play.
> Just apply your lack of knowledge in a systematic way.

By the 1980s, a number of groups in physics and astronomy saw Bayesian analysis as the correct route to resolving inverse problems in the presence of 'incomplete and noisy data' [18]. The advanced state of computational optimization allowed Bayesian MAP estimates to be calculated in large-scale problems, with some notoriety being achieved by the *maximum entropy method* (MEM). The practical properties and limitations of MEM were pointed out by a number of statisticians, most influentially in [12]. In the same period, inverse problems became a 'topic' in statistics, though analysis was limited to regularization *estimators* [38], or Bayesian analyses that used an artificial likelihood conditioned on a regularized solution.

The renewed appreciation of MCMC following the publication of Gelfand and Smith in 1990 influenced those applying Bayesian methods to inverse problems, with the first substantive analyses of inverse problems using MCMC appearing in 1997 [14, 37]. The analysis in [34] of a realistic problem in geophysics also appeared in that year using a Metropolis algorithm, apparently (though somewhat implausibly) unaware of Gelfand and Smith, or Hastings' improvement. That work followed the direction set by Albert Tarantola in formulating inverse problems in a Bayesian framework [41]. The title *Inverse Problems = Quest for Information*, alone, of the 1982 paper by Tarantola and Valette had motivated many in the inverse problems community to explore Bayesian methods.

For Bayesian statisticians the early impact of MCMC was summed up by Peter Clifford when he wrote in 1993,

> from now on we can compare our data with the model we actually want to use rather than with a model which has some mathematical convenient form.

The situation for Bayesian physicists was somewhat different since they were already using physically realistic models (at least for the forward map) but lacked the computational tools for unhindered exploration of the posterior distribution. MCMC provided that tool, though the computational challenges were formidable. Exposure to spatial statistics brought the mid-level and high-level representations [26], that don't fit into a regularization framework, with a clear route for inference having been charted by Grenander and Miller [17].

31.4 Current Bayesian methodology

The methods of regularization and truncation in Section 31.2 provide valid algorithms to tame ill-posed computational problems. They also come close to the Bayesian approach in the sense that a regularization can be interpreted as equivalent to setting prior knowledge—or guess—to some characteristics of the solution. The estimate then will be a compromise produced by the regularization and the measurement data. But a crucial component of the Bayesian approach is still missing: how to produce a proper analysis of the certainty, or rather uncertainty, of the estimates? How much, indeed, can we trust the predictions given by our models, often simulating complex physical systems? Here, we believe, is the main contribution that present day Bayesian Monte Carlo algorithms are able to provide.

In this section we discuss our computational approaches to the statistical aspects of inverse problems, as well as the spirit in which we see ourselves as 'Bayesians'. The discussion is largely influenced by the applied projects from our own experience, and so inevitably is subjective again.

All available data contains measurement errors, so the estimated unknowns are more or less uncertain. A natural question then arises: if measurement noise corrupting the data follows some statistics, what is the *distribution* of the possible solutions after the estimation procedure? Bayesian thinking explicitly allows for the unknown vector x to be interpreted as a *random variable* with a distribution of its own. In addition, the approach typically emphasizes the use of *prior knowledge* in the estimation process, even subjective. As we all know, and we alluded to in our 'history', these questions have been the focus of a longstanding dispute between the two opposing views:

- Frequentists argue that analysis should be driven by the data as much as possible, and that attaching a distribution to a parameter based on one's subjective belief should not be a part of valid statistical analysis. Moreover, parameters indeed are constants without distributions of their own.
- Bayesians argue that treating solutions as random variables is actually more realistic and, by considering different choices for distributions, Bayesian analysis is perfectly valid. Moreover, scientific research most often contains strong *hidden* prior information, such as the choice of model used to explain the phenomena under study.

A practically oriented researcher might find the dispute somewhat academic. In a real modelling project, are we really so concerned about the 'true' interpretation of parameters? In any case we all certainly should be interested in the *reliability* of model predictions. Naturally, the estimates for unknowns should be physically plausible. We have experience in geophysics applications where it is necessary that estimates show a sub-surface structure that is believable to a geologist, before the predictions will be trusted.

But as the solution is estimated from noisy data, some uncertainty always remains, whether we interpret the 'truth' as fixed or random. So, it is essential, in any case, to realize that estimation problems do not have a unique solution. A numerical optimizer may find a true global minimum for a given least squares function with fixed data values. However, a multitude of different solutions may fit the data 'equally well', when we take into account the noise in the measurements. The practical essence of the Bayesian approach, in our experience, is to to find *all* those possible solutions, as well as the respective model predictions, as probability distributions. An added value is also the interpretation of those probabilities in a clear formal perspective (see e.g. [25]), that permit not only useful engineering solutions but valid 'scientific' answers as well.

For many of us, the Bayesian approach is almost synonymous with the use of MCMC methods. The advantages of using MCMC for solving inverse problems are various: full characterization of (non-Gaussian) posterior distributions is possible. We have full freedom in implementing prior information. Even modelling errors can be taken into account in a flexible way. Moreover, we are less likely to get trapped in local minimums than when employing optimization methods to get MAP estimates.

31.4.1 Mathematical formulation

The framework for Bayesian analysis of inverse problems is straightforward in concept; one formulates the likelihood function by modelling the measurement process and errors, specifies a prior distribution over unknowns, and then performs posterior inference (commonly by MCMC).

A general stochastic model for the measurement process is

$$d = G(x, v) \tag{31.11}$$

where x represents the deterministic unknowns, typically physical constants, and v is a random variable accounting for variability between 'identical' experiments. In practice the separation between 'deterministic' and 'random' is a modelling choice, since all effects may be modelled as random. We find that better results are given by modelling as many deterministic processes as possible. However, modelling practicalities often demand that some residual deterministic processes are treated as random.

In the *state space approach*, eqn (31.11) is the *observation equation* in a problem that does not vary with time. The time-varying problem [29, 44] is commonly treated as inference for a hidden Markov model [5], for which sequential Monte Carlo methods are applicable.

In the simplest formulation the stochastic part is attributed to measurement error that is additive and independent of x so that

$$G(x, v) = A(x) + v$$

where $v \sim \pi_{\mathrm{n}}(\cdot)$ comes from the *noise* distribution. Then, when the forward map is treated as certain, the distribution over data conditioned on x is is given by

$$\pi(d|x) = \pi_{\mathrm{n}}(d - A(x)) \tag{31.12}$$

The likelihood function for given data d is the same function considered as a function of the unknown variables x. Hence, formulating the likelihood function requires modelling the forward map as well as the distribution over measurement errors. Evaluation of the likelihood function requires simulation of the forward map, and hence is typically computationally expensive.

Given measurements d, the focus for inference, at least in parameter estimation, is the posterior distribution given by Bayes' theorem

$$\pi(x|d) = \frac{\pi(d|x)\pi(x)}{\pi(d)} \propto \pi(d|x)\pi(x) \tag{31.13}$$

where $\pi(x)$ is the prior distribution and $\pi(d)$ is often called the evidence. Note that we take the usual (and sometimes dangerous) liberty with notation where each density function is implicitly distinguished by the type of argument it takes.

31.4.2 Models for model error

All models are wrong, and particularly so in inverse problems. We are not aware of any inverse problem where the measurement error is greater than the model uncertainty. Perhaps this is because measurements may be made more accurately whereas more accurate physical modelling requires conceptual advances.

It is useful to distinguish between the physical process, the mathematical model and the computational model, that we denote A_p, A_m and A_c, respectively. Kennedy and O'Hagan [31] introduced a model for inadequacy in computer models, writing

$$A_p(x) = A_c(x) + D(x)$$

where the *model inadequacy* $D(x)$ was modelled nonparametrically as a Gaussian process (GP), as was A_c.[41] This approach would be familiar in machine learning. While a nonparametric model for model inadequacy seems very sensible, the use of Gaussian process models is somewhat unsatisfactory for inverse problems. For example, formulating a GP is prohibitive in high dimensions. Instead, modelling D by a Gaussian distribution is feasible, as we will see. Also, building a GP surrogate to the forward map is problematic since the complex input/output structure is effectively only captured in the mean process, but that amounts to tabulating input/output pairs, which is prohibitive. More successful is using a reduced order (computational) model (ROM) A_c^* of the computational forward map A_c that approximately captures that structure with a cheap computation.

The use of ROMs is almost mandatory in large-scale inverse problems, to reduce computational cost of the forward map. There are many schemes for building ROMs, such as local linearization, coarse numerical discretization, or low-order expansions. A systematic approach can be found in [1].

A ROM necessarily introduces a model error that we can analyse as

$$A_c(x) = A_c^*(x) + B(x) \tag{31.14}$$

We call $B(x) = A_c(x) - A_c^*(x)$ the *model reduction error*.

The approximation error model (AEM) of Kaipio and Somersalo [30] has proved effective in mitigating the effects of model reduction error. They modelled the model reduction error as being

[41] A taxonomy for the *arguments* of these functions, that fits well in the inverse problem context, was given by Campbell and McKay in the discussion of [31].

independent of the model parameters and normally distributed. Then the observation process is reduced to

$$d = A_c^*(x) + B + v \tag{31.15}$$

where $B \sim N(\mu_B, \Sigma_B)$, when we assume that the accurate computational model is correct, i.e. $A_p = A_c$, as in [30]. However, it is interesting to note that if the model inadequacy D is also taken to be Gaussian, then eqn (31.15) still holds without the assumption $A_p = A_c$, with the distribution over B also accounting for bias and uncertainty in the mathematical model.

31.4.3 Prior information

Most often, we do not really want to specify a non-trivial prior distribution for the solution. We may just know that the solution components must have some bounded and positive values, leading to *uninformative* or *flat* priors. Naturally, one must remember that 'flatness' depends on the parameterization of the model. For instance, a flat prior for the conductivity σ is non-flat for the resistivity $1/\sigma$, while a model could be equally written in terms of either parameterization.[42]

A practical guide for parameterization is simply to try to write a model that is easily identified by available data. For a given parameterization then, we may just set a box of 'simple bounds', just lower and upper bounds, to constrain the solutions. The analysis is now fully driven by data, supposing that the posterior distribution of the parameters is well inside the given bounds.

If, on the other hand, the posterior does not stay inside any reasonable bounds, we must observe that the available data is *not sufficient to identify the parameters*. This is an important conclusion, and not too unusual! We can then consider a few options:

- *Design of Experiments.* If non-identifiability of parameters is due to lack of data, an obvious remedy is to design new experiments to gain more informative measurements. Several classical linearization-based methods exist. Bayesian analysis and MCMC sampling provide a comprehensive way to design simulation based experiments for, e.g. situations where the classical criteria can not be implemented due to a singular information matrix [35].
- *Model reductions.* Often, however, the non-identifiability is an inherent feature of the forward map and no practically measurable data (or just reparameterization) is able to correct the situation. This occurs when parts of a physics-based model are unobservable due to, e.g. different scales in time or space: fast equilibria of certain parts of chemical kinetics, or negligible diffusion due to small catalyst particles are typical examples. An alternative option for fixing priors for unidentified parameters is then to simplify the model, and thus reduce the list of parameters to be identified. Again, MCMC sampling gives an algorithmic tool here. Instead of 'political decisions' on how to reduce the model, we may create parameter posterior distributions by MCMC to see which model parameters remain unidentified, and reduce the model accordingly. The reduction process itself may require special tools, such as the singular perturbation methods (see [19] for an example).

Typically, Bayes' theorem is seen as the way of putting together a *fixed* prior, data, and model. We may observe that the approaches suggested above rather employ Bayesian sampling techniques as flexible, algorithmic tools for *model development*, that may guide all the relevant steps of modelling: not only the analysis of model parameter identifiability, but also the design of measurements as well as testing different versions of the model for the phenomenon under study. Only if such measures

[42] The Jeffreys' prior for a *scale parameter* works here, that is uniform in $\log(\sigma)$.

are not available, one should carefully seek true prior information to be included as the prior distribution in the estimation process.

Unfortunately, sometimes one is forced to consider prior information in detail. In problems with more than a few unknowns, such as inverse problems, simply setting 'flat' priors over each coordinate direction can result in the prior being highly 'informative' for posterior statistics of interest. We first saw this effect pointed out in the context of estimating occupancy time, or *span*, of archaeological sites from radiocarbon dating of artifacts [36], where a uniform prior over the date of each artifact leads to a strong bias towards larger estimates of span, to the point where a short span is effectively ruled out. We have had to correct for this effect when using electrical capacitance tomography (ECT) to make quantitatively accurate measures of the cross-sectional area of water inclusions in oil pipe lines [44]. Interestingly, in the presence of uncertainties, correcting the prior to give quantitatively accurate estimates of area produces bias in estimates of the boundary length, and reminds us that information is a *relative* concept; uninformative with respect to one question is typically informative with respect to another.

31.4.4 Exploration by sampling

The Metropolis–Hastings (MH) algorithm is the basis of nearly all sampling algorithms that we currently use. This algorithm was originally developed for applications in statistical physics, and was later generalized to allow general proposal distributions [23], and then allowing transitions in state space with differing dimension [16]. Even though we do not always use variable-dimension models, we prefer this Metropolis–Hastings–Green (MHG) 'reversible jump' formulation of MH as it greatly simplifies calculation of acceptance probabilities for the subspace moves that are frequently employed in inverse problems. One step of MHG dynamics can be written as:

Algorithm 1 (MHG)

Let the chain be in state $x_n = x$, then x_{n+1} is determined in the following way:

1. Propose a new candidate state x' from x depending on random numbers γ with density $q(\gamma)$.
2. With probability

$$\alpha(x, x') = \min\left(1, \frac{\pi(x'|d)q(\gamma')}{\pi(x|d)q(\gamma)}\left|\frac{\partial(x', \gamma')}{\partial(x, \gamma)}\right|\right) \quad (31.16)$$

accept the proposed state by setting $x_{n+1} = x'$. Otherwise reject by setting $x_{n+1} = x$.

The last factor in eqn (31.16) denotes the magnitude of the Jacobian determinant of the transformation from (x, γ) to (x', γ'), as implemented in computer code for the proposal. A few details remain to be specified such as the choice of starting state, and the details of the proposal step.

The only choice one has within the MHG algorithm, is *how* to propose a new state x' when at state x. The popular choice of Gibbs sampling is the special case where x' is drawn from a (block) conditional distribution, giving $\alpha(x, x') = 1$. The choice of the proposal density is largely arbitrary, with convergence guaranteed when the resulting chain is irreducible and aperiodic. However, the choice of proposal distribution critically affects efficiency of the resulting sampler. The most common MH variants employ *random walk* proposals that set $x' = x + \gamma$ where γ is a random variable with density $q(\cdot)$, usually centred about zero. In high-dimensional problems, global proposals that attempt to change all components of the state usually have vanishingly small acceptance probability, so are not used. Since ill-posedness results in extremely high correlations, single-component proposals result in slow mixing. Hence, a multi-component update is usually required, that is problem specific.

In problems where the posterior distribution is essentially unimodal, computational cost can be minimized by starting at the MAP estimate computed by computational optimization. Indeed, the optimization step can provide useful input to the MCMC, such as a low rank approximation to the Hessian of the log of the target density when using BFGS optimization. This has been used to seed the proposal covariance in the adaptive Metropolis (AM) algorithm. For multi-modal target distributions, or when debugging code, it is often necessary to start from a randomized starting state drawn from an 'over-dispersed' distribution, though this can be very computationally expensive as the MCMC may require many iterations to find the support of the posterior distribution.

31.4.4.1 *Algorithm performance*

Since many steps of the MHG algorithm are typically required for convergence, and each step requires a computationally expensive evaluation of the forward map, it is important to evaluate and tune on the computational efficiency of a sampling algorithm.

A common measure of *statistical efficiency* is the integrated auto-correlation time (IACT) that measures the number of samples from the chain that have the same variance reducing power as one independent sample. It is desirable to have a small number of steps per IACT, so that estimates evaluated over the chain converge more quickly for a given number of steps.

However, statistical efficiency is not a sufficient measure of algorithmic performance in inverse problems where the CPU time taken per step can vary, such as when using a ROM. For example, in the delayed acceptance algorithms we consider later, the computational cost of a rejection step is much smaller than the computational cost of an acceptance. Hence, it is also necessary to measure the average CPU time per step.

We measure *computational efficiency* as the product of these two terms, to give the *CPU time per IACT*. This then measures the CPU time (sometimes called the wall-clock time) required to reduce the variance in estimates by the same amount that one independent sample would achieve. Clearly, small CPU time per IACT is desirable.

Unfortunately some papers showing new sampling algorithms only report statistical efficiency. We know of several such papers where an 'improved' algorithm is correctly reported as increasing statistical efficiency, but actually decreases computational efficiency, and hence would take longer to produce estimates with a given accuracy compared with the unimproved algorithm.

31.4.5 Atmospheric remote sensing and adaptive Metropolis

As an example of an ill-posed inverse problem, we discuss in some detail the recovery of ozone profiles by satellite measurements. This case study also provides an example on how practical challenges from real-life projects may give us impetus to develop new computational methods.

Remote sensing techniques are today routinely used for atmospheric research. The data processing of these instruments typically involve solving nonlinear inverse problems. GOMOS (Global Ozone Monitoring by Occultation on Stars) is one of the 10 instruments on board the European Space Agency's Envisat satellite which is targeted on studying the Earth's environment. The Envisat satellite was launched on the 1st of March in 2002 to a polar, sun-synchronous orbit at about 800 km above the Earth. It is still fully operational now in 2012. The main objective of GOMOS is to measure the atmospheric composition and especially the ozone concentration in the stratosphere and mesosphere with high vertical resolution. The GOMOS instrument was the first operational instrument that uses the stellar occultation technique to study the Earth's atmosphere. The measurement principle, demonstrated in Figure 31.1, is elegant: the stellar spectrum seen through the atmosphere is compared with the reference spectrum measured above the atmosphere. Due to

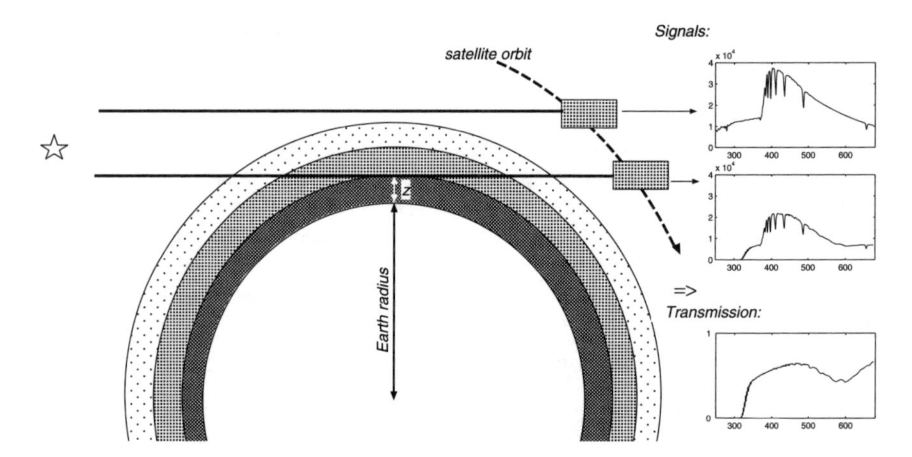

Figure 31.1 GOMOS measurement principle. The horizontal transmission of the atmosphere at tangent altitude z is obtained by dividing the attenuated stellar spectrum with the reference spectrum measured above the atmosphere.

the absorption and scattering in the atmosphere the light measured through the atmosphere is attenuated and the attenuation is proportional to the amount of constituents in the atmosphere. The measurements are repeated at different tangential altitudes to obtain vertical profiles of the concentrations of different atmospheric constituents. The advantages of the GOMOS instrument compared to other instruments measuring ozone are the fairly good global coverage, with 300–400 occultations daily around the Earth combined with the excellent vertical resolution (sampling resolution 0.3–1.7 km). The altitude range which can be covered by GOMOS is large: 15–100 km and the brightest stars can be followed even down to 5 km. Each occultation consists of about 70–100 spectra measured at different tangential altitudes and each UV-vis spectra includes measurements at 1416 different wavelengths. Because of the multitude of stars it is important that the optimal set of stars is selected for each orbit. This optimization was included in the GOMOS mission planning.

In the GOMOS data processing constituent densities are retrieved from stellar spectra attenuated in the atmosphere. The GOMOS inverse problem can be considered as an exterior problem in tomography, but in practice it is solved locally considering only data collected from one occultation at a time. This inverse problem is as follows. By dividing the stellar spectrum measured through the atmosphere with the reference spectrum measured above the atmosphere we obtain a so-called transmission spectrum. The transmission at wavelength λ, measured along the ray path ℓ, includes a term $T^{\text{abs}}_{\lambda,\ell}$ due to absorption and scattering by atmospheric constituents and a term $T^{\text{ref}}_{\lambda,\ell}$ due to refractive attenuation and scintillations, that is, $T_{\lambda,\ell} = T^{\text{abs}}_{\lambda,\ell} T^{\text{ref}}_{\lambda,\ell}$. The dependence of the transmission on the constituent densities along the line of sight ℓ is given by Beer's law:

$$T^{\text{abs}}_{\lambda,\ell} = e^{\left[-\int_\ell \sum_{\text{gas}} \alpha^{\text{gas}}_\lambda (z(s)) \rho^{\text{gas}}(z(s)) ds\right]}$$

where $\rho^{gas}(z)$ gives the constituent density at altitude z and α denotes the cross-sections. Each atmospheric constituent has typical wavelength ranges where the constituent is active either by absorbing, scattering or emitting light. The cross-sections reflect this behaviour and their values are considered to be known from laboratory measurements. In the equation above the sum is over different gases and the integral is taken over the ray path. The problem is ill-posed in the sense that continuous profile is retrieved from a discrete set of measurements. Therefore some additional regularization or prior information is required to make the problem well-posed and solvable. In practice this is done by discretizing the atmosphere into layers and assuming some smoothness prior, or even just constant or linearly varying density inside layers.

The measurements are modelled by

$$y_{\lambda,\ell} = T^{\text{abs}}_{\lambda,\ell} T^{\text{ref}}_{\lambda,\ell} + \epsilon_{\lambda,\ell},$$

$\epsilon_{\lambda,\ell}, \sim N(0, \sigma^2_{\lambda,\ell}), \lambda = \lambda_1, \ldots, \lambda_\Lambda, \ell = \ell_1, \ldots, \ell_M$. The likelihood function for the constituent profiles then reads as

$$P(y|\rho(z)) \propto e^{-\frac{1}{2}(T-y)C^{-1}(T-y)}$$

with $C = \text{diag}(\sigma^2_{\lambda,\ell})$ and $y = (y_{\lambda,\ell})$, $T = (T_{\lambda,\ell})$. The true statistics is Poisson, but can be safely treated as Gaussian. The inverse problem is to estimate the constituent profiles $\rho(z) = (\rho^{\text{gas}}(z))$, gas $= 1, \ldots, n_{\text{gas}}$.

In the operational data processing of GOMOS the problem is divided into two parts. The separation is possible if the measurement noise is independent between successive altitudes and the temperature-dependent cross-sections can be sufficiently well approximated with 'representative' cross-sections (e.g. cross-sections at the temperature of the tangent point of the ray path). In the operational algorithm these simplifications are assumed and the problem is solved in two steps. The *spectral inversion* is given by

$$T^{\text{abs}}_{\lambda,\ell} = \exp\left[-\sum_{\text{gas}} \alpha^{\text{gas}}_{\lambda,\ell} N^{\text{gas}}_{\ell}\right], \; \lambda = \lambda_1, \ldots, \lambda_\Lambda,$$

which is solved for the horizontally integrated line-of-sight densities N^{gas}_{ℓ}. The *vertical inversion*

$$N^{\text{gas}}_{\ell} = \int_{\ell} \rho^{\text{gas}}(z(s))ds, \; \ell = \ell_1, \ldots, \ell_M$$

is solved for local constituent densities ρ^{gas} using the line-of-sight densities from the previous step as the data. Naturally, it is also possible to solve the problem directly in one step by inverting the local densities from the transmission data. This approach is here referred to as the one-step inversion.

The first step of the operational GOMOS data processing, the spectral inversion problem, is nonlinear, with all the usual advantages available if solved using the MCMC technique. At each line-of-sight, the dimension of the problem is small, only some five parameters (horizontally integrated line-of-sight densities of different constituents) to be retrieved. However, the estimation is done repeatedly at each altitude, about 70–100 times for each occultation. The natural way of implementing the MCMC technique is to use random walk MH algorithm. But here we meet the difficulty of tuning the proposal distribution to obtain efficient sampling. The special feature in the GOMOS data processing is that the posterior distributions of the spectral inversion vary strongly. They depend on the tangential altitude and also on the star used for the occultation. The line-of-sight densities vary typically several decades between 15 to 100 km for ozone vertical profile measured by GOMOS. When the star is dim (and hence the signal-to-noise ratio is low) the posterior distributions become many times wider compared with the ones obtained for a bright star. In such a setup it is impossible to find any fixed proposal distribution that would work at all altitudes and for all stars. Therefore, the proposal distributions need to be optimized for each altitude and for each occultation separately. However, any offline manual tuning of the proposal distributions is also impossible to realize because of the huge number of datasets. Automatic algorithms for tuning the proposal distribution were therefore needed.

To overcome these problems of GOMOS spectral inversion problems the adaptive MCMC algorithms were originally developed, AM for the two-step algorithm and adaptive MwG (SCAM) for the one-step inversion. The advantage of these algorithms is that they make the implementation of the MCMC easy; the adaptation can be used in a fully automatic way without increasing the computational time dramatically.

The adaptation only requires a small change in the MHG algorithm. The basic adaptive Metropolis (AM) [21] version uses a Gaussian (and thus symmetric) proposal q, whose covariance is updated by the empirical covariance of the chain:

Algorithm 2 (AM)

At step n, with state $x_n = x$ and covariance C_n, determine x_{n+1} and C_{n+1} in the following way:

1. Generate a proposal $x' \sim \mathrm{N}(x, C_n)$.
2. Accept with probability

$$\alpha(x, x') = \min\left[1, \frac{\pi(x'|d)}{\pi(x|d)}\right]$$

   setting $x_{n+1} = x'$. Otherwise reject by setting $x_{n+1} = x$.
3. Update the proposal covariance by $C_{n+1} = s_d \operatorname{Cov}(x_1, x_2, ..., x_{n+1})$.

The covariance here is scaled down with the parameter s_d with $1/d$ dependence on the dimension d. The adaptation naturally can be started only when there are enough different accepted samples in the chain to compute the covariance. This may be a drawback if the initial proposal is too large; see below the discussion on the DRAM version for a remedy. Also, it may be better, especially in higher-dimensional problems, to keep adapting at some fixed intervals rather than at every step. Note also that the ergodicity does not require of use the *whole* chain but an *increasing* part of it, e.g. the last half.

31.4.6 Cheap MCMC tricks

We now present several other advances to the MHG algorithm that we have developed in response to particular inverse problems. These represent the state-of-the-art for sampling in inverse problems.

31.4.6.1 *Delayed rejection AM*

The delayed rejection (DR) method [16] uses several proposals: when a proposed candidate point in a Metropolis–Hastings chain is rejected, a second stage move is proposed from another proposal distribution. For example, one can use downscaled versions of a 'basic' proposal, with the motive to get acceptance after rejection. Delayed rejection can be combined with AM, as done in [20]. This method (DRAM) has been shown to be efficient in many applications, see e.g. [43]. It is helpful to get the sampler moving, especially in the beginning of the MCMC run, since AM can easily correct a proposal that is too small, but needs accepted points for the adaptation to take place. The DR step can provide such points. In a computationally demanding situation, such as the parameter tuning of a climate model, no standard ways (i.e. preliminary parameter fitting together with the Jacobian-based approximation for the covariance) of getting an initial proposal may be available. In addition, only short chains may be simulated. In such a case, DRAM typically turned out to be a reliable approach to get a reasonably well mixed chain created.

The adaptation in DRAM could be performed in various ways. We have found it enough to keep it simple: only have two proposals, compute the empirical covariance from the chains just as in AM, and keep an identical but down-scaled version of it for the second stage proposal.

31.4.6.2 *Parallel adaptive chains*

Parallelizing the adaptive MCMC algorithms has been studied relatively little. In [4] a parallel MCMC implementation in the context of regeneration was studied. Combining parallel computing and MCMC is inherently difficult, since MCMC is serial by nature. Running many parallel chains independent of each other may not be satisfactory, since it takes time for each single chain to find the mode(s) of the target and for the proposal to adapt. The question whether it is better to run multiple (non-adaptive) short chains or a single long chain has been considered in many studies. In the present case with extremely time-consuming calculations, this question is not relevant, since running a single long chain is simply not possible. Instead, several short chains can be run, and parallel communicating adaptive chains can speed up the mixing of the MCMC chains considerably. For this purpose, we employ a parallel chain version of the AM algorithm. To parallelize AM, we use a simple mechanism called inter-chain adaptation, recently introduced in [8]. In inter-chain adaptation one uses the samples generated by all parallel chains to perform proposal adaptation and the resulting proposal is used for all the chains. This naturally means that one has more points for adaptation and the convergence of every individual MCMC chain is expected to speed up.

The parallel chain approach is rather straightforward to implement. The only difference to running independent parallel AM samplers is that each sampler uses and updates the same joint proposal covariance. Covariance updating can be performed at any given update interval, for instance using the rank-1 covariance update formulas, see [21]. Note that also more advanced adaptation schemes, such as the DRAM and SCAM methods discussed above, can easily be combined with the inter-chain adaptation.

31.4.6.3 *Early rejection*

CPU can also be saved at no cost just by looking closer at the steps of calculations. Suppose the current state in the MH algorithm is x_i. Recall that MH proceeds by proposing a candidate value x' and accepting the proposed value with probability $\alpha = \min(1, \pi(x'|d)/\pi(x)|d)$. In practice, one first evaluates $\pi(x'|d)$, then simulates a uniform random number $u \sim U(0,1)$ and accepts x' if $u < \alpha$. Thus, a point will be rejected if $u > \pi(x'|d)/\pi(x|d)$.

In numerous applications the likelihood can be divided into n independent parts $\pi(d_i|x)$, $i = 1, 2, \ldots, n$. Moreover, the partial unnormalized posterior densities $\tilde{\pi}_k(x|d) = \pi(x)\prod_{i=1}^{k}\pi(d_i|x)$ may be monotonically decreasing with respect to the index k, $k = 1, 2, \ldots, n$. This is the situation, for example, if the likelihood has an exponential form $\pi(d|x) \propto \exp(-l(d|x))$, with $l(d_i|x) \geq 0$, as in the Gaussian case. In these situations, we can reject as soon as $\tilde{\pi}_k(x'|d)/\pi(x|d) < u$ for some value of k. Thus, we can speed up the sampling simply by switching the order of the calculations: generate the random number u first, evaluate the likelihood part by part, and check after each evaluation, if the proposed value will end up being rejected. Naturally, before evaluating any likelihood terms, we can check if the proposed point will be rejected based on the prior only.

The amount of calculation saved by ER depends on the problem (amount of data, properties of the model, shape of the posterior distribution) and on the tuning of the proposal. In cases where the topology of the posterior distribution is complicated (strongly nonlinear, thin 'bananas', or multi-modal), the MH sampler, even if properly tuned, results in low acceptance rates and potentially large performance gains can be achieved through ER. The same is true if the initial proposal covariance is too large: many points are rejected and ER is beneficial again. We have found that this 'cheap trick' may save computational time between around 10% and 80%. In cases with well-posed Gaussian-type posteriors the benefit is lowest. However, these are the situations for which MCMC is not even needed in the first place, as the classical linearization-based Fisher information matrix approach already works quite well.

31.4.6.4 *Delayed acceptance*

The delayed acceptance Metropolis–Hastings [6] (DAMH) algorithm improves computational efficiency of MCMC sampling by taking advantage of approximations to the forward map that are available in many inverse problems. The approximation to the forward map is used to evaluate a computationally fast approximation $\pi_x^*(\cdot)$ to the desired target distribution $\pi(\cdot|d)$, that can depend on the current state x.

Given a proposal drawn from the distribution $q(x, y)$, DAMH first 'tests' the proposal with the approximation $\pi_x^*(y)$ to create a modified proposal distribution $q^*(x, y)$ that is used in a standard MH. DAMH gains computational efficiency by avoiding calculation of $\pi(y|d)$ for poor proposals that are rejected by $\pi_x^*(y)$. One iteration of DAMH is given by:

Algorithm 3 (DAMH)

At step n, with state $x_n = x$, determine x_{n+1} in the following way:

1. Generate a proposal y from $q(x, \cdot)$.
2. When $x \neq y$, with probability
$$\alpha(x, y) = \min\left[1, \frac{\pi_x^*(y)q(y, x)}{\pi_x^*(x)q(x, y)}\right]$$
continue to step 3. Otherwise reject by setting $x_{n+1} = x$ and exit.
3. With probability
$$\beta(x, y) = \min\left[1, \frac{\pi(y|d)q^*(y, x)}{\pi(x|d)q^*(x, y)}\right]$$
accept y setting $x_{n+1} = y$, where $q^*(x, y) = \alpha(x, y)q(x, y)$. Otherwise reject y setting $x_{n+1} = x$.

For a state-dependent approximation we can assume that the approximation is exact when evaluated at the current state, i.e., $\pi_x^*(x) = \pi(x|d)$. Then the second acceptance probability can be simplified to

$$\beta(x, y) = \min\left[1, \frac{\min\left\{\pi(y|d)q(y, x), \pi_y^*(x)q(x, y)\right\}}{\min\left\{\pi(x|d)q(x, y), \pi_x^*(y)q(y, x)\right\}}\right] \tag{31.17}$$

If the approximation does not depend on the current state, we write $\pi^*(\cdot)$ in place of $\pi_x^*(\cdot)$ and the second acceptance probability simplifies to

$$\beta(x, y) = \min\left[1, \frac{\pi(y|d)\pi^*(x)}{\pi(x|d)\pi^*(y)}\right] \tag{31.18}$$

which is exactly the *surrogate transition method* introduced by Liu [33].

DAMH necessarily reduces statistical efficiency, but a good approximation will produce $\beta(x, y) \approx 1$ ([6] Theorem 2) and can increase computational efficiency by up to the inverse of the acceptance ratio. Christen and Fox gave an example in electrical impedance tomography (EIT) using the local linear approximation

$$A_x^*(x + \Delta x) = A(x) + J\Delta x,$$

where J is the Jacobian of A evaluated at state x, that improved computational efficiency by a factor of 25.

31.4.6.5 *Adaptive approximation error*

One way to construct an approximation is to directly replace the forward model A by a reduced-order model (ROM) A^* in evaluating the likelihood function in eqn (31.12). With forward problems that are induced by PDEs, the most obvious approach is to use coarse meshes. These induce a global, or state-independent, approximation. However, as we will see, a substantial improvement in efficiency is achieved by using a local correction that leads to a state-dependent approximation.

Not accounting for model reduction error in eqn (31.15) can give poor results. For example, in an inverse problem in geothermal reservoir modelling [10], we found that simply using a coarse model for A^* in place of A achieved only 17% acceptance in step 3 of DAMH. The reduction in statistical efficiency, by about a factor of 5, nullified any potential gain in computational efficiency.

Kaipio and Somersalo [30] estimated the mean μ_B and covariance Σ_B of the AEM off-line by drawing M samples from the prior distribution over x and used the sample mean and covariance of $\{A(x_i) - A^*(x_i)\}_{i=1}^M$. This AEM will be accurate over the support of the prior distribution, but will not necessarily be accurate over the posterior distribution. Instead, Cui *et al.* [10, 11] constructed the AEM over the posterior distribution adaptively, within the DAMH algorithm. Using this adaptive AEM, and a local correction explained next, resulted in an increase of the second acceptance ratio from 17%, quoted above, to 95%; so the stochastically corrected approximation is effectively perfect.

When implementing a state-independent ROM within DAMH, we have found it is always advantageous to make the zeroth-order local correction

$$A_x^*(y) = A^*(y) + \left[A(x) - A^*(x)\right]$$

which has virtually no computational cost since both $A(x)$ and $A^*(x)$ have been computed when at state x. The resulting approximation $A_x^*(\cdot)$ now depends on the state x, so DAMH is required in eqn (31.17), rather than surrogate transition eqn (31.18). This corrected approximation has the property that AEM has mean of zero [11] and hence the adaptive AEM converges to a zero mean Gaussian. We find in practice that simply setting the mean to zero in the adaptive algorithm gives best results.

One step of the resulting adaptive delayed acceptance Metropolis–Hastings (ADAMH) algorithm is:

Algorithm 4 (ADAMH)

At step n, with state $x_n = x$, approximate target distribution $\pi^*_{x,n}(\cdot)$, and proposal distribution $q_n(x, \cdot)$, determine x_{n+1} and updated distributions in the following way:

1. Generate a proposal y from $q_n(x, \cdot)$.
2. When $x \neq y$, with probability

$$\alpha(x,y) = \min\left[1, \frac{\pi^*_{x,n}(y) q_n(y,x)}{\pi^*_{x,n}(x) q_n(x,y)}\right]$$

   continue to step 3. Otherwise reject by setting $x_{n+1} = x$ and goto step 4.
3. With probability

$$\beta(x,y) = \min\left[1, \frac{\pi(y|d) q^*_n(y,x)}{\pi(x|d) q^*_n(x,y)}\right]$$

   accept y setting $x_{n+1} = y$, where $q^*_n(x,y) = \alpha(x,y) q_n(x,y)$. Otherwise reject y setting $x_{n+1} = x$.
4. Update the AEM covariance by

$$\Sigma_{B,n+1} = \frac{1}{n}\left[(n-1)\,\Sigma_{B,n} + \left[A(x_{n+1}) - A^*_x(x_{n+1})\right]\left[A(x_{n+1}) - A^*_x(x_{n+1})\right]^T\right].$$

5. Update the proposal to $q_{n+1}(x_{n+1}, \cdot)$.

Using this algorithm, Cui *et al.* [11] increased computational efficiency by a factor of 8 in a large-scale nonlinear inverse problem in geothermal modelling with 10^4 continuous unknowns. This reduced computing time from 8 months to 1 month, which is significant. Actually, the performance of ADAMH in that example was remarkable, drawing each *independent* sample from the correct posterior distribution at a cost of only 25 evaluations of the accurate model.

We have not yet given the form of the proposal, yet the choice of proposal distribution is critical in achieving computational feasibility, as with any MH MCMC. While adaptation can remove the need for tuning of proposals, choosing the *structure* of the proposal to adapt to remains something of an art. In high dimensional inverse problems neither of the extremes of single-component proposals (e.g. SCAM) or global proposals (e.g. AM) is optimal; see e.g. [11] for a discussion on this point, and [24] for a demonstration of the failure of AM. Instead, proposing block updates over highly correlated sets of variables, as in [11], can be very effective, although requires some exploration to find a suitable blocking scheme.

31.5 Future directions

Alan Sokal introduced his lecture notes on Monte Carlo methods [40] with the warning,

> Monte Carlo is an extremely bad method; it should be used only when all alternative methods are worse.

We wholeheartedly agree, and add that in practice the situation can be desperate, when we have no decent proposal distribution. Adaptive MCMC methods are useful here by automatically tuning proposals, but even they can never exceed the performance with an optimal proposal. However, sometimes one must sin[43] when there is no alternative route to solving a problem, and we do so nowadays routinely for large classes of models. This leaves a pressing need to improve MCMC sampling.

There are now many options for performing MCMC sampling such as the random-walk MH, hybrid Monte Carlo, proposals based on Langevin diffusions, and many others. A significant issue in inverse problems is not just to rely on algorithms that are provably convergent, but to make sensible algorithmic choices in terms of computational efficiency, and particularly how the algorithm cost scales with problem size.

We expect that lessons learned in computational optimization will be valuable for future improvements in MCMC. In that field many sophisticated algorithms have been developed such as the Krylov space methods that go by the acronyms PCG, Bi-CGSTAB, and GMRES, and the quasi-Newton methods including BFGS. These optimizers navigate high-dimensional surfaces with minimal need to evaluate a complex function, which is a requirement shared by efficient MCMC for inverse problems. There are already sampling algorithms that use these ideas. In [2] LBFGS optimization is used to construct approximate filtering in state spaces that are too high dimensional for the usual extended Kalman filtering. The same approach has been tested for ensemble filtering, and provides a way to high-dimensional MC sampling, without MCMC. The CG sampling algorithm for Gaussian distributions presented in 2001 by Schneider and Willsky, was improved in [39] and characterized for finite precision calculations. The observation that Gibbs samplers are essentially identical to stationary iterative linear solvers that are now considered very slow (see [39] for references) provides a perspective on MCMC in relation to linear solvers, and points towards fundamental improvements.

These algorithms hold the promise of drawing independent samples with the same computational cost as the optimization required for regularized solution. While that would be a dramatic improvement over the current situation, even then the reality is that sample-based inference will only become routine in engineering if the entire cost is no more than a few times the cost of optimization. That means, even with such improvements, that for the foreseeable future 'solutions' will need to be based on at most a handful of samples drawn from the posterior distribution.

If we set aside the goal of accurate estimates of errors on estimates, and set the more modest goal of improving on current practice in inverse problems, we have a chance. As argued in [13], a single sample drawn from the posterior distribution can be better than the regularized solution, in the sense of being more representative. One could then improve substantially by drawing a few samples, since that would at least give some indication of variability in solutions, while a few dozen samples would often be good enough to show the extent of posterior variability (although *which few dozen* might be difficult to determine). This is, especially, true if those few samples already are enough to verify the *negative* conclusion: that our unknown is far from being identified. We should keep in mind that in truly high-dimensional inverse problems the number of samples most likely remains far fewer than the dimension of the unknown, so any discussion on assured convergence of posterior estimates, in the usual sense, remains academic too.

References

[1] Antoulas, A. C. (2005). *Approximation of Large-Scale Dynamical Systems*. SIAM.

[2] Auvinen, H., Bardsley, J., Haario, H. and Kauranne, T. (2010). Variational Kalman filter and an efficient implementation using limited memory BFGS. *International Journal for Numerical Methods in Fluids*, **64**(3), 314–335.

[43] John von Neumann is quoted as saying: 'anyone using Monte Carlo is in a state of sin'.

[3] Birman, M. S. and Solomyak, M. Z. (1977). Estimates of singular numbers of integral operators. *Uspekhi Mat. Nauk*, **32**(1), 17–84. Engl. transl. in: Russian Math. Surveys **32**(1977), no. 1, 15–89.

[4] Brockwell, A. (2006). Parallel Markov chain Monte Carlo simulation by pre-fetching. *Journal of Computational and Graphical Statistics*, **15**(1), 246–260.

[5] Cappé, O., Moulines, E. and Rydén, T. (2005). *Inference in Hidden Markov Models*. Springer Series in Statistics. Springer.

[6] Christen, J. A. and Fox, C. (2005). Markov chain Monte Carlo using an approximation. *Journal of Computational and Graphical Statistics*, **14**(4), 795–810.

[7] Cox, R. T. (1961). *The Algebra of Probable Inference*. Johns Hopkins.

[8] Craiu, R. V., Rosenthal, J. and Yang, C. (2009). Learn from thy neighbor: Parallel-chain and regional adaptive MCMC. *Journal of the American Statistical Association*, **104**(488), 1454–1460.

[9] Cramér, H. (1946). *Mathematical Methods of Statistics* (First US edn). Princeton University Press.

[10] Cui, T., Fox, C. and O'Sullivan, M. J. (2011). Adaptive error modelling in MCMC sampling for large scale inverse problems. Technical Report no. 687, University of Auckland, Faculty of Engineering.

[11] Cui, T., Fox, C. and O'Sullivan, M. J. (2011). Bayesian calibration of a large scale geothermal reservoir model by a new adaptive delayed acceptance Metropolis-Hastings algorithm. *Water Resources Research*, **47**. 26 pp.

[12] Donoho, D. L., Johnstone, I. M., Hoch, J. C. and Stern, A. S. (1992). Maximum entropy and the nearly black object. *Journal of the Royal Statistical Society. Series B*, **54**, 41–81.

[13] Fox, C. (2008). Recent advances in inferential solutions to inverse problems. *Inverse Problems Sci. Eng.*, **16**(6), 797–810.

[14] Fox, C. and Nicholls, G. K. (1997). Sampling conductivity images via MCMC. In *The Art and Science of Bayesian Image Analysis* (ed. K. Mardia, R. Ackroyd, and C. Gills), pp. 91–100. Leeds Annual Statistics Research Workshop: University of Leeds.

[15] Golub, G. H., Heath, M. and Wahba, G. (1979). Generalized cross-validation as a method for choosing a good ridge parameter. *Technometrics*, **21**(2), 215–223.

[16] Green, P. J. and Mira, A. (2001). Delayed rejection in reversible jump Metropolis-Hastings. *Biometrika*, **88**, 1035–1053.

[17] Grenander, U. and Miller, M. (1994). Representations of knowledge in complex systems. *Journal of the Royal Statistical Society. Series B*, **56**(4), 549–603.

[18] Gull, S. F. and Daniell, G. J. (1978). Image reconstruction from incomplete and noisy data. *Nature*, **272**(5655), 686–690.

[19] Haario, H., Kalachev, L. and Laine, M. (2009). Reduced models for algae growth. *Bulletin of Math. Biology*, **71**(7), 1626–1648.

[20] Haario, H., Laine, M., Mira, A. and Saksman, E. (2006). DRAM: Efficient adaptive MCMC. *Statistics and Computing*, **16**(3), 339–354.

[21] Haario, H., Saksman, E. and Tamminen, J. (2001). An adaptive Metropolis algorithm. *Bernoulli*, **7**(2), 223–242.

[22] Hansen, P. C. (1998). *Rank-Deficient and Discrete Ill-Posed Problems. Numerical Aspects of Linear Inversion*. SIAM.

[23] Hastings, W. K. (1970). Monte Carlo sampling methods using Markov chains and their applications. *Biometrika*, **57**(1), 97–109.

[24] Higdon, D., Reese, C. S., Moulton, J. D., Vrugt, J. A. and Fox, C. (2011). Posterior exploration for computationally intensive forward models. In *Handbook of Markov Chain Monte Carlo* (ed. S. Brooks, A. Gelman, G. Jones, and X.-L. Meng), pp. 401–418. Chapman & Hall/CRC.

[25] Howson, C. and Urbach, P. (2005). *Scientific Reasoning: The Bayesian Approach* (3 edn). Open Court.
[26] Hurn, M. A., Husby, O. and Rue, H. (2003). Advances in Bayesian image analysis. In *Highly Structured Stochastic Systems* (ed. P. J. Green, N. Hjort and S. Richardson), pp. 302–322. Oxford: Oxford University Press.
[27] Jaynes, E. T. (1973). The well-posed problem. *Foundations of Physics*, **3**(4), 477–492.
[28] Jaynes, E. T. (1978). Where do we stand on maximun entropy? In *Maximum Entropy Formalism* (ed. R. D. Levine and M. Tribus), p. 16. MIT Press.
[29] Kaipio, J. and Fox, C. (2011). The Bayesian framework for inverse problems in heat transfer. *Heat Transfer Engineering*, **32**(9), 718–753.
[30] Kaipio, J. and Somersalo, E. (2007). Statistical inverse problems: discretization, model reduction and inverse crimes. *J Comput Appl Math*, **198**, 493–504.
[31] Kennedy, M. C. and O'Hagan, A. (2001). Bayesian calibration of computer models (with discussion). *Journal of the Royal Statistical Society: Series B*, **63**, 425–464.
[32] Keynes, J. M. (1921). *A Treatise on Probability*. Macmillan and Co.
[33] Liu, J. S. (2005). *Monte Carlo Strategies in Scientific Computing*. Springer.
[34] Mosegaard, K., Singh, S. C., Snyder, D. and Wagner, H. (1997). Monte Carlo analysis of seismic reflections from Moho and the W-reflector. *Journal of Geophysical Research B*, **102**, 2969–2981.
[35] Müller, P. (1999). Simulation-based optimal design. In *Bayesian Statistics 6* (ed. J. M. Bernardo, J. O. Berger, A. P. Dawid, and A. F. M. Smith), pp. 459–474. Oxford University Press. **6**, 459–474.
[36] Nicholls, G. and Jones, M. (2001). Radiocarbon dating with temporal order constraints. *Journal of the Royal Statistical Society. Series C (Applied Statistics)*, **50**, 503–521.
[37] Oliver, D. S., Cunha, L. B. and Reynolds, A. C. (1997). Markov chain Monte Carlo methods for conditioning a permeability field to pressure data. *Mathematical Geology*, **29**(1), 61–91.
[38] O'Sullivan, F. (1986). A statistical perspective on ill-posed inverse problems. *Statistical Science*, **1**(4), 502–527.
[39] Parker, A. and Fox, C. (2011). Sampling Gaussian distributions in Krylov spaces with conjugate gradients. *SIAM Journal on Scientific Computing*. In the press.
[40] Sokal, A. D. (1996). Monte Carlo methods in statistical mechanics: Foundations and new algorithms. In *Lectures at the Cargése summer school on 'Functional Integration: Basics and Applications'*.
[41] Tarantola, A. (1987). *Inverse Problem Theory: Methods for Data Fitting and Model Parameter Estimation*. Elsevier.
[42] Tikhonov, A. N. and Arsenin, V. Y. (1977). *Solutions of Ill-posed Problems*. Scripta series in mathematics. Winston.
[43] Villagran, A., Huerta, G., Jackson, C. S. and Sen, M. K. (2008). Computational methods for parameter estimation in climate models. *Bayesian Analysis*, **3**(3), 1–27.
[44] Watzenig, D. and Fox, C. (2009). A review of statistical modelling and inference for electrical capacitance tomography. *Measurement Science and Technology*, **20**(5), 22 pp.
[45] Young, N. (1998). *An Introduction to Hilbert Space*. Cambridge University Press.

32 Approximate marginalization over modelling errors and uncertainties in inverse problems

JARI KAIPIO AND VILLE KOLEHMAINEN

In this chapter, we discuss Bayesian modelling of errors that are induced by model uncertainties and practical implementational constraints. Our focus is on problems that are severely resource limited with respect to computational power, memory and time. This is the common case in industrial and biomedical problems in general, and often also in other problems, in which a continuous stream of data is to be processed. Such problems suggest employing highly approximative reduced order models, and often only some of the actual unknowns are of interest. Furthermore, the uncertainties may also be related to such information which traditionally has been considered to be mandatory, for example, the boundary data in problems governed by partial differential equations (PDE).

To put the computational problem in scale, consider the following. In industrial imaging of flows using electromagnetic fields such as electrical impedance tomography, the domain and the unknown(s) are usually modelled as (low level) three-dimensional random fields [23, 49, 60]. Traditional accuracy requirements for the forward model suggest using discretized approximations for the associated PDEs typically with ~1 000–100 000 unknowns. The time frame for the computation of the state, on the other hand, may be of the order of a millisecond. Since the estimates are eventually to be used in (optimal stochastic) control, feasible error estimates (posterior covariance) are also needed [10]. Furthermore, the computations are usually to be computed using standard industrial computers, with processing power comparable to somewhat outdated personal computers.

Such constraints exclude the possibility of (and usually the need for) accurate Bayesian inference. Furthermore, the problems are typically nonlinear, and even the computation of the MAP estimate and highly approximate posterior covariance as spread estimate can be a formidable task. What needs to be done, is to use approximative (simplified) physical models and crude computational approximations, and to neglect many of the actual unknowns. With inverse problems, which can loosely be defined as problems that tolerate measurement and modelling errors poorly, straightforward adherence to such constraints heralds a disaster [6, 24]. Nevertheless, something has to be done to provide feasible estimates.

The approach we take here, is to construct (prior) models for all unknowns and compute a crude likelihood model over the modelling errors that are induced by numerical model reduction and uncertainty in the 'uninteresting' unknowns. The modelling errors are modelled as additive errors in the likelihood model, and thus they can be formally marginalized before the inference. In the sequel, numerical approximation and all modelling errors are referred to as approximation errors.

The rest of the chapter is organized as follows. In Section 32.1, we discuss the characteristics of inverse problems in the Bayesian framework and give a brief review of earlier application of the approximation error approach. In Section 32.2, we give the basic formulation of the approximation error approach, and in Sections 32.3 and 32.4, we discuss general implementational aspects and the high-dimensional observation cases, respectively. In Section 32.5, we discuss the local X-ray tomography problem, in which only a (small) part of the spatial mass attenuation distribution is estimated, and relatively few measurements are acquired. This leads to an unidentifiable problem, with respect to the traditional deterministic framework for inverse problems. This can be modelled as a linear problem with normal posterior model. In Section 32.6, we consider the electrical impedance tomography problem with unknown boundary geometry. We show that the Bayesian approximation error approach facilitates the computation of feasible estimates despite the inaccurately known geometry and, furthermore, also yields feasible approximations for the low-level uncertainties.

32.1 Inverse problems as Bayesian inference problems

In addition to Chapter 31 in this volume, we refer the reader to [6, 24, 55] for treatises of inverse problems in the Bayesian context. Inverse problems are a particular class of Bayesian inference problems due to the following considerations. Firstly, the representations for the unknowns are typically relatively high dimensional. In fact, it is common that the number of unknowns exceeds the number of measurements, sometimes by a factor of ten or more. Secondly, inverse problems tolerate poorly (measurement and modelling) errors, and thus only such problems are usually considered in which the measurements (given an accurate model) are relatively accurate. This means that the likelihoods are typically nonlinear manifolds with very small variances, that is, $\mathrm{std}\langle b, d\rangle$ are small compared to $\mathbb{E}\langle b, d\rangle$, where $d \in \mathbb{R}^m$ is the measurement, $b \in \mathbb{R}^m$ is any vector with $\|b\| = 1$.

Distributed parameter estimation problems induced by partial differential equations and the related initial-boundary value problems constitute perhaps the largest class of inverse problems. Such models always provide a more or less simplified approximation for the physical reality. In particular, the following modelling problems are common:

- Geometry is unknown (biomedical imaging) [16, 17, 33, 45, 46]
- Measurement sensor locations are only approximately known (geophysics, some biomedical imaging modalities) [43]
- Measurement noise statistics are poorly known (most applications) [39, 55]
- Approximate physical measurement models are used (biomedical, geophysical and industrial applications) [34, 56]
- Significant uncertainty in models for the unknown variables (geophysics, industrial flows) [29, 34, 38]
- Boundary conditions are partially unknown (biomedical, geophysics and industrial applications) [33, 36, 49, 59]

As a result, it is almost inevitable that the modelling errors dominate the (effects of the) measurement error. We will be more precise regarding this claim in Section 32.2.

With a fixed parameterization $x = P\bar{x}$ for an unknown physical variable $\bar{x}$, where P is typically a projection operator, the best we can hope for a particular point estimate is $\hat{x} = P\bar{x}$. In terms of probability mass, we can loosely say that unaccounted-for modelling errors move the likelihood and thus posterior probability mass away from the projection $P\bar{x}$. In this chapter, we refer to having $\pi(P\bar{x}|d) \approx 0$ as an infeasible posterior model. In Figure 32.1, we show an example of a posterior marginal density and the actual variable in the case of local X-ray tomography, which is considered in Section 32.5. This is typical behaviour for inverse problems, and gives a probabilistic interpretation for the notion of poor tolerance to modelling errors. In the local tomography problem, the likelihood is normal and the infeasibility of the posterior model is entirely due to the modelling errors in the likelihood. In the case of inverse problems, carrying out proper inference with MCMC methods makes no sense without highly accurate likelihood models.

The Bayesian framework for inverse problems is a natural one when modelling and other uncertainties are to be considered. Sometimes these uncertainties can be parameterized by a small number of unknowns, which suggests using hyperprior models. In the majority of cases, however, the model errors and uncertainties do not allow for a feasible small-dimensional hyperprior model. One example of such cases is that of unknown boundary geometry, which is a common problem in biomedical diffuse tomography. Here, if the geometry were known, the straightforward approach would be to construct the (say, finite element) meshes of the PDE solver to match the geometry. But if the geometry is changing between the measurements (due to breathing, for example), the finite element matrices would have to be recomputed. We are saved from this typically infeasible computational problem by the fact that there is no clinically available measurement method to obtain

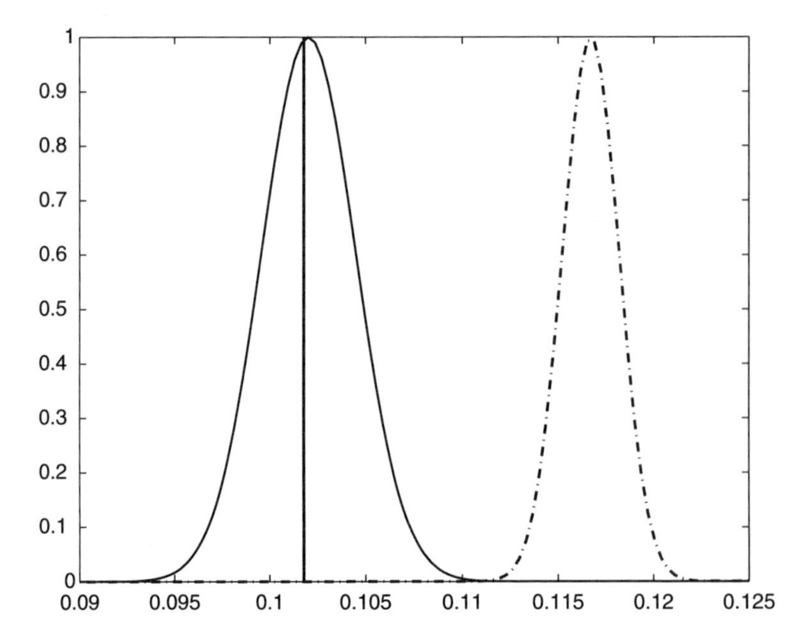

Figure 32.1 A posterior marginal density when the approximation and modelling errors have been modelled (solid line), have not been modelled (dash-dot line); and the actual value of a pixel in a local tomography problem. This infeasibility of the posterior estimates is typical for inverse problems if the approximation and modelling errors are ignored.

the geometry of the body. Thus, we are left with the uncertainty of the boundary geometry. The bad news is that even a relatively small error in boundary geometry usually renders the estimates completely meaningless. We address this problem in Section 32.6.

As explained above, the practical constraints will typically allow a few evaluations of the likelihood and prior only, and the associated models have to be such that even these computations cannot be based on accurate physical models. A further complication is provided by the fact that many inverse problems are time-varying, and the proper formalism is that of sequential Bayesian inference [9]. But again, particle filters are typically out of question, and the best one can do, is to construct a feasible model for observation and state noise processes, and use some form of extended Kalman filters [18–21].

All the above suggests that the actual estimation stage has to be more or less 'quick and dirty'. But this does not usually exclude the possibility of carrying out extensive offline precomputations, possibly with very accurate models. In this chapter, we describe the *Bayesian approximation error approach* [24, 25] which was originally introduced to cope with numerical model reduction in PDE induced distributed parameter estimation problems. The approach is based on constructing a (prior) model for all low-level uncertainties, carrying out simulations with the most accurate available model, and the model that is to be used in the estimation stage. The difference of the prediction of these two models is called the approximation error and is then treated as a (further) additive error term, usually in addition to the measurement errors. A normal approximation to the joint distribution for the primary unknown and the approximation error is constructed and the additive random variables are marginalized before the actual inference. Again, the best that can usually be done within the time and resource constraints in practical applications, is to compute the MAP estimate and an approximate posterior covariance.

As noted above, the approximation error approach was introduced in [24, 25] originally to handle pure model reduction errors. For example, in electrical impedance (resistance) tomography (EIT, ERT) and deconvolution problems, it was shown that significant model reduction is possible without essentially sacrificing the feasibility of estimates. With EIT, for example, this means that very low-dimensional finite element approximations can be used. Later, the approach has also been applied to handle other kinds of approximation and modelling errors as well as other inverse problems: model reduction, domain truncation and unknown anisotropy structures in diffuse optical tomography were treated in [2, 16, 17, 27]. Missing boundary data in the case of image processing and geophysical ERT/EIT were considered in [5] and [33], respectively. Furthermore, in [42, 44] the problem of recovery from simultaneous domain truncation and model reduction was found to be possible, and in [45, 46] recovery from the errors related to inaccurately known body shape was shown feasible.

The approximation error approach was extended to non-stationary inverse problems in [19] in which linear non-stationary (heat transfer) problems were considered, and in [18] and [20] in which nonlinear problems and state space identification problems were considered, respectively. The earliest similar but partial treatment in the framework on non-stationary inverse problems was considered in [50] in which the boundary data that is related to stochastic convection diffusion models was partially unknown. A modification in which the approximation error statistics can be updated with accumulating information was proposed in [21] and an application to hydrogeophysical monitoring in [35].

From pure model reduction and unknown parameters or boundary data, a step forwards was recently considered in [56] in which the physical forward model itself was replaced with a (computationally) much simpler model. In [56], the radiative transfer model (Boltzmann transfer equation) which is considered to be the most accurate PDE model for light transfer in (turbid) media, was replaced with the diffusion approximation. It was found that in this kind of case, the statistical structure of the approximation errors enabled the use of a significantly less complex model, again simultaneously with significant model reduction for the diffusion approximation. But here also, both the absorption and scattering coefficients were estimated simultaneously, while in [30], the scattering coefficient was modelled as a normal Markov random field and was not estimated.

The approximation error approach relies on the Bayesian framework of inverse problems, in which all unknowns are explicitly modelled as random variables [6, 24, 55]. The uncertainty in the unknowns given the models and measurements is reflected in the posterior (probability) distribution. In the Bayesian framework, all unknowns are subject to inference simultaneously, which often results in excessively heavy computational loads. Generally, Markov chain Monte Carlo algorithms have to be used to obtain a representative set of samples from the posterior distribution. Then, after a set of samples has been computed, marginalization over the uninteresting unknowns is trivial. Only in few special but important cases, such as the additive error model, some of the uninteresting unknowns can be eliminated before inference. We refer to such elimination as pre-marginalization.

For Markov chain Monte Carlo inference of inverse problems, see Chapter 31 and the references therein in this volume.

32.2 Approximate marginalization over modelling errors

32.2.1 Marginalization over additive errors

In the approximation error approach, the modelling and other errors as well as uncertainties are propagated to additive errors. With uncertainties, we refer to the random variables (fields etc) that are not estimated, that is, written as variables in the posterior model.

Therefore, we review briefly how the additive errors are formally pre-marginalized [24, 30] and introduce the basic notation. Let the observation model be

$$d = \bar{A}(x) + e \tag{32.1}$$

where e are additive errors and $x \mapsto \bar{A}(x)$ is a deterministic forward model. With deterministic we mean that the model $\bar{A}$ does not contain any uncertainties or other model errors. Let the joint prior model of the unknowns x and e be $\pi(x, e)$. Using the Bayes' theorem repeatedly, the joint distribution of all associated random variables can be decomposed as

$$\pi(d, x, e) = \pi(d \mid x, e)\pi(e \mid x)\pi(x) \tag{32.2}$$

$$= \pi(d, e \mid x)\pi(x) \tag{32.3}$$

In the case that both x and e are fixed, the measurement d in the model (32.1) is completely specified, so the conditional distribution $\pi(d \mid x, e)$ is formally given by

$$\pi(d \mid x, e) = \delta(d - \bar{A}(x) - e) \tag{32.4}$$

where δ is the Dirac delta distribution. Using (32.2)–(32.4), we get the likelihood model

$$\pi(d \mid x) = \int \pi(d, e \mid x)\, \mathrm{d}e = \tag{32.5}$$

$$= \int \delta(d - \bar{A}(x) - e)\pi(e \mid x)\, \mathrm{d}e \tag{32.6}$$

$$= \pi_{e \mid x}(d - \bar{A}(x) \mid x) \tag{32.7}$$

and further, noting that once the measurements have been obtained, $\pi(d) > 0$ is a fixed normalization constant, we have the posterior model

$$\pi(x \mid d) \propto \pi(d \mid x)\pi(x) \tag{32.8}$$

$$= \pi_{e \mid x}(d - \bar{A}(x) \mid x)\pi(x) \tag{32.9}$$

In the quite common case of mutually independent x and e, we have $\pi_{e \mid x}(e \mid x) = \pi_e(e)$. Furthermore, if e and x are normal, we can write $\pi(e) = \mathcal{N}(e_*, \Gamma_e)$ and $\pi(x) = \mathcal{N}(x_*, \Gamma_x)$ and we have the form

$$\pi(x \mid d) \propto \exp\left(-\frac{1}{2}\left(\|L_e(d - \bar{A}(x) - e_*)\|^2 + \|L_x(x - x_*)\|^2\right)\right), \tag{32.10}$$

where $L_e^{\mathrm{T}} L_e = \Gamma_e^{-1}$ and $L_x^{\mathrm{T}} L_x = \Gamma_x^{-1}$, for the posterior model. The MAP estimate for the model (32.10) is obtained by

$$\hat{x} = \arg\min_x \left\{\|L_e(d - \bar{A}(x) - e_*)\|^2 + \|L_x(x - x_*)\|^2\right\} \tag{32.11}$$

An approximate covariance estimate is obtained as

$$\hat{\Gamma}_{x|d} = \left(J^{\mathrm{T}}\Gamma_e^{-1}J + \Gamma_x^{-1}\right)^{-1} \tag{32.12}$$

where the Jacobian matrix J of the mapping A is computed at the MAP estimate.

In the above, the unknown (uninteresting) additive error e was *pre-marginalized*, that is, marginalized before the inference procedure, and e is not present in (32.9) or (32.10).

32.2.2 Approximate pre-marginalization over model reduction related errors and other uncertainties

The problem that is generally related to uninteresting auxiliary unknowns ξ is that we usually cannot perform pre-marginalization such as in (32.6–32.7). In most cases, we have to estimate both x and ξ which may be a considerably more demanding undertaking than estimating just x if ξ was known. For example, if a Markov chain Monte Carlo (MCMC) approach were used, the marginalization over ξ can only be done after running the chain for both x and ξ. Once this is carried out, however, the marginalization over ξ is trivial. For MCMC and inverse problems, see for example [12, 26, 54] for inference in the EIT problem.

In this section, we discuss the computational procedure in more detail in the case in which there are two distributed parameters of which pre-marginalization over the other parameter, together with the additive measurement errors and other uncertainties is to be carried out.

Let now the unknowns be (x, z, ξ, e) where again e represents additive errors while ξ represents auxiliary uncertainties such as unknown boundary data, and (x, z) are two distributed parameters of which only x is of interest. The accurate forward model

$$(x, z, \xi) \mapsto \bar{A}(x, z, \xi) \tag{32.13}$$

is usually a nonlinear one. The uncertainties ξ must sometimes be modelled as mutually dependent with (x, z), especially when ξ represents boundary data on the computational domain boundary and (x, z) are modelled as random fields. On the other hand, if ξ represents unknown boundary shape, ξ might be modelled as mutually independent with (x, z). In the following, we consider the case in which the noise e is additive and the unknowns (x, z, ξ) are not necessarily mutually independent.

Let

$$d = \bar{A}(\bar{x}, z, \xi) + e \in \mathbb{R}^m$$

denote an accurate model for the relation between the measurements and the unknowns[44] and let e be mutually independent with (x, z, ξ).

Below, we approximate the accurate representation of the primary unknown $\bar{x}$ by $x = P\bar{x}$ where P is typically a projection operator. Let $\pi(x, z, \xi, e)$ be a feasible model for the joint distribution of the unknowns. We identify $x = P\bar{x}$ with its coordinates in the associated basis.

In the approximation error approach, we proceed as follows. Instead of using the accurate forward model $(\bar{x}, z, \xi) \mapsto \bar{A}(\bar{x}, z, \xi)$ with $(\bar{x}, z, \xi)$ as the unknowns, we fix the random variables $(z, \xi) \leftarrow (z_0, \xi_0)$ and use a computationally (possibly drastically reduced) approximative model

$$x \mapsto A(x, z_0, \xi_0)$$

Here, the predictions of the two models $\bar{A}(\bar{x}, z, \xi)$ and $A(x, z_0, \xi_0)$ may be drastically different. Thus, we write the measurement model in the form

$$d = \bar{A}(\bar{x}, z, \xi) + e \tag{32.14}$$

$$= A(x, z_0, \xi_0) + \left(\bar{A}(\bar{x}, z, \xi) - A(x, z_0, \xi_0)\right) + e \tag{32.15}$$

$$= A(x, z_0, \xi_0) + \varepsilon + e \tag{32.16}$$

where we define the *approximation error* $\varepsilon = \varphi(\bar{x}, z, \xi) = \bar{A}(\bar{x}, z, \xi) - A(x, z_0, \xi_0)$. Thus, the approximation error is the discrepancy of predictions of the measurements (given the unknowns) when using the accurate model $\bar{A}(\bar{x}, z, \xi)$ and the approximate model $A(x, z_0, \xi_0)$. Note that (32.16) is exact.

Earlier, we referred to the problem of approximation error dominating the measurements errors. In the case of additive measurement errors, we can make this notion quantitative by considering

$$\|e_*\|^2 + \operatorname{trace} \Gamma_e < \|\varepsilon_*\|^2 + \operatorname{trace} \Gamma_\varepsilon \tag{32.17}$$

and

$$e_*(k)^2 + \Gamma_e(k, k) < \varepsilon_*(k)^2 + \Gamma_\varepsilon(k, k), \quad k = 1, \ldots, m \tag{32.18}$$

If (32.17) holds, we say that the approximation errors dominate the measurement errors. Neglecting the approximation errors from the likelihood model will usually result in completely meaningless posterior estimates. But if (32.18) holds for *any* k, this may also be the case. In Section 32.4, we decompose the approximation errors into two components, and (32.18) is used as the decomposition criterion.

Formally, after the models $\bar{A}$ and A are fixed, we have $\pi(\varepsilon \mid \bar{x}, z, \xi) = \delta(\varepsilon - \varphi(\bar{x}, z, \xi))$. We will later, however, employ approximative joint distributions and therefore consider $\pi(\varepsilon, \bar{x}, z, \xi)$ without any special structure. As the first approximation, we approximate $\varphi(\bar{x}, z, \xi) \approx \varphi(Px, z, \xi)$

44 If there are no additive errors, we write $e = 0$ and consider the other types of errors to be included in ξ.

and thus $\pi(\varepsilon \mid \bar{x}, z, \xi) \approx \pi(\varepsilon \mid Px, z, \xi)$. This means that we assume that the model predictions and thus the approximation error is essentially the same for $\bar{x}$ as $x = P\bar{x}$. This assumption holds for inverse problems in most cases. In very severe model reduction for the unknown, such as in the coloured polygon models [41], this may not be the case, and the projection approximation might be needed to be taken into account.

Proceeding as in Section 32.2.1, we use the Bayes' formula repeatedly

$$\begin{aligned}\pi(d, x, z, \xi, e, \varepsilon) &= \pi(d \mid x, z, \xi, e, \varepsilon)\pi(x, z, \xi, e, \varepsilon) \\ &= \delta(d - A(x, z_0, \xi_0) - e - \varepsilon) \\ &\qquad \pi(e, \varepsilon \mid x, z, \xi)\pi(z, \xi \mid x)\pi(x) \\ &= \pi(d, z, \xi, e, \varepsilon \mid x)\pi(x)\end{aligned}$$

Hence

$$\begin{aligned}\pi(d \mid x) &= \iiiint \pi(d, z, \xi, e, \varepsilon \mid x) de\, d\varepsilon\, dz\, d\xi \\ &= \iint \delta(d - A(x, z_0, \xi_0) - e - \varepsilon) \\ &\qquad \cdot \left[\iint \pi(e, \varepsilon \mid x, z, \xi)\pi(z, \xi \mid x) dz\, d\xi\right] de\, d\varepsilon \\ &= \iint \delta(d - A(x, z_0, \xi_0) - e - \varepsilon)\pi(e, \varepsilon|x) de\, d\varepsilon \\ &= \int \pi_e(d - A(x, z_0, \xi_0) - \varepsilon)\pi_{\varepsilon|x}(\varepsilon \mid x)\, d\varepsilon \qquad (32.19)\end{aligned}$$

since e and x are mutually independent, and (32.19) is a convolution integral with respect to ε. In particular, since e is mutually independent with (x, z, ξ), e and ε are also mutually independent.

At this stage, in the approximation error approach, both π_e and $\pi_{\varepsilon|x}$ are approximated with normal distributions. Let the normal approximation for the joint density $\pi(\varepsilon, x)$ be

$$\pi(\varepsilon, x) \propto \exp\left\{-\frac{1}{2}\begin{pmatrix}\varepsilon - \varepsilon_* \\ x - x_*\end{pmatrix}^{\mathrm{T}}\begin{pmatrix}\Gamma_\varepsilon & \Gamma_{\varepsilon x} \\ \Gamma_{x\varepsilon} & \Gamma_x\end{pmatrix}^{-1}\begin{pmatrix}\varepsilon - \varepsilon_* \\ x - x_*\end{pmatrix}\right\} \qquad (32.20)$$

Thus we write

$$e \sim \mathcal{N}(e_*, \Gamma_e), \qquad \varepsilon \mid x \sim \mathcal{N}(\varepsilon_{*,x}, \Gamma_{\varepsilon|x})$$

where

$$\varepsilon_{*,x} = \varepsilon_* + \Gamma_{\varepsilon x}\Gamma_x^{-1}(x - x_*) \qquad (32.21)$$

$$\Gamma_{\varepsilon|x} = \Gamma_\varepsilon - \Gamma_{\varepsilon x}\Gamma_x^{-1}\Gamma_{x\varepsilon} \qquad (32.22)$$

Define the normal random variable ν so that $\nu \mid x = e + \varepsilon \mid x$

$$\nu \mid x \sim \mathcal{N}(\nu_{*|x}, \Gamma_{\nu|x})$$

where

$$\nu_{*|x} = e_* + \varepsilon_* + \Gamma_{\varepsilon x}\Gamma_x^{-1}(x - x_*) \tag{32.23}$$

$$\Gamma_{\nu|x} = \Gamma_e + \Gamma_\varepsilon - \Gamma_{\varepsilon x}\Gamma_x^{-1}\Gamma_{x\varepsilon} \tag{32.24}$$

Thus, we obtain the approximate likelihood model

$$d \mid x \sim \mathcal{N}(d - A(x, z_0, \xi_0) - \nu_{*|x}, \Gamma_{\nu|x})$$

Since we are after computational efficiency, a normal approximation for the prior model would also typically be employed in the construction of the posterior model

$$x \sim \mathcal{N}(x_*, \Gamma_x)$$

Thus, we obtain the approximation for the posterior distribution

$$\pi(x \mid d) \propto \pi(d \mid x)\pi(x) \propto \exp\left(-\frac{1}{2}V(x)\right)$$

where $V(x)$ is the posterior potential

$$\begin{aligned} V(x) &= (d - A(x, z_0, \xi_0) - \nu_{*|x})^{\mathrm{T}}\Gamma_{\nu|x}^{-1}(d - A(x, z_0, \xi_0) - \nu_{*|x}) \\ &\quad + (x - x_*)^{\mathrm{T}}\Gamma_x^{-1}(x - x_*) \end{aligned} \tag{32.25}$$

$$= \|L_{\nu|x}(d - A(x, z_0, \xi_0) - \nu_{*|x})\|^2 + \|L_x(x - x_*)\|^2 \tag{32.26}$$

and where $\Gamma_{\nu \mid x}^{-1} = L_{\nu|x}^{\mathrm{T}}L_{\nu|x}$, $\Gamma_x^{-1} = L_x^{\mathrm{T}}L_x$ and $\nu_{*|x} = \nu_{*|x}(x)$.

The MAP estimate of x with the approximation error model is obtained by

$$\hat{x} = \arg\min_x \left\{\|L_{\nu|x}(d - A(x, z_0, \xi_0) - \nu_{*|x})\|^2 + \|L_x(x - x_*)\|^2\right\} \tag{32.27}$$

Then, the approximate posterior covariance would be computed by linearizing $A(x, z_0, \xi_0)$ at $x = \hat{x}$

$$\hat{\Gamma}_{x|d} \approx \left(\tilde{J}^{\mathrm{T}}\Gamma_{\nu|x}^{-1}\tilde{J} + \Gamma_x^{-1}\right)^{-1} \tag{32.28}$$

where $\tilde{J} = J + \Gamma_{\varepsilon x}\Gamma_x^{-1}$ and J is the Jacobian of $A(x, z_0, \xi_0)$ evaluated at $x = \hat{x}$.

32.3 Computational considerations

32.3.1 Physical versus relatively accurate reference models

Above, we have referred to the model $\bar{A}(\bar{x}, z, \xi)$ as the accurate (physical) model. Of course, we do not have access to such model prediction, but it has turned out that it is adequate to have a 'sufficiently accurate' model as the reference model $\bar{A}$. In practice, one may have to construct a sequence of approximations A_k to determine the 'sufficiently accurate' computational reference model [2, 24, 25].

With respect to uncertainties in general, the key is definitely not to underestimate the uncertainties.

32.3.2 Precomputations versus online computations

The actual computational complexity of the approach can be considerable, since in most cases, a largish number (typically from a few hundred to tens of thousands in the problems studied this far) of accurate predictions $\bar{A}(\bar{x}, z, \xi)$ have to be computed. But the key is that no accurate predictions need to be computed in the inference stage; these predictions are only needed for the computation of the approximate second-order statistics of (x, ε). We also remark that with the typical dimension of the unknowns, the computational complexity is small or comparable to a single MCMC run which usually requires several hundred thousands or millions of evaluations of $\bar{A}(\bar{x}, z, \xi)$.

32.3.3 Sample sizes and the second-order sample statistics

In the case of linear normal cases, the approximation error statistics can be computed analytically. In most cases, however, one needs to compute a number of samples from the joint prior model $\pi(x, z, \xi)$, and then the approximate second-order joint statistics of (x, ε).

Roughly, the required number of samples depends on the mappings $\bar{A}$ and A, as well as the variances of $\pi(x, z, \xi)$ relative to the mean. If the variances are very small in this sense, one needs only a small sample, or could even set $\Gamma_\varepsilon = 0$ and $\varepsilon_* = \bar{A}(\bar{x}_*, z_*, \xi_*) - A(P\bar{x}_*, z_*, \xi_*)$, see for example [33].

32.3.4 Sampling, prior models and atlases

In Section 32.2.2, we wrote the normal approximation (32.20) for the joint distribution of (x, ε). Generally, this approximation is done to make efficient computation of the MAP estimates feasible. If the actual prior model is normal, the marginal distribution of x induced by the (32.20) coincides with the actual prior model. The prior model $\pi(\bar{x}, z, \xi)$ does not, however, have to be jointly normal and neither, in particular, does the marginal prior model $\pi(\bar{x})$. In practice, whatever the prior model $\pi(\bar{x}, z, \xi)$ is, a set of samples $(\bar{x}^{(\ell)}, z^{(\ell)}, \xi^{(\ell)})$ is usually to be drawn and the approximation errors

$$\varepsilon^{(\ell)} = \varphi(\bar{x}^{(\ell)}, z^{(\ell)}, \xi^{(\ell)}) = \bar{A}(\bar{x}^{(\ell)}, z^{(\ell)}, \xi^{(\ell)}) - A(Px^{(\ell)}, z_0, \xi_0)\,, \quad \ell = 1, \ldots, q$$

are then to be computed, where q is the number of draws. The normal approximation for $\pi(\varepsilon, x)$ is then formed by setting $x^{(\ell)} = P\bar{x}^{(\ell)}$ and computing the mean and joint covariance as sample averages over the ensemble.

In Section 32.6, we employ *an (anatomical) atlas* in the construction of the model for the uncertain boundary geometry. If such information is available, a 'closed form' model for $\pi(\xi)$ is not usually needed.

32.3.5 Neglecting the crosscovariances, approximating $x \perp \varepsilon$

Although we have $\varepsilon = \varepsilon(x)$, and thus clearly x and ε could in most cases not be taken as mutually orthogonal, the use of the further (computational) approximation $\Gamma_{x\varepsilon} = 0$ has been shown to lead to very similar estimates with 'full error model' in several problems. With this further approximation, the mean (32.23) and covariance (32.24) become

$$\nu_* = e_* + \varepsilon_* \quad \Gamma_\nu = \Gamma_e + \Gamma_\varepsilon \tag{32.29}$$

In most problems studied thus far, the practical advantage of the further approximation (32.29) has been that it gives feasible estimates with significantly smaller amounts of samples than the full approximation error model. The approximation (32.29) was originally referred to as the *enhanced error model* in [24, 25].

32.4 Dealing with high-dimensional data

If dimension m of the data is very large, such as in 3D X-ray tomography, magnetic resonance imaging, and (inverse) transient wave propagation problems, the computation (and storage) of Γ_ε and especially $L_{e+\varepsilon}$ can be prohibitive tasks. Why these problems don't usually seem to pose a problem without approximation errors, is that the iid assumption or approximation for e is usually adopted $\Gamma_e = \text{diag}\,(\sigma_{e,1}^2, \dots, \sigma_{e,m}^2)$ so that $L_e = \text{diag}\,(\sigma_{e,1}^{-1}, \dots, \sigma_{e,m}^{-1})$. In addition, when the sample covariance Γ_ε is computed, only a smallish number $q \ll m$ of samples may be available.

32.4.1 Eigensystem of the approximation errors

With q samples $\varepsilon^{(\ell)}$ of the approximation error arranged in columns of matrix W, we *could* compute

$$\varepsilon_* = \frac{1}{q}\sum_{\ell=1}^{q} \varepsilon^{(\ell)} \qquad \Gamma_\varepsilon = \frac{1}{q-1} W W^{\mathrm{T}} - \frac{q}{q-1}\varepsilon_*\varepsilon_*^{\mathrm{T}} = \bar{V}\bar{V}^{\mathrm{T}} \tag{32.30}$$

where $\bar{v}_\ell = (q-1)^{-1/2}(\varepsilon^{(\ell)} - \varepsilon_*)$ are the columns of $\bar{V}$. As noted above, we may not be able to either compute or store $\Gamma_\varepsilon \in \mathbb{R}^{m\times m}$ explicitly. Instead, we often need only a small rank approximate covariance $\Gamma_{\varepsilon,p}$ (a small-dimensional principal subspace of the eigensystem), and want to compute it without explicit formation of matrix Γ_ε. In such a case, the *simultaneous or orthogonal iterations* can be employed, see Section 32.4.3.[45] Note that $\bar{V} \in \mathbb{R}^{m\times q}$ and $q < m$ or even $q \ll m$.

Thus, Γ_ε is heavily rank-deficient, and assumes the eigenvalue decomposition

$$\Gamma_\varepsilon = \sum_{k=1}^{m} \lambda_k v_k v_k^{\mathrm{T}} = \sum_{k=1}^{q} \lambda_k v_k v_k^{\mathrm{T}} = V\Lambda V^{\mathrm{T}}$$

since $\lambda_\ell = 0$ at least for $\ell > q$.

Since Γ_ε is positive semidefinite, the eigenvectors $\{v_k,\ k = 1, \dots, q\}$ are (algebraically) orthonormal and

$$\begin{aligned} \text{span}\{\varepsilon^{(\ell)} - \varepsilon_*,\ \ell = 1, \dots, q\} &= \text{span}\{\bar{v}_k,\ k = 1, \dots, q\} \\ &= \text{span}\{v_k,\ k = 1, \dots, q\} \end{aligned}$$

Thus, we can write[46]

$$\varepsilon^{(\ell)} - \varepsilon_* = \sum_{k=1}^{q} \beta_k v_k = \sum_{k=1}^{q} \langle \varepsilon^{(\ell)} - \varepsilon_*, v_k \rangle v_k$$

[45] The idea is not to compute Γ_ε, but to retain it (until further operations) in the form $\bar{V}\bar{V}^{\mathrm{T}}$.

[46] This decomposition is exact for the samples $\varepsilon^{(\ell)}$, but we can also interpret it as a low-dimensional model for the approximation errors in general, as is implicitly done in Section 32.6.

and further

$$\mathbb{E}\beta_k = 0\,, \quad \text{var}\,\beta_k = \lambda_k\,, \quad \mathbb{E}\beta_k\beta_j = 0,\ k \neq j$$

We shift from the linear algebraic interpretation to the statistical interpretation in the following. Of all p-dimensional approximations ε_p for the random vectors ε that are of the form

$$\varepsilon_p = \varepsilon_* + \sum_{k=1}^{p} \langle \varepsilon - \varepsilon_*, w_k \rangle w_k$$

the choice for $\{w_k\}$ that minimizes

$$\mathbb{E}\|\varepsilon - \varepsilon_p\|^2$$

is obtained by setting $w_k = v_k$, where v_k are the eigenvectors of the covariance of ε. We denote the associated subspace by $\mathcal{V}_p = \text{span}\{v_1, \ldots, v_p\}$. This applies both when we consider the actual distributions and covariances, or the sample sets and sample covariances.

This theory is variably known as principal component analysis, proper orthogonal decomposition, Hotelling transform and Karhunen–Loeve decomposition (transform), see for example, [22]. In these theories, the main difference is *how* the decompositions are used and interpreted.

32.4.2 Augmenting the unknowns, approximating $x \perp \varepsilon$

For simplicity, assume that var $e_k = \delta^2$ for all k so that trace $\Gamma_e = \mathbb{E}\|e - e_*\|^2 = m\delta^2$. We write

$$\Gamma_{e+\varepsilon} = \Gamma_e + \Gamma_{\varepsilon,p} + \Gamma_{\varepsilon,-p}$$

where

$$\Gamma_{\varepsilon,p} = \sum_{k=1}^{p} \lambda_k v_k v_k^{\mathrm{T}}\,, \quad \Gamma_{\varepsilon,-p} = \sum_{k=p+1}^{q} \lambda_k v_k v_k^{\mathrm{T}}$$

and we can also write $\varepsilon = \varepsilon_* + \varepsilon_p + \varepsilon_{-p}$. As a rough starting point, choose p so that

$$\sum_{k=p+1}^{q} \lambda_k \lessapprox 0.2m\delta^2$$

With such a choice, we have $\Gamma_e + \Gamma_{\varepsilon,-p} \approx \Gamma_e$, and $\Gamma_{\varepsilon,p}$ is the sought low rank approximation for Γ_ε.

Now, write $\varepsilon_p = \varepsilon_* + \sum_{k=1}^{p} \beta_{p,k} v_k$, $\beta_p \in \mathbb{R}^p$, $V_p = (v_1, \ldots, v_p) \in \mathbb{R}^{m \times p}$ which leads to the computational problem

$$\min_{x,\beta_p} \|L_{e+\varepsilon_{-p}}(y - A(x) - V_p\beta_p - e_* - \varepsilon_*)\|^2 + \|L_x(x - x_*)\|^2 + \|L_p\beta_p\|^2 \tag{32.31}$$

where the last term is brought by the technical approximation of mutual uncorrelatedness of (x, ε_p) and thus (x, β_p), and where we have made the normal approximation for β_p with

$$L_p = \text{diag}\,(\lambda_1^{-1/2}, \ldots, \lambda_p^{-1/2}). \tag{32.32}$$

Furthermore, we can approximate $L_{e+\varepsilon_{-p}} \approx L_e = \delta^{-1} I$

32.4.3 Orthogonal iterations for computation of low rank approximation for the eigensystem

The orthogonal iteration is a Krylov subspace method, and yields p eigenvectors (corresponding to the largest eigenvalues) of a $m \times m$ symmetric positive semidefinite matrix $D = \Gamma_\varepsilon$, see for example [15]. In our case (low rank representation for the covariance), the orthogonal iteration algorithm works as follows.

Define the sequence of matrices $V(k)$ with $V(0) = I_{m \times p}$ or other matrix with linearly independent columns (or, better still, a good guess for the eigenvectors).

```
for k = 1...d
    (Q,R) = qr(V(k-1));
    V(k) = DQ;
end
(V,R) = qr(V(d))
```

Of course, if D were obscenely large and we don't have this, there would be a problem. But in our case, we can again use parentheses (and abuse notations): $V = DQ = \bar{V}(\bar{V}^{\mathrm{T}} Q)$ which is fine, since $\bar{V}^{\mathrm{T}} Q$ is $(q \times m) \times (m \times p) = q \times p$. Here, $\bar{V}$ is the centred and normalized sample matrix as in (32.30). The limit $V(k) \to V_p$ has the p eigenvectors. The eigenvalues are obtained from the diagonal of $V_p^{\mathrm{T}} \Gamma_\varepsilon V_p = V_p^{\mathrm{T}} \bar{V} \bar{V}^{\mathrm{T}} V_p = (V_p^{\mathrm{T}} \bar{V}) \cdot (\bar{V}^{\mathrm{T}} V_p)$.

32.5 Case study 1. Application to local tomography

32.5.1 Local X-ray tomography

The problem of local tomography is to reconstruct a *region of interest* (ROI) inside a target body based on a set of truncated X-ray projection images [40, 52]. In medical imaging, the truncation of the projection images is typically enforced by intentional minimization of the ionizing radiation dose to the patient. The principle of local tomography is illustrated in Figure 32.2. In the conventional (global) computerized tomography (CT) problem, the whole target body is fully visible in all of the X-ray projection images. In the local tomography problem, only part of the target (ROI) is visible in all projections, which is shown as a dark grey patch in Fig. 32.2. For the X-ray tomography problem and the local tomography problem, see [11, 37, 40, 52].

For the construction of the measurement model, we consider the global tomography problem as the starting point. We follow the direct discretization formulation that enables also the further limited-angle limitation of X-ray tomography, see for example [28]. The domain of the unknown attenuation density function is modelled by a bounded subset $\Omega \subset \mathbb{R}^2$ together with a non-negative attenuation coefficient $x(r) : \Omega \to [0, \infty)$. The domain Ω is chosen large enough so that the body under investigation is completely inside the domain Ω, that is, $\text{supp}(x(r)) \subset \Omega$. The X-ray attenuation measurement is commonly modelled by the line integral

$$\int_\gamma x(r)\, \mathrm{d}r = \log I_0 - \log I_1 \tag{32.33}$$

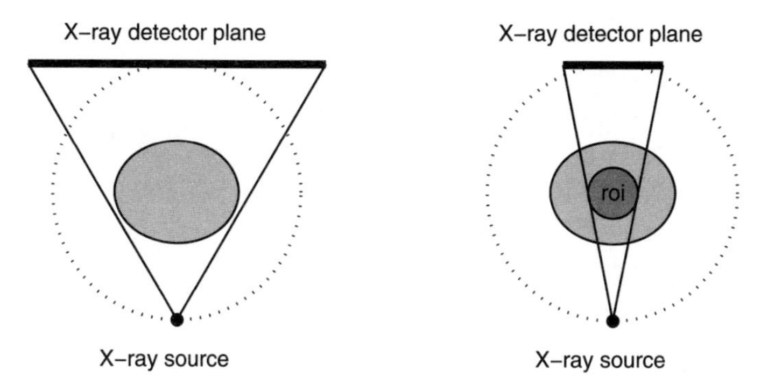

Figure 32.2 Global (left) and local (right) tomography. The source and detector panel rotate around the object along the dotted circle. In the right-hand image, the dark grey patch represents the ROI subdomain that is present in all of the projection images in the local tomography case. In local tomography, only the ROI is to be reconstructed, with a minimal number of projection data.

where γ is the line between the source and detector points, I_0 is the source intensity and I_1 is the attenuated intensity at the detector. In the discretization of the problem, the domain Ω is divided into n disjoint pixels D_i so that $\Omega = \bigcup_{i=1}^{n} D_i$ and the function $x(r)$ is approximated with a piecewise constant function with $x|_{D_i} \equiv x_i$, leading to the approximation

$$\int_{\gamma} x(r)\,\mathrm{d}r \approx \sum_{i=1}^{n} x_i\,|D_i \cap \xi| \tag{32.34}$$

for the line integral in (32.33). Arranging the pixels into a column vector $x = (x_1, x_2, \ldots, x_n)^{\mathrm{T}} \in \mathbb{R}^n$ and all the measured line integrals in the X-ray tomography experiment into $\tilde{d} = (d_1, d_2, \ldots, d_{m_g})^{\mathrm{T}} \in \mathbb{R}^{m_g}$, we get the model

$$\tilde{d} = \tilde{A}x + \tilde{e} \tag{32.35}$$

for the global tomography problem. Here, n is chosen large enough so that (32.35) can be considered an accurate model for the global tomography experiment. The measurement noise e can be taken to be normal mutually independent random variables but with different variances [3, 51].

Consider now the local tomography problem. Decompose the domain into two disjoint subdomains as $\Omega = \Omega_0 + \Omega_1$, where Ω_1 is the region of interest (ROI) and Ω_0 consists of the rest of the image domain, that is, the uninteresting part. Given the global tomography data and model (32.35), a textbook description of local tomography can be stated such that the objective is to estimate $x \in \mathbb{R}^n$ in Ω using a subset $d = \tilde{d}(\mathcal{I}_1) \in \mathbb{R}^m$ of the complete (global) data, where the index set $\mathcal{I}_1 \subset \{1, 2, \ldots, m\}$ indicates the subset of line integrals that intercept the ROI, that is, $j \in \mathcal{I}_1$ if $|\gamma_j \bigcap \Omega_1| > 0$. Thus, we have projection model

$$d = \bar{A}x + e \tag{32.36}$$

where $\bar{A}$ is $m \times n$ block of $\tilde{A}$ such that $\bar{A} := \tilde{A}(\mathcal{I}_1, :)$. Note that model (32.36) is an *accurate* projection model for local tomography in the sense that the unknown attenuation coefficient is modelled in the whole domain Ω. We note that the reconstruction of the entire x from this partial data d is inherently a non-unique problem.

32.5.2 Approximation error model for local tomography

A complication with the model (32.36) is the high computational cost induced by the large dimension of x. Obviously, since in local tomography one is interested in the ROI only, it would be computationally appealing to reduce the model (32.36) by decomposing x into interesting and uninteresting parts as

$$x = x|_{\Omega_0} + x|_{\Omega_1} := x_0 + x_1, \quad x_0 \in \mathbb{R}^{n_0},\, x_1 \in \mathbb{R}^{n_1}, \quad n_0 + n_1 = n$$

and write the accurate measurement model as

$$d = \bar{A}x + e = A_0 x_0 + A_1 x_1 + e$$

and then estimate the interesting part $x_1 \in \mathbb{R}^{n_1}$ from the *truncated* model

$$d \approx A_1 x_1 + e. \tag{32.37}$$

Here, the matrices A_0 and A_1 are blocks of $\bar{A}$ such that $A_0 := \bar{A}(:, \mathcal{J}_0)$ and $A_1 := \bar{A}(:, \mathcal{J}_1)$ where the index sets $\mathcal{J}_0$ and $\mathcal{J}_1$ correspond to the subsets of pixels inside the domains Ω_0 and Ω_1, respectively. However, the model (32.37) employing equality in place of '$\approx$' is severely erroneous, since the attenuation of the rays in Ω_0 is not modelled. Effectively, we would be writing $A_0 x_0 \approx 0$, and as the result, the likelihood and the posterior models would be grossly infeasible, see an example of a marginal posterior of a single pixel in Figure 32.1.

In the following, we marginalize the local tomography problem with respect to the uninteresting variable x_0 using the approximation error approach. For this, we write the accurate model as

$$d = A_1 x_1 + \left[\bar{A}x - A_1 x_1\right] + e = A_1 x_1 + \varepsilon + e \tag{32.38}$$

where the discrepancy ε between the accurate model $\bar{A}x$ and truncated model $A_1 x_1$ is the approximation error. Obviously, we have in this case $\varepsilon = A_0 x_0$.

32.5.3 The employed MRF prior model

We use a proper Gaussian smoothness prior for the unknown image x, constructed as in [24, 27, 30]. In this construction, x is considered in the form

$$x = x_{\mathrm{in}} + x_{\mathrm{bg}}$$

where x_{in} is a spatially inhomogeneous coefficient[47] with zero mean, and x_{bg} is a spatially homogeneous (background) coefficient. For the latter, we can write $x_{\mathrm{bg}}(r) = s\mathbb{I}$, where $\mathbb{I}$ is a vector of ones and s is a scalar random variable with distribution $s \sim \mathcal{N}(x_*, \zeta_{\mathrm{bg}}^2)$. With respect to the basis for x, we have the coordinates $x_{\mathrm{in}} \in \mathbb{R}^n$ and set $x_{\mathrm{in}} \sim \mathcal{N}(0, \Gamma_{\mathrm{in}})$. We model the spatial distributions x_{in} and $s\mathbb{I}$ as mutually independent, that is, the background is mutually independent with the inhomogeneities.

[47] In the sequel, 'in' refers to inhomogeneous, 'bg' to background.

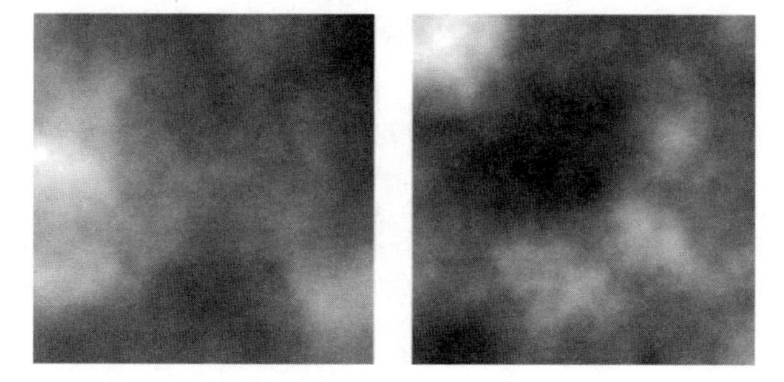

Figure 32.3 Two samples $x^{(\ell)}$ from the prior model $\pi(x)$.

Thus, we have $\Gamma_x = \Gamma_{\text{in}} + \zeta_{\text{bg}}^2 \mathbb{1}\mathbb{1}^{\text{T}}$, and

$$\pi(x) = \mathcal{N}(x_* \mathbb{1}, \Gamma_x) \tag{32.39}$$

In the construction of Γ_{in}, the approximate correlation length can be adjusted to match the size of the expected inhomogeneities. See [24, 27, 30] for details of the construction and [48] for construction of Markov random fields in general.

In the construction of the model for the data considered in this chapter, the prior parameters were set as follows. For the background attenuation, the mean was set as $x_* = 0$ and the standard deviation as $\zeta_{\text{bg}} = 0.005$. For the inhomogeneity part, the target correlation length was set to 7 mm and the standard deviation $\zeta_{\text{in}} = 0.05$. Figure 32.3 shows two random draws from the prior (32.39).

Note that the interesting and uninteresting parts x_1 and x_0 can be modelled as projections $x_1 = P_1 x$ and $x_0 = P_0 x$, and the corresponding marginal priors can be obtained respectively as

$$\pi(x_\ell) = \mathcal{N}(x_{\ell,*}, \Gamma_{x_\ell}), \quad x_{\ell,*} = P_\ell x_*, \quad \Gamma_{x_\ell} = P_\ell \Gamma_x P_\ell^{\text{T}},\ \ell = 0, 1.$$

32.5.4 Experimental setup and projection models

A realization of global tomography X-ray projection data was measured from a tooth specimen using a commercial dental X-ray source, dental CCD X-ray detector and a rotating platform. The measurement equipment was arranged such that the experiment corresponds to the global tomography illustrated in the left image in Figure 32.2. For more details on the experimental setup, see [28]. We took 23 equi-angularly spaced projection images of the tooth specimen from a total view-angle of 187 degrees and extracted one slice of this data to form a 2D tomography problem, resulting in global tomography data $\tilde{d}$ with $m_{\text{g}} = 15\,272$. For the distribution of the measurement noise, we use approximation

$$\pi(e) = \mathcal{N}(0, \Gamma_e), \quad \Gamma_e = \delta^2 I$$

where the standard deviation δ of the noise was estimated using empty (i.e. air) regions around the tooth specimen in the projection images.

For the discretization of the global projection model $\tilde{A}$ in equation (32.35), the domain Ω was chosen as $12 \times 12\text{mm}^2$ square and this domain was divided into $n = 96 \times 96 = 9216$ regular pixels. Thus, the size of $\tilde{A}$ is $15\,272 \times 9216$.

In the construction of the *accurate* local tomography model $\bar{A}$, we chose the region of interest Ω_1 as a square domain that encloses one of the two roots of the tooth specimen, and formed the local tomography data by determining the subset $d = \tilde{d}(\mathcal{I})$ of measured X-rays that intercept the chosen ROI. This resulted in the data $d \in \mathbb{R}^{3964}$ and projection matrix $\bar{A} := \tilde{A}(\mathcal{I}_1, :)$ with size 3964×9216.

For the construction of the truncated model (32.37), we determined the index set $\mathcal{J}_1$ of the pixels that belong to Ω_1, and formed the truncated projection matrix $A_1 := \bar{A}(:, \mathcal{J}_1)$ with size of 3964×1681 (i.e. the number of unknowns is $n_1 = 1681$ in the truncated projection model).

32.5.5 Results

Figure 32.4 shows the MAP estimates with different measurement models for the global and local X-ray tomography example. Top row displays the estimates in the whole image domain Ω. For the estimates (columns 3–5) that are based on the truncated projection model A_1, the (uninteresting) outside of the ROI domain Ω_0 which is not included in the projection model is displayed with constant grey value. The bottom row shows the ROI details Ω_1 from the estimates.

The columns from the left to right show the following MAP estimates:

1. Global tomography MAP estimate using all the available data, i.e. measurement model (32.35):

$$\hat{x} = \arg\min_x \left\{ \| L_{\tilde{e}}(\tilde{d} - \tilde{A}x) \|^2 + \| L_x(x - x_*) \|^2 \right\} \tag{32.40}$$

where $L_{\tilde{e}}^{\mathrm{T}} L_{\tilde{e}} = \Gamma_{\tilde{e}}^{-1}$, $L_x^{\mathrm{T}} L_x = \Gamma_x^{-1}$. This estimate serves for the local tomography estimates as the 'ground truth' reference.

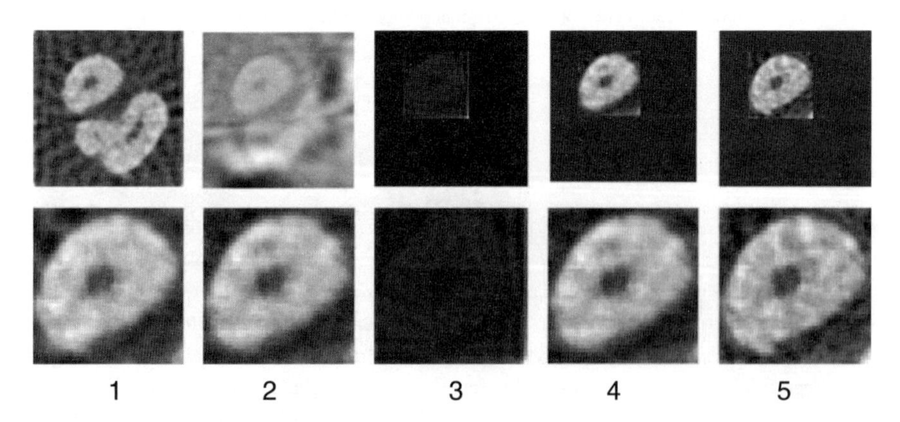

Figure 32.4 Estimates from X-ray projection data from a tooth specimen (23 projections from a view angle of 187 degrees). Columns from left to right: Global tomography, local tomography with accurate model $\bar{A}x$, local tomography with (truncated) model A_1x_1 and conventional measurement error model, local tomography with model A_1x_1 and the approximation error model, local tomography with model A_1x_1 and the high-dimensional modification of the approximation error model. Top row: estimate in whole domain Ω (for the cases with the truncated model A_1x_1, the uninteresting subdomain Ω_0 that is not estimated is displayed with constant value). Bottom row: ROI detail Ω_1 from the estimates.

2. Local tomography with accurate measurement model (32.36):

$$\hat{x} = \arg\min_{x} \left\{ \|L_e(d - \bar{A}x)\|^2 + \|L_x(x - x_*)\|^2 \right\} \tag{32.41}$$

where $L_e^{\mathrm{T}} L_e = \Gamma_e^{-1}$, $L_x^{\mathrm{T}} L_x = \Gamma_x^{-1}$. This estimate serves as reference of local tomography using accurate projection model. Note that the restriction $\hat{x}|_{\Omega_1}$ from this estimate is the exact marginal MAP for x_1.

3. Local tomography with truncated projection model and conventional measurement error model (32.37):

$$\hat{x_1} = \arg\min_{x_1} \left\{ \|L_e(d - A_1 x_1)\|^2 + \|L_{x_1}(x_1 - x_{1,*})\|^2 \right\} \tag{32.42}$$

where $L_{x_1}^{\mathrm{T}} L_{x_1} = \Gamma_{x_1}^{-1}$. This is the straightforward trivial approach corresponding to setting $x_0 = 0$.

4. Local tomography with truncated projection model and approximation error model (32.37):

$$\hat{x_1} = \arg\min_{x_1} \left\{ \|L_{\nu|x}(d - A_1 x_1 - \nu_{*|x})\|^2 + \|L_{x_1}(x_1 - x_{1,*})\|^2 \right\} \tag{32.43}$$

where $\nu_{*|x}$ is given by equation (32.21) and $L_{\nu|x}$ by equation (32.22). Notice that since the problem is linear and Gaussian, all the terms in equations (32.21) and (32.22) have closed-form solutions.

5. Local tomography with (augmented) high-dimensional modification of the approximation error model (32.31):

$$(\hat{x_1}, \hat{\beta_p}) = \arg\min_{x_1, \beta_p} \left\{ \|L_e(y - A_1 x_1 - V_p \beta_p - \varepsilon_*)\|^2 + \|L_{x_1}(x_1 - x_{1,*})\|^2 + \|L_p \beta_p\|^2 \right\} \tag{32.44}$$

where L_p is defined by equation (32.32). To demonstrate the performance of the high-dimensional modification, sample based approximations are used for ε_* and Γ_ε, computed based on 500 samples $\varepsilon^{(\ell)} = \bar{A}x^{(\ell)} - A_1 P_1 x^{(\ell)}$. The dimension p is selected by $\sum_{k=p+1}^{q} \lambda_k \lessapprox 0.2m\delta^2$ and the approximation $\Gamma_e + \Gamma_{\varepsilon,-p} \approx \Gamma_e$ is used.

When the estimate is computed using the truncated projection model A_1 and conventional measurement error model, the estimate (third column) shows stripe-like artefacts in the whole ROI and spiky high amplitude errors near the boundary of the ROI. These errors are due to neglecting the contribution of the tissues outside the ROI to the measured projections. Loosely speaking, discrepancy between the projection model and measurements is compensated in the MAP estimate with spurious details in the ROI domain Ω_1. The estimate (32.43) with the approximation error model (fourth column), on the other hand, is clear of these artefacts and the estimate of the ROI correspond to the global tomography reference (first column) and is equal (within numerical round off errors, relative error $\sim 10^{-10}$) to the MAP of exact marginal posterior (i.e. the ROI part $\hat{x}|_{\Omega_1}$ from the estimate (32.41)). Comparing the estimate (32.44) (fifth column) with the high-dimensional (augmented) modification of the approximation error model to the 'conventional' approximation error estimate (fourth column), one can see that the estimates are very similar despite the fact that the augmented form uses the approximation $\Gamma_{x\varepsilon} = 0$ and is based on using a sample based, low-dimensional approximation for the covariance Γ_ε. This suggests that the augmented formulation in Section 32.4 could be applied also in the very high-dimensional 3D local tomography case in which the explicit construction of the approximation error covariance matrix is computationally not feasible.

32.6 Case study 2. Inaccurately known body shape in electrical impedance tomography

32.6.1 Electrical impedance tomography (EIT)

In EIT, the objective is to estimate the unknown conductivity function inside a body based on electric current and voltage measurements at the exterior boundary of the body. The medical applications of EIT include, for example, the detection of tumours from breast tissue and bedside monitoring of pulmonary function of intensive care patients [4, 13, 58, 61]. For reviews on the EIT problem, see [7, 26].

One of the practical challenges in medical EIT is that the boundary of the body is usually not known. As an example, consider EIT measurements of monitoring pulmonary function from the surface of the thorax. In principle, the shape of the patient's thorax could be obtained from other imaging modalities such as computerized tomography (CT) but such measurement is usually not available (e.g. the patient can not be transferred away from the intensive care). In addition, the shape of the thorax varies in time due to breathing and also depends on the orientation (posture) of the patient. Therefore, the body shape would be inaccurately known even at best and the estimation of the conductivity has to be carried out using an approximate model domain. On the other hand, it is well known that the use of an incorrect model for the body shape leads to serious artefacts in the conductivity images, see for example [14, 31]. In practical biomedical EIT, only so-called difference imaging is usually attempted. In this modality, only a very rough qualitative image can be constructed, but, on the other hand, the unknown boundary shape has less drastic effect, see for example [1].

In the following, we consider the approximate pre-marginalization over the unknown body shape by the approximation error approach. Furthermore, we describe how the high-dimensional modification (augmented form) of the approximation error approach in Section 32.4 can be used to recover an approximation for the unknown body shape based on the low rank estimate ε_p of the approximation error.

32.6.2 Forward model of EIT

Let $\Omega \subset \mathbb{R}^3$ denote the measurement domain and let γ denote a parameterization of the domain boundary $\partial\Omega$. In EIT, an array of N_{el} contact electrodes are attached on the boundary $\partial\Omega$. Using the electrodes, electric currents (called current patterns) are injected into the body (volume conductor) Ω and the resulting voltages are measured using the same electrodes. We model these measurements with the complete electrode model [8, 53]:

$$\nabla \cdot \sigma(x)\nabla u(x) = 0, \quad x \in \Omega \tag{32.45}$$

$$u(x) + z_\ell \sigma(x)\frac{\partial u(x)}{\partial n} = U_\ell, \quad x \in e_\ell \subset \partial\Omega, \tag{32.46}$$

$$\int_{e_\ell} \sigma(x)\frac{\partial u(x)}{\partial n} dS = I_\ell, \quad x \in e_\ell \subset \partial\Omega, \tag{32.47}$$

$$\sigma(x)\frac{\partial u(x)}{\partial n} = 0, \quad x \in \partial\Omega \setminus \bigcup_{l=1}^{N_{\text{el}}} e_\ell. \tag{32.48}$$

where $u(x)$ is the potential distribution in Ω, n is the outward unit normal vector at $\partial\Omega$, $\sigma(x)$ is the conductivity, and z_ℓ is the contact impedance between the object and the electrode e_ℓ. The existence and uniqueness of the solution of the boundary value problem (32.45–32.48) are quaranteed by

$$\sum_{\ell=1}^{N_{\text{el}}} I_\ell = 0, \sum_{\ell=1}^{N_{\text{el}}} U_\ell = 0 \tag{32.49}$$

where the former constraint is the conservation of charge and latter fixes ground level for the voltages.

In the following, the numerical approximation of the boundary value problem (32.45–32.49) is based on the finite element (FEM) approximation, for details of the implementation used in this chapter, see [26, 57]. In the following, we use notation

$$U(\sigma, \gamma) \in \mathbb{R}^m \tag{32.50}$$

for the FEM based forward solution corresponding to single EIT experiment, that is, the vector $U(\sigma, \gamma)$ contains computed voltages for all the current patterns in the measurement paradigm. The dependence of the forward model on the domain Ω is expressed by the parameterization γ of the boundary $\partial\Omega$.

32.6.3 Approximation error model

Let

$$V = U(\bar{\sigma}, \gamma) + e, \tag{32.51}$$

denote a (sufficiently) accurate model between the unknowns and measurements. Here the parameters γ of the boundary $\partial\Omega$ are such that the error in the FEM approximation is smaller than the measurement error (in the sense defined in Section 32.4). The conductivity $\bar{\sigma}$ is a parameterization in the actual Ω and is dense enough in the above sense.

As explained above, in practical clinical measurements one usually lacks accurate knowledge of the shape of the body Ω and therefore the estimation of σ is carried out using an approximate *model domain* $\tilde{\Omega}$. In such a case, the accurate model (32.51) is traditionally replaced by the approximate measurement model:

$$V \approx U(\sigma, \tilde{\gamma}) + e \tag{32.52}$$

where $\tilde{\gamma}$ are the parameters of the boundary $\partial\tilde{\Omega}$ of the model domain, and one hopes that the approximation in (32.52) is a feasible one.

The relation of the representation of the conductivities in (32.51) and (32.52) is of the form $\bar{\sigma}(x) = \sigma(T(x))$, where

$$T : \Omega \mapsto \tilde{\Omega} \tag{32.53}$$

is a mapping that models the deformation of domain Ω to $\tilde{\Omega}$. Obviously, the true deformation T between the measurement domain and model domain is not known and not unique, and one has to choose a model for the deformation. In the numerical examples considered in this chapter, T is chosen such that the angle and relative distance (between the center of the domain and the boundary) of a coordinate point is preserved. The deformation of the conductivity can be numerically implemented by a linear interpolation

$$P\bar{\sigma} = \sigma, \tag{32.54}$$

where P is a matrix that interpolates the conductivity from Ω into $\tilde{\Omega}$ according to the deformation T.

We write the accurate measurement model (32.51) in the form

$$V = U(\sigma, \tilde{\gamma}) + (U(\bar{\sigma}, \gamma) - U(\sigma, \tilde{\gamma})) + e = U(\sigma, \tilde{\gamma}) + \varepsilon(\bar{\sigma}, \gamma) + e, \tag{32.55}$$

where $\varepsilon(\bar{\sigma}, \gamma)$ represents the modelling error due to the incorrect boundary, and we denote $\nu = \varepsilon + e$.

32.6.4 The EIT experiment

The EIT data was measured from a vertically symmetric measurement tank Ω, see Fig. 32.7. Sixteen equally spaced stainless steel electrodes were attached on the boundary $\partial\Omega$ of the tank. The (incorrect) model domain $\tilde{\Omega}$ was a cylinder with diameter equivalent to the diameter of the measurement domain, see Fig. 32.5. The heart and lung shaped inclusions in the target were made of agar and placed in the chest-shaped tank filled with saline of conductivity 3.0 mS cm^{-1}. The conductivity of the lung and heart targets were 0.73 mS cm^{-1} and 5.8 mS cm^{-1}, respectively.

The measurements were carried out with the KIT 4 EIT device using adjacent electrode current patterns measurement paradigm, leading to $m = 256$ voltage measurements, see [32] for details of the measurement system. For the estimation of measurement error statistics, 40 000 realizations were measured and a sample based Gaussian approximation $\pi(e) = \mathcal{N}(e_*, \Gamma_e)$ was constructed, for details see [45].

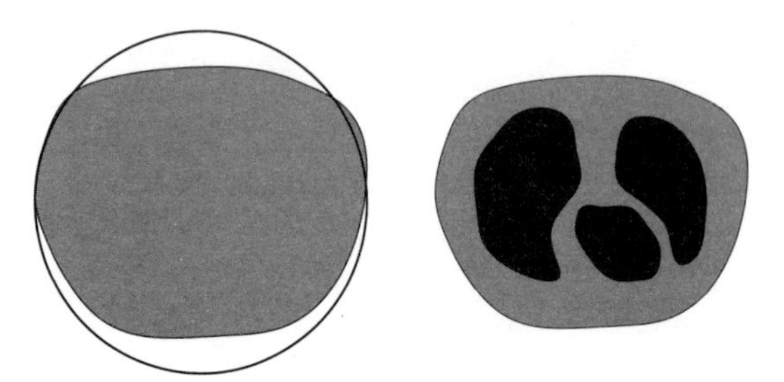

Figure 32.5 Left: The cross-section of the actual domain Ω is shown as a grey patch. The shape of the domain was extracted from a segmented CT image of the thorax. The cross-section of the cylindrical model domain $\tilde{\Omega}$ is shown as a solid line. Right: Arrangement of lung and heart phantoms in the measurement tank.

32.6.5 Prior models and construction of approximation error statistics

As the prior model $\pi(\sigma)$, we use a proper Gaussian smoothness prior $\pi(\sigma) = \mathcal{N}(\sigma_* \mathbb{I}, \Gamma_\sigma)$. with $\Gamma_\sigma = \Gamma_{\text{in}} + \zeta_{\text{bg}}^2 \mathbb{I}\mathbb{I}^{\text{T}}$ constructed similarly to Section 32.5.3. The correlation length in the prior model was set to 7 cm. The other prior parameters were set as follows. The prior mean was set to the conductivity of the saline background, that is, $\sigma_* = 3.0\,\text{mS cm}^{-1}$, $\zeta_{\text{in}} = 0.6\,\text{mS cm}^{-1}$, and $\zeta_{\text{bg}} = 0.15\,\text{mS cm}^{-1}$.

As the prior model $\pi(\gamma)$ we use a sample based Gaussian approximation that is constructed based on an atlas of chest CT images of $n_{\text{pr}} = 150$ different individuals in the population. Since the measurement tank in this example is translationally symmetric, we simplified the implementation procedure by treating the translationally symmetric boundary shape by a 2D parameterization. For this, the CT images were segmented to interior and exterior of the body domain, leading to ensemble of chest domains $\{\Omega^{(\ell)}, \ell = 1, 2, \ldots, n_{\text{pr}}\}$. To obtain the corresponding parametric representations for the chest boundaries $\{\partial\Omega^{(\ell)}\}$, we extracted the boundaries $\{\partial\Omega^{(\ell)}\}$ from the segmented CT images and then computed a Fourier series representation with $n_f = 20$ terms for the boundaries. Thus, we obtained an ensemble $\{\gamma^{(\ell)}, \ell = 1, \ldots, n_{\text{pr}}\}$ of parametric representations of chest shapes. Using the ensemble, we approximate the prior model $\pi(\gamma)$ by a sample based normal distribution $\pi(\gamma) \approx \mathcal{N}(\gamma_*, \Gamma_\gamma)$.

32.6.5.1 *Estimation of the approximation error statistics*

For the estimation of approximation error statistics, we generated a set of $n_{\text{samp}} = 1000$ draws from the prior models $\pi(\gamma)$ and $\pi(\bar{\sigma})$ similarly as explained in Section 32.3.4. Figure 32.6 shows a central cross-section from four samples $\{\bar{\sigma}^{(\ell)}\}$ on the sample domains $\{\gamma^{(\ell)}\}$. The samples were used for the computation of the samples $\varepsilon^{(\ell)}$ of the approximation error as

$$\varepsilon^{(\ell)} = U(\bar{\sigma}^{(\ell)}, \gamma^{(\ell)}) - U(\sigma^{(\ell)}, \tilde{\gamma})$$

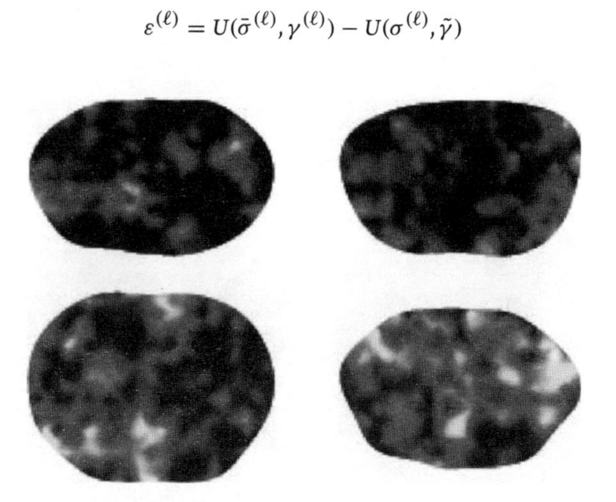

Figure 32.6 Cross-sections from four samples $\{\bar{\sigma}^{(\ell)}, \Omega^{(\ell)}\}$ for the construction of the approximation error model. The sample domains $\{\Omega^{(\ell)}\}$ are from an ensemble of chest CT images of different individuals and the conductivities are drawn from the prior model $\pi(\bar{\sigma})$.

where $\sigma^{(\ell)} = P^{(\ell)}\bar{\sigma}^{(\ell)}$ The second-order statistics of the modelling errors were then estimated as sample based averages to obtain the normal approximation $\pi(\varepsilon) \approx \mathcal{N}(\varepsilon_*, \Gamma_\varepsilon)$.

32.6.6 Results

The MAP estimates with different measurement error models are shown in Fig. 32.7. The images that are shown, are the central horizontal cross-sections of the estimated 3D conductivity. The images are arranged as follows:

- (top left) Photograph of the target conductivity σ_{true}.
- (top right) MAP estimate with conventional measurement error model

$$\hat{\sigma} = \arg\min_{\bar{\sigma}\geq 0} \left\{ \|L_e(V - U(\bar{\sigma},\gamma) - e_*)\|^2 + \|L_{\bar{\sigma}}(\bar{\sigma} - \bar{\sigma}_*)\|^2 \right\} \tag{32.56}$$

  in the correct domain Ω (forward model $U(\bar{\sigma},\gamma)$). This estimate serves as reference when domain boundary is modelled with sufficient accuracy so that the modeling errors are negligible.
- (bottom left) MAP estimate with the conventional measurement error model

$$\hat{\sigma} = \arg\min_{\sigma\geq 0} \left\{ \|L_e(V - U(\sigma,\tilde{\gamma}) - e_*)\|^2 + \|L_\sigma(\sigma - \sigma_*)\|^2 \right\} \tag{32.57}$$

  in the model domain $\tilde{\Omega}$ (forward model $U(\sigma,\tilde{\gamma})$), representing conventional estimate in presence of domain modelling errors.

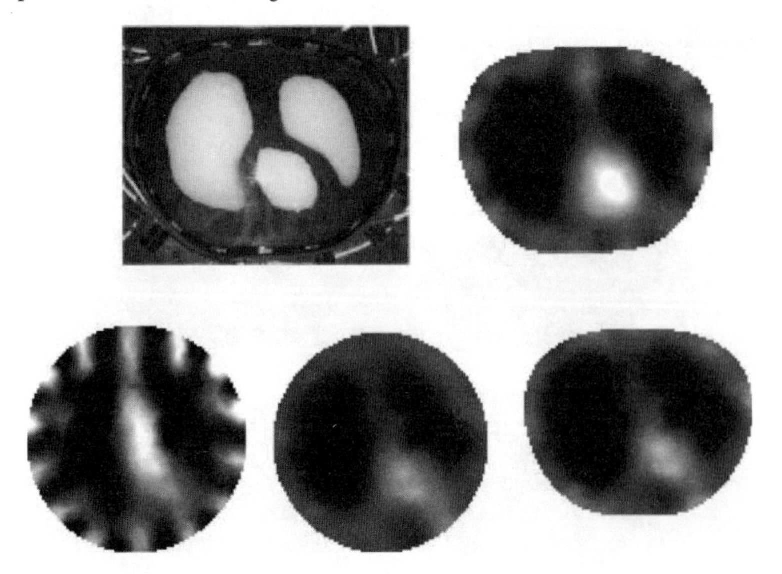

Figure 32.7 Top left: The measurement tank. Top right: MAP estimate using conventional measurement error model and the correct domain Ω. Bottom left: MAP using the conventional measurement error model and incorrect model domain $\tilde{\Omega}$. Bottom middle: MAP using the approximation error model and domain $\tilde{\Omega}$. Bottom right: MAP with the augmented approximation error approach using domain $\tilde{\Omega}$. The images show the central horizontal cross-sections from the 3D reconstructions.

- (bottom middle) MAP with the approximation error model

$$\hat{\sigma} = \arg\min_{\sigma \geq 0} \left\{ \|L_{e+\varepsilon}(V - U(\sigma, \tilde{\gamma}) - e_* - \varepsilon_*)\|^2 + \|L_\sigma(\sigma - \sigma_*)\|^2 \right\} \tag{32.58}$$

  in the model domain $\tilde{\Omega}$ (forward model $U(\sigma, \tilde{\gamma})$). Here we use the approximation (32.29) and set $L_{e+\varepsilon}^{\mathrm{T}} L_{e+\varepsilon} = \Gamma_{\varepsilon+e}^{-1}$.
- (bottom right) MAP with the high-dimensional (augmented) modification of the approximation error model using the model domain $\tilde{\Omega}$. Here the use of the augmented form is not motivated by the dimension of the data, but we instead seek to obtain an approximate recovery of the domain shape from the (low rank) estimate $\hat{\varepsilon}_p$ of the approximation error. For this, we estimate $(\hat{\sigma}, \hat{\beta}_p)$ from

$$(\hat{\sigma}, \hat{\beta}_p) = \arg\min_{\sigma \geq 0, \beta_p} \left\{ \| L_{e+\varepsilon,-p}(V - U(\sigma, \tilde{\gamma}) - V_p\beta_p - \varepsilon_* - e_*) \|^2 \right.$$
$$\left. + \| L_\sigma(\sigma - \sigma_*) \|^2 + \| L_p\beta \|^2 \right\}, \tag{32.59}$$

  Given $\hat{\beta}_p$, the estimate for γ is computed by forming a normal approximation for the joint density of $\varepsilon_p = V_p\beta_p$ and γ and finding the MAP estimate $\hat{\gamma} = \arg\max \tilde{\pi}(\gamma | \hat{\varepsilon}_p)$, given by

$$\hat{\gamma} = \Gamma_{\gamma\varepsilon_p}\Gamma_{\varepsilon,p}^{-1}\hat{\varepsilon}_p + \gamma_*, \tag{32.60}$$

  where $\hat{\varepsilon}_p = V_p\hat{\beta}_p$. Approximate spread estimates for the boundary shape are obtained from the covariance of $\tilde{\pi}(\gamma | \hat{\varepsilon}_p)$, given by

$$\Gamma_{\hat{\gamma}|\hat{\varepsilon}_p} = \Gamma_\gamma - \Gamma_{\gamma\varepsilon,p}\Gamma_{\varepsilon,p}^{-1}\Gamma_{\gamma\varepsilon,p}^{\mathrm{T}}. \tag{32.61}$$

  Note that the estimate (32.59) of the conductivity σ and the projection coefficients β is carried out in the (incorrect) model domain $\tilde{\Omega}$.

  Once the estimates of σ, β and $\hat{\gamma}$ have been computed, the estimated conductivity σ is mapped from the model domain $\tilde{\Omega}$ into the reconstructed domain $\hat{\Omega}$ (that corresponds to estimated $\hat{\gamma}$) by a linear interpolation $\bar{\sigma} = \tilde{P}\hat{\sigma}$, where $\tilde{P}$ implements interpolation from domain to another according to the inverse T^{-1} of the domain deformation model.

The nonlinear minimization problems for the computation of the different MAP estimates were solved with the Gauss–Newton algorithm using a line search algorithm [47].

Figure 32.8 shows the estimated boundary (32.60) and two standard deviation limits for the posterior $\tilde{\pi}(\gamma | \hat{\varepsilon}_p)$ with dashed line and the true measurement domain Ω with grey patch.

The MAP estimate in which the incorrect model domain is used but the approximation error is not accounted for (bottom left), shows infeasible estimation errors. In contrast, the MAP estimate in which the approximation errors are accounted for (bottom middle), is mostly free of these artefacts and gives a deformed image of the conductivity in the model domain $\tilde{\Omega}$. The MAP estimate with the augmented form of the approximation error model (bottom right), on the other hand, is almost as good as the reference estimate in the correct domain (top right). The shape of the domain also has been found quite well, and moreover, the approximate posterior spread estimates for the boundary shape are feasible. In addition to the reconstruction of the shapes of the organs, the actual conductivity values also match the reality quite well.

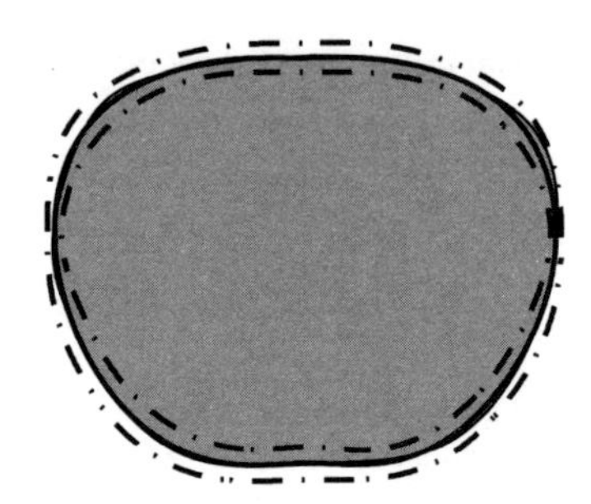

Figure 32.8 The cross-section of the actual domain Ω is shown as a grey patch. The estimated boundary with the augmented approximation error model is shown with a solid line and two posterior (approximate) standard deviation limits for the approximate posterior $\tilde{\pi}(\gamma|\hat{\varepsilon}_D)$ with dashed lines.

32.7 Discussion

Inverse problems are known to tolerate measurement and modelling errors poorly. In terms of statistics, this means that (unaccounted-for) errors and uncertainties shift the entire posterior probability mass away from the actual values, possibly drastically. While the uncertainties can be handled in several ways, most of the approaches require such computational complexity that renders these as academic exercises. In particular, modelling all uncertainties as unknowns and carrying out inference for all these using MCMC, is completely out of the question in most practical inverse problems with continuous data streams.

In this chapter, we have discussed the so-called approximation error approach, which was originally meant to model numerical model reduction only. The approach is based on a number of consecutive approximations for the associated models and densities, based on Bayesian modelling of all uncertainties, approximations and unknowns. The aim is to end up with highly approximative 'dirty but fast' computational models, which are, however, feasible in the sense that the actual unknowns should lie within a couple of approximate credibility intervals.

The approach has been shown to be feasible for a number of modelling errors and uncertainties, including drastic model reduction, unknown geometry and boundary conditions and highly approximate physical models. The approach has also been shown to provide feasible spread estimates, and thus enable, for example, optimal stochastic (feedback) control.

The approach calls for sometimes extensive simulations using relatively accurate computational models and is computationally heavy in this sense. All of these computations are, however, precomputations, and the online real-time calculations are fast and efficient. With a particular application, it is difficult to predict how well the approach will work. The linear normal case is an exception. Thus, it currently seems that the only way to find this out, is to construct all the required models and carry out simulations and then tests with real data.

Acknowledgements

The authors would like to thank Antti Nissinen (PhD) for the figures of the EIT example. The work was supported by the Academy of Finland projects 119270, 140731, 218183 and projects 213476, 250215, Finnish Centre of Excellence in Inverse Problems Research for 2006–2011 and 2012–2017.

References

[1] Adler, A., Guardo, R. and Berthiaume, Y. (1996). Impedance imaging of lung ventilation: Do we need to account for chest expansion? *IEEE Trans. Biomed. Eng.*, **43**, 414–420.

[2] Arridge, S. R., Kaipio, J. P., Kolehmainen, V., Schweiger, M., Somersalo, E., Tarvainen, T. and Vauhkonen, M. (2006). Approximation errors and model reduction with an application in optical diffusion tomography. *Inverse Probl*, **22**, 175–195.

[3] Bouman, C. and Sauer, K. (1993). A generalized Gaussian image model for edge-preserving map estimation. *IEEE Trans. Image Process*, **2**, 296–310.

[4] Boverman, G., Kao, T.-K., Kulkarni, R., Kim, B. S., Isaacson, D., Saulnier, G. J. and Newell, J. C. (2008). Robust linearized image reconstruction for multifrequence eit of the breast. *IEEE Trans. Med. Im.*, **27**, 1439–1448.

[5] Calvetti, D., Kaipio, J. P. and Somersalo, E. (2006). Aristotelian prior boundary conditions. *International Journal of Mathematics*, **1**, 63–81.

[6] Calvetti, D. and Somersalo, E. (2007). *An Introduction to Bayesian Scientific Computing Ten Lectures on Subjective Computing*. Springer. ISBN 978-0-387-73393-7.

[7] Cheney, M., Isaacson, D. and Newell, J. C. (1999). Electrical impedance tomography. *SIAM Rev*, **41**, 85–101.

[8] Cheng, K.-S., Isaacson, D., Newell, J. C. and Gisser, D. G. (1989). Electrode models for electric current computed tomography. *IEEE Trans. Biomed. Eng.*, **36**, 918–924.

[9] Doucet, A., de Freitas, N. and Gordon, N. (2001). *Sequential Monte Carlo Methods in Practice*. Springer.

[10] Duncan, S., Ruuskanen, J. P., Kaipio, A., Malinen, M. and Seppanen, A. (2005). Control systems. In *Handbook of Process Imaging for Automatic Control* (ed. D. Scott and H. McCann), pp. 237–262. CRC Press.

[11] Faridani, A., Finch, D. V., Ritman, E. L. and Smith, K. T. (1997). Local tomography II. *SIAM J. Appl. Math.*, **57**, 1095–1127.

[12] Fox, C. and Nicholls, G. (1997, 1-4 July). Sampling conductivity images via MCMC. In *"The art and science of Bayesian image analysis". Proceedings of the Leeds annual statistics research workshop* (ed. K. V. Mardia, C. A. Gill, and R. G. Aykroyd), Leeds, UK, pp. 91–100. Leeds university press.

[13] Frerichs, I. (2000). Electrical impedance tomography (eit) in applications related to lung and ventilation: a review of experimental and clinical activities. *Physiol. Meas.*, **21**, R1–R21.

[14] Gersing, E., Hoffmann, B. and Osypka, M. (1996). Influence of changing peripheral geometry on electrical impedance tomography measurements. *Med. Biol. Eng. Comput.*, **34**, 359–361.

[15] Golub, G. H. and van Loan, C. F. (1996). *Matrix Computations*, Volume 3rd. The Johns Hopkins University Press, Baltimore, MD.

[16] Heino, J. and Somersalo, E. (2004). A modelling error approach for the estimation of optical absorption in the presence of anisotropies. *Phys Med Biol*, **49**, 4785–4798.

[17] Heino, J., Somersalo, E. and Kaipio, J. P. (2005). Compensation for geometric mismodelling by anisotropies in optical tomography. *Optics Express*, **13**(1), 296–308.

[18] Huttunen, J. M. J. and Kaipio, J. P. (2007). Approximation error analysis in nonlinear state estimation with an application to state-space identification. *Inverse Problems*, **23**, 2141–2157.

[19] Huttunen, J. M. J. and Kaipio, J. P. (2007). Approximation errors in nostationary inverse problems. *Inverse Problem and Imaging*, **1**(1), 77–93.

[20] Huttunen, J. M. J. and Kaipio, J. P. (2009). Model reduction in state identification problems with an application to determination of thermal parameters. *Applied Numerical Mathematics*, **59**, 877–890.

[21] Huttunen, J. M. J., Lehikoinen, A., Hämäläinen, J. and Kaipio, J. P. (2009). Importance filtering approach for the nonstationary approximation error method. *Inverse Problems*. in review.

[22] Jolliffe, I. T. (1986). *Principal Component Analysis*. Springer-Verlag.

[23] Kaipio, J. P., Duncan, S., Seppanen, A., Somersalo, E. and Voutilainen, A. (2005). State estimation. In *Handbook of Process Imaging for Automatic Control* (ed. D. Scott and H. McCann), pp. 207–235. CRC Press.

[24] Kaipio, Jari and Somersalo, Erkki (2005). *Statistical and Computational Inverse Problems*. Springer-Verlag.

[25] Kaipio, J. and Somersalo, E. (2007). Statistical inverse problems: discretization, model reduction and inverse crimes. *J Comput Appl Math*, **198**.

[26] Kaipio, J. P., Kolehmainen, V., Somersalo, E. and Vauhkonen, M. (2000). Statistical inversion and Monte Carlo sampling methods in electrical impedance tomography. *Inverse Probl*, **16**, 1487–1522.

[27] Kolehmainen, V., Schweiger, M., Nissilä, I., Tarvainen, T., Arridge, S. R. and Kaipio, J. P. (2009). Approximation errors and model reduction in three-dimensional diffuse optical tomography. *J. Opt. Soc. Am.*, **26**(10), 2257–2268.

[28] Kolehmainen, V., Siltanen, S., Järvenpää, S., Kaipio, J. P., Koistinen, P., Lassas, M., Pirttilä, J. and Somersalo, E. (2003). Statistical inversion for medical x-ray tomography with few radiographs ii: Application to dental radiology. *Phys. Med. Biol.*, **48**, 1465–1490.

[29] Kolehmainen, V., Tarvainen, T., Arridge, S. R. and Kaipio, J. P. (2010). Marginalization of uninteresting distributed parameters in inverse problems – application to optical tomography. *Int J Uncertainty Quantification*. In review.

[30] Kolehmainen, V., Tarvainen, T., Arridge, S. R. and Kaipio, J. P. (2011). Marginalization of uninteresting distributed parameters in inverse problems – application to optical tomography. *Int. J. Uncertainty Quantification*, **1**(1), 1–17.

[31] Kolehmainen, V., Vauhkonen, M., Karjalainen, P. A. and Kaipio, J. P. (1997). Assessment of errors in static electrical impedance tomography with adjacent and trigonometric current patterns. *Physiol. Meas.*, **18**, 289–303.

[32] Kourunen, J., Savolainen, T., Lehikoinen, A., Vauhkonen, M. and Heikkinen, L. M. (2009). Suitability of a pxi platform for an electrical impedance tomography system. *Meas. Sci. Technol.*, **20**, 015503.

[33] Lehikoinen, A., Finsterle, S., Voutilainen, A., Heikkinen, L. M., Vauhkonen, M. and Kaipio, J. P. (2007). Approximation errors and truncation of computational domains with application to geophysical tomography. *Inverse Probl Imaging*, **1**, 371–389.

[34] Lehikoinen, A., Huttunen, J. M. J., Finsterle, S., Voutilainen, A., Kowalsky, M. B. and Kaipio, J. P. (2010). Dynamic inversion for hydrological process monitoring with electrical resistance tomography under model uncertainties. *Water Resources Res*, **46**, W04513. In press.

[35] Lehikoinen, A., Huttunen, J. M. J., Voutilainen, A., Finsterle, S., Kowalsky, M. B. and Kaipio, J. P. (2009). Dynamic inversion for hydrological process monitoring with electrical resistance tomography under model uncertainties. *Water Resources Res*. In press.

[36] Lipponen, A., Seppänen, A. and Kaipio, J.P. (2009). Nonstationary inversion of convection-diffusion problems – recovery from unknown nonstationary velocity fields. *Inverse Probl*. In review.

[37] Maass, P. (1992). The interior radon transform. *SIAM J. Appl. Math.*, **52**, 710–724.

[38] Mannington, W. O'Sullivan, M. J. and Bullivant, D. P. (2004). Computer modelling of the wairakei-tauhara geothermal system. *Geothermics*, **33**, 401–419.

[39] Mota, C. A. A., Orlande, H. R. B., Carvalho, M. O. M., Kolehmainen, V. and Kaipio, J. P. (2010). Bayesian estimation of temperature-dependent thermophysical properties and boundary heat flux. *Heat Transfer Eng*, **31**, 570–580.

[40] Natterer, F. (1986). *The Mathematics of Computerized Tomography*. Wiley, Philadelphia.

[41] Nicholls, G. K. (1998). Bayesian image analysis with Markov chain Monte Carlo and coloured continuum triangulation models. *J Royal Statistical Society B*, **60**, 643–659.

[42] Nissinen, A., Heikkinen, L. M. and Kaipio, J. P. (2008). The Bayesian approximation error approach for electrical impedance tomography—experimental results. *Meas. Sci. Technol.*, **19**, 015501.

[43] Nissinen, A., Heikkinen, L. M., Kolehmainen, V. and Kaipio, J. P. (2009). Compensation of errors due to discretization, domain truncation and unknown contact impedances in electrical impedance tomography. *Meas Sci Technol*, **20**, doi:10.1088/0957-0233/20/10/105504.

[44] Nissinen, A., Heikkinen, L. M., Kolehmainen, V. and Kaipio, J. P. (2009). Compensation of errors due to discretization, domain truncation and unknown contact impedances in electrical impedance tomography. *Meas. Sci. Technol.*, **20**, 105504.

[45] Nissinen, A., Kolehmainen, V., and Kaipio, J. P. (2011). Compensation of modelling errors due to unknown domain boundary in electrical impedance tomography. *IEEE Trans. Med. Im.*, **30**(2), 231–242.

[46] Nissinen, A., Kolehmainen, V. and Kaipio, J. P. (2011). Reconstruction of domain boundary and conductivity in electrical impedance tomography using the approximation error approach. *Int. J. Uncertainty Quantification.*, **1**(3), 203–222.

[47] Nocedal, J. and Wright, S. J. (2006). *Numerical Optimization (second edition)*. Springer, New York.

[48] Rue, H. and Held, L. (2005). *Gaussian Markov Random Fields: Theory and Applications*. Chapman and Hall.

[49] Seppänen, A., Vauhkonen, M., Vauhkonen, P. J., Somersalo, E. and Kaipio, J. P. (2001). State estimation with fluid dynamical evolution models in process tomography – an application to impedance tomography. *Inverse Probl*, **17**, 467–484.

[50] Seppänen, A., Vauhkonen, M., Vauhkonen, P. J., Somersalo, E. and Kaipio, J. P. (2001). State estimation with fluid dynamical evolution models in process tomography – an application to impedance tomography. *Inverse Probl*, **17**, 467–484.

[51] Siltanen, S., Kolehmainen, V., Järvenpää, S., Kaipio, J. P., Koistinen, P., Lassas, M., Pirttilä, J. and Somersalo, E. (2003). Statistical inversion for medical x-ray tomography with few radiographs i: General theory. *Phys. Med. Biol.*, **48**, 1437–1463.

[52] Smith, K. T. and Keinert, F. (1985). Mathematical foundations of computed tomography. *Appl. Optics*, **24**, 3950–3957.

[53] Somersalo, E., Cheney, M. and Isaacson, D. (1992). Existence and uniqueness for electrode models for electric current computed tomography. *SIAM J. Appl. Math.*, **52**, 1023–1040.

[54] Somersalo, E., Kaipio, J. P., Vauhkonen, M., Baroudi, D. and Järvenpää, S. (1997). Impedance imaging and Markov chain Monte Carlo methods. In *Proc SPIE's 42nd Annual Meeting, Computational, experimental and numerical methods for solving ill-posed inverse imaging problems: medical and nonmedical applications* (ed. R. Barbour, M. Carvlin, and M. Fiddy), San Diego, USA, June 27–August 1, pp. 175–185.

[55] Tarantola, A. (2004). *Inverse Problem Theory and Methods for Model Parameter Estimation*. SIAM, Philadelphia.

[56] Tarvainen, T., Kolehmainen, V., Pulkkinen, A., Vauhkonen, M., Schweiger, M., Arridge, S. R. and Kaipio, J. P. (2010). Approximation error approach for compensating modelling errors between the radiative transfer equation and the diffusion approximation in diffuse optical tomography. *Inv. Probl.*, **26**, 015005 (18pp).

[57] Vauhkonen, P. J., Vauhkonen, M., Savolainen, T. and Kaipio, J. P. (1999). Three-dimensional electrical impedance tomography based on the complete electrode model. *IEEE Trans. Biomed. Eng.*, **46**, 1150–1160.

[58] Victorino, J., Borges, J. B., Okamoto, V. N., Matos, G. F. J., Tucci, M. R., Caramez, M. P. R., Tanaka, H., Sipmann, F. S., Santos, D. C. B., Barbas, C. S. V., Carvalho, C. R. R. and Amato,

M. B. P. (2004). Imbalances in regional lung ventilation. *Am. J. Respir. Crit. Care. Med.*, **169**, 791–800.

[59] Voutilainen, A., Lehikoinen, A., Lipponen, A. and Vauhkonen, M. (2011). A reduced-order filtering approach for 3d dynamical electrical impedance tomography. *Meas Sci Technol*, **22**, 025504.

[60] Williams, R. A. and Beck, M. S. (1995). *Process Tomography: Principles, Techniques and Applications*. Butterworth-Heinemann, Oxford.

[61] Zou, Y. and Guo, Z. (2003). A review of electrical impedance techniques for breast cancer detection. *Med. Eng. Phys.*, **25**, 79–90.

33 Bayesian reconstruction of particle beam phase space

C. NAKHLEH, D. HIGDON, C. K. ALLEN AND R. RYNE

33.1 Introduction

Understanding the physics of charged particle beams (e.g. proton beams) is essential to designing and controlling efficient particle accelerators. The dynamics of a beam are naturally formulated in a six-dimensional phase space (three position and three momentum or velocity dimensions). However, experimental beam profile data taken from accelerators are typically one-dimensional (coordinate) projections of the phase space distribution or image. The objective of this study is to apply Bayesian inversion methodology to reconstruct the initial phase space configuration of the beam using a series of one-dimensional projection datasets (wirescans). Given this initial phase space configuration, the evolution of the phase space down the beam-line can be inferred, providing a key beam diagnostic that is of great interest to accelerator designers. Because of the limited information provided by the wirescan projection data, encoding prior information regarding the high-dimensional initial phase-space configuration of the beam is crucial.

The outline of this chapter is as follows. Firstly, we give a brief description of the proton beam produced by the Low Energy Demonstration Accelerator (LEDA) at the Los Alamos National Laboratory [1]. The proton beam's phase space evolution is affected by various magnets along the beamline, as well as interactions between the charged particles themselves. Two basic strategies exist for modeling the beam's evolution. The first propagation approach uses a computationally demanding forward model [13], using a high-fidelity particle-in-cell approach; the second uses linear transfer matrices to propagate protons along the beamline, sacrificing accuracy for substantial gains in speed. This faster modelling approach proves to be sufficiently accurate, making Markov chain Monte Carlo (MCMC) a feasible approach for posterior exploration, even for a highly parameterized representation of the initial phase space.

Next, we describe our modelling approach, using process convolutions [7] to represent the initial phase space configurations. This process convolution representation gives a parsimonious representation of the particle density as a function of spatial position and momentum, while ensuring positivity. We describe the MCMC scheme to produce posterior draws, producing a reconstruction for the initial phase space. We also outline additional sensitivity studies to assess the impact of the prior smoothness on the resulting initial phase space reconstruction.

33.2 The Low Energy Demonstration Accelerator

LEDA, now decomissioned, was an 11 m-long, 6.7 MeV proton accelerator designed specifically to study continuous, high-current proton beams. The LEDA beamline consisted of 52 focusing/defocusing (FODO) quadrupole magnets; a number of steering magnets; and nine wire scanners. A picture of the actual machine is given in Figure 33.1; the wire scanners are visible in the picture, pointing out of the beamline at a 45 degree angle from straight up.

The data utilized in this chapter come from two separate beamline segments, each containing four pairs of x and y wirescanners—both of these segments are visible in Figure 33.1. At each wire scan station, the proton beam distribution was projected onto orthogonal directions (x and y) at ±45 degrees from vertical. The wirescanner (Figure 33.2) used a 33 μm carbon wire to measure the beam profile via secondary electron emission (because the wire itself was not thick enough to stop the protons themselves).

33.2.1 The experimental setup

For the purposes of this chapter, we focus on a 1.25 m segment of the LEDA beamline, as shown in Figure 33.3. The figure shows a simulated beam with the associated x- and y-direction phase space configurations as the beam passes through six quadrupole magnets and four pairs of wirescanners. The proton beam generally moves along a linear path, influenced by three pairs of quadrupole magnets. Each quadrupole magnet takes up about 7 cm of length along the beamline and is separated by about 14 cm of drift space. A quadrupole magnet acts on the particle beam as a lens acts on a beam of light. A focusing quadrupole causes the beam to converge in the x dimension and to diverge in the y dimension; a defocusing quadrupole causes the beam to diverge in the x dimension and to

Figure 33.1 The Low Energy Demonstration Accelerator at Los Alamos National Laboratory.

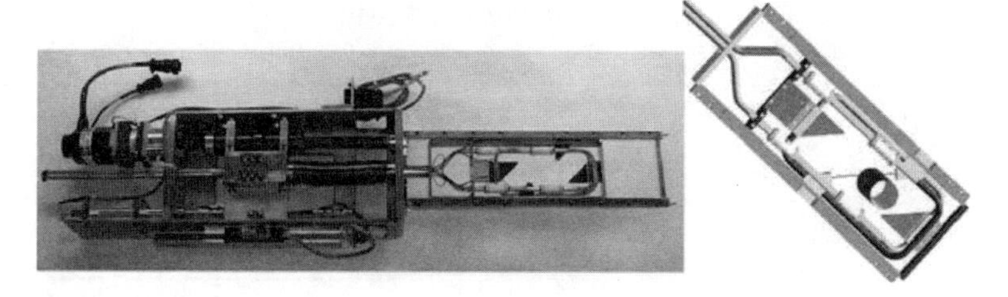

Figure 33.2 A wirescanner used in the LEDA experiments. Groups of four wirescanners are visible in Figure 33.1.

converge in the y dimension. The first of each magnet pair is a defocusing quadrupole, represented by the dark shaded region in the beamline plots of Figure 33.3; the second is a focusing quadrupole, represented by the light shaded region in the beamline plots of Figure 33.3.

As the beam passes the wirescanners, a histogram of the particle density is produced in both the x and y directions, shown by the light dots at the bottom of the phase space diagrams in Figure 33.3. The wirescanners measure a current intensity that is proportional to the beam density projected onto its component axis (either x or y) at four equally spaced positions along the 1.25 m beamline. Hence the experiment produces eight 1-d intensity profiles representing the beam density as a function of distance along the two transverse spatial coordinates.

33.3 The forward model

For the experiments we consider, the beam is in steady state, so its x- and y-phase space configuration is only a function of its initial configuration, and its distance along the beamline. Furthermore, we can expect that correlations between the x and y phase space do not affect how the beam propagates. Hence the x and y phase spaces can be characterized separately, without having to specify additional dependencies between the x and y phase spaces. Note that we use the standard font (e.g. u, θ) to denote statistical parameters, differentiating them from phase space coordinates (e.g. x, $\mathrm{p_x}$).

The initial phase space configuration θ is described by two 2-d point clouds—one over $(\mathrm{x}, \mathrm{p_x})$, and one over $(\mathrm{y}, \mathrm{p_y})$. Alternatively, the initial phase space could be described by two 2-d density images, giving the density of protons over a dense grid of $(\mathrm{x}, \mathrm{p_x})$ or $(\mathrm{y}, \mathrm{p_y})$ points. In Figure 33.3, θ is described by the two leftmost phase space point clouds.

Given the initial phase space configuration θ, the forward model propagates the beam along the 1.25 m beamline, accounting for the effects of the quadrupole magnets on the moving protons, and accounting for charge interactions between the protons themselves. At various positions along the beamline, the proton density can be projected along the x and y coordinates, producing predicted values for the wirescanners. We take $\widehat{d}(\theta)$ to denote the modelled wirescanner output using the initial phase space configuration θ

$$\widehat{d}(\theta) = (\widehat{d}_{1\mathrm{x}}, \widehat{d}_{1\mathrm{y}}, \widehat{d}_{2\mathrm{x}}, \widehat{d}_{2\mathrm{y}}, \widehat{d}_{3\mathrm{x}}, \widehat{d}_{3\mathrm{y}}, \widehat{d}_{4\mathrm{x}}, \widehat{d}_{4\mathrm{y}})$$

The dots in Figure 33.3 show the predicted wirescanner values for that particular simulation.

We considered two candidate forward models for this particular inverse problem. The first is a computational model called MaryLie/IMPACT (ML/I) [13]. This code evolves individual protons according to the electromagnetic forces acting on them. In particular, this code implements a parti-

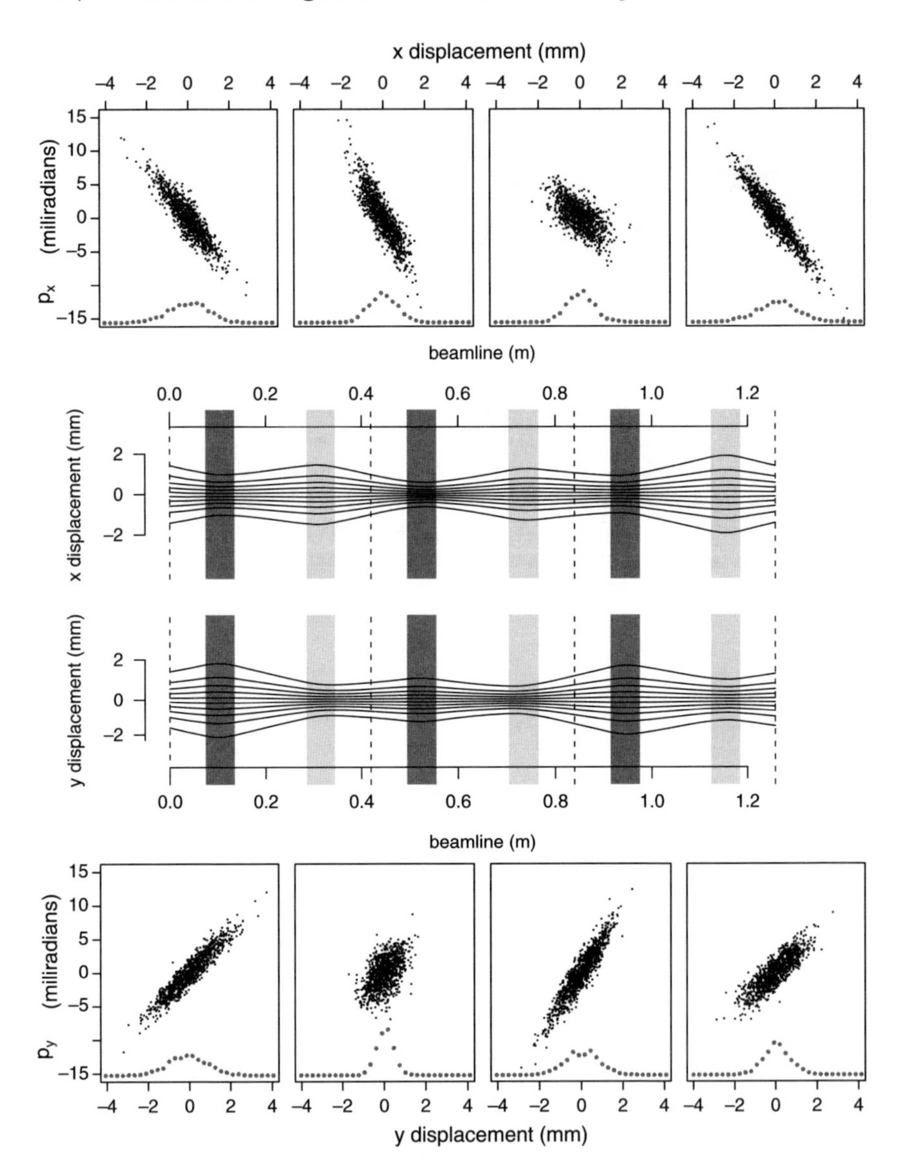

Figure 33.3 Proton beam simulation of a 1.25 m segment of the LEDA beamline. The central diagrams show part of the LEDA beamline consisting of three pairs of focusing and defocusing quadrupole magnets denoted by the dark and light shaded regions respectively. Wirescan locations are given by the four dashed vertical lines. At these four beamline locations, the resulting x and y phase space is shown for this particular simulation. The wirescanners produce a 'histogram' of the phase space projected onto the x- and y-axes (shaded dots at the bottom of the phase space clouds).

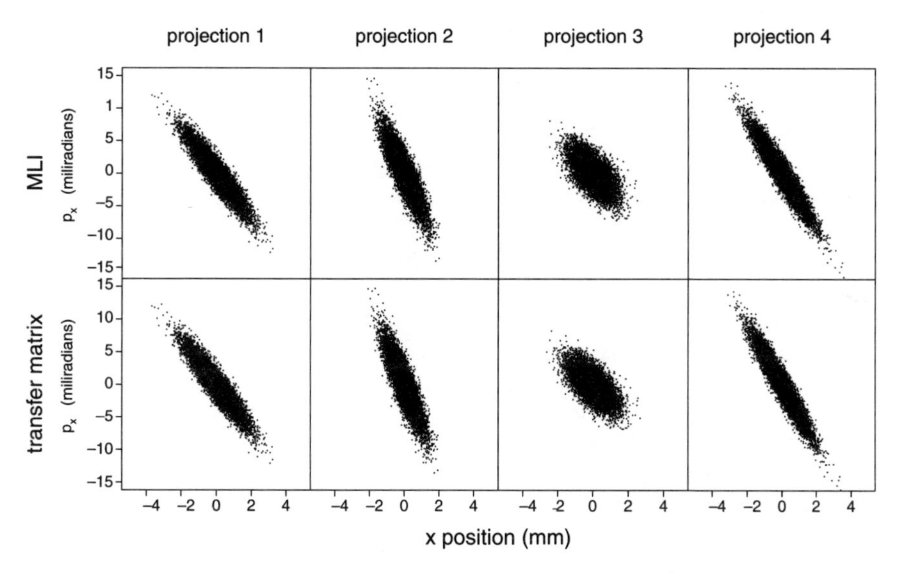

Figure 33.4 A comparison of phase space configurations produced by the computational model ML/I (top) and the transfer matrix approach (bottom). The simulations correspond to the experimental configuration shown in Figure 33.3, using the same initial phase space (leftmost frames labelled projection 1). While the two modelling approaches give nearly identical results for this example, the transfer matrix approach is orders of magnitude faster.

cle in cell algorithm to capture the repulsive effects of the charged particles on one another. While the effect of the quadrupole magnets on the beam is nearly linear, the charge interaction can be highly nonlinear in some cases, making it necessary to use such a high-fidelity code to accurately model the beam dynamics.

An alternative modelling strategy, motivated in part by these LEDA experiments, uses linear transfer matricies to evolve the phase space as the beam propagates down the beamline [2]. This approach accounts for the linear effects of the quadrupole magnets, and uses an approximation to account for the effects of the charge interactions. This approximation has proven to be quite accurate in the conditions produced at the LEDA facility. Figure 33.4 gives a comparison of phase spaces produced by the two forward modelling approaches—the results are nearly identical. Moreover, it can evolve the initial beam θ over the beamline, producing modelled wirescanner output $\widehat{d}(\theta)$ very quickly. This drastic reduction in computing time allows us to consider more highly parameterized process convolution representations [7, 9] of the initial phase space θ—see Section 33.4.2 below. This is a distinct advantage over approaches that use the slower ML/I code [12].

33.4 Statistical formulation

As stated in the introductory chapter on inverse problems, we need to define the sampling model for the data d, a prior for the unknown field to be estimated θ, and possibly additional statistical parameters in the formulations. Each of these terms will be described below, producing a posterior distribution which will be sampled using standard MCMC techniques [4, 14].

33.4.1 Sampling model for the wirescanner data

As mentioned earlier, the wirescanners produce a current that is proportional to the density of the proton beam projected down to the x or y axes, as shown by the light dots in Figure 33.3. Hence the data are eight vectors, each giving an intensity as a function of spatial displacement from the centre of the beam along the x and y axes

$$d = (d_{1x}, d_{1y}, d_{2x}, d_{2y}, d_{3x}, d_{3y}, d_{4x}, d_{4y})$$

Given the fitted wirescanner signal from the forward model $\widehat{d}(\theta)$, each of these components is modelled independently, as multivariate normal draws, after taking a square root transformation. We use the square root transformation to stabilize the variance, improving the normality and making it easier to account for dependence within a single wirescan. So now when we write d or $\widehat{d}(\theta)$, we assume the square root transformation has been carried out.

So, for d_{1x}, using n_{1x} to denote the number of points produced by this particular wirescan, the sampling density can be written

$$L(d_{1x}|\theta, \lambda_d) \propto \lambda_d^{\frac{n_{1x}}{2}} \exp\left\{-\tfrac{1}{2}\lambda_d(d_{1x} - \widehat{d}_{1x}(\theta))' R_{d1x}^{-1}(d_{1x} - \widehat{d}_{1x}(\theta))\right\}.$$

The parameter λ_d is an unknown precision parameter to be estimated in this analysis. Here the error covariance $\lambda_d^{-1} R_{d1x}$ accounts for scatter in the measurements, dependencies induced by the process of producing the wirescanner signal, and by the discrepancy between the model and the actual physical system. More systematic attempts to estimate the appropriate form of R can be found in [8, 11]. Here we use expert judgement and past experience to specify R as a mixture of iid and correlated errors. We use the same R for each of the wirescans.

The same sampling model is used for each of the remaining seven wirescans, using the appropriate forward model prediction $\widehat{d}(\theta)$. This produces a larger normal sampling model for the data

$$L(d|\theta, \lambda_d) \propto \lambda_d^{\frac{n}{2}} \exp\left\{-\tfrac{1}{2}\lambda_d(d - \widehat{d}(\theta))' R_d^{-1}(d - \widehat{d}(\theta))\right\} \tag{33.1}$$

where n is the total number of measurements produced by all eight wirescans and $R_d = \text{diag}(R_{d1x}, \ldots, R_{d4y})$. To complete this specification, we assign a $\Gamma(a_d, b_d)$ prior for the precision parameter

$$\pi(\lambda_d) \propto \lambda_d^{a_d - 1} \exp\{-b_d \lambda_d\}$$

We take $a_d = 1$ and $b_d = 0.001$, giving a rather uninformative prior, allowing the substantial amount of wirescanner data to inform about λ_d.

33.4.2 Process convolution prior for the phase space

We elect to characterize the unknown initial phase space configuration θ using a pair of two-dimensional images θ_x and θ_y, describing the particle density in phase space. Each image is a 200×160, but represented with a lower-dimensional process convolution. Before getting into the specifics of the representation, we first describe the basic approach of using process convolutions on a simple, one-dimensional example that contains some of the features of this accelerator application.

33.4.2.1 *A simple example*

Consider a one-dimensional emission source shown in Figure 33.5. This object, 100 units long, emits objects according to an inhomogeneous Poisson process, whose emission rate is given by the black line. Over a fixed time interval, 382 emissions occur at the locations marked by the dashes at the bottom of Figure 33.5.

The data collected here are the total number of counts occurring in the five intervals $I_1 = [0, 20), \ldots, I_5 = [80, 100)$. These counts, divided by the interval length of 20, are given by the horizontal lines over each of the five segments in Figure 33.5.

The process convolution model represents the intensity function $\theta(s)$, $s \in [0, 100]$ as a simple process $u(s)$ convolved with a smoothing kernel $k(\cdot)$. We take $u(s)$ to be a discrete process, with independent normal values at $m = 7$ fixed knot locations $s_1^*, \ldots, s_m^*$ as shown in the left frame of Figure 33.6. When $u(s)$ is convolved with a normal kernel (a normal density with a standard deviation of 25), the grey line in the right frame of Figure 33.6 is produced. We take $\theta^+(s)$ to be the non-negative part of this smooth process, given by the black line in the right frame of Figure 33.6.

To be more concrete, we specify a normal prior for $u = (u_1, \ldots, u_m)'$, and associate locations $\{s_1^*, \ldots, s_m^*\}$ to each component of u. Thus the model specification becomes

$$\theta(s) = \sum_{j=1}^{m} k(s - s_j^*)u_j$$

$$\pi(u) \propto \lambda_u^{\frac{m}{2}} \exp\{-\tfrac{1}{2}\lambda_u u' u\}$$

$$\pi(\lambda_u) \propto \lambda_u^{a_u - 1} \exp\{-b_u \lambda_u\}$$

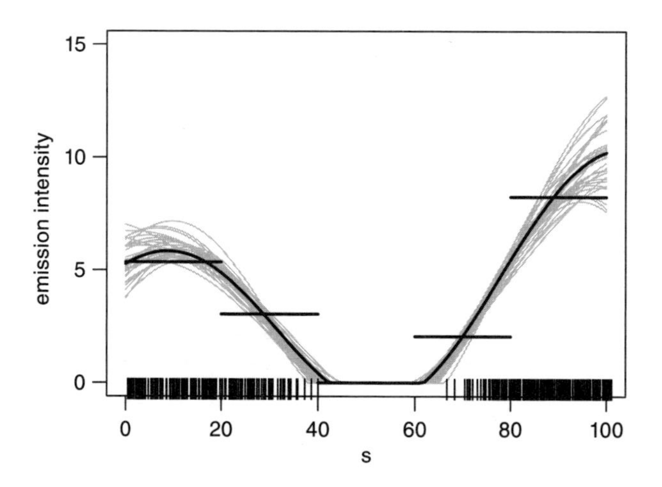

Figure 33.5 A simple example of using process convolutions to estimate the underlying image intensity. A linear region [0,100] is emitting particles at a rate given by the black line. Over a fixed time 382 emissions occur at locations shown by the dark shaded marks at the bottom of the figure. The data are total counts recorded over each of five consecutive intervals of length 20—the dark shaded lines show the total number of counts divided by the interval length (20) for the five aggregated counts. Realizations from the resulting posterior distribution for the image intensity are given by the light shaded lines.

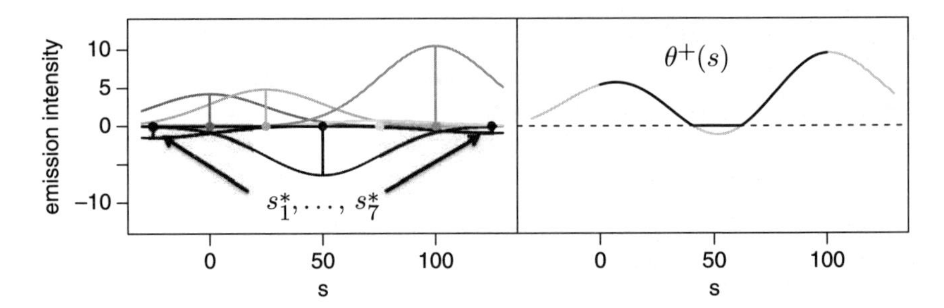

Figure 33.6 Representing a non-negative image with process convolutions.

Note that λ_u, the precision for u, is also treated as a parameter to be estimated; we use the rather uninformative settings $a_u = 1$ and $b_u = 0.001$. Because $\theta(s)$ can go negative, we define the modelled image intensity to be the non-negative part of $\theta(s)$

$$\theta^+(s) = \begin{cases} \theta(s) & \text{if } \theta(s) \geq 0 \\ 0 & \text{if } \theta(s) < 0. \end{cases}$$

Each of the $n = 5$ counts $d = (d_1, \ldots, d_n)'$ follow independent Poisson distributions with rate $\mu_i(\theta)$ given by

$$\mu_i(\theta) = \int_{I_i} \theta^+(s) ds$$

so that

$$L(d|\theta) \propto \prod_{i=1}^{n} \frac{e^{\mu_i(\theta)} \mu_i(\theta)^{d_i}}{d_i!}.$$

The resulting posterior density for this simple image reconstruction problem has the form

$$\pi(u, \lambda_u|d) \propto \prod_{i=1}^{n} e^{\mu_i(\theta)} \mu_i(\theta)^{d_i} \times \lambda_u^{\frac{m}{2}} \exp\{-\tfrac{1}{2}\lambda_u u' u\} \times \lambda_u^{a_u - 1} \exp\{-b_u \lambda_u\}$$

which can be explored using standard single-site MCMC methods. Metropolis updates can be used for the u_is, and a Gibbs step can be used for the precision parameter λ_u. A number of posterior realizations of $\theta^+(s)$ are given by the grey lines in Figure 33.5. Note that the image intensity $\theta^+(s)$ can be evaluated for any s since this function is determined by the knot values u. Also posterior realizations of the model can easily reproduce a 0 emission rate. This will be important for modelling the phase space images where much of the image has 0 particle density. Note also that this process convolution approach shows different behaviour as compared to log-Gaussian fields that are often used to model spatially distributed emissions from a Poisson process [5].

33.4.2.2 *Representing the* x *and* y *phase space images*

Recall the unknown initial phase space configuration θ is described using a pair of 200 × 160 images θ_x and θ_y. As with the previous simple example, we represent each image with a process convolution, but now $s = (x, p_x)$ indexes a two-dimensional phase space. We define the $m = 29 \times 21$

knot locations shown in Figure 33.7. We take the smoothing kernel $k(\cdot)$ to have the tricube [6] form

$$k(s) \propto (1 - |s|^3/r^3)^3 I[|s| \leq r] \tag{33.2}$$

where r is the range of the kernel, and $I[\cdot]$ is the indicator function. This representation also allows for computational savings since it localizes the influence of a particular knot.

Focusing on the x phase space image we have

$$\theta_{\mathrm{x}}(s) = \sum_{j=1}^{m} k(\| s - s_j^* \|) u_j$$

where $\| \cdot \|$ denotes Euclidean distance, the grid points $s_j^* = (\mathrm{x}_j^*, \mathrm{px}_j^*)$ are marked in Figure 33.7, as is the smoothing kernel. The value of r which scales the tricube kernel is chosen to give particle density variation expected by accelerator scientists. Subsequent sensitivity studies have shown this value of r to be good—not oversmoothing, and not producing wildly irregular phase space reconstructions. At the end of Section 33.5, an analysis with r taken to be half its selected value produces overly irregular reconstructions.

The positive part of the phase space images $\theta_{\mathrm{x}}^{+}(s)$ and $\theta_{\mathrm{y}}^{+}(s)$ are then taken as inputs to the forward model, producing fitted wirescans $\widehat{d}(\theta)$ as well as phase space configurations along the beamline. Of particular interest for us are the phase space images at the three locations corresponding to the remaining wirescanners.

33.4.3 Posterior distribution

Putting this all together produces a posterior distribution that depends on the data precision parameter λ_d, the latent knot values u, and the regularizing precision for the knot values λ_u

$$\pi(\lambda_d, u, \lambda_u | d) \propto L(d|u, \lambda_d) \times \pi(\lambda_d) \times \pi(u|\lambda_u) \times \pi(\lambda_u).$$

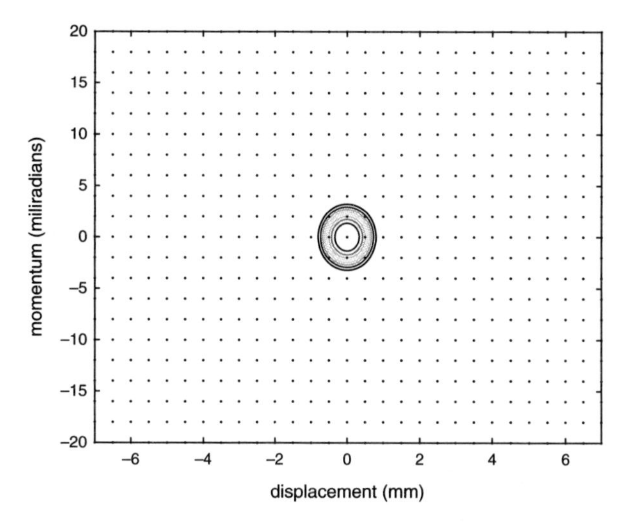

Figure 33.7 Knot locations and kernel used for the process convolution representation of the x and y phase space.

Note that given the values for u, the initial phase space image is completely determined. Mathematically, the posterior can be written

$$\pi(\lambda_d, u, \lambda_u|d) \propto \lambda_d^{\frac{n}{2}} \exp\left\{-\tfrac{1}{2}\lambda_d(d - \widehat{d}(\theta))' R_d^{-1}(d - \widehat{d}(\theta))\right\} \times \quad (33.3)$$
$$\lambda_d^{a_d - 1} \exp\{-b_d\lambda_d\} \times$$
$$\lambda_u^{\frac{m}{2}} \exp\{-\tfrac{1}{2}\lambda_u u' u\} \times \lambda_u^{a_u - 1} \exp\{-b_u\lambda_u\}.$$

The two precision parameters can be updated with Gibbs steps, sampling directly from their Gamma full conditional distributions.

$$\lambda_d|\cdots \sim \Gamma\left(a_d + n/2, b_d + \tfrac{1}{2}(d - \widehat{d}(\theta))' R_d^{-1}(d - \widehat{d}(\theta))\right)$$
$$\lambda_u|\cdots \sim \Gamma\left(a_u + m/2, b_u + \tfrac{1}{2}u' u\right)$$

We update the components of u with standard random walk Metropolis updates. Note that when updating a single u_i, its effect on $\widehat{d}(\theta)$ can be computed very efficiently since its impact on $\theta(s)$ is linear. The final step of computing the impact of changes to $\theta^+(s)$ is also fast, allowing an efficient means for computing changes to the likelihood term (33.1).

33.5 Results

We consider data taken from a single experiment, considering two separate 1.25 m segments of the beamline which conform to Figure 33.3. Results using data from this first segment are shown in Figure 33.8; results from the second segment, about 4 m further down the beamline, are shown in Figure 33.9.

For the x and y phase space, Figure 33.8 shows two rows of four plots. The bottom four plots show the wirescanner data, given by the dots. The top four plots give the posterior mean of the phase space image initially, coincident with the first wirescanner location, and at locations coincident with the remaining three wirescanners. These last three images are produced by propagating the initial phase space estimate through the forward model.

The dark, interior lines in the wirescanner plots show pointwise 90% credible intervals for the beam density projected down to the x or y axes. The width of these intervals is due to uncertainty in the initial phase space images θ. The light, outer lines show pointwise 90% bounds for the wirescanner data, which include additional uncertainty due to the observation process, controlled by λ_d and R.

To assess the impact of our choice of kernel width—controlled by r in equation (33.2)—we first tried altering the prior specification for the process u, replacing the i.i.d. normal prior with a two-dimensional intrinsic Markov random field (MRF) [3, 15]. Such spatial dependence between the components of u can modify the smoothness in the resulting posterior realizations for the image θ [10]. This implementation is simple since it only slightly modifies the posterior distribution, and adds little additional computational complexity. The resulting posterior mean and uncertainty for θ under this alternative formulation is very similar, suggesting the data do not support smoother realizations for θ, and that the choice of kernel width is compatible with the data.

As a final check, we carried out the analysis with our original formulation, but substituting the value of r with one that was half its original value, leading to a much narrorwer kernel $k(\cdot)$ in the process convolution specification. The results are shown in Figure 33.10, using data from the first

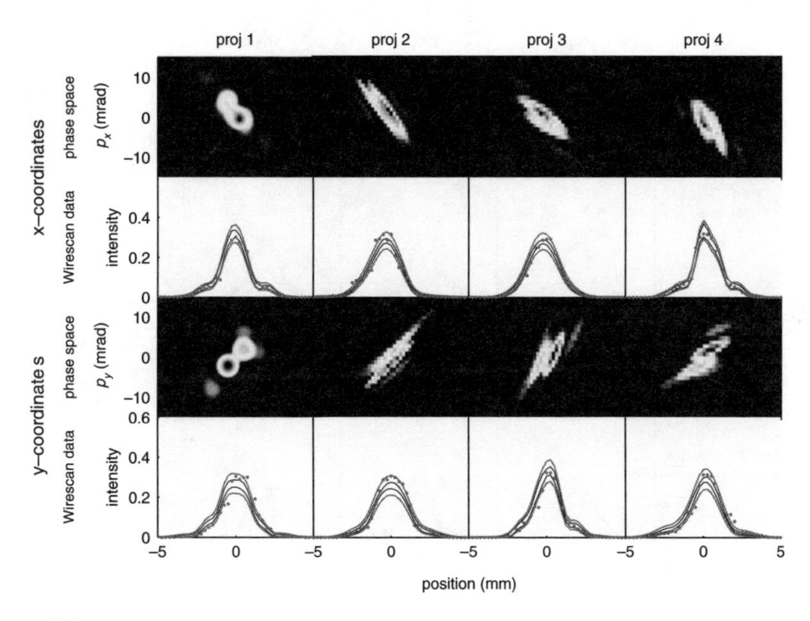

Figure 33.8 Posterior summary of the analysis of wirescanner data from the first LEDA beam segment.

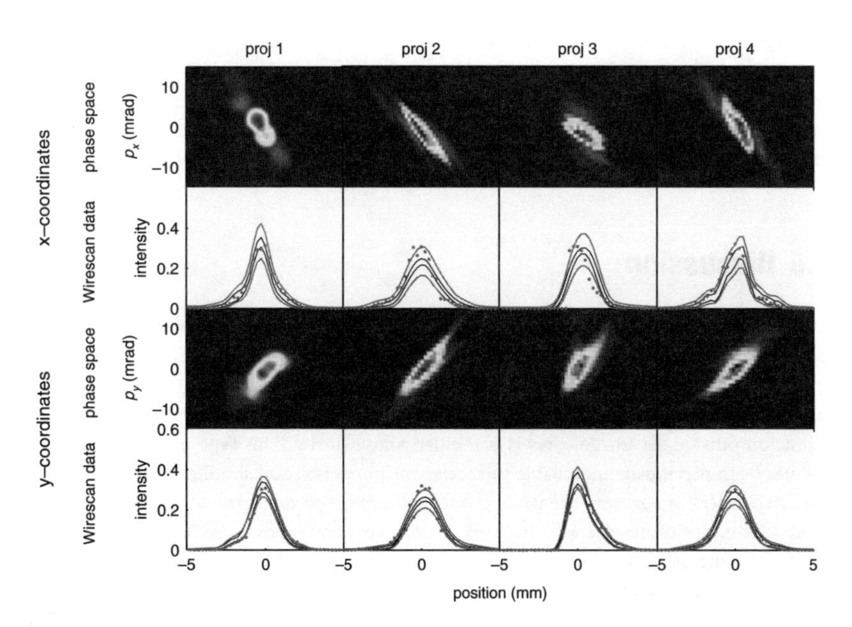

Figure 33.9 Posterior summary of the analysis of wirescanner data from the second LEDA beam segment.

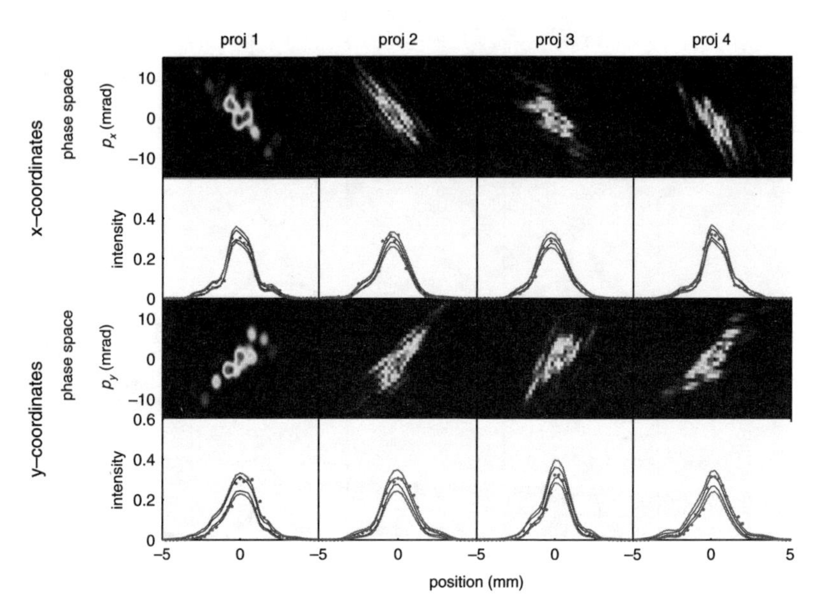

Figure 33.10 Posterior summary of the analysis of wirescanner data from the first LEDA beam segment using a convolution kernel with half the width of the original specification. The resulting phase space reconstructions are unrealistically patchy.

beam segment. Hence this figure corresponds to Figure 33.8 which shows the original results. While this reconstruction produces a posterior mean that gives a slightly better match to the wirescanner data, the phase space reconstruction θ is now unrealistically composed of many small bunches of particles. Again, this sensitivity study argues for our original kernel width choice.

33.6 Discussion

Reconstruction of particle beam phase space images from projection data provides an excellent application of modern computational Bayesian inference. The quality of the reconstructions is quite good, and convincingly demonstrates the ability of this approach to extract physically sensible phase space information from the projection data. The ability to capture complex phase space information outside the image cores is particularly noteworthy. This type of phase space image detail has been previously unavailable to accelerator physicists, and should enhance their ability to predict accelerator performance when changing beamline parameters and to design accelerator upgrades. This type of analysis will furthermore allow accelerator physicists to gain insight into the properties of intense charged particle beams at locations where it might be too expensive or too difficult to make direct measurements.

On the statistical side, the use of the process convolution model for the phase space images θ leads to a number of benefits—easy to control regularization of the image, parameter reduction, ensuring positivity, allowing 0s in the posterior realizations, and leading to a posterior that can be easily explored using standard algorithms. This application might also be adapted to make use of alternative spatial representations such as the integrated nested Laplace approximation [16], alleviating the need for posterior sampling.

The major limiting approximation made in this paper is the use of the transfer matrix method to model the beam dynamics. As we have shown, this approximation is satisfactory for the LEDA accelerator. However, in other applications this approximation is not adequate, so that full particle-in-cell simulations are necessary, making computational expense an important factor. MCMC (or other computationally intensive) techniques place a high premium on finding methods for rapidly obtaining simulator output. This challenge is likely to be met through a combination of advanced algorithms combined with large-scale parallel simulation.

Acknowledgements

This work was supported in part by the US Department of Energy Office of Science, Office of Advanced Scientific Computing Research, Scientific Discovery through Advanced Computing (SciDAC) program. This research also used resources of the National Energy Research Scientific Computing Center, which is supported by the Office of Science of the US Department of Energy under Contract No. DE-AC02-05CH11231.

Sandia National Laboratories is a multi-program laboratory managed and operated by Sandia Corporation, a wholly owned subsidiary of Lockheed Martin Corporation, for the US Department of Energy's National Nuclear Security Administration under contract DE-AC04-94AL85000.

References

[1] Allen, C. K., Chan, K. C. D., Colestock, P. L., Crandall, K. R., Garnett, R. W., Gilpatrick, J. D., Lysenko, W., Qiang, J., Schneider, J. D., Schulze, M. E., Scheffield, R. L., Smith, H. V. and Wangler, T. P. (2002). Beam-halo measurements in high-current proton beams. *Physical Review Letters*, **89**(21), 214802.

[2] Allen, C. K. and Pattengale, N. D. (2002). Theory and technique of beam envelope simulation: Simulation of bunched particle beams with ellipsoidal symmetry and linear space charge forces. Technical Report LA-UR-02-4979, Los Alamos National Laboratory.

[3] Besag, J. (1993). Towards Bayesian image analysis. *Journal of Applied Statistics*, **20**(5–6), 107–119.

[4] Besag, J., Green, P. J., Higdon, D. M. and Mengersen, K. (1995). Bayesian computation and stochastic systems (with discussion). *Statistical Science*, **10**, 3–66.

[5] Brix, A. and Moller, J. (2001). Space-time multi type log Gaussian Cox processes with a view to modelling weeds. *Scandinavian Journal of Statistics*, **28**(3), 471–488.

[6] Cleveland, William S. (1979). Robust locally weighted regression and smoothing scatterplots. *Journal of the American Statistical Association*, **74**, 829–836.

[7] Higdon, Dave (2002). Space and space-time modeling using process convolutions. In *Quantitative Methods for Current Environmental Issues* (ed. C. Anderson, V. Barnett, P. C. Chatwin, and A. H. El-Shaarawi), London, pp. 37–56. Springer Verlag.

[8] Higdon, D., Kennedy, M., Cavendish, J. C., Cafeo, J. A. and Ryne, R. D. (2005). Combining field data and computer simulations for calibration and prediction. *SIAM Journal on Scientific Computing*, **26**(2), 448–466.

[9] Higdon, D. M., Lee, H. and Holloman, C. (2003). Markov chain Monte Carlo-based approaches for inference in computationally intensive inverse problems. In *Bayesian Statistics 7. Proceedings of the Seventh Valencia International Meeting* (ed. J. M. Bernardo, M. J. Bayarri, J. O. Berger, A. P. Dawid, D. Heckerman, A. F. M. Smith and M. West), pp. 181–197. Oxford University Press.

[10] Lee, H. K. H., Higdon, D. M., Calder, C. A. and Holloman, C. H. (2005). Efficient models for correlated data via convolutions of intrinsic processes. *Statistical Modelling*, **5**(1), 53.

[11] Lee, H., Sansó, B., Zhou, W. and Higdon, D. (2006). Inferring particle distribution in a proton accelerator experiment. *Bayesian Analysis*, **1**(2), 249–264.

[12] Lee, H. K. H., Sanso, B., Zhou, W. and Higdon, D. M. (2008). Inference for a proton accelerator using convolution models. *Journal of the American Statistical Association*, **103**(482), 604–613.

[13] Qiang, J., Ryne, R. D. and Habib, S. (2000). Fortran implementation of object-oriented design in parallel beam dynamics simulations. *Computer Physics Communications*, **133**(1), 18–33.

[14] Robert, Christian P. and Casella, George (1999). *Monte Carlo Statistical Methods*. Springer-Verlag Inc.

[15] Rue, H. and Held, L. (2005). *Gaussian Markov Random Fields: Theory and Applications*, Volume 104. Chapman & Hall.

[16] Rue, H., Martino, S. and Chopin, N. (2009). Approximate Bayesian inference for latent Gaussian models by using integrated nested Laplace approximations. *Journal of the Royal Statistical Society: Series B (Statistical Methodology)*, **71**(2), 319–392.

Adrian Smith's research supervision (PhD)

1972–1974	M. Goldstein: Aspects of linear statistical inference, University of Oxford.
1975–1977	D. J. Spiegelhalter: Bayesian inference using finite mixture models, University College London.
1974–1980	B. Booth: Identification of change-points in time series models, University College London.
1975–1983	U. E. Makov: Bayesian approximations to unsupervised learning procedures, University College London.
1977–1983	J. C. Naylor: Some numerical aspects of Bayesian inference, University of Nottingham.
1978–1982	M. West: Aspects of recursive Bayesian estimation, University of Nottingham.
1979–1983	L. I. Pettit: Bayesian approaches to outliers, University of Nottingham.
1980–1985	F. L. Ezzet: Applied sequential procedures: robustness and change-point problems, University of Nottingham.
1982–1985	A. N. Pole: Inference for threshold models, University of Nottingham.
1982–1985	M. Upsdell: Bayesian inference for functions, University of Nottingham.
1981–1986	K. Gordon: Modelling and monitoring of medical time series, University of Nottingham.
1981–1988	J. Marriott: Aspects of Bayesian methodology for ARMA time series, University of Nottingham.
1983–1988	J. E. H. Shaw: Numerical and graphical methods for Bayesian inference, University of Nottingham.
1983–1988	L. D. Sharples: Combining related information, University of Nottingham.
1984–1988	N. G. Polson: Bayesian perspectives on statistical modelling, University of Nottingham.
1984–1992	A. P. Grieve: Applications of Bayesian methods to pharmaceutical problems, University of Nottingham.
1985–1989	S. Hills: Aspects of parameterization in statistical inference, University of Nottingham.
1986–1989	D. Stephens: Applications of change-point methods to image restoration, University of Nottingham.
1986–1992	J. Wakefield: Bayesian analysis of pharamacokinetic models, University of Nottingham.
1989–1994	P. Damien: Some Contributions to Bayesian Nonparametric Inference, Imperial College, London.
1990–1994	N. Gordon: Bayesian methods for tracking, Imperial College, London.
1990–1994	D. Buckle: Bayesian portfolio analysis, Imperial College, London.
1990–1994	L. Foreman: Bayesian analysis of hidden Markov models, Imperial College, London.

1990–1994 D. Phillips: Bayesian analysis of images via templating, Imperial College, London.
1990–1994 P. Vounatsou: Computer graphic techniques and Bayesian inference summaries, Imperial College, London.
1991–1995 E. Gutierrez-Pena: Topics in Bayesian Statistics relating to the exponential family, Imperial College, London.
1991–1995 K. F. Lam: Statistical studies of optimal lending, Imperial College, London.
1992–1996 S. T. B. Choy: Bayesian robustness studies, Imperial College, London.
1992–1996 M-Y. E. Yen: Bayesian classification from images, Imperial College, London.
1992–1996 M. Efstathiou: Laplace approximations and Markov chain Monte Carlo, Imperial College, London.
1992–1997 M. Curtis: Statistical studies of hedging, Imperial College, London.
1993–1996 J. T. Key: Bayesian model choice. Imperial College, London.
1993–1997 A. M. Brink: Bayesian analysis of contingency tables, Imperial College, London.
1994–1998 R. A. Haro-Lopez: Bayesian robustness, Imperial College, London.
1994–1997 D. G. T. Denison: Bayesian curves, CART and MARS, Imperial College, London.
1995–2000 C. J. Hoggart: Bayesian methods in Forensic Science, Imperial College, London.
1995–2000 J. E. Griffin: Bayesian nonlinear design, Imperial College, London.
1995–1999 S. B. Tan: Bayesian clinical trials, Imperial College, London.
1996–2001 N. A. Heard: Bayesian sequential design, Imperial College, London.
1996–2000 J. W. Sandy: Development and validation of an IBS symptom index, Imperial College, London.
1997–2001 R. A. H. J. Al-Jaralla: Bayesian analysis of models where variances depend on covariates, Imperial College, London.

Adrian Smith's publications

Books

1985 (with U. E. Makov and D. M. Titterington) *The Statistical Analysis of Finite Mixture Models.* Wiley.

1994 (with J. M. Bernardo) *Bayesian Theory*. Wiley.

2002 (with D. G. T. Denison, C. C. Holmes and B. K. Mallick) *Bayesian Methods for Nonlinear Classification and Regression.* Wiley

Editing and translation

1974 (co-translator with A. Machi) *Theory of Probability; a critical introductory treatment.* Vol. I, by Bruno de Finetti. Wiley.

1975 (co-translator with A. Machi) *Theory of Probability; a critical introductory treatment.* Vol. II, by Bruno de Finetti. Wiley.

1981 (joint editor with J. M. Bernardo, M. H. DeGroot and D. V. Lindley) Bayesian Statistics: Proceedings of the International Meeting on Bayesian Statistics. University of Valencia Press.

1983 (joint editor with J. P. Florens, M. Mouchart, J. P. Raoult and L. Simar) *Specifying Statistical Models.* Springer-Verlag, New York.

1983 (joint editor with A. P. Dawid) *Practical Bayesian Statistics.* Longman.

1985 (joint editor with J. M. Bernardo, M. H. DeGroot and D. V. Lindley) *Bayesian Statistics 2.* Amsterdam. North-Holland.

1987 (joint editor with A. P. Dawid) *Practical Bayesian Statistics.* Carfax.

1988 (joint editor with J. M. Bernardo, M. H. DeGroot and D. V. Lindley) *Bayesian Statistics 3.* Oxford University Press.

1992 (joint editor with J. Berger, J. M. Bernardo and A. P. Dawid) *Bayesian Statistics 4.* Oxford University Press.

1994 (joint editor with P. R. Freeman) *Aspects of Uncertainty: A Tribute to DV Lindley.* Wiley.

1996 (joint editor with J. Berger, J. M. Bernardo and A. P. Dawid) *Bayesian Statistics 5.* Oxford University Press.

1999 (joint editor with J. Berger, J. M. Bernardo and A. P. Dawid) *Bayesian Statistics 6.* Oxford University Press.

2003 (joint editor with J. M. Bernardo, M. J. Bayarri, J. O. Berger, A. P. Dawid, D. Heckerman and M. West) *Bayesian Statistics 7.* Oxford University Press.

2007 (joint editor with J. M. Bernardo, M. J. Bayarri, J. O. Berger, A. P. Dawid, D. Heckerman and M. West) *Bayesian Statistics 8.* Oxford University Press.

2011 (joint editor with J. M. Bernardo, M. J. Bayarri, J. O. Berger, A. P. Dawid, D. Heckerman and M. West) *Bayesian Statistics 9.* Oxford University Press.

Articles

1972 (with D. V. Lindley) Bayes estimates for the linear model (with discussion). *J R Statist Soc*, B, **34**, 1–41.

1973 Bayes estimates in one-way and two-way models. *Biometrika*, **60**, 319–329.

1973 A general Bayesian linear model. *J R Statist Soc*, B, **35**, 67–75.

1974 (with C. D. Payne) An algorithm for determining Slater's i and all nearest adjoining orders. *Br J Math Statist Psychol*, **27**, 49–52.

1975 (with M. Goldstein) Ridge regression: some comments on a paper of Conniffe and Stone. *Statistician*, **24**, 61–66.

1975 A Bayesian approach to inference about a change-point in a sequence of random variables. *Biometrika*, **62**, 407–416.

1976 (with N. B. Booth) Batch acceptance schemes based on an auto-regressive prior. *Biometrika*, **63**, 133–136.

1977 Bayesian statistics and efficient information processing constrained by probability models. In *Decision Making and Change in Human Affairs* (H. Jungermann and G. de Zeeuw: Editors). Reidel, pp. 479–490.

1977 A Bayesian approach to some time-varying models. Proceedings of the European Congress of Statisticians (J. R. Barra: Editor). North-Holland Publishing Company, pp. 257–267.

1977 (with U. E. Makov) Quasi-Bayes procedures for unsupervised learning. Proceedings of the 1976 IEEE Conference on Decision and Control.

1977 (with U. E. Makov) A quasi-Bayes unsupervised learning procedure for priors. IEEE Transactions on Information Theory, IT-23, 761–764.

1977 A Bayesian note on reliability growth during a development testing program. IEEE Transactions on Reliability, R-26, 346–364.

1977 (Invited Discussant) Comments on "A simulation study of alternatives to ordinary least squares" (Dempster, Schatzoff and Wermuth). *J Amer Statist Ass*, **71**, March.

1978 (with A. P. Dawid, J. I. Galbraith, R. F. Galbraith and M. Stone) A note on forecasting car ownership. *J R Statist Soc*, A, **141**, 64–68.

1978 (with U. E. Makov) A quasi-Bayes sequential procedure for mixtures. *J R Statist Soc*, B, **40**, 106–112.

1980 (with U. E. Makov) Bayesian Detection and Estimation of Jumps in Linear Systems. In *The Analysis and Optimization of Stochastic Systems* (C. Harris and O. Jacobs: Editors). Academic Press, pp. 333–345.

1980 (with D. G. Cook) Switching straight-lines: an analysis of some renal transplant data. *Applied Statistics*, **29**, 180–189.

1980 (with D. J. Spiegelhalter) Bayes factors and choice criteria for linear models. *J R Statistic Soc*, B, **42**, 213–220.

1980 (with I. Verdinelli) A note on Bayesian designs for inference using a hierarchical linear model. *Biometrika*, **67**, 613–619.

1980 Some comments on a paper of D. E. H. Llewelyn, *Clinical Science*, **58**.

1981 Change-point problems: approaches and applications. Proceedings of the International Meeting on Bayesian Statistics (J. M. Bernardo *et al.*: Editors). University of Valencia Press.

1981 (with D. J. Spiegelhalter) Bayes factors in multivariate analysis. Chapter 17 of *Looking at Multivariate Data* (V. Barnett: Editor). Wiley.

1981 On random sequences with centered spherical symmetry. *J R Statist Soc*, B, 208–209.

1981 (with U. E. Makov) Unsupervised learning for signal versus noise. IEEE Trans on Inf Th, 498–500.

1981 (with D. J. Spiegelhalter) Decision analysis and clinical decisions. In *Perspectives in Medical Statistics*. Academic Press, pp. 103–131.

1981 La teoria Bayesiana delle decisioni in medicina. In *Teoria delle decisioni in medicina* (Girelli Bruni: Editor), Bertoni, Verona.

1981 (with R. Dabir, S. J. Ellis, A. Hollingsworth and W. A. Wallace) Comparison of bupivocaine and prilocaine used in Bier block – a double blind trial. *Injury*, 331–336.

1982 (with J. C. Naylor) Applications of a method for the efficient computation of posterior distributions. *Applied Statistics*, **31**, 214–225.

1982 (with D. J. Spiegelhalter) Bayes factors for linear and log-linear models with vague prior information. *J R Statist Soc*, B, **44**, 377–387.

1982 (with N. B. Booth) A Bayesian approach to the retrospective identification of change-points. *J of Econometrics*, **19**, 7–22.

1982 (with M. S. Knapp, I. Trimble, R. Pownall and M. West) The detection of sudden change in renal function by time-series analysis. In *Towards Chronopharmacology* (R. Takalashi *et al.*: Editors). Pergamon Press, Oxford.

1983 (with I. Trimble, M. West, M. S. Knapp and R. Pownall) Detection of renal allograft rejection by computer. *British Medical Journal*, 1695–1699.

1983 Bayesian approaches to outliers and robustness. In *Specifying Statistical Models* (J. P. Florens *et al.*: Editors). Springer-Verlag, New York.

1983 (with M. Knapp, I. Trimble, R. Pownall and K. Gordon) Mathematical and statistical aids to evaluate data from renal patients. *Kidney International*, **24**, 474–486.

1983 (with J. C. Naylor) A contamination model in clinical chemistry. *Statistician*, **32**, 82–87.

1983 (with M. West, K. Gordon, M. S. Knapp and I. Trimble) Monitoring kidney transplant patients. *Statistician*, **34**, 46–54.

1983 (with M. West) Monitoring Renal Transplants: an application of the multi-process Kalman filter. *Biometrics*, **39**, 867–878.

1984 Bayesian Statistics: Chapter 15 of *Handbook of Applicable Mathematics*: Vol. 6; Wiley.

1984 Decision Theory: Chapter 19 of *Handbook of Applicable Mathematics*: Vol. 6; Wiley.

1984 (with L. I. Pettit) Bayesian model comparisons in the presence of outliers. Proceedings of the 44th ISI Meeting, Madrid.

1984 Present position and potential developments – some personal views; Bayesian Statistics. *J R Statist Soc*, A, **147**, 245–259.

1985 (with L. I. Pettit) Outliers and influential observations in linear models. In *Bayesian Statistics 2* (Proceedings of the 2nd Valencia International meeting on Bayesian Statistics). North-Holland.

1985 (with K. Gordon, R. Pownall and M. S. Knapp) The development of new statistical methods for event detection in time series. *The Annual Review of Chronopharmacology*, Vol. **I**, 161–164.

1985 (with A. M. Skene, J. E. H. Shaw, J. C. Naylor and M. Dransfield) The implementation of the Bayesian paradigm. *Commun Statist* **A5**, 1079–1102.

1985 (with J. M. Bernardo and J. R. Ferrandiz) The foundations of decision theory; an intuitive, operational approach with mathematical extensions. *Theory and Decision*, **19**, 127–150.

1985 (with A. N. Pole) A Bayesian approach to some threshold switching models. *J of Econometrics*, **29**, 97–119.

1986 Discussion of Efron's 'Why isn't everyone a Bayesian?' In *American Statistician*, **40**.

1986 Some Bayesian thoughts on modelling and model choice. *Statistician*, **35**, 97–102.

1986 (with A. Racine-Poon, A. P. Grieve and H. Flühler) Bayesian statistics in practice: experiences in the pharmaceutical industry (with discussion). *Appl Statist*, **35**, 93–150.

1986 Observations, Imaginary. In *Encyclopaedia of Statistical Science*, Vol. 7 (S. Kotz and N. L. Johnson: Editors). Wiley.

1987 (with A. M. Skene, J. E. H. Shaw and J. C. Naylor) Progress with numerical and graphical methods for practical Bayesian statistics. *Statistician*, **36**, 75–82.

1987 An overview of some problems relating to finite mixture distributions. *Rassegna di Metodi Statistici*, **5**, 137–150.

1987 Discussion of Hodges' 'Uncertainty, Policy Analysis and Statistics'. In *Statistical Science*, **3**.

1987 (with A. Racine-Poon, A. P. Grieve and H. Flühler) A two-stage procedure for bioequivalence studies. *Biometrics*, **43**, 847–856.

1988 (with J. C. Naylor) Econometric illustrations of novel numerical integration methodology for Bayesian inference. *J of Econometrics*, **38**, 103–125.

1988 (with K. Gordon) Modelling and monitoring discontinuous changes in time series. Chapter 17 of *Bayesian Analysis of Time Series and Dynamic Models* (J. C. Spall: Editor). Marcel Dekker.

1988 (with J. C. Naylor) An archaeological inference problem. *J Amer Statist Ass*, **83**, 588–595.

1988 What should be Bayesian about Bayesian software? In *Bayesian Statistics* 3 (J. M. Bernardo *et al.*: Editors). Oxford University Press.

1988 (with K. Gordon, B. A. Bradley and S. M. Gore) Kalman Filter Monitoring. Chapter 8 of *Renal Transplantation: Sense and Sensitization* (S. M. Gore and B. A. Bradley: Editors). Council of Europe. Kluwer Academic Publishers.

1989 (with A. Racine-Poon) Population Models. Chapter 4 of *Statistical Methodology in the Pharmaceutical Sciences* (D. A. Berry: Editor). Marcel Dekker.

1989 Discussion of Trumbo's 'How to get your first research grant'. *Statistical Science*, **4**, 134–136.

1990 (with J. A. Achcar) Aspects of Reparametrization in Approximate Bayesian Inference. In *Essays in Honour of GA Barnard* (J. Hodges: Editor). North-Holland, pp. 439–452.

1990 (with K. Gordon) Modelling and monitoring biomedical time series. *J Amer Statist Ass*, **85**, 328–337.

1990 (with A. Gelfand) Sampling-based approaches to calculating marginal densities. *J Amer Statist Ass*, **85**, 398–409.

1990 (with A. Gelfand, A. Racine-Poon and S. Hills) Illustration of Bayesian inference in normal data models using Gibbs sampling. In *J Amer Statist Ass*, **85**, 972–985.

1991 (with J. Wakefield and A. Gelfand) Efficient generation of random variates via the ratio-or-uniforms method. *Statistics and Computing*, **1**, 129–133.

1991 (with J. Wakefield, A. Skene and I. Evett) The evaluation of fibre transfer evidence in forensic science: a case study in statistical modelling. *Appl Statist*, **40**, 461–476.

1991 An overview of the Bayesian approach. Chapter 2 of *Bayesian Methods in Reliability* (P. Sander and R. Badoux: Editors). Kluwer Academic Publishers, pp. 15–80.

1991 (with A. Gelfand) Gibbs sampling for posterior expectations. *Commun Statist: Th and Methods*, **20**, 1747–1766.

1991 Bayesian computational methods. *Phil Trans R Soc Lond*, A, **337**, 369–386.

1991 (with C. Buck, J. Kenworthy and C. Litton) Combining archaeological and radiocarbon information: a Bayesian approach to calibration. *Antiquity*, **65**, 808–821.

1991 (with C. Buck and C. Litton) Calibration of radiocarbon results pertaining to related archaeological events. *J Arch Sci*, **19**, 497–512.

1992 (with A. Racine-Poon and C. Weihs) Estimation of relative potency with sequential dilution errors in radioimmunoassay. *Biometrics*, **47**, 1235–1246.

1992 (with J. M. Marriott) Reparametrization aspects of numerical Bayesian methods for ARMA models. *J Time Series Anal*, **13**, 327–343.

1992 (with B. Carlin and A. Gelfand) Hierarchical Bayesian analysis of change point problems. *Appl Statist*, **41**, 389–405.

1992 (with A. Gelfand and T. M. Lee) Bayesian analysis of constrained parameter and truncated data problems via Gibbs sampling. *J Amer Statist Ass*, **87**, 523–532.

1992 (with S. E. Hills) Parameterization issues in Bayesian inference. In *Bayesian Statistics 4* (J. Berger *et al.*: Editors). Oxford University Press, pp. 227–246.

1992 (with L. Kuo) Bayesian computation for survival models via the Gibbs sampler. In *Survival Analysis and Related Topics* (J. P. Klein and P. K. Goel: Editors). Marcel Dekker, pp. 11–24.

1992 (with A. Gelfand) Bayesian statistics without tears: a sampling–resampling perspective. *American Statistician*, **46**, 84–88.

1992 (with L. Pericchi) Exact and approximate posterior moments for a normal location parameter. *J R Statist Soc*, B, **54**, 793–804.

1993 (with I. Verdinelli and F. Ball) Biased coin designs with a Bayesian bias. *J Stat Inf and Plan*, **34**, 403–421.

1993 (with G. O. Roberts) Bayesian computation via the Gibbs sampler and related Markov chain Monte Carlo methods. *J R Statist Soc*, B, **55**, 3–23.

1993 (with D. Stephens) Sampling–resampling techniques for the computation of posterior densities in normal means problems. *Test*, **1**, 1–18.

1993 (with P. Dellaportas) Bayesian inference for generalized linear and proportional hazards models via Gibbs sampling. *Appl Statist*, **42**, 443–459.

1993 (with D. Stephens) Bayesian edge-detection in images via change-point methods. In *Computing Intensive Methods in Statistics* (W. Hardle and L. Simar: Editors). Physica-Verlag, pp. 1–29.

1993 (with N. Gordon and D. J. Salmond) Novel approach to non-linear/non-Gaussian Bayesian state estimation. IEEE Proceedings - F, **140**, 107–113.

1993 (with S. E. Hills) Diagnostic plots for improved parameterization in Bayesian inference. *Biometrika*, **80**, 61–74.

1993 (with E. I. George and U. E. Makov) Conjugate likelihood distributions. *Scand J Statist*, **20**, 147–156.

1993 (with N. Gordon) Approximate non-Gaussian Bayesian estimation and modal consistency. *J R Statist Soc*, B, **55**, 913–918.

1993 (with J. C. Wakefield, A. Racine-Poon and A. E. Gelfand) Bayesian analysis of linear and nonlinear population models using the Gibbs sampler. *Appl Statist*, **43**, 201–222.

1993 (with P. Damien and P. W. Laud) Nonparametric Bayesian bioassay with prior constraints on the shape of the potency curve. *Biometrika*, **80**, 489–498.

1993 (with L. Pericchi and B. Sanso) Posterior cumulant relationships in Bayesian inference involving the exponential family. *J Amer Statist Soc*, **88**, 1419–1426.

1993 (with S. K. Upadhyay) Simulation based Bayesian approaches to the analysis of a lognormal regression model. In *Reliability: A Cutting Edge* (P. J. Van Gestel and W. Roseboom: Editors). Kema, pp. 196–207.

1993 (with P. Damien and P. W. Laud) Random variate generation for D-distributions. *Statist and Comp*, **3**, 109–112.

1993 (with D. Stephens) Bayesian inference in multipoint gene mapping. Ann Hum Genetics, **57**, 65–82.

1993 (with S. K. Upadhyay) Simulation based Bayesian approaches to the analysis of a lognormal regression model. Proc. SRE Symp. "Reliability of a Competitive Edge", The Netherlands, Vol. **1**, 196–207.

1994 (with S. K. Upadhyay) Modelling complexities in reliability, and roles of Bayes simulation. *International Jr Cont Eng Educ*: Sp. Issue on Applied Probability Mod., Vol. **4**, 93–104.

1994 (with G. O. Roberts) Simple conditions for the convergence of the Gibbs sampler and Metropolis–Hastings algorithms. *Stoch Proc and Their Applic*, **49**, 207–216.

1994 (with J. C. Wakefield) The hierarchical Bayesian approach to population pharmacokinetic modelling. *Int J Bio-Med Comp*, **36**, 35–52.

1994 (with E. I. George and U. E. Makov) Fully Bayesian hierarchical analysis for exponential families via Monte Carlo approximation. In *Aspects of Uncertainty* (P. R. Freeman and A. F. M. Smith: Editors). Wiley, pp. 181–199.

1994 (with S. K. Upadhyay) Modelling complexities in reliability, and the role of simulation in Bayesian computation. *Int J Cont Eng Ed*, **4**, 93–104.

1994 (with E. J. Green, F. A. Roesch and W. E. Strawderman) Bayesian estimation for the three-parameter Weibull distribution with tree diameter data. *Biometrics*, **50**, 254–269.

1994 (with D. B. Phillips) Bayesian faces via hierarchical template modelling. *J Amer Statist Assoc*, **89**, 1151–1163.

1994 (with P. Vounatsou) Bayesian analysis of ring-recovery data via Markov chain Monte Carlo simulation. *Biometrics*, **51**, 687–708.

1994 (with J. C. Wakefield, A. Racine-Poon and A. E. Gelfand) Bayesian analysis of linear and non-linear population models by using the Gibbs sampler. *Appl Statist*, **43**, 201–221.

1995 (with E. Gutierrez-Pena) Conjugate parameterizations for natural exponential families. *J Amer Statist Assoc*, **90**.

1995 (with P. Damien and P. W. Laud) Approximate random variate generation from infinitely divisible distributions with applications to Bayesian inference. *J R Statist Soc*, B, **57**, 547–564.

1996 (with S. K. Upadhyay and R. Agrawal) Bayesian analysis of inverse Gaussian non-linear regression by simulation. *Sankhya* Ser. B, **58**, part 3, 363–378.

1996 (with D. Gamerman) Bayesian analysis of longitudinal data studies. In *Bayesian Statistics* 5 (J. M. Bernardo *et al.*: Editors). Oxford University Press, 587–598.

1996 (with E. J. Green and W. E. Strawderman) Construction of thematic maps from satellite imagery. In *Bayesian Statistics* 5 (J. M. Bernardo *et al.*: Editors). Oxford University Press, 181–196.

1996 (with P. Vounatsou) Graphical methods for simulation-based Bayesian inference. In *Bayesian Statistics* 5 (J. M. Bernardo *et al.*: Editors). Oxford University Press, 773–782.

1996 (with P. Damien and P. W. Laud) Implementation of Bayesian non-parametric inference based on beta processes. *Scand J Statist*, **23**, (1), 27–37.

1996 Mad cows and ecstasy: chance and choice in an evidence-based society. *J R Statist Soc*, A, **159**, 367–383.

1996 (with P. Vounatsou) Bayesian analysis of contingency tables: a simulation and graphics-based approach. *Statist Comput*, **6**, 277–287.

1996 (with U. E. Makov and Y-H. Liu) Bayesian methods in actuarial science. *The Statistician*, **45**, 503–515.

1996 (with D. A. Stephens, P. Dellaportas and I. Guttman) A comparative study in perinatal mortality using a two component mixture model. In *Bayesian Biostatistics* (D. A. Berry and D. K. Stangl: Editors). Marcel Dekker, New York.

1996 (with P. Damien and P. W. Laud) A Monte Carlo method approximating a posterior hazard rate process. *Statistics and Computing*, **6**, 77–84.

1997 (with S. T. B. Choy) On robust analysis of a normal location parameter. *J R Statist Soc*, B, **59**, 463–474.

1997 (with R. Moskovic and P. L. Windle) Modeling Charpy impact energy property changes using a Bayesian method. *Metallurgical and Materials Transactions*, A, **28**, 1181–1193.

1997 (with L. A. Foreman and I. W. Evett) A Bayesian approach to validating STR multiplex databases for use in forensic casework. *Int. J Legal Med.* **110**, 244–250.

1997 (with P. Damien and P. W. Laud) Bayesian estimation of unimodal distributions. *Comm in Stat*, **26**, 429–440.

1997 (with P. Damien and P. W. Laud) Bayesian nonparametric and covariate analysis of failure time data. In *Practical Nonparametric Bayes* (D. Dey, D. Sinha and P. Mueller: Editors). Springer-Verlag, New York.

1997 (with D. A. Stephens and R. Moskovic) Charpy impact energy data: A Markov chain Monte Carlo analysis. *Appl Statist*, **46**, (4), 477–492.

1997 (with L. A. Foreman and I. W. Evett) Bayesian analysis of DNA profiling data in forensic identification applications. *J R Statist Soc*, A, **160**, (3), 429–469.

1997 (with E. Gutierrez-Pena) Exponential and Bayesian conjugate families: review and extensions. *Test*, **6**, (1), 1–90.

1998 (with J. J. Deely) Quantitative refinements for comparisons of institutional performance. *J R Statist Soc*, A, **161**, (1), 5–12.

1998 (with N. Lynn and N. Singpurwalla) Bayesian assessment of network reliability. *SIAM Rev*, **40**, (2), 202–227.

1998 (with P. Laud and P. Damien) Bayesian nonparametric and covariate analysis of failure time data. In *Practical Nonparametric Bayes*; D. Dey, D. Sinha and P. Mueller: Editors), Springer-Verlag, New York, 213–225.

1998 (with D. G. T. Denison and B. K. Mallick) Bayesian MARS. *Statistical Comp.*, **8**, 337–346.

1998 (with D. G. T. Denison and B. K. Mallick) A Bayesian CART algorithm. *Biometrika*, **85**, 363–377.

1998 (with D. G. T. Denison and B. K. Mallick) Automatic Bayesian curve fitting. *J R Statist Soc*, B, **60**, 333–350.

1999 (with R. A. Haro-Lopez) On robust Bayesian analysis for location and scale parameters. *Journal of Multivariate Analysis*, **70**, 30–56.

1999 (with S. G. Walker, P. Damien and P. W. Laud) Bayesian nonparametric inference for random distributions and related functions (with discussion). *J R Statist Soc*, B, **61**, 485–516.

1999 (with B. K. Mallick and D. G. T. Denison) Bayesian survival analysis using a MARS model. *Biometrics*, **55**, 1071–1077.

2000 (with S. K. Upadhyay and N. Vasisha) Bayes inference in life testing and reliability via Markov chain Monte Carlo simulation. *Sanhkya*, Series A, **62**, 203–222.

2000 (with D. Ashby) Evidence-based medicine as Bayesian decision-making. *Statistics in Medicine*, **19**, 3291–3305.

2000 (with P. Dellaportas and P. Stavropoulos) Bayesian analysis of mortality data. *J R Statist Soc*, A, 275–292.

2001 (with S. K. Upadhyay and N. Vasisha) Bayes inference in life testing and reliability via Markov chain Monte Carlo simulation. *Sanhkya*, Series A, part I, **63**, 15–40.

2003 (with C. J. Hoggart and S. G. Walker) Bivariate kurtotic distributions of garment fibre data. *Applied Statistics*, **52**, Part 3, 323–335.

Index